THE UNIVERSITY OF ARIZONA SPACE SCIENCE SERIES
Richard P. Binzel, General Editor

Exoplanets
S. Seager, editor, 2010, 526 pages

Europa
Robert T. Pappalardo, William B. McKinnon,
and Krishan K. Khurana, editors, 2009, 727 pages

The Solar System Beyond Neptune
M. Antonietta Barucci, Hermann Boehnhardt, Dale P. Cruikshank,
and Alessandro Morbidelli, editors, 2008, 592 pages

Protostars and Planets V
Bo Reipurth, David Jewitt, and Klaus Keil, editors, 2007, 951 pages

Meteorites and the Early Solar System II
D. S. Lauretta and H. Y. McSween, editors, 2006, 943 pages

Comets II
M. C. Festou, H. U. Keller,
and H. A. Weaver, editors, 2004, 745 pages

Asteroids III
William F. Bottke Jr., Alberto Cellino, Paolo Paolicchi,
and Richard P. Binzel, editors, 2002, 785 pages

Tom Gehrels, General Editor

Origin of the Earth and Moon
R. M. Canup and K. Righter, editors, 2000, 555 pages

Protostars and Planets IV
Vincent Mannings, Alan P. Boss,
and Sara S. Russell, editors, 2000, 1422 pages

Pluto and Charon
S. Alan Stern and David J. Tholen, editors, 1997, 728 pages

**Venus II—Geology, Geophysics, Atmosphere,
and Solar Wind Environment**
S. W. Bougher, D. M. Hunten,
and R. J. Phillips, editors, 1997, 1376 pages

Cosmic Winds and the Heliosphere
J. R. Jokipii, C. P. Sonett,
and M. S. Giampapa, editors, 1997, 1013 pages

Neptune and Triton
Dale P. Cruikshank, editor, 1995, 1249 pages

Hazards Due to Comets and Asteroids
Tom Gehrels, editor, 1994, 1300 pages

Resources of Near-Earth Space
John S. Lewis, Mildred S. Matthews,
and Mary L. Guerrieri, editors, 1993, 977 pages

Protostars and Planets III
Eugene H. Levy and Jonathan I. Lunine, editors, 1993, 1596 pages

Mars
Hugh H. Kieffer, Bruce M. Jakosky, Conway W. Snyder,
and Mildred S. Matthews, editors, 1992, 1498 pages

Solar Interior and Atmosphere
A. N. Cox, W. C. Livingston,
and M. S. Matthews, editors, 1991, 1416 pages

The Sun in Time
C. P. Sonett, M. S. Giampapa,
and M. S. Matthews, editors, 1991, 990 pages

Uranus
Jay T. Bergstralh, Ellis D. Miner,
and Mildred S. Matthews, editors, 1991, 1076 pages

Asteroids II
Richard P. Binzel, Tom Gehrels,
and Mildred S. Matthews, editors, 1989, 1258 pages

Origin and Evolution of Planetary and Satellite Atmospheres
S. K. Atreya, J. B. Pollack,
and Mildred S. Matthews, editors, 1989, 1269 pages

Mercury
Faith Vilas, Clark R. Chapman,
and Mildred S. Matthews, editors, 1988, 794 pages

Meteorites and the Early Solar System
John F. Kerridge and Mildred S. Matthews, editors, 1988, 1269 pages

The Galaxy and the Solar System
Roman Smoluchowski, John N. Bahcall,
and Mildred S. Matthews, editors, 1986, 483 pages

Satellites
Joseph A. Burns and Mildred S. Matthews, editors, 1986, 1021 pages

Protostars and Planets II
David C. Black and Mildred S. Matthews, editors, 1985, 1293 pages

Planetary Rings
Richard Greenberg and André Brahic, editors, 1984, 784 pages

Saturn
Tom Gehrels and Mildred S. Matthews, editors, 1984, 968 pages

Venus
D. M. Hunten, L. Colin, T. M. Donahue,
and V. I. Moroz, editors, 1983, 1143 pages

Satellites of Jupiter
David Morrison, editor, 1982, 972 pages

Comets
Laurel L. Wilkening, editor, 1982, 766 pages

Asteroids
Tom Gehrels, editor, 1979, 1181 pages

Protostars and Planets
Tom Gehrels, editor, 1978, 756 pages

Planetary Satellites
Joseph A. Burns, editor, 1977, 598 pages

Jupiter
Tom Gehrels, editor, 1976, 1254 pages

Planets, Stars and Nebulae, Studied with Photopolarimetry
Tom Gehrels, editor, 1974, 1133 pages

Exoplanets

Exoplanets

Edited by

S. Seager

With the assistance of

Renée Dotson

With 34 collaborating authors

THE UNIVERSITY OF ARIZONA PRESS
Tucson

in collaboration with

LUNAR AND PLANETARY INSTITUTE
Houston

About the front cover:

An imaginary planetary system orbiting a cool M-dwarf star. In the foreground, a giant planet has an Earth-sized satellite, a water-rich body. The giant planet and satellite are orbiting in the star's habitable zone. In the mid-distance is another giant planet (Saturn-sized) close to inferior conjunction with the foreground planet, and a third giant planet is in the far distance, transiting the star. The artist's viewpoint captures the three planets at the time of a fortuitous alignment. Painting by William K. Hartmann, Planetary Science Institute, Tucson, Arizona.

About the back cover:

Image of the HR 8799 planetary system produced by combining the J-, H-, and Ks-band images obtained at the Keck telescope in July (H) and September (J and Ks) 2008. The inner part of the H-band image has been rotated by 1° to compensate for the orbital motion of d between July and September. The planets are listed by lower-case letter in order of chronological discovery. At center is the speckle image of the star. From Marois et al. (2008) *Science, 322,* 1348–1352. Image courtesy of NRC-HIA, C. Marois, and Keck Observatory.

The University of Arizona Press
in collaboration with the Lunar and Planetary Institute
© 2010 The Arizona Board of Regents
All rights reserved
∞ This book is printed on acid-free, archival-quality paper.
Manufactured in the United States of America

17 16 15 14 13   8 7 6 5 4 3

Library of Congress Cataloging-in-Publication Data

Exoplanets / edited by S. Seager ; with the assistance of Renée Dotson ; with 34 collaborating authors.
    p. cm. -- (The University of Arizona space science series)
  Includes bibliographical references and index.
  ISBN 978-0-8165-2945-2 (hardcover : alk. paper)
 1. Extrasolar planets. I. Seager, Sara. II. Dotson, Renée. III. Lunar and Planetary Institute.
QB820.E87 2010
523.2'4--dc22
                      2010042822

*Dedicated to all the people on this planet who have big dreams
and succeed in making them happen.*

Contents

List of Contributing Authors and Scientific Organizing Committee xiii

Foreword xv

Preface xvii

PART I: INTRODUCTION

Introduction to Exoplanets
S. Seager and J. J. Lissauer 3

Keplerian Orbits and Dynamics of Exoplanets
C. D. Murray and A. C. M. Correia 15

PART II: EXOPLANET OBSERVING TECHNIQUES

Radial Velocity Techniques for Exoplanets
C. Lovis and D. Fischer 27

Exoplanet Transits and Occultations
J. N. Winn 55

Microlensing by Exoplanets
B. S. Gaudi 79

Direct Imaging of Exoplanets
W. A. Traub and B. R. Oppenheimer 111

Astrometric Detection and Characterization of Exoplanets
A. Quirrenbach 157

Planets Around Pulsars and Other Evolved Stars: The Fates of Planetary Systems
A. Wolszczan and M. Kuchner 175

Statistical Distribution of Exoplanets
A. Cumming 191

PART III: EXOPLANET DYNAMICS

Non-Keplerian Dynamics of Exoplanets
D. Fabrycky 217

Tidal Evolution of Exoplanets
A. C. M. Correia and J. Laskar 239

PART IV: EXOPLANET FORMATION AND PROTOPLANETARY DISK EVOLUTION

Protoplanetary and Debris Disks
A. Roberge and I. Kamp 269

Terrestrial Planet Formation
J. Chambers 297

Giant Planet Formation
G. D'Angelo, R. H. Durisen, and J. J. Lissauer 319

Planet Migration
S. H. Lubow and S. Ida 347

PART V: EXOPLANET INTERIORS AND ATMOSPHERES

Terrestrial Planet Interiors
C. Sotin, J. M. Jackson, and S. Seager 375

Giant Planet Interior Structure and Thermal Evolution
J. J. Fortney, I. Baraffe, and B. Militzer 397

Giant Planet Atmospheres
A. Burrows and G. Orton 419

Terrestrial Planet Atmospheres and Biosignatures
V. Meadows and S. Seager 441

Atmospheric Circulation of Exoplanets
A. P. Showman, J. Y-K. Cho, and K. Menou 471

Index 517

List of Contributing Authors

Baraffe I.   397
Burrows A.   419
Chambers J.   297
Cho J.Y-K.   471
Correia A. C. M.   15, 239
Cumming A.   191
D'Angelo G.   319
Durisen R. H.   319
Fabrycky D. C.   217
Fischer D.   27
Fortney J. J.   397
Gaudi B. S.   79
Ida S.   347
Jackson J. M.   375
Kamp I.   269
Kuchner M.   175
Laskar J.   239

Lissauer J. J.   319, 3
Lovis C.   27
Lubow S. H.   347
Meadows V.   441
Menou K.   471
Militzer B.   397
Murray C. D.   15
Oppenheimer B. R.   111
Orton G.   419
Quirrenbach A.   157
Roberge A.   269
Seager S.   3, 375, 441
Showman A. P.   471
Sotin C.   375
Traub W. A.   111
Winn J. N.   55
Wolszczan A.   175

Scientific Organizing Committee

Sara Seager, Chair
Alan Boss
Adam Burrows
David Charbonneau
Drake Deming
Debra Fischer
Olivier Grasset

Tristan Guillot
Marc Kuchner
Doug Lin
Jack Lissauer
Vicki Meadows
Wes Traub
Stephane Udry

Foreword

The discovery of exoplanets is arguably the greatest scientific revolution since the time of Copernicus. Simply stated, humanity now knows for the first time as scientific fact: *there actually are planets around other stars*. Yet as profound as this discovery is, the trumpets have not blared and no theorists have been burned at the stake. (An early proponent of other planetary worlds, Giordano Bruno, did not fare as well in 1600.) Acclamation of exoplanets' reality has not penetrated the global consciousness as deeply as it might merit because (I think) the public *already knew* that planets were there. *Star Trek* took us to a new planet every week; *E.T.* had to have some place to phone home to; etc. Even though the transition from *science fiction* to *science fact* may have seemed subtle, historical perspective from decades or centuries henceforth is likely to see this moment as another major turning point in the human perception of our place in the universe.

Now, 400 years after Galileo's observations affirmed the model of Copernicus, we are in the midst of our own golden era of discovery. Just as Galileo first applied the telescope to reveal the details of our own planetary system, that same spirit of inquisitiveness and ingenuity (with the benefit of technological advances) is now revealing details of *other* planetary systems. Thus it is easy to fathom why so many current researchers, and most importantly new students, are drawn so strongly to this field. The attraction of pure discovery, the opportunity to share the same experience as Galileo, is simply too alluring to resist.

Thus sets the stage for this volume, *Exoplanets*, perhaps the most important title yet produced in the Space Science Series. The goals of this book are to merge and embrace the astrophysics of exoplanets within the disciplines of planetary science, to fully expand planetary science beyond the bounds of our own solar system, to serve as a foundation for the interface of these merging and expanding fields, and most importantly, to be the gateway for new students and researchers. To accomplish this, Series authors are each challenged to convey what we know, how we know it, and where we go from here.

I thank my lucky planets that my colleague, Professor Sara Seager, embraced the editorship of this book with such clear vision and conviction for how it should be accomplished, coupled with the requisite determination and stamina. One cannot understate the effort delivered by the authors whose chapters are the cornerstones upon which the field will continue to build. Less visible, but especially deserving of thanks, are Renée Dotson and co-workers of the Lunar and Planetary Institute (LPI), who brought this volume to reality, literally page by page. The support and dedication for the ongoing success of the Space Science Series by LPI Director Dr. Stephen Mackwell and the staff of the University of Arizona Press are paramount. A grant from Dr. Wesley Traub through the NASA Exoplanet Exploration Program has made this gateway accessible to all by supporting costs for this book. No outcome is more desired than that students entering here become the future leaders of the field.

Richard P. Binzel
Space Science Series General Editor
Cambridge, Massachusetts
August 2010

Preface

This is a unique time in human history — for the first time, we are on the technological brink of being able to answer questions that have been around for thousands of years: Are there other planets like Earth? Are they common? Do any have signs of life? The field of exoplanets is rapidly moving toward answering these questions with the discovery of hundreds of exoplanets now pushing toward lower and lower masses; the Kepler Space Telescope with its yield of small planets; plans to use the James Webb Space Telescope (launch date 2014) to study atmospheres of a subset of super Earths; and ongoing development for technology to directly image true Earth analogs. Theoretical studies in dynamics, planet formation, and physical characteristics provide the needed framework for prediction and interpretation.

People working outside of exoplanets often ask if the field of exoplanets is like a dot.com bubble that will burst, deflating excitement and progress. In my opinion, exciting discoveries and theoretical advances will continue indefinitely in the years ahead, albeit at a slower pace than in the first decade. The reason is that observations uncover new kinds and new populations of exoplanets — and these observations rely on technological development that usually takes over a decade to mature. For example, in the early 2000s almost all exoplanets had been discovered by the radial velocity technique. At that time, many groups around the world were working on wide-field transit surveys. But it was not until recently, a decade into the twenty-first century, that the transit technique is responsible for almost one-quarter of known exoplanets. The planet discovery techniques astrometry (as yet to find a planet) and direct imaging have not yet matured; when they do, they will uncover planets within a new parameter space of planet mass and orbital characteristics. In addition, people are working hard to improve the precision for existing planet discovery techniques to detect lower-mass planets and those further from their star. All in all, technology enables slow but sure progress, and this fuels ongoing discovery.

Theory, like observations, also takes time to unfold and mature. We can anticipate an "ultimate" planet formation model similar to the "millenimum simulation" for galaxy formation and evolution. In time, incorporating detailed physics as well as being able to reproduce the generic outcome of planet populations (mass, radius, and orbital characteristics, including period) will enable a deeper understanding of planet formation and migration. Similarly, the ideal exoplanet atmosphere code of the future could be a three-dimensional Monte Carlo code that includes radiative transfer with inhomogeneous cloud coverage and surface features, a code that also solves for the temperature structure and combines with a hydrodynamical simulation to calculate the three-dimensional temperature and wind structure. Classical orbital mechanics, already reinvigorated by interesting exoplanet systems (e.g., planets in resonant orbits, hot Jupiter exoplanets that orbit in the direction opposite to the stellar rotation), also has a role to play in explaining fundamental mechanisms of how planetary system configurations came to be. Orbital dynamics modeling is driving the search for moons and other unseen planet companions by their perturbations on transiting planet signatures.

Exoplanets is a unique science because it involves so many disciplines within and beyond planetary science and astrophysics. The other disciplines include geophysics, high-pressure mineral physics, quantum mechanics, chemistry, and even microbiology. While exoplanet observations clearly belong under the branch of astronomy, for many years the whole discipline of exoplanets lacked a true home. Physics departments have said "Exoplanets: It's interesting, but is it physics?"

Planetary and Earth science departments used to collecting real data in their hands from Earth and *in situ* measurements from solar system planets were, in the early 2000s, reluctant to believe there would ever be enough high-quality data to take the research field of exoplanets seriously. With hundreds of known exoplanets and many fascinating observational and theoretical discoveries, the whole world is embracing exoplanets, a field that can now find its home both in and spanning across planetary science, astronomy and astrophysics, and astrobiology.

We use the term "exoplanets" and not "extrasolar planets" in this book. Exoplanets is the most direct derivative from the Greek language, from which the word "planet" originates. (Note that the word "extrasolar" is derived from Latin.) It is most fitting that the field be named as traditionally as possible, and there is none more traditional than Greek.

The goal of this volume is to cover the range of topics in exoplanet observation and theory at the graduate student level. Each chapter of this volume aims to cover the fundamentals and recent discoveries in such a way as to be a cross between a textbook and a review article. Because the field of exoplanets moves so rapidly, we have tried to arrange each chapter so that the first two sections (the introduction and and an explanation of the basic concepts/fundamental equations) will not go out of date. The next two sections (covering recent highlights and future progress) will go out of date eventually, but serve to capture and review what is going on in a subfield. Some chapters deviate from the standard structure and, by nature of construction, the level of technical detail and writing style will vary in an edited volume with different authors for each chapter. We intend that this book will serve to inspire professional researchers and engineers and students to build a foundation for the next generation of exoplanet observational and theoretical discoveries.

Exoplanets is the 37th book in the University of Arizona Press Space Science Series, the first dedicated to planets beyond our solar system. As editor of this volume, I foremost thank General Editor Richard Binzel for believing in my vision for this book and for his advice and fast turnaround on practical matters. I am extremely grateful to Renée Dotson of the Lunar and Planetary Institute for her patience and careful efforts in compiling and production of the book. The book would not have been possible without the contribution and dedication of each of the chapter authors, often going far beyond what the authors had initially anticipated. The chapter reviewers played an invaluable role in improving the quality of this book. Appreciation to William Hartmann for his original painting that graces the book's cover. Last but not least, I am grateful to Wes Traub of NASA Jet Propulsion Laboratory for his financial support, which has enabled the book to be available for a practical purchase price.

Close to two decades since the first exoplanets were discovered, the field of exoplanets has grown without bound. So many surprising discoveries have been made that by now we know to expect the unexpected. I always like to say that, for exoplanets, anything is possible within the laws of physics and chemistry. And, when asked about which planet or discovery is my favorite one, I like to reply that it is the next one, because in exoplanets, the best is yet to come. I hope you, the reader, will use this book to learn about whichever subdiscipline interests you and to build your knowledge to play a role in the future of exoplanet research.

Sara Seager
September 2010
Cambridge, Massachusetts

Part I:
Introduction

Introduction to Exoplanets

Sara Seager
Massachusetts Institute of Technology

Jack J. Lissauer
NASA Ames Research Center

The discovery of planets around other stars, which we call exoplanets, has emerged over the past two decades as a new, vibrant, fruitful field that spans the disciplines of astrophysics, planetary science, and even parts of biology. The study of exoplanets is now an important part of the curriculum for many graduate students, and this volume is intended to bridge the gap between single-article summaries and specialized reviews. This chapter serves as an introduction both to the study of exoplanets and to this book, providing a starting point to new students and researchers entering the field.

1. A BRIEF HISTORY OF DISCOVERY

The search for our place in the cosmos has fascinated human beings for thousands of years. For the first time in human history we have technological capabilities that put us on the verge of answering a hierarchy of pressing questions: "Do other Earths exist?," "Are they common?," and "Do they have signs of life?" The study of exoplanets seeks to understand how planetary systems formed and evolved, and to understand the diversity of planetary system architectures.

The concept that there may be worlds beyond Earth goes back more than 2000 years, with Epicurus (ca. 300 BCE) asserting, "There are infinite worlds both like and unlike this world of ours . . . We must believe that in all worlds there are living creatures and plants and other things we see in this world." The thirteenth-century scholar and philosopher Albertus Magnus stated, "Do there exist many worlds, or is there but a single world? This is one of the most noble and exalted questions in the study of Nature." Italy's Giordano Bruno asserted that "There are countless suns and countless Earths all rotating around their suns in exactly the same way as the seven planets of our system . . . The countless worlds in the universe are no worse and no less inhabited than our Earth" (*Bruno,* 1584). By the time of Newton, entire books were being written on the topic, with Christopher Huygens asserting in his *Cosmotheoros: Or, Conjectures Concerning the Planetary Worlds* (*Huygens,* 1968) that "the Earth may justly liken'd to the planets . . . [which have] gravity . . . and animals not to be imagin'd too unlike ours . . . and even Men . . . [which] chiefly differ from Beasts in the study of Nature . . . [and who] have Astronomy and its subservient Arts: Geometry, Arithmetick [sic], Writing, Opticks [sic]." These thoughts, however modern sounding, remained speculative (C. Beichman, in preparation).

The contemporary search for exoplanets began with astrometry in the mid-nineteenth century, when a dark companion was suspected to orbit the binary star system 70 Ophiuchi (*Jacob,* 1855; *See,* 1896). Although the 70 Ophiuchi "planet" was soon discredited (*Moulton,* 1899), it likely represents the first published claim of a planet beyond the solar system. A century later, reports of massive planets in the 1940s were controversial and after decades of work were finally discarded as spurious signals. In the 1960s, the detection of the now infamous 24-year-period Jupiter-mass planet around Barnard's star was announced (*van de Kamp,* 1963). This and a companion planet turned out to be instrument systematics. The checkered history of exoplanet detection was born [see *Jayawardhana* (2010) and the chapter by Quirrenbach].

A prescient two-page paper by *Struve* (1952) proposed that Jupiter-like planets might exist in orbits as small as 0.02 AU. Struve pointed out that high-precision radial-velocity measurements could discover such short-period planets and transits of such planets could be found from photometric observations. At the time, and for the following decades, almost all astronomers thought that Jupiter-mass planets should copy our solar system and reside in Jupiter-like orbits at 5 AU from the host star. The key in making radial-velocity observations precise enough to search for Jupiter-mass planet companions in Jupiter-like orbits was to change the reference frame from telluric spectral lines in the Earth's atmosphere to an instrument gas cell. *Campbell and Walker* (1979) were the first to make this advancement, enabling an order-of-magnitude increase in radial-velocity precision, and they embarked on a 12-year search for Jupiter-like planets orbiting 21 bright Sun-like stars. In the late 1980s a substellar object orbiting HD 114762 was detected by *Latham et al.* (1989). Latham et al. referred to this object as a brown dwarf, but if it is very close to the minimum mass obtained by radial-velocity measurements of

HD 11742, then it might be just below the upper mass cutoff for planets. For the history and details of the radial-velocity technique, see the chapter by Lovis and Fischer.

In addition to the radial-velocity planet search method, a technique very different from astrometry and radial velocity was emerging from radio observations of pulsar timing to take advantage of the ultraprecise radiation beaming from millisecond pulsars. Pulsars are neutron stars, remnants of massive stars with ≥8 $M_\odot$. Pulsar timing aims to detect changes as the host star orbits the planet-star common center of mass. An early, tentative claim of planetary companions to pulsars came in 1970 (*Hills, 1970*). The well-publicized announcement of a pulsar planet by *Bailes et al.* (1991) was retracted after the changes in pulsar timing were ascribed to Earth's own motion about the Sun (*Lyne and Bailes, 1992*). Immediately following the retraction of this claim at a scientific conference, the first two bona fide exoplanets were announced to orbit PSR 1257+12 (*Wolszczan and Frail, 1992*). Dubbed "dead worlds" because of the deadly radiation from the host pulsar, pulsar planets are often ignored in favor of planets orbiting main-sequence stars. The chapter by Wolszczan and Kuchner describes the past, present, and future of pulsar planet searches.

Returning to main-sequence stars, *Walker et al.* (1995) published results of their 12-year search for Jupiter-mass companions to nearby stars. In addition to the conclusion of their own study — that none of the 21 Sun-like stars showed reflex motion corresponding to planets with masses less than 1–3 M_{Jup} (Jupiter masses) for orbital periods less than 15 years — they used other evidence to state that no Jupiter-mass planets in short-period circular orbits had been detected around about 45 Sun-like stars.

Later the same year, an exciting, incredible announcement was made of a 0.5 M_{Jup} planet in a 4.2-day period planet orbiting the Sun-like star 51 Peg (*Mayor and Queloz, 1995*). This first planet to orbit a Sun-like star — at about seven times closer to its star than Mercury is to the Sun — shattered the paradigm of our solar system as the model for planetary architecture. 51 Peg and about 100 other so-called hot Jupiters have changed the foundational concepts of planet formation and evolution. Giant planets were not expected to be found so excruciatingly close to the host star, and must have migrated inward after formation. This 51-Peg discovery is perhaps the most significant discovery in exoplanets because it marked the success of the method used to discover hundreds of exoplanets to date. Had jovian-mass planets orbiting within a few AU of their stars been rare, the field of exoplanets would have gone nowhere until surveys had operated long enough uncover the population of true Jupiters in 12-year-period orbits. The existence of short-period exoplanets allowed for a burst of exoplanet discoveries (following within a couple of years), forming the foundation for the now explosive field of exoplanets.

At the present time, five different exoplanet detection techniques have been used to discover exoplanets (Fig. 1). One of the most surprising aspects of the hundreds of known exoplanets is their broad diversity: a seemingly continuous range of masses and orbital parameters. Planet formation gives birth to planets of a wide range of masses in a wide variety of locations in a protoplanetary disk. Planetary migration allows planets to end up very close to the parent star.

It is fair to say that the blank parts of the mass vs. semimajor axis diagram (Fig. 1) are unpopulated because no exoplanet discovery technique can yet reach low-mass planets with modest to large planet-star separations. In particular, any technique has difficulty discovering Earth analogs.

Ultimately, in the future of exoplanets, we would like an image of an Earth twin as beautiful as the Apollo images of Earth (Fig. 2). For our generation, we are instead limited to imaging exoplanets as spatially unresolved, i.e., as point sources. The Voyager 1 spacecraft viewed Earth in such a way from a distance of more than four billion miles (Fig. 2). Earth's features are hidden in a pale blue dot's tiny speck of light. Even to obtain an image of a pale blue dot, the light from the host star, ~10^{10} times as bright as the planet, must be blocked out. This is one of the biggest challenges facing the current generation of astronomers.

This book, *Exoplanets*, aims to introduce principal aspects of exoplanet observation and theory at the graduate student level. The book is organized into five sections. Part I is the introduction, and includes this chapter and a chapter on

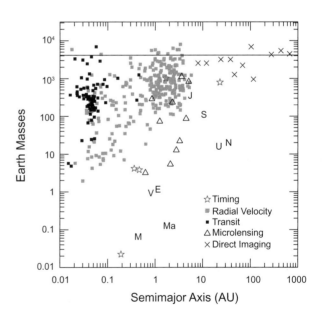

Fig. 1. Known exoplanets as of July 2010. Exoplanets surprisingly are found at a nearly continuous range of masses and semimajor axes. Many different techniques are successful at discovering exoplanets, as indicated by the different symbols. The solar system planets are denoted by the first one or two letters of their name. The horizontal line is the conventional upper limit to a planet mass, 13 M_{Jup} (Jupiter masses). The sloped, lower boundary to the collection of gray squares is due to a selection effect in the radial velocity technique. Data taken from *http://exoplanet.eu/*.

the fundamentals of Keplerian orbits and dynamics. Part II contains exoplanet observing techniques and findings to date, with one chapter per topic and a concluding chapter on the current statistical distribution of exoplanets. The two chapters in Part III discuss orbital dynamics in more detail. Part IV is about exoplanet formation and migration. Part V is about planet interiors and atmospheres. The goal for each chapter is to start with a conceptual introduction followed by a detailed explanation of the foundational equations and methodology. The chapters continue with current research highlights, and conclude with future prospects. The goal of this chapter is to give an introductory overview of the topics covered in *Exoplanets*.

2. WHAT IS A PLANET?

Requirements for membership in the "planet club" received considerable attention in the press when the International Astronomical Union (IAU) "demoted" Pluto during the summer of 2006. But the IAU's still-controversial 2006 definition only covers the four giant planets, the four terrestrial planets, and smaller objects in orbit about our Sun. What about the hundreds of (and rapidly increasing in number) extrasolar "planets" that are the subject of this book?

As larger and more massive objects tend to be easier to detect than smaller ones, the first exoplanets to be found were large, and in many cases more massive than Jupiter. In our solar system, there is a gap of over a factor of 10^3 in mass and more than 10^8 in luminosity between the Sun and its most massive planet. Just as the discoveries of numerous additional small bodies orbiting the Sun have forced astronomers to decide how small an object can be and still be worthy of being classified as a planet, detections of substellar objects orbiting other stars have raised the question of an upper size limit to planethood.

Stars can be defined as objects for which self-sustaining fusion provides sufficient energy for thermal pressure to balance gravity. For solar composition, this yields a lower-mass limit of ≥ 0.075 $M_\odot \approx 80$ M_{Jup}. The smallest-known stars are a bit less than one-tenth as massive as our Sun. One could simply apply the word "planet" to all sizable substellar objects, but the term "brown dwarf" is generally used to denote bodies that are very similar to stars yet too small to reach internal temperatures necessary for thermonuclear fusion of significant quantities of ordinary hydrogen.

Various delineations of the upper boundary of planethood incorporate formation and/or location requirements on the object in question in addition to limits on its physical properties. A few years before the Pluto planethood debate heated up, the IAU's Working Group on Extrasolar Planets decided upon a provisional definition of the attributes required for an object outside of our solar system to be considered to be a planet. This working definition reads as follows:

1. Objects with true masses below the limiting mass for thermonuclear fusion of deuterium (currently calculated to be 13 M_{Jup} for objects of solar metallicity) that orbit stars or stellar remnants are "planets" (no matter how they formed). The minimum mass/size required for an extrasolar object to be considered a planet should be the same as that used in our solar system.

2. Substellar objects with true masses above the limiting mass for thermonuclear fusion of deuterium are "brown dwarfs," no matter how they formed nor where they are located.

3. Free-floating objects in young star clusters with masses below the limiting mass for thermonuclear fusion of deuterium are not "planets," but are "subbrown dwarfs" (or whatever name is most appropriate).

Various hesitant qualifying statements surround the above listing (see *http://www.dtm.ciw.edu/boss/IAU/div3/wgesp/* for details), lest anyone think that the definition wasn't drawn up by a committee.

A few comments are in order: This definition incorporates physical properties and location, but not formation; the theorists on the panel generally objected to the inclusion of mode of formation because of modeling uncertainties, whereas some of the observers argued for including formation-based criteria. 1, 2: Deuterium, which contains one proton and one neutron, is almost five orders of magnitude less abundant than ordinary hydrogen (one proton, no neutrons). Deuterium fuses at a substantially lower temperature than does ordinary hydrogen, but even the fusion of the entire inventory of an object's deuterium does not provide a very large amount of energy because of deuterium's low abundance. There is no fine dividing line dividing no-deuterium fusion from all-deuterium fusion, but the mass at which 50% of deuterium eventually fuses is now estimated at 12 M_{Jup}, down from 13 M_{Jup} estimated several years ago when the IAU exoplanet definition was written. 1: The comment regarding the lower-mass limit shows that the panel was aware of the brewing Pluto controversy, but didn't want to take a position on this matter. Note that the inner "planet" of PSR B1257+12 is less massive than any of the eight planets in our solar system, but more massive than Pluto, and it also has (at least mostly) cleared its orbital zone of debris.

Fig. 2. Earth as viewed from space. **(a)** Image from NASA's Apollo 17 spacecraft in 1972. **(b)** Image from Voyager 1 at a distance of more than four billion miles. Earth lies in the center of a band caused by scattered sunlight in the camera optics.

3. PLANETARY DYNAMICS

By analyzing Tycho Brahe's careful observations of the orbits of the planets, Johannes Kepler deduced the following three laws of planetary motion:

1. All planets move along elliptical paths with the Sun at one focus.
2. A line connecting a planet and the Sun sweeps out equal areas ΔA in equal periods of time Δt

$$\frac{\Delta A}{\Delta t} = \text{constant} \quad (1)$$

Note that the value of this constant differs from one planet to the next.

3. The square of a planet's orbital period about the Sun (in years) is equal to the cube of its semimajor axis.

Although Kepler's laws were originally found from careful observation of planetary motion, they were subsequently shown to be derivable from Newton's laws of motion together with his universal law of gravity.

Newton's analysis corrected some small errors in Kepler's laws, and generalized them to apply to objects not orbiting the Sun. Two mutually gravitating bodies of masses m_1 and m_2 travel on elliptical paths about their mutual center of mass, with an orbital period T given by

$$T^2 = \frac{4\pi^2 a^3}{G(m_1 + m_2)} \quad (2)$$

where a refers to the semimajor axis of relative orbit of the two bodies and G is the universal gravitational constant. The majority of known exoplanets have been found by tracking the small variations in their star's velocity as they moves in response to the planet's gravitational influence (see chapter by Lovis and Fischer).

Newton's laws imply that all massive bodies have influence on one another, so that in systems with more than two bodies, orbits are not perfect ellipses. The planet Neptune was discovered by studying deviations of the path of Uranus from an ellipse that could not be accounted for by the perturbations of the then-known planets. Analogous deviations have been used to precisely specify the masses and orbital inclinations of the two largest planets known to orbit the pulsar PSR B1257+12 (see chapter by Wolszczan and Kuchner), as well as the two resonant giant planets orbiting the small main-sequence star GJ 876 (*Correia et al.*, 2010).

Newton demonstrated that the gravitational force exerted by a spherically symmetric body is equivalent to that of a point particle of the same mass. However, rotation and other processes can cause deviations from spherical symmetry, and these asymmetries affect orbits, especially for bodies whose separations are not substantially larger than their sizes. General relativity implies differences from elliptical trajectories that are most profound for orbits close to massive bodies.

Astrophysical bodies are neither perfectly rigid nor perfectly fluid. Tidal forces can deform a body, and since the strength of the tidal force increases with proximity, so does the amount of deformation. These variations produce flexing in a body traveling on an eccentric orbit, which can dissipate energy as heat within the body at the expense of damping the orbital eccentricity. Moreover, a tidally deformed body exerts a different gravitational force as a result of its nonspherical distribution of mass. For example, the Moon raises tides on Earth, producing a tidal bulge. This bulge points almost along the Earth-Moon line, but as Earth cannot respond instantaneously to the pull of the Moon, Earth's rotation carries the bulge slightly ahead of the Earth-Moon line. The resulting skewed mass distribution within Earth exerts a torque on the Moon, causing its orbit to expand (at the rate of a few centimeters per year) and Earth's rotation to slow.

The basic dynamics of the two-body problem are described in the chapter by Murray and Correia. The chapter by Fabrycky describes planetary perturbations, the principal effects of stellar oblateness, general relativity, and tidal forces as they apply to exoplanet research. The chapter by Correia goes into more detail on tidal evolution.

4. OBSERVATIONAL TECHNIQUES

Exoplanets are being discovered with several different detection techniques (Fig. 1). Each technique has its own selection effects in mass (or radius) and semimajor axis parameter space. Here we introduce and compare the different planet-finding techniques.

Radial velocity: In the presence of a planet, a star orbits the planet-star common center of mass. The star's motion can be described by three components: one along the observer's line-of-sight to the star, and two components on the plane of the sky. The radial-velocity technique measures the star's line-of-sight motion with a selection effect toward more-massive planets close to the star. Radial velocity is the most mature of the planet-finding techniques, giving it the advantage of being able to push to relatively low planetary masses around bright stars. The major disadvantage of the radial-velocity technique is that only an exoplanet's minimum mass (given by $M_p \sin i$, where M_p is the planet mass and i is the orbital inclination) is measured because the orbital inclination of the planet orbit is undetermined. The radial-velocity technique is described in the chapter by Lovis and Fischer.

Astrometry: Astrometry measures the position of stars by the two components of motion in the plane of the sky. The main advantage of astrometric measurements is that they can provide the planetary mass (without the sin i factor inherent in radial-velocity searches) and all planetary orbital parameters. Although the astrometry technique has a long history, with the first claimed exoplanet detections in the mid-nineteenth century, astrometry has yet to discover an exoplanet (as of 2010). For details, see the chapter by Quirrenbach.

Timing: Timing techniques for exoplanet detection include both pulsar timing (*Wolszczan and Frail*, 1992) and time perturbations of stars with stable oscillation periods (e.g., *Silvotti et al.*, 2007) (see the chapter by Wolszczan and

Kuchner for details). Because the planet detection techniques radial velocity, astrometry, and timing infer the presence of the planet from the star's orbital motion, only the planet's mass and orbital elements can be derived.

Gravitational microlensing: Gravitational microlensing occurs when a foreground star happens to pass very close to the observer's line of sight to a more distant background star. The foreground star acts as a lens, magnifying the background source star, as a function of time, by an amount that depends on the angular separation between the lens star and source star. If a planet is orbiting the lens star, the lightcurve may be further perturbed, resulting in a characteristic, short-lived signature of an exoplanet and yielding the planet mass and planet-star physical separation. The major weakness of the microlensing planet-finding technique is that the lensing occurence is a one-time event that will never repeat. The advantages of the current state of microlensing over other planet-finding techniques (except for pulsar timing) is the sensitivity to low planetary masses (potentially to Earth's mass) at relatively large planet-star separation, aiding in understanding the statistical distribution of exoplanets (see Fig. 1). The theoretical and observational foundation for microlensing is laid out in the chapter by Gaudi.

We now turn to the two planet-finding techniques that can also be used to observe physical characteristics beyond planetary mass.

Transits: Exoplanet transits (or primary eclipses) occur when a planet passes in front of its star as seen from Earth. When the planet goes behind the star, the event is called an occultation or secondary eclipse. (The term "transit" is used for small bodies going in front of larger bodies, "occultation" for larger bodies going in front of smaller bodies, and "eclipse" for bodies of about the same size moving in front of each other.) Most transiting planets have been discovered from transit surveys searching large numbers of stars for the characteristic drop in brightness that is indicative of a planet transit. Transiting exoplanets are also detected via follow-up photometric observations to known radial-velocity planets and these tend to be orbiting bright stars, favorable for additional observations. Transits are described in the chapter by Winn.

For an exoplanet to show transits, the exoplanet-star orbit must be aligned with the observer. For a random orientation of stellar inclinations (for zero eccentricity and $R_p \ll R_\star$), the probability, p, to transit is

$$p = R_\star/a \qquad (3)$$

where R_p is the planet radius, $R_\star$ is the stellar radius, and a is the planet's semimajor axis. So, for example, while transits for planets in 1 AU circular orbits are rare (1/215), transits of hot planets in short-period orbits are more common. Transits of short-period planets also occur at shorter time intervals, and thus limited observing programs are more likely to detect them. The major limitation for exoplanet transit searches is, therefore, the huge bias toward short-period planets (Fig. 1).

Transits are not just used for discovering planets. Follow-up observations of transits and eclipses provide a huge amount of information that cannot be obtained from radial-velocity data alone. The size of the planet relative to the size of the star can be measured from the transit lightcurve. The orientation of the planet's orbit relative to the sky plane and relative to the stellar rotation axis can be determined. Transit time anomalies or perturbations in other eclipse properties may indicate the presence of additional planets or moons. A transiting exoplanet's atmosphere can be measured in several ways in the combined light of the planet-star system: during transit, during secondary eclipse, and even around the orbit phase curves. The richness of investigations for transiting exoplanets is described in the chapter by Winn and in the chapters in Part V.

Direct imaging: Direct imaging means taking a snapshot of an exoplanet by spatially separating the planet and star on the sky. More generally, direct detection refers to the ability to distinguish the light emergent from a particular celestial object from that of any other. Stars by virtue of nuclear fusion in their core are much brighter than planets, making the main challenge of direct imaging not the actual spatial separation, but rather the elimination of the scattered or diffracted light from the central star in the telescope optics.

Direct imaging of planets is currently limited to big, bright, young, or massive substellar objects and/or objects located far from their stars (e.g., *Marois et al.*, 2008; *Kalas et al.*, 2008). Four or five or more orders of magnitude improvement in planet-star contrast is needed to reach solar-system-like, solar-system-aged exoplanets. This is the weakness of the direct imaging exoplanet discovery technique — the technical challenges and the complexity of overcoming the large planet-star contrast, the physics of diffracted light, and the engineering needed to mitigate scattered light in the telescope optics.

Direct imaging is advantageous over other exoplanet discovery techniques because of the science it enables. First and foremost, if a direct image (i.e., photometry) can be taken of a planet, if adequate photons are available, so can a spectrum. In addition to exoplanet spectra, the orbit of the planet may be measured, more accurately as combined with astrometry or even radial-velocity measurements. Any circumstellar disks present in a planetary system can also be imaged (see the chapter by Roberge and Kamp). One of the exciting aspects of direct imaging is the fast-paced development of many new coronagraph concepts that have specifically been developed for exoplanet observations, and cannot be found in any optics textbooks. See the chapter by Traub and Oppenheimer for a thorough foundation of the physics and future of direct imaging.

5. PLANET FORMATION

Studies of planetary formation are intimately connected with those of exoplanets. Prevailing views on planet formation have been influential in directing exoplanet studies, although the directions provided have not always been optimal. More than 98% of the planets now known orbit stars other than our Sun, so exoplanet data provide constraints on models of planetary growth. While for exoplanets, we have the numbers, we have much more data on the planets and smaller objects in our solar system, and at present these "local" data remain the primary drivers for planet-formation models. Note that the set of known exoplanets is highly biased based on detectability, whereas our own solar system is not a truly random sample, since we wouldn't reside here if it didn't contain a planet suitable for life to evolve to the point of asking these questions.

Modern theories of star and planet formation are based upon observations of planets and smaller bodies within our own solar system, exoplanets, and young stars and their environments. Terrestrial planets are thought to grow via pairwise accretion of initially small solid bodies known as planetesimals, until the spacing of planetary orbits becomes large enough that the configuration is stable for the age of the system. According to most models, giant planets begin their growth as do terrestrial planets, but they become massive enough that they are able to accumulate substantial amounts of gas before the protoplanetary disk dissipates. These models predict that rocky planets should form in orbit about most single stars. It is uncertain whether or not gas giant planet formation is common, because most protoplanetary disks may dissipate before solid planetary cores can grow large enough to gravitationally trap substantial quantities of gas.

A potential hazard to planetary systems is radial decay of planetary orbits resulting from interactions with material within the disk. Protoplanetary disks are built of the same material as are stars, and thus are initially ~99% hydrogen and helium. In the first few million years of a planetary system's life, some of this gas is still present and provides a large sink/source of angular momentum for (proto)planets. From theoretical consideration, the gravitational interaction between a planet and a massive disk in which it is embedded — Type I migration — ought under most circumstances to bring about a net loss of orbital angular momentum from the planet and decay of its orbit for planets on the order of a few Earth masses. A sufficiently massive planet exerts strong enough torques on the disk to open a gap, thereby locking the planet into the subsequent viscous evolution of the disk in what is called the Type II mode of migration.

Planets more massive than Earth have the potential to suffer the most rapid orbital decay, and may be able to sweep up smaller planets in their path. Planet formation may be an enormously wasteful process, which dumps a steady stream of growing protoplanets onto the primary, and the end result is whatever happens to be left over when the gas fades away.

Significant postformation migration is quite likely to be responsible for the large number of exoplanets detected on close-in orbits. In multiple-planet systems, convergent migration of planets can be invoked to explain resonant capture, and either convergent or divergent migration can lead to eccentricity excitation. As the nebular gas dissipates, it is likely that the tables are eventually turned; the planets, heretofore at the mercy of the gas, assert themselves and serve as anchors to slow down the viscous evolution of the last remains of the disk. Interactions between planets and the planetesimal disk can also lead to significant planetary migration; such migration can account for the orbital distribution of the small bodies within our solar system beyond Neptune. The present observational and theoretical "state of the art" requires us to use a liberal amount of conjecture in attempting to sketch a coherent picture of planet migration. However, it seems equally clear that migration is intimately linked with the formation of the planetary system, and a complete picture of the latter will require a full understanding of the former.

The arrangement of material within the section on circumstellar disks and planet formation is pedagogical rather than chronological. The first chapter, by Roberge and Kamp, covers disks, both of the protoplanetary variety and postformation debris disk. Chambers then reviews terrestrial planet formation, concentrating on data from our solar system and numerical models that attempt to reproduce these observational constraints. As observations of terrestrial exoplanets start to provide significant information on typical planetary system properties, they may well affect the direction of research in this area. Although the formation and most of the migration of giant planets are thought to predate the latter phases of terrestrial planet growth, the physics of the early stages of core-nucleated accretion of gas giants is the same as that of terrestrial planet formation, making the presentation of terrestrial planet formation first more conceptually straightforward. D'Angelo et al. review growth of giant planets, introducing both the prevailing core-nucleated accretion model and the alternative disk instability model. Part IV concludes with a fairly technical chapter by Lubow and Ida reviewing current models of planetary migration.

6. PLANET INTERIORS AND ATMOSPHERES

6.1. Interiors

The diversity of planet interior compositions is large, but can be summarized in a simplified fashion via their bulk composition as a function of rock, ice, or gas components (see Fig. 3). In the solar system, there is a definite relationship between the relative abundances of rock-ice-gas and planet mass. Small planets (M ≤ 1 $M_\oplus$) are rocky. Intermediate-sized planets (~14–18 $M_\oplus$) are thought to be dominated in mass by astrophysical ices and rock with H/He envelopes that are minor components by mass but dominant by volume. Larger planets are predominantly composed of hydrogen and helium. Whether or not exoplanets also follow this pattern is one of the most significant questions of exoplanet formation, migration, and evolution.

Measured exoplanet masses and radii (and hence densities) can be used to estimate the composition of an exoplanet. The mass vs. radius diagram, populated with exoplanets and illustrative theoretical models, provides such a basic picture of exoplanet compositions (Fig. 4). Most prominent are the collection of giant planets in the top right corner of the diagram. These planets are so large for their mass (i.e., have such low densities) that they must be composed mostly of hydrogen and helium gas. The collection of giant planets is interesting because of their spread of masses and radii well beyond Jupiter's. Some Jupiter-mass planets are over

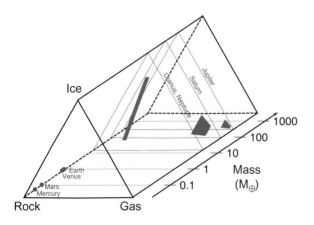

Fig. 3. Schematic diagram illustrating the range of possible planet primordial bulk compositions for exoplanets. In this figure "gas" refers to primordial H and He accreted from the nebula, "ice" refers to ice-forming materials, and "rock" refers to refractory materials. Constraints on the current compositions of the solar system planets are shown and denoted by their first initial. Exoplanets might appear anywhere in this diagram. Adapted from the chapter by Chambers and from *Rogers and Seager* (2010).

1.5 times Jupiter's radius, defying explanation of the large size (see the chapter by Fortney et al. for potential explanations). Some giant planets are up to 10 times more massive than Jupiter but with almost the same size. Such planets are explained as having electron degeneracy pressure in the core, whereby pressure-ionized atoms enable the nuclei to squeeze closer together. This means that a more-massive planet has higher compression in the core and the resulting planet size is similar to that of a less-massive planet.

Moving down in planet mass and radius in Fig. 4, to those with masses between Saturn and Neptune, are planets with no solar system counterparts, exemplifying the diversity of exoplanet interior composition. Planets similar in size and mass to Neptune may be composed mostly of ices plus some rock, with a ~10–15% H/He envelope — or they may alternately have a large rocky interior with a more-massive H/He envelope. Of significant interest are the super Earths, loosely defined as rocky planets significantly more massive than Earth. Super Earths may include planets suitable for life as we know it and are amenable for future observational searches for atmospheric biosignature gases. In the conventional sense, a habitable planet is one with some surface liquid water, because all life on Earth requires liquid water. In contrast to terrestrial planets, giant and Neptune-sized planets enshrouded by gas envelopes have no solid or liquid surfaces to support life as we know it and their temperatures just below the deep atmosphere rapidly become too hot for the complex molecules necessary for life to exist.

To understand planet interiors in more detail, or to use the measured mass and radius to constrain the planet interior composition, models are used. The equations that describe a

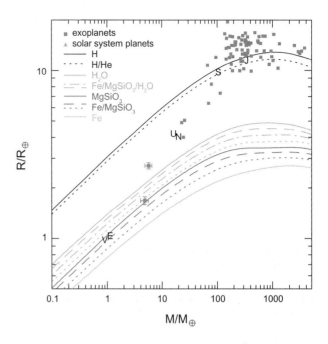

Fig. 4. Mass-radius relationships for cold planets. The solid lines are homogeneous planets. From top to bottom the homogeneous planets are hydrogen, a hydrogen-helium mixture with 25% helium by mass (dotted line), water ice, silicate ($MgSiO_3$ perovskite), and iron (Fe ε). The nonsolid lines are differentiated planets, and are described from top to bottom. The top dashed line is for water planets with 75% water ice, a 22% silicate shell, and a 3% iron core; the dot-dashed line is for water planets with 45% water ice, a 48.5% silicate shell, and a 6.5% iron core (similar to Ganymede); and the dotted line is for water planets with 25% water ice, a 52.5% silicate shell, and a 22.5% iron core. The next lower dashed line is for silicate planets with 32.5% by mass iron cores and 67.5% silicate mantles (similar to Earth), and the adjacent dotted line is for silicate planets with 70% by mass iron core and 30% silicate mantles (similar to Mercury). Solar system planets are shown and denoted by their first intial. The squares designate the transiting exoplanets; mass and radius uncertainties are suppressed for clarity except for the low-mass planets CoRoT 7b and GJ 1214b. Note that electron degeneracy pressure becomes important at high mass, causing the planet radius to become constant and even decrease for increasing mass. Following *Seager et al.* (2007).

planetary interior are conservation of mass; hydrostatic equilibrium (a balance between gravity and pressure gradients); the equation of state (describing the relationship betweeen density, pressure, and temperature for a given material in thermodynamic equilibrium); and energy transport (commonly described by an adiabat). For giant planets, conservation of energy in the form of the change in luminosity as a function of planetary radius is also used. Details of how the equations are used in models, including a description of boundary conditions, are given in the chapters by Fortney et al. and Sotin et al.

Deducing exoplanet interior composition constraints is very difficult. Only two data points are available per planet: mass and radius. Even with perfect measurements there is simply not enough information to uniquely identify a planet's interior composition. Two exceptions are at the density extremes: Giant planets of low density must be composed almost entirely of hydrogen and helium, and any planet of extremely high density must be iron-dominated because iron is the densest cosmically abundant molecular substance.

There are two possible paths for moving forward beyond the limiting degeneracy of interior composition. One path involves observations and interpretation of an exoplanet atmosphere to help break the interior composition degeneracy for a specific exoplanet. Careful work must be done to understand which types of planet interiors can be constrained further with which kind of atmosphere measurements. A second path involves statistics. With enough planets with a measured mass and radius, the hope is that specific planet populations in the mass-radius diagram (Fig. 4) will emerge. With distinct planet populations, characteristics of terrestrial planets in general can be identified, even if the actual composition of individual planets cannot.

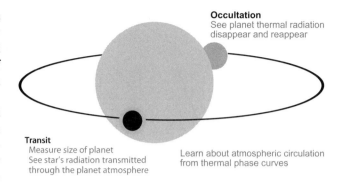

Fig. 5. Schematic of a transiting exoplanet and potential follow-up measurements. Note that the primary eclipse is usually called a transit, and the secondary eclipse is most accurately referred to as an occultation. In this illustration the planet orbit is viewed "edge-on," i.e., sin i = 90°. This figure was adapted from Fig. 5 of *Seager and Deming* (2010).

6.2. Atmospheres

Planetary atmospheres originate either from direct capture of gas from the protoplanetary disk (for massive planets) or from outgassing during planetary accretion (for low-mass planets). The former mechanism is accepted for giant planets like Jupiter and Saturn and the latter is expected for terrestrial planets like Earth and Venus. The final atmospheric mass and composition depends on the net of atmospheric sources vs. atmospheric sinks. The sources, both from direct capture and from outgassing, depend on the planet's location in the protoplanetary disk during formation, due to the compositional gradients in the disk. The atmospheric sinks include atmospheric escape, gas-surface reactions, and sequestering of gases in oceans.

Two ways are available for observing exoplanet atmospheres. Direct imaging is the most natural way to think of observing exoplanet atmospheres (see the chapter by Traub and Oppenheimer). At present, direct imaging is limited to big, bright, young, or massive planets located relatively far from their stars (see Fig. 1).

The second way to observe exoplanet atmospheres is via a collection of techniques used for transiting planets. Transiting planets are observed in the combined light of the planet and star (Fig. 5). The planet and star are not spatially separated as in the direct imaging technique. Instead, the precise on/off nature of the transit and secondary eclipse events provide an intrinsic calibration reference. For more details, see the chapter by Winn.

Over three dozen exoplanet atmospheres have been observed to date, dominated by Spitzer Space Telescope broadband infrared photometric observations of secondary eclipse. A handful of exoplanet atmospheres have been observed by the Hubble Space Telescope and a growing number of hot exoplanets observed with groundbased telescopes. Highlights of exoplanet atmosphere studies include identification of molecules and atoms, constraints on vertical temperature structure (such as the possible identification of thermal inversions), and day-night temperature gradients for the tidally locked hot Jupiters. For a very detailed review of exoplanet atmospheres, see *Seager and Deming* (2010).

The range of possible exoplanet atmospheric mass and composition has yet to be uncovered theoretically and observationally. As such, there is not yet a definitive categorization of atmosphere types (but see *Seager and Deming*, 2010). We do, however, expect atmospheres on different exoplanets to show a wide diversity, just as the orbits, masses, and radii of known exoplanets do. Jupiter (and presumably other massive planets) has retained the gases it formed with, and these gases approximately represent the composition of the Sun. The super Earth atmospheres, on the other hand, could have evolved substantially. In particular, there is an exciting sense of anticipation in observing and studying super Earth atmospheres, simply because we do not know quite what their diversity will be.

In order to interpret, predict, and generally understand exoplanet atmospheres, models are needed. In fact, a variety of different kinds of models are used for different aspects of the atmosphere. To understand the emergent spectrum and to interpret spectral data, radiative transfer codes are used (see the chapters by Burrows and Orton and by Meadows and Seager). But in order to theoretically understand the atmospheric composition for low-mass planets, atmospheric escape and photochemistry models are also needed. Atmospheric circulation describes the hydrodynamic flow in atmospheres and is essential for understanding the three-dimensional temperature structure of tidally locked planets, as well as for getting a handle on surface temperatures for any kind of terrestrial planet (see the chapter by Showman et al.)

Arguably the most exciting investigations that planetary atmosphere observations can enable in the study of exoplanets is the potential to detect biosignatures produced by life. A biosignature gas is one produced by life that accumulates in an exoplanet atmosphere to detectable levels. A surface feature (such as the vegetation red edge) can also be a biosignature. Biosignatures are covered in the final chapter of this volume, by Seager and Meadows.

7. PROSPECTS FOR DETECTION AND CHARACTERIZATION OF HABITABLE PLANETS

We stand on a great divide in the detection and study of exoplanets. On one side of this divide are the hundreds of known massive exoplanets, with measured atmospheric temperatures for a handful of the hottest exoplanets. On the other side of the divide lies the possibility, as yet unrealized, of detecting and characterizing a true Earth analog — an Earth-like planet (a planet of ~1 $M_\oplus$ and 1 $R_\oplus$ orbiting a Sun-like star at a distance of roughly 1 AU).

NASA's Kepler space telescope, currently in an Earth-trailing heliocentric orbit, will determine the frequency of Earth-sized planets (*Borucki et al.*, 2010). Kepler is monitoring 150,000 stars for 3.5 years, seeking the characteristic signature of a transiting Earth and ruling out false positive signatures. Most Kepler stars are too far away and too faint for planet follow-up atmosphere studies, and many of them are out of reach for radial-velocity mass determination.

There is an exciting possibility of a fast track for finding and characterizing habitable exoplanets. This is the search for super Earths orbiting small, cool stars (M dwarfs). M dwarfs are much less luminous than the Sun, so the locations amenable to surface liquid water to support life as we know it are very close to the star, although questions remain regarding the accretion and retention of volatiles by such planets (*Lissauer*, 2007). The relative size and mass of the planet and star are much more favorable for planet detection than the Earth-Sun analog. Two different yet complementary planet-searching techniques (transits and radial velocities) are very sensitive to super Earths orbiting with close separations to M dwarfs. A set of transiting planets will enable average density measurements and hence identification of terrestrial-type planets. Suitable transiting planets can have their atmospheres characterized by the James Webb Space Telescope, now under development, building on the current Spitzer Space Telescope studies of hot Jupiters.

The quest for atmosphere studies for Earth analogs means there will always be a desire for direct imaging to find and characterize exoplanets. Direct imaging at the 10-billion planet-star contrast level needed for Earth analogs will require future spaceborne telescopes.

8. PERSPECTIVES

The field of exoplanet research has blossomed, with hundreds of planets found at very broad ranges of masses and semimajor axes. This points to the diversity of outcomes of planet formation combined with subsequent planet migration. Among the most unexpected discoveries are planets orbiting extremely close to their host stars with periods of only a few days, planets that orbit their star in a direction opposite to the star's rotation (see chapter by Winn), planets with high eccentricities, and (hot) jovian-mass planets with radii significantly larger than that of Jupiter. Regarding exoplanets, we now know that "anything is possible within the laws of physics and chemistry," and we should expect to continue to discover new classes of unpredicted planets.

We do not completely know where the field of exoplanets is heading, and what discoveries will be made. But among the most important results that are emerging, and will continue to emerge, is our sense of perspective: How normal or unusual is our own solar system? How common or how rare are planets analogous to the Earth? Will we find unequivocal biosignatures of life beyond Earth?

At a few special times in history, astronomy has changed the way we see the universe. Hundreds of years ago, humanity believed that Earth was the center of everything — that the known planets and stars all revolved around Earth. In 1453, the Polish astronomer Nicolaus Copernicus advocated the view of the solar system where the Sun was the center, and Earth and the other planets all revolved around it. Gradually, science adopted this "Copernican" theory, but this was only the beginning. Astronomers eventually recognized that our Sun is but one of hundreds of billions of stars in our galaxy, and, in the early twentieth century, that our galaxy is but one of upward of hundreds of billions of galaxies. If and when we find that Earth-like planets are common and see that some of them have signs of life, we will at last complete the Copernican Revolution — a final conceptual move of Earth, and humanity, away from the center of the universe. This is the promise and hope for exoplanets — the detection and characterization of habitable worlds. We are on the verge of, if not in the very midst of, the greatest change in perspective of our place in the universe since the time of Copernicus.

9. APPENDIX: PLANET DEFINITIONS

The word "planet" comes from the Greek word for "wanderer"; planets were originally defined as objects that moved in the night sky with respect to the background of fixed stars. There are no offical definitions for exoplanets or categories of exoplanets. For reference, we nevertheless present commonly accepted definitions here. Some of these definitions are taken from the 2006 *Terrestrial Planet Finder Science and Technology Definition Report* (*Levine et al.*, 2006).

Planet — A planet is an object that is gravitationally bound and supported from gravitational collapse by either electron degeneracy pressure or Coulomb pressure, that is in orbit about a star, and that, during its entire history, never sustains any nuclear fusion reactions in its core (note that in practice the word "never" is too strong). Reliance on theoretical models indicates that such objects are less massive than approximately 13 times the mass of Jupiter (M_{Jup}) for objects with metallicities close to that of the Sun. Between 13 and 75 M_{Jup} objects (known as brown dwarfs) fuse deuterium for a portion of their youth (e.g., *Hubbard et al.*, 2002, and references therein). Objects above 75 M_{Jup} are known as stars. A lower-mass limit to the class of objects called planets has not convincingly been determined.

Solar system planet — The International Astronomical Union has given a specific definition for solar system planets. The exact wording of the official IAU definition is this: "A 'planet' is a celestial body that (a) is in orbit around the Sun, (b) has sufficient mass for its self-gravity to overcome rigid body forces so that it assumes a hydrostatic equilibrium (nearly round) shape, and (c) has cleared the neighbourhood around its orbit." Based on this definition the solar system has eight planets: Mercury, Venus, Earth, Mars, Jupiter, Saturn, Uranus, and Neptune. The IAU also has defined "dwarf planet," bodies such as Pluto that are round but may orbit in a zone that contains many other objects (such as the asteroid or Kuiper belts).

Exoplanet — An exoplanet is a planet orbiting a star other than the Sun. For the working definition of the IAU's Working Group on Extrasolar Planets, see section 2. Note that free-floating planets are not included in the working definition of exoplanets.

Giant planet — Giant planets are those with substantial H/He envelopes. Our solar system retains two types of giant planets ("gas" giants and "ice" giants); other types are likely to exist in exoplanetary systems. Such planets have no solid or liquid surfaces as terrestrial planets do, but their sizes are usually defined at a pressure of roughly 1 bar.

Terrestrial planet — A terrestrial planet, often referred to as a "rocky planet," is a planet that is primarily supported from gravitational collapse through Coulomb pressure, and that has a surface defined by the radial extent of the liquid or solid interior. A gaseous atmosphere may exist above the surface, but this is not a defining feature of a terrestrial planet. Theory suggests that most terrestrial planets will have masses less than about 5–10 $M_\oplus$, as planets larger than this are thought to be likely to capture gas during accretion and develop into giant planets.

Habitable planet — A habitable planet is a terrestrial planet that has surface liquid water. This definition presumes that extraterrestrial life, like Earth life, requires liquid water for its existence. Both the liquid water, and any life that depends on it, must be at the planet's surface in order to be detected remotely. This, in turn, requires the existence of an atmosphere with a surface pressure and temperature suitable for liquid water. (Moons of exoplanets might also be habitable.)

Potentially habitable planet — A potentially habitable planet is one whose orbit lies within the habitable zone, broadly construed, and has a solid or liquid surface. This includes planets that have high eccentricities, but whose semimajor axis is within the habitable zone. A definition for a potentially habitable implies that some planets within the habitable zone may not actually be habitable.

Earth-like Planet — An Earth-like planet or Earth analog is a habitable planet of approximately 1 $M_\oplus$ and 1 $R_\oplus$ in an Earth-like orbit about a Sun-like star. Earth-like planets are not necessarily habitable planets, nor vice-versa; it depends on the context of usage. While an Earth-like planet is used to describe a planet similar in mass, radius, and temperature to Earth, the term Earth twin is usually reserved for an Earth-like planet with liquid water oceans and continental land masses.

Habitable zone and continuously habitable zone — The habitable zone, or HZ, is the region around a star in which a planet may maintain liquid water on its surface, i.e., it is the zone in which Earth-like planets may exist. Its boundaries may be defined empirically (based on the observation that Venus appears to have lost its water some time ago and that Mars appears to have had surface water early in its history) or with models. The continuously habitable zone, or CHZ, is the region that remains habitable over some finite period of time (usually commencing when the star was young) as a star ages. All main-sequence stars brighten with time, and so the HZ moves outward with time.

Exoplanet subcategories — A growing list of terms are being used for exoplanets with no solar system counterparts. Although the following terms may not have long-term stability in the exoplanet community, they are included here for completeness. The term "hot Jupiter" refers to a planet with mass comparable to or greater than Jupiter (but less than the deuterium-burning limit of ~13 M_{Jup}) that is located close to its primary star (e.g., within 0.1 AU). "Super Earths" is a term that has been used to refer to primarily rock worlds (by volume, not just mass) that are significantly larger than our Earth, but is applied by some to other exoplanets with masses <10 $M_\oplus$ and/or radii ≤1.75 $R_\oplus$. Exo-Neptunes might be used for planets between about 10 and 25 $M_\oplus$ whose volume is dominated by a H/He envelope, but mass is dominated by heavier elements. The term "water worlds" has been used for planets that lack large gaseous envelopes and have ~25% or more water by mass. Carbon planets refer to planets that contain more carbon than oxygen, so that rocks are likely to be carbides rather than silicates. At present, observations cannot always distinguish among the above planet types.

Acknowledgments. We thank R. Binzel for very useful comments on this chapter.

REFERENCES

Bailes M., Lyne A. G., and Shemar S. L. (1991) A planet orbiting the neutron star PSR 1829-10. *Nature, 352,* 311–313.

Borucki W., Koch D., Basri G., Batalha N., et al. (2010) Kepler planet-detection mission: Introduction and first results. *Science, 327,* 977–980.

Bruno G. (1584) *De L'infinito Universo E Mondi.*

Campbell B. and Walker G. A. H. (1979) Precision radial velocities with an absorption cell. *Publ. Astron. Soc. Pac., 91,* 540–545.

Correia A. C. M., Couetdic J., Laskar J., Bonfils X., Mayor M., Bertaux J.-L., Bouchy F., Delfosse X., Forveille T., Lovis C., Pepe F., Perrier C., Queloz D., and Udry S. (2010) The HARPS search for southern extra-solar planets. XIX. Characterization and dynamics of the GJ 876 planetary system. *Astron. Astrophys., 511,* 21.

Hills J. G. (1970) Planetary companions of pulsars. *Nature, 226,* 730–731.

Hubbard W. B., Burrows A., and Lunine J. I. (2002) Theory of giant planets. *Annu. Rev. Astron. Astrophys., 40,* 103–136.

Huygens C. (1968) *Cosmotheoros: Or, Conjectures Concerning the Planetary Worlds*. London (reprint; original publication 1698).

Jacob W. S. (1855) On certain anomalies presented by the binary star 70 Ophiuchi. *Mon. Not. R. Astron. Soc., 15,* 228–230.

Jayawardhana R. (2010) *Worlds Beyond: Hot Jupiters, Super-Earths and the Quest for Alien Life*. Princeton Univ., Princeton.

Kalas P., Graham J. R., Chiang E., Fitzgerald M. P., Clampin M., Kite E. S., Stapelfeldt K., Marois C., and Krist J. (2008) Optical images of an exosolar planet 25 light-years from Earth. *Science, 322,* 1345–1348.

Latham D. W., Stefanik R. P., Mazeh T., and Torres G. (1989) Spectroscopic searches for low-mass companions of stars. *Nature, 339,* 38–40.

Levine M., Shaklan S., and Kasting J. (2006) *Terrestrial Planet Finder Science and Technology Definition Team (STDT) Report*. NASA/JPL Document D-34923, Jet Propulsion Laboratory, Pasadena, California.

Lissauer J. J. (2007) Planets formed in habitable zones of M dwarf stars probably are deficient in volatiles. *Astrophys. J. Lett., 660,* L149–L152.

Lyne A. G. and Bailes M. (1992) No planet orbiting PSR 1829-10. *Nature, 355,* 213.

Marois C., Macintosh B., Barman T., Zuckerman B., Song I., Patience J., Lafrenire D., and Doyon R. (2008) Direct imaging of multiple planets orbiting the star HR 8799. *Science, 322,* 1348–1352.

Mayor M. and Queloz D. (1995) A Jupiter-mass companion to a solar-type star. *Nature, 378,* 355–359.

Moulton F. R. (1899) The limits of temporary stability of satellite motion, with an application to the question of the existence of an unseen body in the binary system 70 Ophiuchi. *Astron. J., 20,* 33–37.

Rogers L. A. and Seager S. (2010) Three possible origins for the gas layer on GJ 1214b. *Astrophys. J., 716,* 1208–1216.

Seager S. and Deming D. (2010) Exoplanet atmospheres. *Annu. Rev. Astron. Astrophys., 48,* 631–672.

Seager S., Kuchner M., Hier-Majumder C. A., and Militzer B. (2007) Mass-radius relationships for solid exoplanets. *Astrophys. J., 669,* 1279–1297.

See T. J. J. (1896) Researches on the orbit of F.70 Ophiuchi, and on a periodic perturbation in the motion of the system arising from the action of an unseen body. *Astron. J., 16,* 17–23.

Silvotti R., Schuh S., Janulis R., Solheim J.-E., Bernabei S., et al. (2007) A giant planet orbiting the "extreme horizontal branch" star V391 Pegasi. *Nature, 449,* 189–191.

Struve O. (1952) Proposal for a project of high-precision stellar radial velocity work. *The Observatory, 72,* 199–200.

van de Kamp P. (1963) Astrometric study of Barnard's star from plates taken with the 24-inch Sproul refractor. *Astron. J., 68,* 515–521.

Walker G. A. H., Walker A. R., Irwin A. W., Larson A. M., Yang S. L. S., Richardson D. C. (1995) A search for Jupiter-mass companions to nearby stars. *Icarus, 116,* 359–375.

Wolszczan A. and Frail D. A. (1992) A planetary system around the millisecond pulsar PSR 1257+12. *Nature, 355,* 145–147.

Keplerian Orbits and Dynamics of Exoplanets

Carl D. Murray
Queen Mary University of London

Alexandre C. M. Correia
University of Aveiro

Understanding the consequences of the gravitational interaction between a star and a planet is fundamental to the study of exoplanets. The solution of the two-body problem shows that the planet moves in an elliptical path around the star and that each body moves in an ellipse about the common center of mass. The basic properties of such a system are derived from first principles and described in the context of detecting exoplanets.

1. INTRODUCTION

The motion of a planet around a star can be understood in the context of the two-body problem, where two bodies exert a mutual gravitational effect on each other. The solution to the problem was first presented by *Newton* (1687) in his *Principia*. He was able to show that the observed elliptical path of a planet and the empirical laws of planetary motion derived by *Kepler* (1609, 1619) were a natural consequence of an inverse square law of force acting between a planet and the Sun. According to Newton's universal law of gravitation, the magnitude of the force between any two masses m_1 and m_2 separated by a distance r is given by

$$F = G \frac{m_1 m_2}{r^2} \quad (1)$$

where $G = 6.67260 \times 10^{-11}$ N m² kg⁻² is the universal gravitational constant. The law is applicable in a wide variety of circumstances. For example, the two bodies could be a moon orbiting a planet or a planet orbiting a star. Newton's achievement was to show that motion in an ellipse is the natural consequence of such a law. A more difficult task is to find the position and velocity of an object in the two-body problem; this is commonly referred to as the Kepler problem. In this chapter we derive the basic equations of the two-body problem and solve them to show how elliptical motion arises. We then proceed to solve the Kepler problem showing how motion around the common center of mass of the two-body system can be used to infer the presence of planetary companions to a star. Finally, we give a few representative examples among exoplanets already detected. For the most part we follow the approach of *Murray and Dermott* (1999).

2. BASIC EQUATIONS

Consider a star and a planet of mass m_1 and m_2, respectively, with position vectors $\mathbf{r}_1$ and $\mathbf{r}_2$ referred to an origin O fixed in inertial space (Fig. 1).

The relative motion of the planet with respect to the star is given by the vector $\mathbf{r} = \mathbf{r}_2 - \mathbf{r}_1$. The gravitational forces acting on the star and the planet are

$$\mathbf{F}_1 = m_1 \ddot{\mathbf{r}}_1 = +G \frac{m_1 m_2}{r^3} \mathbf{r} \quad (2)$$

$$\mathbf{F}_2 = m_2 \ddot{\mathbf{r}}_2 = -G \frac{m_1 m_2}{r^3} \mathbf{r} \quad (3)$$

respectively. Now consider the motion of the planet m_2 with respect to the star m_1. If we write $\ddot{\mathbf{r}} = \ddot{\mathbf{r}}_2 - \ddot{\mathbf{r}}_1$ we can use equation (1) to obtain

$$\ddot{\mathbf{r}} + G(m_1 + m_2) \frac{\mathbf{r}}{r^3} = 0 \quad (4)$$

If we take the vector product of $\mathbf{r}$ with equation (4) we have $\mathbf{r} \times \ddot{\mathbf{r}} = 0$, which can be integrated directly to give

$$\mathbf{r} \times \dot{\mathbf{r}} = \mathbf{h} \quad (5)$$

where $\mathbf{h}$ is a constant vector that is simultaneously perpendicular to both $\mathbf{r}$ and $\dot{\mathbf{r}}$. Therefore the motion of the planet about the star lies in a plane (the orbit plane) perpendicular to the direction defined by $\mathbf{h}$. Another consequence of this result is that the position and velocity vectors will always lie in the same plane (see Fig. 2). Equation (5) is often referred to as the angular momentum integral and $\mathbf{h}$ represents a constant of the two-body motion.

In order to solve equation (4) we transform to a polar coordinate system (r, θ) referred to an origin centered on the star with an arbitrary reference line corresponding to $\theta = 0$. In polar coordinates the position, velocity, and acceleration vectors can be written as

$$\mathbf{r} = r \hat{\mathbf{r}} \quad (6)$$

$$\dot{\mathbf{r}} = \dot{r} \hat{\mathbf{r}} + r \dot{\theta} \hat{\boldsymbol{\theta}} \quad (7)$$

$$\ddot{\mathbf{r}} = \left(\ddot{r} - r \dot{\theta}^2 \right) \hat{\mathbf{r}} + \left[\frac{1}{r} \frac{d}{dt} \left(r^2 \dot{\theta} \right) \right] \hat{\boldsymbol{\theta}} \quad (8)$$

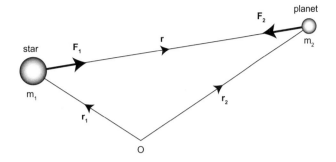

Fig. 1. The forces acting on a star of mass m_1 and a planet of mass m_2 with position vectors $\mathbf{r}_1$ and $\mathbf{r}_2$.

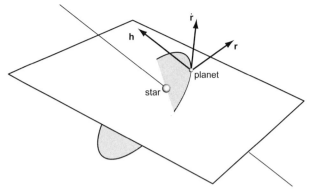

Fig. 2. The motion of m_2 with respect to m_1 defines an orbital plane (shaded region), because $\mathbf{r} \times \dot{\mathbf{r}}$ is a constant vector, $\mathbf{h}$, the angular momentum vector and this is always perpendicular to the orbit plane.

where $\hat{\mathbf{r}}$ and $\hat{\boldsymbol{\theta}}$ denote unit vectors along and perpendicular to the radius vector respectively. Substituting equation (7) into equation (5) gives $\mathbf{h} = r^2\dot{\theta}\hat{\mathbf{z}}$, where $\hat{\mathbf{z}}$ is a unit vector perpendicular to the plane of the orbit forming a righthanded triad with $\hat{\mathbf{r}}$ and $\hat{\boldsymbol{\theta}}$. The magnitude of this vector gives us

$$h = r^2\dot{\theta} \qquad (9)$$

Therefore, although r and θ vary as the planet moves around the star, the quantity $r^2\dot{\theta}$ remains constant. The area element dA swept out by the star-planet radius vector in the time interval dt is given in polar coordinates by

$$dA = \int_0^r r\,dr\,d\theta = \frac{1}{2}r^2 d\theta \qquad (10)$$

and thus

$$\dot{A} = \frac{1}{2}r^2\dot{\theta} = \frac{1}{2}h = \text{constant} \qquad (11)$$

This is equivalent to Kepler's second law of planetary motion, which states that the star-planet line sweeps out equal areas in equal times.

Using equation (6) and comparing the $\hat{\mathbf{r}}$ components of equations (4) and (8) gives the scalar differential equation

$$\ddot{r} - r\dot{\theta}^2 = -\frac{G(m_1 + m_2)}{r^2} \qquad (12)$$

In order to find r as a function of θ we need to make the substitution u = 1/r. By differentiating r with respect to time and making use of equation (9) we can eliminate time in the differential equation. We obtain

$$\ddot{r} = -h\frac{d^2u}{d\theta^2}\dot{\theta} = -h^2u^2\frac{d^2u}{d\theta^2} \qquad (13)$$

and hence equation (12) can be written

$$\frac{d^2u}{d\theta^2} + u = \frac{G(m_1 + m_2)}{h^2} \qquad (14)$$

This is a second-order, linear differential equation, often referred to as Binet's equation, with a general solution

$$u = \frac{G(m_1 + m_2)}{h^2}\left[1 + e\cos(\theta - \varpi)\right] \qquad (15)$$

where e (an amplitude) and ϖ (a phase) are two constants of integration. Substituting back for r gives

$$r = \frac{p}{1 + e\cos(\theta - \varpi)} \qquad (16)$$

where $p = h^2/G(m_1 + m_2)$. This is the general equation in polar coordinates of a set of curves known as conic sections where e is the eccentricity and p is a constant called the semilatus rectum. For a given system the initial conditions will determine the particular conic section (circle, ellipse, parabola, or hyperbola) the planet follows. We consider only elliptical motion for which

$$p = a(1 - e^2) \qquad (17)$$

where a, a constant, is the semimajor axis of the ellipse. The quantities a and e are related by

$$b^2 = a^2(1 - e^2) \qquad (18)$$

where b is the semiminor axis of the ellipse (see Fig. 3).

Therefore, for any given value of θ the radius is calculated using the equation

$$r = \frac{a(1 - e^2)}{1 + e\cos(\theta - \varpi)} \qquad (19)$$

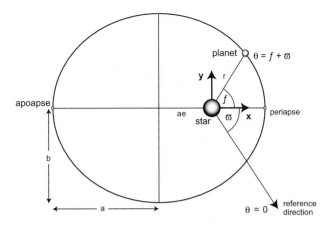

Fig. 3. The geometry of the ellipse of semimajor axis a, semiminor axis b, eccentricity e, and longitude of periapse ϖ.

Hence the path of the planet around the star is an ellipse with the star at one focus; this is Kepler's first law of planetary motion. Note that in the special case where e = 0 (a circular orbit), r = a and the angle ϖ is undefined.

The angle θ is called the true longitude. Equation (19) shows that the minimum and maximum values of r are a(1−e) (at $\theta = \varpi$) and a(1 + e) (at $\theta = \varpi + \pi$), respectively. These points are referred to as the periapse and the apoapse, respectively, although for motion around a star they can also be referred to as the periastron and apastron.

The angle ϖ (pronounced "curly pi") is called the longitude of periapse or longitude of periastron of the planet's orbit and gives the angular location of the closest approach with respect to the reference direction. If we define the true anomaly to be the angle $f = \theta - \varpi$ (see Fig. 3) then f is measured with respect to the periapse direction and equation (19) can be written

$$r = \frac{a(1-e^2)}{1+e\cos f} \quad (20)$$

In this case if we define a cartesian coordinate system centered on the star with the x-axis pointing toward the periapse (see Fig. 3), then the position vector has components

$$x = r\cos f \quad (21)$$
$$y = r\sin f \quad (22)$$

Although we have eliminated the time from our equation of motion, we can relate the orbital period, T, to the semimajor axis, a. The area of an ellipse is A = πab and this is swept out by the star-planet line in a time, T. Hence, from equation (11), A = hT/2 and so

$$T^2 = \frac{4\pi^2}{G(m_1+m_2)}a^3 \quad (23)$$

This is Kepler's third law of planetary motion. It implies that the period of the planet's orbit is independent of e and is purely a function of the sum of the masses and a. If we define the mean motion, n, of the planet's motion as

$$n = \frac{2\pi}{T} \quad (24)$$

then we can write

$$G(m_1+m_2) = n^2 a^3 \quad (25)$$

and hence

$$h = na^2\sqrt{1-e^2} = \sqrt{G(m_1+m_2)a(1-e^2)} \quad (26)$$

There is an additional constant of the two-body motion that is useful in calculating the velocity of the planet. Taking the scalar product of $\dot{\mathbf{r}}$ with equation (4) and using equations (6) and (7) gives the scalar equation

$$\dot{\mathbf{r}} \cdot \ddot{\mathbf{r}} + G(m_1+m_2)\frac{\dot{r}}{r^2} = 0 \quad (27)$$

which can be integrated to give

$$\frac{1}{2}v^2 - \frac{G(m_1+m_2)}{r} = C \quad (28)$$

where $v^2 = \dot{\mathbf{r}} \cdot \dot{\mathbf{r}}$ is the square of the velocity and C is a constant of the motion. Equation (28) is called the vis viva integral. It shows that the orbital energy per unit mass of the system is conserved.

Because ϖ is a constant, $\dot{\theta} = \dot{f}$ and equation (7) gives

$$v^2 = \dot{\mathbf{r}} \cdot \dot{\mathbf{r}} = \dot{r}^2 + r^2\dot{f}^2 \quad (29)$$

By differentiating equation (20) we obtain

$$\dot{r} = \frac{r\dot{f}e\sin f}{1+e\cos f} \quad (30)$$

and hence, using equations (9) and (16), we have

$$\dot{r} = \frac{na}{\sqrt{1-e^2}}e\sin f \quad (31)$$

and

$$r\dot{f} = \frac{na}{\sqrt{1-e^2}}(1+e\cos f) \quad (32)$$

Therefore we can write equation (29) as

$$v^2 = \frac{n^2 a^2}{1-e^2}\left(1+2e\cos f + e^2\right) \quad (33)$$

This shows the dependence of v on f. A little further manipulation gives

$$v^2 = G(m_1 + m_2)\left(\frac{2}{r} - \frac{1}{a}\right) \quad (34)$$

which shows the dependence of v on r.

3. SOLUTION OF THE KEPLER PROBLEM

In the previous section we solved the equation of motion of the two-body problem to show the path of the planet's orbit with respect to the star. However, in the process we eliminated the time and so although we can calculate r for a given value of θ, we have no means of finding r as a function of time. This is the essence of the Kepler problem.

Our starting point is to derive an expression for $\dot{r}$ in terms of r. We can do this by using equations (20), (32), and (34) to rewrite equation (29) as

$$\dot{r}^2 = n^2 a^3 \left(\frac{2}{r} - \frac{1}{a}\right) - \frac{n^2 a^4 (1 - e^2)}{r^2} \quad (35)$$

This simplifies to give

$$\dot{r} = \frac{na}{r}\sqrt{a^2 e^2 - (r - a)^2} \quad (36)$$

In order to solve this differential equation we introduce a new variable, E, the eccentric anomaly, by means of the substitution

$$r = a(1 - e\cos E) \quad (37)$$

The differential equation transforms to

$$\dot{E} = \frac{n}{1 - e\cos E} \quad (38)$$

The solution can be written as

$$n(t - t_0) = E - e\sin E \quad (39)$$

where we have taken t_0 to be the constant of integration and used the boundary condition E = 0 when $t = t_0$. At this point we can define a new quantity, M, as the mean anomaly, such that

$$M = n(t - t_0) \quad (40)$$

where t_0 is a constant called the time of periastron passage. There is no simple geometrical interpretation of M but we note that it has the dimensions of an angle and that it increases linearly with time. Furthermore, $M = f = 0$ when $t = t_0$ or $t = t_0 + T$ (periapse passage) and $M = f = \pi$ when $t = t_0 + T/2$ (apoapse passage). We can write

$$M = E - e\sin E \quad (41)$$

This is Kepler's equation and its solution is fundamental to the problem of finding the orbital position at a given time. For a particular time t we can (1) find M from equation (40), (2) find E by solving Kepler's equation, equation (41), (3) find r using equation (37), and finally (4) find f using equation (20).

The key step is solving Kepler's equation and this is usually done numerically. *Danby* (1988) gives several numerical methods for its solution. For example, if we define the function

$$g(E) = E - e\sin E - M \quad (42)$$

then we can use a Newton-Raphson method to find the root of the nonlinear equation, g(E) = 0. The iteration scheme is

$$E_{i+1} = E_i - \frac{g(E_i)}{g'(E_i)}, \quad i = 0, 1, 2, \ldots \quad (43)$$

where $g'(E_i) = dg(E_i)/dE_i = 1 - e\cos E_i$ and the iterations proceed until convergence is achieved. A reasonable initial value is $E_0 = M$ since E and M differ by a quantity on the order of e (see equation (41)).

Although we cannot have an explicit relation between the angles f and M, from Kepler's second law (equation (9)) and (equation (26)) it is possible to write

$$df = n\sqrt{1 - e^2}\left(\frac{a}{r}\right)^2 dt = \sqrt{1 - e^2}\left(\frac{a}{r}\right)^2 dM \quad (44)$$

The above relation is useful when we want to average any physical quantity over a complete orbit. For instance

$$\left\langle \frac{1}{r^2} \right\rangle = \frac{1}{2\pi}\int_0^{2\pi} \frac{dM}{r^2} = \frac{1}{a^2\sqrt{1 - e^2}} \quad (45)$$

To complete the set of useful angles we define the mean longitude, λ, by

$$\lambda = M + \varpi \quad (46)$$

Therefore λ, like M, is a linear function of time. It is important to note that all longitudes (θ, ϖ, λ) are defined with respect to a common, arbitrary reference direction.

4. THE ORBIT IN THREE DIMENSIONS

Of the orbital elements we have defined so far, two (a and e) are related to the physical dimensions of the orbit and the remaining two (ϖ and f) are related to orientation of the orbit or the location of the planet in its orbit. Note that there are many alternatives to f (e.g., θ, M, and λ) and the time of periapse passage, t_0, can also be used instead of f since the latter can always be calculated from the former. We have already noted that ϖ is the angular location of the periapse direction measured from a reference point on the orbit.

Consider the planet's position vector

$$\mathbf{r} = (x, y, 0) = x\hat{\mathbf{x}} + y\hat{\mathbf{y}} + 0\hat{\mathbf{z}} \qquad (47)$$

in a three-dimensional coordinate system where the x-axis lies along the major (long) axis of the ellipse in the direction of periapse; the y-axis is perpendicular to the x-axis and lies in the orbital plane; and the z-axis is mutually perpendicular to both the x- and y-axes, forming a righthanded triad. By definition the orbital motion is confined to the x-y plane. Consider a standard coordinate system where the direction of the reference line in the reference plane forms the X-axis. The Y-axis is in the reference plane at right angles to the X-axis, while the Z-axis is perpendicular to both the X- and Y-axes, forming a righthanded triad.

Let i denote the inclination, the angle between the orbit plane and the reference plane. The line formed by the intersection of the two planes is called the line of nodes. The ascending node is the point in both planes where the orbit crosses the reference plane moving from below to above the plane. The longitude of ascending node, Ω, is the angle between the reference line and the radius vector to the ascending node. The angle between this same radius vector and the periapse of the orbit is called the argument of periapse, ω. Note that the inclination is always in the range $0 \leq i \leq 180°$. An orbit is said to be prograde if $i < 90°$ while if $i \geq 90°$ the motion is said to be retrograde. We can also define

$$\varpi = \Omega + \omega \qquad (48)$$

where ϖ is the longitude of periapse introduced above but that now, in general, the angles Ω and ω lie in different planes so that ϖ forms a "dog-leg" angle.

The orientation angles i, Ω, and ω are illustrated in Fig. 4. It is clear that coordinates in the (x, y, z) system can be expressed in terms of the (X, Y, Z) system by means of a series of three rotations: (1) a rotation about the z-axis through an angle ω so that the x-axis coincides with the line of nodes, (2) a rotation about the x-axis through an angle i so that the two planes are coincident, and finally (3) a rotation about the z-axis through an angle Ω. We can represent these transformations by two 3 × 3 rotation matrices, denoted by $\mathbf{P}_x(\phi)$ (rotation about the x-axis) and $\mathbf{P}_z(\phi)$ (rotation about the z-axis), with elements

$$\mathbf{P}_x(\phi) = \begin{pmatrix} 1 & 0 & 0 \\ 0 & \cos\phi & -\sin\phi \\ 0 & \sin\phi & \cos\phi \end{pmatrix} \qquad (49)$$

and

$$\mathbf{P}_z(\phi) = \begin{pmatrix} \cos\phi & -\sin\phi & 0 \\ \sin\phi & \cos\phi & 0 \\ 0 & 0 & 1 \end{pmatrix} \qquad (50)$$

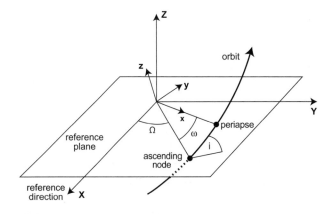

Fig. 4. The relationship between the (x, y, z) and (X, Y, Z) coordinate systems and the angles ω, i, and Ω.

Consequently

$$\begin{pmatrix} X \\ Y \\ Z \end{pmatrix} = \mathbf{P}_z(\Omega)\mathbf{P}_x(i)\mathbf{P}_z(\omega)\begin{pmatrix} x \\ y \\ z \end{pmatrix} \qquad (51)$$

and

$$\begin{pmatrix} x \\ y \\ z \end{pmatrix} = \mathbf{P}_z^{-1}(\omega)\mathbf{P}_x^{-1}(i)\mathbf{P}_z^{-1}(\Omega)\begin{pmatrix} X \\ Y \\ Z \end{pmatrix} \qquad (52)$$

where $\mathbf{P}_x^{-1}(\phi) = \mathbf{P}_x(-\phi)$ and $\mathbf{P}_z^{-1}(\phi) = \mathbf{P}_z(-\phi)$ are the inverse of the matrices of $\mathbf{P}_x(\phi)$ and $\mathbf{P}_z(\phi)$, respectively.

If we now restrict ourselves to coordinates that lie in the orbital plane, we have $x = r\cos f$, $y = r\sin f$, $z = 0$, and

$$X = r(\cos\Omega\cos(\omega+f) - \sin\Omega\sin(\omega+f)\cos i) \qquad (53)$$

$$Y = r(\sin\Omega\cos(\omega+f) + \cos\Omega\sin(\omega+f)\cos i) \qquad (54)$$

$$Z = r\sin(\omega+f)\sin i \qquad (55)$$

5. BARYCENTRIC MOTION

In order to determine the observable effects of an orbiting planet on a star it helps if we consider the motion in the center of mass or barycentric system (see Fig. 5). The position vector of the center of mass of the system is

$$\mathbf{R} = \frac{m_1\mathbf{r}_1 + m_2\mathbf{r}_2}{m_1 + m_2} \qquad (56)$$

From equations (2) and (3) we have

$$\ddot{\mathbf{R}} = \frac{m_1\ddot{\mathbf{r}}_1 + m_2\ddot{\mathbf{r}}_2}{m_1 + m_2} = 0 \qquad (57)$$

and by direct integration $\dot{\mathbf{R}} = \mathbf{V} = $ constant. These equations imply that either (1) the center of mass is stationary (the case when $\mathbf{V} = 0$), or (2) it is moving with a constant velocity (the case when $\mathbf{V} \neq 0$) in a straight line with respect to the origin O. Then, if we write $\mathbf{R}_1 = \mathbf{r}_1 - \mathbf{R}$ and $\mathbf{R}_2 = \mathbf{r}_2 - \mathbf{R}$, we have

$$m_1 \mathbf{R}_1 + m_2 \mathbf{R}_2 = 0 \tag{58}$$

This implies that (1) $\mathbf{R}_1$ is always in the opposite direction to $\mathbf{R}_2$, and hence that (2) the center of mass is always on the line joining m_1 and m_2. Therefore we can write

$$R_1 + R_2 = r \tag{59}$$

where r is the separation of m_1 and m_2, and the distances of the star and planet from their common center of mass are related by $m_1 R_1 = m_2 R_2$ (equation (58)). Hence

$$R_1 = \frac{m_2}{m_1 + m_2} r \quad \text{and} \quad R_2 = \frac{m_1}{m_1 + m_2} r \tag{60}$$

Therefore each object will orbit the center of mass of the system in an ellipse with the same eccentricity but the semimajor axes are reduced in scale by a factor (see Fig. 6)

$$a_1 = \frac{m_2}{m_1 + m_2} a \quad \text{and} \quad a_2 = \frac{m_1}{m_1 + m_2} a \tag{61}$$

The orbital periods of the two objects must each be equal to T and therefore the two mean motions must also be equal ($n_1 = n_2 = n$), although the semimajor axes are not. Each mass then moves on its own elliptical orbit with respect to the common center of mass, and the periapses of their orbits differ by π (see Fig. 6b).

We are now in a position to revisit the expression for the radial velocity of the star, v_r. Observers usually take the reference plane (X, Y) to be the plane of the sky perpendicular to the line of sight, the Z-axis oriented toward the observer (Fig. 7). Thus, the radial velocity of the star is simply given by the projection of the velocity vector on the line of sight. Since $\mathbf{r}_1 = \mathbf{R} + \mathbf{R}_1$ this gives

$$v_r = \dot{\mathbf{r}}_1 \cdot \hat{\mathbf{Z}} = V_Z + \frac{m_2}{m_1 + m_2} \dot{Z} \tag{62}$$

where $V_Z = \mathbf{V} \cdot \hat{\mathbf{Z}}$ is the proper motion of the barycenter and $\dot{Z}$ can be obtained directly from equation (55)

$$\dot{Z} = \dot{r} \sin(\omega + f) \sin i + r \dot{f} \cos(\omega + f) \sin i \tag{63}$$

or, making use of equations (31) and (32)

$$\dot{Z} = \frac{na \sin i}{\sqrt{1 - e^2}} \left(\cos(\omega + f) + e \cos \omega \right) \tag{64}$$

We can now write

$$v_r = V_Z + K \left(\cos(\omega + f) + e \cos \omega \right) \tag{65}$$

where

$$K = \frac{m_2}{m_1 + m_2} \frac{na \sin i}{\sqrt{1 - e^2}} \tag{66}$$

6. APPLICATION TO EXOPLANETS

More than 450 exoplanets are known to date (see *The Extrasolar Planets Encyclopedia* at *http://exoplanet.eu/*), and the number is continuously rising. Looking at this data, we can admire the wide variety of possible orbital parameters and physical properties: central stars of spectral types from F to M, minimum masses from 2 $M_\oplus$ to more than 20 M_{Jup}, orbital periods a little over of 1 day to more than 14 years (the

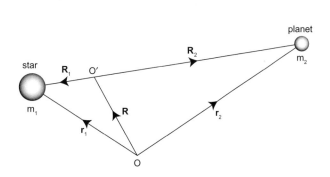

Fig. 5. The position vectors of star and planet with respect to the origin, O, and with respect to the center of mass of the star-planet system, O'.

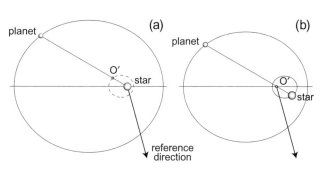

Fig. 6. (a) The motion of the planet m_2 with respect to the star m_1 in the two-body problem; the dashed curve denotes the elliptical path of the center of mass, O'. (b) The motion of the masses m_1 and m_2 with respect to the center of mass, O', for the same system. For the purposes of illustration we used $m_2/m_1 = 0.2$ and $e = 0.5$.

same time as the length of the observations), and eccentricities ranging from perfect circular orbits to extreme values of more than 0.9. There are planets as close as 0.0177 AU and as far as 670 AU from their host stars.

In Table 1 we report two examples of extreme values for the eccentricity, obtained using the radial velocity technique. The first example (HD 156846b) corresponds to a highly eccentric orbit, while the second one (HD 83443b) shows an almost circular orbit. At present, both planets are in single-planet systems, which allows us to apply directly the formulae derived in previous sections to the analysis of their motion.

6.1. HD 156846b

HD 156846 has been observed with the CORALIE spectrograph at La Silla Observatory (ESO) from May 2003 to September 2007. Altogether, 64 radial velocity measurements with a mean uncertainty of 2.8 m/s were gathered. Figure 8a shows the CORALIE radial velocities and the corresponding best-fit Keplerian model. The resulting orbital parameters are T = 359.51 d, e = 0.847, and K = 464 m/s (Table 1). Details on the data analysis using radial velocities are given in the chapter by Lovis and Fischer.

Assuming a stellar mass m_1 = 1.43 $M_\odot$ (*Tamuz et al.*, 2008), equations (25) and (66) can be used to derive a companion minimum mass of m_2 sin i = 10.45 M_{Jup}, orbiting the central star with a semimajor axis a = 0.99 AU. With the radial velocity technique it is impossible to determine the inclination i, and therefore we are unable to describe the orbit in three dimensions and to determine the exact mass of the planet. Nevertheless, the orbit in two dimensions (orbital plane) can be completely characterized. Astrometry is the only observational technique that can provide the full, three-dimensional orbit of the planet, but at present few planets have been observed by this method (chapter by Traub and Oppenheimer).

In Fig. 8b we have drawn the orbit of HD 156846b. Because of its high eccentricity, the orbit is very elongated. As a consequence, the separation between the planet and the star ranges from 0.15 AU at periapse to 1.83 AU at apoapse. In our solar system comets are the only objects that present such large variations in their position relative to the Sun. The origin of such high eccentricities is unknown, but a possible explanation is through close encounters between very massive bodies during the formation process (*Ford and Rasio*, 2008).

The angle ω = 52.2° corresponds to the argument of periapse, which is measured from the nodal line between the plane of the sky and the orbital plane of the planet. For HD 156846b this quantity is well defined, because the orbit is so eccentric. According to equation (34), at periapse the orbital velocity is maximal and therefore it shows an easily identifiable peak in the observational data (Fig. 8).

The planet is at periapse whenever

$$t = t_0 + kT \quad \text{with} \quad k = 0, \pm 1, \pm 2, \ldots \quad (67)$$

where t_0 = JD2453998.1 (September 19, 2006, at 14h 24m UT) is the time of periapse passage. In fact, because the orbit is periodic, any instant of time, t_0, given by equation (67) can be used as the time of periapse passage. This is true for the two-body problem, but no longer valid if additional bodies are

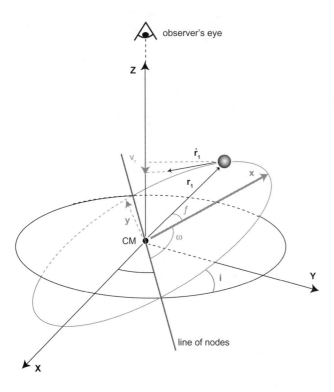

Fig. 7. The relationship between the star's velocity around the center of mass, $\dot{r}_1$, and its radial component along the line of sight, v_r.

TABLE 1. Two examples of extreme orbital eccentricities for exoplanets.

	HD 156846b	HD 83443b
Discovery Year	2007	2002
Data Ref.	*Tamuz et al.* (2008)	*Mayor et al.* (2004)
V_Z (km/s)	−68.540 ± 0.001	29.027 ± 0.001
T(d)	359.51 ± 0.09	2.98565 ± 0.00003
K (km/s)	0.464 ± 0.003	58.1 ± 0.4
e	0.847 ± 0.002	0.013 ± 0.013
ω(°)	52.2 ± 0.4	11 ± 11
t_0 (JD−2.45 × 10^6)	3998.1 ± 0.1	1497.5 ± 0.3
m_1 ($M_\odot$)	1.43	0.90
m_2 sin i (M_{Jup})	10.45	0.38
a (AU)	0.9930	0.03918

present in the system. Indeed, mutual planetary perturbations will disturb the orbits and the time of two successive periapse passages is no longer given exactly given by equation (67) (see chapter by Fabrycky). Although observers often use t_0 as a parameter to characterize orbits, for multiplanet systems it is meaningless. A better option is to use the mean anomaly M (equation (40)) or the mean longitude λ (equation (46)). The observer fixes a date t_f and then provides the value of $M = M_0$ computed for that date (equation (40))

$$M_0 = n\left(t_f - t_0\right) \qquad (68)$$

A practical choice of t_f is to use $t_f = t_0$, since $M_0 = 0$ (and $\lambda = \varpi$). For multiplanetary systems, the planet will still be at periastron whenever $M = 0$, but the time of successive periapse passages will no longer be given by equation (67).

6.2. HD 83443b

HD 83443b is a short period Jupiter-sized planet, therefore belonging to the class of "hot Jupiters." It was first announced at the IAU Symposium 202 in Manchester as a resonant two-planet system with periods $T_1 = 2.986$ d and $T_2 = 29.85$ d, but subsequent observations could not confirm the presence of the companion around 30 d. The origin of the transient signal is not yet clear, but an appealing possibility is to attribute the effect to activity of the star (*Mayor et al., 2004*). As a consequence, the HD 83443 star has been monitored many times. After 257 radial velocity measurements using the CORALIE spectrograph with a mean uncertainty of 8.9 m/s (taken from March 1999 to March 2003), the short-period planet at $T = 2.986$ d was found to revolve alone in an almost circular orbit (e ~ 0) with K = 58.1 km/s (Table 1). In Fig. 9a we show the CORALIE radial velocities and the corresponding best-fit Keplerian model.

Assuming a stellar mass $m_1 = 0.90$ $M_\odot$ (*Mayor et al., 2004*), a companion minimum mass $m_2 \sin i = 0.38$ M_{Jup} and semimajor axis a = 0.039 AU has been derived. Again, it is impossible to determine the inclination i, and the orbit can only be characterized in terms of its orbital plane (Fig. 9b).

Because the orbital eccentricity is small and uncertain (e = 0.013 ± 0.013), so too is the argument of the periapse ($\omega = 11° ± 11°$). Indeed, for perfect circular orbits (e = 0) the argument of the periapse is not defined since the distance from the planet to the star is constant. Therefore, it is also meaningless to

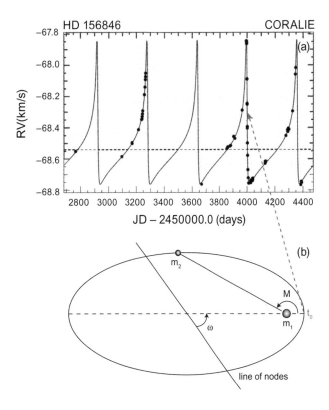

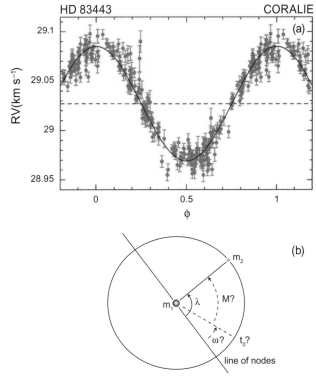

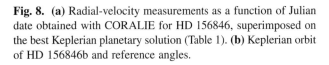

Fig. 8. (a) Radial-velocity measurements as a function of Julian date obtained with CORALIE for HD 156846, superimposed on the best Keplerian planetary solution (Table 1). (b) Keplerian orbit of HD 156846b and reference angles.

Fig. 9. (a) Phase-folded radial-velocity measurements obtained with CORALIE for HD 83443, superimposed on the best Keplerian planetary solution (Table 1). (b) Keplerian orbit of HD 83443b and reference angles.

provide the time of periapse passage (t_0 = JD2451497.5 ± 0.3, i.e., November 15, 1999, at 0h 0m UT). The fact that the error bar is only 0.3 d suggests (erroneously) that it is small. However, since the orbital period of the planet is only 2.986 d, an uncertainty of ±0.3 d is equivalent to a 20% uncertainty in t_0.

For circular orbits the parameters ω and t_0 are not defined (because e = 0) and the same is also true for the mean anomaly M (angle between the periastron and the planet). However, the orbit of the planet is still well determined (Fig. 9a) and we should be able to provide accurate positions of the planet on its orbit. The correct parameter for that purpose is the sum M + ω or the mean longitude λ = M + ϖ (Fig. 9b). Indeed, fixing the date at t_f = JD2453000.0 we obtain λ = 91° ± 1°, which has a relative error of about 0.3%, much better than the 100% uncertainty in ω.

Acknowledgments. This work was partially supported by the Science and Technology Facilities Council (UK) and by the Fundação para a Ciência e a Tecnologia (Portugal).

REFERENCES

Danby J. M. A. (1988) *Fundamentals of Celestial Mechanics*, 2nd edition. Willmann-Bell, Richmond.

Ford E. B. and Rasio F. A. (2008) Origins of eccentric extrasolar planets: Testing the planet-planet scattering model. *Astrophys. J., 686,* 621–636.

Kepler J. (1609) *Astronomia Nova*, Heidelberg.

Kepler J. (1619) *Harmonices Mundi Libri V*, Linz.

Mayor M., Udry S., Naef D., Pepe F., Queloz D., Santos N. C., and Burnet M. (2004) The CORALIE survey for southern extrasolar planets. XII. Orbital solutions for 16 extra-solar planets discovered with CORALIE. *Astron. Astrophys., 415,* 391–402.

Murray C. D. and Dermott S. F. (1999) *Solar System Dynamics.* Cambridge Univ., Cambridge.

Newton I. (1687) *Philosophiae Naturalis Principia Mathematica.* Royal Society, London.

Tamuz O., Ségransan D., Udry S., Mayor M., Eggenberger A., et al. (2008) The CORALIE survey for southern extra-solar planets. *Astron. Astrophys., 480,* L33–L36.

Part II:
Exoplanet Observing Techniques

Radial Velocity Techniques for Exoplanets

Christophe Lovis
Université de Genève

Debra A. Fischer
Yale University

The radial velocity technique was utilized to make the first exoplanet discoveries around Sun-like stars and continues to play a major role in the discovery and characterization of exoplanetary systems. In this chapter we describe how the technique works, and the current precision and limitations. We then review its major successes in the field of exoplanets. With more than 250 planet detections, it is the most prolific technique to date and has led to many milestone discoveries, such as hot Jupiters, multiplanet systems, transiting planets around bright stars, the planet-metallicity correlation, planets around M dwarfs and intermediate-mass stars, and recently, the emergence of a population of low-mass planets: Neptune-mass planets and super Earths. In the near future radial velocities are expected to systematically explore the domain of rocky and icy planets down to a few Earth masses close to the habitable zone of their parent star. Radial velocity measurements will also be used to provide the necessary follow-up observations of transiting candidates detected by space missions. Finally, we also note alternative radial velocity techniques that may play an important role in the future.

1. INTRODUCTION

Since the end of the nineteenth century, radial velocities have been at the heart of many developments and advances in astrophysics. In 1888, Vogel at Potsdam used photography to demonstrate Christian Doppler's 1842 theory that stars in motion along our line of site would exhibit a change in color. This color change, or wavelength shift, is commonly known as a Doppler shift, and has been a powerful tool over the past century, used to measure stellar kinematics, determine orbital parameters for stellar binary systems, and identify stellar pulsations. By 1953, radial velocities had been compiled for more than 15,000 stars in the *General Catalogue of Stellar Radial Velocities* (*Wilson*, 1953) with a typical precision of 750 m s^{-1}, not the precision that is typically associated with planet-hunting. However, at that time, O. Struve proposed that high-precision stellar radial velocity work could be used to search for planets orbiting nearby stars. He made the remarkable assertion that Jupiter-like planets could reside as close as 0.02 AU from their host stars. Furthermore, he noted that if such close-in planets were 10 times the mass of Jupiter, the reflex stellar velocity for an edge-on orbit would be about 2 km s^{-1} and detectable with 1950s Doppler precision (*Struve*, 1952). Two decades later, *Griffin and Griffin* (1973) identified a key weakness in radial velocity techniques of the day; the stellar spectrum was measured with respect to an emission spectrum. However, the calibrating lamps (typically thorium argon) did not illuminate the slit and spectrometer collimator in the same way as the star. Griffin and Griffin outlined a strategy for improving Doppler precision to a remarkable 10 m s^{-1} by differentially measuring stellar line shifts with respect to telluric lines. Assuming that telluric lines are at rest relative to the spectrometer, these absorption lines would trace the stellar light path and illuminate the optics in the same way and at the same time as the star. Although Griffin and Griffin did not obtain this high precision, they had highlighted some of the key challenges that current techniques have overcome.

By 1979, G. Walker and B. Campbell had a version of telluric lines in a bottle: a glass cell containing hydrogen fluoride (HF) that was inserted in the light path at the Canada France Hawaii Telescope (CFHT) (*Campbell and Walker*, 1979). Like telluric lines, the HF absorption lines were imprinted in the stellar spectrum and provided a precise wavelength solution spanning about 50 Å. The spectrum was recorded with a photon-counting Reticon photodiode array. Working from 1980 to 1992, they monitored 17 main-sequence stars and 4 subgiant stars and achieved the unprecedented precision of 15 m s^{-1}. Unfortunately, because of the small sample size, no planets were found. However, upper limits were set on M sin i for orbital periods out to 15 years for the 21 stars that they observed (*Walker et al.*, 1995).

Cross-correlation speedometers were also used to measure radial velocities relative to a stellar template. In 1989, an object with M sin i of 11 M$_{Jup}$ was discovered in an 84-day orbit around HD 114762 (*Latham et al.*, 1989). The velocity amplitude of the star was 600 m s^{-1} and although the single measurement radial velocity precision was only about 400 m s^{-1}, hundreds of observations effectively beat down the noise to permit this first detection of a substellar object.

In 1993, the ELODIE spectrometer was commissioned on the 1.93-m telescope at Observatoire de Haute-Provence

(OHP), France. To bypass the problem of different light paths for stellar and reference lamp sources described by *Griffin and Griffin* (1973), two side-by-side fibers were used at ELODIE. Starlight passed through one fiber and light from a thorium argon lamp illuminated the second fiber. The calibration and stellar spectrum were offset in the cross-dispersion direction on the CCD detector and a velocity precision of about 13 m s^{-1} enabled the detection of a Jupiter-like planet orbiting 51 Pegasi (*Mayor and Queloz*, 1995). In a parallel effort to achieve high Doppler precision, a glass cell containing iodine vapor was employed by *Marcy and Butler* (1995) to confirm the detection of a planetary companion around 51 Peg b. Both Doppler techniques have continued to show remarkable improvements in precision, and have ushered in an era of exoplanet discoveries.

2. DESCRIPTION OF THE TECHNIQUE

2.1. Radial Velocity Signature of Keplerian Motion

The aim of this section is to derive the radial velocity equation, i.e., the relation between the position of a body on its orbit and its radial velocity, in the case of Keplerian motion. We present here a related approach to that of the chapter by Murray and Correia. As shown there, the solutions of the gravitational two-body problem describe elliptical orbits around the common center of mass if the system is bound, with the center of mass located at a focus of the ellipses. Energy and angular momentum are constants of the motion. The semimajor axis of the first body orbit around the center of mass a_1 is related to the semimajor axis of the relative orbit a through

$$a_1 = \frac{m_2}{m_1 + m_2} a \qquad (1)$$

where m_1 and m_2 are the masses of the bodies.

In polar coordinates, the equation of the ellipse described by the first body around the center of mass reads

$$r_1 = \frac{a_1(1-e^2)}{1+e\cos f} = \frac{m_2}{m_1+m_2} \cdot \frac{a(1-e^2)}{1+e\cos f} \qquad (2)$$

where r_1 is the distance of the first body from the center of mass, e is the eccentricity, and f is the true anomaly, i.e., the angle between the periastron direction and the position on the orbit, as measured from the center of mass. The true anomaly as a function of time can be computed via Kepler's equation, which cannot be solved analytically, and therefore numerical methods have to be used (see the chapter by Murray and Correia).

In this section we want to obtain the relation between position on the orbit, given by f, and orbital velocity. More specifically, since our observable will be the radial velocity of the object, we have to project the orbital velocity onto the line of sight linking the observer to the system. In Cartesian coordinates, with the x-axis pointing in the periastron direction and the origin at the center of mass, the position and velocity vectors are given by

$$\mathbf{r}_1 = \begin{pmatrix} r_1 \cos f \\ r_1 \sin f \end{pmatrix} \qquad (3)$$

$$\dot{\mathbf{r}}_1 = \begin{pmatrix} \dot{r}_1 \cos f - r_1 \dot{f} \sin f \\ \dot{r}_1 \sin f + r_1 \dot{f} \cos f \end{pmatrix} \qquad (4)$$

We now need to express $\dot{r}_1$ and $\dot{f}$ as a function of f in order to obtain the velocity as a function of f alone. In a first step, we differentiate equation (2) to obtain $\dot{r}_1$

$$\dot{r}_1 = \frac{a_1 e(1-e^2)\dot{f}\sin f}{(1+e\cos f)^2} = \frac{e r_1^2 \dot{f} \sin f}{a_1(1-e^2)} \qquad (5)$$

Replacing $\dot{r}_1$ in equation (4), we obtain after some algebra

$$\dot{\mathbf{r}}_1 = \frac{r_1^2 \dot{f}}{a_1(1-e^2)} \begin{pmatrix} -\sin f \\ \cos f + e \end{pmatrix} \qquad (6)$$

$$= \frac{h_1}{m_1 a_1(1-e^2)} \begin{pmatrix} -\sin f \\ \cos f + e \end{pmatrix} \qquad (7)$$

where $h_1 = m_1 r_1^2 \dot{f}$ is the angular momentum of the first body, which is a constant of the motion. It can be expressed as a function of the ellipse parameters a and e as (see the chapter by Murray and Correia)

$$h_1 = \frac{m_2}{m_1+m_2} h = \sqrt{\frac{Gm_1^2 m_2^4 a(1-e^2)}{(m_1+m_2)^3}} \qquad (8)$$

Substituting h_1 in equation (6), we finally obtain

$$\dot{\mathbf{r}}_1 = \sqrt{\frac{Gm_2^2}{m_1+m_2} \frac{1}{a(1-e^2)}} \begin{pmatrix} -\sin f \\ \cos f + e \end{pmatrix} \qquad (9)$$

As a final step, the velocity vector has to be projected onto the line of sight of the observer. We define the inclination angle i of the system as the angle between the orbital plane and the plane of the sky (i.e., the perpendicular to the line of sight). We further define the argument of periastron ω as the angle between the line of nodes and the periastron direction (see Fig. 5 of the chapter by Murray and Correia for an overview of the geometry). In a Cartesian coordinate system with x and y axes in the orbital plane as before and the z axis perpendicular to them, the unit vector of the line of sight **k** is given by

$$\mathbf{k} = \begin{pmatrix} \sin\omega \sin i \\ \cos\omega \sin i \\ \cos i \end{pmatrix} \quad (10)$$

The radial velocity equation is obtained by projecting the velocity vector on $\mathbf{k}$

$$v_{r,1} = \dot{\mathbf{r}}_1 \cdot \mathbf{k}$$
$$= \sqrt{\frac{G}{(m_1+m_2)a(1-e^2)}} m_2 \sin i \quad (11)$$
$$\cdot (\cos(\omega+f) + e\cos\omega)$$

This is the fundamental equation relating the radial velocity to the position on the orbit. From there we can derive the radial velocity semiamplitude $K = (v_{r,max} - v_{r,min})/2$

$$K_1 = \sqrt{\frac{G}{(1-e^2)}} m_2 \sin i (m_1+m_2)^{-1/2} a^{-1/2} \quad (12)$$

It is useful to express this formula in more practical units

$$K_1 = \frac{28.4329 \text{ m s}^{-1}}{\sqrt{1-e^2}} \frac{m_2 \sin i}{M_{Jup}} \left(\frac{m_1+m_2}{M_\odot}\right)^{-1/2} \left(\frac{a}{1 \text{ AU}}\right)^{-1/2} \quad (13)$$

Alternatively, one can use Kepler's third law to replace the semimajor axis a with the orbital period P

$$K_1 = \frac{28.4329 \text{ m s}^{-1}}{\sqrt{1-e^2}} \frac{m_2 \sin i}{M_{Jup}} \left(\frac{m_1+m_2}{M_\odot}\right)^{-2/3} \left(\frac{P}{1 \text{ yr}}\right)^{-1/3} \quad (14)$$

In the exoplanet case, only the radial velocity of the parent star is in general observable, because the planet-to-star flux ratio is so small ($\leq 10^{-5}$). Radial velocity observations covering all orbital phases are able to measure the orbital period P, the eccentricity e, and the RV semiamplitude K_1. From these observables, the so-called "minimum mass" $m_2 \sin i$ can be computed, provided the total mass of the system $m_1 + m_2$ is known. In practice, planetary masses are usually negligible compared to the mass of the parent star. The stellar mass can be obtained indirectly via spectroscopic analysis, photometry, parallax measurements, and comparison with stellar evolutionary models. For bright, main-sequence FGKM stars, Hipparcos data and precise spectral synthesis make it possible to estimate stellar masses to ~5%. Another, more precise way of obtaining stellar masses is via asteroseismic observations, since stellar pulsation frequencies are directly related to the basic stellar parameters such as mass, radius, and chemical composition.

The unknown inclination angle i prevents us from measuring the true mass of the companion m_2. While this is an important limitation of the RV technique for individual systems, this fact does not have a large impact on statistical studies of exoplanet populations. Because inclination angles are randomly distributed in space, angles close to 90° (edge-on systems) are much more frequent than pole-on configurations. Indeed, the distribution function for i is given by f(i)di = sin i di. As a consequence, the average value of sin i is equal to $\pi/4$ (0.79). Moreover, the *a priori* probability that sin i is larger than 0.5 is 87%.

Equation (13) gives RV semiamplitudes as a function of orbital parameters. Table 1 lists a few typical examples for planets with different masses and semimajor axes orbiting a solar-mass star. As one can see, the search for exoplanets with the RV technique requires a precision of at least ~30 m s^{-1} to detect giant planets. Toward lower masses, a precision of ~1 m s^{-1} is necessary to access the domain of Neptune-mass planets and super Earths, while measurements at ~0.1 m s^{-1} would be able to reveal Earth. It is interesting to consider here planet searches around lower-mass stars, since the Doppler signal increases as $1/\sqrt{m_1}$ at constant semimajor axis. For a 0.1-$M_\odot$ M dwarf, all RV amplitudes given in Table 1 must be multiplied by 3. Considering that the habitable zone around such stars may be as close as 0.1 AU, an Earth-like planet would induce a RV signal of 0.9 m s^{-1}. While the gain is significant compared to habitable Earths around FGK stars, it must be balanced by the intrinsic faintness of these stars, which makes it difficult to obtain high enough a signal-to-noise ratio to reach the required precision, even with 10-m telescopes.

2.2. Doppler Effect

The fundamental effect on which the radial velocity technique relies is the well-known Doppler effect. In flat space time (special relativity), an emitted photon of wavelength λ_0 in the rest frame of the source will be detected at a different wavelength λ by an observer moving with respect to the emitter, where λ is given by (*Einstein*, 1905)

$$\lambda = \lambda_0 \frac{1 + \frac{1}{c}\mathbf{k}\cdot\mathbf{v}}{\sqrt{1-\frac{v^2}{c^2}}} \quad (15)$$

TABLE 1. Radial velocity signals for different kinds of planets orbiting a solar-mass star.

Planet	a (AU)	K_1 (m s^{-1})
Jupiter	0.1	89.8
Jupiter	1.0	28.4
Jupiter	5.0	12.7
Neptune	0.1	4.8
Neptune	1.0	1.5
Super Earth (5 $M_\oplus$)	0.1	1.4
Super Earth (5 $M_\oplus$)	1.0	0.45
Earth	0.1	0.28
Earth	1.0	0.09

In this equation, **v** is the velocity of the source relative to the observer, **k** is the unit vector pointing from the observer to the source in the rest frame of the observer, and c is the speed of light in vacuum. As one can see, the Doppler shift mainly depends on the projection of the velocity vector along the line of sight, i.e., the radial velocity of the source. There is also a dependence on the magnitude of the velocity vector, which means that the transverse velocity also contributes to the Doppler shift. However, this relativistic effect is often negligible in practical applications where velocities remain small compared to the speed of light.

In the framework of general relativity, the curvature of spacetime, giving rise to the so-called gravitational redshift, has to be taken into account in the derivation of the Doppler shift formula. Neglecting terms on the order of c^{-4} and higher, for an observer at zero gravitational potential (*Lindegren and Dravins*, 2003), the general-relativistic Doppler formula becomes

$$\lambda = \lambda_0 \frac{1 + \frac{1}{c}\mathbf{k}\cdot\mathbf{v}}{1 - \frac{\Phi}{c^2} - \frac{v^2}{2c^2}} \quad (16)$$

where Φ is the Newtonian gravitational potential at the source ($\Phi = GM/r$ at a distance r of a spherically symmetric mass M).

The Doppler effect, being a relation between wavelength (or frequency) of light and velocity of the emitter, introduces the possibility of measuring the (radial) velocity of a star as a function of time, thereby permitting the detection of orbiting companions according to the equations given in the previous section. In practice, the Doppler shift is measured on the numerous spectral lines that are present in stellar spectra. However, Earth-bound Doppler shift measurements must first be corrected from the effects of local motions of the observer, i.e., Earth rotation and revolution around the Sun, before they can be used to study the behavior of the target star. The barycenter of the solar system represents the most obvious local rest frame to which measurements should be referred. A corresponding suitable reference system centered on the solar system barycenter, the International Celestial Reference System (ICRS), has been defined by the International Astronomical Union (IAU) and has become the standard reference system (*Rickman*, 2001). Based on equation (16), the transformation of measured wavelengths at Earth to barycentric wavelengths is called "barycentric correction" and is given by

$$\lambda_B = \lambda_{obs} \frac{1 + \frac{1}{c}\mathbf{k}\cdot\mathbf{v}_{obs}}{1 - \frac{\Phi_{obs}}{c^2} - \frac{v_{obs}^2}{2c^2}} \quad (17)$$

where λ_B is the wavelength that would be measured in the ICRS, λ_{obs} is the measured wavelength in the Earth-bound frame, **k** is the unit vector pointing from the observer to the source in the ICRS, $\mathbf{v}_{obs}$ is the velocity of the observer with respect to the ICRS, and Φ_{obs} is the gravitational potential at the observer.

The main term in this equation is of course the projection of the observer velocity (in the ICRS) in the direction of the source at the time of the observation. It is made of two contributions: Earth's orbital velocity and Earth's rotational velocity at the observer's location. These can be computed using precise solar system ephemerides such as those developed by JPL (*Standish*, 1990) or IMCCE (*Fienga et al.*, 2008). When aiming at radial velocity precisions of 1 m s^{-1} or below, all input parameters must be known to high precision and carefully checked. Target coordinates must be given in the ICRS, corrected from proper motions. The observatory clock must reliably give UTC time, and ideally the photon-weighted midexposure time should be used. Finally, the observatory coordinates and altitude must be known precisely.

The relativistic term appearing in the denominator of equation (17) has a magnitude of a few meters per second in velocity units and, more importantly, exhibits variations at the level of only 0.1 m s^{-1} over the year due to the eccentricity of Earth's orbit. Therefore, if one is interested in radial velocity *variations* at a precision not exceeding 0.1 m s^{-1}, the relativistic term can be neglected and equation (17) reduces to the familiar classical expression

$$\lambda_B = \lambda_{obs}\left(1 + \frac{1}{c}\mathbf{k}\cdot\mathbf{v}_{obs}\right) \quad (18)$$

After correcting the laboratory wavelength scale to the barycentric scale, the radial velocity of the source can finally be obtained from the expression

$$\lambda_B = \lambda_0 \frac{1 + \frac{1}{c}\mathbf{k}\cdot\mathbf{v}_\star}{1 - \frac{\Phi_\star}{c^2} - \frac{v_\star^2}{2c^2}} \quad (19)$$

where $\mathbf{v}_\star$ and $\Phi_\star$ are the source velocity and gravitational potential, respectively. Again, the relativistic term implies that it is in principle necessary to know the transverse velocity of the source to properly derive its radial velocity. However, if one is only interested in low-amplitude radial velocity *variations*, such as those produced by exoplanets, the relativistic term can also be dropped, yielding the familiar expression for the radial velocity

$$\mathbf{k}\cdot\mathbf{v}_\star = c\frac{\lambda_B - \lambda_0}{\lambda_0} \quad (20)$$

We emphasize here that we are focusing on precise (i.e., reproducible), but not accurate (i.e., absolute), radial velocities. Indeed, obtaining absolute stellar radial velocities through the Doppler effect is a very difficult task due to many perturbing systematic effects, such as gravitational redshift, convective blueshifts of stellar lines, line asymmetries, instrumental and wavelength calibration systematics, etc. These effects introduce unknown offsets on the measured radial velocities that may add up to ~1 km s⁻¹ for solar-type stars and high-resolution spectrographs. However, since they remain essentially constant with time, they do not prevent the detection of low-amplitude radial velocity signals. The reader is referred to *Lindegren and Dravins* (2003) for a stringent definition of the concept of radial velocity and an in-depth discussion of all relevant perturbing effects.

2.3. High-Resolution Spectra of Cool Stars

Stellar Doppler shift measurements rely on the presence of spectral features in the emergent spectrum. Since stars emit most of their electromagnetic energy between the UV and mid-IR domains, and since the UV and mid-IR domains themselves are not accessible from the ground, radial velocity work has always focused on spectral lines in the visible and near-IR regions. Stellar spectra vary widely across the different regions of the Hertzsprung-Russell (HR) diagram. Perhaps the single most relevant physical parameter that controls their general properties is the effective temperature of the stellar photosphere. On the hot side ($T_{eff} \gtrsim 10{,}000$ K), all chemical elements are at least partly ionized and the atomic energy levels giving rise to electronic transitions in the visible and near-IR are depopulated (in other words, line opacities in these spectral regions become negligible). Since the spectrum is essentially a continuum, no Doppler shift measurements are possible on these stars. This fact is further reinforced by the usually high rotation rate of hot stars, which smears out spectral lines even more via rotational broadening. On the cool side of the HR diagram ($T_{eff} \lesssim 3500$ K), spectral lines become increasingly densely packed, less contrasted, and overlapping due to the presence of complex molecular bands. From an instrumental point of view, the intrinsic faintness of very cool stars makes it difficult to reach the required signal-to-noise ratio (SNR) for Doppler shift measurements. Moreover, such stars emit most of their flux in the IR, which puts more demanding constraints on groundbased instrumentation (cryogenic parts, background subtraction, removal of telluric lines, etc.).

Cool stars having spectral types from about F5 to M5 are therefore the best suited for precise radial velocity work. On the main sequence, this corresponds to masses between ~1.5 and ~0.1 $M_\odot$, and this represents about the mass range over which the radial velocity technique has been able to find planetary companions. An exception to this rule is the red giant branch and clump region of the HR diagram, where more massive stars spend some time during their post-main-sequence evolution. They are then sufficiently cool and slowly rotating to be targeted by Doppler surveys (see section 3.5).

Solar-type stars and M dwarfs exhibit thousands of absorption lines in their spectra, produced by all kinds of chemical elements. It is clear that one needs as many lines as possible to increase the radial velocity precision. However, strongly saturated stellar lines, such as hydrogen Hα, the calcium H and K lines, or the Na D doublet should be avoided for high-precision radial velocity work because of their very broad wings and potentially variable chromospheric emission in their core. Looking more closely at the thousands of nonsaturated lines present in the spectra of FGK stars, it clearly appears that Fe lines are by far the most numerous. Fe lines therefore represent the necessary basis for all precise Doppler shift measurements in these stars.

A key aspect of radial velocity work is to understand how the velocity precision depends on the shape of spectral lines. Intuitively, it is clear that precision will depend on three main parameters: the SNR in the continuum, the depth, and the width of the spectral line under consideration. The deeper and narrower the line, the better defined its centroid will be. Assuming approximately Gaussian shapes for spectral lines, it can indeed be shown that

$$\sigma_{RV} \sim \frac{\sqrt{FWHM}}{C \cdot SNR} \quad (21)$$

where FWHM is the full width at half maximum of the line, C is its contrast (its depth divided by the continuum level), and SNR is the signal-to-noise ratio in the continuum. From this formula it is clear that the broadening of spectral lines, caused either by stellar rotation or instrumental spectral resolution, can be a killer for radial velocity precision: An increase in rotational velocity, or decrease in resolution, simultaneously increases the FWHM and decreases the contrast since the total equivalent width must be conserved (C ~ 1/FWHM). The RV precision therefore degrades with FWHM³ᐟ². As an example, the achievable RV precision between the Sun ($v_{rot} = 2$ km s⁻¹) and a young G2V star with v sin i = 20 km s⁻¹ decreases by a factor of ~32. Similarly, observing at a spectral resolution R = 100,000 improves the precision by a factor of 2.8 compared to the same observation at R = 50,000, provided the stellar lines are unresolved (this is the case for old solar-type stars) and the observations are made in the photon-limited regime.

More generally, in the presence of photon and detector noise, the Doppler information content of a stellar spectrum can be expressed by the formulae derived by *Connes* (1985) and *Bouchy et al.* (2001)

$$\sigma_{RV,i} = c \frac{\sqrt{A_i + \sigma_D^2}}{\lambda_i \cdot |dA_i/d\lambda|} \quad (22)$$

$$\sigma_{RV} = c \left(\sum_i \frac{\lambda_i^2 \cdot |dA_i/d\lambda|^2}{A_i + \sigma_D^2} \right)^{-1/2} \quad (23)$$

In these equations, A_i is the flux in pixel i expressed in photoelectrons; λ_i is the wavelength of pixel i; σ_D is the detector noise per pixel, also expressed in photoelectrons; and c is the speed of light in vacuum. In these formulae, the shape of spectral lines is hidden in the spectrum derivative $dA_i/d\lambda$. The steeper the spectrum, the higher the Doppler information content will be. Equations (22) and (23) must be used with some caution since they require the numerical computation of this spectrum derivative, which will be systematically overestimated at low SNR and in the continuum regions. As a result, the RV precision will tend to be overestimated as well.

As a practical example, we can now apply these formulae to the spectra of slowly rotating solar-type stars, covering the whole visible range from 3800 to 7000 Å at a resolution R = 100,000. Typically, one obtains a global, photon-limited RV precision of 1 m s^{-1} when the SNR in the continuum at 5500 Å reaches ~145–180 per resolution element of 0.055 Å. The exact numbers depend on the projected rotational velocity of the star v sin i, its metallicity, and its spectral type, with K and M stars having more Doppler information in their spectra than G stars of a given magnitude (*Bouchy et al.,* 2001). One therefore concludes that, as far as photon noise is concerned, the detection of exoplanets is within the capabilities of high-resolution echelle spectrographs on 1-m-class telescopes and above.

As can be seen from equations (21) and (23), the global Doppler information content depends both on the intrinsic line richness (density and depth) and on the SNR of the spectrum. Both parameters have to be taken into account to evaluate the achievable RV precision on a particular type of stars observed with a particular instrument. In general, the spectral richness increases toward shorter wavelengths, and is particularly high in the blue visible region. As a consequence, the visible region is the natural choice for RV measurements of FGK stars, which also emit most of their light at these wavelengths. For M dwarfs, it is tempting to go to the near-IR. However, calculations on real spectra show that for early- to mid-M dwarfs the higher flux in the near-IR does *not* compensate for the large amount of Doppler information in the visible. Only for late-M dwarfs would a near-IR instrument be more efficient than a visible one. However, RV measurements in the near-IR have not reached yet the level of precision achieved by visible instruments. This may change in the near future.

2.4. Stellar Limitations to High-Precision Radial Velocities

Besides the general properties of spectra and spectral lines, other limitations on exoplanet detection around solar-type stars and M dwarfs have to be taken into account. Among these are several phenomena intrinsic to stellar atmospheres, which we will call "stellar noise" or "astrophysical noise" in the following. In this section we review the different sources of stellar noise, classified according to their typical timescales and starting on the high-frequency side.

2.4.1. P-mode oscillations. Stars having an outer convective envelope can stochastically excite p-mode oscillations at their surface through turbulent convection. These so-called solar-like oscillations have typical periods of a few minutes in solar-type stars and typical amplitudes per mode of a few tens of centimeters per second in radial velocity (e.g., *Bouchy and Carrier,* 2001; *Kjeldsen et al.,* 2005). The observed signal is the superposition of a large number of these modes, which may cause RV variations up to several meters per second. P-mode characteristics vary from star to star. The oscillation frequencies scale with the square root of the mean stellar density, while the RV amplitudes scale with the luminosity-to-mass ratio L/M (*Christensen-Dalsgaard,* 2004). As a consequence, oscillation periods become longer toward early-type stars along the main sequence, while they also increase when a star evolves off the main sequence toward the subgiant stage. Similarly, RV amplitudes become larger for early-type and evolved stars due to their higher L/M ratios. From the point of view of exoplanet search, low-mass, nonevolved stars are therefore easier targets because the stellar "noise" due to p-modes is lower. However, even in the most favorable cases, it remains necessary to average out this signal if aiming at the highest RV precision. This can be done by integrating over more than 1–2 typical oscillation periods. Usually, an exposure time of 15 min is sufficient to decrease this source of noise well below 1 m s^{-1} for dwarf stars.

2.4.2. Granulation and supergranulation. Granulation is the photospheric signature of the large-scale convective motions in the outer layers of stars having a convective envelope. The granulation pattern is made of a large number of cells showing bright upflows and darker downflows, tracing the hot matter coming from deeper layers and the matter having cooled at the surface, respectively. On the Sun, the typical velocities of these convective motions are 1–2 km s^{-1} in the vertical direction. Fortunately for planet searches, the large number of granules on the visible stellar surface (~10^6) efficiently averages out these velocity fields. However, the remaining jitter due to granulation is expected to be at the meters per second level for the Sun, probably less for K dwarfs (e.g., *Pallé et al.,* 1995; *Dravins,* 1990). The typical timescale for granulation, i.e., the typical time over which a given granule significantly evolves, is about 10 min for the Sun. On timescales of a few hours to about one day, other similar phenomena occur, called meso- and supergranulation. These are suspected to be larger convective structures in the stellar photosphere that may induce additional stellar noise, similar in amplitude to granulation itself. However, the origin and properties of these structures are still debated. Overall, granulation-related phenomena likely represent a significant noise source when aiming at submeters per second RV precision, and observing strategies to minimize their impact should be envisaged.

2.4.3. Magnetic activity. Magnetic fields at the surface of solar-type stars and M dwarfs are responsible for a number of inhomogeneities in the stellar photosphere that represent yet another source of noise for planet searches via precise RV measurements (e.g., *Saar and Donahue,* 1997; *Santos et al.,* 2000; *Wright,* 2005). Among these are cool spots and bright plages, which may cover a variable fraction of the stellar disk, evolve in time, and are carried across the stellar disk by stellar rotation. Such structures affect RV measurements in several ways, but the most important effect comes from the flux deficit (or excess) in these regions, which moves from the blueshifted to the redshifted part of the stellar disk as the star rotates. This changes the shape of spectral lines and introduces modulations in the measured radial velocities on timescales comparable to the stellar rotation period. The properties of active regions vary widely from star to star depending on their mean activity level, which in turn mainly depends on stellar age for a given spectral type. Solar-type stars undergo a continuous decrease in rotational velocity as they age, starting with typical rotation periods of ~1–2 days at 10 m.y., and slowing down to P_{rot} = 20–50 days at 5 G.y. Correspondingly, the magnetic dynamo significantly weakens with age and active regions become much smaller.

Generally speaking, stellar activity is a major problem for exoplanet searches around young stars (≤1 G.y.), with dark spots causing RV variations larger than 10–100 m s^{-1}. Even hot Jupiters, causing RV amplitudes of hundreds of meters per second, are expected to be difficult to detect around the youngest stars (1–10 m.y.) because stellar rotation periods are then close to the orbital periods. This may preclude investigations of planetary system formation with the RV technique at the stage where the protoplanetary disk is still present. A possible solution to activity-related problems in exoplanet searches is to observe in the near-IR domain, where the photospheric contrast of dark spots is lower than at visible wavelengths. However, this possibility remains to be explored. A broadly used diagnostic of stellar activity is the line bisector, which traces line shape variations due, e.g., to dark spots. In active stars, a clear anticorrelation is often seen between bisector velocity span and radial velocity, which, in certain cases, would make it possible to approximately correct the velocities for the influence of the spot (*Queloz et al.,* 2001). The main problem with bisectors is that they lose most of their sensitivity in stars with a low projected rotational velocity v sin i, because the rotation profile is not resolved any more (e.g., *Desort et al.,* 2007).

The difficulties with young stars have led most planet search surveys to effectively focus on old, slowly rotating solar-type stars. These are usually much more quiet and a large fraction of them show activity levels sufficiently low to permit RV measurements at the 1 m s^{-1} level, and even below in favorable cases.

Detailed studies of the behavior of solar-type stars at this level of precision have yet to be done in order to understand the ultimate limitations of RV measurements set by stellar activity. For example, spectroscopic indicators can be used to monitor activity levels. The most famous is the Ca II H&K chromospheric index, which closely traces the presence of active regions (e.g., *Wilson,* 1978; *Baliunas et al.,* 1995). Precise measurements of this indicator and other spectroscopic diagnostics permits the derivation of stellar rotation periods even for older, chromospherically quiet stars. This analysis helps to distinguish between astrophysical noise sources and dynamical radial velocity signals. The general properties of activity-related noise in quiet stars have to be better quantified. Obviously, the Sun has been extensively studied in this respect and can serve as a prototype to understand other stars. However, solar research and instrumentation have usually pursued different scientific goals and therefore a direct comparison between solar and stellar observations is often not straightforward.

It is clear that the global characteristics of activity related RV noise depend on at least four factors: the area covered by dark spots and plages, the stellar projected rotational velocity v sin i, the stellar rotation period, and the typical lifetime of active regions. While the first two factors directly influence the magnitude of the noise, the latter two determine its characteristic timescales and temporal behavior. Dense sampling, averaging, and binning of the data over timescales similar to the stellar rotation period may allow RV surveys to reach precisions as high as 10 cm s^{-1}, thus making it possible to detect Earth-like planets in the habitable zone around solar-type stars. Indeed, the power spectral density of stellar noise is expected to flatten at frequencies lower than the rotation period, i.e., the noise is likely to become "white" again, instead of "red," when binning the data over sufficient periods of time. Since the timescale of stellar rotation (~30 d) is significantly smaller than the orbital periods of planets at 1 AU (~300 d), such a binning is indeed possible. Accumulating a large number of data points per stellar rotation cycle should then make it possible to efficiently average out stellar noise, although the ultimate attainable levels of precision remain controversial.

2.4.4. Activity cycles. The Sun undergoes a well-known 11-year magnetic cycle, over which the number of active regions present on its surface dramatically changes. At solar minimum, there is sometimes no spot at all to be seen during several months, while at activity maximum about 0.2% of the solar surface may be covered with spots. This obviously impacts the achievable RV precision, although the lack of solar RV data obtained with the "stellar" technique makes it difficult to quantitatively estimate the minimal and maximal RV variability. Other solar-type stars also exhibit similar activity cycles, as shown by Ca II H&K measurements. One can therefore suspect that the RV jitter of solar-type stars varies in time with the activity cycle. Besides varying short-term *scatter* in the radial velocities, there may also be *systematic trends* due to the varying average fraction of the photosphere covered with spots. Such effects may arise, for example, from changes in the granulation pattern that take place in active regions: Solar observations show that granular motions tend to freeze due to strong magnetic fields in these regions. As a result, spectral lines may show reduced (or even vanishing) convective blueshifts, and therefore induce a systematic RV shift in the disk-integrated spectrum (*Saar and Fischer,* 2000; *Lindegren and Dravins,* 2003). However, such effects remain to be better explored and quantified among Doppler planet-search survey stars.

On the other hand, about 10–20% of old solar-type stars seem to be in a very quiet state, having a low mean activity level and showing no cyclic variations. Part of these stars may be in a so-called Maunder-minimum state, as the Sun in the seventeenth century. Such stars represent ideal targets for high-precision RV exoplanet searches. Ongoing Doppler surveys have been accumulating a sufficient number of Ca II H&K measurements to be able to identify this low-activity population in the solar neighborhood, and RV campaigns focusing on these stars are likely to push stellar RV noise down to unprecedented levels. High-quality, densely sampled RV datasets obtained on such stars will eventually answer the questions about the ultimate limits set by stellar activity on the RV technique.

2.5. Instrumental Challenges to High-Precision Radial Velocities

Measuring Doppler shifts with a precision of 1 m s^{-1} is a truly challenging task from an instrumental point of view. Indeed, this corresponds to the measurement of wavelength shifts of a few 10^{-5} Å, which, for a R = 100,000 high-resolution instrument, represent ~1/3000 of the line width or about 1/1000 of a CCD pixel on the detector. In practice, this normally cannot be achieved on a single spectral line because of insufficient SNR. Only the use of thousands of lines in stellar *and* wavelength calibration spectra makes it possible to reach such high levels of precision. These numbers, along with those concerning photon noise (section 2.3), make it obvious that high spectral resolution (R ≥ 50,000), large spectral coverage, and high efficiency are necessary instrumental prerequisites to achieve radial velocity measurements at the meters per second level. This explains why high-resolution, cross-dispersed echelle spectrographs have been systematically chosen for this purpose.

In practice, the problem is that a variety of instrumental effects can potentially cause instrumental shifts 2–3 orders of magnitude larger than the required precision of 1 m s^{-1}. The main effects are described below.

2.5.1. Variations in the index of refraction of air. Changes in ambient temperature and pressure in the spectrograph translate into wavelength shifts due to the varying index of refraction of air. These shifts may amount to several hundreds of meters per second during a single night. Indeed, a temperature change of 0.01 K or a pressure change of 0.01 mbar are sufficient to induce a drift of 1 m s^{-1}. These large-amplitude drifts must therefore be measured and/or corrected for to high accuracy.

2.5.2. Thermal and mechanical effects. Temperature variations of mechanical and optical parts in the spectrograph induce mechanical flexures and optical effects (e.g., PSF changes) that mimic wavelength shifts on the detector. Similarly, moving parts and changes in instrumental configuration may also induce such drifts. Depending on instrumental design and ambient conditions, these effects may easily reach several tens to hundreds of meters per second.

2.5.3. Slit illumination. A spectrograph is basically an optical device producing a dispersed image of the entrance slit on the detector. Therefore, any variations in the slit illumination will translate into variations in the recorded spectrum, including PSF changes and shifts in the main-dispersion direction, i.e., wavelength shifts. Illumination variations arise from changing observing conditions (seeing, air mass), telescope focus, telescope guiding, possible vignetting along the light path, etc. As an order-of-magnitude calculation, a small photocenter shift of 1/100 of the slit width, due for example to guiding corrections, will induce a 30 m s^{-1} radial velocity shift in a R = 100,000 spectrograph. This shows how critical slit illumination is for high-precision radial velocities. Two philosophies have been developed to overcome this major difficulty: a self-referencing method that superimposes iodine lines on the stellar spectrum (see section 2.6), and the use of optical fibers to feed the spectrograph, taking advantage of their high light-scrambling properties (see section 2.7).

2.5.4. Wavelength calibration. The stellar spectrum must be precisely wavelength calibrated using a stable, well-characterized reference having a large number of spectral lines extending over the largest possible wavelength range. We briefly discuss here the intrinsic properties of different calibration sources, and defer the description of how they are used to sections 2.6 and 2.7. Long-term wavelength stability of spectral lines at the level of 1 m s^{-1} (~10^{-9} in relative units) is not a trivial requirement and it is necessary to carefully examine all relevant physical effects that may affect the spectrum of calibration systems at this level of precision.

The two main calibration sources used for high-precision RV measurements have been iodine cells and ThAr hollow cathode lamps. Molecular iodine vapor provides a dense forest of absorption features between 5000 and 6200 Å, while ThAr lamps produce thousands of emission lines over the whole visible and near-IR ranges. Iodine cells are in widespread use in different areas of physics where they serve as absolute wavelength calibrators. In high-precision radial velocity work, they have demonstrated a long-term stability at the ~1–2 m s^{-1} level. However, their intrinsic limitations have to be clarified. Slight short-term or long-term variations in iodine pressure may affect the overall shape of the complex iodine spectrum, made of numerous Doppler-broadened and overlapping molecular transitions. Moreover, the usable wavelength range is limited to ~1000 Å, which makes it impossible to use the stellar Doppler information in other spectral regions. Finally, the Doppler information content of the iodine itself is a function of the achieved SNR in the continuum, which is determined by the stellar flux in the case of star + iodine exposures (see section 2.6). As a consequence, high-SNR observations are required. This sets strong constraints on the target magnitude and/or exposure time.

On the other hand, ThAr lamps also have their advantages and drawbacks. They have also demonstrated a long-term stability at the ~1–2 m s^{-1} level and cover a larger spectral range than iodine. Their spectra show, however, numerous blends and a large dynamic range in line intensities that complicate the precise measurement of line positions. They also have a limited lifetime. Aging effects induce changes in line intensities and small wavelength shifts probably due to slow pressure variations in the lamps. Argon is most sensitive to these effects and Ar lines may drift by several tens of meters per second over the lifetime of a lamp. It is therefore important to avoid using these lines in the wavelength calibration process. Thorium lines are much less sensitive, but drifts of a few meters per second have nonetheless been measured. These can be corrected to high precision by measuring the differential drift of Ar lines with respect to Th lines, knowing the sensitivity ratio between them. With this procedure ThAr lamps have demonstrated a long-term and lamp-to-lamp stability at the ~30 cm s^{-1} level (*Lovis et al.*, 2010, in preparation).

In the near-IR domain, finding an appropriate calibrator is even more difficult. ThAr lamps may be used, but in most cases they do not provide lines with sufficiently high density for a precise calibration. Gas cells with various molecules have been proposed but none of them appears to meet all the requirements such as sufficient wavelength coverage, high line density and depth, nonsuperposition with telluric lines, usability, etc. A straightforward solution is to use the numerous telluric lines available in the near-IR as wavelength references. However, great care has to be taken in disentangling stellar and telluric lines. And, more importantly, telluric lines are known to be variable at the level of ~5–10 m s^{-1} due to atmospheric winds and air mass effects.

Clearly, the ideal calibration source for high-precision RV measurements remains to be invented. A promising new approach is the development of the laser comb technology for use in astronomy (*Murphy et al.*, 2007). Laser frequency combs deliver an extremely regular pattern of equally spaced emission lines (in frequency) over a potentially broad wavelength range. Their decisive advantage comes from the fact that line wavelengths can be precisely controlled by locking the comb to an atomic clock, creating a direct link between the most accurate frequency references ever developed and optical frequencies. The accuracy and stability of comb lines is then basically given by the accuracy of the atomic clock, which can easily exceed 10^{-11} in relative units (corresponding to less than 1 cm s^{-1}). While such systems will definitively solve the reliability problems encountered with other calibration sources, they still have to be adapted to high-resolution astronomical spectrographs. The main difficulty is that current laser repetition rates (~1 GHz) are too low by a factor of ~10 for the resolution of astronomical spectrographs. In other words, consecutive comb peaks are much too close to each other even at R = 100,000 resolution. A possible solution to this problem is to filter out unwanted modes using a Fabry-Pérot cavity (*Li et al.*, 2008; *Steinmetz et al.*, 2008; *Braje et al.*, 2008). In this way, only 1 out of 10 modes, for example, can be transmitted, producing a clean comb spectrum for the spectrograph. Stringent constraints apply to the filtering cavity, however, which should not jeopardize the intrinsic comb accuracy by introducing instabilities or imperfections in the filtering. Another difficulty of laser frequency combs, especially in the visible domain, is the need for a broad wavelength coverage. In brief, these systems have not yet reached a level of reliability and practicality that would make them routinely usable at astronomical observatories. The rapid development of this technology should nevertheless lead to satisfactory solutions to these problems in the near future, making laser combs the calibrator of choice for ultra-high-precision radial velocity measurements.

2.5.5. Detector-related effects. CCD detectors have a number of imperfections that may also impact radial velocity precision. First of all, CCD pixels do not all have the same sensitivity, due partly to intrinsic sensitivity variations and partly to variations in physical size. Flat-fielding cannot make the difference between both effects. Variations in pixel-to-pixel size have an impact on the wavelength solution and may cause spurious small-scale shifts when the stellar spectrum moves on the detector. A promising way to calibrate pixel properties would be to use a tunable wavelength reference, such as a laser frequency comb, that would be able to scan the CCD pixels and measure their response and size.

Thermal stability and control of the detector are also important since thermal dilation and contraction directly affect the dispersion solution. These effects can be corrected with both methods presented in sections 2.6 and 2.7.

Another potentially damaging effect is the charge transfer efficiency (CTE) of the CCD. During CCD readout, charges are transferred from pixel to pixel toward the readout ports, whereby a small fraction of the total charge is inevitably left behind at each step. As a result, line profiles are slightly asymmetrized, producing an apparent shift of spectral lines. Typical fractional losses of modern CCDs are around ~10^{-6} per transfer. Normally, this is just low enough to induce shifts that remain small or comparable to the photon noise on line positions. However, some CCDs seem to be of poorer quality, and in particular a significant degradation of CTE as a function of signal level has been noticed in several cases. This means that at lower signal-to-noise ratios, CTE becomes so low as to induce line shifts that clearly exceed random noise, i.e., systematic effects as a function of SNR are produced. Great care should therefore be taken in selecting and characterizing CCD devices for high-precision radial velocity instruments. If the CTE behavior as a function of flux can be adequately calibrated, CTE-induced effects could in principle be corrected, but it is obviously better to choose a high-quality chip in the first place.

2.5.6. Spectrum contamination. All features appearing in a spectrum that are not related to the target star may induce spurious radial velocity signals. Telluric lines from Earth's atmosphere represent one such example, with a particularly high density of absorption bands in the red and the near-IR. Contaminated spectral regions must be either rejected or properly treated since they are at rest in the observer's frame, contrary to stellar lines. Consequently, the main result of pollution by telluric lines will be a spurious one-year periodic signal in the radial velocities caused by Earth's motion around the Sun. The visible region is much less affected by telluric lines than the near-IR, but several zones starting at ~5500 Å should nonetheless be avoided.

Another important contamination source is moonlight. Diffuse moonlight pollutes the sky background with the reflected solar spectrum, which is superimposed onto the target star spectrum. This obviously also induces spurious radial velocity signals. Their magnitude depends on the relative brightness of the sky with respect to the source in the entrance slit of the spectrograph. Sky transparency is thus an important factor in this respect. Observations close to the Moon or in poor weather conditions when the Moon is present should be avoided.

Finally, the target spectrum may also be contaminated by the light of faint companions or background objects located sufficiently close on the sky to also enter the spectrograph slit. Obvious contaminations due to binary components of similar magnitude are usually easy to identify in the spectrum itself or in the radial velocities (large variations). This is not the case for faint contaminants with magnitude differences larger than ~5 mag. Such objects may induce subtle radial velocity variations both on the short- and long-term. Short-term effects arise from the varying relative mixture of target and contaminant lights due to seeing and telescope pointing. On longer timescales, binary stars in the background may induce RV variations with the period of the binary that could mimic a planetary signal. In most cases, a careful monitoring of line shape variations, e.g., using bisector diagnostics, is able to reveal the presence of contaminants.

2.6. The Iodine Cell Technique

As noted in section 2.5, velocity measurements of 1 m s^{-1} correspond to Doppler wavelength shifts, $d\lambda$, on the order of 1/1000th of a pixel shift. One technique for measuring such tiny shifts makes use of a glass cell containing iodine, which imposes a rich forest of molecular absorption lines into the stellar spectrum. The iodine lines serve as a grid against which the almost imperceptible stellar line shifts can be measured. The molecular transitions span a wavelength range from 5000 Å to 6200 Å. For long-term stability, it is critical that the column density of the reference gas remains constant over the length of the observing program, which may span decades, so once the iodine is in place, the cell is permanently sealed.

An iodine cell can easily be constructed in a chemistry lab. The glass cell is constructed by a glass blower with a small side tube welded onto the cylindrical body of the cell (part of a manifold for importing the iodine gas) and optical flats welded at both ends of the glass cylinder. After construction, antireflective coatings can be applied to the flat surfaces to minimize light losses. The cell should be cleaned with an acidic solution and thoroughly rinsed with water. The cell is then placed in a drying oven for a few hours to ensure that the interior is completely dry. Otherwise, trapped water vapor can condense in the cold night air at the telescope, fogging the inside surface of the exposed optical flats. Once dry, a grain of solid iodine is inserted into the manifold and the system is sealed and pumped to low pressure with a hi-vacuum pump. The entire cell and manifold are then placed into a water bath at a temperature of 37 C to sublimate the iodine, which fills the cell with a lavender pink gas. With the I2 gas now at constant vapor pressure, the glass blower permanently seals off the side tube. The trace amount of iodine is quite nonreactive with glass and the cell has a stable quantity of molecular I2 that will condense into a solid speck if the temperature of the cell is lower than 37 C.

For use at an observatory, a thermocouple is taped to the side of the iodine cell and the cell is wrapped with heat tape. A temperature controller maintains a constant temperature of 55°C throughout the night to ensure that all the enclosed iodine is in a gaseous state at a constant vapor pressure. The iodine cell is inserted in the light path, ideally in front of the slit because it is a heat source. If that is not possible, then the system needs to be thermally isolated to prevent convective currents and temperature variations in the spectrometer. The optical surfaces of the I2 cell and the I2 gas opacity result in about 15% light loss.

In contrast to the well-separated and deep HF molecular lines, the I2 absorption lines are dense and narrow. Figure 1a shows a 2-Å wavelength segment of the iodine spectrum obtained with a Fourier Transform Spectrometer (FTS) obtained with a resolution of about 900,000. Figure 1b shows the spectrum of that same I2 cell, illuminated with a rapidly rotating (featureless) B-type star and observed at Lick Observatory with the Hamilton spectrometer (R = 55,000). The quality of the spectrum is degraded by the lower spectral resolution and by the point spread function (PSF) of the instrument. However, using the FTS spectrum as a starting point, it is possible to find a PSF model that, when convolved with the FTS spectrum (Fig. 1a), will produce the observed spectrum (Fig. 1b). This is a key step in modeling program observations (i.e., stellar observations made with the iodine cell).

Doppler analysis with an iodine reference cell is carried out by forward modeling, rather than cross correlation. The first step is to model an observation of the iodine spectrum (Fig. 1b). It is good practice to obtain an iodine spectrum at the beginning and end of each night to obtain initial guesses for the wavelength solution and the PSF model. A rapidly rotating B-type star is an excellent light source for the iodine spectrum. The stellar spectrum is essentially featureless and the light illuminates the optics in the same manner as the program observations. Because the PSF varies spatially over the detector, the spectrum is divided into smaller chunks, typically 2 or 3 Å wide and a wavelength solution and PSF model is derived for each of these chunks. A Levenberg-Marquardt algorithm drives the model fit for each chunk. To create the model, the matching segment is extracted from the FTS iodine spectrum (containing the

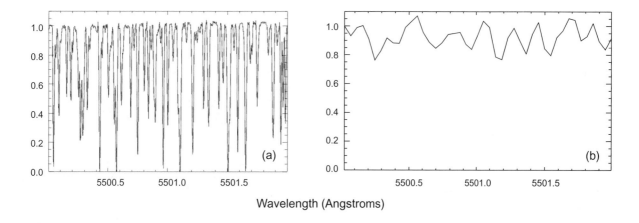

Fig. 1. (a) An R = 900,000 FTS scan of the molecular iodine gas cell spanning only 2 Å. (b) A spectrum of the same cell, illuminated by light from a rapidly rotating B-type star and observed with the Hamilton spectrometer at R = 55,000. The iodine lines contain both wavelength and PSF information.

wavelength information), then splined onto the oversampled wavelength scale of the observation and convolved with a PSF description.

Valenti et al. (1995) describe a flexible technique for modeling the PSF as a sum of Gaussians. A central Gaussian accommodates most of the PSF, but several (five or more) Gaussians are also placed at fixed offsets from the central Gaussian. These flanking Gaussians are allowed to piston in amplitude to model asymmetries in the wings of the PSF. In this first step of modeling an iodine spectrum, there is no Doppler shift to fit, so the free parameters in the Levenberg-Marquardt fitting algorithm consist of the wavelength of the first pixel in the chunk, the dispersion across the chunk, and the PSF parameters (i.e., the FWHM of the central PSF Gaussian and the amplitudes of the flanking Gaussians).

The next step is to model the program observations: stellar observations taken with the iodine cell. The previous analysis of the iodine spectrum provides initial guesses for both the wavelength solution and the PSF model that should be used for the stellar observations and the Doppler shift of the star, $d\lambda = \lambda\, v/c$, is the important additional free parameter. In order to carry out the forward modeling of program observations, three ingredients are needed to model the program observations: the FTS iodine spectrum, a PSF model, and an intrinsic stellar spectrum (ISS).

Ideally, the ISS would have extremely high resolution and high signal-to-noise. In practice, the ISS is generally obtained with the same spectrometer as the program star observations. The starting point for the ISS is a so called template observation, an observation of the star made without the iodine cell and with higher signal-to-noise and higher spectral resolution. Because the template spectrum is smeared by the spectrometer PSF, a deconvolution is carried out to try to recover the true ISS. Observations of featureless B stars through the iodine cell are taken before and after template observations and the PSFs that are recovered from modeling the iodine spectra are used in the deconvolution algorithm. To the extent that the deconvolution fails or introduces ringing in the continuum because of noise, the ISS is less than perfect and compromises Doppler precision.

To model the program observations, the deconvolved stellar template $S(\lambda)$ (our best representation for the true ISS) is multiplied by the FTS iodine spectrum $I2(\lambda + d\lambda)$ and convolved with the PSF description

$$[S(\lambda)\cdot I2(\lambda + d\lambda)]\otimes \text{PSF} = I_{obs}(\lambda) \qquad (24)$$

The free parameters for each chunk of the program observations include the all-important wavelength shift (which gives the velocity of the star), the wavelength of the first pixel, dispersion across the chunk, continuum normalization, and the parameters used to describe the PSF. Figure 2 shows the template observation (Fig. 2a) and the program observation (Fig. 2b) containing iodine absorption lines. The overplotted red dots represent the synthetic model created with equation (24).

With about 80 pixels and 15 free parameters, typical model fits yield a reduced χ^2 for each chunk of about 1.0. Individual chunks typically contain a few spectral lines and provide velocities that are accurate to 30–50 m s^{-1}. The relative change in velocity is always made with respect to the same wavelength segment; barycentric velocities change the location of spectral lines on the physical format, so the chunks vary in pixel space, but are always the same size and contain the same stellar lines. The spectral chunks are independent measurements of the velocity; the uncertainty or precision of the Doppler measurement is the standard deviation of the velocities for all of the chunks and improve over the single chunk precision by the square root of the number of chunks. The velocity for a particular observation is the mean velocity difference in all of the chunks.

The most significant weakness in Doppler modeling is

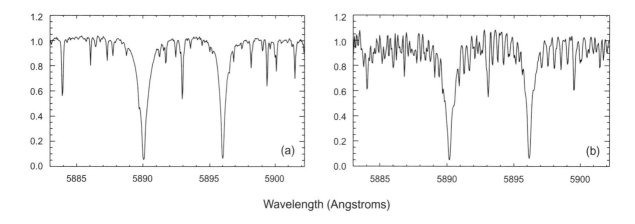

Fig. 2. (a) A template observation is obtained without the iodine cell and deconvolved with a PSF model to produce the intrinsic stellar spectrum. (b) The intrinsic stellar template is multiplied by a high-resolution iodine spectrum and convolved with a PSF to model the program observation of that star. The synthetic model is overplotted with dots on the program observation that contains many narrow iodine absorption lines.

uncertainty in the PSF model. The PSF model is critical to both the deconvolution of the template spectrum and the best-fit model of the program observations. While physical constraints can be included to model the wavelength solution (it should be continuous from chunk to chunk and continuous over the order) and the dispersion (it should also be a smooth function of wavelength across the order and stable throughout the night), the PSF is more problematic. If the PSF only consisted of an instrumental component, then it would vary slowly over the night or season with changing temperature or pressure in the spectrometer. However, the uncontrolled component of the PSF is the variable slit image. Changes in seeing or errors in telescope guiding produce spatial and temporal changes in the cone of light that enters the spectrometer. Because the optical elements in the spectrometer are illuminated differently from observation to observation, the PSF changes from one observation to the next. A fiber optic cable would scramble the light at the slit and overcome this source of error. With a more stable PSF, it would be possible to simultaneously solve for the PSF of a given spectral chunk in stacks of several observations taken during a given night. This would provide a more robust PSF for template deconvolution (to produce the ISS) and a more accurate PSF description for Doppler analysis of program observations.

2.7. The Simultaneous Reference Technique

Another method to achieve high-precision RV measurements without self-referencing the stellar spectra is to build a dedicated instrument in which all possible sources of RV errors are minimized or corrected to high precision. This is the philosophy of the so-called "simultaneous reference" technique, or ThAr technique (*Baranne et al.*, 1996). Going through the different instrumental effects listed in section 2.5, it appears that the three most important problems to deal with are the variations in the index of refraction of air, mechanical flexures, and slit illumination. The first two are internal effects in the spectrograph. To track and correct them, the basic idea is to inject a second, reference spectrum into the spectrograph *simultaneously* with the scientific observation. Normally, a ThAr lamp is used for this purpose. The instrument has therefore two channels: a science and a simultaneous-reference channel. Before scientific observations (at the beginning of the night), wavelength calibration exposures are acquired by injecting the ThAr spectrum into *both* channels. This gives independent dispersion solutions for both channels. Obviously, these wavelength calibrations are only strictly valid at the moment they are taken. The idea is to use the simultaneous-reference channel to monitor instrumental drifts during the night and to correct the wavelength calibration accordingly. The underlying assumption is that the above-mentioned effects (varying index of refraction and mechanical flexures) will perturb the dispersion solutions of both channels in the same way, because both channels follow very similar paths in the spectrograph, from the slit to the detector. The instrumental drift is measured by comparing, in the simultaneous-reference channel, the ThAr spectrum obtained at the beginning of the night to the ThAr spectrum obtained simultaneously with the stellar observation. The measured drift is then subtracted on the science channel, assuming it suffered the same instrumental variations.

The long-term precision of this technique relies on the wavelength calibration obtained at the beginning of each night in the science channel. The derived dispersion solution must be as precise as possible since it determines the "baseline" wavelength calibration for each night. The position of ThAr emission lines must be properly measured, taking care of the blending between neighboring lines and weighting the lines according to their individual Doppler information content. Individual laboratory wavelengths have to be precise enough to avoid introducing excess scatter around dispersion solutions compared to photon noise. To this purpose, an updated list of ThAr reference wavelengths has been published recently (*Lovis and Pepe, 2007*).

Slit illumination is the third major source of instrumental radial velocity jitter. RV shifts induced by atmospheric turbulence and telescope guiding cannot be corrected with the simultaneous-reference channel. In this case, the only way to proceed is to minimize as much as possible illumination variations, i.e., to scramble the light as much as possible before injecting it into the spectrograph, and to stabilize telescope guiding. This is achieved by using optical fibers to link the telescope to the spectrograph, and by developing a high-accuracy guiding system. Optical fibers reduce the inhomogeneities in the light distribution by a factor of typically ~500. Moreover, optical systems exchanging nearfield and farfield ("double scramblers") can be used to further homogenize the light beam. With all these precautions, it has been possible to reduce illumination jitter down to the equivalent of $~10^{-4}$ of the slit width, corresponding to an RV jitter of ~0.3 m s^{-1} for a R = 100,000 spectrograph. Further improvements are certainly possible.

The result of this whole strategy is a precisely wavelength calibrated stellar spectrum, on which the radial velocity can now be measured. This is done with a variant of the cross correlation method. Cross-correlation is an efficient way of computing the global shift between two similar signals. In this case, we would need a reference spectrum for each target star, against which we would correlate the observed spectrum. However, this would require a very large observational effort, since thousands of very high-SNR spectra are needed. A much simpler strategy is to use a binary "mask" containing "holes" at the rest wavelength of stellar lines, i.e., a binary transmission function with ones at the position of stellar lines and zeros elsewhere. Cross-correlating the stellar spectrum with such a mask yields a kind of average stellar "master" line made of the piling up of all lines transmitted through the mask. This procedure is optimal in the sense that it extracts the whole Doppler information from the spectrum using a noise-free template and concentrates it into a single cross-correlation function (CCF). Actually, to really optimize signal extraction, stellar lines have to be weighted by their relative contrast, since Doppler information is proportional to line depth [see equation (21) and *Pepe et al.* (2002)]. Since, for a given star, all lines used in the mask can be assumed to have the same FWHM, line depth is the only relevant parameter to include in the optimal weights (SNR is included by construction). In principle, it would be possible to develop a binary mask for each spectral type. However, differences between close spectral types are not significant from the point of view of signal extraction and experience has shown that a single mask for each main type (G, K, M) is usually sufficient. Finally, the radial velocity is measured by fitting a suitable model to the CCF, usually a Gaussian for slowly rotating stars. In the whole procedure, the important point is to always use the same mask, model, and correlation parameters for a given star, to avoid introducing "algorithm noise" in the radial velocities.

The simultaneous-reference technique has been successfully used since 1993 and the development of the ELODIE spectrograph at Observatoire de Haute-Provence (OHP, France), which led to the discovery of the first exoplanet around a normal star (*Mayor and Queloz, 1995*). From about 10 m s^{-1} in the early days, the precision of the technique has been continuously improved since then, triggered by further instrumental developments such as CORALIE (*Queloz et al., 2000*) at La Silla Observatory (Chile), SOPHIE at OHP (*Bouchy et al., 2006*), and particularly HARPS, also at La Silla (*Mayor et al., 2003*). All these instruments are cross-dispersed echelle spectrographs covering the whole visible range. ELODIE and CORALIE have spectral resolutions R = 50,000 and 60,000 respectively, and are installed in a separate room isolated from the telescope. They are exposed to changing ambient pressure and temperature. Consequently, the typical nightly drifts measured in the simultaneous-reference channel amount to ~100 m s^{-1}. SOPHIE (the successor of ELODIE) has R = 75,000 and lies in a temperature- and pressure-stabilized enclosure, which reduces nightly drifts to a few meters per second. As far as performances are concerned, ELODIE was able to reach a precision of ~7 m s^{-1}, while CORALIE has achieved 3–5 m s^{-1} and SOPHIE ~3 m s^{-1}. The three instruments have been monitoring more than 2000 stars in the solar neighborhood. They have been very successful in discovering exoplanets, with more than 100 giant planets detected and many transiting candidates confirmed. Many statistical properties of the exoplanet population could be derived from these large surveys (see, e.g., *Udry and Santos, 2007*).

In 2003, the installation of the High Accuracy Radial Velocity Planet Searcher (HARPS) on the European Southern Observatory 3.6-m telescope at La Silla marked a further major improvement in radial velocity precision. HARPS was designed from the beginning to minimize all instrumental RV errors, with the goal of achieving 1 m s^{-1}. Most importantly, it is installed in a temperature-controlled enclosure, the spectrograph itself being in a vacuum vessel (see Fig. 3). Temperature is kept constant throughout the year to ±0.01 K and pressure is maintained below 0.01 mbar. As a consequence, nightly drifts as measured in the simultaneous-reference channel never exceed 1 m s^{-1}. Spectral resolution is R = 115,000, which, coupled to the larger telescope aperture, gives a significant gain in radial velocity precision as far as photon noise is concerned. Thanks to all these improvements, HARPS has achieved an unprecedented long-term precision of ~50 cm s^{-1}, and even ~20 cm s^{-1} on the short term (within a night). These performances have created new scientific possibilities: the search for Neptune-mass planets and super Earths down to a few Earth masses (e.g., *Lovis et al., 2006*; *Mayor et al., 2009a*). Indeed, these objects induce radial velocity signals smaller than 3–4 m s^{-1} and were therefore extremely difficult to detect with previous instruments. In a few years of operations, HARPS has revolutionized our knowledge of low-mass planets and will be able to deliver reliable statistics on this new population in the coming years (see sections 3.6 and 3.7).

3. MILESTONE DISCOVERIES AND HIGHLIGHTS

3.1. Overview: Fifteen Years of Discoveries

Figure 4 gives an overview of exoplanet discoveries as a function of time, from 1989 to 2009. The plot shows just how fast this new field of astronomy has developed in recent years and become one of the hot topics of the moment. At the forefront of this breakthrough stands the radial velocity technique, which has detected the majority of the known exoplanets. Individual major discoveries and milestones are discussed in the following sections.

3.2. 51 Pegasi: The First Exoplanet Orbiting a Solar-Type Star, and the First Hot Jupiter

As mentioned in the introduction, major instrumental efforts were being made in the 1980s and the early 1990s to improve radial velocity precision in the hope of finding giant exoplanets and brown dwarfs around nearby solar-type stars. *Campbell et al.* (1988) obtained precise (~10 m s^{-1}) RV measurements of about 20 nearby stars using a hydrogen-fluoride cell as a wavelength reference. Shortly afterward, *Latham et al.* (1989) announced the discovery of an object with a minimum mass of 11 M$_{Jup}$ around the solar-type star HD 114762. Because the unknown inclination angle of the system could only make this object heavier than the deuterium-burning limit, it was considered as a brown dwarf or even a very low-mass star. In 1992, the search for low-mass companions saw an unexpected development: the discovery of three Earth-mass objects orbiting the millisecond pulsar PSR 1257 + 12 using the pulsar timing technique (*Wolszczan and Frail,* 1992; see also the chapter by Wolszczan and Kuchner). Although this demonstrated that planetary bodies may exist in widely diverse environments, the detected objects likely evolved in a different manner than solar system planets.

Finally, new RV surveys of solar-like stars using the two main techniques described in sections 2.6 and 2.7 (*Butler et al.,* 1996; *Baranne et al.,* 1996) started to produce their first results. The achieved precisions of 5–10 m s^{-1} were adequate to easily detect Jupiter-mass planets on close orbits. However, the existence of such objects was not anticipated and observing strategies were really designed to find solar system analogs, i.e., gas giants at 5–10 AU. It came therefore as a huge surprise when *Mayor and Queloz* (1995) announced the discovery of an object with a minimum mass of 0.5 M$_{Jup}$ orbiting the star 51 Peg in only 4.23 d. The detection was made with the ELODIE spectrograph at Observatoire de Haute-Provence (France), where a survey of ~150 nearby stars was being carried out. The original radial velocity curve is shown in Fig. 5. It first had to be demonstrated that the signal indeed was dynamical, and was not related to spurious instrumental or stellar surface phenomena. The confirmation of the reality of the signal had to wait only a few weeks until *Marcy and Butler* (1995)

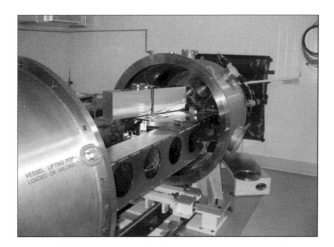

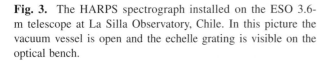

Fig. 3. The HARPS spectrograph installed on the ESO 3.6-m telescope at La Silla Observatory, Chile. In this picture the vacuum vessel is open and the echelle grating is visible on the optical bench.

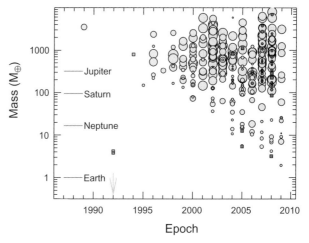

Fig. 4. Timeline of exoplanet detections in a mass vs. year of discovery diagram. Symbol sizes are proportional to orbital eccentricity. The continuously decreasing detection threshold illustrates the progress of radial velocity surveys and shows that planet searches have now entered the Earth-mass domain.

published their own data on 51 Peg. Within one year, this team announced two more distant giant planets (*Marcy and Butler*, 1996; *Butler and Marcy*, 1996) and three other close-in 51 Peg-like planets (*Butler et al.*, 1997). These discoveries marked the beginning of the highly successful search for planets using precise radial velocities. Moreover, a new category of planets, unknown in our solar system, emerged: the hot Jupiters.

Planet formation theories did not foresee the existence of such exotic objects, for the simple reason that it is impossible to form them so close to their parent star. Protoplanetary disks do not contain enough material, hydrogen in particular, in their hot inner regions to possibly form any gas giant. It was thus necessary to invoke a new physical mechanism to explain the existence of hot Jupiters: inward migration of giant planets formed beyond the ice line, caused by interactions with the protoplanetary disk. Migration has remained a major topic of research since then, involving complex physical processes which are not completely understood yet (see the chapter by Lubow and Ida).

3.3. Upsilon Andromedae: The First Multiplanet System

One of the first detected hot Jupiter planets orbits the F8V star Upsilon Andromedae (*Butler et al.*, 1997). The first 36 Doppler measurements from Lick Observatory for Upsilon Andromedae spanned 9 years and revealed a planet in a 4.617-d orbit with a mass of about 0.7 M_{Jup}. The data also showed residual velocity variations with a period of about 2 yr and velocity amplitude of about 50 m s^{-1}. However, announcement of a second planet seemed premature in 1997. After 53 additional measurements from Lick Observatory, spanning 2 more years, it was clear by eye that this second planet suspected by *Butler et al.* (1997) had good phase coverage. The Advanced Fiber Optic Echelle (AFOE) was also being used to carefully monitor this star and in 1999, the Lick and AFOE data were combined for a joint solution. Surprisingly, the double-planet model failed to yield a reasonable χ^2 statistic.

There are many ways to fit multiple planet systems; however, all require a reasonable initial guess for the orbital parameters. In the early days of Keplerian modeling we modeled the systems sequentially with a Levenberg-Marquardt algorithm, first fitting and then subtracting theoretical velocities for the dominant system and then checking for periodicities in the residual velocities. This sequential process was possible for a couple of reasons:

1. Kepler's Laws dictate that the orbital period of the planet is primarily a function of semimajor axis. As a simple first assumption, we expected gas giant planets in stable multiplanet systems to be physically well separated. Therefore the reflex stellar velocities will also be well separated in frequency (i.e., the inverse orbital period).

2. For most planetary systems with gas giant planets, there was typically one dominant planet with a large amplitude and relatively short-period Doppler signal.

In the case of Upsilon Andromedae, the short-period system was modeled in a high-cadence subset of the data that spanned a few months. The 75 m s^{-1} signal from the 4.617-d planet was modulated by a long-period Doppler signal with similar velocity amplitude. Because the fit to the double planet system was inadequate (with a poor χ^2 statistic), the theoretical double-planet velocities were subtracted from the data and the residual noise showed a nearly sinusoidal variation, suggesting the presence of a third planet. This incredible result appeared in the Lick and AFOE data and the independent confirmation was critical for this first detection of a multiplanet system. The velocities for this triple planetary system continue to march along the predicted theoretical curve. Figure 6 shows the latest velocities after subtracting the 4.617-d planet (for clarity).

The challenge of modeling multiplanet systems is greater when the planets are in resonant orbits and orbital periods are commensurate or when velocity amplitudes are smaller (lower signal-to-noise). For multiplanet systems with low-mass (rocky) planets, the highest Doppler precision is required to identify all components in the system. In all multiplanet systems, there are non-Keplerian perturbations from gravitational interactions between planets that can result in slow modifications to a simple superposition of Keplerian models.

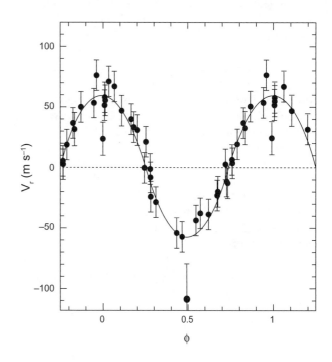

Fig. 5. Original radial velocity curve of the star 51 Peg, phased to a period of 4.23 d, obtained with the ELODIE spectrograph (*Mayor and Queloz*, 1995). The signal is caused by an orbiting companion with a minimum mass of 0.47 M_{Jup}, revealing for the first time an exoplanet around another solar-type star.

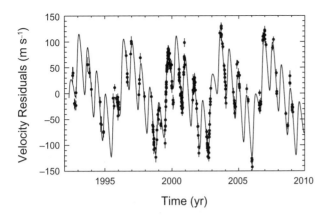

Fig. 6. Radial velocity time series of the Ups And system spanning almost 20 yr (data from Lick Observatory, courtesy of D. Fischer, 2009). The solid line shows the best fit multi-Keplerian model. The signal of the closest planet Ups And b has been subtracted for clarity.

3.4. Transiting Planets Discovered with Radial Velocities

After the detection of the first exoplanets with the Doppler technique, there was skepticism by some astronomers regarding the true nature of these objects. Because the Doppler technique only measures line-of-sight velocity, the orbital inclination is unresolved and M sin i, not M (mass), is actually measured. For small inclinations (nearly face-on orbits), the true mass of the object could even be stellar.

[*An exercise for the student:* Calculate the integrated probability that some of the detected exoplanets are actually stellar binaries in face-on orbits. First, calculate the threshold inclination that would push M sin i to 70 M_{Jup} (the stellar mass threshold) for the ensemble of known exoplanets. Then, calculate the statistical probability of observing an inclination between zero and the threshold inclination: P = 1−cos i_{thresh}. Summing all these probabilities yields a statistical estimate of the number of exoplanets with inclinations likely to move them to stellar masses.]

While the Doppler technique cannot resolve the orbital inclination, it can determine other orbital parameters and predict the time of transit if the inclination is close to 90° (edge-on). Planets are more likely to transit if they are close to the star (see the chapter by Winn). The probability that a close hot Jupiter will transit is relatively large (~10%); therefore, photometric follow-up is carried out for all the short-period planets. By 1999, 10 hot Jupiters had been detected with Doppler observations and none of these exoplanets seemed to transit their host stars. The eleventh hot Jupiter, HD 209458b, finally produced the characteristic dimming expected from a transiting planet at exactly the time predicted with radial velocity data. Two independent groups made the Doppler detection with photometric follow-up (*Charbonneau et al.*, 2000; *Henry et al.*, 2000) that definitively confirmed the interpretation of radial velocity wobbles as exoplanets.

[*An exercise for the student:* Calculate the integrated probability that transiting planets exist in the ensemble of known exoplanets. For a transiting planet, sin $i_{transit}$ must be greater than the ratio of the stellar diameter to the semimajor axis of the orbit (or use the periastron distance). Calculate the probability for this transit inclination for each system. Summing all these probabilities yields a statistical estimate of the number of Doppler-detected exoplanets with transit-favorable inclinations.]

While most transiting exoplanets are now detected with photometric surveys, a few important transiting planets were first detected in Doppler programs and can be given the whimsical designation of "Troppler" planets. Important examples include HD 209458 and HD 189733, which are further discussed in the chapters by Burrows and Orton and by Meadows and Seager.

HD 149026b (*Sato et al.*, 2005) was detected as part of the N2K search for hot Jupiters orbiting metal-rich stars. This star exhibited confusing radial velocities that were difficult to model initially; in retrospect, it was because four of the first seven radial velocities were serendipitously obtained during transit. Because of the Rossiter-McLaughlin effect (see the chapter by Winn), the velocities deviated from Keplerian motion. HD 149026b has a mass that is 10% greater than Saturn and a radius 10% smaller than Saturn. Subsequent models showed that the planet must therefore contain a heavy element core of about 70 $M_\oplus$, more akin to a super Neptune than a hot Jupiter. Because the photometric decrement during transit was only 0.003 mag, this would have been a challenging detection for photometric surveys of the day; the initial Doppler detection was key to demonstrating an unexpected range in the interior structure of gas giant planets.

Doppler observations detected a planet orbiting GJ 436, an M2.5V star (*Butler et al.*, 2004). Subsequent photometric observations at the putative transit ephemeris times showed that the planet orbiting this M-dwarf star was transiting (*Gillon et al.*, 2007). The orbital period of this planet is just 2.64 d and the planet mass is just 22 times that of Earth. The photometric transit depth of 0.009 mag is larger than HD 149026b because of the smaller radius of the host star. This "Troppler" planet remains one of the most intriguing transiting planets because it is the first detected hot Neptune and because it orbits a low-mass M-dwarf star.

The photometric transit probability is greatest for short-period planets, which hug the host star. However, the Doppler technique has been unique in identifying two transiting planets with orbital periods longer than 10 d. Both of these planets are in eccentric orbits and are therefore part-time hot Jupiters during periastron passage. HD 17156b and HD 80606b were both detected by Doppler surveys and both reside in highly eccentric orbits. HD 17156b has an orbital period of 21.2 d and an eccentricity of 0.67. This orbit carries the planet to a close stellar approach of 0.05 AU during periastron passage. HD 80606b is even more extreme;

the orbital period of this planet is 111 d and the eccentricity is an extreme 0.9336. As a result, periastron passage for this planet is only 0.03 AU from the host star. The orbits for HD 17156b (*Fischer et al.*, 2007) and HD 80606b (*Naef et al.*, 2001) were first determined from Doppler observations. This enabled a precise estimate of the transit times for what would have otherwise been a long-shot photometric campaign.

Transit observations measure the ratio of the planet radius to the stellar radius. In order to determine the mass and density of the exoplanet, this technique must be paired with radial velocity measurements. The combination of photometric transit detections and Keplerian modeling of radial velocity data is an example where the sum of two techniques far outweighs the contribution of either individual technique. Finally, it appears that another application of the RV technique is becoming more and more important in the context of transiting planets: the Rossiter-McLaughlin effect (see the chapter by Winn), which gives access to the spin-orbit geometry of the star-planet system and thus has the potential to probe its dynamical history.

3.5. Extending the Primary Mass Spectrum: Planets Around M Dwarfs and Red Giants

The RV technique can also be applied to stars quite different from solar-type stars, provided enough Doppler information is available in their spectrum and stellar noise does not hide planetary signals. This allowed in particular for planet searches around stars with masses significantly different from 1 $M_\odot$. The scientific motivation for this is the study of the formation and properties of exoplanets as a function of stellar mass. Protoplanetary disk masses are expected to scale with the mass of their central star, although the exact dependence is not known. As a consequence, theoretical models based on the core-accretion paradigm predict that gas giants should be rare around M dwarfs, but abundant around intermediate-mass stars (*Laughlin et al.*, 2004; *Ida and Lin*, 2005; *Kennedy and Kenyon*, 2008).

Soon after the first discoveries of planets around FGK stars, new surveys started targeting nearby M dwarfs. For visible spectrographs, early M stars (M0 to M4) in the close solar neighborhood are still bright enough to allow for precise radial velocity measurements at the ~1 m s^{-1} level. About 300–400 M stars with masses ~0.25–0.5 $M_\odot$ have thus been monitored by different groups using instruments like OHP-ELODIE (*Delfosse et al.*, 1998), Keck-HIRES (*Marcy et al.*, 1998), ESO-CES/UVES (*Kuerster et al.*, 1999), HET-HRS (*Endl et al.*, 2006), and ESO-HARPS (*Bonfils et al.*, 2005). These stars are, however, more active on average than FGK stars because the magnetic dynamo weakens more slowly with age compared to solar-type stars. Great care must therefore be taken in analyzing RV signals. Spectroscopic indicators such as Ca II H&K and Hα emission and broadband photometry are useful diagnostics to assess the activity level, measure the stellar rotation period, and disentangle planet-related and activity-related RV signals (e.g., *Bonfils et al.*, 2007).

The first M dwarf found to be orbited by a planet was GJ 876 (*Delfosse et al.*, 1998; *Marcy et al.*, 1998). It later turned out that the GJ 876 system is actually much more complex, with two giant planets in a 2:1 mean-motion resonance (*Marcy et al.*, 2001) and a close-in super Earth (*Rivera et al.*, 2005). However, this peculiar planetary system is not representative of the planet population that has been found around M dwarfs. In fact, only four such stars are known to harbor gas giants, which confirms the rarity of these objects around low-mass stars. On the other hand, several Neptune-mass planets and super Earths have been found over the past few years (see section 3.7), among which a quadruple system around GJ 581 (*Udry et al.*, 2007). This points toward a significant difference between planet populations around Sun-like stars and M dwarfs (*Bonfils et al.*, 2006; *Endl et al.*, 2006; *Johnson et al.*, 2007a). Protoplanetary disk masses are likely to be responsible for these differences, although metallicity may also play a role if M dwarfs had on average a lower metallicity (which is unclear).

On the other side of the stellar mass spectrum, i.e., for intermediate-mass stars, the radial velocity technique encounters several difficulties. In main-sequence A–F stars with M ≥ 1.5 $M_\odot$, the lower number of spectral features and their rotational broadening make it difficult to reach a high RV precision as far as photon noise is concerned. But more importantly, stellar pulsations and activity become the dominant source of noise and may hide planetary signals. A few RV surveys have nevertheless been targeting such stars with some success (e.g., *Galland et al.*, 2005, 2006), but the planet detection limits are usually well above the 1-M_{Jup} threshold.

Another strategy is to observe intermediate-mass stars in a later stage of their evolution, the red giant phase. They have then sufficiently cooled and slowed down to exhibit a rich spectrum with narrow spectral features. However, stellar noise again becomes the limiting factor for RV measurements, since red giants show p-mode oscillations with much larger amplitudes and periods than solar-type stars, have larger granulation noise, and may also exhibit significant levels of RV jitter due to rotational modulations of photospheric features. Great care must therefore be taken in analyzing periodic RV signals in these stars, especially for the more evolved and luminous ones. Photometric and spectroscopic diagnostics should be used to check the planetary origin of RV signals.

Several surveys have been targeting giant stars, but with various sample definitions: nearby bright G and K giants (e.g., *Hatzes and Cochran*, 1993; *Frink et al.*, 2002; *Setiawan et al.*, 2003; *Sato et al.*, 2003; *Doellinger et al.*, 2007; *Niedzielski et al.*, 2007), subgiants (e.g., *Johnson et al.*, 2007b), and open cluster giants (*Lovis and Mayor*, 2007). Stellar masses in these samples vary from ~1 to 4 $M_\odot$, thereby opening a new domain in the parameter space. More than 20 planets have been detected around red giants and numbers are growing fast, showing a probable overabundance of "super Jupiters" and a higher frequency of giant planets compared to solar-type stars (*Lovis and Mayor*, 2007; *Johnson et al.*, 2007a; *Hekker et al.*, 2008). Figure 7 illustrates this by showing the average mass of planetary systems as a function of stellar mass. Although it does not give any details on the underlying planet population, the emerging trend clearly shows that stellar mass (and thus disk mass) plays an important role in the planet formation process. Other interesting properties of this population include a very different semimajor axis distribution, with no planets orbiting below ~0.7 AU (*Johnson et al.*, 2007b), and a possible lack of correlation between planet frequency and stellar metallicity (e.g., *Pasquini et al.*, 2007). This latter point is, however, still debated.

3.6. The First Neptune-Mass Planets

In 2004, exoplanets made primarily of heavy elements instead of hydrogen were discovered for the first time, opening a new field of research and illustrating the capabilities of the radial velocity technique. Based on HARPS data, *Santos et al.* (2004) announced the detection of a hot Neptune orbiting the star μ Ara in 9.6 d, a star already known to host two giant planets. Further measurements revealed yet another giant planet and established the minimum mass of the hot Neptune to be 10.5 $M_\oplus$ (*Pepe et al.*, 2007). Simultaneously, *McArthur et al.* (2004) detected a close-in hot Neptune orbiting in 2.8 d around 55 Cnc, another star already known to harbor several giant planets. *Fischer et al.* (2008) then updated the orbital solutions for this system, adding a fourth gas giant at intermediate distance and fixing the minimum mass of the hot Neptune to 10.8 $M_\oplus$. 55 Cnc thus became the first known system with five planets. Both hot Neptune discoveries highlight the fact that such objects are well detectable with the RV technique, thanks to the improvements in precision and the gathering of a sufficient number of data points, a necessary condition to resolve the complex radial velocity curves of multiplanet systems. Both systems also hint at the fact that planets are rarely single, especially low-mass ones, and may very frequently appear in systems.

A third Neptune-mass object was discovered at the same time: GJ 436b, orbiting an M dwarf on a close-in orbit with a period of 2.64 d and a minimum mass of 23 $M_\oplus$ (*Butler et al.*, 2004). This minimum mass actually also became its true mass when *Gillon et al.* (2007) announced that the planet was transiting its host star. For the first time, the mass and radius of a Neptune-mass planet could be measured, revealing an internal structure close to the one of Uranus and Neptune (mostly water ice with a H/He outer layer of ~10% in mass). More details on this fascinating object are given in section 3.4 and in the chapter by Winn.

3.7. Emergence of a New Population: Super Earths and Ice Giants

The number of RV-detected low-mass planets (Neptune-mass planets or super Earths) has been growing fast since 2004. Table 2 lists all objects with minimum masses below 25 $M_\oplus$ known as of January 2010. As can be seen, this is a rapidly evolving field of research, in which the HARPS spectrograph has made an essential contribution. These discoveries were made possible thanks to the submeters per second precision reached by this instrument. Indeed, most of these objects have RV semiamplitudes K below 3–4 m s^{-1}, and have therefore been very difficult to detect with the previously existing facilities that were limited to ~3 m s^{-1}.

Among these low-mass objects, we note in particular the discovery of three systems containing at least three Neptunes or super Earths, and no gas giants: HD 69830, GJ 581, and HD 40307. The triple-Neptune system HD 69830 was the first to be unveiled (*Lovis et al.*, 2006). The first two planets orbit at 0.08 and 0.18 AU, while the third one is located further away at 0.62 AU. This shows that Neptune-mass objects are now detectable not only on close-in orbits, but all the way out to the habitable zone around solar-type stars. However, this requires very high-precision RV measurements: The total RV dispersion before fitting any planet amounts to only 3.7 m s^{-1}. The postfit residuals have a rms dispersion of 0.8 m s^{-1}, illustrating the unprecedented quality of the HARPS measurements. Figure 8 shows a close-up view of the RV curve for this system, which reveals the complex nature and low amplitude of the RV variations.

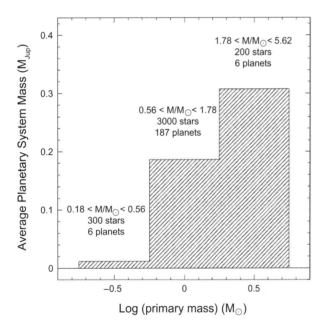

Fig. 7. Average mass of planetary systems as a function of stellar mass, obtained by dividing, in each bin, the total (minimum) mass of all detected planets by the number of survey stars (from *Lovis and Mayor*, 2007). The rising trend toward higher stellar masses shows that giant planets become more frequent and/or more massive with increasing stellar mass, in spite of detection biases toward higher-mass stars.

TABLE 2. RV-detected exoplanets with minimum masses below 25 $M_\oplus$ as of January 2010, ordered according to their minimum mass.

Name	K (m s^{-1})	$m_2 \sin i$ ($M_\oplus$)	P (d)	m_1 ($M_\odot$)	Instrument	Reference
GJ 581e	1.9	1.9	3.15	0.31	HARPS	*Mayor et al.* (2009b)
HD 156668b	1.9	4.2	4.646	0.77	HIRES	Howard et al. (in preparation)
HD 40307b	2.0	4.2	4.31	0.77	HARPS	*Mayor et al.* (2009a)
61 Vir b	2.1	5.1	4.2	0.9	HIRES	*Vogt et al.* (2010)
GJ 581c	3.2	5.4	12.9	0.31	HARPS	*Udry et al.* (2007)
GJ 876d	2.7	5.7	1.94	0.32	HIRES	*Rivera et al.* (2005)
HD 1461b	2.7	7.63	5.77	1.03	HIRES	*Rivera et al.* (2010)
HD 40307c	2.4	6.9	9.62	0.77	HARPS	*Mayor et al.* (2009a)
GJ 581d	2.6	7.1	66.8	0.31	HARPS	*Udry et al.* (2007)
HD 181433b	2.9	7.6	9.37	0.78	HARPS	*Bouchy et al.* (2009)
GJ 176b	4.3	7.8	8.78	0.50	HARPS	*Forveille et al.* (2009)
HD 40307d	2.6	9.2	20.5	0.77	HARPS	*Mayor et al.* (2009a)
HD 7924b	3.9	9.3	5.40	0.83	HIRES	*Howard et al.* (2009)
HD 69830b	3.6	10.5	8.67	0.86	HARPS	*Lovis et al.* (2006)
HD 160691d	3.1	10.5	9.55	1.08	HARPS	*Santos et al.* (2004), *Pepe et al.* (2007)
55 Cnc e	3.7	10.8	2.82	1.03	HRS/HIRES	*McArthur et al.* (2004), *Fischer et al.* (2008)
GJ 674b	8.7	11.8	4.69	0.35	HARPS	*Bonfils et al.* (2007)
HD 69830c	2.9	12.1	31.6	0.86	HARPS	*Lovis et al.* (2006)
HD 4308b	4.0	15.0	15.6	0.83	HARPS	*Udry et al.* (2006)
GJ 581b	12.5	15.7	5.37	0.31	HARPS	*Bonfils et al.* (2005)
HD 190360c	4.6	18.1	17.1	1.04	HIRES	*Vogt et al.* (2005)
HD 69830d	2.2	18.4	197	0.86	HARPS	*Lovis et al.* (2006)
HD 219828b	7.0	21.0	3.83	1.24	HARPS	*Melo et al.* (2007)
HD 16417b	5.0	22.1	17.2	1.20	UCLES	*O'Toole et al.* (2009)
HD 47186b	9.1	22.8	4.08	0.99	HARPS	*Bouchy et al.* (2009)
GJ 436b	18.0	22.9	2.64	0.45	HIRES	*Butler et al.* (2004)
HAT-P-11b	11.6	25.8	4.89	0.81	HAT/HIRES	*Bakos et al.* (2010)

Besides precision, the acquisition of a large number of data points is also critical to resolve the three RV signals. The HD 69830 system appears as a new illustration of the diversity of planetary systems, with several Neptunes on close-in orbits and no gas giants, at least within ~10 AU. Its discovery immediately raised questions about the composition of the Neptunes, whether essentially rocky or icy, depending on where they were formed in the protoplanetary disk (*Alibert et al.*, 2006). The interest in this system is further enhanced by the presence of a warm dust belt at ~1 AU from the star, allowing for in-depth studies of the interactions between planets and debris disks (*Beichman et al.*, 2005; *Lisse et al.*, 2007).

GJ 581 is also a remarkable system, with a close-in Earth-mass planet, a hot Neptune, and two super Earths further away (*Bonfils et al.*, 2005; *Udry et al.*, 2007; *Mayor et al.*, 2009b). With a minimum mass of only 1.94 $M_\oplus$, GJ 581e is presently the lowest-mass object ever discovered around a main-sequence star other than the Sun. The primary is an M dwarf with a mass of only 0.31 $M_\odot$. As a consequence, it is easier to detect a planet of a given mass around this star compared to solar-type stars. Moreover, the habitable zone is expected to be much closer due to the lower stellar luminosity. The two super Earths discovered at 0.07 and 0.22 AU are indeed located close to the inner and outer edges of this zone, which makes them arguably the two most "Earth-like" bodies known outside our solar system. Obviously, too little is presently known about these objects, their atmospheres, and their environment to decide if they more closely resemble Earth, Venus, Mars, or, more likely, something that is unknown in our solar system (see, e.g., *Selsis et al.*, 2007, for a detailed discussion).

Mayor et al. (2009a) discovered yet another triple system, this time containing only super Earths: HD 40307. With minimum masses of 4.2, 6.9, and 9.2 $M_\oplus$, this is the lowest-mass system known. Figure 9 shows the phased radial velocity curves of these objects. The signal is extremely clear despite the very low amplitudes. This again shows that the RV technique has not reached its limits yet on such old and quiet solar-type stars. HD 40307b has one of the lowest minimum masses and RV semiamplitudes (2.0 m s^{-1}) among the known exoplanets. Even if its true mass is not known, this object is expected to be made exclusively of heavy elements (rocks and/or ices), being most probably not massive enough to retain a H/He atmosphere (especially under the intense irradiation from its host star).

Once more, nature has surprised us with the discoveries of these close-in, low-mass systems: Their mere existence was not anticipated and is triggering further theoretical modeling to understand how and where they were formed, and what is their composition. Ice giants do exist in our solar system, but at large orbital distances, while super Earths occupy a completely unexplored mass range, where several mixtures of iron, silicates, ices, and maybe hydrogen are possible.

From the observational point of view, it appears that a new population of low-mass planets is now emerging. Figure 10 shows the mass distribution of all known exoplanets. It is obviously affected by several observational biases, but we are only interested here in its overall shape. Clearly, there seems to be an increase in population below 20–30 $M_\oplus$, despite the strong bias of the RV technique toward low masses. Such an increase is also predicted by some theoretical simulations of planet formation based on the core accretion model (e.g., *Mordasini et al., 2009*), while other models predict a large population only at lower

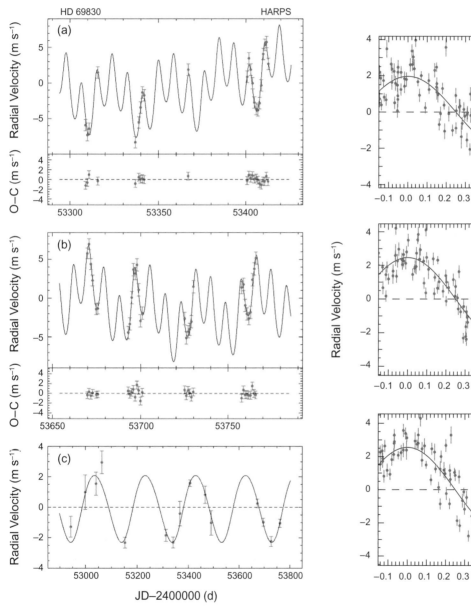

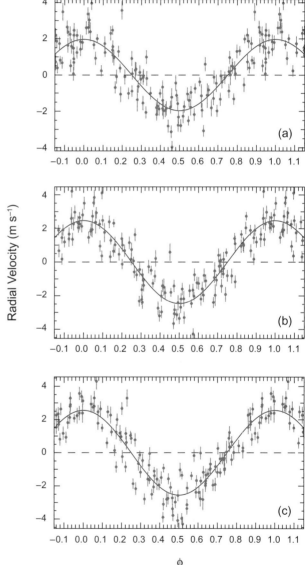

Fig. 8. Close-up views of the HARPS radial velocity curve of HD 69830 as a function of time. The submeters per second precision of the data is critical to reveal the three low-amplitude RV signals caused by the orbiting Neptune-mass planets (from *Lovis et al., 2006*).

Fig. 9. Phase-folded HARPS radial velocity curves of the three super Earths orbiting HD 40307 (from *Mayor et al., 2009a*). In each case, the two other planets have been subtracted. Periods are 4.31, 9.62, and 20.5 d, while minimum masses are 4.2, 6.9, and 9.2 $M_\oplus$, respectively.

masses (e.g., *Ida and Lin*, 2008). In any case, the ongoing HARPS search for low-mass planets has detected many more candidates for which more measurements must be gathered to secure the orbital solutions (*Lovis et al.*, 2009). Several tens of Neptunes and super Earths are expected from this survey, the majority of which are on close-in orbits within ~0.5 AU. Most of them are also in multiplanet systems. Preliminary numbers indicate that the frequency of close-in low-mass planets around solar-type stars may be as high as ~30% (*Mayor et al.*, 2009a). Undoubtedly, the characterization of the low-mass population with the RV technique will bring many new constraints on our understanding of planet formation mechanisms. Moreover, it will allow us to put the Earth into the broader context of telluric planets in the galaxy. Some outstanding issues that can be addressed with the present and near-future discoveries are:

1. What is the precise mass distribution of the low-mass population at short distances? Are super Earths more frequent than Neptunes?

2. Are close-in Neptune-mass planets and super Earths "failed" giant planet cores that could not accrete a massive H/He envelope?

3. Have these objects formed beyond the ice line and subsequently migrated to their present location? Or can some of them have formed *in situ*, like telluric planets in our solar system?

4. Is there a correlation between orbital distance and planet mass? How do migration processes depend on planet mass?

5. What are the relative frequencies of gas giants and low-mass planets? Is there a "desert" at intermediate masses?

6. What is the eccentricity distribution of low-mass objects, and how does it compare to gas giants?

7. Are stars hosting low-mass planets preferentially metal-rich? Is there any correlation between stellar parameters and the presence of low-mass planets?

8. What is the dynamical architecture of multiplanet systems, and what is their dynamical history?

9. Are low-mass planets more or less common around M dwarfs compared to solar-type stars? What is their frequency in the habitable zone of M dwarfs?

The list can be continued indefinitely. The rapidly growing number of known low-mass planets should soon allow us to unveil the main statistical properties of this population. Finally, it has to be noted that RV surveys are targeting nearby bright stars, and every discovered close-in Neptune or super Earth should be considered as a potentially extremely interesting transiting candidate. Transits are rare due to the low geometric probabilities, but the few low-mass objects that do transit a nearby bright star will probably represent "Rosetta stones" for our understanding of exoplanets. They will likely be discovered by RV surveys and follow-up photometry. However, spacebased photometry is necessary to reach the precision required to detect the very small transit depths caused by these objects. Given the wealth of information that can be obtained from transits and antitransits on a bright star (precise density, composition and temperature, in-depth atmospheric studies), the scientific return is certainly worth the effort.

4. FUTURE PROSPECTS

4.1. Alternative Radial Velocity Techniques

4.1.1. Radial velocities in the near-infrared. Radial velocity planet surveys of M-dwarf stars have substantial scientific merit because these are the most common stars in our galaxy. In principle, there is a two-fold enhancement in the detectability of rocky planets at habitable zone distances. Compared with solar-mass stars, the reflex stellar velocity for lower-mass M-dwarf stars is greater by a factor of about 3. The stellar velocity amplitude is further increased because of the proximity of the habitable zone to these low-luminosity stars. However, at optical wavelengths, late-type M dwarfs are faint stars. The flux of M4V stars peaks in the 1–2 µm wavelength range. An M4V star at a distance of 20 parsec has a V-band magnitude of 14, but a J-band magnitude of 9. Therefore, these stars are challenging targets for optical spectroscopy even with 10-m telescopes, but accessible with IR spectroscopy on even moderate-aperture telescopes.

There are still challenges for IR spectroscopy. Telluric absorption lines litter the IR spectrum and vary with the changing column density of water vapor on timescales of hours. Rotational velocity rises for stars later in spectral type than M4V, resulting in broader spectral lines. And

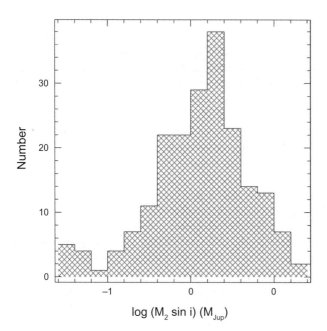

Fig. 10. Observed mass distribution of all known exoplanets. A hint of an increase in population below 20–30 $M_\oplus$ is clearly visible, despite strong observational biases at low masses.

flares are common sources of astrophysical noise for stars at the end of the main sequence. Nevertheless, technology development is ongoing to advance IR spectroscopy to precisions below 10 m s^{-1}, and the field is making rapid progress. Several authors have recently achieved ~5 m s^{-1} precision using the CRIRES instrument at the VLT, a high-resolution near-IR slit spectrograph (*Kaufl et al.,* 2004). On the one hand, *Huelamo et al.* (2008) and *Figueira et al.* (2010) use telluric CO_2 lines in the H band as wavelength references, and select unblended stellar lines to compute radial velocities via the cross-correlation method. On the other hand, *Bean et al.* (2010) have developed a procedure that is similar to the iodine method in the visible. Using an ammonia gas cell in the K band, they model the observed spectrum as a combination of the stellar, ammonia, and telluric spectrum, convolved with the instrument PSF. Both approaches had to deal with broadly similar challenges as in the visible: the varying spectrograph illumination and the need for a suitable wavelength reference. Besides these two aspects, the dense telluric spectrum clearly represents a supplementary difficulty in the near-IR.

These recent results are nevertheless very encouraging and there is little doubt that further progress will occur soon in this field, especially considering the development of several new high-resolution near-IR spectrographs. Radial velocities in the near-IR should become an optimal technique for the detection of planets around stars later than M4V in the near future.

4.1.2. Dispersed fixed-delay interferometry. The dispersed fixed delay interferometer (DFDI) combines a moderate-dispersion spectrograph with a Michelson-type interferometer, adding a graded phase delay perpendicular to the dispersion direction that creates fringes in each resolution element. Doppler shifts cause the fringes in each resolution element to move, allowing a measurement of the radial velocity (*Erskine and Ge,* 2000; *Ge,* 2002; *Ge et al.,* 2002). This Doppler measurement approach is completely different from the echelle approach, which measures the absorption line centroid shifts. The Doppler sensitivity of the DFDI approach weakly depends on the spectral resolution compared to the echelle approach [1/2 power of the spectral resolution vs. 3/2 power of the spectral resolution in the echelle method (*Ge,* 2002)]. This allows the use of a medium-resolution, but high-efficiency spectrometer for dispersing the fringes to boost the overall detection efficiency, while reducing the instrument size and cost. Also, the DFDI approach takes up a small amount of space on a CCD to cover one dispersion order fringing spectrum with a moderate wavelength coverage [e.g., ~600 Å for the Exoplanet Tracker instrument at the Kitt Peak National Observatory 2.1-m telescope (*Ge et al.,* 2006)], which allows either the use of a multiobject fiber feed so that simultaneous, high-throughput, high Doppler sensitivity measurements of many objects may be made with reasonable detector sizes, or the implementation of a cross-dispersion mode for large wavelength coverage, high dispersion, high throughput, and high Doppler sensitivity measurements of a single star (*Ge,* 2002; *Ge et al.,* 2007).

To date, one planet has been discovered by the DFDI instruments, around the star HD 102195 (*Ge et al.,* 2006). Several planet and brown dwarf candidates have been identified from the ongoing planet surveys at the Sloan Digital Sky Survey telescope and the KPNO 2.1-m telescope and are being followed up with additional observations. A next-generation, multiple-object, DFDI Doppler instrument with 60-object capability was commissioned at the SDSS telescope at Apache Point Observatory in September 2008 and is being used for conducting the Multi-object APO Radial-Velocity Exoplanet Large-area Survey (MARVELS), which is one of the four Sloan Digital Sky Survey (SDSS) III survey projects in the period 2008–2014.

4.2. Characterization of Transiting Planets Found by Space Missions

The space missions CoRoT and Kepler (*Baglin et al.,* 2002; *Borucki et al.,* 2003) are both dedicated to the search of transiting exoplanets by collecting long, high-precision photometric time series of thousands of stars. Transit events have to pass first through a series of photometric and spectroscopic tests to exclude alternative scenarios before they can be confirmed as *bona fide* transiting planets. High-precision radial velocity measurements are then necessary to measure the mass of the objects and obtain their mean density. This follow-up work has already been ongoing with great success for CoRoT and several groundbased transit surveys, which are presently discovering a large number of hot Jupiters (e.g., SuperWASP, HAT). CoRoT and particularly Kepler are expected to also find many hot Neptunes and super Earths, whose mass will have to be measured. This task may be challenging due to the faintness of the targets and their activity level. However, for quiet stars with magnitudes V = 11–12, instruments like HARPS and its twin HARPS-North (*Latham,* 2007) will be able to measure the masses of close-in super Earths down to ~5 $M_\oplus$ with an accuracy (~10%) that should be sufficient to usefully constrain their mean density and composition. In the near future, the combination of spacebased transit searches and high-precision RV follow-up is thus likely to considerably improve our knowledge of the close-in low-mass planet population.

4.3. Pushing Down the Limits: Toward the Detection of Earths Around Nearby Stars

What are the ultimate limitations of the radial velocity technique? To answer this difficult question, it is necessary to identify the dominant sources of systematic errors. These fall into two general categories: instrumental and stellar. We first discuss the instrumental aspects, and then examine stellar noise.

Instrumental limitations are presently fairly well understood, and strategies to overcome them are being developed. For HARPS-like instruments, the limiting factors are mainly the guiding system, the light-scrambling efficiency, and the wavelength calibration source. For each of these issues, it appears that significant progress is possible compared to existing facilities like HARPS with relatively modest technological developments and costs. Tests are ongoing on optimized light injection optics and optical fibers to make the spectrograph illumination as homogeneous as possible. As for wavelength calibration, new solutions such as the laser frequency comb (see section 2.5) and stabilized Fabry-Pérot etalons are being developed and tested.

Two major European instrumental projects are now building on the HARPS experience with the goal of improving the overall stability and precision, but also the photon collecting power. Indeed, HARPS is not far from being photon-limited on the typical stars that are presently searched for low-mass planets, so that a larger telescope aperture is needed if performances are to be significantly improved. The first of these projects is ESPRESSO, the successor of HARPS, to be installed on the ESO Very Large Telescope at Paranal Observatory, Chile (*Pepe et al.*, 2010). ESPRESSO is developed with the goal of achieving an instrumental precision of 10 cm s^{-1} on both the short and long term, corresponding to the RV signal induced by Earth orbiting the Sun. Moreover, ESPRESSO will benefit from the large 8.2-m aperture of the VLT. It will even be possible to feed the spectrograph with the four 8.2-m telescopes simultaneously, which represent the equivalent of a single 16-m aperture. In this mode, however, the radial velocity precision will be somewhat degraded compared to the single-telescope mode. According to the typical numbers given in section 2.3, it appears that a SNR of about 700 per pixel will be required (at 550 nm) to achieve a photon limited precision of 10 cm s^{-1} at R = 140,000. This can be obtained in about 20 min of exposure time for stars with V = 7.5 on a 10-m class telescope. ESPRESSO will thus be able to carry out an extremely precise survey of 100–200 nearby bright stars, carefully selected to minimize stellar noise based on HARPS data. It will find all low-mass planets down to a few Earth masses orbiting within ~1 AU of their parent star, and possibly even 1-$M_\oplus$ planets in the habitable zones around K and M dwarfs. This survey should thus yield a comprehensive picture of telluric planets in the solar neighborhood.

The other major European project is CODEX, an ultrastable high-resolution visible spectrograph for the European Extremely Large Telescope (*Pasquini et al.*, 2008). The aimed instrumental precision of CODEX is ~1 cm s^{-1}. Its main science objectives all require extremely high Doppler precision: measurement of the cosmological redshift drift and the dynamics of the universe, search for Earth-like planets, and search for variability in fundamental constants. Although the required performances are quite challenging, there is little doubt that the possibility to address these outstanding scientific issues is well worth the technical effort. The experience gained with HARPS and ESPRESSO is expected to be of great value in this context.

On the shorter term, a twin of the HARPS spectrograph, called HARPS-North, is being built to provide increased follow-up capabilities for Kepler planet candidates. It will be installed in La Palma, Canary Islands, and should see first light in 2012.

The challenges to extreme Doppler precision for reference cell techniques (e.g., the iodine cell technique) are different from those for HARPS-like instruments. The reference spectrum follows the same optical path as the starlight and therefore experiences the same distortions from each of the optical elements (i.e., the instrumental PSF). The reference spectrum provides a measurement of the wavelength solution and Doppler shifts with a precision that is limited by our ability to model the instrumental PSF. As mentioned before, the instrumental PSF would be a slowly varying function of temperature and pressure except for the rapidly varying slit illumination. Therefore, many efforts to improve the Doppler precision from reference cell techniques are focusing on ways to stabilize the slit illumination with fiber optic cables to scramble the incoming light. Work is taking place on fiber optic upgrades at the Hamilton spectrograph at Lick Observatory and for Keck-HIRES. The promising early efforts show an order-of-magnitude improvement in PSF stability with fiber optic feeds instead of traditional slits. With a fiber-optic feed, the 10-m Keck telescope will provide photon-limited precision; ultimately, it will be the systematic astrophysical noise sources that set the true limit to Doppler precision.

A planet search effort at Cerro Tololo Interamerican Observatory began collecting an extensive set of Doppler measurements on α Centauri A and B (V = 0 and V = 1.3 respectively) in January 2009, using a fiber-fed echelle on the 1.5-m telescope. This project is now collecting up to 500 spectra per night during the peak observing season. The stars are monitored all night, every night that they are accessible with a 4-year goal of acquiring 100,000 observations per star. A new R = 80,000 fiber-fed spectrometer will be commissioned for this project at CTIO by the end of 2010. Although this project only focuses on two stars, the extreme cadence allows removal of systematic errors that are common to both stars. An important goal of this project is to empirically check our ability to average down astrophysical noise sources in chromospherically quiet stars.

New facilities in the U.S. for high-precision Doppler planet searches include the 2.4-m Automated Planet Finder at Lick Observatory with a R = 72,000 spectrograph, which will be commissioned in 2010. Spectrometer designs for precision radial velocity searches are also being reviewed for the Thirty Meter Telescope (TMT) and the Giant Magellan Telescope (GMT).

We now turn to the issue of stellar noise in the context of the detection of very low-amplitude radial velocity signals. As described in section 2.4, several physical phenomena in stellar photospheres (oscillations, granulation, active regions) may hide planet-induced RV variations at the meters per second level and below. To address this problem, studies are ongoing to precisely characterize stellar RV noise at the relevant timescales (from minutes for oscillations to about one month for activity). Solar and stellar power density spectra of long, high-precision RV time series can be used to derive the main properties of each noise component and understand how it varies from star to star. From these results it is possible to find and select the most suitable stars for submeters per second RV monitoring. From what is presently known about stellar noise, it clearly appears that F stars and slightly evolved G and K stars are unsuitable because all noise sources become larger in these objects. Between unevolved-G and M stars, oscillation and granulation noise become smaller, but activity levels tend to be higher in late-K and M stars because of the slower decrease of activity with age. However, this is partly compensated by the lower masses of these stars, which make RV signals larger for a given planet mass. Activity-related noise can be minimized by selecting the stars with the lowest Ca II H&K index, and in particular those that do not exhibit activity cycles (or only weak ones). Such objects may be in a Maunder minimum state, an optimal situation to reach unprecedented precision levels on RV measurements. The identification of this low-activity population is presently ongoing, making use of the thousands of spectra that have been gathered by radial velocity surveys over the past years.

Realistic simulations of the stellar noise, based on observed power density spectra of several stars, should be made to derive the optimal observing strategy and compute planet detection limits (e.g., *Dumusque et al.*, in preparation). Preliminary results indicate that a sufficiently dense sampling properly covering the typical timescales of the different noise sources lead to significant improvements in planet detection limits. Moreover, the search for planets in the habitable zone may well benefit from the probably low stellar noise level over timescales of 100–1000 days. As a result, it may well be possible to detect Earth-mass planets in the habitable zone of G, and particularly K, dwarfs. Despite the low RV amplitudes, solar-like stars should therefore also be considered for this kind of searches, together with M dwarfs, which have the obvious advantages of being less massive and having closer habitable zones. Finally, we note that these estimates are not only the result of simulations, but also of real data: As an example, binning the HARPS RV measurements obtained on the star HD 69830 (*Lovis et al.*, 2006) over timescales of ~15 d yields a long-term RV dispersion of only ~30 cm s^{-1} (after subtracting the signal of the three known planets). Given all the instrumental error sources and photon noise that contribute to this number, this observational result shows that we are indeed tantalizingly close to being able to detect an Earth-like planet in the habitable zone of this star.

4.4. Conclusion

In this chapter we have reviewed the history, capabilities, achievements, and present state-of-the-art of the radial velocity technique applied to the search for exoplanets. In summary, over the past 15 years this technique has opened up and developed a new field of astrophysics that has already become one of the most prominent in this discipline, be it in terms of its scientific potential or in terms of the wide-ranging public interest associated with it. After the initial gas giant era, the RV technique has now fully entered the domain of low-mass planets. Given the achieved detection limits, present and near-future RV surveys of nearby bright stars will unveil the statistical properties of the low-mass planet population within ~1 AU. We will then have a global picture of exoplanets, from massive gas giants to telluric bodies, and this will allow us to make decisive steps in our understanding of planet formation and migration. This is all the more true in cases where the RV technique can be coupled to the transit technique. As other detection techniques appear and are developed, radial velocities will remain a fundamental tool for obtaining dynamical masses and characterizing systems. As an important by-product, the obtained high-resolution spectra can be used to characterize the host star. Detecting exoplanets is and will remain an observational challenge requiring significant efforts in instrumental developments. While all possible approaches should certainly be explored, the cost-effectiveness of the RV technique should be particularly emphasized, and thus the development of improved RV facilities should certainly be continued in the future.

Acknowledgments. C.L. would like to thank the Swiss National Science Foundation for its continuous support.

REFERENCES

Alibert Y., Baraffe I., Benz W., Chabrier G., Mordasini C., et al. (2006) Formation and structure of the three Neptune-mass planets system around HD 69830. *Astron. Astrophys., 455,* L25–L28.

Baglin A., Auvergne M., Barge P., Buey J.-T., Catala C., et al. (2002) COROT: Asteroseismology and planet finding. In *Proceedings of the First Eddington Workshop on Stellar Structure and Habitable Planet Finding* (B. Battrick et al., eds.), pp. 17–24. ESA Spec. Publ. 485, Noordwijk, The Netherlands.

Bakos G. Á., Torres G., Pál A., Hartman J., Kovács G., et al. (2010) HAT-P-11b: A super-Neptune planet transiting a bright K star in the Kepler field. *Astrophys. J., 710,* 1724–1745.

Baliunas S. L., Donahue R. A., Soon W. H., Horne J. H., Frazer J., et al. (1995) Chromospheric variations in main-sequence stars. *Astrophys. J., 438,* 269–287.

Baranne A., Queloz D., Mayor M., Adrianzyk G., Knispel G., et al. (1996) ELODIE: A spectrograph for accurate radial velocity measurements. *Astron. Astrophys. Suppl. Ser., 119,* 373–390.

Bean J. L., Seifahrt A., Hartman H., Nilsson H., Wiedemann G., et al. (2010) The CRIRES search for planets around the lowest-mass stars. I. High-precision near-infrared radial velocities with an ammonia gas cell. *Astrophys. J., 713,* 410–422.

Beichman C. A., Bryden G., Gautier T. N., Stapelfeldt K. R., Werner M. W., et al. (2005) An excess due to small grains around the nearby K0 V star HD 69830: Asteroid or cometary debris? *Astrophys. J., 626,* 1061–1069.

Bonfils X., Forveille T., Delfosse X., Udry S., Mayor M., et al. (2005) The HARPS search for southern extra-solar planets. VI. A Neptune-mass planet around the nearby M dwarf Gl 581. *Astron. Astrophys., 443,* L15–L18.

Bonfils X., Delfosse X., Udry S., Forveille T., and Naef D. (2006) Any hot-Jupiter around M dwarfs? In *Tenth Anniversary of 51 Peg-b: Status of and Prospects for Hot Jupiter Studies* (L. Arnold et al., eds.), pp. 111–118. Frontier Group, Paris.

Bonfils X., Mayor M., Delfosse X., Forveille T., Gillon M., et al. (2007) The HARPS search for southern extra-solar planets. X. A m sin i = 11 $M_\oplus$ planet around the nearby spotted M dwarf GJ 674. *Astron. Astrophys., 474*, 293–299.

Borucki W. J., Koch D. G., Lissauer J. J., Basri G. B., Caldwell J. F., et al. (2003) The Kepler mission: A wide-field-of-view photometer designed to determine the frequency of Earth-size planets around solar-like stars. In *Future EUV/UV and Visible Space Astrophysics Missions and Instrumentation* (J. C. Blades et al., eds.), pp. 129–140. SPIE Conf. Series 4854, Bellingham, Washington.

Bouchy F. and Carrier F. (2001) P-mode observations on α Cen A. *Astron. Astrophys., 374*, L5–L8.

Bouchy F., Pepe F., and Queloz D. (2001) Fundamental photon noise limit to radial velocity measurements. *Astron. Astrophys., 374*, 733–739.

Bouchy F. and the Sophie Team (2006) SOPHIE: The successor of the spectrograph ELODIE for extrasolar planet search and characterization. In *Tenth Anniversary of 51 Peg-b: Status of and Prospects for Hot Jupiter Studies* (L. Arnold et al., eds.), pp. 319–325. Frontier Group, Paris.

Bouchy F., Mayor M., Lovis C., Udry S., Benz W., et al. (2009) The HARPS search for southern extra-solar planets. XVII. Super-Earth and Neptune-mass planets in multiple planet systems HD 47 186 and HD 181 433. *Astron. Astrophys., 496*, 527–531.

Braje D. A., Kirchner M. S., Osterman S., Fortier D., and Diddams S. A. (2008) Astronomical spectrograph calibration with broad-spectrum frequency combs. *Eur. Phys. J., D48*, 57–66.

Butler R. P. and Marcy G. W. (1996) A planet orbiting 47 Ursae Majoris. *Astrophys. J. Lett., 464*, L153–L156.

Butler R. P., Marcy G. W., Williams E., McCarthy C., Dosanjh P., and Vogt S. S. (1996) Attaining Doppler precision of 3 m s^{-1}. *Publ. Astron. Soc. Pac., 108*, 500–509.

Butler R. P., Marcy G. W., Williams E., Hauser H., and Shirts P. (1997) Three new "51 Pegasi-type" planets. *Astrophys. J. Lett., 474*, L115–L118.

Butler R. P., Vogt S. S., Marcy G. W., Fischer D. A., Wright J. T., et al. (2004) A Neptune-mass planet orbiting the nearby M dwarf GJ 436. *Astrophys. J., 617*, 580–588.

Campbell B. and Walker G. A. H (1979) Precision radial velocities with an absorption cell. *Publ. Astron. Soc. Pac., 91*, 540–545.

Campbell B., Walker G. A. H., and Yang S. (1988) A search for substellar companions to solar-type stars. *Astrophys. J., 331*, 902–921.

Charbonneau D., Brown T. M., Latham D. W., and Mayor M. (2000) Detection of planetary transits across a Sun-like star. *Astrophys. J. Lett., 529*, L45–L48.

Christensen-Dalsgaard J. (2004) Physics of solar-like oscillations. *Solar Phys., 220(2)*, 137–168.

Connes P. (1985) Absolute astronomical accelerometry. *Astrophys. Space Sci., 110*, 211–255.

Delfosse X., Forveille T., Mayor M., Perrier C., Naef D., and Queloz D. (1998) The closest extrasolar planet. A giant planet around the M4 dwarf GL 876. *Astron. Astrophys., 338*, L67–L70.

Desort M., Lagrange A.-M., Galland F., Udry S., and Mayor M. (2007) Search for exoplanets with the radial-velocity technique: Quantitative diagnostics of stellar activity. *Astron. Astrophys., 473*, 983–993.

Döllinger M. P., Hatzes A. P., Pasquini L., Guenther E. W., Hartmann M., et al. (2007) Discovery of a planet around the K giant star 4 Ursae Majoris. *Astron. Astrophys., 472*, 649–652.

Dravins D. (1990) Stellar granulation. VI — Four-component models and non-solar-type stars. *Astron. Astrophys., 228*, 218–230.

Einstein A. (1905) Zur Elektrodynamik bewegter Körper. *Ann. Phys., 322*, 891–921.

Endl M., Cochran W. D., Kürster M., Paulson D. B., Wittenmyer R. A., et al. (2006) Exploring the frequency of close-in jovian planets around M dwarfs. *Astrophys. J., 649*, 436–443.

Erskine D. J. and Ge J. (2000) A novel interferometer spectrometer for sensitive stellar radial velocimetry. In *Imaging the Universe in Three Dimensions* (W. van Breugel and J. Bland-Hawthorn, eds.), pp. 501–507. ASP Conf. Series 195, Astronomical Society of the Pacific, San Francisco.

Fienga A., Manche H., Laskar J., and Gastineau M. (2008) INPOP06: A new numerical planetary ephemeris. *Astron. Astrophys., 477*, 315–327.

Figueira P., Pepe F., Melo C. H. F., Santos N. C., Lovis C., et al. (2010) Radial velocities with CRIRES: Pushing precision down to 5–10 m/s. *Astron. Astrophys., 511*, A55.

Fischer D. A., Vogt S. S., Marcy G. W., Butler R. P., Sato B., et al. (2007) Five intermediate-period planets from the N2K sample. *Astrophys. J., 669*, 1336–1344.

Fischer D. A., Marcy G. W., Butler R. P., Vogt S. S., Laughlin G., et al. (2008) Five planets orbiting 55 Cancri. *Astrophys. J., 675*, 790–801.

Forveille T., Bonfils X., Delfosse X., Gillon M., Udry S., et al. (2009) The HARPS search for southern extra-solar planets. XIV. Gl 176b, a super-Earth rather than a Neptune, and at a different period. *Astron. Astrophys., 493*, 645–650.

Frink S., Mitchell D. S., Quirrenbach A., Fischer D. A., Marcy G. W., and Butler R. P. (2002) Discovery of a substellar companion to the K2 III Giant ι Draconis. *Astrophys. J., 576*, 478–484.

Galland F., Lagrange A.-M., Udry S., Chelli A., Pepe F., et al. (2005) Extrasolar planets and brown dwarfs around A-F type stars. II. A planet found with ELODIE around the F6V star HD 33564. *Astron. Astrophys., 444*, L21–L24.

Galland F., Lagrange A.-M., Udry S., Beuzit J.-L., Pepe F., et al. (2006) Extrasolar planets and brown dwarfs around A-F type stars. IV. A candidate brown dwarf around the A9V pulsating star HD 180777. *Astron. Astrophys., 452*, 709–714.

Ge J. (2002) Fixed delay interferometry for Doppler extrasolar planet detection. *Astrophys. J. Lett., 571*, L165–L168.

Ge J., Erskine D. J., and Rushford M. (2002) An externally dispersed interferometer for sensitive Doppler extrasolar planet searches. *Publ. Astron. Soc. Pac., 114*, 1016–1028.

Ge J., van Eyken J., Mahadevan S., DeWitt C., Kane S. R., et al. (2006) The first extrasolar planet discovered with a new-generation high-throughput Doppler instrument. *Astrophys. J., 648*, 683–695.

Ge J., van Eyken J. C., Mahadevan S., Wan X., Zhao B., et al. (2007) An all sky extrasolar planet survey with new generation multiple object Doppler instruments at Sloan Telescope. *Rev. Mex. Astron. Astrophys. Conf. Ser., 29*, 30–36.

Gillon M., Pont F., Demory B.-O., Mallmann F., Mayor M., et al. (2007) Detection of transits of the nearby hot Neptune GJ 436b. *Astron. Astrophys., 472*, L13–L16.

Griffin R. and Griffin R. (1973) Accurate wavelengths of stellar and telluric absorption lines near lambda 7000 Angstroms. *Mon. Not. R. Astron. Soc., 162*, 255–260.

Hatzes A. P. and Cochran W. D. (1993) Long-period radial velocity variations in three K giants. *Astrophys. J., 413*, 339–348.

Hekker S., Snellen I. A. G., Aerts C., Quirrenbach A., Reffert S., and Mitchell D. S. (2008) Precise radial velocities of giant stars. IV. A correlation between surface gravity and radial velocity variation and a statistical investigation of companion properties. *Astron. Astrophys., 480*, 215–222.

Henry G. W., Marcy G. W., Butler R. P., and Vogt S. S. (2000) A transiting "51 Peg-like" planet. *Astrophys. J. Lett., 529*, L41–L44.

Howard A. W., Johnson J. A., Marcy G. W., Fischer D. A., Wright J. T., et al. (2009) The NASA-UC Eta-Earth Program. I. A super-Earth orbiting HD 7924. *Astrophys. J., 696*, 75–83.

Huélamo N., Figueira P., Bonfils X., Santos N. C., Pepe F., et al. (2008) TW Hydrae: Evidence of stellar spots instead of a hot Jupiter. *Astron. Astrophys., 489*, L9–L13.

Ida S. and Lin D. N. C. (2005) Toward a deterministic model of planetary formation. III. Mass distribution of short-period planets around stars of various masses. *Astrophys. J., 626*, 1045–1060.

Ida S. and Lin D. N. C. (2008) Toward a deterministic model of planetary formation. V. Accumulation near the ice line and super-Earths. *Astrophys. J., 685*, 584–595.

Johnson J. A., Butler R. P., Marcy G. W., Fischer D. A., Vogt S. S., et al. (2007a) A new planet around an M dwarf: Revealing a correlation between exoplanets and stellar mass. *Astrophys. J., 670*, 833–840.

Johnson J. A., Fischer D. A., Marcy G. W., Wright J. T., Driscoll P., et al. (2007b) Retired A stars and their companions: Exoplanets orbiting three intermediate-mass subgiants. *Astrophys. J., 665*, 785–793.

Kaeufl H.-U., Ballester P., Biereichel P., Delabre B., Donaldson R., et al. (2004) CRIRES: A high-resolution infrared spectrograph for ESO's VLT. In *Ground-based Instrumentation for Astronomy* (A. F. M. Moorwood and M. Iye, eds.), pp. 1218–1227. SPIE Conf. Series 5492, Bellingham, Washington.

Kennedy G. M., and Kenyon S. J. (2008) Planet formation around stars of various masses: The snow line and the frequency of giant planets. *Astrophys. J., 673*, 502–512.

Kjeldsen H., Bedding T. R., Butler R. P., Christensen-Dalsgaard J., Kiss L. L., et al. (2005) Solar-like oscillations in α Centauri B. *Astrophys. J., 635*, 1281–1290.

Kürster M., Hatzes A. P., Cochran W. D., Döbereiner S., Dennerl K., and Endl M. (1999) Precise radial velocities of Proxima Centauri. Strong constraints on a substellar companion. *Astron. Astrophys., 344*, L5–L8.

Latham D. W. (2007) Hot Earths: Prospects for detection by Kepler and characterization with HARPS-NEF. *Bull. Am. Astron. Soc., 38*, 234.

Latham D. W., Stefanik R. P., Mazeh T., Mayor M., and Burki G. (1989) The unseen companion of HD114762 — A probable brown dwarf. *Nature, 339*, 38–40.

Laughlin G., Bodenheimer P., and Adams F. C. (2004) The core accretion model predicts few jovian-mass planets orbiting red dwarfs. *Astrophys. J. Lett., 612*, L73–L76.

Li C.-H., Benedick A. J., Fendel P., Glenday A. G., Kärtner F. X., et al. (2008) A laser frequency comb that enables radial velocity measurements with a precision of cm s^{-1}. *Nature, 452*, 610–612.

Lindegren L. and Dravins D. (2003) The fundamental definition of "radial velocity." *Astron. Astrophys., 401*, 1185–1201.

Lisse C. M., Beichman C. A., Bryden G., and Wyatt M. C. (2007) On the nature of the dust in the debris disk around HD 69830. *Astrophys. J., 658*, 584–592.

Lovis C. and Mayor M. (2007) Planets around evolved intermediate-mass stars. I. Two substellar companions in the open clusters NGC 2423 and NGC 4349. *Astron. Astrophys., 472*, 657–664.

Lovis C. and Pepe F. (2007) A new list of thorium and argon spectral lines in the visible. *Astron. Astrophys., 468*, 1115–1121.

Lovis C., Mayor M., Pepe F., Alibert Y., Benz W., et al. (2006) An extrasolar planetary system with three Neptune-mass planets. *Nature, 441*, 305–309.

Lovis C., Mayor M., Bouchy F., Pepe F., Queloz D., et al. (2009) Towards the characterization of the hot Neptune/super-Earth population around nearby bright stars. In *Transiting Planets* (F. Pont et al., eds.), pp. 502–505. IAU Symp. Proc. 253, Cambridge Univ., Cambridge.

Marcy G. W. and Butler R. P. (1995) The planet around 51 Pegasi. *Bull. Am. Astron. Soc., 27*, 1379.

Marcy G. W. and Butler R. P. (1996) A planetary companion to 70 Virginis. *Astrophys. J. Lett., 464*, L147–L151.

Marcy G. W., Butler R. P., Vogt S. S., Fischer D., and Lissauer J. J. (1998) A planetary companion to a nearby M4 dwarf, Gliese 876. *Astrophys. J. Lett., 505*, L147–L149.

Marcy G. W., Butler R. P., Fischer D., Vogt S. S., Lissauer J. J., and Rivera E. J. (2001) A pair of resonant planets orbiting GJ 876. *Astrophys. J., 556*, 296–301.

Mayor M. and Queloz D. (1995) A Jupiter-mass companion to a solar-type star. *Nature, 378*, 355–359.

Mayor M., Pepe F., Queloz D., Bouchy F., Rupprecht G., et al. (2003) Setting new standards with HARPS. *The Messenger, 114*, 20–24.

Mayor M., Udry S., Lovis C., Pepe F., Queloz D., et al. (2009a) The HARPS search for southern extra-solar planets. XIII. A planetary system with 3 super-Earths (4.2, 6.9, and 9.2 $M_\oplus$). *Astron. Astrophys., 493*, 639–644.

Mayor M., Bonfils X., Forveille T., Delfosse X., Udry S., et al. (2009b) The HARPS search for southern extra-solar planets. XVIII. An Earth-mass planet in the GJ 581 planetary system. *Astron. Astrophys., 507*, 487–494.

McArthur B. E., Endl M., Cochran W. D., Benedict G. F., Fischer D. A., et al. (2004) Detection of a Neptune-mass planet in the ρ^1 Cancri system using the Hobby-Eberly telescope. *Astrophys. J. Lett., 614*, L81–L84.

Melo C., Santos N. C., Gieren W., Pietrzynski G., Ruiz M. T., et al. (2007) A new Neptune-mass planet orbiting HD 219828. *Astron. Astrophys., 467*, 721–727.

Mordasini C., Alibert Y., and Benz W. (2009) Extrasolar planet population synthesis. I. Method, formation tracks, and mass-distance distribution. *Astron. Astrophys., 501*, 1139–1160.

Murphy M. T., Udem T., Holzwarth R., Sizmann A., Pasquini L., et al. (2007) High-precision wavelength calibration of astronomical spectrographs with laser frequency combs. *Mon. Not. R. Astron. Soc., 380*, 839–847.

Naef D., Latham D. W., Mayor M., Mazeh T., Beuzit J. L., et al. (2001) HD 80606b, a planet on an extremely elongated orbit. *Astron. Astrophys., 375*, L27–L30.

Niedzielski A., Konacki M., Wolszczan A., Nowak G., Maciejewski G., et al. (2007) A planetary-mass companion to the K0 giant HD 17092. *Astrophys. J., 669*, 1354–1358.

O'Toole S., Tinney C. G., Butler R. P., Jones H. R. A., Bailey J., et al. (2009) A Neptune-mass planet orbiting the nearby G dwarf HD 16417. *Astrophys. J., 697*, 1263–1268.

Pallé P. L., Jimenez A., Perez Hernandez F., Regulo C., Roca Cortes T., and Sanchez L. (1995) A measurement of the background solar velocity spectrum. *Astrophys. J., 441,* 952–959.

Pasquini L., Döllinger M. P., Weiss A., Girardi L., Chavero C., et al. (2007) Evolved stars suggest an external origin of the enhanced metallicity in planet-hosting stars. *Astron. Astrophys., 473,* 979–982.

Pasquini L., Avila G., Dekker H., Delabre B., D'Odorico S., et al. (2008) CODEX: The high-resolution visual spectrograph for the E-ELT. In *Groundbased and Airborne Instrumentation for Astronomy II* (I. S. McLean and M. M. Casali, eds.), 70141I. SPIE Conf. Series 7014, Bellingham, Washington.

Pepe F., Mayor M., Galland F., Naef D., Queloz D., et al. (2002) The CORALIE survey for southern extra-solar planets VII. Two short-period saturnian companions to HD 108147 and HD 168746. *Astron. Astrophys., 388,* 632–638.

Pepe F., Correia A. C. M., Mayor M., Tamuz O., Couetdic J., et al. (2007) The HARPS search for southern extra-solar planets. VIII. μ Arae, a system with four planets. *Astron. Astrophys., 462,* 769–776.

Pepe F., Cristiani S., Rebolo Lopez R., Santos N. C., Molaro P., et al. (2010) ESPRESSO — Exploring science frontiers with extreme-precision spectroscopy. In *Ground-based and Airborne Instrumentation for Astronomy III* (I .S. McLean et al., eds.), in press. SPIE Conf. Series 7735, Bellingham, Washington.

Queloz D., Mayor M., Weber L., Blécha A., Burnet M., et al. (2000) The CORALIE survey for southern extra-solar planets. I. A planet orbiting the star Gliese 86. *Astron. Astrophys., 354,* 99–102.

Queloz D., Henry G. W., Sivan J. P., Baliunas S. L., Beuzit J. L., et al. (2001) No planet for HD 166435. *Astron. Astrophys., 379,* 279–287.

Rickman H., ed. (2001) *Transactions of the IAU Proceedings of the Twenty-Fourth General Assembly.* IAU Vol. XXIVB, Kluwer, Dordrecht.

Rivera E. J., Lissauer J. J., Butler R. P., Marcy G. W., Vogt S. S., et al. (2005) A 7.5 $M_\oplus$ planet orbiting the nearby star, GJ 876. *Astrophys. J., 634,* 625–640.

Rivera E. J., Butler R. P., Vogt S. S., Laughlin G., Henry G. W., et al. (2010) A super-Earth orbiting the nearby Sun-like star HD 1461. *Astrophys. J., 708,* 1492–1499.

Saar S. H. and Donahue R. A. (1997) Activity-related radial velocity variation in cool stars. *Astrophys. J., 485,* 319–327.

Saar S. H. and Fischer D. (2000) Correcting radial velocities for long-term magnetic activity variations. *Astrophys. J. Lett., 534,* L105–L108.

Santos N. C., Mayor M., Naef D., Pepe F., Queloz D., et al. (2000) The CORALIE survey for southern extra-solar planets. IV. Intrinsic stellar limitations to planet searches with radial-velocity techniques. *Astron. Astrophys., 361,* 265–272.

Santos N. C., Bouchy F., Mayor M., Pepe F., Queloz D., et al. (2004) The HARPS survey for southern extra-solar planets. II. A 14 Earth-masses exoplanet around μ Arae. *Astron. Astrophys., 426,* L19–L23.

Sato B., Ando H., Kambe E., Takeda Y., Izumiura H., et al. (2003) A planetary companion to the G-type giant star HD 104985. *Astrophys. J. Lett., 597,* L157–L160.

Sato B., Fischer D. A., Henry G. W., Laughlin G., Butler R. P., et al. (2005) The N2K Consortium. II. A transiting hot Saturn around HD 149026 with a large dense core. *Astrophys. J., 633,* 465–473.

Selsis F., Kasting J. F., Levrard B., Paillet J., Ribas I., and Delfosse X. (2007) Habitable planets around the star Gliese 581? *Astron. Astrophys., 476,* 1373–1387.

Setiawan J., Hatzes A. P., von der Lühe O., Pasquini L., Naef D., et al. (2003) Evidence of a sub-stellar companion around HD 47536. *Astron. Astrophys., 398,* L19–L23.

Standish E. M. Jr. (1990) The observational basis for JPL's DE 200, the planetary ephemerides of the Astronomical Almanac. *Astron. Astrophys., 233,* 252–271.

Steinmetz T., Wilken T., Araujo-Hauck C., Holzwarth R., Hänsch T. W., et al. (2008) Laser frequency combs for astronomical observations. *Science, 321,* 1335–1337.

Struve O. (1952) Proposal for a project of high-precision stellar radial velocity work. *The Observatory, 72,* 199–200.

Udry S. and Santos N. C. (2007) Statistical properties of exoplanets. *Annu. Rev. Astron. Astrophys., 45,* 397–439.

Udry S., Mayor M., Benz W., Bertaux J.-L., Bouchy F., et al. (2006) The HARPS search for southern extra-solar planets. V. A 14 Earth-masses planet orbiting HD 4308. *Astron. Astrophys., 447,* 361–367.

Udry S., Bonfils X., Delfosse X., Forveille T., Mayor M., et al. (2007) The HARPS search for southern extra-solar planets. XI. Super-Earths (5 and 8 $M_\oplus$) in a 3-planet system. *Astron. Astrophys., 469,* L43–L47.

Valenti J. A., Butler R. P., and Marcy G. W. (1995) Determining spectrometer instrumental profiles using FTS reference spectra. *Publ. Astron. Soc. Pac., 107,* 966–976.

Vogt S. S., Butler R. P., Marcy G. W., Fischer D. A., Henry G. W., et al. (2005) Five new multicomponent planetary systems. *Astrophys. J., 632,* 638–658.

Vogt S. S., Wittenmyer R., Butler R. P., O'Toole S., Henry G. W., et al. (2010) A super Earth and two Neptunes orbiting the nearby Sun-like star 61 Virginis. *Astrophys. J., 708,* 1366–1375.

Walker G. A. H., Walker A. R., Irwin A. W., Larson A. M., Yang S. L. S., and Richardson D. C. (1995) A search for Jupiter-mass companions to nearby stars. *Icarus, 116,* 359–375.

Wilson O. C. (1978) Chromospheric variations in main-sequence stars. *Astrophys. J., 226,* 379–396.

Wilson R. E. (1953) *General Catalogue of Stellar Radial Velocities.* Carnegie Inst. Publ. 601, Washington, DC.

Wolszczan A. and Frail D. A. (1992) A planetary system around the millisecond pulsar PSR1257 + 12. *Nature, 355,* 145–147.

Wright J. T. (2005) Radial velocity jitter in stars from the California and Carnegie Planet Search at Keck Observatory. *Publ. Astron. Soc. Pac., 117,* 657–664.

Exoplanet Transits and Occultations

Joshua N. Winn
Massachusetts Institute of Technology

When we are fortunate enough to view an exoplanetary system nearly edge-on, the star and planet periodically eclipse each other. Observations of eclipses — transits and occultations — provide a bonanza of information that cannot be obtained from radial-velocity data alone, such as the relative dimensions of the planet and its host star, as well as the orientation of the planet's orbit relative to the sky plane and relative to the stellar rotation axis. The wavelength-dependence of the eclipse signal gives clues about the temperature and composition of the planetary atmosphere. Anomalies in the timing or other properties of the eclipses may betray the presence of additional planets or moons. Searching for eclipses is also a productive means of discovering new planets. This chapter reviews the basic geometry and physics of eclipses, and summarizes the knowledge that has been gained through eclipse observations, as well as the information that might be gained in the future.

1. INTRODUCTION

From immemorial antiquity, men have dreamed of a royal road to success — leading directly and easily to some goal that could be reached otherwise only by long approaches and with weary toil. Times beyond number, this dream has proved to be a delusion . . . Nevertheless, there are ways of approach to unknown territory which lead surprisingly far, and repay their followers richly. There is probably no better example of this than eclipses of heavenly bodies.

— Henry Norris Russell (1948)

Vast expanses of scientific territory have been traversed by exploiting the occasions when one astronomical body shadows another. The timing of the eclipses of Jupiter's moons gave the first accurate measure of the speed of light. Observing the passage of Venus across the disk of the Sun provided a highly refined estimate of the astronomical unit. Studying solar eclipses led to the discovery of helium, the recognition that Earth's rotation is slowing down due to tides, and the confirmation of Einstein's prediction for the gravitational deflection of light. The analysis of eclipsing binary stars — the subject Russell had in mind — enabled a precise understanding of stellar structure and evolution.

Continuing in this tradition, eclipses are the "royal road" of exoplanetary science. We can learn intimate details about exoplanets and their parent stars through observations of their combined light, without the weary toil of spatially resolving the planet and the star (see Fig. 1). This chapter shows how eclipse observations are used to gain knowledge of the planet's orbit, mass, radius, temperature, and atmospheric constituents, along with other details that are otherwise hidden. This knowledge, in turn, gives clues about the processes of planet formation and evolution and provides a larger context for understanding the properties of the solar system.

An eclipse is the obscuration of one celestial body by another. When the bodies have very unequal sizes, the passage of the smaller body in front of the larger body is a transit and the passage of the smaller body behind the larger body is an occultation. Formally, transits are cases when the full disk of the smaller body passes completely within that of the larger body, and occultations refer to the complete concealment of the smaller body. We will allow those terms to include the grazing cases in which the bodies' silhouettes do not overlap completely. Please be aware that the exoplanet literature often refers to occultations as secondary eclipses (a more general term that does not connote an extreme size ratio), or by the neologisms "secondary transit" and "antitransit."

This chapter is organized as follows. Section 2 describes the geometry of eclipses and provides the foundational equations, building on the discussion of Keplerian orbits in the chapter by Murray and Correia. Readers seeking a more elementary treatment involving only circular orbits may prefer to start by reading *Sackett* (1999). Section 3 discusses many scientific applications of eclipse data, including the determination of the mass and radius of the planet. Section 4 is a primer on observing the apparent decline in stellar brightness during eclipses (the photometric signal). Section 5 reviews some recent scientific accomplishments, and section 6 offers some thoughts on future prospects.

2. ECLIPSE BASICS

2.1. Geometry of Eclipses

Consider a planet of radius R_p and mass M_p orbiting a star of radius $R_\star$ and mass $M_\star$. The ratio $R_p/R_\star$ occurs frequently enough to deserve its own symbol, for which we will use k, in deference to the literature on eclipsing binary stars. As in the chapter by Murray and Correia, we choose a coordinate system centered on the star, with the sky in the X–Y plane and the +Z axis pointing at the observer (see Fig. 2). Since the orientation of the line of nodes relative to celestial north (or any other externally

56 *Exoplanets*

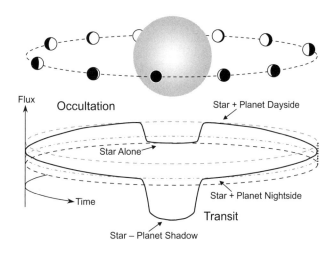

Fig. 1. Illustration of transits and occultations. Only the combined flux of the star and planet is observed. During a transit, the flux drops because the planet blocks a fraction of the starlight. Then the flux rises as the planet's dayside comes into view. The flux drops again when the planet is occulted by the star.

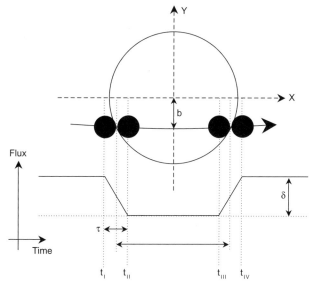

Fig. 2. Illustration of a transit, showing the coordinate system discussed in section 2.1, the four contact points and the quantities T and τ defined in section 2.3, and the idealized lightcurve discussed in section 2.4.

defined axis) is usually unknown and of limited interest, we might as well align the X axis with the line of nodes; we place the descending node of the planet's orbit along the +X axis, giving $\Omega = 180°$.

The distance between the star and planet is given by equation (20) of the chapter by Murray and Correia

$$r = \frac{a(1-e^2)}{1+e\cos f} \quad (1)$$

where a is the semimajor axis of the relative orbit and f is the true anomaly, an implicit function of time depending on the orbital eccentricity e and period P (see section 3 of the chapter by Murray and Correia). This can be resolved into Cartesian coordinates using equations (53)–(55) of the chapter by Murray and Correia, with $\Omega = 180°$

$$X = -r\cos(\omega+f) \quad (2)$$

$$Y = -r\sin(\omega+f)\cos i \quad (3)$$

$$Z = r\sin(\omega+f)\sin i \quad (4)$$

If eclipses occur, they do so when $r_{sky} \equiv \sqrt{X^2+Y^2}$ is a local minimum. Using equations (2) and (3)

$$r_{sky} = \frac{a(1-e^2)}{1+e\cos f}\sqrt{1-\sin^2(\omega+f)\sin^2 i} \quad (5)$$

Minimizing this expression leads to lengthy algebra (*Kipping*, 2008). However, an excellent approximation that we will use throughout this chapter is that eclipses are centered around conjunctions, which are defined by the condition X = 0 and may be inferior (planet in front) or superior (star in front). This gives

$$f_{tra} = +\frac{\pi}{2}-\omega, \quad f_{occ} -\frac{\pi}{2}-\omega \quad (6)$$

where here and elsewhere in this chapter, "tra" refers to transits and "occ" to occultations. This approximation is valid for all cases except extremely eccentric and close-in orbits with grazing eclipses.

The impact parameter b is the sky-projected distance at conjunction, in units of the stellar radius

$$b_{tra} = \frac{a\cos i}{R_\star}\left(\frac{1-e^2}{1+e\sin\omega}\right) \quad (7)$$

$$b_{occ} = \frac{a\cos i}{R_\star}\left(\frac{1-e^2}{1-e\sin\omega}\right) \quad (8)$$

For the common case $R_\star \ll a$, the planet's path across (or behind) the stellar disk is approximately a straight line between the points $X = \pm R_\star\sqrt{1-b^2}$ at $Y = bR_\star$.

2.2. Probability of Eclipses

Eclipses are seen only by privileged observers who view a planet's orbit nearly edge-on. As the planet orbits its star, its shadow describes a cone that sweeps out a band on the celestial sphere, as illustrated in Fig. 3. A distant observer within the shadow band will see transits. The opening angle of the cone, Θ, satisfies the condition $\sin\Theta = (R_\star + R_p)/r$ where r is the instantaneous star-planet distance. This cone is called the penumbra. There is also an interior cone, the antumbra, defined by $\sin\Theta = (R_\star - R_p)/r$, inside of which the transits are full (nongrazing).

A common situation is that e and ω are known and i is unknown, as when a planet is discovered via the Doppler method (see chapter by Lovis and Fischer), but no information is available about eclipses. With reference to Fig. 3, the observer's celestial longitude is specified by ω, but the latitude is unknown. The transit probability is calculated as the shadowed fraction of the line of longitude, or more simply from the requirement $|b| < 1 + k$, using equations (7) and (8) and the knowledge that $\cos i$ is uniformly distributed for a randomly placed observer. Similar logic applies to occultations, leading to the results

$$p_{tra} = \left(\frac{R_\star \pm R_p}{a}\right)\left(\frac{1+e\sin\omega}{1-e^2}\right) \quad (9)$$

$$p_{occ} = \left(\frac{R_\star \pm R_p}{a}\right)\left(\frac{1-e\sin\omega}{1-e^2}\right) \quad (10)$$

where the "+" sign allows grazing eclipses and the "−" sign excludes them. It is worth committing to memory the results for the limiting case $R_p \ll R_\star$ and $e = 0$

$$p_{tra} = p_{occ} = \frac{R_\star}{a} \approx 0.005 \left(\frac{R_\star}{R_\odot}\right)\left(\frac{a}{1\,\text{AU}}\right)^{-1} \quad (11)$$

For a circular orbit, transits and occultations always go together, but for an eccentric orbit it is possible to see transits without occultations or vice versa.

In other situations, one may want to marginalize over all possible values of ω, as when forecasting the expected number of transiting planets to be found in a survey (see section 4.1) or other statistical calculations. Here, one can calculate the solid angle of the entire shadow band and divide by 4π, or average equations (9) and (10) over ω, giving

$$p_{tra} = p_{occ} = \left(\frac{R_\star \pm R_p}{a}\right)\left(\frac{1}{1-e^2}\right) \quad (12)$$

Suppose you want to find a transiting planet at a particular orbital distance around a star of a given radius. If a fraction η of stars have such planets, you must search at least $N \approx (\eta\, p_{tra})^{-1}$ stars before expecting to find a transiting planet. A sample of >200 η^{-1} Sun-like stars is needed to find a transiting planet at 1 AU. Close-in giant planets have an orbital distance of approximately 0.05 AU and $\eta \approx 0.01$, giving $N > 10^3$ stars. In practice, many other factors affect the survey requirements, such as measurement precision, time sampling, and the need for spectroscopic follow-up observations (see section 4.1).

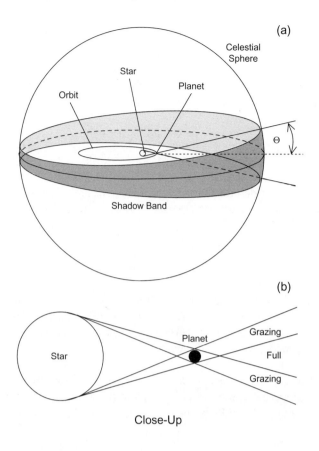

Fig. 3. Calculation of the transit probability. **(a)** Transits are visible by observers within the penumbra of the planet, a cone with opening angle Θ with $\sin\Theta = (R_\star + R_p)/r$, where r is the instantaneous star-planet distance. **(b)** Close-up showing the penumbra (thick lines) as well as the antumbra (thin lines) within which the transits are full, as opposed to grazing.

2.3. Duration of Eclipses

In a nongrazing eclipse, the stellar and planetary disks are tangent at four contact times $t_I - t_{IV}$, illustrated in Fig. 2. (In a grazing eclipse, second and third contact do not occur.) The total duration is $T_{tot} = t_{IV} - t_I$, the full duration is $T_{full} = t_{III} - t_{II}$, the ingress duration is $\tau_{ing} = t_{II} - t_I$, and the egress duration is $\tau_{egr} = t_{IV} - t_{III}$.

Given a set of orbital parameters, the various eclipse durations can be calculated by setting equation (5) equal to $R_\star \pm R_p$ to find the true anomaly at the times of contact, and then integrating equation (44) of the chapter by Murray and Correia, e.g.,

$$t_{III} - t_{II} = \frac{P}{2\pi\sqrt{1-e^2}}\int_{f_{II}}^{f_{III}}\left[\frac{r(f)}{a}\right]^2 df \quad (13)$$

For a circular orbit, some useful results are

$$T_{tot} \equiv t_{IV} - t_I = \frac{P}{\pi}\sin^{-1}\left[\frac{R_\star}{a}\frac{\sqrt{(1+k)^2-b^2}}{\sin i}\right] \quad (14)$$

$$T_{full} \equiv t_{III} - t_{II} = \frac{P}{\pi}\sin^{-1}\left[\frac{R_\star}{a}\frac{\sqrt{(1-k)^2-b^2}}{\sin i}\right] \quad (15)$$

For eccentric orbits, good approximations are obtained by multiplying equations (14) and (15) by

$$\frac{\dot{X}(f_c)[e=0]}{\dot{X}(f_c)} = \frac{\sqrt{1-e^2}}{1\pm e\sin\omega} \quad (16)$$

a dimensionless factor to account for the altered speed of the planet at conjunction. Here, "+" refers to transits and "−" to occultations. One must also compute b using the eccentricity-dependent equations (7) and (8).

For an eccentric orbit, τ_{ing} and τ_{egr} are generally unequal because the projected speed of the planet varies between ingress and egress. In practice the difference is slight; to leading order in $R_\star/a$ and e

$$\frac{\tau_e - \tau_i}{\tau_e + \tau_i} \sim e\cos\omega \left(\frac{R_\star}{a}\right)^3 (1-b^2)^{3/2} \quad (17)$$

which is $<10^{-2}$ e for a close-in planet with $R_\star/a = 0.2$, and even smaller for more distant planets. For this reason we will use a single symbol τ to represent either the ingress or egress duration. Another important timescale is $T \equiv T_{tot} - \tau$, the interval between the halfway points of ingress and egress (sometimes referred to as contact times 1.5 and 3.5).

In the limits $e \to 0$, $R_p \ll R_\star \ll a$, and $b \ll 1-k$ (which excludes near-grazing events), the results are greatly simplified

$$T \approx T_0\sqrt{1-b^2}, \quad \tau \approx \frac{T_0 k}{\sqrt{1-b^2}} \quad (18)$$

where T_0 is the characteristic timescale

$$T_0 \equiv \frac{R_\star P}{\pi a} \approx 13\,h\left(\frac{P}{1\,yr}\right)^{1/3}\left(\frac{\rho_\star}{\rho_\odot}\right)^{-1/3} \quad (19)$$

For eccentric orbits, the additional factor given by equation (16) should be applied. Note that in deriving equation (19), we used Kepler's third law and the approximation $M_p \ll M_\star$ to rewrite the expression in terms of the stellar mean density $\rho_\star$. This is a hint that eclipse observations give a direct measure of $\rho_\star$, a point that is made more explicit in section 3.1.

2.4. Loss of Light During Eclipses

The combined flux F(t) of a planet and star is plotted in Fig. 1. During a transit, the flux drops because the planet blocks a fraction of the starlight. Then the flux rises as the planet's dayside comes into view. The flux drops again when the planet is occulted by the star. Conceptually we may dissect F(t) as

$$F(t) = F_\star(t) + F_p(t) - \begin{cases} k^2\alpha_{tra}(t)F_\star(t) & \text{transits} \\ 0 & \text{outside eclipses} \\ \alpha_{occ}(t)F_p(t) & \text{occultations} \end{cases} \quad (20)$$

where $F_\star$, F_p are the fluxes from the stellar and planetary disks, and α's are dimensionless functions on the order of unity depending on the overlap area between the stellar and planetary disks. In general $F_\star$ may vary in time due to flares, rotation of star spots and plages, rotation of the tidal bulge raised by the planet, or other reasons, but for simplicity of discussion we take it to be a constant. In that case, only the ratio $f(t) \equiv F(t)/F_\star$ is of interest. If we let I_p and $I_\star$ be the disk-averaged intensities of the planet and star, respectively, then $F_p/F_\star = k^2 I_p/I_\star$ and

$$f(t) = 1 + k^2\frac{I_p(t)}{I_\star} - \begin{cases} k^2\alpha_{tra}(t) & \text{transits} \\ 0 & \text{outside eclipses} \\ k^2\frac{I_p(t)}{I_\star}\alpha_{occ}(t) & \text{occultations} \end{cases} \quad (21)$$

Time variations in I_p are caused by the changing illuminated fraction of the planetary disk (its phase function), as well as any changes intrinsic to the planetary atmosphere. To the extent that I_p is constant over the relatively short timespan of a single eclipse, all the observed time variation is from the α functions. As a starting approximation the α's are trapezoids, and f(t) is specified by the depth δ, duration T, ingress or egress duration τ, and time of conjunction t_c (as shown in Fig. 2). For transits the maximum loss of light is

$$\delta_{tra} \approx k^2\left[1 - \frac{I_p(t_{tra})}{I_\star}\right] \quad (22)$$

and in the usual case when the light from the planetary nightside is negligible, $\delta_{tra} \approx k^2$. For occultations

$$\delta_{occ} \approx k^2\frac{I_p(t_{occ})}{I_\star} \quad (23)$$

In the trapezoidal approximation the flux variation during ingress and egress is linear in time. In reality this is not true, partly because of the nonuniform motion of the stellar and planetary disks. More importantly, even with uniform motion the overlap area between the disks is not a linear function of time [see equation (1) of *Mandel and Agol* (2002)]. In addition, the bottom of a transit lightcurve is not flat because real stellar disks do not have uniform intensity, as explained in the next section.

2.5. Limb Darkening

Real stellar disks are brighter in the middle and fainter at the edge (the limb), a phenomenon known as limb darkening. This causes the flux decline during a transit to be larger than k^2 when the planet is near the center of the star, and smaller than k^2 when the planet is near the limb. The effect on the lightcurve is to round off the bottom and blur the second and third contact points, as shown in Fig. 4. Limb darkening is a consequence of variations in temperature and opacity with altitude in the stellar atmosphere. The sightline to the limb follows a highly oblique path into the stellar atmosphere, and therefore an optical depth of unity is reached at a higher altitude, where the temperature is cooler and the radiation is less intense. The resulting intensity profile $I(X,Y)$ is often described with a fitting formula such as

$$I \propto 1 - u_1(1-\mu) - u_2(1-\mu)^2 \quad (24)$$

where $\mu \equiv \sqrt{1-X^2-Y^2}$ and $\{u_1, u_2\}$ are constant coefficients that may be calculated from stellar-atmosphere models or measured from a sufficiently precise transit lightcurve. The decision to use the quadratic function of equation (24) or another of the various limb-darkening "laws" (better described as fitting formulas) is somewhat arbitrary. *Claret* (2004) provides a compilation of theoretical coefficients, and advocates a four-parameter law. *Southworth* (2008) investigates the results of fitting the same dataset with different limb-darkening laws. *Mandel and Agol* (2002) give accurate expressions for $\alpha_{tra}(t)$ for some limb darkening laws, and *Giménez* (2007) shows how to compute $\alpha_{tra}(t)$ for an arbitrary law based on earlier work by *Kopal* (1979).

By using one of these methods to calculate the flux of a limb-darkened disk with a circular obstruction, it is usually possible to model real transit lightcurves to within the measurement errors (see section 4.3). In principle, calculations of occultation lightcurves should take the planetary limb darkening into account, although the precision of current data has not justified this level of detail. Likewise, in exceptional cases it may be necessary to allow for departures from circular shapes, due to rotational or tidal deformation (see section 6). Modelers of eclipsing binary stars have long needed to take into account these and other subtle effects (*Kallrath and Milone*, 2009; *Hilditch*, 2001).

More generally, the loss of light depends on the intensity of the particular patch of the photosphere that is hidden by the planet. The planet provides a raster scan of the stellar intensity across the transit chord. In this manner, star spots and plages can be detected through the flux anomalies that

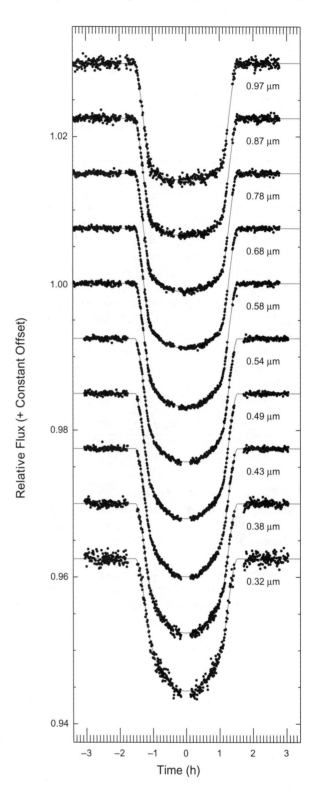

Fig. 4. Transits of the giant planet HD 209458b observed at wavelengths ranging from 0.32 µm (bottom) to 0.97 µm (top). At shorter wavelengths, the limb darkening of the star is more pronounced, and the bottom of the lightcurve is more rounded. The data were collected with the Hubble Space Telescope by *Knutson et al.* (2007a).

are observed when the planet covers them. (An example is given in the top right panel of Fig. 8.) Even spots that are not along the transit chord can produce observable effects, by causing variations in $F_\star$ and thereby causing δ_{tra} to vary from transit to transit.

3. SCIENCE FROM ECLIPSES

3.1. Determining Absolute Dimensions

Chief among the reasons to observe transits is to determine the mass and radius of the planet. Ideally one would like to know the mass in kilograms, and the radius in kilometers, to allow for physical modeling and comparisons with solar system planets. With only a transit lightcurve, this is impossible. The lightcurve by itself reveals the planet-to-star radius ratio $k \equiv R_p/R_\star \approx \sqrt{\delta}$ but not the planetary radius, and says nothing about the planetary mass.

To learn the planetary mass, in addition to the lightcurve one needs the radial-velocity orbit of the host star, and in particular the velocity semiamplitude $K_\star$. Using equation (66) of the chapter by Murray and Correia, and Kepler's third law, we may write

$$\frac{M_p}{(M_p + M_\star)^{2/3}} = \frac{K_\star \sqrt{1-e^2}}{\sin i}\left(\frac{P}{2\pi G}\right)^{1/3} \quad (25)$$

The observation of transits ensures $\sin i \approx 1$, and thereby breaks the usual $M_p \sin i$ degeneracy. However, the planetary mass still cannot be determined independently of the stellar mass. In the usual limit $M_p \ll M_\star$, the data determine $M_p/M_\star^{2/3}$ but not M_p itself.

To determine the absolute dimensions of the planet, one must supplement the transit photometry and radial-velocity orbit with some external information about the star. Depending on the star, the available information may include its luminosity, spectral type, and spectrally derived photospheric properties (effective temperature, surface gravity, and metallicity). A typical approach is to seek consistency between those data and stellar-evolutionary models, which relate the observable properties to the stellar mass, radius, composition, and age. In special cases it may also be possible to pin down the stellar properties using interferometry (*Baines et al.*, 2007), asteroseismology (*Stello et al.*, 2009), or an eclipsing companion star.

Besides δ, the transit lightcurve offers the observables T_{tot} and T_{full} (or T and τ), which can be used to solve for the impact parameter b and the scaled stellar radius $R_\star/a$. For nongrazing transits, in the limit $R_p \ll R_\star \ll a$ we may invert equations (14) and (15) to obtain the approximate formulas

$$b^2 = \frac{(1-\sqrt{\delta})^2 - (T_{full}/T_{tot})^2 (1+\sqrt{\delta})^2}{1 - (T_{full}/T_{tot})^2} \quad (26)$$

$$\frac{R_\star}{a} = \frac{\pi}{2\delta^{1/4}} \frac{\sqrt{T_{tot}^2 - T_{full}^2}}{P}\left(\frac{1+e\sin\omega}{\sqrt{1-e^2}}\right) \quad (27)$$

If in addition $\tau \ll T$, such as is the case for small planets on nongrazing trajectories, the results are simplified still further to

$$b^2 = 1 - \sqrt{\delta}\frac{T}{\tau} \quad (28)$$

$$\frac{R_\star}{a} = \frac{\pi}{\delta^{1/4}} \frac{\sqrt{T\tau}}{P}\left(\frac{1+e\sin\omega}{\sqrt{1-e^2}}\right) \quad (29)$$

The orbital inclination i may then be obtained using equation (7). These approximations are useful for theoretical calculations and for developing an intuition about how the system parameters affect the observable lightcurve. For example, $R_\star/a$ controls the product of T and τ, while b controls their ratio. However, as mentioned earlier, for fitting actual data one needs a realistic limb-darkened model, linked to a Keplerian orbital model.

The dimensionless ratios $R_\star/a$ and R_p/a are important for several reasons: (1) They set the scale of tidal interactions between the star and planet; (2) R_p/a determines what fraction of the stellar luminosity impinges on the planet, as discussed in section 3.4; (3) $R_\star/a$ can be used to determine a particular combination of the stellar mean density $\rho_\star$ and planetary mean density ρ_p

$$\rho_\star + k^3\rho_p = \frac{3\pi}{GP^2}\left(\frac{a}{R_\star}\right)^3 \quad (30)$$

This can be derived from Kepler's third law (*Seager and Mallen-Ornelas*, 2003). Since k^3 is usually small, the second term on the left side of equation (30) is often negligible and $\rho_\star$ can be determined purely from transit photometry.

This method for estimating $\rho_\star$ has proven to be a useful diagnostic in photometric transit surveys: A true transit signal should yield a value of $\rho_\star$ that is consistent with expectations for a star of the given luminosity and spectral type. Furthermore, once a precise lightcurve is available, $\rho_\star$ is a valuable additional constraint on the stellar properties.

Interestingly, it is possible to derive the planetary surface gravity $g_p \equiv GM_p/R_p^2$ independently of the stellar properties

$$g_p = \frac{2\pi}{P}\frac{\sqrt{1-e^2}\,K_\star}{(R_p/a)^2 \sin i} \quad (31)$$

This is derived from equation (66) of the chapter by Murray and Correia and Kepler's third law (*Southworth et al.*, 2007).

In short, precise transit photometry and Doppler velocimetry lead to correspondingly precise values of the stellar mean density $\rho_\star$ and planetary surface gravity g_p. However, the errors in M_p and R_p are ultimately limited by the uncertainties in the stellar properties.

3.2. Timing of Eclipses

The orbital period P can be determined by timing a sequence of transits, or a sequence of occultations, and fitting a linear function

$$t_c[n] = t_c[0] + nP \quad (32)$$

where $t_c[n]$ is the time of conjunction of the nth event. The times must first be corrected to account for Earth's orbital motion and consequent variations in the light travel time. When comparing transit and occultation times, one must further correct for the light travel time across the line-of-sight dimension of the planetary orbit. As long as there is no ambiguity in n, the error in P varies inversely as the total number of eclipses spanned by the observations, making it possible to achieve extraordinary precision in P.

If the orbit does not follow a fixed ellipse — due to forces from additional bodies, tidal or rotational bulges, general relativity, or other non-Keplerian effects — then there will be variations in the interval between successive transits, as well as the interval between transits and occultations and the shape of the transit lightcurve. These variations may be gradual parameter changes due to precession (*Miralda-Escudé*, 2002), or short-term variations due to other planets (*Holman and Murray*, 2005; *Agol et al.*, 2005) or moons (*Kipping*, 2009). The effects can be especially large for bodies in resonant orbits with the transiting planet. By monitoring transits one might hope to detect such bodies, as discussed in the chapter by Fabrycky.

When transits and occultations are both seen, a powerful constraint on the shape of the orbit is available. For a circular orbit, those events are separated in time by P/2, but more generally the time interval depends on e and ω. To first order in e, integrating dt/df between conjunctions gives

$$\Delta t_c \approx \frac{P}{2}\left[1 + \frac{4e\cos\omega}{\pi}\right] \quad (33)$$

In this case, the timing of transits and occultations gives an estimate of $e\cos\omega$. Likewise, the relative durations of the transit and the occultation depend on the complementary parameter $e\sin\omega$

$$\frac{T_{tra}}{T_{occ}} \approx 1 + e\sin\omega \quad (34)$$

Sterne (1940) and *de Kort* (1954) give the lengthy exact results for arbitrary e and i. Because the uncertainty in $\Delta t_c/P$ is typically smaller than that in T_{tra}/T_{occ} (by a factor of P/T), the eclipse data constrain $e\cos\omega$ more powerfully than $e\sin\omega$.

The resulting bounds on e are often valuable. For example, planets on close-in eccentric orbits are internally heated by the friction that accompanies the time-variable tidal distortion of the planet. Empirical constraints on e thereby help to understand the thermal structure of close-in planets. For more distant planets, bounds on e are helpful in the statistical analysis of exoplanetary orbits. As described in the chapter by Cumming and the chapters in Part V of this volume, the observed eccentricity distribution of planetary orbits is a clue about the processes of planet formation and subsequent orbital evolution.

Eclipse-based measurements of t_{tra}, t_{occ}, and P are almost always more precise than those based on spectroscopic or astrometric orbital data. The eclipse-based results can greatly enhance the analysis of those other data. For example, the usual radial-velocity curve has six parameters, but if t_{tra}, t_{occ}, and P are known from eclipses, the number of free parameters is effectively reduced to three, thereby boosting the achievable precision in the other three parameters.

3.3. Transmission Spectroscopy

We have been implicitly assuming that the planetary silhouette has a sharp edge, but in reality the edge is fuzzy. For gas giant planets there is no well-defined surface, and even planets with solid surfaces may have thick atmospheres. During a transit, a small portion of the starlight will be filtered through the upper atmosphere of the planet, where it is only partially absorbed. The absorption will be wavelength dependent due to the scattering properties of atoms and molecules in the planetary atmosphere. At the wavelength of a strong atomic or molecular transition, the atmosphere is more opaque, and the planet's effective silhouette is larger. This raises the prospect of measuring the transmission spectrum of the planet's upper atmosphere and thereby gaining knowledge of its composition.

To calculate the expected signal one must follow the radiative transfer of the incident starlight along a grazing trajectory through the planet's stratified atmosphere. The calculation is rather complicated (see, e.g., *Seager and Sasselov*, 2000; *Brown*, 2001), but the order of magnitude of the effect is easily appreciated. For a strong transition, the effective size of the planet grows by a few atmospheric scale heights H, where

$$H = \frac{k_B T}{\mu_m g} \quad (35)$$

and T is the temperature, μ_m is the mean molecular mass, g is the local gravitational acceleration, and k_B is Boltzmann's constant. Defining R_p as the radius within which the planet is optically thick at all wavelengths, the extra absorption due to the optically thin portion of the atmosphere causes the transit depth to increase by

$$\Delta\delta = \frac{\pi(R_p + N_H H)^2}{\pi R_\star^2} - \frac{\pi R_p^2}{\pi R_\star^2} \approx 2 N_H \delta \left(\frac{H}{R_p}\right) \quad (36)$$

where N_H, the number of scale heights, is on the order of unity. The signal is most readily detectable for planets with large H: low surface gravity, low mean molecular mass, and high temperature. For a "hot Jupiter" around a Sun-like star ($\delta = 0.01$, $T \approx 1300$ K, $g \approx 25$ m s^{-2}, $\mu_m = 2$ amu) the signal is $\Delta\delta \sim 10^{-4}$. For an Earth-like planet around a Sun-like star ($\delta = 10^{-4}$, $T \approx 273$ K, $g \approx 10$ m s^{-2}, $\mu_m = 28$ amu), the signal is $\Delta\delta \sim 10^{-6}$.

The signal can be detected by observing a transit lightcurve at multiple wavelengths, using different filters or a spectrograph. One then fits a limb-darkened lightcurve model to the time series obtained at each wavelength, requiring agreement in the orbital parameters and allowing a value of δ specific to each wavelength. The resulting variations in $\delta(\lambda)$ are expected to be on the order of magnitude given by equation (36). It is best to gather all the data at the same time, because intrinsic stellar variability is also chromatic.

3.4. Occultation Spectroscopy

As discussed in section 2.4, when the planet is completely hidden the starlight declines by a fraction $\delta_{occ} = k^2 I_p/I_\star$, where k is the planet-to-star radius ratio and $I_p/I_\star$ is the ratio of disk-averaged intensities. Observations spanning occultations thereby reveal the relative brightness of the planetary disk, if k is already known from transit observations. The planetary radiation arises from two sources: thermal radiation and reflected starlight. Because the planet is colder than the star, the thermal component emerges at longer wavelengths than the reflected component.

For the moment we suppose that the planet is of uniform brightness, and that the observing wavelength is long enough for thermal emission to dominate. Approximating the planet and star as blackbody radiators

$$\delta_{occ}(\lambda) = k^2 \frac{B_\lambda(T_p)}{B_\lambda(T_\star)} \to k^2 \frac{T_p}{T_\star} \quad (37)$$

where $B_\lambda(T)$ is the Planck function

$$B_\lambda(T) \equiv \frac{2hc^2}{\lambda^5} \frac{1}{e^{hc/(\lambda k_B T)} - 1} \to \frac{2 k_B T}{\lambda^2} \quad (38)$$

in which T is the temperature, λ is the wavelength, h is Planck's constant, and c is the speed of light. The limiting cases are for the "Rayleigh-Jeans" limit $\lambda \gg hc/(k_B T)$. The decrement δ_{occ} that is observed by a given instrument is obtained by integrating equation (37) over the bandpass.

Even when the planetary radiation is not described by the Planck law, one may define a brightness temperature $T_b(\lambda)$ as the equivalent blackbody temperature that would lead to the observed value of $\delta_{occ}(\lambda)$. The brightness temperature is sometimes a convenient way to describe the wavelength dependent intensity even when it is not thermal in origin.

There may be departures from a blackbody spectrum — spectral features — discernible in the variation of brightness temperature with wavelength. In contrast with transmission spectroscopy, which refers to starlight that grazes the planetary limb (terminator), here we are referring to the emission spectrum of the planet averaged over the visible disk of the dayside. Occultation spectroscopy and transit spectroscopy thereby provide different and complementary information about the planetary atmosphere.

It is possible to measure the reflectance spectrum of the planet's dayside by observing at shorter wavelengths, or accurately subtracting the thermal emission. The occultation depth due to reflected light alone is

$$\delta_{occ}(\lambda) = A_\lambda \left(\frac{R_p}{a}\right)^2 \quad (39)$$

where A_λ is the geometric albedo, defined as the flux reflected by the planet when viewed at opposition (full phase), divided by the flux that would be reflected by a flat and perfectly diffusing surface with the same cross-sectional area as the planet. One of the greatest uncertainties in atmospheric modeling is the existence, prevalence, and composition of clouds. Since clouds can produce very large albedo variations, reflectance spectroscopy may help to understand the role of clouds in exoplanetary atmospheres.

For a close-in giant planet, the reflectance signal is $\sim 10^{-4}$ while for an Earth-like planet at 1 AU it is $\sim 10^{-9}$. The detection prospects are better for closer-in planets. However, the closest-in planets are also the hottest, and their radiation may be dominated by thermal emission. Another consideration is that planets with $T > 1500$ K are expected to be so hot that all potentially cloud-forming condensable materials are in gaseous form. The theoretically predicted albedos are very low, on the order of 10^{-3}, due to strong absorption by neutral sodium and potassium.

Real planets do not have uniformly bright disks. Gaseous planets are limb darkened or brightened, and may have latitudinal zones with high contrast, like Jupiter. Rocky planets may have surface features and oceans. To the extent that departures from uniform brightness could be detected, occultation data would provide information on the spatially resolved planetary dayside. Specifically, the lightcurve of the ingress or egress of an occultation gives a one-dimensional cumulative brightness distribution of the planet.

3.5. The Rossiter-McLaughlin Effect

In addition to the spectral variations induced by the planetary atmosphere, there are spectral variations arising from the spatial variation of the stellar spectrum across the stellar disk. The most pronounced of these effects is due to stellar rotation: Light from the approaching half of the stellar disk is blueshifted, and light from the receding half is redshifted. Outside of transits, rotation broadens the spectral lines but does not produce an overall Doppler shift in the disk-integrated starlight. However, when the planet covers part of the blueshifted half of the stellar disk, the integrated starlight appears slightly redshifted, and vice versa.

Thus, the transit produces a time-variable spectral distortion that is usually manifested as an "anomalous" radial velocity, i.e., a Doppler shift that is greater or smaller than the shift expected from only the star's orbital motion. Figure 5 illustrates this effect. It is known as the Rossiter-McLaughlin (RM) effect, after the two astronomers who made the first definitive observations of this kind for binary stars, in 1924.

The maximum amplitude of the anomalous radial velocity is approximately

$$\Delta V_{RM} \approx k^2 \sqrt{1-b^2} \left(v_\star \sin i_\star \right) \quad (40)$$

where δ is the transit depth, b is the impact parameter, and $v_\star \sin i_\star$ is the line-of-sight component of the stellar equatorial rotation velocity. For a Sun-like star ($v_\star \sin i_\star = 2$ km s^{-1}), the maximum amplitude is ~20 m s^{-1} for a jovian planet and ~0.2 m s^{-1} for a terrestrial planet. This amplitude may be comparable to (or even larger than) the amplitude of the spectroscopic orbit of the host star. Furthermore, it is easier to maintain the stability of a spectrograph over the single night of a transit than the longer duration of the orbital period. Hence the RM effect is an effective means of detecting and confirming transits, providing an alternative to photometric detection.

In addition, by monitoring the anomalous Doppler shift throughout a transit, it is possible to measure the angle on the sky between the planetary orbital axis and the stellar rotation axis. Figure 6 shows three trajectories of a transiting planet that have the same impact parameter, and hence produce identical lightcurves, but that have different orientations relative to the stellar spin axis, and hence produce different RM signals. The signal for a well-aligned planet is antisymmetric about the midtransit time (left panels), whereas a strongly misaligned planet that blocks only the receding half of the star will produce only an anomalous blueshift (right panels).

A limitation of this technique is that it is only sensitive to the angle between the sky projections of the spin and orbital angular momentum vectors. The true angle between those vectors is usually poorly constrained because $i_\star$ is unknown. Nevertheless, it may be possible to tell whether the planetary orbit is prograde or retrograde, with respect to the direction of stellar rotation. It is also possible to combine results from different systems to gain statistical knowledge about spin-orbit alignment.

More broadly, just as transit photometry provides a raster scan of the intensity of the stellar photosphere along the transit chord, RM data provide a raster scan of the line-of-sight velocity field of the photosphere. This gives an independent measure of the projected rotation rate $v_\star \sin i_\star$, and reveals the velocity structure of starspots or other features that may exist on the photosphere.

4. OBSERVING ECLIPSES

4.1. Discovering Eclipsing Systems

Figure 7 shows the rate of exoplanet discoveries over the last 20 years, highlighting the subset of planets known to transit. Transits have been discovered in two ways. One way is to find the planet in a Doppler survey (see chapter by Lovis and Fischer), and then check for transits by monitoring the brightness of the star throughout the inferior conjunction. This was the path to discovery for 6 of the 64 known systems, including the first example, HD 209458b (*Charbonneau et al.*, 2000; *Henry et al.*, 2000; *Mazeh et al.*, 2000). The probability that transits will occur is given by equation (9) and the times of inferior conjunction can be calculated from the parameters of the spectroscopic orbit (*Kane*, 2007).

The other way is to conduct photometric surveillance of stars that are not yet known to have any planets, an idea dating back to *Struve* (1952). As an illustration, let us consider

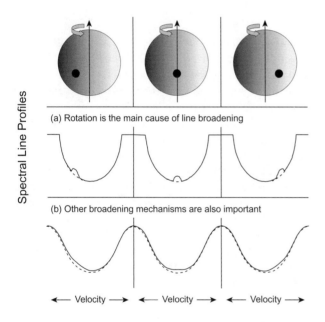

Fig. 5. Illustration of the Rossiter-McLaughlin (RM) effect. The three columns show three successive phases of a transit. The first row shows the stellar disk, with the grayscale representing the projected rotation velocity: The approaching limb is black and the receding limb is white. The second row shows the corresponding stellar absorption line profiles, assuming rotation to be the dominant broadening mechanism. The "bump" occurs because the planet hides a fraction of the light that contributes a particular velocity to the line-broadening kernel. The third row shows the case for which other line-broadening mechanisms are important; here the RM effect is manifested only as an "anomalous Doppler shift." Adapted from *Gaudi and Winn* (2007).

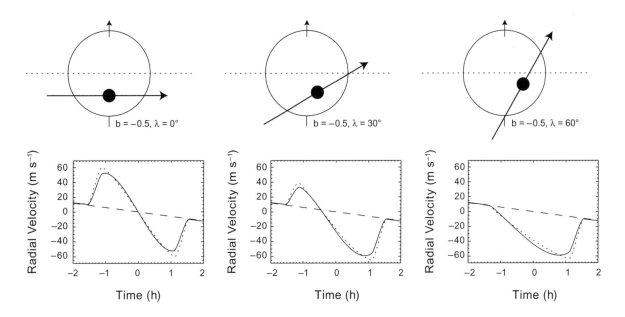

Fig. 6. Using the RM effect to measure the angle λ between the sky projections of the orbital and stellar-rotational axes. Three different possible trajectories of a transiting planet are shown, along with the corresponding RM signal. The trajectories all have the same impact parameter and produce the same lightcurve, but they differ in λ and produce different RM curves. The dotted lines are for the case of no limb darkening, and the solid lines include limb darkening. From *Gaudi and Winn* (2007).

the requirements for a program to discover hot Jupiters. As discussed in section 2.2, a bare minimum of 10^3 Sun-like stars must be examined to find a transiting planet with an orbital distance of 0.05 AU. The observations must be precise enough to detect a 1% flux drop, and must extend for at least several times longer than the three-day orbital period. The difficulty increases dramatically for smaller planets in wider orbits: In an idealized survey of nearby field stars limited by photon noise (see section 4.2), the sensitivity to transiting planets scales approximately as $R_p^6/P^{5/3}$ (*Gaudi*, 2005; *Gaudi et al.*, 2005). More realistic calculations take into account the spread in radius and luminosity among the surveyed stars, the duration and time sampling of observations, the appropriate threshold value of the signal-to-noise ratio, and other factors (see, e.g., *Pepper et al.*, 2003; *Beatty and Gaudi*, 2008).

Photometric surveys are much less efficient than Doppler surveys, in the sense that only a small fraction of planets transit, and even when transits occur they are underway only a small fraction of the time. On the other hand, the starting equipment for a photometric survey is modest, consisting in some cases of amateur-grade telescopes or telephoto lenses and cameras, whereas Doppler surveys require upfront a large telescope and sophisticated spectrograph. For these reasons the royal road of eclipses enticed many astronomers to embark on photometric transit surveys. More than a dozen surveys were undertaken, including a few longitudinally distributed networks to provide more continuous time coverage. Major efforts were made to automate the observations, and to develop algorithms for precise wide-field photometry (*Tamuz et al.*, 2005; *Kovács et al.*, 2005) and transit detection (*Kovács et al.*, 2002). *Horne* (2003) took stock of all the ongoing and planned surveys, and ventured to predict a bounty of 10–100 new planets per month, in an article subtitled "Hot Jupiters Galore."

The royal road turned out to have some potholes. One obstacle was the high rate of "false positives," signals that resemble planetary transits but are actually grazing eclipses of a binary star, or an unresolved combination of an eclipsing binary star and a third star. In the latter case, the deep eclipses of the binary star are diluted to planet-like proportions by the constant light of the third star. In some surveys the false positives outnumbered the planets by 10 to 1. Ruling them out required spectroscopy with large telescopes, which became the bottleneck in the discovery process (see, e.g., *O'Donovan et al.*, 2007). Another obstacle was correlated noise ("red noise") in the survey photometry (see section 4.3). *Pont et al.* (2006) showed that red noise slashed the sensitivity of the search algorithms, thereby providing a quantitative solution to the "Horne problem" of why transiting planets were not being found as rapidly as expected.

Only a few of the surveyors were able to overcome these obstacles. The five most successful surveys were OGLE, which used a 1-m telescope to survey 14–16th-magnitude stars; and the TrES, XO, HAT, and SuperWASP surveys, which used ≈0.1-m lenses to survey 10–12th-magnitude stars. Many of today's transiting planets are named after these surveys; for more details see *Udalski et al.* (2002), *Alonso et al.* (2004), *McCullough et al.* (2005), *Bakos et al.* (2007), and *Pollacco et al.* (2006). The OGLE planets were

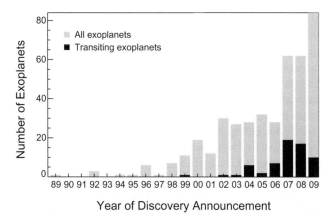

Fig. 7. Rate of exoplanet discovery. The light bars show the number of announcements of newly discovered planets, and the dark bars show the subset of transiting planets. Compiled with data from *exoplanet.eu*, an encyclopedic website maintained by J. Schneider.

found first, although the brighter stars from the wider-field surveys were far more amenable to false-positive rejection and detailed characterization.

On the other end of the cost spectrum are photometric surveys performed from space. By avoiding the deleterious effects of Earth's atmosphere, it is possible to beat the precision of groundbased observations (see section 4.2). It is also possible to avoid the usual interruptions due to the vagaries of the weather, and even to avoid the day-night cycle if the spacecraft is in an orbit far from Earth. The two ongoing spacebased missions CoRoT and Kepler are described in sections 5.1 and 6, respectively.

One spacebased survey that did not discover any planets nevertheless produced an interesting result. *Gilliland et al.* (2000) used the Hubble Space Telescope to seek close-in giant planets in the globular cluster 47 Tucanae. No transits were found, even though 17 were expected based on the survey characteristics and the observed frequency of close-in giant planets around nearby field stars. The absence of close-in planets could be due to the crowded stellar environment (which could inhibit planet formation or disrupt planetary systems) or the cluster's low metallicity (which has been found to be correlated with fewer planets in the Doppler surveys; see the chapters by Lovis and Fischer and by Cumming). The null results of a subsequent survey of the less-crowded outskirts of 47 Tucanae suggest that crowding is not the primary issue (*Weldrake et al.*, 2005).

4.2. Measuring the Photometric Signal

Even after a transiting planet is discovered, it takes a careful hand to measure the transit signal precisely enough to achieve the scientific goals set forth in section 3. The loss of light is only 1% for a Sun-like star crossed by a Jupiter-sized planet, and 0.01% for an Earth-sized planet. Occultations produce still smaller signals. This is the domain of precise time-series differential photometry. The term "differential" applies because only the fractional variations are of interest, as opposed to the actual intensity in Janskys or other standardized units. Eclipses can also be detected via spectroscopy (section 3.5) or polarimetry (*Carciofi and Magalhães*, 2005) but here we focus on the photometric signal, with emphasis on groundbased optical observations.

First you must know when to observe. To observe an eclipse requires a triple coincidence: The eclipse must be happening, the star must be above the horizon, and the Sun must be down. Transit times can be predicted based on a sequence of previously measured transit times, by fitting and extrapolating a straight line (equation (32)). Occultation times can also be predicted from a listing of transit times, but are subject to additional uncertainty due to the dependence on e and ω (equation (33)).

Next you should monitor the flux of the target star along with other nearby stars of comparable brightness. The measured fluxes are affected by short-term variations in atmospheric transparency, as well as the gradual change in the effective atmospheric path length (the airmass) as the star rises and sets. However, the ratios of the fluxes between nearby stars are less affected. As long as some of the comparison stars are constant in brightness, then the relative flux of the target star can be tracked precisely.

This task is usually accomplished with a charge-coupled device (CCD) imaging camera and software for calibrating the images, estimating the sky background level, and counting the photons received from each star in excess of the sky level (aperture photometry). *Howell* (2006) explains the basic principles of CCDs and aperture photometry. For more details on differential aperture photometry using an ensemble of comparison stars, see *Gilliland and Brown* (1988), *Kjeldsen and Frandsen* (1992), and *Everett and Howell* (2001). In short, the fluxes of the comparison stars are combined, and the flux of the target star is divided by the comparison signal, giving a time series of relative flux measurements spanning the eclipse with as little noise as possible. For bright stars observed at optical wavelengths, among the important noise sources are photon noise, scintillation noise, differential extinction, and flat-fielding errors, which we now discuss in turn.

Photon noise refers to the unavoidable fluctuations in the signal due to the quantization of light. It is also called Poisson noise because the photon count rate obeys a Poisson distribution. If a star delivers N photons s^{-1} on average, then the standard deviation in the relative flux due to photon noise is approximately $(N\Delta t)^{-1/2}$ for an exposure lasting Δt seconds. This noise source affects the target and comparison stars independently. The sky background also introduces Poisson noise, which can be troublesome for faint stars or infrared wavelengths. The photon noise in the comparison signal can be reduced by using many bright comparison stars. Beyond that, improvement is possible only by collecting more photons, using a bigger telescope, a more efficient detector, or a wider bandpass.

Scintillation is caused by fluctuations in the index of refraction of air. The more familiar term is "twinkling." For integration times $\Delta t \geq 1$ s, the standard deviation in the relative flux due to scintillation is expected to scale with telescope diameter D, observatory altitude h, integration time, and airmass as

$$\sigma_{scin} = \sigma_0 \frac{(\text{Airmass})^{7/4}}{D^{2/3}(\Delta t)^{1/2}} \exp\left(-\frac{h}{8000 \text{ m}}\right) \qquad (41)$$

based on a theory of atmospheric turbulence by *Reiger* (1963), with empirical support from *Young* (1967) and others. The coefficient σ_0 is often taken to be 0.064 when D is expressed in centimeters and Δt in seconds, but this must be understood to be approximate and dependent on the local meteorology. Scintillation affects both the target and comparison stars, although for closely spaced stars the variations are correlated (*Ryan and Sandler*, 1998). One also expects scintillation noise to decrease with wavelength. Thus, scintillation noise is reduced by employing a large telescope (or combining results from multiple small telescopes), choosing nearby comparison stars, and observing at a long wavelength from a good site.

Differential extinction is used here as a shorthand for "second-order color-dependent differential extinction." To first order, if two stars are observed simultaneously, their fluxes are attenuated by Earth's atmosphere by the same factor and the flux ratio is preserved. To second order, the bluer star is attenuated more, because scattering and absorption are more important at shorter wavelengths. This effect causes the flux ratio to vary with airmass as well as with short-term transparency fluctuations. It can be reduced by choosing comparison stars bracketing the target star in color, or using a narrow bandpass. The advantage of a narrow bandpass must be weighed against the increased photon noise.

Flat fielding is the attempt to correct for nonuniform illumination of the detector and pixel-to-pixel sensitivity variations, usually by dividing the images by a calibration image of a uniformly lit field. If the stars were kept on the same pixels throughout an observation, then flat-fielding errors would not affect the flux ratios. In reality the light from a given star is detected on different pixels at different times, due to pointing errors, focus variations, and seeing variations. Imperfect flat fielding coupled with these variations produce noise in the lightcurve. The impact of flat-fielding errors is reduced by ensuring the calibration images have negligible photon noise, maintaining a consistent pointing, and defocusing the telescope. Defocusing averages down the interpixel variations, reduces the impact of seeing variations, and allows for longer exposure times without saturation, thereby increasing the fraction of the time spent collecting photons as opposed to resetting the detector. Defocusing is good for what ails you, as long as the stars do not blend together.

This discussion of noise sources is not exhaustive; it is in the nature of noise that no such listing can be complete. For example, the gain of the detector may drift with temperature, or scattered moonlight may complicate background subtraction. A general principle is to strive to keep everything about the equipment and the images as consistent as possible. Another good practice is to spend at least as much time observing the star before and after the eclipse as during the eclipse, to establish the baseline signal and to characterize the noise.

It is often advisable to use a long-wavelength bandpass, not only to minimize scintillation and differential extinction, but also to reduce the effects of stellar limb darkening on the transit lightcurve. The degree of limb darkening diminishes with wavelength because the ratio between blackbody spectra of different temperatures is a decreasing function of wavelength. Transit lightcurves observed at longer wavelengths are "boxier," with sharper corners and flatter bottoms. All other things being equal, this reduces the statistical uncertainties in the transit parameters, but other factors should also be considered. For example, at infrared wavelengths, limb darkening may be small, but the sky background is bright and variable.

The spacebased observer need not worry about scintillation and differential extinction, and enjoys a low background level even at infrared wavelengths. Extremely precise photometry is possible, limited only by the size of the telescope and the degree to which the detector is well calibrated. With the Hubble Space Telescope a precision of approximately 10^{-4}-per-minute-long integration has been achieved, a few times better than the best groundbased lightcurves. Figure 8 allows for some side-by-side comparisons of groundbased and spacebased data. However, going to space is not a panacea. When unforeseen calibration issues arise after launch, they are difficult to resolve. In low Earth orbit, there are also unhelpful interruptions due to occultations by Earth, as well as problems with scattered Earthshine and cosmic rays. More remote orbits offer superior observing conditions, but require greater effort and expense to reach.

4.3. Interpreting the Photometric Signal

Once you have an eclipse lightcurve, the task remains to derive the basic parameters $\{\delta, T, \tau, t_c\}$ and their uncertainties, as well as the results for other quantities such as g_p and $\rho_\star$ that can be derived from those parameters. The analytic equations given in section 2 for the eclipse duration, timing, and other properties are rarely used to analyze data. Rather, a parametric model is fitted to the data, based on the numerical integration of Kepler's equation to calculate the relative positions of the star and planet, as well as one of the prescriptions mentioned in section 2.4 for computing the loss of light from the limb-darkened stellar photosphere.

The first task is writing a code that calculates the lightcurve of the star-planet system as a function of the orbital parameters and eclipse parameters. Then, this code is used in conjunction with one of many standard routines to optimize the parameter values, typically by minimizing the sum-of-squares statistic

$$\chi^2 = \sum_{i=1}^{N} \left[\frac{f_i(\text{obs}) - f_i(\text{calc})}{\sigma_i} \right]^2 \qquad (42)$$

where $f_i(\text{obs})$ is the observed value of relative flux at time t_i, $f_i(\text{calc})$ is the calculated flux (depending on the model parameters), and σ_i is the measurement uncertainty. The techniques found in standard works such as *Numerical Recipes* (*Press et al.*, 2007) are applicable here, although a few points of elaboration are warranted for the specific context of eclipse photometry.

It is common to adopt a Bayesian attitude, in which the parameters are viewed as random variables whose probability distributions ("posteriors") are constrained by the data. This can be done in a convenient and elegant fashion using the Monte Carlo Markov Chain (MCMC) method, in which a chain of points is created in parameter space using a few simple rules that ensure the collection of points will converge toward the desired posterior. This method gives the full multidimensional joint probability distribution for all the parameters, rather than merely giving individual error bars, making it easy to visualize any fitting degeneracies and to compute posteriors for any combination of parameters that may be of interest. Although a complete MCMC briefing is beyond the scope of this chapter, the interested reader should consult the textbook by *Gregory* (2005) as well as case studies such as *Holman et al.* (2006), *Collier Cameron et al.* (2007), and *Burke et al.* (2007).

A vexing problem is the presence of correlated noise, as mentioned in section 4.2. The use of equation (42) is based on the premise that the measurement errors are statistically independent. In many cases this is plainly false. Real lightcurves have bumps, wiggles, and slopes spanning many data points. These can be attributed to differential extinction, flat-fielding errors, or astrophysical effects such as starspots. Thus the number of truly independent samples is smaller than the number of data points, and the power to constrain model parameters is correspondingly reduced. Ignoring the correlations leads to false precision in the derived parameters, but accounting for correlations is not straightforward and can be computationally intensive. Some suggestions are given by *Pont et al.* (2006) and *Carter and Winn* (2009).

The treatment of stellar limb darkening presents another unwelcome opportunity to underestimate the parameter uncertainties. It is tempting to adopt one of the standard limb-darkening laws (see section 2.5) and hold the coefficients fixed at values deemed appropriate for the host star, based on stellar-atmosphere models. However, with precise data it is preferable to fit for the coefficients, or at least to allow for some uncertainty in the atmospheric models. In those few cases where the data have been precise enough to test the models, the models have missed the mark (*Claret*, 2009).

Although fitting data is a job for a computer, it is nevertheless useful to have analytic formulas for the achievable precision in the eclipse parameters. The formulas are handy for planning observations, and for order-of-magnitude esti-

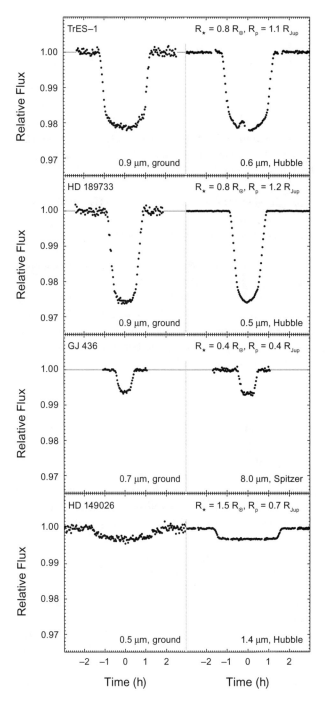

Fig. 8. Examples of transit lightcurves based on groundbased observations (left) and spacebased observations (right). In the top right panel, the "bump" observed just before midtransit is interpreted as the covering of a dark starspot by the planet. From upper left to lower right, the references are *Winn et al.* (2007a), *Rabus et al.* (2009), *Winn et al.* (2007b), *Pont et al.* (2008), *Holman et al.*, in preparation, *Gillon et al.* (2007), *Winn et al.* (2008), and *Carter et al.* (2009).

mates of the observability (or not) of effects such as variations in transit times and durations. The formulas given here are based on the assumptions that the data have uniform time sampling Δt, and independent Gaussian errors σ in the relative flux. A useful figure of merit is $Q \equiv \sqrt{N}\, \delta/\sigma$, where δ is the transit depth and N is the number of data points obtained during the transit. A Fisher information analysis (essentially a glorified error propagation) leads to estimates for the 1σ uncertainties in the transit parameters (*Carter et al.*, 2008)

$$\sigma_\delta \approx Q^{-1}\delta \qquad (43)$$

$$\sigma_{t_c} \approx Q^{-1}T\sqrt{\tau/2T} \qquad (44)$$

$$\sigma_T \approx Q^{-1}T\sqrt{2\tau/T} \qquad (45)$$

$$\sigma_\tau \approx Q^{-1}T\sqrt{6\tau/T} \qquad (46)$$

which are valid when $\delta \ll 1$, limb darkening is weak, and the out-of-transit flux is known precisely. In this case, $\sigma_{t_c} < \sigma_T < \sigma_\tau$. Correlated errors and limb darkening cause these formulas to be underestimates.

5. SUMMARY OF RECENT ACHIEVEMENTS

5.1. Discoveries of transiting planets

Discoveries of exoplanets, and transiting exoplanets in particular, have abounded in recent years (Fig. 7). As of December 2009, approximately 64 transiting planets are known, representing 15% of the total number of exoplanets discovered. Figure 9 shows their masses, radii, and orbital periods. It is important to remember that these planets are *not* a randomly selected subset of exoplanets. The properties of the ensemble have been shaped by powerful selection effects in the surveys that led to their discovery, favoring large planets in short-period orbits.

Despite these selection effects, the known transiting planets exhibit a striking diversity. They span three orders of magnitude in mass, and one order of magnitude in radius. Most are gas giants, comparable in mass and radius to Jupiter. There are also two planets with sizes more like Neptune and Uranus, as well as two even smaller planets with sizes only a few times larger than Earth, a category that has come to be known as "super Earths."

The two transiting super Earths (Fig. 10) were found with completely different strategies. The CoRoT (COnvection, ROtation, and planetary Transits) team uses a satellite equipped with a 0.27-m telescope and CCD cameras to examine fields of ~10,000 stars for a few months at a time, seeking relatively short-period planets (P ≤ 20 d). Along with several giant planets they have found a planet with a radius of 1.7 $R_\oplus$ in a 20-h orbit around a G-dwarf star, producing a transit depth of only 3.4×10^{-4} (*Léger et al.*, 2009). This is smaller than the detection threshold of any of the groundbased surveys, demonstrating the advantage of spacebased photometry. However, it proved difficult to spectroscopically confirm that the signal is indeed due to a planet, because the host star is relatively faint and chromospherically active (*Queloz et al.*, 2009).

Another project, called MEarth (pronounced "mirth"), seeks transits of small planets from the ground by focusing on very small stars (M dwarfs), for which even a super Earth would produce a transit depth on the order of 1%. Because such stars are intrinsically faint, one must search the whole sky to find examples bright enough for follow-up work. That is why MEarth abandoned the usual survey concept in which many stars are monitored within a single telescope's field of view. Instead they monitor M-dwarf stars one at a time, using several 0.4-m telescopes. Using this strategy they found a planet with a radius of 2.7 $R_\oplus$ in a 1.6-d period around a star with a radius of 0.21 $R_\odot$ (*Charbonneau et al.*, 2009). The large transit depth of 1.3% invites follow-up observations to study the planet's atmosphere.

At the other extreme, several transiting objects have masses greater than 10 M_{Jup}, reviving the old debate about what should and should not be considered a planet. The radii of these massive objects are not much larger than Jupiter's radius, in agreement with predictions that between about 1 and 50 M_{Jup} the pressure due to Coulomb forces (which would give $R \propto M^{1/3}$) and electron degeneracy pressure ($R \propto M^{-1/3}$) conspire to mute the mass dependence of the radius.

Almost all the transiting planets have short orbital periods (<10 d), due to the decline in transit probability with orbital distance (equation (12)). Two conspicuous exceptions are HD 17156b with P = 21 d (*Barbieri et al.*, 2007), and HD 80606b with P = 111 d (*Moutou et al.*, 2009; *Garcia-Melendo and McCullough*, 2009; *Fossey et al.*, 2009). Both of those planets were discovered by the Doppler method and found to transit through photometric follow-up observations. They were recognized as high-priority targets because both systems have highly eccentric orbits oriented in such a way as to enhance the probability of eclipses. Thus despite their long orbital periods, their periastron distances are small (<0.1 AU), along with all the other known transiting planets.

5.2. Follow-Up Photometry and Absolute Dimensions

Follow-up observations of transits have allowed the basic transit parameters $\{\delta, T, \tau\}$ to be determined to within 1% or better, and absolute dimensions of planets to within about 5%. Compilations of system parameters are given by *Torres et al.* (2008) and *Southworth* (2009). The orbital periods are known with eight significant digits in some cases. Groundbased observations have achieved a photometric precision of 250 ppm per 1 min sample, through the techniques described in section 4.2.

A pathbreaking achievement was the Hubble Space Telescope lightcurve of HD 209458 by *Brown et al.* (2001), with a time sampling of 80 s and a precision of 110 ppm,

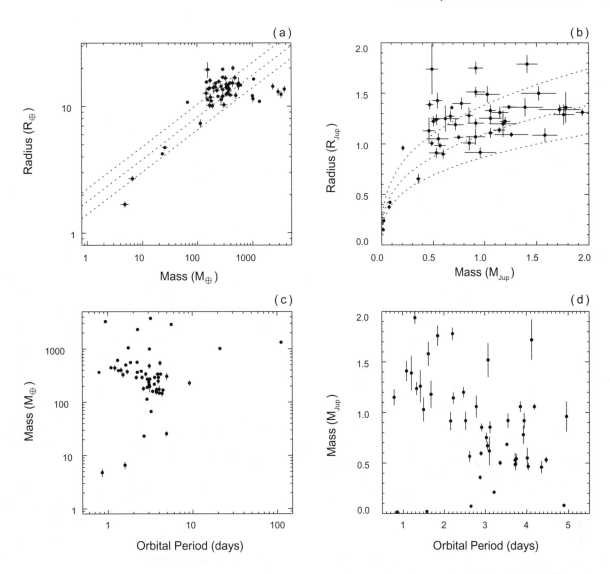

Fig. 9. Masses, radii, and orbital periods of the transiting exoplanets. **(a)** Radius vs. mass, on logarithmic scales. The dotted lines are loci of constant mean density (0.5, 1, and 2 g cm^{-3}, from top to bottom). **(b)** Same, but on linear scales, and with the axes restricted to highlight the gas giants that dominate the sample. **(c)** Mass vs. orbital period, on a logarithmic scale. The two long-period outliers are HD 17156b (P = 21 d) and HD 80606b (P = 111 d). **(d)** Same, but on a linear scale, and with axes restricted to highlight the gas giants. The anticorrelation between mass and orbital period is evident.

without the need for comparison stars. With the Spitzer Space Telescope the photon noise is generally higher, but there are compensatory advantages: There is little limb darkening at at mid-infrared wavelengths, and uninterrupted views of entire events are possible because the satellite is not in a low Earth orbit. Among the most spectacular data yet obtained in this field was the 33-h observation by *Knutson et al.* (2007b) of the K star HD 189733, spanning both a transit and occultation of its giant planet. The data were gathered with Spitzer at a wavelength of 8 µm, and are shown in Fig. 11. They not only provided extremely precise lightcurves of the transit and occultation, but also showed that the combined flux rises gradually in between those two events, demonstrating that the dayside is hotter than the nightside.

Among the properties of the close-in giant planets, a few patterns have emerged. The planetary mass is inversely related to the orbital period (*Mazeh et al.*, 2005). This anticorrelation has been variously attributed to selection effects, tidal interactions, and thermal evaporation of the planetary atmosphere. There is also a positive correlation between the metallicity of the host star and the inferred "core mass" of the planet (*Guillot et al.*, 2006), by which is meant the mass of heavy elements required in models of the planetary interior that agree with the observed mass and radius. It is tempting to interpret this latter correlation as support for the core-accretion theory of planet formation.

A persistent theme in this field, and a source of controversy and speculation, is that several of the transiting giant

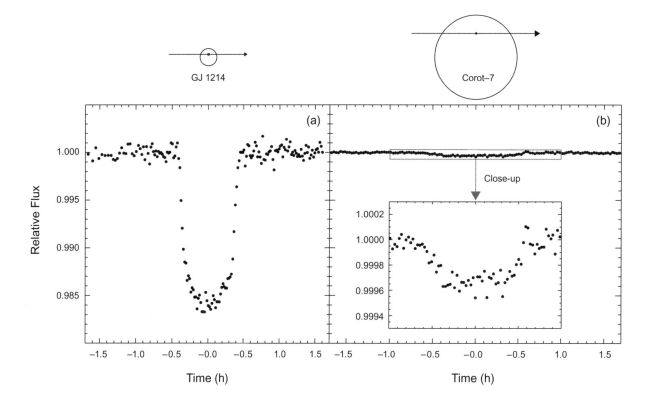

Fig. 10. Transits of two different super Earths: (a) GJ 1214b and (b) CoRoT-7b. The planets are approximately the same size, but because GJ 1214b orbits a small star (spectral type M4.5V), its transit depth is much larger than that of CoRoT-7b, which orbits a larger star (G9V). Data from *Charbonneau et al.* (2009) and *Leger et al.* (2009).

planets have radii that are 10–50% larger than expected from models of hydrogen-helium planets, even after accounting for the intense stellar heating and selection effects (see the chapter by Fortney et al.). Among the possible explanations for these "bloated" planets are tidal heating (see, e.g., *Miller et al.*, 2009); unknown atmospheric constituents that efficiently trap internal heat (*Burrows et al.*, 2007); enhanced downward convection allowing the incident stellar radiation to reach significant depth (*Guillot and Showman*, 2002); and inhibition of convection within the planet that traps internal heat (*Chabrier and Baraffe*, 2007).

Likewise, a few planets are observed to be smaller than expected for a hydrogen-helium planet with the observed mass and degree of irradiation. Some examples are HD 149026b (*Sato et al.*, 2005) and HAT-P-3b (*Torres et al.*, 2007). The favored interpretation is that these planets are enriched in elements heavier than hydrogen and helium, and the increased mean molecular weight leads to a larger overall density. Some degree of enrichment might be expected because Jupiter and Saturn are themselves enriched in hevay elements relative to the Sun. However, the two planets just mentioned would need to be enriched still further; in the case of HD 149026b, theoretical models suggest that the planet is 65% heavy elements by mass.

5.3. Atmospheric Physics

Atmospheric spectra have been obtained for several gas giant planets, especially the two "hot Jupiters" with the brightest host stars, HD 209458b and HD 189733b. Both transmission and emission spectroscopy have been undertaken, mainly with the space telescopes Hubble and Spitzer. Figures 12 and 13 show examples of both transmission (transit) and emission (occultation) spectra. These observations have stimulated much theoretical work on exoplanetary atmospheres (see the chapters by Chambers and by D'Angelo et al. for more details than are presented here).

This enterprise began with the optical detection of neutral sodium in a transmission spectrum of HD 209458b by *Charbonneau et al.* (2002). For that same planet, *Vidal-Madjar et al.* (2003) measured an enormous transit depth of $(15 \pm 4)\%$ within an ultraviolet bandpass bracketing the wavelength of the Lyman-α transition, which they attributed to neutral hydrogen gas being blown off the planet. For HD 189733b, a rise in transit depth toward shorter wavelengths was observed by *Pont et al.* (2008) and interpreted as Rayleigh scattering in the planet's upper atmosphere.

Recent attention has turned to infrared wavelengths, where molecules make their imprint. In the emission spec-

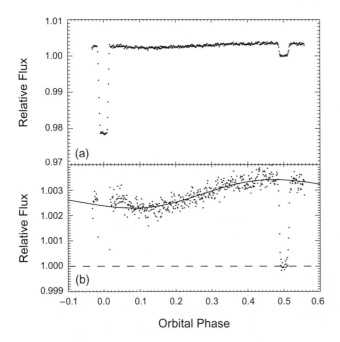

Fig. 11. The combined 8-µm brightness of the K star HD 189733 and its giant planet, over a 33-hr interval including a transit and an occultation. The bottom panel shows the same data as the top panel but with a restricted vertical scale to highlight the gradual rise in brightness as the planet's dayside comes into view. The amplitude of this variation gives the temperature contrast between the dayside (estimated as 1211 ± 11 K) and the nightside (973 ± 33 K). From *Knutson et al.* (2007b).

tra, departures from a blackbody spectrum have been interpreted as arising from water, methane, carbon monoxide, and carbon dioxide (see, e.g., *Grillmair et al.*, 2008; *Swain et al.*, 2009). Absorption features in transmission spectra have been interpreted as arising from water and methane (see, e.g., *Swain et al.*, 2008; *Tinetti et al.*, 2007).

Despite these impressive achievements, many issues regarding the atmospheres of hot Jupiters remain unsettled. Controversies arise because spectral features are observed with a low signal-to-noise ratio and are not always reproducible, as illustrated in Fig. 12. This is understandable, as the observers have pushed the instruments beyond their design specifications to make these demanding measurements. In addition, the theoretical interpretation of the spectra is in a primitive state. A published spectrum is usually accompanied by a model that fits the data, but left unanswered are whether the model is unique and how well constrained are its parameters. Recent work by *Madhusudhan and Seager* (2009) addresses this problem.

One theme that has arisen in the last few years is that some hot Jupiters have an inversion layer in their upper atmospheres, within which the temperature rises with height instead of the usual decline. The evidence for an inversion layer is emission in excess of a blackbody spectrum between 4 and 8 µm, where excess absorption was expected due to water vapor. The interpretation is that water is seen in emission because it exists in a hot, tenuous stratosphere.

Hot stratospheres develop when starlight is strongly absorbed by some species at low pressure (at high altitude) where the atmosphere does not radiate efficiently. The identity of this absorber in hot Jupiter atmospheres has been a topic of debate. Gaseous titanium oxide and vanadium oxide are candidates (*Hubeny et al.*, 2003), as are photochemically produced sulfur compounds (*Zahnle et al.*, 2009). Meanwhile, observers are searching for correlations between the presence or absence of an inversion layer, the degree of stellar irradiation, the magnitude of the day-night temperature difference, and other observable properties.

A few of the known planets have highly eccentric orbits and small periastron distances, and are therefore subject to highly variable stellar irradiation. Observers have monitored the planetary thermal emission following periastron passages, to gauge the amplitude and timescale of the thermal response to stellar heating. In the most extreme case, *Laughlin et al.* (2009) watched the effective temperature of the giant planet HD 80606b rise from about 800 K to 1500 K after periastron passage, and inferred that the characteristic radiative timescale of the upper atmosphere is about 4.5 h.

Although the thermal emission from hot Jupiters has been detected in many cases, at infrared and optical wavelengths, there has been no unambiguous detection of starlight *reflected* from the planetary atmosphere. The best resulting upper limit on the visual planetary albedo is 0.17 (with 3σ confidence), for the case of HD 209458b (*Rowe et al.*, 2008). This rules out highly reflective clouds of the sort that give Jupiter its visual albedo of 0.5. However, the limits have not been of great interest to the theorists, who predicted all along that the visual albedos of hot Jupiters would be very small.

5.4. Tidal Evolution and Migration

Almost all the known transiting planets are close enough to their parent stars for tidal effects to be important. The tidal bulges on the star and planet provide "handles" for the bodies to torque each other, and tidal friction slowly drains energy from the system (see the chapter by Fabrycky). For a typical hot Jupiter, the sequence of events is expected to be: (1) Over $\sim 10^6$ yr, the planet's rotational period is synchronized with its orbital period, and its obliquity (the angle between its rotational and orbital angular momentum vectors) is driven to zero; (2) over $\sim 10^9$ yr, the orbit is circularized; (3) after $\sim 10^{12}$ yr (i.e., not yet), the stellar rotational period is synchronized with the orbital period, the stellar obliquity is driven to zero, and the orbit decays, leading to the engulfment of the planet. The timescales for these processes are highly uncertain, and more complex histories are possible if there are additional planets in the system or if the planet's internal structure is strongly affected by tidal heating.

Tidal circularization is implicated by the fact that planets with orbital periods shorter than ~10 d tend to have smaller

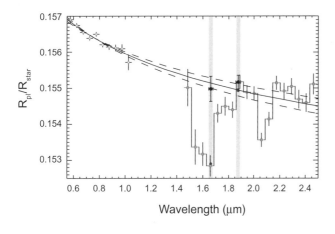

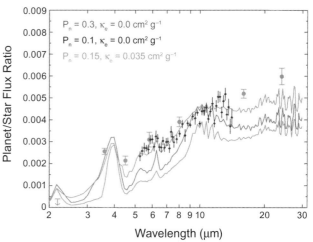

Fig. 12. Transmission (transit) spectroscopy of the gas giant HD 189733b, using the Hubble Space Telescope. The symbols with errors bars are measurements of the effective planet-to-star radius ratio as a function of wavelength. The dip at 1.6 μm was interpreted as evidence for water, and the rise at 2.1 μm as evidence for methane (*Swain et al.*, 2008). However, subsequent observations at 1.7 μm and 1.9 μm, shown with darker symbols and gray bands, disagree with the earlier results and are consistent with a Rayleigh scattering model (solid and dashed curves). From *Sing et al.* (2009).

Fig. 13. Occultation spectroscopy of the gas giant HD 189733b, using the Spitzer Space Telescope. The points show measurements of the flux density ratio of the planet and star as a function of wavelength. The smaller and finer-sampled points are based on observations with a dispersive spectrograph while the larger points are based on broadband filter photometry. The three lines show the outputs of model atmospheres with varying choices for the parameters P_n, specifying the efficiency of heat transfer from dayside to nightside, and κ_e, specifying the opacity of a putative high-altitude absorbing species. From *Grillmair et al.* (2008).

orbital eccentricities than longer-period planets. This fact was already known from Doppler surveys, but for eclipsing planets the eccentricities can be measured more precisely (see section 3.2). It is possible that the "bloating" of some of the close-in giant planets (section 5.3) is related to the heat that was produced during the circularization process, or that may still be ongoing.

As for tidal synchronization, observations with Spitzer have revealed planets with a cold side facing away from the star and a hot side facing the star (see Fig. 11). This could be interpreted as evidence for synchronization, although it is not definitive, because it is also possible that the heat from the star is reradiated too quickly for advection to homogenize the upper atmosphere of the planet.

Tidal decay of the orbit would be observable as a gradual decline in the orbital period (*Sasselov*, 2003). For most systems the theoretical timescale for tidal decay is much longer than the age of the star, and indeed no evidence for this process has been found. However in at least one case (WASP-18b) (*Hellier et al.*, 2009) the theoretical timescale for tidal decay is much *shorter* than the stellar age, because of the large planetary mass and short orbital period. The existence of this system suggests that the theoretical expectations were wrong and dissipation is slower in reality.

Likewise, in most cases one would not expect that enough time has elapsed for tides to modify the star's spin rate or orientation (*Barker and Ogilvie*, 2009). This suggests that the measurements of the projected spin-orbit angle using the RM effect (section 3.5) should be interpreted in the context of planet formation and evolution rather than tides. A close spin-orbit alignment is expected because a star and its planets inherit their angular momentum from a common source: the protostellar disk. However, for hot Jupiters there is the complication that they presumably formed at larger distances and "migrated" inward through processes that are poorly understood. Some of the migration theories predict that the original spin-orbit alignment should be preserved, while others predict occasionally large misalignments. For example, tidal interactions with the protoplanetary disk (see chapter by Showman et al.) should drive the system into close alignment (*Marzari and Nelson*, 2009), while planet-planet scattering or Kozai oscillations with tidal friction (see chapter by Fabrycky) should result in misaligned systems.

The projected spin-orbit angle has been measured for about 20 exoplanets, all of them close-in giants. Some examples of data are shown in Fig. 14. In many cases the results are consistent with good alignment, with measurement precisions ranging from 1° to 20°. However, there are now at least four clear cases of misaligned systems. One such case is XO-3b, a massive planet in a close-in eccentric orbit that is tilted by more than 30° with respect to the stellar equator (*Hébrard et al.*, 2008; *Winn et al.*, 2009a). Even more dramatic is HAT-P-7b, for which the planetary orbit and stellar spin axis are tilted by more than 86° (*Winn et al.*, 2009b; *Narita et al.*, 2009). The planetary orbit is

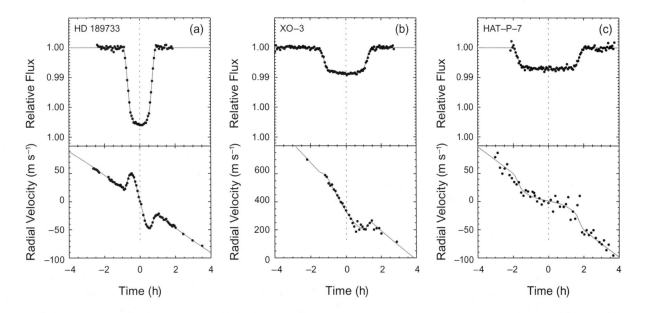

Fig. 14. Examples of data used to measure the projected spin-orbit angle λ. The top panels show transit photometry, and the bottom panels show the apparent radial velocity of the star, including both orbital motion and the anomalous Doppler shift (the Rossiter-McLaughlin effect). (a) Well-aligned system; (b) misaligned system; (c) system for which the stellar and orbital "north poles" are nearly *antiparallel* on the sky, indicating that the planet's orbit is either retrograde or polar (depending on the unknown inclination of the stellar rotation axis). Data from *Winn et al.* (2006, 2009a,b).

either polar (going over the north and south poles of the star) or retrograde (revolving in the opposite direction as the star is rotating). Another possible retrograde system is WASP-7b (*Anderson et al., 2009*).

6. FUTURE PROSPECTS

Eclipses are the "here and now" of exoplanetary science. It seems that every month, someone reports a startling observational feat, describes a creative new application of eclipse data, or proposes an ambitious survey to find ever-larger numbers of ever-smaller transiting planets. Keeping up with the field leaves one breathless, and wary of making predictions.

The only safe bet is that the Kepler space mission will have a major impact. In March 2009, the Kepler satellite was launched into an an Earth-trailing orbit where it will stare at a single field of 10^5 stars for at least 3.5 yr, seeking Earth-like planets in Earth-like orbits around Sun-like stars (*Borucki et al., 2003*). Kepler will end our obsession with close-in, tidally influenced, strongly irradiated giant planets. At last it will be possible to study eclipses of multitudes of longer-period and smaller-radius planets. Stellar heating will not confound models of their atmospheres or interiors, and tidal effects will not confound the interpretation of their orbital properties.

Many of Kepler's discoveries will be small enough to have solid surfaces, making it possible to measure masses and radii of rocky exoplanets. The mission's highest-priority goal is to determine the abundance of terrestrial planets in the "habitable zones" of their parent stars, defined as the range of orbital distances within which water could exist in liquid form on the surface of a rocky planet. The abundance of such planets is a key input to the "Drake equation" as well as to theories of planet formation and to the designs of future missions that will find and study the nearest examples.

The precision of spacebased photometry will also be put to good use to detect a host of subtle effects that have been described in the literature but not yet observed. Already, data from the Kepler and CoRoT satellites have been used to detect occultations in visible light, as opposed to infrared (*Borucki et al., 2009*; *Snellen et al., 2009*), although in both cases the signal is attributed mainly to thermal emission rather than the elusive reflected component.

Many investigators will seek to detect short-term variations in the times, durations, and depths of transits due to gravitational perturbations by additional planets, as mentioned in section 3.2 and discussed further in the chapter by Fabrycky. In particular, Kepler might find habitable planets not only by detecting their transits, but also by observing their effects on other transiting planets. The same principle can be used to detect habitable "exomoons" (*Kipping*, 2009) or habitable planets at the Trojan points (Lagrange L4 and L5) in the orbits of more massive planets (*Ford and Gaudi*, 2006; *Madhusudhan and Winn*, 2009).

Longer-term orbital perturbations should also be measurable, due to additional bodies (*Miralda-Escudé*, 2002) or relativistic precession (*Jordán and Bakos*, 2008; *Pál and*

Kocsis, 2008). For very close-in planets, the orbital precession rate should be dominated by the effects of the planetary tidal bulge, which in turn depends on the deformability of the planet. This raises the prospect of using the measured precession rate to infer some aspects of the planet's interior structure (*Ragozzine and Wolf*, 2009).

In addition, the precise form of the transit lightcurve depends on the shape of the planet's silhouette. With sufficiently precise photometry it may be possible to detect the departures from sphericity due to rings (*Barnes and Fortney*, 2004) or rotation (*Seager and Hui*, 2002; *Barnes and Fortney*, 2003). Already it has been shown that HD 189733b is less oblate than Saturn (*Carter and Winn*, 2010), as expected if the planet's rotation period is synchronized with its 2.2-d orbit (slower than Saturn's 11-h rotation period). Soon we will have a sample of transiting planets with larger orbital distances, for which synchronization is not expected and which may therefore be rotating quickly enough for the oblateness to be detectable.

Another prize that remains to be won is the discovery of a system with more than one transiting planet. This would give empirical constraints on the mutual inclination of exoplanetary orbits (*Nesvorný*, 2009), as well as estimates of the planetary masses that are independent of the stellar mass, through the observable effects of planet-planet gravitational interactions. Already there are a few cases in which there is evidence for a second planet around a star that is known to have a transiting planet (see, e.g., *Bakos et al.*, 2009), but in none of those cases is it known whether the second planet also transits.

Groundbased transit surveys will still play an important role by targeting brighter stars over wider fields than CoRoT and Kepler. With brighter stars it is easier to rule out false positives, and to measure the planetary mass through radial velocity variations of the host star. For the smallest planets around the relatively faint stars in the CoRoT and Kepler fields this will be difficult. Bright stars also offer more photons for the high-precision follow-up investigations that make eclipses so valuable.

Groundbased surveys will continue providing such targets, and in particular will mine the southern sky, which is comparatively unexplored. Two contenders are the Super-WASP and HAT-South surveys, which will aim to improve their photometric precision and discover many Neptune-sized or even smaller planets in addition to the gas giants. The MEarth project, mentioned in section 5.1, is specifically targeting small and low-luminosity stars because of the less stringent requirements on photometric precision and because the habitable zones occur at smaller orbital distances, making transits more likely and more frequent (*Nutzman and Charbonneau*, 2008).

An attractive idea is to design a survey with the fine photometric precision that is possible from space, but that would somehow survey the brightest stars on the sky instead of those within a narrow field of view. Several mission concepts are being studied: The Transiting Exoplanet Survey Satellite (*Deming et al.*, 2009) and the PLATO mission (*Lindberg et al.*, 2009) would tile the sky with the fields-of-view of small cameras, while LEAVITT would have a series of telescopes on a spinning platform (*Olling*, 2007). Another concept is to deploy an armada of "nanosatellites" to target individual stars (S. Seager, personal communication, 2009).

Each time a terrestrial planet is found transiting a bright star, a period of intense anticipation will ensue, as astronomers try to characterize its atmosphere through transmission or emission spectroscopy. The anticipation will be especially keen for any planets in the habitable zone of their parent stars, which may reveal water or other molecules considered important for life. Attaining the necessary signal-to-noise ratio will be excruciating work, and may require the James Webb Space Telescope or an even more capable special-purpose instrument.

Transits and occultations have had a disproportionate impact early in the history of exoplanetary science. This is mainly because of the unexpected existence of close-in giant planets, and their high transit probabilities. In the coming years, other techniques such as astrometry, microlensing, and spatially resolved imaging will mature and give a more complete accounting of exoplanetary systems. Still, in closing, let us recall that most of our most fundamental and precise information about *stars* comes from eclipsing systems, even after more than a century of technological development since eclipses were first observed. The same is likely to be true for exoplanets.

Acknowledgments. The author is grateful to S. Albrecht, J. Carter, D. Fabrycky, D. Kipping, H. Knutson, J. Lissauer, G. Marcy, F. Pont, S. Seager, E. Agol, G. Ricker, as well as the anonymous referees, for helpful comments on this chapter.

REFERENCES

Agol E., Steffen J., Sari R., and Clarkson W. (2005) On detecting terrestrial planets with timing of giant planet transits. *Mon. Not. R. Astron. Soc., 359,* 567–579.

Alonso R. and 11 colleagues (2004) TrES-1: The transiting planet of a bright K0 V star. *Astrophys. J. Lett., 613,* L153–L156.

Anderson D. R. and 20 colleagues (2010) WASP-17b: An ultra-low density planet in a probable retrograde orbit. *Astrophys. J., 709,* 159–167.

Baines E. K., van Belle G. T., ten Brummelaar T. A., McAlister H. A., Swain M., Turner N. H., Sturmann L., and Sturmann J. (2007) Direct measurement of the radius and density of the transiting exoplanet HD 189733b with the CHARA array. *Astrophys. J. Lett., 661,* L195–L198.

Bakos G. Á. and 18 colleagues (2007) HAT-P-1b: A large-radius, low-density exoplanet transiting one member of a stellar binary. *Astrophys. J., 656,* 552–559.

Bakos G. Á. and 18 colleagues (2009) HAT-P-13b,c: A transiting hot Jupiter with a massive outer companion on an eccentric orbit. *Astrophys. J., 707,* 446–456.

Barbieri M. and 10 colleagues (2007) HD 17156b: A transiting planet with a 21.2-day period and an eccentric orbit. *Astron. Astrophys., 476,* L13–L16.

Barker A. J. and Ogilvie G. I. (2009) On the tidal evolution of hot Jupiters on inclined orbits. *Mon. Not. R. Astron. Soc., 395,* 2268–2287.

Barnes J. W. and Fortney J. J. (2003) Measuring the oblateness and rotation of transiting extrasolar giant planets. *Astrophys. J., 588*, 545–556.

Barnes J. W. and Fortney J. J. (2004) Transit detectability of ring systems around extrasolar giant planets. *Astrophys. J., 616*, 1193–1203.

Beatty T. G. and Gaudi B. S. (2008) Predicting the yields of photometric surveys for transiting extrasolar planets. *Astrophys. J., 686*, 1302–1330.

Borucki W. J. and 12 colleagues (2003) The Kepler mission: A wide-field-of-view photometer designed to determine the frequency of Earth-size planets around solar-like stars. In *Future EUV/UV and Visible Space Astrophysics Missions and Instrumentation* (J. C. Blades and O. H. W. Siegmund, eds.), pp. 129–140. SPIE Conf. Series 4854, Bellingham, Washington.

Borucki W. J. and 24 colleagues (2009) Kepler's optical phase curve of the exoplanet HAT-P-7b. *Science, 325*, 709.

Brown T. M. (2001) Transmission spectra as diagnostics of extrasolar giant planet atmospheres. *Astrophys. J., 553*, 1006–1026.

Brown T. M., Charbonneau D., Gilliland R. L., Noyes R. W. and Burrows A. (2001) Hubble Space Telescope time-series photometry of the transiting planet of HD 209458. *Astrophys. J., 552*, 699–709.

Burke C. J. and 17 colleagues (2007) XO-2b: Transiting hot Jupiter in a metal-rich common proper motion binary. *Astrophys. J., 671*, 2115–2128.

Burrows A., Hubeny I., Budaj J., and Hubbard W. B. (2007) Possible solutions to the radius anomalies of transiting giant planets. *Astrophys. J., 661*, 502–514.

Carciofi A. C. and Magalhães A. M. (2005) The polarization signature of extrasolar planet transiting cool dwarfs. *Astrophys. J., 635*, 570–577.

Carter J. A. and Winn J. N. (2009) Parameter estimation from time-series data with correlated errors: A wavelet-based method and its application to transit light curves. *Astrophys. J., 704*, 51–67.

Carter J. A. and Winn J. N. (2010) Empirical constraints on the oblateness of an exoplanet. *Astrophys. J., 709*, 1219–1229.

Carter J. A., Yee J. C., Eastman J., Gaudi B. S., and Winn J. N. (2008) Analytic approximations for transit light-curve observables, uncertainties, and covariances. *Astrophys. J., 689*, 499–512.

Carter J. A., Winn J. N., Gilliland R., and Holman M. J. (2009) Near-infrared transit photometry of the exoplanet HD 149026b. *Astrophys. J., 696*, 241–253.

Chabrier G. and Baraffe I. (2007) Heat transport in giant (exo) planets: A new perspective. *Astrophys. J. Lett., 661*, L81–L84.

Charbonneau D., Brown T. M., Latham D. W., and Mayor M. (2000) Detection of planetary transits across a Sun-like star. *Astrophys. J. Lett., 529*, L45–L48.

Charbonneau D., Brown T. M., Noyes R. W., and Gilliland R. L. (2002) Detection of an extrasolar planet atmosphere. *Astrophys. J., 568*, 377–384.

Charbonneau D. and 18 colleagues (2009) A super-Earth transiting a nearby low-mass star. *Nature, 462*, 891–894.

Claret A. (2004) A new non-linear limb-darkening law for LTE stellar atmosphere models. III. Sloan filters: Calculations for $-5.0 \leq \log[M/H] \leq +1$, $2000 \, K \leq T_{eff} \leq 50000 \, K$ at several surface gravities. *Astron. Astrophys., 428*, 1001–1005.

Claret A. (2009) Does the HD 209458 planetary system pose a challenge to the stellar atmosphere models? *Astron. Astrophys., 506*, 1335–1340.

Collier Cameron A. and 31 colleagues (2007) Efficient identification of exoplanetary transit candidates from SuperWASP light curves. *Mon. Not. R. Astron. Soc., 380*, 1230–1244.

de Kort J. J. M. A. (1954) Upper and lower limits for the eccentricity and longitude of periastron of an eclipsing binary. *Ricerche Astron., 3*, 109–118.

Deming D. and 11 colleagues (2009) Discovery and characterization of transiting super Earths using an all-sky transit survey and follow-up by the James Webb Space Telescope. *Publ. Astron. Soc. Pac., 121*, 952–967.

Everett M. E. and Howell S. B. (2001) A technique for ultrahigh-precision CCD photometry. *Publ. Astron. Soc. Pac., 113*, 1428–1435.

Fabrycky D. C. and Winn J. N. (2009) Exoplanetary spin-orbit alignment: Results from the ensemble of Rossiter-McLaughlin observations. *Astrophys. J., 696*, 1230–1240.

Ford E. B. and Gaudi B. S. (2006) Observational constraints on Trojans of transiting extrasolar planets. *Astrophys. J. Lett., 652*, L137–L140.

Fossey S. J., Waldmann I. P., and Kipping D. M. (2009) Detection of a transit by the planetary companion of HD 80606. *Mon. Not. R. Astron. Soc., 396*, L16–L20.

Garcia-Melendo E. and McCullough P. R. (2009) Photometric detection of a transit of HD 80606b. *Astrophys. J., 698*, 558–561.

Gaudi B. S. (2005) On the size distribution of close-in extrasolar giant planets. *Astrophys. J. Lett., 628*, L73–L76

Gaudi B. S. and Winn J. N. (2007) Prospects for the characterization and confirmation of transiting exoplanets via the Rossiter-McLaughlin effect. *Astrophys. J., 655*, 550–563.

Gaudi B. S., Seager S., and Mallen-Ornelas G. (2005) On the period distribution of close-in extrasolar giant planets. *Astrophys. J., 623*, 472–481.

Gilliland R. L. and Brown T. M. (1988) Time-resolved CCD photometry of an ensemble of stars. *Publ. Astron. Soc. Pac., 100*, 754–765.

Gilliland R. L. and 23 colleagues (2000) A lack of planets in 47 Tucanae from a Hubble Space Telescope search. *Astrophys. J. Lett., 545*, L47–L51.

Gillon M., Demory B.-O., Barman T., Bonfils X., Mazeh T., Pont F., Udry S., Mayor M., and Queloz D. (2007) Accurate Spitzer infrared radius measurement for the hot Neptune GJ 436b. *Astron. Astrophys., 471*, L51–L54.

Giménez A. (2007) Equations for the analysis of the light curves of extra-solar planetary transits. *Astron. Astrophys., 474*, 1049–1049.

Gregory P. C. (2005) *Bayesian Logical Data Analysis for the Physical Sciences: A Comparative Approach with Mathematica Support.* Cambridge Univ., Cambridge.

Grillmair C. J., Burrows A., Charbonneau D., Armus L., Stauffer J., Meadows V., van Cleve J., von Braun K., and Levine D. (2008) Strong water absorption in the dayside emission spectrum of the planet HD 189733b. *Nature, 456*, 767–769.

Guillot T. and Showman A. P. (2002) Evolution of "51 Pegasus b-like" planets. *Astron. Astrophys., 385*, 156–165.

Guillot T., Santos N. C., Pont F., Iro N., Melo C., and Ribas I. (2006) A correlation between the heavy element content of transiting extrasolar planets and the metallicity of their parent stars. *Astron. Astrophys., 453*, L21–L24.

Hébrard G. and 22 colleagues (2008) Misaligned spin-orbit in the XO-3 planetary system? *Astron. Astrophys., 488*, 763–770.

Hellier C. and 22 colleagues (2009) An orbital period of 0.94 days for the hot-Jupiter planet WASP-18b. *Nature, 460*, 1098–1100.

Henry G. W., Marcy G. W., Butler R. P., and Vogt S. S. (2000)

A Transiting "51 Peg-like" planet. *Astrophys. J. Lett., 529,* L41–L44.

Hilditch R. W. (2001) *An Introduction to Close Binary Stars.* Cambridge Univ., Cambridge.

Holman M. J. and Murray N. W. (2005) The use of transit timing to detect terrestrial-mass extrasolar planets. *Science, 307,* 1288–1291.

Holman M. J., Winn J. N., Latham D. W., O'Donovan F. T., Charbonneau D., Bakos G. A., Esquerdo G. A., Hergenrother C., Everett M. E., and Pál A. (2006) The transit light curve project. I. Four consecutive transits of the exoplanet XO-1b. *Astrophys. J., 652,* 1715–1723.

Horne K. (2003) Status and prospects of planetary transit searches: Hot Jupiters galore. In *Scientific Frontiers in Research on Extrasolar Planets* (D. Deming and S. Seager, eds.), pp. 361–370. ASP Conf. Series 294, Astronomical Society of the Pacific, San Francisco.

Howell S. B. (2006) *Handbook of CCD Astronomy,* 2nd edition. Cambridge Univ., Cambridge.

Hubeny I., Burrows A., and Sudarsky D. (2003) A possible bifurcation in atmospheres of strongly irradiated stars and planets. *Astrophys. J., 594,* 1011–1018.

Jordán A. and Bakos G. Á. (2008) Observability of the general relativistic precession of periastra in exoplanets. *Astrophys. J., 685,* 543–552.

Kallrath J. and Milone E. F. (2009) *Eclipsing Binary Stars: Modeling and Analysis,* 2nd edition. Springer, New York, in press.

Kane S. R. (2007) Detectability of exoplanetary transits from radial velocity surveys. *Mon. Not. R. Astron. Soc., 380,* 1488–1496.

Kipping D. M. (2008) Transiting planets — light-curve analysis for eccentric orbits. *Mon. Not. R. Astron. Soc., 389,* 1383–1390.

Kipping D. M. (2009) Transit timing effects due to an exomoon. *Mon. Not. R. Astron. Soc., 392,* 181–189.

Kjeldsen H. and Frandsen S. (1992) High-precision time-resolved CCD photometry. *Publ. Astron. Soc. Pac., 104,* 413–434.

Knutson H. A., Charbonneau D., Noyes R. W., Brown T. M., and Gilliland R. L. (2007a) Using stellar limb-darkening to refine the properties of HD 209458b. *Astrophys. J., 655,* 564–575.

Knutson H. A., Charbonneau D., Allen L. E., Fortney J. J., Agol E., Cowan N. B., Showman A. P., Cooper C. S., and Megeath S. T. (2007b) A map of the day-night contrast of the extrasolar planet HD 189733b. *Nature, 447,* 183–186.

Kopal Z. (1979) *Language of the Stars: Discourse on the Theory of the Light Changes of Eclipsing Variables.* Astrophysics and Space Science Library, Vol. 77, Kluwer, Dordrecht.

Kovács G., Zucker S., and Mazeh T. (2002) A box-fitting algorithm in the search for periodic transits. *Astron. Astrophys., 391,* 369–377.

Kovács G., Bakos G., Noyes R. W. (2005) A trend filtering algorithm for wide-field variability surveys. *Mon. Not. R. Astron. Soc., 356,* 557–567.

Laughlin G., Deming D., Langton J., Kasen D., Vogt S., Butler P., Rivera E., and Meschiari S. (2009) Rapid heating of the atmosphere of an extrasolar planet. *Nature, 457,* 562–564.

Léger A. and 160 colleagues (2009) Transiting exoplanets from the CoRoT space mission. VIII. CoRoT-7b: The first super-Earth with measured radius. *Astron. Astrophys., 506,* 287–302.

Lindberg R., Stankov A., Fridlund M., and Rando N. (2009) Current status of the assessment of the ESA Cosmic Vision mission candidate PLATO. In *Techniques and Instrumentation for Detection of Exoplanets IV* (S. B. Shaklan, ed.), pp. 7440Z-1 to 7440Z-12. SPIE Conf. Series 7440, Bellingham, Washington.

Madhusudhan N. and Seager S. (2009) A temperature and abundance retrieval method for exoplanet atmospheres. *Astrophys. J., 707,* 24–39.

Madhusudhan N. and Winn J. N. (2009) Empirical constraints on Trojan companions and orbital eccentricities in 25 transiting exoplanetary systems. *Astrophys. J., 693,* 784–793.

Mandel K. and Agol E. (2002) Analytic light curves for planetary transit searches. *Astrophys. J. Lett., 580,* L171–L175.

Marzari F. and Nelson A. F. (2009) Interaction of a giant planet in an inclined orbit with a circumstellar disk. *Astrophys. J., 705,* 1575–1583.

Mazeh T. and 19 colleagues (2000) The spectroscopic orbit of the planetary companion transiting HD 209458. *Astrophys. J. Lett., 532,* L55–L58.

Mazeh T., Zucker S., and Pont F. (2005) An intriguing correlation between the masses and periods of the transiting planets. *Mon. Not. R. Astron. Soc., 356,* 955–957.

McCullough P. R., Stys J. E., Valenti J. A., Fleming S. W., Janes K. A., and Heasley J. N. (2005) The XO Project: Searching for transiting extrasolar planet candidates. *Publ. Astron. Soc. Pac., 117,* 783–795.

Miller N., Fortney J. J., and Jackson B. (2009) Inflating and deflating hot Jupiters: Coupled tidal and thermal evolution of known transiting planets. *Astrophys. J., 702,* 1413–1427.

Miralda-Escudé J. (2002) Orbital perturbations of transiting planets: A possible method to measure stellar quadrupoles and to detect Earth-mass planets. *Astrophys. J., 564,* 1019–1023.

Moutou C. and 21 colleagues (2009) Photometric and spectroscopic detection of the primary transit of the 111-day-period planet HD 80606b. *Astron. Astrophys., 498,* L5–L8.

Narita N., Sato B., Hirano T., and Tamura M. (2009) First evidence of a retrograde orbit of a transiting exoplanet HATP-7b. *Publ. Astron. Soc. Japan, 61,* L35–L40.

Nesvorný D. (2009) Transit timing variations for eccentric and inclined exoplanets. *Astrophys. J., 701,* 1116–1122.

Nutzman P. and Charbonneau D. (2008) Design considerations for a ground-based transit search for habitable planets orbiting M dwarfs. *Publ. Astron. Soc. Pac., 120,* 317–327.

O'Donovan F. T., Charbonneau D., Alonso R., Brown T. M., Mandushev G., Dunham E. W., Latham D. W., Stefanik R. P., Torres G., and Everett M. E. (2007) Outcome of six candidate transiting planets from a TrES field in Andromeda. *Astrophys. J., 662,* 658–668.

Olling R. P. (2007) LEAVITT: A MIDEX-class mission for finding and characterizing 10,000 transiting planets in the solar neighborhood. *ArXiv e-prints,* arXiv:0704.3072.

Pál A. and Kocsis B. (2008) Periastron precession measurements in transiting extrasolar planetary systems at the level of general relativity. *Mon. Not. R. Astron. Soc., 389,* 191–198.

Pepper J., Gould A., and Depoy D. L. (2003) Using all-sky surveys to find planetary transits. *Acta Astron., 53,* 213–228.

Pollacco D. L. and 27 colleagues (2006) The WASP project and the SuperWASP cameras. *Publ. Astron. Soc. Pacific, 118,* 1407–1418.

Pont F., Zucker S., and Queloz D. (2006) The effect of red noise on planetary transit detection. *Mon. Not. R. Astron. Soc., 373,* 231–242.

Pont F., Knutson H., Gilliland R. L., Moutou C., and Charbonneau D. (2008) Detection of atmospheric haze on an extrasolar planet: The 0.55–1.05 µm transmission spectrum of HD 189733b with the Hubble Space Telescope. *Mon. Not. R. Astron. Soc., 385,* 109–118.

Press W. H., Teukolsky S. A., Vetterling W. T., and Flannery B. P. (2007) *Numerical Recipes: The Art of Scientific Computing*, 3rd edition. Cambridge Univ., Cambridge.

Queloz D. and 39 colleagues (2009) The CoRoT-7 planetary system: Two orbiting super-Earths. *Astron. Astrophys., 506*, 303–319.

Rabus M., Alonso R., Belmonte J. A., Deeg H. J., Gilliland R. L., Almenara J. M., Brown T. M., Charbonneau D., and Mandushev G. (2009) A cool starspot or a second transiting planet in the TrES-1 system? *Astron. Astrophys., 494*, 391–397.

Ragozzine D. and Wolf A. S. (2009) Probing the interiors of very hot Jupiters using transit light curves. *Astrophys. J., 698*, 1778–1794.

Reiger S. H. (1963) Starlight scintillation and atmospheric turbulence. *Astron. J., 68*, 395.

Rowe J. F. and 10 colleagues (2008) The very low albedo of an extrasolar planet: MOST space-based photometry of HD 209458. *Astrophys. J., 689*, 1345–1353.

Russell H. N. (1948) The royal road of eclipses. *Harvard Coll. Obs. Monograph, 7*, 181–209.

Ryan P. and Sandler D. (1998) Scintillation reduction method for photometric measurements. *Publ. Astron. Soc. Pac., 110*, 1235–1248.

Sackett P. (1999) Searching for unseen planets via occultation and microlensing. In *Planets Outside the Solar System: Theory and Observations* (J.-M. Mariotti and D. Alloin, eds.), p. 189. Kluwer, Dordrecht.

Sasselov D. D. (2003) The new transiting planet OGLE-TR-56b: Orbit and atmosphere. *Astrophys. J., 596*, 1327–1331.

Sato B. and 20 colleagues (2005) The N2K Consortium. II. A transiting hot Saturn around HD 149026 with a large dense core. *Astrophys. J., 633*, 465–473.

Seager S. and Hui L. (2002) Constraining the rotation rate of transiting extrasolar planets by oblateness measurements. *Astrophys. J., 574*, 1004–1010.

Seager S. and Mallén-Ornelas G. (2003) A unique solution of planet and star parameters from an extrasolar planet transit light curve. *Astrophys. J., 585*, 1038–1055.

Seager S. and Sasselov D. D. (2000) Theoretical transmission spectra during extrasolar giant planet transits. *Astrophys. J., 537*, 916–921.

Sing D. K., Désert J.-M., Lecavelier Des Etangs A., Ballester G. E., Vidal-Madjar A., Parmentier V., Hebrard G., and Henry G. W. (2009) Transit spectrophotometry of the exoplanet HD 189733b. I. Searching for water but finding haze with HST NICMOS. *Astron. Astrophys., 505*, 891–899.

Snellen I. A. G., de Mooij E. J. W., and Albrecht S. (2009) The changing phases of extrasolar planet CoRoT-1b. *Nature, 459*, 543–545.

Southworth J. (2008) Homogeneous studies of transiting extrasolar planets. I. Light-curve analyses. *Mon. Not. R. Astron. Soc., 386*, 1644–1666.

Southworth J. (2009) Homogeneous studies of transiting extrasolar planets — II. Physical properties. *Mon. Not. R. Astron. Soc., 394*, 272–294.

Southworth J., Wheatley P. J., and Sams G. (2007) A method for the direct determination of the surface gravities of transiting extrasolar planets. *Mon. Not. R. Astron. Soc., 379*, L11–L15.

Stello D. and 24 colleagues (2009) Radius determination of solar-type stars using asteroseismology: What to expect from the Kepler mission. *Astrophys. J., 700*, 1589–1602.

Sterne T. E. (1940) On the determination of the orbital elements of eccentric eclipsing binaries. *Proc. Natl. Acad. Sci., 26*, 36–40.

Struve O. (1952) Proposal for a project of high-precision stellar radial velocity work. *The Observatory, 72*, 199–200.

Swain M. R., Vasisht G., and Tinetti G. (2008) The presence of methane in the atmosphere of an extrasolar planet. *Nature, 452*, 329–331.

Swain M. R., Vasisht G., Tinetti G., Bouwman J., Chen P., Yung Y., Deming D., and Deroo P. (2009) Molecular signatures in the near-infrared dayside spectrum of HD 189733b. *Astrophys. J. Lett., 690*, L114–L117.

Tamuz O., Mazeh T., and Zucker S. (2005) Correcting systematic effects in a large set of photometric light curves. *Mon. Not. R. Astron. Soc., 356*, 1466–1470.

Tinetti G. and 12 colleagues (2007) Water vapour in the atmosphere of a transiting extrasolar planet. *Nature, 448*, 169–171.

Torres G. and 15 colleagues (2007) HAT-P-3b: A heavy-element-rich planet transiting a K dwarf star. *Astrophys. J. Lett., 666*, L121–L124.

Torres G., Winn J. N., and Holman M. J. (2008) Improved parameters for extrasolar transiting planets. *Astrophys. J., 677*, 1324–1342.

Udalski A., Paczynski B., Zebrun K., Szymanski M., Kubiak M., Soszynski I., Szewczyk O., Wyrzykowski L., and Pietrzynski G. (2002) The optical gravitational lensing experiment. Search for planetary and low-luminosity object transits in the galactic disk. Results of 2001 campaign. *Acta Astron., 52*, 1–37.

Vidal-Madjar A., Lecavelier des Etangs A., Désert J.-M., Ballester G. E., Ferlet R., Hébrard G., and Mayor M. (2003) An extended upper atmosphere around the extrasolar planet HD 209458b. *Nature, 422*, 143–146.

Weldrake D. T. F., Sackett P. D., Bridges T. J., and Freeman K. C. (2005) An absence of hot Jupiter planets in 47 Tucanae: Results of a wide-field transit search. *Astrophys. J., 620*, 1043–1051.

Winn J. N. and 11 colleagues (2006) Measurement of the spin-orbit alignment in the exoplanetary system HD 189733. *Astrophys. J. Lett., 653*, L69–L72.

Winn J. N., Holman M. J., and Roussanova A. (2007a) The transit light curve project. III. Tres transits of TrES-1. *Astrophys. J., 657*, 1098–1106.

Winn J. N. and 10 colleagues (2007b) The transit light curve project. V. System parameters and stellar rotation period of HD 189733. *Astron. J., 133*, 1828–1835.

Winn J. N., Henry G. W., Torres G., and Holman M. J. (2008) Five new transits of the super-Neptune HD 149026b. *Astrophys. J., 675*, 1531–1537.

Winn J. N. and 10 colleagues (2009a) On the spin-orbit misalignment of the XO-3 exoplanetary system. *Astrophys. J., 700*, 302–308.

Winn J. N., Johnson J. A., Albrecht S., Howard A. W., Marcy G. W., Crossfield I. J., and Holman M. J. (2009b) HAT-P-7: A retrograde or polar orbit, and a third body. *Astrophys. J. Lett., 703*, L99–L103.

Young A. T. (1967) Photometric error analysis. VI. Confirmation of Reiger's theory of scintillation. *Astron. J., 72*, 747.

Zahnle K., Marley M. S., Freedman R. S., Lodders K., and Fortney J. J. (2009) Atmospheric sulfur photochemistry on hot Jupiters. *Astrophys. J., 701*, L20–L24.

Microlensing by Exoplanets

B. Scott Gaudi
The Ohio State University

Gravitational microlensing occurs when a foreground star happens to pass very close to our line of sight to a more distant background star. The foreground star acts as a lens, splitting the light from the background source star into two images, which are typically unresolved. However, these images of the source are also magnified, by an amount that depends on the angular separation between the lens and source. The relative motion between the lens and source therefore results in a time-variable magnification of the source: a microlensing event. If the foreground star happens to host a planet with projected separation near the paths of these images, the planet will also act as a lens, further perturbing the images and resulting in a characteristic, short-lived signature of the planet. This chapter provides an introduction to the discovery and characterization of exoplanets with gravitational microlensing. The theoretical foundation of the method is reviewed, focusing in particular on the phenomenology of planetary microlensing perturbations. The strengths and weaknesses of the microlensing technique are discussed, highlighting the fact that it is sensitive to low-mass planetary companions to stars throughout the galactic disk and foreground bulge, and that its sensitivity peaks for planet separations just beyond the snow line. An overview of the practice of microlensing planet searches is given, with a discussion of some of the challenges with detecting and analyzing planetary perturbations. The chapter concludes with a review of the results that have been obtained to date, and a discussion of the near- and long-term prospects for microlensing planet surveys. Ultimately, microlensing is potentially sensitive to multiple-planet systems containing analogs of all the solar system planets except Mercury, as well as to free-floating planets, and will provide a crucial test of planet formation theories by determining the demographics of planets throughout the galaxy.

1. INTRODUCTION

Gravitational lensing generally refers to the bending of light rays of a background light source by a foreground mass. Gravitational microlensing, on the other hand, traditionally refers to the special case when multiple images are created but have separations of less than a few milliarcseconds, and hence are unresolved with current capabilities. Although the idea of the gravitational deflection of light by massive bodies well predates the theory of general relativity, and can be traced as far back as Sir Isaac Newton, the concept of gravitational microlensing appears to be attributable to Einstein himself [see *Schneider et al.* (1992) for a thorough recounting of the history of gravitational lensing]. In 1936 Einstein published a paper in which he derived the equations of microlensing by a foreground star closely aligned to a background star (*Einstein*, 1936).

Indeed, it seems that Einstein had been thinking about this idea as far back as 1912 (*Renn et al.*, 1997), and perhaps had even hoped to use the phenomenon to explain the appearance of Nova Geminorum 1912 (*Sauer*, 2008). However, by 1936 he had dismissed the practical significance of the microlensing effect, concluding that "there is no great chance of observing this phenomenon" (*Einstein*, 1936). Indeed, there is no "great chance" of observing gravitational microlensing. The optical depth to gravitational microlensing, i.e., the probability that any given star is being appreciably lensed at any given time, is on the order of 10^{-6} toward the galactic bulge, and is generally similar or smaller for other lines of sight. (The phenomenon of gravitational lensing of multiply imaged quasars by stars in the foreground galaxy that is creating the multiple images of the quasar is also referred to as microlensing. In this instance, the optical depth to microlensing can be on the order of unity. However, in this chapter we are concerned only with gravitational microlensing of stars within our galaxy or in nearby galaxies, where the optical depths to microlensing are always small.) Thus at least partly due to this low probability, the idea of gravitational microlensing lay mostly dormant for five decades after Einstein's 1936 paper [with some notable exceptions, e.g., *Liebes* (1964) and *Refsdal* (1964)]. It was not until the seminal paper by *Paczynski* (1986) that the idea of gravitational microlensing was resurrected, and then finally put into practice with the initiation of several observational searches for microlensing events toward the large and small Magellenic clouds and galactic bulge in the early 1990s (*Alcock et al.*, 1993; *Aubourg et al.*, 1993; *Udalski et al.*, 1993). The roster of detected microlensing events now numbers in the thousands. These events have been discovered by many different collaborations, toward several lines of sight including the Magellenic clouds (*Alcock et al.*, 1997, 2000; *Palanque-Delabrouille et al.*, 1998), and M31 (*Paulin-Henriksson et al.*, 2002; *de Jong et al.*, 2004; *Uglesich et al.*, 2004; *Calchi Novati et al.*, 2005), with the vast majority found toward the galactic bulge (*Udalski et al.*, 2000; *Sumi et al.*, 2003; *Thomas et al.*, 2005; *Hamadache et al.*, 2006) or nearby fields in the galactic plane (*Derue et al.*, 2001).

Even before the first microlensing events were detected, it was suggested by *Mao and Paczynski* (1991) that gravitational microlensing might also be used to discover planetary companions to the primary microlens stars. The basic idea is illustrated in Fig. 1. As the foreground star passes close to the line of sight to the background source, it splits the source into two images, which sweep out curved trajectories on the sky as the foreground lens star passes close to the line of sight to the source. These images typically have separations on the order of 1 mas, and so are unresolved. However, the total area of these images is also larger than the area of the source, and as a result the background star also exhibits a time-variable magnification, which is referred to as a microlensing event. If the foreground star happens to host a planet with projected separation near the paths of one of these two images, the planet will further perturb the image, resulting in a characteristic, short-lived signature of the planet.

Gould and Loeb (1992) considered this novel method of detecting planets in detail, and laid out the practical requirements for an observational search for planets with microlensing. In particular, they advocated a "two-tier" approach. First, microlensing events are discovered and alerted in real time by a single survey telescope that monitors many square degrees of the galactic bulge on a roughly nightly basis. Second, these ongoing events are then densely monitored with many smaller telescopes to discover the short-lived signatures of planetary perturbations. The search for planets with microlensing began in earnest in 1995 with the formation of several follow-up collaborations dedicated to searching for planetary deviations in ongoing events (*Albrow et al.*, 1998; *Rhie et al.*, 2000). These initial searches adopted the basic approach advocated by *Gould and Loeb* (1992): monitoring ongoing events alerted by several survey collaborations (e.g., *Udalski et al.*, 1994; *Alcock et al.*, 1996), using networks of small telescopes spread throughout the southern hemisphere. Although microlensing planet searches have matured considerably since their initiation, this basic approach is still used to this day, with the important modification that current follow-up collaborations tend to focus on high-magnification events, which individually have higher sensitivity to planets (*Griest and Safizadeh*, 1998), as discussed in detail below.

From 1995 to 2001, no convincing planet detections were made, primarily because the relatively small number of events being alerted each year (~50–100) by the survey collaborations meant that there were only a few events ongoing at any given time, and often these were poorly suited for follow-up. Although interesting upper limits were placed on the frequency of jovian planets (*Gaudi et al.*, 2002; *Snodgrass et al.*, 2004), perhaps the most important result during this period was the development of both the theory and practice of the microlensing method, which resulted in its transformation from a theoretical abstraction to a viable, practical method of searching for planets. In 2001, the OGLE collaboration (*Udalski*, 2003) upgraded to a new camera with a 16× larger field of view and so were able to monitor a larger area of the bulge with a higher cadence. As a result, in 2002 OGLE began alerting nearly an order of magnitude more events per year than previous to the upgrade. These improvements in the alert rate and cadence, combined with improved cooperation and coordination between the survey and follow-up collaborations, led to the first discovery of an exoplanet with microlensing in 2003 (*Bond et al.*, 2004). The MOA collaboration upgraded to a 1.8-m telescope and 2-deg² camera in 2004 (*Sako et al.*, 2008), and in 2007 the OGLE and MOA collaborations started alerting ~850 events per year, thus ushering in the "golden age" of microlensing planet searches.

The first detections of exoplanets with microlensing (*Bond et al.*, 2004; *Udalski et al.*, 2005; *Beaulieu et al.*, 2006; *Gould et al.*, 2006; *Gaudi et al.*, 2008a; *Bennett et al.*, 2008; *Dong et al.*, 2009b; *Janczak et al.*, 2009; *Sumi et al.*, 2010) have provided important lessons about the kinds of information that can be extracted from observed events. When microlensing planet surveys were first being developed, it was thought that the primary virtue of the method would be solely its ability to provide statistics on large-separation and low-mass planets. Individual detections would be of little interest because the only routinely measured property of the planets would be the planet/star mass ratio, and the nature of the host star in any given system would remain unknown because the microlensing event itself provides little information about the properties

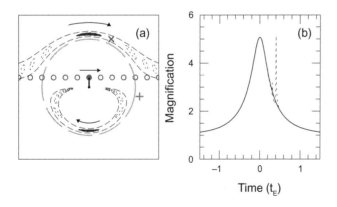

Fig. 1. (a) The images (dotted ovals) are shown for several different positions of the source (solid circles), along with the primary lens (dot) and Einstein ring (long dashed circle). If the primary lens has a planet near the path of one of the images, i.e., within the short-dashed lines, then the planet will perturb the light from the source, creating a deviation to the single lens lightcurve. (b) The magnification as a function of time is shown for the case of a single lens (solid) and accompanying planet (dotted) located at the position of the X in (a). If the planet was located at the + instead, then there would be no detectable perturbation, and the resulting lightcurve would be essentially identical to the solid curve.

of the host star, and follow-up reconnaissance would be difficult or impossible due to the large distances of the detected systems from Earth. In fact, experience with actual events has shown that much more information can typically be gleaned from a combination of a detailed analysis of the lightcurve, and follow-up, high-resolution imaging. As a result, in most cases it is possible to infer the mass of the primary lens and the basic physical properties of a planetary system, including the planet mass and physical separation. In some special cases it is possible to infer considerably more information.

The theoretical foundation of microlensing is highly technical, the practical challenges associated with searching for planets with microlensing are great, and the analysis of observed microlensing datasets is complicated and time consuming. Furthermore, although it is possible to learn far more about individual systems than was originally thought, extracting this information is difficult, and it is certainly the case that the systems will never be as well studied as planets systems found by other methods in the solar neighborhood.

Give the intrinsic limitations and difficulties of the microlensing method, one might therefore wonder: "Why bother?" The primary motivation for going to all this trouble is that microlensing is sensitive to planets in a region of parameter space that is difficult or impossible to probe with other methods, namely low-mass planets beyond the "snow line," the point in the protoplanetary disk beyond which ices can exist. The snow line plays a crucial role in the currently favored model of planet formation. Thus, even the first few microlensing planet detections have provided important constraints on planet-formation theories. The second generation of microlensing surveys will potentially be sensitive to multiple-planet systems containing analogs of all the solar system planets except Mercury, as well as to free-floating planets. Ultimately, when combined with the results from other complementary surveys, microlensing surveys can potentially yield a complete picture of the demographics of essentially *all* planets with masses greater than that of Mars. Thus it is well worth the effort, even given the drawbacks.

The primary goal of this chapter is to provide a general introduction to gravitational microlensing and searches for planets using this method. Currently, the single biggest obstacle to the progress of microlensing searches for planets is simply a lack of human power. There are already more observed binary and planetary events than can be modeled by the handful of researchers in the world with sufficient expertise to do so. This situation is likely to get worse when next-generation microlensing planet surveys come on line. However, because the theoretical foundation of microlensing is quite technical and the practical implementation of microlensing planet searches is fairly complex, it can be very difficult and time consuming for the uninitiated to gain sufficient familiarity with the field to make meaningful contributions. This difficulty is exacerbated by the fact that the explanations of the relevant concepts are scattered over dozens of journal articles. The aim of this chapter is to partially remedy this situation by providing a reasonably comprehensive review, which a beginner can use to gain a least a basic familiarity with the many concepts, tools, and techniques that are required to actively participate and contribute to current microlensing planet searches. Readers are encouraged to peruse reviews by *Paczynski* (1996), *Sackett* (1999), *Mao* (2008), and *Bennett* (2009) for additional background material.

Section 2 provides an overview of the theory of gravitational microlensing searches for planets. This is the heart of the chapter, and is fairly long and detailed. Much of this section may not be of interest to all readers, and some of the material may be too technical for a beginner to the subject. Therefore, the first subsection (section 2.1) provides a primer on the basic properties and features of microlensing. Beginners, or casual readers who are only interested in a basic introduction to the method itself and its features, can read only this section (paying particular attention to Figs. 1, 2, and 3), and then skip to section 4. Section 3 discusses the practical implementation of planetary microlensing, while section 4 reviews the basic advantages and drawbacks of the method. Section 5 provides a summary of the results to date, as well as a brief discussion of their implications. Section 6 discusses future prospects for microlensing planet searches, and in particular the expected yields of a next generation groundbased planet search, and a spacebased mission.

2. FOUNDATIONAL CONCEPTS AND EQUATIONS

2.1. Basic Microlensing

This section provides a general overview of the basic equations, scales, and phenomenology of microlensing by a point mass, and a brief introduction to how microlensing can be used to find planets, and how such planet searches work in practice. It is meant to be self-contained, and therefore the casual reader who is not interested in a detailed discourse on the theory, phenomenology, and practice of planetary microlensing can simply read this section and then skip to section 4 without significant loss of continuity.

A microlensing event occurs when a foreground "lens" happens to pass very close to our line of sight to a more distant background "source." Microlensing is a relatively improbable phenomenon, and so in order to maximize the event rate, microlensing surveys are typically carried about toward dense stellar fields. In particular, the majority of microlensing planet surveys are carried out toward the galactic center. Therefore, for our purposes, the lens is typically a main-sequence star or stellar remnant in the foreground galactic disk or bulge, whereas the source is a main-sequence star or giant typically in the bulge.

Figure 2a shows the basic geometry of microlensing. Light from the source at a distance D_s is deflected by an angle $\hat{\alpha}_d$ by the lens at a distance D_l. For a point lens, $\hat{\alpha}_d = 4GM/(c^2 D_l \theta)$, where M is the mass of the lens, and θ is the angular separation of the images of the source and the lens on the sky. [This form for the bending angle can be derived heuristically by assuming that a photon passing by an object of mass M at a distance b ≡ $D_l\theta$ will experience an impulse given by the Newtonian acceleration GM/b^2 over a time $2b/c$, thereby inducing a velocity perpendicular to the original trajectory of $\delta v = (GM/b^2)(2b/c) = 2GM/(bc)$. The deflection is then $\delta v/c = 2GM/(bc^2)$. The additional factor of 2 cannot be derived classically, and arises from general relativity (see, e.g., *Schneider et al.*, 1992).] The relation between θ, and the angular separation β between the lens and source in the absence of lensing, is called the *lens equation*, and is given trivially by $\beta = \theta - \alpha_d$. From basic geometry and using the small-angle approximation, $\hat{\alpha}_d(D_s - D_l) = \alpha_d D_s$. Therefore, for a point lens

$$\beta = \theta - \frac{4GM}{c^2\theta} \frac{D_s - D_l}{D_s D_l} \qquad (1)$$

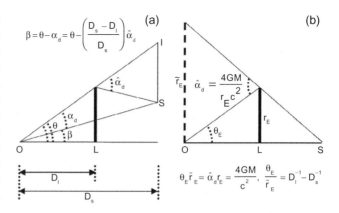

Fig. 2. (a) The lens (L) at a distance D_l from the observer (O) deflects light from the source (S) at distance D_s by the Einstein bending angle $\hat{\alpha}_d$. The angular positions of the images θ and unlensed source β are related by the lens equation, $\beta = \theta - \alpha_d = \theta - (D_s - D_l)/D_s \hat{\alpha}_d$. For a point lens, $\hat{\alpha}_d = 4GM/(c^2 D_l \theta)$. (b) Relation of higher-order observables, the angular (θ_E) and projected ($\tilde{r}_E$) Einstein radii, to physical characteristics of the lensing system. Adapted from *Gould* (2000).

Figure 3a shows the basic source and image configurations for microlensing by a single point mass. From equation (1), if the lens is exactly aligned with the source (β = 0), it images the source into an "Einstein ring" with a radius $\theta_E = \sqrt{\kappa M \pi_{rel}}$, where M is the mass of the lens, $\pi_{rel} = AU/D_{rel}$ is the relative lens-source parallax, $D_{rel} \equiv (D_l^{-1} - D_s^{-1})^{-1}$ is the relative lens-source distance, and $\kappa = 4G/c^2 AU \simeq 8.14$ mas $M_\odot^{-1}$. It is also instructive to note that $\theta_E = \sqrt{2R_{Sch}/D_{rel}}$, where $R_{Sch} \equiv 2GM/c^2$ is the Schwarzschild radius of the lens. Quantitatively

$$\theta_E = 550 \text{ μas} \left(\frac{M}{0.3 M_\odot}\right)^{1/2} \left(\frac{\pi_{rel}}{125 \text{ μas}}\right)^{1/2} \qquad (2)$$

which corresponds to a physical Einstein ring radius at the distance of the lens of

$$r_E \equiv \theta_E D_l \qquad (3)$$

$$\equiv 2.2 \text{ AU} \left(\frac{M}{0.3 M_\odot}\right)^{1/2} \left(\frac{D_s}{8 \text{ kpc}}\right)^{1/2} \left[\frac{x(1-x)}{0.25}\right]^{1/2} \qquad (4)$$

where $x \equiv D_l/D_s$.

Normalizing all angles on the sky by θ_E, we can define $u = \beta/\theta_E$ and $y = \theta/\theta_E$. Using these definitions and the definition for θ_E, the lens equation (1) reduces to the form

$$u = y - y^{-1} \qquad (5)$$

This is equivalent to a quadratic equation in y: $y^2 - uy - 1 = 0$. Thus in the case of imperfect alignment (u ≠ 0), there are two images, with positions

$$y_\pm = \pm \frac{1}{2}\left(\sqrt{u^2 + 4} \pm u\right) \qquad (6)$$

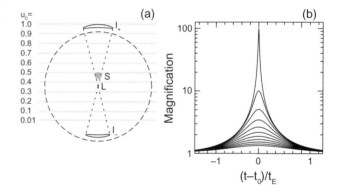

Fig. 3. Basic point-mass microlensing. (a) All angles are normalized by the angular Einstein ring radius θ_E, shown as a dashed circle of radius θ_E. The source (S) is located at an angular separation of u = 0.2 from the lens (L). Two images are created, one image outside the Einstein ring (I_+), on the same side of the lens as the source with position from the lens of $y_+ = 0.5(\sqrt{u^2 + 4} + u)$, and one inside the Einstein ring, on the opposite side of the lens as the source with position from the lens of $y_- = -0.5(\sqrt{u^2 + 4} - u)$. The images are compressed radially but elongated tangentially. Since surface brightness is conserved, the magnification of each image is just the ratio of its area to the area of the source. Since the images are typically unresolved, only the total magnification of the two images is measured, which depends only on u. (b) Magnification as a function of time (lightcurves), for the 10 trajectories shown in the left panel with impact parameters u_0 = 0.01, 0.1, 0.2, . . . , 1.0. Time is relative to the time t_0 of the peak of the event (when $u = u_0$), and in units of the angular Einstein crossing time t_E. Higher magnification implies more elongated images, which leads to increased sensitivity to planetary companions. Adapted from *Paczynski* (1996).

The positive, or "major," image is always outside the Einstein ring, whereas the negative, or "minor," image is always inside the Einstein ring. [The images formed by a gravitational lens can also be found by determining the relative time delay as a function of the (vector) angle θ for hypothetical light rays propagating from the source. This time delay function includes the effect of the difference in the geometric path length, as well the gravitational (Shapiro) delay, as a function of θ. The images are then located at the stationary points (maxima, minima, or saddle points) of the time delay surface. For a point-mass lens, the positive solution in equation (6) corresponds to a minimum in the time delay surface, and so is also referred to as the "minimum" image. The negative solution corresponds to a saddle point, and so is also referred to as the "saddle" image. There is formally a third image, corresponding to the maximum of the time delay surface, but this image is located behind the lens, and is infinitely demagnified (for a true point lens).] As can be seen in Fig. 3, the angular separation between these images at the time of closest alignment is is $\sim 2\theta_E$. Thus for typical lens masses (0.1–1 $M_\odot$) and lens and source distances (1–10 kpc), $\theta_E \lesssim$ mas and so the images are not resolved. However, the images are also distorted, and since surface brightness is conserved, this implies they are also magnified. The magnification of each image is just the ratio of the area of the image to the area of the source. As can be seen in Fig. 3, the images are typically elongated tangentially by an amount $y_\pm/u$, but are compressed radially by an amount $dy_\pm/du$. The magnifications are then

$$A_\pm = \left| \frac{y_\pm}{u} \frac{dy_\pm}{du} \right| = \frac{1}{2} \left[\frac{u^2 + 2}{u\sqrt{u^2 + 4}} \pm 1 \right] \quad (7)$$

and thus the total magnification is

$$A(u) = \frac{u^2 + 2}{u\sqrt{u^2 + 4}} \quad (8)$$

Note that, as $u \to \infty$, $A_+ \to 1$ and $A_- \to 0$. Also, for $u \ll 1$, the magnification takes the form $A \simeq 1/u$.

Because the lens, source, and observer are all in relative motion, the angular separation between the lens and source is a function of time. Therefore, the magnification of the source is also a function of time: a microlensing event. In the simplest case of uniform, rectilinear motion, the relative lens-source motion can be parameterized by

$$u(t) = \left[u_0^2 + \left(\frac{t - t_0}{t_E} \right)^2 \right]^{1/2} \quad (9)$$

where u_0 is the minimum separation between the lens and source (in units of θ_E), t_0 is the time of maximum magnification (corresponding to when $u = u_0$), and t_E is the timescale to cross the angular Einstein ring radius, $t_E \equiv \theta_E/\mu_{rel}$, where μ_{rel} is the proper motion of the source relative to the lens. This form for u gives rise to a smooth, symmetric microlensing event with a characteristic form (often called a "Paczynski curve"), as shown in Fig. 3b. The typical timescales for events toward the galactic bulge are on the order of a month

$$t_E = 19 \text{ d} \left(\frac{M}{0.3 \, M_\odot} \right)^{1/2} \left(\frac{\pi_{rel}}{125 \, \mu as} \right)^{1/2} \left(\frac{\pi_{rel}}{10.5 \, mas/yr} \right)^{-1} \quad (10)$$

but can range from less than a day to years. The magnification is $A > 1.34$ for $u \leq 1$, and so the magnifications are substantial and easily detectable.

If the lens star happens to have a planetary companion, and the planet happens to be located near the paths of one or both of the two images created by the primary lens, the companion can create a short-timescale perturbation on the primary microlensing event when the image(s) sweep by the planet (*Mao and Paczynski*, 1991; *Gould and Loeb*, 1992). There are two conceptually different channels by which planets can be detected with microlensing, delineated by the maximum magnification A_{max} of the primary event in which the planet is detected.

Consider the case of a relatively low-magnification (i.e., $A_{max} \lesssim 10$ or $u_0 \gtrsim 0.1$) primary event, such as illustrated in Fig. 1. Over the course of the primary event, the two images sweep out a path on the sky, with the path of the major (minimum) image entirely contained outside the Einstein ring radius, and the path of the minor (saddle) image entirely contained within the Einstein ring radius. In the main detection channel, the primary hosts a planet that happens to be located near the path of one of these two images for such a low-magnification event. (As can be seen in Fig. 1, for low-magnification events, the two images are well separated, and thus a low-mass-ratio planet will generally only significantly perturb one of the two images.) As the image sweeps by the position of the planet, the planet will further perturb the light from this image and yield a short-timescale deviation (*Gould and Loeb*, 1992). The duration of this deviation is $\sim t_{E,p} = q^{1/2} t_E$, where $q = m_p/M$ is the mass ratio and m_p is the planet mass, and the magnitude of the perturbation depends on how close the perturbed image passes by the planet. Given a range of primary event durations of ~ 10–100 d, the duration of the perturbations range from a few hours for an Earth-mass planet to a few days for jovian-mass planets. As can be seen in Fig. 2, the location of the perturbation relative to the peak of the primary event depends on two parameters: α, the angle of the projected star-planet axis with respect to the source trajectory, and d, the instantaneous angular separation between the planet and host star in units of θ_E. (Conventionally, both the lensing deflection angle and the source trajectory angle are denoted by the symbol α. In order to maintain contact with the literature, this unfortunate convention is maintained here, but to avoid confusion the deflection angle is denoted with the subscript "d.")

Because the orientation of the source trajectory relative to the planet position is random, the time of this perturbation is not predictable and the detection probability is $\sim A(t_{0,p}) \theta_{E,p}/\theta_E$, where $A(t_{0,p})$ is the unperturbed magnification of the image that is being perturbed at the time $t_{0,p}$ of the perturbation (*Horne et al.*, 2009), and $\theta_{E,p} \equiv q^{1/2}\theta_E$ is the Einstein ring radius of the planet. Here the factor A accounts for the fact that the area of the image plane covered by magnified images is larger by their magnification.

Since the planet must be located near one of the two primary images in order to yield a detectable deviation, and these images are always located near the Einstein ring radius when the source is significantly magnified (see Fig. 1), the sensitivity of the microlensing method peaks for planet-star projected separations of $\sim r_E$, i.e., for d ~ 1. However, microlensing can also detect planets well outside the Einstein ring (d ≫ 1), albeit with less sensitivity. Since the magnification of the minor image decreases with position as y_-^4 (see equations (6) and (7)), microlensing is generally not sensitive to planets d ≪ 1, i.e., close-in planets.

Given the existence of a planet with a projected separation within a factor of ~2 of the Einstein ring radius, the detection probabilities range from tens of percent for jovian planets to a few percent for Earth-mass planets (*Gould and Loeb*, 1992; *Bolatto and Falco*, 1994; *Bennett and Rhie*, 1996; *Peale*, 2001). Detecting planets via the main channel requires substantial commitment of resources because the unpredictable nature of the perturbation requires dense, continuous sampling, and furthermore the detection probability per event is relatively low so many events must be monitored.

A useful feature of low-magnification planetary microlensing events such as that shown in Fig. 1 is that it is possible to essentially "read off" the lightcurve parameters from the observed features (*Gould and Loeb*, 1992; *Gaudi and Gould*, 1997b). For the primary event, the three gross observable parameters are the time of maximum magnification t_0, the peak magnification A_{max}, and a measure of the duration of the event such as its full-width half-maximum t_{FWHM}. The latter two observables can then be related to the Einstein timescale t_E and impact parameter u_0 using equations (8) and (9). (For the purposes of exposition, this discussion ignores blending, which is generally important and complicates the interpretation of observed lightcurves; see section 3.1.) For example, for small u_0 we have $u_0 \sim A_{max}^{-1}$ and $t_E \sim 0.5 t_{FWHM} u_0^{-1}$. Unfortunately, only t_E contains any information about the physical properties of the lens, and then only in a degenerate combination of the lens mass, distance, and relative lens-source proper motion. However, as discussed in detail in section 3.3, in many cases it is often possible to obtain additional information that partially or totally breaks this degeneracy. In particular, in those cases where it is possible to isolate the light from the lens itself, this measurement can be used to constrain the lens mass (*Bennett et al.*, 2007). The three parameters that characterize the planetary perturbation are the duration of the perturbation, the time of the perturbation $t_{0,p}$, and the magnitude of the perturbation, δ_p. As mentioned previously, the duration of the perturbation is proportional to $t_{E,p} \equiv q^{1/2} t_E$, and so gives the planet/star mass ratio q. The time and magnitude of the perturbation then specify the projected separation d and angle of the source trajectory with respect to the binary axis α, since $t_{0,p}$ and δ_p depend on the location of the planet relative to the path of the perturbed image (*Gaudi and Gould*, 1997b).

The other channel by which microlensing can detect planets is in high-magnification events (*Griest and Safizadeh*, 1998). (There is no formal definition for "high"-magnification events; however, high magnification typically refers to events with maximum magnification $A_{max} \gtrsim 100$, corresponding to impact parameters $u_0 \lesssim 0.01$.) In addition to perturbing images that happen to pass nearby, planets will also distort the perfect circular symmetry of the Einstein ring. Near the peak of high-magnification events, as the lens passes very close to the observer-source line of sight (i.e., when u ≪ 1), the two primary images are highly elongated and sweep along the Einstein ring, thus probing this distortion. For very-high-magnification events, these images probe nearly the entire Einstein ring radius and so are sensitive to all planets with separations near r_E, regardless of their orientation with respect to the source trajectory. Thus high-magnification events can have nearly 100% sensitivity to planets near the Einstein ring radius, and are very sensitive to low-mass planets (*Griest and Safizadeh*, 1998). However, these events are rare: A fraction $\sim 1/A_{max}$ of events have maximum magnification $\geq A_{max}$. Fortunately, these events can often be predicted several hours to several days ahead of peak, and furthermore the times of high sensitivity to planets are within a full-width half-maximum of the event peak, or roughly a day for typical high-magnification events (*Rattenbury et al.*, 2002). Thus scarce observing resources can be concentrated on these few events and only during the times of maximum sensitivity. Because the source stars are highly magnified, it is also possible to use more common, smaller-aperture telescopes.

2.2. Theory of Microlensing

Gravitational lensing can be thought of as the mapping $\beta \rightarrow \theta$ between the angular position of a source β in the absence of lensing to the angular position(s) θ of the image(s) of the source under the action of the gravitational lens. This mapping is given by the lens equation

$$\beta = \theta - \alpha_d(\theta) \qquad (11)$$

where α_d is deflection of the source due to the lens.

Consider a source at a distance D_s and a lens located at a distance D_l from the observer. Figure 2 shows the lensing geometry. The deflection angle α_d is related to the angle $\hat{\alpha}_d$ by which the lens mass bends the light ray from the source by $\hat{\alpha}_d(D_s - D_l) = \alpha_d D_s$. Assume the lens is a system of N_L

point masses each with mass m_i and position $\theta_{m,i}$. Further assume that the lenses are static (or, more precisely, moving much more slowly than c), and their distribution along the line of sight is small in comparison to D_l, D_s, and $D_s - D_l$. The deflection angle is then

$$\alpha_d(\theta) = \frac{4G}{D_{rel}c^2} \sum_i^{N_l} m_i \frac{\theta - \theta_{m,i}}{|\theta - \theta_{m,i}|^2} \quad (12)$$

where $D_{rel} \equiv (D_l^{-1} - D_s^{-1})^{-1}$. See *Schneider et al.* (1992) and *Petters et al.* (2001) for the expression for α_d for a general mass distribution, as well for the derivation of equations (11) and (12) from the time delay function and ultimately the metric.

It is common practice to normalize all angles to the angular Einstein ring radius

$$\theta_E \equiv (\kappa M \pi_{rel})^{1/2} \quad (13)$$

where $M \equiv \sum_i^{N_l} m_i$ is the total mass of the lens, $\pi_{rel} \equiv AU/D_{rel}$ is the relative lens-source parallax, and $\kappa = 4G/(c^2 AU) \simeq 8.14$ mas $M_\odot^{-1}$. The reason for this convention is clear when considering the single lens (section 2.2.1). The dimensionless vector source position is defined to be $\mathbf{u} \equiv \beta/\theta_E$, and the dimensionless vector image positions are defined to be $\mathbf{y} \equiv \theta/\theta_E$.

It is often convenient to write the lens equation in complex coordinates (*Witt*, 1990). Defining the components of the (dimensionless) source position to be $\mathbf{u} = (u_1, u_2)$ and the image position(s) to be $\mathbf{y} = (y_1, y_2)$, the two dimensional source and images positions can be expressed in complex form as $\zeta = u_1 + iu_2$ and $z = y_1 + iy_2$. The lens equation can now be rewritten

$$\zeta = z - \sum_i^{N_l} \frac{\varepsilon_i}{\bar{z} - \bar{z}_{m,i}} \quad (14)$$

where $z_{m,i}$ is the position of mass i, $\varepsilon \equiv m_i/M$. The overbars denote complex conjugates, which in equation (14) arise from the identity $z/|z|^2 \equiv \bar{z}^{-1}$. This equation can then be solved to find the image positions z_j.

Lensing conserves surface brightness, but because of the mapping the angular area of each image of the source is not necessarily equal to the angular area of the unlensed source. Thus the flux of each image (the area times the surface brightness) is different from the flux of the unlensed source: The source is magnified or demagnified. For a small source, the magnification A_j of each image j is given by the amount the source is "stretched" due to the lens mapping. Mathematically, the amount of stretching is given by the inverse of the determinant of the Jacobian of the mapping equation (14) evaluated at the image position

$$A_j = \frac{1}{\det J}\bigg|_{z=z_j}, \det J \equiv \frac{\partial(x_1, x_2)}{\partial(y_1, y_2)} = 1 - \frac{\partial \zeta}{\partial \bar{z}} \overline{\frac{\partial \zeta}{\partial \bar{z}}} \quad (15)$$

See *Witt* (1990) for a derivation of the rightmost equality. Note that these magnifications can be positive or negative, where the sign corresponds to the parity (handedness) of the image. By definition in microlensing the images are unresolved, and we are interested in the total magnification, which is just the sum of the magnifications of the individual images, $A \equiv \sum_j |A_j|$.

An interesting and critical property of gravitational lenses is that the mapping can be singular for some source positions. At these source positions, $\det J = 0$. In other words, an infinitesimally small displacement in the source position maps to an infinitely large separation in the image position. [This is analogous to the familiar distortion seen in cylindrical map projections of the globe (such as the Mercator projection) for latitudes far from the equator, due to the singular nature of these mappings at the poles.] For point sources at these positions, the magnification is formally infinite, and for sources near these positions, the magnification is large.

From the lens equation (14)

$$\frac{\partial \zeta}{\partial \bar{z}} = \sum_i^{N_l} \frac{\varepsilon_i}{(\bar{z}_{m,i} - \bar{z})^2} \quad (16)$$

and therefore, from equation (15), the image positions where $\det J = 0$ are given by

$$\left| \sum_i^{N_l} \frac{\varepsilon_i}{(\bar{z}_{m,i} - \bar{z})^2} \right|^2 = 1 \quad (17)$$

The set of all such image positions define closed *critical curves*. These can be found by noting that the sum in equation (17) must have a modulus equal to unity. Therefore, we can solve for the critical image positions parametrically by solving the equation

$$\sum_i^{N_l} \frac{\varepsilon_i}{(\bar{z}_{m,i} - \bar{z})^2} = e^{i\phi} \quad (18)$$

for each value of the parameter $\phi = [0, 2\pi)$. The set of source positions corresponding to these image positions define closed curves called *caustics*. By clearing the complex conjugates $\bar{z}$ and fractions (*Witt*, 1990), equation (18) can be written as a complex polynomial of degree $2N_l$. Thus there are at most $2N_l$ critical curves and caustics. Given their importance in planetary microlensing, caustics are discussed in considerably more detail below.

2.2.1. Single lenses. For a single point mass ($N_l = 1$), defining the origin as the position of the lens, the lens equation reduces to

$$\beta = \theta - \frac{\theta_E^2}{\theta} \qquad (19)$$

Since the lens is circularly symmetric, the images are always located along the line connecting the lens and source, and the vector notation has been suppressed, but the convention that positive values of θ are for images on the same side of the lens as the source is kept. If the lens is perfectly aligned with the source, then $\beta = 0$, and $\theta = \theta_E$. In other words, the lens images the source into a ring of radius equal to the angular Einstein ring radius.

The dimensionless single lens equation is

$$u = y - \frac{1}{y} \qquad (20)$$

In the case of imperfect alignment, this becomes a quadratic function of y, and so there are two images of the source, with positions

$$y_\pm = \pm \frac{1}{2}\left(\sqrt{u^2 + 4} \pm u\right) \qquad (21)$$

One of these images (the major image, or minimum) is always outside the Einstein ring radius ($y_+ \geq 1$) on the same side of the lens as the source, and the other image (the minor image, or saddle point) is always inside the Einstein ring radius ($|y_-| \leq 1$) on the opposite side of the lens as the source. The separation between the two images is $|y_+ - y_-| = (u^2 + 4)$, and thus the images are separated by $\sim 2\theta_E$ when both images are significantly magnified (i.e., when $u \lesssim 1$). Since θ_E is on the order of a milliarcsecond for typical lens masses, and source and lens distances in events toward the galactic bulge, the images are unresolved.

The magnifications of each image can be found analytically

$$A_\pm = \frac{1}{2}(A \pm 1) \qquad (22)$$

where the total magnification is

$$A(u) = \frac{u^2 + 2}{u\sqrt{u^2 + 4}} \qquad (23)$$

A few properties are worth noting. First, $u \ll 1$, $A(u) \simeq u^{-1}$. The magnification diverges for $u \to 0$, and the point $u = 0$ defines the caustic in the single lens case. Second, for $u \gg 1$, $A(u) \simeq 1 + 2u^{-4}$, and thus the excess magnification drops rapidly for large source-lens angular separations.

2.2.2. Binary lenses. For a two-point-mass lens ($N_l = 2$), the lens equation is

$$\zeta = z + \frac{\varepsilon_1}{\bar{z}_{m,1} - \bar{z}} + \frac{\varepsilon_2}{\bar{z}_{m,2} - \bar{z}} \qquad (24)$$

This can be written as a fifth-order complex polynomial in z, which cannot be solved analytically. *Witt and Mao* (1995) provide the coefficients of the polynomial, which can easily be solved with standard numerical routines, e.g., Laguerre's method (*Press et al.*, 1992), to yield the image positions z. It is important to note that that the solutions to the fifthorder complex polynomial are not necessarily solutions to the lens equation (equation(24)). Depending on the location of the source with respect to the lens positions, two of the images can be spurious. Thus there are either three or five images.

The boundaries of the three and five image regions are the caustic curves (where $\det J = 0$), and thus the number of images changes by two when the source crosses the caustic. A binary lens has one, two, or three closed and non-self-intersecting caustic curves. Which of these three topologies is exhibited depends on the mass ratio of the lens $q \equiv m_1/m_2$ and on the angular separation of the two lens components in units of Einstein ring radius of binary, $d \equiv |z_{m,1} - z_{m,2}|$. For a given q, the values of d for which the topology changes are given by (*Schneider and Weiss*, 1986; *Dominik*, 1999b)

$$\frac{q}{(1+q)^2} = \frac{(1-d_c)^3}{27 d_c^8}, \quad d_w = \frac{(1+q^{1/3})^{3/2}}{(1+q)^{1/2}} \qquad (25)$$

For $d \leq d_c$, there are three caustic curves; for $d_c \leq d \leq d_w$, there is one caustic curve; and for $d \geq d_w$, there are two caustic curves. These are often referred to as the "close," "intermediate" or "resonant," and "wide" topologies, respectively. Figure 4 plots d_c and d_w as a function of q. For equal-mass binaries, $q = 1$, the critical values of the separation are $d_c = 2^{-1/2}$ and $d_w = 2$.

A useful property of binary lenses is that the total magnification of all the images is always $A \geq 3$ when the source is interior to the caustic curve, i.e., when there are five images of the source (*Witt and Mao*, 1995). Thus if the magnification of a source is observed to be less than 3 during a microlensing event, it can be immediately concluded that either the source is exterior to the caustic, or the source is significantly blended with an unrelated, unlensed star, thus diluting the magnification. [There is a third, less likely possibility that there are additional bodies in the system. In this case, the bound that $A \geq 3$ interior to the caustic can be violated (*Witt and Mao*, 1995).]

2.2.3. Triple lenses and beyond. *Gaudi et al.* (1998) were the first to explore triple lenses in the context of planetary microlensing. For a triple lens ($N_l = 3$), the lens equation can be written as a tenth-order complex polynomial in z, which can be solved using the same techniques as in the binary lens case. *Rhie* (2002) provides the coefficients of the polynomial. There are a maximum of 10 images, and there are a minimum of 4 images, with the number of images changing by a multiple of 2 when the source crosses the caustic. The caustics of triple lenses can exhibit quite complicated topologies, including nested and/

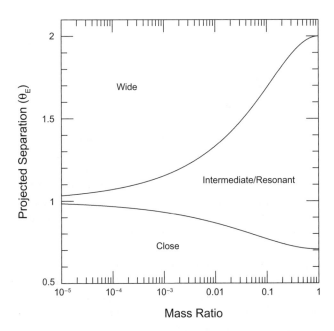

Fig. 4. The critical values of d, the projected separation in units of θ_E, at which the caustic topology (number of caustic curves) of a binary lens changes as a function of the mass ratio q. The upper curve shows d_w, the critical value of d between the wide caustic topology consisting of two disjoint caustics, and the intermediate or resonant caustic topology consisting of a single caustic. The lower curve shows d_c, the critical value between the resonant caustic topology and the close caustic topology consisting of three disjoint caustics.

or self-intersecting caustic curves. The topology depends on five parameters: two mass ratios, two projected separations, and the angle between projected position vectors of the companions and the primary lens.

In general, the lens equation for a system of N_l point lenses can be written as a complex polynomial on the order of $N_l^2 + 1$. However, it has been shown that the maximum number of images is $5(N_l - 1)$ for $N_l \geq 2$ (*Rhie*, 2001, 2003; *Khavinson and Neumann*, 2006). Thus for $N_l > 3$, it is always the case that some of the roots of the polynomial are not solutions to the lens equation.

2.2.4. Magnification near caustics. As mentioned previously, the caustics are the set of source positions for which the magnification of a point source is formally infinite. Caustic curves are characterized by multiple concave segments called *folds*, which meet at points called *cusps*. Folds and cusps are so named because the local lensing properties of sources close to these caustics are equivalent to the generic fold and cusp mapping singularities of mathematical catastrophe theory. For a more precise formulation of this statement and additional discussion, see *Petters et al.* (2001). This leads to the particularly important property of fold and cusps that their local lensing properties are universal, regardless of the global properties of the lens. Thus the positions and magnifications of the critical images of sources near folds and cusps have universal scaling behaviors that can be described essentially analytically, and whose normalization depend only on the local properties of the lens potential. This universal and local behavior of the magnification near caustics has proven quite useful, in that it allows one to analyze lightcurves near caustic crossings separate from and independent of the global lens model (see, e.g., *Gaudi and Gould*, 1999; *Albrow et al.*, 1999b; *Rhie and Bennett*, 1999; *Afonso et al.*, 2001; *Dominik*, 2004a,b).

The lensing behavior near folds and cusps has been discussed in detail by a number of authors (*Schneider and Weiss*, 1986, 1992; *Mao*, 1992; *Zakharov*, 1995, 1999; *Fluke and Webster*, 1999; *Petters et al.*, 2001; *Gaudi and Petters*, 2002a,b; *Pejcha and Heyrovsky*, 2009). The salient properties are briefly reviewed here, but the reader is encouraged to consult these papers for a more in-depth discussion.

For sources close to and interior to a fold caustic, there are two highly magnified images with nearly equal magnification and opposite parity that are nearly equidistant from the corresponding critical curve. (Here, "interior to a fold caustic" is defined such that the caustic curves away from the source.) Neglecting the curvature of the caustic and any changes in the lensing properties parallel to the caustic, the total magnification of these divergent images is (*Schneider and Weiss*, 1986)

$$A_{div}(\Delta u_\perp) = \left(\frac{\Delta u_\perp}{u_r}\right)^{-1/2} \Theta(\Delta u_\perp) \tag{26}$$

where $\Delta u_\perp$ is the perpendicular distance to the caustic, with $\Delta u_\perp > 0$ for sources interior to the caustic, $\Theta(x)$ is the Heaviside step function, and u_r is the characteristic "strength" of the fold caustic locally, and is related to local derivatives of the lens potential. As the source approaches the caustic, the images brighten and merge, disappearing when the source crosses the caustic. The behaviors of the remaining (noncritical) images are continuous as the source crosses the caustic. Thus immediately outside of a fold, the magnification is finite and (typically) modest.

For sources close to and interior to a cusp, there are three highly magnified images. For a source on the axis of symmetry of the cusp, the total magnification of the three images is $\propto \Delta u_c^{-1}$, where Δu_c is the distance of the source from the cusp. As the source approaches the cusp, the magnification of all three images increases and the images merge. Two of the three images disappear as the source exits the cusp. The magnification of the remaining image is continuous as the source exits the cusp, and in particular the image remains highly magnified, also with magnification that is $\propto \Delta u_c^{-1}$. Thus, in contrast to folds, sources immediately exterior to cusps are highly magnified. As with the fold, the behaviors of the noncritical images are continuous as the source crosses the caustic.

Caustic curves are closed, and thus for any given source trajectory (which is simply a continuous path through the source plane), caustic crossings come in pairs. Generally, since the majority of the length of a caustic is made up of fold caustics, both the caustic entry and exit are fold crossings. Since the magnification immediately outside a fold is not divergent, it is usually impossible to predict a caustic entry beforehand. Once one sees a fold caustic entry, a caustic exit is guaranteed, and this is typically a fold exit. Monitoring a caustic exit is useful for two reasons. First, the strong finite source effects (see below) during the crossing can be used to provide additional information about the lens (see section 3.3), and measure the limb-darkening of the source (e.g., *Albrow et al.*, 1999c; *Fields et al.*, 2003). In addition, during the caustic crossing the source is highly magnified and potentially very bright, which allows for otherwise impossible spectroscopic observations to determine properties of the source star, such as its effective temperature and atmospheric abundances (*Minniti et al.*, 1998; *Johnson et al.*, 2008). Unfortunately, it is typically difficult to predict when this exit will happen well before the crossing (*Jaroszyński and Mao*, 2001).

For both fold and cusp caustics, the magnification of a source of finite size begins to deviate by more than a few percent from the point-source approximation when the center of the source is within several source radii of the caustic (*Pejcha and Heyrovský*, 2009). Formally, the magnification of a finite source can be found simply by integrating the point-source magnification over the area of the finite source, weighting by the source surface brightness distribution. Practically, this approach is difficult and costly to implement precisely due to the divergent magnification near the caustic curves. An enormous amount of effort has been put into developing robust and efficient algorithms to compute the magnification for finite sources (*Dominik*, 1995, 2007; *Bennett and Rhie*, 1996; *Wambsganss*, 1997; *Gould and Gaucherel*, 1997; *Griest and Safizadeh*, 1998; *Vermaak*, 2000; *Dong et al.*, 2006; *Gould*, 2008; *Pejcha and Heyrovský*, 2009). The most efficient of these algorithms use a two-pronged approach. First, semianalytic approximations to the finite-source magnification derived from an expansion of the magnification in the vicinity of the source are used where appropriate (*Gould*, 2008; *Pejcha and Heyrovský*, 2009). Second, where necessary a full numerical evaluation of the finite source magnification is performed by integrating in the image plane. Integrating in the image plane removes the difficulties with the divergent behavior of the magnification near caustics, because the surface brightness profiles of the images are smooth and continuous. Thus one simply "shoots" rays in the image plane, and determines which ones "land" on the source using the lens equation. The ratio of the total area of all the images divided by the area of the source (appropriately weighted by the surface brightness profile of the source) gives the total magnification. The devilish details then lie in the manner in which one efficiently samples the image plane (see, e.g., *Rattenbury et al.*, 2002; *Dong et al.*, 2006; *Pejcha and Heyrovský*, 2009).

For planetary microlensing events, efficient routines for evaluating the finite-source magnification are crucial, for two reasons. First, nearly all planetary perturbations are strongly affected by finite source effects. Second, the processes of finding the best-fit model to an observed lightcurve and evaluating the model parameter uncertainties requires calculating tens of thousands of trial model curves (or more), and the majority of the computation time is spent calculating the finite-source magnifications.

2.3. Planetary Microlensing Phenomenology

For binary lenses in which the companion mass ratio is $q \ll 1$ (i.e., planetary companions), the companion will cause a small perturbation to the overall magnification structure of the lens. Thus the majority of source positions will give rise to magnifications that are essentially indistinguishable from a single lens. The source positions for which the magnification deviates significantly from a single lens are all generally confined to a relatively narrow region around the caustics. Thus much of the phenomenology of planetary microlensing can be understood by studying the structure of the caustics and the magnification pattern near the caustics.

Recall there are three different caustic topologies for binary lenses: close, intermediate, and wide (see Fig. 4). For $q \ll 1$, the critical values of the separation where these caustic topologies change can be approximated by $d_c \simeq 1 - 3q^{1/3}/4$ and $d_w \simeq 1 + 3q^{1/3}/2$ (*Dominik*, 1999b). Therefore, for planetary lenses, the intermediate or resonant caustics are confined to a relatively narrow range of separations near $d = 1$, and this range shrinks as $q^{1/3}$. Figure 5 shows the caustics for a Jupiter/Sun mass ratio of $q = 0.001$, and 11 different separations $d = 0.6, 0.7, 0.8, 0.9, 0.95, 1.0, 1.05, 1.11, 1.25, 1.43, 1.67$. For both the close ($d < d_c$; Fig. 5i–k) and wide ($d > d_w$; Fig. 5a–c) topologies, one caustic is always located near the position of the primary (the origin in Fig. 5). This is known as the central caustic. Figure 6 shows an expanded view of these caustics, which have a highly asymmetric "arrow" shape, with one cusp at the arrow tip pointing toward the planet, and three cusps at the "back end" of the caustic pointing away from the planet. The on-axis cusp pointing away from the planet is generally much "weaker" than the other cusps, in the sense that the scale of the gradient in magnification along the cusp axis is smaller for the weaker cusp, so that at fixed distance from the cusp, the excess magnification is smaller for the weaker cusp. The magnification pattern near a central caustic is illustrated in Fig. 7. The lightcurves (one-dimensional slices through the magnification pattern) for sources passing perpendicular to the binary-lens axis close to the back end will exhibit a "U"-shaped double-peaked deviation from the single-lens form (Fig. 8c,d), whereas sources passing perpendicular to the binary-lens axis close to the tip of the central caustic will exhibit a single bump (Fig. 8b,d). Sources passing the caustic parallel to the binary-lens axis will exhibit little deviation from the single lens form, essentially unless they cross the caustic.

In the limit that $q \ll 1$, and $|d-1| \gg q$, it is possible to show that, for fixed d, the size of the central caustic scales as q. Furthermore, the overall shape of the caustic, as quan-

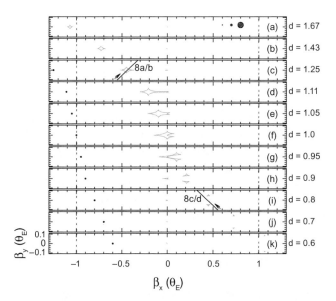

Fig. 5. The gray curves show the caustics for a planetary lens with mass ratio q = 0.001, and various values of d, the projected separation in units of θ_E. The dotted lines show sections of the Einstein ring. The dots show the location of the planet. In panels (c) and (i), an example trajectory is shown that produces a perturbation by the planetary caustic; the resulting lightcurves are shown in Fig. 9. In panel (a), three different representative angular source sizes in units of θ_E are shown: $\rho_* = 0.003$, 0.01, and 0.03. For typical microlensing event parameters, these correspond to stars in the galactic bulge with radii of $\sim R_\odot$, $\sim 3 R_\odot$, and $\sim 10 R_\odot$, i.e., a main-sequence turn-off star, a subgiant, and a clump giant.

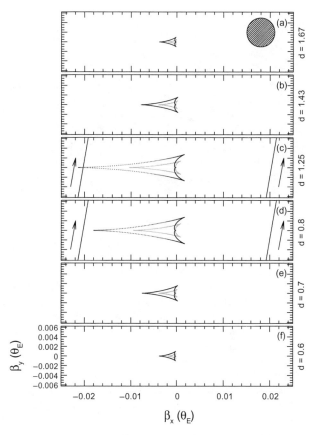

Fig. 6. The black curves show the central caustics for a planetary lens with q = 0.001, and various values of d, the projected separation in units of θ_E. The primary lens is located at the origin, and so trajectories that probe the central caustic correspond to events with small impact parameter u_0, or events with high maximum magnification. The gray curves show the central caustic for a mass ratio of q = 0.0005, demonstrating that the size of the central caustic scales as q. For $q \ll 1$, the central caustic and proximate magnification patterns are essentially identical under the transformation $d \leftrightarrow d^{-1}$. The degree of asymmetry, i.e., the length to width ratio, of the central caustic depends on d, such that the caustic becomes more asymmetric as $d \to 1$. In panels (c) and (d), example trajectories are shown that produce perturbations by the central caustic; the resulting lightcurves are shown in Fig. 8. In panel (a), a representative angular source size in units of θ_E of $\rho_* = 0.003$ is shown. For typical microlensing event parameters, this corresponds to a star in the galactic bulge of radius $\sim R_\odot$, i.e., a main-sequence turn-off star.

tified, e.g., by its length-to-width ratio, depends only on d, such that the caustic becomes more asymmetric (the length-to-width ratio increases) as $d \to 1$. Finally, the central caustic shape and size is invariant under the transformation $d \to d^{-1}$ [see *Chung et al.* (2005) and Fig. 6]. As illustrated in Fig. 8, the $d \to d^{-1}$ duality results in a degeneracy between lightcurves produced by central caustic perturbations due to planetary companions, typically referred to as the close/wide degeneracy (*Griest and Safizadeh*, 1998; *Dominik*, 1999b; *Albrow et al.*, 2002; *An*, 2005).

For the close ($d < d_c$) topology (Fig. 5h–j), there are three caustics, the central caustic and two larger, triangular-shaped caustics with three cusps. The latter caustics are referred to as the planetary caustics, and are centered on the planet/star axis at angular separation of $u_c \simeq |d - d^{-1}|$ from the primary lens, on the opposite side of the primary from the planet. [It is possible to derive the location of the planetary caustic(s) for both the close and wide topologies by noting that, when the source crosses the planetary caustic(s), the planet is perturbing one of the two images of the source created by the primary lens (the major image in the case of the wide topology, and the minor image in the case of the close topology). For a source position u, the locations of the two primary images $y_\pm(u)$ are given by equation (21). The planet must therefore be located at $d \sim y_\pm(u)$ to significantly perturb the image, and thus the center of caustic u_c is located at the solution of the inversion of equation (21), i.e., $u(y_\pm = d)$, which yields $u_c = d - d^{-1}$.] The caustics are symmetrically displaced perpendicular to the planet/star axis, with the separation between the caustics increasing with decreasing d and so increasing u_c. As illustrated in Fig. 7, the magnification pattern near these caustics is characterized by small regions surrounding the caustics where a source exhibits a positive deviation from

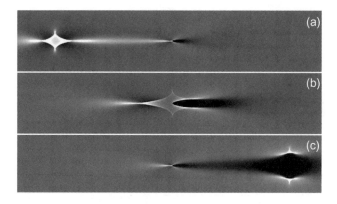

Fig. 7. The magnification pattern as a function of source position for a planetary companion with q = 0.001 and **(a)** d = 1.25, **(b)** d = 1.0, and **(c)** d = 0.8, corresponding to wide, intermediate/resonant, and close topologies, respectively. The grayscale shading denotes $2.5\log(1+\delta)$, where δ is the fractional deviation from the single-lens (i.e., no planet) magnification. White shading corresponds to regions with positive deviation from the single-lens magnification, whereas black shading corresponds to negative deviations. For the wide and close topology, there are two regions of large deviations, corresponding to the central caustics located at the position of the primary (the center of each panel), and the planetary caustics. For the intermediate/resonant topology, there is only one large caustic, which produces relatively weak perturbations for a large fraction of the caustic area.

the single-lens magnification, and a large region between the caustics where a source exhibits a negative deviation from the single-lens magnification. Figure 9c,d shows a representative lightcurve from a source passing near the planetary caustics of a close planetary lens with d = 0.8 and q = 0.001.

For the wide ($d > d_w$) topology (Fig. 5h–j), there are two caustics, the central caustic and a single planetary caustic with four cusps. As for the close planetary caustics, the wide planetary caustic is centered on the planet/star axis at an angular separation of $u_c \simeq |d - d^{-1}|$ from the primary lens, but in this case on the same side of the primary from the planet. The caustic is an asteroid shape, with the length along the planet/star axis being generally longer than the width. The asymmetry (i.e., length-to-width ratio) of the planetary caustic increases as $d \to 1$. The magnification pattern near the wide planetary caustic is characterized by large positive deviations interior to the caustic, and lobes of positive deviation extending outward along the axes of the four cusps, particularly along the planet/star axis in the direction of the primary. There are relatively small regions of slight negative deviation from the single-lens magnification immediately outside the fold caustic between the cusps. Figure 9a,b shows a representative lightcurve from a source passing through the planetary caustic of a wide planetary lens with d = 1.25 and q = 0.001.

In the wide ($d > d_w$) and close ($d < d_c$) cases, the planet is essentially perturbing one of the two images created by the primary lens, and the other image (on the other side of the Einstein ring) is essentially unaffected. In this case, and in the limit that $q \to 0$, the lensing behavior near the planet and the perturbed primary image is equivalent to a single lens with pure external shear (*Gould and Loeb*, 1992; *Dominik*, 1999b; *Gaudi and Gould*, 1997a), i.e., a Chang-Refsdal lens (*Chang and Refsdal*, 1979). The Chang-Refsdal lens has been studied extensively (e.g., *Chang and Refsdal*, 1984; *An*, 2005; *An and Evans*, 2006), and its properties are well understood. The lens equation is

$$\zeta = z - \frac{1}{\bar{z}} - \gamma \bar{z} \qquad (27)$$

where γ is the shear and the origin is taken to be the location of the planet. This is equivalent to a fourth-order complex polynomial in z, which can be solved analytically, or numerically in the same manner as the binary-lens case (see section 2.2). There are two or four images, depending on the source position. The correspondence between the Chang-Refsdal lens and the wide/close planetary case is achieved by setting $\gamma = d^{-2}$ and choosing the origin of the binary lens to be $d - d^{-1}$ from the primary lens, and including in the Chang-Refsdal approximation the magnification of the unperturbed image created by the primary on the other side of the Einstein ring. Note that $\gamma > 1$ corresponds to the planetary caustics for the close topology, whereas $\gamma < 1$ corresponds to the wide topology.

The properties of the caustics of a Chang-Refsdal lens (and thus the planetary caustics of wide/close planetary lenses) can be studied analytically. Based on expressions from *Bozza* (2000), *Han* (2006) has studied the scaling of the planetary caustics. The overall size of the wide (d > 1) planetary caustic is approximately $\propto q^{1/2}d^{-2}$, and the length-to-width ratio is $\sim 1 + d^{-2}$. The shape is independent of q in this approximation. The overall size of the close (d < 1) planetary caustics are approximately $\propto q^{1/2}d^3$, and their shape is also independent of q. The vertical separation between the two planetary caustics in the close topology is $\propto q^{1/2}d^{-1}$.

For the resonant case ($d_c \leq d \leq d_w$) there exists a single, relatively large caustic with six cusps. For fixed q, the resonant caustic is larger than either the central or planetary caustics. The large size of these caustics results in a large cross-section and thus an enhanced detection probability. Indeed, in the first planet detected by microlensing, the source crossed a resonant caustic. The large size also means that the lightcurve deviation can last a significant fraction of the duration of the event. However, resonant caustics are also "weak" in the sense that for a large fraction of the area interior to or immediately outside the caustic, the excess magnification relative to a single lens is small (see Fig. 7). The exceptions to this are source positions in the vicinity of the cusp located on the planet/star axis pointing toward the planet, and source positions near the "back end" of the caustic near the position of the primary, which are characterized by large negative deviations relative to the single-lens magnification. The precise shape of the resonant caustic depends sensitively on d, and thus small changes in the value of d lead to large changes in the caustic morphology, as can be seen in Fig. 5. As a result, the effects of orbital motion, which result in a change in d over the course of the event, are expected to be more important for resonant caustic perturbations. The size of resonant caustics scales as $q^{1/3}$, in contrast to planetary caustics, which scale as $q^{1/2}$, and central caustics, which scale as q.

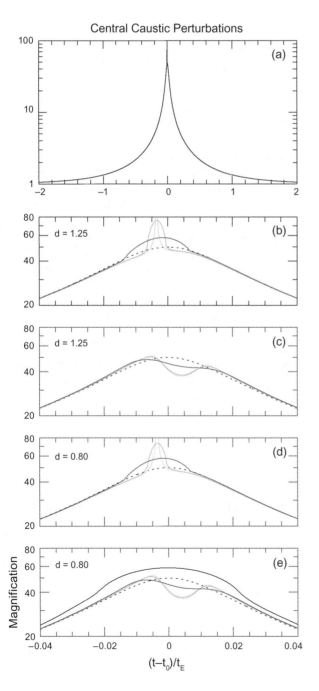

3. PRACTICE OF MICROLENSING

3.1. Lightcurves and Fitting

The apparent relative motion between the lens and the source gives rise to a time-variable magnification of the source: a microlensing event. Is is often (but not always; see section 3.2) a good approximation that the source, lens, and observer are in uniform, rectilinear motion, in which case the angular separation between the lens and source as a function of time can be written as

$$u(t) = \left(\tau^2 + u_0^2\right)^{1/2} \quad (28)$$

where $\tau \equiv (t - t_0)/t_E$, t_0 is the time of closest alignment, which is also the time of maximum magnification, u_0 is the impact parameter of the event, and t_E is the Einstein ring crossing time

$$t_E \equiv \frac{\theta_E}{\mu_{rel}} \quad (29)$$

where μ_{rel} is the relative lens-source proper motion. Figure 3 shows the magnification as a function of time for a microlensing event due to a single lens, with impact parameters of $u_0 = 0.01, 0.1, 0.2, \ldots, 0.9, 1.0$, which serve to illustrate the variety of lightcurve shapes. Of course, what is observed is not the magnification, but the flux of a photometered source as a function of time, which is given by

$$F(t) = F_s A(t) + F_b \quad (30)$$

Here F_s is the flux of the microlensing star, and F_b is the flux of any unresolved light (or "blended light") that is not

Fig. 8. Example lightcurves of planetary perturbations arising from the source passing close to the central caustic in a high-magnification event, for a planet/star mass ratio of q = 0.001. Panel (**a**) shows the overall lightcurve. The impact parameter of the event with respect to the primary lens is $u_0 = 0.02$, corresponding to a peak magnification of $A_{max} \sim u_0^{-1} = 50$. Panels (**b**)–(**e**) show zooms of the lightcurve peak. Two different cases are shown, one case of the wide planetary companion with d = 1.25 [(**b**),(**c**)], and a close planetary companion with d = 0.8 [(**d**),(**e**)]. These two cases satisfy $d \leftrightarrow d^{-1}$ and demonstrate the close/wide degeneracy. The source passes close to the central caustic; two example trajectories are shown in Fig. 6 and the resulting lightcurves including the planetary perturbations are shown in panels (**b**)–(**e**). The dotted line shows the magnification with no planet, whereas the solid lines show the planetary perturbations with source sizes of $\rho_* = 0, 0.003$, and 0.01 (lightest to darkest). In panel (**e**), the lightcurve for $\rho_* = 0.03$ is also shown. In this case, the primary lens transits the source, resulting in a "smoothed" peak. Although the planetary deviation is largely washed out, it is still detectable with sufficiently precise photometry.

being lensed. The latter can include light from a companion to the source, light from unrelated nearby stars, light from a companion to the lens, and (most interestingly) light from the lens itself. Microlensing experiments are typically carried out toward crowded fields in order to maximize the event rate, and therefore one often finds unrelated stars blended with the microlensed source for typical ground-based resolutions of ~1″. Even in the most crowded bulge fields, most unrelated background stars are resolved at the resolution of the Hubble Space Telescope (HST). Figure 10 shows the fields of two events as observed from the ground with typical seeing, with HST, and with groundbased adaptive optics (AO).

The observed flux as a function of time for a microlensing event due to a single lens can be fit by five parameters: t_0, u_0, t_E, F_s, and F_b. It is important to note that several of these parameters tend to be highly degenerate. There are only four gross observable properties of a single-lens curve: t_0, the overall timescale of the event (i.e., t_{FWHM}), and the peak and baseline fluxes. Thus u_0, t_E, F_s, and F_b tend to be highly correlated, and are only differentiated by relatively subtle differences between lightcurves with the same values of the gross observables but different values of u_0, t_E, F_s, and F_b (*Wozniak and Paczynski*, 1997; *Han*, 1999; *Dominik*, 2009). As a result of these degeneracies, when fitting to data it is often useful to employ an alternate parameterization of the single-lens model that is more directly tied to these gross observables, in order to avoid strong covariances between the model parameters.

In practice, several different observatories using several different filter bandpasses typically contribute data to any given observed microlensing event. Since the flux of the source and blend will vary depending the specific bandpass, and furthermore different observatories may have different resolutions and thus different amounts of blended light, one must allow for a different source and blend flux for each filter/observatory combination. Thus the total number of parameters for a generic model fit to an observed dataset is $N_{nl} + 2 \times N_O$, where N_{nl} is the number of (nonlinear) parameters required to specify the magnification as a function of time, and N_O is the total number of independent datasets. Since the observed flux is a linear function of F_s and F_b, for a given set of N_{nl} parameters that specify $A(t)$, the set of source and blend fluxes can be found trivially using a linear least-squares fit. Thus in searching for the best-fit model, one typically uses a hybrid method in which the best-fit nonlinear parameters (e.g., t_0, u_0, t_E for a single lens) are varied using, e.g., a downhill-simplex, Markov Chain Monte Carlo (MCMC), or grid-search method, and the specific best-fit F_s and F_b values for each trial set of nonlinear parameters are determined via linear least squares.

The addition of a second lens component increases the number of model parameters by (at least) three, and enormously increases the complexity. As discussed in section 2.3, the magnification pattern (magnification as a

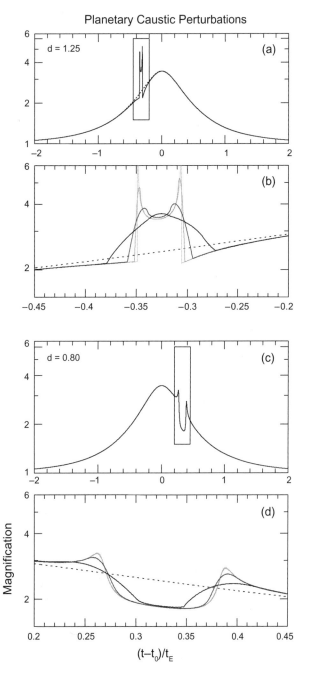

Fig. 9. Example lightcurves of planetary perturbations arising from the source passing close to the planetary caustic for a planet/star mass ratio of q = 0.001. Panels **(a)** and **(c)** show the overall lightcurves, whereas panels **(b)** and **(d)** show zooms of the planetary deviation. Two cases are shown, one case of the wide planetary companion with d = 1.25 [**(a)**,**(b)**], and a close planetary companion with d = 0.8 [**(c)**,**(d)**]. In both cases, the impact parameter of the event with respect to the primary lens is u_0 = 0.3. The trajectories for the lightcurves displayed are shown in Fig. 5. The dotted line shows the magnification with no planet, whereas the solid lines show the planetary perturbations with source sizes of ρ_* = 0, 0.003, 0.01, and 0.03 (lightest to darkest).

function of source position) for a binary lens is described two parameters: d, the separation of lens components in units of the Einstein ring of the lens, and q, the mass ratio of the lens. Four additional parameters specify the trajectory of the source through the magnification patterns as a function of time: t_0, u_0, t_E, and α. The first three are analogous to the single lens case: u_0 is the impact parameter of the trajectory from the origin of the lens in units of θ_E, t_0 is the time when $u = u_0$, and t_E is the Einstein ring crossing time. Finally, α specifies the angle of the trajectory relative to the binary-lens axis. Thus $6 + 2 \times N_O$ parameters are required to specific the lightcurve arising from a generic, static binary lens. There are many possibles choices for the origin of the lens, as well as the mass used for the normalization of θ_E. The optimal choice depends on the particular properties of the lens being considered, i.e., wide stellar binary, close stellar binary, or planetary system (*Dominik*, 1999b). For planetary systems, one typically chooses the location of the primary for the origin and normalizes θ_E the total mass of the system.

Lightcurves arising from binary lenses exhibit an astonishingly diverse and complex phenomenology. While this diversity makes for a rich field of study, it complicates the interpretation of observed lightcurves mightily, for several reasons. First, other than a few important exceptions (i.e., for planetary caustic perturbations; see section 2.1), the salient features of binary and planetary lightcurves have no direct relationship to the canonical parameters of the underlying model. Thus it is often difficult to choose initial guesses for the fit parameters, and even if a trial solution is found, it is difficult to be sure that all possible minima have been located. Second, small changes in the values of the canonical parameters can lead to dramatic changes in the resulting lightcurve. In particular, the sharp changes in the magnification that occur when the source passes close to or crosses a caustic can make any goodness-of-fit statistic such as χ^2 very sensitive to small changes in the underlying parameters. This, combined with the shear size of parameter space, makes brute-force searches difficult and time-consuming.

The complicated and highly corrugated shape of the χ^2 surface also causes many of the usual minimization routines (i.e., downhill simplex) to fail to find the global or even local minimum. These difficulties are compounded by the fact that the magnification of a binary lens is nonanalytic and time-consuming to calculate when finite source effects are important.

Aside from their diverse phenomenology, binary lens lightcurves also have the important property that they can be highly degenerate, in the sense that two very different underlying lens models can produce very similar lightcurves. These degeneracies can be accidental, in the sense that with relatively poor quality data or incomplete coverage of the diagnostic lightcurve features, otherwise distinguishable lightcurves can provide equally good statistical fits to a given dataset (*Dominik and Hirshfeld*, 1996; *Dominik*, 1999a; *Albrow et al.*, 1999b). This is particularly problematic for caustic-crossing binary-lens lightcurves in which only one (i.e., the second) caustic crossing is observed. In this instance, it is typically the case that very different models can fit a given dataset. [An important corollary is that, given an observed first caustic crossing, it is very difficult to predict the time of the second caustic crossing well in advance. For further discussion, see *Albrow et al.* (1999b) and *Jaroszyński and Mao* (2001).] The more insidious degeneracies, however, are those that arise from mathematical symmetries in the lens equation itself (*Dominik*, 1999b). For example, in the limit of a very widely separated ($d_w \gg 1$) or very close ($d_c \ll 1$) binary lenses, Taylor expansion reveals that the lens equations in the two cases are identical (up to an overall coordinate translation) to on the order of d_c^2 or d_w^{-2} for $d_w \leftrightarrow d_c^{-1}(1+q)^{1/2}$ (*Dominik*, 1999b; *Albrow et al.*, 2002; *An*, 2005). Thus the magnifications are also identical to this order in these two cases. Note that for q_c, $q_w \ll 1$, this degeneracy is simply $d_c \leftrightarrow d_w^{-1}$. This degeneracy was discussed in the context of planetary lenses in section 2.3. It is not known if there exist analogous degeneracies for more complex (i.e., triple) lenses. These mathematical degeneracies are insidious because, if the model is deep within

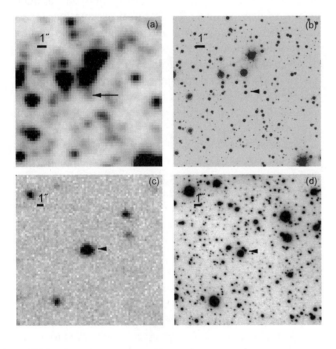

Fig. 10. (a) A 15" × 15" I-band image of the field of planetary microlensing event OGLE-2005-BLG-071 obtained with the OGLE 1.3-m Warsaw telescope at Las Campanas Observatory in Chile. (b) Same field as the top panel, but taken with the Advanced Camera for Surveys instrument on the Hubble Space Telescope in the F814W filter. (c) 22" × 22" H-band image of the field of planetary microlensing event MOA-2008-BLG-310 obtained with the CTIO/SMARTS2 1.3-m telescope at Cerro Tololo InterAmerican Observatory in Chile. (d) Same field and filter as the panel above, but taken with the NACO instrument on VLT. In all panels, the arrow indicates the microlensing target.

the limits where the degeneracies manifest themselves, it is essentially impossible, even in principle, to distinguish between the degenerate models with the photometric data alone [but see *Gould and Han* (2000) for a way to resolve the close/wide degeneracy with astrometry]. Both classes of degeneracies complicate the fitting of observed lightcurves. First, the existence of these degeneracies implies that simply finding a fit does not imply that the fit is unique. Second, these degeneracies may complicate the physical interpretation of observed lightcurves, because the degenerate models generally imply very different lens properties.

A number of authors have developed routines to locate fits to observed binary lens lightcurves. *Mao and DiStefano* (1999) compiled a large library of point-source binary lens lightcurves, and classified these according to their salient features, such as, e.g., the number of peaks and time between peaks. They then match the features in the observed lightcurves to those in the library to find trial solutions for minimization routines. This approach effectively works by establishing a mapping between the lightcurve features and the canonical parameters. *DiStefano and Perna* (1997) adopted a conceptually similar approach, where they decomposed the observed lightcurve into a linear combination of basis functions, then compared the resulting coefficients of this fit to those found for a library of events, again to identify promising trial solutions. Similarly, *Vermaak* (2003) used artificial neural networks to identify promising regions of parameter space. While these methods can in principle be applied to any binary-lens lightcurve, because they are not intrinsically systematic, it is difficult to be sure that all possible fits have been identified.

Albrow et al. (1999b) and *Cassan* (2008) developed algorithms to find a complete set of solutions for binary-lens lightcurves where the source crosses a caustic. These algorithms are robust in the sense that they will identify all possible fits to such lightcurves, and thus uncover all possible degenerate solutions (e.g., *Afonso et al.*, 2000). Unfortunately, they are also fairly user-intensive, and are obviously only applicable to a limited subset of events.

Currently, the most robust and efficient approaches to fitting observed binary-lens lightcurves use some variant of a hybrid approach, e.g., a grid search over those nonlinear parameters that are not simply related to the lightcurve features are thus poorly behaved (such as the mass ratio q, projected separation d, and/or the angle of the trajectory α), combined with a downhill-simplex, steepest-descent, or MCMC fit to the remaining nonlinear parameters that are more directly related to the observed lightcurve features. The linear parameters (such as the source and blend flux) are trivially fit for each trial solution. Often when there is some prior gross knowledge of the approximate model parameters (i.e, a wide or close planetary or binary lens) a judicious choice of parameterization guided by the inherent magnification properties of the underlying lens geometry (i.e., the caustic structure of the lens) can both speed up the fit and improve the robustness of the fitting. Once trial solutions are identified, they can be more carefully explored using, e.g., Markov Chain Monte Carlo techniques (see, e.g., *Gould et al.*, 2006; *Dong et al.*, 2007). Essentially all these approaches must be modified (although sometimes trivially) to include higher-order effects when these provide significant perturbations to the observed lightcurve (see section 3.2).

In general, no robust, practical, universal, and efficient algorithm exists for fitting an arbitrary binary lens lightcurve in an automated way that is not highly user-intensive. For higher-multiplicity (i.e., triple) lenses, the algorithms are generally even less well developed. As mentioned in the introduction, this is likely the current single biggest impediment to the progress of microlensing planet searches. Thus there is a urgent and growing need for the development of (more automated) analysis software.

3.2. Higher-Order Lightcurve Effects

As reviewed in section 3.1, the simplest model of the observed lightcurve arising from an isolated lens can be described by $3 + 2 \times N_O$ parameters: the three parameters that describe the magnification as a function of time (u_0, t_0, t_E), and a blend and source flux for each of the N_O observatory/filter combinations. This model, which accurately describes the vast majority of observed microlensing lightcurves, is derived under a number of assumptions, some of which may break down under certain circumstances. When a lightcurve deviates from this classic "Paczynski" or point-source single-lens form, it is generally classified as "anomalous." Not surprisingly, it is frequently the case that a robust measurement of these "anomalies" allows one to infer additional properties of the primary, planet, or both (see section 3.3).

The most obvious and most common anomaly is when the lens is not isolated. Each additional lens component requires three additional parameters (e.g., mass ratio, projected separation, and angle between the additional component and the primary), and so the lightcurve arising from a lens consisting of N_l masses requires a total of $3 \times N_l + 2 \times N_O$ parameters. (If the additional component is very far away or very close to the primary lens, the effects of the additional component can be described by only two additional parameters: either the shear due to the wide companion or the quadrupole moment of the lens for the close companion, or the angle of the trajectory with respect to the projected binary axis.) Roughly ~10% of all microlensing lightcurves are observed to be due to binaries. In fact, the true fraction of binary lenses is likely considerably higher, since the effects of a binary companion are only apparent when d is on the order of unity.

3.2.1. Finite source effects. The second most common anomaly occurs when the assumption of a point source breaks down (*Gould*, 1994a; *Witt and Mao*, 1994; *Witt*, 1995). As already discussed in section 2.2.4, this assumption breaks down when the curvature of the magnification pattern is significant over the angular size of the source, where "significant" depends on the photometric precision with which the lightcurve is measured. Generally, the center of the source must pass within a few angular source radii θ_* of a caustic in order for finite source effects to be important. The magnitude of finite source effects are parameterized by the angular size of the source in units of the angular Einstein ring radius, $\rho_* \equiv \theta_*/\theta_E$. For typical main-sequence sources in the galactic bulge, ρ_* is on the order of 10^{-3}, whereas for typical giant sources, ρ_* is on the order of 10^{-2}. In addition, the amount of (filter-dependent) limb darkening can also have an important effect on the precise shape of the portions of the lightcurve affected by finite-source effects.

3.2.2. Parallax. Another common anomaly occurs when the assumptions of cospatial and/or nonaccelerating observers break down. For example, the usual expression for u, the angular separation between the lens and source in units of θ_E, presented in equation (28) assumes that the observer (as well as the lens and source) are moving with constant velocity. The assumption that the values of t_0 and u_0 are the same for all observers implies that they are all cospatial. When either of these assumptions break down, this is generally termed (microlens) parallax. This occurs in one of three ways. First, when the duration of the microlensing event is a significant fraction of a year, the acceleration of Earth leads to a significant nonuniform and/or nonrectilinear trajectory of the lens relative to the source, which leads to deviations in the observed lightcurve (*Gould,* 1992). This is referred to as "orbital parallax." Second, for sources very close to a caustic, small parallax displacements due to the differences in the perspective leads to differences in the magnifications seen by observers located at different observatories for fixed time (*Holz and Wald,* 1996). This is commonly known as "terrestrial parallax." Finally, if the observers are separated by a significant fraction of an AU, for example if the event is simultaneously observed from Earth and a satellite in a solar orbit, then the differences in the magnification are large even for moderate magnification (*Refsdal,* 1966; *Gould,* 1994b). This is known as "satellite parallax." In all three cases, the magnitude of the effect depends on the size of relevant length scale that gives rise to the different or changing perspective (i.e., the projected separation between the observatories, the projected size of Earth's orbit, or the projected separation between the satellite and Earth) relative to $\tilde{r}_E \equiv D_{rel}\theta_E$, the angular Einstein ring radius projected on the plane of the observer (see Fig. 2).

3.2.3. Xallarap. Analogously to parallax, if the source undergoes significant acceleration over the course of the event due to a binary companion, this will give rise to deviations from the canonical lightcurve form (*Dominik,* 1998). This effect is commonly known as "xallarap," i.e., parallax spelled backwards, to highlight the symmetry between this effect and orbital parallax. In fact, it is always possible to *exactly* mimic the effects of orbital parallax with a binary source with the appropriate orbital parameters (*Smith et al.,* 2003). For sufficiently precise data, such that the parallax/xallarap parameters are well constrained, these two scenarios can be distinguished because it would be *a priori* unlikely for a binary source to have exactly the correct parameters to mimic the effects of orbital parallax (*Poindexter et al.,* 2005).

The magnitude of xallarap effect depends on the semimajor axis of the binary relative to $\hat{r}_E \equiv D_s\theta_E$, the Einstein ring radius projected on the source plane.

3.2.4. Orbital motion. For binary or higher multiplicity lenses, relative orbital motion of the components can also give rise to deviations from the lightcurve expected under the usual static lens assumption (*Dominik,* 1998). For the general case of a binary lens and a Keplerian orbit, an additional five parameters beyond the usual static-lens parameters are needed to specify the lightcurve (*Dominik,* 1998). Typically, however, only two additional parameters can be measured: the two components of the relative projected velocity of the lenses. These two components can be parameterized by the rate of the change of the binary projected binary axis $\dot{d}$ and the angular rotation rate of the binary ω. The effect of the latter is simply to rotate that magnification pattern of the lens on the sky, whereas the former results in a change in the lens mapping itself, and thus a change in the caustic structure and magnification pattern.

3.2.5. Binary sources. If the source is a binary and the lens happens to pass sufficiently close to both sources, the lightcurve can exhibit a deviation from the generic single-lens, single-source form (*Griest and Hu,* 1992). For static binaries, the morphology of the deviation depends on the flux ratio of the sources, their projected angular separation in units of θ_E, and the impact parameter of the lens from the source companion. By convention, such "binary source" effects are conceptually distinguished from xallarap effects by the nature of the deviation: If the deviation is caused by the source companion being significantly magnified by the lens, is called a binary source effect; if it is caused by the acceleration due to the companion, it is called xallarap. Of course, depending on the parameters of the source and lens, lightcurves can exhibit only binary source effects, only xallarap, or both xallarap and binary-source effects.

3.2.6. Other miscellaneous effects. A number of additional high-order effects have been discussed in the literature, for example, the effects of the finite physical size of the lens (*Bromley,* 1996; *Agol,* 2002). In most cases, these higher-order effects are expected to be unobservable and/or extremely rare.

3.3. Properties of the Detected Systems

For microlensing events due to single, isolated lenses, the parameters that can routinely be measured are the time of maximum magnification t_0, the impact parameter of the event in units of the Einstein ring radius u_0, and the Einstein timescale t_E, along with a source flux F_s and blend flux F_b for each observatory/filter combination. Of these parameters, u_0 and t_0 are simply geometrical parameters and contain no physical information about the lens. The Einstein timescale t_E is a degenerate combination of the lens mass M, the relative lens-source parallax π_{rel}, and the relative lens-source proper motion μ_{rel}. Therefore it is not possible to uniquely determine the mass and distance to the lens from a measurement of t_E alone. The blend flux

F_b contains light from any source that is blended with the source, including light from the lens if it is luminous. Unfortunately, for the typical targets toward the galactic bulge, groundbased images typically contain light from other, unrelated sources, and it is not possible to isolate the light from the lens (see Fig. 10). Therefore, in the vast majority of microlensing events, the mass, distance, and proper motion of the lens are unknown.

As discussed below, for binary and planetary microlensing events it is routinely possible to infer the mass ratio q, and d, the instantaneous projected separation between the planet and star in units of θ_E. However, the mass of the planet is typically not known without a constraint on the primary mass. Furthermore, a measurement of d alone provides very little information about the orbit, since θ_E and the inclination, phase, and ellipticity of the orbit are all unknown *a priori*. For these reasons, when microlensing planet searches were first initiated it was typically believed that detailed information about individual systems would be very limited for planets detected in microlensing lightcurves. This apparent deficiency was exacerbated by the perception that the host stars would typically be too distant and faint for follow-up observations.

Fortunately, in reality much more information can typically be gleaned from a combination of a detailed analysis of the lightcurve and follow-up, high-resolution imaging, using the methods outlined below.

3.3.1. Mass ratio and projected separation. First, the requirements for accurately measuring the minimum three additional parameters needed to describe the lightcurves of binary and planetary microlensing events are discussed. These three additional parameters are the aforementioned q, d, and α.

For planetary caustic perturbations, these parameters can essentially be "read off" of the observed lightcurve. In this case, the gross properties of the planetary perturbation can be characterized by three observable quantities: the time of the planetary perturbation $t_{0,p}$, the timescale of the planetary perturbation $t_{E,p}$, and the magnitude of the perturbation δ_p. These quantities then simply and completely specify the underlying parameters q, d, α, up to a two-fold discrete degeneracy in d, corresponding to whether the planet is perturbing the major or minor image, i.e, $d \leftrightarrow d^{-1}$. Since these two situations result in very different types of perturbations (see Fig. 9), this discrete degeneracy is easily resolved (*Gaudi and Gould*, 1997b). If finite source effects are important but not dominant, then there also exists a continuous degeneracy between q and ρ_*, stemming from the fact that in this regime, both determine the width of the perturbation t_p (*Gaudi and Gould*, 1997b). However, this degeneracy is easily broken by good coverage and reasonably accurate photometry in the wings of the perturbation. Finally, there is also a degeneracy between major image (d > 1) planetary caustic perturbations and a certain class of binary-source events, namely those with extreme flux ratio between the two sources. Specifically, it is always possible to find a binary-source lightcurve that can exactly reproduce the observables $t_{0,p}$, $t_{E,p}$, and δ_p (*Gaudi*, 1998). This degeneracy is also easily broken by good coverage and accurate photometry (*Gaudi*, 1998; *Gaudi and Han*, 2004; *Beaulieu et al.*, 2006).

For central or resonant caustic perturbations in high-magnification events, extracting the parameters q, d, α is typically more complicated due to the fact that there is no simple, general relationship between the salient features of the lightcurve perturbation and these these parameters. Thus fitting these perturbations typically requires a more sophisticated approach, as discussed in section 3.1. In addition, there are a number of degeneracies that plague central caustic perturbations. First, as discussed in section 2.3 and illustrated in Figs. 6 and 8, there is a close/wide duality such that the central caustic shape and associated magnification pattern are highly degenerate under the transformation $d \leftrightarrow d^{-1}$ (*Griest and Safizadeh*, 1998; *Dominik*, 1999b). This degeneracy becomes more severe for very close/very wide planets, and in some cases it is essentially impossible to distinguish between the two solutions, even with extremely accurate photometry and dense coverage of the perturbation (e.g., *Dong et al.*, 2009a). There also exists a degeneracy between central caustic planetary perturbations, and perturbations due to very close or very wide binary lenses. Very close binaries have small, asteroid-shaped caustics located at the center-of-mass of the system, whereas very wide binaries have small, asteroid-shaped caustics near the positions of each of the lenses. The gross features of central caustic perturbations can be reproduced by a source passing by the asteroid-shaped caustic produced by a close/wide binary lens (*Dominik*, 1999a; *Albrow et al.*, 2002; *An*, 2005). Since close/wide binary lenses are themselves degenerate under the transformation $d_w \leftrightarrow d_c^{-1}(1+q)^{1/2}$, there is a fourfold degeneracy for perturbations near the peak of high-magnification events. Fortunately, the degeneracy between planetary central caustic perturbations and close/wide binary lens perturbations can be resolved with good coverage of the perturbation and accurate photometry (*Albrow et al.*, 2002; *Han and Gaudi*, 2008; *Han*, 2009).

3.3.2. Einstein ring radius. The requirement for detecting a planet via microlensing is generally that the source must pass reasonably close to the caustics produced by the planetary companion. However, this is also basically the condition for finite source size effects to be important. Thus for most planetary microlensing events, it is possible to infer the angular size of the source in units of the angular Einstein ring radius, $\rho_* \equiv \theta_*/\theta_E$.

The angular size of the source can be estimated by its de-reddened color and magnitude using empirical color-surface brightness relations determined from angular size measurements of nearby stars (*van Belle*, 1999; *Kervella et al.*, 2004). The source flux F_s is most easily determined by a fit to the microlensing lightcurve of the form $F(t) = F_s A(t) + F_b$, as the variable magnification of the source allows one to "de-blend" the source and blend flux. Determining the color of the source in this manner requires measurements in two passbands, and thus while observations are typically focused on a single passband (typically a far-red visible passband such as R or I), it is important to acquire a few points in a second filter to determine the source color. The extinction toward the source can be approximately determined by comparison to nearby red giant clump stars (*Yoo et al.*, 2004), which have a known and essentially constant luminosity and intrinsic color (e.g., *Paczynski and Stanek*, 1998). An error in the extinction affects both the inferred color and magnitude of the source; fortunately, these have opposite and nearly equal effects on the inferred value of θ_* (*Albrow et al.*, 1999a).

Thus for events in which finite source effects are robustly detected, it is possible to measure θ_E (*Gould*, 1994a). This partially breaks the timescale degeneracy, since

$$\frac{M}{D_{rel}} = \frac{c^2}{4G}\theta_E^2 \qquad (31)$$

The distance to the source is typically known approximately from its color and magnitude (and furthermore the overwhelming majority of sources are in the bulge), and so a measurement of θ_E essentially provides a mass-distance relation for the lens.

3.3.3. Light from the lens. Although the majority of the lenses that give rise to microlensing events are distant and low-mass main-sequence stars, most are nevertheless usually sufficiently bright that their flux can be measured to relative precision of ≤10% with moderate-aperture (1–2 m) telescopes and reasonable (10^2–10^4 s) exposure times, provided that the light from the lens can be isolated. The fit to the microlensing lightcurve gives the flux of the source F_s, and the blend flux F_b. The latter contains the flux from any stars that are not being lensed but are unresolved on the image, i.e., blended with the source star. Generally, this blend flux can be decomposed into several contributions

$$F_b = F_l + F_{l,c} + F_{s,c} + \sum_j F_{u,j} \qquad (32)$$

where F_l is the flux from the lens, $F_{l,c}$ is the flux from any (blended) companions to the lens, $F_{s,c}$ is the flux from any (blended) companions to the source, and $F_{u,j}$ is the flux from each unrelated nearby star j that is blended with the source. As illustrated in Fig. 10, at typical groundbased resolutions of 1″ and in the extremely crowded target fields toward the bulge where microlensing surveys are carried out, it is often the case that there are several unrelated stars blended with the source star. Therefore, the lens light cannot be uniquely identified based on such data alone. At the higher resolutions of 0.05–0.1″ available from HST or groundbased AO imaging, essentially all stars unrelated to the source lens are resolved. Since the source and lens must be aligned to ≤θ_E ~ 1 mas for a microlensing event to occur and the typical relative lens-source proper motions are μ_{rel} ~ 5–10 mas yr^{-1} for microlensing events toward the bulge, the lens and source will be blended in images taken within ~10 years of the event, even at the resolution of HST. However, because the microlensing fit gives F_s, the lens flux can be determined by subtracting this flux from the combined unresolved lens + source flux in the high-resolution image, assuming no blended companions to the lens or source (*Bennett et al.*, 2007).

There are several potential complications to this procedure to determine the lens flux. First, F_s determined from the microlensing fit will generally not be absolutely calibrated. Thus the high-resolution photometric data must be "photometrically aligned" to the microlens dataset. Typically this is done by matching stars common to both sets of images. The accuracy of this alignment is usually limited to ~1% due to the small number of common, isolated stars available (*Dong et al.*, 2009b). It may also be that high-resolution images are taken in a different filter than that for which the source flux F_s is determined, necessitating a (model-dependent) color transformation and introducing additional uncertainties (e.g., *Bennett et al.*, 2008). Note that it is possible to avoid this procedure entirely if the high-resolution images can be taken at two different epochs with substantially different source star magnifications. However, since this means at least one epoch must be taken when the source is significantly magnified during the event, this requires target-of-opportunity observations. Finally, any light in excess of the source detected in the high-resolution images may be attributed to close physical companions to the lens or source. In some favorable cases (i.e., high-magnification events), it is possible to exclude these scenarios by the (lack of) second-order effects the companion would produce in the observed lightcurve. For example, a binary companion to the lens produces a caustic that would be detectable in sufficiently high-magnification events, whereas a sufficiently close companion to the source would give rise to xallarap effects that would be detectable in long-timescale events (*Dong et al.*, 2009b).

A measurement of the flux of the lens in a single passband, along with a model for extinction as a function of distance and a mass-luminosity relationship, gives a mass-distance relationship for the lens (*Bennett et al.*, 2007). A second measurement of the flux in a different passband can provide a unique mass and distance to the lens, subject to the uncertainties in the intrinsic color as a function of mass and the dust extinction properties as a function of wavelength and distance.

3.3.4. Proper motion. For typical values of μ_{rel} ~ 5–10 mas/yr for microlensing events toward the galactic bulge, after a few years, the lens and source will be displaced by ~0.01 arcsec. For luminous lenses, and using space telescope or AO imaging, it is possible measure the relative lens-source proper motion, either by measuring the elongation of the PSF or by measuring the difference in the centroid in several filters if the lens and source have significantly different colors (*Bennett et al.*, 2007). The proper motion can be combined with the timescale to give the angular Einstein ring radius, $\theta_E = \mu_{rel} t_E$.

3.3.5. Microlens parallax. For some classes of events, it is possible to obtain additional information about the lens by measuring the microlensing parallax, π_E, a vector with magnitude $|\pi_E| = AU/\tilde{r}_E$, and direction of the relative lens-source proper motion. Recall $\tilde{r}_E \equiv D_{rel}\theta_E$ is the Einstein ring radius projected onto the observer plane. As discussed in section 3.2, microlens parallax effects arise in one of three varieties: orbital parallax due to the acceleration of Earth during the event, terrestrial parallax in high-magnification events observed by non-cospatial observers, and satellite parallax for events observed from the ground and a satellite in solar orbit.

Orbital parallax deviations are generally only significant for events with timescales that are a significant fraction of a year, and so long as compared to the median timescale of ~20 days. Furthermore, the deviations due to orbital parallax are subject to an array of degeneracies (*Gould et al.*, 1994; *Smith et al.*, 2003; *Gould*, 2004b), which can hamper the ability to extract unique microlens parallax parameters. The severity of these degeneracies depend on the particular parameters of the event in question, but for most events it is the case that the only robustly measured effect in the lightcurve is an overall asymmetry, which only yields one projection of π_E, namely that in the direction perpendicular to the instantaneous Earth-Sun acceleration vector at the time of the event (*Gould et al.*, 1994). If the direction of the relative lens-source proper motion vector μ_{rel} can be independently determined from the proper motion of a luminous lens, then it is possible to determine the full π_E vector, since π_E is parallel to μ_{rel}.

Orbital parallax measurements made from two observatories are also subject to a number of degeneracies, which have been studied by several authors (*Refsdal*, 1966; *Gould*, 1994b, 1995; *Boutreux and Gould*, 1996; *Gaudi and Gould*, 1997a). These can be resolved in a number of ways (*Gould*, 1995, 1999; *Dong et al.*, 2007), including observing from a third observatory, which allows one to uniquely "triangulate" the parallax effects (*Gould*, 1994b). Similarly, terrestrial parallax measurements from only two observatories are subject to degeneracies that can be resolved with simultaneous observations from a third observatory that is not colinear with the other two.

A measurement of the microlens parallax allows one to partially break the timescale degeneracy and provides a mass-distance relation for the lens

$$MD_{rel} = \frac{c^2}{4G}\tilde{r}_E^2 \quad (33)$$

3.3.6. Orbital motion of the planet. In at least two cases, the orbital motion of the planet during the microlensing event has been detected (*Dong et al.*, 2009a; *Bennett et al.*, 2010). The effects of orbital motion generally allow the measurement of the two components of the projected velocity of the planet relative to the primary star. If an external measurement of the mass of the lens is available, and under the assumption of a circular orbit, these two components of the projected velocity completely specify the full orbit of the planet (including inclination), up to a two-fold degeneracy (*Dong et al.*, 2009a). In some cases, higher-order effects of orbital motion can be used to break this degeneracy and even constrain the ellipticity of the orbit (*Bennett et al.*, 2010).

3.3.7. Bayesian analysis. In the cases when only t_E, q, and d can be measured, constraints on the mass and distance to the lens (and therefore mass and semimajor axis of the planet) must rely on a Bayesian analysis that incorporates priors on the distribution of microlens masses, distances, and velocities (e.g., *Dominik*, 2006; *Dong et al.*, 2006).

3.3.8. Complete solutions. In many cases, several of these pieces of information can be measured in the same event, often providing complete or even redundant measurements of the mass, distance, and transverse velocity of the event. For example, a measurement of θ_E from finite source effects, when combined with a measurement of $\tilde{r}_E$ from microlens parallax, yields the lens mass

$$M = \left(\frac{c^2}{4G}\right)\tilde{r}_E \theta_E \quad (34)$$

distance

$$D_l^{-1} = \frac{\theta_E}{\tilde{r}_E} + D_s^{-1} \quad (35)$$

and transverse velocity (*Gould*, 1996).

3.4. Practical Aspects of Current Microlensing Searches

The microlensing event rate toward the galactic bulge is $O(10^{-6})$ events per star per year (*Paczynski*, 1991). In a typical field toward the galactic bulge, the surface density of stars is ~10^7 stars per deg² to I ~ 20 (*Holtzman et al.*, 1998). Thus to detect ~100 events per year, ~10 deg² of the bulge must be monitored. Until relatively recently, large-format CCD cameras typically had fields of view of ~0.25 deg², and thus ~40 pointings were required and so fields could only be monitored once or twice per night. While this cadence is sufficient to detect the primary microlensing events, it is insufficient to detect and characterize planetary perturbations, which last a few days or less.

As a result, microlensing planet searches have operated using a two-stage process. The Optical Gravitational Lens Experiment (OGLE) (*Udalski*, 2003) and the Microlensing Observations in Astrophysics (MOA) (*Sako et al.*, 2008) collaborations monitor several tens of square degrees of the galactic bulge, reducing their data in real time in order to alert a microlensing event in progress. A subset of these alerted events are then monitored by several follow-up collaborations, including the Probing Lensing Anomalies NETwork (PLANET) (*Albrow et al.*, 1998), RoboNet (*Tsapras et al.*, 2009), the Microlensing Network for the Detection of Small Terrestrial Exoplanets (MiNDSTEp), and the Microlensing Follow Up Network (μFUN) (*Yoo et al.*, 2004) collaborations. Since only individual microlensing events are monitored, these teams can achieve the sampling and photometric accuracy necessary to detect planetary perturbations. In fact, the line between the "alert" and "follow-up" collaborations is now somewhat blurry, both because the MOA and OGLE collaborations monitor some fields with sufficient cadence to detect planetary perturbations, and because there is a high level of communication between the collaborations, such that the observing strategies are often altered in real time based on available information about ongoing events.

There are two conceptually different channels by which planets can be detected with microlensing, corresponding to whether the planet is detected through perturbations due to the central caustic (as shown in Fig. 8), or perturbation due to the planetary caustic(s) (as shown in Fig. 9). Because for any given planetary system the planetary caustics are always larger than the central caustics, the majority of planetary perturbations are caused by planetary caustics. Thus searching for planets via the influence of the planetary caustics is termed the "main channel." However, detecting planets via the main channel requires substantial commitment of resources because the unpredictable nature of the perturbation requires dense, continuous sampling, and furthermore the detection probability per event is relatively low so many events must be monitored. Detecting planets via their central caustic perturbations requires monitoring high-magnification primary events near the peak of the event. In this case, the trade-off is that although high-magnification events are rare (a fraction $\sim 1/A_{max}$ of events have maximum magnification and $\geq A_{max}$), they are individually very sensitive to planets. The primary challenge lies with this channel lies with identifying high-magnification events in real time.

Which approach is taken depends on the resources that the individual collaborations have available. The PLANET collaboration has substantial access to 0.6–1.5-m telescopes located in South Africa, Perth, and Tasmania. With these resources, they are able to monitor dozens of events per season, and so are able to search for planets via the main channel. This tactic led to the detection of the first cool rocky/icy exoplanet OGLE-2005-BLG-390Lb (*Beaulieu et al.*, 2006).

On the other hand, the µFUN collaboration uses a single 1-m telescope in Chile to monitor promising alerted events in order try to identify high-magnification events substantially before peak. When likely high-magnification events are identified, the other telescopes in the collaboration are then engaged to obtain continuous coverage of the lightcurve during the high-magnification peak. High-magnification events often reach peak magnitudes of $I \leq 15$, and thus can be monitored with relatively small apertures (0.3–0.4 m). This allows amateur astronomers to contribute to the photometric follow-up. Indeed, over half the members of the µFUN collaboration are amateurs.

4. FEATURES OF THE MICROLENSING METHOD

The unique way in which microlensing finds planets leads to some useful features, as well as some (mostly surmountable) drawbacks. Most of the features of the microlensing method can be understood simply as a result of the fact that planet detection relies on the direct perturbation of images by the gravitational field of the planet, rather than on light from the planet, or the indirect effect of the planet on the parent star.

4.1. Peak Sensitivity Beyond the Snow Line

The peak sensitivity of microlensing is for planet-star separations of $\sim r_E$, which corresponds to equilibrium temperatures of

$$T_{eq} = 287 \text{ K} \left(\frac{L}{L_\odot}\right)^{1/4} \left(\frac{r_E}{AU}\right)^{-1/2} \sim 70 \text{ K} \left(\frac{M}{0.5 M_\odot}\right) \quad (36)$$

where the rightmost expression assumes $L/L_\odot = (M/M_\odot)^5$, $D_l = 4$ kpc, and $D_s = 8$ kpc. Thus microlensing is most sensitive to planets in the regions beyond the "snow line," the point in the protoplanetary disk exterior to which the temperature is less than the condensation temperature of water in a vacuum (*Lecar et al.*, 2006; *Kennedy et al.*, 2007; *Kennedy and Kenyon*, 2008). Giant planets are thought to form in the region immediately beyond the snow line, where the surface density of solids is highest (*Lissauer*, 1987).

Is microlensing sensitive to habitable planets? For assumptions above, and further assuming the habitable zone is centered on $a_{HZ} = AU(L/L_\odot)^{1/2}$, the projected separation of a planet in the habitable zone in units of θ_E is $d_{HZ} \sim 0.25(M/M_\odot)^2$. Thus for typical hosts of $M \leq M_\odot$, the habitable zone is well inside the Einstein ring radius (*Di Stefano*, 1999). Since microlensing is much less sensitive to planets with separations much smaller than the Einstein ring radius as these can only perturb highly demagnified images, it is much less sensitive to planets in the habitable zones of their parent stars (*Park et al.*, 2006) for typical events. However, it is important to note that this is primarily a statistical statement about the typical sizes of the angular Einstein ring radii of microlensing events, rather than a statement about any intrinsic limitations of the microlensing method. A fraction of events have substantially smaller Einstein ring radii, and for these events there is significant sensitivity to habitable planets. In particular, lenses closer to the observer have smaller Einstein ring radii and so microlensing events from nearby stars will have more sensitivity to habitable planets (*Gaudi et al.*, 2008b). Indeed, a spacebased microlensing planet search mission (described in section 6) will have significant sensitivity to habitable planets, primarily due to the large number of events being searched for planets (*Bennett et al.*, 2008).

4.2. Sensitivity to Low-Mass Planets

The amplitudes of the perturbations caused by planets are typically large, $\geq 10\%$. Furthermore, although the durations of the perturbations get shorter with planet mass (as $\sqrt{m_p}$)

and the probability of detection decreases (also roughly as $\sqrt{m_p}$), the amplitude of the perturbations are independent of the planet mass. This holds until the "zone of influence" of the planet, which has a size $\sim \theta_{E,p}$, is smaller than the angular size of the source θ_*. When this happens, the perturbation is "smoothed" over the source size, as demonstrated in Figs. 8 and 9. For typical parameters, $\theta_{E,p} \sim \mu as(m_p/M_\oplus)^{1/2}$, and for a star in the bulge, $\theta_* \sim \mu as(R_*/R_\odot)$. This "finite source" suppression essentially precludes the detection of planets with mass $\lesssim 5\, M_\oplus$ for clump giant sources in the bulge with $R_* \sim 10\, R_\odot$ (*Bennett and Rhie*, 1996). For main-sequence sources ($R \sim R_\odot$), finite source effects become important for planets with the mass of the Earth, but does not completely suppress the perturbations and render then undetectable until masses of $\sim 0.02\, M_\oplus \sim 2\, M_{Moon}$ for main-sequence sources (*Bennett and Rhie*, 1996; *Han et al.*, 2005). Thus microlensing is sensitive to Mars-mass planets and even planets a few times the mass of the Moon, for sufficiently small source sizes.

4.3. Sensitivity to Long-Period and Free-Floating Planets

Since microlensing can "instantaneously" detect planets without waiting for a full orbital period, it is immediately sensitive to planets with very long periods. Although the probability of detecting a planet decreases for planets with separations larger than the Einstein ring radius because the magnifications of the images decline, it does not drop to zero. For events in which the primary star is also detected, the detection probability for very wide planets with $d \gg 1$ is $\sim (\theta_{E,p}/\theta_E)d^{-1} = q^{1/2}d^{-1}$ (*Di Stefano and Scalzo*, 1999b). Indeed, since microlensing is directly sensitive to the planet mass, planets can be detected even without a primary microlensing event (*Di Stefano and Scalzo*, 1999a). Even free-floating planets that are not bound to any host star are detectable in this way (*Han et al.*, 2005). Microlensing is the only method that can detect old, free-floating planets. A significant population of free-floating planets planets is a generic prediction of most planet formation models, particular those that invoke strong dynamical interactions to explain the observed eccentricity distribution of planets (*Goldreich et al.*, 2004; *Juric and Tremaine*, 2008; *Ford and Rasio*, 2008).

4.4. Sensitivity to Planets Orbiting a Wide Range of Host Stars

The hosts probed by microlensing are simply representative of the population of massive objects along the line of sight to the bulge sources, weighted by the lensing probability. Figure 11 shows a model by *Gould* (2000) for the microlensing event rate toward the galactic bulge as a function of the host star mass for all lenses, and also broken down by the type of host, i.e., star, brown dwarf, or remnant. The sensitivity of microlensing is weakly dependent on the host star mass, and has essentially no dependence on the host star luminosity. Thus microlensing is about equally sensitive to planets orbiting stars all along the main sequence, from brown dwarfs to the main-sequence turn-off, as well as planets orbiting white dwarfs, neutron stars, and black holes.

4.5. Sensitivity to Planets Throughout the Galaxy

Because microlensing does not rely on light from the planet or host star, planets can be detecting orbiting stars with distances of several kiloparsecs. The microlensing event rate depends on the intrinsic lensing probability, which peaks for lens distances about halfway to the typical sources in the galactic bulge, and the number density distribution of lenses along the line-of-sight toward the bulge, which peaks at the bulge. The event rate remains substantial for lens distances in the range $D_l \sim 1–8$ kpc. Roughly 40% and 60% of microlensing events toward the bulge are expected to be due to lenses in the disk and bulge, respectively (*Kiraga and Paczynski*, 1994). Specialized surveys may be sensitive to planets with $D_l \lesssim 1$ kpc (*Di Stefano*, 2008; *Gaudi et al.*, 2008b), as well as planets in M31 (*Covone et al.*, 2000; *Chung et al.*, 2006; *Ingrosso et al.*, 2009).

4.6. Sensitivity to Multiple-Planet Systems

For low-magnification events, multiple planets in the same system can be detected only if the source crosses the

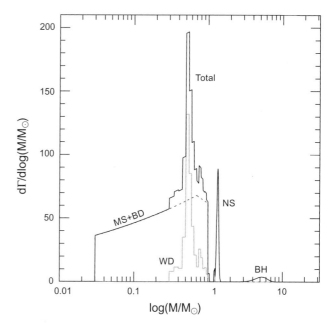

Fig. 11. The rate of microlensing events toward the galactic bulge as a function of the mass of the lens, for main-sequence (MS) stars and brown dwarfs (BD) ($0.03\, M_\odot < M < 1\, M_\odot$) (bold dashed curve) and white dwarfs (WD), neutron stars (NS), and black hole (BH) remnants (solid curves). The total is shown by a bold solid curve. From *Gould* (2000).

planetary caustics of both planets, or equivalently only if both planets happen to have projected positions sufficiently close to the paths of the two images created by the primary lens. The probability of this is simply the product of the individual probabilities, which is typically $O(1\%)$ or less (*Han and Park*, 2002). In high-magnification events, however, individual planets are detected with near-unity probability regardless of the orientation of the planet with respect to the source trajectory (*Griest and Safizadeh*, 1998). This immediately implies all planets sufficiently close to the Einstein ring radius will be revealed in such events (*Gaudi et al.*, 1998). This, along with the fact that high-magnification events are potentially sensitive to very low-mass planets, makes such events excellent probes of planetary systems.

4.7. Sensitivity to Moons of Exoplanets

Because microlensing is potentially sensitive to planets with mass as low as that of a few times the mass of the Moon (*Bennett and Rhie*, 1996), it is potentially sensitive to large moons of exoplanets. A system with a star of mass M, planet of mass m_p, and moon of mass m_m corresponds to a triple lens with the hierarchy $m_m \ll m_p \ll M$. *Bennett and Rhie* (2002), *Han and Han* (2002), and *Han* (2008) have all considered the detectability of moons of various masses, orbiting planets of various masses. Generally, these studies find that massive terrestrial ($m_m \sim M_\oplus$) moons (should they exist) are readily detectable. Less-massive moons, with masses similar to our Moon, are considerably more difficult to detect, but may be detectable in next-generation microlensing planet searches, particularly those from space, in some favorable circumstances (i.e., for sources with small dimensionless source size ρ_*). Analogous to planetary microlensing, moons with projected separation much smaller than the Einstein radius of the planet $r_{E,p} \equiv \theta_{E,p} D_l$ are difficult to detect (*Han*, 2008). Contrary to planetary microlensing, however, the signal of the moon does not diminish for angular separations much greater than $r_{E,p}$ (*Han*, 2008), although of course it becomes less likely that both the planetary and moon signal will be detected simultaneously.

A minimum requirement for a moon to be stable is that its semimajor axis must be less the Hill radius of the planet, which is given by

$$r_H = a_p \left(\frac{m_p}{3M}\right)^{1/3} \qquad (37)$$

where a_p is the semimajor axis of the planet. The ratio of r_H to the Einstein ring radius of the planet is

$$\frac{r_H}{r_{E,p}} \sim d q^{-1/6} \qquad (38)$$

Thus for $q \lesssim 10^{-3}$, detectable moons also (fortuitously) happen to be stable.

5. RECENT HIGHLIGHTS

To date, 10 detections of planets with microlensing have been announced, in 9 systems (*Bond et al.*, 2004; *Udalski et al.*, 2005; *Beaulieu et al.*, 2006; *Gould et al.*, 2006; *Gaudi et al.*, 2008a; *Bennett et al.*, 2008; *Dong et al.*, 2009b; *Janczak et al.*, 2009; *Sumi et al.*, 2010). In addition, there are another seven events with clear, robust planetary signatures that await complete analysis and/or publication. The masses, separations, and equilibrium temperatures of the 10 announced planets are shown in Fig. 12. We can expect a handful of detections per year at the current rate.

Figure 12b demonstrates that the first microlensing planet detections are probing a region of parameter space that has not been previously explored by any method, namely planets beyond the snow line. As a result, although the total number of planets found by microlensing to date is small in comparison to the sample of planets revealed by the radial velocity and transit methods, these discoveries have already provided important empirical constraints on planet formation theories. In particular, the detection of four cold, low-mass (5–20 $M_\oplus$) planets among the sample of microlensing detections indicates that these planets are common (*Beaulieu et al.*, 2006; *Gould et al.*, 2006; *Sumi et al.*, 2010). The detection of a Jupiter/Saturn analog also suggests that solar system analogs are probably not rare (*Gaudi et al.*, 2008a). Finally, the detection of a low-mass planetary companion to a brown-dwarf star suggests that such objects can form planetary systems similar to those around solar-type main-sequence stars (*Bennett et al.*, 2008).

Another important lesson learned from these first few detections is that it is possible to obtain substantially more information about the planetary systems than previously thought. In all 10 cases, finite source effects have been detectable and so it has been possible to measure θ_E, which yields a mass-distance relation for the primary (see section 3.3). Furthermore, for four systems, additional constraints allow for a complete solution for the primary mass and distance, and therefore planet mass.

5.1. Individual Detections

The first two planets found by microlensing, OGLE-2003-BLG-235/MOA-2003-BLG-53Lb (*Bond et al.*, 2004) and OGLE-2005-BLG-071Lb (*Udalski et al.*, 2005), are jovian-mass objects with separations of ~2–4 AU. While the masses and separations of these planets are similar to many of the planets discovered via radial velocity surveys, their host stars are generally less massive and so the planets have substantially lower equilibrium temperatures of ~50–70 K, similar to Saturn and Uranus.

The third and fourth planets discovered by microlensing are significantly lower mass, and indeed inhabit a region of parameter space that was previously unexplored by any method. OGLE-2005-BLG-390Lb is a very-low-mass planet with a planet/star mass ratio of only ~8×10^{-5} (*Beaulieu et al.*, 2006). A Bayesian analysis combined with a mea-

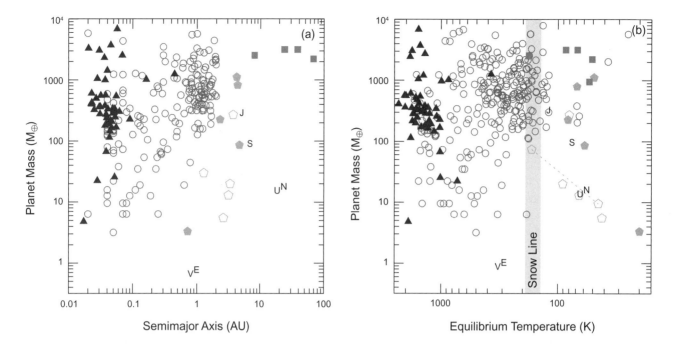

Fig. 12. (a) Mass vs. semimajor axis for known exoplanets. (b) Mass vs. equilibrium temperature for known exoplanets. In both panels, radial velocity detections are indicated by circles, transiting planets are indicated by triangles, microlensing detections are indicated by pentagons, and direct detections are indicated by squares. The letters indicate the locations of the solar system planets. The solid pentagons are those microlensing detections for which the primary mass and distance are measured, and thus planet mass, semimajor axis, and equilibrium temperature are well constrained. The open pentagons indicate those microlensing detections for which only partial information about the primary properties is available, and thus the planet properties are relatively uncertain. An extreme example is shown in (b), where two points connected by a dotted line are plotted for event MOA-2008-BLG-310Lb (*Janczak et al.*, 2009), corresponding to the extrema of the allowed range of planet properties. In (b), the stripe roughly indicates the location of the snow line. These figures also demonstrate the complementarity of the various planet detection techniques: Transit and radial velocity surveys are generally sensitive to planets interior to the snow line, whereas microlensing and direct imaging surveys are sensitive to more distant planets beyond the snow line.

surement of θ_E from finite source effects indicates that the planet likely orbits a low-mass M dwarf with M = $0.22^{+0.21}_{-0.11}$ $M_\odot$, and thus has a mass of only $5.5^{+5.5}_{-2.7}$ $M_\oplus$. Its separation is $2.6^{+1.5}_{-0.6}$ AU, and so has a cool equilibrium temperature of ~50 K. OGLE-2005-BLG-169Lb is another low-mass planet with a mass ratio of 8×10^{-5} (*Gould et al.*, 2006), essentially identical to that of OGLE-2005-BLG-390Lb. A Bayesian analysis indicates a primary mass of $0.52^{+0.19}_{-0.22}$ $M_\odot$, and so a planet of mass ~14^{+5}_{-6} $M_\oplus$, a separation of $3.3^{+1.9}_{-0.9}$ AU, and an equilibrium temperature of ~70 K (*Bennett et al.*, 2007). In terms of its mass and equilibrium temperature, OGLE-2005-BLG-169Lb is very similar to Uranus. OGLE-2005-BLG-169Lb was discovered in a high-magnification A_{max} ~ 800 event; as argued above, such events have significant sensitivity to multiple planets. There is no indication of any additional planetary perturbations in this event, which excludes Jupiter-mass planets with separations between 0.5 and 15 AU, and Saturn-mass planets with separations between 0.8 and 9.5 AU. Thus it appears that this planetary system is likely dominated by the detected Neptune-mass companion.

The fifth and sixth planets discovered by microlensing were detected in what is arguably one of the most information-rich and complex microlensing events ever analyzed, OGLE-2006-BLG-109 (*Gaudi et al.*, 2008a; *Bennett et al.*, 2010). The lightcurve exhibited five distinct features. Four of these features are attributable to the source crossing the resonant caustic of a Saturn-mass planet, and include a short cusp crossing, followed by a pair of caustic crossings, and finally a cusp approach, all spanning roughly two weeks. The fifth feature cannot be explained by the Saturn-mass planet caustic, but rather is due to the source approaching the tip of the central caustic of a inner, Jupiter-mass planet.

The OGLE-2006-BLG-109Lb planetary system bears a remarkable similarity to a scaled version of Jupiter and Saturn. By combining all the available information, it is possible to infer that the primary is an M dwarf with a mass of M = $0.51^{+0.05}_{-0.04}$ $M_\odot$ at distance of D_l = $1.51^{+0.11}_{-0.12}$ kpc. The two planets have masses of 0.73 ± 0.06 M_{Jup} and 0.27 ± 0.02 M_{Jup} and separations of 2.3 ± 0.5 AU and $4.5^{+2.1}_{-1.0}$ AU, respectively. Thus the mass and separation ratios of the two planets are very similar to Jupiter and Saturn. Although the two planets in the OGLE-2006-BLG-109Lb system orbit at about half the distance of their Jupiter/Saturn analogs, their equilibrium temperatures are also similar (although somewhat cooler) to Jupiter and Saturn, because of the lower primary mass and luminosity.

The seventh planet detected by microlensing is also a very low-mass, cool planet (*Bennett et al., 2008*). MOA-2007-BLG-192Lb has a mass ratio of $q = 2 \times 10^{-4}$, somewhat smaller than that of Saturn and the Sun. However, this event exhibits orbital parallax and the best-fit model also shows evidence for finite source effects, allowing a measurement of the primary mass. In this case, the inferred primary mass is quite low: $M = 0.060^{+0.028}_{-0.021}$ $M_\odot$, implying a substellar brown-dwarf host, and very low mass for the planet of $3.3^{+4.9}_{-1.6}$ $M_\oplus$. Remarkably, this planet was detected from data taken entirely by the MOA and OGLE survey collaborations, without the benefit of additional coverage from the follow-up collaborations. This event therefore serves as an early indication of how future microlensing planet surveys will operate, as detailed in the next section. Unfortunately, in this case, coverage of the planetary perturbation was sparse, and as a result the limits on the mass ratio and primary mass are relatively weak. Fortunately, the inferred primary mass can be confirmed and made more precise with follow-up VLT and/or HST observations. If confirmed, this discovery will demonstrate that brown dwarfs can form planetary systems similar to those around solar-type main-sequence stars.

The eighth and ninth microlensing planets, MOA-2007-BLG-400Lb (*Dong et al., 2009b*) and MOA-2008-BLG-310Lb (*Janczak et al., 2009*), were detected in events that bear some remarkable similarities. Both were high-magnification events ($A_{max} \sim 628$ and $A_{max} \sim 400$, respectively) in which the primary lens transited the source, resulting in a dramatic smoothing of the peak of the event (see Fig. 8e for an example). By eye, single lens models provide relatively good matches to the data. Nevertheless, weak but broad and significant residuals to the single-lens model are apparent in both cases, which are well fit by perturbations due to the central caustic of a planetary companion. In both cases, the inferred caustic size is significantly smaller than the source size. Also in both cases, precise constraints on the mass ratio are possible despite the weak perturbation amplitudes. MOA-2007-BLG-400Lb is a cool, jovian-mass planet with mass ratio of $q = 0.0025 \pm 0.0004$, and MOA-2008-BLG-310Lb is a sub-Saturn-mass planet with $q = (3.3 \pm 0.3) \times 10^{-4}$. The angular Einstein ring radius of the primary lens 2008-BLG-310-310L as inferred from finite source effects is quite small ($\theta_E = 0.162 \pm 0.015$ mas), implying that if the primary is a star ($M \geq 0.08$ $M_\odot$), it must have $D_l > 6.0$ kpc and so be in the bulge (*Janczak et al., 2009*). This is the best candidate yet for a bulge planet. Unfortunately, it is not possible to definitely determine if the system is in the bulge with currently available information, and it is possible that the primary is a low-mass star or brown dwarf in the foreground galactic disk. This ambiguity can be resolved with follow-up high-resolution imaging taken immediately, followed by additional observations taken in ~10 years, at which point the source and lens will have separated by ~50 mas.

The tenth planet, OGLE-2007-BLG-368Lb, is another cold Neptune (*Sumi et al., 2010*), making it the fourth low-mass (≤ 20 $M_\oplus$) planet discovered with microlensing. This planet was detected in a relatively low-magnification event with a maximum magnification of only ~13. In this case, the source crossed between the two triangular-shaped planetary caustics of a close (d < 1) topology planetary lens, and thus the lightcurve exhibited a large (~20%) dip, characteristic of the planetary perturbations due to such configurations (see Fig. 9c,d).

5.2. Frequency of Cold Planets

The microlensing detection sensitivity declines with planet mass as $\sim m_p^{1/2}$, and thus the presence of two low-mass planets among the first four detections was an indication that the frequency of cold super Earths/Neptunes (5–15 $M_\oplus$) is substantially higher than that of cold jovian-class planets. *Gould et al.* (2006) performed a quantitative analysis that accounted for the detection sensitivities and Poisson statistics and demonstrated that, at 90% confidence, $38^{+31}_{-22}\%$ of stars host cold super Earths/Neptunes with separations in the range 1.6–4.3 AU. An updated analysis taking into account the planets that have since been detected, as well as those events that did not yield planets, revises this number downward to ~20 ± 10% (D. Bennett, personal communication). Thus, such planets are common, which is ostensibly a confirmation of the core accretion model of planet formation, which predicts that there should exist many more "failed Jupiters" than bonafide jovian-mass planets at such separations, particularly around low-mass primaries (*Laughlin et al., 2004; Ida and Lin, 2005*).

By adopting a simple model for the scaling of the detection efficiency with q, *Sumi et al.* (2010) used the distribution of mass ratios of the 10 microlensing planets discovered to date to derive an intrinsic mass (ratio) function of exoplanets beyond the snow line of $dN/dq \propto q^{-1.7\pm0.2}$. This implies that cold Neptunes/super Earths ($q \sim 5 \times 10^{-5}$) are 7^{+6}_{-3} times more common than cold Jupiters ($q \sim 10^{-3}$), reinforcing the conclusions of *Gould et al.* (2006).

5.3. Properties of the Planetary Systems

As already mentioned, for all of the planet detections, it has been possible to obtain additional information to improve the constraints on the properties of the primaries and planets. For example, for OGLE-2003-BLG-235/MOA-2003-BLG-53Lb, follow-up imaging with HST yielded a detection of light from the lens, which constrains the mass of the primary and planet to ~15%, $M = 0.63^{+0.07}_{-0.09}$ $M_\odot$ and $m_p = 2.6^{+0.8}_{-0.6}$ M_{Jup} (*Bennett et al., 2006*). However, it is the analysis of events OGLE-2006-BLG-109 (*Bennett et al., 2010*) and OGLE-2005-BLG-071 (*Dong et al., 2009a*) that demonstrate most strikingly the detailed information that can be obtained for planets detected via microlensing.

For microlensing event OGLE-2006-BLG-109, besides the basic signatures of the two planets, the lightcurve displays finite source effects and orbital parallax, which allow for a complete solution to the lens system and so measurements of the mass and distance to the primary, as well as the masses of the planets (*Gaudi et al.*, 2008a; *Bennett et al.*, 2010). This primary mass measurement is corroborated by a Keck AO H-band image of the target that yields a measurement of the flux of the lens, which in this case turns out to be brighter than the source. The lens flux is consistent with the mass inferred entirely from the microlensing observables. In addition, the orbital motion of the outer Saturn-mass planet was detected. This was possible because the source crossed or approached the resonant caustic due to this planet at four distinct times. Furthermore, as discussed in section 2.3, the exact shape and size of resonant caustics depend sensitively on d, which changes over the two weeks of the planetary deviations. In fact, four of the six orbital parameters of the Saturn-mass planet are well measured, and a fifth parameter is weakly constrained. This information, combined with an assumption of coplanarity with the Jupiter-mass planet and the requirement of stability, enables a constraint on the orbital eccentricity of the Saturn-mass planet of e = $0.15^{+0.17}_{-0.10}$, and inclination of the system of i = $64^{+4}_{-7}°$ (*Bennett et al.*, 2010).

In the case of OGLE-2005-BLG-071Lb, HST photometry, when combined with information on finite source effects and microlens parallax from the lightcurve, allows for a measurement of the primary mass and distance, and so planet mass and projected separation. This leads to the conclusion that the companion is a massive planet with m_p = 3.8 ± 0.4 M_{Jup} and projected separation $r_\perp$ = 3.6 ± 0.2 AU, orbiting an M-dwarf primary with a mass of M = 0.46 ± 0.04 $M_\odot$ and a distance of D_l = 3.2 ± 0.4 kpc (*Dong et al.*, 2009a). Furthermore, the primary has thick-disk kinematics with a projected velocity relative to the local standard of rest of v = 10^3 ± 15 km s^{-1}, suggesting that it may be metal-poor. Thus, OGLE-2005-BLG-071Lb may be a massive jovian planet orbiting a metal-poor, thick-disk M-dwarf. The existence of such a planet may pose a challenge for core-accretion models of planet formation (e.g., *Laughlin et al.*, 2004; *Ida and Lin*, 2004, 2005).

Interestingly, all four microlensing planet hosts for which it has been possible to measure the distance to the system lie in the foreground disk. As mentioned above, MOA-2008-BLG-310Lb is the most promising candidate for a bulge planet, but in this case the primary could still be a foreground disk brown dwarf. In contrast, roughly 60% of all microlensing events toward the galactic bulge are due to lenses in the bulge (*Kiraga and Paczynski*, 1994). The lack of confirmed microlensing bulge planets could be due to the selection effect that longer events, which are more likely to arise from disk lenses, are preferentially monitored by the follow-up collaborations, or it could reflect a difference between the planet populations in the disk and bulge. Regardless, with a larger sample of planets with well-constrained distances, and a more careful accounting of selection effects, it should be possible to compare the demographics of planets in the galactic disk and bulge.

6. FUTURE PROSPECTS

In the four seasons of 2003–2006 there were five planetary events, containing six detected planets. With the MOA upgrade in 2006 to the MOA-II phase, the rate of planet detections has increased substantially. From the 2007, 2008, and 2009 bulge seasons, there were four, three, and four secure planetary events, respectively (seven of these await publication). Thus even maintaining the current rate, we can expect on the order of a dozen new planet detections over the next several years. In fact, as described below, we can expect the rate of planet detections to increase substantially, as microlensing planet searches transition toward the next generation of surveys. Besides finding more planets, these surveys will also have improved sensitivity to lower-mass, terrestrial planets. Thus we can expect to have robust constraints on the frequency of Earth-mass planets beyond the snow line within the next decade. A spacebased survey would determine determine the demographics of planets with mass greater than that of Mars and semimajor axis ≥ 0.5 AU, determine the frequency of free-floating, Earth-mass planets, and determine the frequency of terrestrial planets in the outer habitable zones of solar-type stars in the galactic disk and bulge.

The transition to the next generation of groundbased surveys is enabled by the advent of large-format cameras with fields-of-view (FOV) of several square degrees. With such large FOVs, it becomes possible to monitor tens of millions of stars every 10–20 minutes, and so discover thousands of microlensing events per year. Furthermore, these events are then simultaneously monitored with the cadence required to detect perturbations due to very low-mass (~$M_\oplus$) planets. Thus next-generation searches will operate in a very different mode than the current alert/follow-up model. In order to obtain around-the-clock coverage and so catch all the perturbations, several such telescopes would be needed, located on three to four continents roughly evenly spread in longitude.

In fact, the transition to the "next generation" is happening already. The MOA-II telescope in New Zealand (1.8 m and 2 deg^2 FOV) already represents one leg of such a survey. The OGLE team has recently upgraded to the OGLEIV phase with a 1.4-deg^2 camera, which will represent the second leg in Chile when it becomes operational in 2010. Although the OGLE telescope has a smaller FOV camera and a smaller aperture (1.3 m) than the MOA telescope, these are mostly compensated by the better site quality in Chile. Finally, the Korean Microlensing Telescope Network (KMTNet) is a project with plans to build three 1.6-m telescopes with 4-deg^2 FOV cameras, one each in South Africa, South America, and Australia. With the completion of the South African leg of this network (planned for 2012), a next-generation survey would effectively be in place. In addition, astronomers from Germany and China are considering initiatives to secure funding to build 1–2-m-class telescopes with wide-FOV cameras in southern Africa or Antarctica.

Detailed simulations of such a next-generation microlensing survey have been performed by several groups (Gaudi et al., unpublished; *Bennett*, 2004). These simulations include models for the galactic population of lenses and sources that match all constraints (*Han and Gould*, 1995, 2003), and account for real-world effects such as weather, variable seeing, moon and sky background, and crowded fields. They reach similar conclusions. Such a survey would increase the planet detection rate at fixed mass by at least an order of magnitude over current surveys. Figure 13 shows the predictions of these simulations for the detection rate of planets of various masses and separations using a survey including MOA-II, OGLE-IV, and a Korean telescope in South Africa. In particular, if Earth-mass planets with semimajor axes of several AU are common around main-sequence stars, a next-generation microlensing survey should detect several such planets per year. This survey would also be sensitive to free-floating planets, and would detect them at a rate of hundreds per year if every star has an ejected Jupiter-mass planet.

Ultimately, however, the true potential of microlensing cannot be realized from the ground. Weather, seeing, crowded fields, and systematic errors all conspire to make the detection of planets with mass less than Earth effectively impossible from the ground (*Bennett*, 2004). As outlined in *Bennett and Rhie* (2002), a spacebased microlensing survey offers several advantages: the main-sequence bulge sources needed to detect sub-Earth-mass planets are resolved from space, the events can be monitored continuously, and it is possible to observe the moderately reddened source stars in the near infrared to improve the photon collection rate. Furthermore, the high spatial resolution afforded by space allows unambiguous identification of light from the primary (lens) stars and so measurements of the primary and planet masses (*Bennett et al.*, 2007).

The expectations from a Discovery-class spacebased microlensing survey are impressive. Such a survey would be sensitive to all planets with mass ≥ 0.1 $M_\oplus$ and separations $a \geq 0.5$ AU, including free-floating planets (*Bennett and Rhie*, 2002; *Bennett et al.*, 2009). This range includes analogs to all the solar system planets except Mercury. If every main-sequence star has an Earth-mass planet in the range 1–2.5 AU, the survey would detect ~500 such planets within its mission lifetime. The survey would also detect a number of habitable Earth-mass planets comparable to that of the Kepler mission (*Borucki et al.*, 2003), and so would provide an important independent measurement of $\eta_\oplus$. When combined with complementary surveys (such as Kepler), a spacebased microlensing planet survey would

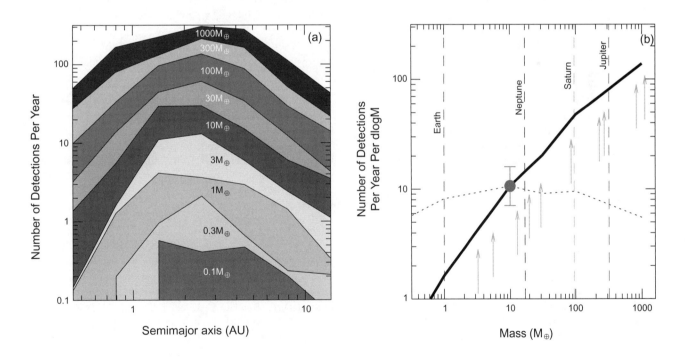

Fig. 13. Expectations from the next generation of groundbased microlensing surveys, including MOA-II, OGLE-IV, and a KMTNet telescope in South Africa. These results represent the average of two independent simulations, which include very different input assumptions but differ in their predictions by only ~0.3 dex. **(a)** Number of planets detected per year as a function of semimajor axis for various masses, assuming every star has a planet of the given mass and semimajor axis. **(b)** Number of planets detected per year as a function of planet mass, normalized by the number of ~10 $M_\oplus$ found to date by microlensing (indicated by the large dot). The solid line is the prediction assuming an equal number of planets per logarithmic mass interval, and the dotted curve assumes that the number of planets per log mass scales as $m_p^{-0.7}$ (*Sumi et al.*, 2010). In both cases, planets are assumed to be distributed uniformly in log(a) between 0.4 and 20 AU. Arrows indicate the locations of the 10 published exoplanets.

determine the demographics of both bound and free-floating planets with masses greater than that of Mars orbiting stars with masses less than that of the Sun.

The basic requirements for a spacebased microlensing planet survey are relatively modest: an aperture of at least 1 m, a large FOV camera (at least ~0.5 deg^2) with optical or near-IR detectors, reasonable image quality of better than ~0.25", and an orbit for which the galactic bulge is continuously visible. The Microlensing Planet Finder (MPF) is an example of a spacebased microlensing survey that can accomplish these objectives, essentially entirely with proven technology, and at a cost of ~$300 million excluding the launch vehicle (*Bennett et al.*, 2009). Interestingly, the requirements for a spacebased microlensing planet survey are very similar to, or less stringent than, the requirements for a number of the proposed dark-energy missions, in particular those that focus on weak lensing measurements. Thus, it may be attractive to consider a combined dark-energy/planet-finding mission that could be accomplished at a substantial savings compared to doing each mission separately (*Gould*, 2009).

Acknowledgments. I would like to thank S. Dong for preparing Fig. 10, A. Udalski for reading over the manuscript, and two anonymous referees for catching some important errors and providing comments that greatly improved the chapter.

REFERENCES

Afonso C., Alard C., Albert J. N., Andersen J., Ansari R., et al. (2000) Combined analysis of the binary lens caustic-crossing event MACHO 98-SMC-1. *Astrophys. J., 532*, 340–352.

Afonso C., Albert J. N., Andersen J., Ansari R., Aubourg É., et al. (2001) Photometric constraints on microlens spectroscopy of EROS-BLG-2000-5. *Astron. Astrophys., 378*, 1014–1023.

Agol E. (2002) Occultation and microlensing. *Astrophys. J., 579*, 430–436.

Albrow M., Beaulieu J.-P., Birch P., Caldwell J. A. R., Kane S., et al. (1998) The 1995 pilot campaign of PLANET: Searching for microlensing anomalies through precise, rapid, round-the-clock monitoring. *Astrophys. J., 509*, 687–702.

Albrow M. D., Beaulieu J.-P., Caldwell J. A. R., Depoy D. L., Dominik M., et al. (1999a) The relative lens-source proper motion in MACHO 98-SMC-1. *Astrophys. J., 512*, 672–677.

Albrow M. D., Beaulieu J.-P., Caldwell J. A. R., Depoy D. L., Dominik M., Gaudi B. S., et al. (1999b) A complete set of solutions for caustic crossing binary microlensing events. *Astrophys. J., 522*, 1022–1036.

Albrow M. D., Beaulieu J.-P., Caldwell J. A. R., Dominik M., Greenhill J., et al. (1999c) Limb darkening of a K giant in the galactic bulge: PLANET photometry of MACHO 97-BLG-28. *Astrophys. J., 522*, 1011–1021.

Albrow M. D., An J., Beaulieu J.-P., Caldwell J. A. R., DePoy D. L., Dominik M., Gaudi B. S., et al. (2002) A short, nonplanetary, microlensing anomaly: Observations and lightcurve analysis of MACHO 99-BLG-47. *Astrophys. J., 572*, 1031–1040.

Alcock C., Akerlof C. W., Allsman R. A., Axelrod T. S., Bennett D. P., et al. (1993) Possible gravitational microlensing of a star in the large Magellanic cloud. *Nature, 365*, 621–623.

Alcock C., Allsman R. A., Alves D., Axelrod T. S., Becker A. C., Bennett D. P., Cook K. H., Freeman K. C., Griest K., Guern J., Lehner M. J., Marshall S. L., Peterson B. A., Pratt M. R., Quinn P. J., Reiss D., Rodgers A. W., Stubbs C. W., Sutherland W., Welch D. L., and The MACHO Collaboration (1996) Real-Time Detection and Multisite Observations of Gravitational Microlensing. *Astrophys. J., 463*, L67–L70.

Alcock C., Allsman R. A., Alves D., Axelrod T. S., Becker A. C., et al. (1997) First detection of a gravitational microlensing candidate toward the small Magellanic cloud. *Astrophys. J., 491*, L11–L13.

Alcock C., Allsman R. A., Alves D. R., Axelrod T. S., Becker A. C., et al. (2000) The MACHO Project: Microlensing results from 5.7 years of large Magellanic cloud observations. *Astrophys. J., 542*, 281–307.

An J. H. (2005) Gravitational lens under perturbations: symmetry of perturbing potentials with invariant caustics. *Mon. Not. R. Astron. Soc., 356*, 1409–1428.

An J. H. and Evans N. W. (2006) The Chang-Refsdal lens revisited. *Mon. Not. R. Astron. Soc., 369*, 317–334.

Aubourg E., Bareyre P., Brehin S., Gros M., Lachieze-Rey M., et al. (1993) Evidence for gravitational microlensing by dark objects in the galactic halo. *Nature, 365*, 623–625.

Beaulieu J.-P., Bennett D. P., Fouqué P., Williams A., Dominik M., et al. (2006) Discovery of a cool planet of 5.5 Earth masses through gravitational microlensing. *Nature, 439*, 437–440.

Bennett D. P. (2004) The detection of terrestrial planets via gravitational microlensing: Space vs. ground-based surveys. In *Extrasolar Planets: Today and Tomorrow* (J.-P. Beaulieu et al., eds.), pp. 59–67. ASP Conf. Proc. 321, Astronomical Society of the Pacific, San Francisco.

Bennett D. P. (2009) Detection of extrasolar planets by gravitational microlensing. *Exoplanets: Detection, Formation, Properties, Habitability* (J. Mason, ed.), pp. 47–88. Springer, Berlin.

Bennett D. P. and Rhie S. H. (1996) Detecting Earth-mass planets with gravitational microlensing. *Astrophys. J., 472*, 660–664.

Bennett D. P. and Rhie S. H. (2002) Simulation of a space-based microlensing survey for terrestrial extrasolar planets. *Astrophys. J., 574*, 985–1003.

Bennett D. P., Anderson J., Bond I. A., Udalski A., and Gould A. (2006) Identification of the OGLE-2003-BLG-235/MOA-2003-BLG-53 planetary host star. *Astrophys. J. Lett., 647*, L171–L174.

Bennett D. P., Anderson J., and Gaudi B. S. (2007) Characterization of gravitational microlensing planetary host stars. *Astrophys. J., 660*, 781–790.

Bennett D. P., Bond I. A., Udalski A., Sumi T., Abe F., et al. (2008) A low-mass planet with a possible sub-stellar-mass host in microlensing event MOA-2007-BLG-192. *Astrophys. J., 684*, 663–683.

Bennett D. P., Anderson J., Beaulieu J. P., Bond I., Cheng E., et al. (2009) A census of exoplanets in orbits beyond 0.5 AU via space-based microlensing. In *Astro2010: The Astronomy and Astrophysics Decadal Survey, Science White Papers*, p. 18.

Bennett D. P., Rhie S. H., Nikolaev S., Gaudi B. S., Udalski A., et al. (2010) Masses and orbital constraints for the OGLE-2006-BLG-109Lb,c Jupiter/Saturn analog planetary system. *Astrophys. J., 713*, 837–855.

Bolatto A. D. and Falco E. E. (1994) The detectability of planetary companions of compact galactic objects from their effects on microlensed light curves of distant stars. *Astrophys. J., 436*, 112–116.

Bond I. A., Udalski A., Jaroszyński M., Rattenbury N. J., Paczyński B., et al. (2004) OGLE 2003-BLG-235/MOA 2003-BLG-53: A planetary microlensing event. *Astrophys. J., 606*, L155–L158.

Borucki W. J., Koch D. G., Lissauer J. J., Basri G B., Caldwell J. F., et al. (2003) The Kepler mission: A wide-field-of-view photometer designed to determine the frequency of Earth-size planets around solar-like stars. *SPIE, 4854*, 129–140.

Boutreux T. and Gould A. (1996) Monte Carlo simulations of MACHO parallaxes from a satellite. *Astrophys. J., 462*, 705.

Bozza V. (2000) Caustics in special multiple lenses. *Astron. Astrophys., 355*, 423–432.

Bromley B. C. (1996) Finite-size gravitational microlenses. *Astrophys. J., 467*, 537–539.

Calchi Novati S., Paulin-Henriksson S., An J., Baillon P., Belokurov V., et al. (2005) POINT-AGAPE pixel lensing survey of M 31. Evidence for a MACHO contribution to galactic halos. *Astron. Astrophys., 443*, 911–928.

Cassan A. (2008) An alternative parameterisation for binary-lens caustic-crossing events. *Astron. Astrophys., 491*, 587–595.

Chang K. and Refsdal S. (1979) Flux variations of QSO 0957+561 A, B and image splitting by stars near the light path. *Nature, 282*, 561–564.

Chang K. and Refsdal S. (1984) Star disturbances in gravitational lens galaxies. *Astron. Astrophys., 132*, 168–178.

Chung S.-J., Han C., Park B.-G., Kim D., Kang S., Ryu Y.-H., Kim K. M., Jeon Y.-B., Lee D.-W., Chang K., Lee W.-B., and Kang Y. H. (2005) Properties of central caustics in planetary microlensing. *Astrophys. J., 630*, 535–542.

Chung S.-J., Kim D., Darnley M. J., Duke J. P., Gould A., Han C., Jeon Y.-B., Kerins E., Newsam A., and Park B.-G. (2006) The possibility of detecting planets in the Andromeda galaxy. *Astrophys. J., 650*, 432–437.

Covone G., de Ritis R., Dominik M., and Marino A. A. (2000) Detecting planets around stars in nearby galaxies. *Astron. Astrophys., 357*, 816–822.

de Jong J. T. A., Kuijken K., Crotts A. P. S., Sackett P. D., Sutherland W. J., et al. (2004) First microlensing candidates from the MEGA survey of M 31. *Astron. Astrophys., 417*, 461–477.

Derue F., Afonso C., Alard C., Albert J.-N., Andersen J., et al. (2001) Observation of microlensing toward the galactic spiral arms. EROS II 3 year survey. *Astron. Astrophys., 373*, 126–138.

Di Stefano R. (1999) Microlensing and the search for extraterrestrial life. *Astrophys. J., 512*, 558–563.

Di Stefano R. (2008) Mesolensing explorations of nearby masses: From planets to black holes. *Astrophys. J., 684*, 59–67.

Di Stefano R. and Perna R. (1997) Identifying microlensing by binaries. *Astrophys. J., 488*, 55–63.

Di Stefano R. and Scalzo R. A. (1999a) A new channel for the detection of planetary systems through microlensing. I. Isolated events due to planet lenses. *Astrophys. J., 512*, 564–578.

Di Stefano R. and Scalzo R. A. (1999b) A new channel for the detection of planetary systems through microlensing. II. Repeating events. *Astrophys. J., 512*, 579–600.

Dominik M. (1995) Improved routines for the inversion of the gravitational lens equation for a set of source points. *Astron. Astrophys., 109*, 597–610.

Dominik M. (1998) Galactic microlensing with rotating binaries. *Astron. Astrophys., 329*, 361–374.

Dominik M. (1999a) Ambiguities in FITS of observed binary lens galactic microlensing events. *Astron. Astrophys., 341*, 943–953.

Dominik M. (1999b) The binary gravitational lens and its extreme cases. *Astron. Astrophys., 349*, 108–125.

Dominik M. (2004a) Theory and practice of microlensing light curves around fold singularities. *Mon. Not. R. Astron. Soc., 353*, 69–86.

Dominik M. (2004b) Revealing stellar brightness profiles by means of microlensing fold caustics. *Mon. Not. R. Astron. Soc., 353*, 118–132.

Dominik M. (2006) Stochastic distributions of lens and source properties for observed galactic microlensing events. *Mon. Not. R. Astron. Soc., 367*, 669–692.

Dominik M. (2007) Adaptive contouring — An efficient way to calculate microlensing light curves of extended sources. *Mon. Not. R. Astron. Soc., 377*, 1679–1688.

Dominik M. (2009) Parameter degeneracies and (un)predictability of gravitational microlensing events. *Mon. Not. R. Astron. Soc., 393*, 816–821.

Dominik M. and Hirshfeld A. C. (1996) Evidence for a binary lens in the MACHO LMC No. 1 microlensing event. *Astron. Astrophys., 313*, 841–850.

Dong S., DePoy D. L., Gaudi B. S., Gould A., Han C., et al. (2006) Planetary detection efficiency of the magnification 3000 microlensing event OGLE-2004-BLG-343. *Astrophys. J., 642*, 842–860.

Dong S., Udalski A., Gould A., Reach W. T., Christie G. W., et al. (2007) First space-based microlens parallax measurement: Spitzer observations of OGLE-2005-SMC-001. *Astrophys. J., 664*, 862–878.

Dong S., Gould A., Udalski A., Anderson J., Christie G. W., Gaudi B. S., et al. (2009a) OGLE-2005-BLG-071Lb, the most massive M dwarf planetary companion? *Astrophys. J., 695*, 970–987.

Dong S., Bond I. A., Gould A., Kozłowski S., Miyake N., Gaudi B. S., et al. (2009b) Microlensing event MOA-2007-BLG-400: Exhuming the buried signature of a cool, jovian-mass planet. *Astrophys. J., 698*, 1826–1837.

Einstein A. (1936) Lens-like action of a star by the deviation of light in the gravitational field. *Science, 84*, 506–507.

Fields D. L., Albrow M. D., An J., Beaulieu J.-P., Caldwell J. A. R., et al. (2003) High-precision limb-darkening measurement of a K3 giant using microlensing. *Astrophys. J., 596*, 1305–1319.

Fluke C. J. and Webster R. L. (1999) Investigating the geometry of quasars with microlensing. *Mon. Not. R. Astron. Soc., 302*, 68–74.

Ford E. B. and Rasio F. A. (2008) Origins of eccentric extrasolar planets: Testing the planet-planet scattering model. *Astrophys. J., 686*, 621–636.

Gaudi B. S. (1998) Distinguishing between binary-source and planetary microlensing perturbations. *Astrophys. J., 506*, 533–539.

Gaudi B. S. and Gould A. (1997a) Satellite parallaxes of lensing events toward the galactic bulge. *Astrophys. J., 477*, 152–162.

Gaudi B. S. and Gould A. (1997b) Planet parameters in microlensing events. *Astrophys. J., 486*, 85–99.

Gaudi B. S. and Gould A. (1999) Spectrophotometric resolution of stellar surfaces with microlensing. *Astrophys. J., 513*, 619–625.

Gaudi B. S. and Han C. (2004) The many possible interpretations of microlensing event OGLE 2002-BLG-055. *Astrophys. J., 611*, 528–536.

Gaudi B. S. and Petters A. O. (2002a) Gravitational microlensing near caustics. I. Folds. *Astrophys. J., 574,* 970–984.

Gaudi B. S. and Petters A. O. (2002b) Gravitational microlensing near caustics. II. Cusps. *Astrophys. J., 580,* 468–489.

Gaudi B. S., Naber R. M., and Sackett P. D. (1998) Microlensing by multiple planets in high-magnification events. *Astrophys. J. Lett., 502,* L33–L37.

Gaudi B. S., Albrow M. D., An J., Beaulieu J.-P., Caldwell J. A. R., et al. (2002) Microlensing constraints on the frequency of Jupiter-mass companions: Analysis of 5 years of PLANET photometry. *Astrophys. J., 566,* 463–499.

Gaudi B. S., Bennett D. P., Udalski A., Gould A., Christie G. W., et al. (2008a) Discovery of a Jupiter/Saturn analog with gravitational microlensing. *Science, 319,* 927–930.

Gaudi B. S., Patterson J., Spiegel D. S., Krajci T., Koff R., et al. (2008b) Discovery of a very bright, nearby gravitational microlensing event. *Astrophys. J., 677,* 1268–1277.

Goldreich P., Lithwick Y., and Sari R. (2004) Final stages of planet formation. *Astrophys. J., 614,* 497–507.

Gould A. (1992) Extending the MACHO search to about 10^6 solar masses. *Astrophys. J., 392,* 442–451.

Gould A. (1994a) Proper motions of MACHOs. *Astrophys. J. Lett., 421,* L71–L74.

Gould A. (1994b) MACHO velocities from satellite-based parallaxes. *Astrophys. J. Lett., 421,* L75–L78.

Gould A. (1995) MACHO parallaxes from a single satellite. *Astrophys. J. Lett., 441,* L21–L24.

Gould A. (1996) Microlensing and the stellar mass function. *Publ. Astron. Soc. Pac., 108,* 465–476.

Gould A. (1999) Microlens parallaxes with SIRTF. *Astrophys. J., 514,* 869–877.

Gould A. (2000) Measuring the remnant mass function of the galactic bulge. *Astrophys. J., 535,* 928–931.

Gould A. (2004) Resolution of the MACHO-LMC-5 puzzle: The jerk-parallax microlens degeneracy. *Astrophys. J., 606,* 319–325.

Gould A. (2008) Hexadecapole approximation in planetary microlensing. *Astrophys. J., 681,* 1593–1598.

Gould A. (2009) Wide field imager in space for dark energy and planets. In *Astro2010: The Astronomy and Astrophysics Decadal Survey, Science White Papers,* p. 100, ArXiv e-prints, arXiv:0902.2211.

Gould A. and Gaucherel C. (1997) Stokes's theorem applied to microlensing of finite sources. *Astrophys. J., 477,* 580–584.

Gould A. and Han C. (2000) Astrometric resolution of severely degenerate binary microlensing events. *Astrophys. J., 538,* 653–656.

Gould A. and Loeb A. (1992) Discovering planetary systems through gravitational microlenses. *Astrophys. J., 396,* 104–114.

Gould A., Miralda-Escude J., and Bahcall J. N. (1994) Microlensing events: Thin disk, thick disk, or halo? *Astrophys. J. Lett., 423,* L105–L108.

Gould A., Udalski A., An D., Bennett D. P., Zhou A.-Y., et al. (2006) Microlens OGLE-2005-BLG-169 implies that cool Neptune-like planets are common. *Astrophys. J., 644,* L37–L40.

Griest K. and Hu W. (1992) Effect of binary sources on the search for massive astrophysical compact halo objects via microlensing. *Astrophys. J., 397,* 362–380.

Griest K. and Safizadeh N. (1998) The use of high-magnification microlensing events in discovering extrasolar planets. *Astrophys. J., 500,* 37–50.

Han C. (1999) Analytic relations between the observed gravitational microlensing parameters with and without the effect of blending. *Mon. Not. R. Astron. Soc., 309,* 373–378.

Han C. (2006) Properties of planetary caustics in gravitational microlensing. *Astrophys. J., 638,* 1080–1085.

Han C. (2008) Microlensing detections of moons of exoplanets. *Astrophys. J., 684,* 684–690.

Han C. (2009) Distinguishing between planetary and binary interpretations of microlensing central perturbations under the severe finite-source effect. *Astrophys. J. Lett., 691,* L9–L12.

Han C. and Gaudi B. S. (2008) A characteristic planetary feature in double-peaked, high-magnification microlensing events. *Astrophys. J., 689,* 53–58.

Han C. and Gould A. (1995) The mass spectrum of MACHOs from parallax measurements. *Astrophys. J., 447,* 53.

Han C. and Gould A. (2003) Stellar contribution to the galactic bulge microlensing optical depth. *Astrophys. J., 592,* 172–175.

Han C. and Han W. (2002) On the feasibility of detecting satellites of extrasolar planets via microlensing. *Astrophys. J., 580,* 490–493.

Han C. and Park M.-G. (2002) A new channel to search for extra-solar systems with multiple planets via gravitational microlensing. *J. Korean Astron. Soc., 35,* 35–40.

Han C., Gaudi B. S., An J. H., and Gould A. (2005) Microlensing detection and characterization of wide-separation planets. *Astrophys. J., 618,* 962–972.

Hamadache C., Le Guillou L., Tisserand P., Afonso C., Albert J. N., et al. (2006) Galactic bulge microlensing optical depth from EROS-2. *Astron. Astrophys., 454,* 185–199.

Holtzman J. A., Watson A. M., Baum W. A., Grillmair C. J., Groth E. J., Light R. M., Lynds R., and O'Neil E. J. Jr. (1998) The luminosity function and initial mass function in the galactic bulge. *Astron. J., 115,* 1946–1957.

Holz D. E. and Wald R. M. (1996) Photon statistics limits for Earth-based parallax measurements of MACHO events. *Astrophys. J., 471,* 64–67.

Horne K., Snodgrass C., and Tsapras Y. (2009) A metric and optimization scheme for microlens planet searches. *Mon. Not. R. Astron. Soc., 396,* 2087–2102.

Ida S. and Lin D. N. C. (2004) Toward a deterministic model of planetary formation. II. The formation and retention of gas giant planets around stars with a range of metallicities. *Astrophys. J., 616,* 567–572.

Ida S. and Lin D. N. C. (2005) Toward a deterministic model of planetary formation. III. Mass distribution of short-period planets around stars of various masses. *Astrophys. J., 626,* 1045–1060.

Ingrosso G., Calchi Novati S., De Paolis F., Jetzer P., Nucita A. A., and Zakharov A. F. (2009) Pixel-lensing as a way to detect extrasolar planets in M31. *Mon. Not. R. Astron. Soc., 399,* 219–228.

Janczak J., Fukui A., Dong S., Monard B., Kozlowski S., Gould A., et al. (2009) Sub-Saturn planet MOA-2008-BLG-310Lb: Likely to be in the galactic bulge. *Astrophys. J., 711,* 731–743.

Jaroszyński M. and Mao S. (2001) Predicting the second caustic crossing in binary microlensing events. *Mon. Not. R. Astron. Soc., 325,* 1546–1552.

Johnson J. A., Gaudi B. S., Sumi T., Bond I. A., and Gould A. (2008) A high-resolution spectrum of the highly magnified bulge G dwarf MOA-2006-BLG-099S. *Astrophys. J., 685,* 508–520.

Jurić M. and Tremaine S. (2008) Dynamical origin of extrasolar planet eccentricity distribution. *Astrophys. J., 686,* 603–620.

Kennedy G. M. and Kenyon S. J. (2008) Planet formation around stars of various masses: The snow line and the frequency of giant planets. *Astrophys. J., 673,* 502–512.

Kennedy G. M., Kenyon S. J., and Bromley B. C. (2007) Planet formation around M-dwarfs: The moving snow line and super-Earths. *Astrophys. Space Sci., 311,* 9–13.

Kervella P., Thévenin F., Di Folco E., and Ségransan D. (2004) The angular sizes of dwarf stars and subgiants. Surface brightness relations calibrated by interferometry. *Astron. Astrophys., 426,* 297–307.

Khavinson D. and Neumann G. (2006) On the number of zeros of certain rational harmonic functions. *Proc. Am. Math. Soc., 134,* 1077–1085.

Kiraga M. and Paczynski B. (1994) Gravitational microlensing of the galactic bulge stars. *Astrophys. J. Lett., 430,* L101–L104.

Laughlin G., Bodenheimer P., and Adams F. C. (2004) The core accretion model predicts few jovian-mass planets orbiting red dwarfs. *Astrophys. J. Lett., 612,* L73–L76.

Lecar M., Podolak M., Sasselov D., and Chiang E. (2006) On the location of the snow line in a protoplanetary disk. *Astrophys. J., 640,* 1115–1118.

Liebes S. (1964) Gravitational lenses. *Phys. Rev., 133,* 835–844.

Lissauer J. J. (1987) Timescales for planetary accretion and the structure of the protoplanetary disk. *Icarus, 69,* 249–265.

Mao S. (1992) Gravitational microlensing by a single star plus external shear. *Astrophys. J., 389,* 63–67.

Mao S. (2008) Introduction to gravitational microlensing. In *Manchester Microlensing Conference, ArXiv e-prints, arXiv:0811.0441.*

Mao S. and Di Stefano R. (1995) Interpretation of gravitational microlensing by binary systems. *Astrophys. J., 440,* 22–27.

Mao S. and Paczynski B. (1991) Gravitational microlensing by double stars and planetary systems. *Astrophys. J. Lett., 374,* L37–L40.

Minniti D., Vandehei T., Cook K. H., Griest K., and Alcock C. (1998) Detection of lithium in a main sequence bulge star using Keck I as a 15 m diameter telescope. *Astrophys. J. Lett., 499,* L175–L178.

Paczynski B. (1986) Gravitational microlensing by the galactic halo. *Astrophys. J., 304,* 1–5.

Paczynski B. (1991) Gravitational microlensing of the galactic bulge stars. *Astrophys. J., 371,* L63–L67.

Paczynski B. (1996) Gravitational microlensing in the local group. *Annu. Rev. Astron. Astrophys., 34,* 419–460.

Paczynski B. and Stanek K. Z. (1998) Galactocentric distance with the optical gravitational lensing experiment and HIPPARCOS red clump stars. *Astrophys. J. Lett., 494,* L219–L222.

Palanque-Delabrouille N., Afonso C., Albert J. N., Andersen J., Ansari R., et al. (1998) Microlensing towards the small Magellanic cloud EROS 2 first year survey. *Astron. Astrophys., 332,* 1–9.

Park B.-G., Jeon Y.-B., Lee C.-U., and Han C. (2006) Microlensing sensitivity to Earth-mass planets in the habitable zone. *Astrophys. J., 643,* 1233–1238.

Paulin-Henriksson S., Baillon P., Bouquet A., Carr B. J., Crézé M., et al. (2002) A candidate M31/M32 intergalactic microlensing event. *Astrophys. J. Lett., 576,* L121–L124.

Peale S. J. (2001) Probability of detecting a planetary companion during a microlensing event. *Astrophys. J., 552,* 889–911.

Pejcha O. and Heyrovský D. (2009) Extended-source effect and chromaticity in two-point-mass microlensing. *Astrophys. J., 690,* 1772–1796.

Petters A. O., Levine H., and Wambsganss J. (2001) *Singularity Theory and Gravitational Lensing.* Birkhäuser, Boston.

Poindexter S., Afonso C., Bennett D. P., Glicenstein J.-F., Gould A., Szymański M. K., and Udalski A. (2005) Systematic analysis of 22 microlensing parallax candidates. *Astrophys. J., 633,* 914–930.

Press W. H., Teukolsky S. A., Vetterling W. T., and Flannery B. P. (1992) *Numerical Recipes in FORTRAN. The Art of Scientific Computing.* Cambridge Univ., Cambridge.

Rattenbury N. J., Bond I. A., Skuljan J., and Yock P. C. M. (2002) Planetary microlensing at high magnification. *Mon. Not. R. Astron. Soc., 335,* 159–169.

Refsdal S. (1964) The gravitational lens effect. *Mon. Not. R. Astron. Soc., 128,* 295–306.

Refsdal S. (1966) On the possibility of determining the distances and masses of stars from the gravitational lens effect. *Mon. Not. R. Astron. Soc., 134,* 315–319.

Renn J., Sauer T., and Stachel J. (1997) The origin of gravitational lensing: A postscript to Einstein's 1936 *Science* paper. *Science, 275,* 184–186.

Rhie S. H. (2001) Can a gravitational quadruple lens produce 17 images?, *ArXiv e-prints, arXiv:astro-ph/0103463.*

Rhie S. H. (2002) How cumbersome is a tenth order polynomial?: The case of gravitational triple lens equation, *ArXiv e-prints, arXiv:astro-ph/0202294.*

Rhie S. H. (2003) n-point gravitational lenses with 5(n-1) images, *ArXiv e-prints, arXiv:astro-ph/0305166.*

Rhie S. H. and Bennett D. P. (1999) Line caustic microlensing and limb darkening, *ArXiv e-prints, arXiv:astro-ph/9912050.*

Rhie S. H., Bennett D. P., Becker A. C., Peterson B. A., Fragile P. C., et al. (2000) On planetary companions to the MACHO 98-BLG-35 microlens star. *Astrophys. J., 533,* 378–391.

Sackett P. D. (1999) Searching for unseen planets via occultation and microlensing. In *Planets Outside the Solar System: Theory and Observations* (J.-M. Mariotti and D. Alloin, eds.), pp. 189–228. Kluwer, Boston.

Sako T., Sekiguchi T., Sasaki M., Okajima K., Abe F., et al. (2008) MOA-cam3: A wide-field mosaic CCD camera for a gravitational microlensing survey in New Zealand. *Experimental Astron., 22,* 51–66.

Sauer T. (2008) Nova Geminorum 1912 and the origin of the idea of gravitational lensing. *Arch. Hist. Exact Sci., 62,* 1–22.

Schneider P. and Weiss A. (1986) The two-point-mass lens — Detailed investigation of a special asymmetric gravitational lens. *Astron. Astrophys., 164,* 237–259.

Schneider P. and Weiss A. (1992) The gravitational lens equation near cusps. *Astron. Astrophys., 260,* 1–2.

Schneider P., Ehlers J., and Falco E. E. (1992) *Gravitational Lenses.* Springer-Verlag, Berlin.

Smith M. C., Mao S., and Paczyński B. (2003) Acceleration and parallax effects in gravitational microlensing. *Mon. Not. R. Astron. Soc., 339,* 925–936.

Snodgrass C., Horne K., and Tsapras Y. (2004) The abundance of galactic planets from OGLE-III 2002 microlensing data. *Mon. Not. R. Astron. Soc., 351,* 967–975.

Sumi T., Abe F., Bond I. A., Dodd R. J., Hearnshaw J. B., et al. (2003) Microlensing optical depth toward the galactic bulge from microlensing observations in astrophysics group observations during 2000 with difference image analysis. *Astrophys. J., 591,* 204–227.

Sumi T., Bennett D. P., Bond I. A., Udalski A., Batista V., et al. (2010) A cold Neptune-mass planet OGLE-2007-BLG-368Lb: Cold Neptunes are common. *Astrophys. J., 710,* 1641–1653.

Thomas C. L., Griest K., Popowski P., Cook K. H., Drake A. J., et al. (2005) Galactic bulge microlensing events from the MACHO collaboration. *Astrophys. J., 631,* 906–934.

Tsapras Y., Street R., Horne K., Snodgrass C., Dominik M., et al. (2009) RoboNet-II: Follow-up observations of microlensing events with a robotic network of telescopes. *Astron. Nachr., 330,* 4–11.

Udalski A. (2003) The optical gravitational lensing experiment. Real time data analysis systems in the OGLE-III survey. *Acta Astron., 53,* 291–305.

Udalski A., Szymanski M., Kaluzny J., Kubiak M., Krzeminski W., Mateo M., Preston G. W., and Paczyński B. (1993) The optical gravitational lensing experiment. Discovery of the first candidate microlensing event in the direction of the galactic sulge. *Acta Astron., 43,* 289–294.

Udalski A., Szymanski M., Kaluzny J., Kubiak M., Mateo M., Krzeminski W., and Paczynski B. (1994) The optical gravitational lensing experiment. The early warning system: Real time microlensing. *Acta Astron., 44,* 227–234.

Udalski A., Zebrun K., Szymanski M., Kubiak M., Pietrzynski G., Soszynski I., and Wozniak P. (2000) The optical gravitational lensing experiment. Catalog of microlensing events in the galactic bulge. *Acta Astron., 50,* 1–65.

Udalski A., Jaroszyński M., Paczyński B., Kubiak M., Szymański M. K., et al. (2005) A jovian-mass planet in microlensing event OGLE-2005-BLG-071. *Astrophys. J. Lett., 628,* L109–L112.

Uglesich R. R., Crotts A. P. S., Baltz E. A., de Jong J., Boyle R. P., and Corbally C. J. (2004) Evidence of halo microlensing in M31. *Astrophys. J., 612,* 877–893.

van Belle G. T. (1999) Predicting stellar angular sizes. *Publ. Astron. Soc. Pacific, 111,* 1515–1523.

Vermaak P. (2000) The effects of resolved sources and blending on the detection of planets via gravitational microlensing. *Mon. Not. R. Astron. Soc., 319,* 1011–1019.

Vermaak P. (2003) Rapid analysis of binary lens gravitational microlensing light curves. *Mon. Not. R. Astron. Soc., 344,* 651–656.

Wambsganss J. (1997) Discovering galactic planets by gravitational microlensing: Magnification patterns and light curves. *Mon. Not. R. Astron. Soc., 284,* 172–188.

Witt H. J. (1990) Investigaion of high amplification events in light curves of gravitationally lensed quasars. *Astron. Astrophys., 236,* 311–322.

Witt H. J. (1995) The effect of the stellar size on microlensing at the Baade window. *Astrophys. J., 449,* 42–46.

Witt H. J. and Mao S. (1994) Can lensed stars be regarded as pointlike for microlensing by MACHOs? *Astrophys. J., 430,* 505–510.

Witt H. J. and Mao S. (1995) On the minimum magnification between caustic crossings for microlensing by binary and multiple stars. *Astrophys. J. Lett., 447,* L105–L108.

Wozniak P. and Paczynski B. (1997) Microlensing of blended stellar images. *Astrophys. J., 487,* 55–60.

Yoo J., DePoy D. L., Gal-Yam A., Gaudi B. S., Gould A., et al. (2004) OGLE-2003-BLG-262: Finite-source effects from a point-mass lens. *Astrophys. J., 603,* 139–151.

Zakharov A. F. (1995) On the magnification of gravitational lens images near cusps. *Astron. Astrophys., 293,* 1–4.

Zakharov A. F. (1999) On the some properties of gravitational lens equation near cusps. *Astron. Astrophys., 18,* 17–25.

Direct Imaging of Exoplanets

Wesley A. Traub
Jet Propulsion Laboratory, California Institute of Technology

Ben R. Oppenheimer
American Museum of Natural History

A direct image of an exoplanet system is a snapshot of the planets and disk around a central star. We can estimate the orbit of a planet from a time series of images, and we can estimate the size, temperature, clouds, atmospheric gases, surface properties, rotation rate, and likelihood of life on a planet from its photometry, colors, and spectra in the visible and infrared. The exoplanets around stars in the solar neighborhood are expected to be bright enough for us to characterize them with direct imaging; however, they are much fainter than their parent star, and separated by very small angles, so conventional imaging techniques are totally inadequate, and new methods are needed. A direct-imaging instrument for exoplanets must (1) suppress the bright star's image and diffraction pattern, and (2) suppress the star's scattered light from imperfections in the telescope. This chapter shows how exoplanets can be imaged by controlling diffraction with a coronagraph or interferometer, and controlling scattered light with deformable mirrors.

1. INTRODUCTION

The first direct images of exoplanets were published in 2008, fully 12 years after exoplanets were discovered, and after more than 300 of them had been measured indirectly by radial velocity, transit, and microlensing techniques. This huge time lag occurred because direct imaging of exoplanets requires extraordinary efforts in order to overcome the barriers imposed by astrophysics (planet-star contrast), physics (diffraction), and engineering (scattering).

The structure of this chapter is as follows. Section 1 discusses the scientific purpose of direct imaging of exoplanets, and includes a glossary of terms. Section 2 discusses basic physical concepts, including brightness, contrast, wavefronts, diffraction, and photons. Section 3 discusses coronagraph and interferometer concepts. Section 4 addresses speckles and adaptive optics. Section 5 sketches recent results from exoplanet imaging and lists current projects. Section 6 outlines future prospects for exoplanet imaging on the ground and in space.

1.1. Exoplanet Images

We illustrate with three examples of direct imaging; as it happens, all three examples are of young, self-luminous objects. Figure 1 shows the dust ring and exoplanet Fomalhaut b by *Kalas et al.* (2008), in the visible. The central star was suppressed using a combination of methods described in this chapter: the rectangular-mask coronagraph (section 3.13) and angular differential imaging (section 4.12). Kalas et al. used the Hubble Space Telescope (HST) Advanced Camera for Surveys (ACS) in coronagraph mode. The planet is about 23 mag fainter than its star, and separated by 12.7 arcsec (98 AU at 7.7 pc distance). It was detected at two epochs, clearly showing common motion as well as orbital motion (see inset).

Figure 2 shows a near-infrared composite image of exoplanets HR 8799 b,c,d by *Marois et al.* (2008). The planets are at angular separations of 1.7, 1.0, and 0.6 arcsec from the star (68, 38, and 24 AU at 40 pc distance). The H-band planet-star contrasts (ratio of planet to star flux) are about 10^{-5}, i.e., roughly 12 mag fainter. The planets would be much

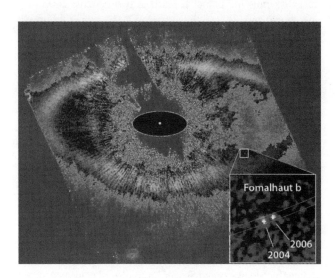

Fig. 1. Visible-wavelength image, from the Hubble Space Telescope, of the exoplanet Fomalhaut b. The planet is located just inside a large dust ring that surrounds the central star. Fomalhaut has been blocked and subtracted to the maximum degree possible.

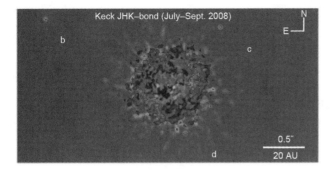

Fig. 2. Near-infrared image, from the Keck telescope, of the exoplanets HR 8799b, c, and d. The central star has been suppressed with angular differential imaging, coupled with adaptive optics. The splatter of dots in the center of this image is simply the small amount of leftover light from the central star that could not be subtracted by ADI, so it is an artifact.

fainter were it not for their youth and consequent internal heat sources, putting their effective temperatures in the 1000 K regime (section 1.3). Marois et al. used the groundbased Gemini and Keck telescopes for these observations. Their techniques included minimizing diffraction using ADI, and minimizing atmospheric speckles using adaptive optics (section 4). If this system were instead the Sun and solar system, then Jupiter would be buried in the inner one-fourth of the speckle field, Earth would be in the inner one-twentieth radius, and both would be 4 to 5 orders of magnitude fainter than the speckles.

Figure 3 shows a near- and mid-infrared composite image of the β Pictoris system, from *Lagrange et al.* (2010). The star itself has been subtracted using a reference star image, and independently with ADI. The composite shows the edge-on dust disk plus the planet β Pic b at two epochs, 2003 (left and above) and 2009 (right and below). The proper motion of β Pic is north, not northwest, so the planet is co-moving, and not a background object. The planet age is ~12 m.y., much younger than the HR 8799 or Fomalhaut planets. The mass is ~9 M_{Jup}, and the semimajor axis in the 8–15 AU range, with a period as short as 17 years. The star is at 19 pc, and the star-planet separations shown are in the 0.3–0.5 arcsec range.

These images set the stage for our goal in this chapter, direct imaging of planets from Earth- to Jupiter-sized around nearby stars.

1.2. Exoplanet Spectra

The spectrum of an exoplanet tells us about its composition, clouds, thermal structure, and variability, as discussed in the chapter by Burrows and Orton for giant planets, and the chapter by Meadows and Seager for terrestrial planets. A direct image of an exoplanet permits us to obtain a spectrum, using a conventional spectrometer or an integral field spectrometer. The resolution will be low, because the planet is faint; however, since many molecular bands are intrinsically low-resolution features, we can still learn much about an atmosphere.

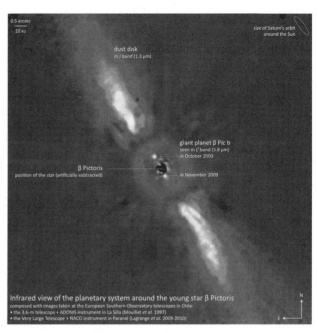

Fig. 3. Near- and mid-infrared composite image of β Pic b and the β Pic dust disk, from the ESO 3.6-m and Very Large Telescopes, with the star subtracted. The planet is shown at two epochs, 2003 and 2009, demonstrating co-moving position with the star as well as orbital motion.

1.3. Hot, Young, and Mature Exoplanets

Hot Jupiters have not yet been directly imaged, but their large thermal flux, 10^{-3} to 10^{-4} times the parent star, means that they will likely be imaged in the future. Their extreme closeness to the parent star requires extreme angular resolution, so the images will come from long-baseline interferometers, not from single-dish telescopes.

Young, self-luminous planets were the first to be directly imaged, because their high temperature and large size give them a strong, detectable flux, and their large distances from their parent stars makes them easier to see in the halo of atmospherically or instrumentally scattered star light. These young, self-luminous planets are likely to continue to be prime targets for detection in the near future, owing to this combination of favorable parameters. However, young planets cool off in a few tens of millions of years, so they will be found only around young stars, and not around nearby (older) stars.

A *mature exoplanet* may be defined here as one with an effective temperature that is roughly comparable to its star-planet equilibrium temperature (section 2.5). These planets, like those in the solar system or around mature nearby stars, will be fainter in the infrared than young, self-luminous planets, and therefore will require more sophisticated techniques to image them. In addition, most of them probably will be closer to their stars than the ones in HR 8799 and Fomalhaut, and will therefore potentially be detectable by single-dish telescopes, but will require the full power of the techniques in this chapter.

1.4. Radial Velocity, Transits, Lensing, and Astrometry

Currently, the techniques of radial velocity (RV), transits, and gravitational lensing are more productive than direct imaging. Remarkably, these techniques, including astrometry and direct imaging, perform nearly independent roles for exoplanet science, so each of them is valuable. Radial velocity has been very successful in measuring masses and periods of planets with masses greater than several Earths and in short-period orbits. Transits have been valuable in measuring the diameters and periods of giant planets, and in combined-light mode have measured temperature distributions, spectral features, and thermal inversions in two gas giant planets. Transits will also be valuable for determining mass and orbit statistics of distant planets, but its geometric bias precludes using it for the vast majority of nearby systems.

Ultimately, exoplanet science will require direct images and spectra of exoplanet systems. For this information, planets around nearby stars will be essential, because these systems will have larger apparent sizes and photon fluxes than more distant systems, and will therefore be relatively accessible to the techniques in this chapter. A combination of astrometry and imaging will provide the mass, period, orbit, and spectroscopic characterization for these planets, down to and including Earth-mass ones.

1.5. Solar System and Exoplanet Systems

There is a strong connection between solar system and exoplanet science. Until exoplanets were discovered in the early 1990s, it was widely thought that exoplanet systems would resemble our solar system. But with the discovery of hot Jupiters, it is now clear that our system is but one of many possible types. There are several points of comparison. Planetary migration and chaotic episodes are now thought to be common to all systems. Self-luminous planets are also common. Dust and debris structures are common as well. Our picture of the evolution of the solar system is strongly influenced by what we are learning about exoplanet systems. In particular, our picture of the evolution of Earth, and of life itself, may well depend on what we learn about habitable-zone terrestrial exoplanets. And for these planets especially, because of their faintness and small angular separations, the techniques in this chapter will be very important.

1.6. Glossary

Some of the terms used in this chapter are briefly defined here for reference:

Visible: wavelength range ~0.3–1.0 μm
Near-infrared: ~1.0–2.5 μm
Mid-infrared: ~2.5–10 μm
Far-infrared: ~10–200 μm
Photometry: broadband (~20%) flux measurement
Color: ratio of two broadband fluxes
Spectrum: narrowband (≤1%) flux measurement
Self-luminous planet: $T_{eff} \gg T_{equil}$
Mature planet: $T_{eff} \approx T_{equil}$
Terrestrial planet: $0.5\ M_\oplus \leq M_p \leq 10\ M_\oplus$
Gas giant planet: $10\ M_\oplus \leq M_p \leq 13\ M_{Jup}$
Habitable zone (HZ): liquid water possible on surface, $0.7\ AU \leq a/L_s^{1/2} \leq 1.5\ AU$
Wavefront: surface of constant phase of a photon
Ray: direction of propagation of photon, always perpendicular to wavefront
Diffraction: bending of wavefront around an obstacle
Scattering: diffraction from polishing or reflectivity errors, a source of speckles
Speckle: light pattern in image plane (coherent with star) from optical path differences in the beam
Coronagraph: telescope with internal amplitude and/or phase masks for imaging faint sources near a bright one
Occulter: coronagraph but with external mask.
Interferometer: two or more telescopes with coherently combined output
Nuller: coronagraph or interferometer using interference of wavefront to suppress a point source

2. FLUX AND PHOTON CONCEPTS

In this section we discuss the underlying equations and concepts needed to calculate flux and photon levels from exoplanets as well as nearby stars and zodi disks. We also discuss the semi-mysterious nature of photons, which are best thought of as waves in some contexts, but must be considered as particles in others; we try to remove the mystery.

2.1. Star Intensity

If we approximate a star as a blackbody of effective temperature T, then its *specific intensity* is the Planck function $B_\nu(T)$ where

$$B_\nu(T) = \frac{2h\nu^3}{c^2 \left(e^{h\nu/kT} - 1\right)} \qquad (1)$$

with units of erg/(s cm² Hz sr), and where, for a star or planet, the unit of area (cm²) is in the plane of the sky, i.e., perpendicular to the line of sight, but not necessarily in the plane of the surface of the object.

It is sometimes convenient to use wavelength units instead of frequency units. From $B_\lambda d\lambda = B_\nu d\nu$ and $\lambda\nu = c$ we get

$$B_\lambda(T) = \frac{2hc^2}{\lambda^5 \left(e^{hc/\lambda kT} - 1\right)} \qquad (2)$$

which has units of erg/(s cm² cm sr).

For the calculation of signal levels and signal to noise ratios we need to know the corresponding specific intensities in units of photons instead of ergs, i.e., $\dot{n}_\nu$ or $\dot{n}_\lambda$. The energy of a photon is $h\nu$, so we get $\dot{n}_\nu = B_\nu/h\nu$, or $\dot{n}_\lambda = B_\lambda \lambda/hc$, which leads to

$$\dot{n}_\nu = \frac{2\nu^2}{c^2\left(e^{h\nu/kT}-1\right)} \quad (3)$$

with units of photons/(s cm² Hz sr), and

$$\dot{n}_\lambda = \frac{2c}{\lambda^4\left(e^{hc/\lambda kT}-1\right)} \quad (4)$$

with units of photons/(s cm² cm sr).

For numerical calculations it is often convenient to insert numerical values of h, c, and k, and to express wavelengths in units of µm instead of cm, indicated by $\lambda_{\mu m}$, where

$$1 \text{ photon} = h\nu = \frac{1.986 \times 10^{-12}}{\lambda_{\mu m}} \text{ erg} \quad (5)$$

Then the specific intensity $\dot{n}_\lambda(T_s)$ in photons is

$$\dot{n}_\lambda(T) = \frac{6 \times 10^{26}}{\lambda_{\mu m}^4 \left(e^{14388/\lambda_{\mu m}T}-1\right)} \quad (6)$$

which has units of photons/(s cm² µm sr).

At a distance d from a star of radius r, such that the star appears to subtend a solid angle $\Omega = \pi(r/d)^2$ steradian (sr), the *photon flux* $\dot{N}_\lambda$ received is

$$\dot{N}_\lambda = \dot{n}_\lambda(T)\Omega \quad (7)$$

with units of photons/(s cm² µm), and likewise for $\dot{N}_\nu$.

Note that for light emitted by an object and subsequently collected by a telescope, or simply for light traversing an optical system, and in the absence of light loss by absorption or blockage, the *etendue*, i.e., the product of area and solid angle, is conserved. Thus the light emitted from the (sky plane) area (A_{star}) of a star, into the solid angle (Ω_{tel}) of a distant telescope, is related to the collecting area (A_{tel}) of the telescope and the solid angle (Ω_{star}) of the distant star

$$A_{star}\Omega_{tel} = A_{tel}\Omega_{star} \quad (8)$$

This explains the switching between the area and solid angle of star and telescope in the above equations.

Stellar flux is often expressed as a *radiant flux* $f_\lambda(m)$, which is a function of *apparent magnitude* m in a *standard spectral band*

$$f_\lambda(m) = 10^{a-0.4m} \quad (9)$$

with units of erg/(s cm² µm), outside Earth's atmosphere. For each standard spectral band there is an effective central wavelength λ_0, an effective bandwidth $\Delta\lambda$ (approximately the full width at half maximum, FWHM), and a corresponding value of a. The latter can differ by up to ±0.03, depending upon the calibration technique. A compilation of these parameters is given in Table 1.

Another common unit of flux density is the Jansky, where

$$1 \text{ Jy} = 10^{-26} \text{ watt}/(m^2 Hz)$$
$$= \frac{3 \times 10^{-9}}{\lambda_{\mu m}^2} \text{erg}/(s\,cm^2\,\mu m) \quad (10)$$
$$= \frac{1.51 \times 10^3}{\lambda_{\mu m}} \text{photon}/(s\,cm^2\,\mu m)$$

Thus a zero-magnitude star has a flux density of about 3750 Jy at V, and 35 Jy at N. The photon densities are 1.03×10^7 at V, and 5.03×10^3 photons/(s cm² µm) at N.

2.2. Angular Separation

Kepler's third law says that a planet with semimajor axis a (AU) and eccentricity e has orbital period P (yr) where

$$P = a^{3/2}/M_s^{1/2} \quad (11)$$

Here M_s is in units of $M_\odot$ and $M_p \ll M_s$. If the distance from star to observer is d(pc), then the maximum angular separation between planet and star is

$$\theta = a(1+e)/d \quad (12)$$

with astronomical units: θ (arcsec), a (AU), and d (pc).

Some examples of angular separations (exoplanet – star) are given in Table 2, along with the required telescope diameters, occulter diameters, and interferometer baselines needed to suppress the star and directly image the exoplanet.

TABLE 1. Standard spectral bands.

Band	λ_0*	$\Delta\lambda$†	a‡
U	0.365	0.068	−4.38
B	0.44	0.098	−4.19
V	0.55	0.089	−4.43
R	0.70	0.22	−4.76
I	0.90	0.24	−5.08
J	1.22	0.26	−5.48
H	1.65	0.29	−5.94
K_s	2.16	0.32	−6.37
L	3.55	0.57	−7.18
M	4.77	0.45	−7.68
N	10.47	5.19	−9.02
Q	20.13	7.8	−10.14

U, B, V, R, and I data is from *Allen* (1991) and *Cox* (2000). J, H, K_s, L, M, N, and Q data is from *Cox* (2000).

*Effective wavelength in µm.
†Effective bandwidth (FWHM) in µm.
‡$\log_{10}(f)$, where f has units of erg/(s cm² µm), at zero magnitude.

For the spectral types AFGKM, the approximate number of stars out to 10 and 30 pc is also noted.

2.3. Contrast of Planet

The spectrum of a planet is the sum of reflected starlight, thermal emission, and nonthermal features, as illustrated in Fig. 4 for the case of the Earth-Sun system as seen from a distance of 10 pc. The reflected and thermal continuum components are discussed in sections 2.4 and 2.5. Background light from zodiacal dust is discussed in section 2.6. A planet's color is discussed in section 2.7, and its absorption line spectrum in section 2.8. Nonthermal features (e.g., auroras) are expected to be faint, and are ignored here.

For direct imaging it is convenient to compare the brightness of a planet to its star, at any wavelength. The *contrast* C is defined to be the ratio of planet (p) to star (s) brightness, so we have

$$C = \frac{f_\lambda(p)}{f_\lambda(s)} = \frac{\dot{N}_\lambda(p)}{\dot{N}_\lambda(s)} \quad (13)$$

where C is a function of wavelength, the properties of the planet, and the apparent geometry of the planet-star system. Here f(p) is the sum of reflected and thermal fluxes.

The expected visible-wavelength contrast of typical Jupiter-like and Earth-like planets around nearby stars is shown in Fig. 5. We see that giant planets beyond the ice line will have typical contrasts on the order of 10^{-9} at visible wavelengths (see section 2.4), and separations of about 0.5 arcsec. Earth-like planets in the habitable zone will have contrasts of about 10^{-10} and separations of about 0.1 arcsec. As suggested by the limiting-case detection lines for several types of groundbased coronagraphs and the HST, these planets cannot be directly imaged by them. However, they could be imaged by a co-

TABLE 2. Angular separation examples.

Distance	10 pc	30 pc
Angular separation of planet at 1 AU (max)	100 mas	33 mas
Telescope diameter (min) at 0.5 μm	3.1 m	6.2 m
Occulter diameter at 0.5 μm	49 m	16 m
Interferometer baseline at 10 μm	21 m	62 m
Number of AFGKM stars	2, 11, 26, 42, 210	27 times greater

Angular separation θ is from equation (12), for the Earth-Sun system. Telescope diameter D is from $\theta = n\lambda/D$, where n = 3 is intermediate between the theoretical minimum for an internal coronagraph (n = 2) and an experimentally demonstrated value (n = 4). Occulter diameter is $D_O = 2\theta d_O$, where the distance between a telescope and its external occulter is $d_O = 50{,}000$ km. Interferometer baseline is $B = \lambda/\theta$. The number of stars is assumed to scale as d^3.

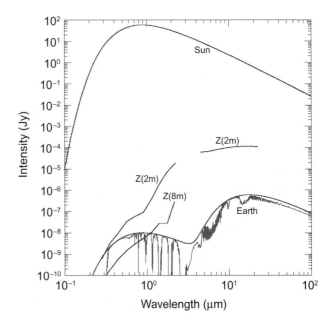

Fig. 4. Schematic spectrum of the Sun and Earth at 10 pc, in the visible and infrared (*Kasting et al.,* 2009). Here Earth at maximum elongation is at 0.1 arcsec (1 AU/10 pc) with a contrast of 10^{-10} in the visible and 10^{-7} in the mid-infrared (~10 μm). The exozodiacal light is sketched for small (2-m) and large (8-m) telescopes in the visible, and an interferometer with 2-m collectors in the infrared.

ronagraph in space, designed for this purpose, as discussed in this chapter.

2.4. Visible Brightness of Planet

Reflected starlight from a planet is often assumed to follow a Lambert law, which states that the light that is incident on a surface, from any direction, is reflected uniformly in all directions, in the sense that the amount of light leaving an element of a surface is proportional to the projected area in the reflected direction. So to an observer, the apparent brightness of any given projected area of the illuminated surface of a planet is proportional to the amount of starlight hitting the surface within that apparent area.

The *phase angle* α of a planet is the planet-centered angle from star to observer. So α = 0 at superior conjunction with the planet behind the star, α = π/2 at quadrature (maximum elongation for a circular orbit), and α = π at inferior conjunction with the planet between the star and observer.

As an example, if the Moon were a Lambert reflector, then the full Moon (α = 0) would appear to be a uniformly bright object, with no limb darkening, but the quarter Moon (α = π/2) would appear to have a bright Sun-facing limb that tapers to zero intensity at the terminator, in proportion to the projected area toward the Sun (i.e., the cosine of the angle between the surface normal and the Sun). In practice, bare-rock bodies like Mars, Earth, and the Moon tend to be more uniformly bright than a Lambert surface, but cloudy planets like Venus and Jupiter tend to be closer to Lambertian.

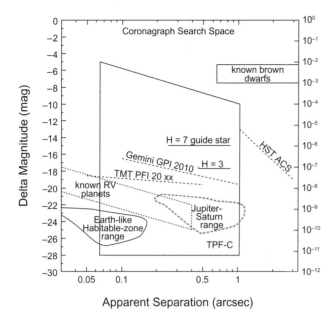

Fig. 5. Contrast vs. separation is shown for several types of companions, along with limits of the TPF-C coronagraph, as once planned. For example, the contrast of Earth and Jupiter twins, for many nearby stars, and for separations from about 20% to maximum elongation are shown.

The *geometric albedo* p of a planet is defined to be the ratio of planet brightness at $\alpha = 0$ to the brightness of a perfectly diffusing disk with the same position and apparent size as the planet. In other words, p is the ratio of the flux reflected toward an observer at zero phase angle to the flux from the star that is incident on the planet. The geometric albedo will in general be wavelength dependent. Numerical values of p_V, for the visible band, are listed in Table 3.

The reflected-light contrast of a planet can be written as

$$C_{vis} = p\phi(\alpha)(r_p/a)^2 \quad (14)$$

where $\phi(\alpha)$ is the phase law, sometimes called the integral phase function, at phase angle α, r_p is the planet radius, and a is the distance from planet to star, here simply written as the semimajor axis. For a Lambert sphere the phase law is

$$\phi(\alpha) = [\sin(\alpha) + (\pi - \alpha)\cos(\alpha)]/\pi \quad (15)$$

For example, in an edge-on system $\phi(0) = 1$ at superior conjunction, $\phi(\pi/2) = 1/\pi$ at maximum elongation, and $\phi(\pi) = 0$ at inferior conjunction.

This law is a good approximation for high-albedo planets such as Venus. There are no convenient expressions for other types of planet surfaces, such as rocky, low-albedo ones, but it is empirically observed that such objects tend to have relatively stronger reflection at zero phase angle, probably from a lack of shadowing on clumpy surfaces, compared to other angles. The net result is that the phase function tends to be smaller than the Lambert law, at angles away from zero. However, for lack of a better version, the Lambert law is often used for exoplanets.

For example, the visible contrasts of the Earth/Sun and Jupiter/Sun systems at maximum elongation, assuming that they reflect as Lambert spheres, are

$$C_{vis}(E) \approx 2.1 \times 10^{-10} \quad (16)$$

$$C_{vis}(J) \approx 1.4 \times 10^{-9} \quad (17)$$

These extreme contrasts, 10^{-10} or Δmag = 25 for Earth, and 10^{-9} or Δmag = 22.5 for Jupiter, are the driving forces behind nearly all the direct imaging discussion in this chapter. These huge brightness ranges, occurring in such close proximity on the sky, mean that scattered light in a telescope system, which is ignored in conventional astronomical imaging, now becomes an important experimental factor in isolating the light of a planet.

2.5. Infrared Brightness of Planet

The *Bond albedo* of a planet, A_{Bond}, is defined to be the ratio of total light reflected to total light incident, where "total" here means bolometric, i.e., integrated over all wavelengths, and the entire planet.

The emittance F of a black-body at *effective temperature* T is the total flow of radiation outward from a unit area of its surface, and is given by

$$F = \int B_\nu d\nu \cos(\vartheta) d\Omega = \int \pi B_\nu d\nu = \sigma T^4 \quad (18)$$

in units of erg/(s cm^2), where ϑ is the angle from the normal to the surface, $d\Omega = \sin(\vartheta)d\vartheta d\varphi$, φ is the azimuth around the

TABLE 3. Albedo and temperature.

Planet	a (AU)	p (visible geom. alb.)	A_{bond} (Bond alb.)	T_{equil}* (K)	T_{eff}† (K)
Mercury	0.387	0.138	0.119	433	433
Venus	0.723	0.84	0.75	231	231
Earth	1.000	0.367	0.306	254	254
Moon	1.000	0.113	0.123	269	269
Mars	1.524	0.15	0.25	210	210
Jupiter	5.203	0.52	0.343	110	124.4
Saturn	9.543	0.47	0.342	81	95.0
Uranus	19.19	0.51	0.290	58	59.1
Neptune	30.07	0.41	0.31	46	59.3

Data adapted from *de Pater and Lissauer* (2001, 2010).

*T_{equil} is calculated from A_{Bond}.
†T_{eff} is set equal to T_{equil} for terrestrial planets, but is measured for gas giants.

normal, and σ is the Stefan-Boltzman constant. For a star, this leads to the *luminosity* L, given by

$$L = 4\pi r_s^2 \sigma T^4 \qquad (19)$$

where r_s is the radius of the star and T the effective temperature.

The flux from a star is diluted by a^{-2} by the time it reaches a planet at distance a. Of this, a fraction $(1-A_{Bond})$ is absorbed by the planet. The resulting *radiative equilibrium temperature* T_{equil} of a planet is determined by setting the incident flux equal to the radiated flux, assuming that the heat from the incident radiation is uniformly distributed over a fraction f of its total surface area, and that it radiates with an emissivity of unity. We find

$$T_{equil} = \left(\frac{1-A_{Bond}}{4f}\right)^{1/4} \left(\frac{r_s}{a}\right)^{1/2} T_s \qquad (20)$$

Here f = 1 for a rapid rotator and f = 0.5 for a tidally locked or slowly rotating planet with no transfer of heat from the hot to cold side. The value of T_{equil} refers to the fraction f of area over which the heat is spread; this simple formulation assumes that none of the incident heat is distributed to the (1–f) of the remaining area, which would therefore be very cold.

The effective temperature T_{eff} of a planet is determined by fitting a blackbody curve to its experimentally measured infrared emission spectrum. If a planet has an internal heat source, then $T_{equil} < T_{eff}$; otherwise, these are equal.

Table 3 lists A_{Bond}, T_{equil}, and T_{eff} for solar system planets. The terrestrial planets have negligible internal heat sources so they have $T_{eff} = T_{equil}$, but the giant planets (save for Uranus) have significant internal heat, as measured in the thermal infrared. For example, Jupiter would have T_{equil} = 110 K from albedo alone, but internal heat pushes the observed effective temperature up to 124 K.

For the Earth we get T_{equil} = 254 K, which is representative of an effective radiating altitude (~40 km). The infrared optical thickness of the atmosphere below this level isolates the radiating level from the surface. The surface of Earth is roughly 288 K, i.e., 34 K warmer owing to the greenhouse effect of H_2O and CO_2, keeping it above the freezing point of water, on average.

The *habitable zone* (HZ) is defined as that range of distances from a star where liquid water can exist on the surface of a planet. For example, Earth has surface oceans and is therefore within its HZ. To extend this to terrestrial exoplanets requires knowing the factor $(1-A_{Bond})/f$ and the greenhouse effect for that planet. Absent this knowledge, we sometimes assume that a planet is like Earth in these respects, in which case the HZ for an Earth-like planet around another star will scale as $L^{1/2}$, giving

$$a(\text{HZ, E-like}) = (1\text{ AU})(L/L_\odot)^{1/2} \qquad (21)$$

The HZ in the solar system is approximately bounded by Venus and Mars. Early in the age of the solar system the luminosity of the Sun was 60% of its present value, so Venus may have been habitable then, before it experienced a runaway greenhouse effect that raised its surface temperature far above the liquid water range. Surprisingly, Mars may have been habitable at early times as well, if it had a sufficiently thick atmosphere with a strong greenhouse effect; however, it has since lost most of that atmosphere and is now well below the liquid water range. On this basis the HZ is empirically defined to be the range 0.7 AU to 1.5 AU, scaled by the square root of stellar luminosity.

The infrared contrast C_{IR} of a planet-star system is estimated by assuming that both bodies are uniformly luminous blackbodies. In this case the contrast depends only on the effective temperatures and radii, and not on the planet phase. We have

$$C_{IR}(\lambda) = \frac{B_\lambda(T_p) r_p^2}{B_\lambda(T_s) r_s^2} \qquad (22)$$

and for a solar system twin we find these contrast values at a reference wavelength of λ = 10 μm

$$C_{IR}(E) \simeq 8.2 \times 10^{-8} \qquad (23)$$

$$C_{IR}(J) \simeq 2.8 \times 10^{-8} \qquad (24)$$

For example, if the Jupiter-Sun system were to be directly imaged from a distance of 10 pc, the intensities of these components would be about as shown in Fig. 6, which is similar to the case of the Earth-Sun system in Fig. 4, except for Jupiter being about an order of magnitude brighter in the visible and infrared. Note also the zodi brightness in both figures, as discussed in the next section.

2.6. Exozodi

The visible, reflected-light surface brightness of a zodiacal disk similar to the solar system disk was calculated by *Kuchner* (2004a) on the basis of visible and infrared observations of the local zodi, for the case of a Hong phase function, for a disk inclined at a median angle of 60°. The tabulated values are closely fit by a simple function given by

$$m_V = 22.1 + 5.6\log(R_{AU}) \qquad (25)$$

where m_V is the apparent brightness in units of V-band magnitudes, for a 1-arcsec² solid angle, and R_{AU} is the radius in the disk with units of AU, in the range R_{AU} = 0.1 to 4.5 AU. The V-band flux is then

$$F_{1-zodi} = 10^{-4.43-0.4m_V} \Omega_{as} \qquad (26)$$

in units of erg = (s cm² μm). Here Ω_{as} is the solid angle of the telescope in square arcseconds, which from equation (50) is

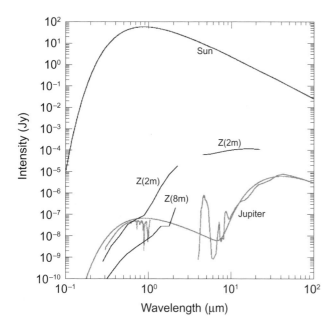

Fig. 6. Schematic spectrum of the Sun and Jupiter at 10 pc (*Kasting et al.*, 2009). Here Jupiter at maximum elongation is at 0.5 arcsec (5 AU/10 pc) with a contrast of 10^{-9} in the visible and 10^{-7} in the mid-infrared (~10 μm). The visible and infrared spectra are roughly approximated by blackbody spectra from reflected and emitted light; the prominent exception is the 4–5-μm peak, which corresponds to a spectral window on Jupiter, allowing us to see deeper and warmer levels compared to the cloud tops. The exozodi for 2-m and 8-m collectors is as in Fig. 4.

$$\Omega_{as} \simeq \frac{\pi}{4}\left(\frac{\lambda}{D} 206{,}000\right)^2 \qquad (27)$$

where the factor of $360 \times 60 \times 60/2\pi \simeq 206{,}000$ converts radians to arcseconds.

As an example, equating the exozodi flux at 1 AU around a solar system twin at 10 pc, to the flux from an Earth at quadrature, and using the fact that the absolute magnitude of the Sun is $M_V = 4.82$, we find that the exozodi signal equals that of Earth for a telescope of diameter $D = 2.4$ m. Thus a telescope of this size or larger is needed to make Earth stand out from a solar-system-like zodi at 10 pc.

The thermal infrared intensity I can be modeled as a dilute blackbody $B_\lambda(T)$, with temperature T and optical depth τ specified empirically as a function of distance R_{AU} from the star

$$I(R_{AU}) = \tau(R_{AU}) B_\lambda\left(T(R_{AU})\right) \qquad (28)$$

where B is given by equation (2), for example. Here we assume the same disk as above, i.e., solar-system-zodi twin around Sun twin at 60° inclination. The optical depth is a weak function of radius, given by

$$\tau = 1.42 \times 10^{-7} R_{AU}^{-0.34} \qquad (29)$$

The local temperature is a somewhat steeper function of radius, given by

$$T(R_{AU}) = 277 \times R_{AU}^{-0.467} \qquad (30)$$

with units of Kelvins. For a nonsolar star, the temperature at 1 AU should be scaled as T_s, per equation (20).

If we ask the same question that we did for the optical range, namely what size telescope diameter D would give an exozodi signal equal to Earth at 10 pc (i.e., $\tau_E B_E \Omega_E = \tau_z B_z \Omega_z$, for $\Omega_z = (\pi/4)(\lambda/D)^2$), we find D = 105 m. This is larger than any future space telescope, and no ground telescope would be relevant because the warm background would be prohibitively large. So we see that the infrared zodi is going to be a bigger problem than the visible zodi. More realistically, we should use smaller (3-m) telescopes in an interferometer configuration (see section 3.18), but in this case the beam pattern of the interferometer will be a central nulling fringe projected onto the full zodi disk. Integrating over these distributions, the configuration of the Terrestrial Planet Finder Interferometer (TPF-I), for example, finds that the exozodi flux is about 100 times stronger than the Earth flux, again a large noise source, but one that might be workable.

References include *Kuchner* (2004a), for the Zodipic algorithm used in this section, and *Beichman et al.* (1999), for the TPF-I concept study.

2.7. Color

Exoplanets are faint, so the first direct images of them may be in broad photometric bands. The ratio of fluxes in two such bands, or equivalently the differences of magnitudes, give *color* information. A color-color diagram is shown in Fig. 7 for planets in the solar system. In the field of stellar astrophysics, color-color diagrams are a useful classification tool, and they could be for exoplanets as well, once we start obtaining direct images in photometric bands.

As an example, some sources of these colors are as follows. A rocky planet with little or no atmosphere tends to be relatively brighter in the red than in the blue, giving these surfaces a slightly red color, and explaining the clustering of points for Mercury, Moon, and Mars in the upper right of this diagram. A cloudy gas-giant planet with a substantial amount of gas-phase methane above its clouds, like Jupiter or Saturn, will be relatively faint in the red owing to the strong absorption bands of methane (see, for example, Table 4 for these band positions), and therefore its color will tend to look slightly blue, thus explaining the cluster of points in the blueward direction of this diagram. In ice giant planets, like Uranus and Neptune, the atmosphere is so cold that the clouds form at a relatively low level in the atmosphere, with a relatively large amount of methane above the clouds, producing almost total absorption of red light, and making the planet look significantly blue-green; this effect explains the extreme positions of these planets in this diagram. A fully cloud-covered terrestrial planet, like Venus, reflects with very little color compared to the Sun, but has a slight absorption at short visible wavelengths, possibly owing to a pigment in the sulfuric-acid cloud droplets. Earth is famously blue owing to strong Rayleigh scattering in its atmosphere (not ocean reflectivity), and therefore occupies a unique position off to the left in this diagram.

2.8. Spectroscopy

The immediate purpose of directly imaging an exoplanet is to measure its photon flux in broad and narrow wavelength bands. From these measurements we can characterize the planet in terms of mass, radius, effective temperature, age, temperature structure, molecular composition, clouds, rotation rate, and atmospheric dynamics. For Earth-like planets we can also search for habitability in terms of a surface temperature and pressure that permits liquid water, as well as signs of life, as evidenced by the presence of disequilibrium species such as coexisting oxygen (or ozone) and methane, and possibly the "red edge" reflective spectral signature from land plants. Small amounts of oxygen can be produced photochemically, and indeed we see oxygen on Mars, for example, but large amounts of oxygen cannot readily be produced (except perhaps in a runaway greenhouse situation where water is photodissociated and the hydrogen escapes, leaving oxygen), so a large amount of oxygen, such as on Earth, is a possible sign of life on a planet.

As examples, Fig. 8 shows Earth's visible spectrum, as seen in Earthshine (light reflected from Earth to the dark side of the Moon, and back again to a groundbased telescope), where the Rayleigh scattering is strong, and bands of oxygen and water are prominent. Figure 9 shows the near-infrared Earthshine spectrum of Earth, in which water bands are very strong. Finally, Fig. 10 shows the thermal infrared spectrum of Earth. In all three of these cases, a simple model of Earth's atmosphere has been used to model the data, successfully reproducing the main features.

Exoplanets are faint, so we may expect that the time sequence of observations will be (1) detection in a convenient broad spectral band; (2) photometry in several broadbands, leading to a characterization by color; and (3) spectroscopy in narrow bands, leading to the identification of molecular bands and strong lines of atomic species. For Earth-like exoplanets,

TABLE 4. Spectral features of Earth.

Species	λ_0 (μm)*	$\Delta\lambda$ (μm)†	Depth‡
O_3	0.32	0.02	0.69
O_3	0.58	0.13	0.20
O_2	0.69	0.01	0.12
H_2O	0.72	0.02	0.37
CH_4	0.73	0.01	0.002
O_2	0.76	0.01	0.47
CH_4	0.79	0.03	0.001
H_2O	0.82	0.02	0.32
CH_4	0.89	0.03	0.002
H_2O	0.94	0.06	0.71
CH_4	1.00	0.05	0.011
CO_2	1.05	0.02	0.0006
H_2O	1.13	0.07	0.80
CO_2	1.21	0.03	0.01
O_2	1.27	0.02	0.15
H_2O	1.41	0.14	0.95
CO_2	1.59	0.14	0.03
CH_4	1.69	0.16	0.012
H_2O	1.88	0.18	0.97
CO_2	2.03	0.12	0.31
CH_4	2.32	0.29	0.009
H_2O	7.00	0.70	0.83
CH_4	7.65	0.59	0.09
N_2O	7.75	0.14	0.10
N_2O	8.52	0.37	0.02
CO_2	9.31	0.49	0.05
O_3	9.65	0.58	0.41
CO_2	10.42	0.65	0.04
CO_2	14.96	3.71	0.52
H_2O	20.49	7.64	0.21

Data adapted from *Des Marais et al.* (2002). Abundances are for present Earth.

*Central wavelength of feature.
†Approximate full-width at half-maximum.
‡Approximate depth of feature (e.g., 0.01 is a weak line, 0.95 strong) for Earth at quadrature, assuming a cloud-free atmosphere; if clouds are present, depths will be somewhat smaller.

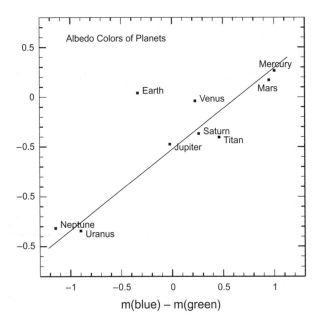

Fig. 7. A color-color diagram for planets in the solar system. The wavelength bands here are blue (0.4–0.6 μm), green (0.6–0.8 μm), and red (0.8–1.0 μm). This diagram shows that low-resolution color information can be valuable in classifying a planet (*Traub*, 2003).

Des Marais et al. (2002) listed all significant spectral features from about 0.3 to 100 μm wavelength, including the width and depth of each for a variety of abundances. They found that the highest resolution needed for an Earth twin is R = $\lambda/\Delta\lambda$ = 70 for the O_2 band at 0.76 μm.

At visible and near-infrared wavelengths a coronagraph can utilize an *integral field spectrometer* (IFS) to provide simultaneous spectra of all pixels in the focal plane. At thermal-infrared wavelengths an interferometer has a single spatial pixel covering the entire star-planet system, so the flux in this pixel must be passed through a spectrometer before detection and image reconstruction.

2.9. Photons as Waves

A photon can be thought of as a particle when it is emitted from an atom on the surface of a star, and again when it is absorbed by a detector, but during its journey through space and through our optical instruments it is necessary to picture it as a wave. This empirical view is an expression of the famous wave-particle duality in quantum mechanics, as applied to photons.

To understand how a photon interacts with a telescope or interferometer, let us first consider how the photon gets from the star to our instruments. It is helpful to think of the light source, here a star, as a collection of many light-emitting atoms, but to visualize only one photon at a time as being emitted from that star, from a random location on the star (weighted by the surface intensity, of course). Each atom emits randomly in time and in phase with respect to each other atom; in other words, there is no coherence between one point and another on the surface of the star. Note that textbooks often speak about the partial coherence of light from a star, but this is an artifact of how we observe the star, not a property of the star itself.

This single photon propagates through space as a spherical expanding shell, with a thickness equal to the coherence length of the photon (speed of light times the lifetime of the emitting state of the atom), but not localized at any particular point or region on the sphere. We often assume that the wave is monochromatic, with a single wavelength and essentially infinite coherence length. The electric field of the photon is proportional to the real part of

$$e^{-i(\omega t - \vec{k}\cdot\vec{r})} \qquad (31)$$

times a constant and times r^{-2} where $\omega = 2\pi/f$, f is the time frequency of oscillation, t is time, $\vec{r}$ is distance from the emitting atom, k = $2\pi/\lambda$, λ is the wavelength, and $\vec{k}/k$ is a direction vector from the atom to any point on the expanding sphere. We could use cos(X) instead of the real part of e^{iX} but the latter is more convenient for calculations. In thinking about diffraction we are entirely concerned with the interaction between the wavelength λ and the spatial dimensions of our apparatus, so we drop the time variation. Also, since the star is very distant compared to our instrument dimensions, we approximate the amplitude A of the electric field of the incident spherical wave as a plane wave

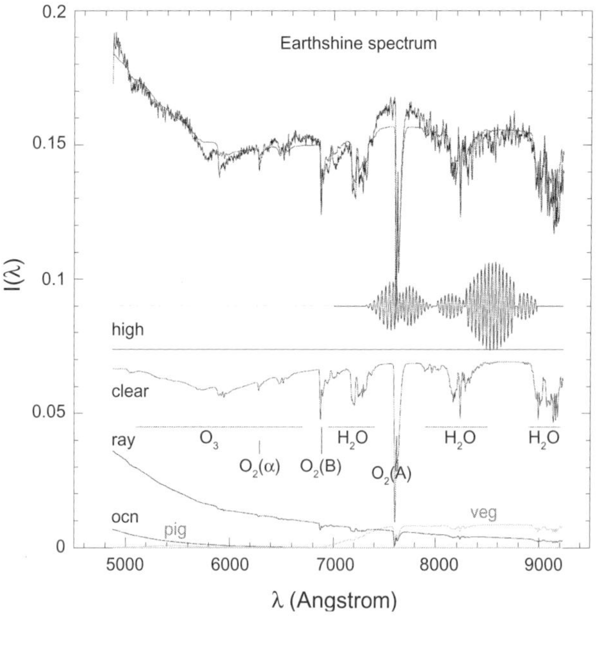

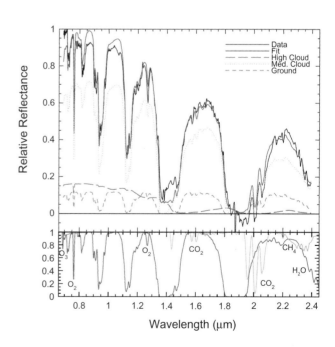

Fig. 8. The visible reflection spectrum of Earth, observed and modeled, along with the contributing spectral components from the clear atmosphere, clouds, Rayleigh scattering, and with weak contributions from the ocean as well as the red edge of land plants (*Woolf et al.*, 2006).

Fig. 9. The near-infrared reflection spectrum of Earth, observed and modeled, along with the contributing spectral components (*Turnbull et al.*, 2006). Gas-phase water is the dominant contributor.

$$A(\vec{k},\vec{x}) = A_0 e^{i\vec{k}\cdot\vec{x}} \qquad (32)$$

where $\vec{x}$ is a local position vector centered in our instrument, and we usually set $A_0 = 1$.

The *wavefront* is defined as the surface containing all contiguous points in space at which the phase of this wave has the same value. Successive wavefronts are separated in space by λ, in phase by 2π, and in time by λ/c. If the medium is not a vacuum, but instead has an index of refraction n, then successive wavefronts are separated in space by λ/n, in phase by 2π, and in time by λ/nc, and the phase delay ϕ along a path of geometrical length z is $\phi = 2\pi n z/\lambda$.

In a medium of index of refraction n, a simple rule is to use a wavelength $\lambda = \lambda_0/n$ where λ_0 is the vacuum wavelength. For example, a wavefront that has passed through an ideal convex lens, thicker in the center than at the edges, will be delayed proportionately to the thickness of glass traversed, thus making it into a spherical converging wavefront.

The trick is now to think of this plane wave as falling on our instrument at all points equally, no matter how large an aperture, or how far apart one element of the aperture is from another. In other words, we can have one or many discrete entrance aperture elements, over as large an area as we wish, and the photon (wave) will somehow take notice of the arrangement and manage to "feel" the entire apparatus.

To visualize the interaction it is helpful to think of the Huygens wavelet picture, which is shown schematically in Fig. 11. Here, at every point on a wavefront, we imagine that little wavelets sprout and propagate outward, each on its own little spherical hemisphere. At a short time thereafter we add up all the electric fields from these propagating points. For a plane wave in free space, these wavelets will tend to cancel each other in all directions except for the single direction that is perpendicular to the wavefront. This direction defines the local *ray*, and is the basic element of geometrical optics.

For a plane wave passing through a finite-size aperture, the wavelets at the center of the aperture will tend to continue onward as before, but the wavelets near the edge will be able to propagate off to the side as well, because there are no canceling wavelets in the "shadow" of the aperture. This concept is the basis for the wave-optics picture that dominates in the following sections.

At any point inside the telescope or interferometer, if we wish to know the total electric field from the incident photon, we need only add up all the wavelets that could have reached that point, or in other words, calculate the sum of wavelet amplitudes *taking into account the phase of each wavelet* according to its distance of travel. We can write this summation as

$$A_{out}(\vec{x}) = \int M(\vec{x}')A_{in}(\vec{x}')e^{i\phi(\vec{x},\vec{x}')}d\vec{x}' \qquad (33)$$

where $A_{out}(\vec{x})$ is the amplitude of the total electric field at point $\vec{x}$ on the detector, $A_{in}(\vec{x}')$ is the incident electric field amplitude on the apparatus at point $\vec{x}'$ in the pupil, $M(\vec{x}')$ is a mask function that modifies the incident wavefront, and

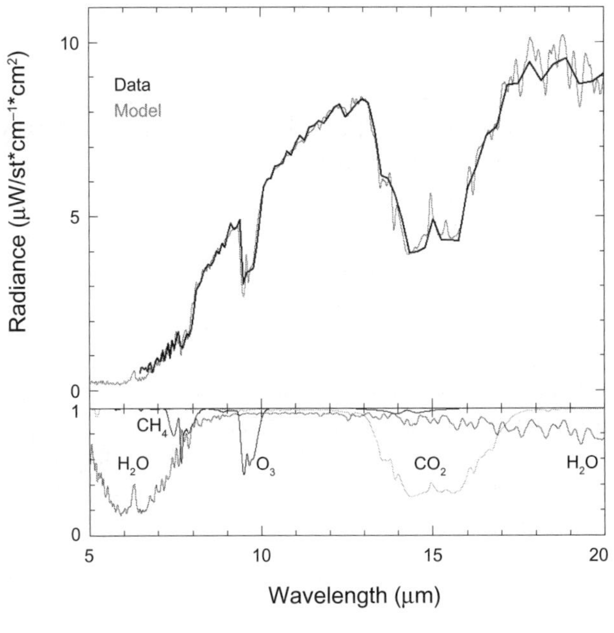

Fig. 10. The far-infrared thermal emission spectrum of Earth, observed and modeled, showing strong contributions from CO_2, O_3, and H_2O. Data (broken heavy line) is from the Thermal Emission Spectrometer, enroute to Mars, and fitted spectrum is from *Kaltenegger et al.* (2007).

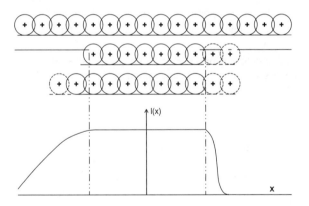

Fig. 11. Huygens' wavelets schematic. The incident wavefront approaches an opening, propagating from top to bottom. Before striking the opening the wave propagates by successive reformation of wavelets emerging from every point on the advancing wavefront, the sum of which, a short distance downstream, recreates a smooth, forward-moving plane wavefront. For illustration, this opening is drawn with a hard edge on the left but a soft semitransparent edge at the right; the geometrical shadow boundary is indicated by vertical dashed lines. At the hard edge, wavelets propagate beyond the geometrical shadow boundary. At the soft edge, wavelets are damped by the presence of adjacent weaker wavelets, with the net effect that the wavefront stays closer to the geometrical shadow boundary than in the hard-edge case.

$\phi(\vec{x}, \vec{x'})$ is the phase difference between the incident electric field at $\vec{x'}$ in the pupil and $\vec{x}$ in the detector, as measured along a minimum-time ray path. For example, a point-source star on the axis of our coordinate system will have $A_{in} = 1$ in the pupil plane, but if it is off-axis at angle $\theta \ll 1$ then $A_{in}(x') = e^{ik\theta x'}$. Also, we often have $M = 1$ in the pupil and $M = 0$ outside, although M can also be partially transparent, and can also have a phase delay of its own.

The intensity at point **x** on the detector is proportional to the magnitude-squared of the electric field

$$I(\vec{x}) = |A(\vec{x})|^2 \quad (34)$$

If the star has a finite angular size, then the intensity (not the electric field) from each point on its surface must be summed. If the detector pixel has a finite size, then the intensity over the area of the pixel must be summed as well. Finally, if the star has multiple wavelengths, then the intensity for each wavelength (weighted by the star spectrum and the transmission of the optics) must be summed.

2.10. Photons as Particles

The net intensity in each detector pixel is proportional to the *probability* that a photon will be detected in that pixel, for example, by the generation of a conduction electron in a CCD or CMOS detector. We can visualize the detection process by picturing all the wavefront segments of a photon collapsing to a single point in on the detector, no matter how large the aperture or how widespread the collection of subapertures. After many photons have passed through the apparatus, the measured intensity pattern will match the shape of the probability pattern.

The detected number of photons in a finite time interval will be given by the *Poisson process*

$$f(n, \bar{n}) = \bar{n}^n e^{-\bar{n}} / n! \quad (35)$$

where f is the probability that there will be exactly n electrons detected in a pixel, for the case where $\bar{n}$ is the expected average number of electrons. Here, of course, $\bar{n}$ is proportional to the calculated intensity distribution of diffracted light. The standard deviation of the number of detected events is $\sigma_n = \bar{n}^{1/2}$; this is called *photon noise* or *shot noise*. For large values of $\bar{n}$, say 10 or more, the probability distribution approaches a Gaussian or *normal process*

$$f(n, \bar{n}, \sigma) = \frac{1}{\sigma\sqrt{2\pi}} e^{-(n-\bar{n})^2/2\sigma^2} \quad (36)$$

where again $\bar{n}$ is the mean and σ is the standard deviation, and in addition $\sigma = \bar{n}^{1/2}$.

These are important relations because directly imaged exoplanets are typically faint. For example, for an Earth twin at 10 pc we expect $\bar{n} \simeq 0.5$ photon/m² s⁻¹ in a 10% bandwidth at visible wavelengths.

2.11. Photons in the Radio and Optical

Photons are photons, whether they have long (e.g., radio) or short (e.g., visible) wavelengths; however, there is a big difference in how they are detected. For example, we know that radio astronomers routinely use heterodyne detection, but optical astronomers do not. Why is this?

One part of the answer is Heisenberg's uncertainty principle, $\Delta E \Delta t \geq h/2\pi$. Suppose that we want to interfere an incident photon with one from a local oscillator. This will fundamentally amount to isolating it in time with an accuracy of a radian, or less, in phase, so $\Delta t \leq (\lambda/2\pi)/c = 1/(2\pi\nu)$. This gives us $\Delta E \geq h\nu$, which tells us that the uncertainty in the number of detected photons is greater than one, even though we have assumed that there is only one incoming photon in the first place. This quandry leads us to the second part of the answer, as follows.

We know from equations (46) and (49) that a wavefront incident on an opening of width D will be diffracted into an emerging beam of angular width approximately λ/D. So in two dimensions the product of area and solid angle is approximately

$$A\Omega = \lambda^2 \quad (37)$$

By time-reversal symmetry, this relationship applies to the emission process as well as the detection process. The conserved quantity, λ^2, defines a single electromagnetic mode. Applying this to the emission process, and using equation (7), we find that the photon rate $\dot{N}_\nu A$ from a blackbody, into solid angle Ω, and from area A, is

$$\dot{N}_\nu A = \dot{n}_\nu A\Omega$$
$$= \frac{2}{e^{h\nu/kT} - 1} \quad (38)$$

photons per second per Hertz into $A\Omega$. Now the uncertainty relation $\Delta p \Delta x \geq h$ for a photon, where its momentum is $p = h\nu/c$ and length is $x = ct$ gives

$$\Delta\nu\Delta t \geq 1 \text{ Hz sec} \quad (39)$$

for the product of a photon's frequency spread and total length in time. In addition, there are two polarization states possible. So using these minimum values we get the number of photons in the minimum area, minimum solid angle, minimum frequency bin, and minimum time interval, per polarization state, i.e., a single electromagnetic mode, as $n_{mode} = \dot{n}_\nu A\Omega\Delta\nu\Delta t/2$

$$n_{mode} = \frac{1}{e^{h\nu/kT} - 1} \quad (40)$$

As an example, suppose we are looking at a star or other object with a brightness temperature of $T = 5000$ K. Then in the visible, say $\lambda < 1$ μm, we get $n_{mode} < 0.1$ photon in a single electromagnetic mode, so we should only expect one photon at a time, on the average. However, at 10 μm we get three identical photons per mode, so heterodyne detection, with its added certainty of one photon, is just barely possible at this wavelength.

At longer wavelengths, say 1 cm, we get 3400 identical photons. This says that this photon is one of a group of 3400 others that are just like it, and are indistinguishable. Therefore we can have multiple radio antennas, each with its own receiver, completely independent of all the others, receiving some of these photons.

This explains why radio interferometers can be analyzed as if they were detecting classical waves, where each antenna can detect a small fraction of the classical wave. This is not how photons are detected, where once a given antenna detects a photon, the other antennas are automatically not allowed to detect that same photon. This is the fundamental reason why radio arrays work, because the array is showered with many identical photons. This is a consequence of stimulated emission in the blackbody source, whereby when one photon is emitted it stimulates many others to be emitted en route to leaving the source.

The Hanbury-Brown Twiss intensity interferometer is based on the above idea that there is a tendency for photons (as bosons) to arrive in pairs. This tendency is weak in the visible, but strong in the radio. The effect is strongest for a point source, as above, and it will be diluted for a source that appears to be resolved by the detecting apparatus. This is the basis for the demonstrated ability of the intensity interferometer to measure the angular diameter of stars.

3. CORONAGRAPH AND INTERFEROMETER CONCEPTS

In this section we discuss how to use coronagraphs and interferometers to observe exoplanets. There are many concepts for coronagraphs, most of which have been invented specifically for exoplanet observations. This is a very exciting field, with new ideas coming along at a fast pace, very little of which can be found in any optics textbook. This is all the more surprising, given that it was once thought that the only way to directly image an exoplanet was with an interferometer in space. Today we know that both approaches are viable, at least in an optical sense. And at the heart of the matter, interferometers and coronagraphs are essentially the same type of machine, balancing the amplitude and phase of one part of a wavefront against that of another part. This section should provide a good understanding of both kinds of telescopes.

3.1. Overview of Types

The types of coronagraphs and interferometers discussed in this section are listed in Table 5. This table is a short, representative list. A much more exhaustive list of types of coronagraphs and their theoretical properties is given in *Guyon et al.* (2006). Column 1 in Table 5 gives the name or names of a class of instrument. Column 2 labels each as primarily a coronagraph or interferometer, although this labeling is largely a matter of taste and history, rather than a fundamentally definable property; for example, in every coronagraph the rays or wavefront segments must be nearly perfectly phased so that they interfere with an extremely high degree of cancelation, so a coronagraph is in fact an interferometer with many contributing elements. If we understand interferometers, then by definition we should be able to understand coronagraphs, and vice versa. In column 3 we designate whether the instrument operates primarily in the pupil plane or the image plane, although here too the label is partially arbitrary, because in all cases some operation (i.e., an amplitude or phase adjustment, or both) is necessary in both planes. In an ideal theoretical view, there are almost no cases in which a deliberately manipulative operation takes place in a plane other than these; the exceptions are in real systems, where slight offsets of the plane of a mask or lens may occur for practical reasons, and in the case of the Talbot effect (section 4.11), where amplitude and phase can be mutually transformed.

3.2. How to Observe Exoplanets: Single Telescope

We now calculate the amplitude and intensity of an incident wavefront as observed by a simple telescope. The phase is calculated across a tilted surface in the pupil, oriented at an angle with respect to the incoming wavefront, at an angle projected onto the sky that corresponds to the point of interest in the focal plane of a perfect lens located in the plane of the pupil. For each such tilted surface there is a corresponding point in the focal plane, on a straight line from the equivalent point in the sky, through the center of the lens, to the focal plane.

The reason for this identification of a tilted surface with a point in the focal plane is that the ideal lens transfers its incident wavelet fronts along a tilted plane, independent of their individual phases, to a converging spherical wave, the convergent center of which is in the focal plane, off-axis by the tangent of the angle times the focal length.

The relative strength of an outgoing wave from one of these surfaces is determined by adding up all the wavelets on that surface. The phase at each point is $2\pi/\lambda$ times the distance between the input and output wavefronts.

3.3. Classical Single Pupil

Our telescope model is as follows. A plane wave from a point on a star is incident on a pupil plane that contains an

TABLE 5. Types of direct imaging instruments.

Name	Type	Main Plane
Pupil-masking	coron.	pupil
Pupil-mapping, PIAA	coron.	pupil
Lyot, Gaussian	coron.	image
Band-limited	coron.	image
Phase, vector vortex	coron.	image
Starshade	coron.	image
Keck Int., TPF-I, Darwin	int.	pupil
Visible nuller	coron/int.	pupil

imbedded ideal lens and is opaque elsewhere; this is shown schematically in Fig. 12. We call this plane number 1. The amplitude of the incident electric field is $A_1(x_1)$, where the subscript denotes this first plane, and the coordinates in this plane are (x_1, y_1). For simplicity of notation, we will work only in the x dimension whenever possible, and we will drop the subscripts wherever the meaning is otherwise clear.

The phase $\phi_1(x_1)$ of a wavelet in plane 1, as seen from a point at position θ_2 in plane 2, and therefore measured with respect to a reference plane tilted at angle θ_2 in the pupil, is

$$\phi_1(x_1) = 2\pi x_1 \sin(\theta_2)/\lambda \simeq 2\pi x_1 \theta_2/\lambda \qquad (41)$$

where $x_1 \sin(\theta_2)$ is the distance between the incoming wavefront from direction $\theta_0 = 0$ and the outgoing direction at angle θ_2, and we assume $\theta_2 \ll 1$.

The imbedded lens will focus the sum of wavelets that exit the pupil at angle θ_2 to a star image at a point in the image plane, i.e., plane number 2. The electric field in this plane is denoted $A_2(\theta_2)$.

The amplitude $A_2(\theta_2)$ in the image plane is the algebraic sum of all wavelets across the pupil

$$A_2(\theta_2) = \sum(\text{wavelets}) = \int_D A_1(x_1) e^{i\phi_1(x_1)} dx_1 \Big/ D \qquad (42)$$

where the integral is over the diameter D of the pupil, and $A_1(x_1)$ is the amplitude of the incoming wave in plane 1. Here the pupil is one-dimensional, but generally it is two-dimensional.

The divisor $\int_D dx_1 = D$ normalizes the righthand side by dividing out the area factor; for simplicity we will usually drop this normalization in most of the rest of this chapter, except where it improves the appearance of the result. Strictly speaking, the units of A_1 and A_2 should be the same, but in this chapter they are not, owing to the integral over the pupil; however, we shall retain this system in order to keep the notation simple. The correct factor in a result can often be calculated by applying conservation of energy between the input pupil and output image plane. Likewise, linear coordinates in each plane should be labeled x_1 for x, and x_2 for θ_2, but we prefer to let the physics dominate the math, and will often use x and θ where it is clear what is meant from the context. Regarding signs, we note that from geometric optics, a linear coordinate in the focal plane x_2 is related to the angle on the sky θ by $x_2 = -f \tan(\theta) \simeq -f\theta$, where f is the focal length of the telescope.

Inserting the approximate expression for $\phi_1(x_1)$ we get

$$A_2(\theta) = \int_D A_1(x_1) e^{i 2\pi \theta x_1/\lambda} dx_1 \qquad (43)$$

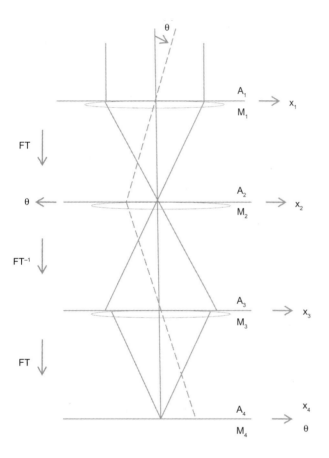

Fig. 12. Four coronagraph planes are shown: 1 is the input pupil; 2 is the image plane; 3 is a reimaged pupil; and 4 is a reimaged image plane. A simple telescope only uses planes 1 and 2, with the detector at plane 2. An internal coronagraph puts the detector at plane 4. The A_i are the incident and propagated electric field amplitudes falling on each plane. The M_i are masks in the planes. The transmitted amplitude just after each plane is $M_i A_i$. There is an ideal lens imbedded in each plane, with focal lengths such that the pupil in plane 1 is imaged on to plane 3, and the sky is imaged onto planes 2 and 4. The mask M_1 is the first pupil stop, and M_3 is often called the Lyot stop. In a conventional coronagraph mask M_2 could be a rectangular or Gaussian dark spot, blocking the central part of the star image. The mask M_4 can be thought of as the assembly of individual pixels in the detector. Linear positions in each plane are positive to the right, but by geometric optics the angle coordinate in plane 2 points left, while in plane 4 it points right. The star here is on-axis, at $\theta_0 = 0$.

This is the basic working equation for much of the imaging calculations that follow. Note that we use a "+" sign in the exponent, owing to our choice of coordinates; it is conventional to use a "–" sign, but the resulting intensities will be the same in either case.

In this case, where we retain only the linear approximation $\sin(\theta) \simeq \theta$ in the pupil plane, we speak of *Fraunhofer diffraction* (*Born and Wolf,* 1999, Chapter 8.3). The more exact, but more difficult to calculate, case is that of *Fresnel diffraction,* in which we retain higher-order terms. If the source and image are at infinity, as in Fig. 11 with the lenses as shown, these give identical results.

Note that the amplitude A(x) can be a real or complex number, where the real part is the magnitude of the electric field and the phase is the phase delay of the wavefront at that point. If the phase is a complex number, then the imaginary part is equivalent to a reduction of the amplitude, i.e., $e^{i(\phi_R(x)+i\phi_I(x))}$ is the same as $e^{-\phi_I(x)} \times e^{i\phi_R(x)}$, a reduced-amplitude wave.

For the case of a one-dimensional pupil, with an input amplitude $A_1(x_1) = 1$, we find

$$A_2(\theta) = \int_{-D/2}^{+D/2} e^{i2\pi x_1 \theta/\lambda} dx_1 = \frac{\sin(\pi\theta D/\lambda)}{\pi\theta D/\lambda} D \qquad (44)$$

The measured intensity is $I = |A|^2$, or

$$I_2(\theta) = \left[\frac{\sin(\pi\theta D/\lambda)}{\pi\theta D/\lambda}\right]^2 D^2 \qquad (45)$$

The intensity pattern is thus the square of a $\text{sinc}(X) \equiv \sin(X)/X$ function, with a strong central peak at the point where the source star would have been imaged with geometrical optics (here $\theta_0 = 0$), and small secondary peaks.

The first zero is the solution of $I_2(\theta_{zero}) = 0$ and is given by

$$\theta_{zero} = \lambda/D \qquad (46)$$

Here are two trivial examples. First, suppose we add a constant phase ϕ_0 across the aperture, for example, with a plane-parallel sheet of glass. The net amplitude in the focal plane is multiplied by $e^{i\phi_0}$ and the intensity is unchanged. Second, suppose the star is off-axis at angle θ_0, or that the telescope is mispointed by the same angle. Then the input wavefront is tilted by angle θ_0, and in the integral θ is replaced by $\theta-\theta_0$, and likewise in the expression for the intensity.

The corresponding results for a two-dimensional circular aperture are roughly similar, but to write the equations correctly we must pay attention to the scaling factor, which for most of this chapter we otherwise ignore. For a circular aperture of diameter D, the physical amplitude in the focal plane (*Born and Wolf*, 1999, Chapter 8.5.2) is

$$A_2(\theta) = \sqrt{I_0} \frac{2J_1(\pi D\theta/\lambda)}{\pi D\theta/\lambda} \qquad (47)$$

and the corresponding intensity is $I_2 = |A_2|^2$, giving

$$I_2(\theta) = I_0 \left[\frac{2J_1(\pi D\theta/\lambda)}{\pi D\theta/\lambda}\right]^2 \qquad (48)$$

Here $J_1(X)$ is the Bessel function of first order, similar to a damped sine function. The first zero-intensity angle is

$$\theta_{zero} \simeq 1.22\lambda/D \qquad (49)$$

which is the famous result for a clear circular aperture. The full-width at half-maximum (FWHM) of the intensity pattern is

$$\theta_{FWHM} \simeq 1.03\lambda/D \simeq \lambda/D \qquad (50)$$

so this value is often referred to as the diameter of the diffraction-limited image of a point source. The relative intensities of the central and first four secondary maxima are 1.0, 0.017, 0.0042, 0.0016, and 0.00078 at respective values of 0.0, 1.6, 2.7, 3.7, 4.7 times λ/D.

The physical part of this result lies in the expression for I_0, which is

$$I_0 = \frac{EA_{pupil}}{\lambda^2 F^2} \qquad (51)$$

where E is the rate of energy per unit area in the pupil plane, for example, $E = f_\lambda \Delta\lambda$ from equation (9), with $\Delta\lambda$ the wavelength range being observed. The term A_{pupil} is the area of the pupil, $\pi D^2/4$ for a circular pupil. The term F is the focal length of the system. Notice that I_0 has units of rate of energy per unit area in the focal plane, similar to E in the pupil plane. The I_0 factor in equation (51) can also be derived by applying conservation of energy in the pupil and image planes, given the shape of the diffraction pattern in the focal plane. As a note of caution, it is not always trivial to convert a one-dimensional diffraction result into a two-dimensional result, but the results in this paragraph show a method that should be useful in other contexts.

3.4. Fourier Optics Approximation

Suppose we write equation (43) above as

$$A_2(\theta) = \int_{-\infty}^{+\infty} M_1(x_1) A_1(x_1) e^{i2\pi\theta x_1/\lambda} dx_1 \qquad (52)$$

where $M_1(x_1)$ is the pupil transmission function, here the top-hat or rectangular function, $M(x) = \text{rect}(x,D)$. The rectangular function is defined here as

$$\begin{aligned}\text{rect}(x,D) &= 1 \quad \text{if } -D/2 < x < +D/2 \\ &= 0 \quad \text{otherwise}\end{aligned} \qquad (53)$$

Here again, $A_1(x_1)$ is the amplitude of the incident wave; however, now it can extend over all values of x_1, as befits an expanding spherical (here nearly flat) wavefront from an atom on a distant star. This gives us the following important result: The amplitude $A_2(\theta)$ of the electric field in the focal plane of a telescope is the Fourier transform of the function $M_1(x_1)A_1(x_1)$, the electric field transmitted by the pupil of the telescope.

Note: Given our physically inspired convention that θ is in the opposite direction of x_2, the above relation is a regular Fourier transform, not an inverse Fourier transform, as this relation is sometimes stated. The difference is not important, as long as it is consistent.

Suppose that the image plane (number 2) is transparent, and is immediately followed by a lens that has a focal length equal to half the distance from the image to pupil plane. From geometric optics we know that the input pupil plane (number 1) will be imaged, one to one, in a conjugate pupil plane (number 3) downstream. Note that in the Fraunhofer approximation, the Fourier-transform integral is a linear operator, and it is reversible, so that the light may be thought of as traveling in either direction; the amplitude in this second pupil plane will be the inverse Fourier transform of the amplitude in the image plane, to within a constant factor. Another way to see this is to notice that in the wavelet picture, the sum (or integral) that propagates from plane number 1 to plane number 2 should be exactly the same in propagating from plane number 2 to plane number 3; however, since $x_3 = -\theta f$, the coordinate in plane 2 is reversed in sign, so the wavelet phases reverse sign, and the Fraunhofer integral becomes an inverse Fourier transform. We denote the amplitude in the third plane by $A_3(x_3)$.

Thus we have the result

$$A_3(x_3) = \int_\theta A_2(\theta) e^{-i2\pi\theta x_3/\lambda} d\theta \qquad (54)$$

Substituting and exchanging the order of integration we get

$$A_3(x_3) = \int_{x_1} M_1(x_1) A_1(x_1) \int_\theta e^{i2\pi\theta(x_1-x_3)/\lambda} d\theta dx_1 \qquad (55)$$

Now use the fact that

$$\int_{-\infty}^{+\infty} e^{i2\pi(x-x')\theta/\lambda} d\theta = \delta((x-x')/\lambda) \qquad (56)$$

where δ is the Dirac delta function, and that

$$\int_{-D/2}^{+D/2} \delta((x-x')/\lambda) dx = \lambda \, \text{rect}(x', D) \qquad (57)$$

We get

$$\begin{aligned} A_3(x_3) &= \lambda \int_{x_1} M_1(x_1) A_1(x_1) \delta(x_1-x_3) dx_1 \\ &= \lambda M_1(x_3) A_1(x_3) \end{aligned} \qquad (58)$$

which is an exact copy of the transmitted amplitude from the input pupil plane, to within a constant. This illustrates a general procedure whereby we can propagate light from one plane to another, using ideal lenses, infinite focal planes, and Fourier transforms.

We have used two simplifications in our picture of diffraction. The first simplification is that we have assumed that the pupil and image spaces are one-dimensional; expanding to the realistic case of two dimensions is in principle straightforward, but in practice often leads to more complex integrals; the net result is an increase in complexity with little gain in understanding. The second simplification is that we use the small angle approximation $\sin(x) \simeq x$, i.e., Fraunhofer diffraction; including the higher-order terms leads to Fresnel diffraction. Both topics are well covered in *Born and Wolf* (1999) and other standard texts.

3.5. Convolution Perspective on Imaging

A conceptually elegant way to view the operation of an imaging system, including a coronagraph, is to take advantage of the Fourier-transform relation between the planes in Fig. 12 and the fact that the Fourier transform of a product of functions is the convolution of the individual Fourier transforms, and also the Fourier transform of a convolution of two functions is the product of the individual Fourier transforms. In other words, $FT(f * g) = FT(f) \cdot FT(g)$, and also $FT(f \cdot g) = FT(f) * FT(g)$.

Referring to plane 1 in Fig. 12, we see that the input amplitude is $A_1(x_1)$, the mask is $M_1(x_1)$, and the output is $M_1 A_1$.

At plane 2, the input amplitude is $FT[M_1 A_1](x_2) = FT(M_1) * [FT(A_1)](x_2)$. The mask is $M_2(x_2)$. And the output is $M_2 \cdot [FT(M_1) * FT(A_1)](x_2)$.

At plane 3 the input is the FT^{-1} of the plane-2 output. We multiply the mask M_3 times that function, and apply the convolution rules again. This gives the output from plane 3 as $M_3 \cdot [FT^{-1}(M_2) * (M_1 \cdot A_1)]$.

At plane 4 the input is the FT of the plane-3 output. Substituting and simplifying we get the field at plane 4 to be $FT(M_3) * [M_2 \cdot FT(M_1 A_1)]$.

The value of this picture will become clear when we look at individual coronagraph designs. The band-limited mask design will show clearly how this picture, and in particular the expression for the output of plane 3, can bypass difficult integrals to give a clear physical picture.

3.6. Imaging Recipes

We summarize the general case of propagation from plane 1 through plane 4 in Fig. 12 as a recipe for later reference, as follows.

The electric field incident on plane 1 is

$$A_1(x_1) = e^{i2\pi x_1 \theta_0/\lambda} \qquad (59)$$

for a point source located at angle θ_0.

A lens in plane 1 produces an electric field $A_2(\theta_2)$ incident on plane 2

$$A_2(\theta_2) = \int_{x_1} M_1(x_1) A_1(x_1) e^{i2\pi\theta_2 x_1/\lambda} dx_1 \qquad (60)$$

where $M_1(x_1)$ is a mask on the output side of plane 1.

Likewise, a lens in plane 2 produces an electric field $A_3(x_3)$ incident on plane 3

$$A_3(x_3) = \int_{\theta_2} M_2(\theta_2) A_2(\theta_2) e^{-i2\pi x_3 \theta_2/\lambda} d\theta_2 \qquad (61)$$

where $M_2(\theta_2)$ is a mask on the output side of plane 2.

Finally, a lens in plane 3 produces an electric field $A_4(\theta_4)$ incident on plane 4

$$A_4(\theta_4) = \int_{x_3} M_3(x_3) A_3(x_3) e^{i2\pi\theta_4 x_3/\lambda} dx_3 \quad (62)$$

where $M_3(x_3)$ is a mask on the output side of plane 3.

3.7. Practical Considerations

Real optics can depart from ideal in several ways. One departure is that opaque baffles will diffract light slightly differently depending on whether the material of the stop is a metal or a dielectric. In practice this effect is mainly noticed in the immediate vicinity (a few wavelengths) of the stop. As an example, subwavelength diameter holes in a screen can have much greater transmission through a metal screen than the corresponding holes in a dielectric screen. A related example is that partially transmitting materials, as are used in some coronagraphs, will always have a phase shift associated with a given level of opacity, as determined by the Kramers-Kronig relation that connects the real (absorbing) and imaginary (phase shifting) parts of the index of refraction of the material.

More mundane considerations include the presence of scattering dust on optics, which can spoil a theoretically low contrast, and atmospheric phase fluctuations in laboratory experiments, which can amount to at least a wavelength or more of time-varying path, especially if there is a heating or cooling air flow nearby.

For coronagraphic telescopes that operate at extreme (planet-detecting) contrasts, the problem of *beam walk* becomes a factor. This arises if the telescope is body-pointed slightly away from a target star, and this error is compensated by the tip-tilt of a subsequent mirror, driven by a star-tracker (for example), thereby slightly shifting the beam transversely across the optics, and encountering a slightly different pattern of surface errors in those optical elements, thereby generating different speckle patterns (cf. section 4).

3.8. Off-Axis Performance

Stars have finite diameters, and telescopes have finite pointing errors. The performance of a coronagraph will be degraded by either effect, because the transmission of a coronagraph will generally increase in both cases, even if it is theoretically zero for an on-axis delta-function source. The degree to which even an ideal coronagraph will leak light is quantified in the concept of the *order of the null*.

The intensity leak is proportional to θ^n where θ is the off-axis angle and n is an integer power (*Kuchner*, 2004b, 2005). Since intensity is proportional to electric field squared, n is always even. Some examples are n = 0 (top-hat, disk phase knife); n = 2 (phase knife, four-quadrant phase mask); n = 4 (notch filter, band-limited mask, Gaussian, achromatic dual zone); n = 8 (band-limited, notch).

Given a functional form of the off-axis transmission of a mask, the effects of a finite-diameter star as well as a slightly mispointed telescope can be directly calculated by integrating the transmission function over the possibly offset disk of the star.

3.9. Effects of Central Obscuration, Spider, Segments

If a circular pupil of diameter D has a central obscuration of diameter d, then the summation of wavelets can be written as the sum over the larger pupil [amplitude $A_2(x_2,D)$] from equation (47) minus the sum over the smaller one [amplitude $A_2(x_2,d)$], and the intensity in plane 2 will be

$$I_2(\theta) = I_0 \left[\frac{2J_1(\pi\theta D/\lambda)}{\pi\theta D/\lambda} A_D - \frac{2J_1(\pi\theta d/\lambda)}{\pi\theta d/\lambda} A_d \right]^2 \quad (63)$$

where A_D is the area of the larger pupil, and A_d is the area of the smaller, obscuring pupil. Notice that the weighting of the respective amplitudes is by area, not diameter. This case is an example of the need to get the correct energy-based coefficients of a diffraction pattern. Comparing this with the nonobscured case in equation (48), we see that the central core of the image is slightly sharper (i.e., slightly better angular resolution), but at the expense of significantly stronger diffraction rings around this core.

If a circular pupil of diameter D has a spider arm of width w placed across its center, or if the pupil is made up of a segmented mirror with a gap of width w, then the intensity pattern of a point source, in the focal plane 2 in a direction perpendicular to the spider or gap, will be

$$I_2(\theta) = I_0 \left[\frac{2J_1(\pi\theta D/\lambda)}{\pi\theta D/\lambda} A_D - \frac{\sin(\pi\theta_w w/\lambda)}{\pi\theta_w w/\lambda} \frac{\sin(\pi\theta_D D/\lambda)}{\pi\theta_D D/\lambda} A_w \right]^2 \quad (64)$$

which is the square of the net amplitude of a circular clear pupil minus the amplitude of a rectangular blocked strip. Here the angular directions in the focal plane are θ_w in a direction parallel to the w dimension of the obscuration, θ_D in a direction parallel to the D dimension of the obscuration, and $\theta = \sqrt{\theta_w^2 + \theta_D^2}$, and $A_D = \pi D^2/4$ is the area of the full pupil, and $A_w = wD$ is the area of the obscuration. The strip obscuration or gap adds a surprisingly large diffracted intensity at large angles.

Equations (63) and (64) are examples of *Babinet's principle*, in which the amplitude resulting from an opaque part in a beam is represented as the negative of the amplitude from a transparent version of the opaque part.

As an example, suppose that we have a segmented primary mirror of total width D, made up from two adjacent segments, each of width D/2, and that there is a thin strip of width w overlying the joint between the segments. Examples are the adjacent segments of the Keck telescopes or the James Webb Space Telescope (JWST). Let us assume that there is a coronagraph that can suppress the central star and its diffraction pattern. Then the contrast of the diffracted spike compared to the (suppressed) central star intensity is $C(\text{spike}) \approx (w/D)^2$. If we want this to be as faint as Earth in brightness, we need

C = 10^{-10}. If the segments are each D = 1 m wide, then we need w ≤ 10 μm in width, about one-eighth the thickness of a human hair. This is much smaller than can be easily accomplished.

These examples show that diffracted light from an obscuration or gap in the pupil can generate a relatively large intensity at angles well away from the diffraction core of λ/D from a point-like star. Only a few types of coronagraphs are immune to these obscuring elements.

3.10. Pupil-Edge Apodization

One way to eliminate the diffraction side lobes of a pupil is to reduce the sharp discontinuity in the transmitted wavefront at the edge of the pupil. As we saw in the discussion of the single pupil with a sharp edge, the Huygens wavelets spread out dramatically at such an edge. An early suggestion was to taper the transmission of the pupil at the edges, to avoid a sharp change of transmission. For example, we could figuratively spray black paint on a telescope mirror so that the center was clear and the edge totally opaque. A more practical (and approximate) method is to surround the perimeter of a pupil with a lot of inward-pointing black triangles or similar pointed spikes, such that the azimuth average transmission drops smoothly from 1 at the center to 0 at the edge; this technique works surprisingly well, and can be implemented with ordinary tools.

Suppose we model this by a Gaussian intensity transmission function $e^{-(x/x_0)^2}$, which corresponds to a Gaussian amplitude function $e^{-(x/x_0)^2/2}$. We want this function to be small at the edge, so we assume that $x_0 < D/2$. The effective diameter D_{eff} is then roughly the FWHM of the intensity distribution, which is $D_{eff} \simeq x_0 2\ln(2)$. Inserting this amplitude into the one-dimensional equation for net amplitude, and making the approximation that $x_0 \ll D/2$, we find the normalized intensity pattern in the focal plane to be

$$I_2(\theta) = e^{-(2\pi\theta x_0/\lambda)^2} \qquad (65)$$

This result shows that tapering the pupil, in the extreme case of strong tapering near the edges, can have a dramatic effect on the image of a point source, namely concentrating it in a tight image with no sidelobes. If we had integrated from 0 to D/2 instead of 0 to ∞ we would have obtained a similarly compact central peak, but with finite sidelobes. In the example shown, the intensity drops to 10^{-10} at an angular distance of about $\theta_{-10} \simeq 2\lambda/D_{eff}$, showing that in principle this is a powerful method of minimizing sidelobes. This technique is generally called *apodizing*, meaning to remove the feet. We could have used a cosine or other similar function, with roughly similar results. Obviously the technique can be extended to a more realistic circular aperture.

In practice, the Gaussian function, which tapers to zero on an infinite range, is replaced by a very similar-looking profile, a *prolate spheroid* function, defined on a finite range. The intensity pattern in the focal plane, with a prolate spheroid tapering of the pupil, can be made to drop to $\theta_{-10} \simeq 4\lambda/D$, which is still a dramatic feat.

3.11. Pupil-Masking Apodization

The concept of *pupil masking* is a practical version of the spray-paint apodization described above. In the pupil-masking apodization method, the pupil is covered by an opaque sheet that has tapered cutouts through which the wavefront can pass, as shown in Fig. 13. The cutouts are designed to transmit more light at the center of the pupil, and less at the (say) left and right edges. The corresponding projected left and right areas on the sky have faint diffracted light in the focal plane, so a planet could be detected in these areas. There are no sharp edges perpendicular to the left and right, so little diffraction. However in the orthogonal direction, say up and down, there are a lot of perpendicular edges, so a lot of light is diffracted in those directions. The search space on the sky is therefore limited to the projected areas with diffraction below a target threshhold, say 10^{-10}. The concept is so simple that it could be tested with paper and scissors at an amateur telescope. These pupil-masking types of stops have been tested in the laboratory, and have achieved dark zones as deep as about 10^{-7}, limited perhaps by minor imperfections in the mask edges. Also their transmission is relatively low, since much of the pupil is covered. Nevertheless these masks stand as a proof of principle that it is possible to beat the iron grip of diffraction, and they have inspired numerous other inventions.

References include *Kasdin et al.* (2003, 2005).

3.12. Pupil-Mapping Apodization

Another way to achieve a Gaussian-like amplitude distribution across a pupil is to rearrange the incoming rays, so to speak, so that they do not uniformly fill the pupil but rather crowd together near the center, and become sparse at the edges. This will make the amplitude of the electric field stronger at the center and weaker at the edges, but for visualization it is easiest to think of rays. Pupil-mapping is illustrated in Fig. 14.

Fig. 13. (a) This optimized pupil mask has six openings within an elliptical envelope designed to match the pupil of TPF-C (section 6.9). (b) The corresponding image plane diffraction pattern, showing a strongly suppressed central star, and dark-hole areas (on the left and right) with residual intensities below a theoretical contrast of 10^{-10}. The IWA is $4\lambda/D$ and the throughput is 30%.

This effect can be created with two aspheric lenses (or mirrors). The first lens is shaped so that it converges the wavefront across most of the aperture but tapers to a flat piece of glass at the extreme edges so that those rays continue on parallel to their input direction. The second lens is placed well before the converging rays cross and shaped so as to make all the output rays parallel again, i.e., a diverging central part tapering to a flat piece of glass at the edges.

The resulting output beam will have the same diameter as the input beam but will be bright near the center and faint near the edge, in a Gaussian-like (prolate spheroid) amplitude distribution. Focusing this beam with a lens will generate a Gaussian-like bright star image with very faint ($\leq 10^{-10}$) sidelobes. The core image can theoretically be contained within a radius of about $2\lambda/D$ sky angle. The pair of lenses can be replaced by a pair of off-axis mirrors, and an off axis paraboloid can be incorporated into the second mirror so that a star image is formed without the need for an additional optic. These lenses or mirrors can also be manufactured at a small diameter and placed at the back end of a large off-axis telescope, at an image of the input pupil. Since this method uses all the input energy collected by the telescope, and forms a star image that is close to the theoretical limit of about $1\lambda/D$ in radius, it is theoretically an optimum type of coronagraph.

The pupil-mapping concept is also called *phase-induced amplitude apodization* (PIAA), a descriptive name reflecting the fact that the electric field amplitude is apodized near the edge of the pupil by manipulating the point-to-point phase of the incoming wavefront. In other words, slowing down the wavefront near the center of the beam causes it to converge locally, while the extreme edge of the beam is not slowed at all, and therefore continues on as if untouched.

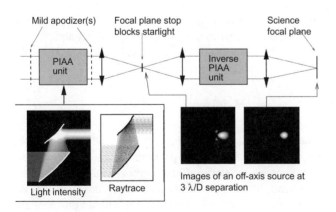

Fig. 14. Pupil-mapping (PIAA) schematic, showing in the lower left how a pair of specially shaped mirrors can reshape a uniformly bright input beam into a Gaussian-like beam (bright in the center, faint at the edges), which when focused by a lens produces a Gaussian-like star image (with extremely weak sidelobes), which in turn can be blocked by a focal-plane stop, so that any adjacent planet images are transmitted. The inverse PIAA unit is needed to reshape the strongly aberrated planet images, giving nearly normal planet images in the science focal plane.

In a practical system, the trick is to shape the outer 1% or so of the radius of the first mirror so as to spray all the light incident on this narrow annulus over a large part of the output pupil, forming the wing of the Gaussian-like distribution; this is difficult because it requires extremely accurate polishing of a relatively sharp radius of curvature into the edge of the optic.

The amplitude near the edge can also be controlled by inserting a thin opaque ring, or a fuzzy outer blocker, at a small loss of light. In practice this step appears to be necessary.

An opaque blocker should be placed at the image of the star, and the surrounding dark field, including any exoplanets, passed on to a detector. The image of an off-axis exoplanet at this point will be aberrated because the pupil mapping mirrors distort any off-axis point source, with the distortion getting worse with distance from the on-axis star; see section 3.23 for a general explanation of the this type of distortion. Fortunately, and surprisingly, this aberration can be eliminated by sending the image through a reversed set of lenses or mirrors.

References include the original paper (*Guyon*, 2003), a ray-trace theory of mirror shapes (*Traub and Vanderbei*, 2003), a proof that the inverse system will restore off-axis image shapes (*Vanderbei and Traub*, 2005), and recent laboratory results (*Guyon et al.*, 2009, 2010).

3.13. Lyot (Hard-Edge) Mask Coronagraph

Suppose that we build the simplest kind of coronagraph, a focusing lens of diameter D in the plane 1, a dark occulting mask of angular diameter θ_r in plane 2, followed by a lens that makes an image of the pupil in plane 3, a mask just after, and another lens that finally images the sky in plane 4. (Equivalently, the dark spot could be replaced by a transmitting hole in a mirror, which could be more convenient.)

The *hard-edge mask* $M_2(\theta_2)$ multiplies the amplitude according to

$$M_2(\theta_2) = 1 - \text{rect}(\theta_2, \theta_r) \qquad (66)$$

so that M is 0 (opaque) in the range $(-\theta_r/2, +\theta_r/2)$ and 1 (transmitting) outside this range. This shape is also known as a *top-hat mask*.

This method was pioneered by B. Lyot for the purpose of imaging the faint ($<10^{-6}$) corona of the Sun. Today any such instrument for observing a faint source near a bright one is called a *coronagraph*, and the particular configuration that uses a hard-edge mask and stop (see below) is called a *Lyot coronagraph*.

We might expect that if the mask covers the central few Airy rings, we might block most of the light, or scatter it off to a large angle where it might be blocked in plane 3, and thus generate a greatly diminished star image (and diffraction pattern) in plane 4. Using equation (61), and an on-axis star, i.e., $A_1(x_1) = 1$, the amplitude $A_3(x_3)$ in plane 3 is

$$A_3(x_3) = \int_{\theta_2} M_2(\theta_2) A_2(\theta_2) e^{-i2\pi x_3 \theta_2/\lambda} d\theta_2 \qquad (67)$$

Then using equation (60) and taking the input pupil to be $M_1(x_1) = \text{rect}(x_1,D)$ we get

$$A_3(x_3) = \int_{\theta_2}\int_D e^{i2\pi(x_1-x_3)\theta_2/\lambda}\left[1-\text{rect}(\theta_2,\theta_r)\right]dx_1 d\theta_2 \quad (68)$$

where the integration range of θ_2 is $\pm\infty$ and the range of x_1 is $\pm D/2$.

Using equation (56) for the delta function and the rectangular function from above, we find the amplitude in the second pupil plane to be

$$A_3(x_3) = \lambda\left[\text{rect}(x_3,D) - \int_{-D/2}^{+D/2}\frac{\sin(\pi(x_1-x_3)\theta_r/\lambda)}{\pi(x_1-x_3)}dx_1\right] \quad (69)$$

There is no simple expression for this amplitude, but by visualizing the terms we can see that the amplitude in plane 3 is a copy of that in plane 1 minus an oscillatory function of the position coordinate x_3. The amplitude is indeed diminished inside the range $\pm D/2$, but there is now amplitude scattered outside this range.

We stop the light at the edge of the x_3 pupil from propagating further by inserting a *Lyot stop* here. This stop is an undersized image of the input pupil, but since the amplitude is small but finite at all radii, the diameter of the stop is a matter of judgement: A small diameter improves the rejection of the sidelobes in the x_4 image plane, but at the expense of overall throughput and angular resolution.

Thus the advantage of the Lyot coronagraph is its simplicity, but the disadvantage is that it will always have a finite leakage, so there is a limit to how small a contrast it can achieve.

Note that the amplitude will have a zero and a sharp discontinuity at $x_3 = \pm D/2$, independent of wavelength, a characteristic of this (and the following) coronagraph.

References include *Lyot* (1933).

3.14. Gaussian Mask Coronagraph

A soft-edge mask to block the first star image should work better than a hard-edge one. An example is the Gaussian-shape amplitude mask

$$M_2(\theta) = 1 - e^{-(\theta/\theta_g)^2} \quad (70)$$

Following through as with the rectangular mask, we find the amplitude in plane 3 to be

$$A_3(x_3) = \lambda\left[\text{rect}(x_3,D) - \frac{1}{\sqrt{\pi}}\int_{z_-}^{z_+}e^{-z^2}dz\right] \quad (71)$$

where the integration limits are $z_\pm = \pi\theta_g(\pm D/2 - x_3)/\lambda$. This can be expressed in term of an error function, but a simple approximation, for illustration here, is to replace the integral by the magnitude of the integrand at the center of the range, multiplied by the width of the integration range (a crude box-car integration). This gives an approximate amplitude

$$A_3(x_3) \approx \lambda\left[\text{rect}(x_3,D) - \frac{\sqrt{\pi}\theta_g D}{\lambda}e^{-(\pi\theta_g x_3/\lambda)^2}\right] \quad (72)$$

Depending on the value of θ_g, which would nominally be in the neighborhood of a few times λ/D, this amplitude also has a small value inside the range $\pm D/2$, a zero at the edge of that range, and more amplitude diffracted out beyond that range, which can then be removed with a hard-edge Lyot stop for M_3.

The diameter of this stop is a free parameter, the tradeoff being that a stop diameter less than D will reduce the background diffracted light in the second focal plane (good), but it will also reduce the light from the off-axis exoplanet (the light of which precisely fills the diameter D) and increase the diffracted diameter of its image (both bad). The reason for the latter is that this stop is now the effective diameter of the system for the exoplanet, and its diameter will determine the image size in plane 4, as can be verified by another Fourier transform.

3.15. Band-Limited Mask Coronagraph

The band-limited mask coronagraph is an evolutionary step beyond the rectangular and Gaussian-mask coronagraphs described above. The band-limited design is the answer to the question; can we find a masking pattern in plane 2 that minimizes the transmitted light from an on axis star, but at the same time allows an off-axis exoplanet image to pass? There are two extremes of answers to this question, absorbing masks (this section) and phase masks (section 3.16), and there are intermediate types that combine absorption and phase.

To illustrate the band-limited concept, we choose an amplitude mask with a periodic modulation

$$M_2(\theta) = c\left[1 - \cos(\theta/\theta_B)\right] \quad (73)$$

where $c = 1/2$ so that $0 \leq M \leq 1$, and θ_B is a scale factor that might chosen to be on the order of one to a few times λ/D so as to suppress the star's diffraction pattern out to the first one or several diffraction sidelobes. Note that this mask has no hard edges, so we might expect that it will not diffract light at large angles; in fact it is periodic, like a diffraction grating, so we might expect it to diffract light at a specific angle. Its spatial frequency range is limited, hence the name. The mask extends over many periods, in principle over $x_2 = \pm\infty$.

We calculate the amplitude of the electric field in the plane 3 using equation (61)

$$A_3(x_3) = \int_D\int_\theta M_2(\theta)e^{i2\pi(x_1-x_3)\theta/\lambda}d\theta dx_1 \quad (74)$$

Substituting, we get

$$A_3(x_3) = c \int_D \int_\theta e^{i2\pi(x_1-x_3)\theta/\lambda} [1-\cos(\theta/\theta_B)] d\theta dx_1 \quad (75)$$

Using $\cos(z) = (e^{iz} + e^{-iz})/2$ and the delta function, and integrating over $x_1 = (-D/2, +D/2)$, we find

$$\begin{aligned}A_3(x_3) = c\lambda\big[&\mathrm{rect}(x_3, D) - \\ &0.5\,\mathrm{rect}(x_3 - \lambda/(2\pi\theta_B), D) - \\ &0.5\,\mathrm{rect}(x_3 + \lambda/(2\pi\theta_B), D)\big]\end{aligned} \quad (76)$$

which is indeed zero over the central range of the pupil, with finite amplitude at the edges, and a zero at the exact edge. We can emphasize this by writing the amplitude as

$$\begin{aligned}A_3(x_3) = c\lambda\big[&0 \times \mathrm{rect}(x_3, D - \lambda/\pi\theta_B) + \\ &\mathrm{wiggle}(x_3 - D, \lambda/\pi\theta_B) + \\ &\mathrm{wiggle}(x_3 + D, \lambda/\pi\theta_B)\big]\end{aligned} \quad (77)$$

where the central part has zero amplitude, and the function called wiggle$(x, \Delta x)$ is defined here by analogy with the rect$(x, \Delta x)$ function.

This technique will work with any band-limited mask function, for example, $\sin^2$ (as above), $\sin^4$, $1-J_0$, $1-\mathrm{sinc}$, $1-\mathrm{sinc}^2$, and $(1-\mathrm{sinc}^2)^2$.

All these masks have greater average transparency away from the central dark region than the 1–cos mask. Here we work out the 1–sinc example

$$M_2(\theta) = c\left[1 - \frac{\sin(\theta/\theta_B)}{\theta/\theta_B}\right] \quad (78)$$

We need this relation first: $\int_0^\infty \sin(z)\cos(mz)/z\,dz$ is 0 if $|m| > 1$ but $\pi/2$ otherwise. The result is

$$\begin{aligned}A_3(x_3) = c\lambda\big[&0 \times \mathrm{rect}(x_3, D - \lambda/\pi\theta_B) + \\ &\mathrm{tilt}(x_3 - D, \lambda/\pi\theta_B) + \\ &\mathrm{tilt}(x_3 + D, \lambda/\pi\theta_B)\big]\end{aligned} \quad (79)$$

where the same central region has zero amplitude, and the ring of scattered light at the edges occupies the same width but has a tilted shape, tilt, similar to the wiggle function above. The amplitude is sketched in Fig. 15.

References include *Kuchner and Traub* (2002) and *Trauger and Traub* (2007).

3.16. Phase and Vector Vortex Mask Coronagraphs

Several types of coronagraphs depend on a pure phase manipulation of the photon in the focal plane.

A *phase mask* in the focal plane is four contiguous quadrants of transparent material onto which the star is focused at the symmetry point, with adjacent quadrants differing in optical thickness by a half-wavelength. The transmitted beam will be nulled out on axis, but an object imaged mainly in one of the quadrants will be transmitted.

A *scalar optical vortex* is a structural helical phase ramp, and generates a longitudinal phase delay by operating on both polarizations. The center of the optical vortex is a phase singularity in an optical field, which generates a point of zero intensity, resulting from a phase screw dislocation of the form $e^{il_p\psi}$, where l_p is called the topological charge, and ψ is the azimuthal coordinate.

A *vector optical vortex* is a space-varying birefringent mask that operates on the orthogonal components of polarization of the photon, such that a star focused at the center of the pattern will be nulled on axis. One version of this design is a set of engraved concentric annular groves in a glass plate, with the groove spacing smaller than a wavelength. The fineness of the grooves ensures that the diffracted light cannot go into side orders, but must continue to propagate in its original direction. However, the groove depth is designed such that the projected part of the photon's electric vector

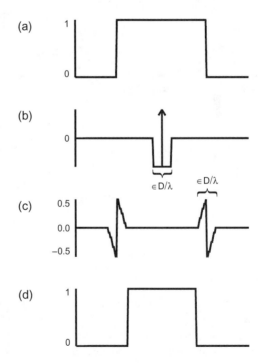

Fig. 15. Band-limited amplitudes for the 1–sinc mask. **(a)** Input pupil amplitude, at plane 1. **(b)** FT of mask, which, when convolved with the pupil, produces the amplitude distribution shown in **(c)**, equivalent to plane 3. **(d)** Transmission of Lyot stop in plane 3, transmitting the amplitude of the star in the central part of the plane (here zero owing to the action of the band-limited mask), and also transmitting the amplitude of any off-axis source, such as a planet. The diffracted light from the star is blocked by the slightly undersized Lyot pupil located just after plane 3.

that is parallel to the local groove direction sees a different optical path than the perpendicular vector component, and by the circularity of the design, a centered star image will effectively have half its amplitude delayed by a half wavelength with respect to the other half. The on-axis light is thus nulled, but of course energy is conserved so the light diffracts off at an angle to the incident beam. These grooves can be noncircularly symmetric, generating total delays of more than one full wavelength per rotation in azimuth. The number of half-wavelengths per turn is called the *topological charge* l_p of the design, with designs ranging from $l_p = 2$ (a minimum) up to $l_p = 6$ (limited somewhat by the complexity of the design). The difficulty of manufacturing subwavelength grooves or index of refraction spiral ramps has encouraged the search for another medium.

Liquid crystal polymers enable another type of vector optical vortex. Here the polymer molecules can be lined up so as to make a locally varying halfwave plate in which the optical axes rotate about the center of symmetry of the plate.

To show the action of a phase mask we illustrate with a one-dimensional example. Suppose that we put a phase mask in plane 2 with a phase discontinuity of π centered on the star image in the image plane. This mask can be written as

$$M_2(\theta_2) = e^{+i\pi/2} \quad \text{if } \theta_2 > 0$$
$$= e^{-i\pi/2} \quad \text{if } \theta_2 < 0 \qquad (80)$$

So with a simple, on-axis star, and a pupil of diameter D in plane 1, equations (60) and (61) give an amplitude in plane 3 which is

$$A_3(x_3) = \frac{2\lambda}{\pi} \int_0^\infty \frac{\sin(z)}{z} \sin(mz) \, dz \qquad (81)$$

where $m = 2x_3/D$, and which has the solution

$$A_3(x_3) = -\frac{\lambda}{2\pi} \ln\left(\frac{m-1}{m+1}\right)^2 \qquad (82)$$

This amplitude has a sharp peak at the edge of the pupil (m = 1), but is not especially small inside the pupil, and therefore fails to be a good coronagraph. It is, however, an illustration of what a point source would do if it were centered on one of the arms of a four-quadrant phase mask, which is one of the reasons that this type is less ideal than the vector vortex type.

To move this calculation into a two-dimensional plane, let us replace x in any plane with (r, ψ), the radial and azimuthal coordinates, with appropriate subscripts for each plane. Following *Mawet* (2005, Appendix C), we find

$$A_3(r_3, \psi_3) = \frac{4e^{i2\psi_3}}{\lambda D} \int_0^\infty J_1(\pi D\theta/\lambda) J_2(2\pi r_3\theta/\lambda) \, d\theta \qquad (83)$$

which illustrates the complexity of working in the full two-dimensional picture, but which fortuitously has a solution in Sonine's integral, giving

$$A_3(r_3, \psi_3) = 0 \quad \text{if } r_3 < D/2$$
$$= \frac{e^{i2\psi_3}}{\pi r_3^2} \quad \text{if } r_3 > D/2 \qquad (84)$$

where the dimensions are arbitrary.

This shows that a point-source star, centered on a vector vortex mask, will have zero amplitude transmitted into the second pupil, following the image plane, and that the star will be diffracted into a bright ring that peaks just outside the second pupil, and falls off quadratically with distance beyond that point. This is therefore an ideal coronagraph, making full use of the collecting pupil, and requiring a Lyot stop that exactly matches the input pupil.

References include *Mawet et al.* (2010).

3.17. External Occulters

An *external occulter* or *star shade* coronagraph is a concept in which a blocking mask is placed between the source and the telescope. The mask is made large enough to cover the star, but not so large as to obscure a nearby planet. In geometrical optics terms, the occulter must be larger than the telescope diameter, but have an angular radius smaller than the planet-star angular separation. For example, if $D_{tel} = 6$ m, and $\theta_{planet} = 0.1$ arcsec, then the occulter must be separated from the telescope by at least $z = D_{tel}/\theta_{planet} = 12,000$ km. Clearly, a telescope in low-Earth orbit will not work, but a drift-away or L2 orbit would be feasible, assuming that the positioning control can be accomplished.

The telescope-facing side of the occulter should look dark compared to the planet, so it must face away from the Sun. Thus we require that the angle between the occulter and the Sun be less than about 90°.

The occulter must move from target to target. For example, if there are about 200 potential targets in about 44,000 deg² of sky, then the average distance between targets is about 15°.

If the mask is circular, then wavelets from the edge will all have an equal optical path to points on the star-occulter axis, and there will be a bright central diffraction spike, the *Arago spot*, sometimes called the *Poisson spot*. To reduce the intensity of the Arago spot, the edge of the occulter must be softened, exactly as with a pupil or image mask for an internal coronagraph. Interestingly, this softening can be in the form of a moderate number of cut-outs around the edge of a circle, the structures between the cut-outs being called *petals*, as in the shape of a flower. Thus the occulter can be fabricated as a connected binary mask.

Suppose that we add a plane 0 in front of plane 1 in Fig. 12, with no imbedded lens. Then the originally proposed "hyper-Gaussian" mask is given by the continuous function

$$M_0(r_0) = 0 \quad \text{if } r_0 < a$$
$$= 1 - \exp\left(-\left[(r_0 - a)/b\right]^n\right) \quad \text{if } r_0 < r_0(\max) \quad (85)$$
$$= 1 \quad \text{if } r_0 > r_0(\max)$$

where M_0 is the rotationally averaged amplitude transmission. The value of $r_0(\max)$ is taken to be that for which the width of the petal is very small, e.g., ~1 mm.

Subsequently, significant improvements to the hyper-Gaussian shape have been made by optimizing the shape function, including accounting for a wide band of wavelengths, and a dark hole over the full diameter of a telescope, with margin for positioning. A typical functional shape for M_0 is a prolate spheroid. An example is shown in Fig. 16.

If a telescope of diameter D_{tel} is placed in the stellar shadow zone of an occulter of tip-to-tip diameter D_{occ}, and their separation is z, then the telescope will have a clear view of a planet that is separated from the star by an angle θ_{IWA} where

$$\theta_{IWA} = (D_{occ} + D_{tel})/2z \quad (86)$$

If diffraction is ignored, so that the shadow diameter is equal to the occulter diameter, and if we wish to have a tolerance margin of, say, ±1 m, for navigating the telescope into the shadow (so that $D_{occ} = D_{tel} + 2$ m here), and if we want an IWA of $\theta_{IWA} \simeq 0.1$ arcsec (to see an Earth around a Sun at 10 pc), and if we assume that the telescope is JWST (so D_{tel} = 6.5 m), then we require $z > (D_{tel} + 1)/\theta_{IWA} \simeq 15,000$ km, and of course $D_{occ} = 8.5$ m.

In reality, diffraction into the shadow region forces these values to be much larger. Since the telescope is not in an image or pupil plane, Fraunhofer diffraction does not apply, and we must use the full power of Fresnel diffraction theory. In this framework, the Fresnel number N_F is conserved, where N_F is given by

$$N_F = D_{occ}^2/\lambda z \quad (87)$$

Fig. 16. An optimally-shaped star shade, from *Vanderbei et al.*, (2007), in which the ideal continuous apodization function is replaced by a 16-petal approximation with discrete (0 or 1) transmission.

The meaning of this is that the relative shape of the diffracted light in the shadow zone is the same if D_{occ} or λ or z are varied, while N_F is held constant.

As a specific example, we show in Table 6 the result of a calculation using an optimized shape for the occulter petals, for cases where the shadow intensity is 10^{-10} or less over the area of a circle of diameter D_{tel} + 2 m (to allow for a ±1 m navigation error). The Fresnel number for these examples is $N_F = 70$. We see that for JWST, for example, with $D_{tel} = 6.5$ m, we will need an occulter with tip-to-tip diameter $D_{occ} = 70$ m, at a distance of z = 140,000 km. For this case we can see a planet relatively close to its star, $\theta_{IWA} = 50$ mas, meaning that a search for Earth-like planets in the habitable zone of nearby stars could be possible. However, to do this, we will need a very large occulter at a very large distance, meaning that fuel for repositioning may become a limiting factor.

Another potential limiting factor is the accuracy requirement on the edge shape of the occulter. Errors in the shape will generate speckles in the focal plane of the telescope. A simulation has shown that edge errors on the order of 0.2 mm RMS can generate focal plane speckles at the 10^{-10} level, so this is approximately the tolerance of manufacturing and deployment of the petals.

References include *Cash* (2006), *Arenberg et al.* (2006), *Lyon et al.* (2007), *Vanderbei et al.* (2007), *Kasdin et al.* (2009), *Shaklan et al.* (2010), and *Glassman et al.* (2010).

3.18. How to Observe Exoplanets: Multiple Pupils

To obtain high angular resolution on a star-planet system we can use two or more separated telescopes instead of a single large one, increasing from a diameter D to a potentially much larger baseline B, and thereby reducing the angular resolution λ/B. A *nulling interferometer* is two or more telescopes arranged so as to collect segments of an incident wavefront and combine them with a half-wavelength path delay, so that the central star is largely canceled by balancing the electric fields and phases. Three examples that are especially relevant to direct imaging of exoplanets are the Keck Interferometer Nuller (KIN), the TPF-I, and the Large Binocular Telescope Interferometer (LBTI). Here we discuss the first two of these.

3.19. Beam Combination with an Interferometer

We distinguish here between nulling interferometers and imaging interferometers.

A nulling interferometer collects segments of a wavefront using several telescopes, sends these segments through delay

TABLE 6. Occulter examples.

D_{occ} (m)	z (km)	D_{tel} (m)	θ_{IWA}
70	140,000	6.5	50
50	72,000	4.0	72
37	39,000	2.4	98

lines to equalize their optical paths from the star, adds a half-wavelength extra delay to one or more of the paths, combines the beams directly on top of each other using semitransparent beam splitters, and focuses all the light onto a single detector. The net effect is that an on-axis star is canceled out by the half-wave delay, but that light from near the star is not canceled because its phase shift differs from π by an amount that increases with distance from the star, reaching 2π or effectively 0 wavelengths at an angle $\theta_0 = 0.5\lambda/B$. The situation can be pictured in terms of a fringe pattern projected on the sky, with the minimum-transmission point of the pattern centered on the star, preventing the star from being detected, while a disk around the star is allowed to be transmitted to the detector, multiplied by the fringe pattern.

An imaging interferometer is similar but adds a repeated and continuously changing delay such that the projected fringe pattern on the sky sweeps back and forth across the star and disk, say, and the transmitted light is recorded and later analyzed to extract the image by a deconvolution or data-fitting algorithm. We do not discuss this type any further in this chapter.

Returning to the nulling interferometer, we note that the examples discussed below are similar in many ways, yet different in that the KIN is designed to measure the zodiacal dust brightness, whereas TPF-I is designed to reject the zodi signal and search instead for point-source planets.

References include *Traub* (2000).

3.20. Interferometric Nulling

The thermal mid-infrared is an attractive spectral range for characterizing exoplanets, because it contains spectral features of H_2O, O_3, CO_2, and potentially CH_4 (see Table 4), and because the planet/star contrast is more favorable than in the visible (see Figs. 3 and 5).

However, for a given angular resolution, longer wavelengths require larger telescopes, or baselines, so for wavelengths 10–20 times greater than visible, conventional (~8 m) telescopes are not sufficient. Fortunately, long-baseline interferometers, with baselines on the order of B ~100 m, are able to do this.

The next two sections discuss an existing groundbased interferometer for exozodi observations, and a proposed space interferometer for exoplanet spectroscopy. The section that follows those is a note on the advantages and disadvantages of rearranging wavefront segments, a topic that is especially relevant to interferometers as well as to the pupil mapping coronagraph.

3.21. Nullers to Measure Zodiacal Light

The KIN was built to measure the zodiacal light in the 8–12 μm wavelength range around nearby stars, in preparation for the Terrestrial Planet Finder Coronagraph (TPF-C) and TPF-I. The KIN uses the two Keck 10-m telescopes, adjusted so that a target star is depressed by a factor of about 100 in intensity, allowing the surrounding zodi to be measured. This task was successfully completed. In operation, the KIN requires that the wavefront segments on the individual telescopes be flattened, using adaptive optics, and that the rapidly varying piston error of each segment be controlled to a fraction of a wavelength, using delay lines. Here we focus on the basic principle of nulling with the KIN.

The KIN has two major pupils (the two telescopes themselves), but four subpupils (the left and right halves of each telescope). The reason for splitting each telescope into two subpupils is that thermal emission from the sky above each telescope can be better suppressed. Figure 17 shows a schematic plan view of the pupils. Figure 18 shows sky coordinates of the target.

The KIN operates in the thermal infrared, so the telescope and sky emission are a huge, fluctuating background that needs to be removed. This is done by rapidly chopping between the three states shown in Table 7. In each state, the optical paths are adjusted to give the phases in this table. In particular, if we chop between states SZB and ZB, we will measure the star flux. Chopping between states ZB and B will give the zodi flux. The ratio of these results gives the contrast zodi/star. The output beams are sent through a prism so that the spectrum is split among 16 wavelength channels, and each is measured separately.

The combining phases here ignore the fact that an ideal beam splitter imparts a $\pi/2$ phase difference. We also treat the chopping as if it is the difference of two discrete states, whereas in fact the delay lines are scanned with a linear ramp that is slightly longer than the longest wavelength, and four measurements are made on the output intensity during this ramp, timed differently for each wavelength channel, from which the amplitude of the signal is extracted.

Numerically, a dust density corresponding to about 100 times the solar system level will produce a mid-infrared contrast at the KIN of about 10^{-2}, and in fact this is about the 1σ level of accuracy.

Each point in the sky, at radial angle θ from the optical axis, and at position angle α from the x axis, is a separate source, and is treated individually. Let us assume that the

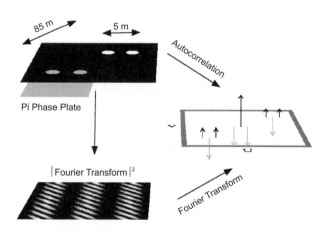

Fig. 17. Keck Nuller input is shown.

amplitude of the electric field from a photon from (θ, α) is $A_1(x_1) = e^{i\phi_1(x_1)}$, i.e., with magnitude unity and phase

$$\phi_1(x_1) = 2\pi \vec{\theta} \cdot \vec{r}/\lambda \tag{88}$$

where $\vec{r}$ is a coordinate in the pupil, projected toward the star.

The four beams are combined with beamsplitters, effectively overlapping the wavefront segments directly on top of each other. The summed amplitudes give an output amplitude A_2 where

$$\begin{aligned}A_2 = \sum_{j=1}^{4} A_1(j,\vec{r}) M_1(j,\vec{r}) = \\ e^{i2\pi[\theta\cos(\alpha)(-B/2)+\theta\sin(\alpha)(+b/2)]/\lambda + \phi_A} + \\ e^{i2\pi[\theta\cos(\alpha)(-B/2)+\theta\sin(\alpha)(-b/2)]/\lambda + \phi_B} + \\ e^{i2\pi[\theta\cos(\alpha)(+B/2)+\theta\sin(\alpha)(+b/2)]/\lambda + \phi_C} + \\ e^{i2\pi[\theta\cos(\alpha)(+B/2)+\theta\sin(\alpha)(-b/2)]/\lambda + \phi_D}\end{aligned} \tag{89}$$

Inserting the phases for each state, collecting terms, squaring to get intensities, and dropping a factor of 16, we find

$$I_2(SZB) = \cos^2(\pi\theta B\cos(\alpha)/\lambda)\cos^2(\pi\theta b\sin(\alpha)/\lambda) \tag{90}$$

$$I_2(ZB) = \sin^2(\pi\theta B\cos(\alpha)/\lambda)\cos^2(\pi\theta b\sin(\alpha)/\lambda) \tag{91}$$

$$I_2(B) = \cos^2(\pi\theta B\cos(\alpha)/\lambda)\sin^2(\pi\theta b\sin(\alpha)/\lambda) \tag{92}$$

A sketch of the resulting fringes, i.e., the transmission pattern projected on the sky, is shown in Fig. 18. We see that the star (at $\theta = 0$) is transmitted in state SZB, but nulled in states ZB and B, so chopping between these gives the star flux as

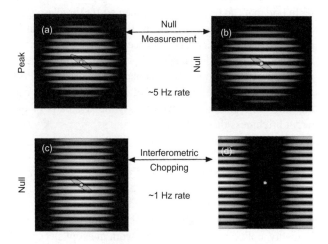

Fig. 18. Keck Nuller output is shown.

TABLE 7. KIN phase states.

State	ϕ_A	ϕ_B	ϕ_C	ϕ_D	Signal
SZB	0	0	0	0	star + zodi + background
ZB	0	0	π	π	zodi + background
B	0	π	π	2π	background

$$I_2(SZB) - I_2(ZB) = I(star) \tag{93}$$

We also see that the zodi (at $\theta > 0$) is transmitted in state ZB, but nulled in state B, so chopping between these gives the zodi, times the transmission pattern, as

$$I_2(ZB) - I_2(B) = \\ I(zodi)\sin^2\left(\frac{\pi\theta B\cos(\alpha)}{\lambda}\right)\cos\left(\frac{2\pi\theta b\sin(\alpha)}{\lambda}\right) \tag{94}$$

The beam pattern is known, and for a zodi disk that extends over several periods of the pattern the effective transmission is about a factor of 1/2. Therefore the contrast, zodi/star, is given by twice the ratio of equations (94) to (93).

References include *Colavita et al.* (2008) and *Barry et al.* (2008).

3.22. Nullers to Measure Exoplanets

The thermal infrared space missions TPF-I and Darwin, in the U.S. and Europe, are designed to detect and characterize exoplanets, down to and including Earth-like ones, around nearby stars. Here we describe the preferred method of operation that these projects have in common.

The TPF-I/Darwin schematic is identical to the KIN schematic. However, in TPF-I/Darwin the four collecting mirrors are mounted on four free-flying separated spacecraft, and the combining unit is another spacecraft at $(x,y) = (0,0)$ but above the plane at $z \gg B$. The key fundamental difference is in the phases added to each beam, which are specified in Table 8.

Inserting these phases into equation (89) and working through as before, we find the following detected intensities, again dropping a factor of 16

$$I_2(\text{left}) = \cos^2\left(\frac{\pi\theta B\cos(\alpha)}{\lambda} + \frac{\pi}{4}\right)\sin^2\left(\frac{\pi\theta b\sin(\alpha)}{\lambda}\right) \tag{95}$$

TABLE 8. TPF-I/Darwin phase states.

State	ϕ_A	ϕ_B	ϕ_C	ϕ_D	Signal
left	0	π	$0 + \pi/2$	$\pi + \pi/2$	null star + left zodi + left planet
right	0	π	$0 - \pi/2$	$\pi - \pi/2$	null star + right zodi + right planet

$$I_2(\text{right}) = \sin^2\left(\frac{\pi\theta B\cos(\alpha)}{\lambda}+\frac{\pi}{4}\right)\sin^2\left(\frac{\pi\theta b\sin(\alpha)}{\lambda}\right) \quad (96)$$

In both states the star at $\theta = 0$ is nulled by the $\sin^2(0) = 0$ term. For a planet at any value of (θ, α) the signal is

$$\begin{aligned}I_2(\text{right}) - I_2(\text{left}) = \\ I(\text{planet})\sin\left(\frac{2\pi\theta B\cos(\alpha)}{\lambda}\right)\sin^2\left(\frac{\pi\theta b\sin(\alpha)}{\lambda}\right)\end{aligned} \quad (97)$$

So the planet will generate a signal given by its intensity times the beam pattern on the sky. As the array rotates about the line of sight, the pattern will sweep across the planet, producing a modulation pattern that is uniquely characteristic of the planet's brightness, and also its radial (θ) and azimuthal (α) position in the sky, so that a map of its position can be unambiguously reconstructed after one-half of a full rotation of the array.

Importantly, any symmetric brightness component will be removed by the chopping, so if the zodi is bright and symmetric, it will drop out of the signal stream. Specifically, if $I_{zodi}(\theta, \alpha) = I_{zodi}(\theta, \alpha + \pi)$, then the zodi signal is

$$I_2(\text{right}) - I_2(\text{left}) = 0 \quad (98)$$

This is helpful in detecting the planet, especially if the zodi is bright. Obviously asymmetries in the zodi will be detected, but since these might be generated by planets in the first place, this will be of value to measure.

Interestingly, by reprogramming the TPF-I chopping sequence we can easily measure either the symmetric part of the target signal or the asymmetric part. And in fact this could be done in a single chopping sequence if desired. Thus a full picture of the target can be built up. The fringe pattern scales with wavelength, so the output beam should be dispersed onto a detector array, the same as for KIN.

As an example, if the individual mirrors are 2 m in diameter, then at 10 μm wavelength the FWHM of each diffraction-limited beam pattern on the sky is $\lambda/D \simeq 1.0$ arcsec FWHM, which just barely will accommodate a Jupiter at 10 pc ($\theta \simeq$ 0.5 arcsec radius). In each wavelength interval, all the light from the system falls on a single pixel. The modulation of that signal gives us a picture of the target at an angular resolution of $\lambda/2B \simeq 0.010$ arcsec, assuming a baseline of up to about B = 100 m, although in principle there is no limit to B.

References include *Beichman et al.* (1999) and *Cockell et al.* (2009).

3.23. Golden Rule

When thinking about an interferometer, and the image that could be formed in its focal plane, there is an important geometrical consideration that should be noticed: the *"golden rule"* of reimaging systems. This rule says that in order to have a wide field of view at a detector, the relative geometry of the input pupil must be preserved at the output pupil, to within a constant magnification factor. This rule was originally formulated for multi-telescope arrays such as the original Multi-Mirror Telescope (MMT) with its six primary mirrors, and it applies to later systems such as TPF-I.

There are three kinds of systems that do not obey this rule, and in each case the focal planes have extremely narrow fields of view. The first kind is the *pupil densification* system of a large array of telescopes in space, covering a baseline of several thousand kilometers, and phased up on a planet around a nearby star, for the purpose of making a true, spatially resolved image of that planet. In this concept the widely spaced collecting telescopes (the input pupil) are reimaged as a close-packed array (the output pupil), followed by an imaging lens. Because the two pupils are not related by a single scaling factor the output image has a very small field of view, which in this case is designed to be slightly larger than the diameter of the planet being imaged.

The second kind is the pupil-mapping coronagraph discussed in section 3.12, in which the output image of the first two mirrors has a field of view on the order of a few times λ/D, not sufficient to image a planetary system. Here the pupils are both circular, but the rays are rearranged, thereby essentially forcing a variable magnification factor between the two pupils, as a function of radial distance. However, in this case the image can be subsequently passed through a reversed set of optics, largely restoring the useful field of view.

The third kind is the family of interferometers discussed in sections 3.21 and 3.22 above. Here, by superposing the output pupils on top of each other, the exit pupil (a single opening) is clearly not a scale copy of the multiple openings in the entrance pupil pattern. This extreme case has a correspondingly tiny field of view, essentially just the diffraction beamwidth of the individual telesopes (λ/D).

References include *Traub* (1986), *Labeyrie* (1996), and *Pedretti et al.* (2000).

3.24. Visible Nuller

The visible nuller (Fig. 19) is a coronagraph-interferometer hybrid that can be used with segmented-mirror telescopes, such as the Thirty Meter Telescope (TMT), and is similar in plan view to the KIN and TPF-I, except that the baselines are more equal because they have to fit within a roughly circular primary mirror footprint. All the equations for KIN and TPF-I can be applied to the visible nuller, and in particular it can measure the symmetric as well as asymmetric parts of the target brightness distribution.

References include *Shao et al.* (2008).

4. SPECKLE CONCEPTS

All the discussions about methods in the preceding sections have dealt with idealized situations using perfect optics. In reality, nothing is perfect, and as work in exoplanet imaging has progressed, the effect of small variations in the photon propagation path lengths across the collected wavefronts has become a significant issue. In imaging, the effects of wavefront errors manifest themselves as *speckles*. Speckles appear as shown in Fig. 20. They are by far the dominant source of noise that must be controlled in order to image exoplanets.

4.1. Speckles from a Phase Step

The basic idea of speckles can be demonstrated with a simple example. Suppose that the wavefront incident on a telescope is advanced by a phase step $\phi/2$ on one half of the pupil, and delayed by $\phi/2$ on the other half. With reference to equation (60), the net amplitude in the focal plane becomes

$$A_2(\theta) = \int_0^{+D/2} e^{+i\phi/2} e^{i2\pi x_1 \theta/\lambda} dx_1 + \int_{-D/2}^{0} e^{-i\phi/2} e^{i2\pi x_1 \theta/\lambda} dx_1 \quad (99)$$

The resulting amplitude in the focal plane is

$$A_2(\theta) = \frac{\sin(\pi D\theta/2\lambda)}{\pi D\theta/2\lambda} \cos(\pi D\theta/2\lambda + \phi/2) D \quad (100)$$

If the wavefront has $\phi = 0$, i.e., no phase jump, then we recover the standard diffraction result

$$A_2(\theta, \phi = 0) = \frac{\sin(\pi D\theta/\lambda)}{\pi D\theta/\lambda} D \quad (101)$$

which is a single peak at the origin. However, if the total phase step is $\phi = \pi$, i.e., a half-wavelength, then we get

$$A_2(\theta, \phi = \pi) = \frac{\sin^2(\pi D\theta/2\lambda)}{\pi D\theta/2\lambda} D \quad (102)$$

which is a pair of peaks (speckles), each similar in width to the original single peak, and separated by about twice that width. In other words, where we once had a single image of the star, we now have two adjacent images. Smaller phase jumps will produce intermediate results, i.e., a pair of speckles but with unequal intensities and smaller separation. Clearly, we could continue to subdivide the pupil into smaller segments, producing about as many speckles as there are distinctive phase patches across the pupil.

4.2. Speckles from Phase and Amplitude Ripples

If we use an ideal telescope to image a star, and the wavefront from the star has been slightly distorted by possibly random phase and amplitude fluctuations, from an intervening atmosphere or from the telescope itself, then the natural result is a weakened star image surrounded by a halo of speckles. If the fluctuations are large, then the speckles will dominate and the star image will become just another speckle. If the telescope is a coronagraph, so the diffraction pattern is suppressed, then the speckles will certainly dominate.

Any continuous wavefront across the pupil can be represented by a sum of sine and cosine waves. For reference, we recall the standard result from Fourier analysis

$$A(x) = \sum_{n=0}^{\infty} \left(a_n \cos(2\pi nx/D) + b_n \sin(2\pi nx/D) \right) \quad (103)$$

where $A(x)$ is any real function on the interval $x = (-D/2, +D/2)$. The $\cos(\ )$ and $\sin(\ )$ functions form an orthogonal basis set, and the coefficients are obtained by projecting $A(x)$ onto this basis set and using the orthogonality. Multiplying both sides by $\cos(2\pi mx/D)$ or $\sin(-)$ and integrating we find

$$a_n = \frac{2}{D} \int_{-D/2}^{+D/2} A(x) \cos(2\pi nx/D) dx \quad (104)$$

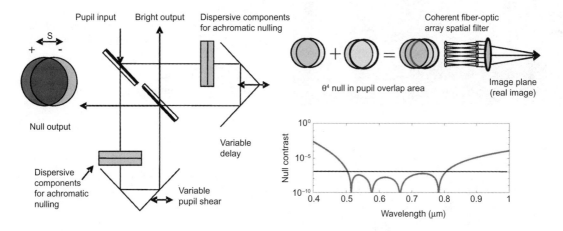

Fig. 19. Visible Nuller schematic is shown. If the input pupil is a segmented mirror, then the shear s is set to a multiple of the segment spacing, in one pass through the interferometer, and set to a multiple along a different axis, in a second interferometer, and only the doubly overlapped segments are used in the fiber output stage.

$$b_n = \frac{2}{D}\int_{-D/2}^{+D/2} A(x)\sin(2\pi nx/D)dx \qquad (105)$$

This method of representing functions in terms of a basis set can be extended to complex functions. Let $A(x)$ be any complex function on the interval $x = (-D/2, +D/2)$. Then $A(x)$ can be expanded as

$$A(x) = \frac{1}{2}\sum_{n=-\infty}^{+\infty} c_n e^{i2\pi nx/D} \qquad (106)$$

where the coefficients are

$$c_n = \frac{2}{D}\int_{-D/2}^{+D/2} A(x) e^{-i2\pi nx/D} dx \qquad (107)$$

and $n = 0, \pm 1, \pm 2, \ldots$

As an example, we could write $A(x) = A_0 e^{i\phi(x)}$ where A_0 is a constant amplitude and $\phi(x)$ is a spatially-varying phase, e.g., linear for a tilted wavefront, etc. If ϕ is complex, then the imaginary part represents the spatial variation of amplitude, which could also be absorbed in the coefficient as $A(x) = A_0 e^{-\mathrm{Im}(\phi)}$. If $\phi(x) \ll 1$ then the exponential can be expanded, giving $A(x) \simeq A_0 \times (1 + \phi(x))$, and in this case the phase $\phi(x)$ itself becomes the function that is expanded in terms of cos and sin basis functions. In this chapter we are dealing with wavefronts that are nearly perfect, e.g., $A(x) \simeq 1$, but have small departures from perfection, and these departures are the cause of speckles. It is for this reason that we view the cos and sin functions as frozen "ripples" on an otherwise flat "ocean" of amplitude.

Both methods can be easily extended to two-dimensional functions $A(x, y)$, where, for example, A could be the wavefront across a two-dimensional pupil. In this case the ripples are two-dimensional, and can be visualized as a set of corrugated surfaces having spatial frequencies from one wave per diameter up to many waves per diameter, and with the corrugations arranged at all possible azimuthal orientations.

Since diffraction operates linearly on the electric field, it operates on each of these ripples independently, and we can sum the resulting amplitudes. We show in this section that the analysis of a single, generalized ripple across the wavefront provides deep insight into the origin of speckles, as well as clues as to how to reduce them in practice.

Recall that $A_1(x_1) = e^{-i\phi_1(x_1)}$ represents the amplitude of the electric field in our one-dimensional pupil, with range $x_1 = (-D/2, +D/2)$. By Fourier analysis, the phase $\phi_1(x_1)$ can be written as the sum of potentially many sinusoidal ripples. Suppose a typical ripple has spatial period x_0, so that the phase of the wavefront can be written

$$\phi_1(x_1) = a\cos(2\pi x_1/x_0 + \alpha) + ib\cos(2\pi x_1/x_0 + \beta) \qquad (108)$$

with units of radians. If the peaks and valleys of the ripple have values $\pm h_0$ (cm), then the corresponding amplitude of phase delay is

$$a = 2\pi h_0/\lambda \qquad (109)$$

If the ripple also represents the patchy nature of scintillation or absorption across the wavefront, from a dark spot on the mirror, for example, then the transmitted field has an intensity ripple $e^{-2b\cos(2\pi x_1/x_0+\beta)}$; to see this, recall that a dark spot in a pupil can be represented by a real function, and is therefore expandable in a Fourier series, as we have shown above.

We find the amplitude in plane 2 to be

$$A_2(\theta) = \int_D e^{i\phi_1(x_1)} e^{i2\pi\theta x_1/\lambda} dx_1 \qquad (110)$$

If we assume that the perturbation is small, $|\phi_1| \ll 1$, then we can expand the first exponential, giving

$$A_2(\theta) \simeq \int_D \Big[1 + i\big(a\cos(2\pi x_1/x_0 + \alpha) + ib\cos(2\pi x_1/x_0 + \beta)\big)\Big] e^{i2\pi\theta x_1/\lambda} dx_1 \qquad (111)$$

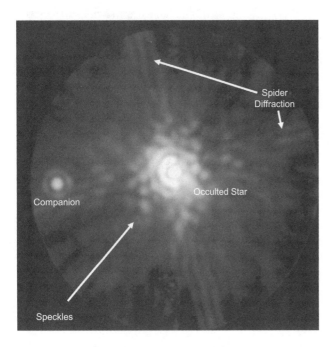

Fig. 20. An example of a classical Lyot-Coronagraph image of a nearby star in the near-IR. The star's PSF is created by occultation by the coronagraph's diffraction-limited rectangular focal plane mask, and a subsequent optimized pupil plane, or Lyot, stop. It shows the effect of spiders in the telescope and the resultant light propagated by a classical coronagraph. In addition, since this is a real observation, assisted by adaptive optics, the speckles have been frozen and are clearly visible. This star has a companion orbiting it, providing a PSF that is not occulted superimposed on the primary star's diffraction and speckle pattern.

Replacing $\cos(z)$ with $(e^{iz} + e^{-iz})/2$ allows us to integrate each term exactly. Then defining the well-known diffracted amplitude of a single star as

$$A_0(\theta) \equiv \frac{\sin(\pi\theta D/\lambda)}{\pi\theta D/\lambda} D \qquad (112)$$

we find that the diffracted amplitude is the sum of a main peak, at the expected ($\theta_{main} = 0$) position of the star, plus two smaller peaks, one on each side, at $\theta_{speckle} = \pm\lambda/x_0$, where

$$A_2(\theta) = A_0(\theta) + \frac{1}{2}\left(iae^{i\alpha} - be^{i\beta}\right)A_0(\theta + \lambda/x_0) + \frac{1}{2}\left(iae^{-i\alpha} - be^{-i\beta}\right)A_0(\theta - \lambda/x_0) \qquad (113)$$

The speckles are diffracted to either side, exactly as would be expected from the first orders of a diffraction grating with rulings spaced by the ripple period. So this is a very physically understandable picture. In a later section we will use this equation to show how a deformable mirror can null out one of these speckles, and indeed, a whole field of them.

The intensity pattern $I_2(\theta) = |A_2(\theta)|^2$ has six terms, and each is a function of θ

$$I_2 = I_0 + I_{(+1)} + I_{(-1)} + I_{(+2)} + I_{(-2)} + I_{(3)} \qquad (114)$$

Here I_0 is the main peak, the central star image, defined above.

The next two terms are *speckles*

$$I_{(\pm 1)} = \frac{1}{4}\left[a^2 + b^2 \pm 2ab\sin(\alpha - \beta)\right]I_0(\theta \pm \lambda/x_0) \qquad (115)$$

These are the symmetrically placed speckles. Their intensities are equal if either phase errors or amplitude errors dominate, but if there is a mixture of these, the intensities can be unequal. Also note that these are exactly equal in shape to translated copies of the central peak, but scaled down.

The next two are *pinned speckles*

$$I_{(\pm 2)} = \left[\mp a\sin(\alpha) - b\cos(\beta)\right]\sqrt{I_0(\theta)I_0(\theta \pm \lambda/x_0)} \qquad (116)$$

These pinned speckles are located at the same place as the ordinary speckles, but they are scaled by the local intensity of the diffraction pattern from the main peak. In other words, they are effectively pinned to the preexisting diffraction rings, and they will show up as a locally enhanced or depressed intensity of the expected ring pattern, varying from point to point. As we will show below, they can be quite strong.

The last term, $I_{(3)}$, is negligible because is a crossproduct of the two speckles, so we drop it.

Here are some numerical examples. Suppose that a telescope in space has a small surface error, from perhaps a residual polishing tool imprint or a quilting from an eggcrate backing structure, so that the error is periodic. If the amplitude of this phase perturbations is $h_0 \simeq \lambda_0/100$, then the phase amplitude is $a = 2\pi/100 \simeq 0.06$ rad. The type 1 speckles will then have an intensity of about $I_{(\pm 1)}/I_0 = a^2/4 \simeq 0.1\%$ times that of the star. The type 2 (pinned) speckle intensity will depend on the preexisting diffraction pattern; for example, if the Airy rings have an intensity that is 0.1% times the main star (around the fourth Airy ring), then the pinned speckles intensity will be about $I_{(\pm 2)}/I_0 = 0.2\%$ times the star, i.e., brighter than the Airy ring itself. In addition, suppose that the reflectivity of the mirror is somewhat patchy, and can be represented by a reflectivity that varies from peak to average by about 1%, giving $b = -(1/2)\ln(0.99) \simeq 0.005$. Then the intensity of the type 1 speckles will be about $ab/2 \simeq 0.01\%$ of the main peak if both a phase and amplitude wave are present, or $b^2/4 \simeq 0.001\%$ if the amplitude ripple is present alone. If the spatial period of either of these ripples is $x_0 \simeq 100$ cm then the speckles will fall at an angle of about ± 0.10 arcsec for visible wavelengths, roughly the location of Earth at 10 pc.

4.3. Speckles from Mirror Surfaces

Real mirrors have surface shape errors that get smaller in proportion to the size of region being considered. Thus the surface errors are not a white noise process, even though, for simplicity, we sometimes make that assumption, as we will do in deriving equations (120)–(125).

A common way to describe the shape error of a mirror surface, which leads directly to phase errors in the reflected wavefront, is through the *power spectral density* function PSD(k). Here k is the spatial frequency of a wave of wavelength Λ on the surface of the mirror. Recall from Fourier analysis that the sum of all such possible wavelengths can reproduce any arbitrary mirror shape error. We have that

$$k(\text{waves per cm}) = \frac{1}{\Lambda(\text{cm})} \qquad (117)$$

Note that k is a two-dimensional quantity, e.g., $k = (k_x, k_y)$, which we will keep in mind as needed.

It is an empirical observation that large mirrors tend to have similar PSD functions, of the form

$$PSD(k) = \frac{A}{1 + (k/k_0)^n} \qquad (118)$$

where $n \simeq 3.0 \pm 0.1$. This form applies to the residual shape of the mirror after low-order terms (typically focus, tip, tilt, coma, and astigmatism) have been subtracted, the reason being that these terms can be largely compensated by better alignment of a primary and secondary mirror system, and are somewhat independent of the intrinsic polished surface. Also, the low-order terms correspond to speckles within roughly $3\lambda/D$, which could be blocked by a coronagraph, but the higher-order terms will contribute directly to speckles at larger angles, where we may expect to be imaging exoplanets.

The rms error of the surface is given by the area under the PSD curve, between specified spatial frequency limits, e.g., the same limits as given by the inner and outer range of a dark hole. We have

$$h_{rms}^2 = \iint PSD(k_x, k_y) dk_x dk_y$$
$$= \iint PSD(k) 2\pi k\, dk \quad (119)$$

The integral can be done analytically or numerically.

An analysis of the residual errors of the HST (2.4-m primary, plus secondary, combined), the Magellan (6.5 m), and a 1.5-m mirror shows that these all have roughly the same PSD shape and value, although it is also clear that more modern mirrors, polished with larger, shape-controlled lapping tools, will have smaller high-spatial-frequency errors.

Numerical values for HST are $A = 6000$ nm^2 cm^2, and $k_0 = 0.040$ cycles/cm. Integrating this over the range 0.02 to 0.28 cm^{-1}, corresponding to $\Lambda = 3.6$–50-cm wavelengths, we find $h_{rms} = 7.5$ nm. Thus the HST mirror surface has a $\lambda/84$ surface at the usual laser wavelength of $\lambda = 633$ nm.

References include *Hull et al.* (2003) and *Borde and Traub* (2006).

4.4. Speckles from the Atmosphere

For a groundbased telescope, observing a star through the atmosphere at visible wavelengths, the incident wavefront is typically distorted on all scales from kilometers to millimeters, with a PSD (see section 4.3) fall-off given by $n = 11/3 \simeq 3.7$, i.e., slightly faster than mirror-polishing errors. The bulk of the distortions take place on scales that lie between the *outer scale* L_0, typically 20 m, and the *inner scale* l_0, typically 0.4 cm.

The length r_0 is the diameter of a region over which the wavefront has an rms variation of about 1 rad (technically, 1.015 rad, by definition in the Kolmogorov model of turbulence). The value of r_0 is about 10 cm in the visible at a typical observatory, and scales as $\lambda^{6/5}$, so it is larger in the infrared. In the "frozen atmosphere" approximation, these patches of density fluctuations are carried along by the wind, so a typical timescale for wavefront change is $\tau_0 \simeq 0.31\, r_0/V$ where V is a typical wind speed in the overlying atmosphere; if $V = 10$ m s^{-1} then τ_0 is about 3 ms. A groundbased image of a star therefore is made up of approximately $(D/r_0)^2$ speckles, churning on a timescale of τ_0, and spread over an angular diameter on the sky of about λ/r_0 or 1 arcsec in the visible, independent of telescope diameter.

The validity of the frozen flow hypothesis has been experimentally investigated by *Poyneer et al.* (2009). They find that speckles at groundbased telescopes can be modeled in terms of about one to three layers in the atmosphere, for over 70% of the time, with a median wind speed of about 10 m s^{-1}, but with a large range from 3 to 40 m s^{-1}. These observations also suggest that frozen flow accounts for about 30($\pm$10)% of the total power that can be controlled by a deformable mirror (DM). These findings, together with the observation that the wind speeds vary by less than 0.5 m s^{-1} over 10-s intervals, suggests that about one-third of the speckle power could be reduced by an predictive algorithm.

References include *Poyneer et al.* (2009).

4.5. Speckle Suppression

Armed with a knowledge of how speckles are formed, we are now prepared to measure and suppress them. The following sections explain wavefront sensing, and how deformable mirrors are used to suppress speckles, singly as well as wholesale, arising from phase and amplitude variations across a pupil. The Talbot effect is examined to show how to suppress speckles from planes before or after the pupil plane. Some of the methods used today at groundbased and space telescopes are discussed, including ADI, SSDI, chromatic speckle suppression, and dual-mode polarimetric imaging. We conclude with a comparison of ground vs. space for direct imaging of exoplanets.

4.6. Wavefront Sensing and Control

At groundbased telescopes the atmosphere drives the incoming wavefront to an rms spatial variation of more than a wavelength, and in addition the wavefront varies rapidly in time, as discussed above. A *wavefront sensor* (WFS) is any device that allows us to measure the wavefront. The term *wavefront sensing and correction* (WFSC) applies when closed-loop corrections are applied.

The six main sources of uncertainty in a WFS are photon noise, chromaticity, aliasing, time delay, scintillation, and non-common-path errors.

Photon noise is obviously fundamental; it can be minimized by using a bright star and large values of r_0 and τ_0 on the ground, or their equivalent in space (e.g., from polishing errors and thermal drift).

Chromaticity arises from WFS systems in which the sensing wavelength is different from the science wavelength, and the wavefront errors are wavelength dependent; it can be minimized by using the same wavelength band for both purposes.

Aliasing arises when the WFS is sensitive to, but cannot distinguish between, a spatial frequency mode of the wavefront and an odd harmonic. For example, a pupil-based wavefront sensor with contiguous detecting elements, each of width w, is sensitive to a spatial wavelengths of size 2w as well as 2w/3.

Time delay occurs because a detector must integrate a signal for a finite length of time before it can be read out, and in addition the servo system has a finite bandwidth, which effectively adds more time delay to the correction signal. The choice of integration time depends on photon rate, the desired signal to noise ratio, and detector read noise.

Scintillation arises through the Talbot effect (section 4.11), in which wavefront phase perturbations generated in the upper atmosphere become intensity perturbations in the lower atmosphere. This causes *shadow bands* to flicker across a telescope pupil, which in turn will generate speckles in the image plane (see section 4.2).

Non-common-path errors arise when the WFS and the science focal plane are separated. For example, it is common for the WFS in a groundbased telescope to be fed by a *beamsplitter* that taps off part of the starlight and sends it to a sensor via an optical path that is different from the optical path to the focal plane. This method assumes that the WFS optics are perfect (or more generally, identical to the science optics), and do not add any wavefront ripples or speckles to the WFS system. Since we always use imperfect optics, this scheme is bound to fail at the level of the quality of the optics.

In principle, and in practice, it is much safer to detect speckles in the science focal plane, because this will avoid all six sources of uncertainty listed above. The decision as to how and where to put a WFS is determined by a combination of practical aspects of a given telescope, the desired level of wavefront correction, and the personal taste of the experimenter.

The Shack-Hartmann WFS is used at many telescopes. It is a simple system, easy to understand, but also probably the least effective. In this system a beamsplitter taps off part of the light, and a lens forms an image of the pupil. This pupil plane is filled with a large number of small lenses, sometimes formed by embossing a sheet of plastic. Each lenslet forms an image of the star onto a position-sensitive detector, e.g., multiple quad cells or a single CCD. A local tilt of the wavefront will produce a shift in the star image. Thus image position measurements can be converted to local wavefront slopes, and by patching together the slopes a full wavefront snapshot can be obtained. This information can then be used to drive a DM, and a closed-loop control established. The Shack-Hartmann WFS is sensitive to aliasing, because it uses a finite number of spatial sensing cells in the pupil plane, as discussed above.

A better WFS would be one that uses some of the bright starlight to interfere with the speckles. This is done in some systems by tapping off the bright star image, passing it through a spatial filter to make it nearly single-mode and therefore a smooth reference wavefront (using a small hole or ideally a single mode fiber), and interfering this with the speckle pattern; unfortunately, this kind of system is susceptible to non-common-path errors, rendering it less than ideal.

An even better WFS would use the star to interfere directly with the speckles, avoiding extra optical paths. This can be done by using a DM to diffract light out of the main star image and onto existing speckles, but adjusting the DM so that the phases cancel. This method is used in sections 4.8, 4.9, and 4.10 below.

References include *Guyon* (2005).

4.7. Contrast in a Dark Hole

The area in the focal plane over which speckles can be suppressed by a DM is called the *dark hole*. If the DM is placed at an image of the primary mirror, and the wavefront errors are entirely due to hills and valleys of that mirror, then it is clear that the DM simply needs to advance or retard the reflected wavefront in order to flatten it.

Suppose that a circular primary mirror, the pupil, is mapped onto a square DM, just filling it. The DM will be controlled to deform in such a way that a bumpy incident wavefront is reflected as a smooth wavefront, to the limit of control of the mirror. A typical DM has a thin glass facesheet, backed by an N × N square array of actuators that push or pull on the facesheet perpendicular to its surface.

In one dimension, it takes a string of four actuators (not two, as is often assumed) to approximate the shape of a single period of a sine or cosine wave, but if there are many such periods in a wave then we can get by with only about two actuators per period. This approximation breaks down in the case of a single period, because two actuators can approximate a cosine, but not a sine, wave, or the reverse, depending on where the actuators are located with respect to the wave.

Thus we can fit up to about N/2 periods of a wave with N actuators in one dimension. Also, there are N/2 whole waves, or modes, that can be approximated by a string of this length, i.e., 1 wave per diameter, 2 waves per diameter, ... N/2 waves per diameter. By analogy, we assume that there are M modes in the full circular area of the pupil, where

$$M = \frac{\pi}{4}\left(\frac{N}{2}\right)^2 \qquad (120)$$

Suppose that the average mode has an amplitude h_0, so that the speckle produced by this mode has relative intensity

$$I(\text{typ. speckle})/I_{\text{star}} \simeq a^2/4 = (\pi h_0/\lambda)^2 \qquad (121)$$

where we used equations (109) and (115).

At each point in the pupil the net amplitude will be the sum of M complex vectors of average length h_0 but with random phases. This is exactly the random walk problem in two dimensions. Therefore the expected average amplitude will be $h_{\text{rms}} \simeq M^{1/2} h_0$, or

$$h_{\text{rms}} = \frac{\sqrt{\pi} N h_0}{4} \qquad (122)$$

So, from equation (121), writing the average contrast as

$$C \equiv \frac{I(\text{ave. speckle})}{I(\text{star})} = \left(\frac{\pi h_0}{\lambda}\right)^2 \qquad (123)$$

we find

$$C = \pi \left(\frac{4 h_{\text{rms}}}{N\lambda}\right)^2 \qquad (124)$$

This says that to achieve a contrast C, with an otherwise perfect coronagraph, we need to control the N × N element DM with an accuracy such that the reflected wavefront has an RMS error of h_{rms} or better. The DM must be controlled to a surface error of $h_{\text{rms}}/2$, of course.

Inverting this relation, we can say that to achieve an average speckle contrast C we need to control the wavefront to a relative accuracy of

$$h_{rms} = \frac{N\lambda\sqrt{C}}{4\sqrt{\pi}} \quad (125)$$

For example, if we desire $C = 10^{-10}$, and we have $N = 64$, then h_{rms} must be about $\lambda/10,000$. Thus the wavefront must be 100 times better than the typically "excellent" $\lambda/100$ wavefront.

In the visible, this means that the DM must control the reflected wavefront to an accuracy of $h_{rms} \simeq 0.5$ Å. This may seem impossible, given that this is about half the radius of a Si or O atom; however, a typical (0.1 to 1.0 mm) DM element averages over many such atoms, and it is the average surface that counts here. In addition, we know from experiment (e.g., *Trauger and Traub*, 2007) that this is perfectly feasible.

It is useful to look at the DM as being a scattering grating device that can be commanded to generate a surface ripple that can diffract starlight to a specific target point in the focal plane. The phase of the controlled scatter can be adjusted by shifting the wave pattern on the DM from sine to cosine, for example. Thus the DM can be used to direct starlight to points in the focal plane, with the desired amplitude and phase so as to cancel starlight that arrived by other means. The DM is thus an extremely powerful device to have in the starlight beam, and it is required in all advanced adaptive optics systems. Note that this only works for light from the star itself; light from an exoplanet is not coherent with starlight, so the DM cannot use starlight to cancel an exoplanet, a fact we will use in section 4.10.

The angular radius of the dark hole is the maximum angle to which the high-frequency spatial period of the DM can scatter light. This angle is λ divided by two pistons of the DM, or D/(N/2), giving the angular radius as

$$\theta(\text{dark hole}) = \pm \frac{N\lambda}{2D} \quad (126)$$

Thus the maximum size of the dark hole is a square of angular size $N\lambda/D$, which is a length of N resolution elements of the pupil. This square is centered on the star.

4.8. Single-Speckle Nulling

We show in this section that it is possible to make a speckle vanish by putting an appropriate pattern on an upstream DM. The method applies to a speckle that is caused by either a phase or amplitude perturbation. The method works for a speckle on one side of a star image, not both sides. To make many speckles vanish, see sections 4.9 and 4.10. To make speckles on both sides vanish, using two DMs, see section 4.11 on the Talbot effect.

Let us start with a simple case in which the one-dimensional pupil has a single ripple across it, formed by either a phase perturbation or an amplitude perturbation (or both).

Recall that arbitrary shapes of such perturbations can be represented by a sum of cosine and sine functions, and that these can be visualized as the basis functions of phase ripples as well as absorption ripples. We assume that a coronagraph is present, so that it suppresses the central star and its diffraction pattern. Thus the telescope diffraction pattern is eliminated, and only scattered light from an imperfect wavefront remains. This is the case that was discussed in section 4.2. For that case we assumed that the pupil had a ripple given by

$$\phi_{pupil}(x_1) = a\cos(2\pi x_1/x_0 + \alpha) + ib\cos(2\pi x_1/x_0 + \beta) \quad (127)$$

and we saw that the resulting amplitude in the image plane was

$$A(\theta) = A_0(\theta) + \frac{1}{2}(iae^{i\alpha} - be^{i\beta})A_0(\theta + \lambda/x_0) + \frac{1}{2}(iae^{-i\alpha} - be^{-i\beta})A_0(\theta - \lambda/x_0) \quad (128)$$

Suppose that we know the values of (a, α, b, β) (we will show how to estimate these later), and we wish to add a ripple to the DM to counteract the existing pupil ripple. If we add a ripple $-a\cos(2\pi x/x_0 + \alpha)$ we will clearly eliminate the a terms in the speckles, but the b terms will still remain; clearly, this is not sufficient.

However, suppose that we decide to eliminate (or null) the speckle at $\theta = \pm\lambda/x_0$, i.e., either the $A_0(\theta-\lambda/x_0)$ or the $A_0(\theta + \lambda/x_0)$ speckle. Let us add a ripple to the DM given by

$$\phi_{DM}(x_1) = a'\cos(2\pi x_1/x_0 + \alpha') \quad (129)$$

Adding this to the existing ripple gives a net speckle amplitude that we set equal to zero

$$iae^{\pm i\alpha} + ia'e^{\pm i\alpha'} - be^{\pm i\beta} = 0 \quad (130)$$

We solve this by setting the real and imaginary parts each equal to zero. The solution for suppressing either of these speckles is

$$a' = -\sqrt{a^2 \pm 2ab\sin(\alpha - \beta) + b^2} \quad (131)$$

$$\tan(\alpha') = \frac{a\sin(\alpha) \pm b\cos(\beta)}{a\cos(\alpha) \mp b\sin(\beta)} \quad (132)$$

This shows that it is possible to suppress a speckle that arises from phase or absorption, or both together, using only the phase ripple of a DM, but only on one side of the star. Speckles originating from phase can be canceled on both sides of the star. Speckles originating from amplitude can be canceled on either one side of the star or the other side, but not both at the same time, since the speckle on the noncanceled side will get larger as the target speckle gets smaller.

For example, if there is no absorption, then b = 0, and we get $a' = -a$, and $\alpha' = \alpha$, which is physically logical. Likewise, if there is pure absorption, then a = 0, and we get $a' = -b$, and $\alpha' = \beta - \pi/2$.

4.9. Multispeckle Nulling

It is tedious to null speckles one by one. Not only are there a lot of speckles to null, but they are all coherent so that the intensity at a given point depends on that at neighboring points. In other words, there is coupling between the speckles, owing to their wings.

In this section we show, in principle, how to estimate the parameters of many speckles simultaneously (as was assumed for a single ripple in section 4.8). With this knowledge the DM can be set to null all speckles on one side of the star, using a method similar to that above. Here again, we assume a perfect coronagraph, but an imperfect residual wavefront.

Suppose that there are multiple speckles present in the pupil, perhaps from an imperfect primary mirror. A DM can null spatial wavelengths as short as two DM actuators in the pupil, so there are up to N/2 waves that need to be measured, in our usual one-dimensional case. (The two-dimensional case is similar.) Let us write the ripple in the pupil as

$$\phi_{pupil}(x_1) = \sum_{n=1}^{N/2} \left[a_n \cos(2\pi n x_1/D + \alpha_n) + i b_n \cos(2\pi n x_1/D + \beta_n) \right] \quad (133)$$

We measure the intensity of this (unknown) starting case at each of N points in the focal plane.

This method requires an extra pupil plane in addition to those shown in Fig. 12, so for this discussion let us assume that plane 1 is an image of the original telescope pupil.

If we now add a new, independent set of ripples (a_n', α_n') to the DM, the added phase will be

$$\phi_{DM1}(x_1) = \sum_{n=1}^{N/2} \left[a_n' \cos(2\pi n x_1/D + \alpha_n') \right] \quad (134)$$

We measure the intensity in the focal plane at N points for this case as well.

We then add yet a different set of ripples

$$\phi_{DM2}(x_1) = \sum_{n=1}^{N/2} \left[a_n'' \cos(2\pi n x_1/D + \alpha_n'') \right] \quad (135)$$

and measure these intensities as well.

In both cases the parameters of the two added sets of ripples can be anything convenient, e.g., totally random ripples or a sharp delta function created by a single actuator. The main point is that the added ripples should be significantly different from the original set.

We now have three sets of intensities in the focal plane, on N pixels, therefore 3N data points. The number of unknowns $(a_n, \alpha_n, b_n, \beta_n)$ for N/2 waves is a total of 2N. Therefore we can use our 3N measurements to solve for 2N parameters.

The reason we need more measurements than parameters is because the intensity is the square of the amplitude, and therefore there are sign ambiguities that need to be resolved. If desired, yet another set of ripples and observations can be made, and a least-squares or singular-value decomposition solution found to the overdetermined set.

Once all the parameters of the original ripples have been measured, then the DM can be set to counteract them, on one side of the star, assuming that both phase and amplitude errors exist.

4.10. Speckle Energy Minimization

In practice, the multispeckle nulling method sketched above has several limitations: (1) intrinsic noise in the system, from photon noise as well as measurement noise, which limits our ability to perfectly measure the intensity pattern; (2) imperfect knowledge of the DM's response to an applied voltage; (3) higher-frequency ripples in the pupil, which can alias down into the dark hole; (4) the pupil phase drifting with time, during a several-minute measurement, from thermal expansion; (5) the need to observe a star over a finite range of wavelengths, e.g., a 10% or 20% band, but the solutions given above are only valid for a single wavelength; (6) the fact that both a sine and cosine ripple in the DM can only be generated out to an angle of half the radius of the dark hole, using the strict rule that four actuators are needed per wave.

Here is the basic idea of energy minimization. We go back to basics for a few steps, to clarify what we are doing. Suppose the amplitude at the input pupil is

$$A_1(x_1) = e^{i\phi_1(x_1)} \simeq 1 + i\phi_1(x_1) \quad (136)$$

where ϕ_1 can be complex. Expanding in a Fourier series we have a useful representation as

$$\phi_1(x_1) = \sum_{n=1}^{\infty} a_n \cos(2\pi n x_1/D + \alpha_n) + i \sum_{n=1}^{\infty} b_n \cos(2\pi n x_1/D + \beta_n) \quad (137)$$

Suppose there is a DM immediately after this pupil, with mask function

$$M_1(x_1) = e^{i\phi_{DM}(x_1)} \simeq 1 + i\phi_{DM}(x_1) \quad (138)$$

and where we expand in a finite Fourier series

$$\phi_{DM}(x_1) = \sum_{n=1}^{N/2} a_n' \cos(2\pi n x_1/D + \alpha_n') \quad (139)$$

Then in plane 2, at the focus of the star, the amplitude $A_2(\theta)$ is

$$A_2(\theta) = \int_D M_1(x_1) A_1(x_1) e^{i2\pi x_1 \theta/\lambda} dx_1 \quad (140)$$

Inserting the expressions for the phases, and keeping only terms to the first order, we get

$$A_2(\theta) = \int_D [1 + \phi_1(x_1) + \phi_{DM}(x_1)] e^{i2\pi x_1 \theta/\lambda} dx_1 \quad (141)$$

Carrying this through will give $A_2(\theta)$ of a single star, as in equation (44), the diffraction pattern of the pupil (Airy rings), and the speckles from both the pupil and DM. Suppose that this is followed by a perfect coronagraph, which in essence allows us to delete the "1" from this expression, and to write the speckle amplitude as

$$A_{2,spec}(\theta) = \int_D \left[\sum_{n=1}^{\infty} a_n \cos(2\pi n x_1/D + \alpha_n) + \right.$$
$$i \sum_{n=1}^{\infty} b_n \cos(2\pi n x_1/D + \beta_n) + \quad (142)$$
$$\left. \sum_{n=1}^{n/2} a'_n \cos(2\pi n x_1/D + \alpha'_n) \right] e^{i2\pi x_1 \theta/\lambda} dx_1$$

Suppose that we wish to calculate the total speckle energy in the range from θ_{min} to θ_{max}, where these might be a target dark hole, i.e., a few times λ/D to $N/2$ times λ/D, for example. This total energy will be E_{spec} where

$$E_{spec} = \int_{\theta_{min}}^{\theta_{max}} |A_{2,spec}(\theta)|^2 d\theta \quad (143)$$

As an extreme example, if we decide to calculate the total energy in the entire focal plane, i.e., θ from $-\infty$ to $+\infty$, after some work we find

$$E_{spec} = \left[\sum_{n=1}^{\infty} a_n^2 + 2 \sum_{n=1}^{N/2} a_n a'_n \cos(\alpha_n - \alpha'_n) + \sum_{n=1}^{\infty} b_n^2 + \sum_{n=1}^{N/2} (a'_n)^2 \right] D/2 \quad (144)$$

If we minimize this with respect to the parameters of the DM, we find that we need to set the DM as

$$a'_n = -a_n$$
$$\alpha''_n = \alpha_n \quad (145)$$

which cancels the wavefront distortion, up to the highest frequency allowed by the DM, but does not affect the absorption part of the wavefront. This is especially clear when we write the value of the minimum energy, which is then

$$E_{spec}(min) = \left[\sum_{n=N/2+1}^{\infty} a_n^2 + \sum_{n=1}^{\infty} b_n^2 \right] D/2 \quad (146)$$

Here the first term is the sum of the power in all the high-frequency errors in the pupil, and the second term is all the absorption error terms, none of which are canceled.

A more useful exercise would be to calculate the total energy in the dark hole on one side of the star, as suggested above, or some other selected area. This would then use the DM to cancel both the delay and absorption terms in a finite window, as shown for a single speckle in equation (132). This is a work in progress.

References include *Borde and Traub* (2006) for the energy minimization method, and *Give'on et al.* (2006) for the electric field conjugation method.

4.11. Exploiting the Talbot Effect

The *Talbot effect* says that if a plane wave is incident on an infinitely wide periodic mask, then the transmitted wave will be periodically replicated downstream. At a distance midway between these replications, a pure phase disturbance will become a pure amplitude disturbance, and vice versa. This seemingly strange curiosity has important consequences in several areas.

Atmospheric scintillation is a familiar example. Suppose that an incident wavefront from a star passes through some turbulence near the boundary between the troposphere and the stratosphere. Breaking up the turbulence into Fourier components, we see that each component will impart a sinusoidal phase ripple onto the incident wave. At a distance Δz lower in the atmosphere (see equation (157)), the phase ripple will become an intensity ripple, or stellar scintillation.

The *Talbot carpet* is another example. Suppose that a plane wave is incident on a one-dimensional mask at $z = 0$ that has many small holes spaced by Λ in the x direction. Then, if we go to a downstream value of z, and ask what the condition is for wave transmitted by every n_Λ-th hole to have a path length that is the minimum distance plus n_λ wavelengths (in other words, an interference-created intensity maximum), from a right-triangle construction, we have the relation

$$z^2 + (n_\Lambda \Lambda)^2 = (z + n_\lambda \lambda)^2 \quad (147)$$

Assuming that $\lambda \ll \Lambda$ we find

$$z = \frac{(n_\Lambda)^2}{n_\lambda} \frac{\Lambda^2}{2\lambda} \quad (148)$$

where n_Λ and n_λ can be 1, 2, 3, ... etc. There will be infinitely many such planes lying between $z = 0$ and $z = z_{TC}$, where

$$z_{TC} \equiv \frac{\Lambda^2}{2\lambda} \quad (149)$$

This pattern of bright points is called the Talbot carpet.

We now ask, can we use this effect to compensate for intensity fluctuations in a pupil, by correcting the resulting downstream phase with a DM? The answer will be yes, but with a caveat. Let us start with an incident plane wave $A_1(x_1) = 1$ in the pupil, and a phase ripple imposed by a mask

$$M_1(x_1) = a\cos(2\pi x_1/\Lambda) \quad (150)$$

We now ask, what is the amplitude in a plane $A_2(x_2)$ downstream a distance z? Note there is no lens at plane 1, since we are allowing the wave to propagate freely in space, i.e., continuing on as a free wave, without a focusing lens. We sum up the wavelet contributions from plane 1, as we did for the Talbot carpet.

$$A_2(x_2,z) = \sum(\text{wavelets}) \\ = \int_{-\infty}^{+\infty} M_1(x_1) A_1(x_1) e^{i2\pi l/\lambda} \frac{1}{\sqrt{l}} dx_1 \quad (151)$$

where l is the distance from x_1 to x_2

$$l = \sqrt{z^2 + (x_1 - x_2)^2} \\ \simeq z + \frac{(x_1 - x_2)^2}{2z} \quad (152)$$

We use $1/\sqrt{l}$ in the integrand to account for the diminished wavelet amplitude with distance, to conserve energy in our two-dimensional space; since amplitude is less important than phase in this integral, we also use $l \simeq z$.

Now, if the phase ripple is weak, i.e., $a \ll 1$, we can expand $e^X \simeq 1 + X$, and use $e^{iX} = \cos X + i \sin X$, and proceed with the integration. We will need the Fresnel cosine integral, defined as

$$C(X) = \int_0^X \cos\left(\frac{\pi}{2}t^2\right) dt \quad (153)$$

where $C(\infty) = 1/2$.

After a bit of work, we find the amplitude in plane 2 to be

$$A_2(x_2,z) = e^{i2\pi z/\lambda} \sqrt{\frac{\lambda}{2}}\left[1 + ia\cos\left(\frac{2\pi x_2}{\Lambda}\right)e^{-i\pi\lambda z/\Lambda^2}\right] \quad (154)$$

The first factor is the familiar plane wave in the z direction. The second factor $(\sqrt{\lambda/2})$ is an artifact of our method of integration and can be ignored. The third term (in brackets) is the same as the input wave except for the additive periodic term in z. The corresponding intensity is

$$I_2(x_2,z) = \frac{\lambda}{2}\left[1 + 2a\cos\left(\frac{2\pi x_2}{\Lambda}\right)\sin\left(\frac{2\pi z}{z_T}\right)\right] \quad (155)$$

where the *Talbot distance* z_T is defined as

$$z_T = \frac{2\Lambda^2}{\lambda} \quad (156)$$

(Confusingly, z_{TC} differs from z_T by a factor of 4, a result of the fact that the former is generated by point sources, and the latter by a continuous wave source.)

We see that the wave that emerges from plane 1 reproduces itself at multiples of z_T, and that at the halfpoint between these reproductions an intensity pattern appears that also reproduces itself. We will have alternating planes of constant intensity (but varying phase), and varying intensity (but constant phase). The distance between adjacent planes of varying and constant intensity is Δz where

$$\Delta z = \frac{z_T}{4} = \frac{\Lambda^2}{2\lambda} \quad (157)$$

So if we go a distance Δz downstream from a pupil image, with no intervening optics, the intensity ripples will become phase ripples. Unfortunately, this distance is chromatic, i.e., it depends on wavelength. But at least we can see a method here to begin to reduce intensity ripples in the pupil.

Here is a numerical example. Suppose that the optics train of a telescope includes a reduced-diameter pupil plane (plane 3a, say), followed by a length Δz with no focusing optics, so that the beam propagates in a nominally parallel fashion to plane 3b. Suppose that there are amplitude variations in plane 3a that can be approximated by ripples of period length Λ. Let us place a DM in plane 3b, where the amplitude ripples will have turned into phase ripples, which we can cancel with the DM. Let us assume that the pupil diameter in plane 3a is such that the period of the disturbance is approximately equal to two actuators of the DM. If each actuator is 1 mm wide, then $\Lambda \approx 2$ mm. If the wavelength is $\lambda \simeq 0.5$ μm, then the separation between 3a and 3b will need to be about $\Delta z \simeq 4$ m, a large but not totally unrealistic length.

4.12. Angular Difference Imaging

If the wavefront sensing occurs in a plane that is different from the science focal plane, and if the speckles from the atmosphere and telescope pupil have been reduced in the wavefront sensing plane, there often remains a residual wavefront deformation in the science plane owing to a *non-common-path* problem. These speckles can be substantial, and since they arise from the telescope optics themselves, they can persist for a long time, typically many minutes or more. Unfortunately, a telescope speckle has the same appearance as an exoplanet, at a given wavelength, so strong, persistent speckles can easily overwhelm a faint exoplanet image.

The *angular differential imaging* (ADI) technique can overcome internal speckles from the telescope by simply rotating the telescope about the line of sight, or at an alt-az telescope by allowing the rotating Earth to rotate the apparent sky (except on the celestial equator). Since the detector remains fixed with respect to the telescope, the non-common-path speckles also remain fixed on the detector. Thus, subtracting a rotated image from a nonrotated one should eliminate the fixed-pattern speckles, allowing the exoplanet to be seen. Another name for this technique is *roll deconvolution*, a method that has had success on the HST.

References include *Marois et al.* (2006), *Hinkley et al.* (2007), and *Artigau et al.* (2008).

4.13. Simultaneous Spectral Differential Imaging

The technique of *simultaneous spectral differential imaging* (SSDI) is based on the fact that speckles are located at an angular distance from the star in proportion to their wavelength. So if images are taken at two or more wavelengths, and they are radially scaled to a common wavelength, then the difference of images should cause the fixed-pattern speckles to drop out. If an exoplanet is in the field it will show up as a radially shifting positive and negative feature.

An additional leverage factor arises if the exoplanet has a strong absorption feature in its spectrum, different from its star. For example, the methane band at 1.7 μm is very deep on some gas giants. The fact that the planet is relatively faint in this band gives it an extra handle for detection.

References include *Racine et al.* (1999), *Marois et al.* (2005), and *Biller et al.* (2006).

4.14. Chromatic Speckle Suppression

As an extension of the SSDI technique, one can use much higher spectral resolution to achieve superior speckle suppression. In this case, a coronagraph is outfitted with a hyperspectral imaging device, also sometimes referred to as an *integral field spectrograph*. Images are obtained at tens to hundreds of wavelengths simultaneously, usually over a single astronomical bandpass. The data forms a cube, with two spatial and one spectral axes. Speckles follow diagonally radiating paths through these data cubes, while real celestial objects will remain at the same spatial separation from the primary star. As such, in principle one can distinguish one wavelength's speckle pattern from another one and effectively remove the speckles without damaging the signal from a bona fide celestial object. The data processing methods are fairly complicated, even though the initial studies of this technique suggested relatively simple solutions for data processing.

References include *Sparks and Ford* (2002).

4.15. Dual-Mode Polarimetric Imaging

Perhaps the most successful of all speckle suppression techniques to date, *dual-mode polarimetric imaging* exploits the fact that in general starlight is very weakly polarized. If one is attempting to image an object or material around a star that exhibits large fractional polarization, the starlight and speckles can be removed with almost arbitrary precision. Images are formed using a Wollaston prism, which sends light with perpendicular polarization vectors in slightly different directions. Two images can formed and sensed simultaneously in this manner. When they are subtracted, only light that is actually polarized remains in the image. If the starlight is not polarized, it will be completely removed. Speckles are formed from unpolarized starlight. This technique has been used very successfully to image disks of dust that polarize light through the scattering process.

References include *Kuhn et al.* (2001), *Perrin et al.* (2004), and *Oppenheimer et al.* (2008).

4.16. Ground Versus Space Direct Imaging

Is it possible to directly image an Earth around a nearby star with a groundbased telescope? Or is it necessary to put an Earth-imaging telescope in space, above the atmosphere? In this section, we show an approach to answering this question.

For a large groundbased telescope, the overlying atmosphere will distort the incident wavefront by several wavelengths. We need to detect this distortion and remove it by reflecting the wavefront from a DM. The distortion may arise from several levels in the atmosphere, but for present purposes let us assume that it arises at a single level, such that we can image that layer on a DM, and remove the phase distortion without suffering any additional error, such as amplitude non-uniformity. Let us also assume that we can do this operation essentially instantaneously, without any time lag due to the measurement interval or the servo system. All these assumptions will be broken in real life, so the current calculation is optimistic in the sense that the real result will always be worse.

Let us start by assuming that we can detect the wavefront error with a Shack-Hartmann device. Suppose that about half the light in the pupil is split off and sent to an array of lenslets, each with diameter r_0. Assume that the local slope of the wavefront is α radians, approximately constant across the patch r_0. Each lenslet will focus the star in an image that has angular size λ/r_0. If there are n detected electrons in that image, we will be able to locate its centroid with an angular accuracy of about $\Delta\alpha = \lambda/(r_0\sqrt{n})$ radian. The uncertainty in the local measured slope of the wavefront will also be $\Delta\alpha$. The uncertainty h_{rms} in the delay of the wavefront over this patch is that error times the width of the patch, so

$$h_{rms} = \Delta\alpha \times r_0 = \frac{\lambda}{\sqrt{n}} \qquad (158)$$

Let us assume that we can correct the wavefront to within this uncertainty.

From equation (125) we see that the resulting wavefront error will generate speckles whose intensity is a factor of C fainter than the star itself, according to

$$h_{rms} = \frac{N_{DM}\lambda\sqrt{C}}{4\sqrt{\pi}} \qquad (159)$$

where N_{DM} is the number of DM elements per diameter D of the telescope. In our case we want to have at least one DM element per r_0 segment, so $N_{DM} = D/r_0$. Collecting terms we find the number of electrons needed is

$$n = \frac{16\pi r_0^2}{CD^2} \quad (160)$$

For a star of magnitude m, the number of electrons that we can collect is limited the number of photons in a volume determined by r_0 and τ_0, which gives

$$n = f_\lambda(m)\Delta\lambda A \tau_0 QE \quad (161)$$

where f is the flux density from equation (9), $\Delta\lambda$ is the bandwidth, A is the collecting area, τ_0 is the collecting time, and QE is the quantum efficiency.

Let us take $\Delta\lambda = 0.20\lambda$, $A = \pi r_0^2/4$, and QE = 0.5, representing the half of the incident light that is split off for the wavefront sensor, and assuming a perfect detector. The integration time is generally given by $\tau_0 = 0.31\, r_0/V$ where $V \simeq 500$ cm s^{-1} is the assumed wind speed, so τ_0 corresponds to the time that it takes the wind to move a patch of air of size r_0 about one-third of the way past a similarly sized collecting lenslet. Collecting terms we get

$$n = 10^{a-0.4m}(0.2\lambda)\left(\frac{\pi}{4}r_0^2\right)\left(0.31\frac{r_0}{V}\right)\left(\frac{QE\times\lambda}{2\times 10^{-12}}\right) \quad (162)$$

where n has units of electrons, λ is in μm, r_0 is in cm, and V is in cm s^{-1}.

Equating the two expressions for n, and using the fact that r_0 scales with wavelength as

$$r_0(\lambda) = r_0(\lambda_V)(\lambda/\lambda_V)^{6/5} \quad (163)$$

and inserting numerical values, we find that the star magnitude m must be at least

$$m = 2.5\left[a + 10.69 + 3.2\log(\lambda_{\mu m}) + \log(CD_m^2)\right] \quad (164)$$

where D_m is the telescope diameter in meters.

Evaluating equation (164) for 30-m and 100-m telescopes, for a contrast $C = 10^{-10}$ appropriate for the Earth/Sun system, and for the BV RI J H K$_s$ bands, we find that the result is nearly independent of band, giving

$$\begin{array}{l} m(10^{-10}, 30\text{ m}) \simeq -4.2 \pm 0.2 \\ m(10^{-10}, 100\text{ m}) \simeq -1.4 \pm 0.2 \end{array} \quad (165)$$

This tells us that there are *no* stars bright enough to drive a servo system to achieve a speckle contrast as small as the ratio of Earth to Sun brightness, in any of these spectral bands, and even for an essentially perfect servo system. The detection might just barely be possible for a 100-m telescope, but even so there would be only a handful of near-infrared targets available. The bottom line is that to detect an Earth, and to characterize it, we will need to go to space, simply to eliminate unavoidably bright speckles from the turbulent atmosphere.

One might ask if a bright nearby guide star could be used for Earth detections instead of the target star. The answer is probably no, because atmospheric turbulence differs enough between stars that this level of compensation would be impossible.

One might also ask if laser guide stars could be used for Earth detections. Here again the answer is no, this time because the brightest laser guide star that has ever been used has an equivalent stellar magnitude of about +5, which, from equation (164), is very far from being bright enough.

If the target is self-luminous young Jupiters, with contrasts around $C \geq 10^{-8}$, then the limiting magnitudes in equation (165) become brighter by about 5 mag, so m(10^{-8}, 30 m) $\simeq$ 0.8, for which a handful of near-infrared target stars might provide sufficiently large signals. Thus large ground-based telescopes should be able to detect self-luminous young Jupiters down to contrasts approaching 10^{-8}.

5. CURRENT PROJECTS

In the previous sections we described the myriad techniques needed to image exoplanets directly. These techniques are employed in many ways, often in combination, in current observational projects. In Table 9 we list many of the currently operating and proposed projects around the world that have exoplanet imaging and spectroscopy as a major part of their scientific justification. Here we briefly compare and contrast these many experiments. Please note that it is impossible in a paper of this nature to include every project in operation or proposed. Instead of being exhaustive and unabridged, our goal here is to provide examples of the applications of the techniques described in detail above. The reader can find additional information about these projects through their associated references.

As an indication of the current status of direct imaging of exoplanets, we note a few recent achievements here. In the laboratory, no one has yet published a coronagraph contrast close to, or better than, the value of 6 × 10^{-10} in monochromatic light, from *Trauger and Traub* (2007), which is rather surprising; this area clearly needs more work. On the bright side, we do have the wonderful images of HR 8799 b,c,d, β Pic b, and Fomalhaut b, as described in the introduction to this chapter. The former triad of young planets was recently directly imaged in the near-infrared with a vector vortex coronagraph using a relatively small (1.5 m) telescope pupil (*Serabyn et al.*, 2010), at K-band contrasts as low as 2 × 10^{-5} and as close as 2λ/D; this is certainly an encouraging step forward.

This section and the next (section 6) are meant to provide a snapshot of the state-of-the-field at the time of publication. Of course the future projects likely will change in some respects. From a very general perspective the observational field of comparative exoplanetary science via direct detection is in a nascent stage. We are just now on the verge of routinely observing such objects with both photometry and spectroscopy. Much in this field will change as observations reveal the advantages or each technique described previously.

TABLE 9. Current, planned, and proposed projects for direct detection and study of exoplanets.

Project Name	First Light (year)	Telescope Diameter (m)	Optimal Wavelength (μm)	AO System Elements	Starlight Suppression Technique	Speckle Suppression Technique
Keck Imaging	2002	10.0 Keck-II	1.0–2.5	249	None	ADI
Gemini NIRI/ Altair Imaging	2004	7.98	1.0–2.5	177	None	ADI, SDI
Lyot Project	2004	3.63	1.2, 1.6	941	Lyot Coronagraph	Polarimetry
VLT/NACO	2005	8.0	4.0	177	Lyot Coronagraph	SDI
HiCIAO	2007	8.2 Subaru	1.2, 1.6, 2.2	188*	Lyot Coronagraph	None
MMT/AO	2008	6.5	5.0		Lyot Coronagraph	ADI
NICI	2008	7.987 Gemini-N	1.2, 1.6, 2.2, 3.8, 4.7	85*	Lyot Coronagraph	SDI or ADI
Project 1640	2008 2011	5.07 Palomar	1.640 1.640	249 3217	APL Coronagraph APL Coronagraph	Chromatic Chromatic Science-Arm
LBTI	2011	2 × 8.1 22.3 eff.	3.0–20.0	2 × 349	Interferometric Nulling	N/A
Gemini Planet Imager	2011	7.798 Gemini-S	0.95, 1.2, 1.6, 2.2	1579	APL Coronagraph	Chromatic- Science-Arm Polarimetry
SPHERE	2011	8.20 VLT	1.2, 1.6, 2.2 0.6, 0.8, 0.9	1312	APL or Phase-Mask Coronagraph	Chromatic Polarimetry
JWST	2015	6.5	5–27	108†	Lyot or Phase-Mask Coronagraph	PSF Subtraction and Chromatic
Planetscope	2015	1.00 Balloon	0.5–1.0	2304	Band-limited Coron.	Science-Arm
30–42 m Telescope	2018	30.0	1.0–2.5	≥3000	APL Coronagraph or Nulling Coronagraph	Science-Arm Chromatic
Probe Missions§	2019‡	1.0–2.0	0.5–1.0	2304	Band-limited, PIAA, Star Shade	Science-Arm
TPF-C	2024	3.5 × 8.0 elliptical	0.5–1.0	2304	PIAA, Band-Limited or Phase Coronagraph	Science-Arm Chromatic
TPF-O	2024	4	0.7–1.0	≈500	Occulter	
DaVINCI	2024¶	4 × 2.5 Single Spacecraft	1.0–13.0	~1000	Interferometric Nulling	N/A
TPF-I/ DARWIN	2028¶	4 × 4 Sep. Spacecraft	5.0–20.0	N/A	Interferometric Nulling	N/A
Large Space Telescope	2035¶	4.0 to 16.0	0.3–20.0	≈4000	Numerous	Unknown

*The HiCIAO and NICI instruments use a curvature-based bimorph mirror so actuator number is not directly comparable to those of the other systems.
†The MIRI instrument has active optics, not adaptive optics, for the JWST primary mirror with 18 segments with 6 actuators per segment.
‡The various proposed space projects have no certain launch dates nor is it known at the time this book was printed which, if any, of these missions would actually be constructed and flown into space.
§There are several different proposed NASA probe-class missions involving star shade occulters or internal coronagraphs and relatively small aperture telescopes with modest AO systems.
¶These projects are highly speculative and are not very well defined at this point.

A summary of the current dynamic range or contrast actually achieved by existing observations can be found in Fig. 21. The projects that have achieved these results are described below.

5.1. HST Coronagraphs

HST has three coronagraphs, one each on these instruments: (1) Advanced Camera for Surveys, High Resolution Camera (ACS-HRC, from 2002 to 2007, now permanently lost); (2) Near Infrared Camera and Multi-Object Spectrometer (NICMOS, from 1997 to 1999, and 2002 to 2008, with a possible restart in the future); and (3) Space Telescope Imaging Spectrograph (STIS, from 1997 to 2004, and 2009 to present).

All three coronagraphs are of the simplest possible type, an opaque top-hat blocker in an image plane, followed by a Lyot mask in a pupil plane, and a detector in a final image plane.

ACS-HRC: The ACS coronagraph is in the uncorrected, aberrated beam. A glass sheet can be inserted near the focus (at the point of least confusion) with a choice between two simple top-hat metalized spots of diameter 1.8 and 3.0 arcsec. This point is upstream of the correcting optics, so the beam is aberrated. The spot sizes were chosen to reduce the diffracted light intensity to be less than the speckle intensity. A thin metal Lyot mask is inserted downstream, just in front of the aberration correction mirror, i.e., close to a pupil plane image. The Lyot mask blocks light from around the edges of the primary, secondary, and spider arms, reducing the throughput to about 48% of its value without the mask. Filters are available. The final image in the HRC is on a CCD detector.

NICMOS: The NICMOS coronagraph is in the corrected beam. It comprises a 0.6-arcsec-diameter circular hole in a focal plane mirror, followed by a cryogenic Lyot mask of the primary, secondary, and spider diffraction in a pupil plane. The sky plane is reimaged on a near-infrared detector. The hole size was chosen to remove 93% of the encircled energy at H band, beyond which point the diffracted and scattered light profile flattens out.

STIS: The STIS coronagraph is in the corrected beam. The image-plane part comprises two orthogonal wedge-shaped blocking masks, where each wedge has a length of 50 arcsec and a width that ranges from 0.5 to 3.0 arcsec, and the wedges overlap at about their 1.25 arcsec width points. The pupil plane part has a Lyot mask of the outer edge of the primary mirror. The final image plane has a CCD detector. No filters are available, so the full 0.2 to 1.03 μm spectrum is imaged.

References include *Krist* (2004), and the HST Instrument Handbook descriptions of the ACS, NICMOS, and STIS coronagraphs, for which up-to-date versions are available on the web.

5.2. Keck and Gemini Imaging

A number of surveys have been conducted using adaptive optics and direct imaging without a coronagraph. In these cases either calibrator stars or an ADI or SDI technique is employed to reduce the starlight in the image. These techniques have been effective, particularly in the case of the star HR 8799 with its three companions that seem to be of planetary mass, one of which was also detected by HST/NICMOS. However, nondetections are in many respects equally important because they provide upper limits to the overall population of planets around nearby stars. For example, Gemini and its Altair AO system were used to conduct the "Gemini Deep Planet Survey" to search for planets at large separations (greater than ~1 arcsec) from their host stars. The survey used the ADI technique. No objects were found around 48 stars that were observed at least two times, and the conclusions are important: The fraction of stars with brown dwarf companions between 25 and 250 AU separations is between 2.2% and 0.4% at the 3σ confidence level. The upper limits on the fraction of stars with at least one planet of mass 0.5–13 M_{Jup} are 28% for the semimajor axis range of 10–25 AU, 13% for 25–50 AU, and 9.3% for 50–250 AU (also with a 3σ confidence).

References include *Marois et al.* (2008), *Janson et al.* (2010), *Herriot et al.* (2000), and *Lafreniere et al.* (2007, 2009).

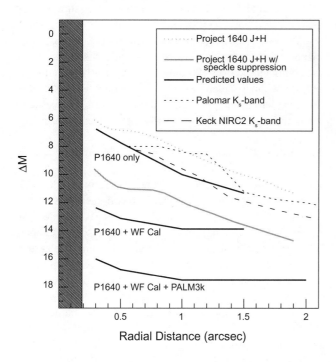

Fig. 21. Actual dynamic range or contrast achieved for existing systems. This is merely a selection of current achieved dynamic ranges, including Keck AO imaging, Palomar AO Lyot Coronagraph imaging, Project 1640 (with and without chromatic speckle suppression), Keck/Gemini ADI (e.g., *Marois et al.*, 2008), and expected performance of Project 1640 with the addition of a science-arm WFS and an extreme AO system (labeled WF Cal and PALM3k), and finally Lyot Project dual-imaging polarimetry. Dynamic range is expressed as the 5σ detection threshold for a point source as a function of radius and as a difference in magnitude with respect to the central star. Generally, ADI sees a gain of about 1 to 2 magnitudes over the speckle noise background, while dual-mode polarimetry can completely remove speckle noise at the level of 10^{-5} or better but only for detection of polarized objects. Image courtesy of S. Hinkley.

5.3. Very Large Telescope NACO and Subaru HiCIAO

A number of surveys of similar nature were undertaken by other investigators using the VLT and Subaru telescopes. In these surveys a classical Lyot coronagraph is fitted behind an AO system. Typically these projects have much smaller fields of view than those described in the previous subsection, but they have more effective starlight suppression due to the presence of a coronagraph. The NACO instrument, which operates at 4 μm with a coronagraph and ADI starlight suppression, has been particularly effective in imaging objects that may be planets or brown dwarfs, not easily distinguishable when very young. The Subaru telescope is outfitted with a new AO coronagraphic imaging device called HiCIAO, which includes speckle suppression modes using dual-mode polarimetry and SDI. This is a very promising instrument currently in operation and producing images of objects in the brown dwarf mass regime (companions to GJ 758) but at temperatures that probably provide spectroscopic opportunities that will be very important for exoplanetary science. HiCIAO will likely find more objects similar to the companions of HR 8799.

References include *Lenzen et al.* (2003), *Rousset et al.* (2003), *Chauvin et al.* (2005), *Neuhauser et al.* (2005), *Tamura et al.* (2006), *Kasper et al.* (2009), *Lagrange et al.* (2009), and *Thalmann et al.* (2009).

5.4. Multiple Mirror Telescope

The MMT AO system operating at 5 μm has also been used to seek warm planets with direct imaging, no coronagraph, and speckle suppression using ADI and SDI. So far only stellar companions have been discovered, but this program is as competitive as the others mentioned so far, due to high-Strehl at the longer wavelengths. A number of experiments with advanced starlight suppression are also being conducted with this system, including phase apodization and nulling interferometry.

References include *Biller et al.* (2007), *Kenworthy et al.* (2007), *Liu et al.* (2007), *Kenworthy et al.* (2009), and *Mamajek et al.* (2010).

5.5. Lyot Project, Project 1640

This is a project that conducted a survey of over 100 nearby stars with the sensitivity to find brown dwarfs and warm, young gas giant planets. Using a Lyot coronagraph along with an extremely high-order AO system on the U.S. Air Force AEOS telescope, it was able to exploit polarimetric speckle suppression to achieve images with a contrast below 10^{-6}, showing a perturbed solar-system-scale disk around AB Aurigae, which may have a planet in formation at 100 AU separation. The project also set constraints on the brown dwarf population of companions through statistical analysis of nondetections. Project 1640, the successor to the Lyot Project at Palomar's 5-m Hale Telescope, combines AO correction with an apodized pupil and hard-edged mask coronagraph as well as an integral field hyperspectral imaging device, simultaneously obtaining coronagraphic images at 30 different wavelengths over the 1.0–1.8-μm range. This system has just begun surveying nearby stars and has found faint stellar companions already. It employs chromatic speckle suppression as well as ADI and SDI, being the first instrument to be able to attempt all three such speckle removal techniques on the same set of data.

References include *Hinkley et al.* (2008), *Oppenheimer et al.* (2008), *Wizinowich* (2008), *LeConte et al.* (2010), *Hinkley et al.* (2010), and *Zimmerman et al.* (2010).

6. FUTURE PROJECTS

Exoplanet imaging and spectroscopy is a growing field of research. Its promise has led to the proposal of many new types of advanced instruments and has been used to justify construction of new major facilities both on the ground and in space. Continuing the path through Table 9, we now describe some of the projects that have not yet begun routine observation, but are already in the process of being built. Finally, at the end of this section we describe a few of the more speculative projects on the decadal time horizon. Since those may substantially change, the discussion provided here is intentionally general. In order to provide a sense of scale in terms of cost, the implementation of all the projects listed in Table 9 is estimated by the authors to be over 100 billion USD. Clearly not all these projects will happen given the level of funding for this kind of research, but this number can be easily compared for scale to many other things that developed societies spend money on.

6.1. Large Binocular Telescope Interferometer

The Large Binocular Telescope Interferometer (LBTI) uses the nulling technique available to two-aperture systems. The stated goals for the project are to detect Jupiter-like planets around younger stars (<2 G.y.) in the solar neighborhood, and 3 M_{Jup} planets around solar-age stars. The system will have a very large field of view in comparison with most current and planned projects. It requires two separate adaptive optics systems as well as the nulling system. The expectation is that nulling at the level of about 10^{-7} will be achieved within a few diffraction elements of the central star. LBTI is expected to begin nulling operations in the fall of 2011.

References include *Hinz et al.* (2008) and *Hinz* (2009).

6.2. Project 1640 Phase II

Project 1640 is due to be upgraded to have a full 3217-actuator AO system for far superior wavefront control, and a second-stage wavefront sensor (called a "science-arm WFS") that obtains the wavefront distortion due to the optical system on a timescale of roughly 1 s. This science-arm WFS is designed to sense and control, through periodic feedback to the AO system, the long-lived speckles that are removed so efficiently by the polarimetric technique, allowing the chromatic speckle suppression to act on much fainter speckles at the 10^{-7} level. The system employs an apodized pupil Lyot coronagraph and the same hyperspectral imaging device as in the current

Phase I. This may allow objects as faint as 10^{-8} at $6\lambda/D$ to be detected and spectra to be extracted, although the field of view is only 4 arcsec wide. This begins to open the exoplanet characterization phase space to the study of many Jupiter-mass exoplanets. First light for this system is expected in 2011.

6.3. Gemini Planet Imager

The Gemini Planet Imager (GPI), with a goal of 10^{-8} contrast at $6\lambda/D$ in raw contrast, is a large instrument consisting of an AO system with 1500 actuators, a science-arm WFS, as well as a hyperspectral imager and a dual-mode polarimetry imager. This is perhaps the most ambitious of the systems currently being developed, employing, like Project 1640, an apodized pupil Lyot coronagraph for starlight suppression. The goals of the project are to provide images and spectra of roughly 100–200 young planets around nearby stars, with a similar field of view as Project 1640. First light is expected in 2011.

References include *Macintosh et al.* (2008).

6.4. Spectro-Polarimetric High-Contrast Exoplanet Research

The Spectro-Polarimetric High-Contrast Exoplanet Research (SPHERE) project at the VLT employs a relatively high-order 1312-actuator AO system with an apodized pupil Lyot coronagraph, and an integral field hyperspectral imager, as well as a separate dual-mode polarimetric imager. It will be possible to conduct both polarimetry in the shorter wavelengths and hyperspectral imaging at longer wavelengths simultaneously over a roughly 4-arcsec field of view. Due to see first light in 2011, this system is predicted to achieve a few times 10^{-7} contrast levels at $6\lambda/D$, enough contrast to image and obtain spectra of several tens of warm, young planets with masses as small as Jupiter around the nearest 100–200 stars.

The four projects above are fully funded and very close to achieving first light around 2011. One can hope that by 2013 or so, low-resolution spectra and orbits of perhaps a hundred exoplanets will be obtained.

References include *Beuzit et al.* (2008).

6.5. James Webb Space Telescope

The JWST is another project that is currently funded and expected to see first light. There are two primary instruments being constructed for JWST that address exoplanet imaging directly. The NIRCAM instrument has several coronagraphic modes operating at 1–5 µm and should achieve contrasts of about 10^{-7} to within a few λ/D, which will allow imaging thermal emission from exoplanets. It employs apodization on the segments of the telescope to reduce speckles pinned to the diffraction pattern as well as several other options for starlight suppression. Grisms permit low-resolution spectroscopy. The TFI instrument has Fabry-Perot etalons operating from 1.5 to 2.4 µm and 3.1 to 5.0 µm, with several hard-edged circular occulting spots and a nonredundant masking mode. Although the inner working angles are larger for this instrument, it should provide performance similar to that of GPI and SPHERE. Another instrument of interest in direct imaging of exoplanets with JWST is the Mid-InfraRed Imager (MIRI), which has several coronagraphic options and images at wavelengths from 9 to 12 µm. This system is predicted to detect planets as cool as 300 K within an arcsecond of a star, using four-quadrant phase masks or a traditional Lyot-style coronagraph. The combination of all these instruments will provide a suite of measurements across a broad range of wavelengths.

References include *Boccaletti et al.* (2005), *Greene et al.* (2007), and *Krist et al.* (2009).

6.6. Planetscope

Planetscope is one of the few proposed balloon experiments that would directly image exoplanets. Using a small-aperture telescope and a coronagraph that includes a low-order AO system (primarily for fine guiding and optical defect correction), this type of project would benefit from getting above more than 99% of atmospheric turbulence. This, and other suborbital projects for exoplanets, hold the promise of delivering science results as well as demonstrating technology for future space missions.

References include *Traub et al.* (2008) and *Chen et al.* (2009).

6.7. Probe-Class Space Missions

At present there are a slew of proposed space missions for direct exoplanet imaging, each costing on the order of 600 to 1000 million USD, and termed *probe-class* (or *medium-class*) missions. These are generally missions that involve relatively small-aperture telescopes, or up to four smaller apertures (to permit a combination of nulling and coronagraphy). Some of these proposed systems use the PIAA technique or band-limited focal plane masks and hyperspectral imaging sensors. These projects typically have the goal of contrasts at the 10^{-9} level, sufficient to detect mature giant planets, but not Earths (unless there is one around a very nearby star). None of these missions will be launched for the next five years or so, but it is possible that one may enter development during that period.

These probe-class projects, in alphabetical order, and the respective lead authors, are ACCESS, a 1.5-m coronagraph (J. T. Trauger); DaVINCI, four 1.1-m visible nullers (M. Shao); EPIC, a 1.65-m visible nuller (M. Clampin); and PECO, a 1.4-m coronagraph (O. Guyon).

References include *Trauger et al.* (2008), *Shao et al.* (2008), *Lyon et al.* (2008), and *Guyon et al.* (2009).

6.8. Large Groundbased Observatories

Moving further into the future, the Thirty Meter Telescope (TMT) project, the Giant Magellan Telescope (GMT), and the European Extremely Large Telescope (E-ELT), all 30–42-m-diameter segmented aperture telescopes, have as core instruments planet detection and characterization projects. These systems face the difficulty of dealing with segmented apertures, which intrinsically diffract light (as do standard spider support structures in on-axis telescope designs). Most of the standard coronagraphic techniques are exceedingly inefficient on segmented mirror telescopes because the optical stops must mask off the segment edges. However, the visible nulling technique overcomes this issue by using a matrix of single-mode fibers mapped to the segments to clean up high-spatial-frequency wavefront errors, and a piston-only deformable mirror to map the segments into one coherent wavefront, as though it were a single aperture. In the process it interferometrically nulls the starlight, giving a peculiar spatially variable throughput. This requires many pointings to achieve sensitivity in the telescope's full field of view, but allows for use of the full aperture of the segmented mirror. Each of these projects expects first light in about 2018; planet detection instrumentation is not likely to be a first-light priority, but it may well be soon thereafter.

References include *Gilmozzi and Spyromilio* (2008), *Johns* (2008), *Nelson and Sanders* (2008), and *Shao et al.* (2008).

6.9. Large Space Missions

On the more distant horizon, several groups around the world have proposed UV/optical space telescopes on the scale of 4–16-m diameter. The most well-studied of these is TPF-C, with an 8-m × 3.5-m oval monolithic clear-aperture primary mirror and an internal coronagraph. TPF-C has had years of research behind its design and science program, and was deemed to be technically feasible. The prime goal of TPF-C is to carry out spectroscopic characterization of planets, at visible wavelengths, for planets down to and including Earth-twins.

Two other large telescopes have recently been suggested: the Advanced Technology Large Aperture Space Telescope (ATLAST), and the eXtrasolar Planet Characterizer (XPC or THEIA). ATLAST actually refers to a family of designs: an 8-m monolithic circular primary, a 9.2-m deployable segmented mirror, and a 16-m deployable segmented version. The 8-m clear-aperture version of ATLAST is similar to TPF-C, and could use its full pupil for an efficient coronagraph, but the segmented versions would require a visible nuller type of coronagraph. THEIA was originally studied as a possible way to combine a 4-m telescope with an internal and external coronagraph, simultaneously, but it was found to be not feasible and this aspect has been set aside in favor of the latter.

The external occulter concept (TPF-O) has been suggested as a way to utilize an existing telescope by adding a distant starshade, with the latter possibly being delivered to orbit (e.g., L2) independently of the former. Potential advantages include a lower cost for the sunshade itself (as compared to an internal coronagraph telescope, and assuming that the companion telescope is not included in the cost), and the ability to image planets closer to their parent star, compared to an internal coronagraph. Disadvantages include being limited to a narrow annulus on the sky, centered near 90° from the Sun, and a long slew time between target observations.

The original Earth-characterizing space instrument was the Terrestrial Planet Finder Interferometer (TPF-I), a thermal-infrared interferometer with formation-flying cryogenic collectors and combiners. A similar concept, Darwin, was developed in parallel, mainly in Europe. These projects have essentially merged in the sense that they are now technically essentially identical. Despite being first to develop, the TPF-I/Darwin concept is currently expected to be the last to be flown, after a TPF-C or similar type is flown, owing to the perceived complexity and cost of a system that requires five free-flying spacecraft, all at cryogenic temperatures.

These large-aperture telescopes would be designed to image planets as small as an Earth-twin, unless the majority of stars have zodiacal disks that are 10–100 times thicker than our own solar system's zodiacal dust. These devices, which in principle could be constructed now and launched within a few years, could begin to tackle the question of life in other planetary systems. The key to this is the detection of chemical abundances in a planet's atmosphere that are not consistent with thermochemical equilibrium and could only be generated by biological activity.

References include *Traub et al.* (2006), *Cash* (2006), *Lawson et al.* (2008), *Cady et al.* (2009), *Cockell et al.* (2009), *Kasting et al.* (2009), *Levine et al.* (2009), *Postman et al.* (2009), and *Soummer et al.* (2009).

6.10. Summary and Outlook

The effort to make direct images of exoplanets is a complicated and rather difficult endeavor. It involves clever optical techniques and exquisite control of the imaging system. Indeed, the systems about to begin collecting data in the next year or so are aiming for $\lambda/1000$ level wavefront control in devices that are dealing with light corrupted by the atmosphere as well as processed by 30–40 optical surfaces. These systems (GPI, SPHERE, P1640) will only be able to study the largest planets around relatively young stars. Yet, the future of this field is exciting. A handful of images have been obtained and spectra of those objects will be available soon.

Next-generation systems will likely provide upwards of 100 or 200 spectra of exoplanets within the following 4 to 5 years. Then the field of comparative planetary science will see a new vitality. For the first time in human history we will be able to compare what many different Jupiter-mass objects have in common, how they evolve through different ages, and whether they even do have anything in common.

In the more distant future, it seems likely that acquiring spectra, orbits, masses, and the other necessary characteristics of the much larger population of older planets in closer orbits to their stars, ones whose signature in our instruments is the result of reflected starlight, not internal radiation, will require spacebased experiments. We showed in section 4.16 that if we were to expand the sorts of groundbased instruments being built now to the regime where they would be sensitive to the majority of planets, they would require guide stars that are brighter than any that exist.

This implies, of course, that moving to space, where one can control wavefronts with far slower cadence, provides the obvious solution. Although we believe that this is true, one of the biggest mistakes a scientist can make is to assume that the primary mode of attacking a particular question is the only one. It is not impossible that the issue of extremely high-contrast imaging will be solved in an alternate way, without adaptive optics, perhaps, or with some other type of optical manipulation.

As such, one must remain optimistic, and in the end, the overwhelmingly compelling nature of the science of exoplanets, and how they directly relate to our own existence, means that the science will get done. Whatever mission, telescope, or technique is eventually used, perhaps even within the next 20 years, other planets similar to Earth with telltale signs of biological forcing of atmospheric chemistry will be discovered.

References include *Oppenheimer and Hinkley (2009)*.

6.11. Epilogue: New Worlds, New Horizons

The Committee for a Decadal Survey issued their report "New Worlds, New Horizons in Astronomy and Astrophysics" in August 2010. This report, also known as Astro2010, specified priority science objectives for the decade 2012–2021 for all of ground- and spacebased astronomy and astrophysics in the U.S., but given the nature of our research, the implications are worldwide.

Astro2010 clearly gives exoplanets a high priority, as high as they can in the face of expected flat future budgets. Specifically, Astro2010 proposes a single flagship mission for the decade with dual science goals: dark energy and exoplanets. The exoplanet data will come from gravitational microlensing of stars toward the galactic bulge, giving us a census of planets with semimajor axes of roughly 1 AU and greater, to complement Kepler's census of planets at roughly 1 AU and smaller, for the purpose of getting the best possible estimate of the frequency of Earth-mass planets in habitable zones. Astro2010 also recommends developing exoplanet technology in the coming 5–10 years to lay the foundation for a future mission to study nearby Earth-like planets, with a possible new start for a flagship in the early 2020s.

Since direct imaging of nearby exoplanets is clearly a potential candidate for this mission, the present chapter is especially relevant. Our hope, as the authors, is that the methods we discuss here will inspire you, the reader, to even newer and better ways of directly imaging nearby exoplanets.

Acknowledgments. W.A.T. thanks NASA's Exoplanet Exploration Program as well as the Kavli Institute for Theoretical Physics (Exoplanet Program) for supporting and hosting part of this work. This research was supported in part by the National Science Foundation under Grant No. PHY05-51164. B.R.O. thanks the AMNH, FUTDI, and the Blavatnik Family Foundation for partial support. Part of the research in this chapter was carried out at the Jet Propulsion Laboratory, California Institute of Technology, under a contract with the National Aeronautics and Space Administration.

REFERENCES

Allen C. W. (1991) *Astrophysical Quantities.* Oxford Univ., Oxford.

Arenberg J. W., Lo A. S., Cash W., and Polidan R. S. (2006) New Worlds Observer: Using occulters to directly observe planets. In *Space Telescopes and Instrumentation I: Optical, Infrared, and Millimeter* (J. Mather et al., eds.), pp. 62651W. SPIE Conf. Series 6265, Bellingham, Washington.

Artigau E., Biller B. A., Wahhaj Z., Hartung M., Hayward T. L., et al. (2008) NICI: Combining coronagraphy, ADI, and SDI. In *Ground-based and Airborne Instrumentation for Astronomy II* (I. McLean and M. Casali, eds.), pp. 70141Z–70141Z–9. SPIE Conf. Series 7014, Bellingham, Washington.

Barry R. K., Danchi W. C., Traub W. A., Sokoloski J. L., Wisniewski J. P., et al. (2008) Milliarcsecond N-band observations of the nova RS Ophiuchi: First science with the Keck Interferometer Nuller. *Astrophys. J., 677,* 1253–1267.

Beichman C. A., Woolf N. J., and Lindensmith C. A. (1999) *The Terrestrial Planet Finder.* JPL Publication 99-3, Jet Propulsion Laboratory, Pasadena, California.

Beuzit J.-L., Feldt M., Dohlen K., Mouillet D., Puget P., et al. (2008) SPHERE: A planet finder instrument for the VLT. In *Ground-based and Airborne Instrumentation for Astronomy II* (I. McLean and M. Casali, eds.), pp. 701418–701418-12. SPIE Conf. Series 7014, Bellingham, Washington.

Biller B. A., Close L. M., Masciadri E., Lenzen R., Brandner W., et al. (2006) Contrast limits with the Simultaneous Differential Extrasolar Planet Imager (SDI) at the VLT and MMT. In *Advances in Adaptive Optics II* (B. Ellerbroek and D. Bonaccini, eds.), pp. 62722D. SPIE Conf. Series 6272, Bellingham, Washington.

Biller B. A., Close L. M., Masciadri E., Nielsen E., Lenzen R., et al. (2007) An imaging survey for extrasolar planets around 45 close, young stars with the Simultaneous Differential Imager at the Very Large Telescope and MMT. *Astrophys. J. Suppl., 173,* 143–165.

Boccaletti A., Baudoz P., Baudrand J., Reess J. M, and Rouan D. (2005) Imaging exoplanets with the coronagraph of JWSR/MIRI. *Adv. Space Res., 36,* 1099–1106.

Borde P. J. and Traub W. A. (2006) High-contrast imaging from space: Speckle nulling in a low-aberration regime. *Astrophys. J., 638,* 488–498.

Born M. and Wolf E. (1999) *Principles of Optics.* Cambridge Univ., Cambridge.

Cady E., Belikov R., Dumont P., Egerman R., Kasdin N. J., et al. (2009) Design of a telescope-occulter system for THEIA. *ArXiV preprints,* arXiv:0912.2938v1.

Cash W. (2006) Detection of Earth-like planets around nearby stars using a petal-shaped occulter. *Nature, 442,* 51–53.

Chauvin G., Lagrange A.-M., Dumas C., Zuckerman B., Mouillet D., et al. (2005) Giant planet companion to 2MASSW J1207334-393254. *Astron. Astrophys., 438,* L25–L28.

Chen P., Traub W. A., Kern B., and Matsuo T. (2009) Seeing in the stratosphere. *Bull. Am. Astron. Soc., 41,* 438.

Cockell C. S., Herbst T., Leger A., Absil O., Beichman C., et al. (2009) Darwin — An experimental astronomy mission to search for habitable planets. *Experimental Astron., 23,* 435–461.

Colavita M. M., Serabyn E., Booth A. J., Crawford S. L., Garcia-Gathright J. I., et al. (2008) Keck interferometer nuller upgrade. In *Optical and Infrared Interferometry* (M. Scholler et al., eds.), pp. 70130A–70130A-14. SPIE Conf. Series 7013, Bellingham, Washington.

Cox A. N., ed. (2000) *Allen's Astrophysical Quantities,* 4th edition. Springer, New York. 719 pp.

de Pater I. and Lissauer J. J. (2001) *Planetary Sciences*. Cambridge Univ., Cambridge.

de Pater I. and Lissauer J. J. (2010) *Planetary Sciences*, 2nd edition. Cambridge Univ., Cambridge.

Des Marais D. J., Harwit M. O., Jucks K. W., Kasting J. F., Lin D. N. C., et al. (2002) Remote sensing of planetary properties and biosignatures on extrasolar terrestrial planets. *Astrobiology, 2*, 153–181.

Gilmozzi R. and Spyromilio J. (2008) The 42-m European ELT: Status. In *Ground-based and Airborne Telescopes II* (L. M. Stepp and R. Gilmozzi, eds.), pp. 701219–701219-10. SPIE Conf. Series 7012, Bellingham, Washington.

Give'on A., Kasdin N. J., Vanderbei R. J., and Avitzour Y. (2006) On representing and correcting wavefront errors in high-contrast imaging systems. *J. Optical Soc. Am., A23*, 1063–1073.

Glassman T., Johnson A., Lo A., Dailey D., Shelton H., and Vogrin J. (2010) Error analysis on the NWO starshade. In *Space Telescopes and Instrumentation 2010: Optical, Infrared, and Millimeter Wave* (J. M. Oschmann Jr. et al., eds.), 773150. SPIE Conf. Series 7731, Bellingham, Washington.

Greene T., Beichman C., Eisenstein D., Horner S., Kelly D., et al. (2007) Observing exoplanets with the JWST NIRCam grisms. In *Techniques and Instrumentation for Detection of Exoplanets III* (D. Coulter, ed.), pp. 66930G–66930G-10. SPIE Conf. Series 6693, Bellingham, Washington.

Guyon O. (2003) Phase-induced amplitude apodization of telescope pupils for extrasolar terrestrial planet imaging. *Astron. Astrophys., 404*, 379–387.

Guyon O. (2005) Limits of adaptive optics for high contrast imaging. *Astrophys. J., 629*, 592–614 (revised version available at arXiv:astroph/0505086v2).

Guyon O., Pluzhnik E. A., Kuchner M. J., Collins B., and Ridgway S. T. (2006) Theoretical limits on extrasolar terrestrial planet detection with coronagraphs. *Astrophys. J., 167*, 81–99.

Guyon O., Angel J. R. P., Belikov R., Egerman R., Gavel D., et al. (2009) Detecting and characterizing exoplanets with a 1.4-m space telescope: The Pupil mapping Exoplanet Coronagraphic Observer (PECO). In *Techniques and Instrumentation for Detection of Exoplanets IV* (S. Shaklan, ed.), pp. 74400F–74400F-10. SPIE Conf. Series 7440, Bellingham, Washington.

Guyon O., Pluzhnik E., Martinache F., Totems J., Tanaka S., et al. (2010) High-contrast imaging and wavefront control with a PIAA coronagraphic laboratory system validation. *Publ. Astron. Soc. Pac., 122*, 71–84.

Herriot G., Morris S., Anthony A., Derdall D., Duncan D., et al. (2000) Progress on Altair: The Gemini North adaptive optics system. In *Adaptive Optical Systems Technology* (P. Wizinowich, ed.), pp. 115–125. SPIE Conf. Series 4007, Bellingham, Washington.

Hinkley S., Oppenheimer B. R., Soummer R., Sivaramakrishnan A., Roberts L. C. Jr., et al. (2007) Temporal evolution of coronagraphic dynamic range and constraints on companions to Vega. *Astrophys. J., 654*, 633–640.

Hinkley S., Oppenheimer B. R., Brenner D., Parry I. R., Sivaramakrishnan A., et al. (2008) A new integral field spectrograph for exoplanetary science at Palomar. In *Adaptive Optics Systems* (N. Hubin, ed.), pp. 701519–701519-10. SPIE Conf. Series 7015, Bellingham, Washington.

Hinkley S., Oppenheimer B. R., Brenner D., Zimmerman N., Roberts L. C. Jr., et al. (2010) Discovery and characterization of a faint stellar companion to the A3V star zeta Virginis. *Astrophys. J., 712*, 421–428.

Hinz P. M. (2009) Detection of debris disks and wide orbit planets with the LBTI. In *Exoplanets and Disks: Their Formation and Diversity*, pp. 313–317. AIP Conf. Series 1158, American Institute of Physics, New York.

Hinz P. M., Solheid E., Durney O., and Hoffmann W. F. (2008) NIC: LBTI's nulling and imaging camera. In *Optical and Infrared Interferometry* (M. Schöller et al., eds.), pp. 701339–701339-12. SPIE Conf. Series 7013, Bellingham, Washington.

Hull T., Trauger J. T., Macenka S. A., Moody D., Olarte G., et al. (2003) Eclipse telescope design factors. In *High-Contrast Imaging for Exo-Planet Detection* (A. Schultz and R. Lyon, eds.), pp. 277–287. SPIE Conf. Series 4860, Bellingham, Washington.

Janson M., Bergfors C, Goto M., Brandner W., and Lafrenière D. (2010) Spatially resolved spectroscopy of the exoplanet HR 8799c. *Astrophys. J. Lett., 710*, L35–L38.

Johns M. (2008) Progress on the GMT. In *Ground-based and Airborne Telescopes II* (L. Stepp and R. Gilmozzi, eds.), pp. 70121B–70121B-15. SPIE Conf. Series 7012, Bellingham, Washington.

Kalas P., Graham J. R., Chiang E., Fitzgerald M. P., Clampin M., et al. (2008) Optical images of an exosolar planet 25 light-years from Earth. *Science, 322*, 1345–1348; erratum 19 Jan. 2009.

Kaltenegger L., Traub W. A., and Jucks K. W. (2007) Spectral evolution of an Earth-like planet. *Astrophys. J., 658*, 598–616.

Kasdin N. J., Vanderbei R. J., Spergel D. N., and Litman M. G. (2003) Extrasolar planet finding via optimal apodized-pupil and shaped-pupil coronagraphs. *Astrophys. J., 582*, 1147–1161.

Kasdin N. J., Belikov R., Beall J., Vanderbei R. J., Littman M. G., et al. (2005) Shaped pupil coronagraphs for planet finding: Optimization, manufacturing, and experimental results. In *Techniques and Instrumentation for Detection of Exoplanets II* (D. Coulter, ed.), pp. 128–136. SPIE Conf. Series 5905, Bellingham, Washington.

Kasdin N. J., Cady E. J., Dumont P. J., Lisman P. D., Shaklan S. B., et al. (2009) Occulter design for THEIA. *Techniques and Instrumentation for Detection of Exoplanets IV* (S. Shaklan, ed.), pp. 744005–744005-8. SPIE Conf. Series 7440, Bellingham, Washington.

Kasper M., Amico P., Pompei E., Ageorges N., Apai D., et al. (2009) Direct imaging of exoplanets and brown dwarfs with the VLT: NACO pupil-stabilized Lyot corongraphy at 4 µm. *Messenger, 137*, 8–13.

Kasting J., Traub W., Roberge A., Leger A., Schwartz A., et al. (2009) Exoplanet characterization and the search for life. *ArXiV preprints*, arXiv:0911.2936v1.

Kenworthy M. A., Codona J. L., Hinz P. M., Angel J. R. P., Heinze A., and Sivanandam S. (2007) First on-sky high-contrast imaging with an apodizing phase plate. *Astrophys. J., 660*, 762–769.

Kenworthy M. A., Mamjek E. E., Hinz P. M., Meyer M. R., Heinze A. N., et al. (2009) MMT/AO 5 µm imaging constraints on the existence of giant planets orbiting Fomalhaut. *Astrophys. J., 697*, 1928–1933.

Krist J. E. (2004) High contrast imaging with the Hubble Space Telescope: Performance and lessons learned. In *Optical, Infrared, and Millimeter Space Telescopes* (J. C. Mather, ed.), pp. 1284–1285. SPIE Conf. Series 5487, Bellingham, Washington.

Krist J. E., Balasubramanian K., Beichman C. A., Echternach P. M., Green J. J., et al. (2009) The JWST/NIRCam coronagraph mask design and fabrication. In *Techniques and Instrumentation for Detection of Exoplanets IV* (S. Shaklan, ed.), pp. 74400W–74400W-10. SPIE Conf. Series 7440, Bellingham, Washington.

Kuchner M. J. (2004a) A minimum-mass extrasolar nebula. *Astrophys. J., 612,* 1147–1151.

Kuchner M. J. (2004b) A unified view of coronagraph image masks. *ArXiV preprints,* arXiv:astro-ph/0401256v1.

Kuchner M. J. and Traub W. A. (2002) A coronagraph with a band-limited mask for finding terrestrial planets. *Astrophys. J., 570,* 900–908.

Kuchner M. J., Crepp J., and Ge J. (2005) Eighth-order image masks for terrestrial planet finding. *Astrophys. J., 628,* 466–473.

Kuhn J. R., Potter D., and Parise B. (2001) Imaging polarimetric observations of a new circumstellar disk system. *Astrophys. J. Lett., 553,* L189–L191.

Labeyrie A. (1996) Resolved imaging of extra-solar planets with future 10–100 km optical interferometric arrays. *Astron. Astrophys. Suppl. Ser., 118,* 517–524.

Lafreniere D., Doyon R., Marois C., Nadeau D., Oppenheimer, B. R., et al. (2007) The Gemini deep planet survey. *Astrophys. J., 670,* 1367–1390.

Lafreniere D., Marois C., Doyon R., and Barman T. (2009) HST/NICMOS detection of HR 8799b in 1998. *Astrophys. J. Lett., 694,* L148–L152.

Lagrange A.-M., Gratadour D., Chauvin G., Fusco T., Ehrenreich D., et al. (2009) A probable giant planet imaged in the beta Pictoris disk. *Astron. Astrophys., 493,* L21–L25.

Lagrange A.-M., Bonnefoy M., Chauvin G., Apai D., et al. (2010) A giant planet imaged in the disk of the young star beta Pictoris. *Science, 329,* 57.

Lawson P. R., Lay O. P., Martin S. R., Peters R. D., Gappinger R. O., et al. (2008) Terrestrial Planet Finder Interferometer: 2007–2008 progress and plans. In *Optical and Infrared Interferometry* (M. Schöller et al., eds.), pp. 70132N–70132N-15. SPIE Conf. Series 7013, Bellingham, Washington.

LeConte J., Soummer R., Hinkley S., Oppenheimer B. R., Sivaramakrishnan A., et al. (2010) The Lyot Project direct imaging survey of substellar companions: Statistical analysis and information from nondetections. *Astrophys. J., 716,* 1551–1565.

Lenzen R., Hartung M., Brandner W., Finger G., Hubin N. N., et al. (2003) NAOS-CONICA first on sky results in a variety of observing modes. In *Instrument Design and Performance for Optical/Infrared Ground-based Telescopes* (M. Iye and A. F. M. Moorwood, eds.), pp. 944–952. SPIE Conf. Series 4841, Bellingham, Washington.

Levine M., Lisman D., Shaklan S., Kastin J., Traub W., et al. (2009) Terrestrial Planet Finder Coronagraph (TPF-C) flight baseline concept. *ArXiV preprints,* arXiv:0911.3200v1.

Liu W. M., Hinz P. M., Meyer M. R., Mamajek E. E., Hoffmann W. F., et al. (2007) Observations of Herbig Ae disks with nulling interferometry. *Astrophys. J., 658,* 1164–1172.

Lyon R. G., Clampin M., Melnick G., Tolls V., Woodruff R., and Vasudevan G. (2008) Extrasolar Planetary Imaging Coronagraph (EPIC): Visible nulling coronagraph testbed results. In *Space Telescopes and Instrumentation 2008: Optical, Infrared, and Millimeter* (J. Oschmann et al., eds.), pp. 701045–701045-7. SPIE Conf. Series 7010, Bellingham, Washington.

Lyot B. (1933) The study of the solar corona without an eclipse. *R. Astron. Soc. Canada, 27,* 225–234, 265–280.

Macintosh B. A., Graham J. R., Palmer D. W., Doyon R., Dunn J., et al. (2008) The Gemini Planet Imager from science to design to construction. In *Adaptive Optics Systems* (N. Hubin et al., eds.), pp. 701518–701518-13. SPIE Conf. Series 7015, Bellingham, Washington.

Mamajek E. E., Kenworthy M. A., Hinz P. M., and Meyer M. R. (2010) Discovery of a faint companion to Alcor using MMT/AO 5 μm imaging. *Astron. J., 139,* 919–925.

Marois C., Doyon R., Nadeau D., Racine R., Riopel M., et al. (2005) TRIDENT: An infrared differential imaging camera optimized for the detection of methanated substellar companions. *Publ. Astron. Soc. Pac., 117,* 745–756.

Marois C., Lafreniere D., Doyon R., Macintosh B., and Nadeau D. (2006) Angular differential imaging: A powerful high-contrast imaging technique. *Astrophys. J., 641,* 556–564.

Marois C., Macintosh B., Barman T., Zuckerman B., Song I., et al. (2008) Direct imaging of multiple planets orbiting the star HR 8799. *Science, 322,* 1348.

Mawet D., Riaud P., Absil O., and Surdej J. (2005) Annular groove phase mask coronagraph. *Astrophys. J., 633,* 1191–1200.

Mawet D., Serabyn E., Liewer K., Burruss R., Hickey J., and Shemo D. (2010) The vector vortex coronagraph: Laboratory results and first light at Palomar Observatory. *Astrophys. J., 709,* 53–57.

Nelson J. and Sanders G. H. (2008) The status of the Thirty Meter Telescope project. In *Ground-based and Airborne Telescopes II* (L. Stepp and R. Gilmozzi, eds.), pp. 70121A–70121A-18. SPIE Conf. Series 7012, Bellingham, Washington.

Neuhauser R., Guenther E. W., Wuchterl G., Mugrauer M., Bedalov A., and Hauschildt P. H. (2005) Evidence for a co-moving substellar companion of GQ Lup. *Astron. Astrophys., 435,* L13–L16.

Oppenheimer B. R. and Hinkley S. (2009) High-contrast imaging in optical and infrared astronomy. *Annu. Rev. Astron. Astrophys., 47,* 253–289.

Oppenheimer B. R., Brenner D., Hinkley S., Zimmerman N., Sivaramakrishnan A., et al. (2008) The solar-system-scale disk around AB Aurigae. *Astrophys. J., 679,* 1574–1581.

Pedretti E., Labeyrie A., Arnold L., Thureau N., Lardiere O., et al. (2000) First images on the sky from a hyper telescope. *Astron. Astrophys. Suppl. Ser., 147,* 285–290.

Perrin M. D., Graham J. R., Kalas P., Lloyd J. P., Max C. E., et al. (2004) Laser guide star adaptive optics imaging polarimitry of Herbig Ae/Be stars. *Science, 303,* 1345–1348.

Postman M., Traub W., Krist J., Stapelfeldt K., Brown R., et al. (2009) Advanced Technology Large-Aperture Space Telescope (ATLAST): Characterizing habitable worlds. *ArXiV preprints,* arXiv:0911.3841v1.

Poyneer L., van Dam M., and Veran J.-P. (2009) Experimental verification of the frozen flow atmospheric turbulence assumption with use of astronomical adaptive optics telemetry. *J. Optical Soc. Am., A26,* 833–846.

Racine R., Walker G. A. H., Nadeau D., Doyon R., and Marois C. (1999) Speckle noise and the detection of faint companions. *Publ. Astron. Soc. Pac., 111,* 587–594.

Rousset G., Lacombe F., Puget P., Hubin N. N., Gendron E., et al. (2003) NAOS, the first AO system of the VLT: On-sky performance. In *Adaptive Optical System Technologies II* (P. L. Wizinowich and D. Bonaccini, eds.), pp. 140–149. SPIE Conf. Series 4839, Bellingham, Washington.

Serabyn E., Mawet D., and Burruss R. (2010) An image of an exoplanet separated by two diffraction beamwidths from a star. *Nature, 464,* 1018–1020.

Shaklan S. B., Noecker M. C., Glassman T., Lo A. S., and Dumont P. J. (2010) Error budgeting and tolerancing of starshades for exoplanet detection. In *Space Telescopes and Instrumentation 2010: Optical, Infrared, and Millimeter Wave* (J. M. Oschmann Jr. et al., eds.), 77312G. SPIE Conf. Series 7731, Bellingham, Washington.

Shao M., Bairstow S., Levine M., Vasisht G., Lane B. F., et al. (2008) DAVINCI, a dilute aperture visible nulling coronagraphic instrument. In *Optical and Infrared Interferometry* (M. Schöller et al., eds.), pp. 70132T–70132T-13. SPIE Conf. Series 7013, Bellingham, Washington.

Soummer R., Cash W., Brown R. A., Jordan I., Roberge A., et al. (2009) A starshade for JWST: Science goals and optimization. In *Techniques and Instrumentation for Detection of Exoplanets IV* (S. Shaklan, ed.), pp. 7440A–7440A-15. SPIE Conf. Series 7440, Bellingham, Washington.

Sparks W. B. and Ford H. C. (2002) Imaging spectroscopy for extrasolar planet detection. *Astrophys. J., 578,* 543–564.

Tamura M., Hodapp K., Takami H., Lyu A., Suto H., et al. (2006) Concept and science of HiCIAO: High contrast instrument for the Subaru next generation adaptive optics. In *Ground-based and Airborne Instrumentation for Astronomy* (I. McLean and M. Iye, eds.), p. 62690V. SPIE Conf. Series 6269, Bellingham, Washington.

Thalmann C., Carson J., Janson M., Goto M., McElwain M., et al. (2009) Discovery of the coldest imaged companion of a Sun-like star. *Astrophys. J. Lett., 707,* L123–L127.

Traub W. A. (1986) Combining beams from separated telescopes. *Appl. Optics, 25,* 528–532.

Traub W. A. (2000) Beam combination and fringe measurement. In *Principles of Long Baseline Interferometry* (P. R. Lawson, ed.), pp. 31–58. JPL Publ. 00-009, Jet Propulsion Laboratory, Pasadena, California. Available online at *http://olbin.jpl.nasa.gov/iss1999/coursenotes.html*.

Traub W. A. (2003) The colors of extrasolar planets. In *Scientific Frontiers in Research on Extrasolar Planets* (D. Deming and S. Seager, eds.), pp. 595–602. ASP Conf. Series 294, Astronomical Society of the Pacific, San Francisco.

Traub W. A. and Vanderbei R. J. (2003) Two-mirror apodization for high-contrast imaging. *Astrophys. J., 599,* 695–701.

Traub W. A., Levine M., Shaklan S., Kasting J., Angel J. R., et al. (2006) TPF-C: Status and recent progress. In *Advances in Stellar Interferometry* (J. Monnier et al., eds.), pp. 62680T. SPIE Conf. Series 6268, Bellingham, Washington.

Traub W., Chen P., Kern B., and Matsuo T. (2008) Planetscope: An exoplanet coronagraph on a balloon platform. In *Space Telescopes and Instrumentation 2008: Optical, Infrared, and Millimeter* (J. Oschmann et al., eds.), pp. 70103S–70103S-12. SPIE Conf. Series 7010, Bellingham, Washington.

Trauger J. T. and Traub W. A. (2007) A laboratory demonstration of the capability to image an Earth-like extrasolar planet. *Nature, 446,* 771–773.

Trauger J. T., Stapelfeldt K., Traub W., Henry C., Krist J., et al. (2008) ACCESS: A NASA mission concept study of an actively corrected coronagraph for exoplanet system studies. In *Space Telescopes and Instrumentation 2008: Optical, Infrared, and Millimeter* (J. Oschmann et al., eds.), pp. 701029–701029-11. SPIE Conf. Series 7010, Bellingham, Washington.

Turnbull M. C., Traub W. A., Jucks K. W., Woolf N. J., Meyer M. R., et al. (2006) Spectrum of a habitable world: Earthshine in the near-infrared. *Astrophys. J., 644,* 551–559.

Vanderbei R. J. and Traub W. A. (2005) Pupil mapping in two dimensions for high-contrast imaging. *Astrophys. J., 626,* 1079–1090.

Vanderbei R. J., Cady E., and Kasdin N. J. (2007) Optimal occulter design for finding extrasolar planets. *Astrophys. J., 665,* 794–798.

Wizinowich P., Dekany R., Gavel D., Max C., Adkins S., et al. (2008) W. M. Keck Observatory's next-generation adaptive optics facility. In *Adaptive Optics Systems* (N. Hubin et al., eds.), pp. 701511–701511-12. SPIE Conf. Series 7015, Bellingham, Washington.

Woolf N. J., Smith P. S., Traub W. A., and Jucks K. W. (2006) The spectrum of Earthshine: A pale blue dot observed from the ground. *Astrophys. J., 574,* 430–433.

Zimmerman N., Oppenheimer B. R., Hinkley S., Brenner D., Parry I. R., et al. (2010) Parallactic motion for companion discovery: An M-dwarf orbiting Alcor. *Astrophys. J., 709,* 733–740.

Astrometric Detection and Characterization of Exoplanets

Andreas Quirrenbach
Landessternwarte, Zentrum für Astronomie der Universität Heidelberg

---

Stars with planetary companions orbit the common center of mass. From astrometric measurements of this motion one can determine the mass and all orbital parameters of the planet. This has been done for a few planets previously known from radial-velocity surveys, using data from the Hipparcos satellite and the Hubble Space Telescope. New techniques are pushing the precision of astrometry from the milliarcsecond level into the microarcsecond regime, enabling more sensitive exoplanet surveys. Astrometric measurements with space interferometers will even be capable of reaching submicroarcsecond precision, as required for conducting a census of Earth analogs in the solar neighborhood.

1. INTRODUCTION

Exoplanets have become the subject of many studies during the past one and a half decades. Monitoring of the radial velocity (RV) and brightness of thousands of stars has revealed the presence of more than 450 planets, which betray their existence through the gravitational interaction with their parent star, or through transits in front of the stellar disk. Contrary to initial expectations, astrometry has not contributed very much so far to this exciting new field of astrophysics. However, technical advances in astrometric instrumentation should push the precision from the present milliarcsecond level into the microarcsecond regime, opening completely new vistas for the detection and characterization of exoplanets with this method.

This chapter provides an overview of the foundations of the astrometric planet detection technique, and summarizes the present state of the art. A more comprehensive overview of the various planet detection methods can be found in the section by Quirrenbach in *Cassen et al.* (2006) and in the chapters in Part II of this book.

1.1. The History of Astrometric Planet Searches

The search for exoplanets has a long and checkered history [see, e.g., *Boss* (1998a) for an easily readable overview]. Because of the enormous brightness contrast between planets and their parent stars, the direct detection of planets by taking images of the vicinity of nearby stars is extremely difficult. Early searches for planets were therefore mostly carried out with the astrometric method. First reports on the detection of massive planets ($\sim 10\,M_{Jup}$) were published during World War II (*Strand*, 1943; *Reuyl and Holmberg*, 1943), but remained controversial, both with regard to the reality of the results and to the question of whether the detected bodies should be called "planets." Much painstaking work over the next few decades led to the realization that these "detections" were spurious. Continued improvements in the astrometric accuracy finally culminated in the announcement of a planet 1.6 times as massive as Jupiter in a 24-year orbit around Barnard's Star (*van de Kamp*, 1963). A decade earlier, O. Struve had written a remarkable paper, in which he noted the possibility that Jupiter-like planets might exist in orbits as small as 0.02 AU, proposed to search for these objects with high-precision RV measurements, and pointed out the feasibility of photometric searches for planets eclipsing their parent stars — all on little more than one journal page (*Struve*, 1952).

By the mid-1960s, the search for exoplanets thus appeared to be a thriving field, with eight planetary companions known from astrometric observations (two of them classified as "existence not completely established"), and a number of potentially promising alternative search methods under consideration (*O'Leary*, 1966). By the same time it had also been recognized that brown dwarfs (termed "black dwarfs" at the time) would form a class of their own, with properties intermediate between those of stars and planets. Both astrometric searches for brown-dwarf companions to low-luminosity stars, and attempts at finding them directly with high-resolution imaging techniques, seemed to be successful (*Harrington et al.*, 1983; *McCarthy et al.*, 1985).

Sadly, none of these early claims for detections of planets and brown dwarfs withstood the test of time. It turned out that systematic instrumental errors had been mistaken for the "planetary companion" of Barnard's Star (*Gatewood and Eichhorn*, 1973), although van de Kamp himself continued to believe in the existence of not only one, but two companions around this star (*van de Kamp*, 1982). What appeared to be the most convincing detection of a brown dwarf, a companion to the star VB8, could never be confirmed (*Perrier and Mariotti*, 1987; *Skrutskie et al.*, 1987). Other putative planets and brown dwarfs did not fare better. By the mid-1990s, all that remained was a candidate brown dwarf companion of HD 114762, detected with the RV method (*Latham et al.*, 1989). (Since the RV technique measures only m sin i, it could not be excluded that HD 114762b is a low-mass star in a nearly face-on orbit.)

This situation changed completely and abruptly with the discovery of 51 Peg b, a Jupiter-like planet in a four-day orbit (*Mayor and Queloz,* 1995). More than 400 planets outside our own solar system are known to date, and new discoveries are announced almost every month. These developments have revolutionized our view of our own place in the universe. We know now that other planetary systems can have a structure that is completely different from that of our solar system, and we have set out to explore their properties and diversity. Consequently, the role of astrometry as a method for planet searches has changed considerably. It is but one tool among several others at the disposal of the observer, to be used judiciously to get information unobtainable in other ways. The motivation and design of modern astrometric programs reflect this more focused approach.

1.2. The Scientific Goals of Astrometric Planet Surveys

The specific strengths of the astrometric method enable it to answer a number of questions that cannot be addressed by any other planet detection method. Among the most prominent goals of astrometric planet surveys are the following:

1.2.1. Mass determination for planets detected in radial velocity surveys (without the sin i factor). The RV method gives only a lower limit to the mass, because the inclination of the orbit with respect to the line-of-sight remains unknown. Astrometry can resolve this ambiguity, because it measures two components of the orbital motion, from which the inclination can be derived. This helps to clarify the status of individual objects, and improves the definition of the planet mass function, especially close to the upper end, where the statistics are poor.

1.2.2. Search for long-period planets around nearby stars of all spectral types. The astrometric precision is independent of the spectral type and rotation velocity of the star, unlike the RV technique, which depends on the presence of many sharp absorption lines in the stellar spectrum. Astrometry can thus contribute to the characterization of planet properties (masses, orbital parameters) as a function of stellar mass, and in this way provide constraints on competing theories of planet formation.

1.2.3. Detection of gas giants around pre-main-sequence stars, signatures of planet formation. Astrometry can detect giant planets around young stars, and thus probe the time of planet formation and migration. Observations of pre-main-sequence stars of different ages can provide a critical test of the formation mechanism of gas giants. Whereas gas accretion on ~10 $M_\oplus$ cores requires ~10 m.y., formation by disk instabilities would proceed rapidly and thus produce an astrometric signature even at very young stellar ages (*Boss,* 1998b). At somewhat larger ages, the planets should still be warm due to the heat trapped during their formation. These will be good targets for follow-up observations with the closure phase method (section 5.6) (see also *Joergens and Quirrenbach,* 2004), which could yield their spectra and thus shine light on the formation process and thermal history of giant planets.

1.2.4. Detection of multiple systems with masses decreasing from the inside out. Whereas the astrometric signal increases linearly with the semimajor axis a of the planetary orbit, the RV signal scales with $1/\sqrt{a}$. This leads to opposite detection biases for the two methods. Systems in which the masses increase with a [e.g., υ And (*Butler et al.,* 1999)] are easily detected by the RV technique because the planets' signatures are of similar amplitudes. Conversely, systems with masses decreasing with a are more easily detected astrometrically.

1.2.5. Determine whether multiple systems are coplanar or not. Many of the known exoplanets have highly eccentric orbits. A plausible origin of these eccentricities is strong gravitational interaction between two or several massive planets (*Lin and Ida,* 1997; *Papaloizou and Terquem,* 2001). This could also lead to orbits that are not aligned with the equatorial plane of the star, and to noncoplanar orbits in multiple systems. The mutual inclination of planetary orbits is thus a tracer of the early history of the system. Masses and mutual inclinations of the planets are also needed for the dynamical modeling of multiple systems, including such fundamental issues as the dynamical stability and orbital resonances.

1.2.6. Search for terrestrial planets orbiting stars in the solar neighborhood. With a precision of ~0.3 µas, a space astrometry mission as exemplified by the Space Interferometry Mission (SIM-Light) will be able to perform a complete census of planets down to a limit of 1 $M_\oplus$ around nearby solar-type stars. Groundbased astrometry can reach rocky planets of a few Earth masses around nearby M dwarfs.

In summary, astrometry is a unique tool for dynamical studies of exoplanetary systems; its capabilities to determine masses and orbits are not matched by any other technique. Astrometric surveys of young and old planetary systems will therefore give unparalleled insight into the mechanisms of planet formation, orbital migration and evolution, orbital resonances, and interaction between planets.

1.3. Synergies with Other Techniques

While astrometric measurements can be used by themselves to detect exoplanets, and to determine their masses and orbital parameters, astrometry can also be employed in conjunction with other techniques to characterize a planetary system more fully. At present, astrometry is "expensive" compared to RV measurements in terms of technical complexity and telescope time. This leads to the strategy of not conducting "blind" astrometric surveys, but rather targeting stars that are already known to host planets. In this case, a relatively small number of astrometric measurements can suffice to determine the "missing" orbital elements, primarily the inclination, which is needed to derive the mass, and the mutual inclination in the case of multiple systems.

As the field of exoplanets is progressing toward a physical characterization of the discovered planets, astrometry is coming into focus as the only technique that can yield the most fundamental parameter of a planet, namely its mass. (For the small fraction of all planets that transit their parent stars, the mass can also be determined from the transit geometry and RV.) Without knowing the mass, one cannot get a handle on the bulk composition or surface gravity, which introduces enormous uncertainties into the analysis of spectroscopic data. Astrometry is therefore an important component of any comprehensive strategy for the characterization of nearby Earth-like planets (e.g., *Lunine et al.*, 2008). Table 1 summarizes the ways in which information from astrometry and spectroscopy in the visible (with a coronograph or external occulter) and/or mid-infrared (with a nulling interferometer) can be combined to derive important properties of terrestrial planets. As an added benefit, an astrometric survey of the solar neighborhood could provide the targets for a spectroscopic mission.

2. PRINCIPLES OF ASTROMETRIC PLANET DETECTION

2.1. The Astrometric Signature of Planets

The existence of exoplanets can be inferred from observations of their parent stars orbiting the common center of mass. This is the foundation of both the RV and astrometric methods. Whereas the RV technique detects the radial component of the orbital motion, astrometry measures the two coordinates in the plane of the sky. For quantitative analyses it is usually assumed that the mass of the parent star $m_\star$ is known from a spectral classification of the star and the general astrophysical knowledge about stellar evolution and fundamental stellar properties. One has to keep in mind that masses of main-sequence stars can be estimated quite accurately in this way, but masses of giant stars are much more difficult to derive, and can be rather uncertain.

From simple geometry and Kepler's Laws it follows immediately that the astrometric signal θ of a planet with mass m_p orbiting a star with mass $m_\star$ at a distance d in a circular orbit of radius a is given by

$$\theta = \frac{m_p}{m_\star} \frac{a}{d} = \left(\frac{G}{4\pi^2}\right)^{1/3} \frac{m_p}{m_\star^{2/3}} \frac{P^{2/3}}{d}$$
$$= 3\,\mu\text{as} \cdot \frac{m_p}{M_\oplus} \cdot \left(\frac{m_\star}{M_\odot}\right)^{-2/3} \left(\frac{P}{\text{yr}}\right)^{2/3} \left(\frac{d}{\text{pc}}\right)^{-1}$$
(1)

Here we have made the approximation $m_p \ll m_\star$; the error introduced thereby is usually much smaller than the measurement uncertainties. The signature θ is shown in Fig. 1 for five sample planets (analogs to Earth, Jupiter, and Sat-

TABLE 1. Summary of measurement synergies for different techniques aimed at characterizing terrestrial exoplanets (adapted from *Beichman et al.*, 2007).

	Astrometry	Visible	Mid-IR
Orbital Parameters			
Stable orbit in habitable zone	Meas	Meas	Meas
Characteristics for Habitability			
Planet temperature	Est	Est	Meas
Temperature variability due to eccentricity	Meas	Meas	Meas
Planet radius	Coop	Coop	Meas
Planet albedo	Coop	Coop	Coop
Planet mass	Meas	Est	Est
Density, bulk composition	Coop	Coop	Coop
Surface gravity	Coop	Coop	Coop
Atmospheric and surface composition	Coop	Meas	Meas
Time variability of composition		Meas	Meas
Presence of water		Meas	Meas
Planetary System Characteristics			
Influence of other planets	Meas	Est	Est
Orbit coplanarity	Meas	Est	Est
Comets, asteroids, and zodiacal dust		Meas	Meas
Indicators of Life			
Atmospheric biomarkers		Meas	Meas
Red absorption edge of vegetation		Meas	

In addition to astrometry, spectroscopic characterization of the planet in the visible wavelength range and in the mid-infrared are considered. "Meas" indicates a directly measured quantity; "Est" indicates a quantity that can be roughly estimated from a single mission; and "Coop" indicates a quantity that is best determined cooperatively using data from several missions.

urn; a "super Earth"; and a "hot Jupiter" with $m_p = 1\ M_{Jup}$ and $P = 4$ d) orbiting a 1 $M_\odot$ star. From this figure, the main strengths and difficulties of astrometric planet detection are readily apparent: (1) The astrometric signature θ is small compared to the precision of "standard" astrometric techniques (≤1 mas). (2) The difficulty of detecting different types of planets varies greatly, with θ ranging from ≤1 μas to ≤1 mas. (3) The sensitivity of astrometry for a given type of planet drops linearly with d (unlike the radial-velocity technique). (4) The detection bias of astrometry with respect to the orbital radius is opposite to that of the RV method, favoring planets at larger separations from their parent stars.

It should be pointed out that for circular orbits the observed astrometric signal is an ellipse with semimajor axis θ independent of the orbital inclination; the mass of the planet can therefore be derived directly from equation (1) if the mass and distance of the parent star are known. The situation is a bit more complicated for noncircular orbits, but even in that case can the orbital inclination be determined from the astrometric data with techniques analogous to those used for fitting orbits of visual binaries (section 2.2) (see also textbooks such as *Binnendijk*, 1960).

2.2. Orbital Fits

Astrometric measurements of a Keplerian orbit yield a time series of positions on the sky, the apparent ellipse. From these data, one wants to determine the orbital elements of the true ellipse: P (period), T (time of periastron passage), e (eccentricity), θ (semimajor axis of the ellipse describing the motion of the star, in angular units), Ω (longitude of the ascending node), ω (argument of the periastron),

and i (orbital inclination). This will be done in practice by performing a nonlinear least-squares fit of a model describing the motion of the star to the data. Including parallax and proper motion, the model looks as follows

$$\xi(t) = \alpha_0^\star + P_{\alpha\star}\varpi + (t - t_0)\mu_{\alpha\star} + BX(t) + GY(t) \quad (2)$$

$$\eta(t) = \delta_0 + P_\delta\varpi + (t - t_0)\mu_\delta + AX(t) + FY(t) \quad (3)$$

Here $\alpha^\star = \alpha \cos\delta$, $(\alpha_0^\star, \delta_0)$ is the reference position of the star, $P_{\alpha\star}$ and P_δ are the parallax factors describing the parallactic ellipse, ϖ is the parallax of the star, and $(\mu_{\alpha\star}, \mu_\delta)$ is the proper motion vector. The so-called Thiele-Innes constants A, B, F, G are related to θ, ω, Ω, and i by the expressions

$$A = \theta(\cos\omega\cos\Omega - \sin\omega\sin\Omega\cos i) \quad (4)$$

$$B = \theta(\cos\omega\sin\Omega + \sin\omega\cos\Omega\cos i) \quad (5)$$

$$F = \theta(-\sin\omega\cos\Omega - \cos\omega\sin\Omega\cos i) \quad (6)$$

$$G = \theta(-\sin\omega\sin\Omega + \cos\omega\cos\Omega\cos i) \quad (7)$$

Finally, (X(t), Y(t)) is the vector describing the motion of the star in its orbit. It is related to the eccentric anomaly E by

$$X(t) = \cos E - e \quad (8)$$

$$Y(t) = \sqrt{1 - e^2}\sin E \quad (9)$$

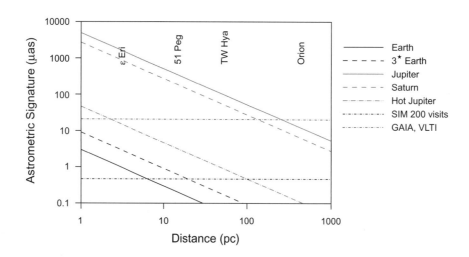

Fig. 1. Astrometric signature (semiamplitude) for five sample planets orbiting a solar-mass star, as a function of distance. Anticipated detection limits for Gaia, VLTI, and the Space Interferometry Mission are also shown. The vertical arrows mark the distances of benchmark objects in the solar neighborhood.

The eccentric anomaly is the solution of Kepler's equation

$$E = \frac{2P}{\pi}(t-T) + e \sin E \qquad (10)$$

A large number of different methods exist to carry out such nonlinear least-squares fits; among the well-known approaches are the Levenberg-Marquardt algorithm, simulated annealing, and genetic algorithms. They can be combined with Monte Carlo techniques to derive not only the best-fit values of the orbital parameters, but also confidence intervals (see, e.g., *Press et al., 2007*).

This 12-parameter model can easily be generalized to multiple systems; a system containing n planets in independent Keplerian orbits is then fully described by 7n + 5 parameters. This approach is justified as long as the gravitational interaction between the planets can be neglected. Capturing planet-planet interactions requires more complicated dynamical models.

An important question in multiple planetary systems is whether all planets orbit (nearly) in one plane or not. The mutual inclination i_{mut} between the orbits of two planets is given by

$$\cos i_{mut} = \cos i_1 \cos i_2 + \sin i_1 \sin i_2 \cos(\Omega_1 - \Omega_2) \qquad (11)$$

This quantity is of fundamental importance for the dynamical modeling of multiple systems, including such important issues as the long-term stability and the occurrence of dynamical resonances. The ability to provide a complete three-dimensional picture of planetary systems is one of the key strengths of the astrometric method.

2.3. Precision, Noise, and Mission Lifetime

The fundamental precision limit σ of any astrometric measurement is determined by the angular resolution of the telescope, and the signal-to-noise ratio of the data. For the case of a Gaussian brightness distribution it is given by

$$\sigma = \frac{1}{SNR} \cdot \frac{FWHM}{2\sqrt{2 \ln 2}} \qquad (12)$$

Similar formulae, with different constants, apply for the cases of diffraction-limited telescopes with circular or rectangular apertures, when the brightness distribution is given by an Airy or $sinc^2$ function, respectively. If the measurement noise is dominated by photon noise, $SNR \propto \sqrt{N}$, and the precision gets better as the integration time T_{int} is increased, $\sigma \propto 1/\sqrt{T_{int}}$. The same scaling with $1/\sqrt{T_{int}}$ applies in the cases where the total noise is dominated by background noise or detector dark current; if detector-read noise dominates, the scaling is $\sigma \propto 1/T_{int}$. It is very important to realize that many astrometric instruments operate in the very high signal-to-noise regime; this means that the astrometric precision can be several orders of magnitude better than the resolution of the telescope.

It is instructive to apply these considerations to a mission that performs an all-sky astrometric survey with a widefield instrument, such as Gaia. (Note that the following general calculations are equally applicable to instruments that scan the sky continuously and to those that stare at each field for a fixed time and then slew to the next; one just has to include the time spent slewing between exposures in the efficiency factor.) Let the collecting area of the instrument be A, the size along the direction in which the astrometric measurements are made be d, the solid angle covered by the survey instrument be Ω, and the total mission time be T_{tot}. Then the average integration time spent on each target is $T_{int} = T_{tot} \times \Omega/4\pi$. The width of the point spread function is $FWHM = k\lambda/d$, with a constant $k \approx 1$ that depends on the exact geometry of the instrument. For a star with photon flux Φ (in photons per collecting area per time), the number of collected photons is $N = \eta \times \Phi \times A \times T_{int}$, where η denotes the overall system efficiency. Inserting everything in equation (12), and setting $\eta^\star = \eta/(4\pi k^2)$, we get

$$\sigma = \frac{\lambda}{d\sqrt{\eta^\star A \Omega T_{tot}}} \times \frac{1}{\sqrt{\Phi}} \qquad (13)$$

This formula has been written such that it can be applied to interferometers (in which case d is the baseline length and A the collecting area of the interferometer elements) and filled-aperture telescopes alike. In the latter case, we can write $A = d \times w$, where w denotes the dimensions of the telescope orthogonal to the measurement direction, and equation (13) becomes

$$\sigma = \frac{\lambda}{\sqrt{\eta^\star d^3 w \Omega T_{tot}}} \times \frac{1}{\sqrt{\Phi}} \qquad (14)$$

From this equation one can see that the astrometric precision for a given stellar brightness can be improved in different ways, but that increasing the telescope dimensions in the direction of the measurement has a particularly high pay-off, because this increases the number of collected photons, and decreases the width of the point spread function at the same time.

An additional important point concerns the dependence of σ on the mission duration T_{tot}. Equations (13) and (14) are applicable to the case where one wants to measure the position of a stationary object, such as a quasar or a very distant star. In contrast, if one wants to determine the proper motion of a star on a linear trajectory, increasing T_{tot} has two effects: the number of photons collected increases $\propto T_{tot}$ leading to an increase in precision as suggested by equations (13) and (14), but in addition the signal, i.e., the difference between the positions at the beginning and at the end of the mission, also increases $\propto T_{tot}$. Together, this means that the precision of a proper motion measurement scales with $\sigma_\mu \propto T_{tot}^{-3/2}$. In real life, parallax complicates the picture by superimposing a periodic motion on the linear trajectory. The exact amount of degeneracy between parallax and proper motion depends on the ecliptic latitude, direction of the proper motion, and timing of the observations, but in general the two terms can be separated well from each other in time series covering at least $T_{tot} \gtrsim 1.5$ yr.

Now consider the effect of exoplanets. For a very long orbital period, the short section of the orbit completed during the mission lifetime will appear almost linear and thus be indistinguishable from proper motion; the planet will be "absorbed" in the parameters $(\mu_{\alpha\star}, \mu_\delta)$. If T_{tot} is increased, the curvature of the orbit becomes visible. The stellar path on the sky can then be described by a parabola characterized by μ and $\dot\mu$. The sensitivity of an astrometric dataset for detecting such planets in long-period orbits increases with $\sigma_{\dot\mu} \propto T_{tot}^{-5/2}$. It is clear, however, that in this regime only lower limits can be given for θ and P, allowing only a rough guess of the companion mass. If the mission time is even longer, a full orbit can be observed. Even in this case, correlations exist between the orbital parameters on the one hand and μ and ϖ on the other hand, most notably when P is close to 1 yr. In addition, there is rarely any *a priori* knowledge about the total number of planets in a system. If a long-period planet exists but is not included in the orbital fit of an inner planet, the orbital parameters derived for the inner planet may be incorrect. For all these reasons, the "value" of astrometric surveys increases much more dramatically than $\propto T_{tot}^{1/2}$, and orbital parameters derived from χ^2 minimization should be taken at face value only when T_{tot} is comfortably larger than P. Careful simulations are needed to assess their reliability in the regime P ~ T_{tot}.

In the preceding discussion we have tacitly assumed that photon noise is the dominant noise source. For the design of the astrometric instrument this means that all instrumental errors have to be kept below this floor, or need to be taken out by calibration procedures to this precision. This is often not possible for bright targets (which have small photon noise errors); in that case the astrometric errors scales with $\sigma \propto \Phi^{-1/2}$ over a certain magnitude range, but approaches a constant level at bright magnitudes. Instrumental errors with zero mean and roughly Gaussian statistics can easily included in the analysis outlined above. Instrumental imperfections producing errors that are correlated between measurements are far more dangerous, especially if the correlation time is long. They can lead to spurious signals that mimic orbits due to companions, and thus lead to false claims of planet discoveries. Since unfortunately there is no foolproof way of identifying such instrumental problems, the history of astrometric planet detection has been plagued with planet announcements that had to be retracted later (see section 1.1). To avoid such mishaps, one should either observe a large sample in a homogeneous way so that the instrument can be characterized quantitatively, or obtain a confirmation of the detections with an independent technique. If neither approach is feasible, at least two full orbital cycles should be observed to reduce the sensitivity to instrumental errors.

The ultimate limit of indirect planet searches is set by jitter induced by stellar variability. Starspots affect photometric, astrometric, and RV measurements, as they rotate into and out of view; the finite lifetime of spots leads to a nonperiodic signal on long timescales. Changes in the convective pattern due to magnetic cycles (analogous to the 11-yr solar activity cycle) can induce an additional spurious signal on the multi-year timescale. Using the simple relation $m_\star a_\star = m_p a_p$, one finds that the orbital motion of the Sun due to the presence of Jupiter is $a_\odot = 740{,}000$ km $= 1.07$ $R_\odot$, i.e., the center of mass of the Sun-Jupiter system lies outside the Sun. From this it is immediately clear that any phenomena on the solar surface that shift the photocenter cannot disturb astrometric measurements aimed at the detection of Jupiter-like planets significantly. This is different for the Sun-Earth system, where $a_\odot = 450$ km $= 6 \cdot 10^{-4}$ $R_\odot$: A spot large enough to shift the photocenter by 1/1000th of the solar radius would give an astrometric signal that is larger than that of Earth. Fortunately, the Sun is rather quiet; the typical jitter due to spots is about a factor of 3 below the astrometric signature of an Earth-like companion, and similar values are expected for typical solar-like dwarf stars (*Makarov et al.,* 2009). In this context it is worth mentioning that the effect of starspots on the RV is about 10 times worse than than that on the astrometric position, each compared to the signal of an Earth-like companion. Astrometry is therefore the ultimate method of choice for searches of low-mass planets in the habitable zones of solar-type stars.

2.4. Atmospheric Limitations and Narrow-Angle Astrometry

Earth's atmosphere imposes serious limitations on the precision that can be achieved with astrometric measurements from the ground. The first-order terms of the atmospheric wavefront distortions (frequently referred to as tip and tilt) are global wavefront gradients, which correspond to a motion of the centroid of the stellar light in the two coordinates. Because most of the power of atmospheric turbulence is in these low-order modes, the amplitude of this image motion is similar to the width of the stellar images, i.e., $\approx \lambda/r_0 \approx 0\overset{''}{.}5 \ldots 1''$. One can obviously reduce this error by taking many exposures and thus averaging over many independent realizations of the atmospheric turbulence, but achieving a precision of a small fraction of a milliarcsecond in this way is clearly not possible.

It helps, however, to make differential measurements over small angles on the sky, i.e., to measure the position of the target star with respect to that of a nearby reference star, or with respect to an ensemble of comparison stars in the field. Because the image motions are correlated in small fields, the atmospheric effects will be much smaller if only differential positions are to be measured. If the angle θ between the target and reference stars is sufficiently small (typically on the order of $1'$), the atmospheric errors are also reduced when large apertures are used. The variance σ_θ^2 of measurements of θ is then approximately given by (*Lindegren,* 1980)

$$\sigma_\theta^2 \approx 6.8 \cdot 10^{-6} d^{-4/3} \theta^2 T^{-1} \quad (15)$$

where d is the telescope size in m, T the integration time in seconds, and θ and σ_θ are measured in arcseconds. The downside is that the information that can be obtained in this way is more restricted, because the local frame may have a motion and rotation of its own. This obviously makes it impossible to measure proper motions. Moreover, all parallax ellipses have the same orientation and axial ratio, which allows only "relative parallaxes" to be measured.

Narrow-angle astrometry is generally sufficient for planet detection, but there are a few caveats. First of all, if only one reference star is used, there is an ambiguity as to which star an astrometric wobble has to be attributed. For example, if one chooses a distant star as reference, and this star happens to be an unrecognized binary, the resulting variation of the position difference could be mistaken for a planetary signature of the much closer target (see equation (1)). This can, of course, be avoided by using multiple reference stars. A somewhat more subtle effect is caused by the rotation of the local reference frame, due to uncertainties in the proper motions of the reference stars. (For illustration purposes, assume that the local frame is defined by two stars. The star in the north has an eastward proper motion, the star in the south a westward one; these proper motions are not known to the observer. The reference frame in which these stars are approximately at rest will have a counterclockwise rotation.) This rotation couples to the proper motion of the target star, and produces a spurious "Coriolis" acceleration, which could be mistaken for the signature of a planet in an orbit with a period longer than the time span covered by the observations. The detection of planets in long-period orbits with narrow-angle astrometry therefore requires accurate knowledge of the proper motions of the reference stars.

2.5. One-Dimensional Versus Two-Dimensional Astrometry

Astrometric data are fundamentally two-dimensional: At each instance, one can determine the two sky coordinates, right ascension and declination (or equivalently, the two coordinates in any other celestial reference frame). When stellar positions are measured on photographic plates or CCD frames, one naturally gets these two components by applying a simple rotation matrix to the raw (x, y) data. However, many astrometric instruments perform only one-dimensional measurements. (Sometimes both coordinates are measured in principle, but the error ellipse has such a large axial ratio that the data can essentially be regarded to be one-dimensional.) The orientation of the data may be constant in time as for a transit instrument or an interferometer with a fixed baseline, or vary from measurement to measurement as in the cases of an interferometer with a rotating baseline, or of scanning satellites such as Hipparcos and Gaia.

For algorithms that perform a χ^2 minimization fit of the model of section 2.2 to astrometric data, it does not matter whether each datum is a two-dimensional vector, or the projection of this vector onto one direction. Nor does it matter whether this direction is always the same, or a different one for each visit. It is of course clear, however, that a series of one-dimensional measurements with a fixed orientation will be blind to certain orbits, and suffer from degeneracies between the orbital parameters, just as inherently one-dimensional RV measurements do. In the case of one-dimensional measurements in which the orientation varies quasirandomly between data points, the main drawbacks are difficulties with the visual representation of the data (cf. Fig. 5) and the necessity to employ algorithms more complex than straightforward periodogram analysis to find starting values for the least-squares model fits.

3. INTERFEROMETRIC ASTROMETRY

3.1. Foundations of Interferometric Astrometry

In interferometry one brings the beams from two telescopes together and and observes the resulting interference pattern. Astrometric observations by interferometry are based on measurements of the delay $D = D_{int} + (\lambda/2\pi)\phi$, where $D_{int} = D_2 - D_1$ is the internal optical path difference measured by a metrology system (see Fig. 2), and ϕ the observed fringe phase, which is 0 for interference maxima and π for interference minima. [Here ϕ has to be unwrapped, i.e., not restricted to the interval $[0, 2\pi)$. In other words, one has to determine which of the sinusoidal fringes was observed. This can, for example, be done with dispersed-fringe techniques (*Quirrenbach*, 2001, and references therein).] D is related to the baseline $\vec{B}$ by

$$D = \vec{B} \cdot \hat{s} = B\cos\theta \qquad (16)$$

where $\hat{s}$ is a unit vector in the direction toward the star, and θ the angle between $\vec{B}$ and $\hat{s}$. Each data point is thus a one-dimensional measurement of the position of the star θ, provided that the length and direction of the baseline are accurately known. The second coordinate can be measured with a separate baseline at a roughly orthogonal orientation. The "trick" contained in Fig. 2 is that the interferometer relates the angle θ to the lengths D_1 and D_2, which can much more easily be measured with high precision. For example, if one assumes a measurement precision of 5 nm (a bit better than 1/100th of the wavelength of a HeNe laser) and B = 100 m, the resulting precision for θ will be 50 prad, which is 10 μas.

To derive D from D_1 and D_2, however, one also has to ensure that one really observes an interference maximum, or if that is not the case, to measure the fringe phase φ. In analogy with equation (12), the photon noise limit for the precision σ of such a measurement is given by

$$\sigma = \frac{1}{SNR} \cdot \frac{\lambda}{2\pi B} \qquad (17)$$

Since high signal-to-noise ratios can be obtained for bright stars, σ can be orders of magnitude smaller than the resolu-

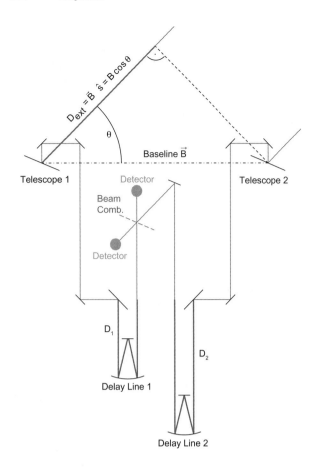

Fig. 2. Schematic drawing of the light path through a two-element interferometer. The external delay $D_{ext} = \vec{B} \cdot \hat{s}$ is compensated by the two delay lines. The path-lengths D_1, D_2 through the delay lines are monitored with laser interferometers. The beams from the two telescopes are combined with a plate that has 50% transmission and 50% reflectivity. The zero-order interference maximum occurs when the delay line positions are such that the internal delay $D_{int} = D_2 - D_1$ is equal to D_{ext}.

tion λ/B of the interferometer. For example, with an SNR ~ 50 in observations at $\lambda = 2$ µm, it is possible to attain a photon noise contribution to the astrometric error of ~10 µas on a 150-m baseline.

3.2. Atmospheric Limitations of Groundbased Interferometry

As explained in section 2.4, Earth's atmosphere imposes serious limitations on the precision that can be achieved with astrometric measurements from the ground. As with single telescopes, it helps for astrometric interferometry to make differential measurements over small angles on the sky, i.e., to measure the position of the target star with respect to that of a nearby reference. It can be shown that the variance σ_θ^2 of measurements of the angle θ is given by (*Shao and Colavita*, 1992)

$$\sigma_\theta^2 \approx \frac{16\pi^2}{B^2 t} \int_0^\infty dh\, v^{-1}(h) \int_0^\infty d\kappa\, \Phi(\kappa,h) \times \\ [1-\cos(B\kappa)] \cdot [1-\cos(\theta h \kappa)] \quad (18)$$

if the integration time $t \gg \max(B, \theta h)/v$. Here $v(h)$ is the wind speed at altitude h, and $\Phi(\kappa, h)$ denotes the three-dimensional spatial power spectrum of the refractive index. It may at first seem surprising that stronger winds should give a smaller measurement error, but within the frozen-turbulence picture a higher wind speed means that one averages faster over independent realizations of the stochastic refractive index fluctuations. Inserting a Kolmogorov power spectrum with $\Phi(\kappa, h) \propto C_N^2(h)\kappa^{-11/3}$ in equation (18) one obtains the two limiting cases

$$\sigma_\theta^2 \approx \begin{cases} 5.25\, B^{-4/3} \theta^2 t^{-1} \int_0^\infty dh\, C_N^2(h) h^2 v^{-1}(h) \\ \quad \text{for } \theta \ll B/h,\ t \gg B/v \\ 5.25\, \theta^{2/3} t^{-1} \int_0^\infty dh\, C_N^2(h) h^{2/3} v^{-1}(h) \\ \quad \text{for } \theta \gg B/h,\ t \gg \theta h/v \end{cases} \quad (19)$$

for long and short baselines, respectively (see also Fig. 3). In particular, one can see that for sufficiently small angles θ the important scaling relations $\sigma_\theta \propto \theta$ and $\sigma_\theta \propto B^{-2/3}$ hold for the astrometric error σ_θ, just as in equation (15). For a good site such as Mauna Kea or Cerro Paranal, astrometric measurements with a precision of ~10 µas are possible over angles of ~10″. It is also apparent from the factor h^2 under the integral in this equation that the astrometric error is dominated by the turbulence at high altitudes. The low level of high-altitude turbulence at Antarctic sites would therefore make an astrometric interferometer at a site on the high Antarctic plateau an attractive possibility (*Lawrence et al.*, 2004).

3.3. Groundbased Dual-Star Interferometry

Because of the short coherence time of the atmosphere, precise astrometry from the ground requires simultaneous observations of the target and astrometric reference. In a dual-star interferometer, each telescope accepts two small fields and sends two separate beams through the delay lines. The delay difference between the two fields is taken out with an additional short-stroke differential delay line; an internal laser metrology system is used to monitor the delay difference (which is equal to the phase difference multiplied with $\lambda/2\pi$, of course). For astrometric observations, this delay difference ΔD is the observable of interest, because it is directly related to the coordinate difference between the target and reference stars; from equation (16) it follows immediately that

$$\Delta D \equiv D_t - D_r = \vec{B} \cdot (\hat{s}_t - \hat{s}_r) = B(\cos\theta_t - \cos\theta_r) \quad (20)$$

where the subscript t is used for the target, and r for the reference, and $\hat{s}_t$ and $\hat{s}_r$ are unit vectors in the directions toward the two stars.

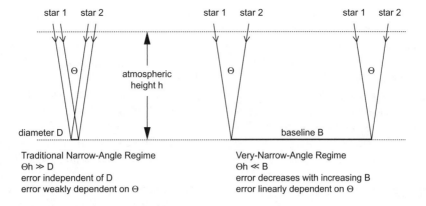

Fig. 3. Schematic of the atmospheric paths for narrow-angle astrometry with short and long baselines B (or telescope diameter D). In each panel, rays from two stars to the two telescopes at the ends of the baseline are shown. The atmosphere is represented by a single layer at height h. In this layer, the two rays originating from the same star to the two telescopes are separated by B; the two rays originating from the two stars to the same telescope are separated by θh. In the lefthand panel the baseline B is short, θh ≫ B, in the righthand panel the baseline is long, θh ≪ B. From *Shao and Colavita* (1992).

The fundamental instrumental requirements can be derived directly from this equation, which can also be written as

$$\Delta D \equiv D_t - D_r = \vec{B} \cdot (\hat{s}_t - \hat{s}_r) \equiv \vec{B} \cdot \Delta \vec{s} \quad (21)$$

The propagation of systematic errors in measurements of the differential delay δΔD and of the baseline vector δB to errors in the derived position difference δΔs can be estimated from the total differential

$$\delta \Delta s \approx \frac{\delta \Delta D}{B} + \frac{\Delta D}{B^2} \delta B = \frac{\delta \Delta D}{B} + \Delta s \frac{\delta B}{B} \quad (22)$$

This formula allows one to draw two important conclusions. First, the systematic astrometric error is inversely proportional to the baseline length. Together with the $B^{-2/3}$ scaling of the atmospheric differential delay r.m.s. this clearly favors longer baselines, up to the limit where the target star gets resolved by the interferometer. The second important conclusion from equation (22) is that the relative error of the baseline measurement gets multiplied with Δs, which is typically on the order of 10^{-4} rad. This means that the requirement regarding the knowledge of the baseline vector is sufficiently relaxed to make calibration schemes possible that rely primarily on the stability of the telescope mount. For a 10 µas (50 prad) contribution to the error budget for a measurement over a 20″ angle, with an interferometer with a 100 m baseline, the metrology system must measure δΔD with a 5 nm precision; the baseline vector has to be known to δB ≈ 50 µm (*Quirrenbach et al.*, 1998). For PRIMA (see section 5.2) it is foreseen that that baseline vector will be determined from repeated observations of stars in the same way that is also customary in radio interferometry.

3.4. Reference Stars for Groundbased Narrow-Angle Astrometry

The discussion in the previous sections has tacitly assumed that there are reference stars against which the motion of the target star can be measured. Identifying suitable references, however, is an important and nontrivial task by itself. One would like to have two such references available for each target, since then any periodic variation in the separation vectors can be uniquely attributed to one of the stars. It is possible to get a feeling for the difficulty of finding reference stars by cross-correlating a list of potential targets with a sufficiently deep catalog of possible reference stars (e.g., USNO-B). The results of one such exercise indicate that a large number of target stars with two astrometric references are available for groundbased astrometric projects (*Quirrenbach*, 2000).

There are a few caveats about the results from such catalog cross-correlations, however, which may lead to an overestimate of the number of potential reference stars. For example, some of the "reference stars" may actually be galaxies that were not recognized as such during the compilation of the reference catalog. In some cases, a target star with high proper motion is found as a potential "reference" for itself because the USNO catalog lists the stellar positions at the epoch of the underlying Schmidt plates. On the other hand, the available catalogs may be incomplete in the vicinity of very bright stars; one therefore has to take high-dynamic range images in the field of each potential target to make sure that one does not miss any available references.

It is also possible to search for planets in bright double stars (*Quirrenbach*, 2000). Most of these are members of wide physical binaries, and searching for planets in these systems is scientifically interesting and technically somewhat less challenging than searches around single stars. The downside of this approach is the difficulty of determining to which of the two system components any detected planet belongs (unless a third star is available as external reference).

The bottom line is that the availability of nearby astrometric references is an important criterion for the selection of suitable targets for groundbased astrometric observations. Optimizing the sensitivity of the faint star channel is clearly very important, because this enhances the chances of finding astrometric references. In addition, one should use as many photons as possible; simultaneous operation in the H and K bands is therefore highly advantageous.

4. CURRENT ASTROMETRIC PLANET HIGHLIGHTS AND PROGRAMS

4.1. Groundbased Surveys

Modern groundbased astrometric planet detection programs are based on the insight that the atmospheric errors decrease strongly with decreasing angle between the target and the reference star(s), and with increasing telescope size (see equation (15)). Test observations of an open cluster with the Palomar 5-m Telescope were used to show that an astrometric precision of ~100 μas can be achieved in practice (*Pravdo and Shaklan*, 1996); even better results have recently been obtained with the FORS1 and FORS2 instruments on the 8.2-m telescopes of the VLT (*Lazorenko et al.*, 2009). Further improvements may be possible with the use of adaptive optics systems (*Cameron et al.*, 2009). At this level of precision it is of course essential to calibrate aberrations of the optical system, as well as errors induced by differential chromatic refraction (DCR). The latter effect is caused by refraction in the atmosphere, which bends starlight by an amount that depends on the zenith angle and on the wavelength, thus displacing bluer stars with respect to redder stars during the course of the night. The amount of DCR can be reduced by employing narrow filters, and corrections can be applied on a star-by-star basis if their colors are known.

A program named STEPS (Stellar Planet Survey) based on these principles was initiated in 1997 at the Palomar 5-m Telescope, with the goal of finding low-mass companions around 30 M dwarfs; a similar survey has recently been initiated with the 2.5-m du Pont Telescope at the Las Campanas Observatory (*Boss et al.*, 2009). In 2009, the apparent discovery of a companion to the star VB10 with m_p = 6.4 M_{Jup} and P = 0.774 yr from STEPS was announced (*Pravdo and Shaklan*, 2009). Follow-up RV observations have shown, however, that this planet does not exist (*Bean et al.*, 2010; *Anglada-Escudé et al.*, 2010). We are thus still waiting for the first confirmed planet discovered by an astrometric search program.

4.2. Results from Hubble and Hipparcos

Looking at planets that are already known from RV surveys is an obvious application of the astrometric technique, because of its ability to determine the planet's mass without sin i ambiguity. Unfortunately, the groundbased instruments described in the previous section are not suited for such observations because of dynamic range problems — the very bright planet host stars completely swamp all potential reference stars within the small field. This is much less of a problem for space missions, which can use reference stars over much larger angles. It is clear from Fig. 1 that the detection of planets is a challenging task with a precision of slightly better than a milliarcsecond, which is currently achievable with Hipparcos data or with the Hubble Fine Guidance Sensors. Nevertheless, observations of Gl 876 with the latter instrument resulted in the detection of the astrometric wobble due to its companion Gl 876b, and thus mark the first secure astrometric detection of an exoplanet (*Benedict et al.*, 2002). Even for a star as close as Gl 876 (d = 4.7 pc) these measurements are at the limit of what is doable with the HST, and therefore the orbital elements can only be determined with rather large error bars. For example, the original publication (*Benedict et al.*, 2002) determined i = 84° ± 6° and derived m_p = 1.89 ± 0.34 M_{Jup}, whereas a more recent analysis that takes into account the information from the interaction between Gl 876b and Gl 876c gives i = $48°.9^{+1.8}_{-1.6}$ and consequently m_p = $2.57^{+0.06}_{-0.08}$ M_{Jup} (*Bean and Seifahrt*, 2009). Hubble data have also been used to confirm the existence of ε Eri b at d = 3.22 pc, and to measure its mass, m_p = 1.55 ± 0.24 M_{Jup} (*Benedict et al.*, 2006). Sometimes astrometric observations can lead to surprises, as in the case of the planet candidate HD 33636b, which turned out to be a 0.14 $M_\odot$ star seen nearly face on at i = $4°.1 ± 0°.1$ (*Bean et al.*, 2007).

The multiple system υ And has been the subject of many observational and theoretical studies. This system, comprising a "hot Jupiter" in a 4.6-day orbit and two additional jovian planets with periods of 241 and 1248 d, is of particular interest because it appears to be on the brink of instability. The astrometric orbits of the two outer planets, υ And c and d, have been detected with the HST Fine Guidance Sensors (*McArthur et al.*, 2010) (see also Fig. 4). The mutual inclination i_{mut} = 30° between these planets has been determined from a combined fit to the astrometry and an extensive set of RV measurements. This provides an important input for dynamical models of the υ And system, and shows that modeling based on the assumption of co-planarity can miss crucial aspects of the dynamical behavior.

In many cases, interesting upper limits on the companion mass can be derived from astrometric observations even if the signature is below the detection limit. A good example is the case of ι Dra b (*Frink et al.*, 2002). The RV observations give P = 536 d, e = 0.70, and m_p sin i = 8.9 M_{Jup} for this object. The nondetection in the Hipparcos data (Fig. 5) places a 3σ upper limit of 45 M_{Jup} on the mass of ι Dra b, and thus firmly establishes its substellar nature. Figure 5 also illustrates the difficulty of visualizing inherently one-dimensional astrometric data. In some favorable cases, the signature of brown dwarf companions is actually discernible in Hipparcos data (*Reffert and Quirrenbach*, 2006); a systematic analysis of the Hipparcos measurements of all RV planet candidates is currently underway (*Reffert and Quirrenbach*, 2010).

4.3. The Prospects of Gaia

The European Gaia mission, whose launch is planned for 2012, will conduct comprehensive astrometric, photometric, and spectroscopic surveys of the whole sky (*Lindegren and Perryman*, 1996; *Perryman et al.*, 2001). With an expected precision of ~30 µas for stars brighter than 13th magnitude (*Mignard and Klioner*, 2009), Gaia will surpass HST and Hipparcos by one to two orders of magnitude. Unfortunately, Gaia's detectors will saturate for very bright stars (V ≤ 6), including the stars closest to the Sun that have highest priority when looking for Earth-like planets. Nevertheless, Gaia will provide masses and orbital parameters for hundreds of giant planets. Since these data will have selection effects that are quite different from those of RV surveys, Gaia will make a substantial contribution to our understanding of the statistical properties of giant planets.

Gaia will perform its observations by scanning the sky with a predetermined regular pattern; on average, each star is visited ~70 times (*Jordan*, 2008). Unlike in targeted observations, the cadence cannot be changed to improve the sampling for "interesting" stars. This is not a major issue for single planets, but may limit the ability to determine the parameters of multiple systems. Extensive simulations indicate that planets can be detected reliably if their astrometric signature exceeds three times the single-measurement precision, and that twice this amplitude is required to derive orbital parameters with 15–20% uncertainties (*Casertano et al.*, 2008).

5. FUTURE PROSPECTS: PLANET SEARCHES WITH ASTROMETRIC INTERFEROMETRY

5.1. PHASES at the Palomar Testbed Interferometer

The Palomar Testbed Interferometer (PTI) has been used to search for low-mass companions around stars in very close pairs with separations ≤1″ (*Lane and Muterspaugh*, 2004; *Muterspaugh et al.*, 2006). The light from both stars enters the interferometer simultaneously, but their fringes are usually separated in delay space. When the delay line is moved, the two fringe packets corresponding to the two binary components appear one after the other (see Fig. 6). The delay difference can be measured from these scans, and converted into the projected separation of the two stars. The fact that the two fringe packets are not measured truly simultaneously introduces a measurement error that can be substantially reduced by stabilizing the fringes. To apply this "phase referencing" technique, a fraction of the light is diverted to a second beam combiner, which produces the error signal for a control loop that keeps the fringes in a fixed position. With this approach, a precision of ~10 µas can be achieved for measurements with the 110-m baseline of the PTI (*Lane and Muterspaugh*, 2004).

The Palomar High-precision Astrometric Search for Exoplanet Systems (PHASES) project has monitored 37 close binary systems with this technique (*Muterspaugh et al.*, 2006). With semimajor axes a ≤ 50 AU, these are systems in which

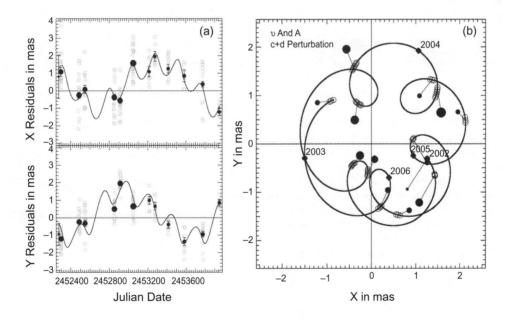

Fig. 4. Astrometric reflex motion of υ And due to υ And c and d. **(a)** x and y coordinates against time. **(b)** Orbital motion on the sky, with times of observations indicated by open circles. The astrometric orbit is shown by the dark line. Dark filled circles are normal points made from the residuals to an astrometric fit of multiple observations [light open circles in **(a)**] at each epoch. Normal point size is proportional to the number of individual measurements that formed the normal point. Error bars represent the 1σ uncertainties of the normal position. Many error bars are smaller than the symbols. From *McArthur et al.* (2010).

the binary dynamics should play a role for planet formation and evolution. No planets were discovered, but meaningful upper limits could be obtained in eight cases. The existence of planets with a few Jupiter masses in orbits with periods from ~20 d to ~3 yr can be excluded for these objects.

5.2. PRIMA at the Very Large Telescope Interferometer

The Phase-Referenced Imaging and Microarcsecond Astrometry (PRIMA) facility (*Quirrenbach et al., 1998*; *Delplancke et al., 2000*) will implement a dual-star capability at the European Southern Observatory's Very Large Telescope Interferometer (VLTI). The purpose of PRIMA is threefold: (1) provide on-axis and off-axis (within the isoplanatic angle) fringe tracking for all VLTI instruments; (2) conduct precise differential astrometry between stars separated by a few tens of arcseconds; and (3) perform phase-referenced imaging of faint sources with off-axis fringe tracking (see Fig. 7).

The present infrastructure of the VLTI consists of the four 8.2-m "unit telescopes" of the VLT, four additional moveable 1.8-m "auxiliary telescopes" (ATs), six long-stroke delay lines (each one with two ports for dual-star operation), beam relay optics, and two focal plane instruments (MIDI and AMBER). The PRIMA hardware consists of four additional major subsystems: (1) star separator systems (sometimes also called "dual star modules") that accept the light from two stars within a 2′ field and transfer it to the two input ports of the long-stroke delay lines; (2) differential delay lines (DDLs) that compensate the delay difference of up to a few centimeters between the two stars; (3) dedicated fringe detection units for the two stars; (4) an end-to-end metrology system that monitors the internal differential delay with high precision.

The first, third, and fourth items from this list have been produced by various suppliers under contracts with ESO. The ESPRI consortium has produced the differential delay lines (*Pepe et al., 2008*), works together with ESO on the analysis of the astrometric error budget and on the development of the observing strategy, and will deliver software tools required to reduce astrometric data (*Elias et al., 2008*).

Each DDL consists of a monolithic cat's eye structure mounted on top of a translation stage that can move the cat's eye mirrors in the longitudinal direction over a distance

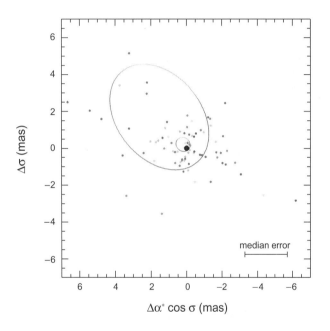

Fig. 5. Illustration of the photocenter motion of ι Dra in the plane of the sky. The effects of proper motion and parallax have been removed. The position of ι Dra in the absence of orbital motion is indicated by the large central dot. The smaller ellipse illustrates the best-fit solution, with an orbital inclination of 38°.7 and a companion mass of 14.2 M_{Jup}. The larger ellipse represents the photocenter motion for the hypothetic case of a companion mass of 0.08 $M_\odot$ (corresponding to an inclination of 6°.1). The small dots represent the abscissa residuals for the case without orbital motion. The color represents orbital phase; abscissa measurements of a certain color correspond to the stretch on the ellipses of the same color. The actual abscissa measurements are only one-dimensional and would be represented by lines perpendicular to the directions connecting the individual small dots with the large central dot. If the companion had been detected, dots with a given color (= orbital phase) would preferentially be found on one side of the central dot. The bar in the lower right shows the median standard error of 2.4 mas for the shown abscissa measurements. The data from both Hipparcos data reduction consortia, FAST and NDAC, are plotted. From *Frink et al.* (2002). A color version of this figure is available at http://iopscience.iop.org/0004-637X/576/1/478/fg3.h.jpg.

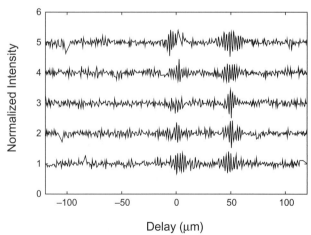

Fig. 6. Five consecutive fringe scans with the Palomar Testbed Interferometer of the binary star HD 171779. The two fringe packets corresponding to the two binary components are visible in each line. Measuring the projected angular separation of these two components corresponds to measuring the delay difference between the two fringe packets. From *Lane and Muterspaugh* (2004).

of 70 mm. Parallel beam sliders with blade spring hinges ensure that a very high accuracy in the lateral directions is maintained. A two-actuator system is chosen; one to provide a long stroke, requiring a very accurate translation mechanism, and one actuator for the high-frequency response over small displacements. The whole system is mounted in a vacuum system to make the differential optical path difference independent of environmental changes in refraction properties of the air. Each pair of delay lines is mounted in a separate vacuum vessel (length × width × height = 1000 mm × 480 mm × 500 mm). A metrology system, based on a He-Ne laser (632 nm), measures the position of the translation stage via the same cat's eye, but along different paths to avoid interfering signals. The resolution of the metrology system is 1 nm. A local control electronics system is responsible for the control of the DDL; it also provides the interface between the DDL and the PRIMA control system.

5.3. PRIMA Observing and Data Reduction Strategy

As explained above, astrometric observations with interferometers are equivalent to measurements of delays, i.e., to measurements of the difference in optical pathlength of light from a star at infinity to the two telescopes forming the interferometer. [There are additional complications if the delay lines are not evacuated (see *Daigne and Lestrade*, 1999).] The accuracy goal of 10 µas = 50 prad corresponding to a total allowable error of 5 nm for a 100-m baseline can only be achieved through a quadruple-differential technique (*Quirrenbach et al.*, 2004; *Elias et al.*, 2008): (1) Two stars with small angular separation are observed simultaneously to reduce the effects of atmospheric turbulence. (2) The optical pathlength within the interferometer is monitored with a laser interferometer. The terms entering the error budgets are thus the differential effects between the starlight and metrology beams, due, e.g., to misalignments or to dispersion between the effective observing wavelength and the wavelength of the metrology system. (3) The paths of the two stars through the instrument are exchanged periodically by rotating the field by 180°. In this way many systematic errors caused by asymmetries are canceled. (4) The orbits of exoplanets are determined from variations of the positions of their parent stars with time; only differences with respect to the position at some reference epoch matter.

It is important to realize that the raw delays have 11 (!) significant digits; astrometric planet detection implies taking differences of large and nearly equal numbers. The implementation of this quadruple-differential technique therefore requires unusual attention to detail in the understanding and calibration of varied astrophysical, atmospheric, and instrumental effects, in the construction of error budgets, in planning the operations, and in specifying and coding the data reduction software.

In particular, the desired accuracy can only be achieved if all systematic sources that can possibly affect the data are properly understood, and removed in a systematic way. While the magnitude of some astrometric errors can be predicted quite reliably (e.g., those related to atmospheric turbulence), others defy simple analysis and may have to be described with parameterized models (e.g., dynamic temperature gradients in the interferometer light ducts). Experience with other forefront astrometric facilities (e.g., the Hipparcos spacecraft, the Mark III Interferometer, and the automated Carlsberg Meridian Circle) also shows that completely unanticipated systematic effects almost inevitably show up in the actual data. The ability to detect, diagnose, and remove such unanticipated effects is of paramount importance for the success of astrometric programs.

It is therefore necessary to perform a careful *a priori* analysis of the errors, and to design and implement systems for *a posteriori* analysis of remaining trends in the data. One further needs an operation and calibration strategy that takes full advantage of, and optimizes the use of, the quadruple-differential technique described above. Finally, one needs software to perform the initial steps of the data reduction, including carrying out said differences with appropriate corrections, and conversion of delays to angles on the sky. This data reduction software has to allow inspection of the residuals and to enable searches for remaining systematic trends over several years. The latter capability is required because the integrity of the data can only be checked after the quadruple-differencing process, and because the residuals are dominated by stellar parallax (which has a period of 1 yr) and proper motion.

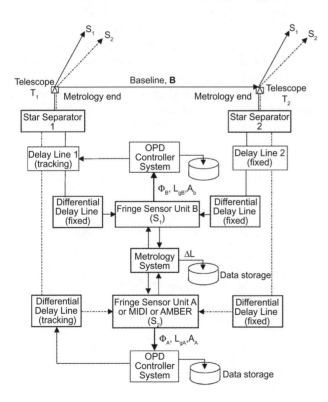

Fig. 7. The subsystems and data flow of PRIMA. Adapted from *Derie et al.* (2003).

5.4. The ESPRI Survey with PRIMA

The ESPRI project (see also *Quirrenbach et al., 2004; Launhardt et al., 2008*) will use the PRIMA facility to conduct a systematic survey of several categories of nearby stars: (1) stars that are already known to have planetary companions from RV surveys, (2) the most nearby stars (d ≤ 15 pc) of any spectral type, and (3) young stars with ages ≤ 300 m.y. within ~100 pc from the Sun. ESPRI will start immediately after the commissioning of PRIMA is completed, likely in late 2010.

Since the targets for the ESPRI survey are rather bright stars (K ≤ 8), they can be used to co-phase the interferometer. Consequently, long coherent integrations are possible for the astrometric references, which therefore can be much fainter; it is expected that stars down to K = 13 . . . 14 should be suitable for PRIMA. As explained in section 3.4, it is necessary to preselect the target stars according to the presence of suitable astrometric references (*Reffert et al., 2005; Geisler et al., 2008*). Therefore a preparatory observing program is being conducted using the SOFI near-infrared camera at the ESO 3.5-m New Technology Telescope on La Silla to obtain high-dynamic-range images of the fields of all potential ESPRI targets. The statistical expectation value of finding a star with K ≤ 14 within 20″ from a given star depends strongly on the galactic latitude, ranging from ~10% at high latitude to almost 100% near the galactic plane. The number of stars in the fields actually found is somewhat larger, because many stars are members of wide physical pairs or multiple systems.

5.5. The Space Interferometry Mission

The astrometric detection of Earth analogs requires a precision of better than 1 µas, which can be achieved only with a spacebased interferometer (see Fig. 1). In fact, a survey with such an instrument is the only proven method at this time with which a census of Earth-like planets in the solar neighborhood could be conducted. This is one of the main scientific drivers for NASA's Space Interferometry Mission (SIM), which is currently under development at the Jet Propulsion Laboratory (*Goulliout et al., 2008*). SIM is a Michelson interferometer just as the groundbased facilities described above (see also Fig. 2). But being above the atmosphere, SIM can achieve much higher precision, and use astrometric reference stars at much larger angular separation from the target.

In space, no convenient stable platform like Earth is available to provide a reference for the baseline orientation. SIM therefore uses an additional interferometer that looks at a guide star to stabilize the spacecraft attitude; this guide interferometer is essentially an extremely precise star tracker. Since the spacecraft structure is not sufficiently stiff on the submicrometer scale, an "optical truss" is formed by laser interferometers and used to monitor the exact position of all important optical elements, including the baseline length and the relative orientation of the main and guide interferometers (see Fig. 8).

The implementation details of SIM have changed several times; the main characteristics of the present "SIM Lite" design are a 6-m baseline, 50-cm siderostats, and a wavelength range from 450 to 900 nm. Since the baseline is an order of magnitude shorter than that of PRIMA, and the intended precision an order of magnitude better, the tolerances on internal errors are two orders of magnitude more stringent. This is only possible with the natural advantages of space (absence of atmospheric turbulence and dispersion, thermal stability in a heliocentric Earth-trailing orbit), but also places extreme requirements on the precision of all mechanisms and optical elements. The strategy to achieve this consists of controlling the interferometer geometry on the nanometer level, but measuring the relevant optical paths in the "optical truss" with picometer accuracy. The design of the components needed has been refined in an extensive technology program at JPL, and the system performance has been verified in a series of demonstrations with laboratory testbeds. The latest results from a 42-h test show that no systematic noise floor was present down to a level that corresponds to an amazing astrometric measurement accuracy of 0.024 µas (*Shao and Nemati, 2009*) — less than the diameter of a human hair at the distance of the Moon.

For any fixed spacecraft orientation, SIM will be able to access stars in a field with a diameter of ~15°. Within each such "tile," a few stars will be selected before the mission to define an astrometric reference grid. A basic observing block will consist of observations of the target object(s) interleaved with observations of the grid stars in the tile; the measured delay differences will thus yield one-dimensional positions of the target(s) relative to the grid stars. The second coordinate can be obtained by rotating the spacecraft around the line of sight by an angle close to 90°. During the course of the mission, the grid stars will be visited regularly, about four to five times per year. By observing overlapping tiles, a full-sky reference grid can be constructed in the same manner as overlapping plates have been used to assemble all-sky astrometric catalogs. The inclusion of quasars in the

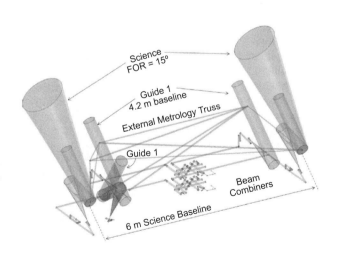

Fig. 8. Design layout of SIM Lite. From *Goulliout et al.* (2008).

grid will ensure that this reference system will represent an inertial frame. The expected performance of SIM in this "global astrometry" mode is 4 µas precision on the derived astrometric parameters (position, parallax) at the end of the nominal 5-yr mission, enabling a rich scientific program addressing many outstanding questions in different fields of astronomy (*Unwin et al.*, 2008).

Many terms in the error budget of SIM measurements depend on the angle between the target and the astrometric references. It is thus possible to improve the accuracy by choosing references close (≤1°) to the object of interest. It is for this "narrow-angle" mode that an accuracy of better than 1 µas for each measurement is envisaged, as discussed in the preceding paragraph. (Note that the term "narrow-angle" as used here denotes the regime where angle-independent terms in the error budget dominate over terms increasing with angle. This is slightly different from the use of this term in section 3.3, where it denoted a break in the scaling laws of atmospheric errors. One has to keep in mind that "narrow-angle" for SIM means ≤1°, whereas for groundbased astrometry it means ≤1′.) Note that this number has to be divided by $\sqrt{N}$ with N = 30 . . . 100 for the number N of visits to obtain the "mission accuracy." This is the fair comparison with the "global" mode and with the figures usually quoted for other missions. These narrow-angle measurements will not be in an inertial reference frame, but only with respect to the selected reference stars, as discussed in section 2.4. The narrow-angle precision will thus not apply to absolute parallaxes and proper motions, but this is the relevant number for determining the astrometric wobble due to unseen companions.

As for Gaia, extensive simulations have been performed to verify that Earth-like planets can indeed be detected in the one-dimensional astrometric data from SIM, even in multiple systems (*Traub et al.*, 2010). There are obvious difficulties if several large bodies are present in the system with periods longer than the mission duration (like the gas and ice giants in the solar system), or if the period of a planet is very close to 1 yr (because of the partial degeneracy with the parallax). Nevertheless, it turned out that terrestrial planets could be detected in a double-blind experiment with a high success rate (i.e., few planets were missed) and good reliability (i.e., there was only a small number of false positives). This means that with SIM the technology is truly at hand to perform a complete census of Earth analogs around ~60 nearby stars.

5.6. Differential Phase and Differential Closure Phase

Differential phase observations with a near-IR interferometer offer a way to obtain spectra of exoplanets. The method makes use of the wavelength dependence of the interferometer phase of the planet/star system, which depends both on the interferometer geometry and on the brightness ratio between the planet and the star (*Quirrenbach*, 2000). In the limit of a separation between star and planet that is much smaller than the resolution of the interferometer, the interferometer phase represents the location of the center-of-light of the system. At a wavelength where the planet is dark, i.e., in an absorption feature in the planetary atmosphere, this location is at the position of the star; at a wavelength where the planet is bright, it is shifted somewhat toward the planet. When longer baselines are used, the system gets more and more resolved, and the phase is given by

$$\phi(\lambda) = \arctan\left[\frac{r(\lambda)\cdot\sin\left(\frac{2\pi}{\lambda}\vec{\theta}\cdot\vec{B}\right)}{1+r(\lambda)\cdot\cos\left(\frac{2\pi}{\lambda}\vec{\theta}\cdot\vec{B}\right)}\right]$$
$$\approx r(\lambda)\cdot\sin\left(\frac{2\pi}{\lambda}\vec{\theta}\cdot\vec{B}\right) \quad (23)$$

Here $\vec{\theta}$ is the separation vector between the star and the planet in the sky plane, and $r(\lambda)$ is the wavelength-dependent brightness ratio between the planet and the star (and thus usually $r(\lambda) \ll 1$). From equation (23) it is evident that very high precision is needed for the measurement of the interferometer phase. For example, taking $r(\lambda) \sim 0.001$ as typical for a hot Jupiter in the infrared, the precision has to be better than 1 mrad.

In practice, one would use the phase at some wavelength λ_0 as a reference and work only with phase differences between the individual wavelength channels. Still, these differential phases are strongly affected by instrumental and atmospheric dispersion effects. Difficulties in calibrating these effects might prevent the application of the differential phase method to systems with a very high contrast, such as exoplanets. However, if the interferometer array consists of three or more telescopes, there is a trick to overcome these obstacles. Rather frequently it is possible to associate the dominant phase errors with the telescopes in the array, i.e., all baselines that involve a certain telescope are affected by the phase error of that telescope in the same way. Examples of such telescope-based errors are atmospheric pathlength variations (for groundbased interferometers), or pathlength variations in the beam train leading from the telescope to the beam combiner. If one considers the baselines formed by three telescopes (i, j, k), the observed phases ϕ'_{ij} are related to the true phases ϕ_{ij} and telescope-based phase errors ψ_i by the following relations

$$\begin{aligned}\phi'_{ij} &= \phi_{ij} + \psi_i - \psi_j \\ \phi'_{jk} &= \phi_{jk} + \psi_j - \psi_k \\ \phi'_{ki} &= \phi_{ki} + \psi_k - \psi_i\end{aligned} \quad (24)$$

If the delay errors are mostly of this type, one can retain some of the phase information in the form of *closure phases* Φ_{ijk}, which are obtained by summing the phases in the closed triangles of the array

$$\Phi_{ijk} = \phi'_{ij} + \phi'_{jk} + \phi'_{ki} = \phi_{ij} + \phi_{jk} + \phi_{ki} \quad (25)$$

Because the phase errors ψ_i cancel each other out in this sum, the closure phases Φ_{ijk} are observables that contain phase information on the target, but are unaffected by telescope-based phase errors.

Figure 9 shows the closure phase as it would be measured with the AMBER instrument at the VLTI, based on a realistic model of the 51 Peg system, and taking into account a theoretical spectrum of the planet as well as the geometry of the VLTI. *Joergens and Quirrenbach* (2004) have presented a strategy to determine the geometry of the planetary system and the spectrum of the exoplanet from such closure phase observations in two steps. First, there is a close relation between the nulls in the closure phase and the nulls in the corresponding single-baseline phases: Every second null of a single-baseline phase is also a null in the closure phase. This means that the nulls in the closure phase do not depend on the spectrum but only on the geometry, so that the geometry of the system can be determined by measuring the nulls in the closure phase at three or more different hour angles. In the second step, the known geometry can then be used to extract the planet spectrum directly from the closure phases.

5.7. Conclusions

Whereas the principle of astrometric planet detection is simple, the technical challenges are formidable. The signatures range from ≤1 mas for super Jupiters to ≤1 µas for Earth analogs and are thus mostly beyond the reach of classical astrometric methods. However, modern developments — large telescopes equipped with CCD cameras, access to space, and interferometry — have made it possible to enter the submilliarcsecond regime; space interferometry should soon even break the microarcsecond barrier. We can thus expect a renaissance of astrometry in general, and rich opportunities for astrometric studies of planetary systems in particular. Astrometry is the only method that can supply the full dynamical information about planetary systems: the planet masses (this assumes that the stellar mass can be determined independently) and all orbital parameters. In addition, its sensitivity does not depend on stellar type, and it is much less susceptible to noise from spots and other disturbances in the stellar atmosphere than the RV technique. Because of these unique capabilities, astrometry will certainly develop into an important tool for comprehensive studies of nearby planetary systems.

Note added in proof: With the decision by the U.S. Decadal Survey committee not to support SIM-Light, it has become unlikely that this mission will fly within this decade. Nevertheless, the science case for an astrometric mission with submicroarcsecond precision remains strong, as this is the only way to determine the masses of nearby Earth-like planets.

REFERENCES

Anglada-Escudé G., Shkolnik E. L., Weinberger A. J., Thompson I. B., Osip D. J., and Debes J. H. (2010) Strong constraints to the putative planet candidate around VB10 using Doppler spectroscopy. *Astrophys. J. Lett., 711,* L24–L29.

Bean J. L. and Seifahrt A. (2009) The architecture of the GJ 876 planetary system. Masses and orbital coplanarity for planets b and c. *Astron. Astrophys., 496,* 249–257.

Bean J. L., McArthur B. E., Benedict G. F., Harrison T. E., Bizyaev D., et al. (2007) The mass of the candidate exoplanet companion to HD 33636 from Hubble Space Telescope astrometry and high-precision radial velocities. *Astron. J., 134,* 749–758.

Bean J. L., Seifahrt A., Hartman H., Nilsson H., Reiners A., et al. (2010) The proposed giant planet orbiting VB10 does not exist. *Astrophys. J. Lett., 711,* L19–L23.

Beichman C. A., Fridlund M., Traub W. A., Stapelfeldt K. R., Quirrenbach A., and Seager S. (2007) Comparative planetology and the search for life beyond the solar system. In *Protostars and Planets V* (B. Reipurth et al., eds), pp. 915–928. Univ. of Arizona, Tucson.

Benedict G. F., McArthur B. E., Forveille T., Delfosse X., Nelan E., et al. (2002) A mass for the extrasolar planet Gl 876 b determined from Hubble Space Telescope Fine Guidance Sensor 3 astrometry and high-precision radial velocities. *Astrophys. J. Lett., 581,* L115–L118.

Benedict G. F., McArthur B. E., Gatewood G., Nelan E., Cochran W. D., et al. (2006) The extrasolar planet ε Eridani b: Orbit and mass. *Astron. J., 132,* 2206–2218.

Binnendijk L. (1960) *Properties of Double Stars.* Univ. of Pennsylvania, Philadelphia.

Boss A. (1998a) *Looking for Earths.* Wiley, New York. 240 pp.

Boss A. P. (1998b) Astrometric signatures of giant-planet formation. *Nature, 393,* 141–143.

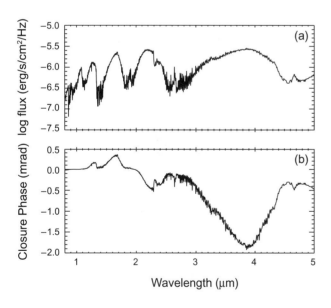

Fig. 9. Spectral signal of planet in the closure phase. **(a)** Theoretical spectrum of the giant irradiated planet orbiting the solar-like star 51 Peg (*Sudarsky et al.,* 2003). Clearly visible are CO and H_2O absorption bands in the near-IR. **(b)** Closure phases in milliradian based on the theoretical spectrum of 51 Peg b as well as of the host star and a simulated observation with the near-IR instrument AMBER at the VLTI using the three telescopes UT1, UT3, and UT4 (baseline lengths 102 m, 62 m, and 130 m). It is evident that the spectrally resolved closure phases encode the spectrum of the planet.

Boss A. P., Weinberger A. J., Anglada-Escudé G., Thompson I. B., Burley G., et al. (2009) The Carnegie astrometric planet search program. *Publ. Astron. Soc. Pacific, 121,* 1218–1231.

Butler R. P., Marcy G. W., Fischer D. A., Brown T. M., Contos A. R., et al. (1999) Evidence for multiple companions to υ Andromedae. *Astrophys. J., 526,* 916–927.

Cameron P. B., Britton M. C., and Kulkarni S. R. (2009) Precision astrometry with adaptive optics. *Astron. J., 137,* 83–93.

Casertano S., Lattanzi M. G., Sozzetti A., Spagna A., Jancart S., et al. (2008) Double-blind test program for astrometric planet detection with Gaia. *Astron. Astrophys., 482,* 699–729.

Cassen P., Guillot T., and Quirrenbach A. (2006) *Extrasolar Planets.* Saas-Fee Advanced Course 31, Springer, New York. 451 pp.

Daigne G. and Lestrade J. F. (1999) Astrometric optical interferometry with non-evacuated delay lines. *Astron. Astrophys., 138,* 355–363.

Delplancke F., Leveque S. A., Kervella P., Glindemann A., and D'Arcio L. (2000) Phase-referenced imaging and microarcsecond astrometry with the VLTI. In *Interferometry in Optical Astronomy* (P. J. Léna and A. Quirrenbach, eds.), pp. 365–376. SPIE Conf. Series 4006, Bellingham, Washington.

Derie F., Delplancke F., Glindemann A., Lévêque S., Ménardi S., et al. (2003) Phase Referenced Imaging and Micro-arcsec Astrometry (PRIMA) technical description and implementation. In *Proceedings of GENIE — DARWIN Workshop — Hunting for Planets* (H. Lacoste, ed.), pp. 6.1–6.13. ESA Spec. Publ. 522, Noordwijk, The Netherlands.

Elias N. M., Köhler R., Stilz I., Reffert S., Geisler R., et al. (2008) The astrometric data-reduction software for exoplanet detection with PRIMA. In *Optical and Infrared Interferometry* (M. Schöller et al., eds.), pp. 1–9. SPIE Conf. Series 7013, Bellingham, Washington.

Frink S., Mitchell D. S., Quirrenbach A., Fischer D. A., Marcy G. W., and Butler R. P. (2002) Discovery of a substellar companion to the K2 III giant ι Draconis. *Astrophys. J., 576,* 478–484.

Gatewood G. and Eichhorn H. (1973) An unsuccessful search for a planetary companion of Barnard's Star (BD + 4°3561) *Astron. J., 78,* 769–776.

Geisler R., Setiawan J., Henning T., Queloz D., Quirrenbach A. et al. (2008) Preparing the exoplanet search with PRIMA: Searching for reference stars and target characterization. In *Exoplanets: Detection, Formation and Dynamics* (Y. S. Sun et al., eds.), pp. 61–63. IAU Symposium 249, Cambridge Univ., Cambridge.

Goullioud R., Catanzarite J. H., Dekens F. G., Shao M., and Marr J. C. (2008) Overview of the SIM PlanetQuest Light mission concept. In *Optical and Infrared Interferometry* (M. Schöller et al., eds.), pp. 1–12, 70134T. SPIE Conf. Proc. Series, Bellingham, Washington.

Harrington R. S., Kallarakal V. V., and Dahn C. C. (1983) Astrometry of the low-luminosity stars VB8 and VB10. *Astron. J., 88,* 1038–1039.

Joergens V. and Quirrenbach A. (2004) Modeling of closure phase measurements with AMBER/VLTI — towards characterization of exoplanetary atmospheres. In *New Frontiers in Stellar Interferometry* (W. A. Traub, ed.), pp. 551–559. SPIE Conf. Series 5491, Bellingham, Washington.

Jordan S. (2008) The Gaia project: Technique, performance and status. *Astron. Nachtr., 329,* 875–880.

Lane B. F. and Muterspaugh M. W. (2004) Differential astrometry of subarcsecond scale binaries at the Palomar Testbed Interferometer. *Astrophys. J., 601,* 1129–1135.

Latham D. W., Mazeh T., Stefanik R. P., Mayor M., and Burki G. (1989) The unseen companion of HD 114762: A probable brown dwarf. *Nature, 339,* 38–40.

Launhardt R., Queloz D., Henning T., Quirrenbach A., Delplancke F., et al. (2008) The ESPRI project: Astrometric exoplanet search with PRIMA. In *Optical and Infrared Interferometry* (M. Schöller et al., eds.), pp. 1–10, 70132I. SPIE Conf. Series 7013, Bellingham, Washington.

Lawrence J. S., Ashley M. C. B., Tokovinin A., and Travouillon T. (2004) Exceptional astronomical seeing conditions above Dome C in Antarctica. *Nature, 431,* 278–281.

Lazorenko P. F., Mayor M., Dominik M., Pepe F., Segransan D., and Udry S. (2009) Precision multi-epoch astrometry with VLT cameras FORS1/2. *Astron. Astrophys., 505,* 903–918.

Lin D. N. C. and Ida S. (1997) On the origin of massive eccentric planets. *Astrophys. J., 477,* 781–791.

Lindegren L. (1980) Atmospheric limitations of narrow-field optical astrometry. *Astron. Astrophys., 89,* 41–47.

Lindegren L. and Perryman M. A. C. (1996) GAIA: Global astrometric interferometer for astrophysics. *Astron. Astrophys., 116,* 579–595.

Lunine J. I., Fischer D., Hammel H. B., Henning T., Hillenbrand L., et al. (2008) Worlds beyond: A strategy for the detection and characterization of exoplanets. Executive summary of a report of the ExoPlanet Task Force Astronomy and Astrophysics Advisory Committee, Washington, DC, June 23, 2008. *Astrobiology, 8,* 875–881.

Makarov V. V., Beichman C. A., Catanzarite J. H., Fischer D. A., Lebreton J., et al. (2009) Starspot jitter in photometry, astrometry, and radial velocity measurements. *Astrophys. J. Lett., 707,* L73–L76.

Mayor M. and Queloz D. (1995) A Jupiter-mass companion to a solar-type star. *Nature, 378,* 355–359.

McArthur B. E., Benedict G. F., Barnes R., Martioli E., Korzennik S., et al. (2010) New observational constraints on the υ Andromedae system with data from the Hubble Space Telescope and Hobby-Eberly Telescope. *Astrophys. J., 715,* 1203–1220.

McCarthy D. W., Probst R. G., and Low F. J. (1985) Infrared detection of a close cool companion to Van Biesbroeck 8. *Astrophys. J. Lett., 290,* L9–L13.

Mignard F. and Klioner S. A. (2009) Gaia: Relativistic modelling and testing. In *Relativity in Fundamental Astronomy* (S. A. Klioner et al., eds.), pp. 306–314. IAU Symposium 261, Cambridge Univ., Cambridge.

Muterspaugh M. W., Lane B. F., Kulkarni S. R., Burke B. F., Colavita M. M., and Shao M. (2006) Limits to tertiary astrometric companions in binary systems. *Astrophys. J., 653,* 1469–1479.

O'Leary B. T. (1966) On the occurrence and nature of planets outside the solar system. *Icarus, 5,* 419–436.

Papaloizou J. C. B. and Terquem C. (2001) Dynamical relaxation and massive extrasolar planets. *Mon. Not. R. Astron. Soc., 325,* 221–230.

Pepe F., Queloz D., Henning T., Quirrenbach A., Delplancke F., et al. (2008) The ESPRI project: Differential delay lines for PRIMA. In *Optical and Infrared Interferometry* (M. Schöller et al., eds.), pp. 1–12, 70130P. SPIE Conf. Series 7013, Bellingham, Washington.

Perrier C. and Mariotti J. M. (1987) On the binary nature of Van Biesbroeck 8. *Astrophys. J. Lett., 312,* L27–L30.

Perryman M. A. C., de Boer K. S., Gilmore G., Høg E., Lattanzi

M. G., et al. (2001) GAIA: Composition, formation and evolution of the galaxy. *Astron. Astrophys., 369,* 339–363.

Pravdo S. H. and Shaklan S. B. (1996) Astrometric detection of extrasolar planets: results of a feasibility study with the Palomar 5 Meter Telescope. *Astrophys. J., 465,* 264–277.

Pravdo S. H. and Shaklan S. B. (2009) An ultracool star's candidate planet. *Astrophys. J., 700,* 623–632.

Press W. H., Teukolsky S. A., Vetterling W. T., and Flannery B. P. (2007) *Numerical Recipes: The Art of Scientific Computing,* 3rd edition. Cambridge Univ., Cambridge.

Quirrenbach A. (2000) Astrometry with the VLT Interferometer. In *From Extrasolar Planets to Cosmology: The VLT Opening Symposium* (J. Bergeron and A. Renzini, eds.), pp. 462–467. Springer, Berlin.

Quirrenbach A. (2001) Optical interferometry. *Annu. Rev. Astron. Astrophys., 39,* 353–401.

Quirrenbach A., Coudé du Foresto V., Daigne G., Hofmann K. H., Hofmann R., et al. (1998) PRIMA — study for a dualbeam instrument for the VLT Interferometer. In *Astronomical Interferometry* (R. D. Reasenberg, ed.), pp. 807–817. SPIE Conf. Series 3350, Bellingham, Washington.

Quirrenbach A., Henning T., Queloz D., Albrecht S., Bakker E. J., et al. (2004) The PRIMA astrometric planet search project. In *New Frontiers in Stellar Interferometry* (W. A. Traub, ed.), pp. 424–432. SPIE Conf. Series 5491, Bellingham, Washington.

Reffert S. and Quirrenbach A. (2006) Hipparcos astrometric orbits for two brown dwarf companions: HD 38529 and HD 168443. *Astron. Astrophys., 449,* 699–702.

Reffert S. and Quirrenbach A. (2010) Mass constraints on known substellar companions from the new Hipparcos Catalogue: 18 confirmed planets and six confirmed brown dwarfs. *Astron. Astrophys.,* in press.

Reffert S., Launhardt R., Hekker S., Henning T., Queloz D., et al. (2005) Choosing suitable target, reference and calibration stars for the PRIMA astrometric planet search. In *Astrometry in the Age of the Next Generation of Large Telescopes* (P. K. Seidelmann and A. K. B. Monet, eds.), pp. 81–89. ASP Conf. Series 338, Astronomical Society of the Pacific, San Francisco.

Reuyl D. and Holmberg E. (1943) On the existence of a third component in the system 70 Ophiuchi. *Astrophys. J., 97,* 41–45.

Shao M. and Colavita M. M. (1992) Potential of long-baseline infrared interferometry for narrow-angle astrometry. *Astron. Astrophys., 262,* 353–358.

Shao M. and Nemati B. (2009) Sub-microarcsecond astrometry with SIM-Lite: A testbed-based performance assessment. *Publ. Astron. Soc. Pacific, 121,* 41–44.

Skrutskie M. F., Forrest W. J., and Shure M. A. (1987) Direct infrared imaging of VB8. *Astrophys. J. Lett., 312,* L55–L58.

Strand K. A. (1943) 61 Cygni as a triple system. *Publ. Astron. Soc. Pacific, 55,* 29–32.

Struve O. (1952) Proposal for a project of high-precision stellar radial velocity work. *The Observatory, 72,* 199–200.

Sudarsky D., Burrows A., and Hubeny I. (2003) Theoretical spectra and atmospheres of extrasolar giant planets. *Astrophys. J., 588,* 1121–1148.

Traub W. A., Beichman C., Boden A. F., Boss A. P., Casertano S., et al. (2010) Detectability of terrestrial planets in multi-planet systems: Preliminary report. In *Extrasolar Planets in Multi-Body Systems: Theory and Observations* (K. Goździewski et al., eds.), pp. 191–199. EAS Publ. Ser. 42, European Astronomical Society, Les Ulis, France.

Unwin S. C., Shao M., Tanner A. M., Allen R. J., Beichman C. A., et al. (2008) Taking the measure of the universe: Precision astrometry with SIM PlanetQuest. *Publ. Astron. Soc. Pacific, 120,* 38–88.

van de Kamp P. (1963) Astrometric study of Barnard's Star from plates taken with the 24-inch Sproul refractor. *Astron. J., 68,* 515–521.

van de Kamp P. (1982) The planetary system of Barnard's star. *Vistas Astron., 26,* 141–157.

Planets Around Pulsars and Other Evolved Stars: The Fates of Planetary Systems

Alex Wolszczan
Pennsylvania State University

Marc J. Kuchner
NASA Goddard Space Flight Center

The first exoplanets ever discovered were found orbiting a pulsar, not a main-sequence star! The pulse timing technique used to detect these first planets has also powered searches for planets around eclipsing binaries, white dwarfs, and other evolved stars. In this chapter, we give a detailed description of this extraordinarily precise planet detection method, discuss the exotic planets orbiting highly evolved stars, and outline possible fates of our solar system and planetary systems in general.

1. INTRODUCTION

"Earth-Sized Planets Confirmed, But They're Dead Worlds," reads one headline. The headline appeared in 2003, more than 11 years after these Earth-sized planets began to emerge in data from the Arecibo radio telescope. The headline reveals the slow turning of the gears of scientific consensus. The world has accepted the existence of pulsar planets as confirmed fact — but perhaps only after declaring them to be dead and uninteresting.

Pulsars are rotating neutron stars, laboratories of extreme physics and astrophysics, laden with mystery. They are the remnants of massive stars, roughly ≥8 $M_\odot$ on the main sequence, that die violent deaths as supernovae. You can find a detailed discussion of all aspects of pulsar astronomy in *Lorimer and Kramer* (2005) and *Lyne and Smith* (2006).

But what amazes planet hunters about pulsars is that for one of these objects, the timing of the arrival of the pulses at Earth contains the signatures of the first exoplanets ever discovered (*Wolszczan and Frail*, 1992). Imagine the reaction to this finding in 1992! Not only were there exoplanets, but they were orbiting a pulsar. No wonder the idea took so long to gain credibility.

What started to change people's minds about these planets — and the existence of exoplanets in general — was the detection, in the timing signal, of the mutual gravitational interaction of the two terrestrial-mass planets (*Rasio et al.*, 1992; *Malhotra et al.*, 1992; *Peale*, 1993; *Wolszczan*, 1994). Luckily, these two larger planets have mean motions close to the 3:2 commensurability, which makes their mutual perturbations relatively large and not too hard to detect. This confirmation paved the way for the discovery of the first 51 Pegasi planet in 1995.

We now know that PSR B1257+12 hosts three planets (*Wolszczan et al.*, 2000). The innermost planet has a mass only twice that of the Moon. The pulsar timing technique is so precise that, in principle, it would allow us to detect planets with masses no greater than that of a large asteroid.

Since 1992, the timing method used to find the first pulsar planets has yielded several other discoveries. Timing observations of the binary pulsar PSR B1620-26 in the M4 globular cluster revealed a Jupiter-mass planet (*Backer et al.*, 1993; *Thorsett et al.*, 1999); we have upper limits on the existence of planets around many other pulsars. The timing method has detected planet-mass bodies around two other highly evolved stellar systems: the helium-burning subdwarf B (sdB) star V391 Peg via pulsation timing (*Silvotti et al.*, 2007) and an sdB eclipsing binary HW Vir via eclipse timing (*Lee et al.*, 2009). Pulsation timing has also helped in placing interesting constraints on the possible presence of planets around white dwarfs (*Kepler et al.*, 2005; *Mullally et al.*, 2008). Table 1 summarizes the properties of all these planets, as described in the above references, and in *Wolszczan* (2008).

It is interesting to note that, as the radial velocity method (see chapter by Lovis and Fischer) provides the most efficient way to detect planets around solar-type stars, the same is true for the timing technique in application to stars in the final stages of their evolution. In this chapter, we use the pulsar timing as a working example to explain, in detail, the technique that led to these discoveries (specifics of the eclipse or transit timing are discussed in the chapter by Winn). Then, we proceed to review the current status of the field and address a selection of questions, some of which are fundamentally important. Where did these exotic planets come from? What is the future of our own solar system? What other kinds of planets might we find one day? We discuss these topics as well.

2. PULSE TIMING FUNDAMENTALS

Finding planets around pulsars does not depend on a detailed understanding of neutron star physics; it relies only on the extreme stability of neutron star rotation, and thereby the pulses we observe. The pulse periods of pulsars, P, have a natural deterministic slowdown rate, $\dot{P}$, that amounts to ~10^{-15} s s^{-1} for normal, middle-aged ones, but becomes

TABLE 1. Planets around evolved stars.

Planet	Orbital Period (d)	M (M_{Jup})	Semimajor Axis (AU)	Eccentricity	Inclination (deg.)
PSR B1257+12b	25.262	$\geq 7 \times 10^{-5}$	0.19	0	—
PSR B1257+12c	66.5419	0.013	0.36	0.0186	53
PSR B1257+12d	98.2114	0.012	0.46	0.0252	47
PSR B1620-26b	36526	2.5	23	0	55
V391 Peg b	1170	≥ 3.2	1.7	0	—
HW Vir b	3321	≥ 8.5	3.6	0.31	—
HW Vir c	5767	≥ 19.2	5.3	0.46	—

only ~10^{-20} s s^{-1} for the rapidly spinning millisecond pulsars (MSPs). This extraordinary property of the MSPs, combined with our ability to measure the arrival times of their pulses with microsecond precision, makes them excellent celestial clocks, rivaling the stability of atomic clocks on Earth (e.g., *Matsakis et al.*, 1997).

The pulsar clocks, whose stability and precision are governed by the huge angular momentum of their spins and rotational energy loss through magnetic dipole radiation and relativistic particle winds, are not entirely unique. Similar clock properties, albeit at much lower levels of stability and precision, are exhibited by pulsating white dwarfs. The g-mode oscillations of the hydrogen atmosphere (DAV) white dwarfs have periods on the order of hundreds of seconds and cooling-generated slowdown at rates as small as 10^{-15} s s^{-1}. These white dwarf pulsations can be timed with a ~ 1 s precision (e.g., *Kepler et al.*, 2005).

Stellar eclipses generated by orbital motion and geometry represent yet another, less stable but still useful astrophysical clock mechanism. Eclipsing binary stars are characterized by $\dot{P}$ values in the range of -10^{-10} to -10^{-12} s s^{-1}, dictated by the loss of orbital angular momentum of the secondary star due to a magnetized wind from the primary (*Demircan et al.*, 2006; *Lee et al.*, 2009). A $\leq$10-s eclipse timing precision has been demonstrated for these objects by *Deeg et al.* (2008) and *Lee et al.* (2009).

In what follows, we discuss the basics of the standard procedures employed in the timing of pulsars with binary or multiple companions and in the modeling of Keplerian orbits from both pulse period and arrival time measurements. As noted in section 1, a more in-depth presentation of these and other aspects of pulsar astronomy can be found in *Lorimer and Kramer* (2005) and *Lyne and Smith* (2006). The most recent implementations of these procedures in the timing of white dwarf oscillations and stellar eclipses are described in *Mullally et al.* (2008) and *Lee et al.* (2009).

2.1. Measurement of Pulse Arrival Times

High-precision pulsar timing measurements are carried out by synchronously averaging the detected pulsar signal at a period that is Doppler-corrected for Earth's orbital motion, to form a series of high signal-to-noise (S/N) integrated pulse profiles sampled at intervals of about a milliperiod (Fig. 1). Each integration is time-tagged using the observatory's atomic time standard and stored for further processing.

The precision of pulse time-of-arrival (TOA) measurements can be estimated from the radiometer equation suitably modified for use with a pulsed signal (*Lorimer and Kramer*, 2005)

$$\sigma_{TOA} \approx \frac{S_{sys} P \delta_c^{3/2}}{S_m \sqrt{t_{int} \Delta f}} \quad (1)$$

where S_{sys} is the system flux density, which is a measure of the system noise, $\delta_c = W/P$ is the pulse duty cycle (W is the pulse width), S_m is the mean flux density of the pulsar, t_{int} is the integration time, and Δf is the receiver bandwidth. Obviously, it is advantageous to time strong, short-period, narrow-pulse pulsars with large telescopes equipped with low-noise receivers. Ultimately, with all the instrumental and propagation factors affecting the timing accuracy properly accounted for, the final TOA precision limit is set by causes intrinsic to the pulsar, such as rotation noise, single pulse jitter, and intrinsic width of the pulse.

With a proper choice of the instrumental time resolution, one still has to combat the effect of pulse smearing by a frequency-dependent, dispersive delay of the pulsar signal, as it travels through the ionized interstellar gas. This phenomenon is adequately described by the dispersion law for a cold, low-density, weakly magnetized plasma in which case the refractive index μ at radio frequency f is given in terms of the plasma frequency, $f_p = \sqrt{n_e e^2 / \pi m_e}$, and the cyclotron frequency, $f_B = eB_\parallel / 2\pi m_e c$, as

$$\mu = \sqrt{1 - \frac{f_p^2}{f^2} \mp \frac{f_p^2 f_B}{f^3}} \quad (2)$$

where $n_e \sim 0.03$ cm^{-3} is the electron density, $B_\parallel \sim 1$ μG is the line-of-sight component of the magnetic field, e and m_e are the electron charge and mass, respectively, and c is the speed of light. Given the above typical values of n_e and $B_\parallel$ for the interstellar medium, one obtains $f_p \sim 1.5$ kHz and $f_B \sim 3$ Hz. A reduction of the refractive index of the interstellar plasma by the presence of the magnetic field is very small and can be ignored. However, the plasma frequency is typically high enough to cause a sufficient frequency dependent change of the group velocity $v_g = c\mu$ of the wave to produce a measurable propagation delay of the pulsar signal.

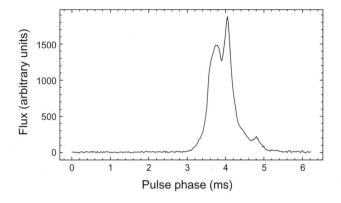

Fig. 1. An integrated pulse profile of PSR B1257+12 observed with the Arecibo telescope at 430 MHz.

In order to quantify this effect, one can drop the f_B term in equation (2), assume $f_p \ll f$, expand the remaining f_p term in Taylor series, and integrate the dispersive delay $dt = dl/v_g$ along the line of sight to obtain a time delay, Δt (in seconds) between the two observing frequencies f_1 and f_2

$$\Delta t = D \times DM \left[f_1^{-2} - f_2^{-2} \right] \qquad (3)$$

where $D = e^2/2\pi m_e c = 4.148808 \times 10^3 \text{MHz}^2 \text{ pc}^{-1} \text{ cm}^3 \text{ s}$ is the dispersion constant

$$DM = \int_0^d n_e \, dl \qquad (4)$$

is the dispersion measure expressed in units of pc cm^{-3}, and d is the pulsar distance.

Using DM = 10.1 pc cm^{-3} and W ~ 0.5 ms of the planet's pulsar, PSR B1257+12, as an example, assuming a typical observing frequency f = 1.4 GHz, and the receiver bandwidth Δf = 100 MHz, one gets Δt ~ 3 ms from equation (3). This means that, with the above observing parameters, the interstellar dispersion would smear the pulses from the pulsar over almost half its 6.2-ms rotation period! As is clear from equation (1), such a broadening of the pulse would seriously impact its timing precision, which requires possibly narrow pulses to be observed over large bandwidths.

The simplest way to compensate for the dispersion smearing is to integrate the incoming pulses in a number of narrow, adjacent subbands (frequency channels) to obtain the final, dispersion-corrected profile by an appropriate phase shifting and adding of the single-channel profiles. This so-called incoherent dedispersion scheme is practically realized with the aid of analog or digital filterbank receivers. A more precise, coherent dedispersion method relies on the fact that the dispersion effect modifies the phase of the radio signal as it propagates through the interstellar plasma. This frequency-dependent phase change can be conveniently modeled in frequency domain as a transfer function

$$H(f_0 + f) = e^{-ik(f_0 + f)d} \qquad (5)$$

where $k(f) = (2\pi/c)\mu f$ is the wavenumber, f_0 is the center frequency, and $|f| < \Delta f/2$ (the signal is band-limited). Substituting for μ from equation (2), after some algebra, one gets an expression for the transfer function in terms of the observables

$$H(f_0 + f) = e^{i \frac{2\pi D}{(f+f_0)f_0^2} DM f^2} \qquad (6)$$

This function can be treated as an interstellar dispersion filter that can be deconvolved out of the sampled, complex voltage output of the receiver to yield an almost dispersion-free pulse profile, depending on the accuracy of the DM determination.

A practical method of determining the TOAs is dictated by another very useful property of pulsars, which is the stability of their pulse profiles. A sufficiently long averaging (typically a few thousand pulse periods) leads to stable pulse shapes for a given pulsar, differing only by a deterministic bias, a scale factor, an additive random noise, and an arrival time-dependent time offset, which must be measured and added to the start time of the observation, to determine a topocentric TOA. A commonly used method to measure this offset is to cross-correlate the observed pulses with a high S/N template profile. This technique, when applied to observations of the millisecond pulsars, results in a microsecond precision of TOA measurements.

2.2. Pulse Timing Analysis

The pulsar timing technique, in its application to planet detection, relies on modeling of planetary orbits on the basis of precise measurements of a varying Römer delay of times-of-arrival (TOAs) of the pulsar pulses (think of them as of ticks of the pulsar clock), $t_R = \Delta z/c$, due to the changing line-of-sight projection, Δz, of the reflex motion of a pulsar. In the modeling process, the number of pulses N_i, received over a time interval $t_i = t_{obs} - t_0$ between some initial epoch t_0 and the topocentric arrival time at the telescope, t_{obs}, is a measure of the accumulated phase, which can be expressed as a continuous variable

$$\phi_i = \int_{t_0}^{t} \nu(t) dt \qquad (7)$$

where P(t) is the instantaneous period of the pulsar and $\nu(t) = 1/P(t)$ is the corresponding rotation frequency.

For most of the old MSPs, there is no evidence that processes other than the spindown caused by a gradual loss of the neutron star's rotational energy have a significant effect on their rotational stability. This is not necessarily true for young pulsars with strong magnetic fields and large spindown rates, in which case the second frequency derivative may be measurable, and the stability of a pulsar clock may be affected by the timing noise.

The pulsar spindown process is adequately described in terms of a magnetic dipole, rotating at the angular frequency $\Omega = 2\pi\nu$ and emitting electromagnetic radiation at frequency Ω with a total power $2/3 M_\perp^2 \Omega^4 c^{-3}$, where $M_\perp$ is the component of the magnetic moment perpendicular to the rotation axis. The energy for this "magnetic braking" is provided by the pulsar's rotational energy loss, $\dot{E}_{rot} = d(1/2 I \Omega^2)/dt = I\Omega\dot{\Omega}$, which means that

$$\dot{E}_{rot} = -\frac{2}{3} M_\perp^2 \Omega^4 c^{-3} \qquad (8)$$

where I is the moment of inertia.

In reality, the spindown energy of a pulsar is not only carried away by dipole radiation, but a large part of it is converted into a wind of relativistic particles and high-energy radiation. Nevertheless, because the behavior of the upper magnetosphere is governed by the magnetic field, equation (8) represents an entirely adequate description of the deterministic process of the pulsar slowdown. This allows the phase of equation (7) to be parameterized in terms of the rotation frequency and its time derivatives as a Taylor series

$$\phi_i = \phi_0 + \nu t_i + \frac{1}{2}\dot{\nu}t_i^2 + \frac{1}{6}\ddot{\nu}t_i^3 + \cdots \qquad (9)$$

In most cases, only the first two terms in this equation are measurable in the old, "field" millisecond pulsars. The exceptions are the globular cluster pulsars, whose spin period derivatives, $\dot{P}$, can be significantly affected by accelerations, $a = c\dot{P}/P$, caused by the mean cluster potential and distant encounters with other stars. Another possible exception is exemplified by the binary pulsar PSR B1620-26 in the M4 cluster, which has a third, long-period, Jupiter-mass companion (*Sigurdsson and Thorsett*, 2005). In this case, the presence of a planet in this hierarchical triple system has been deduced from a precise timing analysis of the dynamical "contamination" of the higher-order period derivatives (see also *Joshi and Rasio*, 1997).

Yet another potentially measurable contribution to the observed spindown rate, called the "Shklovskii effect" (*Shklovskii*, 1970), arises from the pulsar's transverse motion, which changes the apparent spin period by

$$\dot{P}/P = \frac{1}{c}\frac{v_t^2}{d} \qquad (10)$$

where v_t is the transverse velocity of the pulsar. The Shklovskii effect has to be taken into account in the case of nearby millisecond pulsars with small values of $\dot{P}$.

The topocentric TOA measurements are made at a telescope that takes part in Earth's rotation and in its motion within the solar system. To correct for the effects of these motions, the TOAs are transformed to the solar system barycenter (SSB), which represents a very good inertial reference frame. For the measured topocentric TOAs, the general form of their transformation to the corresponding barycentric pulse arrival times, t_B, is

$$t_B = t_{obs} + t_{clk} - (\mathbf{r}\times\mathbf{n})/c - D'/f^2 + t_{S\odot} + t_{E\odot} + t_R \qquad (11)$$

where t_{clk} represents the net clock correction that has to be applied to t_{obs}; r is the net vector from the observatory to the barycenter; n is the unit vector in the direction to the pulsar; D' is a constant related to a column density of interstellar electrons along the line of sight; f is the observing frequency, Doppler-shifted to the barycenter; $t_{E\odot}$ is the combined effect of gravitational redshift and time dilation in the solar system ("Einstein delay"); and $t_{S\odot}$ is the delay acquired by the signal while propagating through the Schwarzschild space in the solar system ("Shapiro delay"). The product $(\mathbf{r} \times \mathbf{n})/c$ represents the Römer delay of the signal, as it propagates between the telescope and the SSB, whereas t_R is the additional Römer delay due to a Keplerian orbital motion.

2.2.1. Clock corrections. The net clock correction, t_{clk}, in equation (10) consists of several steps that convert the topocentric TOA, t_{obs}, measured with an atomic clock at the observatory, to the Terrestrial Time (TT), which represents the geocentric time kept by a perfect clock on the terrestrial geoid. Corrections to the local time at the observatory are determined by comparing it to the Coordinated Universal Time (UTC), which is distributed by the U.S. National Institute of Standards and Technology (NIST) via the Global Positioning System (GPS). UTC itself is expressed as an integer number of seconds of the International Atomic Time (TAI), which is determined from an average of a number of terrestrial atomic clocks by the Bureau International des Poids et Mesures (BIPM) and remains within one second of UT1, the timescale based on the nonuniform Earth rotation. Finally, TT, the time the observed TOAs are corrected to, is provided by the BIPM as a retroactive, uniform time standard expressed in SI seconds and related to TAI by TT = TAI + 32.184 s.

2.2.2. Römer delay. The classical Römer delay, which is simply a light-travel time of the signal through the solar system, is taken into account in the process of transforming the observed TOAs to the SSB. This is done with the solar system ephemeris, which gives the distance **r** between the observer and the SSB, and with the assumed pulsar position, in ecliptic coordinates, which defines the vector **n**. The vector **r** is the sum of three vectors pointing from the observer to Earth's center, from there to the center of the Sun, and then to the SSB.

The first of these vectors depends on the Earth's radius and the elevation of the telescope, H. Given the mean radius of Earth, this correction amounts to 21.2 sin H ms.

To compute it with the required accuracy, one also has to include the effect of nonuniform Earth rotation, which is accomplished with the use of the published UT1 corrections.

The determination of the remaining two vectors requires a precise knowledge of masses and positions of the Sun and all the major bodies in the solar system. The most commonly used solar system ephemerides that provide these data are the DE200 and DE405 published by the Jet Propulsion Laboratory (JPL) (e.g., *Standish*, 1990). The vector from Earth to the Sun center is computed by taking into account both the motion of the Earth-Moon barycenter and the Earth motion in that system, whereas the vector R_B pointing from the center of the Sun to the SSB is derived from

$$R_B = \frac{1}{1+\Sigma m_i^{-1}} \Sigma \left(R_i m_i^{-1} \right) \quad (12)$$

by summing over all the relevant masses, m_i, and positions, R_i, of bodies in the solar system.

2.2.3. Dispersion delay. As discussed in section 2.1, the measured pulse TOAs are affected by dispersive (i.e., frequency dependent) delays induced by the presence of ionized gas along the line of sight to the pulsar. The way to compensate for this effect in the timing analysis is to shift the TOAs to an infinitely high frequency and remove the delay given by the term D'/f^2 in equation (10) (note that, from equation (3), $D' = D \times DM$).

Precision timing observations must take into account the fact that dispersion measure changes over time, as the line of sight to the moving pulsar traverses the interstellar medium. This is accomplished by measuring the TOAs at two or more frequencies, calculating DM corrections from equation (3), and applying them to TOAs either by modeling the DM changes or using appropriate time averages obtained from direct DM measurements. An example of a practical application of DM corrections to the TOA measurements of PSR B1257+12 is shown in Fig. 4 of section 3.

2.2.4. Einstein and Shapiro delays. The remaining two relativistic time delays that have to be corrected for, in order to refer the topocentric TOAs to the SSB, arise from the presence of masses in the solar system. These and other relativistic effects that influence the pulsar timing are discussed in detail by *Backer and Hellings* (1986).

The Shapiro delay is acquired by the signal that propagates through space-time, which is curved by a presence of sufficiently massive bodies. Ignoring the eccentricity of Earth's orbit and contributions to the delay by planets, the correction can be computed from

$$t_{S\odot} = \frac{2GM_\odot}{c^3} \ln(1+\cos\theta) \quad (13)$$

where G is the gravitational constant, $M_\odot$ is the mass of the Sun, and θ is the pulsar-Sun-Earth angle that is computed from the solar system ephemeris. The largest contribution to the Shapiro delay obviously comes from the Sun and amounts to ~120 μs, whereas Jupiter can contribute as much as ~200 ns to it.

The Einstein delay is a combined effect of time dilation and gravitational redshift of the signal due to the annual variation of a terrestrial atomic clock as Earth moves around the Sun on its eccentric orbit and to the presence of other masses in the solar system. Again, restricting the problem to contributions from the Sun, the delay is calculated from an integral of the equation

$$\frac{dt_{E\odot}}{dt} = \frac{2GM_\odot}{c^2}\left(\frac{1}{r}-\frac{1}{4a}\right) - \text{const} \quad (14)$$

where r is the Earth-Sun distance, and a is the semimajor axis of Earth's orbit. The maximum amplitude of this time varying correction is ~1.66 ms.

2.2.5. Position and proper motion. Another factor affecting the pulse arrival times that must be taken into account is the accuracy of the pulsar position. If the unit vector in the pulsar's direction is **n**′ rather than **n**, as in equation (10), then an annual sinusoidal term (**n**′−**n**) × r/c, which has a maximum amplitude of (1 AU/c) cos β ≈ 500 cos β s, adds to the arrival times. The effect depends on the ecliptic latitude, β, of the pulsar and may become quite large for objects located close to the ecliptic plane.

In addition, particularly in the case of high spatial velocity pulsars, their proper motion, μ_t, due to the transverse velocity

$$v_t = 4.74 \text{ km s}^{-1} \left(\frac{\mu_t}{\text{mas yr}^{-1}}\right)\left(\frac{d}{\text{kpc}}\right) \quad (15)$$

will make the vector **n** change in time, adding to the observed TOAs a corresponding sinusoidal variation with a gradually growing amplitude.

2.2.6. Keplerian orbital motion. The process of modeling the Keplerian orbits by means of the pulse timing observations of a varying Römer delay, t_R, in equation (10) begins with measurements of Doppler shifts of the pulsar period in response to the star's orbital motion. This approach, which is analogous to the single-line Doppler spectroscopy used in searches for planets around normal stars, is employed to obtain initial values of model parameters. The model orbits are then gradually refined using the much more precise pulse timing method.

As described in the chapter by Murray and Correia, in the case of a Keplerian binary orbit, a radial (line-of-sight) component, v_r, of the relative orbital velocity $V_r - V_\oplus$ between the pulsar and Earth varies as

$$v_r = \frac{2\pi a_1 \sin i}{P_b (1-e^2)^{1/2}} \left[\cos(\omega+\vartheta)+e\cos\omega\right] \quad (16)$$

where P_b is the orbital period, $a_1 \sin i$ is the projected semimajor axis, e is the eccentricity of the orbit, ω is the longitude of periastron, and ϑ is the true anomaly. The observed period of an orbiting pulsar will be Doppler-shifted according to

$$P = P_0 \left(1 + \frac{V_r}{c}\right) \quad (17)$$

where P_0 is the intrinsic pulsar period. Measuring P over a sufficient number of orbital periods allows a determination of orbital elements from equations (15) and (16). One can then make the standard assumption that the pulsar mass, $m_1 \sim 1.35\ M_\odot$ (*Thorsett and Chakrabarty,* 1999), and proceed to estimate the mass of a planetary companion from the mass function

$$f_1(m_1, m_2 \sin i) = \frac{(m_2 \sin i)^3}{(m_1 + m_2)^2} = \frac{4\pi^2}{G} \frac{(a_1 \sin i)^3}{P_b^2} \quad (18)$$

where G is the gravitational constant, and m_2 is the companion mass.

Because the pulsar timing involves counting of the individual pulses to make accurate and unambiguous predictions of the accumulated pulse phase (equations (7)–(8)), the estimated orbital parameters must be close enough to their true values to avoid phase ambiguities. With that accomplished, a Keplerian orbit of the pulsar is modeled in terms of its parameters as (*Blandford and Teukolsky,* 1976)

$$t_R = x(\cos E - e)\sin\omega + x \sin E \sqrt{1 - e^2} \cos\omega \quad (19)$$

where $x = a_1 \sin(i)/c$, and E, the eccentric anomaly, is related to the mean anomaly, $M = (2\pi/P_b)(t - T_0)$, through the equation

$$E - e \sin E = M \quad (20)$$

where t is the epoch of observation.

The extraordinary precision of the pulse timing method can be illustrated with the simple scaling relationship for a circular orbit, $i = 90°$, and $m_1 = 1.35\ M_\odot$

$$t_R \simeq 0.5\,s \left(\frac{a_2}{1\ AU}\right) \left(\frac{m_2}{M_{Jup}}\right) \left(\frac{m_1}{M_\odot}\right)^{-1} \quad (21)$$

where M_{Jup} is the Jupiter mass, a_2 is the semimajor axis of the planetary orbit, and $m_2 \ll m_1$. Indeed, using this equation, it is easy to verify that large asteroid-mass bodies in long-period orbits would be detectable around a pulsar that could be timed with microsecond precision (Fig. 2). Also note that, as in the case of planet detection with astrometry, and in contrast with the Doppler velocity method, sensitivity of the timing technique increases with the orbital size.

2.3. Time of Arrival Modeling in Practice

In summary of the discussion so far, the timing model of a pulsar consists of two spin parameters, ν and $\dot\nu$, four astrometric parameters, which are corrections to the pulsar position, $\Delta\alpha$ and $\Delta\delta$, proper motion parameters, μ_α and μ_δ, and the propagation parameter D'. The pulsar's orbital motion adds one or more Keplerian sets of orbital parameters to the model.

Practical translation of the topocentric TOAs to the arrival times at the barycenter, and all other necessary corrections defined in equation (10) and described in the following discussion, are accomplished by a timing modeling software that determines the model parameters as corrections to their initial values in the process of a linearized least-squares fit of the timing model to the arrival time data. A goodness of this fit is assessed from a χ^2 statistic given by

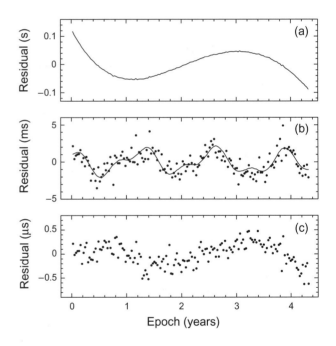

Fig. 2. Simulated TOA residuals for the solar system planets around a 1.35 $M_\odot$ pulsar after subtraction of the best-fit timing model including the initial phase, the pulsar spin period, and its slowdown rate. **(a)** All planets present: The residuals are dominated by Jupiter. Note that most of the Keplerian TOA variation due to Jupiter is absorbed in the linear and parabolic terms of equation (8). The remaining cubic term, not included in the model, represents the leftover contribution of this planet to TOA residuals in the form of a nonzero value of $\dddot\nu$. **(b)** Outer planets fitted out: The residuals for a slow pulsar (filled circles, timing accuracy 0.5 ms) show the presence of Venus and Earth. The solid line represents the same detection with a slow pulsar replaced with a millisecond pulsar (0.1-μs timing precision). **(c)** All major plants removed: The residuals from a millisecond pulsar (defined as above) reveal the presence of Ceres, the largest asteroid.

$$\chi^2 = \sum_{i=1}^{N} \left[\frac{\phi_i - N_i}{\sigma_i/P} \right]^2 \qquad (22)$$

where N_i is the closest integer number of pulses corresponding to each computed ϕ_i, and σ_i is the uncertainty of the arrival times, t_i. The most widely used computer code to perform this analysis, called TEMPO, is available at the Australia Telescope National Facility (ATNF) website (*www.atnf.csiro.au/research/pulsar/tempo*).

An application of the above procedure to pulsar planet detection is illustrated in Fig. 3. Note that the best-fit rms residual for a two-planet, initial Keplerian model based on Doppler shift measurements of the pulsar period (equation (16), Fig. 3a) amounts to only ~3 cm s^{-1}, which is still a factor of ~100 better than the typical radial velocity precision of planet searches in the optical domain! Further refinements of the model with the aid of the timing method (equations (11) and (19), Figs. 3b,c) resulted in a postfit residual of only 5 µs, which is equivalent to a S/N ≈ 300 detection of each of the two ~4 $M_\oplus$ planets orbiting the pulsar.

3. PULSAR PLANETS AND DISKS

3.1. Discovery

The story of the discovery of the first exoplanets began in early 1990, when a routine inspection of the 1000-ft Arecibo radio telescope revealed structural defects, which had developed over time as the result of material fatigue. With the memories of the recent dramatic collapse of the 300-ft Green Bank telescope still very much alive in the minds of the community, the only logical decision was to shut down the routine Arecibo operations to repair the damages. The process was estimated to last several weeks, and normal observations were obviously out of the question during that period.

The nature of the repairs had ruled out using the telescope in its normal tracking mode, but once a day, it could be "parked" at a desired azimuth and zenith angle and utilized as a transit instrument, which is a very efficient way to conduct large surveys under any circumstances. At the observing frequency of 430 MHz, a typical choice for pulsar searches at Arecibo in the 1980s, the 10-arcmin beamwidth of the telescope was equivalent to about 40 s integration time (depending on source declination), which was long enough to deliver an excellent ~1 mJy sensitivity of the detection of a pulsed radio signal.

One of us (A.W.), the resident Arecibo pulsar astronomer at the time, had long been interested in conducting a large survey to search for millisecond pulsars away from the galactic plane to test the idea that the local population of old neutron stars should be isotropically distributed over the sky (*Kulkarni and Narayan*, 1988). With the telescope operating normally, he would have very little chance to get enough observing time to carry out such a project. But, with the telescope practically unavailable to outside observers, his suggestion that a significant part of the available telescope time should be devoted to a large pulsar search was easily accepted. This is how he was granted about one-third of the time of the world's largest radio telescope for almost a month.

This fortunate circumstance had triggered a sequence of events starting with the discovery of a new millisecond pulsar, PSR B1257+12, and culminating in a surprising realization that this old neutron star is orbited by a system of three terrestrial-mass planets (*Wolszczan and Frail*, 1992; *Wolszczan*, 1994). For the most recent review of the pulsar planet research, the reader is referred to *Wolszczan* (2008).

3.2. PSR B1257+12 Planetary System

The 6.2-ms radio pulsar, PSR B1257+12, is orbited by three terrestrial-mass planets forming a compact system that is not much larger than the orbit of Mercury. Relative sizes of the orbits and a distribution of masses of planets b, c, and d are strikingly similar to those of the three inner planets in the solar system. The near 3:2 mean-motion resonance (MMR) between planets c and d and the existence of detectable gravitational perturbations between the two planets provide the mechanism to derive their masses without an *a priori* knowledge of orbital inclinations. The measured respective masses of 4.3 $M_\oplus$ and 3.9 $M_\oplus$ imply that the orbits of these planets are almost coplanar, which strongly suggests their disk origin (*Konacki and Wolszczan*, 2003).

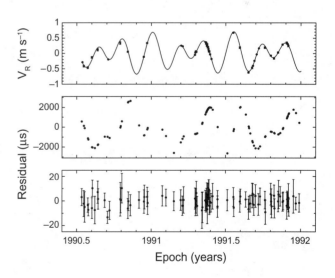

Fig. 3. A practical application of the timing method to detect pulsar planets. The process is illustrated with the first 18 months of observations of planets around PSR B1257+12 with the Arecibo radiotelescope at 430 MHz. **(a)** Period variations converted to the equivalent changes of the pulsar's radial velocity with the superimposed best-fit, two-planet model (solid curve). **(b)** The best-fit TOA residuals for a standard pulsar timing model without planets. **(c)** As above, with the two terrestrial-mass planets included in the fit.

Orbital parameters of the PSR B1257+12 planets are listed in Table 1. Postfit residuals for a complete timing model, including the pulsar spin parameters, time-varying interstellar dispersion, astrometric parameters, three planetary orbits, and the MMR-generated perturbations between planets c and d are shown in Fig. 4. Together with the true mass measurements of the planets, these results offer a fairly complete dynamical characterization of the first exoplanetary system known to contain terrestrial-mass planets. Its existence provides a demonstration that protoplanetary disks beyond the Sun can evolve Earth-mass planets to a dynamical configuration that is similar to that of the planets in the inner solar system.

Many possible formation pathways for this system have been proposed; *Phinney and Hansen* (1993) and *Podsiadlowski* (1993) summarize the vast variety of explanations that have been put forth. Some of these scenarios have been discarded after further theoretical investigation. But a few promising candidates remain.

Perhaps the supernova that engendered the pulsar helped form the planets. Some material from the supernova ejecta may have fallen back onto the neutron star, forming a protoplantary disk. Scenarios that envoke this possibility are called fallback scenarios. A simple fallback scenario, however, begs the question of why PSR B1257+12 is a millisecond pulsar, not a typical radio pulsar. Also popular are supernova recoil scenarios, in which the neutron star receives a kick that knocks it into a companion star, capturing matter from the star that then becomes a protplanetary disk. This kick geometry is obviously rare — but then, so are pulsar planets.

Other currently favored scenarios are the disrupted companion scenarios, where the progenitor of the system is a close binary containing the pulsar and a low-mass main-sequence star or a white dwarf. As the binary shrinks and the companion star overflows its Roche lobe, the material from the companion forms a protoplanetary disk. Fully degenerate or convective stars expand when they lose mass, so these stars can be disrupted completely by this process.

Theoretical investigations of the conditions of survival and evolution of pulsar protoplanetary disks (e.g., *Currie and Hansen*, 2007) suggest that unless they have exceptionally low viscosity, gas disks created by the above mechanisms should dissipate in $\sim 10^5$ yr. This timescale is probably too short for formation of gas giant planets. However, this time is sufficient to sediment solid grains that decouple from the gas, leaving solids behind to accrete into planetesimals and then planets, and conceivably to be detectable as a debris disk. If the disk is heavy in carbon and oxygen, e.g., from a disrupted white dwarf companion, that speeds the planet-formation process.

3.3. Jovian Planet Around PSR B1620-26

The second pulsar found to be orbited by a planet-mass body is a neutron star-white dwarf binary: PSR B1620-26 in the globular cluster M4. This pulsar has an outer, substellar-mass companion on a wide, moderately eccentric orbit. The substellar companion was detected because its dynamical influence on the pulsar induces accelerations that are measurable in the form of higher-order derivatives of its spin period, as shown in Fig. 5.

The inferred mass of the outer companion to PSR B1620-26 has been informed by detection of the pulsar's inner, white dwarf companion with the Hubble Space Telescope. The low mass (0.34 ± 0.04 M$_\odot$) and a relatively young age ($4.8 \times 10^8 \pm 1.4 \times 10^8$ yr) of the white dwarf support a formation scenario for the PSR B1620-26 system in which the planet starts out orbiting a main-sequence star in the cluster core and then survives an exchange interaction with a neutron star binary (*Sigurdsson et al.*, 2003).

The pulsar cannot be the original member of the system, because a substantial natal kick at the pulsar's birth would be required to prevent disruption of the observed tight binary. Such a kick would most likely lead to the ejection of the system from the cluster. More probably, the pulsar originated in a binary that had formed together with the cluster and evolved into a system of a spunup millisecond pulsar and a heavy white dwarf remaining in the cluster core.

In the exchange reaction, the original white dwarf companion to the neutron star was replaced by the planet's parent

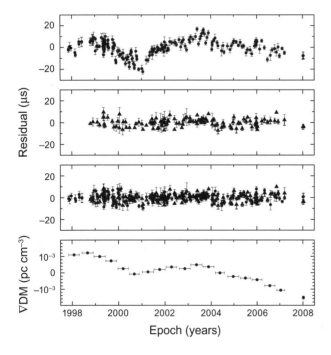

Fig. 4. The best-fit residuals for the timing model of PSR B1257+12 including the standard pulsar parameters, the three planets, and the perturbations between planets c and d. From top to bottom: residuals at 430 MHz (circles), residuals at 1400 MHz (triangles), and the best-fit to dual frequency data (circles and triangles) with the long-term dispersion measure variations (bottom panel) corrected for.

star — which then evolved into the white dwarf we see in the system today. Since PSR B1620-26 is quite bright and young, and so is its white dwarf companion, the system must have undergone an exchange interaction relatively recently (1–2 G.y. ago). The acquisition of a new companion, most likely a primordial (12.7 G.y.) main-sequence star near the turnoff mass, and a subsequent second spinup episode of the pulsar, must have caused a premature termination of the star's RGB evolution and the formation of an undermassive, relatively bright, young white dwarf, which is precisely as observed.

Given the mass of the pulsar's white dwarf companion one can estimate the orbital inclination of the inner binary $(55^{+14°}_{-8})$. This constrains the semimajor axis and the mass of the outer companion to be 23 AU and 2.5 ± 1 M_{Jup}, respectively. Since this mass is obviously in the giant planet range, the PSR B1620-26 planet is not only the oldest one detected so far, but it also represents a rare case of a planet born in a metal-poor environment.

3.4. Neutron Star Disks

The planetary system around PSR B1257+12 appears to have formed from an evolving disk similar to that invoked to describe planet formation around normal stars; the three scenarios for the formation of the planetary system around PSR B1257+12 that we discussed above are all essentially disk formation scenarios. The necessity of a disk follows from the low eccentricities and near coplanarity of the orbits of the planets. Disks around neutron stars are also invoked to explain the high spin rates of millisecond pulsars, which could have accreted angular momentum through a disk. But can we observe these putative neutron star disks?

To set the basic constraints on the observability of such a disk with a minimum number of assumptions, one can assume that a fraction f of the pulsar spindown luminosity L_{sd} heats N dust grains of size a to a temperature T_{dust}, and calculate the expected infrared flux as a function of disk parameters and pulsar distance from $fL_{sd} \sim 4\pi a^2 N\sigma T^4_{dust}$ (*Foster and Fischer*, 1996). These estimates can be compared with the upper flux limits derived from the attempts to detect the hypothetical circumpulsar disks with both the space- and the groundbased telescopes.

Another, less arbitrary, approach is to exploit the fact that, given the dust temperature, the ratio of dust luminosity, L_{dust}, to L_{sd} can be constrained from flux measurements (*Bryden et al.*, 2006). Of course, both methods must treat T as a free parameter, because in most cases, the primary source of energy to heat the dust is the relativistic pulsar wind, whose interaction with dust in a circumpulsar disk is not well understood. An illustration of the latter approach is given in Fig. 6, which shows upper limits to the L_{dust}/L_{sd} ratio over a range of temperatures for PSR B1257+12, derived from the existing observations. These limits rule out the presence of a dense, massive disk, but they do not exclude a possibility of an optically thin debris disk around the pulsar.

So far, the most credible evidence for a formation and survival of disks around neutron stars has come from the recent, still tentative detection of a cool, 10 $M_\oplus$ disk around a young X-ray pulsar, 4U 0142+61 (*Wang et al.*, 2006). The Spitzer telescope observation of a mid-infrared emission at the pulsar's position has been interpreted as the result of a passively illuminated dust orbiting the star at a distance of $\geq R_\odot$ (Fig. 7). This object is an anomalous X-ray pulsar (AXP), one of a class of young neutron stars whose X-ray luminosities greatly exceed their rates of rotational energy loss. These objects have pulsation periods of 5–12 s, much longer than those of millisecond pulsars like PSR 1257+12. The 4U 0142+61 disk could be related to the AXP phe-

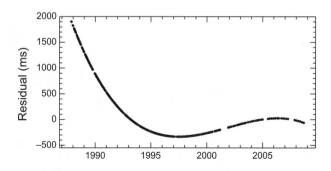

Fig. 5. The timing residuals for PSR B1620-26 after a fit for the spin period and its first derivative. The astrometric parameters and those of the inner orbit were held fixed at their best-fit values. Observations were made with the Jodrell Bank Lovell, Mark II, and Mark III telescopes, the VLA, and the Green Bank 140-ft and GBT telescopes. Courtesy of I. Stairs.

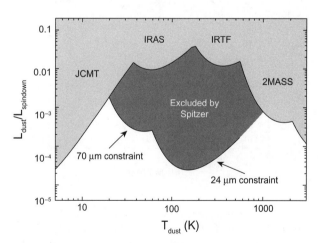

Fig. 6. Observational limits on the temperature and dust luminosity in the PSR B1257+12 planetary system. The light shaded region corresponds to limits set by observations previous to the most recent ones described in *Bryden et al.* (2006) and represented by the dark shaded region. Courtesy of G. Bryden.

nomenon. Nonetheless, the existence of this candidate disk suggests supports the idea that pulsars in general could form planets after the death of their progenitor stars.

4. PULSATING STARS AND CLOSE BINARIES

Pulsars are not the only degenerate objects in the universe that emit periodic signals; certain white dwarfs pulsate in nonradial g-modes with the stability needed to permit planet detection via the pulsation timing technique. These pulsating white dwarfs, called DAVs or ZZ Ceti stars, fall in a region of the Hertzprung-Russell diagram called the ZZ Ceti Instability strip, analogous to the instabililty strip containing Cepheids and RR Lyrae stars. The DAV pulsations have typical periods of 100–1500 s, amplitudes of a few percent, and pulsation period decay rates of $\dot{P} \sim$ a few × 10^{-15} (e.g., *Kepler et al.*, 2005).

So far, no planet has yet been detected around a white dwarf via this technique, although a promising candidate exists (*Mullally et al.*, 2008). There are also other signs of planetary systems around white dwarfs, like the planet orbiting PSR B1620-26 (see above). Moreover, roughly 10% of white dwarfs are now known to host circumstellar disks containing metal gas (*Gänsicke et al.*, 2006) or small silicate dust grains. These cool white dwarfs are thought to be remnants of planetary systems like the solar system; we will discuss them further below.

Meanwhile, the pulsation timing technique has revealed a planet orbiting yet a third kind of pulsating object; the sdB star V391 Pegasi (Fig. 8). This star hosts a planetary-mass body (M sin i = 3.2 M_{Jup}) orbiting at a semimajor axis of about 1.7 AU with a period of 3.2 yr (*Silvotti et al.*, 2007). The nonradial pulsations of sdB stars typically have periods of minutes to hours and a stability of $\dot{P} \sim 10^{-12}$.

V391 Pegasi is a kind of "extreme horizontal branch" star. Extreme horizontal branch stars are thought to be stars that evolved through the red giant branch, then somehow lost most of their hydrogen envelopes; they burn helium. This helium burning and lack of a substantial envelope make them extremely hot, with temperatures up to 35,000 K. Subdwarf B stars are thought to be low-mass extreme horizontal branch stars in which helium burning has just ignited. V391 Pegasi currently has a mass of about 0.5 $M_\odot$, but it burns blue-white with a luminosity of 15 $L_\odot$.

Possibly, V391 Pegasi shows us the future of an ordinary planetary system around an ordinary low-mass star. In this case, the relatively short-period orbit of the planet gives hope for the survival of planets near main-sequence stars; the orbit of the planet probably started at roughly 1 AU and expanded as the star lost mass (see below). Or maybe V391 Pegasi represents something much more exotic. The missing envelopes of sdB stars have led some authors to suggest that these objects form by losing mass to a binary

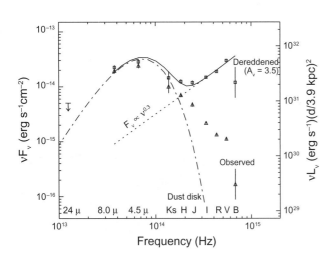

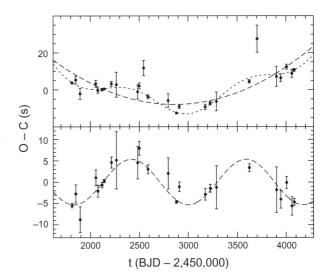

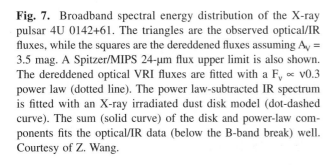

Fig. 7. Broadband spectral energy distribution of the X-ray pulsar 4U 0142+61. The triangles are the observed optical/IR fluxes, while the squares are the dereddened fluxes assuming A_V = 3.5 mag. A Spitzer/MIPS 24-μm flux upper limit is also shown. The dereddened optical VRI fluxes are fitted with a $F_\nu \propto \nu^{0.3}$ power law (dotted line). The power law-subtracted IR spectrum is fitted with an X-ray irradiated dust disk model (dot-dashed curve). The sum (solid curve) of the disk and power-law components fits the optical/IR data (below the B-band break) well. Courtesy of Z. Wang.

Fig. 8. Modeling of the pulsation timing measurements of the sdB subdwarf V391 Peg at the main period of 349.5 s. The upper panel shows the best fit to the data of a model consisting of a parabolic term to describe the long-term increase of the period due to evolutionary changes of the stellar structure, and a Keplerian, sinusoidal term to account for a wobble of the star around the barycenter induced by the presence of a planetary companion ($m_2 \sin i$ = 3.2 M_J). The lower panel shows the best-fit, sinusoidal residual after the removal of the long-term lengthening of the pulsation period. Courtesy of R. Silvotti.

companion, or perhaps from the merger of two helium white dwarfs (e.g., *Han et al.*, 2002).

Finally, there is another technique closely related to pulsation timing that has yielded candidate planet detections around some kinds of special binary stars. In a close binary system, stellar eclipses can serve as a frequent regular signal, like stellar pulsations, that can indicate the time variation of the light-travel time to the object. Two candidate planetary systems have been inferred so far via this method of eclipse timing: a pair of low-mass companions around the sdB eclipsing binary HW Virginis (*Lee et al.*, 2009) and a planet around the M4.5/M4.5 binary, CM Draconis (*Deeg et al.*, 2008).

The CM Draconis companion may have a period of 18.5 ± 4.5 yr and mass of about 3 M_{Jup}; a more massive companion with a longer period is another possible fit to the data. The HW Virginis companions have masses of m sin i = 19.2 M_{Jup} m sin i = 8.5 M_{Jup} and periods of 15.84 ± 0.14 and 9.08 ± 0.22 yr respectively. The presence of these candidates around close binaries reinforces one of the central lessons from the last 20 years of exoplanet observations; planetary systems are common and diverse and they inhabit all kinds of stellar systems.

Close binary stars are also known to host circumstellar dust. The RS CVn class of close binaries have an IR excess occurrence rate of about 40% based on an IRAS survey twice as high as that of main-sequence stars and 34 times higher than giants (*Busso et al.*, 1988). This class of binaries typically consists of a mid-F through mid-K giant plus a lower-mass main-sequence star; one of the components must be chromospherically active. The full definition of the class is given by *Hall* (1976).

The infrared excesses seen around these binaries are hot, typically ~1500 K. The dust in these systems could conceivably represent circumstellar envelopes built by magnetically driven stellar winds (*Scaltriti et al.*, 1993). Or perhaps these dusty close binaries host debris disks recently revived by the evolving binary, similar to the debris disks thought to orbit some white dwarfs (see below). Modeling and recent Spitzer observations of these systems suggest that the size of the central holes in these disks is consistent with dynamical clearing of debris by the binary (*Matranga et al.*, 2010).

5. FATES OF PLANETARY SYSTEMS

Pulsars, white dwarfs, and sdB stars all used to be main-sequence stars. So finding planetary systems around these objects begs us to ask, what happens to planetary systems as their stars evolve? And more specifically, what is the future and fate of our solar system?

5.1. Solar System

Let us start by looking at the Sun, a relatively low-mass star that will evolve into a white dwarf. The Sun will remain on the main sequence until it reaches an age of about 10 G.y., roughly 5.4 G.y. from now. At that age, it will have a radius of about 1.4 $R_\odot$, and a luminosity of about 1.84 $L_\odot$. *Gough* (1981) provides the following rule for the time variation of the solar luminosity, $L_{Sun}(t)$, while the Sun is in the main sequence

$$L_{Sun}(t) = \left(1 + 0.4(1 - t/t_0)\right)^{-1} L_\odot \quad (23)$$

where t_0 represents today, and $L_\odot$ refers strictly to today's solar luminosity.

Next, the Sun will switch from central hydrogen burning to hydrogen shell burning, and start climbing up the RGB branch. At an age of about 12.2 G.y., the Sun will reach the tip of the RGB, with a luminosity of 2730 $L_\odot$ and a radius of 256 $R_\odot$ or 1.19 AU. Clearly this expansion poses a potential problem for the survival of the planets of the solar system.

Long-term integrations of the orbits of the giant planets of the solar system suggest that they could survive for at least 10 G.y. from the present (*Duncan and Lissauer*, 1998). But whether or not the Earth becomes engulfed as the Sun expands is a subtle question. Solar mass loss causes the orbits of the planets to expand, while tides raised by the planets on the RGB Sun cause their orbits to decay. Because of the competition between these two phenomena, the survival of Earth has been a subject of debate.

If the mass loss is adiabatic, i.e., slow compared to the orbital time, the orbits of the planets expand in such a way to conserve the action of the planet's orbit, $\oint \dot{r} \, dr$, where r is the planet-star distance, as the star loses mass. The semimajor axis of the orbit, a, evolves as

$$\frac{a(t)}{a(t=0)} = \frac{M_\star(t=0)}{M_\star(t)} \quad (24)$$

Schröder and Cuntz (2007) provide a convenient expression for the mass loss from an RGB star assuming the winds are not dust-driven

$$\dot{M} = \eta \frac{L(t)_\star R(t)_\star}{M_\star(t)} \left(\frac{T_{eff}}{4000 \, K}\right)^{3.5} \left(\frac{1 + g_\odot}{4300 \, g_\star(t)}\right) \quad (25)$$

where $\eta \approx 8 \times 10^{-14} \, M_\odot \, y^{-1}$, $g_\odot$ is the gravitational acceleration at the solar surface, and $L_\star$, $M_\star$, $R_\star$, and $g_\star$ are in solar units. Assuming this relationship holds for the Sun, and neglecting tides, *Schröder and Smith* (2008) calculated that the Sun loses 0.332 $M_\odot$ by the time it reaches the tip of the RGB, and Earth's orbit expands to a semimajor axis of 1.50 AU.

On the other hand, the Sun will have virtually stopped spinning by the time it expands to the tip of the RGB. So the tidal bulge raised on the RGB Sun by Earth will lag behind Earth in its orbit, exerting a torque on it with a typical value of $\Gamma = -3.3 \times 10^{26}$ kg m^2 s^{-2}. This torque, integrated over the RGB evolution of the Sun, seems likely to be enough to dump Earth into the Sun's expanded envelope, despite the Sun's mass loss.

In general, although planets orbiting closer than ~5 AU from a solar-type star seem likely to be lost during post-main-sequence evolution, planets beyond ~5 AU should be able to survive — or at least some of the planets can (*Debes and Sigurdsson*, 2002). Multiple-planet systems can become destabilized during stellar mass loss. Because of chaos in the n-body problem, the exact consequences of this instability for a given planet cannot be predicted. But typically, one or more of the surviving planets will be ejected, or sent crashing into the Sun or into another planet, until relative stability returns.

Exactly which solar system planets become engulfed by the Sun may not really matter; life as we know it will end long before all these dramatic events occur. A mere 10% increase in the solar flux will likely cause rapid loss of Earth's oceans via photolysis of water in the stratosphere followed by the escape of the hydrogen to space (*Kasting*, 1988). Using the rule above, we find that catastrophic ocean loss could occur in about 200 m.y. Moreover, even if humans manage to survive this calamity, they will probably find it uncomfortable when the continued increase in the Sun's luminosity tips Earth's atmosphere into a runaway greenhouse state shortly thereafter, and the former biosphere of Earth comes to resemble the surface of Venus.

5.2. Disks Around White Dwarfs

Besides the search for planets around white dwarfs via pulsation timing, there have been many searches for brown dwarfs and planets aiming to directly detect light from low-mass companions. Some surveys use infrared photometry to search for unresolved thermal emission in excess of emission from a white dwarf's photosphere (*Mullally et al.*, 2007). Other surveys have tried to directly image faint point sources orbiting white dwarfs using modern point-source detection algorithms (e.g., *Hogan et al.*, 2009), or coronagraphy (e.g., *Debes et al.*, 2005).

Although these surveys have not yet turned up any planets, they have uncovered a handful of brown dwarf companions to white dwarfs; about 0.5% of white dwarfs seem to have brown dwarf companions (*Farihi et al.*, 2005). We recognize these brown dwarf companions by the distinctive bump in their spectra around 4–5 μm between absorption bands of methane and water (e.g., *Sudarsky et al.*, 2003). The spectra of jovian planets also contain this feature, but planets would create less infrared excess than brown dwarfs at a given age — unless they have been reheated, e.g., via collisions during the star's post-main-sequence evolution.

Some of these searches have also turned up white dwarfs that apparently host circumstellar dust disks. The first discovered white dwarf with infrared excess, G 29-38, was originally thought to host a brown dwarf (*Zuckerman and Becklin*, 1987). But now we know from spectroscopy of this object with the Spitzer Space Telescope that the 4–5-μm bump characteristic of a brown dwarf is absent, and instead the spectrum shows a prominent bump at 10 μm, indicating much of the excess radiation from G 29-38 comes from small silicate grains (*Reach et al.*, 2005). Figure 9 shows the spectrum of this infrared excess, compared to a model that combines an optically thick, physically thin disk with an optically thin cloud of olivine and bronzite (*Reach et al.*, 2009).

Twelve similar cool (<20,000 K) white dwarfs with infrared excesses consistent with circumstellar dust are now known. Five of these (including G 29-38) are now known to show 10-μm silicate emission features. In the case of G 29-38, there is evidence pointing to a disk geometry for this dust. Some of the white dwarf's pulsation modes are seen strongly in the near-IR flux from this source — radiation that is reprocessed by the dust. Other modes are almost absent in the radiation from the dust, suggesting that the signal in these modes averages out over the dust cloud. This evidence shows that the dust geometry is very different from a spherical cloud (*Graham et al.*, 1990).

What is the origin of these dusty white dwarf disks? Many of the small bodies — comets and Kuiper belt objects — in the solar system should survive the post-main-sequence evolution of the Sun, at first. But as the Sun loses mass, and the orbits of the planets become unstable, planets can become scattered into highly eccentric orbits. Such scattered planets can, in turn, destabilize the Kuiper belt, the asteroid belt, and the Oort cloud, leading to a high flux of comets into the inner solar system, a kind of second late heavy bombardment starting 10^7–10^8 yr after the mass loss phase (*Debes and Sigurdsson*, 2002). When they approach the white dwarf, comets and asteroids should be shredded by the strong tides near the white dwarf, just as Comet Shoemaker-Levy 9 was tidally shredded when as it approached Jupiter in 1994.

Such tidally disrupted asteroids or comets may be the origin of the disks observed around white dwarfs. These

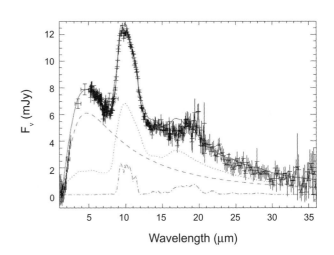

Fig. 9. The infrared spectrum of the white dwarf G29-38 minus the photospheric emission from 2MASS, IRTF, and Spitzer. The solid line shows a best-fitting model, with a physically thin, optically thick disk plus an optically thin cloud. Dashed, dotted, and dot-dash lines show contributions to the model from the optically thick component (dashed), amorphous olivine (dotted), and crystalline bronzite (dash-dot). Courtesy of W. Reach.

disks typically extend outward to roughly the Roche radius of the white dwarf (*Jura*, 2003), the radius interior to which an asteroid would be torn apart by tides. For an asteroid of density ρ_a near a star of density $\rho_\star$ and radius $R_\star$, this radius is (*Davidsson*, 1999)

$$R_{tide} = C_{tide}\left(\frac{\rho_\star}{\rho_a}\right)R_\star \qquad (26)$$

where C_{tide} is a numerical constant on the order of unity that depends on the orbital parameters of the asteroid, its rotation, and its composition. For an asteroid with $\rho_a = 3$ g cm^{-3} around G 29-38, R_{tide} works out to about 1.5 $C_{tide}R_\odot$.

For more luminous white dwarfs, the sublimation radius for silicate dust exceeds R_{tide}. These objects have been observed to mostly lack detectable circumstellar dust (*Kilic et al.*, 2006), as one might expect. Instead, four of these moderately hot ($T_{eff} \geq 15,000$ K) white dwarfs show double-peaked calcium lines indicative of rotating *gaseous* circumstellar disks. One unusual object, the white dwarf SDSSJ1228+1040, seems to have both circumstellar gas and dust coexisting in a disk extending out to about 1.2 $R_\odot$ (*Brinkworth et al.*, 2009).

Besides these cool white dwarfs with detectable circumstellar material, an even larger fraction of white dwarfs than the subset seen to have infrared excesses have atmospheres polluted by metals. These metals are expected to sink out of a white dwarf's atmosphere on timescales of 1 yr, implying that whatever metals we see were recently accreted, perhaps from comets or planetesimals. Perhaps all these metal-rich white dwarfs host remnants of planetary systems, although accretion of metals from the ISM is another possible source for these photospheric contaminants.

Another scenario to generate disks around massive white dwarfs is the merger of two white dwarfs (*Livio et al.*, 2005). This mechanism could produce a gas disk formed from white dwarf material orbiting a single, massive white dwarf. So far, all the white dwarfs thought to have disks have masses <0.9 M$_\odot$ (*Kilic et al.*, 2008); none is massive enough to point to this mechanism for disk formation. But such massive white dwarfs are rare, and few have been searched for circumstellar dust; perhaps future infrared surveys will spot white dwarf merger disks.

5.3. Planetary Systems Around More Massive Stars

The fate of the solar system, gloomy as it is, pales in contrast to the fate of planetary systems around more massive stars, the kind that die as supernovae, leaving pulsars as remnants (*Thorsett and Dewey*, 1993). The asymmetry of supernova explosions means pulsars are born with a kick, a velocity of typically ~200 km s^{-1} compared to their original frame. This kick occurs over a timescale of minutes, much less than any planet's orbital period, and it is much greater than the orbital velocity of any ordinary planet; it would probably unbind any planetary system originally orbiting the star. So any planets that survived the post-main-sequence swelling of the massive star and the blast of the supernova wind would either be lost or kicked into highly eccentric orbits. The pulsar planets, for this reason, are thought to be second-generation planets, formed in a disk of material that accreted onto the pulsar after the supernova.

But even planets that form around a neutron star live in a dangerous neighborhood, as the "dead worlds" headline alludes to. Photon luminosity produced by accretion and high-energy particles produced by pulsar spindown can ablate and evaporate planetesimals and planets. Forming any planets may require a disk with at least 2–4 M$_\oplus$ of material to shield planetesimals from these relativistic particles (*Miller and Hamilton*, 2001). The presence of the innermost planet in the PSR 1257+12 system tells us that the accretion activity that spun up the pulsar must have occurred before the planet formed — otherwise the accretion energy likely would have evaporated it!

5.4. Exotic Planet Chemistries

Whatever their formation mechanism, disks around pulsars probably have chemical compositions quite different from the compositions of ordinary protoplanetary disks around young stars. For example, a pulsar disk formed from a shredded white dwarf would likely be hydrogen-poor but rich in helium, carbon, oxygen, etc., depending on the mass of the white dwarf. Since planets form out of disk material, this possibility has led some to speculate about the possibility of gas planets made of carbon dioxide (*Livio et al.*, 2005), or rocky planets made of carbides and diamond (*Kuchner and Seager*, 2005).

The possibility of carbon-rich disks essentially splits planet formation into two domains. With a dissociation energy of 11.108 eV (for ^{12}C^{16}O) (*Huber and Herzberg*, 1979), carbon monoxide is the most tightly bound molecule in astrophysics; in protoplanetary disks, carbon monoxide tends to form from all the available carbon and oxygen, and whatever is left over — carbon or oxygen — goes into making solids. Because of this effect, the chemistry of planet formation contains a kind of switch: When the C:O ratio is <1, solid planets should form substantially from oxygen compounds, like silicates, but when the C:O ratio is >1, carbon compounds should begin to dominate.

What other kinds of exotic planet chemistries could there be? The solar system contains planets mostly made of hydrogen (Jupiter), planets made mostly of silicates (Earth), and planets made substantially of water and other ices (Neptune). If you could strip away the mantle of a terrestrial planet, leaving its core, you would have a planet made mostly of iron; this process may have been the origin of the high uncompressed density of Mercury. One could also imagine planets formed from the remnants of supergiant stars that burn via the CNO cycle. This process produces copious amounts of nitrogen.

Seager et al. (2007) have modeled the mass-radius relationships for planets of a wide variety of compositions.

Extremely dense (e.g., Fe) planets are easy to recognize from their masses and radii, which can be determined directly via transit and Doppler observations. Other bulk compositions, or mixed compositions, are impossible to discern exactly, and can only be inferred from context. For planets in the solar system and for exoplanets as well, it is hard to tell what is in the center of a planet when you can only see the outside. Nonetheless, the range of possibilities is entertaining to contemplate, especially in a discussion about the chemical origins of life.

6. FUTURE PROSPECTS

Millisecond pulsars are rare, which, aside from the physics of planet formation, may contribute to the observed rarity of pulsar planets. Except for PSR B1257+12, no planetary companions have emerged from the precision timing of the other 65 known field MSPs that, unlike PSR B1620-26 discussed in section 3.3, are not located in globular clusters (*Lorimer*, 2008). Among these, 16 pulsars (including PSR B1257+12) are solitary in the sense that they do not have binary stellar companions. If this is the requirement for a millisecond pulsar to have planets, because, for example, the material for a protoplanetary disk is typically supplied by an obliterated stellar companion, the frequency of pulsar planets is most probably lower that 5%. Similar estimates appear to emerge from the recent H-band imaging search for substellar companions to young, isolated neutron stars with the Very Large Telescope (*Posselt et al.*, 2009).

The statistics of planets around pulsars should eventually improve as continuing pulsar surveys, such as the Arecibo PALFA project (*Cordes et al.*, 2006), discover new pulsars. Better yet, the proposed Square Kilometer Array (SKA) promises to be a pulsar powerhouse. By reasonable estimates, the SKA should be able to find 15,000 new pulsars; this number would be limited mostly by the computational power required to form the synthesized telescope beam (*Smits et al.*, 2009). Assuming again that the isolated MSPs are the best candidates for planetary companions, and using the existing statistics, one can expect that the SKA will discover some 250 isolated MSPs, out of which ~10 or so may have planets around them. The discovery process may get accelerated by the two precursors to the SKA that are presently under construction — South Africa's Karoo Array Telescope, known as MeerKAT, and the Australian Square Kilometre Array Pathfinder (ASKAP) — and should also find many new pulsars.

In the absence of new detections of planets around pulsars, further searches for dust around these objects are a logical way to further constrain the models of a creation and evolution of pulsar protoplanetary disks and pulsar planets themselves. The existing IR flux limits for PSR B1257+12 (and a few other pulsars) rule out massive, ~0.01 $M_\odot$ disks similar to those thought to give rise to planets around normal stars. However, observational constraints do not contradict the possibility that some pulsars may be accompanied by much less massive disks with masses ranging from a fraction of the asteroid belt mass to a few hundred $M_\oplus$ over a wide range of temperatures and grain sizes. This conclusion is consistent with the masses of the PSR B1257+12 planets and their disk origin deduced from timing observations of the pulsar, and with the theoretical constraints on neutron star planet formation scenarios outlined above. Searches for disks around pulsars should benefit greatly from the high sensitivity and angular resolution of the James Webb Space Telescope (JWST) and the Atacama Large Millimeter Array (ALMA), which are under construction.

The use of pulsation timing to search for planets around white dwarfs and other evolved stars is a new trick, bound to yield more discoveries soon. These techniques require special patience, since planets are thought to be excluded from short-period orbits around these evolved stars. The Kepler mission, recently launched, will be one important tool for studying planets around evolved stars via transit photomery. The Kepler field is expected to contain roughly 350 eclipsing binaries. Kepler will be sensitive to both transit timing variations of the eclipsing binaries, and also transits of planets across the stellar disks in these systems.

Direct imaging searches for planets around white dwarfs will benefit greatly from the next generation of adaptive optics instruments on large groundbased telescopes. For example, the SEEDS survey using the High Contrast Instrument for the Subaru Next Generation Adaptive Optics (HiCIAO) instrument on the Subaru telescope plans to include a substantial white dwarf component. Studies of dust around white dwarfs should also benefit greatly from JWST, while ongoing studies of white dwarf photospheres with optical spectroscopy from existing large telescopes should help us understand the chemistry and variability of white dwarf accretion disks. In the more distant future, the Laser Interferometer Space Antenna (LISA) could conceivably detect planets around double white dwarf binaries via the phase modulation they should induce in the gravitational waves from these systems (*Seto*, 2008).

Acknowledgments. This work was partially supported by the NASA Astrobiology Institute. The Center for Exoplanets and Habitable Worlds is supported by the Pennsylvania State University, the Eberly College of Science, and the Pennsylvania Space Grant Consortium. Thanks to J. Debes, T. Currie, and two anonymous referees for reading drafts.

REFERENCES

Backer D. C. and Hellings R. W. (1986) Pulsar timing and general relativity. *Annu. Rev. Astron. Astrophys., 24,* 537–575.

Blandford R. and Teukolsky S. A. (1976) Arrival-time analysis for a pulsar in a binary system. *Astrophys. J., 205,* 580–591.

Brinkworth C. S., Gänsicke B. T., Marsh T. R., Hoard D. W., and Tappert (2009) A dusty component to the gaseous debris disk around the white dwarf SDSS J1228+1040. *Astrophys. J., 696,* 1402–1406.

Bryden G., Beichman C. A., Rieke G. H., Stansberry J. A., Stapelfeldt K. R., et al. (2006) Spitzer MIPS limits on asteroidal dust in the pulsar planetary system PSR B1257+12. *Astrophys. J., 646,* 1038–1042.

Busso M., Scaltriti F., Origlia L., Persi P., Ferrari-Toniolo M. (1988) IRAS observations and IR excesses of RS CVn-type binaries. *Mon. Not. R. Astron. Soc., 234,* 445–457.

Cordes J. M., Freire P. C. C., Lorimer D. R., Camilo F., Champion D. J., et al. (2006) Arecibo pulsar survey using ALFA. I. Survey strategy and first discoveries. *Astrophys. J., 637,* 446–455.

Currie T. and Hansen B. (2007) The evolution of protoplanetary disks around millisecond pulsars: The PSR 1257+12 system. *Astrophys. J., 666,* 1232–1244.

Davidsson B. J. R. (1999) Tidal splitting and rotational breakup of solid spheres. *Icarus, 142,* 525.

Debes J. H. and Sigurdsson S. (2002) Are there unstable planetary systems around white dwarfs? *Astrophys. J., 572,* 556–565.

Debes J. H., Sigurdsson S., and Woodgate B. E. (2005) Cool customers in the stellar graveyard. II. Limits to substellar objects around nearby DAZ white dwarfs. *Astron. J., 130,* 1221–1230.

Deeg H. J., Ocaña B., Kozhevnikov V. P., Charbonneau D., O'Donovan F. T., and Doyle L. R. (2008) Extrasolar planet detection by binary stellar eclipse timing: Evidence for a third body around CM Draconis. *Astron. Astrophys., 480,* 563–571.

Demircan O., Eker Z., Karatas Y., and Bilir S. (2006) Mass loss and orbital period decrease in detached chromospherically active binaries. *Mon. Not. R. Astron. Soc., 366,* 1511–1519.

Duncan M. J. and Lissauer J. J. (1998) The effects of post-main-sequence solar mass loss on the stability of our planetary system. *Icarus, 134,* 303–310.

Farihi J., Becklin E. E., and Zuckerman B. (2005) Low-luminosity companions to white dwarfs. *Astrophys. J. Suppl., 161,* 394–428.

Foster R. S. and Fischer J. (1996) Search for protoplanetary and debris disks around millisecond pulsars. *Astrophys. J., 460,* 902.

Gänsicke B. T., Marsh T. R., Southworth J. and Rebassa-Mansergas A. (2006) A gaseous metal disk around a white dwarf. *Science, 314,* 1908.

Gough D. O. (1981) Solar interior structure and luminosity variations. *Solar Phys., 74,* 21–34.

Graham J. R., Matthews K., Neugebauer G., and Soifer B. T. (1990) The infrared dxcess of G29-38 — A brown dwarf or dust? *Astrophys. J., 357,* 216–223.

Hall D. S. (1976) The RS CVn binaries and binaries with similar PROPERTIS. In *Multiple Periodic Variable Stars* (W. S. Fitch, ed.), p. 287. IAU Colloquium No. 29, Astrophysics and Space Science Library, Vol. 60.

Han Z., Podsiadlowski Ph., Maxted P. F. L., Marsh T. R., and Ivanova N. (2002) The origin of subdwarf B stars — I. The formation channels. *Mon. Not. R. Astron. Soc., 336,* 449–466.

Hogan E., Burleigh M. R., and Clarke F. J. (2009) The DODO Survey II: A Gemini direct imaging search for substellar and planetary mass companions around nearby equatorial and northern hemisphere white dwarfs. *Mon. Not. R. Astron. Soc., 396,* 2074.

Huber K. and Herzberg G. (1979) *Molecular Spectra and Molecular Structure IV: Constants of Diatomic Molecules.* Van Nostrand, New York.

Joshi K. J. and Rasio F. A. (1997) Distant companions and planets around millisecond pulsars. *Astrophys. J., 479,* 948–959.

Jura M. (2003) A tidally disrupted asteroid around the white dwarf G29-38. *Astrophys. J. Lett., 584,* L91–L94.

Kasting J. F. (1988) Runaway and moist greenhouse atmospheres and the evolution of Earth and Venus. *Icarus, 74,* 472–494.

Kepler S. O., Costa J. E. S., Castanheira B. G., Winget D. E., Mullally F., et al. (2005) Measuring the evolution of the most stable optical clock G117-B15A. *Astrophys. J., 634,* 1311–1318.

Kilic M., von Hippel T., Leggett S. K., and Winget D. E. (2006) Debris disks around white dwarfs: The DAZ connection. *Astrophys. J., 646,* 474–479.

Kilic M., Thorstensen J. R., and Koester D. (2008) Direct distance measurement to the dusty white dwarf GD 362. *Astrophys. J. Lett., 689,* L45–L47.

Konacki M. and Wolszczan A. (2003) Masses and orbital inclinations of planets in the PSR B1257+12 system. *Astrophys. J. Lett., 591,* L147–L150.

Kuchner M. and Seager S. (2005) Extrasolar carbon planets. *ArXiv Preprints,* astro-ph/0504214.

Kulkarni S. R. and Narayan R. (1988) Birthrates of low mass binary pulsars and low-mass X-ray binaries. *Astrophys. J., 335,* 755–768.

Lee J. W., Kim S.-L., Kim C.-H., Koch R. H., Lee C.-Uk, et al. (2009) The sdB+M eclipsing system HW Virginis and its circumbinary planets. *Astron. J., 137,* 3181–3190.

Livio M., Pringle J. E., and Wood K. (2005) Disks and planets around massive white dwarfs. *Astrophys. J. Lett., 632,* L37–L39.

Lorimer D. R. (2008) Binary and millisecond pulsars. In *Living Reviews in Relativity,* available online at *relativity.livingreviews.org/Articles/lrr-2008-8/.*

Lorimer D. R. and Kramer M. (2001) *Handbook of Pulsar Astronomy.* Cambridge Univ., Cambridge.

Lyne A. G. and Smith F. G. (2006) *Pulsar Astronomy.* Cambridge Univ., Cambridge.

Malhotra R., Black D., Eck A., and Jackson A. (1992) Resonant orbital evolution in the putative planetary system of PSR1257+12. *Nature, 356,* 583.

Matranga M., Drake J. J., Kashyap V. L., Mareng M., and Kuchner M. J. (2010) Close binaries with infrared excess: Destroyers of worlds? *Astrophys. J.,* in press.

Matsakis D. N., Taylor J. H., and Eubanks T. M. (1997) A statistic for describing pulsar and clock stabilities. *Astron. Astrophys., 326,* 924–928.

Miller M. C. and Hamilton D. P. (2001) Implications of the PSR 1257+12 planetary system for isolated millisecond pulsars. *Astrophys. J., 550,* 863–870.

Mullally F., Kilic M., Reach W. T., Kuchner M. J., von Hippel T., Burrows A., and Winget D. E. (2007) A Spitzer white dwarf infrared survey. *Astrophys. J. Suppl., 171,* 206–218.

Mullally F., Winget D. E., De Gennaro S., Jeffery E., Thompson S. E., et al. (2008) Limits on planets around pulsating white dwarf stars. *Astrophys. J., 676,* 573–583.

Peale S. J. (1993) On the verification of the planetary system around PSR 1257+12. *Astron. J., 105,* 1562–1570.

Phinney E. S. and Hansen B. M. S. (1993) The pulsar planet production process. In *Planets Around Pulsars* (J. A. Phillips et al., eds.), p. 371. ASP Conference Series 36, Astronomical Society of the Pacific, San Francisco.

Podsiadlowski P. (1993) Planet formation scenarios. In *Planets Around Pulsars* (J. A. Phillips et al., eds.), p. 149. ASP Conference Series 36, Astronomical Society of the Pacific, San Francisco.

Posselt B., Neuhäuser R., and Haberl F. (2009) Searching for substellar companions of young isolated neutron stars. *Astron. Astrophys., 496,* 533–545.

Rasio F. A., Nicholson P. D., Shapiro S. L., and Teukolsky S. A. (1992) An observational test for the existence of a planetary system orbiting PSR1257+12. *Nature, 355,* 325.

Reach W. T., Kuchner M. J., von Hippel T., Burrows A., Mullally F., Kilic M., and Winget D. E. (2005) The dust cloud around the white dwarf G29-38. *Astrophys. J. Lett., 635,* L161–L164.

Reach W. T., Lisse C., von Hippel T., and Mullally F. (2009) The dust cloud around the white dwarf G 29-38. II. Spectrum from 5 to 40 μm and mid-infrared photometric variability. *Astrophys. J., 693,* 697–712.

Scaltriti F., Busso M., Ferrari-Toniolo M., Origlia L., Persi P., Robberto M., and Silvestro G. (1993) Evidence from infrared observations of circumstellar matter around chromospherically active binaries. *Mon. Not. R. Astron. Soc., 264,* 5–15.

Schröder K.-P. and Connon Smith R. (2008) Distant future of the Sun and Earth revisited. *Mon. Not. R. Astron. Soc., 386,* 155–163.

Schröder K.-P. and Cuntz M. (2007) A critical test of empirical mass loss formulas applied to individual giants and supergiants. *Astron. Astrophys., 465,* 593–601.

Seager S., Kuchner M., Hier-Majumder C. A., and Militzer B. (2007) Mass-radius relationships for solid exoplanets. *Astrophys. J., 669,* 1279–1297.

Seto N. (2008) Detecting planets around compact binaries with gravitational wave detectors in space. *Astrophys. J., 677,* 55.

Shklovskii I. S. (1970) Possible causes of the secular increase in pulsar periods. *Soviet Astron., 13,* 562.

Sigurdsson S. and Thorsett S. E. (2005) Update on pulsar B1620-26 in M4: Observations, models, and implications. In *Binary Radio Pulsars* (F. A. Rasio and I. H. Stairs, eds.), p. 213. ASP Conference Series 328, Astronomical Society of the Pacific, San Francisco.

Sigurdsson S., Richer H. B., Hansen B. M., Stairs I. H., and Thorsett S. E. (2003) A young white dwarf companion to pulsar B1620-26: Evidence for early planet formation. *Science, 301,* 193–196.

Silvotti R., Schuh S., Janulis R., Solheim J.-E., Bernabei S., et al. (2007) A giant planet orbiting the "extreme horizontal branch" star V391 Pegasi. *Nature, 449,* 189–191.

Smits R., Kramer M., Stappers B., Lorimer D. R., Cordes J., and Faulkner A. (2009) Pulsar searches and timing with the square kilometre array. *Astron. Astrophys., 493,* 1161–1170.

Standish E. M. Jr. (1990) The observational basis for JPL's DE 200, the planetary ephemerides of the Astronomical Almanac. *Astron. Astrophys., 233,* 252–271.

Sudarsky D., Burrows A., and Hubeny I. (2003) Theoretical spectra and atmospheres of extrasolar giant planets. *Astrophys. J., 588,* 1121–1148.

Thorsett S. E. and Chakrabarty D. (1999) Neutron star mass measurements. I. Radio pulsars. *Astrophys. J., 512,* 288–299.

Thorsett S. E. and Dewey R. J. (1993) Limits on planets orbiting massive stars from radio pulsar timing. *Astrophys. J. Lett., 419,* L65.

Thorsett S. E., Arzoumanian Z., Camilo F., and Lyne A. G. (1999) The triple pular system PSR B1620-26 in M4. *Astrophys. J., 523,* 763–770.

Wang Z., Chakrabarty D., and Kaplan D. L. (2006) A debris disk around an isolated young neutron star. *Nature, 440,* 772–775.

Wolszczan A. (1994) Confirmation of Earth-mass planets orbiting the millisecond pulsar PSR B1257+12. *Science, 264,* 5.

Wolszczan A. (2008) Fifteen years of the neutron star planet research. *Phys. Scripta, 130,* 014005.

Wolszczan A. and Frail D. A. (1992) A planetary system around the millisecond pulsar PSR1257+12. *Nature, 355,* 145–147.

Wolszczan A., Hoffman I. M., Konacki M., Anderson S. B., and Xilouris K. M. (2000) A 25.3 day periodicity in the timing of the pulsar PSR B1257+12: A planet or a heliospheric propagation effect? *Astrophys. J. Lett., 540,* L41–L44.

Zuckerman B. and Becklin E. E. (1987) Excess infrared radiation from a white dwarf — an orbiting brown dwarf? *Nature, 330,* 138–140.

Statistical Distribution of Exoplanets

Andrew Cumming
McGill University

This chapter discusses the current statistical sample of exoplanets. We discuss the selection effects in radial velocity and transit surveys, followed by a brief introduction to statistical techniques for characterizing the orbital properties of planets and how to include completeness corrections in population studies. We then highlight the major features of the planet population discovered so far, discuss some of the implications for planet formation theories, and the future prospects for increasing the sample of known planets.

1. INTRODUCTION

The last 15 years have seen a tremendous rate of discovery of exoplanetary systems. At the time of writing in September 2010, the Extrasolar Planets Encyclopedia website reports 490 planets in 413 planetary systems (*http://exoplanet.eu*, maintained by J. Schneider, Paris Observatory, accessed on September 1, 2010). These planets have masses that range from 3 $M_\oplus$ and up, orbital periods from close to one day to several years, and a wide range of eccentricities. The number of confirmed multiple planet systems is 49, with many other single planet systems showing evidence for an additional companion at long orbital periods. This large number of planets has allowed the first studies of the distribution of orbital periods, eccentricities, and planet masses, as well as the dependence on host star properties. For example, it is now well established that the planet occurrence rate increases strongly with the metallicity of the host star. The number of transiting planets is over 100, allowing studies of the mass-radius relation for hot Jupiters, and the first glimpses of exoplanet atmospheres.

One reason to be interested in the statistical properties of exoplanets is to answer the question of how common planetary systems are, particularly those with habitable planets. We now know that at least 10% of solar type stars, perhaps up to 20%, harbor gas giant planets, and information about lower masses will follow. Already, lower-mass planets with masses ~10 $M_\oplus$ comparable to Neptune are being studied in close orbits. Systems similar to our own solar system, dominated in mass by Jupiter at 5 AU from the Sun, are beginning to be found, including a Jupiter/Saturn analog recently detected in a microlensing event. The frequency of planets determined from current data is an important input for future surveys, such as astrometric and direct searches.

Another reason for studying the statistical distributions of planet properties is that they offer us a tremendous amount of information about the process of planet formation. Planets likely form in the outer regions of protoplanetary disks and migrate inward. The radial distribution of observed planets offers clues to the physics of migration and the process by which it stops close to the star. The increasing planet occurrence rate with stellar metallicity and stellar mass is likely directly related to the amount of planet-building materials in the protoplanetary disk. The nonzero eccentricities of most exoplanets perhaps point to planet-planet scattering or other gravitational interactions as an important process early in the life of most planetary systems. Part of the puzzle is to understand how our solar system, in which the planets have nearly circular orbits and the gas giants orbit at larger distances than most of the exoplanets so far discovered, fits into this picture.

When studying the statistical distributions of planet properties, we must be careful to account for selection effects. For example, in radial velocity searches, the velocity precision (typically $\sigma \approx 1$–3 m/s) sets a lower limit on the amplitude of velocity variations that can be detected. The result is that low-mass planets in wide orbits are much harder to detect than massive planets in close orbits. We must include this fact when trying to estimate the fraction of stars with planets, and when interpreting the distribution of planetary masses and orbital periods.

This chapter is an overview of the statistical properties of the known planets as of September 2010. We focus on the planets detected by radial velocity, transit searches, and microlensing techniques. Radial velocity and transit searches have both led to large samples of planets. Although microlensing has so far uncovered only a handful of planets, the results already constrain the occurrence rates of planets with different properties than those detected by the radial velocity and transit methods. We start in section 2 by describing the selection effects in each of these search techniques, and how to understand them. We review the main statistical techniques that are used to characterize the properties of the planet from the data, and infer properties of the population from large surveys. In section 3, we present the observations and highlight some of the lessons so far about planet formation from the distribution of planet properties. We conclude in section 4 with a discussion of future prospects for this field. The number of planets promises to increase dramatically in future years, making this a particularly exciting time for the study of exoplanet statistics.

2. STATISTICAL TECHNIQUES AND SELECTION EFFECTS

In this section, we first discuss how to set the detection threshold for a radial velocity or transit observation (section 2.1), and then how to understand the resulting selection effects in radial velocity (section 2.2), transit (section 2.3), and microlensing surveys (section 2.4). In section 2.5, we discuss the importance of understanding the properties of the host star and population of stars in the survey. In section 2.6, we briefly introduce some of the techniques that have been developed to characterize the orbital properties of a planet from the data. Finally, in section 2.7, we discuss how to use our knowledge of the selection effects to infer properties of the planet population.

2.1. Setting a Detection Threshold

A typical set of radial velocity measurements is shown in Fig. 1, taken from the paper by *Robinson et al.* (2007) that announced the discovery of a planet with minimum mass 1.9 M_{Jup} in a 675-d low-eccentricity orbit around the star HD 5319. The velocity variations due to the planet can be clearly seen in this dataset, as they have a much larger amplitude than the measurement errors. But it is interesting to consider what would happen for smaller-mass planets, which induce a smaller amplitude of velocity variations. When this amplitude becomes comparable to the measurement errors, there must be a point at which we can no longer say with confidence that we are seeing an orbiting planet in the data. How to calculate this detection threshold is the subject of this section.

We start by considering radial velocity data, but similar ideas apply to analysis of lightcurves for transits. Imagine we are given a set of N measured velocities v_i where i labels each individual measurement, i = 1 to N, along with the observation time t_i and the measurement error σ_i. The simplest approach is to carry out a χ^2 fit of a Keplerian orbit to the data. For each trial orbit, we calculate

$$\chi^2 = \sum_{i=1}^{N} \left(\frac{v_i - V(t_i)}{e_i} \right)^2 \quad (1)$$

where $V(t_i)$ is the predicted velocity at time t_i for the orbital parameters being considered, and the estimated error e_i for each measurement includes the Doppler measurement errors σ_i, but could also be augmented by other sources of noise such as intrinsic stellar variability. The best fitting orbit is the one that minimizes the value of χ^2. [We show how to calculate $V(t_i)$ in section 2.2. We need five parameters to specify $V(t_i)$ for each planet in the model: the orbital period P, the planet minimum mass $M_p \sin i$, the eccentricity e, longitude of periastron ω, and the time of periastron t_P. In addition, we need to include the systemic velocity, and in some cases (such as in Fig. 1) add a long-term velocity trend.]

Minimizing χ^2 in this way tells us the set of best-fitting

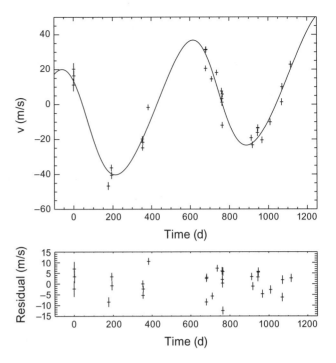

Fig. 1. An example radial velocity dataset for the star HD 5319 (*Robinson et al.*, 2007). The error bars correspond to the measurement errors. The best fit model is shown, which has a planet in a 1.84 yr orbit with K = 33.7 m/s and e = 0.12. A long term acceleration is included in the fit indicating a massive long period companion is also present in the system. The lower panel shows the measured velocities after subtracting the best fit Keplerian orbit. The scatter is significantly larger than the measurement errors, but can be accounted for by intrinsic stellar jitter.

orbital parameters, i.e., characterizes the orbit. (In addition, how quickly χ^2 grows as the parameters are changed from their best-fitting values gives a measure of the error in each parameter. We will say more about this in section 2.5.) But the question of whether we have a detection is a different one. A model that includes a planet will always fit the data better, in the sense that it will always lower the minimum χ^2, because we have more free parameters to adjust. The important question to ask is whether the reduction in χ^2 is significant. If we were to model the velocities in Fig. 1 without the planet, the value of χ^2 would be significantly larger. Because we can lower χ^2 by such a large amount with only five additional parameters, the planet model is strongly preferred. The principle at work here is similar to Occam's razor; we want to keep the model as simple as possible while adequately describing the data.

This idea underlies the Lomb-Scargle (LS) periodogram, a commonly used tool for searching for periodicities in measurements that are unevenly sampled in time (*Lomb*, 1976; *Scargle*, 1982). The LS periodogram power is calculated for each trial period P as

$$z(P) = \frac{\Delta\chi^2}{\chi^2} = \frac{\chi_0^2 - \chi^2(P)}{\chi^2(P)} \quad (2)$$

where χ_0^2 is the minimum value of χ^2 from a fit of a constant to the data, and $\chi^2(P)$ is the minimum χ^2 from a fit of a sinusoid with period P plus constant. If including the sinusoid reduces χ^2 significantly, the power z will be large. The procedure is then to calculate z(P) for a number of different trial periods and look for the period that gives the largest power, z_{max}. Figure 2 shows the LS periodogram for the data shown in Fig. 1. There is a large peak in the power at a period just under 700 days, which matches nicely the final orbital period determined by *Robinson et al.* (2007).

The LS periodogram is often used in planet searches to quickly identify likely periods for planets. For circular orbits, the velocity variations are sinusoidal, and so the LS periodogram fits the correct model, but even for eccentric orbits, where the velocity curve becomes significantly non-sinusoidal, the LS peridogram often gives a good initial estimate for the orbital period. (To do better for noncircular orbits, *Cumming* (2004) defined a "Keplerian periodogram" in an analogous way to equation (2) but fitting a Keplerian orbit at each period.)

We can use the values of z_{max} to decide whether we have detected a planet: Only if the power exceeds a threshold value, $z_{max} > z_{th}$, do we count the fit as significant and say that we have a detection. The detection threshold z_{th} can be determined with Monte Carlo simulations. By generating many synthetic datasets consisting of a constant velocity plus random noise chosen to represent the measurement errors and stellar variability, the distribution of z_{max} that arises due to noise fluctuations can be determined. [In the Monte Carlo simulations, the noise can be generated by assuming a Gaussian distribution, or by using the measured velocities themselves to estimate the noise distribution (so-called "bootstrapping") (see *Press et al.*, 1992). For example, *Marcy et al.* (2005a) discuss the calculation of the false alarm probability by scrambling the order of the detected velocities.] The threshold z_{th} is then set to be the value of z_{max} that is exceeded in some small fraction F of the Monte Carlo simulations. If we observe $z_{max} > z_{th}$, then we know that we have a detection, with a small false alarm probability F that the signal is due to noise fluctuations. The choice of F is determined by our willingness to tolerate false alarms. The higher we set z_{th}, the lower the false alarm rate F, but we also lose the ability to detect low-amplitude signals. This trade off is something that always has to be dealt with when deciding where to set the detection threshold (e.g., *Wainstein and Zubakov*, 1962). Figure 2 of *Cumming* (2004) gives the false alarm probability for Keplerian orbit fits for different values of N and $\Delta\chi^2/\chi^2$.

An important quantity is the number of statistical trials that we make in our search of parameter space. The more different combinations of parameters we try, the more likely it is that noise fluctuations will mimic a planetary signal and give a large value of z. Therefore the value of z_{th} increases

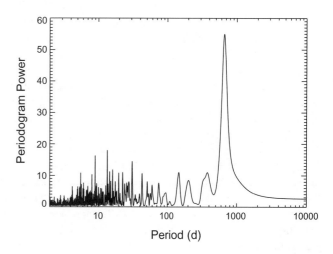

Fig. 2. Lomb-Scargle periodogram of the data for HD 5319 shown in Fig. 1, which measures the improvement in χ^2 when the data are modeled by a sinusoidal velocity variation plus constant rather than a constant alone. There is a peak at 1.8 yr, which matches well with the final derived orbital period when a long-term trend and nonzero eccentricity are included. The periodogram shows many peaks, spaced equally in frequency with separation ~1/T where T is the timespan of the dataset. The peaks are equally spaced in frequency, so that on the log scale used here, the density of peaks increases toward shorter periods.

as we search a larger range of parameters. An example is the number of independent frequencies N_f (or periods) that are searched. Monte Carlo simulations must be used to determine N_f, but a simple analytic estimate gives $N_f \approx$ 3000 for typical radial velocity datasets (*Cumming*, 2004). This means that to detect a signal with a false alarm probability of 10^{-3}, the periodogram power for that signal must be large enough, or the Keplerian fit good enough, that the false alarm probability for a search at a single frequency would be $\sim 10^{-6}$.

For transits, the same approach applies, of defining a detection statistic, and setting a threshold that takes into account the noise level and number of trials. There are a variety of detection statistics that have been proposed for detecting the signature of a transiting planet. These detection statistics are essentially based on χ^2 fitting of a box-shaped dip to the stellar lightcurve. *Moutou et al.* (2005) carry out a blind test comparison of five different detection methods on simulated lightcurves, which include a matched filter method (*Jenkins et al.*, 1996), box-fitting least squares (*Kovács et al.*, 2002), and a Bayesian approach (*Defaÿ et al.*, 2001), finding that no one technique offers an advantage in all situations. Indeed, *Schwarzenberg-Czerny and Beaulieu* (2006) point out the equivalence of some of these methods for Gaussian noise, but this may not be the case if the data have a red noise component coming from systematic errors that introduce correlations over time (*Pont et al.*, 2006).

It is important to stress that we have been discussing the statistical detection threshold, but that may in fact not be the criterion used to determine detections, especially early in a survey when systematic effects are not fully understood, and the same detection threshold may not be applied uniformly in a given survey. For example, the first planets discovered by Doppler surveys had large signal-to-noise ratios, and are essentially detectable by eye. *Marcy and Butler* (1998) note that "experience shows that a confident detection requires that the amplitude be ~4 times the Doppler error." Recent radial velocity detections have velocity amplitudes much closer to the noise level, and statistical analysis has become more important, with the planet search teams often reporting false alarm probabilities for their detections (*Marcy et al.*, 2005a). Another example is the OGLE-III transit survey, in which the transit candidates were selected by eye from a list of "pre-candidates." *Gould et al.* (2006a) pass a number of simulated transits through the same by-eye selection procedure in an attempt to objectively quantify the corresponding detection threshold as a function of the signal-to-noise ratio.

The other point to note is that the methods discussed here identify significant periodic signals in radial velocity data or dips in stellar lightcurves, but it may be the case that the detected signal is not due to a planet. For example, in radial velocity searches, a periodic signal could be caused by stellar magnetic activity, and be related to the rotation period of the star. In transit surveys, typically only a small fraction of identified candidates turn out to be transiting planets; most are other kinds of systems such as transits by low-mass stars.

2.2. Selection Effects in Radial Velocity Searches

Having set a detection threshold, we can ask what orbital parameters lead to signals that are detectable. The radial velocity of a star with an orbiting planet is

$$V(t) = V_Z + K\left[\cos(f(t) + \omega) + e\cos\omega\right] \quad (3)$$

where V_Z is the systemic velocity, K is the semiamplitude, and the remaining factor in square brackets sets the shape of the velocity curve. The amplitude is

$$K = \frac{28.4 \text{m s}^{-1}}{\sqrt{1-e^2}} \left(\frac{M_P \sin i}{M_J}\right) \left(\frac{P}{1 \text{ yr}}\right)^{-1/3} \left(\frac{M_\star}{M_\odot}\right)^{-2/3} \quad (4)$$

where the mass of the star is $M_\star$, the planet mass is M_P, the orbital period is P, and the inclination of the orbit to the line of sight is i (where i = 90° if we are looking at the orbit edge-on). The shape of the velocity curve depends on the longitude of periastron ω and how the true anomaly f varies with time, given by the relations

$$\tan\left(\frac{f(t)}{2}\right) = \left(\frac{1+e}{1-e}\right)^{1/2} \tan\left(\frac{E(t)}{2}\right)$$

$$E(t) - e\sin E(t) = M(t) \quad M(t) = \frac{2\pi}{P}(t - t_P) \quad (5)$$

where E(t) is the eccentric anomaly, and M(t) is the mean anomaly. For a circular orbit, f(t) = E(t) = M(t), and the velocity variations are sinusoidal.

There are three main factors that determine whether the velocity variations induced by a planet will be detectable in a radial velocity dataset. The first is the signal-to-noise ratio. A larger-amplitude K leads to a larger-power z_{max} on average, and so increases the likelihood of detection. Using analytic results for Gaussian noise, the signal-to-noise ratio required for a 50% detection rate is

$$\frac{K_0}{\sqrt{2}s} \approx \frac{1}{\sqrt{N}} \left[2\ln\left(\frac{N_f}{F}\right)\right]^{1/2} \quad (6)$$

(*Cumming*, 2004) (here we give the large N limit), where N_f is the number of independent frequencies that were searched, and F is the desired false alarm probability. The noise level s includes contributions from measurement errors and stellar jitter. The $1/\sqrt{N}$ factor should be no surprise: If we take N measurements each with an error s, then the uncertainty in the mean of those measurements we expect to be $s/\sqrt{N}$, and this gives a measure of what velocity amplitude could be detected. For example, for typical values $N_f \approx 1000$ and $F \approx 10^{-3}$ (so that on the order of one false alarm would be expected in a survey of 1000 stars), we find $K_0 \approx 7 \text{ s}/\sqrt{N}$. Using equation (4), the planet mass that can be detected 50% of the time is

$$M_{p,50} = \frac{0.14 \, M_J}{\sin i} \left(\frac{P}{1 \text{ yr}}\right)^{1/3} \left(\frac{M_\star}{M_\odot}\right)^{2/3} \left(\frac{s}{3 \text{ m/s}}\right) \quad (7)$$

or

$$M_{p,50} = \frac{1.0 \, M_\oplus}{\sin i} \left(\frac{P}{1 \text{ d}}\right)^{1/3} \left(\frac{M_\star}{0.3 \, M_\odot}\right)^{2/3} \left(\frac{s}{1 \text{ m/s}}\right) \quad (8)$$

where we assume N = 30 (with $M_p \propto 1/\sqrt{N}$ for different N).

The second factor is the time span of the observations, T. For long orbital periods P > T, we see only part of the orbit, and the velocity variation is smaller than the full velocity amplitude K. Therefore the velocity amplitude needed for detection K_{det} increases with period for P > T. The rate at which it increases depends on the phase of the orbit being detected: When the orbit is observed near a zero-crossing rather than a maximum or minimum, the velocity variation is larger. Averaging over phase, *Cumming* (2004) shows that for a 50% detection threshold (in which case we can rely on the zero-crossing phases only), $K_{det} \propto P$ whereas for a larger detection efficiency, the scaling is $\propto P^2$. Of course, even if a long-period orbit is detected, it is not possible to determine whether the companion is of planetary mass: Characterization of the orbit is possible only after a whole orbit has been observed. For example, the velocities in Fig. 1 show a clear upward trend over the time span of the data, presumably due to an additional companion with a long-period orbit. However, only after continued monitoring will its orbital period and mass be securely identified.

The third factor is the shape of the velocity curve, which depends on the eccentricity and the orientation of the orbit. The orbit shown in Fig. 1 has a low eccentricity, giving a radial velocity curve that is close to sinusoidal. A planet on a very eccentric orbit on the other hand has a much more distorted velocity curve (Fig. 3), with large velocities during periastron passage, but smaller velocities during most of the orbit. Unless we are lucky enough to observe the star while the planet is close to periastron, the orbit could go undetected. The two velocity curves in Fig. 3 are drawn from a large number of simulated observations using the same observation times and velocity amplitude, but random phases, of an orbit with e = 0.5. The upper panel in the figure shows a case that was detected by the detection algorithm, whereas the lower panel was not. The effect of eccentricity on detectability has been calculated by several authors (*Endl et al.*, 2002; *Cumming*, 2004; *Wittenmyer et al.*, 2006). There is good agreement that the detectability falls off for e ≥ 0.5–0.6.

In equations (7) and (8) above, we use values s = 1 and 3 m/s, corresponding to the long-term precision currently available. For example, the error bars in Fig. 1 show the measurement errors from the HIRES spectrometer at the Keck telescope, and range from 1.5 to 3.6 m/s. A long-term accuracy of $\sigma \approx 1$ m/s has been achieved at the HARPS spectrograph at La Silla Observatory, Chile (*Lovis et al.*, 2006). The ability to track stellar radial velocities at the ~1 m/s level on timescales of 10 yr is a remarkable achievement. In fact, the precision is good enough that the scatter in the residuals to orbital solutions usually has a significant component from stellar variability or "jitter" (added in quadrature to the measurement errors), which we discuss in section 2.4.

The observed detection threshold of radial velocity surveys matches these expectations quite well. For example, *Cumming et al.* (2008) show that equation (6) provides a good description of the K-N relation for announced planets from the Keck Planet Search, if the noise s for each star is estimated from the residuals to the best-fitting orbital solution. However, for N ≥ 60 there is an apparent floor in the signal-to-noise ratio of announced planets of K/s ≈ 2. The likely cause of this is that at small signal-to-noise ratio it is more difficult to rule out other explanations such as stellar variability and thereby confirm a statistically significant signal is indeed due to a planet.

2.3. Selection Effects in Transit Searches

Searches for transiting planets look for dips in the stellar lightcurve as the planet repeatedly transits its host star. *Mandel and Agol* (2002) and *Seager and Mallén-Ornelas* (2003) give analytic expressions for transit lightcurves. Four main parameters can be determined from the lightcurve. The time between transits gives the orbital period. The depth of the transit depends on the ratio of the areas of the disk of the planet and star, $\delta \approx (R_p/R_\star)^2$, or $\delta \approx 0.01$ for Jupiter. The other two observables are the duration of the

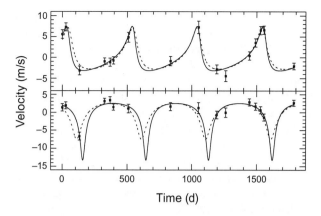

Fig. 3. Examples of velocity curves with e = 0.5 that are (top panel) and are not (bottom panel) detected. The dotted line in each case shows the true orbit, the points are the observed velocities, and the solid curve shows the best-fitting orbit. In both cases, the solid curve gives a lower χ^2 than the dotted curve. The lower panel has only a single measurement during the periastron passage, and is not a significant detection.

transit and the time for ingress or egress, which allow the orbital radius and inclination to be determined. The detailed shape of the transit lightcurve also depends on the stellar limb darkening.

The observed transiting planets have radii ranging from ≈0.15 to 1.8 R_{Jup}, masses ≈5 $M_\oplus$ to >10 M_{Jup} (radial velocity follow-up is required to determine the mass of a transiting planet), and short orbital periods, concentrated at ~1 to several days. These properties are consistent with the fact that detectability falls off strongly with increasing orbital period and decreasing planet radius. To see this (e.g., *Horne*, 2003; *Gaudi et al.*, 2005; *Gaudi*, 2005), the first step is to write down the expected signal-to noise ratio. If there are N_{obs} observations made of a given star, the number of observations during transit will be on average $N_t \approx N_{obs}(R_\star/\pi a)$, since the duration of the transit is roughly a fraction $\approx R_\star/\pi a$ of the orbital period (a planet orbiting farther out spends a smaller fraction of its time in transit than one orbiting close in). The signal-to-noise ratio will then be

$$S/N \approx \sqrt{N_t}\,\frac{\delta}{\sigma} \qquad (9)$$

where σ is the photometric precision per observation. By analogy with equation (6) for radial velocities, we expect that there will be a critical value of S/N required for a detection. For example, *Tingley* (2003) finds that S/N ≈ 16 for a 50% detection rate with 1% false alarm probability, the exact number depending on the particular detection statistic adopted (although note that a FAP of much less than 1% will be typically necessary given the large numbers of stars surveyed).

The sensitivity to orbital period and planet radius comes about because the search volume, or the number of stars surveyed, is different for different planet parameters (*Gaudi*, 2005). If we assume that the noise level is set by source dominated photon noise, then $\sigma \propto 1/\sqrt{N_\gamma} \propto d$ where d is the distance to the star, since the stellar flux falls off as $1/d^2$. Therefore, if we fix the signal-to-noise threshold and stellar properties, equation (9) gives a search volume $\propto d^3 \propto N_t^{3/2} \delta^3 \propto R_p^6/P$. The probability that a planetary orbit has an inclination leading to a transit is $\approx R_\star/a$. Multiplying the transit probability by the search volume, and assuming a constant space density of stars, we find that the number of detections is $\propto R_p^6/P^{5/3}$ (*Gaudi*, 2005). Therefore, the number of planets discovered should increase strongly with planet radius, and fall off strongly with orbital period. Based on this simple formula, we see that going from an orbital period of 1 to 3 d gives a factor of 6 suppression in detectability, and going from a radius of 1 R_{Jup} to 1.3 R_{Jup} increases detectability be a factor of 5.

This estimate predicts the general fall off with period well, but in detail the detection probability as a function of period shows a series of sharp dips around this trend set by the spacing of the observations [see Fig. 1 of *Beatty and Gaudi* (2008) for an example of the detectability as a function of orbital period, known as the window function]. This aliasing arises because the observations are taken during the same ≈8-h window each day. Depending on the phase of the orbit at the time of observation, this can make orbits with periods close to an integer number of days easier or harder to detect. For example, if the phase is such that the transit occurs in every observing window, then the detectability is greater than a noninteger period, but the opposite could also be true, that the phase is such that transits happen outside the observing window! *Gould et al.* (2006a) show that for a signal-to-noise limited survey the first effect wins out, leading to a net enhancement of detectability at integer periods. The finite spacing of the observations has other consequences. For example, scheduling radial velocity follow up to confirm the mass of a candidate transiting planet requires that two transits be observed, giving the orbital period. For a given observing strategy, this becomes less likely for longer orbital periods (*Gaudi et al.*, 2005).

Another important complication is the presence of red noise (*Pont et al.*, 2006). Equation (9) for the signal-to-noise ratio assumes that the photometric errors are uncorrelated, i.e., that the noise was white. In fact, systematic effects such as changing atmospheric or telescope conditions over time lead to correlations in the errors in transit surveys, so that equation (9) overestimates the effective signal-to-noise. Several recent papers attempt to untangle the effects of red noise on detection thresholds in transit surveys; see, e.g., *Pont et al.* (2006) and *von Braun et al.* (2009).

Understanding the selection effects in transit surveys in detail has proved to be complex. Although there are now more than 100 transiting planets, the discoveries at first did not come as quickly as expected based on the known occurrence rate of hot Jupiters from radial velocity surveys. It became apparent that the yield of transit surveys was initially overestimated by large factors (e.g., *Horne*, 2003). The yields are now much better understood. *Beatty and Gaudi* (2008) consider a number of different transit surveys that have adopted different observing strategies, and carefully estimate the predicted yields, finding numbers that are close to the actual number of reported detections.

2.4. Selection Effects in Microlensing Surveys

Ten planets discovered using microlensing had been published at the time of writing, with masses in the range 3 $M_\oplus$ to 3.5 M_{Jup}, and orbital radii from 0.6 to 5.1 AU. This range of planet mass and orbital radius can be understood by considering that the planet must cause a significant deviation from the microlensing lightcurve that would otherwise result from lensing of the background star by the lens star. This requires that the orbital radius of the planet lies close to the Einstein radius

$$R_E \approx 2.9 \mathrm{AU} (M_L/M_\odot)^{1/2} (D/\mathrm{kpc})^{1/2} \qquad (10)$$

where $R_E^2 = (4 GM_L/c^2)D$, M_L is the mass of the lens (planet host star), and D is the reduced distance ($1/D = 1/D_{LS} + 1/D_{OL}$ with D_{LS} being the lens source distance, and D_{OL} the observer lens distance).

Given a planet in the lensing zone (typically a within about a factor of 2 of R_E), the probability of detection can be close to 100% for high-magnification central caustic events, in which the source image sweeps around the Einstein ring (*Griest and Safizadeh*, 1998). This is a powerful way to detect multiple planet systems, since there is a high probability of seeing all detectable planets within the lensing zone. For lower-magnification planetary caustic events, the detection probability depends on the relative path of the source and lens, but is typically tens of percent for Jupiter-mass planets to a few percent for Earth-mass planets (*Gould and Loeb*, 1992; *Bennett and Rhie*, 1996).

A limit on the detectable planet mass is set by the finite angular size of the source star, which washes out the perturbation due to low-mass planets if the angular size of the source star is greater than the Einstein radius of the planet. (However, note that detection of finite size effects in the lightcurve is important, as it enables a measurement of the Einstein radius and therefore the lens mass.) The detectable mass can therefore be roughly estimated by setting the angular size of the planet's Einstein radius $\theta_E = R_{E,p}/D_L = q^{1/2}R_E/D_L$, where $q = M_p/M_\star$ is the mass ratio, to the angular radius of the source $\theta_\star = R_\star/D_S$. This gives

$$M_P \geq 1 M_\oplus \left(\frac{R_\star}{R_\odot}\right)^2 \left(\frac{D_L}{D_S}\right)^2 \left(\frac{D}{1 \mathrm{kpc}}\right)^{-1} \qquad (11)$$

where the stellar radius depends on the type of star, $R_\star \sim R_\odot$ for a main-sequence star or could be ~10 $R_\odot$ for a giant. Such low planet masses at orbital radii of several AU are not possible to detect with other techniques, so microlensing provides a unique and important probe of the planet distribution in this region.

2.5. Stellar Properties

Alongside the rapid growth in the number of known exoplanets has been a large amount of work to understand the properties of their host stars. Understanding stellar properties is important for detecting and characterizing the planets in a given system, and for interpreting the statistical properties of the planet sample.

For transiting planets, knowledge of the planet properties is limited by how well the properties of the host star can be determined. For example, the geometric measurements of the planet radius R_P and orbital radius a are in units of the stellar radius. Differences in the way stellar properties are determined from star to star are important to take into account when looking at the sample of transiting planets as a whole. *Torres et al.* (2008) and *Southworth* (2008) present subsamples of transiting planets for which the stellar parameters have been obtained in a uniform way.

For radial velocity searches, the intrinsic stellar radial velocity variability, or stellar jitter, provides an important source of noise. Stellar jitter is believed to correlate with magnetic activity, and can arise from a few different sources. For example, a magnetic spot that covers a fraction f_{spot} of the area of a rotating star would give a shift in the observed line centroids of roughly f_{spot} v sin i ≈ 10 m/s (f_{spot}/0.01) (v sin i/1 km/s) (*Saar and Donahue*, 1997) (to set the scale, note that $2\pi R_\odot/10$ d = 5 km/s). An old inactive star like the Sun has $f_{spot} \sim 10^{-3}$; active stars can have f_{spot} of several percent. Other sources of stellar jitter are spatial variations in convective velocities across the surface, changes in line profiles with time, or stellar oscillations. The effect of stellar jitter can be seen in the lower panel of Fig. 1, in which the residuals have a scatter ≈6 m s^{-1}, greater than the measurement errors, but consistent with the predicted jitter for this star. [For F, G, and K dwarfs in radial velocity searches, the level of jitter has been empirically calibrated (*Wright et al.*, 2004; *Wright*, 2005; see also *Saar and Donahue*, 1997; *Saar et al.*, 1998) in terms of observables absolute magnitude M_V, color B–V, and activity level S. The quantity S measures the flux in Ca H and K emission lines (an indicator of magnetic activity) relative to the neighboring continuum.] As well as hindering detection as an extra noise source, time-dependent jitter can also mimic planetary orbits. Timescales range from the stellar rotation period (~10 d) to magnetic cycle timescales of ~10 yr. Monitoring magnetic activity indicators over time is important to check that a planetary signal is in fact real [e.g., see *Queloz* (2001) for a study of the magnetically active star HD 166435].

As well as individual stellar properties, the properties of the sample of stars surveyed is important to understand when trying to draw conclusions about the planet population. For example, the selection of stars in the Keck Planet Search is described by *Wright et al.* (2004) and *Marcy et al.* (2005a). The ≈975 stars are selected to be chromospherically quiet, to lie at most 3 magnitudes above the main sequence (thereby excluding giant stars), to have no companion within 2", a color selection B–V > 0.55, and are magnitude limited to V = 8. As discussed by *Marcy et al.* (2005a), the fact that the survey is magnitude limited introduces some interesting biases. The first is a Malmquist bias, that brighter stars such as subgiants can be seen to a larger distance, and so they are overrepresented in the sample. Second, there is a metallicity bias introduced by the fact that at a fixed B–V color, metal-rich stars are brighter than metal-poor stars (e.g., see *Santos et al.*, 2004; also see discussion in *Gould et al.*, 2006a). This comes about because metal-rich stars are redder than metal-poor stars of the same mass, due to increased line blanketing. Therefore in a given B–V bin, there are more-massive, brighter, metal-rich stars, and less-massive, fainter, metal-poor stars. Because the sample is magnitude-limited, metal-rich stars are seen out to a larger volume and therefore overrepresented. The same effect leads to a correlation between stellar metallicity and mass for stars in the sample (*Fischer and Valenti*, 2005).

A second example is the CORALIE survey of 1650 dwarfs (*Udry et al.*, 2000), which is selected using Hipparcos parallaxes to be volume limited (d < 50 pc) rather than magnitude limited, in which case the Malmquist biases are absent. However, the fainter cool dwarfs are removed from the sample by implementing a color-dependent distance cutoff for K dwarfs, and so the search volume depends on the B–V bin. As a final example, the N2K consortium deliberately targets metal-rich stars in a search for closely orbiting (P ≤ 14 d) planets (*Fischer et al.*, 2005). It is important to keep these different sample biases in mind when comparing the results of different surveys, or looking at differences between the planet populations around stars with different metallicities or masses.

2.6. Exploring Parameter Space: Fitting and Uncertainties

We now turn to techniques for exploring parameter space to find the best-fitting orbital solution, and to assess the errors in the fitted parameters. Consider fitting Keplerian curves (equation (3)) to radial velocity measurements. A common approach to finding the best-fitting solution is to use a Levenberg-Marquardt algorithm (e.g., *Press et al.*, 1992). This is an efficient algorithm that marches downhill from a starting guess to the χ^2 minimum. The difficulty is that in general there isn't a single χ^2 minimum in parameter space, but rather a complicated χ^2 surface with many local minima. For example, the many peaks in periodogram power in Fig. 2 reflect the many local minima in χ^2 as a function of period for a circular orbit fit. For nonzero eccentricities, there are also many peaks as a function of the phase (time of periastron t_P), since there can be many values of t_P for which the radial velocity peak at periastron passage intercepts the sparsely spaced data points. The problem becomes more complicated for multiple planet systems since each additional planet adds an extra five parameters to the model. A grid search through parameter space with a resolution of 100 in each direction would require an additional factor of 10^{10} evaluations for each additional planet added.

Often, Levenberg-Marquardt is used with many different starting points in parameter space to determine which leads to the lowest value of χ^2. The LS periodogram can be used to give initial guesses for the orbital period. For multiple planet systems, the data can be analyzed by subtracting successive planet orbits from the data and analyzing the residuals for further companions, before carrying out a full multiple-planet fit.

A related question is determining the uncertainties in fitted parameters. The probability distribution of certain parameters such as eccentricity is not necessarily Gaussian (e.g., *Shen and Turner,* 2008), so that the usual estimate of uncertainty from the covariance matrix, roughly $\delta_a \approx (\partial^2 \chi^2 / \partial a^2)^{-1/2}$ for parameter a (e.g., *Press et al.,* 1992), does not give a good measure of the error.

A good framework for systematically calculating the probability distributions of the different parameters of the model is a Bayesian approach. [We only have room here to give the basic idea. For a short but thorough introduction to the subject that will get you up and calculating quickly, we recommend the book by *Sivia* (1996).] The goal is to calculate the probability distribution of model parameters given the data, Prob($\bar{a}$|data), where, for example, $\bar{a} = (V_0, P, K, e, \omega, t_P)$ for a single Keplerian orbit. Bayes' theorem

$$\mathrm{Prob}(\bar{a}|\mathrm{data}) \propto \mathrm{Prob}(\bar{a}) \mathrm{Prob}(\mathrm{data}|\bar{a}) \quad (12)$$

gives this probability in terms of something we can calculate directly, the probability of obtaining the data given the parameters $\bar{a}$, Prob(data|$\bar{a}$). If the noise is Gaussian and uncorrelated from observation to observation, the probability of obtaining a set of measurements $\{t_i, v_i, \sigma_i\}$ if the underlying velocities are $V(t_i)$ is Prob(data|$\bar{a}$) =

$$\begin{aligned}\mathrm{Prob}(\mathrm{data}|\bar{a}) &= \prod_{i=1}^{N} \frac{\exp\left(-(v_i - V(t_i))^2 / \sigma_i^2\right)}{\sqrt{2\pi\sigma_i^2}} \\ &= \frac{1}{\sqrt{2\pi}\prod \sigma_i} \exp\left(-\frac{1}{2}\sum_{i=1}^{N}\left(\frac{v_i - V(t_i)}{\sigma_i}\right)^2\right) \\ &= \frac{1}{\sqrt{2\pi}\prod \sigma_i} \exp\left(-\frac{\chi^2(\bar{a})}{2}\right)\end{aligned} \quad (13)$$

where in the last step, we see that the probability is related in a simple way to the χ^2 for that choice of parameters. The remaining factor in equation (12) is the prior probability of the parameters $\bar{a}$, Prob($\bar{a}$), which allows us to specify any prior information about the parameters.

Equations (12) and (13) give the joint probability densities for all parameters $\bar{a}$. However, often we are interested in the probability distribution of some subset of the parameters of the model. For example, consider circular orbits, for which the model has four parameters $\bar{a} = (P, K, \phi, V_0)$,

$$V(t_i) = V_0 + K \sin\left(\frac{2\pi t_i}{P} + \phi\right) \quad (14)$$

Often we are interested in the period P and velocity amplitude K of the orbit (since they relate to physical properties of the system), but not ϕ and V_0. The probability distribution of P and K can be obtained by marginalizing, or integrating over the uninteresting parameters

$$\mathrm{Prob}(K, P|\mathrm{data}) = \int dV_0 d\phi \, \mathrm{Prob}(K, P, \phi, V_0|\mathrm{data}) \quad (15)$$

A good way to think of this integral is as a weighted average of the probability over the parameters V_0 and ϕ. Therefore, all possible values of ϕ and V_0 are taken into account, in contrast to traditional χ^2 minimization, in which one would find the best fit V_0 and ϕ at each K and P. For eccentric orbits, similar integrals are needed, but over different choices of the Keplerian parameters.

There has been a lot of work recently on Markov Chain Monte Carlo (MCMC) techniques as a method for evaluating marginalization integrals for Keplerian fits to radial velocity data (e.g., *Ford,* 2006; *Gregory,* 2007). The basic algorithm is very simple. One starts off at some point in the parameter space $\bar{a}$, and then proposes jumping to a new location $\bar{a}' = \bar{a} + \delta\bar{a}$. If χ^2 for the new set of parameters at $\bar{a}'$ is lower than at $\bar{a}$, then we accept the jump. In this way, we always accept a move to a better-fitting model. If the χ^2 is larger at the new location, we accept the jump with a probability $\exp(-\Delta\chi^2/2)$. Every jump that is accepted generates a new "link" in the chain. After following this simple algorithm many times (typically at least several thousands of accepted jumps are required), we end up with a sample of points in parameter space. The amazing property of this algorithm is that the density of points in parameter space is proportional to the probability density in equation (12). [For this to be the case, certain rules must be followed, for example, the jumps we make in parameter space $\delta\bar{a}$ must have certain properties, e.g., the probability to jump from point A to point B in parameter space should be the same as B to A, and the jumps should be uniform in the prior distribution. The size of the jumps is generally adjusted so that the jump is accepted ≈25% of the time. The choice of parameters in which to make jumps can be tuned to make this efficient (*Ford,* 2006).] Marginalization is then trivial. Say we generate a series of points in the parameter space (P, K, ϕ, V_0) for a circular orbit; to find Prob(P), we simply plot a histogram of the P values of all the points that were visited.

As an example of the kind of constraints that are obtained, Fig. 4b shows the joint probability distribution of P and e derived from the radial velocity data for the star

HD 72659 from *Butler et al.* (2003). The radial velocity data for this star (Fig. 4a) show clear long-period variations. However, because the orbit has not yet closed, a range of orbital periods are possible, with longer periods corresponding to a larger eccentricity. This plot is based on Fig. 4 of *Ford* (2005), but calculated by direct integration (*Cumming and Dragomir*, 2010) rather than using a MCMC technique.

There are alternative methods for investigating the uncertainties in derived parameters. *Marcy et al.* (2005a) calculate uncertainties by a Monte Carlo method in which they make a large number of radial velocity curves consisting of the best-fitting model plus "noise" drawn from the residuals to the best fit. The resulting distribution of measured parameters gives an estimate of the uncertainties.

There are also many other methods for searching complicated parameter spaces that have been applied in the literature. For example, *Lovis et al.* (2006) use a genetic algorithm in which sets of parameter values are bred (along with occasional mutations) such that only the fittest (the best-fitting models) survive. *Gregory* (2007) has developed a parallel tempering technique in which MCMC chains with different temperatures are run simultaneously, and exchange information. A "hot" chain, which has artificially increased error bars, explores the broad parameter space, jumping large distances and exploring many local minima; a "cold" chain takes small steps locally and explores the local minima. This technique has been applied with success to a number of multiple-planet systems.

In some cases, more than one method may be applied. A good example is fitting microlensing lightcurves. The χ^2 surface in that problem is extremely complex. Small changes in parameters that affect the location of caustics can lead to dramatic changes in the lightcurve and therefore χ^2. This makes it nontrivial to explore the parameter space and find the best-fitting solutions. *Gould et al.* (2006b) adopt a brute force scan of parameter space in some nonlinear parameters, and use a minimization algorithm for others. *Bennett et al.* (2008) use MCMC to explore the complex parameter space.

A different application of Bayesian techniques is in microlensing and transit searches, when a statistical model of the stellar population must be considered. For example, in microlensing events, constraints on the lens mass and distance can be derived using as input priors on the distances and masses of lenses and sources [see *Beaulieu et al.* (2006) for an application of this approach to one of the detected microlensing planets].

2.7. Completeness Corrections: Determining Planet Occurrence Rates

Understanding selection effects is crucial for population studies, as it allows us to correct for incompleteness, i.e., to determine the effective size of the stellar sample. For each set of planet parameters, e.g., mass, orbital period, and eccentricity, the idea is to work out which stars in the survey have data that would allow a planet with those

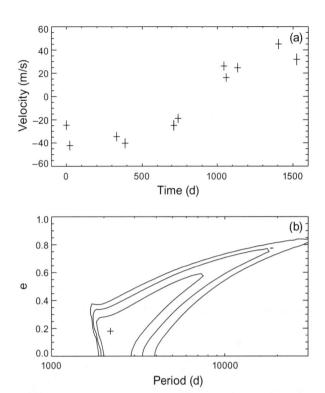

Fig. 4. Constraints on the period and eccentricity of the companion to HD 72659 (data from *Butler et al.*, 2003), for a companion with mass less than 10 M_{Jup}. The best fit values given in that paper are e = 0.18 and P = 5.98 yr, shown by the cross. From *Cumming and Dragomir* (2010), based on Fig. 4 of *Ford* (2005).

parameters to be detected. Only those stars are useful in constraining the population. To see this, consider a simple example. Imagine surveying 100 stars, detecting 10 planets, and being able to rule out the presence of planets around the remaining 90 stars. Then the best estimate of the fraction of stars with planets is 10%, since we detected 10 planets out of 100 stars surveyed. However, it could be that the data for 80 of those stars are not good enough to say whether a planet is present or not. In that case, only 20 stars have data good enough to constrain the presence of a planet, and since planets were detected around 10 of these, the best estimate of the planet fraction is 50%. In practice, the ability to detect or rule out the presence of a planet depends on the orbital parameters, and so when doing this, we must account for every combination of orbital parameters of interest.

We can write this argument down mathematically using a maximum likelihood method. Assume that a survey of $N_\star$ stars detects N_p planets, and rules out planetary companions for the remaining $N_\star - N_p$ stars. If the fraction of stars with planets is f, then the probability that a given star has zero or one planet can be written e^{-f} or $f e^{-f}$ (considering each star as an independent Poisson trial). The total likelihood is the product of these probabilities for all stars

$$L = \left(fe^{-f}\right)^{N_d} \left(e^{-f}\right)^{N_\star - N_d} = f^{N_d} e^{-N_\star f} \quad (16)$$

Maximizing the likelihood with respect to f by setting $\partial L/\partial f = 0$ gives $f = N_d/N_\star$, exactly what we expect: The fraction of stars with planets is given by the number of detections divided by the total number of stars. We can introduce selection effects into this calculation by including the fact that for each star i there is some probability p_i of detecting a planet. Then

$$L = \prod_{i=1}^{N_d} p_i f e^{-p_i f} \prod_{j=1}^{N_\star - N_d} e^{-p_j f}$$
$$= f^{N_d} \left(\prod_{i=1}^{N_d} p_i\right) \exp\left(-f \sum_{k=1}^{N_\star} p_k\right) \quad (17)$$

where the sum labeled i is over stars with a detected planet, the sum labeled j is over stars with no detection, and the sum labeled k is over all stars. The value of f that maximizes the likelihood L is now $f = N_d/N_{eff}$ where the effective number of stars $N_{eff} = \sum_{i=1}^{N_\star} p_i$. This is the mathematical statement of our previous argument. We are guaranteed to detect a planet around stars with $p_i = 1$, and so these stars contribute fully to N_{eff}; stars with lower quality or fewer data points that have small p_i do not contribute as much to N_{eff}. In practice, the detection probability depends on the orbital parameters of the planet, and varies from star to star depending on the number of data points, measurement errors, and stellar properties.

There have been several detailed calculations of detection probabilities and thresholds for radial velocity surveys (*Walker et al.*, 1995; *Cumming et al.*, 1999, 2008; *Endl et al.*, 2002; *Naef et al.*, 2005; *Wittenmyer et al.*, 2006; *O'Toole et al.*, 2009). These papers essentially use the same method for determining detection probabilities or upper limits on the planet mass at a given orbital period. The idea is to use a Monte Carlo method to find the value of velocity amplitude K as a function of period that results in a particular detection efficiency. The technique for modeling the noise differs in different studies, ranging from assuming Gaussian noise with amplitude set by the observed variability, to sampling directly from the observed residuals to the best-fit orbit.

Early investigations of the planet mass and orbital period distributions did not take these detailed star-by-star mass limits into account (*Tabachnik and Tremaine*, 2002; *Udry et al.*, 2003; *Lineweaver and Grether*, 2003), but recent work has included mass limits calculated for each star. *Naef et al.* (2005) estimate planet occurrence rates in the ELODIE Planet Search, *Cumming et al.* (2008) use their results from the Keck Planet Search to constrain the minimum mass and orbital period distribution of planets, and *O'Toole et al.* (2009) use observations from the Anglo-Australian Planet Search to focus on the frequency and mass distribution of low-mass planets. *Cumming et al.* (2008) use a method similar to *Tabachnik and Tremaine* (2002), but including the upper limits derived separately for each star, rather than assuming a constant velocity limit for the whole survey. The method is a generalization of equations (16) and (17) to bins in the mass-period plane [based on the method of *Avni et al.* (1980) for one-dimensional distributions with upper limits].

Calculations of the incidence of planets or limits on the planet fraction in transit surveys include *Gilliland et al.* (2000), *Brown* (2003), *Mochejska et al.* (2005), *Bramich and Horne* (2006), *Gould et al.* (2006a), *Beatty and Gaudi* (2008), and *Weldrake et al.* (2008). Monte Carlo methods are used in which fake planetary transits are injected into the observations, and subject to the same analysis procedure as the real data. As discussed in section 2.3, the effective size of the stellar sample in a transit survey depends sensitively on the planet radius and orbit. Determining this number accurately requires a model of galactic structure, the stellar mass function, and accounting for the effects of extinction. For example, *Gould et al.* (2006a) give a detailed discussion of the selection effects and stellar sample size for the OGLE-III transit surveys, and derive the frequencies of hot and very hot Jupiters.

Examples of calculations of detection sensitivities and occurrence rates in microlensing surveys are *Gaudi et al.* (2002) and *Snodgrass et al.* (2004). Exactly as we discuss above, the challenge is to identify a subset of microlensing events for which planetary systems could have been identified, e.g., free from nonplanetary features and with good enough signal-to-noise data. Detection sensitivities are then calculated as a function of planet-lens mass ratio and separation [see Fig. 8 of *Gaudi et al.* (2002) for examples] and used to infer occurrence rates of planets.

3. PROPERTIES OF OBSERVED PLANETS

In this section, we discuss the statistical properties of observed planets, highlighting the major features of the population discovered so far, and some of the implications for planet formation theories. As we go, we will point out the places where the selection effects discussed in section 2 play a role.

3.1. Catalogs of Exoplanet Properties

There are a few different catalogs of exoplanet properties. The Extrasolar Planets Encyclopedia website (*http://exoplanet.eu*), maintained by J. Schneider of the Paris Observatory, gives properties of all planets announced to date, organized by discovery method. In this section, we use data from the Extrasolar Planets Encyclopedia for planets discovered by radial velocity measurements, transits, or microlensing, accessed on September 1, 2010, a total of 469 planets with 101 transiting planets and 10 microlensing planets. *Butler et al.* (2006a) published a catalog of 172 exoplanets within 200 pc with M sin i < 24 M_{Jup}. An updated version of this catalog is available (*http://exoplanets.org/exotable/exoTable.html*), which at the time of writing lists 372 planets. This catalog includes radial velocity data for those stars with planets observed at the Keck, Lick, or Anglo-Australian telescopes. When showing plots of planet properties in this section, we use data from the Extrasolar Planets Encyclopedia, but our conclusions would not be changed had we used the *Butler et al.* (2006a) sample.

For transiting planets, there are several other sources of information. A website maintained by F. Pont (*http://www.in science.ch/transits/*) gives properties and references for all known transiting planets. As we mentioned in section 2.4, *Torres et al.* (2008) and *Southworth* (2008) have published data for subsamples of the transiting planets in which the data have been analyzed in a uniform way.

There is also a lot of information available about planet host stars. For example, the Spectroscopic Properties of Cool Stars (SPOCS) survey (*Valenti and Fischer*, 2005; *Takeda et al.*, 2007) provides a catalog of stellar properties for 1040 F, G, and K stars that have radial velocity measurements as part of the Keck, Lick, or Anglo-Australian Planet Search programs.

A new resource is the NASA/IPAC/NExScI Star and Exoplanet Database (NStED) available at *http://nsted.ipac.caltech.edu/*. Designed to support NASA's planet-finding and characterization activities, this is a comprehensive database of stellar and planet properties, including radial velocity data and lightcurves.

3.2. Fraction of Stars with Planets

The fraction of F, G, and K stars with giant planets is fairly well understood, out to orbital periods of several years that can be probed with the current data, with good agreement between different surveys. The simplest estimate that can be made is to divide the number of detected planets by the number of stars surveyed. *Marcy et al.* (2005b) did this for the Lick, Keck, and Anglo-Australian planet surveys. They found that 16/1330 = 1.2 ± 0.3% of stars have a hot Jupiter (a < 0.1 AU), 88/1330 = 6.6% of stars have a gas giant within 5 AU, and extrapolating the observed orbital period distribution to longer periods, 12% of F, G, and K stars have a gas giant within 20 AU. *Udry and Santos* (2007) give the corresponding numbers for the CORALIE Planet Search, finding 0.8% of stars with hot Jupiters, and 63/1120 = 5.6% of stars having giant planets within 5 AU.

These estimates are lower limits to the true planet fraction for these stellar samples, because they do not account for selection effects. Including completeness corrections for the Keck Planet Search, *Cumming et al.* (2008) concluded that 11 ± 1.7% of stars harbor gas giants within 5 AU, or 17% within 20 AU (Saturn mass and larger), using a flat extrapolation of the period distribution observed for a < 3 AU. *Naef et al.* (2005) find 7.3 ± 1.5% of stars have gas giants (M > 0.5 M_{Jup}) within 5 AU from the ELODIE survey. (Taking into account the different lower mass limits, these numbers are consistent at the 1σ level.)

It is important to be aware of the selection effects in the stellar sample when discussing the planet occurrence rate. The planet fractions we mentioned previously are for the particular samples of F, G, and K stars studied, which are biased selections of stars. For example, *Beatty and Gaudi* (2008) argue that a comparison between the 0.8 ± 0.3% hot Jupiter occurrence rate reported by *Udry and Santos* (2007) for the volume-limited CORALIE survey and the 1.5 ± 0.6% reported by *Cumming et al.* (2008) for the magnitude-limited Keck sample should take into account metallicity bias in the magnitude-limited sample. Accurately knowing the hot Jupiter occurrence rate is necessary to understand the yields of transit searches (*Beatty and Gaudi*, 2008). *Gould et al.* (2006a) calculated a hot Jupiter frequency of $0.3^{+0.4}_{-0.2}\%$ (P < 5 days) from the OGLE-III transit survey results, which appears consistent with the radial velocity numbers (see also *Gaudi et al.*, 2005).

At larger separations, beyond the reach of transit and current radial velocity surveys, there are indications from microlensing that low-mass planets are common. *Gould et al.* (2006b) detected a Neptune-mass 13 $M_⊕$ planetary companion to a 0.5 $M_⊙$ star with orbital separation 2.7 AU. They were able to infer that such cool Neptune planets (~10 $M_⊕$ in orbits beyond the ice line, ~1–4 AU) are common based on the detection of two planets from four events sensitive to them, with a derived frequency and 90% upper and lower limits of $0.38^{+0.31}_{-0.22}$ (see also *Beaulieu et al.*, 2006). This is significantly larger than the frequency of Jupiter-mass planets at similar distances from solar-mass stars, found to be 2–5% (for Jupiter mass and above between 2–5 AU) by *Cumming et al.* (2008).

3.3. Mass and Orbital Period Distributions

The mass-orbital period distribution is shown in Fig. 5. The two different symbols distinguish the transiting and radial velocity planets. The gray solid curves show the expected 50% and 99% detection thresholds for a radial velocity survey of eight years with σ = 3 m/s for solar mass stars. These curves match the lower envelope of detected planets quite well. Figure 6 shows the mass-semimajor axis distribution, now including the planets discovered by microlensing.

Perhaps the most striking feature is that the distribution of orbital periods is bimodal, with a population of "hot" planets with orbital periods close to 3 d, and a population of long-period planets with periods of ≥200 d. Figure 7 shows the distribution of semimajor axes. *Udry et al.* (2003) and *Jones et al.* (2003) emphasized this "period valley" in the period distribution.

The rising number of gas giants with orbital period has been fit with a power-law distribution. *Cumming et al.* (2008) fit the orbital period and M sin i distribution from the Keck Planet Search jointly with a power law, and found that dn ∝ $M^α P^β$ d ln M d ln P with α = –0.31 ± 0.2 and β = 0.26 ± 0.1 for a range of masses and periods M sin i = 0.3–10 M_{Jup} and P = 2–2000 d [similar values for α and β were found by *Tabachnik and Tremaine* (2002), who included detections from several surveys but with a simplified treatment of detection thresholds]. In terms of semimajor axis, this corresponds to dn/d ln a ∝ $a^{0.39}$, assuming equal stellar masses. *Cumming et al.* (2008) pointed out that an equally good description of the distribution is a step in the number of planets per decade beyond orbital periods of ≈300 d or semimajor axis ≈ 1 AU. For the Keck sample of F, G, and K dwarfs, they found the fraction of stars with planets per decade in orbital period dn/d log$_{10}$ P = 6.5 ± 1.4% at long periods, compared with dn/d log$_{10}$ P = 1.3 ± 0.4% at short periods.

The detection curves in Fig. 5 rise steeply at long periods, as discussed in section 2.2, but the data show a much sharper cutoff at P ≈ 2000–3000 d. This is because even though a companion can be detected statistically, it cannot be confirmed as a planet, as opposed to a low-mass star, for example, until a full orbit has been observed. There is a lot of information in the radial velocity datasets about orbits at long periods. Analysis of these long-period candidate planets is beginning to appear in the literature (e.g., *Fischer et al.*, 2001; *Wright et al.*, 2007; *Patel et al.*, 2007).

The group of hot Jupiters at short orbital periods shows a "pileup" in their orbital periods at P ≈ 3 d, with a tail that stretches out to longer periods. Figure 8 shows the distribution of radial velocity and transiting planets at short periods. There are planets with orbits inside 3 d, particularly those found by transit surveys, which survey a larger number of stars and whose sensitivity depends strongly on orbital period. Initially there were questions about why the "very hot" planets found by transit surveys were not being found by radial velocity surveys, but in fact the likely explanation is that they are rare (*Gaudi et al.*, 2005; *Gould et al.*, 2006a). Interestingly, the distribution of transiting planets is showing increasing numbers at 3–4-d orbital periods.

Marcy et al. (2005b, following *Marcy and Butler*, 2000) fit a power law to the observed minimum mass distribution, finding $dn/dM_p \sin i \propto 1/(M_p \sin i)^{1.05}$. In other words, the planets are distributed almost uniformly in log mass, rising slightly more steeply than that toward lower masses. The power law fits of *Tabachnik and Tremaine* (2002) and *Cumming et al.* (2008) given earlier rise more quickly to lower masses than the *Marcy et al.* (2005b) fit, which is as expected since the completeness corrections account for undetected low-mass planets.

The observed $M_p \sin i$ distribution of planets is shown in Fig. 9, for all observed planets and for those in close orbits with P < 100 d. For orbital periods beyond P ≈ 1 yr, the low-velocity amplitude makes it difficult to constrain the mass distribution below a Saturn mass. However, in close orbits, particularly the "hot" orbits within 10-d orbital periods, Fig. 5 shows that there are now many examples of Neptune-mass planets and below (mass range ~10 $M_\oplus$). Indeed, *Lovis et al.* (2006) discovered a three-Neptune planetary system around HD 69830. The long-term radial velocity precision of only ≈1 m/s of the HARPS spectrometer at La Silla Observatory allowed the detection of these signals from this 0.86 $M_\odot$ star, each of which has an amplitude of less than 3 m/s. Many, but not all, of the known hot Neptunes orbit M dwarfs, which helps with their detection since a less-massive star has a larger-velocity amplitude for a given planet mass (in equation (8) we scale to a typical M_{dwarf} mass).

The distribution of planet periods and masses has been contrasted with that of low-mass stars and stellar binaries. The roughly $dn/dM \propto 1/M$ mass distribution is significant because it clearly separates the distribution of planet masses from that of low-mass stars. These objects are not an extension of the low-mass stars/brown dwarfs to lower masses. [Indeed, there is a paucity of brown dwarfs at semimajor axes <3 AU, the so-called "brown dwarf desert" (*Marcy and Benitz*, 1989)]. This has been used to argue for differ-

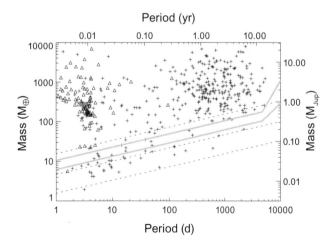

Fig. 5. Mass and orbital periods of known exoplanets. Crosses show planets detected by radial velocity only; triangles show transiting planets. The dotted lines correspond to stellar radial velocity amplitudes of K = 1, 3, and 10 m/s for a solar mass star. The gray curves give the 50% and 99% detection threshold for a survey of 8 yr with σ = 3 m/s, assuming solar-mass stars (see equation (7)).

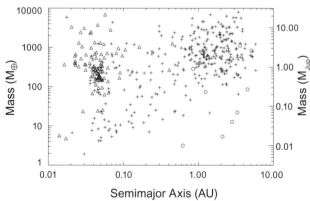

Fig. 6. Mass and semimajor axis of known exoplanets. Crosses show planets detected by radial velocity only; triangles show transiting planets; circles show planets discovered by microlensing.

ent formation mechanisms for brown dwarfs and planets. *Jones et al.* (2003) contrasted the distribution of planetary orbital periods with the stellar binary orbital period distribution, which shows no dip at intermediate orbital periods, being well fitted by a log-Gaussian (*Duquennoy and Mayor*, 1991), again consistent with the different formation mechanism of giant planets in a gas disk through which migration occurs.

In general, the distribution of planet masses and orbital periods can be described by core accretion models of planet formation including inward migration (e.g., *Ida and Lin*, 2004a,b, 2005, 2008a,b; *Armitage*, 2007; *Mordasini et al.*, 2009a,b). Models based on type II migration of gas giants predict a smooth increase in the number of gas giants with orbital radius, because of the increasing migration rate as the planet moves inward. In principle, therefore, how quickly the observed distribution increases to long periods can be used to determine the ratio of migration to formation/disk-depletion timescales. The most recent models by *Ida and Lin* (2008a,b) include type I migration, and propose that the upturn in planet frequency at ~1 AU corresponds to retention of solids near the ice line, which gives a preferred radius in the protoplanetary disk. *Mordasini et al.* (2009a,b) also find an upturn, although further out than observed by about a factor of 2, coming from two different evolutionary tracks: one in which cores migrate inward before reaching the condition for runaway accretion of gas, the other from cores that run away *in situ*. In addition, they predict a distribution of ≈20 $M_\oplus$ planets ranging in semimajor axis from ~0.1 to 5 AU, which is consistent so far with the Neptunes being found within a wide range of semimajor axis both by radial velocity surveys and microlensing.

3.4. Eccentricity Distribution

Exoplanets have a large range of eccentricities. This goes against what we might have expected in comparison with the solar system, in which the orbits are almost circular (e.g., Jupiter has an eccentricity of 0.05; Mercury's eccentricity is significantly larger than the other planets, with e = 0.21). The eccentricity distribution is shown in Fig. 10 for planets with orbital periods greater than 10 d, for which tidal damping of eccentricity is not expected to have occurred. The distribution is consistent with being flat at low eccentricity, continuing up to eccentricities beyond the median value of 0.25–0.3 and then drops sharply above e = 0.5. The reason for the drop at high eccentricity may be partly due to selection effects. Figure 11 illustrates this by comparing the K–e distribution against the detection curves calculated by *Cumming* (2004) (this is an updated version of Fig. 13 in that paper). The planets with large eccentricities all have large K values, K > 20 m/s for e ≥ 0.7. This suggests that the tail of eccentricities in Fig. 10 may be underestimated due to the finite sampling of radial velocity surveys, although a detailed analysis taking into account the selection effects on a star-by-star basis in different surveys is needed.

Marcy et al. (2005b) point out that the most massive planets have nonzero eccentricities. This is not due to a selection effect, since more massive planets should be easier to detect at all eccentricities than low-mass planets, yet only lower-mass planets are found with circular orbits. Figure 12 shows the eccentricity against mass for planets with P > 10 d (to avoid circularization effects). The median eccentricity for $M_P \sin i \geq 5$ M_{Jup} is clearly greater than the median eccentricity for M < 5 M_{Jup}. This could indicate that the most massive planets have a different formation mechanism (*Ribas and Miralda-Escudé*, 2007; *Ford and Rasio*, 2008).

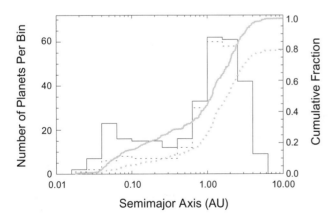

Fig. 7. The distribution of semimajor axes for nontransiting planets (radial velocity detection only). The solid curve is all planets; the dotted curves are for $M_P \sin i > 0.5$ M_{Jup}. The gray curves show the cumulative distribution in each case. The cutoff at a ≈ 5 AU is due to the finite length of the radial velocity surveys (orbital periods ~10 yr and greater).

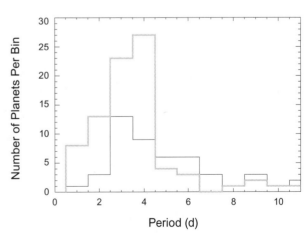

Fig. 8. Orbital periods of "hot" planets. The solid curve is for radial velocity detected planets; the gray curve is for transiting planets.

The eccentricity-semimajor axis distribution is shown in Fig. 13. The orbits of short-period planets are expected to be circularized by tides, and the eccentricity-semimajor axis distribution provides evidence for this. The circularization timescale for small e due to tides raised on the planet is

$$\tau_e = \frac{4Q}{63\Omega}\left(\frac{M_P}{M_\star}\right)\left(\frac{a}{R_P}\right)^5 \quad (18)$$

where the parameter Q describes the tidal dissipation inside the planet. For a Jupiter-mass and radius planet orbiting a solar-mass star, this is $\tau_e \approx 1$ G.y. $(Q/10^6)(a/0.05 \text{ AU})^{13/2}$. The value of Q is uncertain; values somewhere between ~10^5 to 10^6 are likely (*Wu*, 2003). Therefore we see that within a periastron distance on the order of 0.05 AU, the circularization time is smaller than 1 G.y., and it is reasonable to expect that the planet's orbit will have circularized over the lifetime of the star. In Fig. 13 the dotted curve shows a constant value of a(1−e) = 0.05 AU. Indeed, this roughly separates the occupied and unoccupied regions, suggesting that tidal dissipation sets the upper envelope of the distribution. This cannot be the only factor, however, as there is a significant group of planets with nonzero eccentricity at short orbital periods. In some cases, a yet-undetected additional companion may excite the eccentricity of the short-period planet through gravitational interactions.

Earlier, we discussed the differences in the mass and orbital period distributions of exoplanets compared to stellar binaries, and how these differences have been used to argue that the planets are a distinct population with a different formation mechanism. For the eccentricity distribution, the opposite is the case. *Stepinski and Black* (2001) pointed out that the period-eccentricity distribution for planets is very similar to that for binary stars, and suggested that this may result from a similar formation mechanism. More recently, *Halbwachs et al.* (2005) point out that there are differences, e.g., there are several long-period planets with almost circular orbits (analogous to our solar system), which is not the case in the stellar binaries.

The origin of the observed exoplanet eccentricities is still not understood. Since gas giants are believed to form in close to circular orbits in a gas disk, most ideas rely on dynamical interactions following formation of the planet to increase its eccentricity. Building on their earlier work, *Ford and Rasio* (2008) find that two-body scattering with a range of planet mass ratios can explain the observed eccentricity distribution of most planets, but that there should be a well-defined maximum eccentricity of ≈0.8 (for initially circular orbits of both planets), so that the most eccentric planets require some other mechanism to excite their eccentricity. *Jurić and Tremaine* (2008) simulate systems of several planets and find that the eccentricity distribution for a wide range of initial conditions relaxes to the form dn ∝ e exp$\left[-\frac{1}{2}(e/0.3)^2\right]$de. This distribution is shown as the dotted curve in Fig. 10, and matches the shape of the observed distribution well at larger eccentricities, but not for small eccentricities. Therefore, an extra mechanism is needed to make the population of planets on circular orbits (perhaps these are planets that formed in a disk and did not undergo gravitational scattering).

One problem with these models is that the distribution of semimajor axis tends to be smooth with increasing a, and does not match the factor of 5 observed increase in the number of planets orbiting beyond ~1 AU compared to within 1 AU (see Fig. 7 of *Jurić and Tremaine*, 2008). This increase is present even when planets with circular orbits are excluded (see Fig. 13). Other proposals for producing the observed eccentricities are due to the interaction of a

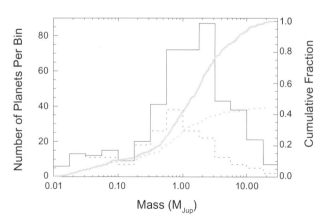

Fig. 9. Observed mass distribution of exoplanets. The gray curve shows the cumulative fraction. The dotted curves are for planets with P < 100 d.

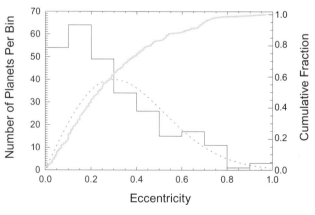

Fig. 10. Eccentricity distribution for planets with orbital periods greater than 10 d. The gray curve shows the cumulative distribution. The dotted curve shows the theoretical distribution from planet-planet scattering derived by *Jurić and Tremaine* (2008). The normalization has been adjusted to roughly fit the tail of the observed distribution.

migrating planet with the disk itself (*Goldreich and Sari*, 2003), and interactions with a passing star (*Zakamska and Tremaine*, 2004), or companion star (*Holman et al.*, 1997; *Wu and Murray*, 2003; *Fabrycky and Tremaine*, 2007; *Wu et al.*, 2007). The observed eccentricity distribution is likely due to a combination of these mechanisms.

3.5. Planet Incidence Versus Stellar Metallicity

It is now well established that the occurrence rate of giant planets increases dramatically with increasing metallicity of the host star. Evidence that this is the case emerged from the first planet discoveries. *Gonzalez* (1997) performed abundance analyses of four of the early planet discoveries, and noted that they were all metal rich relative to the Sun. This trend has been confirmed with the many planet detections since then: There are many more planets known around metal-rich than metal-poor stars. One might worry that this could be an effect of the stellar selection. We have already discussed the possible metallicity biases possible in stellar samples; e.g., metal-rich stars are overrepresented in a magnitude-limited sample with a cutoff in B–V (section 2.4). To eliminate these selection effects, *Santos et al.* (2004) and *Fischer and Valenti* (2005) measured metallicities in a uniform way for samples of stars with and without planets, and so could directly address the likelihood that a star in a particularly metallicity bin would have a planet.

Figures 14 and 15 show the results from *Fischer and Valenti* (2005), who determined metallicities and other stellar parameters for more than 1000 stars observed as part of the Keck, Lick, and Anglo-Australian Planet Searches. They identified a subset of 850 stars that they estimated had similar radial velocity detectability (more than 10 observations spanning four or more years). [You may also wonder whether the typical velocity precision would change with metallicity and could introduce a bias. In fact, the mean velocity precision seems to be constant for [Fe/H] in the range –0.5 to 0.5 (see *Valenti and Fischer*, 2008).] Figure 14 compares the metallicity distributions for the stars with and without planets, and Fig. 15 shows the fraction of stars with planets in each metallicity bin. *Fischer and Valenti* (2005) find that the probability that a star has a planet is well fit by the expression

$$\text{Prob(planet)} = 0.03 \times 10^{2.0[\text{Fe/H}]}$$
$$= 0.03 \left[\frac{N_{Fe}/N_H}{(N_{Fe}/N_H)_\odot} \right]^2 \qquad (19)$$

(shown by the gray curve in Fig. 15).

For hot Jupiters, another way to address the incidence of planets at low metallicities is transit searches of globular clusters. *Gilliland et al.* (2000) found that the frequency of hot Jupiters in the globular cluster 47 Tuc is at least an order of magnitude below that found in radial velocity surveys of stars in the solar neighborhood. They found no transiting hot Jupiters in their sample of ~34,000 main-sequence stars, whereas if the occurrence rate is 1%, 17 detections would have been expected. More recently, *Weldrake et al.* (2008) placed limits on the occurrence rate of planets at periods between 1 and 5 d in the globular cluster ω Cen. The majority of stars studied are members of the low-metallicity population of the cluster, which has [Fe/H] = –1.7. The combined rate of hot and very hot Jupiters that they find is approximately <1/600. We should note that both of these studies are for planets with radii larger than $R_p = 1.3$ R_{Jup} or 1.5 R_{Jup} respectively, which decreases the expected rate compared to that for all hot Jupiters (see the distribution of transiting planet radii in Fig. 18).

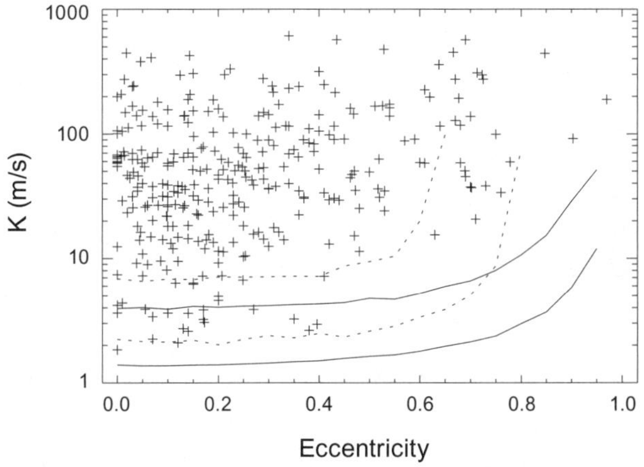

Fig. 11. Radial velocity amplitude against eccentricity. The curves show 50% (solid) and 99% (dotted) detection thresholds calculated by *Cumming* (2004), assuming 16 (top curve in each pair) and 39 (bottom curve in each pair) measurements, and σ = 5 m/s.

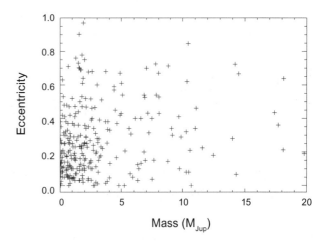

Fig. 12. Planet mass against eccentricity. More massive planets have a larger median eccentricity than low-mass planets.

The origin of the metallicity correlation could be either that planet formation or migration are enhanced around metal-rich stars, or that the metallicity of the star is enhanced by planet formation itself. Following the suggestion of *Lin et al.* (1996) that the hot Jupiters migrated inward along with the remains of the protoplanetary disk, *Gonzalez* (1997) proposed that the addition of heavy elements to the outer convection zone of the star would be enough to raise the star's metallicity by a significant amount. The thickness of the convection zone at the time of the accretion event, which depends on the mass of the star (more massive stars have thinner outer convective envelopes) and the timing of the accretion event (the mass of the convection zone decreases as the star settles onto the main sequence), is critical since the added material is mixed throughout the convection zone and therefore diluted. However, studies of the metallicities of stars as a function of their effective temperature, or in subgiants compared to main-sequence stars, do not show any clear evidence that the accretion of solids gives rise to the observed metallicity dependence of the planet occurrence rate (*Pinsonneault et al.*, 2001).

The alternative is that metallicity affects the migration process, which brings planets closer to their host star where they are easier to detect, or that metallicity directly affects planet formation. In the context of the core-accretion model for forming Jupiters, this would not be surprising. We might naively expect the probability of forming a planet to scale with N^2 (equation (19)) if planet formation involved binary collisions of solids. Core-accretion models by *Ida and Lin* (2004b) or *Mordasini et al.* (2009b) do indeed show an increasing abundance of planets with metallicity.

The behavior at low metallicity is still being debated. The planet abundance against [Fe/H] appears consistent with going to a constant at low values of [Fe/H] (Fig. 15), as suggested by *Santos et al.* (2004). However, *Cochran et al.* (2008) find that many of the stars with low [Fe/H] are members of the thick disk population rather than the thin disk, and have a larger fraction of α-elements than thin disk stars. This means that [Fe/H] actually underestimates their total metallicity if measured as the total mass fraction of elements heavier than helium, $Z/Z_\odot$. It may be that the planet fraction does indeed go to zero below some value of $Z/Z_\odot$, as expected as the raw materials for planet building become rarer. There is clearly a lot of interesting work to do studying the relationships between stellar abundances and planet occurrence rates and properties.

3.6. Planet Incidence Versus Stellar Mass

Stellar mass is another parameter that could affect the likelihood of forming planets, and their masses and orbital properties. The distribution of masses and metallicities of known planet-bearing stars is shown in Fig. 16. The range of stellar masses goes from ≈0.3 $M_\odot$ (main-sequence M dwarfs) to almost 3 $M_\odot$ (evolved stars). However, most planet detections are concentrated in the stellar mass range close to 1 $M_\odot$ (mostly G and K spectral types). This is because radial velocity surveys have traditionally focused on this narrow range of stellar masses, where the stellar spectra are most amenable to radial velocity measurements. At higher stellar masses, main-sequence stars earlier than late F-type are generally not included in radial velocity surveys, because of the low radial velocity precision that can be achieved (~50–100 m/s). The reason for this is that these stars have hot atmospheres giving fewer spectral lines, and are rapid rotators giving high levels of jitter and rotationally broadened lines. At lower masses, because M dwarfs are faint they require more observing time to achieve the same precision as brighter solar-type stars.

Observational programs have begun to enlarge the sample

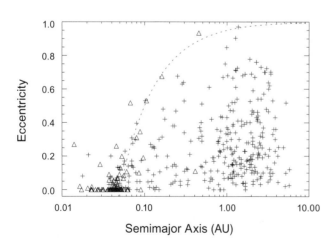

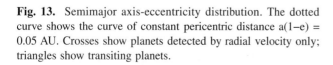

Fig. 13. Semimajor axis-eccentricity distribution. The dotted curve shows the curve of constant pericentric distance a(1−e) = 0.05 AU. Crosses show planets detected by radial velocity only; triangles show transiting planets.

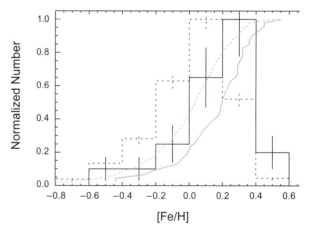

Fig. 14. Distribution of stellar metallicities of stars with uniform radial velocity detectability (dotted curves), and those with planets (solid curves). The histograms have been normalized so that the peak value is 1. Data are from *Fischer and Valenti* (2005) and *Valenti and Fischer* (2005). There are 850 stars in total, of which 47 have planets.

of low- and high-mass stars surveyed for planets, and evidence is emerging that the planet occurrence rate and even the distribution of orbital properties depends on stellar mass. One potential selection effect that must be considered carefully is the mass-metallicity correlation introduced into a stellar sample by making a cut in B–V (section 2.4).

At the low-mass end, there is now good evidence for a low occurrence rate of giant planets around M dwarfs. For example, *Butler et al.* (2006b) estimated the occurrence rate of giant planets within 2.5 AU as 2/114 = 1.8 ± 1.2% for planet masses ≥0.4 M_{Jup}. The equivalent occurrence rate for F, G, and K stars is between 6% and 7% (*Cumming et al.*, 2008). Taking into account differences in selection effects between the F, G, and K stars and the M dwarfs, *Cumming et al.* (2008) found that M dwarfs were 3–10 times less likely to harbor a gas giant planet with an orbital period less than 2000 d (set by the duration of the Keck survey).

A low occurrence rate is predicted by core accretion models for planet formation, which find that Jupiter-mass planets should be rare around M dwarfs, with the mass function of planets shifted toward lower masses (*Laughlin et al.*, 2004; *Ida and Lin*, 2005; *Kennedy and Kenyon*, 2008). *Kennedy and Kenyon* (2008) assume that the mass of the protoplanetary disk scales $\propto M_\star$, and include a detailed calculation of the position of the snow line. Their Fig. 7 shows that the probability of having at least one giant planet is six times lower for 0.4 $M_\odot$ star than a 1 $M_\odot$ star, in good agreement with the observed fraction from radial velocity surveys. A challenge for these models is to still be able to produce massive planets, e.g., the recent microlensing discovery of a 3.8 M_{Jup} planet at 3.6 AU from a 0.46 $M_\odot$ M dwarf (*Dong et al.*, 2009).

At higher stellar masses, the evidence is that giant planets are more common. The most massive stars in Fig. 16 are evolved stars, either subgiants or giants. Whereas it is not possible to achieve high-precision velocities for massive stars on the main sequence, once these stars evolve off the main sequence and move to later spectral types, it again becomes possible to look for planets. *Johnson et al.* (2007, 2008) describe the first results from survey of 159 subgiants with Lick and Keck, for which they can achieve a precision of ≈2–5 m/s. They use these results to investigate the stellar mass dependence of the planet occurrence rate. To ensure uniform detectability between stars of different masses, they select stars that have at least eight observations with an observing time necessary to detect a companion at a = 2.5 AU, and consider only $M_p \sin i > 0.8$ M_{Jup} (this mass limit is set by the need to be able to detect companion around subgiants, which have lower radial velocity precision, and being more massive, lower stellar velocity amplitudes for a given planet mass). They find planet occurrence rates of 1.8 ± 1.0% for $M_\star < 0.7$ $M_\odot$, 4.2 ± 0.7% for 0.7 $M_\odot < M_\star < 1.3$ $M_\odot$, and 8.9 ± 2.9% for $M_\star > 1.3$ $M_\odot$. The increasing trend of planet occurrence rate with mass remains after correcting for the mean metallicity of each mass bin.

This trend is consistent with the core accretion models of *Kennedy and Kenyon* (2008). They find that the planet occurrence rate is sensitive to the assumed dependence of snow line location with stellar mass. For example, *Ida and Lin* (2005), with a steeper dependence of snow line position on mass, predict a decrease in the planet fraction for $M_\star \geq 1$ $M_\odot$.

Johnson et al. (2007, 2008) emphasize the different semimajor axis distribution of the planets discovered around subgiants and giants compared to the sample of F, G, and K stars. This is illustrated in Fig. 17, which shows the mass semimajor axis distribution of known planets, with stellar masses <0.5 $M_\odot$ (corresponding to M dwarfs) and >1.5 $M_\odot$ (evolved stars) highlighted. [Note that the ≈3 $M_\oplus$ microlensing planet MOA-2007-BLG-192-L b may orbit a substellar object. The best-fit mass for the host star is well

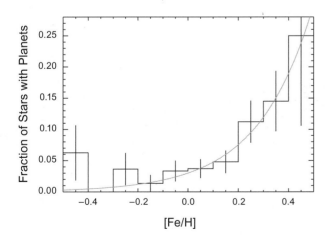

Fig. 15. Fraction of stars with planets as a function of [Fe/H], calculated using the same data as in Fig. 14. The gray curve shows the relation $0.03 \times 10^{2[Fe/H]}$ from *Fischer and Valenti* (2005).

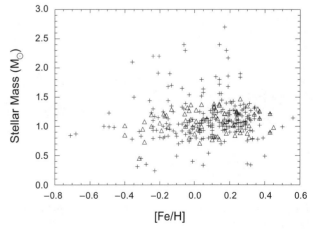

Fig. 16. Distribution of stellar metallicities and masses for transiting planet host stars (triangles) and stars with RV detected planets (crosses).

below an M dwarf mass, $M_\star = 0.06^{+0.028}_{-0.021} M_\odot$ (*Bennett et al.*, 2008).] The planets orbiting evolved stars are concentrated in long-period orbits, out beyond 0.6 AU [the exception in Fig. 17 is the hot Jupiter orbiting HD 102956 (*Johnson et al.*, 2010)]. This cannot be due to selection effects, since shorter orbital periods would be easier to detect. Similarly, no planet with mass ~1 M_{Jup} has been detected within 0.1 AU of an M dwarf. Continued observations of evolved stars and M dwarfs will clarify this trend. Particularly for giants, tidal effects on close orbits must be included when trying to infer how much of this trend is due to differences in planet formation or migration between stars of different masses (*Sato et al.*, 2008).

3.7. Mass-Radius Relation of Transiting Planets

The statistical study of the properties of transiting planets is just beginning. Transiting planets are particularly interesting because we are not restricted to measuring the orbital elements, but can study planet properties such as radius and atmospheres. Some interesting trends in the population are already emerging.

The masses and radii of the transiting planets are shown in Fig. 18. At a given mass, the transiting planets show a range of radii, or alternatively a range of densities. Given the uncertainties in the physics of giant planet interiors (e.g., *Saumon and Guillot*, 2004), the study of a large sample of gas giant planets is of great interest. The first transiting planet HD 209458 had a larger radius than expected (1.3 M_{Jup}) (*Charbonneau et al.*, 2000), which has been attributed to tidal heating caused by damping of eccentricity driven by gravitational interactions with a second planet in the system (*Bodenheimer et al.*, 2003, *Laughlin et al.*,

2005b), or the insulating effect of stellar irradiation as the planet migrates inward, which keeps the internal entropy of the planet at a larger value than a planet cooling in isolation (*Burrows et al.*, 2003). A smaller radius than expected can be explained by the presence of a rocky core. For example, the Saturn-mass planet orbiting HD 149026 has a radius of only 0.73 R_{Jup}, and is inferred to have a core of mass ≈70 $M_\oplus$ (*Sato et al.*, 2005). *Guillot et al.* (2006) propose that the radii of all transiting planets can be accommodated by having different size cores, with the core mass increasing with stellar metallicity (see also *Burrows et al.*, 2007).

Figure 19 shows the masses and orbital periods of closely orbiting planets. Inspection of this figure shows that the planet mass is anticorrelated with the orbital period (for planet masses ≥0.1 M_{Jup}). One proposal is that this is evidence that planet-planet scattering is responsible for producing hot Jupiters. Some fraction of planets will scatter into orbits in which they approach very close to the star, and could tidally circularize there. In this picture, there should be a limit on the orbital radius of hot Jupiters of twice the Roche limit (*Rasio and Ford*, 1996; *Ford and Rasio*, 2006), since a highly eccentric orbit circularizes to twice its periastron distance. [To see this, conserve angular momentum $\propto \sqrt{GMa(1-e^2)}$ during the circularization, so that $a_{final} = a_{initial}(1-e^2) \approx 2a_{periastron}$ for e ≈ 1.] Writing the Roche limit a_R as $R_p = 0.462 \ a_R (M_p/M_\star)^{1/3}$, then $a_{circ} = 2a_R$ gives $M_p = 1.1 \ M_{Jup} \ (R_p/R_{Jup})3(P/1 \ d)^{-2}$, which is plotted as a dotted line in Fig. 19. This is quite close to the lower envelope of the planets with $M_p > 0.1 \ M_{Jup}$.

Other correlations between parameters have been discussed in the literature. *Hansen and Barman* (2007) point out that the transiting planets appear to fall into two classes based on Safronov number, $\Theta = (1/2)(v_{esc}/v_{orb})^2 = (a/R_p) (M_p/M_\star)$, where v_{esc} is the escape speed from the planet, and v_{orb} is the planet's orbital velocity. The size of the Safronov

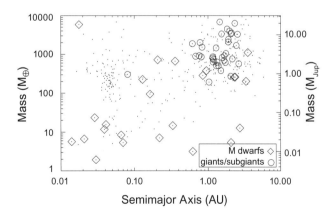

Fig. 17. Mass and orbital period distribution as a function of stellar mass. The dots show all the planets plotted in Fig. 6. The circled points are planets around stars more massive than 1.5 $M_\odot$ (evolved stars: giants and subgiants); the diamonds show stars less massive than 0.5 $M_\odot$ (M dwarfs). The dotted lines correspond to stellar radial velocity amplitudes of K = 1, 3, and 10 m/s for a solar-mass star.

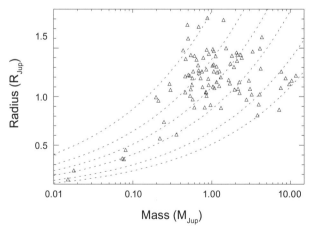

Fig. 18. Mass-radius relation for observed transiting planets. The dotted curves show constant mean density $\langle \rho \rangle = 3 M_p/4\pi R_p^3 = 0.2$, 0.5, 1.0, 2.0, 5.0, and 10.0 g cm^{-3}.

determines whether the planet captures or scatters other objects in neighboring orbits, and so this difference could point to differences in migration or stopping mechanisms. However, the statistical significance of the correlation has been questioned (*Southworth*, 2008). As the number of transiting planets increases further, we are sure to learn a tremendous amount about their internal structure and formation and evolution histories.

3.8. Multiple Planet Systems

There are 46 multiple-planet systems in the set of planets considered here, about 10% of the 397 planetary systems. Of these, 15 have three or more planets and 5 have four or more planets, including the star 55 Cnc, which has five companions (*Fischer et al.*, 2008). Note that the chance of having additional planets is much greater for a two-planet system (15 out of 46) than for a single-planet system (46 out of 397). This may be partially due to a selection effect, in that two-planet systems will have been observed more carefully than many single planet systems, making it more likely to detect additional companions. Analysis of the multiplicity of planets taking into account selection effects has not been carried out.

Several multiple-planet systems show orbital period ratios that are integers or close to integers. For example, the two planets in GJ 876 have P ≈ 30 and 60 d, and are believed to be in a 2:1 mean-motion resonance. A natural way to form such an arrangement is by migration. As planets migrate inward, they can be trapped in resonant configurations (e.g., *Kley et al.*, 2005). The occurrence of resonant configurations is a strong argument that migration does take place.

Planets in resonant configurations are particularly interesting since over time they will show correlated changes in orbital elements. An example is the GJ 876 system (*Laughlin et al.*, 2005a; *Rivera et al.*, 2005; *Bean and Seifahrt*, 2009). The fit to the radial velocity measurements for this system is significantly improved if the gravitational interactions between the planets is included. Instead of fitting Keplerians, the equations of motion for the multiple body system are integrated directly to predict the expected radial velocity evolution. Because the dynamical interactions depend on the planet masses, whereas the Keplerian fits determine $M_P \sin i$, the interactions can be used to determine the inclination of the orbits, and whether the orbits are coplanar. Another example is the three-planet system orbiting υ And (the first multiple-planet system discovered orbiting a main-sequence star). The outer two planets orbit with periods of 241 d and 1301 d (the third planet is a hot Jupiter), and are expected to show evolution of their eccentricities and longitudes of periastron on long timescales of thousands of years (*Ford et al.*, 2005).

In Fig. 20, we show the known planets in the mass orbital period plane, highlighting those planets that are known to be in multiple systems. Note that this does not include stars with single planets that have an underlying trend in their radial velocity, many of which could have a planet at long orbital periods (*Fischer et al.*, 2001; *Wright et al.*, 2007). Quite striking in this figure is that the majority of the super Earths and Neptunes ($M_P \sin i \lesssim 30\ M_\oplus$) are in multiple-planet systems (see, e.g., *Mayor et al.*, 2009). However, one could imagine that this is a selection effect. Neptune-mass planets are detected with small signal-to-noise values, requiring many observations. Multiple-planet systems are likely to have been subject to intense observational scrutiny, accumulating many observations that could make them amenable to detecting planets with low masses.

Microlensing allows planetary systems at large orbital separations to be probed. There has been one detection so far; *Gaudi et al.* (2008) report the detection of a planetary

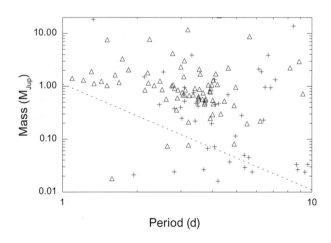

Fig. 19. Mass and orbital periods of closely orbiting exoplanets. Crosses show planets detected by radial velocity only; triangles show transiting planets. The dotted line shows twice the Roche limit, assuming $R_p = R_{Jup}$.

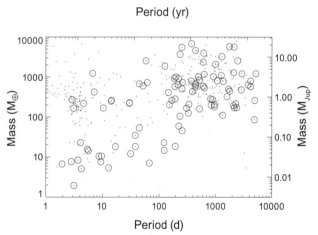

Fig. 20. Orbital period and mass distribution of planets, with members of multiple planet systems circled. Announced multiple planet systems only. This does not include those planets that show a long-term trend indicating a planet further out.

system that is a scaled version of our Jupiter/Saturn system, with 0.71 M_{Jup} and 0.27 M_{Jup} planets with orbital separations of 2.3 and 4.6 AU from a 0.5 $M_\odot$ host star. Based on the fact that the two other Jupiter-mass planets detected by microlensing were not very sensitive to multiple-planet systems, they suggested that the occurrence rate of multiple Jupiters is likely to be high, which promises more of these systems will be discovered in the future.

4. FUTURE PROSPECTS

This is an exciting time in exoplanet statistics. The current sample of planets is now large enough to yield statistically meaningful constraints on orbital properties, giving input for planet formation theories. In the next few years, we should learn much more as the sample increases further and new regions of the mass-orbital period plane and a wider range of stellar properties are explored.

Doppler surveys are moving forward in a few different ways. Doppler precision continues to improve beyond the 1 m s^{-1} level. *Pepe and Lovis* (2008) discuss the prospects for lowering the Doppler precision from the current ≈1 m s^{-1} level to the cm s^{-1} level. They conclude that $\sigma \approx 10$ cm s^{-1} should be possible for some stars that are "quiet" enough, although it is not yet clear for how long such a precision could be maintained. Precise radial velocity measurements so far have been made at optical wavelengths, but *Ramsey et al.* (2008) demonstrate the first measurements in the near-infrared at the 10 m s^{-1} level. This has the potential to allow studies of M dwarfs that are not currently possible, moving the mass function of hot planets to smaller masses. As part of the Sloan Digital Sky Survey III, the Multi-Object APO Radial Velocity Exoplanet Large-Area Survey (MARVELS) will monitor 11,000 bright stars with a precision ~10 m s^{-1}. This survey makes use of a dispersed fixed-delay interferometer to make simultaneous observations of ≈60 stars in a given exposure, and ≈500 stars per night (*Mahadevan et al.*, 2008). *Johnson et al.* (2008) have added 300 new subgiants to their program to look for planets around massive stars, which should enlarge the sample by 20–30 more planets in the next few years. Continued monitoring of stars included in current surveys will reveal planets in long-period orbits. *Wright et al.* (2008) announced the discovery of a Jupiter twin around the star HD 154345. This planet has a mass of 0.95 M_{Jup}, and a 9.2-yr circular orbit, with no evidence so far for other planets in the system.

For transits, the number of planets has grown tremendously in the last few years, and will continue to increase in the near future with observations from the ground and from space (the CoRoT and Kepler missions). Whereas photometric precision from the ground is limited to ≈0.1%, spacebased observations allow a photometric precision adequate to detect Earth-sized planets [recall that $\delta \approx (R_P/R_\star)^2 \approx 10^{-4}$ for $R_P = R_\oplus$]. The NASA Kepler mission, launched in 2009, will continuously monitor ≈10^5 stars for transits over the 3.5-yr mission lifetime. It should find hundreds of terrestrial-mass planets if they are common around solar-type stars (scaling from the ≈10% probability of observing a transit for a hot Jupiter at 0.05 AU, we see that the probability is ≈0.5% at 1 AU for a solar-type star), out to orbits of 1 AU and beyond, and increase the sample of hot Jupiters by tens (*Basri et al.*, 2005). This will provide a census of Earths and super Earths in close orbits.

Even with only a handful of detections, microlensing has already provided interesting constraints on the planet population beyond the snow line, complementing the discoveries made by radial velocity and transit surveys, e.g., indicating that cold Neptunes are common at orbital radii of a few AU. Microlensing offers some unique views of the planet population: It is potentially sensitive to Earth-mass planets beyond the snow line; it can detect old, free-floating planets; it surveys a different population of stars, being sensitive to planets in the bulge and disk of the Milky Way at large distances, and probes a wide range of host stars. Next-generation microlensing searches are currently being put together or planned (including the MOA-II and OGLE-IV upgrades, and the South Korean KMTNet project undergoing construction over the next few years), involving continuous wide-field large area surveys that would survey ~10^7 stars on ~10-min timescales, giving thousands of microlensing events per year that would all be monitored for planetary perturbations (rather than a small subsample as in current alert-follow-up surveys). Simulations indicate a detection rate of ≈10 super Earths per year and ≈1 Earth-mass planet per year (e.g., *Gaudi*, 2008), an order of magnitude improvement on the current detection rate. A spacebased mission such as the Microlensing Planet Finder [a 1-m-class telescope, imaging the galactic bulge continuously for several months (*Bennett et al.*, 2007)] would be 100 times as effective at finding Earth-mass planets (*Bennett*, 2004).

Direct searches are beginning to detect planetary companions, e.g., the HR 8799 system with three massive gas giants at ≈20–70 AU (*Marois et al.*, 2008), and had already placed interesting statistical constraints on giant planet occurrence rates at large orbital radii. The Gemini Deep Planet Survey (*Lafrenière et al.*, 2007), a near-infrared survey of 85 nearby young stars using adaptive optics at the Gemini North telescope, was able to place 95% upper limits on the fraction of stars with a gas giant between 10 and 25 AU of 0.28, 25 and 50 AU of 0.13, and 50 and 250 AU of 0.09. This compares to the fraction 0.17–0.2 for gas giants within 20 AU from the Keck radial velocity survey (*Cumming et al.*, 2008). Follow-on projects in the next few years such as the Gemini Planet Imager (*Macintosh et al.*, 2006) should increase the detection rate of gas giants at separations ≥5 AU, informing our understanding of the early evolution of gas giants, and when and how planets migrate to large orbital radii.

Future astrometry space missions should allow detection of a significant sample of planets at AU separations. Depending on the bright star single measurement precision achievable (they assume 8 µas), *Casertano et al.* (2008) find that Gaia could potentially find several thousands of gas giant planets at orbital radii 1–4 AU around stars within 200 pc, with hundreds of multiple-planet systems. The flux-limited stellar sample would consist of a range of spectral types, ages, and metallicities. The Space Interferometry Mission, with a single measurement accuracy of 1 µas, could uncover tens of terrestrial planets (~3 M$_\oplus$) in habitable zones around stars within 30 pc (*Ford and Tremaine*, 2003; *Catanzarite et al.*, 2006).

Theoretical models of core accretion can reproduce many of the statistical properties of exoplanets; recent papers include *Ida and Lin* (2008b), *Kennedy and Kenyon* (2008), and *Mordasini et al.* (2009b). As the observational sample grows to include a wider range of orbital, planet, and stellar properties, a challenge for theorists is to identify which properties of the observed distributions constrain key aspects of the theoretical models, and can distinguish between alternative scenarios. In this way, the full potential of the upcoming observational discoveries can be realized.

Acknowledgments. I thank the referees for a careful reading of this chapter and constructive suggestions that improved the text. I am grateful for support from the National Sciences and Engineering Research Council of Canada (NSERC) and the Canadian Institute for Advanced Research (CIFAR).

REFERENCES

Armitage P. J. (2007) Massive planet migration: Theoretical predictions and comparison with observations. *Astrophys. J., 665*, 1381–1390.

Avni Y., Soltan A., Tananbaum H., and Zamorani G. (1980) A method for determining luminosity functions incorporating both flux measurements and flux upper limits, with applications to the average X-ray to optical luminosity ratio for quasars. *Astrophys. J., 238*, 800–807.

Basri G., Borucki W. J., and Koch D. (2005) The Kepler mission: A wide-field transit search for terrestrial planets. *New Astron. Rev., 49*, 478–485.

Bean J. L. and Seifahrt A. (2009) The architecture of the GJ 876 planetary system. Masses and orbital coplanarity for planets b and c. *Astron. Astrophys., 496*, 249.

Beatty T. G. and Gaudi B. S. (2008) Predicting the yields of photometric surveys for transiting extrasolar planets. *Astrophys. J., 686*, 1302–1330.

Beaulieu J.-P. et al. (2006) Discovery of a cool planet of 5.5 Earth masses through gravitational microlensing. *Nature, 439*, 437.

Bennett D. P. (2004) The detection of terrestrial planets via gravitational microlensing: Space vs. ground-based surveys. In *Extrasolar Planets: Today and Tomorrow* (J.-P. Beaulieu et al., eds.), p. 59. ASP Conf. Proc. 321, San Francisco.

Bennett D. P. and Rhie S. H. (1996) Detecting Earth-mass planets with gravitational microlensing. *Astrophys J., 472*, 660.

Bennett D. P. et al. (2007) An extrasolar planet census with a space-based microlensing survey. White paper submitted to the NASA/NSF ExoPlanet Task Force. *ArXiv e-prints, arXiv:0704.0454.*

Bennett D. P. et al. (2008) A low-mass planet with a possible substellar-mass host in microlensing event MOA-2007-BLG-192. *Astrophys. J., 684*, 663.

Bodenheimer P., Laughlin G., and Lin D. N. C. (2003) On the radii of extrasolar giant planets. *Astrophys. J., 592*, 555.

Bramich D. M. and Horne K. (2006). Upper limits on the hot Jupiter fraction in the field of NGC 7789. *Mon. Not. R. Astron. Soc., 367*, 1677.

Brown T. M. (2003) Expected detection and false alarm rates for transiting jovian planets. *Astrophys J. Lett., 593*, L125.

Burrows A., Sudarsky D., and Hubbard W. B. (2003) A theory for the radius of the transiting giant planet HD 209458b. *Astrophys. J., 594*, 545–551.

Burrows A., Hubeny I., Budaj J., and Hubbard W. B. (2007) Possible solutions to the radius anomalies of transiting giant planets. *Astrophys. J., 661*, 502–514.

Butler R. P., Marcy G. W., Vogt S. S., Fischer D. A., Henry G. W., Laughlin G., and Wright J. T. (2003) Seven new Keck planets orbiting G and K dwarfs. *Astrophys. J., 582*, 455.

Butler R. P., Wright J. T., Marcy G. W., Fischer D. A., Vogt S. S., Tinney C. G., Jones H. R. A., Carter B. D., Johnson J. A., McCarthy C., and Penny A. J. (2006a) Catalog of nearby exoplanets. *Astrophys. J., 646*, 505–522.

Butler R. P., Johnson J. A., Marcy G. W., Wright J. T., Vogt S. S., and Fischer D. A. (2006b) A long-period Jupiter-mass planet orbiting the nearby M dwarf GJ 849. *Publ. Astron. Soc. Pacific, 118*, 1685–1689.

Casertano S. et al. (2008) Double-blind test program for astrometric planet detection with Gaia. *Astron. Astrophys., 482*, 699–729.

Catanzarite J., Shao M., Tanner A., Unwin S., and Yu J. (2006) Astrometric detection of terrestrial planets in the habitable zones of nearby stars with SIM PlanetQuest. *Publ.. Astron. Soc. Pacific, 118*, 1319.

Charbonneau D., Brown T. M., Latham D. W., and Mayor M. (2000) Detection of planetary transits across a Sun-like star. *Astrophys. J. Lett., 529*, L45–L48.

Cochran W. D., Endl M., MacQueen P. J., and Barnes S. (2008) The planet-metallicity correlation. *American Astronomical Society, DPS meeting #40, #4.05.*

Cumming A. (2004) Detectability of extrasolar planets in radial velocity surveys. *Mon. Not. R. Astron. Soc., 354*, 1165.

Cumming A. and Dragomir D. (2010) An integrated analysis of radial velocities in planet searches. *Mon. Not. R. Astron. Soc., 401*, 1029.

Cumming A., Marcy G. W., and Butler R. P. (1999) The Lick Planet Search: Detectability and mass thresholds. *Astrophys. J., 526*, 890.

Cumming A, Butler R. P., Marcy G. W., Vogt S. S., Wright J. T., and Fischer D. A. (2008) The Keck Planet Search: Detectability and the minimum mass and orbital period distribution of extrasolar planets. *Publ. Astron. Soc. Pacific, 120*, 531.

Defäy C., Deleuil M., and Barge P. (2001) A Bayesian method for the detection of planetary transits. *Astron. Astrophys., 365*, 330–340.

Dong S., Gould A., Udalski A., Anderson J., Christie G. W., et al. (2009) OGLE-2005-BLG-071Lb, the most massive M dwarf planetary companion? *Astrophys. J., 695*, 970.

Duquennoy A. and Mayor M. (1991) Multiplicity among solartype stars in the solar neighbourhood. II — Distribution of the orbital elements in an unbiased sample. *Astron. Astrophys., 248*, 485–524.

Endl M., Krster M., Els S., Hatzes A. P., Cochran W. D., Dennerl K., and Dbereiner S. (2002) The Planet Search Program at the ESO Coud Echelle spectrometer. III. The complete Long Camera survey results. *Astron. Astrophys., 392*, 671–690.

Fabrycky D. and Tremaine S. (2007) Shrinking binary and planetary orbits by Kozai cycles with tidal friction. *Astrophys. J., 669*, 1298.

Fischer D. A. and Valenti J. (2005) The planet-metallicity correlation. *Astrophys. J., 622,* 1102–1117.

Fischer D. A., Marcy G. W., Butler R. P., Vogt S. S., Frink S., and Apps K. (2001) Planetary companions to HD 12661, HD 92788, and HD 38529 and variations in Keplerian residuals of extrasolar planets. *Astrophys. J., 551,* 1107–1118.

Fischer D. A. et al. (2005) The N2K consortium. I. A hot Saturn planet orbiting HD 88133. *Astrophys. J., 620,* 481–486.

Fischer D. A. et al. (2008) Five planets orbiting 55 Cancri. *Astrophys. J., 675,* 790–801.

Ford E. B. (2005) Quantifying the uncertainty in the orbits of extrasolar planets. *Astron. J., 129,* 1706–1717.

Ford E. B. (2006) Improving the efficiency of Markov Chain Monte Carlo for analyzing the orbits of extrasolar planets. *Astrophys. J., 642,* 505–522.

Ford E. B. and Rasio F. A. (2006) On the relation between hot Jupiters and the Roche limit. *Astrophys. J. Lett., 638,* L45–L48.

Ford E. B. and Rasio F. A. (2008) Origins of eccentric extrasolar planets: Testing the planet-planet scattering model. *Astrophys. J., 686,* 621–636.

Ford E. B. and Tremaine S. (2003) Planet-finding prospects for the Space Interferometry mission. *Publ. Astron. Soc. Pacific, 115,* 1171–1186.

Ford E. B., Lystad V., and Rasio F. A. (2005) Planet-planet scattering in the upsilon Andromedae system. *Nature, 434,* 873–876.

Gaudi B. S. (2005) On the size distribution of close-in extrasolar giant planets. *Astrophys. J. Lett., 628,* L73–L76.

Gaudi B. S. (2008) Microlensing searches for planets: Results and future prospects. In *Extreme Solar Systems* (D. Fischer et al., eds.), p. 479. ASP Conf. Ser. 398, San Francisco.

Gaudi B. S. et al. (2002) Microlensing constraints on the frequency of Jupiter-mass companions: Analysis of 5 years of PLANET photometry. *Astrophys. J., 566,* 463.

Gaudi B. S., Seager S., and Mallen-Ornelas G. (2005) On the period distribution of close-in extrasolar giant planets. *Astrophys. J., 623,* 472–481.

Gaudi B. S. et al. (2008) Discovery of a Jupiter/Saturn analog with gravitational microlensing. *Science, 319,* 927.

Gilliland R. L. et al. (2000) A lack of planets in 47 Tucanae from a Hubble Space Telescope search. *Astrophys. J. Lett., 545,* L47.

Goldreich P. and Sari R. (2003) Eccentricity evolution for planets in gaseous disks. *Astrophys. J., 585,* 1024–1037.

Gonzalez G. (1997) The stellar metallicity-giant planet connection. *Mon. Not. R. Astron. Soc., 285,* 403–412.

Gould A. and Loeb A. (1992) Discovering planetary systems through gravitational microlenses. *Astrophys J., 396,* 104.

Gould A., Dorsher S., Gaudi B. S., and Udalski A. (2006a) Frequency of hot Jupiters and very hot Jupiters from the OGLEIII transit surveys toward the galactic bulge and Carina. *Acta Astron., 56,* 1–50.

Gould A., Udalski A., An D., Bennett D. P., Zhou A.-Y., et al. (2006b) Microlens OGLE-2005-BLG-169 implies that cool Neptune-like planets are common. *Astrophys. J. Lett., 644,* L37–L40.

Gregory P. C. (2007) A Bayesian Kepler periodogram detects a second planet in HD 208487. *Mon. Not. R. Astron. Soc., 374,* 1321–1333.

Griest K. and Safizadeh N. (1998) The use of high-magnification microlensing events in discovering extrasolar planets. *Astrophys. J., 500,* 37.

Guillot S., Guillot T., Santos N. C., Pont F., Iro N., Melo C., and Ribas I. (2006) A correlation between the heavy element content of transiting extrasolar planets and the metallicity of their parent stars. *Astron. Astrophys., 453,* L21–L24.

Halbwachs J. L., Mayor M., and Udry S. (2005) Statistical properties of exoplanets. IV. The period-eccentricity relations of exoplanets and of binary stars. *Astron. Astrophys., 431,* 1129–1137.

Hansen B. M. S. and Barman T. (2007) Two classes of hot Jupiters. *Astrophys. J., 671,* 861-871.

Holman M., Touma J., and Tremaine S. (1997) Chaotic variations in the eccentricity of the planet orbiting 16 Cygni B. *Nature, 386,* 254.

Horne K. (2003) Status and prospects of planetary transit searches: Hot Jupiters galore. In *Scientific Frontiers in Research on Extrasolar Planets* (D. Deming and S. Seager, eds.), pp. 361–369. ASP Conf. Ser. 294, San Francisco.

Ida S. and Lin D. N. C. (2004a) Toward a deterministic model of planetary formation. I. A desert in the mass and semimajor axis distributions of extrasolar planets. *Astrophys. J., 604,* 388–413.

Ida S. and Lin D. N. C. (2004b) Toward a deterministic model of planetary formation. II. The formation and retention of gas giant planets around stars with a range of metallicities. *Astrophys. J., 616,* 567–572.

Ida S. and Lin D. N. C. (2005) Toward a deterministic model of planetary formation. III. Mass distribution of short-period planets around stars of various masses. *Astrophys. J., 626,* 1045–1060.

Ida S. and Lin D. N. C. (2008a) Toward a deterministic model of planetary formation. IV. Effects of type I migration. *Astrophys. J., 673,* 487–501.

Ida S. and Lin D. N. C. (2008b) Toward a deterministic model of planetary formation. V. Accumulation near the ice line and super-Earths. *Astrophys. J., 685,* 584–595.

Jenkins J. M., Doyle L. R., and Cullers D. K. (1996) A matched filter method for ground-based sub-noise detection of terrestrial extrasolar planets in eclipsing binaries: Application to CM Draconis. *Icarus, 119,* 244–260.

Johnson J. A., Fischer D. A., Marcy G. W., Wright J. T., Driscoll P., et al. (2007) Retired A stars and their companions: Exoplanets orbiting three intermediate-mass subgiants. *Astrophys. J., 665,* 785–793.

Johnson J. A., Marcy G. W., Fischer D. A., Wright J. T., Reffert S., Kregenow J. M., Williams P. K. G. and Peek K. M. G. (2008) Retired A stars and their companions. II. Jovian planets orbiting κ CrB and HD 167042. *Astrophys. J., 675,* 784–789.

Johnson J. A., Bowler B. P., Howard A. W., Henry G. W., Marcy G. W., et al. (2010) Retired A stars and their companions V. A hot Jupiter orbiting the 1.7 Msun subgiant HD 102956. *ArXiv e-prints,* arXiv:1007.4555.

Jones H. R. A., Butler R. P., Tinney C. G., Marcy G. W., Penny A. J., McCarthy C., and Carter B. D. (2003) An exoplanet in orbit around τ^1 Gruis. *Mon. Not. R. Astron. Soc., 341,* 948–952.

Jurić M. and Tremaine S. (2008) Dynamical origin of extrasolar planet eccentricity distribution. *Astrophys J., 686,* 603–620.

Kennedy G. M. and Kenyon S. J. (2008) Planet formation around stars of various masses: The snow line and the frequency of giant planets. *Astrophys. J., 673,* 502–512.

Kley W., Lee M. H., Murray N., and Peale S. J. (2005) Modeling the resonant planetary system GJ 876. *Astron. Astrophys., 437,* 727–742.

Kovács G., Zucker S., and Mazeh T. (2002) A box-fitting algorithm in the search for periodic transits. *Astron. Astrophys., 391,* 369–377.

Lafrenière D., Doyon R., Marois C., Nadeau D., Oppenheimer B. R., et al. (2007) The Gemini Deep Planet Survey — GDPS. *Astrophys. J., 670,* 1367–1390.

Laughlin G., Bodenheimer P., and Adams F. C. (2004) The core accretion model predicts few jovian-mass planets orbiting red dwarfs. *Astrophys. J. Lett., 612,* L73–L76.

Laughlin G., Butler R. P., Fischer D. A., Marcy G. W., Vogt S. S., and Wolf A. S. (2005a) The GJ 876 planetary system: A progress report. *Astrophys. J., 622,* 1182–1190.

Laughlin G., Wolf A., Vanmunster T., Bodenheimer P., Fischer D., Marcy G., Butler P., and Vogt S. (2005b) A comparison of observationally determined radii with theoretical radius predictions for short-period transiting extrasolar planets. *Astrophys. J., 621,* 1072–1078.

Lin D. N. C., Bodenheimer P., and Richardson D. C. (1996) Orbital migration of the planetary companion of 51 Pegasi to its present location. *Nature, 380,* 606–607.

Lineweaver C. H. and Grether D. (2003) What fraction of Sun-like stars have planets? *Astrophys. J., 598,* 1350–1360.

Lomb N. R. (1976) Least-squares frequency analysis of unequally spaced data. *Astrophys. Space Sci., 39,* 447–462.

Lovis C. et al. (2006) An extrasolar planetary system with three Neptune-mass planets. *Nature, 441,* 305–309.

Macintosh B., Graham J., Palmer D., Doyon R., Gavel D., et al. (2006) The Gemini Planet Imager. In *Advances in Adaptive Optics II* (B. L. Ellerbroek and D. B. Calia, eds.), 62720. Proc. SPIE, Vol. 6272, DOI: 10.1117/12.672430.

Mahadevan S., van Eyken J., Ge J., DeWitt C., Fleming S. W., Cohen R., Crepp J., and Vanden Heuvel A. (2008) Measuring stellar radial velocities with a dispersed fixed-delay interferometer. *Astrophys. J., 678,* 1505–1510.

Marcy G. W. and Benitz K. J. (1989) A search for substellar companions to low-mass stars. *Astrophys. J., 344,* 441–453.

Marcy G. W. and Butler R. P. (1998) Detection of extrasolar giant planets. *Annu. Rev. Astron. Astrophys., 36,* 57–97.

Marcy G. W., Butler R. P., Vogt S. S., Fischer D. A., Henry G. W., Laughlin G., Wright J. T., and Johnson J. A. (2005a) Five new extrasolar planets. *Astrophys. J., 619,* 570–584.

Marcy G., Butler R. P., Fischer D., Vogt S., Wright J. T., Tinney C. G., and Jones H. R. A. (2005b) Observed properties of exoplanets: Masses, orbits and metallicities. *Progr. Theor. Phys. Suppl., 158,* 24–42.

Mandel K. and Agol E. (2002) Analytic light curves for planetary transit searches. *Astrophys. J. Lett., 580,* L171–L175.

Marois C., Macintosh B., Barman T., Zuckerman B., Song I., Patience J., Lafrenière D., and Doyon R. (2008) Direct imaging of multiple planets orbiting the star HR 8799. *Science, 322,* 1348.

Mayor M., Bonfils X., Forveille T., Delfosse X., Udry S., Bertaux J.-L., Beust H., Bouchy F., Lovis C., Pepe F., Perrier C., Queloz D., and Santos N. C. (2009) The HARPS search for southern extra-solar planets. XVIII. An Earth-mass planet in the GJ 581 planetary system. *Astron. Astrophys., 507,* 487.

Mochejska B. J. et al. (2005) Planets in stellar clusters extensive search. III. A search for transiting planets in the metal-rich open cluster NGC 6791. *Astron. J., 129,* 2856.

Mordasini C., Alibert Y., and Benz W. (2009a) Extrasolar planet population synthesis. I. Method, formation tracks, and mass-distance distribution. *Astron. Astrophys., 501,* 1139.

Mordasini C., Alibert Y., Benz W., and Naef D. (2009b) Extrasolar planet population synthesis. II. Statistical comparison with observations. *Astron. Astrophys., 501,* 1161.

Moutou C., Pont F., Barge P., Aigrain S., Auvergne M., et al. (2005) Comparative blind test of five planetary transit detection algorithms on realistic synthetic light curves. *Astron. Astrophys., 437,* 355–368.

Naef D., Mayor M., Beuzit J.-L., Perrier C., Queloz,D., Sivan J.-P., and Udry S. (2005) The ELODIE Planet Search: Synthetic view of the survey and its global detection threshold. In *Proc. 13th Cambridge Workshop on Cool Stars, Stellar Systems and the Sun* (F. Favata et al., eds.), pp. 833–836. ESA SP-560, Noordwijk, The Netherlands.

O'Toole S. J., Jones H. R. A., Tinney C. G., Butler R. P., Marcy G. W., Carter B., Bailey J., and Wittenmyer R. A. (2009) The frequency of low-mass exoplanets. *Astrophys. J., 701,* 1732.

Patel S. G., Vogt S. S., Marcy G. W., Johnson J. A., Fischer D. A., Wright J. T., and Butler R. P. (2007) Fourteen new companions from the Keck and Lick radial velocity survey including five brown dwarf candidates. *Astrophys. J., 665,* 744–753.

Pepe F. A. and Lovis C. (2008) From HARPS to CODEX: Exploring the limits of Doppler measurements. *Phys. Scripta T, 130,* 014007.

Pinsonneault M. H., DePoy D. L. and Coffee M. (2001) The mass of the convective zone in FGK main-sequence stars and the effect of accreted planetary material on apparent metallicity determinations. *Astrophys. J. Lett., 556,* L59.

Pont F., Zucker S., and Queloz D. (2006) The effect of red noise on planetary transit detection. *Mon. Not. R. Astron. Soc., 373,* 231–242.

Press W. H., Teukolsky S. A., Vetterling W. T., and Flannery B. P. (1992) *Numerical Recipes: The Art of Scientic Computing,* 2nd edition, Cambridge Univ., Cambridge.

Queloz D. et al. (2001) No planet for HD 166435. *Astron. Astrophys., 379,* 279–287.

Ramsey L. W., Barnes J., Redman S. L., Jones H. R. A., Wolszczan A., Bongiorno S., Engel L., and Jenkins J. (2008) A Pathfinder instrument for precision radial velocities in the near-infrared. *Publ. Astron. Soc. Pacific, 120,* 887–894.

Rasio F. A. and Ford E. B. (1996) Dynamical instabilities and the formation of extrasolar planetary systems. *Science, 274,* 954–956.

Ribas I. and Miralda-Escudé J. (2007) The eccentricity-mass distribution of exoplanets: Signatures of different formation mechanisms? *Astron. Astrophys., 464,* 779–785.

Rivera E. J., Lissauer J. J., Butler R. P., Marcy G. W., Vogt S. S., Fischer D. A., Brown T. M., Laughlin G., and Henry G. W. (2005) A 7.5 Earth mass planet orbiting the nearby star, GJ 876. *Astrophys. J., 634,* 625.

Robinson S. E., Laughlin G., Vogt S. S., Fischer D. A., Butler R. P., Marcy G. W., Henry G. W., Driscoll P., Takeda G., and Johnson J. A. (2007) Two jovian-mass planets in Earthlike orbits. *Astrophys. J., 670,* 1391.

Saar S. H. and Donahue R. A. (1997) Activity-related radial velocity variation in cool stars. *Astrophys. J., 485,* 319–327.

Saar S. H., Butler R. P., and Marcy G. W. (1998) Magnetic activity-related radial velocity variations in cool stars: First results from the Lick extrasolar planet survey. *Astrophys. J. Lett., 498,* L153–L157.

Santos N. C., Israelian G., and Mayor M. (2004) Spectroscopic [Fe/H] for 98 extra-solar planet-host stars: Exploring the probability of planet formation. *Astron. Astrophys., 415,* 1153.

Sato B., Fischer D. A., Henry G. W., Laughlin G., Butler R. P., et al. (2005) The N2K consortium. II. A Transiting hot Saturn around HD 149026 with a large dense core. *Astrophys. J., 633,* 465–473.

Sato B., Izumiura H., Toyota E., Kambe E., Ikoma M., et al. (2008) Planetary companions around three intermediate-mass G and K giants: 18 Delphini, ξ Aquilae, and HD 81688. *Proc. Astron. Soc. Japan, 60,* 539.

Saumon D. and Guillot T. (2004) Shock compression of deuterium and the interiors of Jupiter and Saturn. *Astrophys. J., 609*, 1170–1180.

Scargle J. D. (1982) Studies in astronomical time series analysis. II — Statistical aspects of spectral analysis of unevenly spaced data. *Astrophys. J., 263*, 835–853.

Schwarzenberg-Czerny A. and Beaulieu J.-P. (2006) Efficient analysis in planet transit surveys. *Mon. Not. R. Astron. Soc., 365*, 165–170.

Seager S. and Mallén-Ornelas G. (2003) A unique solution of planet and star parameters from an extrasolar planet transit light curve. *Astrophys. J., 585*, 1038–1055.

Shen Y. and Turner E. L. (2008) On the eccentricity distribution of exoplanets from radial velocity surveys. *Astrophys. J., 685*, 553–559.

Sivia D. S. (1996) *Data Analysis: A Bayesian Tutorial*. Oxford Univ., New York.

Snodgrass C., Horne K., and Tsapras Y. (2004) The abundance of galactic planets from OGLE-III 2002 microlensing data. *Mon. Not. R. Astron. Soc., 351*, 967.

Southworth J. (2008) Homogeneous studies of transiting extrasolar planets. II. Physical properties. *Mon. Not. R. Astron. Soc., 394*, 272.

Stepinski T. F. and Black D. C. (2001) On orbital elements of extrasolar planetary candidates and spectroscopic binaries. *Astron. Astrophys., 371*, 250–259.

Tabachnik S. and Tremaine S. (2002) Maximum-likelihood method for estimating the mass and period distributions of extrasolar planets. *Mon. Not. R. Astron. Soc., 335*, 151–158.

Takeda G., Ford E. B., Sills A., Rasio F. A., Fischer D. A., and Valenti Jeff A. (2007) Structure and evolution of nearby stars with planets. II. Physical properties of 1000 cool stars from the SPOCS catalog. *Astrophys. J. Suppl., 168*, 297.

Tingley B. (2003) Improvements to existing transit detection algorithms and their comparison. *Astron. Astrophys., 408*, L5–L7.

Torres G., Winn J. N., and Holman M. J. (2008) Improved parameters for extrasolar transiting planets. *Astrophys. J., 677*, 1324–1342.

Udry S. and Santos N. C. (2007) Statistical properties of exoplanets. *Annu. Rev. Astron. Astrophys., 45*, 397–439.

Udry S., Mayor M., Naef D., Pepe F., Queloz D., Santos N. C., Burnet M., Conno B., and Melo C. (2000) The CORALIE survey for southern extra-solar planets II. The short-period planetary companions to HD 75289 and HD 130322. *Astron. Astrophys., 356*, 590–598.

Udry S., Mayor M., and Santos N. C. (2003) Statistical properties of exoplanets I. The period distribution: Constraints for the migration scenario. *Astron. Astrophys., 407*, 369–376.

Valenti J. A. and Fischer D. A. (2005) Spectroscopic properties of cool stars (SPOCS). I. 1040 F, G, and K dwarfs from Keck, Lick, and AAT Planet Search programs. *Astrophys. J. Suppl., 159*, 141–166.

von Braun K., Kane S. R. and Ciardi D. R. (2009) Observational window functions in planet transit surveys. *Astrophys. J., 702*, 779.

Wainstein L. A. and Zubakov V. D. (1962) *Extraction of Signals from Noise*. Prentice-Hall, New Jersey.

Walker G. A. H., Walker A. R., Irwin A. W., Larson A. M., Yang S. L. S., and Richardson D. C. (1995) A search for Jupiter-mass companions to nearby stars. *Icarus, 116*, 359–375.

Weldrake D. T. F., Sackett P. D., and Bridges T. J. (2008) The frequency of large-radius hot and very hot Jupiters in ω Centauri. *Astrophys. J., 674*, 1117.

Wittenmyer R. A., Endl M., Cochran W. D., Hatzes A. P., Walker G. A. H., Yang S. L. S., and Paulson D. B. (2006) Detection limits from the McDonald Observatory Planet Search program. *Astron. J., 132*, 177–188.

Wright J. T. (2005) Radial velocity jitter in stars from the California and Carnegie Planet Search at Keck observatory. *Publ. Astron. Soc. Pacific, 117*, 657–664.

Wright J. T., Marcy G. W., Butler R. P., and Vogt S. S. (2004) Chromospheric Ca II emission in nearby F, G, K, and M stars. *Astrophys. J. Suppl., 152*, 261–295.

Wright J. T., Marcy G. W., Fischer D. A., Butler R. P., Vogt S. S., et al. (2007) Four new exoplanets and hints of additional substellar companions to exoplanet host stars. *Astrophys. J., 657*, 533–545.

Wright J. T., Marcy G. W., Butler R. P., Vogt S. S., Henry G. W., Isaacson H., and Howard A. W. (2008) The Jupiter twin HD 154345b. *Astrophys. J. Lett., 683*, L63–L66.

Wu Y. (2003) Tidal circularization and Kozai migration. In *Scientific Frontiers in Research on Extrasolar Planets* (D. Deming and S. Seager, eds.), pp. 213–216. ASP Conf. Ser. 294, San Francisco.

Wu Y. and Murray N. (2003) Planet migration and binary companions: The case of HD 80606b. *Astrophys. J., 589*, 605–614.

Wu Y., Murray N. W., and Ramsahai J. M. (2007) Hot Jupiters in binary star systems. *Astrophys. J., 670*, 820–825.

Zakamska N. L. and Tremaine S. (2004) Excitation and propagation of eccentricity disturbances in planetary systems. *Astron. J., 128*, 869–877.

Part III:
Exoplanet Dynamics

Non-Keplerian Dynamics of Exoplanets

Daniel C. Fabrycky
Harvard-Smithsonian Center for Astrophysics

Exoplanets are often found with short periods or high eccentricities, and multiple-planet systems are often in resonance. They require dynamical theories that describe more extreme motions than those of the relatively placid planetary orbits of the solar system. We describe the most important dynamical processes in fully formed planetary systems and how they are modeled. Such methods have been applied to detect the evolution of exoplanet orbits in action and to infer dramatic histories from the dynamical properties of planetary systems.

1. INTRODUCTION

After a planet has formed via giant impacts and has outlasted migration torques from the gaseous disk, perils still await. Dynamical instabilities among the planets of the system, long-term orbital changes due to a companion star, and tidal interactions with the host star could all eject the planet or toss it into the host star. None of these would threaten planets if they continued on Keplerian orbits (as described in the chapter by Murray and Correia), in which the planet and star are considered as point masses, orbiting each other according to Newton's approximate theory of gravity, in isolation from other bodies. To model these more interesting interactions, we explore non-Keplerian dynamics, an extension of orbital theory to (1) the astrophysical two-body problem in more detail, including nonspherical bodies and relativity theory, and (2) true planetary systems: systems with more than just one planet and one star.

Although most dynamical concepts were originally designed to describe the solar system, the discovery of exoplanets has channeled research in new directions. Previously most analytical and even numerical techniques required certain quantities (mass ratios, eccentricities, mutual inclinations) to be small. These approximations are appropriate for applications in the solar system, but more general methods are needed to describe the bewildering variety of exoplanetary systems. Here we describe these methods — both new directions of analytic theory and an introduction to numerical work.

Orbital motion departs from fixed Keplerian ellipses on a wide range of timescales, from orbital periods to stellar lifetimes. These variations have a wide range of magnitudes, from slight deflections to complete reorientations and ejections. Slight deflections, due to planets passing one another, may be too small to detect, apart from sensitive transit-timing measurements. Secular interactions and dissipative effects may be too slow to detect in individual systems, but they may be inferred by statistical studies of populations. Ejections may be too rare to see in action, but they are expected from numerical modeling. Resonant interactions, however, can produce substantially non-Keplerian motion over a timescale suitable for observation, and resonant systems can stably persist for the star's entire lifetime.

The organization of this chapter is as follows. In section 2, we begin with equations of motion for planets that include tidal distortion, rotational oblateness, effects of relativity, and gravitational interactions among planets. The framework for solving these equations is presented, including introductions to both coordinate systems and numerical algorithms. In section 3, the phenomena that arise from these equations are discussed. We emphasize phenomena that are important for known systems, as well as types of orbits that could exist but are not yet observed. Next, in section 4, we turn to observational highlights that have used and continue to challenge these concepts, showing how theories and systems in nature have enjoyed a symbiotic relationship. Finally, section 5 discusses observations that would be particularly useful for constraining theories of dynamical evolution. With this structure, some individual topics in dynamics reappear in several different sections; so we provide Table 1 as a topical guide.

2. EQUATIONS OF MOTION AND NUMERICAL METHODS

2.1. Astrophysical Two-Body Problem

First, let us examine the astrophysical two-body problem including non-Keplerian effects. The masses of the star and planet are $m_\star$ and m_p, respectively, and the orbital elements refer to the displacement vector $\mathbf{r}$ (of magnitude r) of the planet relative to the star, as in the chapter by Seager and Lissauer. The equation of motion is

$$\ddot{\mathbf{r}} = -G(m_\star + m_p)\frac{\mathbf{r}}{r^3} + \mathbf{f} \qquad (1)$$

where $\mathbf{f}$ is a force other than that of mutual gravity of point masses, so $\mathbf{f} = 0$ yields Keplerian motion (see equation (4) in the chapter by Murray and Correia).

2.1.1. Relativistic effects. The lowest-order post-Newtonian effects may be implemented using the force

$$\mathbf{f}_{GR} = -\frac{G(m_\star + m_p)}{r^2 c^2} \times \left(-2(2-\eta)\dot{r}\dot{\mathbf{r}} \right.$$
$$\left. + \left[(1+3\eta)\dot{\mathbf{r}}\cdot\dot{\mathbf{r}} - \frac{3}{2}\eta\dot{r}^2 \right. \right. \quad (2)$$
$$\left. \left. -2(2+\eta)\frac{G(m_\star + m_p)}{r} \right]\hat{\mathbf{r}} \right)$$

[where $\eta = m_\star m_p/(m_\star + m_p)^2$; *Kidder* (1995); *Mardling and Lin* (2002)] for $\mathbf{f}$ in equation (1). We will examine one dynamical consequence of this force, apsidal motion, in section 3.1.1. Alternatively, a potential that mimics lowest-order relativistic effects, which is especially suitable for analysis within a Hamiltonian framework, was given by equation (31) of *Saha and Tremaine* (1992).

2.1.2. Effects of nonspherical bodies. The orbital parameters change when we consider the bodies not as point masses, but as physical objects capable of distortion and internal energy dissipation. Tidal effects on the orbit become more and more pronounced as two gravitating bodies of finite extent get closer to one another. Tides have apparently caused many exoplanet orbits to become circular within about 0.1 AU of their main-sequence stars.

The tidal force of each body distorts the potential energy surfaces of its companion. For stars and gaseous planets in hydrostatic equilibrium, the surfaces of constant density will settle to these new equipotential surfaces, and the distortion of the body itself modifies them further, until a self-consistent solution is obtained. For solid planets, the rigidity of the material also comes into play. These properties determine the planet's tidal deformability and are summarized by a Love number k_L (*Love*, 1911), which is the amplitude ratio of the quadrupolar potential due to the deformed body to the tidal potential imposed on the body, evaluated at the surface of the body; a table of typical values of k_L may be found in *Mardling and Lin* (2004). Below we shall also denote the Love number of the star as $k_{L,\star}$, which is *twice* the apsidal motion constant, the conventional parameter in the literature regarding eclipsing binaries (*Sterne*, 1939).

First, the star raises a tidal bulge of the planet with a size $\propto r^{-3}$. This bulge creates its own external field, which falls off like r^{-3}. Acting back on the star, these radial scalings combine to augment the radial gravitational force (per unit mass) with a term having steep radial dependence

$$\mathbf{f}_T = -3k_L \frac{Gm_\star^2}{m_p} \frac{R_p^5}{r^7}\hat{\mathbf{r}} \quad (3)$$

TABLE 1. Guide to topics.

Topic	Sections
Single-Planet Orbital Evolution	
Periastron advance	3.1.1
Tides	3.1.2
Miscellaneous	3.1.3
Few-Body Orbital Interactions	
Short-term orbit fluctuations	3.2, 5.2
Resonances	3.2, 4.1, 4.2, 5.1
Secular effects	3.2, 3.3, 4.3, 4.4
Chaos	3.4
Dynamical Niches	
Resonance protection	3.2, 4.1
Satellites	3.5, 5.1
Trojans and horseshoes	3.5, 5.1
Interlocking orbits	3.3
Habitable zones	3.5
Data Analysis	
Astrometry and direct imaging	2.4
Radial velocity	2.4
Transits	2.5, 5.1, 5.2
Individual Systems	
GJ 876	4.1
Pulsar 1257+12	4.2
16 Cyg B	3.5, 4.4
HD 80606	3.3, 4.4
Resonant systems	Table 3

Second, the rotation of the star causes it to become oblate. Its degree of oblateness depends on the square of the stellar angular rotation rate $\Omega_\star$ divided by its breakup angular rate, $\sqrt{Gm_\star/R_\star^3}$. The quadrupolar potential outside of the star scales as r^{-3}. Thus the figure of the rotating star induces an extra force

$$\mathbf{f}_R = -\frac{1}{2}k_{L,\star}\Omega_\star^2 \frac{R_\star^5}{r^4}\hat{\mathbf{r}} \quad (4)$$

The forces of the foregoing two subsections may be included in numerical integrations to simulate more realistic orbital behavior than Keplerian ellipses. They may also be used in analytic calculations to determine orbit changes on a longer timescale, results of which we quote in section 3.1.

2.2. N-Body Equations and Coordinates

When a star hosts N (>1) planets, gravitational interactions among the planets can affect their orbits in complex ways. Neglecting the effects discussed in section 2.1, the equation of motion of planet planet i (of mass m_i) is

$$\ddot{\mathbf{r}}_i = -G(m_0 + m_i)\frac{\mathbf{r}_i}{r_i^3} + G\sum_{j=1;j\neq i}^{N} m_j\left(\frac{\mathbf{r}_j - \mathbf{r}_i}{|\mathbf{r}_j - \mathbf{r}_i|^3} - \frac{\mathbf{r}_j}{r_j^3}\right) \quad (5)$$

where each of the coordinates $\mathbf{r}_i$ is referred to the central star, of mass $m_0 \equiv m_\star$. The interaction terms in the sum over each of the other planets are the *direct* gravitational force (first term) and the *indirect* effective force due to the bodies causing the star, and thus the reference frame, to accelerate (second term). In numerical work, these N second-order differential equations are most often transformed into a system of 2N first-order differential equations in the quantities $\mathbf{r}_i$ and $\dot{\mathbf{r}}_i$. In response to the motions of the planets, the star's position and velocity with respect to the barycenter of the system (the center of mass) are

$$\mathbf{R}_0 = -\left(\sum_{i=1}^{N} m_i \mathbf{r}_i\right) \bigg/ \left(\sum_{i=0}^{N} m_i\right) \quad (6)$$

$$\dot{\mathbf{R}}_0 = -\left(\sum_{i=1}^{N} m_i \dot{\mathbf{r}}_i\right) \bigg/ \left(\sum_{i=0}^{N} m_i\right) \quad (7)$$

which are useful for self-consistent fits of data (see section 2.4).

The coordinates of equation (5) are called astrocentric, because they refer each planet to the (moving) position of the star. There are several other possible coordinate systems — barycentric, Jacobian, or Poincaré coordinates — which are defined as follows. Barycentric coordinates refer all positions and velocities to the center of mass, and the only force is Newton's attractive force between bodies (no indirect term). Jacobian coordinates are hierarchical, in which the positions and velocities of each of the planets is referred to the center of mass of all interior bodies. Poincaré coordinates [*Laskar and Robutel* (1995), also called democratic heliocentric coordinates (*Duncan et al.*, 1998), or canonical heliocentric (*Sussman and Wisdom*, 2001, p. 435)], on the other hand, take the positions of the planets to be astrocentric, but take the velocities of the planets to be barycentric.

The various coordinate systems have different strengths and weaknesses, in both numerical integration and in analytical studies, as follows.

Astrocentric coordinates have a few drawbacks. In numerical integrations, all planets must be integrated with timesteps smaller than the innermost planet's period, because of the high-frequency forcing it introduces. Also, for theoretical studies in a Hamiltonian framework, the astrocentric coordinates are rather cumbersome (*Beaugé et al.*, 2007).

Although barycentric coordinates are very simple, they do not take advantage of the fact that the center of mass is invariant. Therefore the position of the star must either be numerically integrated or updated at each timestep according to equation (6). A reduction by those three degrees of freedom is achieved using Jacobian or Poincaré coordinates.

Lee and Peale (2003) have advocated using Jacobian coordinates, instead of astrocentric coordinates, when turning the solutions of multiple Keplerian radial velocity fits into a self-consistent N-body realization (section 2.4). Particularly for systems that are hierarchical, these are the coordinates in which the planets perturb each other minimally on orbital timescales, so the independent-Keplerian model is satisfied best. Once an integration is set up in Jacobian coordinates, one might want to do the integration in astrocentric coordinates, so that, e.g., transit times of the outer planet can be easily calculated (section 2.5). The hierarchical structure of Jacobian coordinates is not always physically well motivated, e.g., for systems with Trojan orbits. Most authors take the relevant mass binding the jth planet of mass m_j to the mass interior to it $m_{interior,j}$ to be $m_j + m_{interior,j}$ (in generalization of the mass term equation (1)), but in symplectic integrations (described in section 2.3) it can be more efficient to use $\tilde{M}_j \equiv m_\star m_{interior,j}/m_{interior,j-1}$ (*Wisdom and Holman*, 1991; *Murray and Dermott*, 1999).

Poincaré coordinates have positions and velocities with different origins, so they are not conceptually based on elliptical motion. However, analytic studies can be cleaner, as the resulting Hamiltonian has rather clear symmetry.

For each coordinate system, each planet has "osculating" orbital elements that correspond to the orbit the planet would have if all the interaction forces are ignored. For instance, orbital elements based on the astrocentric coordinates have a simple physical interpretation: If (N−1) planets of the system suddenly disappeared, a single planet would be left orbiting the star with those orbital elements. Osculating elements have a drawback, which is that the physical, observed properties, which include the continual perturbations of the other bodies, do not correspond to the osculating value. For instance, the orbital period observed with transits or radial velocities is not the osculating period in an astrocentric coordinate system (for a conversion, see *Ferraz-Mello et al.*, 2005).

For explicit forms of these coordinate systems, and a fuller discussion, see *Beaugé et al.* (2007) and *Morbidelli* (2002).

2.3. Numerical Integration Techniques

Here we summarize some of the numerical techniques and codes that are used in exoplanet dynamics.

Currently, the fastest algorithms that are reliable for long-term studies are symplectic integrators, in which separate integrations delimiting a certain volume of phase space continue to delimit the same volume as the integration proceeds. This property automatically respects the Hamiltonian nature of the gravitational problem; e.g., energy is not secularly lost or gained. The *Wisdom and Holman* (1991) integrator is the most well known of the symplectic integrators. It has been supplemented with algorithms that can handle close approaches and collisions, which forms the backbone of the publicly available codes Mercury (*http://www.arm.ac.uk/~jec/Ihome.html*) (*Chambers*, 1999), Swift (*http://www.boulder.swri.edu/~hal/swift.html*) (*Levison and Duncan*, 1994), and SyMBA (available upon request from Hal Levison) (*Duncan et al.*, 1998). A symplectic algorithm was also built specifically to handle stellar mass companions (*Chambers et al.*, 2002), either with planets orbiting one star of the pair (satellite-type) or both stars (planetary-type). The speed of symplectic integrators allowed some of the first long-term integrations of all the planets of the solar system, showing their orbits to be chaotic (e.g., *Sussman and Wisdom*, 1992). High-order symplectic schemes have been used to follow the whole solar system for its lifetime (*Laskar and Gastineau*, 2009).

There are several frequently-used, but nonsymplectic, algorithms. RADAU (15) by *Everhart* (1985) is a very high-order method of integration by Radau quadrature — a prescription for times to evaluate forces and weights to apply to them — and it is included in Mercury. General-purpose integrators from *Numerical Recipes* (*Press et al.*, 1992) (e.g., the popular Burlisch-Stoer) and the GNU Scientific Library [e.g., Embedded Runge-Kutta Prince-Dormand (8/9 order), as advocated by *Fregeau et al.* (2004)] have also been set to work on problems in exoplanet dynamics. These integrators are particularly useful for short-term, high-precision work.

If mean-motion resonances are unimportant, one can derive approximate secular equations of motion by analytically averaging each planets' gravitational effect over its orbit. This procedure results in differential equations for the orbital elements, which can be numerically integrated much faster than the positions themselves. *Mardling and Lin* (2002) have presented equations to evolve three-planet systems, as well as separate equations for the system once the inner planet's orbit is averaged. *Eggleton and Kiseleva-Eggleton* (2001) have presented equations for the secular problem for three-body systems where the semimajor axis ratio is large, and the outermost orbit dominates the angular momentum budget; this approach is applicable to planets in binaries. Higher-order expansions in the ratio of semimajor axes, with an emphasis on planetary systems, have also been derived (*Ford et al.*, 2000; *Lee and Peale*, 2003; *Migaszewski and Góźdźiewski*, 2008). Taking no restrictions on relative size of orbits or orbital elements, one can compute by brute force the time-averaged torque acting among a set of nearly Keplerian orbits, which is called Gauss' method (e.g., *Touma et al.*, 2009).

Finally, a general consideration, which is surprising at first, is that codes are usually much slower at printing data to files than computing data. For instance, if the position of the planets is printed once per orbit, the printouts will generally completely dominate the runtime. It is generally true that performance can be limited by the weakest link (the slowest part of the algorithm), so one must think about the program as a whole, including the reads and writes, when trying to speed up an algorithm.

2.4. N-Body Fits to Data

To directly detect non-Keplerian motion, a framework must be constructed for comparing the outputs of numerical algorithms to data. We describe the conventional approach here.

The astrocentric coordinates for planets are $\mathbf{r}_i = (x_i, y_i, z_i)$; the barycentric coordinates of the planets are $\mathbf{R}_i = (X_i, Y_i, Z_i)$, and the barycentric coordinates of the host star are $\mathbf{R}_0 = (X_0, Y_0, Z_0)$, which obey equation (7). The standard coordinate system (e.g., chapter by Murray and Correia) has the Z axis pointed away from the observer. Let the X–Y plane be the plane of the sky, which passes through the barycenter. Then aligning the X axis with the north direction and the Y axis with the east direction forms a righthanded coordinate system with Z. The ascending node Ω is measured east from north, and the argument of periastron ω is measured from the sky plane. These coordinates are useful for fitting data from essentially all techniques (neglecting the rectilinear motion of the system as a whole): (1) radial velocity of the host = $-\dot{Z}_0$; (2) astrometric position of the host = (X_0, Y_0); (3) transits and secondary eclipses may be modeled using the lightcurve equations of *Mandel and Agol* (2002), setting the projected distance between the centers, called d, equal to $\sqrt{x_i^2 + y_i^2}$ (see also section 2.5 below); (4) direct images of planets have a sky-offset from their host of (x_i, y_i); and (5) the times of arrival (TOA) of pulses of pulsars, or phases of pulsating stars, or eclipses of close binaries, that are orbited by planets are delayed by $\delta_t = -Z_0/c$, etc.

Therefore a common coordinate system can be used to fit different types of datasets simultaneously, as has already been done for radial velocity plus astrometry (*Bean and Seifahrt*, 2009; *Wright and Howard*, 2009) and for radial velocity plus transits (e.g., *Winn et al.*, 2005).

The values of G and $m_\odot$ are not known to nearly the precision of their combination. A result is that dynamical fits reported in the literature are typically not reproducible if the masses are reported in physical units (e.g., grams). One solution is to report all masses in units of solar masses. Alternatively, one could abide by the International Astronomical Union's conventional value of $k \equiv \sqrt{Gm_\odot} = 0.01720209895$ AU$^{3/2}$ d^{-1}.

There are numerous methods for converging on the solution and evaluating uncertainties. Markov Chain Monte Carlo (*Ford*, 2005, 2006), Levenberg-Marquardt (*Laughlin and Chambers*, 2001; *Rivera and Lissauer*, 2001), genetic algorithm (*Góźdźiewski et al.*, 2005), and multiplexed simulated annealing [also known as parallel tempering (*Gregory*, 2007)] have all been used extensively, and these references describe the implementations and advantages of each method.

The magnitude and/or timescale of perturbations scale as powers of the mass ratios, so the true planetary masses in systems with measurable non-Keplerian motion are accessible (sections 4.1 and 4.2). Moreover, often only a subset of orbital parameters that fit the observations actually result in long-term stability, so additional information about the system is accessible by requiring long-term stability. This idea was pursued for the first multiplanet system by *Rivera and Lissauer* (2000) and *Stepinski et al.* (2000), and it has been particularly useful for resonant systems; see *Góźdźiewski et al.* (2008b) for an overview.

2.5. Focus on Transit Variations

Now let us focus on dynamical fits to transit data, which have the potential to discover small planets and to reveal the detailed dynamical properties of multiplanet systems (see section 5). The data begin as flux measurements of a star as a function of time; the star is dimmed as the transiting planet covers some of the stellar surface (see chapter by Winn). Encoded in this time series is information about the position of the planet relative to the star, so integrators based on astrocentric coordinates are the most natural for this application. The data allow for precise positions and times to be observed, so high precision in coordinates and arbitrary timesteps is required. Typically, the flux measurements are used to derive transit parameters — midtransit times and durations (or impact parameters or inclinations) — as a function of transit number, which may be compared directly to a numerical integration.

Theoretical midtransit and contact times for the ith planet may be computed by integrating forward and backward in time by the Newton-Raphson method, seeking roots of functions of $\mathbf{r}_{s,i} \equiv (x_i, y_i)$, the relative separation vector of the planet and star on the plane of the sky (see Fig. 1). This method is described next. The only requirement for the following algorithm is that the sky-projected trajectory has a radius of curvature larger than $R_\star + R_p$, which is easily fulfilled in practice.

For each transit the midtime is found first, as follows. The first step is to advance the integrator in time to the vicinity of an observed transit, e.g., by stopping an integration once an observed transit time has been passed, or advancing the appropriate number of nominal orbital periods from a previous transit. Midtransit times may be found by minimizing $|\mathbf{r}_{s,i}|$. Taking the derivative of this quantity (expressed in x_i and y_i) and setting it to zero, we find that minimizing $|\mathbf{r}_{s,i}|$ amounts to solving

$$g(x_i, \dot{x}_i, y_i, \dot{y}_i) \equiv \mathbf{r}_{s,i} \cdot \dot{\mathbf{r}}_{s,i} = x_i \dot{x}_i + y_i \dot{y}_i = 0 \quad (8)$$

The first guess at the time of transit must be within about one-eighth of a period, or the algorithm may converge to an adjacent local maximum of $|\mathbf{r}_{s,i}|$. The solution of equation (8) is found by 5–10 iterations of moving the integrator by

$$\delta t = -g \left(\frac{\partial g}{\partial t} \right)^{-1} \quad (9)$$

where

$$\frac{\partial g}{\partial t} = \dot{x}_i^2 + x_i \ddot{x}_i + \dot{y}_i^2 + y_i \ddot{y}_i \quad (10)$$

according to equation (8). Once δt is below the required accuracy, the midtime t_{mid} and position $\mathbf{r}_{s,i}^{mid}$ of that transit have been found. This method is computationally considerably faster than searching for a minimum of $|\mathbf{r}_{s,i}|$ directly, as root-finding is a simpler operation than minimum finding.

Next, before solving for times of contact, we can determine if they exist for this particular transit. (See the chapter by Winn for the definitions of points of contact and for the numbering scheme.) If $|\mathbf{r}_{s,i}^{mid}| < R_\star - R_p$, then all four

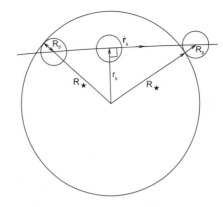

Fig. 1. Diagram of a transiting planet at the contact and midtransit points of its orbit.

times of contact exist; the trajectory is not grazing. If $R_\star - R_p < |\mathbf{r}_{s,i}^{mid}| < R_\star + R_p$, then the trajectory is grazing and only first and fourth contact exist. If $|\mathbf{r}_{s,i}^{mid}| > R_\star + R_p$, the planet is not transiting at all and searching for times of contact is a waste. To search for times of contact, first advance the integrator to $t_{mid} \mp R_\star/|\dot{\mathbf{r}}_{s,i}^{mid}|$, where − is taken for first and second contact, and + is taken for third and fourth contact. Next, solve the equation

$$h(x_i, y_i) = |\mathbf{r}_{s,i}|^2 - (R_\star \pm R_p)^2 = 0 \quad (11)$$

where + is taken for first and fourth contact, and − is taken for second and third contact. As before, this is done by iteratively driving the integrator by

$$\delta t = -h \left(\frac{\partial h}{\partial t} \right)^{-1} \quad (12)$$

where now

$$\frac{\partial h}{\partial t} = 2 x_i \dot{x}_i + 2 y_i \dot{y}_i \quad (13)$$

It is important not to start the search near midtransit, otherwise $\frac{\partial h}{\partial t} = 2g \approx 0$ and equation (12) will ask for an enormous jump: Preemptively advancing to near or beyond the times of contact (as above) avoids that fate. As usual with Newton-Raphson's method, sensible bounds should be placed on the search for the root.

After times of contact and midtransit are computed, they need to be corrected for the finite travel time of light. Thus the apparent time of an event depends on the distance between the bodies and the center of mass of the system, along the line of sight (see, e.g., *Knutson et al.*, 2007; *Bakos et al.*, 2009). Suppose, for instance, that the ephemeris of midtransit times for a transiting planet is well established, and the values of eccentricity e and longitude of periastron ω are known. Then the time of secondary eclipse could be predicted, but it will actually be observed a time Δt later due to the finite speed of light, where

$$\Delta t = \frac{2a}{c} \frac{m_\star^2 - m_p^2}{(m_\star + m_p)^2} \frac{1-e^2}{1-e^2 \sin^2 \omega} \quad (14)$$

This equation is derived by delaying the image of each body according to its line-of-sight distance, then solving for the time at which these images cross, i.e., the midtime of secondary eclipse as observed. This delay can be large for transiting planets, e.g., Δt ≈ 160 s for HD 80606b, a systematic effect that is comparable in magnitude to the measurement error of ~260 s for an individual secondary eclipse (*Laughlin et al.*, 2009). The reason this effect has an observable magnitude is because it originates within special relativity, and so it scales as 1/c. Effects of general relativity, such as those embodied in equation (2) or the effect of curvature on light propagation (*Shapiro*, 1964), scale as $1/c^2$, and are thus harder to detect.

A final step is needed to compare these theoretical event times to the observed data. The most important effect of a second body on transit times is to change the true period (as mentioned in section 2.2), so in practice the times are scaled to match the observed period and shifted to match the observed epoch of transit. Then variations on top of that simple linear ephemeris can be used, e.g., to detect or set limits on the presence of perturbing bodies (section 5.1). Care needs to be taken if radial velocities are being simultaneously fit, because such data specify a particular *velocity* of the star corresponding to a particular *time*, whereas transit data specify a particular *time* corresponding to a particular relative *position* of the planet to the star.

3. DYNAMICAL PHENOMENA

3.1. Astrophysical Two-Body Problem

3.1.1. Periastron advance. In section 2.1 we encountered several extra forces that modify realistic two-body motion from Keplerian ellipses. Let us take a perturbation theory approach, in which the force is calculated over a Keplerian orbit, to find how that orbit itself evolves (*Burns*, 1976). Now that the effective force of gravity is no longer of the form $1/r^2$, the ellipse does not close, and the periastron advances, which amounts to a reorientation of the orbit within its own plane. We shall calculate this precession rate due to the perturbing forces of relativity, tides, and rotational oblateness.

Periastron advance is the relativistic effect that changes orbits on the shortest timescale, and averaging equation (2) over an orbit, we determine its angular rate to be

$$\dot{\omega}_{GR} = \frac{3G^{3/2}(m_\star + m_p)^{3/2}}{a^{5/2} c^2 (1-e^2)} \quad (15)$$

This effect causes an additional 43 arcsec per century of precession for Mercury (in addition to the precession caused by the other planets), which was the famous first hint that nature obeyed Einstein's equations. Higher-order corrections to this precession rate, precession due to the star's spin (the Lenz-Thirring effect), orbital decay due to gravitational wave emission, and other relativistic effects are all negligible for exoplanets.

The force due to tidal distortion of the planet (equation (3)) causes apsidal motion at the rate

$$\dot{\omega}_T = \frac{15}{2} n k_L \frac{m_\star}{m_p} \frac{1+(3/2)e^2 + (1/8)e^4}{(1-e^2)^5} \left(\frac{R_p}{a}\right)^5 \quad (16)$$

where n ≡ 2π/P is the mean motion (*Sterne*, 1939). This effect is generally much bigger than that of the tide raised on the star by the planet, and for hot Jupiters with periods less than three days, it can dominate all other precessional effects (*Ragozzine and Wolf*, 2009). For physically smaller planets, and for Jupiter-type planets with periods ≥3 d, relativistic precession (equation (15)) typically dominates.

Rotational distortion gives rise to a force (equation (4)) that causes a periastron advance rate

$$\dot{\omega}_R = \frac{n k_{L,\star}}{2} \frac{1+m_p/m_\star}{(1-e^2)^2} \left(\frac{\Omega_\star}{n}\right)^2 \left(\frac{R_\star}{a}\right)^5 \quad (17)$$

Around fast-rotating and large stars (i.e., young or early type), this effect can dominate the others. If the stellar spin is misaligned with the orbit by an angle ψ, equation (17) requires an extra term (5 cos² ψ −1)/4; for spin-orbit angles satisfying 63.4° < ψ < 116.6°, the apsidal motion is retrograde. With spin-orbit misalignment, the nodal angle also precesses; equations for the coupled spin and orbital motion are given by *Eggleton and Kiseleva-Eggleton* (2001).

Of course, the star has a tidal bulge and the planet has a rotational bulge as well, but these never contribute substantially to the total precession.

3.1.2. Tidal dissipation. Tidal energy is converted to heat when a tidal bulge rotates through a body or varies in amplitude, due to the material's resistance to shearing motion. First, the dissipative torque changes the rotation of the planet to a rate at which the time average of that torque vanishes. At this spin rate the time average of the shear, and the energy dissipation rate, is minimized. In a fixed, circular orbit, the spin angular velocity equals the orbital angular velocity and the obliquity is zero, so in the frame corotating with the perturber, the tide is no longer time variable, stopping energy loss. In an eccentric orbit, the spin will either settle at a pseudosynchronous state (*Peale and Gold*, 1965; *Hut*, 1981; *Levrard et al.*, 2007), or be trapped in a spin-orbit resonance (of which Mercury is the prototype); the latter is only possible if the body has a permanent quadrupole moment due to its rigidity, and is therefore not expected for gas giants. For rocky planets with dynamically important atmospheres (of which Venus is the prototype), the picture can be qualitatively different, including up to four stable rotation states (*Correia et al., 2008*).

On a longer timescale, the eccentricity damps. The correlation between eccentricity and orbital distance (or period) is the main constraint on tidal theory for exoplanets (see chapter by Cumming). This damping can in principle be due to either dissipation in the planet or the star. If dissipation in the star is important, eventually the planetary orbit will decay into the star (e.g., *Rasio et al., 1996; Jackson et al., 2009; Barker and Ogilvie, 2009*). If dissipation in the planet is important, then it may have ingested more tidal energy than its own binding energy. In that case, gas giants could inflate or even disrupt (*Gu et al., 2003*), and such heating on terrestrial planets would have significant geophysical consequences (*Wisdom, 2008*).

The physical causes of tidal damping for giant planets are still poorly known; as yet no first-principle theory is efficient enough to damp the eccentricity of hot Jupiters, or to generate the inferred histories of satellite systems around the four solar system giants. *Lin et al.* (2000) and *Ogilvie and Lin* (2004) discuss these matters and review the literature. In the absence of such a theory, a phenomenological approach has gained currency (*Goldreich and Soter, 1966*). A fraction 1/Q of the tidal energy is dissipated per tidal forcing cycle (or per orbit, depending on the author). This allows differential equations for tidal damping to be derived, in which damping times scale with Q (*Mardling and Lin, 2002; Matsumura et al. 2008*). Empirical constraints on Q for close-in gas giants have been worked out (*Wu, 2003; Jackson et al., 2008; Matsumura et al., 2008*).

3.1.3. Miscellaneous orbital evolution. There are numerous other effects that can modify a planet's orbit about its star. Here we simply list some of these effects, referring the reader to work that describes them in detail.

Close in to the star, the planet may be tidally stripped of mass. As the mass leaves the planet, it applies a torque on its orbit. The reaction of the orbit has been calculated for circular orbits as the planet finishes migration due to a gas disk (*Trilling et al., 1998*), for moderate eccentricities as the planet tidally circularizes and perhaps inflates (*Gu et al., 2003*), and for eccentricities near 1 when the planet is shot near the star by either a dynamical instability or a chance flyby (*Faber et al., 2005*).

Once close to the star, the planet's atmosphere absorbs and reradiates photons in preferential directions, which can lead to at most a 5% change in semimajor axis — enough to influence resonant configurations with more distant planets (*Fabrycky, 2008*).

A planet or planets may scatter and eject a sea of small bodies (planetesimals) after the main formation phase, which leads to planetary migration. This effect was first worked out for the giant planets of the solar system (*Fernandez and Ip, 1984; Malhotra, 1993a, 1995*), and has since been applied to exoplanets (*Murray et al., 1998; Morbidelli et al., 2007; Thommes et al., 2008*). For more on migration, particularly in a gas disk, see the chapter by Lubow and Ida.

Finally, far from the host star, passing stars may perturb planetary orbits (see, e.g., *Spurzem et al., 2009*).

3.2. Short-Period, Secular, and Resonant Interactions

The interaction terms in the equations of motion (equation (5)) lead to all the interesting behavior of N-planet systems, and here we show heuristically how short-period, secular, and resonant behaviors arise. [For a traditional expansion in terms of orbital elements, using the so-called disturbing function, see *Murray and Dermott* (1999).] Figure 2 shows how the Jacobian orbital elements of two planets, initially on circular orbits, evolve as the planets move from opposition (being on opposite sides of the star), through conjunction (lined up with the star on the same side), then back to opposition. As the planets approach each other, the mutual gravity moves them onto slightly different orbits. At conjunction, the inner planet has been torqued forward, to a more distant, slower orbit; the outer planet has been torqued backward, to a closer-in, faster orbit. As the planets move through and recede from conjunction, these orbit changes are mostly reversed.

Now let us introduce moderate eccentricity to the orbits, and follow the system for many conjunctions. Due to the eccentricity, the paths of the planets at various conjunctions are either converging or diverging, and the changes in orbital elements on either side of conjunction do not cancel as completely. At a single conjunction, this causes the orbits to transfer energy (the semimajor axes change) and angular momentum (the eccentricities and orbit orientations change). After multiple conjunctions, the behavior of the system depends on whether the periods are near a ratio of small integers.

First, consider Fig. 3, which shows a hypothetical system with a period ratio far from any ratio of small integers. Because of this property, the conjunctions sample all parts of both orbits rather equally. The semimajor axes exhibit no long-term changes, which means the energy of each orbit is conserved. However, the angular momentum of each orbit is exchanged on long timescales, resulting in eccentricity variations. One way of seeing why this happens is by considering the time average of a planet over its orbit, so that its gravitational effect is that of an elliptical wire weighted inversely to the Keplerian velocity at each position. Each of the planets respond to the other planets as if they were such rings. Because the potential from such a ring is not time dependent, it produces a conservative force, and no energy can be exchanged: Semimajor axes and periods may not change. However, the lopsided rings do torque each other, and this corresponds to angular momentum (and thus eccentricity and inclination) changes. For instance, two eccentric, coplanar planets will undergo periodic oscillations in eccentricity that are 180° out of phase from each other. For more on this topic, called secular evolution, see section 3.3 and section 4.3.

Next, consider Fig. 4, which is a hypothetical system with periods very close to a ratio of small integers (2.01:1). In this situation, called a mean-motion resonance, conjunc-

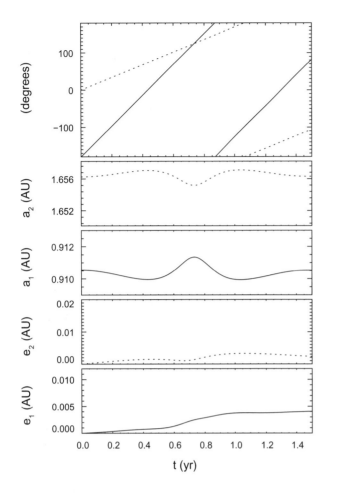

Fig. 2. Orbital element changes on the timescale of conjunctions, in a hypothetical system. The stellar mass is $m_\odot$, both planets have mass 10^{-3} $m_\odot$, their orbits are coplanar and initially circular, and they start on opposite sides of the star. Here and elsewhere, planets are numbered by increasing semimajor axis. *Top panel*: The mean longitudes of each planet (the two lines cross at conjunction); *second and third panels*: the planets' semimajor axes, which vary symmetrically about the conjunction; *fourth and fifth panels*: the planetary eccentricities, which receive a small kick.

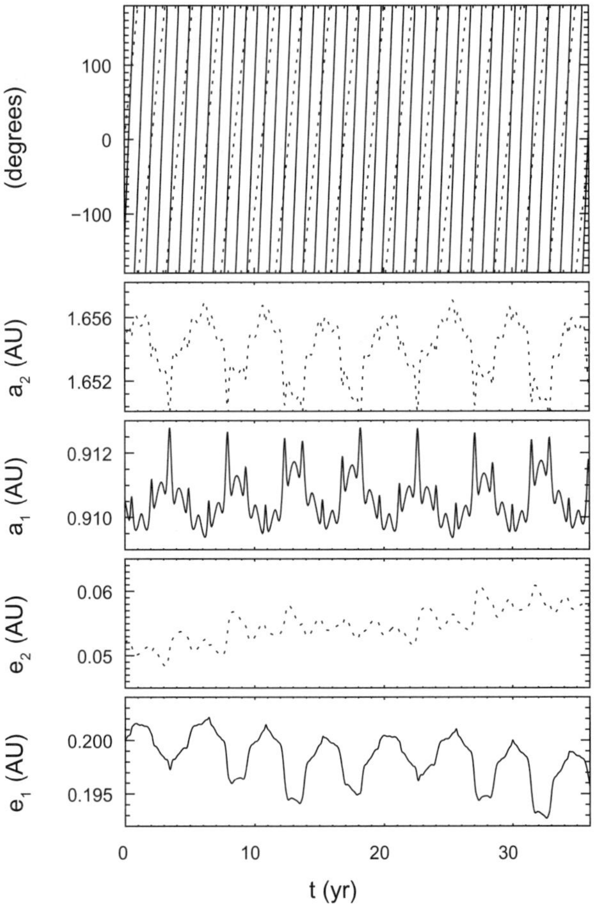

Fig. 3. Orbital element changes after many conjunctions. The hypothetical system is the same as in Fig. 2, except both planets start with eccentricity, with the inner planet's argument of pericenter 45° ahead of the outer planet's. The vertical axis on each panel has the same scaling as the corresponding panel in Fig. 2, emphasizing that the semimajor axes experience no net drift, but the eccentricities do.

tions occur at the same part of the orbit many times in a row, and the change in orbital elements builds. One may consider a changing period the hallmark of a mean-motion resonance. Along with period changes come eccentricity oscillations, which can be rather large over only tens of orbits. For noncoplanar planets, the distance between the location of conjunctions and the intersection of the orbital planes affects their dynamics, so resonances can also involve inclinations and not only eccentricities. In general, the angles that dictate the behavior of the resonance are called critical angles, and they have the form

$$\phi = j_1\lambda_1 + j_2\lambda_2 + j_3\varpi_1 + j_4\varpi_2 + j_5\Omega_1 + j_6\Omega_2 \qquad (18)$$

where each planet has a mean motion of λ, a longitude of ascending node Ω, and a longitude of periastron ϖ. The j values are integers obeying $\Sigma j_i = 0$ (called the d'Alembert relation), which is required by the invariance of the system's behavior to the arbitrary reference direction from which angles are measured. At the very center of each resonance, where the planets come to conjunction at exactly the same point in their orbits, the periods are constant and precession rate is constant. In this case, there exists a slowly rotating frame in which the motion of each planet is perfectly periodic, yet not perfectly elliptical.

Individual resonances can help keep a system stable. For instance, when the critical argument for the interior 2:1 resonance has zero libration amplitude ($\theta = 2\lambda_2 - \lambda_1 - \varpi_1 = 0$, $\dot\theta =$

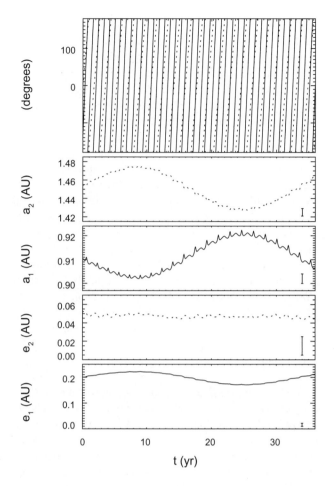

Fig. 4. Orbital element changes induced by a mean-motion resonance. The system is the same as in Fig. 3, except the outer planet starts at a different period, with a ratio of osculating periods of 2.01:1. Note that successive conjunctions (where the lines intersect in the top panel) occur at nearly the same longitude, which causes the period change to grow. The bar shows the vertical scale of each corresponding panel in Figs. 2 and 3, emphasizing that the semimajor axes and eccentricities are experiencing a large oscillation.

0), we may rearrange the equation to read $\lambda_1 - \lambda_2 = \lambda_2 - \varpi_1$, which shows that when the two planets are at conjunction ($\lambda_1 - \lambda_2 = 0$), then the inner body is also at pericenter ($\lambda_1 - \varpi_1 = 0$), so a close approach is avoided. Conversely, whenever the outer body is at the azimuthal location of the inner body's apocenter ($\lambda_2 - \varpi_1 = \pi$), the two bodies are farthest apart ($\lambda_2 - \lambda_1 = \pi$). This argument, applied generally to mean-motion resonances, is called a resonance protection mechanism for otherwise unstable systems.

The role and behavior of resonances during planetary migration is beyond the scope of this chapter. However, the reader is referred to *Peale* (1976) and *Malhotra* (1998) for Hamiltonian descriptions of resonances, which can cleanly treat migration. For a recent applications to exoplanets, see *Lee and Peale* (2002) for how two planets can capture into a resonance if their migration converges and *Chiang* (2003) for how two planets can excite each other's eccentricities as they pass through a resonance while their orbits diverge.

3.3. Advanced Interactions

Having surveyed the basic interactions between two planets in section 3.2, we now introduce several more advanced topics.

We previously saw that the eccentricities of planets outside mean-motion resonance can change on a long timescale. We now extend that concept to systems with three or more planets, systems in which the orbital elements of each planet vary on many different timescales. Resonances between these timescales can excite eccentricities to very high values [see *Moro-Martín et al.* (2007) for an example in an exoplanetary system]. For the three-planet system Upsilon Andromeda (υ And) in the Newtonian approximation, the innermost planet's eccentricity periodically reaches ~0.4, compared to ~0.06 in the absence of the outer planet or ~0.025 in the absence of the middle planet (*Barnes*, 2008). Also important in determining the qualitative behavior of the secular dynamics is extra precession, e.g., that supplied by relativistic or tidal effects (section 2.1); see *Wu and Goldreich* (2002) and *Migaszewski and Góździewski* (2009b). In Fig. 5 we plot the long-term behavior of the eccentricities in the υ And system considering (1) only Newtonian point masses, and (2) an extra force modeling the tidal bulge raised on the inner planet. The behavior of the outer planets is not much different, but the effect of the extra precession on the inner planet is to detune its pericenter precession rate away from an eccentricity-exciting secular resonance. Thus, the low current value of e_b argues an additional precession is active (*Adams and Laughlin*, 2006); both relativity and tides probably contribute with roughly equal precession rates of $\dot{\omega} \simeq 10^{-11}$ s^{-1} each.

An extension of secular evolution theory can be made into the regime of high inclination and eccentricity, in which they are strongly coupled (*Kozai*, 1962; *Lidov*, 1962). In a system with a planet on an initially circular orbit, a third body on a distant exterior orbit will periodically pump the planetary eccentricity to a maximum of

$$e_{max} \approx \sqrt{1 - (5/3)\cos^2 i} \qquad (19)$$

where i is the mutual inclination, which must initially be in the range 39.2°–140.8°. [For a detailed description and a derivation of this behavior, called Kozai oscillations, see *Fabrycky and Tremaine* (2007).] Note that an initially perpendicular orbit (i = 90°) leads to an eccentricity of unity. In some systems, relativistic precession would suppress this behavior, but in others, tidal dissipation would take hold at the eccentricity maximum and circularize the planet at a period of a few days (*Fabrycky and Tremaine*, 2007; *Wu et al.*, 2007; *Nagasawa et al.*, 2008). An important

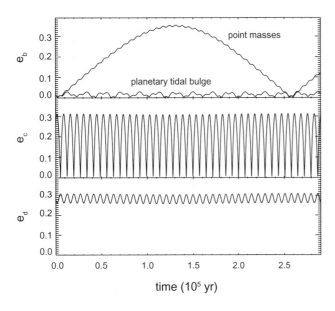

Fig. 5. Eccentricities as a function of time for the υ And system. Orbital elements of *Ford et al.* (2005), and edge-on, coplanar orbits were assumed. We use $m_\star = 1.27\ m_\odot$ and $(m_b, m_c, m_d) = (0.6777, 1.943, 3.943)\ m_{Jup}$. The outer planets, c ($P_c = 241.32$ d) and d ($P_d = 1301.0$ d), were initialized using Jacobian coordinates, and the plotted eccentricities are also Jacobian. Two integrations were performed, one in which the planets were point masses and followed the Newtonian equations of motion (equation (5)), and one in which $\mathbf{f} = \mathbf{f}_T$ of equations (1) and (3) to model the tidal bulge of planet b ($P_b = 4.61714$ d), with $R_p = 1.3 R_{Jup}$, $k_L = 0.34$ assumed. In the point-mass model, e_b quasiperiodically reaches ~0.36 (top curve of top panel). Including the tidal model significantly suppresses the induced e_b (bottom curve of top panel), and including relativistic precession as well (not shown) suppresses it still further. The evolution of e_c and e_d are almost identical in the two cases, so they are plotted only once.

indicator of whether close-in planets have undergone such an event is their orbital orientation relative to the stellar spin. With transiting planets, this can be measured via the spectroscopic Rossiter-McLaughlin effect (see section 4.4 and the chapter by Winn).

Another secular effect was found by *Migaszewski and Góździewski* (2009a), who used a numerical Hamiltonian approach to classify stationary solutions for two-planet, noncoplanar systems. They found that *interlocking* planetary orbits could stably exist even if no mean-motion resonance protection is active. Instead, each orbit nodally precesses because of the other orbit, and they do so at the same rate, so the two orbits remain in the same configuration with respect to each other as the whole system precesses. However, in numerical tests of the full equations of motion, we find that a specific example of this class (Fig. 17 of *Migaszewski and Góździewski,* 2009a) becomes chaotic within several million years unless the mass ratios are smaller than ~10^{-4}. A system of Neptune-mass planets could thus exist around a solar-like star in this interlocked configuration.

The high eccentricities of exoplanets call for a mixture of numerical and analytical work to understand the nonlinear dynamics. The typical method of expanding in small eccentricities and inclinations has been fruitfully supplemented with numerical methods (e.g., *Michtchenko and Malhotra*, 2004) to average over short-timescale effects like those of Fig. 3. Resonant orbits at high eccentricity can have novel properties (*Lee and Peale*, 2002; *Beaugé et al.*, 2003; *Lee*, 2004). One way exoplanets can be dynamically simpler than solar system planets is that they are often hierarchical, i.e., the period ratios can be large. When this is the case, an expansion in the semimajor axis ratio becomes a useful analytic method (*Ford et al.*, 2000) (see also section 2.3).

Mean-motion resonances may be shared among more than two planets with critical angles that are extensions of equation (18); these are called three-body mean-motion resonances. They have been shown to destabilize asteroids (*Nesvorný and Morbidelli*, 1998), the outer planets of the solar system (*Murray and Holman*, 1999), and potentially exoplanetary systems (*Góździewski et al.*, 2008a).

3.4. Chaos

Nothing ensures that planets emerging from a protoplanetary disk — in which collisions were frequent — will be on nearly Keplerian, stable orbits. On the contrary, models of planet formation suggest that the orbits of planets should be closely packed together, with the timescale of collisions or ejections being comparable to a system's current age (e.g., *Laskar*, 2000). Indeed, observed multiplanet systems are often close to instability (*Barnes et al.*, 2008). A dynamical system is said to be chaotic if trajectories that are initially separated by an infinitesimal amount diverge exponentially. This "microscopic" definition is often used as a computational tool to detect whether and when "macroscopic" events (i.e., ejections and collisions) are likely to occur (*Lecar et al.*, 1992; *Morbidelli and Froeschlé*, 1995); these usually occur on a timescale orders of magnitude longer.

Resonance overlap is a general condition for strong chaos (*Chirikov*, 1979). Qualitatively, each resonance allows the system to explore a finite zone of semimajor axes, called the width of the resonance. If two resonances overlap, the system has access to a wider swath of semimajor axes. This region might even connect to infinity, allowing an ejection. *Wisdom* (1980) found that resonance overlap could account for the orbits of massless particles becoming chaotic near a planet on a circular orbit. For a planet with a semimajor axis of 1 and a mass (relative to the star) of μ, he found that for semimajor axes within $|\Delta a|/a < 1.3\ \mu^{2/7}$ of the planet, a particle's orbit is chaotic: Its fate is either a planetary collision or an ejection. The mass scaling of this limit is found by comparing the widths of the resonances to their spacing.

Duncan et al. (1989) presented a complementary understanding of the chaotic zone. Suppose that two planets follow Keplerian orbits except at their mutual conjunction, where they give each other a kick. The kick leads to a change in eccentricity that is second-order in the orbital separation, and a change in orbital period that is even higher order. If the kick in orbital period is strong enough such that the next conjunction occurs more than half an orbital period earlier or later than it would have without the kick, the result of successive kicks is uncorrelated and the system executes a random walk, eventually leading to orbital instability. This argument leads to nearly the same scaling law of the chaotic region: $|\Delta a| \lesssim 1.24\ \mu^{2/7}$. This concept can be readily extended to N-planet systems. *Chambers et al.* (1996) applied it to terrestrial planet formation (rather small μ values), found that there is no critical separation beyond which stability is assured, and mapped out numerical timescales to instability as a function of orbital separation. *Zhou et al.* (2007) extended these arguments to the mass ratios of most known exoplanets ($\mu \sim 10^{-3}$), finding that kicks at conjunction can explain the instability of planetary systems, but that empirical corrections are needed for quantitative agreement.

At the relatively larger masses of most known exoplanets ($\mu \sim 10^{-3}$), both these methods start breaking down: (1) Their chaotic zones extend to low p among the (p + 1):p resonances, so the peculiarities of individual strong resonances need to be taken into account in the resonance overlap picture; and (2) the physics of interactions between planets is not nearly as localized to conjunctions, and this derivation, based on the limit $\mu \to 0$, breaks down. Interestingly, the method of resonance overlap no longer breaks down when the concept is applied to even higher mass ratios: *Mardling* (2008) has shown that for comparable-mass triple stars, n:1 resonances are the only ones relevant for stability. This method has been used to find the instability boundary for planets orbiting one star of a binary star system (*Mudryk and Wu*, 2006).

A more general framework for the stability of three-body systems is called Hill stability, in which the orbits of two planets can never cross if a particular inequality is satisfied (*Marchal and Bozis,* 1982; *Gladman,* 1993; *Veras and Armitage,* 2004). There is no known sufficient condition for *instability*, unfortunately, but *Barnes and Greenberg* (2007b) have shown that Hill's stability criterion and practical stability are rather close, numerically.

Chaotic trajectories are particularly important during planet formation, in which a given surface density of material is converted into fewer numbers of larger bodies. Chaotic zones around a given body scale as only a shallow function of its mass, so by putting the same surface density into more widely spaced bodies, the system becomes more stable. This concept underlies the derivation of the isolation mass — the mass at which a protoplanet has cleared its feeding zone and no longer accretes — laid out in section 2.4 of the chapter by Chambers.

One popular application of chaotic trajectories is the origin of eccentricities. Exoplanets tend to have larger eccentricities than the solar system planets. Perhaps they typically start in systems of several planets of comparable mass, which perturb each other into crossing orbits, resulting in ejection or accretion (*Rasio and Ford,* 1996; *Chatterjee et al.,* 2008; *Jurić and Tremaine,* 2008). If indeed this mechanism explains the observed distribution, as discussed in section 3.4 of the chapter by Cumming, it implies that most systems of giant planets will self-destruct and many free-floating planets exist. See *Ford and Rasio* (2008) for a thorough review of the work on this hypothesis and references to competing hypotheses for the origin of eccentricities.

3.5. Stable Orbits Near Planets and in Habitable Zones

We have seen in section 3.4 that close to a planet, orbits become unstable. However, there are particular configurations that allow for long-term stability. Although none of the following configurations have been observed yet in exoplanetary systems, it is worthwhile to discuss here their existence (their potential detectability is discussed in section 5.1). For each type of configuration, we consider stable orbits of test particles in orbits similar to a planet, then extend the notion to a pair of planets.

Particles could orbit the planet on the orbit of a satellite, within its Hill sphere. Consider a particle between the star and a planet on a circular orbit, in a frame that corotates with the planet. The Hill sphere is the region within which the gravity from the planet ($Gm_p r_p^{-2}$, where r_p is the distance to the planet) is comparable to the tidal gravity from the star ($2Gm_\star r_p a^{-3}$, in the radial direction). Accounting for a differential centrifugal force ($Gm_\star r_p a^{-3}$) yields the customary definition for the Hill radius

$$r_H = a \left(\frac{m_p}{3 m_\star} \right)^{1/3} \quad (20)$$

Bodies orbiting within some fraction of the Hill radius orbit stably as satellites (*Domingos et al.,* 2006). The same types of orbits exist even if the satellite is massive, even up to the mass equal to that of the planet. In this case, the planetary mass m_p used to define the Hill radius (equation (20)) would be the sum of the two bodies' masses (*Henon and Petit,* 1986).

Orbits can also stay near the planet if they are in a 1:1 resonance, of which there are three kinds.

The first is an extension of satellite orbits to outside the Hill sphere, called quasisatellite orbits (e.g., *Shen and Tremaine,* 2008). From the planet's perspective, the particle would look like a very distant satellite, orbiting the planet in a direction retrograde to their common motion around the star. From an astrocentric perspective, the particle's orbital phase is similar to the planet's orbit, but its eccentricity is different, which carries it between the planet and the star at periastron and to the far side of the planet from the star at apastron. If that body is endowed with mass as a second planet, then both orbits will evolve due to their mutual perturbation; the eccentricity can be passed back and forth between the planets (*Laughlin and Chambers,* 2002). *Hadjidemetriou et al.* (2009) have explored the relationship between this type of 1:1 resonance to satellite-type mutual orbits.

The second type of 1:1 resonance is a Trojan orbit, named after the asteroids that inhabit this resonance with Jupiter.

Their average position is 60° ahead of or 60° behind the planet in its orbit; these are the stable Lagrange points labeled L_4 and L_5 (*Dvorak et al.,* 2004). However, the orbits may also wander stably around those points, tracing out a shape of a tadpole, so they are sometimes called tadpole orbits. Second planets of any mass relative to the more massive planet can exist in these points, but $\mu = (m_1 + m_2)/m_\star$ must not exceed ~0.038 (*Laughlin and Chambers,* 2002), or the system will be unstable. In this case, both planets will trace out tadpole shapes in the frame rotating with the long-term mean angular velocity, with the size of the tadpole inversely proportional to each planet's mass.

The width of the stable tadpole region scales as $\mu^{1/2}$, which is steeper than the Hill sphere's scaling $\mu^{1/3}$ (equation (20)), so for low planetary masses there is a region between them. In this region lies the third type of 1:1 resonance: the horseshoe orbits. Such orbits trace out a horseshoe shape in the frame rotating with the orbit of the massive planet, which encompass both L_4 and L_5. As with the other resonances, any relative mass of two planets may be in this resonance. In fact, the solar system furnishes an example of a ~4:1 mass ratio in this resonance: the saturnian satellites Janus and Epimetheus (*Dermott and Murray,* 1981).

In systems with known giant planets, the orbital stability of hypothetical terrestrial planets has been studied extensively, particularly in the region that allows for habitable climates (see the chapter by Fortney et al. regarding habitable zones). An appropriate approximation is that the terrestrial planet is too small to affect the orbits of the giants: They are treated as test particles (e.g., *Sándor et al.,* 2007). Habitable planets might also reside in the dynamical niches described above: satellites (*Williams et al.,* 1997) or Trojans (*Dvorak et al.,* 2004). In systems already known to have multiple giant planets, the zones of stability can be quite complicated, and numerical integrations are indispensable.

Terrestrial planets that avoid ejection may still be subject to oscillating orbital and spin properties (see respectively *Menou and Tabachnik,* 2003; *Laskar and Robutel,* 1993), and these oscillations may cause climate changes [a much enhanced form of Milankovitch cycles (*Hays et al.,* 1976)]. The planet may not need to stay in the habitable zone for all of its orbit, depending how well the atmosphere buffers seasonal temperature changes (*Williams and Pollard,* 2002), so the upper limit on a habitable eccentricity is a function of planetary properties. It has been argued that moons of planets with eccentricities as high as 0.69 (16 Cyg B b) might still be habitable. Therefore, ejection may be the only dynamical effect that will spoil a habitable world.

4. HIGHLIGHTS: DYNAMICS IN NATURE

4.1. GJ 876 and Mean-Motion Resonances

The only exoplanet system hosted by a main-sequence star for which non-Keplerian motion has been conclusively detected is GJ 876 (*Marcy et al.,* 2001; *Laughlin and Chambers,* 2001; *Rivera and Lissauer,* 2001). The M-dwarf primary hosts three planets, whose properties are listed in Table 2.

These numbers come from a self-consistent Newtonian fit of the radial velocity data (Fig. 6), in which a coplanar configuration is assumed and the best-fitting common sky inclination is found to be i = 50° (*Rivera et al.,* 2005). For typical planets discovered by radial velocity, which are not found to transit, the data are only sensitive to the gravitational influence of the planet in the line-of-sight direction, which yields the quantity $m_p \sin i$, not m_p or i independently. In this system, the periastron of each planet rotates at a rate proportional to the mass of the other planet, and that rate is well determined by the data. Therefore m_p is independently measured, breaking the $m_p \sin i$ degeneracy to yield i.

However, in this system the dynamically derived inclination is in tension with the astrometric orbit of planet b (*Benedict et al.,* 2002). Nevertheless, *Bean and Seifahrt* (2009) have fit both datasets simultaneously, and they even determined a mutual inclination of $\Phi_{bc} = 5.0°^{+3.9°}_{-2.3°}$, close but marginally inconsistent with coplanar.

Almost no dynamical constraint can currently be given on the orbital orientation of planet d, from either stability considerations or fits to radial velocity or astrometry data. Its short period means it is only weakly coupled to the outer planets on the timescale of the data.

The dynamically interesting aspect of the system is that the outer two planets are deeply engaged in a 2:1 resonance, with small-amplitude libration of both critical arguments $\theta_1 \equiv \lambda_c - 2\lambda_b - \varpi_c$ and $\theta_2 \equiv \lambda_c - 2\lambda_b - \varpi_b$ about 0° (compare equation (18)). As a consequence, $\Delta\varpi \equiv \varpi_c - \varpi_b = \theta_2 - \theta_1$, also librates around 0°. The three-planet, Newtonian, coplanar model with i = 50°, for which system parameters were quoted above, yields $|\theta_1|_{max} = 5.4° \pm 0.9°$, $|\theta_2|_{max} = 19.5° \pm 3.8°$, and $|\Delta\varpi|_{max} = 19.4° \pm 4.3°$ (*Rivera et al.,* 2005).

Apart from GJ 876, about seven other planetary systems have been shown to be in resonance (Table 3). This is usually accomplished, for planets of lower-quality data or fewer dynamical times, by noticing the system would be unstable if not for the resonance (*Vogt et al.,* 2005; *Correia et al.,* 2005; *Lee et al.,* 2006; *Fabrycky and Murray-Clay,* 2010) [for an overview, see *Barnes and Greenberg* (2007b) and Table 3]. This logic is needed because (1) not enough orbital periods have been observed for them to be measured to a precision such that nonresonant systems are inconsistent with the radial

TABLE 2. The GJ 876 system.

Planet	m_p	P(days)	e
b	2.530 ± 0.008 m_{Jup}	60.83(2)	0.0338 ± 0.0025
c	0.790 ± 0.006 m_{Jup}	30.46(2)	0.2632 ± 0.0013
d	7.53 ± 0.70 $m_\oplus$	1.93774(6)	0.0 (assumed)

Three-planet Newtonian solution from *Rivera et al.* (2005), which optimally fits the radial velocities (Fig. 6). Uses $Gm_\star = 0.32\ Gm_\odot$.

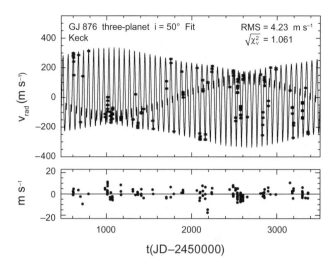

Fig. 6. Radial velocity data and best three-planet Newtonian fit to the planetary system GJ 876. From *Rivera et al.* (2005).

velocity data; and (2) usually only a small fraction of the resonance libration cycle, or precession cycle, is completed during the observation span, so the orbits appear Keplerian.

4.2. Planets Around Pulsar 1257+12

The discovery of a planetary system around a pulsar (see chapter by Wolszczan and Kuchner) preceded the first solid radial velocity detections around main-sequence stars (*Wolszczan and Frail*, 1992). The discovery technique was to infer the gravitational influence of the planets through tracking the line-of-sight motion of the host star, as in the radial velocity technique. In this case, however, the light time effect delays pulse profiles. Nevertheless, with only Keplerian motion detected, the typical $m_p \sin i$ degeneracy held sway.

The theorists set to work quickly, showing that (1) the perturbations could be detected with more data, (2) they would confirm the planetary nature of the timing residuals, and (3) the amplitude and the character of the detected perturbations would determine the masses of the objects (*Rasio et al.*, 1992; *Malhotra et al.*, 1992). The final point is analogous to how the $m_p \sin i$ degeneracy is broken for the GJ 876 system (section 4.1). However, here the main observable effect was shown to be period variations (*Malhotra*, 1993b; *Peale*, 1993), which build up to a sinusoidally varying phase shift with a period of $2\pi/[n_B-(3/2)n_C] \approx 5.5$ yr (see Fig. 7). The same type of near-resonant behavior is famously active in the solar system between Jupiter and Saturn, which has a large effect on their orbital phases known classically as the Great Inequality.

The observers answered the challenge (see the chapter by Wolszczan and Kuchner), not only detecting perturbations to constrain the masses of the planets (*Wolszczan*, 1994), but showing their orbits are consistent with coplanar to within ~13° (*Konacki and Wolszczan*, 2003). The data and the timing model are shown in Fig. 8.

4.3. Secular Apsidal Alignment

Eccentricities oscillate and apses precess due to their planets' secular interaction. If two planets have apses precessing at the same rate, on average, they are said to be in *apsidal lock*, with a critical angle $\Delta\varpi \equiv \varpi_1 - \varpi_2$. This angle can librate around either 0° (apses aligned) or 180° (apses antialigned), depending on the masses and initial orbital elements, with the restoring torque supplied by the secular terms. [Mean-motion resonance terms can also result in libration of $\Delta\varpi$; e.g., *Beaugé et al.* (2003) and section 4.1, but this phenomenon is not our current focus.] If one planet

TABLE 3. Resonant systems.

System	Planets	Resonance	P_{in} (days)	N Inner Orbits Observed	$m_{p,in}$, $m_{p,out}$ (m_{Jup}/sin i)	Reference
HD 45364	b–c	3:2	227	7	0.19, 0.66	[1]
GJ 876	c-d	2:1	30	~80	0.56, 1.94	[2]
HD 82943	b–c	2:1	441.2	8	2.01, 1.75	[3]
HD 128311	b–c	2:1	448.6	6	2.18, 3.21	[4]
HD 73526	b–c	2:1	188.3	13	2.9, 2.5	[5]
HD 160691 = μ Arae	d–b	2:1	310.5	9	0.52, 1.68	[6]
HD 60532	b–c	3:1	201	4.5	3.15, 7.46	[7]
HD 202206	b–c	5:1	255.87	8	17.4, 2.4	[8]
HD 108874	b–c	4:1	395	6	1.36, 1.02	[9]
55 Cnc	b–c near	3:1	14.65	~300	0.82, 0.17	[10]

References: [1] Surrounded by chaos (*Correia et al.*, 2009); [2] king of the resonant planets (section 4.1); [3] *Mayor et al.* (2004), *Ferraz-Mello et al.* (2005), *Lee et al.* (2006), *Góździewski and Konacki* (2006), *Beaugé et al.* (2008b); [4] *Vogt et al.* (2005), *Góździewski and Konacki* (2006); [5] *Tinney et al.* (2006); [6] also known as HD 160691; *Pepe et al.* (2007), *Short et al.* (2008); [7] *Desort et al.* (2008), *Laskar and Correia* (2009). [8] *Correia et al.* (2005); [9] not necessarily resonant (*Vogt et al.*, 2005); [10] according to *Fischer et al.* (2008), it is actually just outside this resonance.

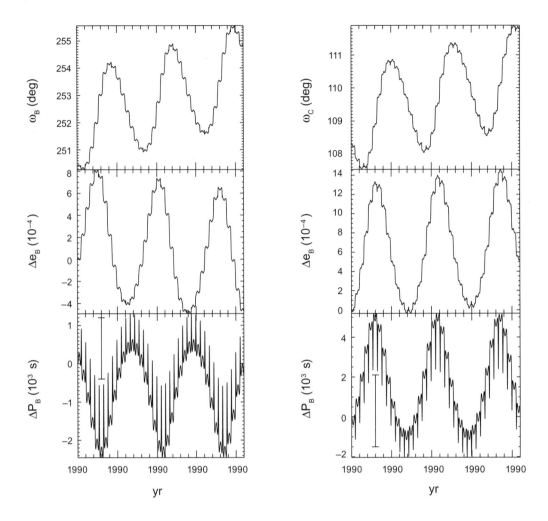

Fig. 7. Evolution of the orbital elements of planets B and C around Pulsar 1257+12. The latest orbital elements (*Konacki and Wolszczan*, 2003), taking the planets to be coplanar, are used, and planet A is not included. Following *Rasio et al.* (1992), the elements displayed are argument of pericenter (ω), change in eccentricity (Δe), and change in period (ΔP) of the Jacobian coordinates of each planet. Error bars of each planet's P from the discovery paper (*Wolszczan and Frail*, 1992) are indicated in the bottom boxes, showing period changes were nearly detectable (but the changes in ω and e were far from detectable). With equally precise data spread over three years, the period variations were detected (*Wolszczan*, 1994).

periodically reaches e = 0, we follow *Ford* (2008) in calling such a system borderline: For such systems, $\Delta\varpi$ is on the border between librating and circulating. We note that there is little dynamical significance to this border, and systems on either side of it remain close to each other in phase space. In polar coordinates (e cos $\Delta\varpi$, e sin $\Delta\varpi$), the difference is just whether the trajectory contains the origin or not.

Recently, there have been several attempts to use the libration amplitude, or the proximity to the borderline state, to shed light on earlier epochs in multiplanet systems. The idea is that if all the planets of the system start out in circular orbits, but one is forced to an eccentric orbit, the system will have a small libration amplitude if this forcing is much slower than the secular timescale (*Chiang and Murray*, 2002), or will be left near the borderline state if this forcing is much faster than the secular timescale (*Malhotra*, 2002). The agent imparting the initial eccentricity might be the protoplanetary disk or an additional planet, which is subsequently ejected. *Ford et al.* (2005) presented the borderline behavior of the more massive planets in υ And (see Fig. 5) as evidence of the latter. Later work showed that, while borderline behavior is surprisingly common among multiplanet systems (*Barnes and Greenberg*, 2006), it is perhaps even *too* common for simple models of scattering among planets to be the explanation (*Barnes and Greenberg*, 2007a). The final answer awaits a rigorous statistical comparison between the secular behavior resulting from scattering simulations (e.g., *Chatterjee et al.*, 2008) and the secular behavior inferred from data (e.g., *Veras et al.*, 2009).

4.4. Kozai Oscillations

One of the first very eccentric exoplanets, 16 Cyg B b, caused excitement because its orbit is quite unlike the nearly

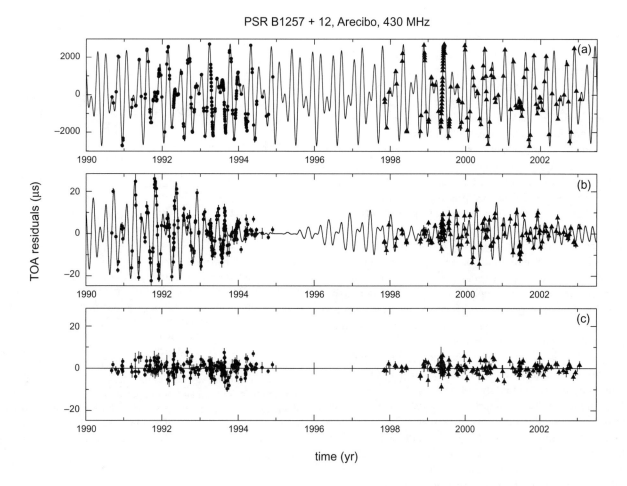

Fig. 8. Time-of-arrival (TOA) of pulses for the pulsar that hosts a three planet system. **(a)** The daily-averaged data with the best triple-Keplerian model as a solid line; **(b)** the residuals of the Keplerian fit and the difference between the best Newtonian and Keplerian models as a solid line; **(c)** the residuals from the Newtonian model. From *Konacki and Wolszczan* (2003).

circular orbits of the giant planets of the solar system, and it was unclear how a giant planet could form on an eccentric orbit. It also suggested that the low eccentricities of the previous discoveries were perhaps not primordial, but tidally damped. This particular system has a distant companion star, which could cause Kozai eccentricity oscillations [section 2.3, and *Holman et al.* (1997), *Mazeh et al.* (1997)], resulting in the high eccentricity.

As pointed out by *Takeda et al.* (2008), the four planets with the highest eccentricities (e > 0.8), and ~50% of the 18 planets with e > 0.6, have confirmed stellar companions; this correlation is statistically significant, considering that only ~16% of exoplanet hosts have stellar companions (*Mugrauer and Neuhaeuser*, 2008) due to survey biases. These numbers may be interpreted as a statistical detection of Kozai oscillations, which have much too long a timescale to be directly observed in the current data of any particular system. However, there is additional evidence that one of these four systems, HD 80606b with e = 0.93, has indeed experienced Kozai cycles. It was shown that a natural prediction of the Kozai scenario (as described by *Wu and Murray*, 2003) is that the stellar spin is currently misaligned from the planetary orbit (*Fabrycky and Tremaine*, 2007). This prediction of misalignment was recently verified (*Moutou et al.*, 2009; *Winn et al.*, 2009).

5. FUTURE PROSPECTS

5.1. Searching for Small Planets

One of the applications of dynamical calculations, which has yet to be realized, is the detection of previously unknown small planets via their dynamical effect on known planets. This potential is particularly ripe for the transit-timing method.

Transiting planets offer a wealth of information that is inaccessible for a usual radial-velocity-detected planet (see chapter by Winn). An exciting opportunity is afforded by the extreme phase sensitivity of a transit lightcurve. With high-quality data, a timing precision of ~10 s is achievable, which translates to a phase measurement of ~0.01° along a four-day orbit. Only very slight perturbations lead to along track variations of that order. In comparison, for a Jupiter-mass planet on a four-day orbit around a solar-mass star, a radial velocity datum with precision of 3 m s^{-1} translates to a phase measurement of ~1°.

Although it has potential, no compelling examples of transit-time variations have yet been published. However, *Steffen and Agol* (2005) — and many authors following their example — have published stringent upper limits on hypothetical second planets, which are required by constant transit times. *Agol and Steffen* (2007) have analyzed transits of the first-discovered transiting planet HD 209458b using the Hubble Space Telescope, searching for other companions. They were able to put stringent upper limits on the mass of a hypothetical second planet, as a function of its period (see Fig. 9). The sensitivity of transit-time measurements is orders of magnitude better within resonances than midway between them (*Agol et al.*, 2005; *Holman and Murray*, 2005). This is simply a restatement of our earlier identification of resonances as effective locations for a and P variations. Thus they provide complementary information to radial velocity observations, which are not particularly good at discovering planets in resonance [especially the 2:1 resonance (*Anglada-Escudé et al.*, 2010)], but they have sensitivity over a wide range of periods.

Although midtransit times are the most sensitive characteristic to period changes over timescales short compared to the observations, transit *duration* variations (TDV) have been recognized as more sensitive to variations of much longer timescale (*Miralda-Escudé*, 2002; *Heyl and Gladman*, 2007; *Pál and Kocsis*, 2008), and they may eliminate degeneracies inherent to transit-time variation (TTV) measurements (*Kipping*, 2009).

Transit-timing measurements are a possible way to find objects in qualitatively different orbits than those available to the radial velocity method. Trojans are hard to pull out of the radial velocity data because their orbits have the same harmonics as the main planet (*Góździewski and Konacki*, 2006). However, if the two planets have low eccentricity and a large libration amplitude, the signal would be a single sinusoid with an amplitude that slowly oscillates (every ~10 orbital periods), markedly different from the signal of a single planet (*Laughlin and Chambers*, 2002). Such a libration could be easily seen in transit data (*Ford and Holman*, 2007). Trojan planets that make a perfect equilateral triangle with the star would not librate at all. However, a combination of transit and radial velocity data could still detect it (*Ford and Gaudi*, 2006), which has been used to place upper limits of varying sensitivity in 25 systems (*Madhusudhan and Winn*, 2009). Moons of giant planets could also be searched for by transit timing (*Simon et al.*, 2007; *Kipping*, 2009), but the conceivable limits are well above the mass of Earth's moon or of the moons of giant planets in the solar system. The masses of moons may be limited by their formation mechanism (*Canup and Ward*, 2006) or by the requirement that tidal evolution has not destroyed them (*Barnes and O'Brien*, 2002; *Cassidy et al.*, 2009). Nevertheless, many of the giant exoplanets are in the habitable zone of their stars, so the first such planet to

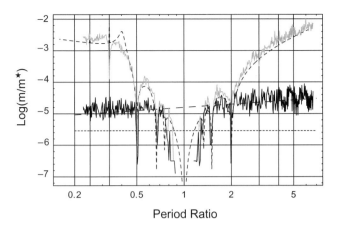

Fig. 9. The 3σ upper limits on the mass of a hypothetical second planet (with assumed eccentricity 0.02) relative to its host star in the HD 209458 system, based on the precise and nonvariable transit times of planet b (solid, gray curve) and that data combined with the radial velocity time series (solid, black curve). A perturbation theory calculation (thin, dashed curve) matches the transit time constraints between resonances, and an analytic expression (large dots) matches within resonances. The most sensitive constraints are within resonances, and extend an order of magnitude lower than both upper limits from radial velocity (thick dashed line) and the mass of Earth (thin dashed horizontal line). Near the period ratio of 1, orbits of small bodies are within the chaotic zone (section 3.4) and are thus unstable. From *Agol and Steffen* (2007).

transit will be carefully scrutinized for both habitable moons (*Williams et al.*, 1997) and habitable Trojan companions (*Dvorak et al.*, 2004).

5.2. System Architectures Through Detecting Perturbations

In this section we shall examine how more observations of non-Keperian motion will contribute to our understanding of the architectures of planetary systems, some aspects of which we currently know very little. We shall see that (1) resonant orbits and (2) mutual inclinations of planetary systems will likely enjoy fundamental advances in the coming years via detected perturbations.

First, a historical look at the two well-observed non-Keplerian systems shows their enormous contribution to our understanding of planets. In 1992, detecting perturbations for the planets of PSR 1257 + 12 was decisive in demonstrating the orbital and planetary nature of the signal, rather than an unforeseen pulsar oscillation (*Rasio et al.*, 1992; *Wolszczan*, 1994). Thus the confirmation of the first exoplanets, and the detection of the only sub-Earth-mass exoplanet known, required the modeling of non-Keplerian motion. In 2001, detecting perturbations for the planetary system GJ 876 was important for demonstrating the true masses of planets orbiting main-sequence stars; they are smaller than masses of brown dwarfs (*Marcy et al.*, 2001). Only later did true masses become known for many planets thanks to transit measurements. This history suggests we can expect perturbations to reveal new aspects of planetary systems.

One aspect of system architectures that is ripe for observational input is resonant orbits. The resonance of GJ 876bc has already been mapped out in detail, but orbital changes due to resonance could be detectable on decade timescales in other systems as well (*Ferraz-Mello et al.*, 2005; *Correia et al.*, 2005, 2009). It might seem that 55 Cnc, with five planets, two of which are close to resonance, would be a good candidate. But neither *Marcy et al.* (2002) nor *Fischer et al.* (2008) found improved fits when the system was modeled with Newtonian equations (equation (5)) rather than independent Keplerians. Although perturbations have not been detected in most systems, see Table 3 for a summary of the systems that have been plausibly claimed to be resonant. Determining whether this conclusion is robust, on a case-by-case basis, and finding the libration amplitude, has not received much observational attention. However, the frequency of resonances and the expected libration amplitude of the critical angles, or their circulation, has been a frequent topic of theoretical research. Currently observed resonances have the potential to constrain nebular conditions that lead to their formation and/or destruction (*Lee and Peale*, 2002; *Beaugé et al.*, 2003; *Adams et al.*, 2008; *Lecoanet et al.*, 2009), perturbative events in the system's history (*Sándor and Kley*, 2006; *Rein and Papaloizou*, 2009; *Lee et al.*, 2009), and tidal dissipation for close-in inner planets (*Novak et al.*, 2003; *Terquem and Papaloizou*, 2007). Therefore, with some additional high-quality observational input, we stand to learn fundamental information about the architecture of resonances, which in turn informs several classes of theories.

Just outside resonance, an oscillation in the orbital periods and eccentricities is expected, as shown in section 4.2. Depending on the proximity from resonance, it may be small enough to have eluded detection even in well-studied systems, e.g., in 55 Cnc. Conversely, since no perturbations are detected there, a constraint may be placed on the true masses, as resonance widths grow with mass (*Malhotra et al.*, 1992). Thus the *absence* of detected non-Keplerian motion can yield important constraints.

A second aspect of system architectures that has received rather little attention is mutual inclination. Although mutual inclination is recognized as a fundamental quantity for planet formation theory (e.g., it was a key inspiration for Laplace's nebular hypothesis), it cannot be measured by radial velocity if planets stay on Keplerian orbits.

Secular precession due to noncoplanarity has been detected in triple stars by radial velocity (e.g., *Jha et al.*, 2000) and by eclipses (e.g., *Torres and Stefanik*, 2000), and it has led to constraints on their mutual inclination. In a noncoplanar two-planet system, each orbit exerts secular torques on the other, and they each precess like a top. The precession period in the low-eccentricity, nearly coplanar case, with $P_{out} \gg P_{in}$, is

$$P_{sec} = \frac{8\pi}{3} \frac{P_{out}^2}{P_{in}} \frac{m_\star}{m_{p,out} + m_{p,in}(P_{in}/P_{out})^{1/3}} \quad (21)$$

where "in" and "out" refer to the inner and outer planets, respectively (*Murray and Dermott*, 1999) (see also section 7.2). This period is also the right order of magnitude for moderate mutual inclinations and eccentricities, and it also roughly describes the periapse precession of such a system (*Lee and Peale*, 2003). In systems for which this period is small, one might hope to observe precession directly and probe mutual inclination. The shortest precession period due to secular terms among known systems is that of GJ 876bc, with $P_{sec} \gtrsim 100$ yr. Although this timescale is much longer than a manageable observing program, even a small part of this cycle could produce observable effects, because over the complete cycle the orbit can entirely reorient. So far mutual inclination in planetary systems has only been measured in two systems, based on resonant perturbations (*Konacki and Wolszczan*, 2003; *Bean and Seifahrt*, 2009).

Transit-timing variations on orbital timescales may be a much faster route to mutual inclinations. Theoretical evidence has been building that the pattern made by transit times can reveal mutual inclination in a single passage of an external, eccentric planet (*Borkovits et al.*, 2003; *Bakos et al.*, 2009) or the detailed signal of short-term interactions (*Agol et al.*, 2005; *Nesvorný and Beaugé*, 2010). Therefore, as soon as transit time variations are detected, a constraint can be put on mutual inclination.

The transit-timing method may prove to be crucial in interpreting aspects of the HAT-P-13 planetary system, for which planet b, the inner planet, transits. According to *Batygin et al.* (2009), following *Mardling* (2007), planet b should have damped to a calculable, nonzero eccentricity due to forcing by the outer planet, which is massive and eccentric. The value of this forced eccentricity depends on the planet's tidal deformability (the Love number k_L) and thus its interior structure. Thus, for the first time, we could have a constraint on the mass distribution interior to an exoplanet. However, this chain of logic assumes that the two planets are coplanar, so it is crucial to establish that fact. *Bakos et al.* (2009) showed in the discovery paper that ~5 s transit time variations are expected if the system is coplanar, and even larger variations (of a different shape) are expected if the system is noncoplanar. Therefore, a tidal bulge may be soon inferred, resting on two aspects of non-Keplerian dynamics: (1) orbital timescale perturbations for observers to determine the mutual inclination, and (2) secular timescale and tidal evolution, which causes the interior structure of the planet to feed back on its observable eccentricity.

5.3. Summary

The study of non-Keplerian motion in exoplanets has played a critical role in understanding their nature and histories, from the first known exoplanets (around a pulsar, of all places) to systems of planets in dynamical configurations that shed light on their early history. The relatively simple equations hold a wealth of complexity, the extent of which is still being mapped in parallel with the continued discovery of planets unlike those of our solar system.

For further reading, the standard reference textbook for planetary dynamics is *Solar System Dynamics* by *Murray and Dermott* (1999). It contains much of the material here, with all the required mathematical detail, but it was written too early to include many results on exoplanets. Two recent reviews on the dynamics of exoplanet systems fill in many of the details of this chapter: *Michtchenko et al.* (2007) treat secular and resonant effects in two-planet systems, and *Beaugé et al.* (2008a) discuss resonant dynamics (particularly of the 2:1 resonance) and orbital fits to such systems.

Acknowledgments. I acknowledge support from the Michelson Fellowship, supported by the National Aeronautics and Space Administration and administered by the Michelson Science Center, and I thank an anonymous referee, E. Agol, J. Bean, R. Dawson, E. Ford, K. Gozdziewski, M. Holman, R. Malhotra, R. Murray-Clay, D. Ragozzine, J. Steffen, S. Tremaine, D. Veras, J. Winn, and J. Wisdom for discussions and comments on a draft of this chapter that considerably improved it.

REFERENCES

Adams F. C. and Laughlin G. (2006) Effects of secular interactions in extrasolar planetary systems. *Astrophys. J., 649,* 992.

Adams F. C., Laughlin G., and Bloch A. M. (2008) Turbulence implies that mean motion resonances are rare. *Astrophys. J., 683,* 1117.

Agol E. and Steffen J. H. (2007) A limit on the presence of Earth-mass planets around a Sun-like star. *Mon. Not. R. Astron. Soc., 374,* 941.

Agol E., Steffen J., Sari R., and Clarkson W. (2005) On detecting terrestrial planets with timing of giant planet transits. *Mon. Not. R. Astron. Soc., 359,* 567.

Anglada-Escudé G., López-Morales M., and Chambers J. E. (2010) How eccentric orbital solutions can hide planetary systems in 2:1 resonant orbits. *Astrophys. J., 709,* 168.

Bakos G. Á., Howard A. W., Noyes R. W., Hartman J., Torres G., Kovács G., Fischer D. A., Latham D. W., Johnson J. A., Marcy G. W., Sasselov D. D., Stefanik R. P., Sipöcz B., Kovács G., Esquerdo G. A., Pál A., Lázár J., Papp I., and Sári P. (2009) HAT-P-13b,c: A transiting hot Jupiter with a massive outer companion on an eccentric orbit. *Astrophys. J., 707,* 446.

Barker A. J. and Ogilvie G. I. (2009) On the tidal evolution of hot Jupiters on inclined orbits. *Mon. Not. R. Astron. Soc., 395,* 2268.

Barnes J. W. and O'Brien D. P. (2002) Stability of satellites around close-in extrasolar giant planets. *Astrophys. J., 575,* 1087.

Barnes R. (2008) Dynamics of multiple planet systems. In *Exoplanets: Detection, Formation, Properties, Habitability* (J. W. Mason, ed.), pp. 177–208. Praxis, Chichester.

Barnes R. and Greenberg R. (2006) Behavior of apsidal orientations in planetary systems. *Astrophys. J. Lett., 652,* L53.

Barnes R. and Greenberg R. (2007a) Apsidal behavior among planetary orbits: Testing the planet-planet scattering model. *Astrophys. J. Lett., 659,* L53.

Barnes R. and Greenberg R. (2007b) Stability limits in resonant planetary systems. *Astrophys. J. Lett., 665,* L67.

Barnes R., Gózdziewski K., and Raymond S. N. (2008) The successful prediction of the extrasolar planet HD 74156d. *Astrophys. J. Lett., 680,* L57.

Batygin K., Bodenheimer P., and Laughlin G. (2009) Determination of the interior structure of transiting planets in multiple-planet systems. *Astrophys. J. Lett., 704,* L49.

Bean J. L. and Seifahrt A. (2009) The architecture of the GJ 876 planetary system. Masses and orbital coplanarity for planets b and c. *Astron. Astrophys., 496,* 249.

Beaugé C., Ferraz-Mello S., and Michtchenko T. A. (2003) Extrasolar planets in mean-motion resonance: Apses alignment and asymmetric stationary solutions. *Astrophys. J., 593,* 1124.

Beaugé C., Ferraz-Mello S., and Michtchenko T. A. (2007) Planetary masses and orbital parameters from radial velocity measurements. In *Exoplanets: Formation, Detection and Dynamics* (R. Dvorak, ed.), p. 1. Wiley, New York.

Beaugé C., Ferraz-Mello S., Michtchenko T. A., and Giuppone C. A. (2008a) Orbital determination and dynamics of resonant extrasolar planetary systems. In *Exoplanets: Detection, Formation and Dynamics* (Y.-S. Sun et al., eds.), pp. 427–440. IAU Symposium No. 249, Cambridge Univ., Cambridge.

Beaugé C., Giuppone C. A., Ferraz-Mello S., and Michtchenko T. A. (2008b) Reliability of orbital fits for resonant extrasolar planetary systems: The case of HD 82943. *Mon. Not. R. Astron. Soc., 385,* 2151.

Benedict G. F., McArthur B. E., Forveille T., Delfosse X., Nelan E., Butler R. P., Spiesman W., Marcy G., Goldman B., Perrier C., Jefferys W. H., and Mayor M. (2002) A mass for the extrasolar planet Gliese 876b determined from Hubble Space Telescope Fine Guidance Sensor 3 astrometry and high-precision radial velocities. *Astrophys. J. Lett., 581,* L115.

Borkovits T., Érdi B., Forgács-Dajka E., and Kovács T. (2003) On the detectability of long period perturbations in close hierarchical triple stellar systems. *Astron. Astrophys., 398,* 1091.

Burns J. A. (1976) Elementary derivation of the perturbation equations of celestial mechanics. *Am. J. Phys., 44,* 944.

Canup R. M. and Ward W. R. (2006) A common mass scaling for satellite systems of gaseous planets. *Nature, 441,* 834.

Cassidy T. A., Mendez R., Arras P., Johnson R. E., and Skrutskie M. F. (2009) Massive satellites of close-in gas giant exoplanets. *Astrophys. J., 704,* 1341.

Chambers J. E. (1999) A hybrid symplectic integrator that permits close encounters between massive bodies. *Mon. Not. R. Astron. Soc., 304,* 793.

Chambers J. E., Wetherill G. W., and Boss A. P. (1996) The stability of multi-planet systems. *Icarus, 119,* 261.

Chambers J. E., Quintana E. V., Duncan M. J., and Lissauer J. J. (2002) Symplectic integrator algorithms for modeling planetary accretion in binary star systems. *Astron. J., 123,* 2884.

Chatterjee S., Ford E. B., Matsumura S., and Rasio F. A. (2008) Dynamical outcomes of planet-planet scattering. *Astrophys. J., 686,* 580.

Chiang E. I. (2003) Excitation of orbital eccentricities by repeated resonance crossings: Requirements. *Astrophys. J., 584,* 465.

Chiang E. I. and Murray N. (2002) Eccentricity excitation and apsidal resonance capture in the planetary system υ Andromedae. *Astrophys. J., 576,* 473.

Chirikov B. V. (1979) A universal instability of many-dimensional oscillator systems. *Phys. Rept., 52,* 263.

Correia A. C. M., Udry S., Mayor M., Laskar J., Naef D., Pepe F., Queloz D., and Santos N. C. (2005) The CORALIE survey for southern extra-solar planets. XIII. A pair of planets around

HD 202206 or a circumbinary planet? *Astron. Astrophys., 440,* 751.

Correia A. C. M., Levrard B., and Laskar J. (2008) On the equilibrium rotation of Earth-like extra-solar planets. *Astron. Astrophys., 488,* L63.

Correia A. C. M., Udry S., Mayor M., Benz W., Bertaux J.-L., Bouchy F., Laskar J., Lovis C., Mordasini C., Pepe F., and Queloz D. (2009) The HARPS search for southern extra-solar planets. XVI. HD 45364, a pair of planets in a 3:2 mean motion resonance. *Astron. Astrophys., 496,* 521.

Dermott S. F. and Murray C. D. (1981) The dynamics of tadpole and horseshoe orbits II. The coorbital satellites of Saturn. *Icarus, 48,* 12.

Desort M., Lagrange A.-M., Galland F., Beust H., Udry S., Mayor M., and Lo Curto G. (2008) Extrasolar planets and brown dwarfs around A–F type stars. V. A planetary system found with HARPS around the F6IV-V star HD 60532. *Astron. Astrophys., 491,* 883.

Domingos R. C., Winter O. C., and Yokoyama T. (2006) Stable satellites around extrasolar giant planets. *Mon. Not. R. Astron. Soc., 373,* 1227.

Duncan M., Quinn T., and Tremaine S. (1989) The long-term evolution of orbits in the solar system — A mapping approach. *Icarus, 82,* 402.

Duncan M. J., Levison H. F., and Lee M. H. (1998) A multiple time step symplectic algorithm for integrating close encounters. *Astron. J., 116,* 2067.

Dvorak R., Pilat-Lohinger E., Schwarz R., and Freistetter F. (2004) Extrasolar Trojan planets close to habitable zones. *Astron. Astrophys., 426,* L37.

Eggleton P. P. and Kiseleva-Eggleton L. (2001) Orbital evolution in binary and triple stars, with an application to SS Lacertae. *Astrophys. J., 562,* 1012.

Everhart E. (1985) An efficient integrator that uses Gauss-Radau spacings. In *Dynamics of Comets: Their Origin and Evolution* (A. Carusi and G. B. Valsecchi, eds.), p. 185. IAU Colloquium 83, Reidel, Dordrecht.

Faber J. A., Rasio F. A., and Willems B. (2005) Tidal interactions and disruptions of giant planets on highly eccentric orbits. *Icarus, 175,* 248.

Fabrycky D. (2008) Radiative thrusters on close-in extrasolar planets. *Astrophys. J. Lett., 677,* L117.

Fabrycky D. C. and Murray-Clay R. A. (2010) Stability of the directly imaged multiplanet system HR 8799: Resonance and masses. *Astrophys. J., 710,* 1408.

Fabrycky D. and Tremaine S. (2007) Shrinking binary and planetary orbits by Kozai cycles with tidal friction. *Astrophys. J., 669,* 1298.

Fernandez J. A. and Ip W. (1984) Some dynamical aspects of the accretion of Uranus and Neptune — The exchange of orbital angular momentum with planetesimals. *Icarus, 58,* 109.

Ferraz-Mello S., Michtchenko T. A., and Beaugé C. (2005) The orbits of the extrasolar planets HD 82943c and b. *Astrophys. J., 621,* 473.

Fischer D. A., Marcy G. W., Butler R. P., Vogt S. S., Laughlin G., Henry G. W., Abouav D., Peek K. M. G., Wright J. T., Johnson J. A., McCarthy C., and Isaacson H. (2008) Five planets orbiting 55 Cancri. *Astrophys. J., 675,* 790.

Ford E. B. (2005) Quantifying the uncertainty in the orbits of extrasolar planets. *Astron. J., 129,* 1706.

Ford E. B. (2006) Improving the efficiency of Markov Chain Monte Carlo for analyzing the orbits of extrasolar planets. *Astrophys. J., 642,* 505.

Ford E. B. (2008) Dynamics and instabilities in exoplanetary systems. In *Exoplanets: Detection, Formation and Dynamics* (Y.-S. Sun et al., eds.), pp. 441–446. IAU Symposium No. 249, Cambridge Univ., Cambridge.

Ford E. B. and Gaudi B. S. (2006) Observational constraints on Trojans of transiting extrasolar planets. *Astrophys. J. Lett., 652,* L137.

Ford E. B. and Holman M. J. (2007) Using transit timing observations to search for trojans of transiting extrasolar planets. *Astrophys. J. Lett., 664,* L51.

Ford E. B. and Rasio F. A. (2008) Origins of eccentric extrasolar planets: Testing the planet-planet scattering model. *Astrophys. J., 686,* 621.

Ford E. B., Kozinsky B., and Rasio F. A. (2000) Secular evolution of hierarchical triple star systems. *Astrophys. J., 535,* 385.

Ford E. B., Lystad V., and Rasio F. A. (2005) Planet-planet scattering in the upsilon Andromedae system. *Nature, 434,* 873.

Fregeau J. M., Cheung P., Portegies Zwart S. F., and Rasio F. A. (2004) Stellar collisions during binary-binary and binary-single star interactions. *Mon. Not. R. Astron. Soc., 352,* 1.

Gladman B. (1993) Dynamics of systems of two close planets. *Icarus, 106,* 247.

Goldreich P. and Soter S. (1966) Q in the solar system. *Icarus, 5,* 375.

Góździewski K. and Konacki M. (2006) Trojan pairs in the HD 128311 and HD 82943 planetary systems. *Astrophys. J., 647,* 573.

Góździewski K., Konacki M., and Maciejewski A. J. (2005) Orbital solutions to the HD 160691 μ Arae doppler signal. *Astrophys. J., 622,* 1136.

Góździewski K., Breiter S., and Borczyk W. (2008a) The long-term stability of extrasolar system HD 37124. Numerical study of resonance effects. *Mon. Not. R. Astron. Soc., 383,* 989.

Góździewski K., Migaszewski C., and Musieli ski A. (2008b) Stability constraints in modeling of multi-planet extrasolar systems. In *Exoplanets: Detection, Formation and Dynamics* (Y.-S. Sun et al., eds.), pp. 447–460. IAU Symposium No. 249, Cambridge Univ., Cambridge.

Gregory P. C. (2007) A Bayesian Kepler periodogram detects a second planet in HD 208487. *Mon. Not. R. Astron. Soc., 374,* 1321.

Gu P.-G., Lin D. N. C., and Bodenheimer P. H. (2003) The effect of tidal inflation instability on the mass and dynamical evolution of extrasolar planets with ultrashort periods. *Astrophys. J., 588,* 509.

Hadjidemetriou J. D., Psychoyos D., and Voyatzis G. (2009) The 1/1 resonance in extrasolar planetary systems. *Cel. Mech. Dyn. Astron., 104,* 23.

Hays J. D., Imbrie J., and Shackleton N. J. (1976) Variations in the Earth's orbit: Pacemaker of the Ice Ages. *Science, 194,* 1121.

Henon M. and Petit J. (1986) Series expansion for encounter-type solutions of Hill's problem. *Cel. Mech., 38,* 67.

Heyl J. S. and Gladman B. J. (2007) Using long-term transit timing to detect terrestrial planets. *Mon. Not. R. Astron. Soc., 377,* 1511.

Holman M. J. and Murray N. W. (2005) The use of transit timing to detect terrestrial-mass extrasolar planets. *Science, 307,* 1288.

Holman M., Touma J., and Tremaine S. (1997) Chaotic variations in the eccentricity of the planet orbiting 16 Cyg B. *Nature, 386,* 254.

Hut P. (1981) Tidal evolution in close binary systems. *Astron. Astrophys., 99,* 126.

Jackson B., Greenberg R., and Barnes R. (2008) Tidal evolution of close-in extrasolar planets. *Astrophys. J., 678,* 1396.

Jackson B., Barnes R., and Greenberg R. (2009) Observational evidence for tidal destruction of exoplanets. *Astrophys. J., 698,* 1357.

Jha S., Torres G., Stefanik R. P., Latham D. W., and Mazeh T. (2000) Studies of multiple stellar systems — III. Modulation of orbital elements in the triple-lined system HD 109648. *Mon. Not. R. Astron. Soc., 317,* 375.

Jurić M. and Tremaine S. (2008) Dynamical origin of extrasolar planet eccentricity distribution. *Astrophys. J., 686,* 603.

Kidder L. E. (1995) Coalescing binary systems of compact objects to (post)$^{5/2}$-Newtonian order. V. Spin effects. *Phys. Rev. D, 52,* 821.

Kipping D. M. (2009) Transit timing effects due to an exomoon. *Mon. Not. R. Astron. Soc., 392,* 181.

Knutson H. A., Charbonneau D., Allen L. E., Fortney J. J., Agol E., Cowan N. B., Showman A. P., Cooper C. S., and Megeath S. T. (2007) A map of the day-night contrast of the extrasolar planet HD 189733b. *Nature, 447,* 183.

Konacki M. and Wolszczan A. (2003) Masses and orbital inclinations of planets in the PSR B1257+12 system. *Astrophys. J. Lett., 591,* L147.

Kozai Y. (1962) Secular perturbations of asteroids with high inclination and eccentricity. *Astron. J., 67,* 591.

Laskar J. (2000) On the spacing of planetary systems. *Phys. Rev. Lett., 84,* 3240.

Laskar J. and Correia A. C. M. (2009) HD 60532, a planetary system in a 3:1 mean motion resonance. *Astron. Astrophys., 496,* L5.

Laskar J. and Gastineau M. (2009) Existence of collisional trajectories of Mercury, Mars and Venus with the Earth. *Nature, 459,* 817.

Laskar J. and Robutel P. (1993) The chaotic obliquity of the planets. *Nature, 361,* 608.

Laskar J. and Robutel P. (1995) Stability of the planetary three-body problem. I. Expansion of the planetary Hamiltonian. *Cel. Mech. Dyn. Astron., 62,* 193.

Laughlin G. and Chambers J. E. (2001) Short-term dynamical interactions among extrasolar planets. *Astrophys. J. Lett., 551,* L109.

Laughlin G. and Chambers J. E. (2002) Extrasolar Trojans: The viability and detectability of planets in the 1:1 resonance. *Astron. J., 124,* 592.

Laughlin G., Deming D., Langton J., Kasen D., Vogt S., Butler P., Rivera E., and Meschiari S. (2009) Rapid heating of the atmosphere of an extrasolar planet. *Nature, 457,* 562.

Lecar M., Franklin F., and Murison M. (1992) On predicting long-term orbital instability — A relation between the Lyapunov time and sudden orbital transitions. *Astron. J., 104,* 1230.

Lecoanet D., Adams F. C., and Bloch A. M. (2009) Mean motion resonances in extrasolar planetary systems with turbulence, interactions, and damping. *Astrophys. J., 692,* 659.

Lee A. T., Thommes E. W., and Rasio F. A. (2009) Resonance trapping in protoplanetary disks. I. Coplanar systems. *Astrophys. J., 691,* 1684.

Lee M. H. (2004) Diversity and origin of 2:1 orbital resonances in extrasolar planetary systems. *Astrophys. J., 611,* 517.

Lee M. H. and Peale S. J. (2002) Dynamics and origin of the 2:1 orbital resonances of the GJ 876 planets. *Astrophys. J., 567,* 596.

Lee M. H. and Peale S. J. (2003) Secular evolution of hierarchical planetary systems. *Astrophys. J., 592,* 1201.

Lee M. H., Butler R. P., Fischer D. A., Marcy G. W., and Vogt S. S. (2006) On the 2:1 orbital resonance in the HD 82943 planetary system. *Astrophys. J., 641,* 1178.

Levison H. F. and Duncan M. J. (1994) The long-term dynamical behavior of short-period comets. *Icarus, 108,* 18.

Levrard B., Correia A. C. M., Chabrier G., Baraffe I., Selsis F., and Laskar J. (2007) Tidal dissipation within hot Jupiters: A new appraisal. *Astron. Astrophys., 462,* L5.

Lidov M. L. (1962) The evolution of orbits of artificial satellites of planets under the action of gravitational perturbations of external bodies. *Planet. Space Sci., 9,* 719.

Lin D. N. C., Papaloizou J. C. B., Terquem C., Bryden G., and Ida S. (2000) Orbital evolution and planet-star tidal interaction. In *Protostars and Planets IV* (V. Mannings et al., eds.), p. 1111. Univ. of Arizona, Tucson.

Love A. E. H. (1911) *Some Problems of Geodynamics.* Cambridge Univ., Cambridge.

Madhusudhan N. and Winn J. N. (2009) Empirical constraints on Trojan companions and orbital eccentricities in 25 transiting exoplanetary systems. *Astrophys. J., 693,* 784.

Malhotra R. (1993a) The origin of Pluto's peculiar orbit. *Nature, 365,* 819.

Malhotra R. (1993b) Three-body effects in the PSR 1257+12 planetary system. *Astrophys. J., 407,* 266.

Malhotra R. (1995) The origin of Pluto's orbit: Implications for the solar system beyond Neptune. *Astron. J., 110,* 420.

Malhotra R. (1998) Orbital resonances and chaos in the solar system. In *Solar System Formation and Evolution* (D. Lazzaro et al., eds), p. 37. ASP Conference Series 149, Astronomical Society of the Pacific, San Francisco.

Malhotra R. (2002) A dynamical mechanism for establishing apsidal resonance. *Astrophys. J. Lett., 575,* L33.

Malhotra R., Black D., Eck A., and Jackson A. (1992) Resonant orbital evolution in the putative planetary system of PSR 1257+12. *Nature, 356,* 583.

Mandel K. and Agol E. (2002) Analytic light curves for planetary transit searches. *Astrophys. J. Lett., 580,* L171.

Marchal C. and Bozis G. (1982) Hill stability and distance curves for the general three-body problem. *Cel. Mech., 26,* 311.

Marcy G. W., Butler R. P., Fischer D., Vogt S. S., Lissauer J. J., and Rivera E. J. (2001) A pair of resonant planets orbiting GJ 876. *Astrophys. J., 556,* 296.

Marcy G. W., Butler R. P., Fischer D. A., Laughlin G., Vogt S. S., Henry G. W., and Pourbaix D. (2002) A planet at 5 AU around 55 Cancri. *Astrophys. J., 581,* 1375.

Mardling R. A. (2007) Long-term tidal evolution of short-period planets with companions. *Mon. Not. R. Astron. Soc., 382,* 1768.

Mardling R. A. (2008) Resonance, chaos and stability in the general three-body problem. In *Dynamical Evolution of Dense Stellar Systems* (E. Vesperini et al., eds.), pp. 199–208. IAU Symposium No. 246, Cambridge Univ., Cambridge.

Mardling R. A. and Lin D. N. C. (2002) Calculating the tidal, spin, and dynamical evolution of extrasolar planetary systems. *Astrophys. J., 573,* 829.

Mardling R. A. and Lin D. N. C. (2004) On the survival of short-period terrestrial planets. *Astrophys. J., 614,* 955.

Matsumura S., Takeda G., and Rasio F. A. (2008) On the origins of eccentric close-in planets. *Astrophys. J. Lett., 686,* L29.

Mayor M., Udry S., Naef D., Pepe F., Queloz D., Santos N. C., and Burnet M. (2004) The CORALIE survey for southern extrasolar planets. XII. Orbital solutions for 16 extra-solar planets discovered with CORALIE. *Astron. Astrophys., 415,* 391.

Mazeh T., Krymolowski Y., and Rosenfeld G. (1997) The high eccentricity of the planet orbiting 16 Cygni B. *Astrophys. J. Lett., 477,* L103.

Menou K. and Tabachnik S. (2003) Dynamical habitability of known extrasolar planetary systems. *Astrophys. J., 583,* 473.

Michtchenko T. A. and Malhotra R. (2004) Secular dynamics of the three-body problem: Application to the υ Andromedae planetary system. *Icarus, 168,* 237.

Michtchenko T. A., Ferraz-Mello S., and Beaugé C. (2007) Dynamics of the extrasolar planetary systems. In *Extrasolar Planets: Formation, Detection and Dynamics* (R. Dvorak, ed.), p. 151. Wiley, New York.

Migaszewski C. and Góźdiewski K. (2008) A secular theory of coplanar, non-resonant planetary system. *Mon. Not. R. Astron. Soc., 388,* 789.

Migaszewski C. and Góździewski K. (2009a) Equilibria in the secular, non-co-planar two-planet problem. *Mon. Not. R. Astron. Soc., 395,* 1777.

Migaszewski C. and Góździewski K. (2009b) Secular dynamics of a coplanar, non-resonant planetary system under the general relativity and quadrupole moment perturbations. *Mon. Not. R. Astron. Soc., 392,* 2.

Miralda-Escudé J. (2002) Orbital perturbations of transiting planets: A possible method to measure stellar quadrupoles and to detect Earth-mass planets. *Astrophys. J., 564,* 1019.

Morbidelli A. (2002) *Modern Celestial Mechanics: Aspects of Solar System Dynamics.* Taylor and Francis, London.

Morbidelli A. and Froeschlé C. (1995) On the relationship between Lyapunov times and macroscopic instability times. *Cel. Mech. Dyn. Astron., 63,* 227.

Morbidelli A., Tsiganis K., Crida A., Levison H. F., and Gomes R. (2007) Dynamics of the giant planets of the solar system in the gaseous protoplanetary disk and their relationship to the current orbital architecture. *Astron. J., 134,* 1790.

Moro-Martín A., Malhotra R., Carpenter J. M., Hillenbrand L. A., Wolf S., Meyer M. R., Hollenbach D., Najita J., and Henning T. (2007) The dust, planetesimals, and planets of HD 38529. *Astrophys. J., 668,* 1165.

Moutou C., Hébrard G., Bouchy F., Eggenberger A., Boisse I., Bonfils X., Gravallon D., Ehrenreich D., Forveille T., Delfosse X., Desort M., Lagrange A., Lovis C., Mayor M., Pepe F., Perrier C., Pont F., Queloz D., Santos N. C., Ségransan D., Udry S., and Vidal-Madjar A. (2009) Photometric and spectroscopic detection of the primary transit of the 111-day-period planet HD 80606b. *Astron. Astrophys., 498,* L5.

Mudryk L. R. and Wu Y. (2006) Resonance overlap is responsible for ejecting planets in binary systems. *Astrophys. J., 639,* 423.

Mugrauer M. and Neuhaeuser R. (2008) The multiplicity of exoplanet host stars. New low-mass stellar companions of the exoplanet host stars HD 125612 and HD 212301. *Astron. Astrophys., 494,* 373.

Murray C. D. and Dermott S. F. (1999) *Solar System Dynamics.* Cambridge Univ., Cambridge.

Murray N. and Holman M. (1999) The origin of chaos in the outer solar system. *Science, 283,* 1877.

Murray N., Hansen B., Holman M. and Tremaine S. (1998) Migrating planets. *Science, 279,* 69.

Nagasawa M., Ida S., and Bessho T. (2008) Formation of hot planets by a combination of planet scattering, tidal circularization, and the Kozai mechanism. *Astrophys. J., 678,* 498.

Nesvorný D. and Beaugé C. (2010) Fast inversion method for determination of planetary parameters from transit timing variations. *Astrophys. J. Lett., 709,* L44.

Nesvorný D. and Morbidelli A. (1998) Three-body mean motion resonances and the chaotic structure of the asteroid belt. *Astron. J., 116,* 3029.

Novak G. S., Lai D. and Lin D. N. C. (2003) The interesting dynamics of the 55 Cancri system. In *Scientific Frontiers in Research on Extrasolar Planets* (D. Deming and S. Seager, eds.), pp. 177–180. ASP Conf. Series 294, Astronomical Society of the Pacific, San Francisco.

Ogilvie G. I. and Lin D. N. C. (2004) Tidal dissipation in rotating giant planets. *Astrophys. J., 610,* 477.

Pál A. and Kocsis B. (2008) Periastron precession measurements in transiting extrasolar planetary systems at the level of general relativity. *Mon. Not. R. Astron. Soc., 389,* 191.

Peale S. J. (1976) Orbital resonances in the solar system. *Annu. Rev. Astron. Astrophys., 14,* 215.

Peale S. J. (1993) On the verification of the planetary system around PSR 1257+12. *Astron. J., 105,* 1562.

Peale S. J. and Gold T. (1965) Rotation of the planet Mercury. *Nature, 206,* 1240.

Pepe F., Correia A. C. M., Mayor M., Tamuz O., Couetdic J., Benz W., Bertaux J.-L., Bouchy F., Laskar J., Lovis C., Naef D., Queloz D., Santos N. C., Sivan J.-P., Sosnowska D., and Udry S. (2007) The HARPS search for southern extra-solar planets. VIII. μ Arae, a system with four planets. *Astron. Astrophys., 462,* 769.

Press W. H., Teukolsky S. A., Vetterling W. T., and Flannery B. P. (1992) *Numerical Recipes in FORTRAN. The Art of Scientific Computing, 2nd edition.* Cambridge Univ., Cambridge.

Ragozzine D. and Wolf A. S. (2009) Probing the interiors of very hot Jupiters using transit light curves. *Astrophys. J., 698,* 1778.

Rasio F. A. and Ford E. B. (1996) Dynamical instabilities and the formation of extrasolar planetary systems. *Science, 274,* 954.

Rasio F. A., Nicholson P. D., Shapiro S. L., and Teukolsky S. A. (1992) A observational test for the existence of a planetary system orbiting PSR 1257+12. *Nature, 355,* 325.

Rasio F. A., Tout C. A., Lubow S. H., and Livio M. (1996) Tidal decay of close planetary orbits. *Astrophys. J., 470,* 1187.

Rein H. and Papaloizou J. C. B. (2009) On the evolution of mean motion resonances through stochastic forcing: Fast and slow libration modes and the origin of HD 128311. *Astron. Astrophys., 497,* 595–609.

Rivera E. J. and Lissauer J. J. (2000) Stability analysis of the planetary system orbiting υ Andromedae. *Astrophys. J., 530,* 454.

Rivera E. J. and Lissauer J. J. (2001) Dynamical models of the resonant pair of planets orbiting the star GJ 876. *Astrophys. J., 558,* 392.

Rivera E. J., Lissauer J. J., Butler R. P., Marcy G. W., Vogt S. S., Fischer D. A., Brown T. M., Laughlin G., and Henry G. W. (2005) A ~7.5 $M_\oplus$ planet orbiting the nearby star, GJ 876. *Astrophys. J., 634,* 625.

Saha P. and Tremaine S. (1992) Symplectic integrators for solar system dynamics. *Astron. J., 104,* 1633.

Sándor Z. and Kley W. (2006) On the evolution of the resonant planetary system HD 128311. *Astron. Astrophys., 451,* L31.

Sándor Z., Süli Á., Érdi B., Pilat-Lohinger E., and Dvorak R. (2007) A stability catalogue of the habitable zones in extrasolar planetary systems. *Mon. Not. R. Astron. Soc., 375,* 1495.

Shapiro I. I. (1964) Fourth test of general relativity. *Phys. Rev. Lett., 13,* 789.

Shen Y. and Tremaine S. (2008) Stability of the distant satellites of the giant planets in the solar system. *Astron. J., 136,* 2453.

Short D., Windmiller G., and Orosz J. A. (2008) New solutions for the planetary dynamics in HD 160691 using a Newtonian model and latest data. *Mon. Not. R. Astron. Soc., 386,* L43.

Simon A., Szatmáry K., and Szabó G. M. (2007) Determination of the size, mass, and density of "exomoons" from photometric transit timing variations. *Astron. Astrophys., 470,* 727.

Spurzem R., Giersz M., Heggie D. C., and Lin D. N. C. (2009) Dynamics of planetary systems in star clusters. *Astrophys. J., 697,* 458.

Steffen J. H. and Agol E. (2005) An analysis of the transit times of TrES-1b. *Mon. Not. R. Astron. Soc., 364,* L96.

Stepinski T. F., Malhotra R., and Black D. C. (2000) The Andromedae system: Models and stability. *Astrophys. J., 545,* 1044.

Sterne T. E. (1939) Apsidal motion in binary stars. *Mon. Not. R. Astron. Soc., 99,* 451.

Sussman G. J. and Wisdom J. (1992) Chaotic evolution of the solar system. *Science, 257,* 56.

Sussman G. J. and Wisdom J. (2001) *Structure and Interpretation of Classical Mechanics.* MIT, Cambridge.

Takeda G., Kita R., and Rasio F. A. (2008) Planetary systems in binaries. I. Dynamical classification. *Astrophys. J., 683,* 1063.

Terquem C. and Papaloizou J. C. B. (2007) Migration and the formation of systems of hot super-Earths and Neptunes. *Astrophys. J., 654,* 1110.

Thommes E. W., Bryden G., Wu Y., and Rasio F. A. (2008) From mean motion resonances to scattered planets: Producing the solar system, eccentric exoplanets, and late heavy bombardments. *Astrophys. J., 675,* 1538.

Tinney C. G., Butler R. P., Marcy G. W., Jones H. R. A., Laughlin G., Carter B. D., Bailey J. A., and O'Toole S. (2006) The 2:1 resonant exoplanetary system orbiting HD 73526. *Astrophys. J., 647,* 594.

Torres G. and Stefanik R. P. (2000) The cessation of eclipses in SS Lacertae: The mystery solved. *Astron. J., 119,* 1914.

Touma J. R., Tremaine S., and Kazandjian M. V. (2009) Gauss's method for secular dynamics, softened. *Mon. Not. R. Astron. Soc., 394,* 1085.

Trilling D. E., Benz W., Guillot T., Lunine J. I., Hubbard W. B., and Burrows A. (1998) Orbital evolution and migration of giant planets: Modeling extrasolar planets. *Astrophys. J., 500,* 428.

Veras D. and Armitage P. J. (2004) The dynamics of two massive planets on inclined orbits. *Icarus, 172,* 349.

Veras D., Crepp J. R., and Ford E. B. (2009) Formation, survival, and detectability of planets beyond 100 AU. *Astrophys. J., 696,* 1600.

Vogt S. S., Butler R. P., Marcy G. W., Fischer D. A., Henry G. W., Laughlin G., Wright J. T., and Johnson J. A. (2005) Five new multicomponent planetary systems. *Astrophys. J., 632,* 638.

Williams D. M. and Pollard D. (2002) Habitable planets on eccentric orbits. In *The Evolving Sun and its Influence on Planetary Environments* (B. Montesinos et al., eds.), p. 201. ASP Conf. Series 269, Astronomical Society of the Pacific, San Francisco.

Williams D. M., Kasting J. F., and Wade R. A. (1997) Habitable moons around extrasolar giant planets. *Nature, 385,* 234.

Winn J. N., Noyes R. W., Holman M. J., Charbonneau D., Ohta Y., Taruya A., Suto Y., Narita N., Turner E. L., Johnson J. A., Marcy G. W., Butler R. P., and Vogt S. S. (2005) GJ 436 TLC. *Astrophys. J., 631,* 1215.

Winn J. N., Howard A. W., Johnson J. A., Marcy G. W., Gazak J. Z., Starkey D., Ford E. B., Colón K. D., Reyes F., Nortmann L., Dreizler S., Odewahn S., Welsh W. F., Kadakia S., Vanderbei R. J., Adams E. R., Lockhart M., Crossfield I. J., Valenti J. A., Dantowitz R., and Carter J. A. (2009) The transit ingress and the tilted orbit of the extraordinarily eccentric exoplanet HD 80606b. *Astrophys. J., 703,* 2091.

Wisdom J. (1980) The resonance overlap criterion and the onset of stochastic behavior in the restricted three-body problem. *Astron. J., 85,* 1122.

Wisdom J. (2008) Tidal dissipation at arbitrary eccentricity and obliquity. *Icarus, 193,* 637.

Wisdom J. and Holman M. (1991) Symplectic maps for the n-body problem. *Astron. J., 102,* 1528.

Wolszczan A. (1994) Confirmation of Earth mass planets orbiting the millisecond pulsar PSR:B1257+12. *Science, 264,* 538.

Wolszczan A. and Frail D. A. (1992) A planetary system around the millisecond pulsar PSR1257+12. *Nature, 355,* 145.

Wright J. T. and Howard A. W. (2009) Efficient fitting of multiplanet Keplerian models to radial velocity and astrometry data. *Astrophys. J. Suppl., 182,* 205.

Wu Y. (2003) Tidal circularization and Kozai migration. In *Scientific Frontiers in Research on Extrasolar Planets* (D. Deming and S. Seager, eds.), pp. 213–216. ASP Conf. Series 294, Astronomical Society of the Pacific, San Francisco.

Wu Y. and Goldreich P. (2002) Tidal evolution of the planetary system around HD 83443. *Astrophys. J., 564,* 1024.

Wu Y. and Murray N. (2003) Planet migration and binary companions: The case of HD 80606b. *Astrophys. J., 589,* 605.

Wu Y., Murray N. W., and Ramsahai J. M. (2007) Hot Jupiters in binary star systems. *Astrophys. J., 670,* 820.

Zhou J.-L., Lin D. N. C., and Sun Y.-S. (2007) Post-oligarchic evolution of protoplanetary embryos and the stability of planetary systems. *Astrophys. J., 666,* 423.

Tidal Evolution of Exoplanets

Alexandre C. M. Correia
University of Aveiro

Jacques Laskar
Paris Observatory

Tidal effects arise from differential and inelastic deformation of a planet by a perturbing body. The continuous action of tides modify the rotation of the planet together with its orbit until an equilibrium situation is reached. It is often believed that synchronous motion is the most probable outcome of the tidal evolution process, since synchronous rotation is observed for the majority of the satellites in the solar system. However, in the nineteenth century, Schiaparelli also assumed synchronous motion for the rotations of Mercury and Venus, and was later proven wrong. Rather, for planets in eccentric orbits, synchronous rotation is very unlikely. The rotation period and axial tilt of exoplanets is still unknown, but a large number of planets have been detected close to the parent star and should have evolved to a final equilibrium situation. Therefore, based on the well-studied cases in the solar system, we can make some predictions for exoplanets. Here we describe in detail the main tidal effects that modify the secular evolution of the spin and the orbit of a planet. We then apply our knowledge acquired from solar system situations to exoplanet cases. In particular, we will focus on two classes of planets, hot Jupiters (fluid) and super Earths (rocky with atmosphere).

1. INTRODUCTION

The occurrence on most open ocean coasts of high sea tide at about the time of the Moon's passage across the meridian long ago prompted the idea that Earth's satellite exerts an attraction on the water. The occurrence of a second high tide when the Moon is on the opposite meridian was a great puzzle, but the correct explanation of the tidal phenomena was given by Newton in *Philosophiæ Naturalis Principia Mathematica*. Tides are a consequence of the lunar and solar gravitational forces acting in accordance with laws of mechanics. Newton realized that the tidal forces also must affect the atmosphere, but he assumed that the atmospheric tides would be too small to be detected, because changes in weather would introduce large irregular variations upon barometric measurements.

However, the semidiurnal oscillations of the atmospheric surface pressure has proven to be one of the most regular of all meteorological phenomena. It is readily detectable by harmonic analysis at any station in the world (e.g., *Chapman and Lindzen*, 1970). The main difference in respect to ocean tides is that atmospheric tides follow the Sun and not the Moon, as the atmosphere is essentially excited by solar heat. Even though tides of gravitational origin are present in the atmosphere, the thermal tides are more important as the pressure variations on the ground are more sensitive to the temperature gradients than to the gravitational ones.

The inner planets of the solar system as well as the majority of the main satellites present today a spin that is different from what is believed to have been the initial one (e.g., *Goldreich and Soter*, 1966; *Goldreich and Peale*, 1968). Planets and satellites are supposed to rotate much faster in the beginning, and any orientation of the spin axis may be allowed (e.g., *Dones and Tremaine*, 1993; *Kokubo and Ida*, 2007). However, tidal dissipation within the internal layers give rise to secular evolution of planetary spins and orbits. In the case of the satellites, spin and orbital evolution is mainly driven by tidal interactions with the central planet, whereas for the inner planets the main source of tidal dissipation is the Sun (in the case of Earth, tides raised by the Moon are also important).

Orbital and spin evolution cannot be dissociated because the total angular momentum must be conserved. As a consequence, a reduction in the rotation rate of a body implies an increment of the orbit semimajor axis and vice-versa. For instance, the Earth's rotation period is increasing about 2 ms/century (e.g., *Williams*, 1990), and the Moon is consequently moving away about 3.8 cm/yr (e.g., *Dickey et al.*, 1994). On the other hand, Neptune's moon, Triton, and the martian moon, Phobos, are spiraling down into the planet, clearly indicating that the present orbits are not primordial, and may have undergone a long evolving process from a previous capture from a heliocentric orbit (e.g., *Mignard*, 1981; *Goldreich et al.*, 1989; *Correia*, 2009). Both the Earth's Moon and Pluto's moon, Charon, have a significant fraction of the mass of their systems, and therefore could be classified as double planets rather than as satellites. The protoplanetary disk is unlikely to produce double-planet systems whose origin seems to be due to a catastrophic impact of the initial planet with a body of comparable dimensions (e.g., *Canup and Asphaug*, 2001; *Canup*, 2005). The resulting orbits after collision are most likely eccentric, but the present orbits are almost circular, suggesting that tidal evolution subsequently occurred.

The ultimate stage for tidal evolution corresponds to the synchronous rotation, a configuration where the rotation rate coincides with the orbital mean motion, since synchronous equilibrium corresponds to the minimum of dissipation of energy. However, when the eccentricity is different from zero, some other configurations are possible, such as the 3/2 spin-orbit resonance observed for the planet Mercury (*Colombo*, 1965; *Goldreich and Peale*, 1966; *Correia and Laskar*, 2004) or the chaotic rotation of Hyperion (*Wisdom et al.*, 1984). When a dense atmosphere is present, thermal atmospheric tides may counterbalance the gravitational tidal effect and nonresonant equilibrium configurations are also possible, as illustrated by the retrograde rotation of Venus (*Correia and Laskar*, 2001). Additional effects may also contribute to the final evolution of the spin, such as planetary perturbations or core-mantle friction.

Despite the proximity of Mercury and Venus to Earth, the determination of their rotational periods has only been achieved in the second half of the twentieth century, when it became possible to use radar ranging on the planets (*Pettengill and Dyce*, 1965; *Goldstein*, 1964; *Carpenter*, 1964). We thus do not expect that it will be easy to observe the rotation of the recently discovered exoplanets. Nevertheless, many of the exoplanets are close to their host star, and we can assume that exoplanets' spin and orbit have already undergone enough dissipation and evolved into a final equilibrium possibility. An identical assumption has been done before for Mercury and Venus by *Schiaparelli* (1889), who made predictions for their rotations based on *Darwin*'s work (1880). Schiaparelli's predictions were later proven to be wrong, but were nevertheless much closer to the true rotation periods than most values derived from observations in the two previous centuries. Like Schiaparelli, we may dare to establish predictions for the rotation periods of some already known exoplanets. We hope that the additional knowledge that we gained from a better understanding of the rotation of Mercury and Venus will help us to be at least as close to reality as Schiaparelli. Indeed, observations also show that many of the exoplanets have highly eccentric orbits. In some cases eccentricities larger than 0.9 are found (e.g., *Naef et al.*, 2001; *Jones et al.*, 2006; *Tamuz et al.*, 2008), which introduces the possibility of a wide variety of final tidal equilibrium positions, different from what we observe around the Sun.

In this chapter we will describe the tidal effects that modify the secular evolution of the spin and orbit of a planet. We then apply our knowledge acquired from solar system situations to exoplanet cases. In particular, we will focus on two classes of planets, hot Jupiters (fluid) and super Earths (rocky), which are close to the star and therefore more susceptible to having arrived in a final equilibrium situation.

2. MODEL DESCRIPTION

We will first omit the tidal effects, and describe the spin motion of the planet in a conservative framework. The motion equations will be obtained from a Hamiltonian formalism (e.g., *Goldstein*, 1950) of the total gravitational energy of the planet (section 2.1). Gravitational tides (section 2.2) and thermal atmospheric tides (section 2.3) will be described later. We also discuss the impact of spin-orbit resonances (section 2.4) and planetary perturbations (section 2.5).

2.1. Conservative Motion

The planet is considered here as a rigid body with mass m and moments of inertia A ≤ B < C, supported by the reference frame (**I**, **J**, **K**), fixed with respect to the planet's figure. Let **L** be the total rotational angular momentum and (**i**, **j**, **k**) a reference frame linked to the orbital plane (where **k** is the normal for this plane). As we are interested in the long-term behavior of the spin axis, we merge the axis of figure **K** with the direction of the angular momentum **L**. Indeed, the average of **K** coincides with L/L up to J^2, where cos J = **L** · **K** (*Boué and Laskar*, 2006). J is extremely small for large rocky planets (J ≈ 7 × 10^{-7} for Earth), being even smaller for Jupiter-like planets that behave as fluids. The angle between **K** and **k** is the obliquity, ε, and thus, cos ε = **k** · **K** (Fig. 1).

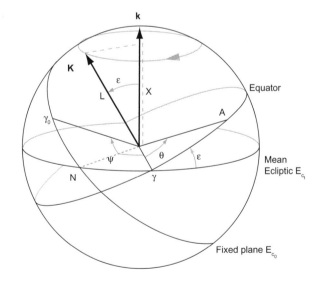

Fig. 1. Andoyer's canonical variables. L is the projection of the total rotational angular momentum vector **L** on the principal axis of inertia **K**, and X the projection of the angular momentum vector on the normal to the orbit (or ecliptic) **k**. The angle between the equinox of date γ and a fixed point of the equator A is the hour angle θ, and ψ = γN + N γ$_0$ is the general precession angle. The direction of γ$_0$ is on a fixed plane E$_{c0}$, while γ is on the mean orbital (or ecliptic) E$_{ct}$ of date t.

The Hamiltonian of the motion can be written using canonical Andoyer's action variables (L, X) and their conjugate angles (θ, –ψ) (*Andoyer*, 1923; *Kinoshita*, 1977). L = **L** · **K** = Cω is the projection of the angular momentum on the C axis, with rotation rate ω = θ̇–ψ̇ cos ε, and X = **L** · **k** is the projection of the angular momentum on the normal to the ecliptic; θ is the hour angle between the equinox of date and a fixed point of the equator, and ψ is the general precession angle, an angle that simultaneously accounts for the precession of the spin axis and the orbit (Fig. 1).

2.1.1. Gravitational potential. The gravitational potential V (energy per unit mass) generated by the planet at a generic point of the space **r**, expanded in degree two of R/r, where R is the planet's radius, is given by (e.g., *Tisserand*, 1891; *Smart*, 1953)

$$V(\mathbf{r}) = -\frac{Gm}{r} + \frac{G(B-A)}{r^3}P_2(\hat{\mathbf{r}} \cdot \mathbf{J}) \qquad (1)$$
$$+ \frac{G(C-A)}{r^3}P_2(\hat{\mathbf{r}} \cdot \mathbf{K})$$

where $\hat{\mathbf{r}} = \mathbf{r}/r$, G is the gravitational constant, and $P_2(x) = (3x^2-1)/2$ are the Legendre polynomials of degree two. The potential energy U when orbiting a central star of mass $m_\star$ is then

$$U = m_\star V(r) \qquad (2)$$

For a planet evolving in a nonperturbed Keplerian orbit, we write

$$\hat{\mathbf{r}} = \cos(\varpi + v)\mathbf{i} + \sin(\varpi + v)\mathbf{j} \qquad (3)$$

where ϖ is the longitude of the periapse and v the true anomaly (see chapter by Murray and Correia). Thus, transforming the body equatorial frame (**I**, **J**, **K**) into the orbital frame (**i**, **j**, **k**), we obtain (Fig. 1)

$$\begin{cases} \hat{\mathbf{r}} \cdot \mathbf{J} = -\cos w \sin\theta + \sin w \cos\theta \cos\varepsilon \\ \hat{\mathbf{r}} \cdot \mathbf{K} = -\sin w \sin\varepsilon \end{cases} \qquad (4)$$

where $w = \varpi + \psi + v$ is the true longitude to date. The expression for the potential energy (equation (2)) becomes (e.g., *Correia*, 2006)

$$U = -\frac{Gmm_\star}{r} + \frac{GCm_\star}{r^3}E_d P_2(\sin w \sin\varepsilon) \qquad (5)$$
$$- \frac{3Gm_\star}{8r^3}(B-A)F(\theta, w, \varepsilon)$$

where

$$F(\theta, w, \varepsilon) = 2\cos(2\theta - 2w)\cos^4\left(\frac{\varepsilon}{2}\right) \qquad (6)$$
$$+ 2\cos(2\theta + 2w)\sin^4\left(\frac{\varepsilon}{2}\right) + \cos(2\theta)\sin^2\varepsilon$$

and

$$E_d = \frac{C - \tfrac{1}{2}(A+B)}{C} = \frac{k_f R^5}{3GC}\omega^2 + \delta E_d \qquad (7)$$

where E_d is the dynamical ellipticity, and k_f is the fluid Love number (pertaining to a perfectly fluid body with the same mass distribution as the actual planet). The first part of E_d (equation (7)) corresponds to the flattening in hydrostatic equilibrium (*Lambeck*, 1980), and δE_d to the departure from this equilibrium.

2.1.2. Averaged potential. Since we are only interested in the study of the long-term motion, we will average the potential energy U over the rotation angle θ and the mean anomaly M

$$\bar{U} = \frac{1}{4\pi^2}\int_0^{2\pi}\int_0^{2\pi} U \, dM \, d\theta \qquad (8)$$

However, when the rotation frequency ω ≈ θ̇ and the mean motion n = Ṁ are close to resonance (ω ≈ pn, for a semi-integer value p), the terms with argument 2(θ–pM) vary slowly and must be retained in the expansions (e.g., *Murray and Dermott*, 1999)

$$\frac{\cos(2\theta)}{r^3} = \frac{1}{a^3}\sum_{p=-\infty}^{+\infty} G(p,e)\cos 2(\theta - pM) \qquad (9)$$

and

$$\frac{\cos(2\theta - 2w)}{r^3} = \frac{1}{a^3}\sum_{p=-\infty}^{+\infty} H(p,e)\cos 2(\theta - pM) \qquad (10)$$

where a and e are the semimajor axis and the eccentricity of the planet's orbit, respectively. The functions G(p, e) and H(p, e) can be expressed in power series in e (Table 1). The averaged nonconstant part of the potential $\bar{U}$ becomes

$$\frac{\bar{U}}{C} = -\alpha\frac{\omega x^2}{2} - \frac{\beta}{4}\Big[(1-x^2)G(p,e)\cos 2(\theta - pM) \qquad (11)$$
$$+ \frac{(1+x)^2}{2}H(p,e)\cos 2(\theta - pM - \phi)$$
$$+ \frac{(1-x)^2}{2}H(-p,e)\cos 2(\theta - pM + \phi)\Big]$$

where x = X/L = cos ε, φ = ϖ + ψ

$$\alpha = \frac{3Gm_\star}{2a^3(1-e^2)^{3/2}}\frac{E_d}{\omega} \approx \frac{3}{2}\frac{n^2}{\omega}(1-e^2)^{-3/2}E_d \qquad (12)$$

is the "precession constant" and

$$\beta = \frac{3Gm_\star}{2a^3}\frac{B-A}{C} \approx \frac{3}{2}n^2\frac{B-A}{C} \qquad (13)$$

TABLE 1. Coefficients of G (p,e) and H (p,e) to e^4.

p	G(p,e)	H(p,e)
−1	$\frac{9}{4}e^2 + \frac{7}{4}e^4$	$\frac{1}{24}e^4$
−1/2	$\frac{3}{2}e + \frac{27}{16}e^3$	$\frac{1}{48}e^3$
0	$1 + \frac{3}{2}e^2 + \frac{15}{8}e^4$	0
1/2	$\frac{3}{2}e + \frac{27}{16}e^3$	$-\frac{1}{2}e + \frac{1}{16}e^3$
1	$\frac{9}{4}e^2 + \frac{7}{4}e^4$	$1 - \frac{5}{2}e^2 + \frac{13}{16}e^4$
3/2	$\frac{53}{16}e^3$	$\frac{7}{2}e - \frac{123}{16}e^3$
2	$\frac{77}{16}e^4$	$\frac{17}{2}e^2 - \frac{115}{6}e^4$
5/2		$\frac{845}{48}e^3$
3		$\frac{533}{16}e^4$

The exact expression of the coefficients is given by $G(p,e) = \frac{1}{\pi}\int_0^\pi \left(\frac{a}{r}\right)^3 \exp(i2pM)\, dM$ and $H(p,e) = \frac{1}{\pi}\int_0^\pi \left(\frac{a}{r}\right)^3 \exp(i2\nu)\exp(i2pM)\, dM$.

For nonresonant motion, that is, when $(B-A)/C \approx 0$ (e.g., gaseous planets) or $|\omega| \gg pn$, we can simplify expression (11) as

$$\frac{\bar{U}}{C} = -\alpha\frac{\omega x^2}{2} \qquad (14)$$

2.1.3. Equations of motion. The Andoyer variables (L, θ) and (X, −ψ) are canonically conjugated and thus (e.g., *Goldstein,* 1950; *Kinoshita,* 1977)

$$\frac{dL}{dt} = -\frac{\partial \bar{U}}{\partial \theta}, \quad \frac{dX}{dt} = \frac{\partial \bar{U}}{\partial \psi}, \quad \frac{d\psi}{dt} = -\frac{\partial \bar{U}}{\partial X} \qquad (15)$$

Andoyer's variables do not give a clear view of the spin variations, despite their practical use. Since $\omega = L/C$ and $\cos\varepsilon = x = X/L$ the spin variations can be obtained as

$$\frac{d\omega}{dt} = -\frac{\partial}{\partial\theta}\left(\frac{\bar{U}}{C}\right), \quad \frac{d\psi}{dt} = -\frac{1}{\omega}\frac{\partial}{\partial x}\left(\frac{\bar{U}}{C}\right) \qquad (16)$$

and

$$\frac{dx}{dt} = -\frac{1}{L}\left(\frac{X}{L}\frac{dL}{dt} - \frac{dX}{dt}\right)$$
$$= \frac{1}{\omega}\left[x\frac{\partial}{\partial\theta} + \frac{\partial}{\partial\psi}\right]\left(\frac{\bar{U}}{C}\right) \qquad (17)$$

For nonresonant motion, we get from equation (14)

$$\frac{d\omega}{dt} = \frac{dx}{dt} = 0 \quad \text{and} \quad \frac{d\psi}{dt} = \alpha x \qquad (18)$$

The spin motion reduces to the precession of the spin vector about the normal to the orbital plane with rate αx.

2.2. Gravitational Tides

Gravitational tides arise from differential and inelastic deformations of the planet due to the gravitational effect of a perturbing body (which can be the central star or a satellite). Tidal contributions to the planet evolution are based on a very general formulation of the tidal potential, initiated by *Darwin* (1880). The attraction of a body with mass $m_\star$ at a distance r from the center of mass of the planet can be expressed as the gradient of a scalar potential V′, which is a sum of Legendre polynomials (e.g., *Kaula,* 1964; *Efroimsky and Williams,* 2009)

$$V' = \sum_{l=2}^{\infty} V'_l = -\frac{Gm_\star}{r}\sum_{l=2}^{\infty}\left(\frac{r'}{r}\right)^l P_l(\cos S) \qquad (19)$$

where r′ is the radial distance from the planet's center, and S the angle between **r** and **r**′. The distortion of the planet by the potential V′ gives rise to a tidal potential

$$V_g = \sum_{l=2}^{\infty} (V_g)_l \qquad (20)$$

where $(V_g)_l = k_l V'_l$ at the planet's surface and k_l is the Love number for potential (Fig. 2). Typically, $k_2 \sim 0.25$ for Earth-like planets, and $k_2 \sim 0.40$ for giant planets (*Yoder*, 1995). Since the tidal potential $(V_g)_l$ is an lth degree harmonic, it is a solution of a Dirichlet problem, and exterior to the planet it must be proportional to r^{-l-1} (e.g., *Abramowitz and Stegun*, 1972; *Lambeck*, 1980). Furthermore, as upon the surface $r' = R \ll r$, we can retain in expression (20) only the first term, $l = 2$

$$V_g = -k_2 \frac{Gm_\star}{R} \left(\frac{R}{r}\right)^3 \left(\frac{R}{r'}\right)^3 P_2(\cos S) \qquad (21)$$

In general, imperfect elasticity will cause the phase angle of V_g to lag behind that of V' (*Kaula*, 1964) by an angle $\delta_g(\sigma)$ such that

$$2\delta_g(\sigma) = \sigma \Delta t_g(\sigma) \qquad (22)$$

$\Delta t_g(\sigma)$ being the time lag associated with the tidal frequency σ (a linear combination of the inertial rotation rate ω and the mean orbital motion n) (Fig. 3).

2.2.1. Equations of motion. Expressing the tidal potential given by expression (21) in terms of Andoyer angles (θ, ψ), we can obtain the contribution to the spin evolution from expression (15) using $U_g = m'V_g$ at the place of $\bar{U}$, where m' is the mass of the interacting body. As we are interested here in the study of the secular evolution of the spin, we also average U_g over the periods of mean anomaly and longitude of the periapse of the orbit. When the interacting body is the same as the perturbing body ($m' = m_\star$), we obtain

$$\frac{d\omega}{dt} = -\frac{Gm_\star^2 R^5}{Ca^6} \sum_\sigma b_g(\sigma) \Omega_\sigma^g(x,e) \qquad (23)$$

$$\frac{d\varepsilon}{dt} = -\frac{Gm_\star^2 R^5}{Ca^6} \frac{\sin\varepsilon}{\omega} \sum_\sigma b_g(\sigma) \mathcal{E}_\sigma^g(x,e) \qquad (24)$$

where the coefficients $\Omega_\sigma^g(x, e)$ and $\mathcal{E}_\sigma^g(x, e)$ are polynomials in the eccentricity (*Kaula*, 1964). When the eccentricity is small, we can neglect the terms in e^2, and we have

$$\begin{aligned}
\Sigma_\sigma b_\tau(\sigma) \Omega_\sigma^\tau =\; & b_\tau(\omega) \tfrac{3}{4} x^2 (1-x^2) \\
& + b_\tau(\omega-2n) \tfrac{3}{16}(1+x)^2 (1-x^2) \\
& + b_\tau(\omega+2n) \tfrac{3}{16}(1-x)^2 (1-x^2) \\
& + b_\tau(2\omega) \tfrac{3}{8}(1+x^2)^2 \\
& + b_\tau(2\omega-2n) \tfrac{3}{32}(1+x)^4 \\
& + b_\tau(2\omega+2n) \tfrac{3}{32}(1-x)^4
\end{aligned} \qquad (25)$$

and

$$\begin{aligned}
\Sigma_\sigma b_\tau(\sigma) \mathcal{E}_\sigma^\tau =\; & b_\tau(2n) \tfrac{9}{16}(1-x^2) \\
& + b_\tau(\omega) \tfrac{3}{4} x^3 \\
& - b_\tau(\omega-2n) \tfrac{3}{16}(1+x)^2 (2-x) \\
& + b_\tau(\omega+2n) \tfrac{3}{16}(1-x)^2 (2+x) \\
& + b_\tau(2\omega) \tfrac{3}{8} x (1-x^2) \\
& - b_\tau(2\omega-2n) \tfrac{3}{32}(1+x)^3
\end{aligned} \qquad (26)$$

The coefficients $b_\tau(\sigma)$ are related to the dissipation of the mechanical energy of tides in the planet's interior, responsible for the time delay $\Delta t_g(\sigma)$ between the position of "maximal tide" and the substellar point. They are related to the phase lag $\delta_g(\sigma)$ as

$$b_g(\sigma) = k_2 \sin 2\delta_g(\sigma) = k_2 \sin(\sigma \Delta t_g(\sigma)) \qquad (27)$$

where $\tau = g$ for gravitational tides. Dissipation equations (23) and (24) must be invariant under the change (ω, x) by $(-\omega, -x)$, which imposes that $b(\sigma) = -b(-\sigma)$, that is, $b(\sigma)$ is an odd function of σ. Although mathematically equivalent, the couples (ω, x) and $(-\omega, -x)$ correspond to two different physical situations (*Correia and Laskar*, 2001).

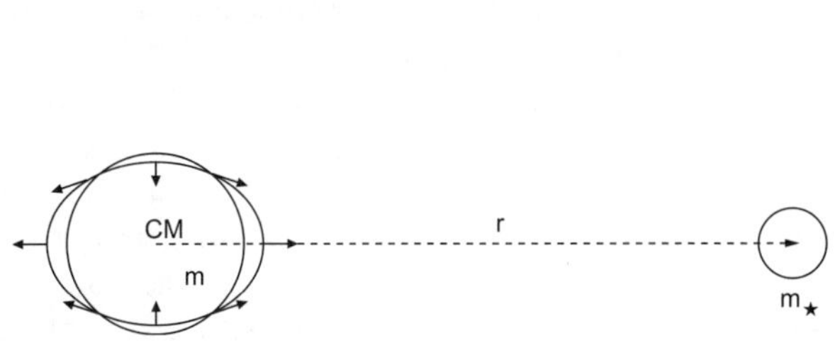

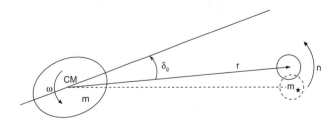

Fig. 2. Gravitational tides. The difference between the gravitational force exerted by the mass m on a point of the surface and the center of mass is schematized by the arrows. The planet will deform following the equipotential of all present forces.

Fig. 3. Phase lag for gravitational tides. The tidal deformation takes a delay time Δt_g to attain the equilibrium. During the time Δt_g, the planet turns by an angle $\omega \Delta t_g$ and the star by $n \Delta t_g$. For $\varepsilon = 0$, the bulge phase lag is given by $\delta_g \approx (\omega - n) \Delta t_g$.

The tidal potential given by expression (21) can also be directly used to compute the orbital evolution due to tides. Indeed, it can be seen as a perturbation of the gravitational potential (equation (1)), and the contributions to the orbit are computed using Lagrange planetary equations (e.g., *Brouwer and Clemence*, 1961; *Kaula* 1964)

$$\frac{da}{dt} = \frac{2}{mna} \frac{\partial U}{\partial M} \quad (28)$$

$$\frac{de}{dt} = \frac{\sqrt{1-e^2}}{mna^2 e} \left[\sqrt{1-e^2} \frac{\partial U}{\partial M} - \frac{\partial U}{\partial \varpi} \right] \quad (29)$$

We then find for the orbital evolution of the planet

$$\frac{da}{dt} = -\frac{6Gm_\star^2 R^5}{mna^7} \sum_\sigma b_g(\sigma) A_\sigma^g(x,e) \quad (30)$$

$$\frac{de}{dt} = -e \frac{3Gm_\star^2 R^5}{mna^8} \sum_\sigma b_g(\sigma) E_\sigma^g(x,e) \quad (31)$$

where the coefficients $A_\sigma^g(x, e)$ and $E_\sigma^g(x, e)$ are again polynomials in the eccentricity. When the eccentricity is small, we can neglect the terms in e^2, and we have

$$\begin{aligned}\Sigma_\sigma b_\tau(\sigma) A_\sigma^\tau = \ &b_\tau(2n)\tfrac{9}{16}(1-x^2)^2 \\ &- b_\tau(\omega-2n)\tfrac{3}{8}(1-x^2)(1+x)^2 \\ &+ b_\tau(\omega+2n)\tfrac{3}{8}(1-x^2)(1-x)^2 \\ &- b_\tau(2\omega-2n)\tfrac{3}{32}(1+x)^4 \\ &+ b_\tau(2\omega+2n)\tfrac{3}{32}(1-x)^4\end{aligned} \quad (32)$$

and

$$\begin{aligned}\Sigma_\sigma b_\tau(\sigma) E_\sigma^\tau = \ &b_\tau(n)\tfrac{9}{128}(5x^2-1)(7x^2-3) \\ &- b_\tau(2n)\tfrac{9}{32}(1-x^2)^2 \\ &+ b_\tau(3n)\tfrac{441}{128}(1-x^2)^2 \\ &- b_\tau(\omega-n)\tfrac{3}{64}(5x-1)(7x+1)(1-x^2) \\ &+ b_\tau(\omega+n)\tfrac{3}{64}(5x+1)(7x-1)(1-x^2) \\ &+ b_\tau(\omega-2n)\tfrac{3}{16}(1-x^2)(1+x)^2 \\ &- b_\tau(\omega+2n)\tfrac{3}{16}(1-x^2)(1-x)^2 \\ &- b_\tau(\omega-3n)\tfrac{3}{64}(1-x^2)(1+x)^2 \\ &+ b_\tau(\omega+3n)\tfrac{3}{64}(1-x^2)(1-x)^2 \\ &- b_\tau(2\omega-n)\tfrac{3}{256}(5x-7)(7x-5)(1+x)^2 \\ &+ b_\tau(2\omega+n)\tfrac{3}{256}(5x+7)(7x+5)(1-x)^2 \\ &+ b_\tau(2\omega-2n)\tfrac{3}{64}(1+x)^4 \\ &- b_\tau(2\omega+2n)\tfrac{3}{64}(1-x)^4 \\ &- b_\tau(2\omega-3n)\tfrac{147}{256}(1+x)^4 \\ &+ b_\tau(2\omega+3n)\tfrac{147}{256}(1-x)^4\end{aligned} \quad (33)$$

2.2.2. Dissipation models. The dissipation of the mechanical energy of tides in the planet's interior is responsible for the phase lag $\delta(\sigma)$. A commonly used dimensionless measure of tidal damping is the quality factor Q (*Munk and MacDonald*, 1960), defined as the inverse of the "specific" dissipation and related to the phase lags by

$$Q(\sigma) = \frac{2\pi E}{\Delta E} = \cot 2\delta(\sigma) \quad (34)$$

where E is the total tidal energy stored in the planet, and ΔE the energy dissipated per cycle. We can rewrite expression (27) as

$$b_g(\sigma) = \frac{k_2 \mathrm{sign}(\sigma)}{\sqrt{Q^2(\sigma)+1}} \approx \mathrm{sign}(\sigma) \frac{k_2}{Q(\sigma)} \quad (35)$$

The present Q value for the planets in the solar system can be estimated from orbital measurements, but as the rheology of the planets is poorly known, the exact dependence of $b_\tau(\sigma)$ on the tidal frequency σ is unknown. Many different authors have studied the problem and several models for $b_\tau(\sigma)$ have been developed so far, from the simplest to the more complex (for a review see *Efroimsky and Williams*, 2009). The huge problem in declaring one model to be more valid than the others is the difficulty of comparing theoretical results with the observations, as the effect of tides are very small and can only be efficiently detected after long periods of time. Therefore, here we will only describe a few simplified models that are commonly used:

The viscoelastic model. Darwin (1908) assumed that the planet behaves like a Maxwell solid, i.e., the planet responds to stresses like a massless, damped harmonic oscillator. It is characterized by a rigidity (or shear modulus) μ_e and by a viscosity υ_e. A Maxwell solid behaves like an elastic solid over short timescales, but flows like a fluid over long periods of time. This behavior is also known as elasticoviscosity. For a constant density ρ, we have

$$b_g(\sigma) = k_f \frac{\tau_b - \tau_a}{1+(\tau_b \sigma)^2} \sigma \quad (36)$$

where k_f is the fluid Love number (equation (7)). $\tau_a = \upsilon_e/\mu_e$ and $\tau_b = \tau_a(1 + 19\mu_e R/2Gm\rho)$ are time constants for the damping of gravitational tides.

The viscoelastic model is a realistic approximation of the planet's deformation with the tidal frequency (e.g., *Escribano et al.*, 2008). However, when substituting expression (36) into the dynamical equations (23) and (24) we get an infinite sum of terms, which is not practical. As a consequence, simplified versions of the viscoelastic model for specific values of the tidal frequency σ are often used. For instance, when σ is small, $(\tau_b \sigma)^2$ can be neglected in expression (36) and $b_g(\sigma)$ becomes proportional to σ.

The viscous or linear model. In the viscous model, it is assumed that the response time delay to the perturbation is independent of the tidal frequency, i.e., the position of the "maximal tide" is shifted from the substellar point by a constant time lag Δt_g (*Mignard*, 1979, 1980). As we usually have $\sigma \Delta t_g \ll 1$, the viscous model becomes linear

$$b_g(\sigma) = k_2 \sin(\sigma \Delta t_g) \approx k_2 \sigma \Delta t_g \quad (37)$$

The viscous model is a particular case of the viscoelastic model and is specially adapted to describe the behavior of planets in slow rotating regimes ($\omega \sim n$).

The constant-Q model. Since for Earth, Q changes by less than an order of magnitude between the Chandler wobble period (about 440 d) and seismic periods of a few seconds (*Munk and MacDonald*, 1960), it is also common to treat the specific dissipation as independent of frequency. Thus

$$b_g(\sigma) = \text{sign}(\sigma) k_2 / Q \quad (38)$$

The constant-Q model can be used for periods of time where the tidal frequency does not change much, as is the case for fast rotating planets. However, for long-term evolutions and slow-rotating planets, the constant-Q model is not appropriate as it gives rise to discontinuities for $\sigma = 0$.

2.2.3. Consequences for the spin. Although both linear and constant models have some limitations, for reasons of simplicity they are the most widely used in literature. The linear model nevertheless has an important advantage over the constant model: It is appropriate to describe the behavior of the planet near the equilibrium positions, since the linear model closely follows the realistic viscoelastic model for slow rotation rates. The equations of motion can also be expressed in an elegant way, so we will adopt the viscous model for the remainder of this chapter, without loss of generality concerning the main consequences of tidal effects.

Using the approximation (37) in expressions (23) and (24), we simplify the spin equations as (*Correia and Laskar*, 2010, Appendix B)

$$\dot{\omega} = -\frac{Kn}{C}\left(f_1(e)\frac{1+\cos^2\varepsilon}{2}\frac{\omega}{n} - f_2(e)\cos\varepsilon\right) \quad (39)$$

and

$$\dot{\varepsilon} \approx \frac{Kn}{C\omega}\sin\varepsilon\left(f_1(e)\cos\varepsilon\frac{\omega}{2n} - f_2(e)\right) \quad (40)$$

where

$$f_1(e) = \frac{1+3e^2+3e^4/8}{(1-e^2)^{9/2}} \quad (41)$$

$$f_2(e) = \frac{1+15e^2/2+45e^4/8+5e^6/16}{(1-e^2)^6} \quad (42)$$

and

$$K = \Delta t \frac{3k_2 G m_\star^2 R^5}{a^6} \quad (43)$$

Because of the factor $1/\omega$ in the magnitude of the obliquity variations (equation (40)), for an initial fast-rotating planet the timescale for the obliquity evolution will be longer than the timescale for the rotation rate evolution (equation (39)). As a consequence, it is to be expected that the rotation rate reaches an equilibrium value earlier than the obliquity. For a given obliquity and eccentricity, the equilibrium rotation rate, obtained when $\dot{\omega} = 0$, is then attained (see Fig. 4) for

$$\frac{\omega_e}{n} = \frac{f_2(e)}{f_1(e)}\frac{2\cos\varepsilon}{1+\cos^2\varepsilon} \quad (44)$$

Replacing the previous equation in the expression for obliquity variations (equation (40)), we find

$$\dot{\varepsilon} \approx -\frac{Kn}{C\omega}f_2(e)\frac{\sin\varepsilon}{1+\cos^2\varepsilon} \quad (45)$$

We then conclude that the obliquity can only decrease by tidal effect, since $\dot{\varepsilon} \leq 0$, and the final obliquity always tends to zero.

2.2.4. Consequences for the orbit. As for the spin, the semimajor axis and the eccentricity evolution can be obtained using approximation (37) in expressions (32) and (33), respectively (*Correia*, 2009)

$$\dot{a} = \frac{2K}{ma}\left(f_2(e)\cos\varepsilon\frac{\omega}{n} - f_3(e)\right) \quad (46)$$

and

$$\dot{e} = \frac{9K}{ma^2}\left(\frac{11}{18}f_4(e)\cos\varepsilon\frac{\omega}{n} - f_5(e)\right)e \quad (47)$$

where

$$f_3(e) = \frac{1+31e^2/2+255e^4/8+185e^6/16+25e^8/64}{(1-e^2)^{15/2}} \quad (48)$$

$$f_4(e) = \frac{1+3e^2/2+e^4/8}{(1-e^2)^5} \quad (49)$$

$$f_5(e) = \frac{1+15e^2/4+15e^4/8+5e^6/64}{(1-e^2)^{13/2}} \quad (50)$$

The ratio between orbital and spin evolution timescales is roughly given by $C/(ma^2) \ll 1$, meaning that the spin achieves an equilibrium position much faster than the orbit.

Replacing the equilibrium rotation rate (equation (44)) with $\varepsilon = 0$ (for simplicity) in equations (46) and (47) gives

$$\dot{a} = -\frac{7K}{ma} f_6(e) e^2 \quad (51)$$

$$\dot{e} = -\frac{7K}{2ma^2} f_6(e)(1-e^2) e^2 \quad (52)$$

where $f_6(e) = (1 + 45e^2/14 + 8e^4 + 685e^6/224 + 255e^8/448 + 25e^{10}/1792)(1-e^2)^{-15/2}/(1 + 3e^2 + 3e^4/8)$. Thus, we always have $\dot{a} \le 0$ and $\dot{e} \le 0$, and the final eccentricity is zero. Another consequence is that the quantity $a(1-e^2)$ is conserved (equation (101)). The final equilibrium semimajor axis is then given by

$$a_f = a(1-e^2) \quad (53)$$

which is a natural consequence of the orbital angular momentum conservation (since the rotational angular momentum of the planet is much smaller). Notice, however, that once the equilibrium semimajor axis a_f is attained, the tidal effects on the star cannot be neglected, and they govern the future evolution of the planet's orbit.

2.3. Thermal Atmospheric Tides

The differential absorption of the solar heat by the planet's atmosphere gives rise to local variations of temperature and consequently to pressure gradients. The mass of the atmosphere is then permanently redistributed, adjusting for an equilibrium position. More precisely, the particles of the atmosphere move from the high-temperature zone (at the substellar point) to the low-temperature areas. Indeed, observations on Earth show that the pressure redistribution is essentially a superposition of two pressure waves (see *Chapman and Lindzen*, 1970): a daily (or diurnal) tide of small amplitude (the pressure is minimal at the substellar point and maximal at the antipode) and a strong half-daily (semidiurnal) tide (the pressure is minimal at the substellar point and at the antipode) (Fig. 5).

The gravitational potential generated by all the particles in the atmosphere at a generic point of the space $\mathbf{r}$ is given by

$$V_a = -G \int_{(\mathcal{M})} \frac{d\mathcal{M}}{|\mathbf{r}-\mathbf{r}'|} \quad (54)$$

where $\mathbf{r}' = (r', \theta', \varphi')$ is the position of the atmosphere mass element $d\mathcal{M}$ with density $\rho_a(\mathbf{r}')$ and

$$d\mathcal{M} = \rho_a(\mathbf{r}') r'^2 \sin\theta' dr' d\theta' d\varphi' \quad (55)$$

Assuming that the radius of the planet is constant and that the height of the atmosphere can be neglected, we approximate expression (55) as

$$d\mathcal{M} = \frac{R^2}{g} p_s(\theta', \varphi') \sin\theta' d\theta' d\varphi' \quad (56)$$

where g is the mean surface gravity acceleration and p_s the surface pressure, which depends on the stellar insolation. Thus, p_s depends on S, the angle between the direction of the Sun and the normal to the surface

$$p_s(\theta', \varphi') = p_s(S) = \sum_{l=0}^{+\infty} \tilde{p}_l P_l(\cos S) \quad (57)$$

where P_l are the Legendre polynomials on the order of l and $\tilde{p}_l$ its coefficients. Developing also $|\mathbf{r}-\mathbf{r}'|^{-1}$ in Legendre polynomials we rewrite expression (54) as

$$V_a = -\frac{1}{\bar{\rho}} \sum_{l=0}^{+\infty} \frac{3}{2l+1} \tilde{p}_l \left(\frac{R}{r}\right)^{l+1} P_l(\cos S) \quad (58)$$

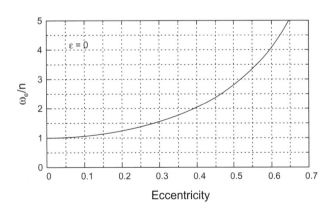

Fig. 4. Evolution of the equilibrium rotation rate $\omega_e/n = f_2(e)/f_1(e)$ with the eccentricity when $\varepsilon = 0°$ using the viscous model (equation (44)). As the eccentricity increases, ω_e also increases.

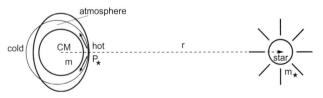

Fig. 5. Thermal atmospheric tides. The atmosphere's heating decreases with the distance to the substellar point $P_\star$. The atmospheric mass redistribution is essentially decomposed in a weak daily tide (round shape) and in a strong half-daily tide (oval shape).

where $\bar{\rho}$ is the mean density of the planet. Since we are only interested in pressure oscillations, we must subtract the term of constant pressure ($l = 0$) in order to obtain the tidal potential. We also eliminate the diurnal terms ($l = 1$) because they correspond to a displacement of the center of mass of the atmosphere bulge, which has no dynamical implications. Thus, since we usually have r ≫ R, retaining only the semidiurnal terms ($l = 2$), we write

$$V_a = -\frac{3}{5}\frac{\tilde{p}_2}{\bar{\rho}}\left(\frac{R}{r}\right)^3 P_2(\cos S) \quad (59)$$

2.3.1. Equations of motion. Using the same methodology of previous sections, the contributions of thermal atmospheric tides to the spin evolution are obtained from expression (15) using $U_a = m_\star V_a$ in the place of $\bar{U}$

$$\frac{d\omega}{dt} = -\frac{3m_\star R^3}{5C\bar{\rho}a^3}\sum_\sigma b_a(\sigma)\Omega^a_\sigma(x,e) \quad (60)$$

$$\frac{d\varepsilon}{dt} = -\frac{3m_\star R^3}{5C\bar{\rho}a^3}\frac{\sin\varepsilon}{\omega}\sum_\sigma b_a(\sigma)\mathcal{E}^a_\sigma(x,e) \quad (61)$$

where the terms $\Omega^a_\sigma(x, e)$ and $\mathcal{E}^a_\sigma(x, e)$ are also polynomials in the eccentricity, but different from their analogs for gravitational tides (equations (23) and (24)). Nevertheless, when neglecting the terms in e^2, they become equal and are given by expressions (25) and (26), respectively (with $\tau = a$).

For thermal atmospheric tides there is also a delay before the response of the atmosphere to the excitation (Fig. 6). We name the time delay $\Delta t_a(\sigma)$ and the corresponding phase angle $\delta_a(\sigma)$ (equation (22)). The dissipation factor $b_a(\sigma)$ is given here by

$$b_a(\sigma) = \tilde{p}_2(\sigma)\sin 2\delta_a(\sigma) = \tilde{p}_2(\sigma)\sin(\sigma\Delta t_a(\sigma)) \quad (62)$$

Siebert (1961) and *Chapman and Lindzen* (1970) have shown that when

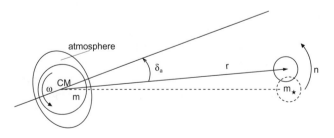

Fig. 6. Phase lag for thermal atmospheric tides. During the time Δt_a the planet turns by an angle $\omega\Delta t_a$ and the star by $n\Delta t_a$. For $\varepsilon = 0$, the bulge phase lag is given by $\delta_a \approx (\omega-n)\Delta t_a$.

$$|\tilde{p}_2(\sigma)| \ll \tilde{p}_0 \quad (63)$$

the amplitudes of the pressure variations on the ground are given by

$$\tilde{p}_2(\sigma) = i\frac{\gamma}{\sigma}\tilde{p}_0\left(\nabla\cdot\mathbf{v}_\sigma - \frac{\gamma-1}{\gamma}\frac{J_\sigma}{gH_0}\right) \quad (64)$$

where $\gamma = 7/5$ for a perfect gas, **v** is the velocity of tidal winds, J_σ is the amount of heat absorbed or emitted by a unit mass of air per unit time, and H_0 is the scale height at the surface. We can rewrite expression (64) as

$$\tilde{p}_2(\sigma) = \frac{\gamma}{|\sigma|}\tilde{p}_0\left|\nabla\cdot\mathbf{v}_\sigma - \frac{\gamma-1}{\gamma}\frac{J_\sigma}{gH_0}\right|e^{\pm i\frac{\pi}{2}} \quad (65)$$
$$= |\tilde{p}_2(\sigma)|e^{\pm i\frac{\pi}{2}}$$

where the factor $e^{\pm i\frac{\pi}{2}}$ can be seen as a supplementary phase lag of $\pm\pi/2$

$$b_a(\sigma) = |\tilde{p}_2(\sigma)|\sin 2\left(\delta_a(\sigma) \pm \frac{\pi}{2}\right) \quad (66)$$
$$= -|\tilde{p}_2(\sigma)|\sin 2\delta_a(\sigma)$$

The minus sign above causes pressure variations to lead the Sun whenever $\delta_a(\sigma) < \pi/2$ (*Chapman and Lindzen*, 1970; *Dobrovolskis and Ingersoll*, 1980) (Fig. 6).

2.3.2. Dissipation models. Unfortunately, our knowledge of the atmospheric response to thermal excitation is still very incomplete. As for the gravitational tides, models are developed to deal with the unknowns. *Dobrovolskis and Ingersoll* (1980) adopted a model called "heating at the ground," where they suppose that all the stellar flux absorbed by the ground F_s is immediately deposited in a thin layer of atmosphere at the surface. The heating distributing may then be written as a δ function just above the ground

$$J(r) = \frac{g}{\tilde{p}_0}F_s\delta(r-0^+) \quad (67)$$

Neglecting **v** over the thin heated layer, expression (64) simplifies as

$$|\tilde{p}_2(\sigma)| = \frac{5}{16}\frac{\gamma-1}{|\sigma|}\frac{F_s}{H_0} = \frac{5}{16}\frac{\gamma}{|\sigma|}\frac{gF_s}{c_p\bar{T}_s} \quad (68)$$

where the factor 5/16 represents the second-degree harmonic component of the insolation contribution (*Dobrovolskis and Ingersoll*, 1980), c_p is the specific heat at constant pressure, and $\bar{T}_s$ the mean surface temperature.

Nevertheless, according to expression (68), if $\sigma = 0$, the amplitude of the pressure variations $\tilde{p}_2(\sigma)$ becomes infinite. The amplitude cannot grow infinitely, as for a tidal frequency equal to zero, a steady distribution is attained. Indeed, expression (64) is not valid when $\sigma \approx 0$ because condition (63) is no longer verified. Using the typically accepted values for the venusian atmosphere, $c_p \approx 1000$ K kg^{-1}s^{-1}, $\bar{T}_s \approx 730$ K, and $F_s \approx 100$ Wm^{-2} (*Avduevskii et al.*, 1976), we compute

$$\left|\tilde{p}_2(\sigma)\right| \approx 10^{-4}\tilde{p}_0 \frac{n}{|\sigma|} \tag{69}$$

which means that for $\sigma \sim n$, the "heating at the ground" model of *Dobrovolskis and Ingersoll* (1980) can still be applied. Since we are only interested in long-term behaviors we can set $\tilde{p}_2(\sigma) = 0$ whenever $|\sigma| \ll n/100$, because for tidal frequencies $\sigma \sim 0$ the dissipation lag $\sin(\sigma\Delta t_a(\sigma)) \approx \sigma\Delta t_a(\sigma)$ also goes to zero. We expect that further studies about the atmospheres of synchronous exoplanets (e.g., *Joshi et al.*, 1997; *Arras and Socrates*, 2010) may provide a more accurate solution for the case $\sigma \approx 0$.

In the presence of a dense atmosphere, another type of tides can arise: The atmosphere pressure upon the surface gives rise to a deformation, a pressure bulge, that will also be affected by the stellar torque. At the same time, the atmosphere itself exerts a torque over the planet's bulges (gravitational and pressure bulge). Nevertheless, we do not need to take into account additional tidal effects as their consequences upon the dynamical equations can be neglected (*Hinderer et al.*, 1987; *Correia and Laskar*, 2003a).

2.4. Spin-Orbit Resonances

A spin-orbit resonance occurs when there is a commensurability between the rotation rate ω and the mean motion of the orbit n (equations (9) and (10)). The synchronous rotation of the Moon is the most common example. After the discovery of the 3/2 spin-orbit resonance of Mercury (*Colombo*, 1965), spin-orbit resonances were studied in great detail (*Colombo and Shapiro*, 1966; *Goldreich and Peale*, 1966; *Counselman and Shapiro*, 1970; *Correia and Laskar*, 2004, 2009, 2010). When resonant motion is present we cannot neglect the terms in β in expression (11). Assuming for simplicity a low obliquity ($x \approx 1$), we obtain a nonzero contribution for the rotation rate (equation (16))

$$\frac{d\omega}{dt} = -\beta H(p,e)\sin 2\gamma \tag{70}$$

where $\gamma = \theta - pM - \phi$. The rotation of the planet will therefore present oscillations around a mean value. The width of the corresponding resonance, centered at $\omega = pn$, is

$$\Delta\omega = \sqrt{2\beta H(p,e)} \tag{71}$$

Due to the tidal torque (equation (23)), here denoted by $\bar{T}$, the mean rotation rate does not remain constant and may therefore cross and be captured in a spin-orbit resonance. *Goldreich and Peale* (1966) computed a simple estimation of the capture probability P_{cap}, and subsequent more detailed studies proved their expression to be essentially correct (for a review, see *Henrard*, 1993). Since the tidal torques can usually be described by means of the torques considered by *Goldreich and Peale* (1966), we will adopt here the same notations. Let

$$\bar{T} = -K\left(V + \frac{\dot{\gamma}}{n}\right) \tag{72}$$

where K and V are positive constant torques, and $\dot{\gamma} = \omega - pn$. The probability of capture into resonance is then given by (*Goldreich and Peale*, 1966)

$$P_{cap} = \frac{2}{1+\pi Vn/\Delta\omega} \tag{73}$$

where $\Delta\omega$ is the resonance width (equation (71)). In the slow rotation regime ($\omega \sim n$), where the spin encounters spin-orbit resonances and capture may occur, we compute for the viscous tidal model (equation (39))

$$P_{cap} = 2\left[1+\left(p-\frac{2x}{1+x^2}\frac{f_2(e)}{f_1(e)}\right)\frac{n\pi}{2\Delta\omega}\right]^{-1} \tag{74}$$

2.5. Planetary Perturbations

As is the case for planets in the solar system, many exoplanets are not alone in their orbits, but belong to multiplanet systems. Because of mutual planetary perturbations the orbital parameters of the planets do not remain constant and undergo secular variations in time (see chapter by Fabrycky). An important consequence for the spin of the planets is that the reference orbital plane (to which the obliquity and the precession were defined) will also present variations. We can track the orbital plane variations by the inclination to an inertial reference plane, I, and by the longitude of the line of nodes, Ω. Under the assumption of principal axis rotation, the energy perturbation attached to an inertial frame can be written (*Kinoshita*, 1977; *Néron de Surgy and Laskar*, 1997)

$$U_{pp} = \left[X(1-\cos I) - L\sin\varepsilon\sin I\cos\varphi\right]\frac{d\Omega}{dt} \\ + L\sin\varepsilon\sin\varphi\frac{dI}{dt} \tag{75}$$

where $\varphi = -\Omega - \psi$.

Although the solar system motion is chaotic (*Laskar*, 1989, 1990), the motion can be approximated over several million of years by quasiperiodic series. In particular, for the orbital elements that are involved in the precession-driving terms (equation (75)), we have (*Laskar and Robutel*, 1993)

$$\left(\frac{dI}{dt}+i\frac{d\Omega}{dt}\sin I\right)e^{i\Omega} = \sum_k J_k e^{i(\nu_k t+\phi_k)} \tag{76}$$

and

$$(1-\cos I)\frac{d\Omega}{dt} = \sum_k \mathcal{L}_k \cos(v_k t + \varphi_k) \quad (77)$$

where v_k are secular frequencies of the orbital motion with amplitude J_k and phase ϕ_k, and $i = \sqrt{-1}$. We may then rewrite expression (75) as

$$U_{pp} = L\sum_k \Big[\mathcal{L}_k \cos(v_k t + \varphi_k) x \\ - J_k \sqrt{1-x^2} \sin(v_k t + \psi + \phi_k)\Big] \quad (78)$$

Assuming nonresonant motion, from equations (16) and (17) we get for the spin motion

$$\frac{d\varepsilon}{dt} = \sum_k J_k \cos(v_k t + \psi + \phi_k) \quad (79)$$

and

$$\frac{d\psi}{dt} = \alpha\cos\varepsilon - \sum_k \mathcal{L}_k \cos(v_k t + \varphi_k) \\ - \cot\varepsilon \sum_k J_k \sin(v_k t + \psi + \phi_k) \quad (80)$$

For planetary systems like the solar system, the mutual inclinations remain small (*Laskar*, 1990; *Correia et al.*, 2010), and it follows from expressions (76) and (77) that the amplitudes of J_k and $\mathcal{L}_k$ are bounded respectively by

$$J_k \sim v_k I_{max}, \quad \mathcal{L}_k \sim v_k I_{max}^2/2 \quad (81)$$

The term in $\mathcal{L}_k$ in expression (80) for the precession variations can then be neglected for small inclinations.

From expression (79) it is clear that a resonance can occur whenever the precession frequency is equal to the opposite of a secular frequency v_k (that is, $\dot\psi = -v_k$). Retaining only the terms in k, the problem becomes integrable. We can search for the equilibrium positions by setting the obliquity variations equal to zero ($\dot\varepsilon = 0$). It follows then from expression (79) that $\psi + v_k t + \phi_k = \pm\pi/2$, and replacing it in expression (80) with $\dot\psi = -v_k$, we get

$$\alpha\cos\varepsilon\sin\varepsilon + v_k \sin\varepsilon \approx J_k \cos\varepsilon \quad (82)$$

which gives the equilibrium positions for the spin of the planet, generally known as "Cassini states" (e.g., *Henrard and Murigande*, 1987). Since $J_k/v_k \ll 1$ (equation (81)), the equilibrium positions for the obliquity are then

$$\tan\varepsilon \approx \frac{J_k}{v_k \pm \alpha}, \quad \cos\varepsilon \approx -\frac{v_k}{\alpha} \quad (83)$$

When $|\alpha/v_k| \ll |\alpha/v|_{crit}$, the first expression gives states 2 and 3, while the second expression has no real roots (states 1 and 4 do not exist). When $|\alpha/v_k| \gg |\alpha/v|_{crit}$, the first expression approximates Cassini states 1 and 3, while the second one gives states 2 and 4. States 1, 2 and 3 are stable, while state 4 is unstable. Although gravitational tides always decrease the obliquity (equation (45)), the ultimate stage of the obliquity evolution is to be captured into a Cassini resonant state, similar to the capture of the rotation in a spin-orbit resonance (equation (73)).

The complete system (equations (79) and (80)) is usually not integrable as there are several terms in expression (76), but we can look individually to the location of each resonance. When the resonances are far apart, the motion will behave locally as in the integrable case, with the addition of supplementary small oscillations. However, if several resonances overlap, the motion is no longer regular and becomes chaotic (*Chirikov*, 1979; *Laskar*, 1996). For instance, the present obliquity variations on Mars are chaotic and can vary from 0° to nearly 60° (*Laskar and Robutel*, 1993; *Touma and Wisdom*, 1993; *Laskar et al.*, 2004b).

3. APPLICATION TO THE PLANETS

The orbital parameters of exoplanets are reasonably well determined from radial velocity, transit, or astrometry techniques, but the spins of exoplanets remain a mystery. The same applies to the primordial spins of the terrestrial planets in the solar system, since very few constraints can be derived from the present planetary formation models. Indeed, a small number of large impacts at the end of the formation process will not average, and can change the spin direction. The angular velocities are also unpredictable, but they are usually high, $\omega \gg n$ (*Dones and Tremaine*, 1993; *Agnor et al.*, 1999; *Kokubo and Ida*, 2007), although impacts can also form a slow-rotating planet ($\omega \sim n$) if the size of the typical accreting bodies is much smaller than the protoplanet (e.g., *Schlichting and Sari*, 2007). The critical angular velocity for rotational instability is (*Kokubo and Ida*, 2007)

$$\omega_{cr} \approx 3.3 \left(\frac{\rho}{3 g\,cm^{-3}}\right)^{1/2} hr^{-1} \quad (84)$$

which sets a maximum initial rotation periods of about 1.4 h, for the inner planets of the solar system.

For the jovian planets in the solar system no important mechanism capable of altering the rotation rate is known, but the orientation of the axis may also change by secular resonance with the planets (e.g., *Correia and Laskar*, 2003b; *Ward and Hamilton*, 2004; *Boué and Laskar*, 2010). The fact that all jovian planets rotate fast (Table 3) seems to be in agreement with theoretical predictions (e.g., *Takata and Stevenson*, 1996).

An empirical relation derived by *MacDonald* (1964) based on the present rotation rates of planets from Mars to Neptune (assumed almost unchanged) gives for the initial rotation rates

$$\omega_0 \propto m^{4/5} R^{-2} \quad (85)$$

Extrapolating for the remaining inner planets, we get initial rotation periods of about 18.9 h, 13.5 h, and 12.7 h for Mercury, Venus, and Earth, respectively, much faster than today's values, which are in agreement with the present formation theories.

The above considerations and expressions can also be extended to exoplanets. However, since many of the exoplanets are close to their host stars, it is believed that the spins have undergone significant tidal dissipation and eventually reached some equilibrium positions, as happens for Mercury and Venus in the solar system. Therefore, in this section we will first review the rotation of the terrestrial planets, and then look at the already known exoplanets. In particular, we will focus our attention on two classes of exoplanets, the hot Jupiters (fluid) and the super Earths (rocky with atmosphere), for which tidal effects may play an important role in orbital and spin evolution.

3.1. Solar System Examples

3.1.1. Mercury. The present spin of Mercury is very peculiar: The planet rotates three times around its axis in the same time as it completes two orbital revolutions (*Pettengill and Dyce*, 1965). Within a year of the discovery, the stability of this 3/2 spin-orbit resonance became understood as the result of the solar torque on Mercury's quadrupolar moment of inertia combined with an eccentric orbit (equation (70)) (*Colombo and Shapiro*, 1966; *Goldreich and Peale*, 1966). The way the planet evolved into the 3/2 configuration remained a mystery for long time, but can be explained as the result of tidal evolution combined with the eccentricity variations due to planetary perturbations (*Correia and Laskar*, 2004, 2009). Mercury has no atmosphere, and the spin evolution of the planet is therefore controlled by gravitational tidal interactions with the Sun. Tidal effects drive the final obliquity of Mercury close to zero (equation (45)), and the averaged equation for the rotation motion near the p resonance can be written combining expressions (70) and (39) as

$$\frac{d\omega}{dt} = -\beta' \sin 2\gamma - K' \left[\frac{\omega}{n} - \frac{f_2(e)}{f_1(e)} \right] \quad (86)$$

where $\gamma = \theta - pM - \phi$, $\beta' = \beta H(p, e)$, and $K' = Knf_1(e)/C$. Note that near the p resonance a contribution from core-mantle friction may also be present, but we will neglect it here (for a full description see *Correia and Laskar*, 2009, 2010). The tidal equilibrium is achieved when $d\omega/dt = 0$, i.e., for a constant eccentricity e, when $\omega/n = f_2(e)/f_1(e)$. In a circular orbit (e = 0) the tidal equilibrium coincides with synchronization, while the equilibrium rotation rate $\omega/n = 3/2$ is achieved for $e_{3/2} = 0.284927$ (Fig. 4).

For the present value of Mercury's eccentricity (e ≈ 0.206), the capture probability in the 3/2 spin-orbit resonance is only about 7% (equation (74)). However, as the eccentricity of Mercury suffers strong chaotic variations in time due to planetary secular perturbations, the eccentricity can vary from nearly 0 to more than 0.45, and thus reach values higher than the critical value $e_{3/2} = 0.284927$. Additional capture into resonance can then occur at any time during the planet's history (*Correia and Laskar*, 2004).

In order to check the past evolution of Mercury's spin, it is not possible to use a single orbital solution; because of the chaotic behavior the motion cannot be predicted precisely beyond a few tens of millions of years. A statistical study of the past evolutions of Mercury's orbit was then performed, with the integration of 1000 orbits over 4 G.y. in the past, starting with very close initial conditions, within the uncertainty of the present determinations (*Correia and Laskar*, 2004, 2009).

For each of the 1000 orbital motions of Mercury, the rotational motion (equation (86)) was integrated numerically with planetary perturbations. As e(t) is not constant, $\omega(t)$ will tend toward a limit value $\tilde{\omega}(t)$ that is similar to an averaged value of $(f_2/f_1)(e(t))$, and capture into resonance can occur more often (Fig. 7). Globally, only 38.8% of the solutions did not end in resonance, and the final capture probability distribution was (*Correia and Laskar*, 2004) $P_{1/1} = 2.2\%$, $P_{3/2} = 55.4\%$, $P_{2/1} = 3.6\%$.

With the consideration of the chaotic evolution of the eccentricity of Mercury, the present 3/2 resonant state becomes the most probable outcome for the spin evolution. The largest unknown remains the dissipation factor $k_2\Delta t$ in the expression of K (equation (43)). A stronger dissipation increases the probability of capture into the 3/2 resonance, as ω/n would follow more closely $f_2(e)/f_1(e)$ (Fig. 7), while lower dissipa-

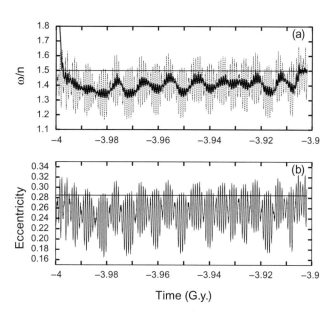

Fig. 7. Simultaneous evolution of **(a)** the rotation rate and **(b)** the eccentricity of Mercury. In this example, there is no capture at the first encounter with the 3/2 resonance (at t ≈ −3.9974 G.y.). About 100 m.y. later, as the mean eccentricity increases, additional crossing of the 3/2 resonance occurs, leading to capture with damping of the libration (*Correia and Laskar*, 2006).

tion slightly decreases the capture probability. The inclusion of core-mantle friction also increases the chances of capture for all resonances (*Correia and Laskar*, 2009, 2010).

3.1.2. Venus. Venus is a unique case in the solar system: It presents a slow retrograde rotation, with an obliquity close to 180° and a 243-d period (*Smith*, 1963; *Goldstein*, 1964; *Carpenter*, 1970). According to planetary formation theories it is highly improbable that the present spin of Venus is primordial, since we would expect a lower obliquity and a fast-rotating planet (equation (85)).

The present rotation of Venus is believed to represent a steady state resulting from a balance between gravitational tides, which drives the planet toward synchronous rotation, and thermally driven atmospheric tides, which drives the rotation away (e.g., *Gold and Soter*, 1969). The conjugated effect of tides and core-mantle friction can tilt Venus down during the planet's past evolution, but requires high values of the initial obliquity (e.g., *Dobrovolskis*, 1980; *Yoder*, 1997). However, the crossing in the past of a large chaotic zone for the spin, resulting from secular planetary perturbations (*Laskar and Robutel*, 1993), can lead Venus to the present retrograde configuration for most initial conditions (*Correia and Laskar*, 2001, 2003b).

Venus has a dense atmosphere and the planet is also close enough to the Sun to undergo significant tidal dissipation. Venus' spin evolution is then controlled by tidal effects (both gravitational and thermal). Tidal effects combined can drive the obliquity either to $\varepsilon = 0°$ or $\varepsilon = 180°$ (*Correia et al.*, 2003). For the two final obliquity possibilities, the tidal components become very simplified, with (at second order in the planetary eccentricity) a single term of tidal frequency $\sigma = 2\omega - 2n$ for $\varepsilon = 0$ and $\sigma = 2\omega + 2n$ for $\varepsilon = \pi$ (equation (25)). Combining expressions (23) and (60), for the rotation rate we can write

$$\left.\frac{d\omega}{dt}\right|_0 = -\frac{3}{2}\left[K_g b_g (2\omega - 2n) + K_a b_a (2\omega - 2n)\right]$$
$$\left.\frac{d\omega}{dt}\right|_\pi = -\frac{3}{2}\left[K_g b_g (2\omega + 2n) + K_a b_a (2\omega + 2n)\right] \quad (87)$$

where K_g and K_a are given by the constant part of expressions (23) and (60), respectively. Let $f(\sigma)$ be defined as

$$f(\sigma) = \frac{b_a(2\sigma)}{b_g(2\sigma)} \quad (88)$$

As $b_\tau(\sigma)$ is an odd function of σ (equation (25)), $f(\sigma)$ is an even function of σ of the form $f(|\sigma|)$. Thus, at equilibrium, with $d\omega/dt = 0$, we obtain an equilibrium condition

$$f(|\omega - xn|) = -\frac{K_g}{K_a} \quad (89)$$

where $x = +1$ for $\varepsilon = 0$ and $x = -1$ for $\varepsilon = \pi$. Moreover, for all commonly used dissipation models f is monotonic and decreasing for slow rotation rates. There are thus only four possible values for the final rotation rate ω_f of Venus, given by

$$|\omega - xn| = f^{-1}\left(-\frac{K_g}{K_a}\right) = \omega_s \quad (90)$$

Assuming that the present rotation of Venus corresponds to a stable retrograde rotation, since $\omega_s > 0$ the only possibilities for the present rotation are $\varepsilon = 0$ and $\omega_{obs} = n - \omega_s$, or $\varepsilon = \pi$ and $\omega_{obs} = \omega_s - n$. In both cases, $\omega_s = n + |\omega_{obs}|$ (ω_s is thus the synodic frequency). With

$$\omega_{obs} = 2\pi/243.0185 \text{ day}; \quad n = 2\pi/224.701 \text{ day} \quad (91)$$

we have

$$\omega_s = 2\pi/116.751 \text{ day} \quad (92)$$

We can then determine all four final states for Venus (Table 2). There are two retrograde states (F_0^- and F_π^-) and two prograde states (F_0^+ and F_π^+). The two retrograde states correspond to the observed present retrograde state of Venus with a period of 243.02 d, while the two other states have a prograde rotation period of 76.83 d. Looking to the present rotation state of the planet, it is impossible to distinguish between the two states with the same angular momentum (Fig. 8).

In order to obtain a global view of the possible final evolutions of Venus' spin, numerical integrations of the equations of motion for the dissipative effects (equations (23), (24), (60), and (61)), with the addition of planetary perturbations (equations (79) and (80)), were performed (*Correia and Laskar*, 2001, 2003b). In Fig. 9 we show the possible final evolutions for a planet starting with an initial period ranging from 3 to 12 d, with an increment of 0.25 d, and initial obliquity from 0° to 180°, with an increment of 2.5° (rotation periods faster than 3 d are excluded as they do not allow the planet to reach a final rotation state within the age of the solar system). Each color represents one of the possible final states. For high initial obliquities, the spin of Venus always evolves into the retrograde final state F_π^-. It is essentially the same evolution as without planetary

TABLE 2. Possible final spin states of Venus.

State	ε	ω	P (days)	P_s (days)
F_0^+	0°	$n + \omega_s$	76.83	116.75
F_0^-	0°	$n - \omega_s$	−243.02	−116.75
F_π^+	180°	$-n - \omega_s$	−76.83	116.75
F_π^-	180°	$-n + \omega_s$	243.02	−116.75

There are two retrograde states (F_0^- and F_π^-) and two prograde states (F_0^+ and F_π^+). In all cases the synodic period P_s is the same (*Correia and Laskar*, 2001).

perturbations, since none of the trajectories encounters a chaotic zone for the obliquity (*Laskar and Robutel*, 1993). However, for evolutionary paths starting with low initial obliquities, we can distinguish two different zones: one zone corresponding to slow initial rotation periods ($P_i > 8$ d) where the prograde rotation final state F_0^- is prevailing, and another zone for faster initial rotation periods ($P_i < 8$ d), where we find a mixture of the three attainable final states, F_0^+, F_0^-, and F_π^-. To emphasize the chaotic behavior, we integrated twice more the zone with $P_i < 8$ d, with a difference of 10^{-9} in the initial eccentricity of Mars (Fig. 9b), and with a difference of 10^{-9} in the initial eccentricity of Neptune (Fig. 9c). The passage through the chaotic zone is reflected by the scattering of the final states in the lefthand side of the figure.

3.1.3. Earth and Mars. Contrary to Mercury and Venus, Earth and Mars are not tidally evolved. For Mars, the tidal dissipation from the Sun is negligible. For Earth, tidal dissipation is noticeable due to the presence of the Moon, but Earth's spin is still far from the equilibrium (e.g., *Néron de Surgy and Laskar*, 1997). Nevertheless, the spin axis of both planets is subjected to planetary perturbations and thus present some significant variations (section 2.5).

In the case of Mars, the presence of numerous secular resonances of the kind $\dot\psi = -v_k$ (equation (79)) induce large chaotic variations in the obliquity, which can evolve between 0° and 60° (Fig. 10). At present, the obliquity of Mars is very similar to the obliquity of Earth, which is a mere coincidence. Indeed, the obliquity of Mars has most certainly reached values larger than 45° in the past (*Laskar et al.*, 2004a). The periods of high obliquity led to large climatic changes on Mars, with the possible occurrence of large-scale ice cycles where the polar caps are sublimated during high-obliquity stages and the ice is deposited in the equatorial regions (*Laskar et al.*, 2002; *Levrard et al.*, 2004, 2007b).

In the case of Earth, the precession frequency is not in resonance with any orbital secular frequency ($\dot\psi \neq v_k$).

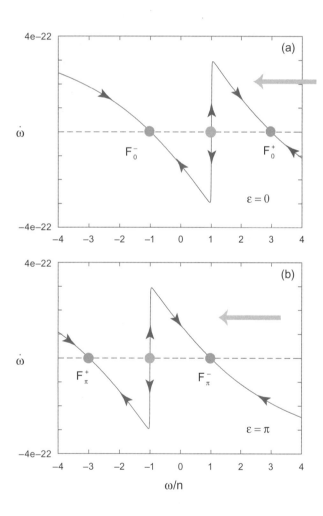

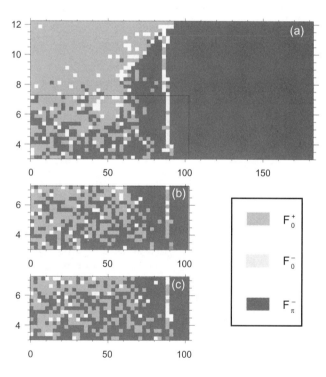

Fig. 8. Final states for a planet with strong atmospheric thermal tides. The original equilibrium point obtained at synchronization ($\omega/n = 1$), when considering uniquely the gravitational tides, becomes unstable, and bifurcates at $\varepsilon = 0$ into two new stable fixed points F_0^- and F_0^+, and at $\varepsilon = \pi$ into F_π^- and F_π^+ (*Correia and Laskar*, 2001, 2003b).

Fig. 9. Final states of Venus' spin with planetary perturbations for initial obliquity [$\varepsilon_i \in (0°, 180°)$] and period [$P_i \in (3$ d, 12 d$)$]. For high initial obliquities, the final evolution of Venus remains essentially unchanged since none of the trajectories crossed the chaotic zone. The passage through the chaotic zone is reflected by the scattering of the final states in the left side of the picture. To emphasize the chaotic behavior, in the bottom left corner of **(a)**, additional integrations were done with the same initial conditions, but with a difference of 10^{-9} in the initial eccentricity of **(b)** Mars and **(c)** Neptune (*Correia and Laskar*, 2003b).

The obliquity of Earth is then only subject to small oscillations of about 1.3° around the mean value (23.3°) with main periodicities around 40,000 yr (*Laskar et al.*, 2004b). The small obliquity variations are nevertheless sufficiently important to induce substantial changes in the insolation received in summer in high-latitude regions on Earth, and they are imprinted in the geological stratigraphic sequences (*Hays et al.*, 1976; *Imbrie*, 1982).

Due to tidal dissipation in the Earth-Moon system, the Moon is moving away from Earth at a rate of 3.8 cm/yr (*Dickey et al.*, 1994), and the rotation rate of Earth is slowing down (equations (46) and (39), respectively). As a consequence, the torque exerted on the equatorial bulge of Earth decreases and thus the Earth's precession frequency (equations (7) and (12)) as well. Using the present dissipation parameters of Earth, *Néron de Surgy and Laskar* (1997) found that after 1.5 G.y., the spin of Earth will enter a large chaotic zone of overlapping orbital secular resonances. From then, Earth's spin axis will evolve in a wildly chaotic way, with a possible range from 0° to nearly 90° (Fig. 11).

The main difference between Earth and Mars is thus due to the presence of the Moon, whose gravitational torque on the equatorial bulge of Earth prevents the spin axis from evolving in a largely chaotic state. Without the presence of the Moon, the behavior of the spin axis of Earth and Mars would be identical (*Laskar et al.*, 1993; *Laskar and Robutel*, 1993). Depending on the orbital configuration of exoplanetary systems, we thus expect to find planets that would be either in a chaotic state, as Mars or the moonless Earth, or in a regular state, as Earth with the Moon. It should be stressed, however, that the presence of a large satellite is not mandatory in order to stabilize the spin axis. Since the stability of the axis is very important for the exoplanet climate, planetary perturbations should be taken into consideration when searching for other Earth-like environments.

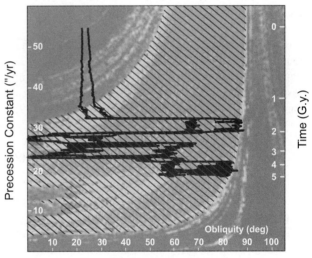

Fig. 11. Example of possible evolution of Earth's obliquity for 5 G.y. in the future, due to tidal dissipation in the Earth-Moon system. The background of the figure is obtained as the result of a stability analysis on about 250,000 numerical integrations of the obliquity of Earth under planetary perturbations for 36 m.y., for various values of the initial obliquity of Earth (x-axis) and various values of the precession constant (left y-axis). We observe a very large chaotic zone (with stripes on the figure) resulting from overlap of orbital secular resonances (*Laskar et al.*, 1993; *Laskar and Robutel*, 1993). The numerical integration is then conducted over 5 G.y. in the future for the obliquity of Earth, including tidal dissipative effects. The two bold curves correspond to the minimum and maximum values reached by the obliquity and the timescale is given in the right y-axis. As long as the orbits stay in the regular region, the motion suffers only small (and regular) variations. As soon as the orbit enters the chaotic zone, very strong variations of the obliquity are observed, which wanders throughout the chaotic zone, and very high values (close to 90°) are reached (*Néron de Surgy and Laskar*, 1997).

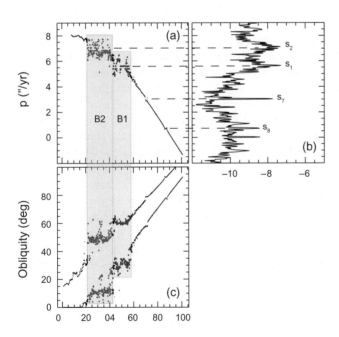

Fig. 10. Frequency analysis of Mars' obliquity. **(a)** The frequency map is obtained by reporting in the ordinate the value of the precession frequency obtained for 1000 integrations over 56 m.y. for the different values of the initial obliquity (abscissa). A large chaotic zone is visible, ranging from 0° to about 60°, with two distinct zones of large chaos, B1 and B2. **(b)** Power spectrum of the orbital forcing term (equation (76)) given in logarithmic scale, showing the correspondence of the chaotic zone with the main secular frequencies s_1, s_2, s_7, s_8 (*Laskar et al.*, 2004a). **(c)** Maximum and minimum values of the obliquity reached over 56 m.y.

3.2. Hot Jupiters

One of the most surprising findings concerning exoplanets was the discovery of several giant planets with periods down to 3 d, designated as "hot Jupiters" (e.g., *Santos et al.,* 2005). Many of the hot Jupiters were simultaneously detected by transit method and radial Doppler shift, which allows the direct and accurate determination of both mass and radius of the exoplanet. Therefore, hot Jupiters are among the better characterized planets outside the solar system.

Due to the proximity of the host star, hot Jupiters are almost certainly tidally evolved. Given the large mass of hot Jupiters, they may essentially be composed of an extensive hydrogen atmosphere, similar to Jupiter or Saturn. As a consequence, despite the presence of an inner metallic core, hot Jupiters can be treated as fluid planets, and we may adopt a viscous model for the tidal dissipation (equations (39) and (40)). Thermal atmospheric tides may also be present (e.g., *Arras and Socrates,* 2010, 2010; *Goodman,* 2010), but we did not take thermal tides into account, as gravitational tides are so strong for hot Jupiters that they probably rule over all the remaining effects (see section 3.3).

3.2.1. Rotational synchronization. The effect of gravitational tides over the obliquity is to straighten the spin axis (equation (45)), so we will adopt $\varepsilon = 0°$ for the obliquity. Assuming that the eccentricity and semimajor axis of the planet are constant, we can derive from expression (39)

$$\frac{\omega}{n} = \frac{f_2(e)}{f_1(e)} + \left(\frac{\omega_0}{n} - \frac{f_2(e)}{f_1(e)}\right) \exp(-t/\tau_{eq}) \qquad (93)$$

where $\tau_{eq}^{-1} = Kf_1(e)/C$ is the characteristic timescale for fully despinning the planet.

As for Mercury, the final equilibrium rotation driven by tides (t → +∞) is given by the equilibrium position $\omega_e/n = f_2(e)/f_1(e)$, which is different from synchronous rotation if the eccentricity is not zero (Fig. 4). Unlike Mercury, because hot Jupiters are assumed to be fluid, they should not present many irregularities in the internal structures. Therefore (B−A)/C ≈ 0 and we do not expect hot Jupiters to be captured in spin-orbit resonances. Indeed, determination of second-degree harmonics of the gravity fields of Jupiter and Saturn from Pioneer and Voyager tracking data (*Campbell and Anderson,* 1989) provided a crude estimate of the (B−A)/C value lower than ~10^{-5} for Saturn and ~10^{-7} for Jupiter. The (B−A)/C values for Jupiter and Saturn are more than one order of magnitude smaller than the Moon's or Mercury's value, leading to insignificant chances of capture. Furthermore, the detection of an equatorial asymmetry is questionable. If an equatorial bulge originates from local mass inhomogeneities driven by convection, it is probably not permanent and must have a more negligible effect if averaged spatially and temporally.

The time required for dampening the rotation of the planet depends on the dissipation factor $k_2\Delta t$ (equation (43)). Assuming that hot Jupiters are similar to the solar system giant planets, we can adopt $k_2 = 0.4$, and a range for Q from 10^4 to 10^5 (Table 3). The Q factor and the time lag Δt can be relayed using expressions (22) and (34):

$$Q^{-1} \approx \sigma\Delta t \qquad (94)$$

Since we are using a viscous model, for which Δt is made constant, Q will be modified across the evolution as Q is inversely proportional to the tidal frequency σ. The Q factor for the solar system gaseous planets is measured for their present rotation states, which correspond to less than 1 d (Table 3). We may then assume that exoplanets should present identical Q values when they were rotating as rapidly as Jupiter, i.e., $Q_0^{-1} = \omega_0 \Delta t$. For $Q_0 = 10^4$ and $\omega_0 = 2\pi/10$ h, we compute a constant $\Delta t \approx 0.57$ s.

In Fig. 12 we have plotted all known exoplanets, taken from *The Extrasolar Planets Encyclopedia* (http://exoplanet.eu/), that could have been tidally evolved. We consider that exoplanets are fully evolved if their rotation rate, starting with an initial period of 10 h, is dampened to a value such that $|\omega/n - f_2(e)/f_1(e)| < 0.01$. The curves represent the planets that are tidally evolved in a given time interval ranging from 0.001 G.y. to 10 G.y. Figure 12 allows us to check whether the planet should be fully evolved. For a solar-type star, we can expect that all exoplanets that are above the 1 G.y. curve have already reached the equilibrium rotation ω_e. On the other hand, exoplanets that are below the 10 G.y. curve are probably not yet fully tidally evolved. As expected, all planets in circular orbits with a < 0.05 AU are tidally evolved. However, we are more interested in exoplanets further from the star with nonzero eccentricity that are tidally evolved, since the rotation period is not synchronous, but given by expression (44). For instance, for the planet around HD 80606, the orbital period is 111.7 d, but since e = 0.92 we predict a rotation period of about 1.9 d.

3.2.2. Cassini states. Until now we have assumed that the final obliquity of the planet is 0°. However, *Winn and Holman* (2005) suggested that high-obliquity values could be maintained if the planet has been trapped in a Cassini state resonance (equation (82)) since the early despinning process. For small-amplitude variations of the eccentricity

TABLE 3. Constants for the solar system outer planets.

Quantity	Jupiter	Saturn	Uranus	Neptune
P_0(h)	9.92	10.66	17.24	16.11
ρ(g/cm^3)	1.33	0.69	1.32	1.64
C/mR^2	0.25	0.21	0.23	0.24
k_2	0.49	0.32	0.36	0.41
Q (×10^4)	~3	~2	1~3	1~30

Data from *Yoder* (1995); *Veeder et al.* (1994); *Dermott et al.* (1988); *Tittemore and Wisdom* (1990); *Banfield and Murray* (1992).

and inclination, the equilibrium positions for the obliquity are given by expression (83). Unless $\alpha = |v_k|$, state 1 is close to 0°, and state 3 is close to 180°. We thus focus only on state 2, which may maintain a significant obliquity.

To test the possibility of capture in the high-obliquity Cassini state 2, we can consider a simple scenario where a hot Jupiter forms at a large orbital distance (approximately several AU) and migrates inward to the current position (~0.05 AU). Before the planet reaches typically ~0.5 AU, tidal effects do not affect the spin evolution, but the reduction in the semimajor axis increases the precession constant (equation (12)), so that the precession frequency $\dot\psi$ may become resonant with some orbital frequencies v_k (equation (80)). The passage through resonance generally causes the obliquity to change (*Ward, 1975; Ward and Hamilton, 2004; Hamilton and Ward, 2004; Boué et al., 2009*), raising the possibility that the obliquity has a somewhat arbitrary value when the semimajor axis attains ~0.5 AU.

Tidal effects become efficient for a < 0.5 AU and drive the obliquity to an equilibrium value $\cos\varepsilon \approx 2n(1 + 6e^2)/\omega$ (equation (40)). For initial fast rotation rates ($\omega \gg n$), the equilibrium obliquity tends to 90°. As the rotation rate is decreased by tides, the equilibrium obliquity is reduced to 0° (equation (45)). It is then possible that the obliquity crosses several resonances (one for each frequency v_k) in both ways (increasing and decreasing obliquity), and that a capture occurs. Inside the resonance island, the restoring torque causes the obliquity to librate with amplitude (*Correia and Laskar, 2003b*)

$$\cos\varepsilon_2 \pm \Delta\cos\varepsilon_2 \approx -\frac{v_k}{\alpha} \pm 2\sqrt{\frac{J_k}{\alpha}}\sqrt{1-\frac{v_k^2}{\alpha^2}} \quad (95)$$

Using a linear approximation of the tidal torque (equation (40)) around the resonant obliquity ε_2, the probability of capture in the Cassini state 2 can be estimated from the analytical approach for spin-orbit resonances (equation (73)), with (*Levrard et al., 2007a*)

$$\frac{\Delta\omega}{\pi V n} = \left[\frac{\left(1-3\cos^2\varepsilon_2\right)\frac{\omega}{n} + 2\cos\varepsilon_2}{\pi\sin^2\varepsilon_2\left(2-\cos\varepsilon_2\,\omega/n\right)}\right]\Delta\cos\varepsilon_2 \quad (96)$$

In Fig. 13, we plotted the capture probabilities at 0.05, 0.1, and 0.5 AU as a function of the rotation period for different amplitudes (J_k) and frequencies (v_k) characteristic of the solar system (*Laskar and Robutel, 1993*). As a reasonable example, we choose $v_k = -10''/\text{yr}$, but the results are not affected by changes on this value. Assuming an initial rotation period of 12 h, the capture is possible and even unavoidable if $J_k > 0.1''/\text{yr}$ at 0.5 AU. On the contrary, the chances of capture at 0.05 AU are negligible (<1%) because a decrease in the semimajor axis leads to an increase in the precession constant and reduces the width of the resonance (equation (95)). Theoretical estimations can be compared with numerical simulations (*Levrard et al., 2007a*). To that purpose, the spin equations (equations (79) and (80)) were integrated in the presence of tidal effects (equations (39) and (40)) considering 1000 initial precession angles equally distributed over $0-2\pi$ for each initial obliquity. Statistics of capture were found to be in good agreement with previous theoretical estimates.

To test the influence of migration on the capture stability, additional numerical simulations were performed for various initial obliquities and secular perturbations over typically ~5 × 10⁷ yr. The migration process was simulated by exponentially decreasing the semimajor axis toward 0.05 AU with a 10⁵–10⁷ yr timescale. The obliquity librations were found to be significantly shorter than spin-down and migration timescales so that the spin trajectory follows an "adiabatic invariant" in the phase space. Nevertheless, expression (40) indicates that the tidal torque dramatically increases with both spin-down and inward migration processes ($d\varepsilon/dt \propto a^{-15/2}\omega^{-1}$). If the tidal torque exceeds the maximum possible restoring torque (equation (79)), the resonant equilibrium is destroyed (the evolution is no longer adiabatic). For a given semimajor axis, the stability condition requires then that the rotation rate must always be larger than a threshold value ω_{crit}, which is always verified if $\omega_{\text{crit}} < f_2(e)/f_1(e)$. The stability condition can be simply written as (*Levrard et al., 2007a*)

$$\tan(\varepsilon) < J_k \times \tau_{\text{eq}} \quad (97)$$

where τ_{eq} is the timescale of tidal despinning (equation (93)). It then follows that the final obliquity of the planet cannot be too large, otherwise the planet would quit the resonance.

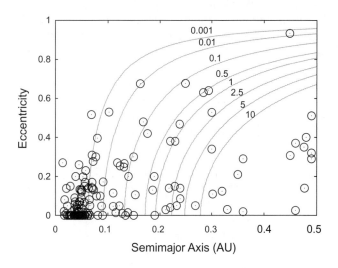

Fig. 12. Tidally evolved exospins with $Q_0 = 10^4$ and initial rotation period $P_0 = 10$ h. The labeled curves denote (in G.y.) the time needed by the rotation to reach the equilibrium (timescales are linearly proportional to Q_0). We assumed Jupiter's geophysical parameters for all planets (Table 3) (updated from *Laskar and Correia, 2004*).

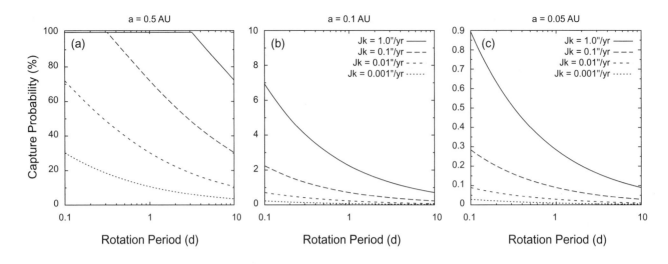

Fig. 13. Obliquity capture probabilities in resonance under the effect of gravitational tides, as a function of the rotation period at (a) 0.5 AU, (b) 0.1 AU, (c) 0.05 AU. J_k is the amplitude of the secular orbital perturbations and $\nu_k = -10.0''/\text{yr}$ (*Levrard et al.*, 2007a).

For instance, taking $J_k = 1''/\text{yr}$ at 0.05 AU (the highest value in Fig. 13), we need an obliquity $\varepsilon < 21°$. For the more realistic amplitude $J_k = 0.1''/\text{yr}$, the resonant obliquity drops to $\varepsilon < 2°$. Such a low resonant obliquity at 0.05 AU is highly unlikely because, according to expression (83), the resonant state requires very high values of the orbital secular frequencies ($|\nu_k| > 7.2 \times 10^{6}{''}/\text{yr}$). At 0.5 and 0.1 AU, critical obliquity values are respectively 83° and 41° and require more reasonable orbital secular frequencies so that a stable capture is possible. In numerical simulations, the stability criteria for the final obliquity (equation (97)) is empirically retrieved with excellent agreement (*Levrard et al.*, 2007a). When the obliquity leaves the resonance, the obliquity ultimately rapidly switches to the resonant stable Cassini state 1, which tends to 0° (equation (83)). We then conclude that locking a hot Jupiter in an oblique Cassini state seems to be a very unlikely scenario.

3.2.3. Energy balance. Tidal energy is dissipated in the planet at the expense of the rotational and orbital energy so that $\dot{E} = -C\omega\dot{\omega} - \dot{a}(Gm_\star m)/(2a^2)$. Replacing the equilibrium rotation given by equation (44) in expression (46) we obtain for the tidal energy

$$\dot{E} = Kn^2 \left[f_3(e) - \frac{f_2^2(e)}{f_1(e)} \frac{2x^2}{1+x^2} \right] \quad (98)$$

or, at second order in eccentricity

$$\dot{E} = \frac{Kn^2}{1+\cos^2\varepsilon} \left[\sin^2\varepsilon + e^2\left(7 + 16\sin^2\varepsilon\right) \right] \quad (99)$$

which is always larger than in the synchronous case (e.g., *Wisdom*, 2004). In Fig. 14 the rate of tidal heating within a nonsynchronous and synchronous planet as a function of the eccentricity ($0 < e < 0.25$) is compared for two different obliquities (*Levrard et al.*, 2007a). The ratio between the tidal heating in the two situations is an increasing function of both eccentricity and obliquity. For $e \approx 0$, as observed for hot Jupiters, the ratio may reach ~1.3 and 2.0 at 45° and 90° obliquity respectively, not being significantly modified at larger eccentricity.

We then conclude that planets in eccentric orbits and/or with high obliquity dissipate more energy than planets in synchronous circular orbits. This may explain why some planets appear to be more inflated than initially expected (e.g., *Knutson et al.*, 2007). A correct tidal energy balance must then take into account the present spin and eccentricity of the orbit.

3.2.4. Orbital circularization. In section 2.2.4 we saw that under tidal friction the spin of the planet attains an equilibrium position faster than the orbit. As a consequence, we can use the expression of the equilibrium rotation rate (equation (44)) in the semimajor axis and eccentricity variations and find simplified expressions (equations (51) and (52)). Combining the two equations we get

$$\frac{da}{a} = \frac{2e\,de}{(1-e^2)} \quad (100)$$

whose solution is given by

$$a = a_f\left(1-e^2\right)^{-1} \quad (101)$$

Replacing the above relation in expression (52) we find a differential equation that rules the eccentricity evolution

$$\dot{e} = -K_0 f_6(e)\left(1-e^2\right)^9 e \quad (102)$$

where K_0 is a constant parameter

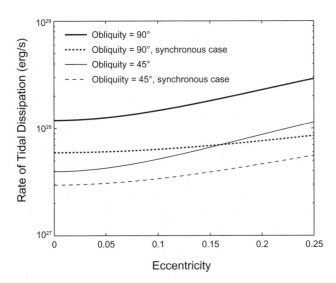
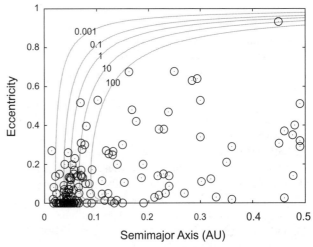

Fig. 14. Rate of tidal dissipation within HD 209458b as a function of the eccentricity for 45° (solid thin line) and 90° (solid thick line) obliquity. The synchronous case is plotted with dashed lines for comparison. The dissipation factor Q_0/k_2 is set to 10^6 (*Levrard et al.*, 2007a).

Fig. 15. Tidally evolved orbits of exoplanets with $Q_0 = 10^4$. The labeled curves denote (in G.y.) the time needed to circularize the orbits of the planets (e < 0.01) (timescales are linearly proportional to Q_0). We assumed Jupiter's geophysical parameters for all planets (Table 3).

$$K_0 = \Delta t \frac{21 k_2 G m_\star^2 R^5}{2 m a_f^8} \quad (103)$$

The solution of the above equation is given by

$$F(e) = F(e_0) e^{-K_0 t} \quad (104)$$

where F(e) is an implicit function of e, which converges to zero as t → +∞. For small eccentricities, we can neglect terms in e^4 and $F(e) = e|7-9e^2|^{-1/2}$. The characteristic timescale for fully dampening the eccentricity of the orbit is then $\tau_{orb} \sim 1/K_0$, and the ratio between the spin and orbital timescales

$$\frac{\tau_{eq}}{\tau_{orb}} \sim \left(\frac{R}{a_f}\right)^2 \quad (105)$$

Since $a_f = a(1-e^2)$, for initial very eccentric orbits the two timescales become comparable.

In Fig. 15 we have plotted all known exoplanets, taken from *The Extrasolar Planets Encyclopedia* (*http://exoplanet.eu/*), whose orbits could have been tidally evolved. We consider that they are fully evolved if the eccentricity is dampened to a value e < 0.01. The curves represent the time needed to damp the eccentricity starting with the present orbital parameters, for time intervals ranging from 0.001 G.y. to 100 G.y. Figure 15 allows us to simultaneously check whether the planet is tidally evolved, and the time needed to fully damp the present eccentricity. All planets in eccentric orbits experience stronger tidal effects because the planet is close to the star at the periapse. As a consequence, tidal friction can, under certain conditions, be an important mechanism for the formation of hot Jupiters (see section 3.2.5).

According to Fig. 15, a significant fraction of exoplanets that are close to the host star (a < 0.1 AU) still present eccentricities up to 0.4, although tidal effects should have already damped the eccentricity to zero. Observational errors and/or weaker tidal dissipation ($Q_0 \gg 10^4$) can be a possible explanation, but they will hardly justify all the observed situations. A more plausible explanation is that the eccentricity of the exoplanet is being excited by gravitational perturbations from an outer planetary companion (section 2.5). Indeed, the eccentricity of a inner short-period planet can be excited as long as its (nonresonant) outer companion's eccentricity is nonzero. *Mardling* (2007) has shown that the eccentricity of the outer planet will decay on a timescale that depends on the structure of the inner planet, and that the eccentricities of both planets are damped at the same rate, controlled by the outer planet (Fig. 16). The mechanism is so efficient that the outer planet may be an Earth-mass planet in the "habitable zones" of some stars. As a consequence, the evolution timescale for both eccentricities can be as long as gigayears instead of millions of years, which could explain the current observations of nonzero eccentricity for some hot Jupiters.

3.2.5. Kozai migration. In current theories of planetary formation, the region within 0.1 AU of a protostar is too hot and rarefied for a Jupiter-mass planet to form, so hot Jupiters likely form further away and then migrate inward. A significant fraction of hot Jupiters have been found in systems of binary stars (e.g., *Eggenberger et al.*, 2004), suggesting that the stellar companion may play an important role in the shrinkage of the planetary orbits. In addition, close binary star systems (separation comparable to the stellar radius) are also often accompanied by a third star. For instance, *Tokovinin et al.* (2006) found that 96% of a sample of spectroscopic binaries with periods less than 3 d have a tertiary component. Indeed, in some circumstances the distant companion enhances tidal interactions in the inner binary, causing the binary orbital period to shrink to the currently observed values. Three-body systems can be stable for long timescales provided that the system is hierarchical, i.e., if the system is formed by an inner binary (star and planet) in a nearly Keplerian orbit with a semimajor axis a, and a outer star also in a nearly Keplerian orbit about the center of mass of the inner system with semimajor a′ ≫ a. An additional requirement is that the eccentricity e′ of the outer orbit is not too large, in order to prevent close encounters with the inner system. In this situation, perturbations on the inner planetary orbit are weak, but can have important long-term effects (see chapter by Fabrycky).

The most striking effect is known as the Lidov-Kozai mechanism (*Kozai*, 1962; *Lidov*, 1962), which allows the inner orbit to periodically exchange eccentricity with inclination. Even at large distances (a′ >1000 AU), the outer star can significantly perturb the planetary orbit as long as the two orbital planes are initially inclined to each other more than I > 39.2° (i.e., for cos I < (3/5)$^{1/2}$). When I < 39.2° there is little variation in the planet's inclination and eccentricity. Secular effects of the Lidov-Kozai type can then produce large cyclic variations in the planet's eccentricity e as a result of angular momentum exchange with the companion orbit. Since the z component of the planet's angular momentum must be conserved and a is not modified by the secular perturbations, the Kozai integral

$$L_K = \left(1-e^2\right)^{1/2} \cos I \qquad (106)$$

is conserved during the oscillations (*Lidov and Ziglin*, 1976). Maxima in e occur with minima in I, and vice versa.

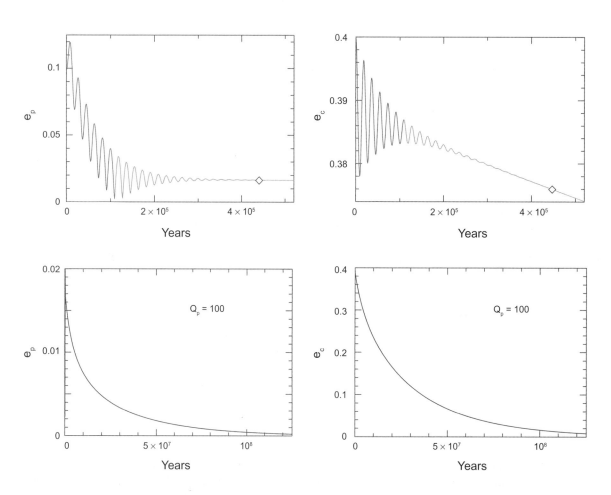

Fig. 16. Tidal evolution of the eccentricities of planet HD 209458b (e$_p$) perturbed by a 0.1 M$_{Jup}$ companion at 0.4 AU (e$_c$). The dissipation factor Q$_p$ = 100 of the observed planet is set artificially low in order to illustrate the damping process (timescales are linearly proportional to Q$_p$). The top figures show the first stages of the evolution. The change in grayscale shows the transition of the eccentricity from the circulation phase to the libration phase. The diamond represents the moment where the eccentricity librations are damped. Bottom figures show the final evolution of the eccentricities. The result of the presence of a companion is that e$_p$ decays at the same rate as e$_c$, while the dissipation rate for e$_c$ is controlled by Q$_p$ and not by Q$_c$ (*Mardling*, 2007).

If the inner orbit is initially circular, the maximum eccentricity achieved in a Kozai cycle is $e_{max} = (1-(5/3)\cos^2 I)^{1/2}$ and the oscillation period of a cycle is approximately P'^2/P (*Kiseleva et al.*, 1998). The maximum eccentricity of the inner orbit in the Kozai cycle will remain fixed for different masses and distances of the outer star, but the period of the Kozai cycle will grow with a'^3. Kozai cycles persist as long as the perturbation from the outer star is the dominant cause of periapse precession in the planetary orbit. However, small additional sources of periapse precession such as the quadrupole moments, additional companions, general relativity, or even tides can compensate the Kozai precession and suppress the eccentricity/inclination oscillations (e.g., *Migaszewski and Góździewski*, 2009).

Because the Lidov-Kozai mechanism is able to induce large eccentricity excitations, a planet in an initial nearly circular orbit (for instance, a Jupiter-like planet at 5 AU around a Sun-like star) can experiment close approaches to the host star at the periapse when the eccentricity increases to very high values. As a consequence, tidal effects increase by several orders of magnitude, and according to expression (51) the semimajor axis of the orbit will decrease and the planet migrate inward. At some point of the evolution, the periapse precession will be dominated by other effects and the eccentricity oscillations suppressed. From that moment on, the eccentricity is damped according to expression (52) and the final semimajor axis given by $a_f = a(1-e^2)$. *Ford and Rasio* (2006) have derived that tidal evolution of high eccentric orbits would end at a semimajor axis a_f equal to about twice the Roche limit R_L. Indeed, at the closest periapse distance, attained for $e \approx 1$, we will have $a(1-e) = R_L$, and thus $a_f = (1 + e)R_L \approx 2R_L$.

In Fig. 17 we plot an example of combined Kozai-tidal migration of the planet HD 80606b. The planet is initially set in an orbit with a = 5 AU, e = 0.1, and I = 85.6°. The stellar companion is supposed to be a Sun-like star at a′ = 1000 AU, and e′ = 0.5 (*Wu and Murray*, 2003; *Fabrycky and Tremaine*, 2007). Prominent eccentricity oscillations are seen from the very beginning and the energy in the planet's spin is transferred to the orbit, increasing the semimajor axis for the first 0.1 G.y. (equation (39)). As the equilibrium rotation is achieved (equation (44)) the orbital evolution is essentially controlled by equations (51) and (52), whose contributions are enhanced when the eccentricity reaches high values. The semimajor axis evolution is executed by apparent "discontinuous" transitions precisely because the tidal dissipation is only efficient during periods of high eccentricity. As dissipation shrinks the semimajor axis, periapse precession becomes gradually dominated by relativity rather than by the third body, and the periapse starts circulating as the eccentricity passes close to 0 at 0.7 G.y. Tidal evolution stops when the orbit is completely circularized. The present semimajor axis and eccentricity of planet HD 80606b are a = 0.45 AU and e = 0.92, respectively, meaning that the tidal evolution on HD 80606b is still underway (Fig. 15). The final semimajor axis is estimated to about $a_f = 0.07$ AU, which corresponds to a regular hot Jupiter.

3.3. Super Earths

After a significant number of discoveries of gaseous giant exoplanets, a new barrier has been passed with the detections of several exoplanets in the Neptune and even Earth-mass ($M_\oplus$) regime: 2–12 $M_\oplus$ (*Rivera et al.*, 2005; *Lovis et al.*, 2006; *Udry et al.*, 2007; *Bonfils et al.*, 2007); these exoplanets are commonly designated as "super Earths." If the commonly accepted core-accretion model can account for the formation of super Earths, resulting in a mainly icy/rocky composition, the fraction of the residual He-H$_2$ atmospheric envelope accreted during the planet migration is not tightly constrained for planets more massive than Earth (e.g., *Alibert et al.*, 2006). A minimum mass of below 10 $M_\oplus$ is usually considered to be the boundary between terrestrial and giant planets, but *Rafikov* (2006) found that planets more massive than 6 $M_\oplus$ could have retained more than 1 $M_\oplus$ of the He-H$_2$ gaseous envelope. For comparison, masses of Earth's and Venus' atmosphere are respectively $\sim 10^{-6}$ and 10^{-4} times the planet's mass. Despite significant uncertainties, the discoveries of super Earths provide an opportunity to test some properties that could be similar to those of the more familiar terrestrial planets of the solar system.

Because some of the super Earths are potentially in the "habitable zone" (*Udry et al.*, 2007; *Selsis et al.*, 2007), the present spin state is an important factor to constrain the climates. As for Venus, thermal atmospheric tides may have a profound influence on the spin of super Earths. However, the small eccentricity approximation calculated for Venus (equation (87)) may no longer be adequate for super Earths, which exhibit a wide range of eccentricities, orbital distances, or central star types. Although our knowledge of super Earths is restricted to their orbital parameters and minimum masses, we can attempt to place new constraints on the surface rotation rate, assuming that super Earths have a dense atmosphere.

As for Venus, the combined effect of tides is to set the final obliquity at 0° or 180° (*Correia et al.*, 2003). Adopting a viscous dissipation model for tidal effects (equation (37)) and the "heating at the ground" model (*Dobrovolskis and Ingersoll*, 1980) for surface pressure variations (equation (68)), the average evolution of the rotation rate is then obtained by adding the effects of both tidal torques acting on the planet. From expressions (23) and (60) we get for $\varepsilon = 0°$ and to the second order in the eccentricity

$$\frac{\dot{\omega}}{\tau_{eq}^{-1}} = \omega - (1+6e^2)n - \omega_s \left[\left(1 - \tfrac{21}{2}e^2\right) \text{sign}(\omega - n) \right. \\ \left. - e^2 \text{sign}(2\omega - n) + 9e^2 \text{sign}(2\omega - 3n) \right] \quad (107)$$

where

$$\omega_s = \frac{F_s}{16 H_0 k_2} \frac{K_a \Delta t_a}{K_g \Delta t_g} \propto \frac{L_\star}{m_\star} \frac{R}{m} a \quad (108)$$

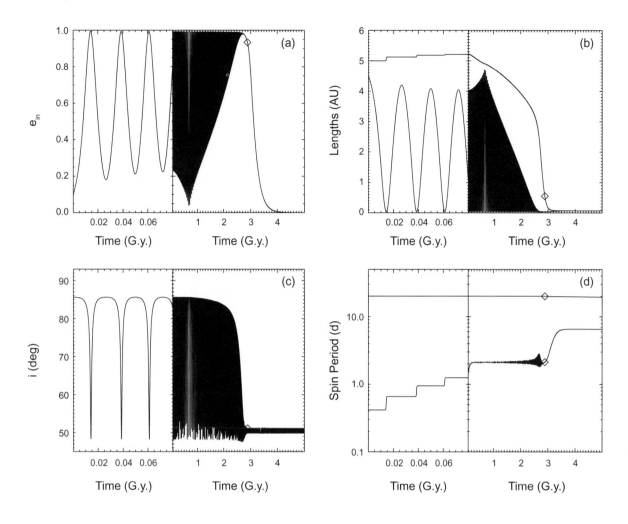

Fig. 17. Possible evolution of the planet HD 80606b initially in an orbit with a = 5 AU, e = 0.1, and I = 85.6°. The stellar companion is supposed to be a Sun-like star at a' = 1000 AU, and e' = 0.5. The diamonds mark the current position of HD 80606b along this possible evolution (*Wu and Murray*, 2003; *Fabrycky and Tremaine*, 2007).

When e = 0, we saw in the case of Venus that final positions of the rotation rate at zero obliquity are given by (equation (90))

$$|\omega - n| = \omega_s \quad (109)$$

i.e., there are two final possibilities for the equilibrium rotation of the planet, given by $\omega^\pm = n \pm \omega_s$. When $e \neq 0$, the expression (109) above is no longer valid and additional equilibrium positions for the rotation rate may occur. For moderate values of the eccentricity, from expression (107) we have that the effect of the eccentricity is to eventually split each previous equilibrium rotation rate into two new equilibrium values. Thus, four final equilibrium positions for the rotation rate are possible (eight if we consider the case $\varepsilon = 180°$), obtained with $\dot\omega = 0$ (Fig. 18)

$$\omega_{1,2}^\pm = n \pm \omega_s + e^2 \delta_{1,2}^\pm \quad (110)$$

with

$$\delta_1^- = 6n + \frac{1}{2}\omega_s, \quad \delta_1^+ = 6n - \frac{41}{2}\omega_s \quad (111)$$

and

$$\delta_2^- = 6n + \frac{5}{2}\omega_s, \quad \delta_2^+ = 6n - \frac{5}{2}\omega_s \quad (112)$$

Because the set of $\omega_{1,2}^\pm$ values must verify the additional condition

$$\omega_2^- < n/2 < \omega_1^- < n < \omega_1^+ < 3n/2 < \omega_2^+ \quad (113)$$

the four equilibrium rotation states cannot, in general, exist simultaneously, depending on the values of ω_s and e. In particular, the final states ω_1^- and ω_1^+ can never coexist with ω_2^-. At most, three different equilibrium states are therefore possible, obtained when ω_s/n is close to 1/2, or more precisely, when $1/2 - 19\,e^2/4 < \omega_s/n < 1/2 + 17\,e^2/4$. Conversely, we find that one single final state $\omega_1^+ = (1 + 6e^2)\,n + (1 - 41e^2/2)\,\omega_s$ exists when $\omega_s/n < 6e^2(1 + e^2/2)$.

Correia and Laskar: Tidal Evolution of Exoplanets   261

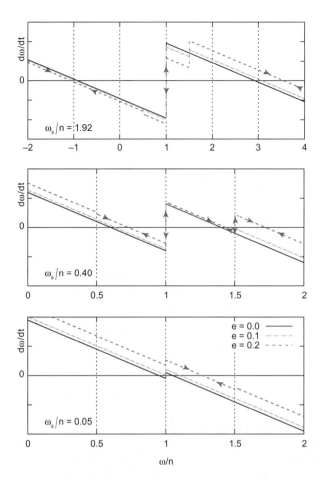

Fig. 18. Evolution of $\dot{\omega}$ (equation (107)) with ω/n for different atmospheric strengths ($\omega_s/n = 1.92, 0.40, 0.05$) and eccentricities ($e = 0.0, 0.1, 0.2$). The top picture with $e = 0$ is the same as Fig. 8 for Venus. The equilibrium rotation rates are given by $\dot{\omega} = 0$ and the arrows indicate whether the equilibrium position is stable or unstable. For $\omega_s/n > 1$, we have two equilibrium possibilities, $\omega_2^{\pm}$, one of which corresponds to a retrograde rotation (as for Venus). For $\omega_s/n < 1$, retrograde states are not possible, but we can still observe final rotation rates $\omega^- < n$. For eccentric orbits, because of the harmonics in $\sigma = 2\omega - n$ and $\sigma = 2\omega - 3n$, we may have at most four different final possibilities (equation (110)). When ω_s/n becomes extremely small, which is the case for the present observed exoplanets with some eccentricity (Table 4), a single final equilibrium is possible for ω_1^+ (*Correia et al.*, 2008).

Earth and Venus are the only planets that can be included in the category of super Earths for which the atmosphere and spin are known. Only Venus is tidally evolved and therefore suitable for applying the above expressions for tidal equilibrium. We can nevertheless investigate the final equilibrium rotation states of the already detected super Earths. For that purpose, we considered only exoplanets with masses smaller than 12 $M_{\oplus}$ that we classified as rocky planets with a dense atmosphere, although we stress that this mass boundary is quite arbitrary.

Using the empirical mass-luminosity relation $L_{\star} \propto m_{\star}^4$ (e.g., *Cester et al.*, 1983) and the mass-radius relationship for terrestrial planets $R \propto m^{0.274}$ (*Sotin et al.*, 2007), expression (108) can be written as

$$\omega_s/n = \kappa (am_{\star})^{2.5} m^{-0.726} \qquad (114)$$

where κ is a proportionality coefficient that contains all the constant parameters, but also the parameters that we are unable to constrain such as H_0, k_2, Δt_g, or Δt_a. In this context, as a first-order approximation we consider that for all super Earths the parameter κ has the same value as for Venus. Assuming that the rotation of Venus is presently stabilized in the ω^- final state, i.e., $2\pi/\omega^- = -243$ d (*Carpenter*, 1970), we compute $2\pi/\omega_s = 116.7$ d. Replacing the present rotation in expression (114), we find for Venus that $\kappa = 3.723 \, M_{\oplus}^{0.726} \, M_{\odot}^{-2.5} \, AU^{-2.5}$. We can then estimate the ratio ω_s/n for all considered super Earths in order to derive their respective equilibrium rotation rates (Table 4).

The number and values of the allowed equilibrium rotation states are plotted as a function of $am_{\star}$ for different eccentricities in Fig. 19. All eccentric planets have a ratio ω_s/n that is lower than 6×10^{-3} (Table 4), which verifies the condition $\omega_s/n < 6e^2(1 + e^2/2)$. As a consequence, only one single final state exists, $\omega_1^+/n \approx (1 + 6e^2)$, corresponding to the equilibrium rotation resulting from gravitational tides (equation (44)). The main reason is that the effect of atmospheric tides is clearly disfavored relative to the effect of gravitational tides on super Earths discovered orbiting M-dwarf stars: The short orbital periods strengthens the effect of gravitational tides, which are proportional to $1/a^6$, while the effect of thermal tides varies as $1/a^5$. Moreover, the small mass of the central star also strongly affects the luminosity received by the planet and hence the size of the atmospheric bulge driven by thermal contrasts.

For the planets with nearly zero eccentricity (GJ 581e, HD 40307b,c,d, and GJ 176b), two equilibrium rotation states $\omega_1^{\pm}$ are possible. However, the two final states $\omega_1^{\pm}$ are so close to the mean motion n, that the quadrupole moment of inertia $(B-A)/C$ will probably capture the rotation of the planet in the synchronous resonance. We then conclude that super Earths orbiting close to their host stars (in particular M dwarfs) will be dominated by gravitational tides and present a final equilibrium rotation rate given by $\omega_e/n \approx f_2(e)/f_1(e)$ (Fig. 4), or present spin-orbit resonances like Mercury.

4. FUTURE PROSPECTS

The classical theory of tides initiated by *Darwin* (1880, 1908) is sufficient to understand the main effects of tidal friction upon planetary evolution. However, the exact mechanism for how tidal energy is dissipated within the internal layers of the planet remains a challenge for planetary scientists. *Kaula* (1964) derived a generalization of Darwin's work, with consideration of higher-order tides and without the adoption of any dissipation model. The tidal potential is

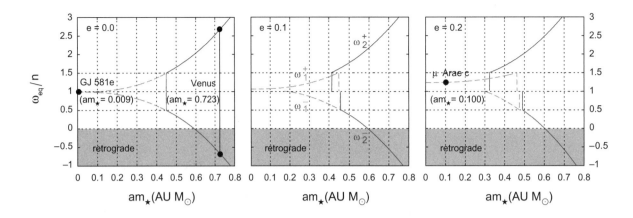

Fig. 19. Equilibrium positions of the rotation rate for super Earths as a function of the product $am_\star$ for three different values of the eccentricity (e = 0.0, 0.1, 0.2). Each curve corresponds to a different final state (dotted lines for $\omega_1^\pm$ and solid lines for $\omega_2^\pm$). For $e \approx 0$ (case of Venus), we always count two final states that are symmetrical about n. For small values of $am_\star$, the two equilibrium possibilities are so close to n that the most likely scenario for the planet is to be captured in the synchronous resonance (case of GJ 581e). As we increase the eccentricity, we can count at most three final equilibrium rotations, depending on the value of ω_s/n (computed from equation (114)). When $e \approx 0.2$, only one equilibrium state exists for $am_\star < 0.3$, resulting from $\omega_s/n < 6e^2(1 + e^2/2)$. This is the present situation of μ Arae c and most of the super Earths listed in Table 4 (*Correia et al.*, 2008).

described using infinite series in eccentricity and inclination, which is not practical and can only be correctly handled by computers. Ever since, many efforts have been made in order to either simplify the tidal equations, or to correctly model the tidal dissipation (for a review see *Ferraz-Mello et al.*, 2008; *Efroimsky and Williams*, 2009). Many solar system phenomena have been successfully explained using the existent tidal models, so we expect that they are suitable to describe the tidal evolution of exoplanets.

Nevertheless, many exoplanets are totally different from the solar system cases, and we cannot exclude the observation of some unexpected behaviors. For instance, it is likely that dissipation within hot Jupiters is closer to dissipation within stars (e.g., *Zahn*, 1975), while dissipation within super Earths is closer to dissipation observed for rocky planets (e.g., *Henning et al.*, 2009). It is then necessary to continue improving tidal models in order to get a more realistic description for each planetary system. In particular, a correct description of the tidal dissipation and how it evolves with the tidal frequency is critical for the evolution timescale.

The orbital architecture of exoplanetary systems is relatively well determined from the present observational techniques. However, the spins of the exoplanets are not easy to measure, as the lightcurve coming from the planet is always dimmed by the light from the star. The continuous improvements that have been made in photometry and spectrography let us believe that the determination of exoplanets' spins can be a true possibility in the near future. In particular, infrared spectrographs are being developed that will allow acquisition of the spectra of the planets if we manage to subtract the stellar contribution (e.g., *Barnes et al.*, 2010).

Some additional methods for detecting the rotation and/or obliquity of exoplanets have also been tested and suggested so far. For instance, indirect sensing of the planetary gravitational quadrupole and shape, which is linked to both spin rate and obliquity (e.g., *Seager and Hui*, 2002; *Ragozzine and Wolf*, 2009), or transient heating of one face of the planet, which then spins into and out of view, has been attempted for the system HD 80606 (*Laughlin et al.*, 2009). The effect of planetary rotation on the transit spectrum of a giant exoplanet is another possibility. During ingress and egress, absorption features arising from the planet's atmosphere are Doppler shifted by a factor on the order of the planet's rotational velocity (~1–2 km s^{-1}) relative to where they would be if the planet were not rotating (e.g., *Spiegel et al.*, 2007). Finally, for planets whose light is spatially separated from the star, variations may be discernible in the lightcurve obtained by low-precision photometry due to meteorological variability, composition of the surface, or spots (e.g., *Ford et al.*, 2001).

Although the spin states of exoplanets cannot be measured, for exoplanets that are tidally evolved we can still try to make predictions for the rotation rates. When the eccentricity is large, the rotation of many of the observed exoplanets can still be tidally evolved even if the planets are not very close to their central stars (Fig. 12). For tidally evolved hot Jupiters, we can conjecture that the rotation periods are the limit values $P_{orb} \times f_1(e)/f_2(e)$ (Fig. 4). It becomes a new challenge for the observers to be able to confirm these predictions.

Thermal atmospheric tides may very well destabilize the tidal equilibrium from gravitational tides and create additional possible stable limit values, with the possibility of retrograde rotations, as for planet Venus (Fig. 8).

TABLE 4. Characteristics and equilibrium rotations of some "super Earths" with masses lower than 12 $M_\oplus$.

Name	Ref.	$m_\star$ ($M_\odot$)	Age (G.y.)	τ_{eq} (G.y.)	m sin i ($M_\oplus$)	a (AU)	e	ω_s/n	$2\pi/n$ (d)	$2\pi/\omega_2^-$ (d)	$2\pi/\omega_1^-$ (d)	$2\pi/\omega_1^+$ (d)	$2\pi/\omega_2^+$ (d)
Venus		1.00	4.5	2.3	0.82	0.723	0.007	1.92	224.7	−243.0			76.8
Earth*		1.00	4.5	16	1.00	1.000	0.017	3.75	365.3	−132.9			77.1
GJ 581e	[1]	0.31	7–11	10^{-7}	1.94	0.03	0	10^{-5}	3.4087		3.4088	3.4087	
HD 40307b	[2]	0.77	—	10^{-7}	4.2	0.047	0	0.0003	4.2413		4.2427	4.240	
GJ 581c	[1]	0.31	7–11	10^{-5}	5.36	0.07	0.17	10^{-4}	12.14			10.6335	
GJ 876d	[3]	0.32	9.9	10^{-8}	6.3	0.021	0.14	10^{-6}	1.9649			1.7822	
HD 40307c	[2]	0.77	—	10^{-5}	6.9	0.081	0	0.0009	9.5956		9.6042	9.5871	
GJ 581d	[1]	0.31	7–11	0.02	7.09	0.22	0.38	0.0011	67.6918			47.8226	
HD 181433b	[4]	0.78	—	10^{-5}	7.5	0.08	0.396	0.0008	9.3579			6.535	
GJ 176b	[5]	0.5	—	10^{-5}	8.4	0.066	0	0.0002	8.7583		8.7596	8.7568	
HD 40307d	[2]	0.77	—	10^{-4}	9.2	0.134	0	0.0025	20.4175		20.4696	20.3656	
HD 7924b	[6]	0.83	—	10^{-6}	9.26	0.057	0.17	0.0004	5.4493			4.7688	
HD 69830b	[7]	0.86	4–10	10^{-5}	10.2	0.079	0.10	0.0008	8.6625			8.1995	
µ Arae c	[8]	1.1	6.41	10^{-5}	10.6	0.091	0.172	0.0021	9.5505			8.13313	
55 Cnc e	[9]	1.03	5.5	10^{-7}	10.8	0.038	0.07	0.0002	2.6659			2.592	
GJ 674b	[10]	0.35	0.1–1	10^{-7}	11.09	0.039	0.2	10^{-5}	4.7549			4.0138	
HD 69830c	[7]	0.86	4–10	10^{-3}	11.8	0.186	0.13	0.0064	31.5943			28.8691	

*Moon tidal effects were not included. τ_{eq} was computed with $k_2 = 1/3$ and $\Delta t_g = 640$ s (Earth's values).
References: [1] *Mayor et al.* (2009a); [2] *Mayor et al.* (2009b); [3] *Correia et al.* (2010); [4] *Bouchy et al.* (2009); [5] *Forveille et al.* (2009); [6] *Howard et al.* (2009); [7] *Lovis et al.* (2006); [8] *Pepe et al.* (2007); [9] *Fischer et al.* (2008); [10] *Bonfils et al.* (2007).

Thermal tides should be particularly important for super Earths, which are expected to have a distinct rocky body surrounded by a dense atmosphere. In a paradoxical way, the final rotation rate of super Earths are the most difficult to predict, as the equilibrium configurations depend on the composition of the atmospheres. Thermal tides are nevertheless more relevant for exoplanets that orbit Sun-like stars at some distance, like Venus (Fig. 19).

We also assumed that the final obliquity of exoplanets is either 0° or 180°, as the two values represent the final outcome of tidal evolution. However, each planetary system has its own architecture, and planetary perturbations on the spin can lead to resonant capture in high oblique Cassini states or even to chaotic motion. Thus, the final spin evolution of a planet cannot be dissociated from its environment, and a more realistic description of the rotation of exoplanets can only be achieved with the full knowledge of the system orbital dynamics.

Acknowledgments. We thank an anonymous referee for valuable suggestions that helped to improve this work. We acknowledge support from the Fundação para a Ciência e a Tecnologia (Portugal) and PNP-CNRS (France).

REFERENCES

Abramowitz M. and Stegun I. A. (1972) *Handbook of Mathematical Functions.* Dover, New York.

Agnor C. B., Canup R. M., and Levison H. F. (1999) On the character and consequences of large impacts in the late stage of terrestrial planet formation. *Icarus, 142,* 219–237.

Alibert Y., Baraffe I., Benz W., Chabrier G., Mordasini C., Lovis C., Mayor M., Pepe F., Bouchy F., Queloz D., and Udry S. (2006) Formation and structure of the three Neptune-mass planets system around HD 69830. *Astron. Astrophys., 455,* L25–L28.

Andoyer H. (1923) *Cours de Mécanique Céleste.* Gauthier-Villars, Paris.

Arras P. and Socrates A. (2010) Thermal tides in short period exoplanets. *ArXiv e-prints,* arXiv:0901.0735v1.

Arras P. and Socrates A. (2010) Thermal tides in fluid extrasolar planets. *Astrophys. J., 714,* 1–12.

Avduevskii V. S., Golovin I. M., Zavelevich F. S., Likhushin V. I., Marov M. I., Melnikov D. A., Merson I. I., Moshkin B. E., Razin K. A., and Chernoshchekov L. I. (1976) Preliminary results of an investigation of the light regime in the atmosphere and on the surface of Venus. *Kosmich. Issled., 14,* 735–742.

Banfield D. and Murray N. (1992) A dynamical history of the inner neptunian satellites. *Icarus, 99,* 390–401.

Barnes J. R., Barman T. S., Jones H. R. A., Barber R. J., Hansen B. M. S., Prato L., Rice E. L., Leigh C. J., Cameron A. C., and Pinfield D. J. (2010) A search for molecules in the atmosphere of HD 189733b. *Mon. Not. R. Astron. Soc., 401,* 445–454.

Bonfils X., Mayor M., Delfosse X., Forveille T., Gillon M., Perrier C., Udry S., Bouchy F., Lovis C., Pepe F., Queloz D., Santos N. C., and Bertaux J.-L. (2007) The HARPS search for southern extra-solar planets. X. A m sin i = 11 $M_\oplus$ planet around the nearby spotted M dwarf GJ 674. *Astron. Astrophys., 474,* 293–299.

Bouchy F., Mayor M., Lovis C., Udry S., Benz W., Bertaux J., Delfosse X., Mordasini C., Pepe F., Queloz D., and Segransan D. (2009) The HARPS search for southern extra-solar planets. XVII. Super-Earth and Neptune-mass planets in multiple planet systems HD 47186 and HD 181433. *Astron. Astrophys., 496,* 527–531.

Boué G. and Laskar J. (2006) Precession of a planet with a satellite. *Icarus, 185,* 312–330.

Boué G. and Laskar J. (2010) A collisionless scenario for Uranus tilting. *Astrophys. J. Lett., 712,* L44–L47.

Boué G., Laskar J., and Kuchynka P. (2009) Speed limit on Neptune migration imposed by Saturn tilting. *Astrophys. J. Lett., 702,* L19–L22.

Brouwer D. and Clemence G. M. (1961) *Methods of Celestial Mechanics.* Academic, New York.

Campbell J. K. and Anderson J. D. (1989) Gravity field of the saturnian system from Pioneer and Voyager tracking data. *Astron. J., 97,* 1485–1495.

Canup R. M. (2005) A giant impact origin of Pluto-Charon. *Science, 307,* 546–550.

Canup R. M. and Asphaug E. (2001) Origin of the Moon in a giant impact near the end of the Earth's formation. *Nature, 412,* 708–712.

Carpenter R. L. (1964) Symposium on radar and radiometric observations of Venus during the 1962 conjunction: Study of Venus by CW radar. *Astron. J., 69,* 2–11.

Carpenter R. L. (1970) A radar determination of the rotation of Venus. *Astron. J., 75,* 61–66.

Cester B., Ferluga S., and Boehm C. (1983) The empirical mass-luminosity relation. *Astrophys. Space Sci., 96,* 125–140.

Chapman S. and Lindzen R. (1970) *Atmospheric Tides: Thermal and Gravitational.* Reidel, Dordrecht.

Chirikov B. V. (1979) A universal instability of many dimensional oscillator systems. *Phys. Rept., 52,* 263–379.

Colombo G. (1965) Rotational period of the planet Mercury. *Nature, 208,* 575–578.

Colombo G. and Shapiro I. I. (1966) The rotation of the planet Mercury. *Astrophys. J., 145,* 296–307.

Correia A. C. M. (2006) The core-mantle friction effect on the secular spin evolution of terrestrial planets. *Earth Planet. Sci. Lett., 252,* 398–412.

Correia A. C. M. (2009) Secular evolution of a satellite by tidal effect: Application to Triton. *Astrophys. J. Lett., 704,* L1–L4.

Correia A. C. M. and Laskar J. (2001) The four final rotation states of Venus. *Nature, 411,* 767–770.

Correia A. C. M. and Laskar J. (2003a) Different tidal torques on a planet with a dense atmosphere and consequences to the spin dynamics. *J. Geophys. Res.–Planets, 108(E11),* 5123-10.

Correia A. C. M. and Laskar J. (2003b) Long-term evolution of the spin of Venus II. Numerical simulations. *Icarus, 163,* 24–45.

Correia A. C. M. and Laskar J. (2004) Mercury's capture into the 3/2 spin-orbit resonance as a result of its chaotic dynamics. *Nature, 429,* 848–850.

Correia A. C. M. and Laskar J. (2006) Evolution of the spin of Mercury and its capture into the 3/2 spin-orbit resonance. In *Past Meets Present in Astronomy and Astrophysics, Proceedings of the 15th Portuguese National Meeting,* pp. 1–4. World Scientific, Singapore.

Correia A. C. M. and Laskar J. (2009) Mercury's capture into the 3/2 spin-orbit resonance including the effect of core-mantle friction. *Icarus, 201,* 1–11.

Correia A. C. M. and Laskar J. (2010) Long-term evolution of the spin of Mercury. I. Effect of the obliquity and core-mantle friction. *Icarus, 205,* 338–355.

Correia A. C. M., Laskar J., and Néron de Surgy O. (2003) Long-term evolution of the spin of Venus I. Theory. *Icarus, 163,* 1–23.

Correia A. C. M., Levrard B., and Laskar J. (2008) On the equilibrium rotation of Earth-like extra-solar planets. *Astron. Astrophys., 488,* L63–L66.

Correia A. C. M., Couetdic J., Laskar J., Bonfils X., Mayor M., Bertaux J., Bouchy F., Delfosse X., Forveille T., Lovis C., Pepe F., Perrier C., Queloz D., and Udry S. (2010) The HARPS search for southern extra-solar planets. XIX. Characterization and dynamics of the GJ 876 planetary system. *Astron. Astrophys., 511,* A21.

Counselman C. C. and Shapiro I. I. (1970) Spin-orbit resonance of Mercury. *Symp. Math., 3,* 121–169.

Darwin G. H. (1880) On the secular change in the elements of a satellite revolving around a tidally distorted planet. *Philos. Trans. R. Soc. London, 171,* 713–891.

Darwin G. H. (1908) *Scientific Papers.* Cambridge Univ., Cambridge.

Dermott S. F., Malhotra R., and Murray C. D. (1988) Dynamics of the uranian and saturnian satellite systems — A chaotic route to melting Miranda? *Icarus, 76,* 295–334.

Dickey J. O., Bender P. L., Faller J. E., Newhall X. X., Ricklefs R. L., Ries J. G., Shelus P. J., Veillet C., Whipple A. L., Wiant J. R., Williams J. G., and Yoder C. F. (1994) Lunar laser ranging — a continuing legacy of the Apollo program. *Science, 265,* 482–490.

Dobrovolskis A. R. (1980) Atmospheric tides and the rotation of Venus. II — Spin evolution. *Icarus, 41,* 18–35.

Dobrovolskis A. R. and Ingersoll A. P. (1980) Atmospheric tides and the rotation of Venus. I — Tidal theory and the balance of torques. *Icarus, 41,* 1–17.

Dones L. and Tremaine S. (1993) On the origin of planetary spins. *Icarus, 103,* 67–92.

Efroimsky M. and Williams J. G. (2009) Tidal torques: A critical review of some techniques. *Cel. Mech. Dyn. Astron., 104,* 257–289.

Eggenberger A., Udry S., and Mayor M. (2004) Statistical properties of exoplanets. III. Planet properties and stellar multiplicity. *Astron. Astrophys., 417,* 353–360.

Escribano B., Vanyo J., Tuval I., Cartwright J. H. E., González D. L., Piro O., and Tél T. (2008) Dynamics of tidal synchronization and orbit circularization of celestial bodies. *Phys. Rev., E78(3),* 036216.

Fabrycky D. and Tremaine S. (2007) Shrinking binary and planetary orbits by Kozai cycles with tidal friction. *Astrophys. J., 669,* 1298–1315.

Ferraz-Mello S., Rodríguez A., and Hussmann H. (2008) Tidal friction in close-in satellites and exoplanets: The Darwin theory re-visited. *Cel. Mech. Dyn. Astron., 101,* 171–201.

Fischer D. A., Marcy G. W., Butler R. P., Vogt S. S., Laughlin G., Henry G. W., Abouav D., Peek K. M. G., Wright J. T., Johnson J. A., McCarthy C., and Isaacson H. (2008) Five planets orbiting 55 Cancri. *Astrophys. J., 675,* 790–801.

Ford E. B. and Rasio F. A. (2006) On the relation between hot Jupiters and the Roche limit. *Astrophys. J. Lett., 638,* L45–L48.

Ford E. B., Seager S., and Turner E. L. (2001) Characterization of extrasolar terrestrial planets from diurnal photometric variability. *Nature, 412,* 885–887.

Forveille T., Bonfils X., Delfosse X., Gillon M., Udry S., Bouchy F., Lovis C., Mayor M., Pepe F., Perrier C., Queloz D., Santos N., and Bertaux J.-L. (2009) The HARPS search for southern extra-solar planets. XIV. Gl 176b, a super-Earth rather than a Neptune, and at a different period. *Astron. Astrophys., 493,* 645–650.

Gold T. and Soter S. (1969) Atmospheric tides and the resonant rotation of Venus. *Icarus, 11,* 356–366.

Goldreich P. and Peale S. (1966) Spin-orbit coupling in the solar system. *Astron. J., 71,* 425–438.

Goldreich P. and Peale S. J. (1968) The dynamics of planetary rotations. *Annu. Rev. Astron. Astrophys., 6,* 287–320.

Goldreich P. and Soter S. (1966) Q in the solar system. *Icarus, 5,* 375–389.

Goldreich P., Murray N., Longaretti P. Y., and Banfield D. (1989) Neptune's story. *Science, 245,* 500–504.

Goldstein H. (1950) *Classical Mechanics.* Addison-Wesley, Reading.

Goldstein R. M. (1964) Symposium on radar and radiometric observations of Venus during the 1962 conjunction: Venus characteristics by Earth-based radar. *Astron. J., 69,* 12–19.

Goodman J. (2010) Concerning thermal tides on hot Jupiters. *ArXiv e-prints,* arXiv:0901.3279v1.

Hamilton D. P. and Ward W. R. (2004) Tilting Saturn. II. Numerical model. *Astron. J., 128,* 2510–2517.

Hays J. D., Imbrie J., and Shackleton N. J. (1976) Variations in the Earth's orbit: Pacemaker of the ice ages. *Science, 194,* 1121–1132.

Henning W. G., O'Connell R. J., and Sasselov D. D. (2009) Tidally heated terrestrial exoplanets: Viscoelastic response models. *Astrophys. J., 707,* 1000–1015.

Henrard J. (1993) The adiabatic invariant in classical dynamics. In *Dynamics Reported,* pp. 117–235. Springer Verlag, New York.

Henrard J. and Murigande C. (1987) Colombo's top. *Cel. Mech., 40,* 345–366.

Hinderer J., Legros H., and Pedotti G. (1987) Atmospheric pressure torque and axial rotation of Venus. *Adv. Space Res., 7,* 311–314.

Howard A. W., Johnson J. A., Marcy G. W., Fischer D. A., Wright J. T., Henry G. W., Giguere M. J., Isaacson H., Valenti J. A., Anderson J., and Piskunov N. E. (2009) The NASA-UC Eta-Earth Program. I. A super-Earth orbiting HD 7924. *Astrophys. J., 696,* 75–83.

Imbrie J. (1982) Astronomical theory of the Pleistocene ice ages — A brief historical review. *Icarus, 50,* 408–422.

Jones H. R. A., Butler R. P., Tinney C. G., Marcy G. W., Carter B. D., Penny A. J., McCarthy C., and Bailey J. (2006) High-eccentricity planets from the Anglo-Australian Planet Search. *Mon. Not. R. Astron. Soc., 369,* 249–256.

Joshi M. M., Haberle R. M., and Reynolds R. T. (1997) Simulations of the atmospheres of synchronously rotating terrestrial planets orbiting M dwarfs: Conditions for atmospheric collapse and the implications for habitability. *Icarus, 129,* 450–465.

Kaula W. M. (1964) Tidal dissipation by solid friction and the resulting orbital evolution. *Rev. Geophys., 2,* 661–685.

Kinoshita H. (1977) Theory of the rotation of the rigid Earth. *Cel. Mech., 15,* 277–326.

Kiseleva L. G., Eggleton P. P., and Mikkola S. (1998) Tidal friction in triple stars. *Mon. Not. R. Astron. Soc., 300,* 292–302.

Knutson H. A., Charbonneau D., Noyes R. W., Brown T. M., and Gilliland R. L. (2007) Using stellar limb-darkening to refine the properties of HD 209458b. *Astrophys. J., 655,* 564–575.

Kokubo E. and Ida S. (2007) Formation of terrestrial planets from protoplanets. II. Statistics of planetary spin. *Astrophys. J., 671,* 2082–2090.

Kozai Y. (1962) Secular perturbations of asteroids with high inclination and eccentricity. *Astron. J., 67,* 591–598.

Lambeck K. (1980) *The Earth's Variable Rotation: Geophysical Causes and Consequences.* Cambridge Univ., Cambridge.

Laskar J. (1989) A numerical experiment on the chaotic behaviour of the solar system. *Nature, 338,* 237– 238.

Laskar J. (1990) The chaotic motion of the solar system — A numerical estimate of the size of the chaotic zones. *Icarus, 88,* 266–291.

Laskar J. (1996) Large scale chaos and marginal stability in the solar system. *Cel. Mech. Dyn. Astron., 64,* 115–162.

Laskar J. and Correia A. C. M. (2004) The rotation of extra-solar planets. In *Extrasolar Planets: Today and Tomorrow* (J.-P. Beaulieu et al., eds.), pp. 401–409. ASP Conf. Ser. 321, Astronomical Society of the Pacific, San Francisco.

Laskar J. and Robutel P. (1993) The chaotic obliquity of the planets. *Nature, 361,* 608–612.

Laskar J., Joutel, F., and Robutel P. (1993) Stabilization of the Earth's obliquity by the Moon. *Nature, 361,* 615–617.

Laskar J., Levrard B., and Mustard J. F. (2002) Orbital forcing of the martian polar layered deposits. *Nature, 419,* 375–377.

Laskar J., Correia A. C. M., Gastineau M., Joutel F., Levrard B., and Robutel P. (2004a) Long term evolution and chaotic diffusion of the insolation quantities of Mars. *Icarus, 170,* 343–364.

Laskar J., Robutel P., Joutel F., Gastineau M., Correia A. C. M., and Levrard B. (2004b) A long-term numerical solution for the insolation quantities of the Earth. *Astron. Astrophys., 428,* 261–285.

Laughlin G., Deming D., Langton J., Kasen D., Vogt S., Butler P., Rivera E., and Meschiari S. (2009) Rapid heating of the atmosphere of an extrasolar planet. *Nature, 457,* 562–564.

Levrard B., Forget F., Montmessin F., and Laskar J. (2004) Recent ice-rich deposits formed at high latitudes on Mars by sublimation of unstable equatorial ice during low obliquity. *Nature, 431,* 1072–1075.

Levrard B., Correia A. C. M., Chabrier G., Baraffe I., Selsis F., and Laskar J. (2007a) Tidal dissipation within hot Jupiters: A new appraisal. *Astron. Astrophys., 462,* L5–L8.

Levrard B., Forget F., Montmessin F., and Laskar J. (2007b) Recent formation and evolution of northern martian polar layered deposits as inferred from a global climate model. *J. Geophys. Res.–Planets, 112(E11),* 6012.

Lidov M. L. (1962) The evolution of orbits of artificial satellites of planets under the action of gravitational perturbations of external bodies. *Planet. Space Sci., 9,* 719–759.

Lidov M. L. and Ziglin S. L. (1976) Non-restricted double-averaged three body problem in Hill's case. *Cel. Mech., 13,* 471–489.

Lovis C., Mayor M., Pepe F., Alibert Y., Benz W., Bouchy F., Correia A. C. M., Laskar J., Mordasini C., Queloz D., Santos N. C., Udry S., Bertaux J.-L., Sivan J.-P., (2006) An extrasolar planetary system with three Neptune-mass planets. *Nature, 441,* 305–309.

MacDonald G. J. F. (1964) Tidal friction. *Rev. Geophys., 2,* 467–541.

Mardling R. A. (2007) Long-term tidal evolution of short-period planets with companions. *Mon. Not. R. Astron. Soc., 382,* 1768–1790.

Mayor M., Bonfils X., Forveille T., Delfosse X., Udry S., Bertaux J., Beust H., Bouchy F., Lovis C., Pepe F., Perrier C., Queloz D., and Santos N. C. (2009a) The HARPS search for southern extra-solar planets. XVIII. An Earth-mass planet in the GJ 581 planetary system. *Astron. Astrophys., 507,* 487–494.

Mayor M., Udry S., Lovis C., Pepe F., Queloz D., Benz W., Bertaux J.-L., Bouchy F., Mordasini C., and Segransan D. (2009b) The HARPS search for southern extra-solar planets. XIII. A planetary system with 3 super-Earths (4.2, 6.9, and 9.2 $M_\oplus$). *Astron. Astrophys., 493,* 639–644.

Migaszewski C. and Goździewski K. (2009) Secular dynamics of a coplanar, non-resonant planetary system under the general relativity and quadrupole moment perturbations. *Mon. Not. R. Astron. Soc., 392,* 2–18.

Mignard F. (1979) The evolution of the lunar orbit revisited. I. *Moon Planets, 20,* 301–315.

Mignard F. (1980) The evolution of the lunar orbit revisited. II. *Moon Planets, 23,* 185–201.

Mignard F. (1981) Evolution of the martian satellites. *Mon. Not. R. Astron. Soc., 194,* 365–379.

Munk W. H. and MacDonald G. J. F. (1960) *The Rotation of the Earth; A Geophysical Discussion.* Cambridge Univ., Cambridge.

Murray C. D. and Dermott S. F. (1999) *Solar System Dynamics.* Cambridge Univ., Cambridge.

Naef D., Latham D. W., Mayor M., Mazeh T., Beuzit J. L., Drukier G. A., Perrier-Bellet C., Queloz D., Sivan J. P., Torres G., Udry S., and Zucker S. (2001) HD 80606b, a planet on an extremely elongated orbit. *Astron. Astrophys., 375,* L27–L30.

Néron de Surgy O. and Laskar J. (1997) On the long term evolution of the spin of the Earth. *Astron. Astrophys., 318,* 975–989.

Pepe F., Correia A. C. M., Mayor M., Tamuz O., Couetdic J., Benz W., Bertaux J.-L., Bouchy F., Laskar J., Lovis C., Naef D., Queloz D., Santos N. C., Sivan J.-P., Sosnowska D., and Udry S. (2007) The HARPS search for southern extra-solar planets. VIII. μ Arae, a system with four planets. *Astron. Astrophys., 462,* 769–776.

Pettengill G. H. and Dyce R. B. (1965) A radar determination of the rotation of the planet Mercury. *Nature, 206,* 1240–1241.

Rafikov R. R. (2006) Atmospheres of protoplanetary cores: Critical mass for nucleated instability. *Astrophys. J., 648,* 666–682.

Ragozzine D. and Wolf A. S. (2009) Probing the interiors of very hot Jupiters using transit light curves. *Astrophys. J., 698,* 1778–1794.

Rivera E. J., Lissauer J. J., Butler R. P., Marcy G. W., Vogt S. S., Fischer D. A., Brown T. M., Laughlin G., and Henry G. W. (2005) A ~ 7.5 $M_\oplus$ planet orbiting the nearby star, GJ 876. *Astrophys. J., 634,* 625–640.

Santos N. C., Benz W., and Mayor M. (2005) Extrasolar planets: Constraints for planet formation models. *Science, 310,* 251–255.

Schiaparelli G. V. (1889) Sulla rotazione di Mercurio. *Astronom. Nach., 123,* 241–250.

Schlichting H. E. and Sari R. (2007) The effect of semicollisional accretion on planetary spins. *Astrophys. J., 658,* 593–597.

Seager S. and Hui L. (2002) Constraining the rotation rate of transiting extrasolar planets by oblateness measurements. *Astrophys. J., 574,* 1004–1010.

Selsis F., Kasting J., Levrard B., Paillet J., Ribas I., and Delfosse X. (2007) Habitable planets around the star Gliese 581? *Astron. Astrophys., 476,* 1373–1387.

Siebert M. (1961) Atmospheric tides. *Adv. Geophys., 7,* 105–187.

Smart W. M. (1953) *Celestial Mechanics.* London, New York.

Smith W. B. (1963) Radar observations of Venus, 1961 and 1959. *Astron. J., 68,* 15–21.

Sotin C., Grasset A., and Mocquet A. (2007) Mass-radius curve for extrasolar Earth-like planets and ocean planets. *Icarus, 191,* 337–351.

Spiegel D. S., Haiman Z., and Gaudi B. S. (2007) On constraining a transiting exoplanet's rotation rate with its transit spectrum. *Astrophys. J., 669,* 1324–1335.

Takata T. and Stevenson D. J. (1996) Despin mechanism for protogiant planets and ionization state of protogiant planetary disks. *Icarus, 123,* 404–421.

Tamuz O., Ségransan D., Udry S., Mayor M., Eggenberger A., Naef D., Pepe F., Queloz D., Santos N. C., Demory B., Figuera P., Marmier M., and Montagnier G. (2008) The CORALIE survey for southern extrasolar planets. XV. Discovery of two eccentric planets orbiting HD 4113 and HD 156846. *Astron. Astrophys., 480,* L33–L36.

Tisserand F. (1891) *Traité de Mécanique Céleste (Tome II).* Gauthier-Villars, Paris.

Tittemore W. C. and Wisdom J. (1990) Tidal evolution of the uranian satellites. III — Evolution through the Miranda-Umbriel 3:1, Miranda-Ariel 5:3, and Ariel-Umbriel 2:1 mean-motion commensurabilities. *Icarus, 85,* 394–443.

Tokovinin A., Thomas S., Sterzik M., and Udry S. (2006) Tertiary companions to close spectroscopic binaries. *Astron. Astrophys., 450,* 681–693.

Touma J. and Wisdom J. (1993) The chaotic obliquity of Mars. *Science, 259,* 1294–1297.

Udry S., Bonfils X., Delfosse X., Forveille T., Mayor M., Perrier C., Bouchy F., Lovis C., Pepe F., Queloz D., and Bertaux J.-L. (2007) The HARPS search for southern extra-solar planets. XI. Super-Earths (5 and 8 $M_\oplus$) in a 3-planet system. *Astron. Astrophys., 469,* L43–L47.

Veeder G. J., Matson D. L., Johnson T. V., Blaney D. L., and Goguen J. D. (1994) Io's heat flow from infrared radiometry: 1983–1993. *J. Geophys. Res., 99,* 17095–17162.

Ward W. R. (1975) Tidal friction and generalized Cassini's laws in the solar system. *Astron. J., 80,* 64–70.

Ward W. R. and Hamilton D. P. (2004) Tilting Saturn. I. Analytic model. *Astron. J., 128,* 2501–2509.

Williams G. E. (1990) Precambrian cyclic rhythmites: Solar-climatic or tidal signatures? *Phil. Trans. R. Soc. London A, 330,* 445–457.

Winn J. N. and Holman M. J. (2005) Obliquity tides on hot Jupiters. *Astrophys. J. Lett., 628,* L159–L162.

Wisdom J. (2004) Spin-orbit secondary resonance dynamics of Enceladus. *Astron. J., 128,* 484–491.

Wisdom J., Peale S. J., and Mignard F. (1984) The chaotic rotation of Hyperion. *Icarus, 58,* 137–152.

Wu Y. and Murray N. (2003) Planet migration and binary companions: The Case of HD 80606b. *Astrophys. J., 589,* 605–614.

Yoder C. F. (1995) Venus' free obliquity. *Icarus, 117,* 250–286.

Yoder C. F. (1997) Venusian spin dynamics. In *Venus II — Geology, Geophysics, Atmosphere, and Solar Wind Environment* (S. W. Bougher et al., eds.), pp. 1087–1124. Univ. of Arizona, Tucson.

Zahn J. (1975) The dynamical tide in close binaries. *Astron. Astrophys., 41,* 329–344.

Part IV:
Exoplanet Formation and Protoplanetary Disk Evolution

Protoplanetary and Debris Disks

Aki Roberge
NASA Goddard Space Flight Center

Inga Kamp
Kapteyn Astronomical Institute

The discovery of protoplanetary and debris disks around young stars, together with the discovery of exoplanets, confirmed the longstanding notion that planet formation in rotating circumstellar disks is a natural consequence of the star formation process. Disk studies both challenge and inform modern planet formation theories, providing insights on the birth of the solar system and the origins of unexpectedly varied exoplanets. The older disks, debris disks, can also help or hinder observations of mature exoplanets. We introduce here the various disk classes and connect them to theories of planetary system formation. The fundamental theoretical aspects of disks — structure, heating/cooling, chemistry, dynamics, and dissipation — are examined, as well as the challenges associated with developing realistic disk models. Then we discuss the basic observational parameters most relevant to both disk theory and planet formation theory — disk masses, geometries, gas-to-dust ratios, composition, and lifetimes — and describe how they are determined (poorly, in some cases). Finally, the next decade promises to be a particularly fruitful one for disk studies, with new opportunities to probe vital but previously veiled characteristics (most importantly, the gas component).

1. INTRODUCTION

Beginning in the seventeenth century, scientists surmised that the solar system planets formed out of a rotating envelope of gas surrounding the young Sun (e.g., René Descartes' theory of vortices). Before the era of modern astronomy began, they reasoned that such an envelope would contract under its own gravity and that conservation of angular momentum would flatten it into a disk, explaining the planets' co-planar, nearly circular orbits. Although the mathematical and observational basis for these theories was weak at the time, the general conclusions have been born out over the last 25 years, through images of gas and dust disks around nearby young stars and spectra that have proved their rotation (e.g., *Weinberger et al.,* 1999; *Kawabe et al.,* 1993).

For the first time, observations of these disks allow us to study the planet formation process in operation. The discovery of hundreds of nearby exoplanets has taught us that the process is both common and robust, capable of forming diverse planets in widely varying environments. Both of these developments have tested modern theories of planet formation, which were largely created over the second half of the twentieth century to explain the solar system's present-day characteristics (e.g., *Isaacman and Sagan,* 1977).

In this chapter, we will summarize the theoretical and observational advances in disk studies that have changed our understanding of planet formation. These studies are also important for efforts to detect and characterize mature exoplanetary systems. Planets can imprint their dynamical signatures on circumstellar (CS) dust disks that are present even after billions of years, just as Earth produces a ring and clumps in the zodiacal dust of the inner solar system (*Dermott et al.,* 1994). Various dust structures are commonly seen in these older CS disks, the debris disks. These structures provide indirect evidence of exoplanets that are inaccessible to other detection techniques, such as radial velocity or transits, because of their large distances from the central stars.

A dramatic example is the direct imaging of a gas giant planet orbiting Fomalhaut at a distance of 119 AU, shown in Fig. 1 (*Kalas et al.,* 2008). For many years, the characteristics of the Fomalhaut CS dust ring strongly indicated the presence of a planet near the location of the one found. Those ring characteristics helped constrain the mass of the imaged planet (<3 M_{Jup}) (*Chiang et al.,* 2009). This first direct connection between a CS dust structure and the exoplanet causing it opens the door to many more such discoveries.

While CS dust disks provide indirect evidence of planets right now, such dust will very likely be the largest source of noise in direct imaging and spectroscopy of habitable exoplanets. In addition, dust structures caused by the exoplanets themselves will probably be the most troublesome source of confusion in direct imaging, since they orbit the stars just as the planets do. At the end of this chapter, we briefly discuss these problems and the progress toward solving them expected in the next several years.

1.1. Formation of Planetary Systems

In order to place protoplanetary and debris disks in the context of planet formation theory, we begin by sketching the

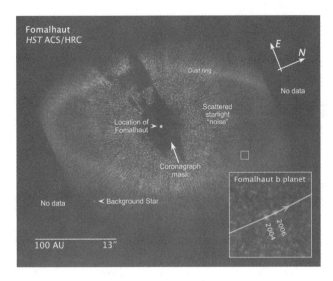

Fig. 1. Direct imaging of an exoplanet in a debris disk (*Kalas et al.*, 2008). This optical wavelength coronagraphic image from the Hubble Space Telescope (HST) shows the debris dust ring around the nearby ~200-m.y.-old A3V star Fomalhaut. A possible gas giant planet is also seen (inset panel). Images taken at two epochs show that the planet candidate orbits the star. Image credit: NASA, ESA, and Z. Levay (STScI).

three main phases in the formation of a mature planetary system. Detailed reviews of planet formation theory may be found in other chapters of this volume. Our simple description of planetary system formation implies a rather orderly progression from one phase to the next. However, it is important to emphasize that the different phases may be occurring simultaneously in different parts of a disk. Other complications in this simple picture will become clear later.

The first phase is the formation of gas giant planets. The standard theory begins with the accretion of kilometer-sized planetesimals, otherwise known as asteroids and comets, into a solid core. When this core reaches ~10 $M_\oplus$, it may gravitationally accrete gas. Fairly rapidly, the bulk of the final planet mass is acquired as a gaseous envelope. This formation scenario is usually referred to as core accretion. Recent core-accretion models can form gas giant planets within a few million years around solar-mass stars, assuming disk masses a few times the estimated minimum mass of the initial solar protoplanetary disk (e.g., *Alibert et al.*, 2005). From here on, we will refer to the initial solar protoplanetary disk as the solar nebula. An alternate theory proposes that gas giant planets form by direct collapse of self-gravitating clumps in unstable protoplanetary disks, without initial massive cores (for a review, see *Durisen et al.*, 2007). While we do not extensively discuss unstable disks in this chapter, it appears that some gas giant planets may form by gravitational instability (see section 3.1).

The next phase is the formation of solid terrestrial planets. In the most general terms, terrestrial planets form through inelastic collisions between planetesimals, as do giant planet cores. There are many unsolved theoretical problems involved with building up tiny submicrometer dust grains into kilometer-sized planetesimals (for details, see *Dominik et al.*, 2007). However, evidence from solar system meteorites indicates that planetesimals grew rapidly, within about 1 m.y. after condensation of the most refractory grains in the solar nebula (*Wadhwa et al.*, 2007). These refractory grains, the calcium-aluminum-rich inclusions (CAIs), are the oldest dated solar system materials. It is worth noting that solar system timelines put "time zero" at CAI formation. On the other hand, stellar ages are typically anchored at the beginning of the pre-main-sequence phase, when stellar evolution models assume that the central protostar begins to be in hydrostatic equilibrium (*Palla and Stahler*, 1993). In a recent examination of the issue, *Pascucci and Tachibana* (2010) argue that the CAIs formed within the first 1 m.y. after the Sun's "stellar time zero."

Therefore, kilometer-sized planetesimals probably formed within about 2 m.y. after the birth of the Sun. Models of the later stages of terrestrial planet formation around solar-type stars suggest that Mars-sized bodies form about 1 m.y. after planetesimal formation (*Kokubo and Ida*, 1998), while Earth-sized bodies take 10–100 m.y. (*Kenyon and Bromley*, 2006). In total, one might expect that it should take no more than about 3 m.y. to build up a Mars-mass planet around a solar-type star and tens of millions of years to reach Earth-mass. The observed lifetimes of protoplanetary disks (discussed in section 3.5) indicate that while terrestrial planet formation may continue after the bulk of the disk gas has been dissipated, it likely begins while significant gas remains.

The final stage in formation of a mature planetary system is removal of most leftover planetesimals that were not incorporated into giant or terrestrial planets. Of course, planetesimals survive for billions of years in the asteroid and Kuiper belts of the solar system. But the present total mass of bodies in these regions constitutes a small fraction of the initial solid mass (e.g., *Weidenschilling*, 1977; *Kenyon and Luu*, 1999). The mechanisms for removing planetesimals will be discussed in section 2.5. The planetesimal clearing phase may also be considered the terrestrial atmosphere formation phase. Most models for the formation of Earth indicate that it was initially very dry (e.g., *Raymond et al.*, 2006). The bulk of Earth's current surface water and other volatiles was probably delivered after the planet was largely complete, by the impact of water-rich planetesimals formed further from the Sun (e.g., *Morbidelli et al.*, 2000).

1.2. General Disk Characteristics

Circumstellar disks around nearby stars were first conclusively identified through unresolved photometry of infrared (IR) emission orders of magnitude brighter than expected from the central stars alone (*Gillett*, 1986). This excess emission arises when short wavelength light from the star is absorbed by CS dust and reradiated at thermal wavelengths. Figure 2 shows theoretical disk spectral energy distributions (SEDs), displaying characteristic excess emission at long wavelengths. Disk images and spectra have confirmed the early conclusion that most stars showing unresolved IR excess harbor protoplanetary disks. Hundreds of disk systems have been found through photometry of IR excess emission and it remains the primary disk detection method (e.g., *Moór et al.*, 2006). However, IR excess in a star's SED only indicates the presence of CS dust and provides relatively little information on the dust's spatial distribution.

A disk SED may be approximated by summing the fluxes from annuli emitting as blackbodies at the local dust temperature. Using basic radiative transfer, the flux from an axisymmetric, spatially thin disk is

$$F_\nu = \frac{\cos\theta}{D^2} \int_{r_{inner}}^{r_{outer}} B_\nu(T_d)\left(1-e^{-\tau_\nu}\right) 2\pi r\, dr \qquad (1)$$

where θ is disk's inclination to the line of sight, D is the distance to the system, r_{inner} and r_{outer} are the disk's inner and outer radii, and $B_\nu(T_d)$ is the Planck function for blackbody dust at temperature T_d (*Beckwith et al.*, 1990). The dust optical depth is

$$\tau_\nu = \kappa_\nu \Sigma_d(r)/\cos\theta \qquad (2)$$

where κ_ν is the dust opacity per unit mass and $\Sigma_d(r)$ is the mass surface density of dust. Therefore, the dust emission from a disk is determined by its geometry (inner/outer radii, inclination), dust properties (κ_ν), and temperature and density structure (T_d and Σ_d). While these formulae may seem straightforward, the parameters are correlated and fitting an SED is highly degenerate. Theoretical prescriptions for Σ_d and T_d are discussed in section 2.1. Spatially resolved images of dust emission (discussed in section 3.2) or spectrally resolved observations of emission from rotating gas are often needed to constrain disk geometries.

As will be discussed in section 3.1, turning an observed IR excess into a total dust mass is difficult. Therefore, observers often quantify the amount of dust in a disk using the system's fractional IR luminosity, $L_{IR}/L_\star$, which is the light absorbed by the dust and reemitted at IR wavelengths relative to the stellar luminosity. This parameter is determined by integrating the total IR excess flux seen in a disk SED over frequency. $L_{IR}/L_\star$ is not a unit of optical depth, dust mass, or surface brightness. In the optically thin case, it is proportional to the dust mass but is affected by the grain properties (see equation (21)).

1.3. Disk Classes

Circumstellar disks may be divided into three classes that appear to roughly correspond to the theoretical phases in planet formation outlined above. In section 3, observed disk characteristics will be discussed in detail. Here, we identify the classes and briefly describe them. To clarify a point of possible confusion, the term "protoplanetary" is typically only applied to the two youngest disk classes, the gas-rich

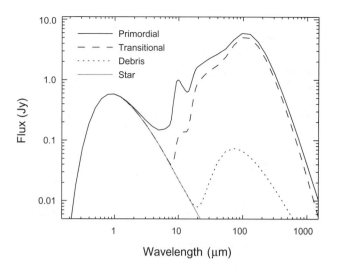

Fig. 2. Theoretical spectral energy distributions for a primordial disk (solid line), a transitional disk (dashed line), and a debris disk (dotted line). The flux contribution from the star is shown with a solid gray line. The primordial and transitional SEDs were produced using the CGPlus code (*Dullemond et al.*, 2001); the debris disk SED is a simple 70 K blackbody model. These SEDs illustrate the general characteristics of the disk classes, but are not models of actual disks. The $L_{IR}/L_\star$ values are 0.2 (primordial), 0.1 (transitional), and 0.0015 (debris). Compared to the primordial SED, the transitional SED has a deficit of near-IR emission, caused by a paucity of warm grains in the inner region of the disk (a central cavity). The primordial and transitional SEDs show an emission feature near 10 μm originating from small, warm silicate grains in optically thin regions of the disks. Debris disks are colder than primordial/transitional disks and their masses are lower, which is why the IR excess emission in the debris disk SED begins at longer wavelengths and is overall fainter.

primordial and transitional disks.

The first class is primordial disks, which are generally found around pre-main-sequence stars located in or near the interstellar (IS) molecular clouds where they were born. In most cases, accretion of disk material onto the star is still occurring, indicating that star formation is not quite finished. Strong submillimeter CO emission from these massive, optically thick disks shows that they are gas-rich (e.g., *Thi et al.*, 2001). Presumably, their gas-to-dust mass ratios are close to the typical value seen in IS clouds (100:1) (e.g., *Knapp and Kerr*, 1974). Thus, they appear to correspond to the solar nebula and are the likely sites of gas giant planet formation. Primordial disks intercept and reradiate a significant fraction of the total stellar light and have $L_{IR}/L_\star$ values around 10–20%.

The next class, transitional disks, was first identified as a possible class in the late 1980s but has gained much more attention in recent years (*Strom et al.*, 1989). The lack of near-IR excess emission in the SEDs of these disks indicates that their inner regions are optically thin, with few or no small dust grains, while large mid- to far-IR excesses demonstrate that their outer disks are optically thick. Transitional disks often display submillimeter CO emission, like the primordial disks, showing that their outer disks contain substantial amounts of gas (e.g., *Qi et al.*, 2004). The central dust clearings have radii ranging from around 0.5 AU to tens of AU (*Brown et al.*, 2007). In some cases, gas is present within the dust clearing, as shown by direct observation of warm gas (e.g., *Salyk et al.*, 2007) and implied by the fact that some transitional disk systems show weak accretion of disk gas onto the central stars (*Najita et al.*, 2007).

Various scenarios have been suggested to explain the apparent removal of primordial material from the inner regions of the disks. A newly formed gas giant planet might inhibit or prevent outer disk material from moving past the planet to resupply the inner disk after material there accretes onto the central star (e.g., *Skrutskie et al.*, 1990). The coagulation of submicrometer dust grains to larger sizes, starting at small disk radii, could lead to reduction of the dust optical depth in the inner disk (e.g., *Strom et al.*, 1989). This scenario may not be consistent with the observation that transitional disks have lower accretion rates than disks without central dust clearings, since grain growth should not affect the gas component (*Najita et al.*, 2007). Photoevaporation of the disk by the central star might clear the disk from the inside out (e.g., *Alexander et al.*, 2006), although this mechanism would remove gas as well as dust. And finally, in at least three cases (CS Cha, HH 30, and CoKu Tau/4), the central cavity is dynamically cleared by a close stellar companion (*Guenther et al.*, 2007; *Anglada et al.*, 2007; *Ireland and Kraus*, 2008). It appears that no single scenario can explain the diversity of transitional disk systems.

Transitional disks have some characteristics in common with primordial disks and some in common with the next disk class (debris disks). It is tempting to surmise that transitional disks are evolving from one class to the other and that they may be in the terrestrial planet formation phase. However, transitional disks are generally classified using a single observational parameter, i.e., a deficit of near-IR excess in the disk SED. This can occur for various reasons that may or may not be related to disk evolution, possibly including some phenomena not yet appreciated as well as the ones discussed above. At present, our understanding of the exact nature of transitional disks is very incomplete.

The final class of disks is very different from the previous two classes. Despite that, debris disks naturally fit into the scenario for formation of a mature planetary system. These disks are found around main-sequence stars with a wide range of ages, from about 5 to 10 m.y. to older than the solar system, although most have ages less than about 1 G.y. (e.g., *Su et al.*, 2006). They have modest amounts of dust; their $L_{IR}/L_\star$ values range from a few × 10^{-3} down to the current detection limits (~10^{-5} for Sun-like stars) (*Beichman et al.*, 2006). Unlike the previous two classes, debris disks are gas-poor. But as will be discussed later, at least some debris disks contain small amounts of gas, which has very different characteristics from the gas in primordial and transitional disks.

Very soon after the discovery of debris disks, it was realized that the dynamical lifetimes of the disks' dust grains are much shorter than the stellar ages (*Gillett*, 1986). Therefore, the bulk of the dust cannot be primordial IS grains left over from star formation. It must have been recently produced by collisions between and evaporation of planetesimals. Over the last two decades, this conclusion has been borne out by many lines of evidence. Debris disks are basically extrasolar versions of the solar system's asteroid and Kuiper belts. Given the ages of debris disks, some of the young ones are likely in the late stages of terrestrial planet formation. The older ones correspond to the planetesimal clearing phase discussed above.

While the nature of the youngest and oldest disk classes seems fairly well understood, there are some weaknesses in the disk classification scheme outlined here. The transitional disk class is particularly problematic, since we cannot discriminate between the different causes for a central dust cavity solely from observations of the dust component alone. It is also possible that such cavities are transient phenomena and that there are disks "in transition" between the primordial and debris disk phases that do not show them.

Many classification problems hinge on the fact that disk gas abundances are difficult to determine. Since primordial and transitional disks are largely made of gas, this makes it difficult to determine accurate total disk masses. As will be seen, the gas abundance in a disk is critical for understanding its place in the planet formation process. Primordial disks obviously have high gas abundances and debris disks obviously have low ones, but the precise evolution of the gas-to-dust ratio is not known. New opportunities to study the gas component in disks will become available in the next several years (discussed in section 5.1). A disk classification scheme based on the gas-to-dust ratio may soon be possible. Such a scheme would more accurately capture the whole-disk evolution and should help us figure out what is happening in transitional disks.

2. MODELING CONCEPTS

From a theoretical viewpoint, protoplanetary disks can be fully described by hydrodynamics. However, the equation of state is strongly related to the chemical structure of the gas in the disk. Hence, dynamics, radiative transfer, and chemistry are intertwined and solving them self-consistently remains a challenge. For that reason, various approximations exist that focus on particular aspects of protoplanetary disks.

Steady-state dust disk models adopt a radial density profile for the disk and derive the vertical structure (perpendicular to the midplane) from hydrostatic equilibrium and continuum radiative transfer. Steady-state gas disk models add line radiative transfer (often approximated) and a more-or-less detailed chemical network to determine the gas composition and temperature. Sometimes complex chemistry calculations are simply performed on the background of a detailed dust disk model, assuming that the gas temperature equals that of the dust everywhere in the disk.

Hydrodynamic gas models focus on the impact of transport processes such as turbulent mixing and diffusion, as well as the chemistry and ionization degree of disks. They are also employed to study the long-term evolution of disks, considering processes such as photoevaporation in the presence of viscous evolution. Two-fluid hydrodynamic models describe the friction and mutual dynamical impact of gas and dust. They are most often used to model dust grain growth, the formation of protoplanetesimals, and the formation of dust structures in debris disks (e.g., rings).

2.1. Disk Structure

Star and disk formation begins with the gravitational collapse of dense cores in IS molecular clouds. A rotating disk of gas and dust rapidly develops around the young star, which grows by accreting cloud material that falls onto the disk and moves inward through it. Detailed reviews of these earliest phases may be found in *Reipurth et al.* (2007). We begin our discussion at the end of the core collapse phase, when the central star has acquired most of its final mass and its protoplanetary disk is fairly stable.

The disk initially contains relatively unmodified material from its parental cloud, which typically has a gas-to-dust mass ratio of 100:1. Disk material still accreting onto the central star is continuously replenished from the outer disk. At this time, the stellar mass accretion rate is typically $\dot{M} = 10^{-7}$–10^{-8} $M_\odot$/yr (*Muzerolle et al.*, 1998). The disk can be approximated as a steady-state rotating disk with no transport of material in the vertical direction.

If protoplanetary disks are massive and cool enough, they could be gravitationally unstable (e.g., *Boss*, 1998). However, detailed modeling of the structure of typical T Tauri disks suggests that stellar irradiation will provide enough heat to stabilize them (*D'Alessio et al.*, 1998). Although a few protoplanetary disks appear massive enough to be gravitationally unstable (discussed in section 3.1), we will not examine the theory of such disks in detail (for a review, see *Durisen et al.*, 2007).

2.1.1. Surface and vertical density structure. To move orbiting material farther into the star's gravitational potential well (i.e., accretion), energy and angular momentum must be removed. Gas may readily lose energy by radiative cooling (see section 2.1.3); losing angular momentum is more problematic. Viscous torques damp shearing motions in rotating gas, transporting angular momentum to larger radii. Most of the disk material can then move inward, while a smaller portion moves out, carrying away angular momentum. This causes protoplanetary disks to spread out to sizes of a few hundred AU in the first million years (*Hueso and Guillot*, 2005).

However, the familiar molecular viscosity is too weak to explain the observed stellar mass accretion rates. The physics of angular momentum transport in protoplanetary disks is still hotly debated. Possible mechanisms to generate anomalously strong viscosity are convection, gravitational instabilities, and the magnetorotational instability (see *Dullemond et al.*, 2007, and references therein). In the absence of a well-understood mechanism, angular momentum transport is conceptualized using the kinematic — or turbulent — viscosity (ν), typically parameterized as an α-viscosity

$$\nu = \alpha \, c_s H_{gas} \tag{3}$$

where α is a dimensionless scaling factor, c_s is the gas sound speed, and H_{gas} is the vertical scale height of the gas disk (*Shakura and Syunyaev*, 1973). The gas sound speed is defined as $c_s^2 = \partial P / \partial \rho_g$, where P is the gas pressure and ρ_g is the gas density. For an ideal gas, $c_s = \sqrt{kT_g/(\mu m_p)}$. Here, k is the Boltzmann constant, T_g is the gas temperature, μ is the mean molecular weight of the gas, and m_p is the proton mass.

The disk structure follows from radial and angular momentum conservation and the assumption that the vertical component of gravity from the star is balanced by the vertical gas pressure gradient (vertical hydrostatic equilibrium). A detailed examination of accretion disk theory appears in *Pringle* (1981). Here, a few key points are discussed. The radial momentum conservation equation for a steady-state flow is

$$v_r \frac{\partial v_r}{\partial r} - \frac{v_\phi^2}{r} + \frac{1}{\rho_g} \frac{\partial P}{\partial r} + \frac{GM_\star}{r^2} = 0 \tag{4}$$

where v_r is the radial velocity of the gas, v_ϕ is the circular velocity, and $M_\star$ is the mass of the central star. The four terms in this equation arise from radial mass flow, centrifugal force, gas pressure, and gravity. Since pressure typically decreases with increasing radius, the third term is nearly always negative; effectively, gas pressure resists the gravitational force, resulting in gas rotating at sub-Keplerian orbital velocities.

The angular momentum of a thin annulus of the disk is $2\pi r \, \Delta r \, \Sigma r^2 \Omega$, where Δr is the width of the annulus, Σ is the disk surface density, and Ω is the angular velocity of the gas. Therefore, the steady-state angular momentum conservation equation is

$$\frac{\partial}{\partial r}\left(r \, v_r \, \Sigma r^2 \Omega \right) = \frac{\partial}{\partial r}\left(\nu \Sigma r^3 \frac{\partial \Omega}{\partial r} \right) \tag{5}$$

where the lefthand side is the radial change in angular momentum and the righthand side arises from viscous torques. Integrating this equation yields the disk surface density

$$\Sigma = \frac{\dot{M}}{3\pi \nu} \tag{6}$$

for radii much larger than the stellar radius.

The vertical (z-direction) disk structure is found by solving the equation of hydrostatic equilibrium

$$\frac{1}{\rho_g} \frac{\partial P}{\partial z} = \frac{\partial}{\partial z}\left(\frac{GM_\star}{\sqrt{r^2 + z^2}} \right) \tag{7}$$

If the disk is vertically thin (z ≪ r) and the gas temperature does not depend on z (i.e., a vertically isothermal disk), equation (7) can be integrated to give the vertical density structure

$$\rho_g(r,z) = \rho_c(r) e^{-z^2/(2H_{gas}^2)} \qquad (8)$$

where $\rho_c(r)$ is the density at the disk midplane. The gas scale height is

$$H_{gas} = \sqrt{kT_c r^3/(\mu m_p G M_\star)} \qquad (9)$$

where T_c is the midplane gas temperature. This equation shows that the gas scale height is the ratio of the gas sound speed to the angular velocity ($H_{gas} = c_s/\Omega$).

Having found an expression for the scale height, we return to the radial disk structure and write equation (6) as

$$\Sigma = \frac{\mu m_p \sqrt{GM_\star}}{3\pi k} \frac{\dot{M}}{\alpha T_c r^{3/2}} \qquad (10)$$

For a disk with a simple power-law midplane temperature profile, $T_c \propto r^{-q}$, the surface density is proportional to $r^{q-3/2}$. Further, the midplane density may be written as

$$\rho_c \simeq \Sigma/H_{gas} = \rho_{in} (r/r_{in})^\varepsilon \qquad (11)$$

where ρ_{in} is the density at the inner disk radius and the exponent ε equals $3/2q-3$.

The next step is determining the actual disk temperature profile (i.e., the exponent of the temperature power law, q). Energy transport in disks is dominated by radiation processes. *D'Alessio et al.* (1998) and *Nomura* (2002) find that typical T Tauri disks are convectively stable and turbulence plays only a minor role, mostly in the midplane regions of the inner disk (r ≲ 10 AU). Hence, even though some sort of turbulence is responsible for viscosity in the disk and thus for its long-term evolution (spreading and accretion), the disk's temperature balance at any point in time is largely determined by the radiation field. The energy equation, used to calculate gas and dust temperatures, is generally expressed in terms of a detailed balance of heating and cooling processes. Dust heating and cooling is a continuum process, while gas heating and cooling mostly occurs at discrete wavelengths (line absorption/emission) except in a few cases like H⁻.

2.1.2. Dust temperature. Protoplanetary disks rapidly become optically thick in radial direction. However, stellar photons can efficiently irradiate the disk surface if the gas scale height increases with radius (a flaring disk). In this case, the disk is heated even at large radii where the material would otherwise be extremely cool.

Kenyon and Hartmann (1987) found that the efficient reprocessing of stellar irradiation by a flaring disk leads to a strong far-IR excess, as frequently seen in the SEDs of young stars. *Chiang and Goldreich* (1997) modeled the temperature structure of a passive reprocessing disk, one without a high intrinsic luminosity, using a simple two-layer approach. They assumed an optically thin disk surface plus an optically thick midplane, dust continuum radiative transfer, and vertical hydrostatic equilibrium. A fraction of the stellar radiation gets scattered into the disk by dust grains in the surface, creating a diffuse stellar radiation field in the disk interior. Another fraction is absorbed then reemitted according to the local surface dust temperature. Under these conditions, the gas scale height increases with distance from the star ($H_{gas} \propto r^{9/7}$) and the disk midplane temperature scales as $r^{-3/7}$ (*Chiang et al.*, 2001). Inserting this temperature profile into equation (10) yields the surface density profile for a standard flaring disk

$$\Sigma \approx \Sigma_{in} (r/r_{in})^{-1} \qquad (12)$$

where Σ_{in} is the surface density at the disk inner radius.

In the disk's inner regions, the optical depth is so high that radiation cannot efficiently penetrate to the midplane. These dense layers are instead primarily heated by local viscous dissipation (*Frank et al.*, 1992). The viscous heating rate is

$$\Gamma_{vis} = \frac{9}{4} \rho_g \nu \Omega^2 \qquad (13)$$

For r ≲ 2 AU, viscous dissipation can dominate the dust temperature balance (*D'Alessio et al.*, 1998).

Disk structure modeling was initially driven by dust observations and thus the dust temperature determination played a key role. Simple two-layer disk models, like those of *Chiang and Goldreich* (1997) and *D'Alessio et al.* (1998), have been further developed with the addition of more detailed radiative transfer prescriptions iteratively coupled to the disk vertical structure given by hydrostatic equilibrium (e.g., *Dullemond et al.*, 2002). In all these models, the underlying assumption is that gas and dust are well-coupled, both thermally and dynamically. However, under realistic conditions, the gas temperature can differ from the dust temperature. The disk vertical structure will be given by the gas scale height and the dust will be forced to follow if it is dynamically coupled.

2.1.3. Gas temperature. The gas heating/cooling balance is far more complex, due to the multitude of atomic and molecular emission lines that can carry away energy (cooling lines). In addition, the gas chemical and thermal balance are intricately linked. Collisions between gas particles and dust grains thermally couple the gas to the dust ($T_g = T_d$) only if radiative processes become ineffective due to high optical depth ($\tau \gg 1$) and/or densities become very large ($n \gtrsim 10^7$ cm^{-3}).

The detailed gas energy balance has been modeled by several groups: *Kamp and Dullemond* (2004), *Jonkheid et al.* (2004), *Gorti and Hollenbach* (2004, 2008), *Nomura and Millar* (2005); *Nomura et al.* (2007), *Meijerink et al.* (2008), and *Woods and Willacy* (2008). These approaches differ in the number of heating/cooling processes included, the size of the chemical network used, and the irradiation considered [X-rays and/or ultraviolet (UV) stellar radiation]. Recently, *Woitke et al.* (2009) extended these models to include cooling by electronic transitions — important at higher gas temperatures and in the inner disk — and a continuum background radiation field for excitation of atoms and molecules, which was derived from detailed radiative transfer calculations.

These models indicate that in the disk surface, down to a continuum optical depth of ~1, the gas is thermally decoupled from the dust. Here, the gas temperature ranges from a few thousand to several tens of Kelvins (*Kamp and Dullemond*, 2004; *Jonkheid et al.*, 2004). The dominant heating source in the presence of UV radiation is the photoelectric effect on grains, either small polycyclic aromatic hydrocarbons (PAHs) or larger dust grains. The electrons ejected from the grains transfer their kinetic energy to the gas through collisions.

X-rays also heat the gas via production of energetic electrons during photoionization of atoms and molecules. While far-UV radiation (6 eV < hν < 13.6 eV) typically gets absorbed within gas column densities of 10^{21} cm^{-2}, X-rays can penetrate roughly a factor of 100 deeper and can start to dominate the heating in layers where far-UV-driven processes are ineffective. This heating plays an important role at fairly high X-ray luminosities ($L_X \sim 10^{30}$–10^{31}) (*Glassgold et al.*, 2004). At lower luminosities (approximately a few 10^{29}), their impact is restricted to intermediate depths at larger radii (r > 10 AU).

Figure 3 illustrates the two-dimensional temperature structure of a gas disk. In the optically thick deeper layers of the disk, collisions efficiently couple the gas and dust temperatures. This produces a temperature profile that is roughly isothermal in the vertical direction and declines approximately as $r^{-0.5}$ in the radial direction, as predicted by the simple two-layer flared disk model of *Chiang and Goldreich* (1997). The temperature profiles in the upper disk layers, however, are more complex. Cooling of the surface layers proceeds through atomic emission, mainly the [C II] and [O I] fine structure lines (for gas temperatures of a few hundred Kelvins), Fe II electronic transitions, Lyman α, and O I 6300 Å (for temperatures of a few thousand Kelvins). While photoelectric heating scales roughly as the density squared, the cooling lines often come from species that are not in local thermodynamic equilibrium (LTE) and thus scale linearly with n. Combined with the weak dependence of the photoelectric heating process on gas temperature, this can lead to strong vertical temperature gradients.

2.2. Disk Chemistry

Gas chemistry in protoplanetary disks is studied using a variety of approaches. Stationary chemistry assumes that the chemical timescales are generally much smaller than dynamic timescales. Time-dependent chemistry follows the advection of gas parcels as they move through the disk along an accretion trajectory (e.g., *Aikawa et al.*, 1999). Some approaches also include global diffusion and turbulent mixing (e.g., *Ilgner et al.*, 2004).

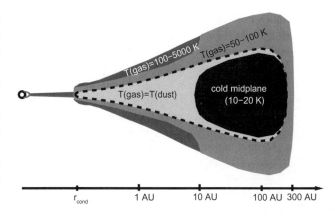

Fig. 3. Schematic gas temperature structure of a protoplanetary disk. The x-axis shows the approximate distance from the star. The inner edge of the dust disk is at r_{cond}, where the gas temperature surpasses the dust vaporization temperature (~1500 K). Interior to this radius, only gas exists. The disk surface is heated by the photoelectric effect on grains and/or by X-ray ionization. As the incoming radiation is attenuated by optical depth and/or distance from the star, the gas becomes cooler. Below a continuum optical depth of τ ~ 1, shown by the dashed line, gas and dust can efficiently thermally couple through collisions.

The complexity of the chemical networks used varies substantially, ranging from simple gas-phase chemistry to more complex gas-grain chemistry. The latter includes ice formation on cold dust grains, standard desorption processes such as thermal desorption or nonthermal processes such as cosmic ray-induced desorption (*Hasegawa and Herbst*, 1993), photodesorption (*Willacy and Langer*, 2000), and exothermic surface reactions (*Garrod et al.*, 2007). Including the formation of complex molecules on grain surfaces, along with gas chemistry and ice formation, remains numerically challenging (*Stantcheva and Herbst*, 2004).

Gas-phase chemistry and, to some extent, ice formation can be described with a rate equation approach, where the time variation in the average density of a species is given by the sum of various production and loss rates. A generic rate equation for species i is

$$\frac{dn_i}{dt} = \mathcal{P}_i - \mathcal{L}_i$$
$$= \sum_{jl} k_{ijl}(T_g) n_j n_l + \sum_{j} \left(\Gamma_{ij}(r,z) + \varsigma_{ij}(r,z)\right) \quad (14)$$
$$- n_i \left[\sum_{jl} k_{jil}(T_g) n_l + \sum_{j} \left(\Gamma_{ji}(r,z) + \varsigma_{ji}(r,z)\right) \right]$$

where n_i is the average volume density of species i and $\mathcal{P}_i$ and $\mathcal{L}_i$ are the chemical production and loss rates for that species. The first and third sums are the production and loss terms for species i through chemical reactions with species j and l, at rates k_{ijl} and k_{jil}. The second and fourth sums are the production and loss terms for species i through photoreactions involving species j (at rates Γ_{ij} and Γ_{ji}) and cosmic-ray reactions involving species j (at rates ζ_{ij} and ζ_{ji}).

On the other hand, grain surface chemistry is stochastic in nature and strongly depends on the surface coverage of the species involved in the reaction at any instant. Hence, it is much better described using a master equation approach

$$\frac{d}{dt} P(i_1 \ldots i_N) = \sum (\text{accretion} + \text{evaporation} + \text{surface reactions}) \qquad (15)$$

where P is the probability that i particles of species 1 to N are on the grain surface. *Caselli et al.* (1998, 2002) find that the rate equation approach can be modified for grain surface chemistry by taking into account the accretion timescale for a species as well as its migration timescale on the grain surface. This enables efficient simultaneous treatment of gas- and dust-phase chemistry.

After more than a decade of chemical studies, the picture of a layered disk chemical structure has emerged. The surface is generally ionized by stellar and IS UV radiation. Photoionization and photodissociation strongly depend on the shape of the incident radiation field and the nature and wavelength range of the ionization/dissociation processes. Molecules such as H_2 and CO are dissociated through discrete band absorption, while ionization of atoms is a continuum process. A compilation of photodissociation cross sections appears in *van Dishoeck et al.* (2006).

Lyman α emission plays a crucial role in photodissociation and ionization of certain species such as CH, CO_2, H_2O, and HCN (*Bergin et al.*, 2003). Since other molecules (e.g., CN, CO, and H_2) are unaffected by Lyman α emission, the relative strength of line vs. continuum irridation strongly affects certain abundance ratios such as CN/HCN. *van Zadelhoff et al.* (2003) employed two-dimensional radiative transfer to show that UV radiation is efficiently scattered into the disk, enhancing the amount of radicals like CN and C_2H. For a recent review of the chemical evolution of protoplanetary disks, see *Bergin et al.* (2007).

The disk thermal balance and chemical composition are intimately coupled. In moderately optically thick regions (medium gray areas in Fig. 3), a large fraction of material is shielded from incoming dissociating and ionizing UV flux. Here, gas temperatures are around 100 K and molecules can form. Due to X-rays and cosmic rays, the ionization fraction in these layers can still be up to 10^{-6}, driving rich ion-molecule chemistry. Cosmic rays also induce a nonnegligible degree of ionization deep in the disk, thereby driving CO and N_2 into CO_2, CH_4, NH_3, and HCN via efficient ion-molecule chemistry (e.g., *Semenov et al.*, 2004). The formation of more complex organic molecules is especially interesting for astrobiology, as these molecules — in the gas or the ice phase — could "seed" planetary atmospheres at later evolutionary stages.

2.2.1. Snow line(s). The position in a protoplanetary disk beyond which the temperature is low enough for water ice to condense is generally called the "snow line." The location of the snow line in the solar nebula (at 3–5 AU) apparently defined the transition between terrestrial planets and gas giant planets. Outside the snow line, the surface density of solids is larger by a factor of ~4, allowing coagulation of much more massive planetary cores that can gravitationally accrete massive gaseous envelopes.

The location of the snow line depends strongly on grain opacities and hence on grain composition. It occurs at different temperatures in the midplane (170 K) and the disk photosphere (150 K), mainly because evaporation and sublimation of ice depend on the amount of irradiation (*Podolak and Zucker*, 2004). Disk evolution models suggest that the snow line is not static. It moves inward as the disk cools and the dust opacity decreases due to grain growth and lowered surface densities (e.g., *Ruden and Lin*, 1986).

In more general terms, a "snow line" exists for any molecule that freezes out below a certain temperature. The various snow lines thus follow from the disk thermal structure and chemical composition. Radial diffusion and turbulent mixing of material in a disk probably turns the canonical water ice snow line into something more like a "slush zone" (see section 2.3).

2.2.2. Interstellar heritage. The initial chemical composition of a protoplanetary disk resembles the composition of the IS molecular cloud from which it formed. Accretion shocks occur as IS material rains onto the forming disk. Postshock gas temperatures can be well above 100 K inward of 10 AU from the central star (*Wood*, 1984). However, these shocks are not strong enough to substantially erase the original chemical pattern except in the innermost region of the disk. The postshock temperature will not be high enough to melt refractory dust grains at radii greater than a few AU (*Iida et al.*, 2001). The high D/H ratio found in cometary refractory grains points to a pristine IS heritage of some material in the solar nebula (*Wooden*, 2008).

A significant fraction of IS ices in the outer disk (particularly water ice) remains intact and is incorporated into mantles on grains. However, the large D/H ratios observed in comet volatiles are within the range of deuteration achieved in chemical models of cold outer disks and suggest some processing of volatile material (*Aikawa et al.*, 1999). *Woods and Willacy* (2007) find that the physical conditions in the inner disk (r < 3 AU) are favorable for the formation of larger carbon chains and even rings such as benzene (c-C_6H_6). This result has revived debate over whether the PAHs in solar system planetesimals are IS or were formed in the solar nebula.

2.3. Protoplanetary Disk Dynamics

A complete discussion of hydrodynamical disk models is beyond the scope of this chapter. Instead, we refer the reader to the excellent summary of instabilities, turbulence, and transport processes in accretion disks by *Balbus and Hawley* (1998). Dust grain growth, settling, and dynamical interactions between protoplanetesimals lead eventually to the formation of terrestrial planets inside the snow line and the cores of gas giant planets outside it. The details of the core-accretion scenario and the relevant dynamics are reviewed in the chapters by Chambers and by D'Angelo et al.

As planets form in disks, they can dynamically interact with the gas, leading to radial migration both inward or outward. However, planet formation can also disturb the disk structure by locally inverting pressure gradients and opening density gaps. Further details of the underlying physics are discussed in the chapter by Lubow and Ida. In the following, we concentrate on two dynamical processes — turbulent mixing and diffusion — that are strongly linked to observations of protoplanetary disks. Debris disk dynamics involves very different physics and will be discussed separately in section 2.5.

Turbulent radial/vertical mixing and diffusion can affect chemical fractionation and deuteration in disks. Comparison between models and observations supply important information on the amount and efficiency of such mixing processes (e.g., *Woods and Willacy*, 2008). Chemical diffusion (transport due to chemical gradients) and advection (transport due to global particle flows) can be added to the general rate equation approach (equation (14))

$$\frac{\partial n_i}{\partial t} + \nabla \Phi_i + n_i \nabla v = \mathcal{P}_i - \mathcal{L}_i \qquad (16)$$

where Φ_i is the diffusion flux of species i and v is the flow velocity. In the presence of vertical abundance gradients, the diffusion flux can be written as

$$\Phi_i = \alpha c_s H_{gas} n_{tot} \frac{d\varepsilon_i}{dz} \qquad (17)$$

where n_{tot} is the total hydrogen particle density and ε_i is the abundance of species i with respect to n_{tot}. The advection term follows directly from the flow velocity v.

Several papers have studied the role of vertical mixing, advection, and radial diffusion on the chemical structure of protoplanetary disks (e.g., *Aikawa*, 2007). Vertical mixing can lead to destruction of molecules by exposing them to unfavorable conditions, such as dissociating radiation and/or high temperatures. These conditions occur at the disk surface (*Willacy et al.*, 2006) and in the inner disk (r < 5 AU) (*Ilgner et al.*, 2004), where mixing timescales are shorter than reformation timescales. The overall layered chemical structure is not greatly affected by these transport processes, but the extent of the intermediate molecular layer generally widens (*Willacy et al.*, 2006). This is caused by upward mixing of ices, which then thermally desorb, and downward mixing of atoms and ions, which then form molecules.

Turbulence has another impact on the gas-grain chemistry, specifically, on the freeze-out of molecules. Transport processes can cause molecules to remain in the gas phase even below their sublimation temperature (see section 3.3), due to the finite timescale for freezing onto dust grains (*Semenov et al.*, 2006; *Aikawa*, 2007). It is clearly important to include all mixing and transport processes, both radially and vertically. But more work is needed to understand the complex interplay between chemistry, temperature, and dynamics, especially when gas-grain chemistry is involved. None of the above models take into account the reaction of the gas temperature to changes in chemical composition. This can be important for the intermediate molecular layer, where gas and dust are partially decoupled.

2.4. Disk Dispersal

Over millions of years, protoplanetary disks are subject to viscous evolution. The angular momentum of disk material accreting onto the star is transported outward, leading to large-scale radial spreading of the disk. Especially during the first 10^5 years, stellar outflows and disk winds also carry away part of the disk mass and angular momentum (for a recent review, see *Ferreira*, 2008). For T Tauri stars, typical ratios between mass outflow and mass accretion rates are of the order of 10^{-2} (*Hartigan et al.*, 1995). While dynamical interactions such as tidal stripping by a passing star dominates the disk dispersal in a small fraction of young systems, other processes such as photoevaporation by the central star or a source of external illumination (e.g., a nearby O/B star) are considered the most general means of disk dispersal.

A detailed study of the timescales for various disk dispersal processes reveals that viscous accretion onto the star prevails in the inner ~10 AU of the disk, while photoevaporation dominates outside the gravitational radius

$$r_g = \frac{G M_\star}{c_s^2} \qquad (18)$$

determined by the ratio between the gravitational and thermal energy of the gas (*Hollenbach et al.*, 2000). Current hydrodynamical models that combine viscous disk evolution with photoevaporation by ionizing extreme-UV (hν < 13.6 eV) stellar radiation show a multistage clearing process (for a recent review, see *Alexander*, 2008). After a long period of viscous spreading, lasting a few million years, the mass accretion rate drops to a point where photoevaporation can cut off resupply of the inner disk from the outer regions. A gap opens near r_g and the inner disk rapidly accretes onto the star. After this, the outer disk photoevaporates from the inside out, on a slightly longer timescale. These models indicate that the mass loss rate due to extreme-UV photoevaporation is

$$\frac{dM}{dt} \approx 4.4 \times 10^{-10} \left(\frac{\Phi}{10^{41} s^{-1}}\right)^{1/2} \left(\frac{M_\star}{M_\odot}\right)^{1/2} M_\odot yr^{-1} \qquad (19)$$

where Φ is the stellar ionizing flux (*Alexander*, 2008).

Since far-UV photons can penetrate much larger column densities, they affect protoplanetary disks at earlier stages, when stellar extreme-UV photons cannot pass through the dense gas accreting onto the star (i.e., when $\dot{M}$ is large). *Gorti and Hollenbach* (2009) show that far-UV photoevaporation depletes the outer disk (r ~ 100 AU), where the viscous and photoevaporation timescales are equal. This reduces the mass accretion rate, speeding the onset of the dispersal scenario outlined in the previous paragraph. They find that far-UV photoevaporation of the outer disk sets the total disk lifetime at ~2 m.y., in fairly good agreement with observed disk dispersal timescales (see section 3.5).

2.5. Debris Disk Dynamics

In contrast to protoplanetary disks, debris disks are optically thin and largely gas-free, making different dynamical processes important. Most of the dust in debris disks is probably produced in planetesimal collisions, rather than outgassing of comet-like bodies. If the planetesimal velocity dispersion is high enough, collisions between them become destructive and they are broken down into smaller and smaller pieces (collisional cascade) (e.g., *Wyatt et al.*, 2007). In most if not all of the currently known debris disks, the dust densities are high, leading to short collision times between grains (e.g., *Artymowicz*, 1988). They are rapidly reduced to small sizes, then blown out of the system by the radial force of stellar photons (radiation pressure). These disks are described as collision-dominated.

In low-density debris disks (like the solar system), the collision timescales between grains are long, leading to a low production rate of grains small enough to be blown out by radiation pressure. Instead, the major dust removal processes bring grains inward. The processes are Poynting-Robertson (P-R) drag (effectively, the headwind of stellar photons seen by an orbiting grain) and stellar wind drag, both of which are slower than radiation pressure and act on larger grains. In both collision- and P-R drag-dominated debris disks, the dynamics of grains is largely controlled by their interactions with stellar radiation.

However, collisional cascade alone probably cannot account for the current low masses of the solar system asteroid and Kuiper belts. Many planetesimals must have been ejected by dynamical interactions with planets (for a recent review of this process, see *Levison et al.*, 2007). Evidence from the heavily cratered surfaces of the Moon and other solar system solid bodies indicates that a sudden burst in this dynamical ejection process, called the late heavy bombardment (LHB), occurred about 700 m.y. after planetesimal formation (e.g., *Kring and Cohen*, 2002).

A new explanation for this phenomenon has been suggested (*Gomes et al.*, 2005, and references therein). This model, called the Nice Model after the city in France, describes the coupled dynamical evolution of the gas giant planets and a remaining outer reservoir of planetesimals in the solar system. The Nice Model indicates that the LHB was caused by Saturn crossing the 2:1 mean-motion resonance (MMR) with Jupiter, which stirred up the remnant planetesimal population, injecting many of them into the inner solar system. If the planetesimal disk is placed outside of 15.3 AU, the time at which the MMR is crossed in the model agrees well with the time of the LHB.

Debris disk images often show dust structures, for example, axisymmetric rings as well as clumps, warps, and other asymmetrical features. Most of the dust seen in debris disks must be spread far from the parent planetesimals. But in collision-dominated disks, observed dust structures probably trace the locations of planetesimal populations (e.g., *Wyatt*, 2008a). This is not necessarily so in P-R drag-dominated disks (e.g., *Kuchner and Holman*, 2003).

Returning to collision-dominated disks, rings could arise from planetesimals that are either themselves confined to a belt or experiencing an enhanced collision rate due to stirring by a protoplanet (e.g., *Kenyon and Bromley*, 2008). Very narrow or noncircular rings (see Fig. 1) suggest perturbation by a planet (e.g., *Quillen*, 2006). Clumps might be produced from planetesimals trapped into mean-motion resonances with a planet or by a recent collision between large planetesimals (*Wyatt*, 2003; *Kenyon and Bromley*, 2005). Recent high-quality images of "warps" in debris disks show they are actually sub-disks inclined with respect to the main disk (*Golimowski et al.*, 2006). This might indicate a population of planetesimals perturbed by a planet on an inclined orbit. Suffice it to say that the diverse morphology of debris disks may arise from many causes, most of which are linked to planets.

3. OBSERVATIONAL CONSTRAINTS

In this section, we explore the observed disk characteristics most relevant to the theories discussed in the various subsections of section 2. Both the disk parameters and the common techniques for measuring them will be examined. As will become clear, some basic disk attributes are poorly known at present, hindering further development of generic planet formation models.

3.1. Masses

The most fundamental parameter of a protoplanetary disk is its total mass, which dictates whether there is enough material to form particular types of planets, the speed with which they form, and the mode of their formation. Unfortunately, total disk masses are difficult to measure for several reasons. Here we discuss three of the most important ones. First, dust grains emit most efficiently at wavelengths comparable to the grain size (for an explanation, see Chapter 7 in *Spitzer*, 1998). Therefore, observations of dust continuum emission at some wavelength are insensitive to grains that are much smaller and larger. The longest wavelengths we can realistically observe from protoplanetary disks are millimeters to a few centimeters; once grains grow larger than these sizes, they are basically invisible although they contain most of the solid mass.

Initially, the dust in the very youngest disks will resemble IS grains, which have radii around 0.2–0.3 µm or less (*Natta et al.*, 2007). A complete discussion of grain growth mechanisms appears in *Dominik et al.* (2007). Suffice it to say, it appears that grain growth proceeds very rapidly. Millimeter- and centimeter-wavelength photometry has provided evidence for centimeter-sized grains in disks with ages ranging from about 10 m.y. (e.g., *Calvet et al.*, 2002; *Wilner et al.*, 2005) to as little as ~1 m.y. (*Greaves et al.*, 2008). However, recall that solar system planetesimals apparently grew to kilometer sizes within roughly 2 m.y. after the birth of the Sun. It is very possible that a significant fraction of the solid mass in protoplanetary disks is already hidden in pebbles (centimeter sizes) and boulders (meter sizes) at fairly young ages.

The second problem relates to optical depth effects. When the dust density and therefore optical depth are low, the term $(1-e^{-\tau_v})$ in equation (1) can be approximated as τ_v. In this optically thin case, assuming all the dust is at the same temperature, the emitted flux may be written as

$$F_v \approx \frac{\kappa_v}{D^2} B_v(T_d) \int_{r_{inner}}^{r_{outer}} \Sigma_d(r) \, 2\pi r \, dr \quad (20)$$

$$\approx \frac{\kappa_v}{D^2} B_v(T_d) M_d \quad (21)$$

where M_d is the total disk dust mass. Equation (21) can be used to quickly estimate total disk dust masses if the observed emission is optically thin.

On the other hand, examination of equation (1) shows that the emitted flux approaches a constant value as the dust optical depth increases. The observed flux mostly comes from lower density regions like the disk surface and no longer accurately reflects the total dust density. Emission from protoplanetary disks is typically optically thick at wavelengths shorter than about 100 μm and only becomes optically thin in the submillimeter or millimeter (*Beckwith et al.*, 1990). However, even at those wavelengths, the innermost region of the disk may be optically thick. Spatially resolved millimeter-wavelength images, which constrain the disk outer radius and surface density profile, are needed to ensure that the whole disk is optically thin (e.g., *Testi et al.*, 2003).

Finally, the third problem is the difficulty in measuring disk gas masses. Turning a dust mass into a total disk mass requires an assumption about the disk gas-to-dust ratio. The mass ratio seen in IS molecular clouds (100:1) is generally used for protoplanetary disks. However, while this value is probably appropriate for the youngest primordial disks, the gas-to-dust ratio rapidly declines to nearly zero in debris disks. At the moment, we know very little about the evolution of disk gas-to-dust ratios. Since dust makes up only a small fraction of the total mass (at least in the primordial and transitional disks), small errors in the measured dust mass translate into large errors in the total disk mass. In section 3.3, we will discuss the problems associated with measuring gas masses and what little is known about disk gas-to-dust ratios.

As might be expected from the preceding discussion, observed total disk masses are very uncertain. The vast majority of measurements depend on unresolved submillimeter and millimeter fluxes, assuming the canonical gas-to-dust mass ratio of 100:1. An early 1.3-mm survey of T Tauri stars in the Taurus star-forming region by *Beckwith et al.* (1990) found total disk masses ranging from about 0.001 to 0.5 $M_\odot$, with an average mass of ~0.02 $M_\odot$. A more recent submillimeter survey of Taurus sources also found a wide range of masses but somewhat lower values, ranging from about 0.0004 $M_\odot$ to 0.2 $M_\odot$, with an average mass of ~0.005 $M_\odot$ (*Andrews and Williams*, 2005). Similar results have been found for disks in the Orion and ρ Ophiuchius star-forming regions (*Williams et al.*, 2005; *Andrews and Williams*, 2007).

What are the implications of these estimated disk masses for planet formation? The short lifetimes of protoplanetary disks, discussed in section 3.5, challenged early models of giant planet formation through the core-accretion process. Newer models form giant planets faster partly by increasing the initial disk masses, providing sufficient solid surface density to form massive cores within a few million years around solar-mass stars. Disk surface densities a few times that of the minimum mass solar nebula (MMSN) are required (e.g., *Hubickyj et al.*, 2005; *Alibert et al.*, 2005). A commonly used formulation for the total (gas + solids) surface density of the MMSN, determined by *Hayashi* (1981), is

$$\Sigma(r) \approx 1700 \left(\frac{r}{1 \, AU}\right)^{-3/2} \, g \, cm^{-2} \quad (22)$$

Integrating equation (22) from 0.3 to 30 AU gives a total MMSN mass of 0.013 $M_\odot$. This value is well within the range of disk masses estimated from submillimeter photometry. As long as giant planets are not found around a very high fraction of stars, it appears that increasing the initial disk masses in core-accretion models is justified for solar-mass stars. However, it is very difficult to form gas giant planets around low-mass stars (M stars) in a reasonable time with a core-accretion model (*Laughlin et al.*, 2004).

Another response to short disk lifetimes was to postulate a different mode of giant planet formation through direct collapse of clumps in gravitationally unstable disks, which might possibly form giant planets very rapidly (e.g., *Boss*, 1998). Some models of this process, generally called gravitational instability (GI), suggest that a disk more massive than ~0.1 $M_\odot$ around a roughly solar-mass star should be unstable (*Durisen et al.*, 2007). Other analyses apparently preclude giant planet formation through GI within several tens of AU of a solar-mass star (e.g., *Rafikov*, 2007). It is important to note that detailed, time-dependent heating and cooling calculations, which are of critical importance, have not yet been incorporated into any GI models. To summarize the current state of affairs, it seems that giant planet formation through GI might occur in particular situations: in the most massive disks and in the outermost regions of disks (several tens to ~200 AU).

Debris disks have much lower masses than primordial and transitional disks. Even the brightest ones only contain on the order of a few lunar masses of dust (e.g., *Dent et al.*, 2000). They are optically thin at all radii, both in the vertical and radial directions. Debris disks also have very low gas masses (discussed further in section 3.3). The nondetection of submillimeter CO emission from the vast majority of debris disks suggests that primordial gas has completely dissipated by this stage (e.g., *Zuckerman et al.,* 1995). Upper limits on H_2 in the midplanes of two debris disks support this conclusion (β Pic and AU Mic) (*Lecavelier des Etangs et al.,* 2001; *Roberge et al.,* 2005).

3.2. Observed Structure

As mentioned before, unresolved SEDs provide relatively little information on disk geometries. They have, however, revealed that primordial and transitional disks often appear to be flared, which led to the development of the disk structure models discussed in section 2.1. Recall that if the disks were flat, their outer regions would not intercept much stellar radiation, would be cooler, and would reemit less long-wavelength flux than observed in their SEDs. Images of a few edge-on protoplanetary disks support this scenario; an example is the HH 30 disk shown in Fig. 4 (*Burrows et al.,* 1996).

However, there are additional complexities in disk structure. First, there is some temperature above which dust must be vaporized (~1500 K for silicate grains) (*Nagahara et al.,* 1988). In disks around Herbig Ae/Be stars, this sublimation temperature should be reached at a radius between roughly 0.5 to 1 AU (*Dullemond,* 2002). There will be a dust-free inner cavity and an inner rim in the dust disk, which is directly illuminated by the central star (as long as there is no optically thick gas between the star and the rim). The rim will be hotter than expected, producing additional near-IR flux in the disk SED (*Dullemond et al.,* 2001). Note that a central cavity caused by dust vaporization is an expected feature of primordial disks and is not the kind of cavity seen in transitional disks; those are too large to be explained in this way.

This effect explains the "near-IR" bump seen in the SEDs of Herbig Ae/Be stars with disks (*Natta et al.,* 2001). More recently, *Cieza et al.* (2005) found evidence for this phenomenon in the disks of T Tauri stars, although the rims appear to be at larger radii than expected based on the dust sublimation temperature and calculated disk temperature profiles. This may be a consequence of neglecting additional rim heating by excess stellar UV radiation associated with accretion onto the star (*Muzerolle et al.,* 2003). The errors introduced by neglecting UV excess will be more severe for stars with cooler photospheres that do not produce significant UV flux (i.e., T Tauri stars).

Due to its increased temperature, the disk rim will also have a larger scale height than expected. Such a "puffed up" rim could shadow areas of the disk further from the star, reducing the amount of stellar irradiation reaching the outer disk and lowering the outer disk temperatures. This effect tends to make the disk surface brightness profile more like that of a flat disk, which has steeper temperature and density profiles (*Dullemond,* 2002). At large radii, the disk might emerge from the rim shadow and become flaring again, or the self-shadowing might extend over the whole disk.

With all these possible complexities, even for the simplest primordial disks, it is not surprising that the relatively

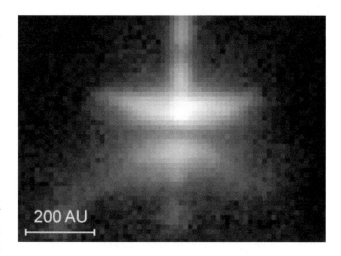

Fig. 4. The edge-on, flaring disk around HH 30 (*Burrows et al.,* 1996). Although it is not apparent in this image, this transitional disk has an inner dust cavity (r ≈ 37 AU), which was cleared by a companion star (*Guilloteau et al.,* 2008). The vertical linear feature is emission from hot gas in a bipolar jet driven by accretion of disk material onto the pre-main-sequence star. Image credit: C. Burrows (STScI), the WFPC2 Science Team, and NASA.

few disks imaged at high spatial resolution appear diverse (*Watson et al.,* 2007, and references therein). They show power-law surface brightness profiles with a range of exponents, from shallow profiles indicative of flared disks to steep ones that might indicate self-shadowing. Also, the exponent of the surface brightness power-law often changes with radius over a single disk.

However, images of disks have provided vital information on their basic geometries. At present, the most detailed disk images come from coronagraphic imaging of scattered light at optical and near-IR wavelengths. These images only detect light scattered from small grains with sizes comparable to the wavelength of observation, but are not sensitive to grain temperatures. This technique has been particularly fruitful for debris disks. They are not typically located in crowded regions of high IS extinction, i.e., star-forming regions, so they are not obscured at short wavelengths.

Disks have also been imaged in thermal emission at IR, submillimeter, and millimeter wavelengths. These images generally have low spatial resolution compared to scattered light images, but can detect emission from larger grains and probe deeper layers of the disks. Thermal images at different wavelengths probe grains with different temperatures; the longer the wavelength, the colder the grain.

Many primordial and transitional disks extend to large radii, typically on the order of hundreds of AU. In a few cases, the disk extends over 1000 AU from the central star (e.g., AB Aur outer radius = 1300 AU) (*Grady et al.,* 1999). Scattered light from debris disks has been detected out to even larger radii (e.g., β Pic outer radius = 1800 U) (*Larwood and Kalas,* 2001). There is obviously a selection bias toward imaging large disks. But there are also some small imaged disks, with outer radii on the order of tens of AU. Many (but not all) of these small disks are located in regions of high-mass star formation. It appears that these disks are being evaporated from the outside by intense UV irradiation from nearby O/B stars (e.g., *Richling and Yorke, 2000*). In the Orion nebula, this phenomenon has been directly observed (the Orion "proplyds") (e.g., *Bally et al., 1998*).

3.3. Gas-to-Dust Ratios

As difficult as it is to accurately measure dust masses in protoplanetary and debris disks, it is even more difficult to measure the gas component. Primordial disks resemble IS molecular clouds in various ways; therefore, we generally assume that they have similar compositions and gas-to-dust mass ratios. However, as will become clear, disk gas-to-dust ratios apparently decline to nearly zero quite rapidly, within roughly 10 m.y. or so. The exact nature of this decline and the factors that affect it are not currently understood. Here, we discuss what is known about the bulk gas in protoplanetary and debris disks.

3.3.1. Molecular hydrogen. If protoplanetary disks really do resemble molecular clouds, then most of the gas and total mass in primordial disks is molecular hydrogen. This symmetrical homonuclear diatomic molecule is notoriously difficult to observe, since it has no permanent electric or magnetic dipole moment. Therefore, the only strong, dipole-allowed transitions of H_2 are the transitions between electronic levels, which lie in the UV. The bulk of the gas in protoplanetary disks is far too cold to emit in these transitions, barring the special case of fluorescence. In this process, H_2 molecules are pumped into high electronic levels by absorption of UV photons, then cascade down to lower levels, emitting in a number of UV electronic transitions (and some near-IR transitions too; see below). The far-UV spectra of some classical T Tauri stars (CTTS) show fluorescent H_2 emission lines (e.g., *Herczeg et al., 2006*). However, the emission appears to be associated with disk surfaces or nearby diffuse IS cloud material, low-density environments where stellar UV photons can penetrate, and does not probe the bulk of the disk mass.

In a few cases, H_2 absorption lines have been observed in far-UV spectra of young stars (e.g., *Roberge et al., 2001*). Since these lines arise from many low-lying energy levels, they can sensitively probe gas at essentially any temperature, but only in relatively optically thin material lying along the line of sight to a UV-bright background source (i.e., the central star). The H_2 observed in absorption near Herbig Be stars appears to be mainly associated with large-scale CS envelopes, displaying characteristics similar to those of IS photodissociation regions (PDRs) (*Martin-Zaidi et al., 2008*). However, the H_2 absorption seen toward Herbig Ae stars is not consistent with PDR models. The gas appears to be associated with warm/hot material located close to the central stars; in at least two cases, the gas may lie in puffed-up inner disk rims (*Lecavelier des Etangs et al., 2003*). While these results are intriguing, UV absorption spectroscopy of H_2 does not probe the bulk of the disk mass, as in the case of H_2 fluorescence.

The mid-IR pure rotational and near-IR rovibrational emission lines of H_2 probe gas at temperatures of a few hundred and a few thousand Kelvins, respectively. Unfortunately, they are electric quadrupole transitions and are roughly a billion times weaker than the UV transitions. Nonetheless, near-IR H_2 emission is commonly seen in star-formation regions, typically either UV-pumped fluorescent emission from diffuse IS cloud material or shocked H_2 emission associated with Herbig-Haro flows (for a review, see *Bally et al., 2007*). In only a few cases has near-IR H_2 emission been firmly associated with protoplanetary disks (e.g., *Bary et al., 2003*).

Mid-infrared pure rotational H_2 emission lines probe gas in the warm layers of protoplanetary disks (the light gray regions in Fig. 3). If the emission is optically thin, then the integrated flux in a rotational line with upper level u and lower level l is

$$F_{ul} = \frac{hc}{4\pi\lambda} N_{H_2} A_{ul} x_u \Omega \quad (23)$$

where λ is the wavelength of the emission line, N_{H_2} is the column density of H_2 molecules, A_{ul} is the spontaneous transition probability (or Einstein A coefficient), x_u is the fraction of molecules in the upper level, and Ω is the angular source size. Assuming LTE, the population of the upper level is given by

$$x_u = \frac{(2J_u + 1)g_N \, e^{-E_u/kT_{ex}}}{Q_{H_2}(T_{ex})} \quad (24)$$

where J_u is the angular momentum quantum number of the upper level, g_N is nuclear statistical weight (1 for even J_u and 3 for odd J_u), E_u is the energy of the upper level, T_{ex} is the excitation temperature, and $Q_{H_2}(T_{ex})$ is the partition function for the given excitation temperature. If more than one rotational line is observed, T_{ex} may be calculated; if not, it must be assumed. With slight modifications, these equations are generally applicable to other gas emission lines. Measurement of F_{ul} allows one to calculate N_{H_2}, which may be turned into a total H_2 mass if the source size is known. In practice, the source is not usually resolved and assumptions about its shape and size must be made.

Unfortunately, mid-IR pure rotational H_2 emission is especially hard to detect from protoplanetary disks, where the weak lines are superimposed on strong mid-IR dust emission, resulting in low line-to-continuum contrast. Detections of this emission in Infrared Space Observatory spectra of young stars implied large amounts of warm H_2 gas in protoplanetary disks (*Thi et al., 2001*). However, follow-up observations either failed to confirm those detections or indicated that the gas actually lies in large-scale CS envelopes (e.g., *Richter et al., 2002*). Again, mid-IR H_2 emission has been firmly associated with protoplanetary disks in only a few cases (e.g., *Bitner et al.,* 2008). Nondetections of mid-IR H_2 emission from protoplanetary disks with the Spitzer Space Telescope have provided upper limits of tens to a few Jupiter masses of warm (100–200 K) H_2 gas (*Lahuis et al.,* 2007). Taking into account the expected temperature structure of a protoplanetary disk, these upper limits are too high to usefully constrain total disk masses and gas-to-dust ratios. Molecular hydrogen is well-termed "the dark matter of planet formation" (M. Kuchner).

3.3.2. Carbon monoxide. The most commonly used method of detecting bulk gas in protoplanetary disks is therefore submillimeter and millimeter observations of pure rotational emission from CO, the second most abundant molecule in disks after H_2. These strong, electric dipole allowed lines probe gas at temperatures of a few tens of Kelvins to about 100 K. In principle, they provide access to the disk midplane and cool molecular layer. In many cases, the CO emission is spatially resolved, showing that gas disks can extend to several hundred AU from both CTTS and Herbig Ae/Be stars (e.g., *Qi et al.,* 2004). Spectrally resolved CO observations also show that the disks are typically in Keplerian rotation (a notable exception is the disk around the Herbig Ae star AB Aur) (*Piétu et al.,* 2005). For a recent review of submillimeter/millimeter observations of CO in protoplanetary disks, see *Dutrey et al.* (2007).

In practice, submillimeter/millimeter ^{12}CO emission is often optically thick throughout the disk and can only provide a lower limit on the total CO gas mass. The various CO rotational lines are more useful for probing the three-dimensional disk structure. Observations of the ^{12}CO J = 2–1 transition typically show a disk temperature gradient of $T(r) \propto r^{-0.6}$, roughly in agreement with the predictions of flared disk models (*Dutrey et al.,* 2007). Deeper disk layers can be probed by rotational lines of CO isotopologs (e.g., ^{13}CO), since they are sometimes optically thin in the outer regions of disks. Such observations have revealed vertical gas temperature gradients in a few disks, as predicted by the gas disk models discussed in section 2.1 (e.g., *Piétu et al.,* 2007). Furthermore, it appears that the surfaces and midplanes of disks around Herbig Ae/Be stars are hotter than those of CTTS disks; in one case (AB Aur), the gas temperature never goes below the point at which CO can freeze out on dust grains (17 K). In CTTS disks, the temperature beyond ~150 AU from the central star is below the CO freeze-out temperature; however, there is still some observable gas phase CO in these outer regions (see section 2.3 for an explanation).

This leads us to the difficulties involved with determining total disk gas masses from submillimeter/millimeter CO emission. In addition to saturation of the emission lines, the possibility that CO may freeze out on grains makes it difficult to extrapolate a CO mass to a total gas mass using an assumed CO-to-H_2 ratio. In some Herbig Ae/Be disks, the temperatures can be high enough that CO depletion onto grains cannot occur and a CO-to-H_2 ratio typical of a star-forming region may be appropriate. This is obviously not the case for disks around cooler stars. Another source of uncertainty in the CO-to-H_2 ratio is CO photodissociation by stellar and IS UV radiation in optically thin outer regions of disks (e.g., *Kamp and Bertoldi,* 2000).

One point does appear clear. Bona fide weak-line T Tauri stars (WTTS) do not display submillimeter CO emission (e.g., *Duvert et al.,* 2000). By definition, WTTS spectra do not show the broad and bright high-temperature emission lines that signal strong accretion of disk gas onto the central stars. Their SEDs show no near-IR excess emission, indicating that small dust has been cleared from their inner regions. The lack of cold CO emission suggests that outer disk gas has also been largely dispersed, on a timescale not terribly different from that of the inner small dust; this issue is discussed further in section 3.5.

At this time, observations of gas in disks only provide relative gas abundances at best. Total disk gas masses cannot be accurately determined from spectra of a single gas species, although observations of several gas species in conjunction with improved thermochemical disk models may provide answers soon (see section 5.1). In sum, there is little to do at the moment but assume a gas-to-dust mass ratio of 100:1 for primordial and transitional disks.

3.3.3. Debris gas. As mentioned earlier, submillimeter CO emission has been detected from only one debris disk (*Dent et al.,* 2005). The lack of CO emission has been interpreted as a sign that the primordial gas has been completely removed from debris disks and that their gas-to-dust ratios are effectively zero (e.g., *Zuckerman et al.,* 1995). Nondetections of far-UV H_2 absorption in two edge-on debris disks support the first part of the preceding statement. The upper limits on the H_2-to-dust ratios in β Pictoris and AU Microscopii are 3:1 and 6:1, respectively (*Lecavelier des Etangs et al.,* 2001; *Roberge et al.,* 2005).

While the gas-to-dust ratios in debris disks are certainly low, at least some do contain gas. The gas is usually detected through absorption lines superimposed on optical/UV spectra of the central stars. This technique is very sensitive to small amounts of cold gas in optically thin environments like debris disks but only probes the line of sight to the central star. Therefore, it is important to be sure that the absorption lines arise in CS (not IS) material, which is the case for all the debris disks with firm gas detections. Either the absorption lines are too strong to be IS given the known distance to the star or lines arising from excited energy levels are present. Fairly high gas densities are needed to populate these levels through collisions; such absorption lines are not seen in the low-density diffuse interstellar medium (ISM).

Currently, there are eight debris disks known to have CS gas: 49 Cet (*Dent et al.,* 2005), β Pic (e.g., *Lagrange et al.,* 1998), 51 Ophiuchi (e.g., *Roberge et al.,* 2002), σ Herculis (*Chen and Jura,* 2003), HD 32297 (*Redfield,* 2007), HD 158352, HD 118232, and HD 21620 (*Roberge and Weinberger,* 2008). With the possible exception of 49 Cet, the gas seen is primarily atomic and ionic, in contrast to the molecular gas in primordial and transitional disks. This is because the low dust masses in debris disks make them completely optically thin to dissociating stellar and IS UV radiation.

Several lines of evidence indicate that the gas in these systems is secondary material produced from comets and asteroids, just like the dust. Many of the species seen have short photoionization lifetimes in optically thin environments, showing that the gas has been recently released from dust grains and/or larger solid bodies. Stellar radiation pressure on some observed species should blow them out of the disks on fairly short timescales, again indicating recent production. Finally, in two well-studied cases, the measured gas abundances are very unusual (51 Oph and β Pic) (*Roberge et al.*, 2002, 2006). The primary production mechanisms for debris gas are currently unknown, but may include photon-stimulated desorption from dust grains (*Chen et al.*, 2007) and/or grain-grain collisions (*Czechowski and Mann*, 2007).

The only debris disk whose gas is well-characterized is β Pic, one of the first four debris disks discovered (*Gillett*, 1986). This disk is very close to edge-on, so the line of sight to the central star probes the disk midplane. Circumstellar absorption lines in spectra of β Pic were studied even before it was known to have a dust disk (*Slettebak*, 1982). The absorption features fall into one of two categories. Every line observed shows an unvarying narrow absorption at the velocity of the star. This component contains the bulk of the CS gas and is called the stable component. On the wings of most absorption lines are broad, variable absorption features, which are typically redshifted with respect to the star. These features arise from gas falling toward the star at high velocity, and are produced by vaporization of star-grazing planetesimals [for more about this interesting phenomenon, see *Beust et al.* (1990)].

Over the last two decades, the line-of-sight column densities of many gaseous species in the β Pic stable gas have been measured [an inventory appears in the online Supplementary Information for *Roberge et al.* (2006)]. Using images of scattered disk gas emission, the three-dimensional structure of the gas disk has been determined (*Brandeker et al.*, 2004). The column densities and the three-dimensional structure may be combined to estimate the total disk gas mass. Preliminary work suggests that the gas-to-dust mass ratio in β Pic is between 0.01 and 1 (A. Roberge, in preparation). Most models of dust dynamics in debris disks do not include gas drag. Some that do suggest that fairly small amounts of gas have an important effect on the grain spatial distributions and the resulting disk scattered light profiles (e.g., *Thebault and Augereau*, 2005); others do not (*Krivov et al.*, 2009).

3.4. Composition

3.4.1. Solids in protoplanetary disks. The most prominent feature of protoplanetary disk spectra is IR dust emission; most is simply continuum emission that provides little information on the grain compositions. However, the spectra also typically show a 10-μm emission feature produced by the Si-O stretching mode in small silicate grains (e.g., *Kessler-Silacci et al.*, 2006). Unlike the smooth triangular 10-μm silicate feature seen in the diffuse ISM, this feature in protoplanetary disk spectra often has a broader trapezoidal shape. The ISM feature is well modeled with submicrometer-sized amorphous olivine and pyroxene grains and the upper limit on the fraction of crystalline silicates is about 2.2% (*Kemper et al.*, 2004). Apparently, some process in the ISM turns crystalline silicates produced by evolved stars into amorphous ones, although dilution by amorphous silicate dust produced in supernovae remains marginally possible. Regardless, one expects that silicate grains in primordial disks are initially amorphous.

The broad, flat shape of the 10-μm feature in protoplanetary disk spectra has been interpreted as evidence for crystalline silicates (e.g., *Bouwman et al.*, 2001). The crystalline Mg-rich silicate forsterite produces an emission feature at 11.3 μm; in low-resolution spectra, this feature blends with the amorphous feature and broadens the combined emission. However, it has been recognized that growth of amorphous silicate grains above submicrometer sizes also broadens and flattens the 10-μm emission feature (e.g., *Przygodda et al.*, 2003). Therefore, detection of unblended features at longer mid-IR wavelengths is needed to determine the presence of crystalline silicates. Such features are more difficult to observe, but have been seen from many disks around T Tauri and Herbig Ae/Be stars (e.g., *Kessler-Silacci et al.*, 2006).

Temperatures above ~800 K are needed to make crystalline silicates, either through recondensation of vaporized silicates or annealing of amorphous grains (e.g., *Gail*, 2001). Dust grains reach these high temperatures only in the innermost region of a viscously heated disk (at $r \lesssim 1$ AU for a solar-type star) or in shocks associated with spiral density waves occurring at $r \lesssim 10$ AU (e.g., *D'Alessio et al.*, 1998; *Harker and Desch*, 2002). The fact that crystalline silicates appear to be largely confined within ~10 AU is consistent with either scenario, although the former would also require large-scale radial mixing (*van Boekel et al.*, 2004; *Watson et al.*, 2009). However, the lack of correlations between the crystalline fraction and stellar mass, luminosity, stellar accretion rate, disk mass, or disk/star mass ratio indicates that both scenarios are incomplete (*Watson et al.*, 2009). Other materials signifying high-temperature processing of IS dust have been identified in protoplanetary disks (e.g., silica) (*Honda et al.*, 2003; *Sargent et al.*, 2009).

Another common feature of protoplanetary disk spectra is IR emission from PAHs, the smallest "grains" in disks. Since particular conditions must be met for PAH formation, it is generally thought that they cannot form or grow in protoplanetary disks and therefore originate in the parent IS molecular clouds (e.g., *Visser et al.*, 2007), although the issue has been reopened for discussion (*Woods and Willacy*, 2007). At any rate, PAHs can be photodissociated in protoplanetary disks by absorption of multiple UV photons. However, to produce IR emission features at all, PAHs must absorb UV photons to become excited and cool through emission in C–H and C–C stretching and bending modes. Therefore, IR PAH emission must arise in optically thin regions of disks (like the surface layers and outermost regions) and is very sensitive to the radiation field. This is why Herbig Ae/Be disks show IR PAH emission more often and more strongly than disks around the cooler T Tauri stars (*Acke and van den Ancker*, 2004; *Geers et al.*, 2006). Apparently, UV-excess emission from stellar accretion in CTTS does not compensate for their faint UV continua.

Given the cold temperatures in large parts of protoplanetary disks, a great deal of material should be present as ices, probably coating the surfaces of dust grains. Ice absorption features are often seen in spectra of young stellar objects; however, the ices apparently lie in the ambient molecular cloud material and CS envelopes (e.g., *Pontoppidan et al.*, 2008). Ices in disks are harder to detect, although water ice emission and absorption features have been seen in a few cases (e.g., *Terada et al.*, 2007). No clear evidence for snow lines in disks has yet been seen, either through changes in grain albedo at some radius in coronagraphic scattered light images or through spatially resolved spectroscopy of ice emission.

3.4.2. Gas in protoplanetary disks. As discussed above, gas observations are intrinsically difficult and have focused until recently on strong rotational lines of abundant molecules such as CO, HCN, HCO^+ (e.g., *Koerner et al.*, 1993; *Dutrey et al.*, 1997; *Thi et al.*, 2004). These lines generally trace the outer cooler regions of protoplanetary disks and/or layers at intermediate heights, where the stellar UV radiation is sufficiently shielded to suppress photodissociation, but still provides enough ionization to drive a rich ion-molecule chemistry (*Bergin et al.*, 2007). However, the sensitivity of current radio telescopes only allows observations of small samples (*Dent et al.*, 2005) and detailed studies of a few individual objects (e.g., *Qi et al.*, 2003; *Semenov et al.*, 2005; *Qi et al.*, 2008).

The optically thin surfaces and hot inner regions of disks can produce rovibrational line emission in the near-IR, either through fluorescence and/or thermal excitation (e.g., *Brittain et al.*, 2007). More recently, near- and mid-IR gas emission lines have been seen in Spitzer spectra of protoplanetary disks. Water, OH, and simple organic molecules were detected in the terrestrial planet regions, suggesting that ices from cold regions have been transported inward or mixed upward, then vaporized (*Salyk et al.*, 2008; *Carr and Najita*, 2008). The spectra also commonly show surprisingly strong atomic and ionic gas emission from disk surfaces; the presence of Ne II highlights the importance of stellar X-rays (*Pascucci et al.*, 2007; *Lahuis et al.*, 2007).

3.4.3. Debris disk dust. Mid-infrared spectra of a few debris disks show silicate dust emission features superimposed on the dust continuum emission; the features have been modeled using grains similar to solar system comet dust (e.g., *Beichman et al.*, 2005). Although the detailed grain composition is somewhat ambiguous, it does seem clear that a mixture of small amorphous and crystalline silicates is required. However, the mid-IR spectra of most debris disks are featureless (*Chen et al.*, 2006). For some disks, this may be because they do not have grains warm enough to emit at mid-IR wavelengths. But even disks with mid-IR excess emission may not show spectral features.

This is due to the fact that most grains in these systems are larger than about 10 μm and radiate basically like blackbodies. Smaller grains that might produce spectral features are probably rapidly removed by radiation pressure. No matter the reason, mid-IR spectra of debris disks have provided little information about the typical grain composition.

Another way to get information on dust composition is to look at the broadband colors of debris disks in scattered light images. Figure 5 shows the albedos of some materials in the optical and near-IR. There are at least seven debris disks that have been imaged in scattered light in more than one bandpass: HD 32297 (*Schneider et al.*, 2005), AU Mic (*Krist et al.*, 2005), HD 15115 (*Kalas et al.*, 2007), HD 92945 (*Golimowski et al.*, 2007), HR 4796A (*Debes et al.*, 2008), β Pic (*Golimowski et al.*, 2006), HD 181327 (*Schneider et al.*, 2006), and HD 107146 (*Ardila et al.*, 2004). HD 32297, AU Mic, and HD 15115 show the blue color expected for very small silicate grains. HD 92945 displays neutral scattering characteristic of large silicate grains. Somewhat surprisingly, the others show red colors suggestive of organic material. Icy bodies in the outer solar system commonly have red scattered light colors indicative of complex organics produced by UV and cosmic-ray reactions in ices (e.g., *Cruikshank and Dalle Ore*, 2003). However, slightly red scattered light colors might be produced by intermediate-sized silicate grains, as seen in Fig. 5.

There is more information available for the HR 4796A disk, which has been imaged in scattered light at seven optical and near-IR wavelengths. The disk's scattering coefficient has a steep red slope inconsistent with any size distribution of pure silicate grains but consistent with complex organic molecules on the grain surfaces (i.e., tholins) (*Debes et al.*, 2008). However, the color of the HR 4796A disk might also be explained by porous grains composed of amorphous silicates, amorphous carbon, and water ice (*Kohler et al.*, 2008). In sum, debris disk grain compositions are not known for certain, although there are intriguing hints of organic material as well as silicates.

3.4.4. Debris disk gas. Again, β Pic is the only debris disk with a fairly complete measured gas composition. Far-UV absorption spectroscopy has revealed a modest amount of cold CO in the stable gas (*Roberge et al.*, 2000). The lower limit on the CO-to-H_2 ratio shows that the CO is overabundant, rather than depleted as is common in primordial and transitional disks (*Lecavelier des Etangs et al.*, 2001). This strongly suggests that the CO is produced by evaporation of icy comet-like material; solar system comets contain abundant CO, but most hydrogen is in water, not H_2.

Most measured elements in the β Pic stable gas have roughly solar abundance relative to each other, as does the central star (*Roberge et al.*, 2006). The exception is carbon, which is extremely overabundant relative to every other measured element (e.g., C/O = 18× solar). Little of the carbon comes from photodissociation of CO. The carbon overabundance suggests that a previously unsuspected and currently unknown process operates in the disk, during which the planetesimals preferentially lose volatile carbo-

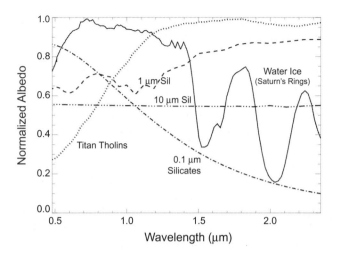

Fig. 5. Optical and near-IR albedo spectra for various grains. The solid line shows the albedo spectrum of Saturn's rings, displaying strong absorption bands of water ice. Silicate grains have basically featureless albedo spectra at these wavelengths. Very small grains in an optically thin disk will produce blue scattered light colors due to Rayleigh scattering (dot-dashed line). Grains larger than about 2 μm will scatter neutrally (dot-dot-dot-dashed line). The calculated albedo for 1-μm astronomical silicates from *Draine and Lee* (1984) is shown by the dashed line, which indicates that grains with intermediate sizes produce slightly red scattering. However, some other silicate grain models show neutral-blue scattering for intermediate-sized grains. The dotted line shows the albedo spectrum of tholins, which are complex organic molecules produced from simple carbon-bearing species (e.g., methane) by UV and cosmic ray reactions. Tholins appear abundant in the atmosphere of Saturn's moon Titan. Image credit: A. Weinberger (Carnegie DTM).

naceous material. In addition, radiation pressure from the central star is weaker for carbon than for most atomic and ionic species. Therefore, the carbon overabundance allows the whole β Pic gas disk to resist stellar radiation pressure and remain in Keplerian rotation out to hundreds of AU (for details, see *Roberge et al.,* 2006; *Fernández et al.,* 2006).

3.5. Lifetimes

The timescale for dissipation of abundant primordial material left over from star formation is critical for models of planetary system formation. First, the primordial gas dissipation timescale limits the time available for formation and migration of giant planets. Second, the amount of gas remaining during terrestrial planet formation affects the final dynamical arrangement of the planets. While the primordial gas in the disk appears to be largely dissipated by the time Earth-mass planets are fully formed (approximately tens of millions of years), there is likely to be significant gas present during the formation of Mars-sized bodies (less than approximately a few million years). These bodies may excite density waves in the gas disk, leading to a net torque on the planets, loss of their orbital angular momentum,

and their inward migration. Models indicate that it takes a Mars-mass body less than ~1 m.y. to spiral into the central star (*Kominami et al.,* 2005). Obviously, this is a serious problem for terrestrial planet formation. It is even more serious for the formation of giant planet cores, which definitely form in a dense gas disk. Many people are working on theoretical ways to halt or slow planet migration (see chapter by Lubow and Ida).

On the other hand, a small amount of gas present at the late stages of formation helps damp the eccentricities and inclinations of terrestrial planets to produce circular, coplanar orbits (e.g., *Nagasawa et al.,* 2005). So if there is too much gas present during terrestrial planet formation, the protoplanets are rapidly destroyed by migrating onto the central star. But if there is too little, the final planet orbits may not resemble those seen in the solar system. The current terrestrial planet formation models predict that the details of gas dissipation between 1 and 10 m.y. have an important effect on the final system architectures.

3.5.1. Protoplanetary disk fractions with age. While primordial gas is what really affects planet formation and dynamics, the aforementioned difficulties associated with measuring gas masses mean that current observed disk lifetimes are largely based on disappearance of dust signatures. An early estimate of disk lifetimes used measurements of near-IR excess (i.e., L-band photometry at 3.45 μm) from stars in young clusters of various ages; such excess emission arises from hot dust within the innermost regions (r ≲ 0.1 AU) of disks (*Haisch et al.,* 2001). This study showed a high initial disk fraction (fraction of stars with a detectable near-IR excess), which rapidly declined to about 50% within about 3 m.y. and nearly zero within an average age of about 6 m.y. Mid-infrared and submillimeter photometric surveys of T Tauri stars indicate that, in general, disks lacking near-IR excess also lack cooler outer dust (r ~ 1–100 AU) (*Cieza et al.,* 2005; *Andrews and Williams,* 2005).

Over the last several years, IR photometry has been obtained for larger numbers of stars and other young clusters, primarily using Spitzer. These observations were at longer near- and mid-IR wavelengths, making them more sensitive to cooler dust in the presumed planet-forming regions of disks [for a recent review of disk dissipation, see *Pascucci and Tachibana* (2010)]. A compilation of recent results is shown in Fig. 6. The primordial disk fraction declines to ~1% for roughly solar-mass stars (0.8 $M_\odot$ < $M_\star$ < 1.8 $M_\odot$) with ages ≥8 m.y. There is evidence for more rapid disk dissipation around high-mass stars ($M_\star$ ≥ 1.8 $M_\odot$). But the picture is still far from clear, with a negligible fraction of transitional disks found around roughly solar-mass stars and conflicting evidence about the evolution timescale for disks around low-mass stars, possibly due to incompleteness of the samples. As pointed out by *Hillenbrand* (2005), the dispersion in apparent disk lifetimes within particular clusters is at least a factor of a few, possibly an order of magnitude. Nonetheless, roughly solar-mass single stars appear to disperse their primordial dust disks within about 10 m.y., with the caveat that particular disks may have much shorter or somewhat longer lifetimes.

Do these dust lifetimes actually reflect gas dissipation timescales? Or more generally, do the gas and dust evolve in concert? To some extent, we expect that they do not. As protoplanetary disk grains grow, becoming invisible to observers, they should eventually settle out of the gas and collect in the disk midplane, accumulating into planetesimals, then planets. Most such material should be retained by the star as the gas dissipates. As mentioned earlier, there is observational evidence of up to centimeter-sized grains in protoplanetary disks. Various observations of both dust and some gas species in a few disks show signs of gas-dust decoupling and dust settling (e.g., *Rettig et al.*, 2006; *Fedele et al.*, 2008). However, little is known about this sort of vertical stratification or how emission from small dust grains reflects the gas abundance at any particular time (the evolution of the gas-to-dust ratio, again).

However, there is indirect evidence that the disappearance of small dust in inner disks does say something about the gas evolution. By definition, WTTS spectra lack emission lines arising from accretion, indicating that disk gas is no longer falling onto the stars. Gas has probably been removed from their inner disks. Recall that WTTS also do not show submillimeter/millimeter CO emission, suggesting low total disk gas masses. These two items imply that inner and outer disk gas dissipate on similar timescales. Furthermore, for T Tauri stars, the presence of accretion signatures and the detection of near-IR excess emission (indicative of inner disk dust) are strongly correlated (*Hartigan et al.*, 1995). So there appears to be a connection between disk dust (traced by IR excess), inner disk gas (traced by accretion signatures), and outer disk gas (traced by submillimeter/millimeter CO emission). The similar timescales for disappearance of all these observables suggests that once disk evolution begins, it progresses rapidly throughout the whole disk within about 10 m.y. Tighter constraints on gas dissipation await sensitive far-IR and submillimeter/millimeter spectroscopy (see section 5.1).

Another important complication is the effect of binarity on disk evolution. First, protoplanetary disks have been observed around binary stars, both as disks around the individual stars and as circumbinary disks. In addition, mature planets have been found in multiple stellar systems. The circumbinary disks around very close binaries and the CS disks around wide binaries are not noticeably different from those around single stars (e.g., *Mathieu et al.*, 2000). However, binaries with separations between about 5 and 30 AU have a lower disk frequency (*Kraus et al.*, 2008). Apparently, the circumbinary disks in these medium-separation systems disperse more quickly, within about 1–2 m.y., probably because the secondary stars have orbits that can dynamically disrupt the disks. Most photometric disk surveys contain both single and binary stars. Therefore, the disk lifetimes for single stars may be somewhat underestimated.

3.5.2. Debris disk fractions and current detection limits. Debris disks have a wide range of ages, although younger disks tend to have greater amounts of dust than older ones. The 24- and 70-μm excess emission from a large number of A-type main-sequence stars with various ages was measured using Spitzer (e.g., *Su et al.*, 2006). The 24-μm emission comes from warm dust located about 5 to 50 AU from the central stars, while the 70-μm emission comes from cold dust at larger distances (between about 50 and 200 AU). The upper envelope of the excess distribution as a function of age falls off as $t_\star^{-1}$, with a characteristic decay time around 150 m.y. for warm dust and a slower decay for cold dust. However, there is a great deal of scatter in the excesses seen from disks of nearly the same age. This was initially explained by stochastic dust production through recent large collisions between planetesimals (*Rieke et al.*, 2005). However, newer modeling indicates that the power-law decay is consistent with steady collisional evolution (except for a few unusually bright disks); the dispersion in dust excesses is largely due to the expected spread in initial disk masses and radii (*Wyatt et al.*, 2007).

Overall, 32% ± 5% of field A-stars have warm dust detectable with Spitzer, while at least 33% ± 5% have cold dust (*Su et al.*, 2006). Both these values are really lower limits, since Spitzer's sensitivity to 24-μm excess emission was limited by its photometric accuracy. These results imply quite high

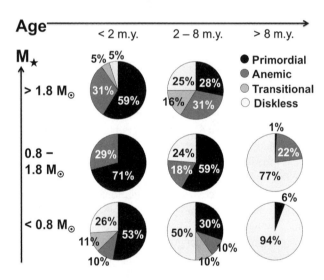

Fig. 6. Percentages of disk types as functions of stellar age and mass. Age increases from left to right and stellar mass decreases from top to bottom. These results are from several large Spitzer surveys of star-forming regions and stellar associations with different mean ages and masses (*Merín et al.*, 2008; *Harvey et al.*, 2008; *Evans et al.*, 2009; *Hartmann et al.*, 2005; *Lada et al.*, 2006; *Silverstone et al.*, 2006; *Carpenter et al.*, 2009). In these charts, primordial and transitional disk fractions are indicated in black and light gray. An additional disk class, "anemic," is indicated in dark gray. This class is defined as having an SED shape similar to that of a primordial disk, but considerably fainter at all IR wavelengths. Stars with no detectable IR excess emission are called "diskless" and are indicated in white. However, some of these stars may have debris disks that are below the survey detection levels. Image credit: B. Merín (ESA).

planetesimal formation efficiency around early type stars. On the other hand, Spitzer surveys of smaller sets of F-, G-, K-, and M-type stars showed somewhat lower disk fractions and only weak evidence of declining dust abundance with increasing age (e.g., *Bryden et al.*, 2006; *Gautier et al.*, 2007; *Carpenter et al.*, 2009). This may not reflect the true disk fraction around late-type stars, however, since photometry of IR excess emission can detect lower dust abundances around higher-luminosity stars. Assuming the stellar spectrum is well-described by a Rayleigh-Jeans law (as it typically is at IR wavelengths) and the dust emission is single-temperature blackbody radiation, the fractional IR luminosity may be expressed as

$$\frac{L_{IR}}{L_\star} = \left(\frac{F_{IR}}{F_\star}\right) \frac{kT_d^4 \left(e^{h\nu/kT_d} - 1\right)}{h\nu T_\star^3} \quad (25)$$

where F_{IR} and $F_\star$ are the dust and stellar fluxes at some frequency ν. The brighter and hotter the star, the lower the $L_{IR}/L_\star$ value and dust mass that can be detected. All else being equal (T_d and the S/N of the observations), the minimum $L_{IR}/L_\star$ value that can be detected for an A5 star is about 35% of the minimum value for a G5 star.

In the solar system, interplanetary dust interior to the astroid belt is called the zodiacal dust, which comes from asteroid collisions and comet comae, just like the dust in any debris disk. The measured $L_{IR}/L_\star$ value for the zodiacal dust is about 10^{-7} (*Dermott et al.*, 2002). In the context of direct detection of exoplanets, debris dust around other stars is called exozodiacal dust and its abundance typically described in units of "zodis." If the exozodiacal dust has exactly the same properties as the zodiacal dust, a one-zodi disk has the same total mass and surface brightness as the zodiacal dust. In practice, one zodi simply corresponds to a dust disk with $L_{IR}/L_\star = 10^{-7}$. Measurements of ^{3}He in seafloor sediments indicate that the amount of zodiacal dust impacting Earth has varied over the last ~80 m.y., with some spikes in the impact rate linked to particular collisions between asteroids (e.g., *Farley et al.*, 2006).

For warm dust located around 1 AU from solar-type stars, the lowest $L_{IR}/L_\star$ value that could be detected with Spitzer is about 10^{-4} (*Beichman et al.*, 2006). This limit corresponds to about 1000 zodis. For cooler dust between about 5 and 10 AU from solar-type stars, Spitzer could detect $L_{IR}/L_\star \geq 10^{-5}$ (≥ 100 zodis). Spitzer's sensitivity to debris dust around nearby solar-type stars is shown in Fig. 7. Given these high detection thresholds, the lower disk fractions for late-type stars compared to A-type stars do not necessarily indicate lower planetesimal formation efficiency in the terrestrial planet regions of solar-type stars. The limits do suggest that asteroid belts more than about 10× more massive than the solar system's are rare around solar-type stars with ages greater than 1 G.y. (*Beichman et al.*, 2006).

So far, there is no significant correlation between the presence of a debris disk and the presence of a gas giant exoplanet or the stellar metallicity (*Bryden et al.*, 2006;

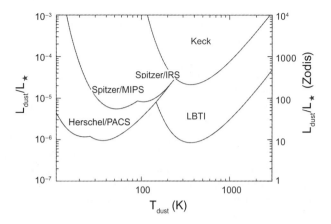

Fig. 7. Sensitivity limits for detection of debris dust around nearby solar-type stars, for various current (Spitzer, the Keck Interferometer, and the Herschel Space Observatory) and upcoming facilities (the Large Binocular Telescope Interferometer). The Herschel curve is based on the prelaunch sensitivity estimate for the PACS instrument. The curves show 3σ detection limits in terms of the dust's fractional IR luminosity vs. its temperature. On the right y-axis, $L_{IR}/L_\star$ is given in units of zodis. The assumed 1σ accuracies used for these curves are 20% of the stellar flux for Spitzer/MIPS at 70 μm, 2.5% for Spitzer/IRS at 32 μm, S/N = 10 for Herschel/PACS at 100 μm, S/N = 2 for Herschel/PACS at 160 μm, 0.5% null for the Keck Interferometer at 10 μm, and starlight removal to 0.01% for LBTI at 10 μm. Image credit: G. Bryden (JPL).

Greaves et al., 2006). This suggests that planetesimals can readily form in disks with relatively low masses of solids, in contrast to gas giant planets, which are more often found around stars with high metallicity. However, the exact implications of this result for planet formation theory are not yet clear.

4. RECENT HIGHLIGHTS

Important recent advances and outstanding questions in protoplanetary and debris disk studies, which were discussed in the previous sections, are summarized here.

1. The transitional disks, with their signs of inner disk clearing, raise the exciting possibility of observing sudden disk changes that may be related to planet formation (section 1.3). But diverse explanations for their observed SEDs calls for intensive study to firmly determine their evolutionary status.

2. Detailed new steady-state thermochemical models of protoplanetary disks show complex, layered gas temperature and chemical structures (sections 2.1 and 2.2). However, the interactions between dynamics, chemistry, and temperature have not yet been fully captured in any disk models.

3. Theoretical studies of protoplanetary disk dispersal now suggest that viscous evolution followed by UV pho-

toevaporation is the dominant mechanism, although other processes must certainly play a role (section 2.4).

4. Images of dust structures in debris disks have finally begun to fulfill their promise as an indirect exoplanet detection technique (section 2.5).

5. The observed dust masses of many protoplanetary disks imply total disk masses large enough for gas giant planets to form through core accretion in reasonable times (section 3.1). Some massive primordial disks may be gravitationally unstable, possibly allowing a different mode of gas giant planet formation (gravitational instability).

6. Poor knowledge of the gas component in protoplanetary and debris disks hampers efforts to model planet formation and migration, as well as debris disk dynamics (section 3.3).

7. The scattered light colors of debris disks (and the gas composition in one debris disk) show intriguing signs of carbon-rich material, possibly complex organics on the grain surfaces (section 3.4).

8. Protoplanetary disk lifetimes seem to have an inherently large scatter, although nearly all disks apparently dissipate within ~10 m.y. (section 3.5). However, this conclusion is largely based on the disappearance of dust rather than declining total disk masses.

5. FUTURE PROSPECTS

The opportunities for new disk studies in the next decade are almost too numerous to mention. Three major facilities will be available: the Herschel Space Observatory, the James Webb Space Telescope (JWST), and the Atacama Large Millimeter Array (ALMA). SOFIA and the Large Binocular Telescope Interferometer (LBTI) will also make significant contributions. Here we highlight two particularly exciting areas where we expect tremendous progress in the near future.

5.1. Disk Gas Observations

The successful launch of Herschel on May 14, 2009, will revolutionize the study of gas in protoplanetary disks. This far-IR/submillimeter facility can sensitively observe the dominant cooling lines from the disk surface, [O I] and [C II], as well as many additional molecular tracers of the warmer inner disk such as water and CO. Protoplanetary gas observations are the focus of several guaranteed and open-time key programs, such as "Water in Star Forming Regions with Herschel" (WISH) (P.I.: E. van Dishoeck), "HIFI Spectral Surveys of Star Forming Regions" (P.I.: C. Ceccarelli), and "Gas in Protoplanetary Systems" (GASPS) (P.I.: W. Dent).

The JWST will provide medium-resolution (up to R = 3000) spectroscopy between 5 and ~28 μm. With its much higher sensitivities, spectral studies of faint objects will be possible, such as disks around brown dwarfs. In addition, JWST will reveal weaker emission lines that were hard to detect above the continuum in low-resolution Spitzer spectra. Some improvements in disk models will be needed, since the ones described in section 2.1 do not yet fully contain the physics and chemistry appropriate for the densities and temperatures found at the inner disk rim or in the gas within the dust-free zone, $n_g > 10^{12}$ cm^{-3} and $T_g \gtrsim$ few 1000 K.

At much longer wavelengths, ALMA will study the warm (up to several 100 K) and cold molecular gas in disks of all ages (*van Dishoeck and Jørgensen*, 2008). *Semenov et al.* (2008) discuss the observational requirements needed to unravel the thermal and chemical structure of young protoplanetary disks. *Kamp et al.* (2005) show that using the standard CO rotational lines, ALMA will be sensitive down to gas masses of ~1 $M_⊕$; this sensitivity should allow studies of tenuous gas in many debris disks.

Combining Herschel, JWST, and ALMA results with modeling advances, we may finally compose a complete picture of the gas component in disks of all ages and arrive at a better understanding of the interpretation and diagnostic power of individual lines. In addition, these observations will test and challenge the current generation of thermochemical disk models, providing crucial feedback for improvement. A suitable combination of gas emission lines, in combination with advanced two-dimensional disk models, will hopefully enable reliable disk gas mass determinations at any stage of evolution. Together with complementary dust observations, gas-to-dust mass ratios in disks may be measured for the first time, eventually leading to a new disk classification scheme based on evolution of the gas-to-dust mass ratio.

5.2. Planets in Debris Disks

New or revitalized high-contrast imaging capabilities have opened the door to a long-discussed method for indirection detection of planets. Ever since the first images of debris disk structures, astronomers have suspected that unseen planets were responsible. Such planets were (and are) inaccessible to other techniques that use the central star, since they would be located at large distances from young, mostly early-type stars. However, until a disk structure and the planet causing it were both observed, this method remained somewhat theoretical. The detection of a planet candidate in the Fomalhaut disk has changed this situation (see Fig. 1). Three planet candidates were also recently imaged around HR 8799 using a high-contrast groundbased adaptive optics system (*Marois et al.*, 2008). The star is only about 30–160 m.y. old, even younger than Fomalhaut, and has an IR excess indicative of a fairly dense debris disk. Unfortunately, the disk has not yet been imaged as of this writing.

These discoveries highlight the complementary investigations needed to develop studies of planets in debris disks, some observational, some theoretical. First, more disks must be imaged. The most sensitive high-resolution images of debris disks to date were obtained with the Advanced Camera for Surveys (ACS) on HST; unfortunately, this instrument no longer has coronagraphic imaging capability. Happily, a new generation of groundbased high-contrast imaging instruments will become available in the next several years (e.g., HiCIAO on the Subaru Telescope, GPI on the Gemini

Telescope, and SPHERE on the VLT). Second, we need more confidence that debris disk dynamical models can uniquely determine the parameters of a planet producing an observed structure. Additional test cases where the disk is imaged *and* the planet detected are required. Since the dust structures that can be seen are at large distances, the planets responsible are best detected with direct imaging. Multi-epoch observations are crucial, to show that the unresolved planet candidate is orbiting the star.

Third, some advances in disk dynamical theory are called for, as detailed in *Wyatt* (2008b). Suffice it to say, models of structures created by planets are typically designed for low-density, P-R drag-dominated systems. These models need to incorporate mutual collisions between dust grains more thoroughly to realistically model the currently observed debris disks. In addition, stellar winds (corpuscular drag) and debris gas must be included. Some steps have been taken to integrate collisions (e.g., *Stark and Kuchner*, 2009) and to estimate the effects of gas on grains in optically thin disks (e.g., *Besla and Wu*, 2007). But much remains to be done, as the numerous explanations for dust structures described in section 2.5 might suggest.

Finally, we return to the impact of debris disks on future imaging and characterization of habitable terrestrial planets, which cannot be underestimated. Although tenuous, the solar system's zodiacal and Kuiper belt dust covers a huge surface area. The magnitude of the exozodiacal dust emission within the point-spread function of a 4-m telescope is $V \approx 28$, assuming a one-zodi disk viewed at 60° inclination. This is about two magnitudes brighter than Earth viewed at quadrature from 10 pc away.

There are three obvious sources of background flux in imaging of habitable terrestrial planets: the local zodiacal dust, the exozodiacal dust in the target system, and unsuppressed light from the target star. Calculations by groups designing future telescopes for direct observations of habitable exoplanets show that the first two are much more important than the third (e.g., *Defrère et al.*, 2008; Turnbull et al., in preparation). In both images and spectra, light from the zodiacal and exozodiacal dust will likely dominate the planet signal. As much as want to know the fraction of stars with habitable planets ($\eta_\oplus$), we also want to know the exozodiacal dust levels around nearby stars, at least down to the level of tens to 100 zodis. This is more than an order of magnitude below our current detection limits for debris dust in the habitable zones of solar-type stars (see section 3.5).

Herschel will dramatically improve our sensitivity to cold dust at Kuiper-belt-like distances. There are two open-time key programs planning sensitive far-IR disk surveys: "Dust Disks Around Nearby Stars" (DUNES) (P.I.: C. Eiroa) and "Disc Emission via a Bias-Free Reconnaissance in the Infrared/Submillimetre" (DEBRIS) (P.I.: B. Matthews). To probe dust in terrestrial planet zones, we must observe at shorter wavelengths. A mid-IR nulling interferometer designed for the Large Binocular Telescope should reach ~10 zodis in these zones for a set of northern hemisphere stars (*Hinz et al.*, 2009). Currently, LBTI is expected to begin operation in 2010. The expected debris dust sensitivity curves for Herschel and LBTI are shown in Fig. 7.

The problem of confusion between a planet and a dust clump, when both are spatially unresolved, has not yet been thoroughly addressed. When a planet causes a structure, it will also orbit the star, although perhaps not with the same period as the planet (e.g., *Kuchner and Holman*, 2003). So images at different epochs will not eliminate dust clumps by themselves. Another suggestion for ruling out dust clumps is multicolor imaging. However, at optical wavelengths, the emission from exozodiacal dust is largely reflected starlight, as is the light from the planet. Clump colors may not be strikingly different from those of planets. In addition, debris disks have diverse scattered light colors (see section 3.4), which will probably be the case for terrestrial planets as well. In the thermal-IR, the clump emission will reflect the dust temperature. This would possibly distinguish between dust and a planet with an atmosphere, if the orbit was known and one had a great deal of confidence in the atmosphere model.

In the end, a spectrum will likely allow us to easily distinguish a dust clump, which should have a featureless optical/IR spectrum, from a habitable exoplanet with an atmosphere. However, any method of ruling out dust clumps before taking a time-consuming spectrum would be valuable. These issues will be studied in the coming years, through development of more realistic exoplanet detection and characterization simulations incorporating improved models of exozodiacal disks and planets.

Acknowledgments. We thank J. Greaves and an anonymous reviewer for their very helpful comments. This work was partially supported by the Goddard Center for Astrobiology, a part of the NASA Astrobiology Institute, and the NASA Herschel Science Center.

REFERENCES

Acke B. and van den Ancker M. E. (2004) ISO spectroscopy of disks around Herbig Ae/Be stars. *Astron. Astrophys., 426,* 151–170.

Aikawa Y. (2007) Cold CO gas in protoplanetary disks. *Astrophys. J. Lett., 656,* L93–L96.

Aikawa Y., Umebayashi T., Nakano T., and Miyama S. M. (1999) Evolution of molecular abundances in protoplanetary disks with accretion flow. *Astrophys. J., 519,* 705–725.

Alexander R. (2008) From discs to planetesimals: Evolution of gas and dust discs. *New Astronomy Review, 52,* 60–77.

Alexander R. D., Clarke C. J., and Pringle J. E. (2006) Photoevaporation of protoplanetary discs — II. Evolutionary models and observable properties. *Mon. Not. R. Astron. Soc., 369,* 229–239.

Alibert Y., Mordasini C., Benz W., and Winisdoerffer C. (2005) Models of giant planet formation with migration and disc evolution. *Astron. Astrophys., 434,* 343–353.

Andrews S. M. and Williams J. P. (2005) Circumstellar dust disks in Taurus-Auriga: The sub-millimeter perspective. *Astrophys. J., 631,* 1134–1160.

Andrews S. M. and Williams J. P. (2007) A submillimeter view of circumstellar dust disks in ρ Ophiuchi. *Astrophys. J., 671,* 1800–1812.

Anglada G., López R., Estalella R., Masegosa J., Riera A., et al. (2007) Proper motions of the jets in the region of HH 30 and HL/XZ Tau: Evidence for a binary exciting source of the HH 30 jet. *Astron. J., 133,* 2799–2814.

Ardila D. R., Golimowski D. A., Krist J. E., Clampin M., Williams J. P., et al. (2004) A resolved debris disk around the G2V star HD 107146. *Astrophys. J. Lett., 617,* L147–L150.

Artymowicz P. (1988) Radiation pressure forces on particles in the beta Pictoris system. *Astrophys. J. Lett., 335,* L79–L82.

Balbus S. A. and Hawley J. F. (1998) Instability, turbulence, and enhanced transport in accretion disks. *Rev. Mod. Phys., 70,* 1–53.

Bally J., Sutherland R. S., Devine D., and Johnstone D. (1998) Externally illuminated young stellar environments in the Orion nebula: Hubble Space Telescope Planetary Camera and ultraviolet observations. *Astron. J., 116,* 293–321.

Bally J., Reipurth B., and Davis C. J. (2007) Observations of jets and outflows from young stars. In *Protostars and Planets V* (B. Reipurth et al., eds.), pp. 215–230. Univ. of Arizona, Tucson.

Bary J. S., Weintraub D. A., and Kastner J. H. (2003) Detections of rovibrational H_2 emission from the disks of T Tauri stars. *Astrophys. J., 586,* 1136–1147.

Beckwith S. V. W., Sargent A. I., Chini R. S., and Guesten R. (1990) A survey for circumstellar disks around young stellar objects. *Astron. J., 99,* 924–945.

Beichman C. A., Bryden G., Gautier T. N., Stapelfeldt K. R., Werner M. W., et al. (2005) An excess due to small grains around the nearby K0 V star HD 69830: Asteroid or cometary debris? *Astrophys. J., 626,* 1061–1069.

Beichman C. A., Tanner A., Bryden G., Stapelfeldt K. R., Werner M. W., et al. (2006) IRS spectra of solar-type stars: A search for asteroid belt analogs. *Astrophys. J., 639,* 1166–1176.

Bergin E., Calvet N., D'Alessio P., and Herczeg G. J. (2003) The effects of UV continuum and Lyα radiation on the chemical equilibrium of T Tauri disks. *Astrophys. J. Lett., 591,* L159–L162.

Bergin E. A., Aikawa Y., Blake G. A., and van Dishoeck E. F. (2007) The chemical evolution of protoplanetary disks. In *Protostars and Planets V* (B. Reipurth et al., eds.), pp. 751–766. Univ. of Arizona, Tucson.

Besla G. and Wu Y. (2007) Formation of narrow dust rings in circumstellar debris disks. *Astrophys. J., 655,* 528–540.

Beust H., Vidal-Madjar A., Ferlet R., and Lagrange-Henri A. M. (1990) The Beta Pictoris circumstellar disk. X — Numerical simulations of infalling evaporating bodies. *Astron. Astrophys., 236,* 202–216.

Bitner M. A., Richter M. J., Lacy J. H., Herczeg G. J., Greathouse T. K., et al. (2008) The TEXES survey for H_2 emission from protoplanetary disks. *Astrophys. J., 688,* 1326–1344.

Boss A. P. (1998) Evolution of the solar nebula. IV. Giant gaseous protoplanet formation. *Astrophys. J., 503,* 923.

Bouwman J., Meeus G., de Koter A., Hony S., Dominik C., et al. (2001) Processing of silicate dust grains in Herbig Ae/Be systems. *Astron. Astrophys., 375,* 950–962.

Brandeker A., Liseau R., Olofsson G., and Fridlund M. (2004) The spatial structure of the β Pictoris gas disk. *Astron. Astrophys., 413,* 681–691.

Brittain S. D., Simon T., Najita J. R., and Rettig T. W. (2007) Warm gas in the inner disks around young intermediate-mass stars. *Astrophys. J., 659,* 685–704.

Brown J. M., Blake G. A., Dullemond C. P., Merín B., Augereau J. C., et al. (2007) Cold disks: Spitzer spectroscopy of disks around young stars with large gaps. *Astrophys. J. Lett., 664,* L107–L110.

Bryden G., Beichman C. A., Trilling D. E., Rieke G. H., Holmes E. K., et al. (2006) Frequency of debris disks around solar-type stars: First results from a Spitzer MIPS survey. *Astrophys. J., 636,* 1098–1113.

Burrows C. J., Stapelfeldt K. R., Watson A. M., Krist J. E., Ballester G. E., et al. (1996) Hubble Space Telescope observations of the disk and jet of HH 30. *Astrophys. J., 473,* 437.

Calvet N., D'Alessio P., Hartmann L., Wilner D., Walsh A., et al. (2002) Evidence for a developing gap in a 10 m.y. old protoplanetary disk. *Astrophys. J., 568,* 1008–1016.

Carpenter J. M., Bouwman J., Mamajek E. E., Meyer M. R., Hillenbrand L. A., et al. (2009) Formation and evolution of planetary systems: Properties of debris dust around solar-type stars. *Astrophys. J. Suppl., 181,* 197–226.

Carr J. S. and Najita J. R. (2008) Organic molecules and water in the planet formation region of young circumstellar disks. *Science, 319,* 1504.

Caselli P., Hasegawa T. I., and Herbst E. (1998) A proposed modification of the rate equations for reactions on grain surfaces. *Astrophys. J., 495,* 309.

Caselli P., Stantcheva T., Shalabiea O., Shematovich V. I., and Herbst E. (2002) Deuterium fractionation on interstellar grains studied with modified rate equations and a Monte Carlo approach. *Planet. Space Sci., 50,* 1257–1266.

Chen C. H. and Jura M. (2003) The low-velocity wind from the circumstellar matter around the B9 V star σ Herculis. *Astrophys. J., 582,* 443–448.

Chen C. H., Sargent B. A., Bohac C., Kim K. H., Leibensperger E., et al. (2006) Spitzer IRS spectroscopy of IRAS-discovered debris disks. *Astrophys. J. Suppl., 166,* 351–377.

Chen C. H., Li A., Bohac C., Kim K. H., Watson D. M., et al. (2007) The dust and gas around β Pictoris. *Astrophys. J., 666,* 466–474.

Chiang E. I. and Goldreich P. (1997) Spectral energy distributions of T Tauri stars with passive circumstellar disks. *Astrophys. J., 490,* 368.

Chiang E. I., Joung M. K., Creech-Eakman M. J., Qi C., Kessler J. E., et al. (2001) Spectral energy distributions of passive T Tauri and Herbig Ae disks: Grain mineralogy, parameter dependences, and comparison with Infrared Space Observatory LWS observations. *Astrophys. J., 547,* 1077–1089.

Chiang E., Kite E., Kalas P., Graham J. R., and Clampin M. (2009) Fomalhaut's debris disk and planet: Constraining the mass of Fomalhaut b from disk morphology. *Astrophys. J., 693,* 734–749.

Cieza L. A., Kessler-Silacci J. E., Jaffe D. T., Harvey P. M., and Evans N. J. II (2005) Evidence for J- and H-band excess in classical T Tauri stars and the implications for disk structure and estimated ages. *Astrophys. J., 635,* 422–441.

Cruikshank D. P. and Dalle Ore C. M. (2003) Spectral models of Kuiper belt objects and Centaurs. *Earth Moon Planets, 92,* 315–330.

Czechowski A. and Mann I. (2007) Collisional vaporization of dust and production of gas in the β Pictoris dust disk. *Astrophys. J., 660,* 1541–1555.

D'Alessio P., Canto J., Calvet N., and Lizano S. (1998) Accretion disks around young objects. I. The detailed vertical structure. *Astrophys. J., 500,* 411.

Debes J. H., Weinberger A. J., and Schneider G. (2008) Complex organic materials in the circumstellar disk of HR 4796A. *Astrophys. J. Lett., 673,* L191–L194.

Defrère D., Lay O., den Hartog R., and Absil O. (2008) Earthlike planets: Science performance predictions for future nulling interferometry missions. In *Optical and Infrared Interferometry* (M. Schöller et al., eds.), pp. 701321 to 701321-12 . SPIE Conference Series 7013, Bellingham, Washington.

Dent W. R. F., Walker H. J., Holland W. S., and Greaves J. S. (2000) Models of the dust structures around Vega-excess stars. *Mon. Not. R. Astron. Soc., 314,* 702–712.

Dent W. R. F., Greaves J. S., and Coulson I. M. (2005) CO emission from discs around isolated HAeBe and Vega-excess stars. *Mon. Not. R. Astron. Soc., 359,* 663–676.

Dermott S. F., Jayaraman S., Xu Y. L., Gustafson B. A. S., and Liou J. C. (1994) A circumsolar ring of asteroidal dust in resonant lock with the Earth. *Nature, 369,* 719–723.

Dermott S. F., Kehoe T. J. J., Durda D. D., Grogan K., and Nesvorny D. (2002) Recent rubble-pile origin of asteroidal solar system dust bands and asteroidal interplanetary dust particles. In *Asteroids, Comets, and Meteors: ACM 2002* (B. Warmbein, ed.), pp. 319–322. ESA SP-500, Noordwijk, The Netherlands.

Dominik C., Blum J., Cuzzi J. N., and Wurm G. (2007) Growth of dust as the initial step toward planet formation. In *Protostars and Planets V* (B. Reipurth et al., eds.), pp. 783–800. Univ. of Arizona, Tucson.

Draine B. T. and Lee H. M. (1984) Optical properties of interstellar graphite and silicate grains. *Astrophys. J., 285,* 89–108.

Dullemond C. P. (2002) The 2-D structure of dusty disks around Herbig Ae/Be stars. I. Models with grey opacities. *Astron. Astrophys., 395,* 853–862.

Dullemond C. P., Dominik C., and Natta A. (2001) Passive irradiated circumstellar disks with an inner hole. *Astrophys. J., 560,* 957–969.

Dullemond C. P., van Zadelhoff G. J., and Natta A. (2002) Vertical structure models of T Tauri and Herbig Ae/Be disks. *Astron. Astrophys., 389,* 464–474.

Dullemond C. P., Hollenbach D., Kamp I., and D'Alessio P. (2007) Models of the structure and evolution of protoplanetary disks. In *Protostars and Planets V* (B. Reipurth et al., eds.), pp. 555–572. Univ. of Arizona, Tucson.

Durisen R. H., Boss A. P., Mayer L., Nelson A. F., Quinn T., et al. (2007) Gravitational instabilities in gaseous protoplanetary disks and implications for giant planet formation. In *Protostars and Planets V* (B. Reipurth et al., eds.), pp. 607–622. Univ. of Arizona, Tucson.

Dutrey A., Guilloteau S., and Guelin M. (1997) Chemistry of protosolar-like nebulae: The molecular content of the DM Tau and GG Tau disks. *Astron. Astrophys., 317,* L55–L58.

Dutrey A., Guilloteau S., and Ho P. (2007) Interferometric spectroimaging of molecular gas in protoplanetary disks. In *Protostars and Planets V* (B. Reipurth et al., eds.), pp. 495–506. Univ. of Arizona, Tucson.

Duvert G., Guilloteau S., Ménard F., Simon M., and Dutrey A. (2000) A search for extended disks around weak-lined T Tauri stars. *Astron. Astrophys., 355,* 165–170.

Evans N. J. II, Dunham M. M., Jørgensen J. K., Enoch M. L., Merín B., et al. (2009) The Spitzer c2d legacy results: Star formation rates and efficiencies; evolution and lifetimes. *Astrophys. J. Suppl., 181,* 321–350, DOI: 10.1088/0067-0049/181/2/321.

Farley K. A., Vokrouhlický D., Bottke W. F., and Nesvorný D. (2006) A late Miocene dust shower from the break-up of an asteroid in the main belt. *Nature, 439,* 295–297.

Fedele D., van den Ancker M. E., Acke B., van der Plas G., van Boekel R., et al. (2008) The structure of the protoplanetary disk surrounding three young intermediate mass stars. II. Spatially resolved dust and gas distribution. *Astron. Astrophys., 491,* 809–820.

Fernández R., Brandeker A., and Wu Y. (2006) Braking the gas in the β Pictoris disk. *Astrophys. J., 643,* 509–522.

Ferreira J. (2008) The inner magnetized regions of circumstellar accretion discs. *New Astron. Rev., 52,* 42–59.

Frank J., King A., and Raine D. (1992) *Accretion Power in Astrophysics.* Cambridge Univ., Cambridge.

Gail H.-P. (2001) Radial mixing in protoplanetary accretion disks. I. Stationary disc models with annealing and carbon combustion. *Astron. Astrophys., 378,* 192–213.

Garrod R. T., Wakelam V., and Herbst E. (2007) Non-thermal desorption from interstellar dust grains via exothermic surface reactions. *Astron. Astrophys., 467,* 1103–1115.

Gautier T. N. III, Rieke G. H., Stansberry J., Bryden G. C., Stapelfeldt K. R., et al. (2007) Far-infrared properties of M dwarfs. *Astrophys. J., 667,* 527–536.

Geers V. C., Augereau J.-C., Pontoppidan K. M., Dullemond C. P., Visser R., et al. (2006) C2D Spitzer-IRS spectra of disks around T Tauri stars. II. PAH emission features. *Astron. Astrophys., 459,* 545–556.

Gillett F. C. (1986) IRAS Observations of cool excess around main sequence stars. In *Light on Dark Matter* (F. P. Israel, ed.), pp. 61–69. ASSL Vol. 124, Reidel, Dordrecht.

Glassgold A. E., Najita J., and Igea J. (2004) Heating protoplanetary disk atmospheres. *Astrophys. J., 615,* 972–990.

Golimowski D. A., Ardila D. R., Krist J. E., Clampin M., Ford H. C., et al. (2006) Hubble Space Telescope ACS multiband coronagraphic imaging of the debris disk around β Pictoris. *Astron. J., 131,* 3109–3130.

Golimowski D., John Krist J., Chen C., Stapelfeldt K., Ardila D., et al. (2007) Observations and models of the debris disk around the K dwarf HD 92945. In *In the Spirit of Bernard Lyot: The Direct Detection of Planets and Circumstellar Disks in the 21st Century* (P. Kalas, ed.), Univ. of California, Berkeley.

Gomes R., Levison H. F., Tsiganis K., and Morbidelli A. (2005) Origin of the cataclysmic late heavy bombardment period of the terrestrial planets. *Nature, 435,* 466–469.

Gorti U. and Hollenbach D. (2004) Models of chemistry, thermal balance, and infrared spectra from intermediate-aged disks around G and K stars. *Astrophys. J., 613,* 424–447.

Gorti U. and Hollenbach D. (2008) Line emission from gas in optically thick dust disks around young stars. *Astrophys. J., 683,* 287–303.

Gorti U. and Hollenbach D. (2009) Photoevaporation of circumstellar disks by far-ultraviolet, extreme-ultraviolet and x-ray radiation from the central star. *Astrophys. J., 690,* 1539–1552.

Grady C. A., Woodgate B., Bruhweiler F. C., Boggess A., Plait P., et al. (1999) Hubble Space Telescope imaging spectrograph coronagraphic imaging of the Herbig Ae star AB Aurigae. *Astrophys. J. Lett., 523,* L151–L154.

Greaves J. S., Fischer D. A., and Wyatt M. C. (2006) Metallicity, debris discs and planets. *Mon. Not. R. Astron. Soc., 366,* 283–286.

Greaves J. S., Richards A. M. S., Rice W. K. M., and Muxlow T. W. B. (2008) Enhanced dust emission in the HL Tau disc: A low-mass companion in formation? *Mon. Not. R. Astron. Soc., 391,* L74–L78.

Guenther E. W., Esposito M., Mundt R., Covino E., Alcalá J. M., et al. (2007) Pre-main sequence spectroscopic binaries suitable for VLTI observations. *Astron. Astrophys., 467,* 1147–1155.

Guilloteau S., Dutrey A., Pety J., and Gueth F. (2008) Resolving the circumbinary dust disk surrounding HH 30. *Astron. Astrophys., 478,* L31–L34.

Haisch K. E., Lada E. A., and Lada C. J. (2001) Disk frequencies and lifetimes in young clusters. *Astrophys. J. Lett., 553,* L153–L156.

Harker D. E. and Desch S. J. (2002) Annealing of silicate dust by nebular shocks at 10 AU. *Astrophys. J. Lett., 565,* L109–L112.

Hartigan P., Edwards S., and Ghandour L. (1995) Disk accretion and mass loss from young stars. *Astrophys. J., 452,* 736.

Hartmann L., Megeath S. T., Allen L., Luhman K., Calvet N., et al. (2005) IRAC observations of Taurus pre-main-sequence stars. *Astrophys. J., 629,* 881–896.

Harvey P. M., Huard T. L., Jørgensen J. K., Gutermuth R. A., Mamajek E. E., et al. (2008) The Spitzer survey of interstellar clouds in the Gould belt. I. IC 5146 observed with IRAC and MIPS. *Astrophys. J., 680,* 495–516.

Hasegawa T. I. and Herbst E. (1993) New gas-grain chemical models of quiescent dense interstellar clouds — The effects of H_2 tunnelling reactions and cosmic ray induced desorption. *Mon. Not. R. Astron. Soc., 261,* 83–102.

Hayashi C. (1981) Structure of the solar nebula, growth and decay of magnetic fields and effects of magnetic and turbulent viscosities on the nebula. *Progr. Theor. Phys. Suppl., 70,* 35–53.

Herczeg G. J., Linsky J. L., Walter F. M., Gahm G. F., and Johns-Krull C. M. (2006) The origins of fluorescent H_2 emission from T Tauri stars. *Astrophys. J. Suppl., 165,* 256–282.

Hillenbrand L. A. (2005) Observational constraints on dust disk lifetimes: Implications for planet formation. In *A Decade of Discovery: Planets Around Other Stars* (M. Livio, ed.), STScI Symposium Series 19, in press, arXiv:astro-ph/0511083v1.

Hinz P., Millan-Gabet R., and the Exoplanet Forum 2008 Exozodiacal Disk Group (2009) Exozodiacal disk characterization. In *2008 Exoplanet Forum Report* (P. R. Lawson et al., eds.), pp. 113–140. Jet Propulsion Laboratory, Pasadena.

Hollenbach D. J., Yorke H. W., and Johnstone D. (2000) Disk dispersal around young stars. In *Protostars and Planets IV* (V. Mannings et al., eds.), pp. 401–428. Univ. of Arizona, Tucson.

Honda M., Kataza H., Okamoto Y. K., Miyata T., Yamashita T., et al. (2003) Detection of crystalline silicates around the T Tauri star Hen 3-600A. *Astrophys. J. Lett., 585,* L59–L63.

Hubickyj O., Bodenheimer P., and Lissauer J. J. (2005) Accretion of the gaseous envelope of Jupiter around a 5–10 Earth-mass core. *Icarus, 179,* 415–431.

Hueso R. and Guillot T. (2005) Evolution of protoplanetary disks: Constraints from DM Tauri and GM Aurigae. *Astron. Astrophys., 442,* 703–725.

Iida A., Nakamoto T., Susa H., and Nakagawa Y. (2001) A shock heating model for chondrule formation in a protoplanetary disk. *Icarus, 153,* 430–450.

Ilgner M., Henning T., Markwick A. J., and Millar T. J. (2004) Transport processes and chemical evolution in steady accretion disk flows. *Astron. Astrophys., 415,* 643–659.

Ireland M. J. and Kraus A. L. (2008) The disk around CoKu Tauri/4: Circumbinary, not transitional. *Astrophys. J. Lett., 678,* L59–L62.

Isaacman R. and Sagan C. (1977) Computer simulations of planetary accretion dynamics — Sensitivity to initial conditions. *Icarus, 31,* 510–533.

Jonkheid B., Faas F. G. A., van Zadelhoff G.-J., and van Dishoeck E. F. (2004) The gas temperature in flaring disks around pre-main sequence stars. *Astron. Astrophys., 428,* 511–521.

Kalas P., Fitzgerald M. P., and Graham J. R. (2007) Discovery of extreme asymmetry in the debris disk surrounding HD 15115. *Astrophys. J. Lett., 661,* L85–L88.

Kalas P., Graham J. R., Chiang E., Fitzgerald M. P., Clampin M., et al. (2008) Optical images of an exosolar planet 25 light-years from Earth. *Science, 322,* 1345.

Kamp I. and Bertoldi F. (2000) CO in the circumstellar disks of Vega and Beta Pictoris. *Astron. Astrophys., 353,* 276–286.

Kamp I. and Dullemond C. P. (2004) The gas temperature in the surface layers of protoplanetary disks. *Astrophys. J., 615,* 991–999.

Kamp I., Dullemond C. P., Hogerheijde M., and Enriquez J. E. (2005) Chemistry and line emission of outer protoplanetary disks. In *Astrochemistry: Recent Successes and Current Challenges* (D. C. Lis et al., eds.), pp. 377–386. IAU Symposium No. 231, Astronomical Society of the Pacific, San Francisco.

Kawabe R., Ishiguro M., Omodaka T., Kitamura Y., and Miyama S. M. (1993) Discovery of a rotating protoplanetary gas disk around the young star GG Tauri. *Astrophys. J. Lett., 404,* L63–L66.

Kemper F., Vriend W. J., and Tielens A. G. G. M. (2004) The absence of crystalline silicates in the diffuse interstellar medium. *Astrophys. J., 609,* 826–837.

Kenyon S. J. and Bromley B. C. (2005) Prospects for detection of catastrophic collisions in debris disks. *Astron. J., 130,* 269–279.

Kenyon S. J. and Bromley B. C. (2006) Terrestrial planet formation. I. The transition from oligarchic growth to chaotic growth. *Astron. J., 131,* 1837–1850.

Kenyon S. J. and Bromley B. C. (2008) Variations on debris disks: Icy planet formation at 30–150 AU for 1–3 $M_\odot$ main-sequence stars. *Astrophys. J. Suppl., 179,* 451–483.

Kenyon S. J. and Hartmann L. (1987) Spectral energy distributions of T Tauri stars — Disk flaring and limits on accretion. *Astrophys. J., 323,* 714–733.

Kenyon S. J. and Luu J. X. (1999) Accretion in the early Kuiper belt. II. Fragmentation. *Astron. J., 118,* 1101–1119.

Kessler-Silacci J., Augereau J.-C., Dullemond C. P., Geers V., Lahuis F., et al. (2006) c2d Spitzer IRS spectra of disks around T Tauri stars. I. Silicate emission and grain growth. *Astrophys. J., 639,* 275–291.

Knapp G. R. and Kerr F. J. (1974) The galactic gas-to-dust ratio from observations of eighty-one globular clusters. *Astron. Astrophys., 35,* 361–379.

Koerner D. W., Sargent A. I., and Beckwith S. V. W. (1993) A rotating gaseous disk around the T Tauri star GM Aurigae. *Icarus, 106,* 2.

Kohler M., Mann I., and Li A. (2008) Complex organic materials in the HR 4796A disk? *Astrophys. J. Lett., 686,* L95–L98.

Kokubo E. and Ida S. (1998) Oligarchic growth of protoplanets. *Icarus, 131,* 171–178.

Kominami J., Tanaka H., and Ida S. (2005) Orbital evolution and accretion of protoplanets tidally interacting with a gas disk. I. Effects of interaction with planetesimals and other protoplanets. *Icarus, 178,* 540–552.

Kraus A. L., Ireland M. J., Martinache F., Lloyd J. P., and Hillenbrand L. A. (2008) The role of multiplicity in protoplanetary disk evolution. In *New Light on Young Stars: Spitzer's View of Circumstellar Disks*, p. 71. Infrared Processing and Analysis Center, California Institute of Technology, Pasadena.

Kring D. A. and Cohen B. A. (2002) Cataclysmic bombardment throughout the inner solar system 3.9–4.0 Ga. *J. Geophys. Res.–Planets, 107,* 5009.

Krist J. E., Ardila D. R., Golimowski D. A., Clampin M., Ford H. C., et al. (2005) Hubble Space Telescope Advanced Camera for surveys coronagraphic imaging of the AU Microscopii debris disk. *Astron. J., 129,* 1008–1017.

Krivov A. V., Herrmann F., Brandeker A., and Thébault P. (2009) Can gas in young debris disks be constrained by their radial brightness profiles? *Astron. Astrophy., 507,* 1503–1516.

Kuchner M. J. and Holman M. J. (2003) The geometry of resonant signatures in debris disks with planets. *Astrophys. J., 588,* 1110–1120.

Lada C. J., Muench A. A., Luhman K. L., Allen L., Hartmann L., et al. (2006) Spitzer observations of IC 348: The disk population at 2–3 million years. *Astron. J., 131,* 1574–1607.

Lagrange A.-M., Beust H., Mouillet D., Deleuil M., Feldman P. D., et al. (1998) The Beta Pictoris circumstellar disk. XXIV. Clues to the origin of the stable gas. *Astron. Astrophys., 330,* 1091–1108.

Lahuis F., van Dishoeck E. F., Blake G. A., Evans N. J. II, Kessler-Silacci J. E., et al. (2007) c2d Spitzer IRS spectra of disks around T Tauri stars. III. [Ne II], [Fe I], and H_2 gas-phase lines. *Astrophys. J., 665,* 492–511.

Larwood J. D. and Kalas P. G. (2001) Close stellar encounters with planetesimal discs: The dynamics of asymmetry in β Pictoris. *Mon. Not. R. Astron. Soc., 323,* 402–416.

Laughlin G., Bodenheimer P., and Adams F. C. (2004) The core accretion model predicts few jovian-mass planets orbiting red dwarfs. *Astrophys. J. Lett., 612,* L73–L76.

Lecavelier des Etangs A., Vidal-Madjar A., Roberge A., Feldman P. D., Deleuil M., et al. (2001) Deficiency of molecular hydrogen in the disk of β Pictoris. *Nature, 412,* 706–708.

Lecavelier des Etangs A., Deleuil M., Vidal-Madjar A., Roberge A., Le Petit F., et al. (2003) FUSE observations of H_2 around the Herbig AeBe stars HD 100546 and HD 163296. *Astron. Astrophys., 407,* 935–939.

Levison H. F., Morbidelli A., Gomes R., and Backman D. (2007) Planet migration in planetesimal disks. In *Protostars and Planets V* (B. Reipurth et al., eds.), pp. 669–684. Univ. of Arizona, Tucson.

Marois C., Macintosh B., Barman T., Zuckerman B., Song I., et al. (2008) Direct imaging of multiple planets orbiting the star HR 8799. *Science, 322,* 1348–1352, DOI: 10.1126/science.1166585.

Martin-Zaidi C., Deleuil M., Le Bourlot J., Bouret J.-C., Roberge A., et al. (2008) Molecular hydrogen in the circumstellar environments of Herbig Ae/Be stars probed by FUSE. *Astron. Astrophys., 484,* 225–239.

Mathieu R. D., Ghez A. M., Jensen E. L. N., and Simon M. (2000) Young binary stars and associated disks. In *Protostars and Planets IV* (V. Mannings et al., eds.), p. 703. Univ. of Arizona, Tucson.

Meijerink R., Glassgold A. E., and Najita J. R. (2008) Atomic diagnostics of X-ray-irradiated protoplanetary disks. *Astrophys. J., 676,* 518–531.

Merín B., Jørgensen J., Spezzi L., Alcalá J. M., Evans N. J., et al. (2008) The Spitzer c2d survey of large, nearby, interstellar clouds. XI. Lupus observed with IRAC and MIPS. *Astrophys. J. Suppl. Ser., 177,* 551–583.

Moór A., Ábrahám P., Derekas A., Kiss C., Kiss L. L., et al. (2006) Nearby debris disk systems with high fractional luminosity reconsidered. *Astrophys. J., 644,* 525–542.

Morbidelli A., Chambers J., Lunine J. I., Petit J. M., Robert F., et al. (2000) Source regions and time scales for the delivery of water to Earth. *Meteoritics & Planet. Sci., 35,* 1309–1320.

Muzerolle J., Hartmann L., and Calvet N. (1998) A Br–γ probe of disk accretion in T Tauri stars and embedded young stellar objects. *Astron. J., 116,* 2965–2974.

Muzerolle J., Calvet N., Hartmann L., and D'Alessio P. (2003) Unveiling the inner disk structure of T Tauri stars. *Astrophys. J. Lett., 597,* L149–L152.

Nagahara H., Kushiro I., Mori H., and Mysen B. O. (1988) Experimental vaporization and condensation of olivine solid solution. *Nature, 331,* 516–518.

Nagasawa M., Lin D. N. C., and Thommes E. (2005) Dynamical shake-up of planetary systems. I. Embryo trapping and induced collisions by the sweeping secular resonance and embryo-disk tidal interaction. *Astrophys. J., 635,* 578–598.

Najita J. R., Strom S. E., and Muzerolle J. (2007) Demographics of transition objects. *Mon. Not. R. Astron. Soc., 378,* 369–378.

Natta A., Prusti T., Neri R., Wooden D., Grinin V. P., et al. (2001) A reconsideration of disk properties in Herbig Ae stars. *Astron. Astrophys., 371,* 186–197.

Natta A., Testi L., Calvet N., Henning T., Waters R., et al. (2007) Dust in protoplanetary disks: Properties and evolution. In *Protostars and Planets V* (B. Reipurth et al., eds.), pp. 767–781. Univ. of Arizona, Tucson.

Nomura H. (2002) Structure and instabilities of an irradiated viscous protoplanetary disk. *Astrophys. J., 567,* 587–595.

Nomura H. and Millar T. J. (2005) Molecular hydrogen emission from protoplanetary disks. *Astron. Astrophys., 438,* 923–938.

Nomura H., Aikawa Y., Tsujimoto M., Nakagawa Y., and Millar T. J. (2007) Molecular hydrogen emission from protoplanetary disks. II. Effects of X-ray irradiation and dust evolution. *Astrophys. J., 661,* 334–353.

Palla F. and Stahler S. W. (1993) The pre-main-sequence evolution of intermediate-mass stars. *Astrophys. J., 418,* 414.

Pascucci I. and Tachibana S. (2010) The clearing of protoplanetary disks and of the protosolar nebula. In *Protoplanetary Dust: Astrophysical and Cosmochemical Perspectives* (D. A. Apai and D. S. Lauretta, eds.), pp. 263–298. Cambridge Univ., New York.

Pascucci I., Hollenbach D., Najita J., Muzerolle J., Gorti U., et al. (2007) Detection of [Ne II] emission from young circumstellar disks. *Astrophys. J., 663,* 383–393.

Piétu V., Guilloteau S., and Dutrey A. (2005) Sub-arcsec imaging of the AB Aur molecular disk and envelope at millimeter wavelengths: A non Keplerian disk. *Astron. Astrophys., 443,* 945–954.

Piétu V., Dutrey A., and Guilloteau S. (2007) Probing the structure of protoplanetary disks: A comparative study of DM Tau, LkCa 15, and MWC 480. *Astron. Astrophys., 467,* 163–178.

Podolak M. and Zucker S. (2004) A note on the snow line in protostellar accretion disks. *Meteoritics & Planet. Sci., 39,* 1859–1868.

Pontoppidan K. M., Boogert A. C. A., Fraser H. J., van Dishoeck E. F., Blake G. A., et al. (2008) The c2d Spitzer spectroscopic survey of ices around low-mass young stellar objects. II. CO_2. *Astrophys. J., 678,* 1005–1031.

Pringle J. E. (1981) Accretion discs in astrophysics. *Annu. Rev. Astron. Astrophys., 19,* 137–162.

Przygodda F., van Boekel R., Ábrahám P., Melnikov S. Y., Waters L. B. F. M., et al. (2003) Evidence for grain growth in T Tauri disks. *Astron. Astrophys., 412,* L43–L46.

Qi C., Kessler J. E., Koerner D. W., Sargent A. I., and Blake G. A. (2003) Continuum and CO/HCO^+ emission from the disk around the T Tauri star LkCa 15. *Astrophys. J., 597,* 986–997.

Qi C., Ho P. T. P., Wilner D. J., Takakuwa S., Hirano N., et al. (2004) Imaging the disk around TW Hydrae with the Submillimeter Array. *Astrophys. J. Lett., 616,* L11–L14.

Qi C., Wilner D. J., Aikawa Y., Blake G. A., and Hogerheijde M. R. (2008) Resolving the chemistry in the disk of TW Hydrae. I. Deuterated Species. *Astrophys. J., 681,* 1396–1407.

Quillen A. C. (2006) Predictions for a planet just inside Fomalhaut's eccentric ring. *Mon. Not. R. Astron. Soc., 372,* L14–L18.

Rafikov R. R. (2007) Convective cooling and fragmentation of gravitationally unstable disks. *Astrophys. J., 662,* 642–650.

Raymond S. N., Quinn T., and Lunine J. I. (2006) High resolution simulations of the final assembly of Earth-like planets I. Terrestrial accretion and dynamics. *Icarus, 183,* 265–282.

Redfield S. (2007) Gas absorption detected from the edge-on debris disk surrounding HD 32297. *Astrophys. J. Lett., 656,* L97–L100.

Reipurth B., Jewitt D., and Keil K., eds. (2007) *Protostars and Planets V.* Univ. of Arizona, Tucson. 951 pp.

Rettig T., Brittain S., Simon T., Gibb E., Balsara D. S., et al. (2006) Dust stratification in young circumstellar disks. *Astrophys. J., 646,* 342–350.

Richling S. and Yorke H. W. (2000) Photoevaporation of protostellar disks. V. Circumstellar disks under the influence of both extreme-ultraviolet and far-ultraviolet radiation. *Astrophys. J., 539,* 258–272.

Richter M. J., Jaffe D. T., Blake G. A., and Lacy J. H. (2002) Looking for pure rotational H_2 emission from protoplanetary disks. *Astrophys. J. Lett., 572,* L161–L164.

Rieke G. H., Su K. Y. L., Stansberry J. A., Trilling D., Bryden G., et al. (2005) Decay of planetary debris disks. *Astrophys. J., 620,* 1010–1026.

Roberge A. and Weinberger A. J. (2008) Debris disks around nearby stars with circumstellar gas. *Astrophys. J., 676,* 509–517.

Roberge A., Feldman P. D., Lagrange A. M., Vidal-Madjar A., Ferlet R., et al. (2000) High-resolution Hubble Space Telescope STIS spectra of C I and CO in the Beta Pictoris circumstellar disk. *Astrophys. J., 538,* 904–910.

Roberge A., Lecavelier des Etangs A., Grady C. A., Vidal-Madjar A., Bouret J.-C., et al. (2001) FUSE and Hubble Space Telescope/STIS observations of hot and cold gas in the AB Aurigae system. *Astrophys. J. Lett., 551,* L97–L100.

Roberge A., Feldman P. D., Lecavelier des Etangs A., Vidal-Madjar A., Deleuil M., et al. (2002) FUSE observations of possible infalling planetesimals in the 51 Ophiuchi circumstellar disk. *Astrophys. J., 568,* 343–351.

Roberge A., Weinberger A. J., Redfield S., and Feldman P. D. (2005) Rapid dissipation of primordial gas from the AU Microscopii debris disk. *Astrophys. J. Lett., 626,* L105–L108.

Roberge A., Feldman P. D., Weinberger A. J., Deleuil M., and Bouret J.-C. (2006) Stabilization of the disk around β Pictoris by extremely carbon-rich gas. *Nature, 441,* 724–726.

Ruden S. P. and Lin D. N. C. (1986) The global evolution of the primordial solar nebula. *Astrophys. J., 308,* 883–901.

Salyk C., Blake G. A., Boogert A. C. A., and Brown J. M. (2007) Molecular gas in the inner 1 AU of the TW Hya and GM Aur transitional disks. *Astrophys. J. Lett., 655,* L105–L108.

Salyk C., Pontoppidan K. M., Blake G. A., Lahuis F., van Dishoeck E. F., et al. (2008) H_2O and OH gas in the terrestrial planet-forming zones of protoplanetary disks. *Astrophys. J. Lett., 676,* L49–L52.

Sargent B. A., Forrest W. J., Tayrien C., McClure M. K., Li A., et al. (2009) Silica in protoplanetary disks. *Astrophys. J., 690,* 1193–1207.

Schneider G., Silverstone M. D., and Hines D. C. (2005) Discovery of a nearly edge-on disk around HD 32297. *Astrophys. J. Lett., 629,* L117–L120.

Schneider G., Silverstone M. D., Hines D. C., Augereau J.-C., Pinte C., et al. (2006) Discovery of an 86 AU radius debris ring around HD 181327. *Astrophys. J., 650,* 414–431.

Semenov D., Wiebe D., and Henning T. (2004) Reduction of chemical networks. II. Analysis of the fractional ionisation in protoplanetary discs. *Astron. Astrophys., 417,* 93–106.

Semenov D., Pavlyuchenkov Y., Schreyer K., Henning T., Dullemond C., et al. (2005) Millimeter observations and modeling of the AB Aurigae system. *Astrophys. J., 621,* 853–874.

Semenov D., Wiebe D., and Henning T. (2006) Gas-phase CO in protoplanetary disks: A challenge for turbulent mixing. *Astrophys. J. Lett., 647,* L57–L60.

Semenov D., Pavlyuchenkov Y., Henning T., Wolf S., and Launhardt R. (2008) Chemical and thermal structure of protoplanetary disks as observed with ALMA. *Astrophys. J. Lett., 673,* L195–L198.

Shakura N. I. and Syunyaev R. A. (1973) Black holes in binary systems. Observational appearance. *Astron. Astrophys., 24,* 337–355.

Silverstone M. D., Meyer M. R., Mamajek E. E., Hines D. C., Hillenbrand L. A., et al. (2006) Formation and evolution of planetary systems (FEPS): Primordial warm dust evolution from 3 to 30 m.y. around Sun-like stars. *Astrophys. J., 639,* 1138–1146.

Skrutskie M. F., Dutkevitch D., Strom S. E., Edwards S., Strom K. M., et al. (1990) A sensitive 10-micron search for emission arising from circumstellar dust associated with solar-type pre-main-sequence stars. *Astron. J., 99,* 1187–1195.

Slettebak A. (1982) Spectral types and rotational velocities of the brighter Be stars and A-F type shell stars. *Astrophys. J. Suppl., 50,* 55–83.

Spitzer L. (1998) *Physical Processes in the Interstellar Medium.* Wiley and Sons, New York.

Stantcheva T. and Herbst E. (2004) Models of gas-grain chemistry in interstellar cloud cores with a stochastic approach to surface chemistry. *Astron. Astrophys., 423,* 241–251.

Stark C. C. and Kuchner M. J. (2009) A new algorithm for self-consistent three-dimensional modeling of collisions in dusty debris disks. *Astrophys. J., 707,* 543–553.

Strom K. M., Strom S. E., Edwards S., Cabrit S., and Skrutskie M. F. (1989) Circumstellar material associated with solar-type pre-main-sequence stars — A possible constraint on the timescale for planet building. *Astron. J., 97,* 1451–1470.

Su K. Y. L., Rieke G. H., Stansberry J. A., Bryden G., Stapelfeldt K. R., et al. (2006) Debris disk evolution around A stars. *Astrophys. J., 653,* 675–689.

Terada H., Tokunaga A. T., Kobayashi N., Takato N., Hayano Y., et al. (2007) Detection of water ice in edge-on protoplanetary disks: HK Tauri B and HV Tauri C. *Astrophys. J., 667,* 303–307.

Testi L., Natta A., Shepherd D. S., and Wilner D. J. (2003) Large grains in the disk of CQ Tau. *Astron. Astrophys., 403,* 323–328.

Thebault P. and Augereau J.-C. (2005) Upper limit on the gas density in the β Pictoris system. *Astron. Astrophys., 437,* 141–148.

Thi W. F., van Dishoeck E. F., Blake G. A., van Zadelhoff G. J., Horn J., et al. (2001) H_2 and CO emission from disks around T Tauri and Herbig Ae pre-main-sequence stars and from debris disks around young stars: Warm and cold circumstellar gas. *Astrophys. J., 561,* 1074–1094.

Thi W.-F., van Zadelhoff G.-J., and van Dishoeck E. F. (2004) Organic molecules in protoplanetary disks around T Tauri and

Herbig Ae stars. *Astron. Astrophys., 425,* 955–972.
van Boekel R., Min M., Leinert C., Waters L. B. F. M., Richichi A., et al. (2004) The building blocks of planets within the "terrestrial" region of protoplanetary disks. *Nature, 432,* 479–482.
van Dishoeck E. F. and Jørgensen J. K. (2008) Star and planet-formation with ALMA: An overview. *Astrophys. Space Sci., 313,* 15–22.
van Dishoeck E., Jonkheid B., and Hemert M. (2006) Photoprocesses in protoplanetary disks. *Faraday Discussions, 133,* 231–243.
van Zadelhoff G.-J., Aikawa Y., Hogerheijde M. R., and van Dishoeck E. F. (2003) Axi-symmetric models of ultraviolet radiative transfer with applications to circumstellar disk chemistry. *Astron. Astrophys., 397,* 789–802.
Visser R., Geers V. C., Dullemond C. P., Augereau J.-C., Pontoppidan K. M., et al. (2007) PAH chemistry and IR emission from circumstellar disks. *Astron. Astrophys., 466,* 229–241.
Wadhwa M., Amelin Y., Davis A. M., Lugmair G. W., Meyer B., et al. (2007) From dust to planetesimals: Implications for the solar protoplanetary disk from short-lived radionuclides. In *Protostars and Planets V* (B. Reipurth et al., eds.), pp. 835–848. Univ. of Arizona, Tucson.
Watson A. M., Stapelfeldt K. R., Wood K., and Ménard F. (2007) Multiwavelength imaging of young stellar object disks: Toward an understanding of disk structure and dust evolution. In *Protostars and Planets V* (B. Reipurth et al., eds.), pp. 523–538. Univ. of Arizona, Tucson.
Watson D. M., Leisenring J. M., Furlan E., Bohac C. J., Sargent B., et al. (2009) Crystalline silicates and dust processing in the protoplanetary disks of the Taurus young cluster. *Astrophys. J. Suppl., 180,* 84–101.
Weidenschilling S. J. (1977) The distribution of mass in the planetary system and solar nebula. *Astrophys. Space Sci., 51,* 153–158.
Weinberger A. J., Becklin E. E., Schneider G., Smith B. A., Lowrance P. J., et al. (1999) The circumstellar disk of HD 141569 imaged with NICMOS. *Astrophys. J. Lett., 525,* L53–L56.
Willacy K. and Langer W. D. (2000) The importance of photoprocessing in protoplanetary disks. *Astrophys. J., 544,* 903–920.
Willacy K., Langer W., Allen M., and Bryden G. (2006) Turbulence-driven diffusion in protoplanetary disks: Chemical effects in the outer regions. *Astrophys. J., 644,* 1202–1213.
Williams J. P., Andrews S. M., and Wilner D. J. (2005) The masses of the Orion proplyds from submillimeter dust emission. *Astrophys. J., 634,* 495–500.
Wilner D. J., D'Alessio P., Calvet N., Claussen M. J., and Hartmann L. (2005) Toward planetesimals in the disk around TW Hydrae: 3.5 centimeter dust emission. *Astrophys. J. Lett., 626,* L109–L112.
Woitke P., Kamp I., and Thi W.-F. (2009) Radiation thermochemical models of protoplanetary disks I. Hydrostatic disk structure and inner rim. *Astron. Astrophys.,* in press, arXiv:0904.0334v1.
Wood J. A. (1984) *Calculation of Infall Heating at the Surfaces of Model Solar Nebulae.* SAO Special Report No. 394, Smithsonian Astrophysical Observatory.
Wooden D. H. (2008) Cometary refractory grains: Interstellar and nebular sources. *Space Sci. Rev., 138,* 75–108.
Woods P. M. and Willacy K. (2007) Benzene formation in the inner regions of protostellar disks. *Astrophys. J. Lett., 655,* L49–L52.
Woods P. M. and Willacy K. (2008) Carbon isotope fractionation in protoplanetary disks. *Astrophys. J.,* in press, arXiv:0812.0269.
Wyatt M. C. (2003) Resonant trapping of planetesimals by planet migration: Debris disk clumps and Vega's similarity to the solar system. *Astrophys. J., 598,* 1321–1340.
Wyatt M. C. (2008a) Dynamics of small bodies in planetary systems. *Lect. Notes Phys., 758,* in press, arXiv:0807.1272.
Wyatt M. C. (2008b) Evolution of debris disks. *Annu. Rev. Astron. Astrophys., 46,* 339–383.
Wyatt M. C., Smith R., Su K. Y. L., Rieke G. H., Greaves J. S., et al. (2007) Steady state evolution of debris disks around A stars. *Astrophys. J., 663,* 365–382.
Zuckerman B., Forveille T., and Kastner J. H. (1995) Inhibition of giant planet formation by rapid gas depletion around young stars. *Nature, 373,* 494–496.

Terrestrial Planet Formation

John Chambers
Carnegie Institution for Science

The standard planetesimal model of terrestrial planet formation is based on astronomical and cosmochemical observations, and the results of laboratory experiments and numerical simulations. In this model, planets grow in a series of stages beginning with the micrometer-sized dust grains observed in protoplanetary disks. Dust grains readily stick together to form millimeter- to centimeter-sized aggregates, some of which are heated to form chondrules. Growth beyond meter size via pairwise sticking is problematic, especially in a turbulent disk. Turbulence also prevents the direct formation of planetesimals in a gravitationally unstable dust layer. Turbulent concentration can lead to the formation of gravitationally bound clumps that become 10–1000-km planetesimals. Dynamical interactions between planetsimals give the largest objects the most favorable orbits for further growth, leading to runaway and oligarchic growth and the formation of Moon- to Mars-sized planetary embryos. Large embryos acquire substantial atmospheres, speeding up planetesimal capture. Embryos also interact tidally with the gas disk, leading to orbit modification and migration. Oligarchic growth ceases when planetesimals become depleted. Embryos develop crossing orbits, and occasionally collide, leaving a handful of terrestrial planets on widely spaced orbits. The Moon probably formed via one such collision. Most stages of planet formation probably took place in the asteroid belt, but dynamical perturbations from the giant planets removed the great majority of embryos and planetesimals from this region.

1. INTRODUCTION

When we think of a planet, our first conception is a body like Earth with an atmosphere, continents, and oceans. However, the Sun's planets are a diverse group of objects that come in several varieties, and exoplanets are more diverse still (Fig. 1). In this chapter I will focus on terrestrial planets — planets that are mostly composed of refractory materials such as silicates and metal. These objects are large enough to be roughly spherical due to their own gravity. They have a solid surface one could walk around on. They may have an atmosphere, but gases make up a negligible fraction of their total mass.

Terrestrial planets are the only place we know for certain that life can exist. While living organisms might survive on icy satellites like Titan, or in the atmospheres of giant planets, terrestrial planets can provide a number of benefits for life. These include a solid substrate, access to abundant sunlight, and the possibility of liquid water at or near the surface.

Clues to the origin of the solar system and its planets come from several sources including astronomical observations, data returned from spacecraft, and cosmochemical analysis of samples from planets and asteroids. Numerical simulations are increasingly used to try to make sense of these data and examine different theoretical models for planet formation.

The Sun's planets all orbit in the same direction and have roughly coplanar orbits, suggesting the solar system formed from a disk-shaped structure. Many young stars are surrounded by solar-system-sized disks of gas and dust, and these structures are commonly referred to as protoplanetary disks. A typical protoplanetary disk is mostly composed of hydrogen and helium gas. Submicrometer- to centimeter-sized dust grains composed of silicates and water ice have been observed in these disks using infrared- and radiotelescopes.

The Sun's protoplanetary disk, also called the solar nebula, must have had a mass of at least 1–2% that of the Sun in order to provide the heavy elements seen in the planets today. However, this "minimum mass nebula" is only a lower limit — the solar nebula could have been an order of magnitude more massive than this. The solar nebula probably had a composition similar to the Sun itself since most primitive meteorites have fairly similar elemental abundances to the Sun (normalized to silicon) except for highly volatile elements such as hydrogen and the noble gases.

Samples from Earth, the Moon, Mars, and most asteroids have roughly uniform isotopic abundances, after physical processes leading to mass-dependent fractionation are taken into account. (Oxygen is a notable exception.) The solar nebula was probably made up of a mixture of material from different stellar sources, which suggests widescale mixing took place, perhaps during an early hot phase. The great depletion of ice-forming elements on the terrestrial planets compared to the outer planets and their satellites suggests the inner nebula was too hot for ices to condense while the inner planets were forming, while the outer nebula must have been cooler.

There are several indications that planets form shortly after stars themselves, and that planet formation is a natural part of star formation. Protoplanetary disks are only seen around stars that are thought to be younger than about 10 m.y. Gas-giant planets like Jupiter, which must acquire most of their mass in gaseous form, have to form while a disk is still present. Disks also provide a way to damp radial and vertical motions that can arise during planet formation, and that would otherwise slow or halt planetary growth. Age measurements using radioactive isotopes suggest the terrestrial planets and asteroids in the solar system took less than 200 m.y. to form, and are roughly 4.5 G.y. old, which matches estimates for the age of the Sun based on stellar evolution models.

The ancient surfaces of the Moon, Mercury, Mars, and some of the satellites of the giant planets are covered in impact craters, which suggests collisions played a much more important role in the early solar system than they do today. This observation and the abundance of dust grains seen in protoplanetary disks has given rise to the planetesimal model for planet formation, which implies that planets formed as a result of numerous collisions between small objects to form larger ones. The asteroid belts between Mars and Jupiter, and beyond the orbit of Neptune, can be thought of as leftover material from this collisional epoch that was prevented from growing into additional planets.

The planetesimal hypothesis is widely accepted today as the basis of terrestrial planet formation. It remains unclear whether gas-giant planets also form this way. The standard version of this model may require some modification in light of the growing variety of exoplanetary systems that are being discovered. In particular, the fact that many exoplanets lie close to their parent star or have highly eccentric orbits argues that a planet's orbit can change substantially during or after its formation. It is plausible that extrasolar systems will contain new types of planet that do not exist in the solar system, and these objects may have rather different formation histories than Earth.

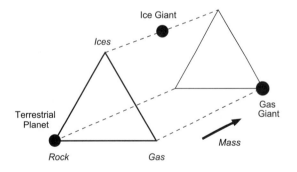

Fig. 1. Schematic diagram showing different planetary classes seen in the solar system. Here, "rock" indicates refractory materials such as silicates, metal, and sulfides; "ices" refers to ice-forming materials such as water, methane, and ammonia, in either solid or fluid form. Other types of planets may exist elsewhere, such as "super Earths" — bodies with masses comparable to Uranus and Neptune but composed almost entirely of rock.

2. FOUNDATIONAL CONCEPTS AND EQUATIONS

2.1. Dust Dynamics

In a protoplanetary disk with a composition similar to the Sun, roughly 99% of the initial mass is gas, primarily hydrogen and helium. The remaining 1% of the mass exists in solid grains that are ~1 μm in size. These dust grains provide the starting point for the formation of terrestrial planets.

As dust grains orbit the star, they interact with the gas in the disk, experiencing aerodynamic drag forces. The strength of this interaction can be characterized by the stopping time

$$t_{stop} \equiv \frac{M|v - v_{gas}|}{F_{drag}} \qquad (1)$$

where M is the mass of a dust grain, and v and v_{gas} are the velocity of the grain and the gas respectively.

The drag force F_{drag} varies depending on the size and shape of a particle. For a spherical particle with radius R < 9λ/4, where λ is the mean free path of the gas molecules, the particle undergoes Epstein drag

$$F_{drag} = -\left(\frac{\rho_{gas} c_s}{\rho R}\right) M (v - v_{gas}) \qquad (2)$$

where ρ_{gas} and c_s are the density and sound speed of the gas, and ρ is the density of the particle. For larger spherical objects, the drag force is

$$F_{drag} = -\frac{3 C_D \rho_{gas}}{8 \rho R} M (v - v_{gas}) |v - v_{gas}| \qquad (3)$$

The drag coefficient C_D depends on the Reynolds number Re, and is approximately given by

$$\begin{aligned} C_D &\simeq 24 \, Re^{-1} & Re < 1 \\ &\simeq 24 \, Re^{-0.6} & 1 < Re < 800 \\ &\simeq 0.44 & Re > 800 \end{aligned} \qquad (4)$$

where

$$Re \equiv \frac{2R|v - v_{gas}|}{v_m} \qquad (5)$$

and $v_m \simeq \lambda c_s/2$ is the molecular viscosity (*Weidenschilling*, 1977; *Cuzzi et al.*, 1993). Note that the last of equation (4) corresponds to Stokes drag. The stopping time increases with particle size, so small particles are more tightly coupled to the motion of the gas than large ones.

Very small dust grains undergo Brownian motion due to collisions with individual gas molecules. The typical relative velocity for grains of mass M can be determined from approximate equipartition of kinetic energy

$$v_{rel} \simeq \left(\frac{M_m}{M}\right)^{1/2} c_s \qquad (6)$$

where M_m is the mass of a gas molecule.

Grains also sediment toward the midplane of the disk due to the vertical component of the star's gravity, moving at terminal vertical velocity v_{settle} determined by a balance between gravity and gas drag

$$v_{settle} = -\Omega^2 z t_{stop} \qquad (7)$$

where z is the distance from the midplane and Ω is the orbital angular frequency. Large particles settle faster than small ones, allowing them to sweep up material as they fall.

Gas density, pressure, and temperature in a disk all tend to decrease radially with distance from the star. The outward pressure gradient means that gas orbits the star at slightly less than the Keplerian orbital velocity of a solid body

$$v_{gas} = \Omega r (1 - \eta) \qquad (8)$$

where r is the distance from the star. The factor η can be calculated by balancing the gravitational, centrifugal, and pressure forces on a parcel of gas, giving

$$\eta = -\frac{1}{2\rho_{gas}\Omega^2 r}\frac{dP}{dr} \qquad (9)$$

The reduced orbital velocity of the gas means that dust particles experience a headwind and lose angular momentum. As a result, particles drift radially through the disk toward the star or the nearest pressure maximum. Particles drift inward at a velocity determined by a balance between stellar gravity, gas pressure, and centrifugal and Coriollis forces, where

$$\frac{dr}{dt} \simeq -\frac{2\eta r}{t_{stop}}\left[\frac{(\Omega t_{stop})^2}{1+(\Omega t_{stop})^2}\right] \qquad (10)$$

(*Weidenschilling*, 1977). Drift rates are highest when $\Omega t_{stop} = 1$, which corresponds to roughly meter-sized particles. In this case, particles move inward by 1 AU every 10^2–10^3 yr, and this rapid drift poses a challenge for models of planet formation. Particles continue drifting until they collide with another object or reach a region that is hot enough for them to evaporate.

Protoplanetary disks are likely to be turbulent, at least in some locations. Turbulence provides a way to drive the viscous evolution of disks and the observed accretion of material onto the central star. However, the mechanism that sustains turbulence is unclear. The turbulent viscosity ν is often assumed to have the form

$$\nu = \alpha c_s H_{gas} \qquad (11)$$

where $H_{gas} \simeq c_s/\Omega$ is the vertical scale height of the gas disk, and $\alpha \leq 1$ is a parameter that depends on the mechanism driving the turbulence (*Shakura and Sunyaev*, 1973). This is equivalent to saying that the horizontal shear stress on a parcel of gas in the disk is approximately α times the gas pressure. The form of equation (11) was originally justified on the assumption that turbulence in disks has a hydrodynamic origin such as thermal convection. It is less clear whether the same form will apply for other sources of turbulence. As a result, α probably varies in time and space in real disks. Observed accretion rates in protoplanetary disks suggest $\alpha \sim 10^{-2}$–10^{-3} on average (*Hartmann et al.*, 1998; *Hueso and Guillot*, 2005).

The largest turbulent eddies in a disk probably rotate at an angular frequency Ω_L that is no slower than Ω, due to the Coriolis force. The size L and rotational velocity V_L of the largest eddies are related via the viscosity, such that $\nu \simeq LV_L$. This implies that $V_L \sim c_s\sqrt{\alpha}$ and $L \sim H_{gas}\sqrt{\alpha}$ (*Cuzzi et al.*, 2001). Turbulence typically follows a Kolmogorov energy spectrum, which can be derived from dimensional analysis, such that an eddy with radius l has a rotational velocity

$$v_l = l\Omega_l \sim V_L\left(\frac{l}{L}\right)^{1/3} \qquad (12)$$

where Ω_l is the angular frequency of the eddy (*Cuzzi and Weidenschilling*, 2006). As a result, gas moves more slowly in small eddies than in large ones. The size of the smallest eddy is set by the point at which collisions between gas molecules smooth out turbulent motions.

Particles couple strongly to large eddies with $\Omega_l \ll 1/t_{stop}$. Neighboring particles moving in the same eddy have similar velocities. Particles couple poorly to small eddies with $\Omega_l \gtrsim 1/t_{stop}$, and undergo a randomly fluctuating acceleration instead. In this case, neighboring particles will have different velocities even if the particles are the same size. Since v_l increases with l, the relative velocities between equally sized particles will be greatest when $\Omega_L t_{stop} = 1$. These objects will have $R \sim 1$ m if $\Omega_L \sim \Omega$.

Larger objects will also be affected by turbulent fluctuations, but the accelerations will be smaller due to their greater inertia. For all sizes, the velocity of a particle due to turbulence, relative to a body moving on a circular Keplerian orbit, is roughly

$$v_{turb} \sim \frac{V_L}{(1+t_{stop}/t_L)^{1/2}} \qquad (13)$$

where $t_L = 2\pi/\Omega_L$ (*Cuzzi et al.*, 1993).

2.2. Dust Grain Collisions

Most of what we know about collisions between dust grains comes from laboratory experiments. *Poppe et al.* (2000) have found that individual, uncharged, spherical silica grains stick together due to electrostatic forces when they collide at low speeds. At high speeds, grains collide and rebound. There is a fairly sharp transition between these two regimes. The dividing line depends on particle radius, and lies at ~1 m/s for micrometer-sized grains. Irregularly shaped grains behave differently: The probability of sticking vs. rebound declines with increasing collision speed, but there is no sharp transition. Some collisions can lead to sticking at speeds of up to 50 m/s (*Poppe et al.*, 2000).

Dust collision experiments in microgravity show that low-velocity collisions (≪1 m/s) lead to the formation of loose fractal aggregates with fractal dimension ~2 (*Blum and Wurm*, 2000). At larger impact speeds, sticking still occurs but growth is accompanied by compaction as grain-grain bonds break and grains start to roll over one another. At collision speeds ≥1 m/s, individual grains are ejected from loose aggregates. Still more energetic collisions catastrophically disrupt an existing aggregate (*Blum and Wurm*, 2000).

Collisions involving compacted aggregates of dust grains are somewhat different. When millimeter- to centimeter-sized aggregates of micrometer-sized grains collide at speeds of a few meters per second, they rebound with the loss of some individual grains. However, above ~10 m/s, small aggregates embed themselves in larger ones. Some fragments escape from the larger aggregate but it gains mass overall (*Wurm et al.*, 2005).

Many meteorites are primarily composed of rounded millimeter-sized particles called chondrules. These appear to be dust aggregates that were heated in the solar nebula and partially or completely melted. Experiments designed to reproduce chondrule textures suggest that they cooled slowly over a period of hours (*Connelly and Jones*, 2005). These slow cooling rates suggest chondrules formed in dense regions of the nebula, with high solid-to-gas ratios, possibly in shocks.

Collisions between monolithic objects like chondrules differ from those between loosely bound dust aggregates. Typically, low-speed collisions lead to rebound, while collisions above ~20 m/s lead to some fragmentation (*Ueda et al.*, 2001). The presence of fine dust grains in chondrule-forming regions allowed chondrules to accrete thick dust rims in 10^2–10^3 yr (*Cuzzi*, 2004). These porous dust rims help chondrules stick together during collisions, as compaction of the rims absorbs impact energy (*Ormel et al.*, 2008). This process ceases once all the dust is accreted and the rims become compacted, at which point the largest aggregates are likely to be <1 m in radius.

Numerical simulations show that in the absence of turbulence micrometer-sized dust grains aggregate into meter-sized bodies in ~10^4 orbital periods for plausible sticking probabilities (*Weidenschilling*, 1997). Growth mainly occurs when large aggregates sweep up small ones. Without turbulence, similarly sized objects have low relative velocities, so mutual collisions are not disruptive, even though they may not lead to growth.

In a turbulent nebula, meter-sized objects collide with one another at substantial speeds, probably leading to fragmentation rather than growth. Meter-sized particles also have short drift lifetimes, severely limiting the amount of time available to grow into larger objects. These difficulties suggest that growth may stall when objects reach ~1 m in diameter, a problem referred to as the "meter-size barrier." This issue is discussed further in section 3.1.

2.3. Gravitational Instability

In the absence of turbulence, dust grains will gradually sediment toward a thin layer at the disk midplane. Perturbations in this layer can potentially grow in size leading to gravitational instability. This has long been considered as a possible mechanism for the formation of planetesimals (*Goldreich and Ward*, 1973), although this issue remains controversial.

A perturbation in the density of the particle layer with frequency ω satisfies a dispersion relation given by

$$\omega^2 = k^2 v_{rel}^2 + \Omega^2 - 2\pi G \Sigma_{solid} k \qquad (14)$$

where Σ_{solid} is the surface density of particles, v_{rel} is their velocity dispersion, and $k = 2\pi/\lambda$ where λ is the wavelength of the perturbation (*Goldreich and Ward*, 1973). The dust layer becomes unstable when $\omega^2 \leq 0$, which occurs for some values of λ when

$$v_{rel} \leq \frac{\pi G \Sigma_{solid}}{\Omega} \qquad (15)$$

The largest region that can become unstable (the case where $v_{rel} \simeq 0$) is

$$\lambda_{max} \simeq \frac{4\pi^2 G \Sigma_{solid}}{\Omega^2} \qquad (16)$$

If the particles in this clump collapse to form a solid body, this object will have a mass given by

$$M_{max} \sim \Sigma_{solid} \lambda_{max}^2 \qquad (17)$$

However, this is only a very rough guide to the size of object that may form by gravitational instability. Complete collapse may be prevented by random motions or rotation, so that only part of a gravitationally unstable clump collapses. For particles with $\Omega t_{stop} \ll 1$, it may take many orbital periods for the particles to sediment to the center of a clump (*Cuzzi and Weidenschilling*, 2006), and neighboring clumps may merge during this time.

Settling of dust grains to the midplane is opposed by intrinsic turbulence in the disk. As a vertical concentration gradient develops, turbulent diffusion tends to move particles away from the midplane again. A rough estimate for the scale height H_{solid} of the particle layer comes from assuming settling and turbulence are in equilibrium

$$\left(\frac{H_{solid}}{H_{gas}}\right)^2 \simeq \frac{\alpha}{\Omega t_{stop}\left(1+\Omega t_{stop}\right)} \quad (18)$$

(*Cuzzi and Weidenschilling*, 2006). Intrinsic turbulence generally prevents gravitational instability unless the particles are large or α is very small.

Dust settling can generate turbulence in an otherwise laminar disk. When the mass of solid particles near the midplane exceeds that of the gas, the particles begin to drag the gas along at the Keplerian velocity $v_{kep} = \Omega r$. Gas above and below the midplane orbits at sub-Keplerian speeds $v_{kep}(1-\eta)$ due to the radial pressure gradient in the disk. This vertical velocity shear generates turbulence that opposes further dust settling (*Weidenschilling*, 1980). Gravitational instability is probably prevented by shear induced turbulence except in regions where $\Sigma_{solid} \sim \Sigma_{gas}$ (*Garaud and Lin*, 2004) or in parts of the disk where there is no radial pressure gradient.

2.4. Runaway and Oligarchic Growth

Although it is currently unclear how planetesimals formed, it is assumed they did so in large numbers at an early stage in the solar nebula. Two new mechanisms that may generate planeteimals in a turbulent disk are described in section 3.1.

Once planetesimals are present, mutual collisions can lead to growth or fragmentation depending on the strength of the objects and the impact velocity. Collisions can take place in one of two size regimes depending on whether gravity is an important factor.

Monolithic bodies with radius $R \leq 100$ m have negligible gravitational fields. These objects tend to grow weaker with increasing size since big bodies typically contain larger flaws than small bodies. The tensile stress required to activate a flaw decreases with the length of the flaw, so large bodies break apart at lower impact speeds than smaller bodies. Collisions involving large bodies also last longer, so there is more time for flaws to grow and coalesce (*Housen and Holsapple*, 1999).

Impact strength is often characterized using $Q_S^\star$, which is the kinetic energy per unit target mass required to break up a body such that the largest fragment contains half the original mass of the target. Experiments and numerical simulations find that

$$Q_S^\star \propto R^{-b_s} \quad (19)$$

where $b_s \simeq 0.4-0.6$ for rock and water ice, while the constant of proportionality depends on the material (*Housen and Holsapple*, 1999; *Benz and Asphaug*, 1999). The outcome of a collision also depends on the mass ratio of the projectile and target. For a given projectile kinetic energy, low velocity collisions involving large projectiles are more disruptive than high-velocity impacts by small projectiles (*Benz*, 2000).

Bodies with $R \geq 100$ m become harder to disrupt with increasing size because fragments need to be ejected rapidly enough to escape from the object's gravitational field. The impact strength in this regime is characterized using $Q_D^\star$, which is the energy per unit target mass required to break up and disperse the body so that the largest reaccumulated object contains half the mass of the target. Numerical simulations suggest that

$$Q_D^\star \propto \rho R^{b_g} v^c \quad (20)$$

where $b_g \simeq 1.2-1.5$ for rock and water ice, and $c \sim 0.5$ (*Melosh and Ryan*, 1997; *Benz and Asphaug*, 1999; *Benz*, 2000). Again there is a dependence on the impact velocity v, and hence on the projectile to target mass ratio. Objects are most easily disrupted when they are hit by large, slow moving projectiles.

Initially, planetesimals were probably loosely compacted aggregates rather than monolithic bodies. In the strength regime (≤ 100 m in size), aggregates are weaker than monolithic bodies since energy that would go into breaking the object instead goes into dispersing the preexisting pieces (*Benz*, 2000). Aggregates can actually be stronger than monoliths in the gravity regime, in some cases, since shockwaves propagate less effectively through aggregates (*Asphaug et al.*, 1998). Planetesimals that formed within the first 2 m.y. in the solar nebula, and grew larger than a few tens of kilometers, probably melted and differentiated due to heat released by short-lived radioactive isotopes such as ^{26}Al (*Woolum and Cassen*, 1999). These objects would have behaved as monoliths rather than aggregates, although energetic impacts may subsequently have converted them into gravitationally bound "rubble piles." The values of $Q_S^\star$ and $Q_D^\star$ for real planetesimals are poorly constrained at present.

In both the strength and gravity regimes, the mass of the largest surviving body (which may consist of gravitationally reaccumulated fragments) is

$$\frac{M_{largest}}{\left(M_{target}+M_{projectile}\right)} \simeq 0.5 + s\left(1-\frac{Q}{Q_D^\star}\right) \quad (21)$$

where Q is the kinetic energy of the projectile per unit target mass and $s \sim 0.5$ (*Benz and Asphaug*, 1999). Thus, low-energy collisions lead to net growth while high-energy collisions lead to erosion. In a size distribution of planetesimals, large objects are more likely to grow via collisions while small planetesimals may be disrupted.

When planetesimals are large enough to have appreciable gravitational fields, they focus the trajectories of other passing objects toward them, increasing the chance of a collision. From conservation of energy and angular momentum, the collision probability is increased by a gravitational focusing factor

$$F_{grav} = 1 + \left(\frac{v_{esc}}{v_{rel}}\right)^2 \quad (22)$$

where v_{esc} is the escape velocity and v_{rel} is the relative velocity of the objects when they are far apart.

The growth rate of a planetesimal of mass M and radius R, moving through a population of smaller bodies of radius R_{small} can be calculated by determining the volume of space that the planetesimal moves through, modifying for gravitational focusing

$$\frac{dM}{dt} = \frac{\pi(R+R_{small})^2 \Sigma_{solid} v_{rel}}{2H_{solid}} F_{grav} F_{3B} \qquad (23)$$

where we have assumed that the mass of escaping fragments is negligible. Here Σ_{solid} and H_{solid} are the surface density and vertical scale height of the planetesimal disk. In deriving this equation, we have neglected the fact that planetesimals actually travel on curved orbits about the central star. For an ensemble of planetesimals with a range of eccentricities e and inclinations i, the presence of the star increases the collision probability for those objects with small e and i. This can be accounted for with a correction factor $F_{3B} \sim 3$ for the mean growth rate (*Greenzweig and Lissauer*, 1992).

When e and i are very small, the gravitational reach of a planetesimal can become larger than the scale height of the particle disk. In this case, the problem is essentially two-dimensional, and the growth rate becomes

$$\frac{dM}{dt} = 2(R+R_{small})\Sigma_{solid} v_{rel} \left(1 + \frac{v_{esc}^2}{v_{rel}^2}\right)^{1/2} \qquad (24)$$

Gravitational interactions between planetesimals with different masses causes dynamical friction, which tends to equipartition the kinetic energy associated with radial and vertical motions. If equipartition of energy goes to completion

$$\begin{aligned} Me^2 &\simeq constant \\ Mi^2 &\simeq constant \end{aligned} \qquad (25)$$

so large bodies have small e and i, and vice versa. In practice, equipartition is not reached, partly because close encounters tend to increase e and i for all bodies (referred to as viscous stirring), and also because gas drag continually damps e and i at rates that depend on an object's size.

Gas drag damping rates are roughly

$$\begin{aligned} \frac{de}{dt} &\simeq \frac{e}{t_{stop}}\left(\eta^2 + \frac{5}{8}e^2 + \frac{1}{2}i^2\right)^{1/2} \\ \frac{di}{dt} &\simeq \frac{i}{2t_{stop}}\left(\eta^2 + \frac{5}{8}e^2 + \frac{1}{2}i^2\right)^{1/2} \end{aligned} \qquad (26)$$

where the gas drag time is

$$t_{stop} = \frac{8\rho R}{3C_D \rho_{gas} v_{rel}} \qquad (27)$$

where v_{rel} is the velocity of the object with respect to the gas, and $C_D \sim 1$ for planetesimal-sized bodies (*Adachi et al.*, 1976).

During the early stages of growth, viscous stirring rates are determined by the mean planetesimal mass, which changes slowly over time. As a result, v_{rel} also changes slowly. Eventually F_{grav} becomes large for some objects since $v_{esc} \propto M^{1/3}$. Neglecting the first term on the righthand side of equation (22), the growth rate for the largest planetesimals becomes

$$\frac{dM}{dt} \propto \Sigma_{solid} M^{4/3} \qquad (28)$$

from equation (23), where we have assumed that $H_{solid} \propto v_{rel}$. The corresponding growth timescale is

$$t_{grow} \sim M \Big/ \frac{dM}{dt} \propto \frac{1}{M^{1/3}} \qquad (29)$$

Under these circumstances, large planetesimals grow faster than small ones, a situation called runaway growth. Dynamical friction gives large bodies nearly circular, coplanar orbits, so these bodies have large gravitational focussing factors. Conversely, F_{grav} is smaller for low-mass bodies and these grow more slowly.

Runaway growth ceases when the largest bodies become massive enough to control the velocity distribution of the smaller planetesimals. This happens when the mass of the largest objects M_{large} satisfies

$$2M_{large}\Sigma_{Large} \geq \bar{M}\Sigma_{solid} \qquad (30)$$

where $\bar{M}$ is the mean planetesimal mass (*Ida and Makino*, 1993).

At this point, an approximate balance between viscous stirring and gas drag means that v_{rel} for the small planetesimals depends on the mass of the largest body in their vicinity, such that $v_{rel} \propto M_{large}^{1/3}$ (*Thommes et al.*, 2003). The growth rate of the largest planetesimals now becomes

$$\frac{dM}{dt} \propto \Sigma_{solid} M^{2/3} \qquad (31)$$

from equation (23), and the corresponding growth timescale is

$$t_{grow} \sim M \Big/ \frac{dM}{dt} \propto M^{1/3} \qquad (32)$$

A new regime called oligarchic growth is established, in which each region of the disk tends to be dominated by a single large body called a planetary embryo (Fig. 2). Equation (32) shows that the masses of neighboring embryos tend to converge over time. All embryos grow faster than a typical planetesimal because they have larger gravitational focussing factors. Embryos typically maintain nonoverlapping orbits with a separation bR_H, where the Hill radius

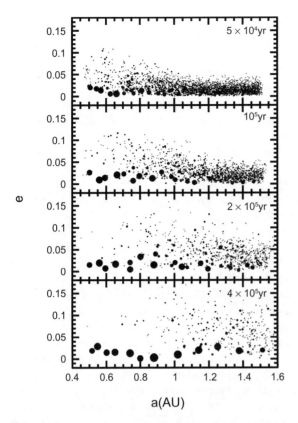

Fig. 2. A numerical simulation of oligarchic growth beginning with 10,000 equal-mass planetesimals with a total mass 2.5 times that of Earth. The panels show four snapshots in time. Each circular symbol shows the orbital semimajor axis a and eccentricity e of a planetesimal, with symbol radius proportional to the planetesimal's radius. Figure kindly supplied by Eiichiro Kokubo.

R_H is a measure of the gravitational reach of the embryo, given by

$$R_H = a\left(\frac{M}{3M_\star}\right)^{1/3} \quad (33)$$

where a is the semimajor axis of the orbit, and b ~ 10 (*Kokubo and Ida*, 1998). Embryos typically collide with one another when b ≪ 10 so that their wide spacing is maintained.

Embryos primarily accrete planetesimals from an annulus a few Hill radii wide centered on their orbit, called a feeding zone. The width of a feeding zone is proportional to $M^{1/3}$, while the surface density of planetesimals in the feeding zone declines with M. As a result, the maximum mass of an embryo is finite, even if no other embryos are present. An embryo that accretes all the mass in its feeding zone reaches its isolation mass, given by

$$M_{iso} = \left(\frac{8\pi^3 \Sigma_{solid}^3 a^6 b^3}{3M_\star}\right)^{1/2} \quad (34)$$

The isolation mass will be modified if solid material moves radially through the disk, e.g., via gas drag for plan-etesimals, or planetary migration for embryos. Oligarchic growth usually ceases before the embryos reach M_{iso} since dynamical friction weakens as planetesimals are consumed, and embryos no longer have nearly circular, nonoverlapping orbits (*Kenyon and Bromley*, 2006). In the solar system, the runaway and oligarchic growth stages probably lasted 10^5–10^6 yr after planetesimals formed (*Wetherill and Stewart*, 1993; *Kokubo and Ida*, 2000). At the end of the oligarchic growth stage, embryos at 1 AU from the Sun would have had masses of 0.01–0.1 $M_\oplus$ for plausible values of Σ_{solid}.

2.5. Late-Stage Growth

When planetary embryos reach roughly the mass of Mars, they acquire thick atmospheres of gas captured from the surrounding disk. Planetesimals passing through an atmosphere are slowed, increasing the probability of capture. Using a rough estimate for the energy lost due to gas drag, the atmospheric density needed to capture a planetesimal of radius R and density ρ is

$$\rho_{gas}(r) \simeq \frac{2R\rho}{3r_H}\left(\frac{24 + 5e^2/h^2}{24}\right) \quad (35)$$

where r_H is the Hill radius of the embryo, $h = r_H/a$, and e is the orbital eccentricity of the planetesimal (*Inaba and Ikoma*, 2003; *Chambers*, 2006).

An accurate determination of the enhanced collision cross section of an embryo requires a detailed atmospheric model. However, in the inner regions of a purely radiative atmosphere, with a mass that is negligible compared to the embryo, the density at a distance r from the center of the embryo is roughly

$$\rho_{gas} \sim \frac{\rho_0}{W_0}\left(\frac{V_0}{4}\right)^3\left(\frac{R_0}{r}\right)^3 \quad (36)$$

where

$$V_0 = \frac{GM\rho_0}{R_0 P_0}$$
$$W_0 = \frac{3\kappa L P_0}{4\pi a_r c G M T_0^4} \quad (37)$$

where R_0 is the outer radius of the atmosphere where it meets the surrounding nebula; T_0, P_0, and ρ_0 are the temperature, pressure, and density at this outer radius; a_r is the radiation density constant; κ is the atmospheric opacity due mainly to dust grains; and L is the luminosity due mainly to energy released from impacting planetesimals (*Inaba and Ikoma*, 2003).

Above a critical mass, an embryo's gravity is too strong to maintain a static atmosphere. This mass is approximately

$$M_{crit} \sim 7\left(\frac{\dot{M}}{1\times 10^{-7} M_\oplus/y}\right)^{s_1}\left(\frac{\kappa}{1\ cm^2/g}\right)^{s_2} M_\oplus \quad (38)$$

where κ is the opacity of the atmosphere, $\dot{M}$ is the rate at which the embryo is accreting mass in the form of planetesimals, and $s_1 \simeq 1/4$ and $s_2 \simeq 1/4$ (*Ikoma et al., 2000*). Gas from the disk flows onto objects with $M > M_{crit}$, providing one pathway to giant-planet formation. For small values of $\dot{M}$ and κ, Earth-mass bodies can exceed the critical mass; however, gas accretion rates are extremely low for such bodies (*Ikoma et al., 2000*).

A planet interacts tidally with gas in the disk, generating torques that can alter the planet's orbit. Interactions are particular important at Lindblad resonances, where the orbital frequency of the gas Ω_{gas} is related to that of the planet by

$$m\left(\Omega_{gas} - \Omega_{planet}\right) = \pm\kappa_e \quad (39)$$

where $\kappa_e \simeq \Omega_{gas}$ is the epicyclic frequency, and m is an integer. At a Lindblad resonance, gas moving on an eccentric orbit always has conjunctions with the planet at the same phase in its orbit, enhancing the planetary perturbation. Interactions can also be important at the corotation resonance, where $\Omega_{gas} = \Omega_{planet}$, although this tends to be less important than the Lindblad resonances for terrestrial-mass planets.

Tidal interactions with the gas disk have two effects: They damp a planet's orbital eccentricity and inclination, and they alter the semimajor axis of the orbit. The latter effect is commonly referred to as "type-I" planetary migration (another kind of migration, "type-II," affects giant planets). In a nonmagnetic, vertically isothermal disk, these changes are

$$\begin{aligned}
\frac{da}{dt} &= -(2.7 - 1.1x)\left(\frac{c_s}{v_{kep}}\right)^2 \frac{a}{t_{tidal}} \\
\frac{de}{dt} &= -0.780 \frac{e}{t_{tidal}} \quad (40) \\
\frac{di}{dt} &= -0.544 \frac{i}{t_{tidal}}
\end{aligned}$$

where

$$t_{tidal} = \left(\frac{M_\star}{M}\right)\left(\frac{M_\star}{\Sigma_{gas}a^2}\right)\left(\frac{c_s}{v_{kep}}\right)^4 \quad (41)$$

and $\Sigma_{gas} \propto a^x$ (*Tanaka et al., 2002, Tanaka and Ward, 2004*).

Damping of e and i happens on a timescale that is 2–3 orders of magnitude shorter than that for changes in a. Even so, inward migration can be rapid: An Earth-mass planet at 1 AU has an inward migration timescale of $\sim 10^5$ yr in a minimum mass nebula.

Equations (40) were derived for an idealized case. The magnitude and direction of migration in real disks are highly uncertain at present. Migration rates may be reduced or reversed in nonisothermal disks where the planetary perturbation on the gas is taken into account (*Paardekooper and Mellema, 2006*). Inward migration is also slowed in regions where the disk opacity changes rapidly (*Menou and Goodman, 2004*), or in the presence of a toroidal magnetic field (*Fromang et al., 2005*). Migration is likely to be outward rather than inward in regions containing steep, positive surface density gradients, such as those at the edge of an inner cavity in the disk (*Masset et al., 2006*).

In a turbulent disk, turbulent density fluctuations in the gas can generate torques on a planet's orbit that change a, e, and i. These fluctuations typically change on timescales comparable to the orbital period, so the long-term effect is that the orbit undergoes a random walk, sometimes referred to as stochastic migration (*Nelson, 2005*). Unlike smooth tidal torques, stochastic migration is potentially important for low-mass objects like planetesimals. In particular, it is likely to raise the relative velocities of planetesimals, potentially reducing growth rates and increasing collisional fragmentation.

Protoplanetary disks typically disperse after a few million years, probably due to a combination of viscous accretion onto the star and photoevaporation (*Haish et al., 2001; Alexander et al., 2006*). At this point, tidal damping of e and i ceases. Dynamical friction with planetesimals also becomes much less effective as planetesimals are removed and oligarchic growth ends. At this point, the terrestrial planet region of the solar nebula probably still contained 10–100 embryos. Perturbations between neighboring embryos increase e and i, causing their orbits to cross. The number of embryos is gradually reduced due to mutual collisions, while residual planetesimals are swept up. This process continues until the remaining objects have noncrossing orbits that are stable for the age of the star.

The time required to form Earth can be estimated from equation (23) noting that $H \simeq v_{rel}/\Omega$ and $F_{grav} \simeq F_{3B} \simeq 1$ after oligarchic growth ceases, so that

$$t_{late} = M / \left(\frac{dM}{dt}\right) \sim \frac{4R\rho}{3\Sigma_{solid}\Omega} \quad (42)$$

At 1 AU from the Sun, assuming that $\Sigma_{solid} \sim 4$ g/cm² when oligarchic growth ceased, gives $t_{late} \sim 2 \times 10^8$ yr. This is roughly consistent with estimates based on radioactive dating (see section 3.2).

A system of two planets of mass M_1 and M_2 orbiting a star of mass $M_\star$ is stable against collisions if the energy and angular momentum satisfy

$$-\frac{2M_{tot}EJ^2}{G^2M_{prod}^3} > 1 + 3^{4/3}\frac{M_1M_2}{M^{2/3}(M_1+M_2)^{4/3}} + \cdots \quad (43)$$

where M_{tot} is the total mass of the system, E is the total energy, J is the total angular momentum, and $M_{prod} = M_\star(M_1 + M_2) + M_1M_2$ (*Gladman, 1993*).

For three or more planets, there is no known analytic stability criterion. However, numerical simulations have provided an approximate empirical way to gauge instability. Planets with mass M, on initially circular orbits, develop crossing orbits on a timescale t_{cross} given by

$$\log t_{cross} \sim Bb + C \qquad (44)$$

where b is the mean orbital separation in Hill radii, and $B \propto M^{1/12}$ and C are constants that depend on the number of planets N and planetary mass, although B is roughly independent of N for $N \geq 5$ (*Chambers et al., 1996*). Extrapolating this relation would suggest that a system of four terrestrial-mass planets would be stable for the age of the solar system if their mean orbital separation was $b \sim 12$. The inner planets are actually spaced further apart than this by factors of 3–5. This may reflect evolutionary processes during their formation, but it must also partly be due to the fact that the orbits of the planets are not circular.

An approximate way to gauge the stability of eccentric orbits is by considering their angular momentum deficit (AMD), which is the difference between the angular momentum and that of a circular orbit with the same semimajor axis

$$AMD = M(GM_\star a)^{1/2}\left(1 - \sqrt{1-e^2}\cos i\right) \qquad (45)$$

For a system of planets on widely spaced orbits, AMD is approximately constant since the total angular momentum is conserved and the semimajor axes change only slightly. In the solar system, angular momentum transfer between the inner and outer planets is relatively inefficient, so to a first approximation, the AMD of the inner planets is roughly constant over gigayear timescales (*Laskar, 1997*). However, AMD is readily exchanged between the inner planets, resulting in changes in e and i. The maximum value of e for each planet occurs when it absorbs the entire AMD for the system. These values of e are shown in Table 1, together with the resulting perihelion and aphelion distances q and Q respectively. It appears that the inner planets will probably avoid a collision over the age of the solar system. However, this is not a rigorous constraint since some AMD exchange with the outer planets does take place.

Collisions between embryos do not always result in a merger. Oblique impacts between similarly sized embryos can have a high specific angular momentum. If the embryos merged to form a single body the rotation frequency may exceed the critical value at which rotational breakup occurs

$$\omega_{crit} \simeq \left(\frac{4\pi G\rho}{3}\right)^{1/2} \qquad (46)$$

Numerical simulations show that pairs of embryos involved in oblique collisions often separate again at greater than their mutual escape velocity, exchanging some material during the collision (*Agnor and Asphaug, 2004*). Head-on collisions do not suffer from this problem, but at high impact speeds, embryo-embryo collisions can lead to net erosion rather than growth. Mercury may have experienced such a collision that stripped away much of its rocky mantle, leaving a high-density planet with a relatively large iron-rich core (*Benz et al., 1988*).

TABLE 1. Maximum eccentricities of the inner planets from conservation of angular momentum deficit.

Planet	e_{max}	q_{min} (AU)	Q_{max} (AU)
Mercury	0.409	0.229	0.545
Venus	0.093	0.656	0.790
Earth	0.077	0.923	1.077
Mars	0.212	1.201	1.847

The Moon probably formed as a result of an oblique impact between a Mars-mass embryo and Earth toward the end of its accretion. Numerical simulations show that if the two bodies were already differentiated, their iron cores would coalesce. However, tidal torques from the nonspherical distribution of mass would have ejected a substantial amount of rocky material, mostly from the impactor, into orbit around Earth forming a disk (*Canup and Asphaug, 1999*; *Canup, 2004*). Unlike many bodies in the solar system, Earth and the Moon have identical oxygen isotope ratios (*Wiechert et al., 2001*), which suggests there was substantial exchange of material between Earth and the protolunar disk (*Pahlevan and Stevenson, 2007*). Mass in the disk would have quickly coalesced into the Moon provided it was beyond the Roche radius, given by

$$R_{Roche} \simeq 2.44\left(\frac{\rho_\oplus}{\rho_{disk}}\right)^{1/3} \qquad (47)$$

where ρ_{disk} is the mean density of material in the disk. Subsequent tidal evolution rapidly expanded the Moon's orbit and slowed Earth's rotation.

Runaway and oligarchic growth probably took place in the asteroid belt but no planet exists there today. Compositional differences between meteorites show they come from at least several dozen different parent asteroids (*Meibom and Clark, 1999*), so the asteroids do not represent the remains of a single planet. The asteroid belt is highly depleted in mass compared to other regions of the solar system with a total mass of $\sim 1/2000 M_\oplus$. This suggests most of the primordial mass in this region was removed either before or after it formed into planets. Collisional fragmentation probably played only a minor role in removing this mass for several reasons. These include the preservation of an almost intact basaltic crust on asteroid 4 Vesta, the small number of impact-generated satellites around large asteroids (*Durda et al., 2004*), and the paucity of impact events between the formation of the asteroid belt and the onset of the late heavy bombardment ~4 G.y. ago as recorded in meteorites (*Bogard, 1995*).

The asteroid belt probably lost most of its mass dynamically due to gravitational perturbations from the giant planets. Perturbations are especially effective at resonances. A mean-motion resonance (MMR) between two objects occurs when

$$p_1 n_1 + p_2 n_2 + p_3 \dot{\varpi}_1 + p_4 \dot{\varpi}_2 = 0 \qquad (48)$$

where $n_i = 2\pi/P_i$, where P is the orbital period, ϖ_i is the longitude of periapse of planet i, and the p_j are integers such that $\Sigma p_j = 0$. Since the orbital precession period is usually much longer than the orbital period, an approximate condition for a MMR is that the two periods are related by

$$\frac{P_1}{P_2} \simeq -\frac{p_1}{p_2} \qquad (49)$$

A secular resonance occurs when the periapse or nodal precession frequency of an object matches one of the eigenfrequencies ν of the linearized secular equations for the planetary system. In the asteroid belt, the most important secular resonances are

$$\begin{aligned}\dot\varpi &= \nu_5 \simeq \dot\varpi_J \\ \dot\varpi &= \nu_6 \simeq \dot\varpi_S \\ \dot\Omega &= \nu_{16} \simeq \dot\Omega_S\end{aligned} \qquad (50)$$

where in this instance Ω refers to the longitude of the ascending node of an orbit, and ϖ_J, ϖ_S, and Ω_S are the longitudes of perihelion of Jupiter and Saturn and the longitude of the ascending node of Saturn respectively (*Murray and Dermott*, 2000).

An asteroid located in a MMR typically has a highly chaotic orbit because multiple subresonances (different values of p_3 and p_4) are located close to one another and these resonances overlap. For these orbits, e increases stochastically on a timescale of ~1 m.y., until the object collides with the Sun or a planet, or has a close encounter with Jupiter, leading to hyperbolic ejection (*Gladman et al.*, 1997). Secular resonances cause a monotonic increase in e and lead to the same outcome.

Resonances currently occupy a small fraction of the asteroid belt (Fig. 3), but their importance in the early solar system was increased by two factors. While the nebula was still present, its gravity modified the resonance locations. As the nebula dispersed, the resonances moved radially, sweeping across the asteroid belt, and potentially removing a large fraction of the planetesimals and planetary embryos that were present (*Lecar and Franklin*, 1997; *Nagasawa et al.*, 2000). Second, if planetary embryos were present in the asteroid belt, their mutual perturbations would have occasionally nudged one of them into a resonance, increasing the fraction of objects removed (*Chambers and Wetherill*, 2001).

2.6. Chemical Evolution

Models for the solar nebula suggest the inner few AU may initially have been hot enough to vaporize most minerals (*Boss*, 1996). However, protoplanetary disks cool over time as energy released by viscous accretion and the luminosity of the central star both decline. Table 2 shows some of the materials that would condense as the temperature decreased in the solar nebula. This condensation sequence assumes a C/O number density ratio $\simeq 0.5$ equivalent to that in the Sun. In a disk with C/O > 1, oxides and silicates would be replaced by carbides and nitrides. At low temperatures, chemical reactions may be kinetically inhibited so the mixture of solids will not necessarily reach an equilibrium (*Lewis and Prinn*, 1980). As a result, it is unclear what the dominant C- and N-bearing species will be in cool regions of a disk. In the inner solar nebula, these elements mainly existed as gases.

The inner planets and all meteorites, except CI chondrites, are depleted in moderately volatile elements such as sodium and sulfur, relative to silicon, compared to the Sun. In chondritic meteorites, the degree of depletion is roughly correlated with an element's volatility, but the pattern differs from one group of meteorites to another. These depletion patterns may be a signature of incomplete condensation of these elements from a gradually cooling nebula (*Cassen*, 1996). As the nebula cooled, it continued to lose gas, so the more volatile elements were underrepresented when planetesimals formed. Alternatively, the depletions may be a signature of thermal processing of solids in the nebula, perhaps associated with chondrule formation. The lack of isotopic fractionation in elements such as potassium (*Humayun and Clayton*, 1995), and the partial retention of elements such as sodium, argue that chondrule formation must have occurred in regions with high dust/gas densities in this case (*Alexander et al.*, 2008).

Beyond a certain distance from the star, temperatures in a protoplanetary disk are cold enough for water ice to condense. This location is called the ice line. Over time, the ice line and other condensation fronts move closer to the star as the disk cools, so the composition of solid material at a given location changes (see Fig. 4). Chondrule ages measured using radioactive isotope systems show that chondrule formation spanned several million years in the solar nebula, which suggests that planetesimals also formed over a range of times. Planetesimals that formed in the same location at different times would have had different compo-

TABLE 2. Selected nebula condensation temperatures at 10^{-4} bar (following *Lodders*, 2003; *Lewis and Prinn*, 1980).

Mineral	Temperature (K)
Re	1821
ZrO_2	1764
Al_2O_3	1677
$CaTiO_3$	1593
$MgAl_2O_4$	1397
Fe	1357
Mg_2SiO_4	1354
$MgSiO_3$	1316
FeS	704
H_2O	182
$NH_3 \cdot H_2O$	131
$CH_4 \cdot 7H_2O$	78
CH_4	41
CO	25
N_2	22

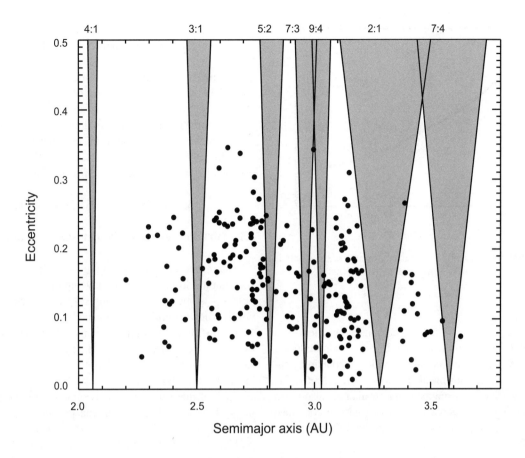

Fig. 3. Orbits of main-belt asteroids larger than 100 km in diameter, and approximate locations of the major mean-motion resonances with Jupiter. The resonance regions are typically empty. The ν_6 secular resonance removes asteroids with semimajor axes $2.0 \leq a \leq 2.2$ AU. Adapted from *Nesvorný et al.* (2002).

sitions because local nebula conditions would have changed. Similarly, planetesimals forming concurrently at different distances from the star would have different compositions.

Main-belt asteroids show a gradation in spectral properties with distance from the Sun (*Gradie and Tedesco*, 1982) suggesting differences in composition or degree of thermal processing. Asteroids in the inner belt tend to be S types, which are thought to be dry and rich in iron-magnesium silicates, or M types, some of which may be the parent bodies of iron meteorites. The middle belt is dominated by C types and related classes, many of which have spectral features suggesting they contain hydrated silicates (*Rivkin et al.*, 2003), which probably formed by reactions between water ice and dry rock. The outer belt and Trojan asteroids are P and D types showing no signs of hydrated silicates, although they may contain water ice. Several asteroids in the outer belt display comet-like activity, suggesting they contain some ice (*Hsieh and Jewitt*, 2006).

Meteorites have undergone various degrees of thermal processing and aqueous alteration due to reactions with water. Some carbonaceous chondrites contain up to 10% water by mass in the form of hydrated silicates. Ordinary chondrites contain little water and are more depleted in moderately volatile elements, but there are hints that water was once present (*Grossman et al.*, 2000). Ordinary chondrites have also undergone thermal metamorphism. Primitive achondrite meteorites come from parent bodies that have partially melted, but many were clearly once similar to chondrites. Finally, most iron meteorites and some achondrites come from asteroids that have completely melted and differentiated. The wide range of thermal histories seen in meteorites probably reflects differences in the time of formation and the corresponding degree of heating due to short-lived radioactive isotopes, as well as differences in the initial amount of water ice they contained.

The oxidation state of the solar nebula probably changed over time and with location. At early times, boulder-sized bodies would have drifted inward across the ice line, evaporating, and increasing the O/H ratio of the gas (*Ciesla and Cuzzi*, 2006). In a turbulent disk, water vapor from the inner disk would have diffused outward across the ice line, potentially depositing large amounts of water ice at a "cold trap" (*Stevenson and Lunine*, 1988). Eventually, much of the water became locked up in large planetesimals and embryos, shutting off the flow of boulders and making the inner disk chemically reducing. This may explain differences in chon-

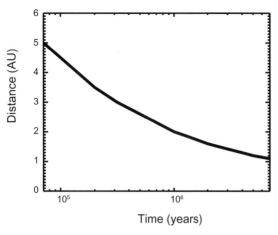

Fig. 4. Approximate location of the ice line in the solar nebula over time. Adapted from *Kennedy and Kenyon* (2008).

drite chemistry: The CM and CI carbonaceous chondrites are relatively oxidized, ordinary chondrites are more reduced, while enstatite chondrites are highly reduced, containing nitrides and silicon-bearing metal (*Weisberg et al.*, 2006). The removal of water vapor from the inner nebula may also mean that late-forming planetesimals in the asteroid belt contained little water ice, even though they were beyond the ice line at this point (Fig. 4).

Terrestrial planets are more depleted in moderately volatile elements than chondrites, while some achondrites are more highly depleted still (*Halliday and Porcelli*, 2001). This suggests that volatile materials escaped as planetary embryos were heated during impacts and by radioactive decay of short-lived isotopes. The isotopic ratios of some noble gases in Earth's atmosphere are enriched in heavier isotopes, arguing that much of the original complement of these elements escaped into space, preferentially leaving heavy isotopes behind (*Pepin*, 2006). Earth may have lost other volatile materials at the same time. The Moon is highly depleted in volatiles, which argues in favor of an impact origin.

Planetary embryos the size of Mars or larger are likely to melt and differentiate as a result of the kinetic energy released during impacts with other large bodies (*Tonks and Melosh*, 1992). Iron and other siderophile elements sink to the center of these bodies, leaving a silicate-mantle surrounding a metal-rich core. Highly siderophile elements like iridium should have been almost entirely extracted into the core as Earth differentiated. However, the highly siderophile elements are still present in small amounts in the mantle, and in roughly chondritic ratios (*Drake and Righter*, 2002). This suggests that Earth accreted the last 1% of its mass from a primitive source after core formation was complete, referred to as the "late veneer."

The principal source of Earth's water is uncertain. Earth's water has a D/H ratio six times that of the Sun (*Robert*, 2001), suggesting that this hydrogen was not captured directly from the nebula. The ice line in the solar nebula may have been within 1 AU of the Sun near the end of nebula's lifetime, allowing planetesimals to incorporate water ice. However, it seems likely that planetesimals had already finished forming by this point. If water ice was not present in planetesimals at 1 AU it must have come from further out in the disk. Impacts with planetesimals and planetary embryos originating in the asteroid belt and outer solar system are possibilities. The former source is plausible due to the high collision probability of asteroids with Earth, and the fact that the D/H ratio of water in carbonaceous chondrites matches the ratio in Earth's oceans (*Morbidelli et al.*, 2000; *Robert*, 2001).

3. RESEARCH HIGHLIGHTS

3.1. Turbulence-Driven Planetesimal Formation

The formation of planetesimals — objects 1–1000 km in size that are large enough to have appreciable gravitational fields — is one of the main unresolved problems associated with planet formation. Traditionally there have been two schools of thought: Planetesimals either formed gradually by pairwise collisions between dust grains, or planetesimals formed rapidly by gravitational instability in a thin dust layer at the midplane of the disk.

Each of these models has encountered severe difficulties in the face of turbulence. In the particle-sticking model, growth is especially difficult for boulder-sized bodies, a problem referred to as the "meter-size barrier." For plausible disk turbulence levels, meter-sized particles collide with one another at speeds of tens of meters per second. These collisions probably lead to erosion rather than growth. Differential radial drift rates also mean that meter-sized bodies collide with smaller objects at speeds of up to 100 m/s. Experiments show that small aggregates of micrometer-sized dust particles can embed themselves in larger dust aggregates at speeds of at least 25 m/s (*Wurm et al.*, 2005). However, many asteroids are primarily composed of dense millimeter-sized chondrules. Chondrule aggregates are unlikely to merge at such high speeds, even when coated with dust rims (*Ormel et al.*, 2008).

Turbulence poses an even more severe challenge to the conventional model for gravitational instability. Turbulence levels equivalent to $\alpha \sim 10^{-7}$ and 10^{-4} are sufficient to prevent millimeter- and meter-sized particles respectively from settling to a layer thin enough to become gravitationally unstable (*Cuzzi and Weidenschilling*, 2006). Even in an intrinsically laminar nebula, shear instabilities associated with dust settling will generate local turbulence, and this is sufficient to prevent gravitationally instability unless the solid-to-gas ratio is enhanced by two orders of magnitude compared to that in the solar nebula (*Garaud and Lin*, 2004).

In light of these problems, the hunt is on to find a new mechanism for planetesimal formation that can operate in a turbulent environment. Here we will examine two models that have recently been developed that consider the dynamics of millimeter- and meter-sized particles respectively.

In a turbulent disk, small particles tend to become concentrated in stagnant regions between eddies (*Cuzzi et al.*, 2001). Concentration is most efficient for the smallest eddies whose size is determined by the point at which turbulence is damped by molecular viscosity. These eddies primarily affect millimeter-sized particles, which have stopping times comparable to the eddy turnover time.

Turbulent concentration is a stochastic process. Eddies continually form and break up, and particles enter and leave dense clumps repeatedly. However, a particle tends to spend more time in regions of high particle density than in the rarified regions in between. High-density clumps form less often than low-density clumps. In regions of very high particle density, the particles tend to damp down turbulent motions, preventing further concentration. Calculations show that the local solid-to-gas density ratio is ~100 in the densest clumps, which is roughly four orders of magnitude higher than the average for material in the solar nebula (*Cuzzi et al.*, 2008).

Large clumps of particles can be gravitationally bound but they will still have a very low density compared to a planetesimal. Particles are accelerated toward the center of the clump by gravity. However, particles are also tightly coupled to the gas, so the gas becomes compressed as the particles move inward, opposing further contraction. Calculations suggest it may take hundreds of orbital periods for millimeter-sized particles to sediment to the center of a clump at roughly terminal velocity (*Cuzzi et al.*, 2008).

Particles in dense clumps drag along the gas so that the clump as a whole orbits the star at roughly the Keplerian velocity. Gas in the surrounding region typically moves at sub-Keplerian speeds due to the outward radial pressure gradient in the disk. As a result, a clump is potentially vulnerable to disruption by ram pressure associated with this difference in velocity. Ram pressure increases with the surface area of the clump, while the gravitational binding energy increases with its mass. This implies there is likely to be a minimum mass for a clump that can survive, equivalent to a solid planetesimal with radius of 10–100 km (*Cuzzi et al.*, 2008).

The turbulent concentration model is still at a relatively early stage of development, but it has several appealing features in light of the observed characteristics of primitive meteorites. Surviving clumps should eventually shrink to form planetesimals composed of loosely compacted millimeter-sized particles similar to chondrite parent bodies. These particles are size sorted during the turbulent concentration phase and are likely to have a narrow size distribution, another property shared by chondritic meteorites (*Cuzzi et al.*, 2001). Finally, the stochastic nature of turbulence, and the fact that high-density clumps have a relatively low probability of formation, means that turbulent concentration can naturally explain why planetesimal formation continued in the asteroid belt for several million years.

If a large fraction of the dust in a disk becomes incorporated into meter-sized boulders, a second possible route to planetesimal formation arises. Meter-sized objects are strongly affected by gas drag, so they quickly move radially toward temporary pressure maxima in a turbulently evolving disk (*Johansen et al.*, 2006). The short radial drift timescale means that these objects can undergo significant concentration before the pressure maximum disappears.

Once the density of boulders is enhanced in a particular region, the particles begin to affect the motion of the surrounding gas. In general, gas orbits the star more slowly than a large solid body due to its outward pressure gradient. As the solid particles become concentrated they begin to drag the gas along at higher speeds. As a result, the relative velocity between the particles and the gas is reduced, and radial drift rates decline. Boulders further out in the disk continue to drift inward rapidly as before, so solids tend to accumulate in regions where the solid/gas density ratio is already high. This positive feedback effect, referred to as the "streaming instability," further enhances the solid-to-gas ratio until gravitationally bound clumps form (Fig. 5). Numerical simulations show that such clumps are likely to grow in mass and contract over a few orbital periods due to collisions between particles and gas drag (*Johansen et al.*, 2007).

The streaming instability model is attractive in that it provides a way to form bodies comparable in size to the largest asteroids in a vigorously turbulent disk. A necessary prerequisite is that a substantial fraction of the solid material in the disk forms into meter-sized objects. Whether or not this happens, and whether these objects survive mutual collisions in the turbulent disk depends on their mechanical properties, which are poorly constrained at present. An obvious shortcoming of the model is that primitive meteorites are typically composed of millimeter-sized particles rather than larger objects, although it is conceivable that chondrules first aggregated into roughly meter-sized boulders and these objects subsequently formed planetesimals.

3.2. Radioisotope Dating of Planetary Growth

Theories for planet formation can be tested or constrained using data from radioactive isotope systems to measure timescales. Useful isotope systems fall into two categories. Long-lived isotopes such as ^{238}U were present when rocks formed early in the solar system, and their decay products can be used to measure the absolute time that has elapsed since these rocks formed. Short-lived isotopes such as ^{26}Al have half-lives comparable to timescales involved in planet formation, and they were present in the solar nebula. These isotopes have all decayed now, but the distribution of their daughter isotopes can tell us about the relative timing of different events during planet formation.

For a long-lived isotope system such as Rb-Sr, the number of atoms of the daughter isotope ^{87}Sr increases over time t as the number of atoms of the parent ^{87}Rb declines

$$\frac{^{87}Sr(t)}{^{86}Sr} = \frac{^{87}Sr(0)}{^{86}Sr} + \frac{^{87}Rb(t)}{^{86}Sr}\left[\exp(\lambda t)-1\right] \qquad (51)$$

where λ is the decay constant that is related to the half-life. Here, ^{86}Sr is the number of atoms of another stable isotope of the daughter element not involved in radioactive decay, which is used for comparison.

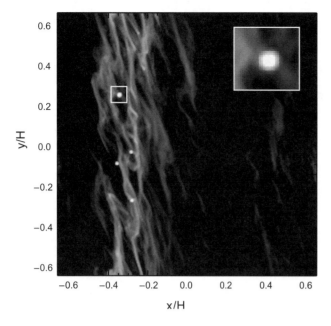

Fig. 5. A simulation showing the streaming instability involving meter-sized particles in a turbulent disk. The x and y axes refer to the radial and azimuthal directions in the disk respectively. Brighter shades of gray indicate locations with high particle surface density. The region highlighted by the square box shows a location where a gravitationally bound clump has formed. Figure kindly supplied by A. Johansen.

This equation contains two unknowns: the time that has elapsed since the sample formed, and the original amount of the daughter isotope. However, by plotting the current isotopic ratios for different minerals in the same sample, having different elemental ratios, it is possible to determine both quantities. An implicit assumption is that the rocks have remained closed systems. If the system has been disturbed by mixing with other reservoirs, the accuracy of the age estimate will be compromised.

Among long-lived isotopes, the U-Pb system is most widely used. Uranium has two long-lived isotopes, ^{235}U and ^{238}U, that each decay into a different isotope of lead. The corresponding half-lives are 0.7 and 4.5 G.y., respectively, which are convenient for dating events comparable to the age of the solar system. In addition, the presence of two isotopic clocks running concurrently means it is possible to obtain useful dates even in systems that have been disturbed. A variant of this technique is the Pb-Pb method in which the age of a sample is deduced from measurements of lead isotopes alone, making use of the fact that the two uranium isotopes decay into lead at different rates.

The U-Pb system was the first to be used to measure the age of bodies in the solar system. More recent measurements have obtained very precise ages for calcium-aluminum-rich inclusions (CAIs), the oldest known objects that formed in the solar system: 4567.2 ± 0.1 m.y. (*Amelin*, 2006). Lead-lead measurements of rocks on Earth show that it must have formed roughly 100 m.y. later (*Allégre et al.*, 1995).

A number of short-lived isotopes were present early in the solar system (see Table 3). The source of these isotopes has been the subject of much debate. Light isotopes such as ^{10}Be may have been synthesized in the solar nebula as dust grains close to the Sun were bombarded with energetic particles. Most isotopes, especially heavy, neutron-rich isotopes, were probably generated in nearby massive stars before being ejected into the interstellar medium, ultimately ending up in the solar nebula.

To make use of these systems, it is necessary to assume that the isotopes were uniformly distributed in the solar nebula so that the isotopic ratios for each element depend only on time. The Mg-isotope ratios in bulk chondrites, Earth, the Moon, and Mars suggest that ^{26}Al, at least, was uniformly distributed (*Thrane et al.*, 2006). In addition, it appears that ^{60}Fe was uniformly distributed and injected at an early stage (*Dauphas et al.*, 2008). The situation is less clear for some other isotopes, and ^{53}Mn may have been unevenly distributed (*Lugmair and Shukolyukov*, 2001).

The ^{26}Al-^{26}Mg system is particularly useful for chronology since the parent isotope has a half-life of only 0.7 m.y., which makes high precision dating possible, and both parent and daughter isotopes are refractory, reducing the probability that the system will be disturbed. From conservation of ^{26}Al + ^{26}Mg, the isotopic ratios that are observed today are related to those at the time a rock formed by

$$\left(\frac{^{26}\text{Mg}}{^{24}\text{Mg}}\right)_{\text{today}} = \left(\frac{^{26}\text{Mg}}{^{24}\text{Mg}}\right)_{\text{original}} + \left(\frac{^{26}\text{Al}}{^{27}\text{Al}}\right)_{\text{original}} \times \left(\frac{^{27}\text{Al}}{^{24}\text{Mg}}\right)_{\text{today}} \quad (52)$$

A plot of the isotope ratios measured today, for different minerals in the same sample, gives the ^{26}Al/^{27}Al ratio at the time the rock formed. By comparing this ratio to the ^{26}Al/^{27}Al ratio in other objects, it is possible to deduce their relative formation times.

On the basis of the Al-Mg system, the oldest solid objects known to have formed in the solar system are CAIs, which are found in most chondritic meteorites (Fig. 6). Many CAIs had similar ^{26}Al/^{27}Al ratios when they formed, suggesting they formed over a narrow time interval, possibly as short as 20,000 yr (*Thrane et al.*, 2006). The ^{26}Al/^{27}Al ratio is $\simeq 5 \times 10^{-5}$ for these objects, and this is often assumed to be the ratio shortly after ^{26}Al was synthesized or injected into the solar nebula. Typically, other timescales in cosmochemistry are measured with respect to the time at which these CAIs formed. A few CAIs apparently had little or no ^{26}Al when they formed, which suggests they actually formed even earlier, at a time before ^{26}Al had appeared in the solar nebula (*Krot et al.*, 2008).

Chondrule ages measured using the Al-Mg system are generally 1–2 m.y. younger than CAIs (*Mostefaoui et al.,* 2002), a result that is in excellent agreement with absolute Pb-Pb ages for these objects (*Amelin,* 2006; *Connelly et al.,* 2008). This implies that the chondrite parent bodies formed several million years after CAIs. However, the presence of CAIs in most chondritic meteorites implies that some CAIs survived in the nebula for this length of time.

The Al-Mg and Hf-W systems have been applied to the HED meteorites, which are thought to come from asteroid 4 Vesta. These analyses show that Vesta must have formed within the first 5 m.y. of the solar system (*Srinivasan et al.,* 1999; *Touboul et al.,* 2008). Similarly, the angrites, which come from an unidentified differentiated body, formed within 5 m.y. of CAIs (*Markowski et al.,* 2007). The parent bodies of the iron meteorites formed within the first 1.5 m.y. of the solar system, according to the Hf-W system (*Kleine et al.,* 2005), which means their formation probably predates that of most chondrules.

The Hf-W system is particularly useful for studying differentiated bodies because the parent isotope ^{182}Hf is lithophile, partitioning into the mantle, while the daughter isotope ^{182}W is siderophile, partitioning into the core. The Hf-W system provides a way to date the timing of core formation, and is most easily applied to objects where core formation occurs over a short space of time. For example, the lack of a ^{182}W excess in lunar samples indicates that the giant impact leading to the formation of the Moon happened >50 m.y. after the start of the solar system (*Touboul et al.,* 2007).

Caution is required when the Hf-W system is applied to objects like Earth and Mars, where growth and core formation took place concurrently over extended periods of time. A ^{182}W excess has been measured in Earth's mantle (*Yin et al.,* 2002; *Kleine et al.,* 2002). Assuming an exponentially declining rate of growth and complete mixing after each impact event, Earth would have grown to 63% of its final mass in ~11 m.y. This is substantially shorter than calculations based on the Pb-Pb and I-Xe systems (*Allégre et al.,* 1995; *Halliday,* 2004) and estimates based on numerical simulations of planet formation. These numbers can probably be reconciled if it is assumed that only partial mixing occurs during giant impacts (*Halliday,* 2004), although this is still a matter of debate. Similar issues affect age estimates for Mars, in addition to uncertainties about the Hf/W elemental ratio of the planet. However, it appears that Mars formed more rapidly than Earth, within 10 m.y. of CAIs (*Nimmo and Kleine,* 2007).

3.3. N-Body Simulations of Planet Formation

The final stage of terrestrial planet formation involved a few tens to a few hundred embryos moving on crossing orbits, and lasted ~10^8 yr according to data from radioactive isotopes. The small number of bodies involved and their extensive dynamical interactions means that the kind of statistical analysis often used to study runaway and oligarchic growth is inappropriate here. Instead, this stage is ideally suited to N-body simulations where each object is followed explicitly.

TABLE 3. Short-lived isotopes in the solar nebula (adapted from *Wadhwa et al.*, 2007).

Isotope	Half Life (m.y.)	Daughter Isotope
^{41}Ca	0.1	^{41}K
^{26}Al	0.72	^{26}Mg
^{60}Fe	1.5	^{60}Ni
^{10}Be	1.5	^{10}B
^{53}Mn	3.7	^{53}Cr
^{107}Pd	6.5	^{107}Ag
^{182}Hf	8.9	^{182}W
^{129}I	15.7	^{129}Xe
^{244}Pu	82	various
^{146}Sm	103	^{142}Nd

Early studies considered systems of 30–150 embryos, and showed that these typically evolved via dynamical interactions and collisions into a system of two to four terrestrial planets on a timescale 100–200 m.y. (*Chambers and Wetherill,* 1998; *Agnor et al.,* 1999; *Chambers,* 2001) (Fig. 7). The evolution in these simulations is highly stochastic. Many outcomes are possible for similar sets of initial conditions, including differences in the number of planets and their orbits and masses. The prolonged evolution, and large number of close encounters per collision, lead to substantial radial mixing, so that planets like Earth are probably a composite of embryos from throughout the inner solar system. Potential moon-forming impacts are also commonly seen in these simulations (*Agnor et al.,* 1999).

The results of these early simulations differ from the observed terrestrial planets in two significant ways. First, they tend to produce planets on moderately eccentric orbits, with e ~ 0.1, compared to the lower time-averaged values e ~ 0.03 for Earth and Venus. Second, they fail to reproduce the low mass of Mars and the high concentration of mass seen in the region between 0.7 and 1.0 AU, which contains 90% of the planetary mass interior to Jupiter. Newer simulations involving ~1000 planetesimals and embryos have produced terrestrial planets with low eccentricities, more like Earth and Venus (*O'Brien et al.,* 2006; *Raymond et al.,* 2006a). Although planetesimals are continually depleted in these simulations, enough survive for long intervals that dynamical friction with the larger embryos reduces the eccentricities of the latter, resulting in final planets with nearly circular orbits.

N-body simulations of the asteroid-belt region have confirmed the hypothesis that perturbations from the giant planets would dynamically deplete the belt on timescales that are short compared to the age of the solar system. Simulations involving 15–200 lunar- to Mars-sized embryos find that almost all these objects typically enter resonances due to their mutual interactions, and are subsequently removed from the asteroid belt (*Chambers and Wetherill,* 2001). In roughly two-thirds of the cases, no embryos survive. Complementary simulations that include test particles find that the loss

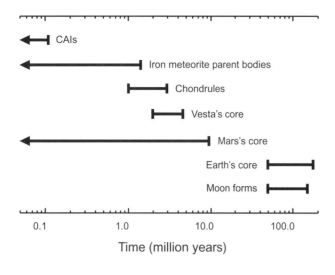

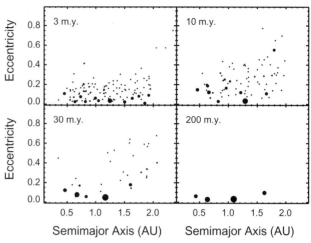

Fig. 6. Timescales for various events in the early solar system determined by radioactive isotope dating. Data from *Kleine et al.* (2002), *Halliday* (2004), *Kleine et al.* (2005), *Mostefaoui et al.* (2002), *Nimmo and Kleine* (2007), *Scott* (2007), and *Touboul et al.* (2007, 2008).

Fig. 7. A simulation of the final stage of terrestrial planet formation by the author. The simulation begins with 150 lunar-to-Mars-mass planetary embryos in the presence of fully formed Jupiter and Saturn. The figure shows four snapshots in time. Each circle represents a single planetary embryo with symbol radius proportional to that of the embryo. The largest planet at 200 m.y. has a mass almost identical to that of Earth.

of embryos is likely to be accompanied by the removal of smaller asteroids, with ~99% of test particles being removed from the asteroid belt in these simulations (*Petit et al.*, 2001).

Perturbations from giant planets also influence the growth of terrestrial planets outside the asteroid belt. In particular, orbital eccentricities are excited at secular resonances, and this excitation can then be transmitted to neighboring regions by interactions between embryos. The final number of terrestrial planets declines with increasing excitation from the giant planets, with a corresponding increase in the mean planetary mass (*Levison and Agnor*, 2003).

A side effect of the dynamical clearing of the asteroid belt is that planetesimals and embryos develop highly eccentric orbits that cross those of the inner planets. Collisions between these two populations provide one way to deliver water and other volatiles to the terrestrial planets (*Morbidelli et al.*, 2000; *Raymond et al.*, 2006a), although this issue remains controversial. The small number of embryos involved means that the amount of water received by each planet from embryos will vary stochastically from one planet to another. Water delivered by planetesimals should be more uniformly distributed (*Raymond et al.*, 2007). Overall large planets like Earth probably receive more volatiles than smaller planets like Mars (*Lunine et al.*, 2003).

N-body simulations also show that the efficiency of volatile delivery is highly sensitive to the orbital eccentricities of the giant planets, which are poorly constrained at present (*Chambers and Wetherill*, 2001; *Chambers and Cassen*, 2002; *O'Brien et al.*, 2006). When Jupiter and Saturn have eccentric orbits, resonances occupy a large fraction of orbital phase space and these strongly excite the eccentricities of asteroids. As a result, objects in the asteroid belt have short dynamical lifetimes and low collision probabilities with the terrestrial planets. When the giant planets have nearly circular orbits, their perturbations are weaker. The orbits of some embryos in the belt gradually diffuse into the inner solar system where the chance of hitting one of the inner planets is much higher. Terrestrial planets are likely to be more water rich in this latter case (*O'Brien et al.*, 2006).

Simulations that include smooth type-I planetary migration find that this can have a profound effect on planet formation. Many of the embryos initially in the terrestrial planet region are lost, to be replaced by material from further out in the disk (*McNeil et al.*, 2005; *Daisaka et al.*, 2006). Large embryos migrating inward can capture smaller objects at an interior MMR, so that both objects migrate inward together at a rate determined by their average mass (*McNeil et al.*, 2005; *Cresswell and Nelson*, 2008). As a result, small objects can be indirectly affected by migration. Migration rates decline over time as the gas dissipates, so late-forming embryos and those that migrate inward from outside the terrestrial planet region can survive. Migration rates are higher in massive disks, so these may actually be less likely to form terrestrial planets despite having more solid material available.

In the absence of smooth migration, stochastic tidal torques caused by turbulent density fluctuations tend to excite eccentricities. Tidal damping reduces e while keeping the angular momentum roughly constant, so that $a(1-e^2)$ is unchanged. As a result, a decreases, and embryos move slowly toward the star, even in the absence of conventional type-I migration (*Ogihara et al.*, 2007).

Similarly, embryos probably moved inward in the solar system as secular resonances swept across the asteroid belt during the dissipation of the gas disk (*Nagasawa et al.*, 2005). The v_5 resonance was initially located in the outer main belt when the nebula was present. As it moved inward, embryos' eccentricities increased, which combined with tidal damping caused the embryos to move inward. Simulations show that embryos often tend to move inward at the same rate as the sweeping resonance, transporting large amounts of mass from the asteroid belt to the terrestrial planet region. As embryos gradually escape from the resonance, their inward motion ceases. Eventually the v_5 resonance stopped moving when the nebula was completely dispersed, coming to rest in the region now occupied by the terrestrial planets. This "dynamical shake-up" model predicts a pile-up of mass near this final location. It may explain the current concentration of mass between 0.7 and 1.0 AU (*Nagasawa et al.*, 2005; *Thommes et al.*, 2008), but only if the giant planets had their current orbits at this time (*O'Brien et al.*, 2007).

One of the most interesting scenarios examined recently is the possibility that terrestrial planets can form and survive in systems that contain a "hot Jupiter" — a giant planet orbiting within ~0.1 AU of its star. Hot Jupiters are observed in a number of exoplanetary systems. They probably formed in the outer parts of their star's protoplanetary disk and then migrated inward to their current location due to tidal interactions with the disk. During this migration, planetesimals and planetary embryos in the terrestrial planet region are either captured into MMRs and shepherded inward by the giant planet, or they are scattered onto eccentric orbits crossing the outer disk (*Raymond et al.*, 2006b; *Fogg and Nelson*, 2007). Dynamical friction and gas drag subsequently damp the eccentricities of the embryos and planetesimals, respectively. The simulations show that a new disk of material forms beyond the orbit of the migrating giant, and this subsequently evolves into one or more terrestrial planets (Fig. 8). These second-generation terrestrial planets typically contain substantial amounts of volatile-rich material from the outer disk, and may be unlike anything in the solar system.

4. FUTURE PROSPECTS

The next decade should see advances in several areas related to the study of planet formation. Continuing discoveries by the Spitzer Space Telescope and the upcoming Atacama Large Millimeter Array (ALMA) should greatly improve our knowledge of protoplanetary disks, including their compositions and temporal evolution. The very high

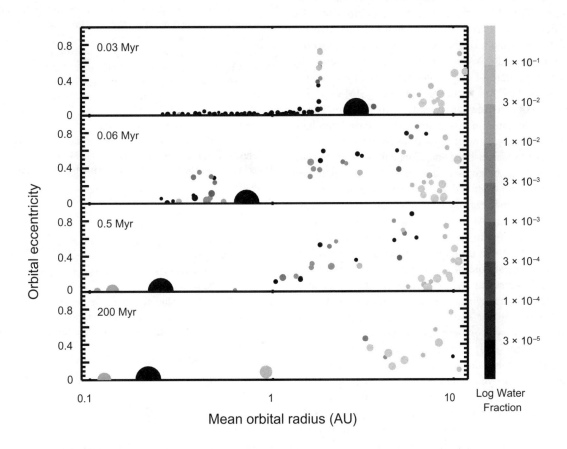

Fig. 8. Simulation of terrestrial planet formation in the presence of a migrating gas-giant planet. A Jupiter-mass planet, indicated by the large circle, begins at 5 AU and migrates to 0.25 AU in 10^5 yr. The small symbols show planetary embryos, with the embryo's radius proportional to the symbol radius, and the embryo's water mass fraction indicated by the gray scale on the righthand side of the figure. Smaller planetesimals were also included in the simulation but are not shown here. Adapted from *Raymond* (2006b). Data kindly supplied by S. Raymond.

spatial resolution of ALMA will provide detailed snapshots of the dust distribution within protoplanetary disks, and should help answer questions about how planets interact with their disks and how disks disperse. The ongoing CoRoT mission, and recently launched Kepler mission, should provide the first detections of Earth-like planets orbiting other stars like the Sun, and possibly provide compositional constraints. At the same time, further improvements in groundbased radial-velocity and transit surveys should continue the inexorable trend toward the discovery of ever lower-mass planets. Searches for planets orbiting low-mass M stars are likely to be particularly fruitful in this respect.

Several relevant space missions are en route to objects in the solar system. These include the ongoing MESSENGER mission to explore Mercury, which is probably the least understood of the terrestrial planets in the solar system. The Dawn mission is scheduled to rendezvous with the two largest main-belt asteroids, Ceres and Vesta, in the coming decade. This will provide an opportunity to compare what we think we know about Vesta on the basis of the HED meteorites with close-up observations of their parent body. For Ceres, the prospect is even more enticing since little is known about this dwarf planet, including major uncertainties about its formation and composition. At the far end of the solar system, the New Horizons mission will visit several Kuiper-belt objects including Pluto. While these objects are compositionally very different than Earth, studying them will improve our understanding of the formation of solid planets in general.

Astronomical discoveries are likely to be matched by progress in the field of cosmochemistry. The new generation of secondary ion mass spectrometer (SIMS) probes makes it possible to analyze extremely small samples from meteorites, including matrix grains, which have been hard to study previously, as well as samples from Comet Wild 2 collected by the Stardust mission. This will improve our understanding of the behavior of dust grains in protoplanetary disks, and when combined with advances in radiometric dating, should provide strong constraints on the early stages of planet formation. At the same time, ongoing programs to collect meteorites in Antarctica and other remote locations will expand our sampling of primitive and evolved bodies from the asteroid belt. The likely discovery of new "ungrouped" meteorites will continue to test our theories of what actually happened early in the solar system.

There is scope for substantial breakthroughs on the theoretical front too. New and rapidly evolving models for the evolution of small particles in turbulent gas suggest we may finally gain a detailed understanding of how planetesimals form, 40 years after Safronov first proposed his planetesimal model for planet formation. At the same time, our understanding of what drives the evolution and dispersal of protoplanetary disks is advancing rapidly, and this field is likely to mature substantially over the next 10 years. Planetary migration represents a major area of uncertainty at present, but a better understanding of the physical processes involved, and continuing improvements in the resolution of hydrodynamical simulations, offer the prospect that we may yet get to grips with this tricky issue. Finally, improved numerical simulations of planet formation, coupled with models for the physical and chemical evolution of planets and protoplanetary disks, should eventually make it clear exactly where Earth's water originated, and what it takes to make a habitable planet.

Acknowledgments. I would like to thank S. Raymond, S. Seager, and an anonymous reviewer for helpful comments that helped to improve this article. This work was partially supported by NASA's Origins of Solar Systems program.

REFERENCES

Adachi I., Hayashi C., and Nakazawa K. (1976) The gas drag effect on the elliptic motion of a solid body in the primordial solar nebula. *Prog. Theor. Phys., 56,* 1756–1771.

Agnor C. and Asphaug E. (2004) Accretion efficiency during planetary collisions. *Astrophys. J. Lett., 613,* L157–L160.

Agnor C. B., Canup R. M., and Levison H. F. (1999) On the character and consequences of large impacts in the late stage of terrestrial planet formation. *Icarus, 142,* 219–237.

Alexander C. M. O'D, Grossman J. N., Ebel D. S., and Ciesla F. J. (2008) Formation conditions of chondrules and chondrites. *Science, 320,* 1617–1619.

Alexander R. D., Clarke C. J., and Pringle J. E. (2006) Photoevaporation of protoplanetary discs II. Evolutionary models and observable properties. *Mon. Not. R. Astron. Soc., 369,* 229–239.

Allégre C. J., Manhés G., and Gopel C. (1995) The age of Earth. *Geochim. Cosmochim. Acta, 59,* 1445–1456.

Amelin Y. (2006) The prospect of high precision Pb isotopic dating of meteorites. *Meteoritics & Planet. Sci., 41,* 7–17.

Asphaug E., Ostro S. J., Hudson R. S., Scheeres D. J., and Benz W. (1998) Disruption of kilometer-sized asteroids by energetic collisions. *Nature, 393,* 437–440.

Benz W. (2000) Low velocity collisions and the growth of planetesimals. *Space Sci. Rev., 92,* 279–294.

Benz W. and Asphaug E. (1999) Catastrophic disruptions revisited. *Icarus, 142,* 5–20.

Benz W., Slattery W. L., and Cameron A. G. W. (1988) Collisional stripping of Mercury's mantle. *Icarus, 74,* 516–528.

Blum J. and Wurm G. (2000) Experiments on sticking, restructuring and fragmentation of preplanetary dust aggregates. *Icarus, 143,* 138–146.

Bogard D. D. (1995) Impact ages of meteorites: A synthesis. *Meteoritics, 30,* 244–268.

Boss A. P. (1996) Evolution of the solar nebula III. Protoplanetary disks undergoing mass accretion. *Astrophys. J., 469,* 906–920.

Canup R. M. (2004) Simulations of a late lunar-forming impact. *Icarus, 168,* 433–456.

Canup R. M. and Asphaug E. (1999) Origin of the Moon in a giant impact near the end of the Earth's formation. *Nature, 412,* 708–712.

Cassen P. (1996) Models for the fractionation of moderately volatile elements in the solar nebula. *Meteoritics & Planet. Sci., 31,* 793–806.

Chambers J. E. (2001) Making more terrestrial planets. *Icarus, 152,* 205–224.

Chambers J. E. (2006) A semi-analytic model for oligarchic growth. *Icarus, 180,* 496–513.

Chambers J. E. and Cassen P. (2002) The effects of nebula surface density profile and giant-planet eccentricities on planetary accretion in the inner solar system. *Meteoritics & Planet. Sci., 37*, 1523–1540.

Chambers J. E. and Wetherill G. W. (1998) Making the terrestrial planets: N-body integrations of planetary embryos in three dimensions. *Icarus, 136*, 304–327.

Chambers J. E. and Wetherill G. W. (2001) Planets in the asteroid belt. *Meteoritics & Planet. Sci., 36*, 381–399.

Chambers J. E., Wetherill G. W., and Boss A. P. (1996) The stability of multi-planet systems. *Icarus, 119*, 261–268.

Ciesla F. J. and Cuzzi J. N. (2006) The evolution of the water distribution in a viscous protoplanetary disk. *Icarus, 181*, 178–204.

Connelly H. C. and Jones R. H. (2005) Understanding the cooling rates experienced by type II porphyritic chondrules. In *Lunar and Planetary Science XXXVI*, Abstract #1881. Lunar and Planetary Institute, Houston (CD-ROM).

Connelly J. N., Amelin Y., Krot A. N., and Bizzarro M. (2008) Chronology of the solar system's oldest solids. *Astrophys. J. Lett., 675*, L121–L124.

Creswell P. and Nelson R. P. (2008) Three-dimensional simulations of multiple protoplanets embedded in a protostellar disc. *Astron. Astrophys., 482*, 677–690.

Cuzzi J. N. (2004) Blowing in the wind III. Accretion of dust rims by chondrule-sized particles in a turbulent protoplanetary nebula. *Icarus, 168*, 484–497.

Cuzzi J. N. and Weidenschilling S. J. (2006) Particle-gas dynamics and primary accretion. In *Meteorites and the Early Solar System II* (D. S.Lauretta and H. Y. McSween Jr., eds.), pp. 353–382. Univ. of Arizona, Tucson.

Cuzzi J. N., Dobrovolskis A. R., and Champney J. M. (1993) Particle-gas dynamics in the midplane of a protoplanetary nebula. *Icarus, 106*, 102–134.

Cuzzi J. N., Hogan R. C., Paque J. M., and Dobrovolskis A. R. (2001) Size-selective concentration of chondrules and other small particles in protoplanetary nebula turbulence. *Astrophys. J., 546*, 496–508.

Cuzzi J. N., Hogan R. C., and Shariff K. (2008) Towards planetesimals: Dense chondrule clumps in the protoplanetary nebula. *Astrophys. J., 687*, 1432–1447.

Daisaka J. K., Tanaka H., and Ida S. (2006) Orbital evolution and accretion of protoplanets tidally interacting with a gas disk II. Solid surface density evolution with type-I migration. *Icarus, 185*, 492–507.

Dauphas N., Cook D. L., Sacarabany C., Frohlich C., Davis A. M., Wadhwa M., Pourmand A., Rauscher T., and Gallino R. (2008) Iron 60 evidence for early injection and efficient mixing of stellar debris in the protosolar nebula. *Astrophys. J., 686*, 560–569.

Drake M. J. and Righter K. (2002) Determining the composition of the Earth. *Nature, 416*, 39–44.

Durda D. D., Bottke W. F., Enke B. L., Merline W. J., Asphaug E., Richardson D. C., and Leinhardt Z. M. (2004) The formation of asteroid satellites in large impacts: Results from numerical simulations. *Icarus, 170*, 243–257.

Fogg M. J. and Nelson R. P. (2007) On the formation of terrestrial planets in hot-Jupiter systems. *Astron. Astrophys., 461*, 1195–1208.

Fromang S., Terquem C., and Nelson R. P. (2005) Numerical simulations of type I planetary migration in non-turbulent magnetized discs. *Mon. Not. R. Astron. Soc., 363*, 943–953.

Garaud P. and Lin D. N. C. (2004) On the evolution and stability of a protoplanetary disk dust layer. *Astrophys. J., 608*, 1050–1075.

Gladman B. (1993) Dynamics of systems of two close planets. *Icarus, 106*, 247–263.

Gladman B. J., Miglorini F., Morbidelli A., Zappala V, Michel P., Cellino A., Froeschle C, Levison H. F., Bailey M., and Duncan M. (1997) Dynamical lifetimes of objects injected into asteroid belt resonances. *Science, 277*, 197–201.

Goldreich P. and Ward W. R. (1973) The formation of planetesimals. *Astrophys. J., 183*, 1051–1061.

Gradie J. and Tedesco E. (1982) Compositional structure of the asteroid belt. *Science, 216*, 1405–1407.

Greenzweig Y. and Lissauer J. J. (1992) Accretion rates of protoplanets. *Icarus, 100*, 440–463.

Grossman J. N., Alexander C. M. O'D., Wang J., and Brearley A. J. (2000) Bleached chondrules: Evidence for widespread aqueous processes on the parent asteroids of ordinary chondrites. *Meteoritics & Planet. Sci., 35*, 467–486.

Haisch K. E., Lada E. A., and Lada C. J. (2001) Disk frequencies and lifetimes in young clusters. *Astrophys. J. Lett., 553*, L153–L156.

Halliday A. N. (2004) Mixing, volatile loss and compositional change during impact-driven accretion of the Earth. *Nature, 427*, 505–509.

Halliday A. N. and Porcelli D. (2001) In search of lost planets — the paleocosmochemistry of the inner solar system. *Earth Planet. Sci. Lett., 192*, 545–559.

Hartmann L., Calvet N., Gullbring E., and D'Alessio P. (1998) Accretion and the evolution of T Tauri disks. *Astrophys. J., 495*, 385–400.

Housen K. R. and Holsapple K. A. (1999) Scale effects in strength-dominated collisions of rocky asteroids. *Icarus, 142*, 21–33.

Hsieh H. H. and Jewitt J. (2006) A population of comets in the main asteroid belt. *Science, 312*, 561–563.

Hueso R. and Guillot T. (2005) Evolution of protoplanetary disks: Constraints from DM Tauri and GM Aurigae. *Astron. Astrophys., 442*, 703–725.

Humayun M. and Clayton R .N. (1995) Potassium isotope cosmochemistry: Genetic implications of volatile element depletion. *Geochim. Cosmochim. Acta., 59*, 2131–2148.

Ida S. and Makino J. (1993) Scattering of planetesimals by a protoplanet: Slowing down of runaway growth. *Icarus, 106*, 210–227.

Ikoma M., Nakazawa K., and Emori H. (2000) Formation of giant planets: Dependences on core accretion rate and grain opacity. *Astrophys. J., 537*, 1013–1025.

Inaba S. and Ikoma M. (2003) Enhanced collisional growth of a protoplanet that has an atmosphere. *Astron. Astrophys., 410*, 711–723.

Johansen A., Klahr H., and Henning Th. (2006) Gravoturbulent formation of planetesimals. *Astrophys. J., 636*, 1121–1134.

Johansen A., Oishi J. S., Mac Low M. M., Klahr H., Henning Th., and Youdin A. (2007) Rapid planetesimal formation in turbulent circumstellar disks. *Nature, 448*, 1022–1025.

Kennedy G. M. and Kenyon S. J. (2008) Planet formation around stars of various masses: The snow line and the frequency of giant planets. *Astrophys. J., 673*, 502–512.

Kenyon S. J. and Bromley B. C. (2006) Terrestrial planet formation I. The transition from oligarchic growth to chaotic growth. *Astron. J., 131*, 1837–1850.

Kleine T., Münker C., Mezger K., and Palme H. (2002) Rapid accretion and early core formation on asteroids and the terrestrial planets from Hf-W chronometry. *Nature, 418*, 952–955.

Kleine T., Mezger K., Palme H., Scherer E., and Münker C. (2005) Early core formation in asteroids and late accretion of chondrite parent bodies: Evidence from ^{182}Hf-^{182}W in CAIs, metal-rich chondrites and iron meteorites. *Geochim. Cosmochim. Acta, 69,* 5805–5818.

Kokubo E. and Ida S. (1998) Oligarchic growth of protoplanets. *Icarus, 131,* 171–178.

Kokubo E. and Ida S. (2000) Formation of protoplanets from planetesimals in the solar nebula. *Icarus, 143,* 15–27.

Krot A. N., Nagashima K., Bizzarro M., Huss G. R., Davis A. M., Meyer B. S., and Ulyanov A. A. (2008) Multiple generations of refractory inclusions in the metal-rich carbonaceous chondrites Acfer 182/214 and Isheyevo. *Astrophys. J., 672,* 713–721.

Laskar J. (1997) Large scale chaos and the spacings of the inner planets. *Astron. Astrophys., 317,* L75–L78.

Lecar M. and Franklin F. (1997) The solar nebula, secular resonances, gas drag and the asteroid belt. *Icarus, 129,* 134–146.

Levison H. F. and Agnor C. (2003) The role of giant planets in terrestrial planet formation. *Astron. J., 125,* 2692–2713.

Lewis J. S. and Prinn R. G. (1980) Kinetic inhibition of CO and N_2 reduction in the solar nebula. *Astrophys. J., 238,* 357–364.

Lodders K. (2003) Solar system abundances and condensation temperatures of the elements. *Astrophys. J., 591,* 1220–1247.

Lugmair G. W. and Shukolyukov A. (2001) Early solar system events and timescales. *Meteoritics & Planet. Sci., 36,* 1017–1026.

Lunine J. I., Chambers J., Morbidelli A., and Leshin L. A. (2003) The origin of water on Mars. *Icarus, 165,* 1–8.

Markowski A., Quitté G., Kleine T., Halliday A. N., Bizzarro M., and Irving A. J. (2007) Hafnium tungsten chronometry of angrites and the earliest evolution of planetary objects. *Earth Planet. Sci. Lett., 262,* 214–229.

Masset F. S., Morbidelli A., Crida A., and Ferreira J. (2006) Disk surface density transitions as protoplanet traps. *Astrophys. J., 642,* 478–487.

McNeil D., Duncan M., and Levison H. F. (2005) Effects of type-I migration on terrestrial planet formation. *Astron. J., 130,* 2884–2899.

Meibom A. and Clark B. E. (1999) Invited review: Evidence for the insignificance of ordinary chondrite material in the asteroid belt. *Meteoritics & Planet. Sci., 34,* 7–24.

Melosh H. J. and Ryan E. V. (1997) Note: Asteroids shattered but not dispersed. *Icarus, 129,* 562–564.

Menou K. and Goodman J. (2004) Low-mass protoplanet migration in T Tauri alpha disks. *Astrophys. J., 606,* 520–531.

Morbidelli A., Chambers J., Lunine J. I., Petit J. M., Robert F., Valsecchi G. B., and Cyr K. E. (2000) Source regions and timescales for the delivery of water to the Earth. *Meteoritics & Planet. Sci., 35,* 1309–1320.

Mostefaoui S., Kita N. T., Togashi S., Tachibana S., Nagahara H., and Morishita Y. (2002) The relative formation ages of ferromagnesian chondrules inferred from their initial aluminum 26/aluminum 27 ratios. *Meteoritics & Planet. Sci., 37,* 421–438.

Murray C. D. and Dermott S. F. (2000) *Solar System Dynamics.* Cambridge Univ., Cambridge.

Nagasawa M., Tanaka H., and Ida S. (2000) Orbital evolution of asteroids during depletion of the solar nebula. *Astron. J., 119,* 1480–1497.

Nagasawa M., Lin D. N. C., and Thommes E. (2005) Dynamical shake-up of planetary systems I. Embryo trapping and induced collisions by the sweeping secular resonance and embryo-disk tidal interaction. *Astrophys. J., 635,* 578–598.

Nelson R. P. (2005) On the orbital evolution of low mass protoplanets in turbulent, magnetised disks. *Astron. Astrophys., 443,* 1067–1085.

Nesvorný D., Ferraz-Mello S., Holman M., and Morbidelli A. (2002) Regular and chaotic dynamics in the mean-motion resonances: Implications for the structure and evolution of the asteroid belt. In *Asteroids III* (W. F. Bottke et al., eds.), pp. 379–394. Univ. of Arizona, Tucson.

Nimmo F. and Kleine T. (2007) How rapidly did Mars accrete? Uncertainties in the Hf-W timing of core formation. *Icarus, 191,* 497–504.

O'Brien D. P., Morbidelli A., and Levison H. F. (2006) Terrestrial planet formation with strong dynamical friction. *Icarus, 184,* 39–58.

O'Brien D. P., Morbidelli A., and Bottke W. F. (2007) The primordial excitation and clearing of the asteroid belt — revisited. *Icarus, 191,* 434–452.

Ogihara M., Ida S., and Morbidelli A. (2007) Accretion of terrestrial planets from oligarchs in a turbulent disk. *Icarus, 188,* 522–534.

Ormel C. W., Cuzzi J. N., and Tielens A. G. G. M. (2008) Co-accretion of chondrules and dust in the solar nebula. *Astrophys. J., 679,* 1588–1610.

Paardekooper S. J. and Mellema G. (2006) Halting type I planet migration in non-isothermal disks. *Astron. Astrophys., 459,* L17–L20.

Pahlevan K. and Stevenson D. J. (2007) Equilibration in the aftermath of the lunar-forming giant impact. *Earth Planet. Sci. Lett., 262,* 438–449.

Pepin R. O. (2006) Atmospheres on the terrestrial planets: Clues to origin and evolution. *Earth Planet. Sci. Lett., 252,* 1–14.

Petit J. M., Morbidelli A., and Chambers J. (2001) The primordial excitation and clearing of the asteroid belt. *Icarus, 153,* 338–347.

Poppe T., Blum J., and Henning T. (2000) Analogous experiments on the stickiness of micron-sized preplanetary dust. *Astrophys. J., 533,* 454–471.

Raymond S. N., Quinn T., and Lunine J. I. (2006a) High-resolution simulations of the final assembly of Earth-like planets I. Terrestrial accretion and dynamics. *Icarus, 183,* 265–282.

Raymond S. N., Mandell A. M., and Sigurdsson S. (2006b) Exotic Earths: Forming habitable worlds with giant planet migration. *Science, 313,* 1413–1416.

Raymond S. N., Quinn T., and Lunine J. I. (2007) High-resolution simulations of the final assembly of Earth-like planets. Water delivery and planetary habitability. *Astrobiology, 7,* 66–84.

Rivkin A. S., Davies J. K., Johnson J. R., Ellison S. L., Trilling D. E., Brown R. H., and Lebofsky L. A. (2003) Hydrogen concentrations on C-class asteroids derived from remote sensing. *Meteoritics & Planet. Sci., 38,* 1383–1398.

Robert F. (2001) The origin of water on Earth. *Science, 293,* 1056–1058.

Scott E. R. D. (2007) Chondrites and the protoplanetary disk. *Annu. Rev. Earth Planet. Sci., 35,* 577–620.

Shakura N. I. and Sunyaev R. A. (1973) Black holes in binary systems. Observational appearance. *Astron. Astrophys., 24,* 337–355.

Srinivasan G., Goswami J. N., and Bhandari N. (1999) ^{26}Al eucrite Piplia Kalan: Plausible heat source and formation chronology. *Science, 284,* 1348–1350.

Stevenson D. J. and Lunine J. I. (1988) Rapid formation of Jupiter by diffusive redistribution of water vapor in the solar nebula. *Icarus, 75,* 146–155.

Tanaka H. and Ward W. R. (2004) Three-dimensional interaction between a planet and an isothermal gaseous disk II. Eccentricity waves and bending waves. *Astrophys. J., 602,* 388–395.

Tanaka H., Takeuchi T., and Ward W. R. (2002) Three-dimensional interaction between a planet and an isothermal gaseous disk I. Corotation and Lindblad torques and planet migration. *Astrophys. J., 565*, 1257–1274.

Thommes E. W., Duncan M. J., and Levison H. F. (2003) Oligarchic growth of giant planets. *Icarus, 161*, 431–455.

Thommes E., Nagasawa M., and Lin D. N. C. (2008) Dynamical shake-up of planetary systems II. N-body simulations of solar system terrestrial planet formation induced by secular resonance sweeping. *Astrophys. J., 676*, 728–739.

Thrane K., Bizzarro M., and Baker J. A. (2006) Extremely brief formation interval for refractory inclusions and uniform distribution of ^{26}Al in the early solar system. *Astrophys. J. Lett., 646*, L159–L162.

Tonks W. B. and Melosh H. J. (1992) Core formation by giant impacts. *Icarus, 100*, 326–346.

Touboul M., Kleine T., Bourdon B., Palme H., and Wieler R. (2007) Late formation and prolonged differentiation of the Moon inferred from W isotopes in lunar metals. *Nature, 450*, 1206–1209.

Touboul M., Kleine T., and Bourdon B. (2008) Hf-W systematics of cumulate eucrites and the chronology of the eucrite parent body. In *Lunar and Planetary Science XIX*, Abstract #2336. Lunar and Planetary Institute, Houston (CD-ROM).

Ueda T., Murakami Y., Ishitsu N., Kawabe H., Inoue R., Nakamura T., Sekiya M., and Takaoka N. (2001) Collisional destruction experiment of chondrules and formation of fragments in the solar nebula. *Earth Planets Space, 53*, 927–935.

Wadhwa M., Amelin Y., Davis A. M., Lugmair G. W., Meyer B., Gounelle M., and Desch S. J. (2007) From dust to planetesimals: Implications for the solar protoplanetary disk from short-lived radionuclides. In *Protostars and Planets V* (B. Reipurth et al., eds.), pp. 835–848. Univ. of Arizona, Tucson.

Weidenschilling S. J. (1977) Aerodynamics of solid bodies in the solar nebula. *Mon. Not. R. Astron. Soc., 180*, 57–70.

Weidenschilling S. J. (1980) Dust to planetesimals: Settling and coagulation in the solar nebula. *Icarus, 44*, 172–189.

Weidenschilling S. J. (1997) The origin of comets in the solar nebula: A unified model. *Icarus, 127*, 290–306.

Weisberg M. K., McCoy T. J., and Krot A. N. (2006) Systematics and evaluation of meteorite classification. In *Meteorites and the Early Solar System II* (D. S. Lauretta and H. Y. McSween Jr., eds.), pp. 353–382. Univ. of Arizona, Tucson.

Wetherill G. W. and Stewart G. R. (1993) Formation of planetary embryos: Effects of fragmentation, low relative velocity, and independent variation of eccentricity and inclination. *Icarus, 106*, 190–209.

Wiechert U., Halliday A. N., Lee D. C., Snyder G. A., Taylor L. A., and Rumble D. (2001) Oxygen isotopes and the Moon-forming giant impact. *Science, 294*, 345–348.

Woolum D. S. and Cassen P. (1999) Astronomical constraints on nebular temperatures: Implications for planetesimal formation. *Meteoritics & Planet. Sci., 34*, 897–907.

Wurm G., Paraskov G., and Krauss O. (2005) Growth of planetesimals by impacts at ~25 m/s. *Icarus, 178*, 253–263.

Yin Q., Jacobsen S. B., Yamashita K., Blichert-Toft J., Télouk P., and Albaréde F. (2002) A short timescale for terrestrial planet formation from Hf-W chronometry of meteorites. *Nature, 418*, 949–952.

Giant Planet Formation

Gennaro D'Angelo
NASA Ames Research Center and University of California, Santa Cruz

Richard H. Durisen
Indiana University

Jack J. Lissauer
NASA Ames Research Center

Gas giant planets play a fundamental role in shaping the orbital architecture of planetary systems and in affecting the delivery of volatile materials to terrestrial planets in the habitable zones. Current theories of gas giant planet formation rely on either of two mechanisms: the core accretion model and the disk instability model. In this chapter, we describe the essential principles upon which these models are built and discuss the successes and limitations of each model in explaining observational data of giant planets orbiting the Sun and other stars.

1. INTRODUCTION

Jupiter and Saturn are composed predominantly of hydrogen and helium and are therefore referred to as gas giants, although most of these elements are not in gaseous form at the high pressures of these planets' interiors. The majority of exoplanets discovered so far have masses in excess of about one-fourth of Jupiter's mass (M_{Jup}) and are known to be or suspected of being gas giants. Since helium and molecular hydrogen do not condense under conditions typically found in star-forming regions and in protoplanetary disks, giant planets must have accumulated them as gases. Therefore, giant planets must form prior to the dissipation of protoplanetary disks. Optically thick dust disks typically survive for only a few million years and protoplanetary disks lose all of their gaseous contents by an age of $\leq 10^7$ yr (see the chapter by Roberge and Kamp), hence giant planets must form on this timescale or less.

Giant planets contain most of the mass and angular momentum of our planetary system, and thus they must have played a dominant role in influencing the orbital properties of smaller planets. Gas giants may also affect the timing and efficiency of the delivery of volatile materials (*Chambers and Wetherill*, 2001), such as water and carbon compounds, to the habitable zones of planetary systems, where liquid water can exist on a (rocky) planet's surface. In the solar system, Jupiter is also believed to have reduced the impact rate of minor bodies (such as comets) on Earth (*Horner et al.*, 2010). Therefore, understanding giant planet formation is essential for the formulation of theories describing the origins and evolution of terrestrial planets capable of sustaining life in the form of complex organisms.

Observations of exoplanets have vastly expanded our database by increasing the number of known planets by over 1.5 orders of magnitude. The distribution of observed exoplanets is highly biased toward those objects that are most easily detectable using the Doppler radial velocity and transit photometry techniques (see chapters by Lovis and Fischer and by Winn), which have been by far the most effective methods of discovering planetary-mass objects orbiting other stars. Although these exoplanetary systems are generally different from the solar system, it is not yet known whether our planetary system is the norm, quite atypical, or somewhere in between. Nonetheless, some unbiased statistical information can be distilled from available exoplanet data (see the chapter by Cumming).

The mass distribution function of young compact objects in star-forming regions (e.g., *Zapatero Osorio et al.*, 2000) extends down through the brown dwarf mass range to below the deuterium-burning limit (12–14 M_{Jup}). This observed continuity implies that most isolated brown dwarfs and isolated high planetary-mass objects form via the same collapse process as do stars.

However, star-like direct quasispherical collapse is not considered a viable mechanism for the formation of Jupiter-mass planets, because of both theoretical arguments and observational evidence. The brown dwarf desert, a profound dip over the range from ~15 M_{Jup} to ~60 M_{Jup} in the mass distribution function of companions orbiting within several AU of Sun-like stars, suggests that the vast majority of gas giants form via a mechanism different from that of stars.

A theory based on a unified formation scenario for rocky planets, ice giants much as Uranus and Neptune, and gas giants is the core accretion model, in which the initial phases of a giant planet's growth resemble those of a terrestrial planet's. The only alternative formation scenario receiving significant attention is the disk instability model, in which a giant planet forms directly from the contraction of a clump of gas produced via a gravitational instability in the protoplanetary disk.

In this chapter, we introduce the basic physical concepts of gas giant planet formation according to core accretion models (section 2) and disk instability models (section 3). In section 4, we present theoretical arguments suggesting that most of the giant planets known to date formed via core accretion and pose some of the outstanding questions that still need an answer. Future prospects that may settle some of the lingering issues are discussed in section 5.

2. CORE ACCRETION MODELS

The initial phases of a giant planet's growth by core nucleated accretion proceed through an accumulation process of solid material, in the same fashion as terrestrial planets form. Dust and small solid grains, which are entrained in the predominant gas component of a protoplanetary disk, coagulate into larger particles. Centimeter-sized particles tend to settle toward the disk midplane, aggregating and eventually forming kilometer-sized agglomerates, referred to as planetesimals.

Planetesimals grow larger via pairwise collisions, leading to the formation of a planetary embryo. An embryo may have the ability to grow at a rate that increases as its mass increases, eventually consuming nearly all planetesimals in the neighborhood, while rapidly gaining mass and becoming a planetary core or protoplanet. Once a planetary core grows large enough and the escape velocity from its surface exceeds the thermal speed of the surrounding gas, a tenuous envelope of gas begins to accumulate around the solid core.

For most of the following planet's growth history, thermal pressure effects within the envelope regulate the accretion of gas. The ability of the envelope to radiate away the gravitational energy released by incoming planetesimals and by contraction limits the amount of gas that can be accreted by the planet. A slow contraction phase ensues in which the accretion rates of both solid material and gas are small. However, as the protoplanet grows and its total mass exceeds the value beyond which the pressure gradient (in the envelope) can no longer balance the gravitational force, the envelope undergoes a phase of rapid contraction, which allows more gas to be accreted. The augmented mass of the envelope triggers further contraction and gas can be thereby accreted at an ever-increasing rate.

During this epoch, known as runaway gas accretion phase, the gas accretion rate is regulated by the ability of the surrounding disk to supply gas to the planet's vicinity. In this stage, once the planet's mass exceeds a few tenths of Jupiter's mass (for typical temperature and viscosity conditions in a protoplanetary disk around a solar-mass star), the gas accretion rate decreases as the planet mass increases due to tidal interactions between the planet and the disk. The giant planet continues to grow at an ever-decreasing rate until there is no gas available within the planet's gravitational reach.

2.1. From Dust to Planetesimals

The formation of a heavy element core is an essential part of any core accretion model. Hence, for completeness, we present a summary of the basic elements involved in this process. An in-depth discussion of these concepts can be found in the chapter by Chambers.

The core formation starts at some distance from the star, most likely beyond the "snow line," where disk temperatures allow for condensation of water ice and solid material. This process begins from (sub)micrometer-sized dust particles, which may have originated from the interstellar medium and/or condensed within the disk's gas. Such small solid particles are well-coupled to the gas, on account of their large surface area to mass ratio, and therefore move with it.

As particles collide and stick together, they can grow larger and start to decouple from the gas and interact with it because of differential rotation between solids and gas. The relative velocity stems from the fact that the gas rotates about the star slower than do solids due to the radial pressure gradient that partially counteracts the gravitational attraction of the gas toward the star. The resulting interaction can be described in terms of friction via gas drag on the particle, producing an acceleration proportional to and in the opposite direction of the relative velocity between the particle and the gas. This friction generally removes orbital angular momentum from solid particles, causing them to drift toward the star.

Along with the radial drift, small solids also experience a friction in the direction perpendicular to the disk midplane. Assuming that the vertical motion of the gas is negligible (compared to that of solids), the vertical motion of a solid particle is obtained by applying the second law of dynamics

$$\frac{d^2 z}{dt^2} + \frac{1}{\tau_f}\frac{dz}{dt} + z\Omega^2 = 0 \quad (1)$$

in which z represents the distance over the disk midplane, τ_f is the friction timescale (also referred to as "stopping time"), and Ω is the particle's angular velocity about the star. The second term on the lefthand side of equation (1) is the opposite of the frictional force per unit mass exerted by the gas and the third term is the opposite of the vertical gravitational acceleration imposed by the star. In the limit $\tau_f \to 0$, dust is perfectly coupled to gas and dz/dt is equal to the vertical gas velocity, which is zero by assumption. Hence z is a constant and no sedimentation would occur. Notice that for the gaseous part of the disk, equation (1) is replaced by an equation for hydrostatic equilibrium in the vertical direction, with the vertical component of the gas pressure gradient balancing the vertical gravitational force (see equation (3)). In the limit $\tau_f \to \infty$, equation (1) assumes the form of a harmonic oscillator, hence particles oscillate about the disk midplane with a period equal to $2\pi/\Omega$.

For solid particles whose size is shorter than the mean free path of gas molecules and whose velocity relative to the gas is slower than the gas sound speed, c, the friction time is (*Epstein*, 1924)

$$\tau_f = \left(\frac{\rho_{sp}}{\rho}\right)\left(\frac{R_{sp}}{c}\right) \quad (2)$$

where ρ_{sp} and R_{sp} are the solid particle's volume density and radius, respectively, while ρ is the gas density. Indicating with H the vertical scale-height of the gaseous component of the disk and with a the orbital distance to the star, hydrostatic equilibrium in the vertical direction requires that

$$\frac{1}{\rho}\frac{\partial p}{\partial z} + \left(\frac{GM_\star}{a^2}\right)\left(\frac{z}{a}\right) = 0 \quad (3)$$

in which $p = p(a, z)$ is the gas pressure and $M_\star$ is the mass of the star. Writing the pressure as $p = c^2\rho$, approximating c as being independent of height, and integrating over the disk thickness, one finds that $c^2 \approx H^2\Omega^2$, where $\Omega = \sqrt{GM_\star/a^3}$ is the Keplerian angular velocity of the gas. (A more accurate determination of the disk's rotation rate can be derived from imposing hydrostatic equilibrium in the radial direction, which results in $\Omega^2 \simeq (GM_\star/a^3)[1-(H/a)^2]$). Thus, equation (2) gives

$$\Omega\tau_f = \left(\frac{a}{H}\right)\left(\frac{\rho_{sp}}{\rho}\right)\left(\frac{R_{sp}}{a}\right) \quad (4)$$

Assuming typical values at $a \approx 5$ AU for ρ and H/a around a solar-mass star of 10^{-10} g cm^{-3} and 0.05, respectively, and expressing both the density and radius of the particle in cgs units, $\Omega\tau_f \sim 10^{-3}\rho_{sp} R_{sp}$. This estimate is applicable up to values of R_{sp} less than tens of centimeters, i.e., on the order of the mean free path of gas molecules ($\propto 1/\rho$) under the adopted gas conditions.

Equation (1) has the form of a damped harmonic oscillator and can be integrated once initial conditions are provided for position and velocity. Setting $z = z_0$ and $\dot{z} = dz/dt = 0$ at time zero and taking into account the inequality $\Omega\tau_f \ll 1$ found above, the solution can be approximated as

$$z \simeq z_0\left(e^{-t\Omega^2\tau_f} - \Omega^2\tau_f^2 e^{-t/\tau_f}\right) \quad (5)$$

The second term in the above solution is a fast transient that rapidly decays to zero, hence we can further approximate the solution as $z \approx z_0 e^{-t\Omega^2\tau_f}$. Therefore, in order for the altitude above the midplane to reduce by more than 99% of its initial value, $\Omega t > 5/(\Omega\tau_f) \sim 5 \times 10^3/(\rho_{sp} R_{sp})$, in which again ρ_{sp} and R_{sp} are expressed in cgs units. Micrometer-sized particles would take millions of orbital periods ($\tau_{rot} = 2\pi/\Omega$)

to settle, which suggests that they first need to aggregate and grow into larger particles. Centimeter-sized icy/rocky aggregates would require only on the order of thousands of orbits to approach the disk's midplane.

A thin layer of solid material may thus accumulate at the disk's midplane in a relatively short amount of time compared to disk lifetimes. If this layer is sufficiently dense, clumps may form through gravitational instabilities within the layer (e.g., *Goldreich and Ward*, 1973). Such a process may produce kilometer-sized bodies, known as planetesimals.

The assumption that the gas vertical velocity is small compared to the particle vertical velocity is, however, only valid in the absence of sustained turbulent motions. In general, protoplanetary disks are expected to be somewhat turbulent and moderate amounts of turbulence could affect the settling timescales of small grains ($R_{sp} \lesssim 1$ cm). Ignoring the second term in equation (5), particle speeds are $\dot{z} \approx -z\Omega^2\tau_f$. Assuming a turbulent kinematic viscosity of the form $\nu = \alpha Hc$ (*Shakura and Syunyaev*, 1973) and indicating with λ_{turb} the vertical mixing length (the typical size of eddies), vertical gas speeds due to turbulence can be estimated as $|\dot{z}_{turb}| \sim \nu/\lambda_{turb}$ or $|\dot{z}_{turb}| \sim \alpha Hc/\lambda_{turb} = \alpha H^2\Omega/\lambda_{turb}$, and thus

$$\left|\frac{\dot{z}}{\dot{z}_{turb}}\right| \sim \left(\frac{z\lambda_{turb}}{H^2}\right)\left(\frac{\Omega\tau_f}{\alpha}\right) \quad (6)$$

Since $z\lambda_{turb}/H^2 \leq 1$ (eddies cannot be larger than the disk's thickness), we have that $|\dot{z}/\dot{z}_{turb}| < \Omega\tau_f/\alpha$. Notice that in order for turbulent motions to be subsonic (i.e., $|\dot{z}_{turb}| \leq c$), the turbulent kinematic viscosity $\nu \leq c\lambda_{turb} \leq cH$ and thus $\alpha \leq 1$. Under the disk's conditions adopted above, the ratio of the particle's settling velocity to the gas vertical turbulent speed is then $|\dot{z}/\dot{z}_{turb}| < 10^{-3} \rho_{sp} R_{sp}/\alpha$ (ρ_{sp} and R_{sp} are in cgs units).

Therefore, if the turbulence parameter $\alpha \gtrsim 10^{-3}$, the settling time of centimeter-sized particles may be affected since vertical turbulent mixing could influence their vertical transport. [Based on observed gas accretion rates and other properties of young stellar objects with disks, typical values of α are in the range from $\sim 10^{-4}$ to ~ 0.1 (e.g., *Hueso and Guillot*, 2005; *Isella et al.*, 2009)]. Studies of grain settling in the presence of turbulent motions indicate that particles tend to concentrate in stagnant regions of the flow and that concentrations are size-dependent, which may lead to the accumulation of subcentimeter-sized particles (*Cuzzi et al.*, 2001). Previous growth of these agglomerates would still rely on sticking collisions of smaller particles.

The growth from centimeter- to kilometer-sized bodies is still a poorly understood process and an active field of both theoretical and experimental research. Nonetheless, there is observational evidence that it does occur in nature: Dust particles are observed in debris disks around other stars and small bodies of tens to hundreds of kilometers in size are observed in the Kuiper belt around the Sun.

2.2. From Planetesimals to Planetary Cores

We shall now assume that the solid component of the protoplanetary disk has had time to agglomerate into planetesimals, rocky/icy bodies of a kilometer (or larger) in size. These objects are massive enough ($\geq 10^{15}$–10^{16} g ~ 10^{-12} $M_\oplus$) that they may gravitationally interact with their neighbors and perturb their velocities. As a result of these interactions, planetesimals become prone to collisions.

Although the outcome of a collision between two planetesimals depends upon their relative velocity, we assume that collisions among such objects result in mergers rather than fragmentation. Under this assumption, the growth rate of a planetesimal of mass M_s can be written as (*Safronov*, 1969)

$$\frac{dM_s}{dt} = \pi R^2 v_{rel} \rho_s F_g \qquad (7)$$

where R is the planetesimal radius, v_{rel} is the relative velocity between the two impacting bodies, and ρ_s is the volume density of the solid component of the disk. The product $v_{rel} \rho_s$ represents a mass flux, i.e., the amount of solid material sweeping across the target planetesimal per unit time and unit area. The quantity $F_g = (R_{eff}/R)^2$ is a gravitational enhancement factor, which is the ratio of the effective cross-section (πR^2_{eff}) of the accreting planetesimal to its geometrical cross-section (πR^2). This factor accounts for the ability of the growing body to bend toward itself the trajectories of other, sufficiently close, planetesimals.

It is often useful to express the accretion rate dM_s/dt in terms of the surface density of the solid material $\Sigma_s = \int \rho_s dz \sim H_s \rho_s$, with H_s being the vertical thickness of the planetesimal disk. In order to do so, one can assume that the gravitational force exerted by the star is the major component of the force acting on planetesimals in the vertical direction of the disk (as in equations (1) and (3)) and that relative velocities between planetesimals are isotropic. Thus, H_s is on the order of v_{rel}/Ω, where Ω is the Keplerian angular velocity of the growing planetesimal. Equation (7) can then be cast in the form $dM_s/dt = \pi R^2 \Omega \Sigma_s F_g$. Since the angular velocity along a Keplerian orbit is $\Omega \propto a^{-3/2}$, the accretion rate of planetesimals is slower farther from the star (neglecting variations of $\Sigma_s F_g$ with distance). This would imply that dM_s/dt at the current location of Uranus (19.2 AU) was about 7 times as small as it was at the current location of Jupiter (5.2 AU).

If we neglect the collective gravitational action of the other planetesimals and that of the star during an encounter, two interacting bodies can be described in the framework of a two-body problem. Assuming that the target planetesimal has already grown somewhat larger than the neighboring bodies, hence becoming a planetary embryo, the impacting body can be thought of as a projectile. We can therefore use the approximation that the embryo sits on the center of mass of the two-body system. In the rest frame of the embryo, conservation of the specific angular momentum (i.e., angular momentum per unit mass) reads $R_{eff} v_{rel} \simeq R v_{ta}$, where v_{ta} is the relative velocity for a tangential approach (when the projectile grazes the embryo). Conservation of specific energy during the collision requires that $v^2_{rel} \simeq v^2_{ta} - 2GM_s/R$ and therefore the gravitational enhancement factor, i.e., the ratio $(R_{eff}/R)^2$, can be cast in the form

$$F_g = 1 + \frac{v^2_{esc}}{v^2_{rel}} \qquad (8)$$

in which $v_{esc} = \sqrt{2GM_s/R}$ is the escape velocity from the surface of the target planetesimal. Notice that, if the planetesimal radius is not negligible compared to the embryo radius, the radius R in equation (7) and in v_{esc} should be replaced by the sum of the two radii.

If relative velocities are high and $v_{rel} \gg v_{esc}$, then $F_g \approx 1$ (i.e., $R_{eff} \approx R$) and equation (7) yields $dM_s/dt \propto R^2 \propto M_s^{2/3}$. If relative velocities are low and $v_{rel} \ll v_{esc}$, then $F_g \approx v^2_{esc}/v^2_{rel} \propto R^2$ and $dM_s/dt \propto R^4 \propto M_s^{4/3}$. If the escape velocity is very much larger than the relative velocity, then three-body effects (star, planetary embryo, and planetesimal), neglected in deriving equation (8), must be taken into account to compute F_g (*Greenzweig and Lissauer*, 1992).

In the high-relative-velocity regime, the growth timescale $\tau_s = M_s(dM_s/dt)^{-1}$ (i.e., the time it takes for the mass of the embryo to increase by a factor e ≈ 2.7), is proportional to $M_s^{1/3}$ and therefore the growth rate, $1/\tau_s$, of an embryo reduces as it grows larger. This implies an orderly growth of large planetesimals, which tend to attain similar masses. From equation (7) cast in terms of solids' surface density, one obtains that the timescale τ_{og} necessary to build a large embryo, or planetary core, of mass M can be estimated as

$$\Omega \tau_{og} \sim \frac{1}{\pi} \left(\frac{4\pi}{3} \right)^{2/3} \frac{\left(\rho^2_{sp} M \right)^{1/3}}{\Sigma_s} \qquad (9)$$

To assemble a body of mass M ~ 1.6×10^{-4} $M_\oplus$ (about as massive as Ceres, the largest object in the asteroid belt) within a solids' disk of density 10 g cm^{-2}, orderly growth would require a few 10^6 orbital periods, which already represents a fairly long timescale compared to lifetime of protoplanetary disks around solar-mass stars ($\leq 10^7$ yr).

In the low-relative-velocity regime, the growth rate of an embryo is $1/\tau_s \propto M_s^{1/3}$, and thus the larger the planetary embryo the faster it grows, a process known as runaway growth. During runaway growth, the largest embryo grows faster than any other embryo within its accretion region. Although a two-body approximation (equation (8)) yields an unlimited gravitational enhancement factor as the ratio $v_{rel}/v_{esc} \to 0$, gravitational scattering due to three-body effects set a limit to F_g, which cannot exceed values much beyond several thousands (*Lissauer*, 1993).

During the assembly of a planetary core through the growth of an embryo, relative velocities among planetesimals play a crucial role in determining the accretion rates, as indicated by equations (7) and (8). The velocity distribution of a swarm of planetesimals is affected by a number of physical processes, such as elastic and inelastic collisions, gravitational scattering, and frictional drag by the gas, the results of which can be highly stochastic.

If the orbit radial excursion of planetesimals, relative to the orbit of an embryo, is on the order of the embryo's Hill radius

$$R_H = a \left(\frac{M_s}{3 M_\star} \right)^{1/3} \quad (10)$$

or smaller, then the accretion is said to be shear-dominated because growth is dictated by Keplerian shear in the disk, rather than by planetesimals' random velocities. [The Hill radius, R_H, represents the distance of the Lagrange point L_1 from the secondary in the circular restricted three-body problem (e.g., *Murray and Dermott*, 2000). It provides a rough measure of the distance from the secondary beyond which the gravitational attraction of the primary and centrifugal effects prevail over the gravity of the secondary. Note, however, that this region, which identifies the Roche lobe, is not a sphere and its volume is about one-third that of a sphere with a radius R_H (e.g., *Eggleton*, 1983).] If the embryo's orbit is nearly circular, this situation requires that orbital eccentricities and inclinations of planetesimals should be $\sim R_H/a$, or smaller. The relative velocity between an embryo and a planetesimal traveling on a circular Keplerian orbit with radii a and a + Δa ($\Delta a/a \ll 1$), respectively, is $v_{rel} \simeq a|\Omega(a + \Delta a) - \Omega(a)| \approx a\Omega(a)|[1 - 3\Delta a/(2a)] - 1|$, hence $|v_{rel}| \approx 3\Omega\Delta a/2$. If we approximate Δa as the halfwidth of the region within which the gravity field of the embryo dominates over that of nearby embryos, the accretion rate is $dM_s/dt \propto M_s^{2/3} \Omega \Delta a \rho_s F_g$. We will see in section 2.3 that Δa is generally proportional to the embryo's Hill radius, R_H. Since $|v_{rel}| \sim \Omega R_H$ and $\rho_s \sim \Sigma_s/H_s \sim \Sigma_s/R_H$, the planetesimal accretion rate becomes $dM_s/dt \propto M_s^{2/3} \Omega \Sigma_s F_g$.

This result is formally the same as the accretion rate in the orderly regime (in which $v_{esc}^2/v_{rel}^2 \ll 1$ and $F_g \approx 1$), except that now $v_{esc}^2/v_{rel}^2 \sim 6R_H/R \gg 1$, and therefore $F_g \gg 1$. Accordingly, the growth timescale (in units of $1/\Omega$) is given by the righthand side of equation (9) divided by the gravitational enhancement factor, F_g. The growth rate of an embryo is much larger than it is during the orderly growth phase (since $F_g \gg 1$), but it reduces as the embryo mass grows larger (see equation (9)). This phase of growth, often referred to as oligarchic, may lead to the formation of massive embryos at regular intervals in semimajor axis. Notice that, neglecting variations of $\Sigma_s F_g$ with orbital distance, as in the orderly growth regime $dM_s/dt \propto M_\star^{1/2} a^{-3/2}$. This implies that the accretion rate of planetesimals reduces as the stellar mass decreases or as the distance from the star increases.

2.3. Isolation Mass of Planetary Cores

The rapid (runaway/oligarchic) growth of a planetary embryo continues until its neighborhood, or feeding zone, has been substantially cleared of planetesimals (*Lissauer*, 1987). The feeding zone represents the domain within which the dominant embryo is able to significantly deflect the paths of other planetesimals toward itself. The radial extent of this region is several Hill radii, R_H, which can be understood by recalling the definition of R_H. This is the distance of the equilibrium point from mass M_s (the embryo), on the line connecting (and in between) masses $M_\star$ and M_s, in a reference frame rotating at angular velocity Ω. The force balance for a planetesimal of mass much smaller than both $M_\star$ and M_s requires that $M_s R_H^{-2} - M_\star (a - R_H)^{-2} + \Omega^2 (a - R_H) = 0$. Hence, to leading order in R_H/a, $M_s R_H^{-2} - M_\star a^{-2}(1 + 2R_H/a) + M_\star a^{-2}(1 - R_H/a) = 0$, whose solution is the righthand side of equation (10).

Once the feeding zone has been severely depleted, the planetary embryo becomes nearly isolated. Numerical N-body simulations suggest that isolation of an embryo occurs once a region of width $\Delta a \sim b R_H$, where $b \approx 4$, on each side of the embryo's orbit becomes nearly emptied of planetesimals (*Kokubo and Ida*, 2000). The isolation mass, M_{iso}, of a planetary embryo can therefore be calculated as $M_{iso} \sim 4\pi a \Delta a \Sigma_s \sim 4\pi a b R_H \Sigma_s$. Notice that here Σ_s refers to the *initial* value of the solids' surface density. By using equation (10), the isolation mass can be written as

$$M_{iso} \sim \sqrt{\frac{\left(4\pi a^2 b \Sigma_s\right)^3}{3 M_\star}} \quad (11)$$

According to equation (11), a planetary core of mass $\sim 11 M_\oplus$ (if $b \approx 4$) would become nearly isolated at 5.2 AU from a solar-mass star if the local surface density of solids were equal to 10 g cm^{-2}. This value for Σ_s is a couple of times as large as that predicted for a minimum mass (proto) solar nebula (MMSN). [This is defined as the amount of heavy elements (heavier than helium) observed in the planets and minor bodies of the solar system (mostly contained in giant planets) augmented by an amount of gas such to render the protosolar nebula composition equal to that of the young Sun. Such definition constrains the total mass, $\sim 0.02 M_\odot$, and, to a lesser extent, the surface density distribution of the MMSN (see, e.g., *Davis*, 2005).] Additionally, the isolation mass increases with distance from the star, although this increase can be somewhat compensated for by a reduced surface density of solids, which is expected to decrease as a increases.

An order-of-magnitude estimate of the timescale, τ_{iso}, required to reach isolation can be obtained by taking the ratio of equation (11) to the planetesimal accretion rate (expressed in terms of Σ_s, see section 2.2), which yields

$$\Omega \tau_{iso} \sim \frac{C_{iso}}{F_g} \sqrt{\frac{M_\star}{a^2 \Sigma_s}} \left(\frac{a^3 \rho_{sp}}{M_\star} \right)^{2/3} \quad (12)$$

where $C_{iso} = (2\pi)^{2/3} \left(2/\sqrt{3}\right)^{5/3} \sqrt{b/\pi} \simeq 2.44 \sqrt{b}$. Notice that for given values of F_g and Σ_s, this timescale increases with increasing distance from the star and decreases with increasing stellar mass: $\tau_{iso} \propto a^{5/2} M_\star^{-2/3}$. At the current location of Jupiter around a solar-mass star, $\Omega \tau_{iso} \sim 3 \times 10^8/F_g$, if Σ_s is 10 g cm^{-2}. Assuming a situation in which the gravitational enhancement factor is on the order of 1000 over the entire accretion epoch, τ_{iso} is several tens of thousands of orbital periods or several 10^5 yr.

In deriving equation (12) we assumed that the surface density of solids, Σ_s, is comparable to its initial value and the geometrical cross-section of the planet only depends on the core radius, R. Yet, the surface density of solids drops if accreted planetesimals are not replaced by others arriving from outside the core's feeding zone, which would operate toward increasing τ_{iso}. Moreover, during its growth toward isolation, a planetary core also accretes gas from the disk, although at a much smaller rate than it accretes solids. Once the atmospheric envelope becomes massive enough to dissipate the kinetic energy of incoming planetesimals via drag friction, the geometrical cross-section of the protoplanet in equation (7) becomes substantially larger than πR^2. This effect would operate toward reducing τ_{iso}.

A planetary core does not necessarily stop growing once attaining the mass given by equation (11). Perturbations among planetesimals and other embryos can supply additional solid material to the core's feeding zone via scattering. Furthermore, a planet in excess of a Mars mass can exert gravitational torques on the surrounding gas. The disk responds by exerting the same amount of torques onto the planet, which modify the planet's orbital angular momentum, forcing it to radially migrate within the disk. As a result of this radial displacement, a planet may reach disk regions that still contain planetesimals. However, as the planet's mass grows, so does its ability to scatter planetesimals away from the orbital path or to trap them into mean-motion resonances (so that the ratio of a planetesimal's orbital period to the planet's orbital period is a rational number).

The order-of-magnitude estimates given above for the isolation mass and the timescale to reach it neglect many aspects of the physical processes involved in the growth of a planetary core. Nonetheless, detailed calculations of giant planet formation by core accretion and gas capture indicate that those estimates are valid under appropriate conditions. Calculations that start from a planetary embryo of $M_s \sim 0.1\ M_\oplus$ (about equal to the mass of Mars) orbiting a solar-mass star at 5.2 AU and undergoing rapid growth within a planetesimal disk of initial surface density $\Sigma_s \approx 10$ g cm^{-2} show that a planetary core becomes nearly isolated within less than half a million years when $M_{iso} \approx 11\ M_\oplus$ (*Pollack et al.*, 1996), as also indicated by the solid line in Fig. 1a (see caption for further details).

2.4. Growth of Thermally Regulated Envelopes

Gas can accrete onto a planetary embryo when the thermal energy is smaller than the gravitational energy binding the gas to the embryo. This condition is satisfied when the escape velocity from the surface of the embryo, v_{esc} (see section 2.2), exceeds the local thermal speed of the disk's gas $\left(\sqrt{8/\pi} c\right)$, which occurs when $M_s \geq 4(H/a)^2 M_\star R/(\pi a)$, or

$$M_s \geq \sqrt{\frac{M_\star^3}{a^3 \rho_{sp}}} \left(\frac{H}{a}\right)^3 \qquad (13)$$

At distances of several AU from a solar-mass star, relatively small bodies ($M_s \sim 0.01\ M_\oplus$) can retain an atmosphere. In these early phases, the atmospheric gas is optically thin and thermal energy released by impacting planetesimals can be readily radiated away, allowing for contraction of the envelope and for additional gas to be accreted.

Prior to achieving isolation from the planetesimal disk, the accretion rate of solids is much larger than that of gas. The growing core mass forces the envelope to contract and more gas can be collected from the surrounding gaseous disk. However, as the envelope grows more massive, it becomes increasingly optically thick to its own radiation, which therefore cannot escape to outer space as easily as in earlier phases. As a consequence, the envelope's temperature and density exceed those of the local disk's gas.

The pressure gradient that builds up in the envelope opposes gravitational contraction, preventing accretion of large amounts of gas. On the one hand, ongoing accretion of solid material and growth of the core help envelope contraction; on the other hand, the gravitational energy released in the envelope by gas compression and that supplied by accreted planetesimals help maintain a relatively large pressure gradient. Therefore, contraction becomes self-regulated. Once the planet achieves (near) isolation from planetesimals, the accretion rate of solids becomes small and accretion of gas continues to the extent allowed by envelope compression.

The envelope enters a stage of quasistatic contraction that can be characterized by long evolution timescales. If the accretion of solids and gas was negligibly small, the timescale over which the envelope contracts would be related to the ratio of the gravitational energy released by contraction, $|E_{grav}|$, to the envelope's luminosity, L. This ratio defines the Kelvin-Helmholtz timescale $\tau_{KH} = |E_{grav}|/L$. Indicating the protoplanet's core and envelope masses with M_c and M_e, respectively, $|E_{grav}| \sim GM_c M_e/\bar{R}$, $\bar{R}$ being the average radius comprising most of the protoplanet's mass, and thus the contraction timescale is

$$\tau_{KH} \sim \frac{GM_c M_e}{\bar{R} L} \qquad (14)$$

Using values from Figs. 1 and 2 around the middle of the planetesimal isolation phase, τ_{KH} would be on the order of 10^5 yr. However, for the case illustrated in the Figs. 1 and 2, most of the luminosity produced during this slow growth phase is due to gravitational energy generated by accretion rather than by contraction of the envelope.

The length of this epoch depends on several factors, but principally on the opacity of the envelope and on the

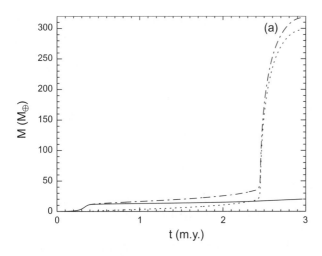
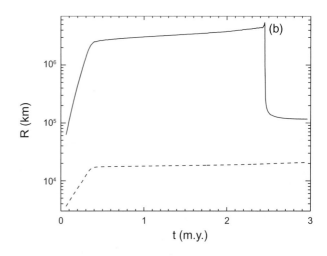

Fig. 1. Formation models of a giant planet by core nucleated accretion need to take into account many physical effects, including (1) calculation of the planetesimal accretion rate onto the protoplanet; (2) calculation of the interaction via gas drag of impacting planetesimals with the protoplanet's gaseous envelope; (3) thermodynamics calculation of the envelope structure; (4) calculation of the gas accretion rate from the protoplanetary disk onto the protoplanet during the phase of slow envelope growth; and (5) hydrodynamics calculation of constraints on envelope size and disk-limited gas accretion rates during the phase of runaway gas accretion. Some results from one such calculation are reported here for a model of Jupiter's formation (*Lissauer et al.*, 2009). This model assumes that the planetesimal disk has $\Sigma_s \approx 10$ g cm^{-2} and the grain opacity in the protoplanet's envelope is 2% the value of the interstellar medium. The gaseous disk is assumed to dissipate within 3 m.y. **(a)** Mass of solids (solid line), gas (dotted line), and total mass of the planet (dot-dashed line) as functions of time. **(b)** Radius of the planet (solid line) and the planet's solid core (dashed line) from the same model.

solids' accretion rate. The more opaque the gas to outgoing radiation, the less able the envelope to dissipate the energy provided by gas compression and/or by continued planetesimal accretion. Moreover, accreted solids can be consumed by ablation during their atmospheric entry, releasing dust and increasing the envelope's opacity. Therefore, a reduced accretion of solids shortens this epoch. As shown in Fig. 1a, the slow contraction phase, from the time the core mass reaches near-isolation from surrounding planetesimals to the time the rapid gas accretion phase begins, lasts about 2 m.y. The assumption made for this particular model is that the grain opacity in the protoplanet's envelope is 2% that of the interstellar medium and that planetesimals continue to accrete throughout the slow contraction phase. All else being equal, this time-span increases to 6 m.y. when the full interstellar opacity is adopted and decreases to ≈ 1 m.y. if planetesimal accretion ceases entirely once the isolation mass is achieved (*Hubickyj et al.*, 2005).

It thus appears that the duration of the slow contraction phase represents the main uncertainty in determining whether a giant planet can form before the gaseous component of the disk dissipates (i.e., within a few to several millions of years; see the chapter by Roberge and Kamp). However, it offers a natural explanation for the formation of planets such as Uranus and Neptune in our solar system, which do not possess very massive envelopes. These planets may have not been able, because of core and envelope conditions, or may have not had enough time, because of a dissipating disk, to acquire a large envelope. In addition, static atmospheric models indicate that the removal of grains by growth and settling from the radiative zone of a planet's envelope may significantly reduce their contribution to the opacity (*Movshovitz and Podolak*, 2008), hence shortening the contraction timescale of the envelope.

Quasistatic calculations of the envelope thermal structure suggest that the growth timescale of a protoplanet's envelope $\tau_e = M(dM_e/dt)^{-1}$ can be cast in the following parametric form (e.g., *Ida and Lin*, 2008)

$$\tau_e = \bar{\tau}_e \left(\frac{M_\oplus}{M}\right)^\xi \quad (15)$$

in which $M = M_c + M_e$ is the total planet's mass and $M_c \geq M_{iso}$. The timescale $\bar{\tau}_e$ depends on several factors, but mostly on the opacity of the envelope. For the model reported in Fig. 1, $\bar{\tau}_e \sim 10^{10}$ yr and $\xi \sim 3$. Notice, though, that many factors affect the values of parameters ξ, especially $\bar{\tau}_e$. In general, applicability of equation (15) requires some previous knowledge of the envelope thermal conditions, e.g., through theoretical arguments or numerical modeling. When $M_e \geq M_c$, the inner (densest) parts of the envelope, which contain most of the mass, effectively contribute to the compression of the outer parts. Yet, around this stage, the envelope may become gravitationally unstable and collapse (see Fig. 1).

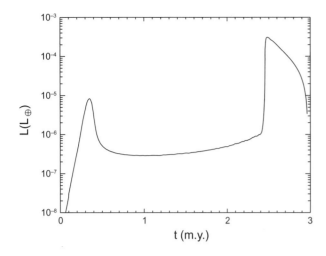

Fig. 2. Luminosity as a function of time from the model reported in Fig. 1. From the time of the first luminosity peak (t ≈ 0.35 m.y.) to the middle of the planetesimal isolation phase (t ~ 1.6 m.y.), the luminosity of the protoplanet decreases by more than a factor of 25. This is mainly due to the reduced accretion rate of solids.

2.5. Critical Core Mass and Envelope Collapse

In order to gain some insight into the conditions leading to envelope collapse, we can construct simplified analytical models that qualitatively describe the physical state of the envelope. The first assumption we shall make is that a slowly contracting envelope evolves through stages of quasi-hydrostatic equilibria. Early studies of a static envelope surrounding a planetary core (e.g., *Perri and Cameron*, 1974) suggested that there may not be any static structure beyond some value of the core mass. Past such a critical core mass, the extended outer portions of the gaseous envelope become hydrodynamically unstable and collapse onto the core. Simplified models of hydrostatic envelopes can shed some light over the concept of critical core mass. One such model was used by *Stevenson* (1982) to interpret results of static envelope calculations performed by *Mizuno* (1980).

Let us assume that a gaseous envelope of mass M_e surrounds a core of mass M_c and that the envelope is spherically symmetric around the core. Hydrostatic equilibrium in an envelope shell of radius r and thickness dr implies that there is a balance between gravitational force and pressure acting at the shell surface boundaries. Therefore, the condition for hydrostatic equilibrium reads

$$\frac{dp}{dr} = -\frac{Gm\rho}{r^2} \qquad (16)$$

where p and ρ are, respectively, the pressure and density at radius r, and m is the mass within r.

An important assumption we shall make is that energy is transported through the envelope via radiative diffusion (i.e., this model applies to a radiative envelope) and that the envelope is stable against convection. In general, this may be true only in the outer layers of the envelope (*Bodenheimer and Pollack*, 1986).

The amount of radiation energy transported through an envelope shell per unit surface area and unit time, i.e., the radiation flux, is $F_{rad} = -\mathcal{D} dE_{rad}/dr$, where the radiation energy density $E_{rad} \propto T^4$ and T is the temperature. The radiation diffusion coefficient, $\mathcal{D}$, is proportional to the speed of light times the mean free path of photons, $1/(\bar{\kappa}\rho)$, where $\bar{\kappa}$ is a frequency-integrated opacity (or mass absorption coefficient). Hence, for an envelope shell of mass dm, the radiation flux is also equal to (e.g., *Kippenhahn and Weigert*, 1990) $F_{rad} = -16\pi r^2 \sigma_{rad}/(3\bar{\kappa})(dT^4/dm)$, where σ_{rad} is the Stefan-Boltzmann constant. If the luminosity through the envelope, L, is nearly constant, then one can write $F_{rad} = 4\pi r^2 L$, and thus

$$\frac{dT^4}{dm} = -\frac{3\bar{\kappa} L}{64\pi^2 \sigma_{rad} r^4} \qquad (17)$$

Another assumption we shall rely upon is that the gradient $\nabla = d \ln T/d \ln p$ is nearly constant in the envelope. The temperature gradient can thus be cast in the form

$$\frac{dT}{dr} = \left(\frac{T}{p}\right)\left(\frac{dp}{dr}\right)\nabla \qquad (18)$$

In order for energy to be transported via radiation alone, the gradient ∇ must be smaller than its adiabatic value, ∇_{ad}. For a perfect monoatomic gas and negligible radiation pressure, $\nabla_{ad} = 2/5$. If $\nabla \geq \nabla_{ad}$, the envelope becomes convectively unstable. By substituting equation (16) in place of the pressure gradient in equation (18) and using the equation of state for a perfect gas $p/\rho = \mathcal{R}T/\mu$ ($\mathcal{R}$ is the gas constant), we have that $dT/dr \sim -\mu GM\nabla/(\mathcal{R}r^2)$, where $M = M_c + M_e$, and hence $T \sim \mu GM\nabla/(\mathcal{R}r)$. In the above relations, μ indicates the mean-molecular weight, i.e., the average number of atomic mass units per gas particle. Notice that the approximation m ~ M made above determines how deep in the envelope this temperature stratification is applicable.

Equation (17) can be written as

$$\frac{dm}{dr} = -\frac{4^4 \pi^2 \sigma_{rad}}{3\bar{\kappa} L} r^4 T^3 \frac{dT}{dr} \qquad (19)$$

from which it follows that

$$\frac{dm}{dr} = \frac{4^4 \pi^2 \sigma_{rad}}{3\bar{\kappa} L}\left(\frac{\mu GM\nabla}{\mathcal{R}}\right)^4 \frac{1}{r} \qquad (20)$$

Integrating both sides of this equation along r, from the core radius, R_c, to the envelope radius, R, and neglecting opacity variations, one obtains

$$M - M_c = \mathcal{A}_0 \frac{(\mu\nabla)^4}{\bar{\kappa} L} M^4 \qquad (21)$$

in which $\mathcal{A}_0 = (\pi^2\sigma_{rad}/3)(4G/\mathcal{R})^4 \ln(R/R_c)$. [A more general opacity law, given by the product of some power of p and some power of T, would result in the same relation as equation (21), but with a modified form of coefficient $\mathcal{A}_0$ (*Wuchterl et al., 2000*).]

We seek for the largest value of the core mass that can have a fully radiative envelope in hydrostatic equilibrium. This can be obtained from equation (21), which gives an explicit function $M_c = M_c(M)$, if L is independent of both M and M_c. The value of M for which M_c is maximum is given by

$$M^{cr} = \left[\frac{1}{4\mathcal{A}_0}\frac{\overline{\kappa}L}{(\mu\nabla)^4}\right]^{1/3} \quad (22)$$

Physically significant solutions of equation (21) can only be obtained for $M/M^{cr} < 4^{1/3}$, and therefore the mass of *strictly* hydrostatic envelopes does not exceed $4^{1/3}$ M^{cr}. The critical core mass for radiative envelopes is found by substituting equation (22) in equation (21), which gives

$$M_c^{cr} = \frac{3}{4}M^{cr} \quad (23)$$

According to these simple analytical arguments, the critical core mass, M_c^{cr}, is independent of the local density and temperature of the protoplanetary disk, it is weakly dependent on envelope opacity and luminosity, but more strongly dependent on μ and ∇. Using radii and luminosity values before collapse from Figs. 1 and 2, $\mu = 2.25$, $\nabla = 0.235$ (*Kippenhahn and Weigert*, 1990), and $\overline{\kappa} = 0.02$ cm^2 g^{-1}, one obtains $M_c^{cr} \sim 16$ $M_\oplus$ and a total mass of ~ 21 $M_\oplus$ (recall that these values would apply to a strictly hydrostatic and fully radiative envelope).

As mentioned in section 2.4, most of the protoplanet's luminosity during the slow growth phase may be due to gravitational energy released by accreted planetesimals. In these circumstances the luminosity can be approximated as $L \sim (GM_c/R_c)(dM_c/dt)$, in which dM_c/dt is the planetesimal accretion rate, thus $L \propto M_c^{2/3}(dM_c/dt)$. Substituting in equation (21), one obtains M_c as an implicit function of M. By differentiation of this function, one can readily show that again $M_c^{cr} = (3/4)$ M^{cr}.

The deepest and densest layers of an envelope are typically convective (e.g., *Bodenheimer and Pollack*, 1986). Simple analytical models of hydrostatic envelopes can also be constructed under the hypothesis that energy is transported only via convection, as that proposed by *Wuchterl* (1993). By employing the adiabatic gradient ∇_{ad}, the temperature gradient becomes $dT/dr \sim -\mu GM\nabla_{ad}/(\mathcal{R}r^2)$. Instead of an energy equation, in a fully convective envelope one can use the polytropic law $T/T_D = (\rho/\rho_D)^\Gamma$, in which T_D and ρ_D indicate values where the envelope merges with the disk. In a perfect monoatomic gas with negligible radiation pressure, $\Gamma = (5/3)\nabla_{ad}$. By differentiation of the polytropic law, it follows that

$$\frac{dT}{d\rho} = \Gamma\left(\frac{T_D}{\rho_D}\right)\left(\frac{\rho}{\rho_D}\right)^{(\Gamma-1)} \quad (24)$$

which combined with the temperature gradient above yields

$$\Gamma\left(\frac{T_D}{\rho_D}\right)\left(\frac{\rho}{\rho_D}\right)^{(\Gamma-1)}\frac{d\rho}{dr} = -\left(\frac{\mu GM\nabla_{ad}}{\mathcal{R}r^2}\right) \quad (25)$$

Equation (25) can be integrated in radius to obtain the envelope's density stratification

$$\left(\frac{\rho}{\rho_D}\right)^\Gamma = -\left(\frac{\mu\nabla_{ad}}{\mathcal{R}T_D}\right)\left(\frac{GM}{r}\right) \quad (26)$$

The approximation m ~ M, introduced above to derive the temperature gradient, again determines the envelope depths to which this density stratification applies. In equation (26) the constant of integration is set to zero by choosing as outer envelope radius $R = \mu GM\nabla_{ad}/(\mathcal{R}T_D)$. By integrating the quantity $4\pi r^2\rho dr$ from the core radius, R_c, to R we get

$$M - M_c = \mathcal{B}_0\rho_D\left(\frac{\Gamma}{3\Gamma-1}\right)\left(\frac{\mu\nabla_{ad}}{T_D}\right)^3 M^3 \quad (27)$$

where $\mathcal{B}_0 = 4\pi(G/\mathcal{R})^3 [1-(R_c/R)^{(3-1/\Gamma)}]$, which can be approximated to $4\pi(G/\mathcal{R})^3$ since $R_c \ll R$ and $3 - 1/\Gamma > 1$ when radiation pressure can be ignored.

The largest core mass that bears a fully convective envelope in hydrostatic equilibrium corresponds to a total mass

$$M^{cr} = \left[\frac{1}{12\pi}\left(\frac{3\Gamma-1}{\Gamma}\right)\left(\frac{\mathcal{R}}{\mu G\nabla_{ad}}\right)^3 \frac{T_D^3}{\rho_D}\right]^{1/2} \quad (28)$$

In order for equation (27) to admit physically significant solutions, $M/M^{cr} < \sqrt{3}$. According to these simple arguments, strictly hydrostatic and convective envelopes are limited in mass to $\sqrt{3}M^{cr}$. From equations (27) and (28), one finds that the critical core mass of a convective envelope is

$$M_c^{cr} = \frac{2}{3}M^{cr} \quad (29)$$

Unlike the solution for radiative envelopes, the critical core mass for convective envelopes explicitly depends on the disk density and temperature. Using values $\rho_D \sim 10^{-10}$ g cm^{-3} and $T_D \sim 100$ K, which would apply to a protoplanetary disk at ~5 AU from a solar-mass star if there was no planet, equation (28) gives $M^{cr} \sim 170$ $M_\oplus$. However, once M achieves several tens of Earth masses, both ρ_D and T_D are affected by disk-planet interactions and hence depend on M. In these cases, because of the gravitational perturbation produced by the protoplanet, local disk densities as well as temperatures can be smaller than the corresponding values in absence of the protoplanet (*D'Angelo et al., 2003a*), if effects of stellar irradiation can be neglected.

The two analytic solutions for fully radiative and fully convective envelopes can be combined to obtain an estimate of the critical core mass for composite convective-radiative envelopes. Following the same line of argument, we consider a two-layer model in which the "convective" solution applies to the inner convective layer that extends from the core radius, R_c, to the radius R_{clb}, which plays the role of envelope radius (R) in the convective solution given above. Accordingly, temperature T_{clb} and density ρ_{clb} at the convective layer boundary, R_{clb}, replace the disk values T_D and ρ_D, respectively. Then, indicating with M_{clb} the mass within R_{clb} (core mass plus mass of the convective layer), equation (29) gives a critical value $M_{clb}^{cr} = (3/2)\,M_c^{cr}$.

The "radiative" solution applies to the outer radiative layer, whose boundaries are at the convective layer (outer) radius R_{clb}, which plays the role of core radius (R_c) in the radiative solution above, and at the envelope radius R. Thus, the mass within the convective layer boundary, M_{clb}, replaces the core mass (M_c) in the radiative solution above. Equation (23) gives the critical value of the total mass $M^{cr} = (4/3)\,M_{clb}^{cr}$. Therefore, the largest planet mass for which both the convective and radiative layers, and hence the entire envelope, can be in hydrostatic equilibrium is

$$M^{cr} = \frac{4}{3} M_{clb}^{cr} = 2 M_c^{cr} \qquad (30)$$

According to equation (30), a composite convective radiative envelope collapses once the envelope mass is equal to the core mass. This prediction of the simple analytic model is in agreement with the results from the detailed calculation shown in Fig. 1a. Equation (22) can still be used to estimate M^{cr}, where now $\mathcal{A}_0 \propto \ln(R/R_{clb})$. At the epoch of envelope collapse, the calculation in Fig. 1 gives a ratio $R/R_{clb} \approx 4.74$, which, together with the other values adopted above, yields $M^{cr} \sim 32\,M_\oplus$ and $M_c^{cr} \sim 16\,M_\oplus$, consistent with the results illustrated in Fig. 1a.

2.6. Disk-Limited Gas Accretion Rates

During the collapse phase, large amounts of gas can be accreted from the protoplanetary disk. The rate at which the protoplanet's envelope grows soon becomes very large. When the total mass is $M \sim 20\,M_\oplus$, equation (15) yields an envelope growth timescale on the order of 10^6 yr. But once the envelope mass, M_e, exceeds the core mass (equation (30)), this length of time can shorten by orders of magnitude. In the model illustrated in Fig. 1, when $M_e \sim 2\,M_c$ the growth timescale becomes on the order of 10^4 yr.

The protoplanetary disk feeds gas to the planet's vicinity. Yet, the rate at which such gas supply occurs is limited. In this stage, since a protoplanet can basically accrete gas at arbitrarily large rates, gas accretion is governed by hydrodynamical factors that involve tidal (i.e., gravitational) interactions between the growing planet and the disk. To derive simple estimates of disk-limited gas accretion rates, dM/dt, we can assume that a gas parcel orbits about the star, at a distance $a + \Delta a$, with an orbital velocity relative to the planet on the order of $\Omega\Delta a$ (see section 2.2), where Ω is the angular velocity of the planet and a its orbital radius. (In this phase, we designate the accretion rate of gas simply as dM/dt since total mass variations are overwhelmingly determined by accretion of gas, regardless of whether there is continued accretion of solids.) The protoplanet is able to capture gas within an effective radius R_{gc} (i.e., πR_{gc}^2 is its effective cross-section for gas capture) smaller than the disk's local thickness, H. If ρ is the gas volume density, then the mass flux through the planet's cross-section is $\rho\Omega R_{gc}$. The gas accretion rate is given by the mass flux times the planet's cross-section, that is

$$\frac{dM}{dt} \sim \frac{\Sigma}{H}\Omega R_{gc}^3 \qquad (31)$$

in which the volume density is expressed in terms of surface density ($\rho \sim \Sigma/H$).

There are two relevant length-scales over which a protoplanet can attract gas: the Hill and Bondi radii. The Hill radius, R_H, is defined in equation (10) (in which M_s is replaced with the total mass M) and represents a measure of the distance past which centrifugal forces and gravitational forces by the star dominate over the gravitational force exerted on the gas by the planet. The Bondi radius is

$$R_B = \frac{GM}{c^2} \qquad (32)$$

($c = H\Omega$ is the gas sound speed) and marks the distance beyond which the thermal energy of the gas is larger than the gravitational energy binding that gas to the planet. In the case that $R_B < R_H$, forces due to pressure gradients within the Hill sphere prevent gas from becoming bound to the protoplanet. Then, the Bondi radius is the relevant distance over which gas can be accreted. We will refer to this regime as Bondi-type gas accretion. The Hill radius becomes the relevant distance for gas accretion if $R_H < R_B$, which will be designated as Hill-type gas accretion regime. Therefore, in the general case, the effective radius for gas capture of a protoplanet embedded in a gaseous disk is

$$R_{gc} = \min(R_B, R_H) \qquad (33)$$

By substituting equation (32) in equation (31), one finds that the protoplanet's growth rate, (dM/dt)/M, in the Bondi-type gas accretion regime is

$$\frac{1}{\tau_B} C_B \Omega \left(\frac{a^2\Sigma}{M_\star}\right)\left(\frac{a}{H}\right)^7 \left(\frac{M}{M_\star}\right)^2 \qquad (34)$$

whereas, in the Hill-type gas accretion regime, the growth rate is found by using equation (10) (with M in place of M_s) and equation (31), resulting in

$$\frac{1}{\tau_H} = \frac{1}{3} C_H \Omega \left(\frac{a^2 \Sigma}{M_\star}\right)\left(\frac{a}{H}\right) \qquad (35)$$

The dimensionless constants C_B and C_H account for small corrections due to effects neglected in deriving this simple model and should be on the order of unity. They can be obtained from direct numerical simulations. Notice that $dM/dt \propto M^3$ for Bondi-type gas accretion (implying a faster than exponential growth), while $dM/dt \propto M$ for Hill-type gas accretion (implying an exponential growth).

Bondi-type gas accretion is a very steep function of the disk thickness, i.e., of the local gas sound speed. If the disk is locally very warm, high thermal energy prevents gas from becoming bound and hence accreted (the planet's cross-section for gas accretion decreases for increasing temperature). Hill-type gas accretion is less sensitive on disk temperature because in this regime gravitational factors dominate over thermal ones.

In general, the mass growth rate of the protoplanet, $1/\tau_D$, which is limited by disk hydrodynamics, is given by

$$\frac{1}{\tau_D} = \begin{cases} 1/\tau_B & \text{for } M < M_{tr} \\ 1/\tau_H & \text{for } M \leq M_{tr} \end{cases} \qquad (36)$$

where the transition between the two regimes occurs at the transition mass M_{tr}, for which $\tau_H = \tau_B$, and thus

$$M_{tr} = \frac{M_\star}{\sqrt{3}} \sqrt{\frac{C_H}{C_B}} \left(\frac{H}{a}\right)^3 \qquad (37)$$

This simple gas accretion model can be compared against results from three-dimensional hydrodynamical simulations of a planet gravitationally interacting with a disk and growing in mass at a disk-limited gas accretion rate. Figure 3a shows numerical results (solid line) and the growth rate given by equation (36) (dashed line). Agreement is found for values of the constants $C_B \approx 2.6$ and $C_H \approx 0.9$. Note that the mass scaling, i.e., the slopes of the two dashed line segments, is correctly predicted by the simple model and so is the transition mass between the two regimes ($M_{tr} \approx 4 \times 10^{-5} M_\star$, or $\approx 14 M_\oplus$ if $M_\star = 1 M_\odot$). The disk thickness in the example shown in Fig. 3 is H/a = 0.05, which corresponds, for typical disk properties, to a local temperature T ~ 100 K. Since the gas sound speed $c \propto \sqrt{T}$, a factor of 2 increase in the local temperature would reduce the planet's growth rate by an order of magnitude during the Bondi-type accretion regime.

As the planet's mass increases, the effective radius for gas capture may exceed the disk thickness. Moreover, density perturbations due to tidal interactions of the protoplanet with the disk are no longer negligible. The gas density along the planet's orbit starts to be affected (see Fig. 3b), and the simple accretion model becomes inapplicable beyond some value of the protoplanet mass. In the example shown in Fig. 3, this occurs for masses $M \gtrsim 10^{-4} M_\star$, when disk-planet interactions have changed the local surface density by more than 20% of its unperturbed value (i.e., when there is no planet in the disk), as illustrated in Fig. 3b.

2.7. Gap Formation

A protoplanet embedded in a disk exerts a gravitational torque on the gas interior of $a - |\Delta a|$ and exterior of $a + |\Delta a|$ whose magnitude is on the order of

$$\mathcal{T}_g \approx \frac{\Sigma}{f} a^4 \Omega^2 \left(\frac{a}{\Delta a}\right)^3 \left(\frac{M}{M_\star}\right)^2 \qquad (38)$$

which leads to exchange of orbital angular momentum between the planet and the disk (see the chapter by Lubow and Ida for a derivation of this torque). The factor f is typically on the order of unity. It can be shown that $|\Delta a|$ is the larger between the Hill radius, R_H, and the disk scale-height, H. The sign of $\mathcal{T}_g$ is positive for material lying outside the planet's orbit ($\Delta a > 0$) and negative for material inside it ($\Delta a < 0$). Hence, material orbiting outside of the planet's orbit gains angular momentum, moving toward larger radii, whereas material inside of the planet's orbit loses angular momentum, moving toward smaller radii. This process tends to deplete the disk of gas along the planet's orbit, thus forming an annular gap in the local density distribution (e.g., *Lin and Papaloizou*, 1986). However, in a viscous (Keplerian) disk, because of differential rotation gas is also subject to a viscous torque generated by viscous friction between adjacent disk rings. The viscous torque exerted by material inside of the orbital radius a on material outside of a is (*Lynden-Bell and Pringle*, 1974)

$$\mathcal{T}_v \approx 3\pi v a^2 \Sigma \Omega \qquad (39)$$

This torque tends to smooth out density gradients and redistribute material across the planet's orbit, filling in the gap.

A condition for gap formation requires a net gain of orbital angular momentum for material outside the radius a, i.e., $\mathcal{T}_g - \mathcal{T}_v > 0$, and a net loss of orbital angular momentum for material inside a, $\mathcal{T}_g + \mathcal{T}_v < 0$. If $H \geq R_H$, the condition $|\mathcal{T}_g| > \mathcal{T}_v$ translates into a condition for the minimum planet's mass necessary to open up a density gap in a disk

$$\left(\frac{M}{M_\star}\right)^2 \geq 3\pi f \alpha \left(\frac{H}{a}\right)^5 \qquad (40)$$

where the turbulence parameter α is defined in section 2.1. In the case of very cold disks or very massive planets, $R_H > H$ and the torque inequality gives $(M/M_\star) \geq \pi f \alpha (H/a)^2$. For conditions adopted in the simulations reported in Fig. 3, substantial gas depletion is expected for $M \gtrsim 2 \times 10^{-4}\, M_\star$ (or $M \gtrsim 60\, M_\oplus$ for a disk around a solar-mass star). In accord with this estimate, Fig. 3b indicates that the average surface density along the planet's orbit is reduced by 40% of its unperturbed value when $M \gtrsim 2 \times 10^{-4}\, M_\star$.

The timescale for gap formation can be estimated from the equation of motion of a thin viscous Keplerian disk, which also evolves under the action of gravitational torques exerted by an embedded object (*Lin and Papaloizou*, 1986)

$$\pi a \frac{\partial \Sigma}{\partial t} = \frac{\partial}{\partial a}\left[\frac{1}{a\Omega}\frac{\partial}{\partial a}(\mathcal{T}_v - \mathcal{T}_g)\right] \quad (41)$$

where radius a is now interpreted as a variable indicating the distance from the star. Equation (41) can be derived from imposing conservation of mass and angular momentum in an axisymmetric and flat disk. Ignoring viscous torques ($\mathcal{T}_v < |\mathcal{T}_g|$) and taking the approximation $\partial/\partial a \to 1/\Delta a$ in equation (41), defining the gap formation timescale as $\tau_{\mathrm{gap}} = \Sigma |\partial \Sigma/\partial t|^{-1}$ and using equation (38), it follows that

$$\Omega \tau_{\mathrm{gap}} \sim \pi f \left(\frac{M_\star}{M}\right)^2 \left(\frac{|\Delta a|}{a}\right)^5 \quad (42)$$

with $|\Delta a|$ equal to the larger of H and R_H. This timescale is typically short, on the order of tens of orbital periods.

Once gap formation starts, accretion rates drop because of gas depletion around the planet. According to condition (40), lower disk viscosity allows for gas depletion around smaller mass planets. By using equation (37), one finds that for

$$\alpha \lesssim \frac{C_H}{9\pi f\, C_B}\left(\frac{H}{a}\right) \quad (43)$$

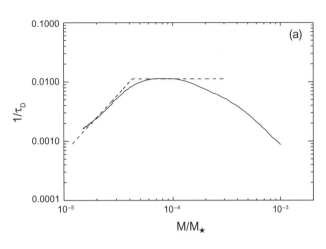

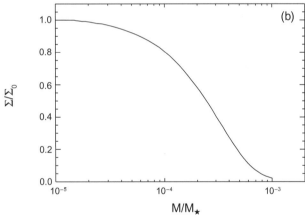

Fig. 3. (a) Mass growth rate, $1/\tau_D = (dM/dt)/M$, vs. mass of a protoplanet accreting gas at a limiting rate provided by a protoplanetary disk. These results were obtained from a three-dimensional, high-resolution hydrodynamical model of a growing protoplanet interacting with a disk of initial surface density at $a = 5.2$ AU of $100\, \mathrm{g\, cm^{-2}}$, relative thickness $H/a = 0.05$, and turbulence parameter $\alpha = 4 \times 10^{-3}$ (*D'Angelo and Lubow*, 2008). The timescale τ_D is in units of orbital periods. The dashed line represents the growth rate given by equation (36): The slanted portion corresponds to the rate in the Bondi-type gas accretion regime ($1/\tau_B$ with $C_B \simeq 2.6$) and the horizontal portion corresponds to the Hill-type gas accretion rate ($1/\tau_H$ with $C_H \simeq 0.9$). At large planet masses, the growth rates drop due to the formation of the tidally produced gap in the local density distribution. (b) Average surface density distribution of the disk's gas, Σ, near the planet relative to the local unperturbed (i.e., without a planet) value, Σ_0, as a function of the planet's mass.

gas depletion begins before the transition to a Hill-type gas accretion and the growth rate starts to decline for smaller mass planets. In a disk of thickness $H/a \approx 0.05$, this may occur for α less than few times 10^{-4}.

Effects of gap formation can be seen in Fig. 4. The figure shows disk-limited gas accretion rates, derived from hydrodynamical simulations (for details, see *D'Angelo et al.*, 2003b), for protoplanets embedded in a moderately viscous ($\alpha = 4 \times 10^{-3}$) and in a low-viscosity ($\alpha = 4 \times 10^{-4}$) disk. Notice, though, that if $\tau_{\mathrm{gap}} > \tau_H$, i.e., $f C_H (a^2 \Sigma/M_\star) \gtrsim (M/M_\star)^2 (a/H)^4$, Hill-type gas accretion may persist until $R_H \sim H$.

Disk-limited gas accretion rates can be quite large, unless the disk's gas density is low. A Saturn-mass planet orbiting a solar-mass star at ~5 AU in a disk, whose gas surface density just outside the gap region is on the order of $100\, \mathrm{g\, cm^{-2}}$, may accrete gas at rates $\sim 10^{-3}$–$10^{-2}\, M_\oplus$ per year, thereby reaching a Jupiter's mass in 10^4–10^5 yr.

2.7. Final Masses of Giant Planets

It is not yet entirely clear what all the factors are that determine the final mass of a giant planet. Yet, tidal truncation of the disk by gravitational interactions with the planet is likely one of the main factors. As indicated in Fig. 3b, gaps can become quite deep around a Jupiter-mass planet, even at moderate values of the turbulence parameter α. However, gas can filter through the tidal barrier (i.e., the gap) and continue to accrete toward the planet.

This can be seen in Fig. 5, which shows the gap in the surface density distribution around a giant planet together with trajectories of material moving toward the planet along the inner and outer edges of the gap. The trajectories are drawn in a frame co-moving with the planet. This material becomes trapped in the inner parts of the planet's Roche lobe and is eventually accreted. As mentioned in the previous section, accretion through a gap can be quite efficient. Assuming a gas surface density of ~100 g cm^{-2} just outside the gap region at ~5 AU, the growth timescale (due to disk-limited gas accretion), $\tau_D = M(dM/dt)^{-1}$, of a Jupiter-mass planet in a disk with $\alpha \sim 10^{-3}$ and $H/a \approx 0.05$ would be several thousands of orbital periods (see Fig. 4). Around a solar-mass star, the tidal barrier alone would constrain the planet's final mass to about 6–7 M_{Jup}.

A lower disk temperature, i.e., a smaller disk thickness, can increase the planet's tidal barrier and reduce gas accretion. However, it is unclear whether accretion can be stopped around $M \sim 1\ M_{Jup}$ for reasonable values of H/a (*Lubow and D'Angelo*, 2006). Lower disk viscosity also helps reduce disk-limited gas accretion rates, as indicated in Fig. 4. Assuming a disk around a solar-mass star with $\alpha \sim 10^{-4}$, for the same example made above, the tidal barrier would limit the planet's final mass to ~2 M_{Jup}.

The considerations above do not take into account the fact that a protoplanetary disk evaporates due to irradiation from the central star and, possibly, from other external sources (*Hollenbach et al.*, 2000). Both observations and theory suggest that photoevaporation timescales are on the order of a few million years. Therefore, disk consumption by photoevaporation may become a relevant process in the final stages of a giant planet's growth. It is likely that a combination of these three factors, i.e., a reducing disk thickness H/a, a decaying turbulence α, and gas depletion by photoevaporation, all of which may take place as a protoplanetary disk ages, plays a significant role in determining final masses of gas giant planets.

Once gas accretion stops, a giant planet evolves at a constant mass. The planet gradually cools down and contracts on a Kelvin-Helmholtz timescale (equation (14)). The luminosity is mostly due to gravitational energy released in the contraction.

2.8. Orbital Migration of Protoplanets

Gravitational interactions between a protoplanetary disk and an embedded planet result in an exchange of orbital angular momentum (see the chapter by Lubow and Ida for a thorough discussion on this topic). Assuming that $H \geq R_H$ and that condition (40) for gap formation is not satisfied, the disk's surface density is largely unperturbed and equation (38) applies.

If the planet exerts a torque $\mathcal{T}_g$ on the gas outside its orbit, then conservation of angular momentum dictates that this gas exerts the same torque (with an opposite sign) on the planet. Therefore, the planet loses orbital angular momentum to material exterior to its orbit. By the same principle, the planet gains orbital angular momentum from material interior to its orbit. The torque expression in equation (38) is symmetrical with respect to the planet's radial

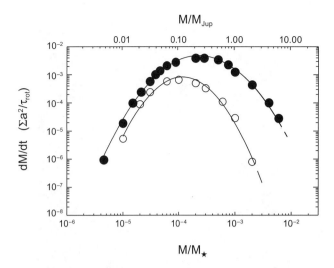

Fig. 4. Disk-limited gas accretion rates as a function of the planet mass obtained from three-dimensional, high-resolution hydrodynamical calculations of a planet interacting with a protoplanetary disk. Accretion rates are in units of the unperturbed surface density at the planet's orbital radius, a, and the planet's orbital period, τ_{rot}. The top axis uses $M_\star = 1\ M_\odot$. Filled circles correspond to results for a disk with a turbulence parameter $\alpha = 4 \times 10^{-3}$ (see section 2.1). Empty circles are for a disk with $\alpha = 4 \times 10^{-4}$. The disk's aspect ratio is $H/a = 0.05$. The curves represent fits to the data.

position, which would result in a zero net torque acting on the planet. However, one can show that, because of global variations of the disk properties across the planet's orbital radius a, the net torque exerted on the planet by disk material is on the order of $|\mathcal{T}_p| \sim (H/a)\,|\mathcal{T}_g|$ (*Goldreich and Tremaine*, 1980; *Ward*, 1997), hence

$$\mathcal{T}_p \approx C_I \Sigma a^4 \Omega^2 \left(\frac{a}{H}\right)^2 \left(\frac{M}{M_\star}\right)^2 \qquad (44)$$

Quantity C_I depends on the radial gradients of surface density, temperature, and pressure of the disk across the planet's orbit. In typical circumstances, this quantity is negative and on the order of unity (*Tanaka et al.*, 2002).

The orbital angular momentum of a planet on a circular orbit is $L_p = M a^2 \Omega$. If the planet is acted upon by a torque $\mathcal{T}_p$, conservation of angular momentum imposes that $dL_p/dt - \mathcal{T}_p = 0$, and thus

$$\frac{1}{M}\frac{dM}{dt} + \frac{1}{2a}\frac{da}{dt} = \frac{\mathcal{T}_p}{M a^2 \Omega} \qquad (45)$$

If the first term on the lefthand side of equation (45) is negligible compared to the second, then $da/dt = 2\ \mathcal{T}_p/(M a \Omega)$. This phenomenon is referred to as orbital migration. Since the torque experienced by the planet is generally negative, the orbit shrinks. When the torque acting on a (non-gap-

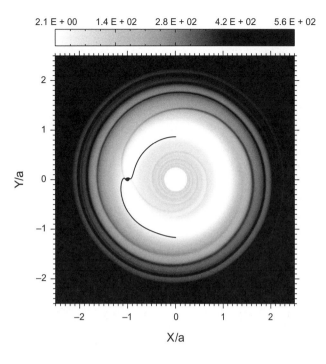

Fig. 5. Surface density of a gaseous disk containing a protoplanet whose mass is $M = 10^{-3} M_\star$ (i.e., 1 M_{Jup} for a solar-mass star). The relative disk thickness is $H/a = 0.05$ and the turbulence parameter is $\alpha = 4 \times 10^{-3}$. The planet is located at $(-1, 0)$, the star at $(0, 0)$, and the disk is rotating in the counterclockwise direction. The grayscale bar is in cgs units. The plot shows the density gap along the planet's orbit and the wave pattern generated by disk-planet interactions. The two black lines represent trajectories of gas in the co-moving frame of the planet. These gas parcels move along the inner and outer gap edges, become gravitationally bound to the planet, and are eventually accreted.

opening) planet is of the type in equation (44), migration is said to be of type I. The migration timescale, $a|da/dt|^{-1}$, in these circumstances is

$$\Omega \tau_I \sim \left(\frac{M_\star}{a^2 \Sigma}\right)\left(\frac{M_\star}{M}\right)\left(\frac{H}{a}\right)^2 \quad (46)$$

Orbital decay due to type I migration can be quite rapid. A 10 $M_\oplus$ ($M = 3 \times 10^{-5} M_\star$) planetary core interacting with a disk, whose surface density is $\Sigma \sim 100$ g cm^{-2} and $H/a \sim 0.05$, would migrate on a characteristic timescale of $\sim 4 \times 10^4$ orbital periods. For a planet orbiting at ~ 5 AU around a solar-mass star this period would amount to several times 10^5 yr, a factor of a few shorter than the thermally regulated envelope phase (see section 2.4) shown in Fig. 1a. Such rapid migration rates may be difficult to reconcile with formation via core accretion of giant planets that orbit at large distances from their stars, unless the time spent by a planet in the slow growth stage and the migration timescale τ_I are actually closer (see, e.g., *Alibert et al.*, 2005).

Once a planet becomes a gas giant, a density gap forms along its orbit (because condition (40) is typically satisfied) where the surface density is drastically depleted (see Figs. 3b and 5). In the limit of a very clean gap, there is a balance between viscous torques and gravitational torques at gap edges (interior and exterior to the planet's orbit) and the planet remains locked in the gap. This argument is valid as long as the disk inside the orbit of the planet is not significantly depleted. The planet drifts toward the star carried by the viscous diffusion of the disk, if the local mass of the disk is comparable to the planet's mass, or larger. This is referred to as type II migration. Viscous diffusion through the disk occurs on a timescale $a|da/dt|^{-1} \sim a^2/\nu$, and therefore

$$\Omega \tau_{II} \sim \frac{1}{\alpha}\left(\frac{a}{H}\right)^2 \quad (47)$$

In typical circumstances, there is residual gas in the gap region and there may not be an exact balance between viscous and gravitational torques at gap edges. Nonetheless, results from multidimensional simulations indicate that migration occurs on a timescale on the order given by equation (47) in the presence of a sufficiently deep gap (e.g., *D'Angelo and Lubow*, 2008). In a disk with $\alpha \sim 10^{-3}$ and $H/a \sim 0.05$, a gap-opening planet would undergo orbital migration on timescales of many tens of thousands of orbital periods, or many 10^5 yr at ~ 5 AU from a solar-mass star.

According to equation (47), $\tau_{II} \propto a^{3/2}$ (if H/a is approximately constant) and thus the migration timescale becomes shorter as a giant planet approaches the star. However, efficient exchange of orbital angular momentum between a gas giant and a protoplanetary disk requires that the local disk mass is at least comparable with the planet's mass, i.e., $\pi a^2 \Sigma \gtrsim M$. If this is not the case, planet's inertia acts to slow down migration. Strict type II migration of a 1 M_{Jup} planet would therefore necessitate a gas surface density Σ at both gap edges greater than ~ 100 g cm^{-2} at 5 AU from the star and $\gtrsim 2.8 \times 10^3$ g cm^{-2} at 1 AU.

If the disk inside the planet's orbit is significantly depleted (due to the tidal barrier of a massive planet), conservation of angular momentum at the gap edge exterior of the planet's orbit requires that $dL_p/dt + \mathcal{T}_v = 0$ (e.g., *Syer and Clarke*, 1995), where the first term on the lefthand side is the rate of change of the planet's orbital angular momentum and the second term is the (positive) viscous torque exerted on the disk, given by equation (39). Thus, for a nearly truncated disk at the gap edge exterior to the planet's orbit $da/dt = -2\, \mathcal{T}_v/(Ma\Omega)$, and the migration timescale is

$$\Omega \tau_{ED} \sim \frac{1}{6\pi\alpha}\left(\frac{a}{H}\right)^2 \left(\frac{M}{a^2 \Sigma}\right) \quad (48)$$

Inertia effects are expected to be important at small orbital radii, when a disk is sufficiently depleted. Migration of giant planets provides a natural explanation for the occurrence of

Jupiter-mass planets orbiting within a few tenth of an AU from their parent stars, and whose existence would be otherwise difficult to explain by means of *in situ* formation via core accretion or disk instability (as we shall see in section 3).

2.9. Examples of Giant Planet Evolution Tracks

An illustrative example of models combining planet's growth and orbital migration is displayed in Fig. 6. These results are obtained from calculations that encapsulate in a simplified manner all the basic aspects of a core accretion model and standard migration theory discussed thus far. We consider here the case of a planetary embryo embedded in an axisymmetric (around the star) disk, which evolves under the action of viscous torques (equation (39)), gravitational torques, and photoevaporation induced by the central star. All three effects are accounted for in equation (41) once the lefthand side of that equation is replaced with $\pi a \partial(\Sigma + \Sigma_{pe})/\partial t$, in which $\partial \Sigma_{pe}/\partial t$ is the mass loss flux due to photoevaporation.

Tidal torques exerted by the planet on the disk are incorporated in equation (41) through a set of torque density distributions (torque per unit disk mass as a function of radius), which cover the relevant range of planet masses and are obtained from three-dimensional hydrodynamical simulations of disk-planet interactions (see *D'Angelo and Lubow*, 2008, for details). This procedure allows us to encompass the different regimes of orbital migration experienced by the planet as it grows and discussed in section 2.8.

We use a simplified form of the mass loss rate from the disk that is zero within the "gravitational" radius $a_g \approx 10\,(M_\star/M_\odot)$ AU, where the thermal energy of the gas equals in magnitude its gravitational energy, otherwise $\partial \Sigma_{pe}/\partial t = \dot{\Sigma}^g_{pe}(a_g/a)^{5/2}$ (see the chapter by Roberge and Kamp). The mass loss flux at radius a_g is $\dot{\Sigma}^g_{pe} = 3.7 \times 10^{-13}\,(\Phi^{41}_{pe})^{1/2}\,(M_\star/M_\odot)^{-3/2}$ in units of $M_\odot \mathrm{AU}^{-2}$ per year. The constant Φ^{41}_{pe} is the rate of ionizing photons emitted by the star, Φ_{pe}, in units of $10^{41}\,\mathrm{s}^{-1}$.

At time $t = 0$, the disk contains $\approx 0.022\,M_\odot$ of gas within about 40 AU (*Davis*, 2005). The disk evolution equation is solved numerically, adopting a turbulence parameter $\alpha \approx 4 \times 10^{-3}$ and a disk scale-height $H/a = 0.05$. The evolution begins at 13 AU from a solar-mass star.

A planet embryo, of initial mass $0.1\,M_\oplus$, accretes solids at an oligarchic rate ($\propto M^{2/3}$, see section 2.2) and becomes isolated when $M \approx 13\,M_\oplus$. The gas accretion rate, dM/dt, is given by the smaller of M/τ_e (equation (15)) and the upper fit to data in Fig. 4. Three values of Φ^{41}_{pe} are used: 1, 10, and 50. Thicker and thinner lines represent cases in which the planet's evolution starts at time $t = 0$ and 1.5 m.y., respectively. The evolution tracks illustrated in Fig. 6 highlight the importance of the disk's initial conditions in which a giant planet's core begins to evolve and on the details of disk photoevaporation. It also supports the contention that early core accumulation may lead to substantial amounts of radial migration, whereas late-stage formation scenarios may imply much less orbital decay.

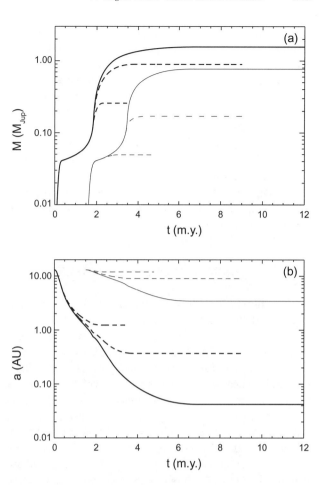

Fig. 6. Evolution tracks of giant planet formation via core accretion obtained from models that account for orbital migration by tidal interactions with the protoplanetary disk (see section 2.9 for details). **(a)** Total mass and **(b)** orbital radius, both shown as a function of time. Two sets of tracks represent cases in which the planet evolution begins at time $t = 0$ (thicker lines) and at time $t = 1.5$ m.y. (thinner lines). Within each set of tracks, three different line types represent different values of the applied rate of ionizing photons emitted by the star: $\Phi_{pe} = 10^{41}\,\mathrm{s}^{-1}$ (solid lines), $10^{42}\,\mathrm{s}^{-1}$ (long-dashed lines), and $5 \times 10^{42}\,\mathrm{s}^{-1}$ (short-dashed lines).

2.10. Summary of Core Accretion Models

The formation of a gas giant via core nucleated accretion may take about one to a few million years and is initiated by a solid planetary core of at least a few Earth masses. Core formation requires that the heavy elements of the disk have condensed out and coagulated into planetesimals. Although the process of planetesimal formation is still under scrutiny, small bodies of ~10 to ~100 km in size are observed in the Kuiper belt and Neptune-mass planets are observed around the Sun and other stars. Core accretion models provide an explanation for the existence of not only the four giant planets in the solar system, but also the majority of giant planets that have been observed around other stars.

Giant planets contain large amounts of hydrogen and helium, which are acquired in gaseous form from protoplanetary disks. Therefore, gas dispersal timescales of disks set a strict upper limit on the time available to form these planets (a few to ~10 m.y.). The phase of slow envelope contraction that precedes runaway gas accretion seems to represent the greatest hurdle to overcome since it may last longer than the disk's gases, at least in some situations. Orbital decay by tidal torques may require formation in evolved rather than young disks to avoid amounts of radial migration that are too large. Additionally, since the oligarchic growth timescale is $\propto a^{3/2} M_\star^{-1/2}$, core formation at orbital distances of many tens of AU or around low-mass stars (e.g., red dwarfs) may take too long to produce a gas giant.

3. DISK INSTABILITY MODELS

Stars form through the gravitational collapse of interstellar clouds, and *Boss* (1997) has championed the idea (*Kuiper*, 1951; *Cameron*, 1978) that gas giant planets might form from protoplanetary disks in a similar way. This is commonly referred to as the disk instability theory of gas giant planet formation, because the mechanism by which the planets form is the onset of gas-phase gravitational instabilities (GIs) leading to fragmentation of a protoplanetary disk into bound, self-gravitating clumps. An important distinction with core accretion is that, if the conditions prevail for disk instability to work, then planets can form directly out of the gas phase within only a few to tens of disk orbit periods ($\tau_{rot} = 2\pi/\Omega$). The planets gain most of their complement of gas immediately and subsequently sediment heavy element cores and/or sweep up planetesimals over a longer timescale. Disk instability is top down and initially rapid; core accretion is bottom up and initially slow. The critical issue for disk instability is when and where the conditions necessary for fragmentation might actually occur in protoplanetary disks.

We will begin by discussing how GIs operate and what is required for the outcome to be fragmentation. To decide whether real disks fragment, we must rely heavily on detailed numerical simulations.

3.1. Gravitational Instabilities: Linear Instability

Gravitational instabilities are caused by the self-gravity of the disk, but, unlike star formation, the onset of planet formation via disk instability is much more strongly influenced by rotation, and it happens in a different thermodynamic environment.

To understand how GIs work, consider first the basic equilibrium of a gas disk orbiting a central star for the case where disk self-gravity cannot be ignored. For the vertical equilibrium in a thin disk where parameters vary smoothly and slowly with orbital distance a, equation (3) becomes

$$\frac{1}{\rho}\frac{\partial p}{\partial z} + \left(\frac{GM_\star}{a^2}\right)\left(\frac{z}{a}\right) + 2\pi G\Sigma(a,z) = 0 \qquad (49)$$

where $\Sigma(a, z) = \int_{-z}^{z} \rho\, dz'$ is the surface density of gas at radial position a between −z and z, with z = 0 being the midplane. In what follows, we use $\Sigma = \Sigma(a)$ to refer to $\Sigma(a,\infty)$. The additional (third) term on the lefthand side of equation (49) characterizes the vertical component of the gravitational field due to the gas disk. The radial equilibrium for a thin disk is then

$$\frac{1}{\rho}\frac{\partial p}{\partial a} - a\Omega^2 + \left(\frac{GM_\star}{a^2}\right) - g_a^D = 0 \qquad (50)$$

where g_a^D is the additional radial component of the gravitational field due to disk self-gravity, which depends on the detailed distribution of the disk surface density $\Sigma(a)$. When H/a is small, disk self-gravity can be important in equation (49) without necessarily being a major contribution in equation (50). So, to derive simple relations in later sections, we will sometimes approximate Ω by using the Keplerian angular speed, $\sqrt{GM_\star/a^3}$.

GIs arise through perturbation of the equilibrium quantified by equations (49) and (50). A classic linear stability analysis by *Toomre* (1981) for axisymmetric radial ripples is very instructive. Equations (49) and (50) define the zero-order equilibrium state of the disk, which we denote by a subscript zero, e.g., p_0, ρ_0, etc. Physical variables in the perturbed disk can then be represented by these zero-order quantities plus small perturbations, i.e., $p = p_0 + \delta p$, with $|\delta p| \ll p_0$. To simplify the analysis, the radial wavelength λ of the perturbations is taken to be small compared with both a and the scale length of radial changes in Σ, but large compared with the vertical scale-height H. The thin vertical structure of the disk is assumed to remain in hydrostatic equilibrium according to equation (49) even when perturbed. So, after the acceleration term $\partial v_a/\partial t \simeq \partial(\delta v_a)/\partial t$ is introduced into the lefthand side of equation (50) in order to describe the dynamics of the perturbations, the radial equation is integrated over z. Equation (50) allows the zero-order terms to be canceled, and only terms to first order in the perturbed quantities are retained in the linear analysis (for more details, see *Binney and Tremaine*, 1987).

By looking for wave-like perturbations such that, e.g., $\delta p \propto \exp(i\omega t \pm 2\pi a/\lambda)$, one can derive the dispersion relation

$$\omega^2 = \kappa^2 - \frac{4\pi^2 G\Sigma}{\lambda} + \frac{4\pi^2 c^2}{\lambda^2} \qquad (51)$$

where λ is positive-definite and c^2 is the sound speed squared. The quantity κ in equation (51) is the epicyclic frequency at which a perturbed fluid element oscillates about its equilibrium position (*Binney and Tremaine*, 1987). For a strictly Keplerian disk, $\kappa = \Omega$, because, when the radial equilibrium is dominated by the central force of the star, a slightly perturbed fluid element follows an elliptical orbit with the same period as τ_{rot}. As viewed from its unperturbed equilibrium position, an oscillating fluid element appears to execute a small elliptical "epicycle." The epicycle has the opposite or retrograde sense of rotation compared with the orbital motion of the disk.

Equation (51) reveals a great deal about the behavior of perturbations in self-gravitating disks. As long as the righthand side (RHS) of equation (51) is positive, $i\omega t \pm 2\pi a/\lambda$ is a complex number and so the ripples are waves that oscillate and propagate radially. However, when the RHS is negative, there are stationary solutions that grow exponentially (the argument of the exponential is real). The disk is then unstable and tends to break apart into dense concentric rings.

An examination of the individual terms on the RHS of equation (51) reveals the stabilizing and destabilizing influences. The first and third terms are always positive and represent stabilization by rotation and gas pressure, respectively. As λ gets large, the first term dominates the RHS of equation (51); in other words, rotation stabilizes long wavelengths. Because the central gravitational force of the star appears in equation (51) through κ, it is also a stabilizing influence. When λ is small, the third term dominates; in other words, gas pressure stabilizes small wavelengths. In contrast to the first and third terms, the self-gravity of the disk represented by the second term on the RHS, which is only important at intermediate wavelengths, is always negative, and hence is a destabilizing influence. If the acceleration imparted by disk gas, $G\Sigma$, is large enough relative to $\kappa^2\lambda$ and c^2/λ, some middle range of wavelengths will be unstable. Because the destabilizing term is due to disk self-gravity, we refer to such instabilities as gravitational instabilities.

Setting $\omega = 0$ in equation (51) and solving for λ, one can see by simple algebra that there is a range of unstable wavelengths when

$$Q = \frac{c\kappa}{\pi G \Sigma} < 1 \qquad (52)$$

Notice that the numerator of the Toomre stability parameter Q is, in effect, a geometric mean of the stabilizing influences of pressure and rotation, while the denominator represents the destabilizing influence of disk self-gravity. Thus, fixing the mass of the star, a protoplanetary gas disk becomes unstable to GIs if it is sufficiently cold (low c) or massive (high Σ). The critical wavelength λ_{cr} that first becomes unstable when Q dips below unity is

$$\lambda_{cr} = \frac{2\pi^2 G \Sigma}{\kappa^2} \approx \frac{2\pi H}{Q} \qquad (53)$$

as can be derived by setting $\omega = 0$ and $Q = 1$ in equation (51). For the approximate relation on the right in equation (53), we use $c \approx H\Omega$, which still holds to within a factor on the order of unity for a self-gravitating disk.

As summarized in various reviews (e.g., *Durisen et al.*, 2007), multidimensional simulations of disks show that nonaxisymmetric perturbations, i.e., spiral waves, become unstable for

$$Q = \frac{c\kappa}{\pi G \Sigma} < Q_{cr} \qquad (54)$$

where Q_{cr} is between about 1.5 and 2.0, depending on the detailed structure of the gas disk. Because spiral waves can be unstable in disks that are stable against growth of the axially symmetric ripples we just discussed, spiral disturbances are considered to be more unstable, i.e., a disk that transitions slowly from stable to unstable conditions will first manifest GIs in the form of nonaxisymmetric waves. In fact, in simulations, the linear growth of GIs from noise is characterized by the appearance of multiarmed spirals, which are typically trailing due to the Keplerian shear. These waves grow exponentially from noise in the linear regime on the dynamic timescale τ_{rot} (e.g., *Nelson et al.*, 1998; *Pickett et al.*, 1998).

With a little algebra, Q in equation (52) can be rewritten approximately as twice the ratio of the star's vertical gravitational field to the disk's vertical gravitational field, namely, $2g_z^\star/g_z^D$, where both $g_z^\star$ and g_z^D are evaluated at $z = H$. So, $Q < 2$ corresponds roughly to $g_z^D > g_z^\star$. As expected, GIs appear when the self-gravity of the gas disk becomes sufficiently important, i.e., when the second and third terms on the lefthand side of equation (49) become comparable.

The discussion associated with equation (51) makes it clear how self-gravity could break a disk into ringlets, but how do spiral waves grow? The likely mechanism is explained in *Toomre* (1981) and *Binney and Tremaine* (1987). Consider part of a leading spiral wave spanning its own corotation radius (CR), the radial location in the disk where the angular pattern speed of the wave equals Ω, i.e., where one arm of the wave goes around the disk in exactly τ_{rot}. Shear in the disk causes the arc to "swing" from leading to trailing. Because $\kappa \approx \Omega$ in a nearly Keplerian disk, the fluid elements of the disk in the crest of the wave near the CR are executing epicyclic oscillations that keep them in the crest. This allows self-gravity to amplify the density maximum at the crest of the wave. Numerical studies of protoplanetary disks in which spiral waves grow show that the conditions are indeed right for the swing amplification mechanism to operate (*Pickett et al.*, 1998; *Mayer et al.*, 2004), and the spiral waves that grow always straddle their CR (see, e.g., *Mejía et al.*, 2005).

3.2. Occurrence of Gravitational Instabilities

To get a feeling for when GIs are expected in protoplanetary disks, we can evaluate equation (54) for the typical parameters used in earlier sections, namely $H/a \approx 0.05$ at $a = 5.2$ AU for a disk orbiting a solar-mass star (corresponding to a temperature $T \sim 75$ K if c is the adiabatic sound speed). Let us assume that $Q_{cr} \approx 1.7$. Then

$$Q \approx Q_{cr} \left(\frac{3 \times 10^3 \, \text{g cm}^{-2}}{\Sigma} \right) \qquad (55)$$

The typical solids' surface density $\Sigma_s \sim 10$ g cm^{-2} discussed in section 2 for core accretion models scales to $\Sigma \sim 10^3$ g cm^{-2} for gas with solar composition when half the heavy elements are assumed to be in the solid phase. A surface density $\Sigma = 10^3$ g cm^{-2} at 5.2 AU is three times the standard Hayashi MMSN (*Hayashi*, 1981), but is only a factor of about 1.5 larger than more recent MMSN models (*Davis*, 2005; *Desch*, 2007). For a disk around a solar-type star to be unstable to gas-phase GIs at Jupiter's orbit radius, i.e., $Q < Q_{cr}$, equation (55) tells us that Σ must be $\geq 3 \times 10^3$ g cm^{-2}, which is 10 times higher than the Hayashi MMSN and 4 or 5 times higher than the newer MMSN models.

There are two ways that such a high Σ might occur at 5.2 AU: Either the disk is massive everywhere or there is a localized concentration of mass near this radius. In the first case, where $\Sigma(a)$ is relatively smooth, gas-phase GIs at the orbit radii of the gas giants in the solar system require a total protoplanetary gas disk mass of about 0.1 M$_\odot$ or more (*Boss*, 2002a), and disk masses of this order are typical in hydrodynamics simulations used to study GIs in protoplanetary disks. Disks may well have masses this large or larger during the accretion phase, when gas is falling onto the young star/disk system at rates of 10^{-5} M$_\odot$ per year or higher (e.g., *Vorobyov and Basu*, 2006). Alternatively, a disk might have a localized enhancement of surface density due to accumulation of mass in or at the edges of a dead zone (*Gammie*, 1996), where radial transport of mass becomes inefficient. This could lead to episodic eruptive phenomena related to FU Orionis outbursts (e.g., *Armitage et al.*, 2001; *Boley and Durisen*, 2008; *Zhu et al.*, 2009).

Although we have not yet discussed the conditions for fragmentation, it is instructive to estimate how much mass is involved in the GIs. Using equation (53) as an estimate for the radial extent of the instability, we expect the mass in one arm of a k-armed unstable wave at 5.2 AU to be about

$$M_{frag} \approx 2\pi \frac{a\lambda_{cr}\Sigma}{k} \sim \frac{11}{k} M_J \qquad (56)$$

Simulations show that k is typically four or five at the onset of GIs (*Mejía et al.*, 2005; *Boley and Durisen*, 2008) when $M_D \sim 0.14 M_\star$, and therefore the arms of the spirals would indeed have a gas inventory sufficient to produce a giant planet.

Even if a disk is stable ($Q > Q_{cr}$) at the radius of Jupiter's orbit, it may be susceptible to GIs at larger radii. Suppose that the disk temperature $T \propto a^{-1/2}$ (i.e., $H/a \propto a^{1/4}$, a "flared" disk). We know that $\kappa \approx \Omega \propto a^{-3/2}$, so if $\Sigma \propto a^{-s}$,

$$Q = \frac{c\kappa}{\pi G\Sigma} \propto \frac{T^{1/2}\Omega}{\Sigma} \propto a^{s-7/4} \qquad (57)$$

Thus, for any disk with s < 7/4, Q becomes less than Q_{cr} if the disk extends to a sufficiently large radius. For a disk with s < 2, $M_D(a) = \int_0^a 2\pi a'\Sigma da' \sim a^{2-s}$, and so the disk mass $M_D(a)$ interior of a diverges as a increases. As a result, if a disk with s < 7/4 is extended enough to have GIs in its outer regions, it is likely to have a mass that is a significant fraction of the star's mass. Again, this is most likely to occur during the early accretion phase of the star/disk system. If we put in parameters appropriate for this situation at a few hundred AU around a solar-type star, M_{frag} can become tens of Jupiter masses or more (*Stamatellos and Whitworth*, 2009).

The above arguments are not very sensitive to the temperature distribution T(a). If we use $T(a) \propto a^0$ or $\propto a^{-1}$ instead, then we get $H/a \propto a^{1/2}$ or H/a equal to a constant, respectively, and $Q \propto a^{s-3/2}$ or $Q \propto a^{s-2}$, respectively. Parameter Q will still become less than Q_{cr} at large enough radii for surface densities that do not fall off too rapidly. During the early accretion phase, one expects that, in spatially extended disks, the gas temperature levels off to a constant value ($T \propto a^0$) at large radii due to envelope irradiation. At small radii, some disks may not be irradiated by their central stars due to shadowing. Such disks would have steeper temperature fall-offs than $a^{-1/2}$ and may not be flared. For more detailed analytic arguments about gravitational instabilities in disks during the accretion phase, see *Clarke* (2009) and *Rafikov* (2009).

3.3. Nonlinear Growth of Gravitational Instabilities and Fragmentation

There are two possible nonlinear outcomes for GIs: steady-state balance of heating and cooling, or fragmentation into dense clumps. As the spiral waves grow in amplitude and steepen into shocks, they turn some of the ordered rotational energy of the disk into heat. Moreover, gravitational torques due to the predominantly trailing character of the spirals lead to net outward transport of angular momentum and inward transport of mass. This transport causes gravity to do work on fluid elements as they sink into the central gravitational potential and is another important source of heat for the gas. If the disk does not shed this heat, its temperature and hence c inevitably becomes larger, which raises Q and shuts off the instability. Disks do radiate, and so GIs can be sustained in an quasisteady state with an average Q near but somewhat below Q_{cr} where the heating by GIs is balanced by radiative cooling (*Gammie*, 2001; *Lodato and Rice*, 2004; *Mejía et al.*, 2005). Mejía et al. refer to this as the *asymptotic* state. GIs grow on a dynamical timescale τ_{rot} for Q well below Q_{cr}, but the growth times become long as $Q \rightarrow Q_{cr}$ from the unstable side. So the heating rate due to GIs can adjust to the prevailing radiative cooling rate. In this sense, the thermal physics of the disk controls the limiting amplitude of GIs (*Tomley et al.*, 1991; *Pickett et al.*, 2000).

Let us define a radiative cooling time in a column-wise sense by

$$\tau_{cool} = \frac{\int_{-\infty}^{\infty} E_{int} dz}{2\sigma_{rad} T_{eff}^4} \qquad (58)$$

where E_{int} is the gas internal energy density and the effective temperature, T_{eff}, characterizes the radiant flux (energy per unit area per unit time) $\sigma_{rad}T_{eff}^4$ out the top of a column through the disk. The factor of 2 in equation (58) accounts for radiation from both the top and bottom of the disk. This equation makes the approximation that all energy loss from the disk is vertical, which need not be true in complicated dynamical situations.

When τ_{cool} is comparable to or less than τ_{rot}, the balance of heating and cooling discussed above may not be possible. The gas could cool profoundly between the spiral arms before it is reheated by the shock in the next arm. Another way to say the same thing is that, because the growth time of GIs is τ_{rot}, it is the shortest timescale on which GIs can replenish thermal energy in the gas. There will then be large compressions behind the shocks where the self-gravity of the gas could win out entirely. The arms may then break up into dense clumps, which become bound in the sense that the magnitude of the clump's self-gravitational energy exceeds the clump's own internal and rotational energies.

The precise value of τ_{cool} for which dense clumps form must be determined numerically. The first systematic treatment by *Gammie* (2001) for a simple equation of state in a local patch of a razor-thin disk showed that fragmentation occurs for $\Omega\tau_{cool} \lesssim 3$, or equivalently $\tau_{cool}/\tau_{rot} \lesssim 1/2$. Global three-dimensional simulations by several groups using a variety of simulation techniques have verified that the fragmentation criterion is

$$\frac{\tau_{cool}}{\tau_{rot}} < f_{frag} \quad (59)$$

where f_{frag} depends on the gas equation of state and, if τ_{cool} is a function of time, on the thermal history of the gas (*Rice et al.*, 2003, 2005; *Mejía et al.*, 2005; *Clarke et al.*, 2007). For ideal gases with ratios of specific heats $\Gamma_1 = 5/3$ and 7/5, $f_{frag} \approx 1$ and 2, respectively. In other words, a "softer" equation of state in the sense of lower Γ_1 makes a disk more unstable. This may be relevant, because 5.2 AU is close to where the effective Γ_1 for molecular hydrogen is expected to change from 5/3 at large disk radii to 7/5 at intermediate radii. When τ_{cool} is time-dependent and approaches the fragmentation condition slowly, f_{frag} becomes about half as large as when τ_{cool} is constant, and so the disks become more stable and harder to fragment.

For the special case of gas disks that behave isothermally when perturbed, τ_{cool} is essentially zero, and, instead of a fragmentation criterion based on the ratio τ_{cool}/τ_{rot}, one finds that all disks fragment when $Q < Q_{frag}$ where $Q_{frag} \approx 1.4–1.5 < Q_{cr}$ (*Boss*, 2000; *Mayer et al.*, 2002; *Johnson and Gammie*, 2003). A classic example of fragmentation in an isothermally evolved disk is shown in Fig. 7.

The masses of planets that form in fragmenting simulations are roughly compatible with equation (56), but the process of fragmentation can be violent and chaotic, often leading to multiple interacting protoplanetary clumps. Some clumps merge or shear out, and new clumps can form. No one has yet integrated a large number of fragmenting simulations over long times, but a moderately broad spectrum of final clump masses would probably result in a mass range between tenths to multiples of M_{Jup} for disks with $M_D \sim 0.1\,M_\star$ that are a few tens of AU in radius (see, e.g., *Mayer et al.*, 2004). A discussion of fragmentation spectra for disks with more extended radii can be found in *Stamatellos and Whitworth* (2009).

3.4. Realistic Radiative Cooling

The pertinent question now is whether τ_{cool} for realistic dust opacities is actually short enough to cause fragmentation. The vertical optical depth down into a disk is defined by

$$\tau(z) = \int_z^\infty \bar{\kappa}\rho\, dz' \quad (60)$$

where $\bar{\kappa}$, with units of cross-section per unit mass, is the mass absorption coefficient for radiation (opacity) averaged in an appropriate way over frequency. If $\bar{\kappa}$ does not vary much with z, then the midplane optical depth is $\tau(0) \approx \bar{\kappa}\Sigma/2$. This is essentially the number of photon mean free paths between the midplane and $z = \infty$.

If the radiation emitted by the gas and dust is thermal and if a frequency-independent (gray) opacity characterized by $\bar{\kappa}$ is a good approximation, then momentum conservation for radiation gives (e.g., *Gray*, 1992)

$$\frac{dp_{rad}}{d\tau} = \left(\frac{a_{rad}}{4\sigma_{rad}}\right)F_{rad} \quad (61)$$

for plane-parallel geometry, where p_{rad} is the radiation pressure, F_{rad} is the net upward flux of radiant energy, and $a_{rad} \approx 7.57 \times 10^{-15}$ erg cm^{-3} K^{-4} is the radiation-density constant (notice that the ratio in parentheses in equation (61) is equal to the inverse of the speed of light).

For no radiation shining down on the disk, for a constant upward flux ($F_{rad} = \sigma_{rad}T_{eff}^4$), and for a radiation field that is well described by black-body properties, e.g., $p_{rad} = a_{rad}T^4/3$, the standard gray atmosphere solution to equation (61) is

$$T^4 = \frac{3}{4}T_{eff}^4(\tau + q) \quad (62)$$

where q depends on boundary conditions and approximations but is roughly 2/3. Equation (62) strictly applies only to a semi-infinite atmosphere with a constant upward flux. A disk is finite and has a flux that varies with τ (*Hubeny*, 1990). Nevertheless, within a factor on the order of unity, equation (62) is a useful estimate relating the midplane temperature $T_{mid} = T(z = 0)$, i.e., $T(\tau)$ at $\tau = \tau(0)$, to the expected effective temperature T_{eff} in radiative equilibrium given a midplane optical depth $\tau(0)$.

When $\tau(0) \gg q$, the disk is optically thick. In equation (58), one can approximate the integral of E_{int} to be $3\Sigma\mathcal{R}T_{mid}/(2\mu)$, where $\mathcal{R}$ is the gas constant and μ is the mean-molecular weight (see section 2.5). Then

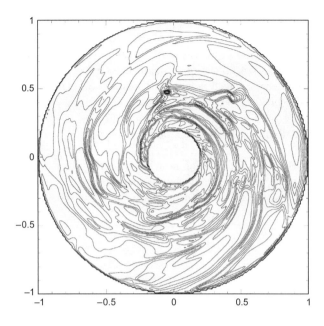

Fig. 7. Midplane density contours after 374 yr for an isothermal simulation of a protoplanetary gas disk spanning 4 to 20 AU around a solar-mass star with $M_D = 0.09\ M_\odot$ and an initial $Q_{min} = 1.3$. The dense clump near twelve o'clock is gravitationally bound and has about 5 M_{Jup}. Adapted from *Boss* (2000).

$$\tau_{cool} \sim \frac{9}{32}\left(\frac{\mathcal{R}}{\sigma_{rad}}\right)\left(\frac{\bar{\kappa}}{\mu}\right)\left(\frac{\Sigma^2}{T_{mid}^3}\right) \propto a^{(3-4s)/2} \quad (63)$$

where, for simplicity, the final dependence on a is based on assuming that $\bar{\kappa}$ is constant and $T_{mid} \propto a^{-1/2}$. (If $T_{mid} \propto a^0$ or $\propto a^{-1}$ then $\tau_{cool} \propto a^{-2s}$ or $\tau_{cool} \propto a^{3-2s}$, respectively.) At 5.2 AU for the usual parameters, equations (60) and (63) give

$$\tau(0) \sim 10^3 \left(\frac{\bar{\kappa}}{1\ cm^2\ g^{-1}}\right)\left(\frac{\Sigma}{3\times 10^3\ g\ cm^{-2}}\right) \quad (64)$$

and

$$\tau_{cool} \sim 10^5\ yr \left(\frac{\bar{\kappa}}{1\ cm^2\ g^{-1}}\right)\left(\frac{\Sigma}{3\times 10^3\ g\ cm^{-2}}\right)^2 \quad (65)$$

Note that, for H/a = 0.05 at 5.2 AU, $T_{mid} \approx 75$ K if we use an adiabatic sound speed with $\Gamma_1 = 5/3$ and $\mu \approx 2.4$. In an unstable disk at 5.2 AU, where $\tau_{rot} \approx 12$ yr, equation (65) gives $\tau_{cool}/\tau_{rot} \approx 10^4$, which is orders of magnitude greater than f_{frag} for any reasonable value of Γ_1.

We have just seen that, for our standard parameters, equations (54), (55), (59), and (63) together imply that, at 5.2 AU, a disk that is dense enough to be gravitationally unstable will cool too slowly to fragment. *Rafikov* (2005) first pointed out this conundrum. Although it depends on detailed assumptions about how disk properties vary with a, it can reasonably be argued that this problem extends out to about 40 AU or so, exactly the region where one would like to form the giant planets in our own solar system. Moreover, if we relax our choice of H/a = 0.05, then lowering τ_{cool} at fixed $\bar{\kappa}$ requires a lower Σ and/or a higher T_{mid}, both of which would tend to make the disk more stable.

One can instead imagine lowering $\bar{\kappa}$ itself because dust grows and settles to the midplane, but there is a limit to the efficacy of this ploy. The most efficient cooling by thermal emission occurs for $\tau(0) \approx 1$. For lower $\tau(0)$, the disk is a poor absorber and, hence, by Kirchhoff's law, a poor thermal emitter. All else being equal, equation (64) shows that a reduction of $\tau(0)$ to unity requires a reduction of $\bar{\kappa}$ by a factor of 10^3. With this reduction, τ_{cool}/τ_{rot} becomes about 10, which is still larger than the largest expected value of f_{frag}.

These analytic arguments suggest that gas giant planet formation by disk instability probably does not work inside several tens of AU. On the other hand, equations (57) and (63) further suggest that the instability criterion (equation (54)) and the fragmentation criterion (equation (59)) may be satisfied simultaneously at large enough radii, roughly 100 AU and beyond, when $0 < s < 7/4$ (see also *Clarke*, 2009; *Rafikov*, 2009).

3.5. Computational Results for Disk Instability

3.5.1. Simulations with realistic radiative cooling. The assumptions that go into the results of section 3.4 are rather simplistic and the details are subject to debate. For instance, $\bar{\kappa}$ is a complicated function of particle size and temperature and T_{mid} at very large orbital radii may reach a constant value. It can also be tricky to define τ_{cool} properly for the fragmentation criterion in a radiatively cooled disk evolving under the action of GIs (*Johnson and Gammie*, 2003). Disk instability must be tested by detailed multidimensional radiative hydrodynamics simulations. The numerical techniques for such simulations remain difficult, and results to date are disparate and controversial. A detailed discussion of numerical schemes goes beyond the scope of this chapter. Here we only summarize the current situation regarding global simulations that attempt to treat the cooling correctly with realistic opacities. This section deals with simulations of disk instability in regions extending out to tens of AU.

The analytic arguments of section 3.4 suggest that fragmentation should not occur inside 40 AU or so. Evidence supporting this conclusion comes from simulations using both smoothed particle hydrodynamics (SPH) (*Nelson et al.*, 2000; *Stamatellos and Whitworth*, 2008; *Forgan et al.*, 2009) and grid-based hydrodynamics (*Cai et al.*, 2006; *Boley et al.*, 2006, 2007; *Boley and Durisen*, 2008). These papers employ a variety of radiative algorithms. Figure 8 illustrates a typical midplane density structure for a nonfragmenting simulation. These studies treat 0.5–1 $M_\odot$ stars with a disk mass $M_D \sim 0.1\ M_\star$ and with outer disk radii of 10 to tens of AU. The disks do not cool rapidly enough to fragment. In the cases where τ_{cool} is explicitly calculated, nonfragmentation is consistent with the application of equation (59) for the relevant equation of state. Some efforts have been made by those who do not see fragmentation to evolve similar disks and to test their radiative schemes against analytic expectations, e.g., compare *Boley et al.* (2007) and *Stamatellos and Whitworth* (2008).

Evidence for fragmentation comes from two groups, one using grid-based simulations (*Boss*, 2001, 2002a,b, 2007, 2008) and the other using SPH (*Mayer et al.*, 2007). The disk parameters are similar to those of the nonfragmenting simulations, but these researchers find that their disks fragment into dense, bound clumps, with masses of typically a few M_{Jup} (see equation (56)). These groups do not agree in detail on the conditions necessary for fragmentation. *Boss* (2001, 2002a,b, 2007, 2008) finds that fragmentation is robust, i.e., fairly independent of many physical and numerical parameters, while *Mayer et al.* (2007) find that fragmentation requires somewhat higher disk masses and high mean molecular weight and that it is sensitive to some choices of numerical parameters. *Cai et al.* (2010), using an improved radiative cooling scheme similar to that of *Boley et al.* (2006) but with higher numerical resolution, have now evolved the same initial disk and perturbation as in one of the *Boss* (2007) simulations. No fragmentation occurs. The physical conditions are almost identical, so *Cai et al.* (2010) conclude that the difference really seems to be a matter of radiative algorithms.

The different outcomes are not due to inadequate resolution in the nonfragmenting calculations. For instance, some of the simulations in *Boley and Durisen* (2008) have extremely large grids, achieving resolutions of 0.025 AU over a radial range of about 13 AU. This resolution is more than adequate to detect fragmentation when conditions are artificially forced to satisfy equation (59).

Both *Boss* (2004) and *Mayer et al.* (2007) recognize that purely radiative cooling is not fast enough to satisfy equation (59). Both attribute the fragmentation they see to rapid cooling due to convective transport in the vertical direction and cite upwellings at the sound speed associated with spiral arms in their disks as evidence for thermal convection. However, numerical studies (*Boley et al.*, 2006, 2007) combined with analytic arguments (*Rafikov*, 2007) suggest that the grid-based codes that do not see fragmentation are fully capable of modeling thermal convection and that thermal convection in a disk is unlikely to carry more than tens of percent of the vertical flux, insufficient to reduce cooling times by orders of magnitude. *Boley and Durisen* (2006) explain the upwellings associated with spiral shocks, which are also seen in nonfragmenting simulations, as hydraulic jumps, a dynamic phenomenon very different from thermal convection.

The disagreement about cooling times leads to disagreement about how the disk metallicity affects fragmentation. *Boss* (2002b) studies variations in metallicity by varying $\bar{\kappa}$ up and down by factors of 10 relative to solar composition and finds that fragmentation is not sensitive to $\bar{\kappa}$. On the other hand, *Cai et al.* (2006), using simulation techniques similar to those in *Boley et al.* (2006), find that GIs are

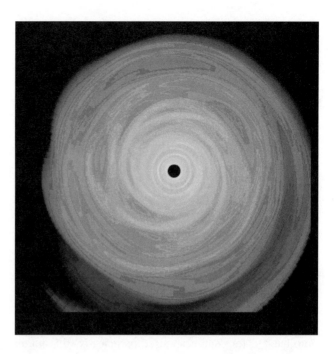

Fig. 8. Midplane density grayscale of a nonfragmenting and radiatively cooled disk with $M_{\star} = 0.5\ M_{\odot}$, $M_D = 0.07\ M_{\odot}$, and an initial $Q_{min} = 1.5$ after about 2000 yr. The box shown is 80 AU on a side. Adapted from *Boley et al.* (2006).

sensitive to metallicity in simulations where $\bar{\kappa}$ is varied only from 1/4 to 2 times the solar opacity. No simulation in *Cai et al.* (2006) produces fragmentation, but the GIs in the lower metallicity disks are stronger, which is expected if τ_{cool} is mostly determined by radiative processes.

Finally, *Boss* (2002a), *Cai et al.* (2008), and *Stamatellos and Whitworth* (2008) do agree with analytic arguments by *Matzner and Levin* (2005) that sufficiently strong external irradiation of a disk can weaken and suppress GIs. Strongly irradiated disks should not fragment. An easy way to estimate the necessary flux of irradiation $F_{irr} = \sigma_{rad} T_{irr}^4$ to stabilize a disk is to compute the irradiation temperature, T_{irr}, such that the resulting sound speed $c(T_{irr})$ satisfies the inequality in equation (54). At 5.2 AU, $T_{irr} \sim 75$ K suffices to suppress GIs altogether for $M_D \approx 0.1\ M_{\odot}$. No matter what else one might believe about disk instability in the region within tens of AU, it seems pretty clear that irradiation on this order or higher prevents planets from forming by GIs.

3.5.2. Enrichment of heavy elements. In the standard disk instability model, heavy element cores form *after* the bulk of the gas in the planet becomes a gravitationally bound equilibrium object. Although such protoplanets are likely to be rotating (see, e.g., *Mayer et al.*, 2004), let us approximate them here as hydrostatic equilibrium spheres, with the gas pressure gradient balancing gravity. As discussed in section 2.1, solid particles respond to gas pressure forces only indirectly through gas drag. Thus, in the absence of turbulence, small solid particles slowly sink to the center of a gas giant protoplanet at a drift speed $v_d \approx g\tau_f$ (see section 2.1), where g is the magnitude of the gravitational acceleration inside the protoplanet and τ_f is the friction time. For a newly formed 1 M_{Jup} protoplanet with a radius R $\approx$ 0.5 AU, approximately half λ_{cr} in equation (53) for our standard parameters, $g \approx GM_{Jup}/R^2 \simeq 2.3 \times 10^{-3}$ cm s^{-2}, and the internal temperature would be T ~ 100 K. Using equation (2), one gets a sedimentation time

$$\tau_{sed} \approx \frac{R}{v_d} \sim 6 \times 10^3 \, yr \left(\frac{1 \, g \, cm^{-3}}{\rho_{sp}}\right)\left(\frac{1 \, cm}{R_{sp}}\right) \quad (66)$$

in rough agreement with an estimate of *Boss* (1997). Detailed calculations by *Helled et al.* (2008), starting with micrometer-sized particles but including aggregation and growth, give similar times. The sedimentation timescale, τ_{sed}, is short compared with the Kelvin-Helmholtz contraction timescale, τ_{KH}, of the protoplanet, which is a few times 10^5 yr (*Helled et al.*, 2006). Turbulence slows sedimentation, but τ_{sed} for centimeter-sized particles is still shorter than τ_{KH}. Therefore, gas giants formed by disk instability probably are able to produce heavy element cores out of their initial complements of heavy elements. A 1 M_{Jup} mass of solar composition could form a ~6 $M_\oplus$ core.

Jupiter and Saturn are strongly enhanced in heavy elements compared with solar composition, as are some transiting exoplanets (e.g., HD 149026b). There are two ways that gas giants formed by disk instability could end up with complements of heavy elements that are fractionally larger than in their natal gas disk: accreting planetesimals from the surrounding disk after formation or forming in a gas clump already enhanced in solids above the original gas disk composition.

Detailed computations of planetesimal accretion (*Helled et al.*, 2006; *Helled and Schubert*, 2009) for a 1 M_{Jup} protoplanet formed by disk instability show that total masses of accreted planetesimals can vary anywhere from 1 to over 100 $M_\oplus$, depending upon the initial location of the planet and upon the assumed mass and radial density distribution of the parental disk. Large accretion rates of heavy elements are made possible by the appreciable starting radii of gas giants in the disk instability model coupled with their moderately long contraction timescale. Such calculations are of course sensitive to assumptions, and a great deal more work is needed, but it appears that heavy element enrichment of planets may not be a strong discriminator between disk instability and core accretion models.

The other intriguing possibility is that, at the time of fragmentation, spiral arms may already be enhanced in solids relative to the ambient disk due to gas drag effects. How might this occur? We have seen that solid particles respond to pressure gradient forces by drifting relative to the gas in a direction opposite to the gradient. For example, particles drift or sediment toward the pressure maximum at the disk midplane (section 2.1) or at the center of a protoplanet (as discussed above). The same thing applies to radial drift of solid particles in the protoplanetary disk: They drift radially in a direction opposite to the direction of the radial pressure gradient (*Weidenschilling*, 1977), which affects the gas orbital velocity, and concentrate at a radial pressure maximum (*Haghighipour and Boss*, 2003).

The physics in this case is complicated by angular momentum exchange. Equation (50) determines the equilibrium rotation rate of the gas, Ω, which includes the effect of the gas pressure term. Centrifugal balance against gravity in the absence of gas pressure requires a different angular speed for solids, Ω_s, such that $\Omega_s^2 = GM_\star/a^3 - g_a^D/a$. From equation (50) we have that if $\partial p/\partial a < 0$, then $\Omega < \Omega_s$, while if $\partial p/\partial a > 0$, then $\Omega > \Omega_s$. As a consequence, gas drag exerts a torque (per unit mass) on solid particles equal to $(\Omega-\Omega_s) a^2/\tau_f$. Then, by conservation of angular momentum, the radial drift velocity is $da/dt = (a/\pi)(\Omega-\Omega_s)(\tau_{rot}/\tau_f)$ (see section 2.8). In disk regions with a negative radial pressure gradient, an orbiting particle feels a headwind and drifts inward due to loss of orbital angular momentum. In disk regions with a positive radial pressure gradient, a solid particle feels a tailwind and drifts outward due to gain of orbital angular momentum. If there is a pressure maximum at some value of a, solids drift toward the pressure maximum from both sides and become concentrated relative to the gas.

The magnitude of the radial drift speed depends on the ratio τ_f/τ_{rot}. For small particles, $\tau_f/\tau_{rot} \ll 1$ (see equation (4)), and particles are essentially entrained with the gas. The head- and tailwinds are very small ($\Omega_s \simeq \Omega$), and so the radial drift is slow. Drift speeds increase with particle size in the small particle regime. Very large particles, for which $\tau_f/\tau_{rot} \gg 1$, orbit at angular speed $\Omega_s \neq \Omega$. They can experience considerable head- and tailwinds, but radial drift speeds are again small because of the long friction times. Drift speeds decrease with particle size in this regime. Maximum radial drift speeds (~$a|\Omega - \Omega_s|$) occur for $\tau_f/\tau_{rot} \sim 1$, corresponding to particles ~1 m in radius (*Weidenschilling*, 1977). Such particles can be swept into radial pressure maxima on timescales of only a few hundred years (*Haghighipour and Boss*, 2003).

It is not obvious *a priori* that this phenomenon generalizes to the dynamic environment of a GI-active disk with spiral waves, but, in fact, solid particles of this optimal size do drift into the pressure maxima of spiral arms on similarly short timescales (*Rice et al.*, 2003). This concentration of solids relative to gas in spiral arms can trigger GIs in the solids themselves (*Rice et al.*, 2006) and could even increase the tendency of the gaseous arms to fragment (*Mayer et al.*, 2007). In the latter case, a protoplanet formed by disk instability could initially be enhanced in heavy elements relative to the disk gas.

3.6. Bimodal and Hybrid Planet Formation Models

As argued analytically in section 3.4 and illustrated in Fig. 9, even codes that do not produce fragmentation in disks within tens of AU around solar-type stars confirm that fragmentation can and does occur in radially extended and massive disks, beyond about 100 AU (*Boley*, 2009; *Stamatellos and Whitworth*, 2009). Appropriate scaling of equation (56) to large orbital distance and very massive disks leads us to expect the pieces to be super Jupiters, brown dwarfs, or low-mass stars. This is exciting in light of recent discoveries, discussed elsewhere in this volume (see the chapter by Traub and Oppenheimer), that massive gas giant planets seem to exist at large radii around relatively young stars. Standard core accretion models do not form planets at large orbital distances because it may take too long to form a solid core, whereas violent GIs in an outer disk may occur during the early disk accretion phase for a range of plausible parameters (*Vorobyov and Basu*, 2006; *Clarke*, 2009; *Rafikov*, 2009).

Unless planets formed by core accretion can migrate or be scattered to large distances, it seems reasonable to conjecture that planets found in large orbits are formed *in situ* by disk instability. But, if nonfragmentation of GI-active disks in the inner tens of AU prevails as the consensus view, we still have to rely on core accretion to explain the bulk of gas giant exoplanets discovered so far, whose orbits are within a few AU of their stars. This leads to the notion (*Boley*, 2009) that gas giant planet formation may actually be bimodal, with different mechanisms dominating in different regions of protoplanetary disks. Even in this case, it is possible that gas-phase GIs play a role in core accretion by accelerating the formation of solid planetesimals through migration of solids into structures formed in the disk by GIs (e.g., *Haghighipour and Boss*, 2003; *Rice et al.*, 2004, 2006; *Durisen et al.*, 2005).

3.7. Summary of Disk Instability Models

Once the necessary conditions exist, namely $Q \lesssim 1.5\text{--}1.7$ and $\tau_{cool}/\tau_{rot} < f_{frag} \sim 1$, disk fragmentation into bound clumps of gas is a robust dynamic process that requires only a few to tens of orbital periods. Gas giants formed by disk instability can subsequently produce heavy element cores by sedimentation and can become enriched in heavy elements relative to the background gas disk by accreting planetesimals. Disk instability is expected to occur in the outer regions of extended massive disks during the accretion phase, and it may be the only way to understand the existence of high planetary-mass objects at large distances from their stars, such as Fomalhaut b (*Kalas et al.*, 2008) and HR 8799b, c, and d (*Marois et al.*, 2008).

Although there is still some controversy, analytic arguments and a number of simulations indicate that disks are unlikely to fragment within a few tens of AU, because the conditions of low Q and a low τ_{cool}/τ_{rot} cannot be simultaneously satisfied. The inclusion of stellar irradiation, which tends to stabilize these regions by maintaining a high Q, makes the problem worse. Disk instability is most likely to occur in the outer regions of disks relatively early in disk evolution, when disks are massive and still accreting from their protostellar cloud at a high rate. The ultimate fate and survivability of planets formed at such an early phase of disk evolution is unclear. The principle of parsimonious explanations makes appeal to more than one formation mechanism for gas giants unappealing.

4. HIGHLIGHTS AND OUTSTANDING QUESTIONS

4.1. Solar Planets and Exoplanets

Any theoretical model of giant planet formation has to provide a quantitative explanation for the existence of the four outer planets of the solar system. Although Jupiter's core mass remains uncertain, there is theoretical evidence that Saturn, Uranus, and Neptune have cores of 10–20 $M_\oplus$ (*Podolak et al.*, 1995; *Saumon and Guillot*, 2004). All four planets have massive hydrogen/helium envelopes, which represent most of the mass of Jupiter and Saturn and more than 10% of the mass of Uranus and Neptune (e.g., *Guillot*, 2005).

Core accretion models offer a natural explanation for both the formation of heavy element cores and the accretion of massive gaseous envelopes, such as those of Jupiter and Saturn. In this scenario, Uranus and Neptune reached their current masses in an environment that may have been deprived of gases because their cores took too long to form, and never underwent a runaway gas accretion phase. It seems unlikely, instead, that any of these planets formed via GIs in the early solar nebula, principally because theoretical arguments show that fragmentation may only occur beyond many tens of AU from a solar-type star.

Any formation theory also needs to explain the wide range of physical and orbital properties of giant planets

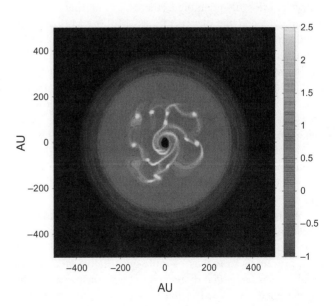

Fig. 9. Logarithmic surface density (in cgs units) for a radiatively cooled simulation of a fragmenting protoplanetary gas disk around a solar-mass star that is accreting at a rate of 10^{-4} $M_\odot$ per year. The dense clumps have masses ranging from 4 M_{Jup} to 14 M_{Jup}. Adapted from *Boley* (2009).

around other stars (see the chapter by Cumming). About 6% of Sun-like stars have giant planets orbiting within 4 AU. Jupiter-mass planets in orbit within several AU of Sun-like stars are more common than planets of several Jupiter masses, and substellar companions more massive than ~10 M_{Jup} are rare. Neptune-mass planets may be rather common compared to planets a few times the mass of Neptune. These observational results are broadly predicted by core accretion models, provided that gas in protoplanetary disks survives for at least a few million years. Disk instability models, instead, would predict an abundance of massive objects (≥10 M_{Jup}) relative to Jupiter-mass planets, since formation begins at very early times in massive disks. GIs, though, provide an appealing prospect for the formation of ~10 M_{Jup} objects observed at distances ≥100 AU from their stars, where core formation timescales are probably too long.

Exoplanet data show a strong correlation between stellar metallicity and frequency of giant planet detections (e.g., *Udry and Santos*, 2007): ~25% of stars whose metallicity is twice that of the Sun host a gas giant (within a few AU). This percentage reduces to only ~5% for solar-metallicity stars. Disk instability models do not seem to predict this trend. In fact, there is some evidence that GIs are stronger in lower-metallicity disks (*Cai et al.*, 2006). Core accretion models do predict this trend (*Kornet et al.*, 2005; *Mordasini et al.*, 2009), provided that the density of solids in a disk is proportional to the metallicity of the central star.

Planet searches around M dwarf stars ($M_\star \leq 0.5$ $M_\odot$) seem to suggest that the occurrence of giant planets is less likely than it is around more massive stars (*Johnson et al.*, 2007; *Eggenberger and Udry*, 2010). Although these surveys are still incomplete, there is mounting observational evidence that planets of mass M ≤ 0.1 M_{Jup} orbiting M dwarfs may be common (*Forveille et al.*, 2009). If confirmed by more complete statistics, a correlation between stellar masses and the masses of hosted planets may provide valuable information on the process of planet formation. According to core accretion models, gas giants orbiting low-mass stars should be rare (*Kennedy and Kenyon*, 2008), whereas Neptune-mass planets should be common (*Laughlin et al.*, 2004). According to disk instability models, if conditions are appropriate for disk fragmentation, the occurrence rate of giant planets should not depend strongly on the stellar mass (*Boss*, 2006).

Interpretations of observational data based on either of these mechanisms remain difficult and predictions are still rather qualitative. There is a number of unresolved issues that prevent us from making more quantitative predictions. Such issues are generally related to poor constraints on the initial conditions used in the formation models. Some of these open questions are outlined below.

What factors determined the final masses of Jupiter and other gas giants? The mass of Saturn may have been influenced by the presence of Jupiter and similar conditions may have operated in some exoplanet systems with multiple gas giants. However, a more general answer to this question will require a much deeper understanding of the physical conditions that existed in the solar nebula and other protoplanetary disks.

Are there peaks in the mass distribution function of massive planets? During the final phases of growth, giant planets are in a symbiotic relationship with their parent disks. Therefore, we need better constraints on masses, temperatures, and lifetimes of protoplanetary disks to make predictions on the relative abundances of giant planet masses.

What is the occurrence rate of gas giants and Neptune-mass planets as a function of the stellar mass? What is the percentage of red dwarfs that host massive planets? Currently, a handful of cases are known, as the two Jupiter-mass planets orbiting Gliese 876 (see *http://exoplanet.eu*), but conclusions cannot be drawn from such small-number statistics.

Is there a numerous population of gas giants orbiting at large separations from their host stars, as GIs models seem to predict? If so, how do their orbital properties and envelope composition differ from those of other known gas giants revolving on smaller orbits? What is the likelihood that they formed closer to the star (perhaps via core accretion) and were later scattered or migrated to larger distances?

4.2. Impact of Orbital Migration

Planetary cores of ~10 $M_\oplus$ that undergo type I migration may be subject to rapid orbital decay. One of the main advantages of disk instability models is that they would bypass this phase of planetary growth altogether and avoid this potential threat. Planets formed via GIs, instead, may be subject to prolonged periods of type II migration. But how relevant is planetary migration during planet formation?

Although a complete answer is not presently known, there is evidence that substantial migration may occur during the formation process. A significant fraction of multiplanet exoplanet systems are close to or in a mean-motion resonance (*Udry and Santos*, 2007). Since it is unlikely that such orbital configurations can be produced by *in situ* formation models, these systems may lend support to convergent migration hypotheses. Recent studies aimed at explaining the structure of the outer solar system (e.g., *Morbidelli et al.*, 2009) suggest that Jupiter and Saturn may have undergone convergent migration, as a result of which Saturn became temporarily captured into a mean motion resonance with Jupiter.

The presence of Neptune-mass planets and gas giants orbiting stars within a few tenths of AU may also constitute evidence of orbital migration, although alternative interpretations are possible. Planet-planet scattering and Kozai cycles (see the chapter by Lubow and Ida) can, in principle, drive gas giants very close to their host stars. However, while these last two mechanisms could produce a large misalignment between the stellar spin axis and the planet's orbit axis, disk-induced orbital migration would preserve spin-orbit alignment. Some of the transiting planets allow for the measurement of the sky-projected angle between stellar spins and orbit axes (see the chapter by Winn). Currently, about two-thirds of these measurements are consistent with a close spin-orbit alignment (*Simpson et al.*, 2010), suggesting that migration by tidal interaction with the parent disk drove these planets close to their stars.

The standard theory of orbital migration by disk-planet interactions seems able to reproduce the main features of the orbital period distribution of exoplanets (e.g., *Armitage*, 2007), which suggest that considerable migration may have occurred. Yet, the question of whether or not substantial migration is the norm can only be assessed once observations of long-period planets become available. A more complete distribution function of orbital periods can also provide quantitative indications on whether current estimates of the migration rates of Neptune-mass planets are too rapid and, if so, to what extent they are.

5. FUTURE PROSPECTS

To achieve a more comprehensive understanding of the formation of planetary systems, at least some of the questions highlighted in section 4 need to be addressed. We envision that answers to those questions can indeed be found in the near future, as further progress is made along three main avenues: data collection from the solar system; observations of star-forming regions, protoplanetary disks, and exoplanets; theoretical studies of star and planet formation.

Interplanetary spacecraft (e.g., Voyager, Galileo, and Cassini) have provided a wealth of information about the origins and evolution of the outer solar system. The Juno mission may finally reveal the mass of Jupiter's core by direct measurements of the planet's gravity field. Although it seems implausible that such information alone will provide definitive evidence for one or the other formation scenario, it is essential for refining interior models of the gas giant nearest to us.

To better constrain current estimates of sizes, masses, dust contents, and lifetimes of protoplanetary disks, it is important to know the detailed physical conditions that exist at various stages of a disk's evolution, starting from the latter phases of stellar formation. The observation of star-forming regions and disks surrounding young stars plays a pivotal role in establishing more comprehensive and accurate models of disk evolution. These models can be then applied in the context of increasingly sophisticated calculations of planet formation that include disk dispersal.

Space observatories, like the Hubble and Spitzer Space Telescopes, have provided valuable data that has significantly improved our knowledge of star-forming regions and planet-forming environments. Over the next decade, a lot more is expected to be learned about protoplanetary disks with both existing and new-generation instruments, such as the Atacama Large Millimeter Array (*Tarenghi*, 2008), the Stratospheric Observatory for Infrared Astronomy airborne observatory (*Gehrz et al.*, 2009), and the James Webb Space Telescope (*http://www.jwst.nasa.gov*). These facilities, among others, will improve in completeness the mass distribution function of Neptune-mass and larger planets. They will likely improve our knowledge on the relationships between stellar metallicity and dust content of young disks. They should also allow direct imaging of high-planetary-mass objects and low-mass brown dwarfs, orbiting at several tens of AU from stars, and provide precious data on the abundances of heavy elements and on the chemical compositions of their envelopes.

Groundbased interferometers (e.g., Very Large Telescope and Keck telescope) and astrometry space observatories, such as Gaia (*http://sci.esa.int/GAIA*) and the Space Interferometry Mission (*http://sim.jpl.nasa.gov*), will extend the orbital period distribution function of detected exoplanets toward longer-period orbits. Correlations between planet detection frequency and stellar metallicity and between stellar mass and planet mass are also expected to become more accurate and statistically significant. If a significant fraction of M dwarf stars possess gas giants, observational data should allow us to compare their orbital and physical properties with those of planets orbiting more massive stars. If a large population of high-mass planets is found at large separations, their atmospheric composition can be compared to those of transiting planets, whose number should also grow significantly thanks to space missions such as CoRoT (*http://smsc.cnes.fr/COROT/*), Kepler (*http://kepler.nasa.gov/*), and TESS (*http://space.mit.edu/TESS*). Within the next 5 to 10 years, we should be able to test the bimodal distribution hypothesis for gas giant formation based on statistical analysis of orbital properties, mass distribution functions, and envelope compositions.

The trend of exoplanet detections and characterization witnessed over the past decade will carry forward to the next decade. Theoretical studies need to tackle still-unsolved problems on the timescales and physical conditions required for gas giant formation. Core accretion models must be extended to include proper feedback on the protoplanetary disk and tidally induced orbital migration in a self-consistent manner. Disk instability models need to be refined in order to make quantitative predictions on survivability, final orbital radii, and final masses of planetary mass objects formed after disk fragmentation. As data of better and better quality become available, it may be challenging, from a theoretical standpoint, to combine all pieces in a coherent framework. Some of the pieces now regarded as important may be discarded and replaced by others. Yet, the prospects are in place to turn a blurred picture into one with much higher definition within the next 10 years.

Acknowledgments. We thank A. C. Boley, U. Gorti, and two anonymous reviewers for useful comments. G.D. acknowledges support from NASA Origins of Solar Systems (OSS) Program grants NNX08AH82G and NNX07AI72G. J.J.L. was supported by NASA OSS grant 811073.02.07.02.4. R.H.D. was partially supported during manuscript preparation by NASA OSS grants NNG05GN11G and NNX08AK36G. G.D. acknowledges computational resources provided by the NASA High-End Computing (HEC) Program through the NASA Advanced Supercomputing (NAS) Division at Ames Research Center.

REFERENCES

Alibert Y., Mordasini C., Benz W., and Winisdoerffer C. (2005) Models of giant planet formation with migration and disc evolution. *Astron. Astrophys., 434,* 343–353.

Armitage P. J. (2007) Massive planet migration: theoretical predictions and comparison with observations. *Astrophys. J., 665,* 1381–1390.

Armitage P. J., Livio M., and Pringle J. E. (2001) Episodic accretion in magnetically layered protoplanetary discs. *Mon. Not. R. Astron. Soc., 324,* 705–711.

Binney J. and Tremaine S. (1987) *Galactic Dynamics.* Princeton Univ., Princeton, New Jersey.

Bodenheimer P. and Pollack J. B. (1986) Calculations of the accretion and evolution of giant planets: the effects of solid cores. *Icarus, 67,* 391–408.

Boley A. C. (2009) The two modes of gas giant planet formation. *Astrophys. J. Lett., 695,* L53–L57.

Boley A. C. and Durisen R. H. (2006) Hydraulic/shock jumps in protoplanetary disks. *Astrophys. J., 641,* 534–546.

Boley A. C. and Durisen R. H. (2008) Gravitational instabilities, chondrule formation, and the FU Orionis phenomenon. *Astrophys. J., 685,* 1193–1209.

Boley A. C., Mejía A. C., Durisen R. H., Cai K., Pickett M. K., and D'Alessio P. (2006) The thermal regulation of gravitational instabilities in protoplanetary disks. III. Simulations with radiative cooling and realistic opacities. *Astrophys. J., 651,* 517–534.

Boley A. C., Durisen R. H., Nordlund Å., and Lord J. (2007) Three-dimensional radiative hydrodynamics for disk stability simulations: A proposed testing standard and new results. *Astrophys. J., 665,* 1254–1267.

Boss A. P. (1997) Giant planet formation by gravitational instability. *Science, 276,* 1836–1839.

Boss A. P. (2000) Possible rapid gas giant planet formation in the solar nebula and other protoplanetary disks. *Astrophys. J. Lett., 536,* L101–L104.

Boss A. P. (2001) Gas giant protoplanet formation: Disk instability models with thermodynamics and radiative transfer. *Astrophys. J., 563,* 367–373.

Boss A. P. (2002a) Evolution of the solar nebula. V. Disk instabilities with varied thermodynamics. *Astrophys. J., 576,* 462–472.

Boss A. P. (2002b) Stellar metallicity and the formation of extrasolar gas giant planets. *Astrophys. J. Lett., 567,* L149–L153.

Boss A. P. (2004) Convective cooling of protoplanetary disks and rapid giant planet formation. *Astrophys. J., 610,* 456–463.

Boss A. P. (2006) Rapid formation of gas giant planets around M dwarf stars. *Astrophys. J., 643,* 501–508.

Boss A. P. (2007) Testing disk instability models for giant planet formation. *Astrophys. J. Lett., 661,* L73–L76.

Boss A. P. (2008) Flux-limited diffusion approximation models of giant planet formation by disk instability. *Astrophys. J., 677,* 607–615.

Cai K., Durisen R. H., Michael S., Boley A. C., Mejía A. C., Pickett M. K., and D'Alessio P. (2006) The effects of metallicity and grain size on gravitational instabilities in protoplanetary disks. *Astrophys. J. Lett., 636,* L149–L152.

Cai K., Durisen R. H., Boley A. C., Pickett M. K., and Mejía A. C. (2008) The thermal regulation of gravitational instabilities in protoplanetary disks. IV. Simulations with envelope irradiation. *Astrophys. J., 673,* 1138–1153.

Cai K., Pickett M. K., Durisen R. H., and Milne A. M. (2010) Giant planet formation by disk instability: A comparison simulation with an improved radiative scheme. *Astrophys. J. Lett., 716,* 176.

Cameron A. G. W. (1978) Physics of the primitive solar accretion disk. *Moon and Planets, 18,* 5–40.

Chambers J. E. and Wetherill G. W. (2001) Planets in the asteroid belt. *Meteoritics & Planet. Sci., 36,* 381–399.

Clarke C. J. (2009) Pseudo-viscous modelling of self-gravitating discs and the formation of low mass ratio binaries. *Mon. Not. R. Astron. Soc., 396,* 1066–1074.

Clarke C. J., Harper-Clark E., and Lodato G. (2007) The response of self-gravitating protostellar discs to slow reduction in cooling time-scale: The fragmentation boundary revisited. *Mon. Not. R. Astron. Soc., 381,* 1543–1547.

Cuzzi J. N., Hogan R. C., Paque J. M., and Dobrovolskis A. R. (2001) Size-selective concentration of chondrules and other small particles in protoplanetary nebula turbulence. *Astrophys. J., 546,* 496–508.

D'Angelo G. and Lubow S. H. (2008) Evolution of migrating planets undergoing gas accretion. *Astrophys. J., 685,* 560–583.

D'Angelo G., Henning T., and Kley W. (2003a) Thermohydrodynamics of circumstellar disks with high-mass planets. *Astrophys. J., 599,* 548–576.

D'Angelo G., Kley W., and Henning T. (2003b) Orbital migration and mass accretion of protoplanets in three-dimensional global computations with nested grids. *Astrophys. J., 586,* 540–561.

Davis S. S. (2005) The surface density distribution in the solar nebula. *Astrophys. J. Lett., 627,* L153–L155.

Desch S. J. (2007) Mass distribution and planet formation in the solar nebula. *Astrophys. J., 671,* 878–893.

Durisen R. H., Cai K., Mejía A. C., and Pickett M. K. (2005) A hybrid scenario for gas giant planet formation in rings. *Icarus, 173,* 417–424.

Durisen R. H., Boss A. P., Mayer L., Nelson A. F., Quinn T., and Rice W. K. M. (2007) Gravitational instabilities in gaseous protoplanetary disks and implications for giant planet formation. In *Protostars and Planets V* (B. Reipurth et al., eds.), pp. 607–622. Univ. of Arizona, Tucson.

Eggenberger A. and Udry S. (2010) Detection and characterization of extrasolar planets through doppler spectroscopy. In *Physics and Astrophysics of Planetary Systems* (T. Montmerle et al, eds.), pp. 27–75. EAS Publications Series, Vol. 41, European Astronomical Society.

Eggleton P. P. (1983) Approximations to the radii of Roche lobes. *Astrophys. J., 268,* 368–369.

Epstein P. S. (1924) On the resistance experienced by spheres in their motion through gases. *Phys. Rev., 23,* 710–733.

Forgan D., Rice K., Stamatellos D., and Whitworth A. (2009) Introducing a hybrid radiative transfer method for smoothed particle hydrodynamics. *Mon. Not. R. Astron. Soc., 394,* 882–891.

Forveille T., Bonfils X., Delfosse X., Gillon M., Udry S., Bouchy F., Lovis C., Mayor M., Pepe F., Perrier C., Queloz D., Santos N., and Bertaux J. (2009) The HARPS search for southern extra-solar planets. XIV. Gl 176b, a super-Earth rather than a Neptune, and at a different period. *Astron. Astrophys., 493,* 645–650.

Gammie C. F. (1996) Layered accretion in T Tauri disks. *Astrophys. J., 457,* 355–362.

Gammie C. F. (2001) Nonlinear outcome of gravitational instability in cooling, gaseous disks. *Astrophys. J., 553,* 174–183.

Gehrz R. D., Becklin E. E., de Pater I., Lester D. F., Roellig T. L., and Woodward C. E. (2009) A new window on the cosmos: The Stratospheric Observatory for Infrared Astronomy (SOFIA). *Adv. Space Res., 44,* 413–432.

Guillot T. (2005) The interiors of giant planets: Models and outstanding questions. *Annu. Rev. Earth Planet. Sci., 33,* 493–530.

Goldreich P. and Tremaine S. (1980) Disk-satellite interactions. *Astrophys. J., 241,* 425–441.

Goldreich P. and Ward W. R. (1973) The formation of planetesimals. *Astrophys. J., 183,* 1051–1062.

Gray D. F. (1992) *The Observation and Analysis of Stellar Photospheres.* Cambridge Univ., Cambridge.

Greenzweig Y. and Lissauer J. J. (1992) Accretion rates of protoplanets. II — Gaussian distributions of planetesimal velocities. *Icarus, 100,* 440–463.

Haghighipour N. and Boss A. P. (2003) On pressure gradients and rapid migration of solids in a nonuniform solar nebula. *Astrophys. J., 583,* 996–1003.

Hayashi C. (1981) Structure of the solar nebula, growth and decay of magnetic fields and effects of magnetic and turbulent viscosities on the nebula. *Progr. Theor. Phys. Suppl., 70,* 35–53.

Helled R. and Schubert G. (2009) Heavy-element enrichment of a Jupiter-mass protoplanet as a function of orbital location. *Astrophys. J., 697,* 1256–1262.

Helled R., Podolak M., and Kovetz A. (2006) Planetesimal capture in the disk instability model. *Icarus, 185,* 64–71.

Helled R., Podolak M., and Kovetz A. (2008) Grain sedimentation in a giant gaseous protoplanet. *Icarus, 195,* 863–870.

Hollenbach D. J., Yorke H. W., and Johnstone D. (2000) Disk dispersal around young stars. In *Protostars and Planets IV* (V. Mannings et al. eds.), pp. 401–428. Univ. of Arizona, Tucson.

Horner J., Jones B. W., and Chambers J. (2010) Jupiter — friend or foe? III: The Oort cloud comets. *Intl. J. Astrobiol., 9,* 1–10.

Hubeny I. (1990) Vertical structure of accretion disks — A simplified analytical model. *Astrophys. J., 351,* 632–641.

Hubickyj O., Bodenheimer P., and Lissauer J. J. (2005) Accretion of the gaseous envelope of jupiter around a 5–10 Earth-mass core. *Icarus, 179,* 415–431.

Hueso R. and Guillot T. (2005) Evolution of protoplanetary disks: Constraints from DM Tauri and GM Aurigae. *Astron. Astrophys., 442,* 703–725.

Ida S. and Lin D. N. C. (2008) Toward a deterministic model of planetary formation. IV. Effects of type I migration. *Astrophys. J., 673,* 487–501.

Isella A., Carpenter J. M., and Sargent A. I. (2009) Structure and evolution of pre-main-sequence circumstellar disks. *Astrophys. J., 701,* 260–282.

Johnson B. M. and Gammie C. F. (2003) Nonlinear outcome of gravitational instability in disks with realistic cooling. *Astrophys. J., 597,* 131–141.

Johnson J. A., Butler R. P., Marcy G. W., Fischer D. A., Vogt S. S., Wright J. T., and Peek K. M. G. (2007) A new planet around an M dwarf: Revealing a correlation between exoplanets and stellar mass. *Astrophys. J., 670,* 833–840.

Kalas P., Graham J. R., Chiang E., Fitzgerald M. P., Clampin M., Kite E. S., Stapelfeldt K., Marois C., and Krist J. (2008) Optical images of an exosolar planet 25 light-years from Earth. *Science, 322,* 1345–1348.

Kennedy G. M. and Kenyon S. J. (2008) Planet formation around stars of various masses: The snow line and the frequency of giant planets. *Astrophys. J., 673,* 502–512.

Kippenhahn R. and Weigert A. (1990) *Stellar Structure and Evolution.* Springer-Verlag, Berlin.

Kokubo E. and Ida S. (2000) Formation of protoplanets from planetesimals in the solar nebula. *Icarus, 143,* 15–27.

Kornet K., Bodenheimer P., Różyczka M., and Stepinski T. F. (2005) Formation of giant planets in disks with different metallicities. *Astron. Astrophys., 430,* 1133–1138.

Kuiper G. P. (1951) On the origin of the solar system. *Proc. Natl. Acad. Sci., 37,* 1–14.

Laughlin G., Bodenheimer P., and Adams F. C. (2004) The core accretion model predicts few jovian-mass planets orbiting red dwarfs. *Astrophys. J. Lett., 612,* L73–L76.

Lin D. N. C. and Papaloizou J. (1986) On the tidal interaction between protoplanets and the protoplanetary disk. III — Orbital migration of protoplanets. *Astrophys. J., 309,* 846–857.

Lissauer J. J. (1987) Timescales for planetary accretion and the structure of the protoplanetary disk. *Icarus, 69,* 249–265.

Lissauer J. J. (1993) Planet formation. *Annu. Rev. Astron. Astrophys., 31,* 129–174.

Lissauer J. J., Hubickyj O., D'Angelo G., and Bodenheimer P. (2009) Models of Jupiter's growth incorporating thermal and hydrodynamic constraints. *Icarus, 199,* 338–350.

Lodato G. and Rice W. K. M. (2004) Testing the locality of transport in self-gravitating accretion discs. *Mon. Not. R. Astron. Soc., 351,* 630–642.

Lubow S. H. and D'Angelo G. (2006) Gas flow across gaps in protoplanetary disks. *Astrophys. J., 641,* 526–533.

Lynden-Bell D. and Pringle J. E. (1974) The evolution of viscous discs and the origin of the nebular variables. *Mon. Not. R. Astron. Soc., 168,* 603–637.

Marois C., Macintosh B., Barman T., Zuckerman B., Song I., Patience J., Lafrenière D., and Doyon R. (2008) Direct imaging of multiple planets orbiting the star HR 8799. *Science, 322,* 1348–1352.

Matzner C. D. and Levin Y. (2005) Protostellar disks: Formation, fragmentation, and the brown dwarf desert. *Astrophys. J., 628,* 817–831.

Mayer L., Quinn T., Wadsley J., and Stadel J. (2002) Formation of giant planets by fragmentation of protoplanetary disks. *Science, 298,* 1756–1759.

Mayer L., Quinn T., Wadsley J., and Stadel J. (2004) The evolution of gravitationally unstable protoplanetary disks: Fragmentation and possible giant planet formation. *Astrophys. J., 609,* 1045–1064.

Mayer L., Lufkin G., Quinn T., and Wadsley J. (2007) Fragmentation of gravitationally unstable gaseous protoplanetary disks with radiative transfer. *Astrophys. J. Lett., 661,* L77–L80.

Mejía A. C., Durisen R. H., Pickett M. K., and Cai K. (2005) The thermal regulation of gravitational instabilities in protoplanetary disks. II. Extended simulations with varied cooling rates. *Astrophys. J., 619,* 1098–1113.

Mizuno H. (1980) Formation of the giant planets. *Progr. Theor. Phys., 64,* 544–557.

Morbidelli A., Brasser R., Tsiganis K., Gomes R., and Levison H. F. (2009) Constructing the secular architecture of the solar system. I. The giant planets. *Astron. Astrophys., 507,* 1041–1052.

Mordasini C., Alibert Y., Benz W., and Naef D. (2009) Extrasolar planet population synthesis. II. Statistical comparison with observations. *Astron. Astrophys., 501,* 1161–1184.

Movshovitz N. and Podolak M. (2008) The opacity of grains in protoplanetary atmospheres. *Icarus, 194,* 368–378.

Murray C. D. and Dermott S. F. (2000) *Solar System Dynamics.* Cambridge Univ., Cambridge.

Nelson A. F., Benz W., Adams F. C., and Arnett D. (1998) Dynamics of circumstellar disks. *Astrophys. J., 502,* 342–371.

Nelson A. F., Benz W., and Ruzmaikina T. V. (2000) Dynamics of circumstellar disks. II. Heating and cooling. *Astrophys. J., 529,* 357–390.

Perri F. and Cameron A. G. W. (1974) Hydrodynamic instability of the solar nebula in the presence of a planetary core. *Icarus, 22,* 416–425.

Pickett B. K., Cassen P., Durisen R. H., and Link R. (1998) The effects of thermal energetics on three-dimensional hydrodynamic instabilities in massive protostellar disks. *Astrophys. J., 504,* 468–491.

Pickett B. K., Cassen P., Durisen R. H., and Link R. (2000) The effects of thermal energetics on three-dimensional hydrodynamic instabilities in massive protostellar disks. II. High-resolution and adiabatic evolutions. *Astrophys. J., 529,* 1034–1053.

Podolak M., Weizman A., and Marley M. (1995) Comparative models of Uranus and Neptune. *Planet. Space Sci., 43,* 1517–1522.

Pollack J. B., Hubickyj O., Bodenheimer P., Lissauer J. J., Podolak M., and Greenzweig Y. (1996) Formation of the giant planets by concurrent accretion of solids and gas. *Icarus, 124,* 62–85.

Rafikov R. R. (2005) Can giant planets form by direct gravitational instability? *Astrophys. J. Lett., 621,* L69–L72.

Rafikov R. R. (2007) Convective cooling and fragmentation of gravitationally unstable disks. *Astrophys. J., 662,* 642–650.

Rafikov R. R. (2009) Properties of gravitoturbulent accretion disks. *Astrophys. J., 704,* 281–291.

Rice W. K. M., Armitage P. J., Bate M. R., and Bonnell I. A. (2003) The effect of cooling on the global stability of self-gravitating protoplanetary discs. *Mon. Not. R. Astron. Soc., 339,* 1025–1030.

Rice W. K. M., Lodato G., Pringle J. E., Armitage P. J., and Bonnell I. A. (2004) Accelerated planetesimal growth in self-gravitating protoplanetary discs. *Mon. Not. R. Astron. Soc., 355,* 543–552.

Rice W. K. M., Lodato G., and Armitage P. J. (2005) Investigating fragmentation conditions in self-gravitating accretion discs. *Mon. Not. R. Astron. Soc., 364,* L56–L60.

Rice W. K. M., Lodato G., Pringle J. E., Armitage P. J., and Bonnell I. A. (2006) Planetesimal formation via fragmentation in self-gravitating protoplanetary discs. *Mon. Not. R. Astron. Soc., 372,* L9–L13.

Safronov V. S. (1969) *Evolution of the Protoplanetary Cloud and Formation of the Earth and Planets.* Moscow, Nauka Press. Translated in English as NASA TTF-677 (1972).

Saumon D. and Guillot T. (2004) Shock compression of deuterium and the interiors of Jupiter and Saturn. *Astrophys. J., 609,* 1170–1180.

Shakura N. I. and Syunyaev R. A. (1973) Black holes in binary systems. Observational appearance. *Astron. Astrophys., 24,* 337–355.

Simpson E. K., Pollacco D., Hebrard G., Gibson N. P., Barros S. C. C., Bouchy F., Collier Cameron A., Boisse I., Watson C. A., and Keenan F. P. (2010) The spin-orbit alignment of the transiting exoplanet WASP-3b from Rossiter-McLaughlin observations. *Mon. Not. R. Astron. Soc., 405,* 1867.

Stamatellos D. and Whitworth A. P. (2008) Can giant planets form by gravitational fragmentation of discs? *Astron. Astrophys., 480,* 879–887.

Stamatellos D. and Whitworth A. P. (2009) The properties of brown dwarfs and low-mass hydrogen-burning stars formed by disc fragmentation. *Mon. Not. R. Astron. Soc., 392,* 413–427.

Stevenson D. J. (1982) Formation of the giant planets. *Planet. Space Sci., 30,* 755–764.

Syer D. and Clarke C. J. (1995) Satellites in discs: Regulating the accretion luminosity. *Mon. Not. R. Astron. Soc., 277,* 758–766.

Tanaka H., Takeuchi T., and Ward W. R. (2002) Three dimensional interaction between a planet and an isothermal gaseous disk. I. Corotation and lindblad torques and planet migration. *Astrophys. J., 565,* 1257–1274.

Tarenghi M. (2008) The Atacama Large Millimeter/Submillimeter Array: Overview and status. *Astrophys. Space Sci., 313,* 1–7.

Tomley L., Cassen P., and Steiman-Cameron T. (1991) On the evolution of gravitationally unstable protostellar disks. *Astrophys. J., 382,* 530–543.

Toomre A. (1981) What amplifies the spirals. In *Structure and Evolution of Normal Galaxies* (S. M. Fall and D. Lynden-Bell, eds.), pp. 111–136. Cambridge Univ., Cambridge.

Udry S. and Santos N. C. (2007) Statistical properties of exoplanets. *Annu. Rev. Astron. Astrophys., 45,* 397–439.

Vorobyov E. I. and Basu S. (2006) The burst mode of protostellar accretion. *Astrophys. J., 650,* 956–969.

Ward W. R. (1997) Protoplanet migration by nebula tides. *Icarus, 126,* 261–281.

Weidenschilling S. J. (1977) Aerodynamics of solid bodies in the solar nebula. *Mon. Not. R. Astron. Soc., 180,* 57–70.

Wuchterl G. (1993) The critical mass for protoplanets revisited — Massive envelopes through convection. *Icarus, 106,* 323–334.

Wuchterl G., Guillot T., and Lissauer J. J. (2000) Giant planet formation. In *Protostars and Planets IV* (V. Mannings et al., eds.), pp. 1081–1109. Univ. of Arizona, Tucson.

Zapatero Osorio, M. R., Béjar V. J. S., Martín E. L., Rebolo R., Barrado y Navascués D., Bailer-Jones C. A. L., and Mundt R. (2000) Discovery of young, isolated planetary mass objects in the σ Orionis star cluster. *Science, 290,* 103–107.

Zhu Z., Hartmann L., Gammie C., and McKinney J. C. (2009) Two-dimensional simulations of FU Orionis disk outbursts. *Astrophys. J., 701,* 620–634.

Planet Migration

Stephen H. Lubow
Space Telescope Science Institute

Shigeru Ida
Tokyo Institute of Technology

Planet migration is the process by which a planet's orbital radius changes in time. The main agent for causing gas giant planet migration is the gravitational interaction of the young planet with the gaseous disk from which it forms. We describe the migration rates resulting from these interactions based on a simple model for disk properties. These migration rates are higher than is reasonable for planet survival. We discuss some proposed models for which the migration rates are lower. There are major uncertainties in migration rates due to a lack of knowledge about the detailed physical properties of disks. We also describe some additional forms of migration.

1. INTRODUCTION

Planet formation occurs in disks of material that orbit young stars (see chapters by Chambers and by D'Angelo et al.). At early stages of evolution, the disks are largely gaseous and have masses of typically a few percent or more of the stellar masses. At later times, after several 10^6 yr, the gas disperses and disks largely consist of solid material. Such disks last for timescales of ~10^8 yr. The interactions of a young planet with its surrounding disk affects the planet's orbital energy and angular momentum. One consequence of such interactions is that a planet may move radially through the disk (toward or away from the central star), a process called migration that is the topic of this chapter. In this chapter, we mainly concentrate on migration caused by gaseous disks. Although gaseous disks survive only a small fraction of the lifetime of the star and planet (~10^{10} yr), calculations of disk-planet interactions described in this chapter show them to be strong enough to have caused substantial migration. In fact, a major problem is that the effects of migration are typically predicted to be too strong to have allowed the formation of gas giant planets to complete. That is, the typical timescales for migration are found to be shorter than both the lifetimes of gaseous disks and the predicted formation timescales of gas giant planets in the so-called core accretion model of ~10^6 yr (see chapter by D'Angelo et al.). Migration models applied some simple disk structures predict that a forming planet should fall into (or very close to) the central star before it grows in mass to become a gas giant (*Ward*, 1997).

Early theoretical studies of planet migration, such as *Goldreich and Tremaine* (1980), understandably concentrated on the role of migration for the planets in our solar system. These studies suggested that there was substantial migration of Jupiter while it was immersed in the solar nebula. Migration should have occurred both at Jupiter's early stages of formation when it was a solid core and at later stages when it was a fully developed gas giant planet interacting with the solar nebula. The migration timescales were estimated to be much shorter than lifetime of the nebula, although the direction of migration was not determined. However, the evidence for migration was not apparent. The location of Jupiter is just outside the snow line, where conditions for rapid core formation are most favorable (see chapter by D'Angelo et al.). This situation seemed to suggest that Jupiter formed and remained near its current location. Therefore, a plausible conclusion was that migration did not play an important role for Jupiter and therefore perhaps for all planets.

Evidence for the importance of migration changed dramatically in 1995 with the discovery of the first giant planet, 51 Peg b, which has an orbital period of only about four days (*Mayor and Queloz*, 1995). Subsequent discoveries of many giant planets with orbital radii substantially smaller than Jupiter's added to the evidence (e.g., *Butler et al.*, 2006). Their existence suggested that migration had occurred, since giant planet formation close to a star is not likely to occur (*Bodenheimer et al.*, 2000; *Ida and Lin*, 2004).

We might crudely think of a planet orbiting in a gaseous disk around a star as being similar to a satellite in orbit about Earth. Atmospheric gas drag on the satellite causes it to lose angular momentum and spiral down (migrate inward) toward Earth. However, the disk-planet situation is somewhat different. The planet and disk are both in orbit about the star, and there is a relatively small difference between the rotational speeds of the planet and its neighboring gas. Although the velocity difference can produce an important level of gas drag for small mass solid objects, gas drag does not produce the dominant torque for objects as large as planets, as is discussed further in section 2.1. For disk-planet interactions, gravitational torques play the dominant role.

The gravitational interactions between a planet and a gaseous disk are difficult to analyze in a precise way. But a general picture has developed. The results depend somewhat on the disk structure. We consider the disk to be what is called an accretion disk. Accretion disks occur in many astronomical objects, such as binary star systems and active galactic nuclei. Such disks typically arise as gas flows toward a central body with some rotation about that body. Due to its angular momentum, the gas cannot fall directly onto the central object and instead forms a disk in near centrifugal balance. Observations of such systems, including disks around young stars, reveal evidence that the gas is not simply orbiting about the star, but also flowing inward (accreting) toward the central object. The inflow velocity within the disk is very small compared to the orbital velocity. But it is sufficient to reveal various observational signatures as a consequence of this mass flow toward and onto the central object. In particular, the inflow can result in the emission of radiation from the accreting gas as it moves deeper in the gravitational potential of the central object. In the case of interest to planet migration, the gaseous disk orbits about a young central star. These so-called protoplanetary disks are described in the chapter by Roberge and Kamp.

Turbulence within the disk is thought to be the main driver of the accretion. The disk does not rotate about the star at a constant rate with distance from the star. Instead, as a consequence of centrifugal balance, it undergoes differential rotation, and the gas rotation rate decreases with distance from the star. The gas orbits at (nearly) the so-called Keplerian orbital frequency at orbital radius r from the star of mass M_s given by

$$\Omega(r) = \sqrt{\frac{GM_s}{r^3}} \qquad (1)$$

In a simple model, the disk turbulence can be considered to cause an effective friction. Neighboring rings of gas in the disk interact by frictional forces as they rub against each other due to the differential rotation. This friction produces orbital energy losses and a torque that causes nearly all the material within the disk to move in (accrete), while angular momentum is transported outward (*Lynden-Bell and Pringle*, 1974). Instabilities within the disk are the cause of the turbulence. The most likely candidates for causing the instabilities are disk self-gravity (e.g., *Paczynski*, 1978; *Lodato and Rice*, 2004) and magnetic fields (*Balbus and Hawley*, 1991). Disk self-gravity instabilities are more likely to be important at the earliest stages of disk evolution, while it is more massive.

For small mass planets, the density structure of the disk is largely unaffected by the presence of the planet. This situation leads to the so-called Type I regime of planet migration. In this simple model, the disk density structure is determined by the disk turbulence. The disk differential rotation implies that gas that lies outside the orbit of the planet moves more slowly than gas inside the orbit of the planet. The planet disturbs the disk by its gravity as gas passes near it. The disk disturbances act back on the planet and cause a net torque to be exerted on the planet. This torque causes the planet to migrate. An analysis, such as in section 2.2, shows that the migration timescale (orbit decay timescale) for a 10 $M_\oplus$ (Earth mass) planet embedded in a disk that is somewhat similar to the minimum mass solar nebula (*Hayashi*, 1981) is only about 10^5 yr (see Fig. 1). Furthermore, the migration timescale varies inversely with planet mass. This result has the consequence that planets migrate even faster as they grow in mass. The shortness of this timescale poses a challenge to our understanding of migration, as discussed earlier.

For a sufficiently high mass planet, its tidal torques can overpower the viscous torques associated with disk turbulence, and the planet can alter the disk density in its vicinity. The result is that the planet opens a gap in the disk about its orbit. In this situation, we have the so-called Type II regime of planet migration. The disk-planet interactions in this case are still gravitational. But the gravitational torque on the planet is reduced because there is less material close to the planet compared to the Type I regime. In the limiting case that the planet mass is small compared to the disk mass, the planet moves inward like a test particle along with the accretion inflow of the gas. The disk, being much more massive than the planet, sets the migration rate of the planet to its inflow rate. The disk density near the gap region adjusts so that the torque on the planet matches the disk's inflow speed.

The mass required for the planet to open a gap, as well as the migration rate, depends on the level of disk turbulence. For typically adopted disk parameters, the transition to Type II migration occurs at a few tenths of a Jupiter mass. The migration timescales are about 10^5 yr. This timescale is also short compared to the disk lifetimes, but is much longer than it would have been if the planet had continued to undergo Type I migration (see Fig. 1). In this idealized Type II regime of high disk-to-planet mass ratio, the planet migration rate is independent of planet mass, since it is determined by the disk only. But for less extreme disk-to planet mass ratios, the inertia of the planet plays a role and the migration rate decreases somewhat with planet mass.

The determination of planet migration rates is not easy and various aspects remain controversial. There are several reasons for this. Although the interaction between a planet and its surrounding disk is gravitational, the interaction behaves in a way that is reminiscent of friction (dynamical friction). Gas that lies somewhat exterior (interior) to the planet's orbit rotates slower (faster) than the planet and provides an inward (outward) torque on the planet as a consequence of the interaction. The net torque therefore involves a competition between nearly equal and opposite torques. Accurate calculations of the inward and outward torques are then required to determine the migration rate and the direction of migration. Another complication is that the disk-planet interactions generally get stronger closer to the planet. The results therefore depend on the details of how these torques become limited near the planet.

Another issue involves the effects of material that lies very close to the planet. This so-called co-orbital region involves

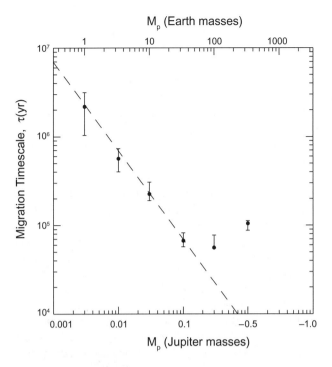

Fig. 1. Migration timescales vs. planet mass for a planet embedded in a three-dimensional disk of mass ~0.02M_s with surface density $\Sigma \propto r^{-1/2}$ and H/r = 0.05. The planets are on fixed circular orbits and have fixed masses and orbit a solar-mass star. The dots with error bars denote results of three-dimensional numerical simulations with the same disk parameters (*Bate et al.*, 2003). The dashed line plots equation (41) based on linear theory (*Tanaka et al.*, 2002). Above about 0.1 M_J, the planet opens a gap in the disk, Type I theory becomes invalid, and Type II migration occurs.

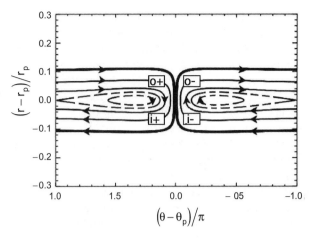

Fig. 2. Coorbital streamlines near a Saturn-mass planet that orbits a solar mass star, $M_p = 3 \times 10^{-4}\, M_s$. The origin is centered on the planet and corotates with it. Azimuthal angle θ about the star increases to the left. The outermost streamlines are at the edge of the coorbital region. Solid streamlines are the horseshoe orbits. Locations i∓ and o∓ label the positions near the encounter with the planet.

gas that does not circulate past the planet (see Fig. 2). Instead, the gas undergoes abrupt changes in its angular momentum due to encounters with the planet that take it from interior of the planet's orbit radius to exterior of the planet's orbital radius and vice versa. The torques resulting from this region are quite different from those involving more weakly perturbed material that lies outside this region. The two regions, the weakly deflected and strongly deflected regions, are dynamically quite different and require separate descriptions. Yet another issue is that the torques depend on the detailed conditions within the disk. The disk structural variations, i.e., how the disk density and temperature vary in radius, can play an important role in determining the torque. The detailed structural properties of disks around young stars are not well understood and could be complicated (e.g., *Armitage et al.*, 2001; *Terquem*, 2008). The resulting structures are likely more complicated than the simple power-law density distributions of the minimum mass solar nebula model. The level of disk turbulence in a protostellar disk influences the disk structure and the rate of planet migration. The amount of turbulence can vary with distance from the star and in time, but it is not well understood.

In section 2 we describe calculations of migration rates due to disk-planet interactions. Section 2.1 discusses the effects of gas drag. Section 2.2 describes the migration rate of a low-mass planet interacting with a disk, modeled simply as a set of ballistic particles that undergo mild or weak interactions with the planet. This leads to an estimate of the migration rate, but the approximations are not accurate enough to determine the direction of migration. Section 2.3 discusses the torques due to material that comes very close to the planet, torques in the co-orbital region. Section 2.4 discusses the more accurate determination of Type I migration rates for a gaseous disk and the factors affecting migration. Section 3 describes some of the outstanding issues: how planet migration may be slower than estimated in section 2, some effects in planet migration that are missing from the idealized models in section 2, some forms of migration not involving a gaseous disk, and some issues about the techniques used in carrying out numerical simulations. Section 4 describes some future prospects.

2. MIGRATION RATES DUE TO DISK-PLANET INTERACTIONS

2.1. Aerodynamic Gas Drag

Aerodynamic gas drag provides the dominant influence on the orbital evolution of low mass solids embedded in a gaseous disk that orbits a star (*Weidenschilling*, 1977; *Cuzzi and Weidenschilling*, 2006). The gas is subject to a radial pressure force, in addition to the gravitational force of the star. For a disk with smooth structural variations in radius, this force induces slight departures from Keplerian speeds on the order of $(H/r)^2 \Omega r$, where H is the disk thickness and Ω is the angular speed of the disk at radius r from the star. For a thin disk (H $\ll$ r) whose pressure declines in radius, as is typically expected, the pressure force acts radially outward. The gas rotation rate required to achieve centrifugal balance is then slightly below the Keplerian rate. Sufficiently high mass solids are largely dynamically decoupled from the gas and orbit at nearly the Keplerian rate, but are subject to drag forces from the more slowly rotating gas. The drag leads to their orbit decay. Both the drag force and the inertia increase with the size of the solid, R. The drag force on the object increases with its area ($\sim R^2$), while its inertia increases with its volume ($\sim R^3$). In the high-mass solid regime, the dominance of inertia over drag causes the orbital decay rate to decrease with object size as 1/R or with object mass as $1/M^{1/3}$.

The orbital migration rate due to gravitational interactions between an object and a disk increases with the mass of the object, as we will see in the next section. There is then a crossover mass above which the orbital changes due to disk gravitational forces dominate over those due to drag forces. For typical parameters, this value is much less than 1 $M_\oplus$ (*Hourigan and Ward*, 1984). Consequently, for the purposes of planet migration, we will ignore the effects of aerodynamic gas drag.

2.2. Torques Due to Mildly Perturbed Particles

As an initial description of gravitational disk-planet interactions, we consider the disk to consist of noninteracting particles, each having mass M much less than the mass of the planet M_p. The disk is assumed to extend smoothly across the orbit of the planet without a gap. The planet is then in the Type I regime of planet migration, as discussed in section 1. The model leads to insights about several issues concerning planet migration. In particular, it explains why the migration rate increases with planet mass and allows us to estimate the magnitude of the migration rate. This estimated migration rate is numerically close to what more detailed calculations reveal for simple disk structures. The approximations fall short of allowing a determination of whether the migration is inward or outward and the determination of the detailed dependence of the migration rate on disk properties, such as the density and temperature distributions. The treatment describes material that orbits about the star and undergoes a small deflection by the planet. It does not describe the effects of material that lies very close to the planet in the so-called co-orbital region, where the deflections are large (see Fig. 2). We will consider a model for that region in section 2.3.

In this model, the particles are considered to encounter and pass by the vicinity of a planet that is on a fixed circular orbit of radius a about the star of mass M_s. As a result of their interactions, the planet and disk exchange energy and angular momentum. This situation is a special case of the famous three-body problem in celestial mechanics. The tidal or Hill radius of the planet where planetary gravitational forces dominate over stellar and centrifugal forces is given by

$$R_H = a\left(\frac{M_p}{3M_s}\right)^{1/3} \quad (2)$$

where radius R_H is measured from the center of the planet.

Consider a particle that approaches the planet on a circular orbit about the star with orbital radius r sufficiently different from a to allow it to freely pass by the planet with a small deflection. This condition requires that the closest approach between the particle and planet $\sim |r-a|$ to be somewhat greater than a few times R_H. The solid line in Fig. 3 shows the path of a particle deflected by a planet whose mass is $10^{-6} M_s$ (corresponding to 0.3 $M_\oplus$ for a planet that orbits a solar-mass star), Hill radius $R_H \simeq 0.007a$, and orbital separation r – a $\simeq 3.5 R_H$. The planet and particle orbit counterclockwise in the inertial frame with angular speeds Ω_p and $\Omega(r)$, respectively. In the frame of the planet, the particle moves downward in the figure, since its angular speed is slower than that of the planet [$\Omega(r) < \Omega_p$ for r > a].

To estimate the angular momentum change of a particle like that in Fig. 3, we consider a Cartesian coordinate system that is centered on the planet and co-rotates with it. The x axis lies along a line between the star and planet and points away from the star. We consider in detail the fate of a particle that starts its approach toward the planet on a circular orbit of orbital radius r that lies outside the orbit of the planet, x = r – a > 0. The planet and the preencounter particle are on circular Keplerian orbits whose orbital frequencies are given by equation (1). We later discuss what happens to particles that begin their encounter at orbital radii smaller than the planet's (x < 0). The dashed line in Fig. 3 traces the path that the particle would take in the absence of the planet, while the solid line shows the path in the presence of the planet. In both cases, the particle paths are generally along the negative y direction. The two paths are nearly identical prior to the encounter with the planet (for y > 0).

The particle velocity in the frame that co-rotates with the planet is then approximately given by

$$\mathbf{v} \simeq r\left(\Omega(r)-\Omega_p\right)\mathbf{e}_y \simeq xa\frac{d\Omega}{dr}\mathbf{e}_y \simeq -\frac{3}{2}\Omega_p x\mathbf{e}_y \quad (3)$$

where x = r – a, $\Omega(r)$ is given by equation (1), and $\Omega_p = \Omega(a)$. Upon interaction with the planet, the particle is deflected slightly toward it. The particle of mass M experiences a force $F_x = -GMM_p x/(x^2 + y(t)^2)^{3/2}$. This force mainly acts over a time t when $|y(t)| \lesssim x$ and is $F_x \sim -GMM_p/x^2$. From equation (3), it follows that the encounter time $\Delta t \sim x/v \sim 1/\Omega_p$, on the order of the orbital period of the planet, independent of x.

To proceed, we apply the so-called impulse approximation, as employed by *Lin and Papaloizou* (1979). The approximation involves the assumption that the duration of the interaction is much shorter than the orbital period. Since the duration of the encounter is on the order of the orbital period, the approximation is only marginally satisfied. Consequently, the expressions we obtain cannot be determined

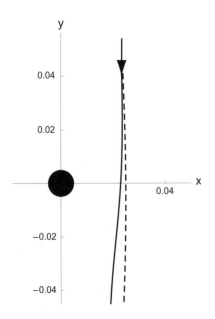

Fig. 3. Path of a particle that passes by a planet of mass $M_p = 10^{-6} M_s$ (0.3 $M_\oplus$ for a planet that orbits a solar-mass star). The coordinates are in units of the orbital radius of the planet a. The planet lies at the origin, while the star lies at (–1, 0). The dashed line follows the path that is undisturbed by the planet with x = 0.025a, while the solid line follows the path resulting from the interaction with the planet. In the frame of the planet, the particle moves in the negative y direction.

with high accuracy in this approximation. They contain the proper dependences on various physical quantities. But the approximation does not lead to the correct dimensionless numerical coefficients of proportionality for the angular momentum change of a particle. Therefore, we suppress the numerical coefficients in the analysis. An exact treatment in the limit of weak perturbations exerted by a small-mass planet is given in *Goldreich and Tremaine* (1982).

As a result of the encounter, the particle with x > 0 acquires an x (radial) velocity

$$v_x \sim \frac{F_x \Delta t}{M} \sim -\frac{GM_p}{x^2 \Omega_p} \quad (4)$$

The particle is then deflected by an angle

$$\delta \sim \frac{v_x}{v} \sim \left(\frac{M_p}{M_s}\right)\left(\frac{a}{x}\right)^3 \sim \left(\frac{R_H}{x}\right)^3 \quad (5)$$

after the encounter (see Fig. 3).

To determine the change in angular momentum of the particle, we need to determine its change in velocity along the azimuthal direction about the star, i.e., the same as the y direction near the planet. To determine this velocity change, Δv_y, we ignore the effects of the star during the encounter, and apply conservation of kinetic energy between the start and end of the encounter in the frame of the planet. The velocity magnitude v is then the same before and after the encounter, although the direction changes by angle δ. Since the particle in Fig. 3 moves in the negative y direction, its preencounter y velocity is –v and its postencounter y velocity is –v cos δ. We then have that the change of the velocity of the particle along the y direction is

$$\Delta v_y = -v \cos\delta + v \quad (6)$$

where $v \simeq 1.5\Omega_p x$ is the magnitude of the velocity before the encounter (see equation (3)). We assume that the perturbation is weak, $\delta \ll 1$, and obtain from equations (3), (5), and (6)

$$\Delta v_y \sim v\delta^2 \sim a\Omega_p \left(\frac{M_p}{M_s}\right)^2 \left(\frac{a}{x}\right)^5 \quad (7)$$

It then follows that the change in angular momentum of the particle is given by

$$\Delta J \sim Ma\Delta v_y \sim Ma^2\Omega_p \left(\frac{M_p}{M_s}\right)^2 \left(\frac{a}{x}\right)^5 \quad (8)$$

Figure 4 plots the orbital radius as a function of time for the particle plotted in Fig. 3, where t = 0 is the time of closest approach to the planet. Notice that after the encounter, the particle orbit acquires an eccentricity, as seen by the radial oscillations, and an increased angular momentum, as seen by the mean shift of the radius for the oscillations. Although the particle is initially deflected inward (negative x direction, toward the star), as expected by equation (4), it rebounds after the encounter to an increased time-averaged radius and higher angular momentum, as expected by equation (8).

Figure 5 plots the torque on a particle due to the planet as a function of time, where t = 0 is the time of closest approach between the particle and the planet. The solid curves are for particles that follow their actual paths (similar to the solid line in Fig. 3). The dashed curves are for particles that are made to follow the unperturbed paths (similar to the dashed line in Fig. 3). The net angular momentum change is the time-integrated torque. The torque on a particle along an unperturbed path (such as the dashed curve in Fig. 3) is antisymmetric in the time t. Consequently, there is no net change in angular momentum accumulated along this path. The change in angular momentum along the unperturbed path from t = –∞ to t = 0 is linear in the planet mass, since the force due to the planet is proportional to its mass. The departures from antisymmetry of the torque vs. time result in the net angular momentum change. These departures are a consequence of the path deflection (solid curve in Fig. 3).

For the case of the closer encounter plotted in Fig. 5a, the departures from the unperturbed case are substantial. But for a slightly larger orbit (Fig. 5b), the torque along the perturbed path differs only slightly from the torque along

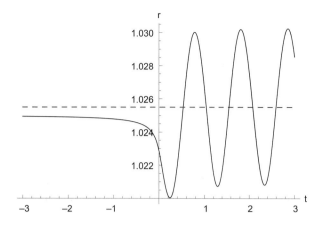

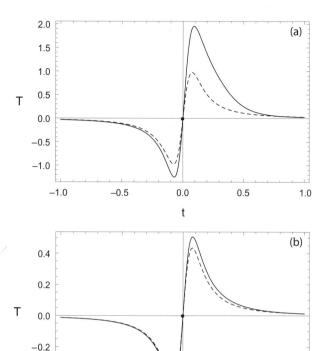

Fig. 4. The solid line plots the particle distance from the star r in units of a as a function of time in units of planet orbit periods for the particle that follows the perturbed path in Fig. 3. The particle passes the planet at time t = 0. Immediately after passage by the planet, the particle is deflected toward smaller radii, toward the planet, and acquires an eccentricity, as indicated by the radial oscillations. The dashed line plots the mean radius of these oscillations. Since the mean radius of the oscillations is larger than the initial orbital radius (compare dashed line with solid line at t < −1), the particle gained energy and angular momentum, as a consequence of its interaction with the planet.

Fig. 5. Torque on the particle of mass M that is normalized by 10^3 $M\Omega_p^2 a^2$ as a function of time in units of the planet's orbit period along the unperturbed (dashed lines) and perturbed paths (solid lines). The planet mass $M_p = 10^{-6} M_s$ (0.3 $M_\oplus$ for a planet that orbits a solar-mass star). **(a)** Particle having x = 0.02a, the case in Fig. 3; **(b)** particle with x = 0.03a.

the unperturbed path that integrates to zero. This behavior is a consequence of the steep decline of ΔJ with x in equation (8). For x greater than a few R_H, the angular momentum change acquired before the encounter is nearly equal and opposite to the angular momentum change at later times t > 0. The net angular momentum change is given by equation (8) and is quadratic in the planet mass. The quadratic dependence on M_p is a consequence of the deviations in torque between the perturbed and the unperturbed paths, since the angular momentum change along the unperturbed path is zero. These deviations involve the product of the linear dependence of the force on planet mass with the linear dependence of the path deflection on planet mass (equation (5)) and so are quadratic in M_p. This quadratic dependence of torque on planet mass is important. It leads to the conclusion we obtain later that the migration rate of a planet is proportional to its mass.

The particle in Fig. 3 gains angular momentum, as predicted by equation (8). The reason is that the path deflection occurs mainly after the particle passes the planet, i.e., for y < 0. The deflection takes the particle closer to the planet than would be the case along the unperturbed path. The planet then pulls the particle toward positive y, causing it to gain angular momentum. Just the opposite would happen for a particle with r < a, for x < 0 in equation (8). The particle would approach the planet in the positive y direction, be deflected toward the planet for y > 0, and be pulled by the planet in the negative y direction, causing it to lose angular momentum. The interaction behaves somewhat like friction.

The particle gains (loses) angular momentum if it moves slower (faster) than the planet. The angular momentum then flows outward as a result of the interactions. That is, for a decreasing angular velocity with radius ($d\Omega/dr < 0$) as in the Keplerian case, a particle whose orbit lies interior to the planet gives angular momentum to the planet, since the planet has a lower angular speed than the particle. The planet in turn gives angular momentum to a particle whose orbit lies exterior to it.

The planet in effect pushes material away from its orbit. A particle outside the orbit of the planet is forced outward (in a time-averaged sense) as it gains angular momentum, and a particle inside is forced inward as it loses angular momentum. From this point of view, the gravitational effects of the planet behave in a repulsive manner.

Figure 6 shows the results of numerical tests of equation (8). It verifies the dependence of ΔJ on x and M_p. Departures of the expected dependences (solid lines) occur when $x \simeq 3 R_H$. At somewhat smaller values of x, the particle orbits do not pass smoothly by the planet. Instead, they lie within the co-orbital region where they periodically undergo strong deflections, as seen in Fig. 2. We omit this region from current consideration and consider it later in section 2.3.

We now determine the torque on the planet for a set of particles that form a continuous disk (of zero thickness) that lies in the orbit plane of the planet with surface density (mass per unit area) Σ that we take to be constant in the region near the planet. The particle disk provides a flux of mass (defined to be positive) past the planet between x and $x + dx$

$$d\dot{M} \sim |v_y|\Sigma dx \sim \Sigma\Omega_p x dx \quad (9)$$

We evaluate the torque T_{out} on the planet due to disk material that extends outside the orbit of the planet from $r = a + \Delta r$ to ∞ or x from Δr to ∞, where $\Delta r > 0$. We use the fact that the torque the planet exerts on the disk is equal and opposite to the torque the disk exerts on the planet. We then have

$$T_{out} \sim -\int_{\Delta r}^{\infty} \frac{\Delta J}{M} \frac{d\dot{M}}{dx} dx \quad (10)$$

$$T_{out} = -C_T \Sigma \Omega_p^2 a^4 \left(\frac{M_p}{M_s}\right)^2 \left(\frac{a}{\Delta r}\right)^3 \quad (11)$$

where C_T is a dimensionless positive constant on the order of unity and ΔJ is evaluated through equation (8). The torque on the planet due to the disk interior to the orbit of the planet from $x = -\infty$ to $x = \Delta r$ with $\Delta r < 0$ evaluates to $T_{in} = -T_{out}$ or

$$T_{in} = C_T \Sigma \Omega_p^2 a^4 \left(\frac{M_p}{M_s}\right)^2 \left(\frac{a}{|\Delta r|}\right)^3 \quad (12)$$

The equations of motion for particles subject to only gravitational forces are time-reversible. We saw that particle in Fig. 3 gains angular momentum from its interaction with the planet. But, if we time-reverse the particle planet encounter in Fig. 3, we see that the eccentric orbit particle would approach the planet (both on clockwise orbits) and then lose angular momentum (apply $t \to -t$ in Fig. 4). What determines whether a disk particle gains or loses angular momentum? We have assumed that the particles always approach the perturber on circular orbits. The particles periodically encounter the gravitational effects of the planet. The closer the particle orbits to the planet, the smaller the relative orbital speeds and the longer the time between encounters. From equation (3), it follows that the time between encounters $\tau \simeq 2\pi a/v$ is estimated as

$$\tau \sim \frac{aP}{|x|} \quad (13)$$

where P is the planet orbital period. Since the encounters are close $|x| \ll a$, the time between encounters is long compared to the encounter time $\sim P$, i.e., $\tau \gg P$. But we

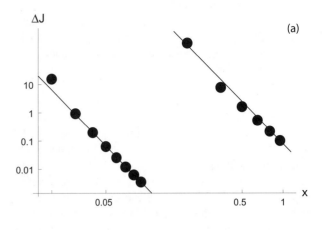

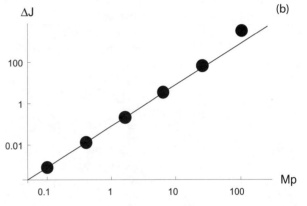

Fig. 6. Numerical test of equation (8) based on orbit integrations. (a) Log-log plot of $10^4 \Delta J/(Ma^2\Omega_p)$ as a function of preimpact distance from the planet's orbit x in units of a. The lower set of points is for a planet of mass $M_p = 10^{-6} M_s$ (0.3 $M_\oplus$ for a planet that orbits a solar-mass star). The upper set is for $M_p = 10^{-3} M_s$ (a Jupiter mass for a planet that orbits a solar-mass star). The solid lines are for $\Delta J \propto x^{-5}$ that pass through the respective rightmost points. (b) Log-log plot of $10^5 \Delta J/(Ma^2\Omega_p)$ as a function of planet mass in Earth masses for a planet that orbits a solar-mass star. The points are the results of numerical simulations for a fixed value of $x = 0.2a$. The solid line is for $\Delta J \propto (M_p/M_s)^2$ that passes through the leftmost point.

saw in Fig. 4 that the particles acquire eccentricity after the encounter. For this model to be physically consistent, we require that this eccentricity damp between encounters with the planet. The eccentricity damping produces an arrow of time for the angular momentum exchange process that favors circular orbits ahead of the encounter as shown in Fig. 3, resulting in a gain of angular momentum for particles that lie outside the orbit of the planet.

Equations (11) and (12) have important consequences. The torques on the planet arising from the inner and outer disks are quite powerful and oppose each other. This does not mean that the net torque on the planet is zero because we have assumed perfect symmetry across $r = a$. The symmetry is broken by higher-order considerations. For this reason, migration torques are often referred to as differential torques.

Since the torques are singular in Δr, they are dominated by material that comes close to the planet. Consequently, the asymmetries occur through differences in physical quantities at radial distances Δr from the planet. In a gaseous disk, the effects of both temperature and density variations in radius can play a role in the asymmetry, as well as asymmetries associated with the differential rotation of the disk. We cannot describe the asymmetries due to temperature in this model because temperature is not described by the ballistic particles. For this simple model of a (pressure-free) particle disk, we consider for example the effect of the density variation in radius that we have ignored up to this point, although the other asymmetries are important. It does not matter which quantity is considered for the purposes of obtaining a rough expression for the net torque.

We expand the disk density in a Taylor expansion about the orbit of the planet and obtain that

$$\Sigma_{out} - \Sigma_{in} \simeq 2\Delta r \frac{d\Sigma}{dr} \quad (14)$$

where Σ_{in} and Σ_{out} are the surface densities at $r = a - \Delta r$ and $r = a + \Delta r$, respectively. Consequently, for $|d\Sigma/dr| \sim \Sigma_p/a$, we expect that the sum of the inner and outer torques to be smaller than their individual values by an amount on the order of $\Delta r/a$. Similar considerations apply to variations in other quantities. That is, we have that the absolute value of the net torque T on the planet is approximately given by

$$|T| = |T_{in} + T_{out}| \quad (15)$$

$$\sim |T_{in}| \frac{|\Delta r|}{a} \quad (16)$$

$$\sim \Sigma \Omega_p^2 a^4 \left(\frac{M_p}{M_s}\right)^2 \left(\frac{a}{|\Delta r|}\right)^2 \quad (17)$$

The above equation for the torque (equation (17)) must break down for small Δr, in order to yield a finite result. For values of $|\Delta r| \lesssim 3 R_H$, the particles become trapped in closed orbits in the so-called co-orbital region (see Fig. 2). We exclude this region from current considerations, since the torque derivation we considered here does not apply in this region. In particular, the assumption that particles pass by the planet with a small deflection is invalid in this region. Using equation (17) with $|\Delta r| \sim R_H$, we obtain a torque

$$|T| \sim \Sigma \Omega_p^2 a^4 \left(\frac{M_p}{M_s}\right)^{4/3} \quad (18)$$

Another limit on Δr comes about due to gas pressure. One effect of gas pressure is to cause the disk to have a nonzero thickness H. The disk thickness is measured by its scale height out of the orbital plane. The disk scale height is determined by the force balance in the direction perpendicular to the orbit plane, the z or vertical direction. Gas pressure forces act to spread the gas in this direction, while gravitational forces act to confine it. We consider the vertical structure of an axisymmetric disk, unperturbed by a planet. There are no motions in the vertical direction and the disk is said to be in hydrostatic balance. The vertical hydrostatic balance condition can be written as

$$\frac{\partial p(r,z)}{\partial z} = \rho g_z \quad (19)$$

where $p(r, z)$ is the gas pressure, $\rho(r, z)$ is the gas density, and $g_z(r, z)$ is the vertical gravity. The vertical gravity is the z component of the gravitational force per unit disk mass due to the star that is equal to

$$g_z(r,z) = -\frac{GM_s z}{r^3} = -\Omega^2(r)z \quad (20)$$

where Ω is given by equation (1). For a vertically isothermal disk, a disk whose temperature and therefore sound speed $c(r)$ is independent of z, we have

$$p(r,z) = \rho(r,z)c^2(r) \quad (21)$$

Substituting equations (20) and (21) into equation (19), we obtain

$$\rho(r,z) = \rho(r,0)\exp\left(-\frac{z^2}{2H^2}\right) \quad (22)$$

and

$$H(r) = \frac{c(r)}{\Omega(r)} \quad (23)$$

where c is the gas sound speed that is a function of disk temperature. Disk thickness H is a measure of the importance of gas pressure. For conditions in protostellar disks, the disk thickness to radius ratio (disk aspect ratio) is typically $0.03 \lesssim H/r \lesssim 0.1$, corresponding to a gas sound speed of about 1.5 km s^{-1} at 1 AU from a solar-mass star.

For an axisymmetric disk of nonzero thickness, the surface density $\Sigma = \int_{-\infty}^{\infty} \rho dz$ evaluates to

$$\Sigma(r) = \int_{-\infty}^{\infty} \rho(r,z)dz = \sqrt{2\pi}H(r)\rho(r,0) \quad (24)$$

where $\rho(r,z)$ is defined by equation (22). The gas density is in effect smeared over distance H out of the orbit plane. Near the planet, the gas gravitational effects are then smoothed over distance H. Distance Δr is in effect limited to be no smaller than $\sim H$. For $\Delta r \sim H$, the torque expression (17) is then estimated as

$$|T| \sim \Sigma \Omega_p^2 a^4 \left(\frac{M_p}{M_s}\right)^2 \left(\frac{a}{H}\right)^2 \quad (25)$$

Which form of the torque applies (equation (18) or (25)) to a particular system depends on the importance of gas pressure. For $H < R_H$, gas pressure effects are small compared to the gravitational effects of the planet near the Hill radius. Consequently, we expect equation (18) to be applicable in the case of relatively weak gas pressure, $H < R_H$, and equation (25) to be applicable otherwise. For typically expected conditions in gaseous protostellar disks, it turns out that equation (25) is the relevant one for planets undergoing Type I migration.

The migration rate $\dot{r}_p/a = T/J_p$ for a planet with angular momentum J_p is then linear in planet mass, since T is quadratic while J_p is linear in planet mass. Therefore, the Type I migration rate increases with planet mass, as asserted in section 2.1. This somewhat surprising result that more massive planets migrate faster is in turn a consequence of the quadratic variation of ΔJ with planet mass in equation (8). This quadratic dependence occurs because the possible linear dependence of ΔJ on planet mass vanishes due to the antisymmetry of the torque as a function of time along the unperturbed particle path, as discussed after equation (8).

Based on equation (25) with typical parameters for the minimum mass solar nebula at the location of Jupiter $a = 5$ AU ($\Sigma = 150$ g cm^{-2}, $\Omega = 1.8 \times 10^{-8}$ s^{-1}, and $H = 0.05a$), we estimate the planetary migration timescale J_p/T for a planetary core of 10 $M_\oplus$ embedded in a minimum mass solar nebula as 4×10^5 yr. This timescale is short compared to the disk lifetime, estimated as several times 10^6 yr, or the Jupiter formation timescale of $\geq 10^6$ yr in the core accretion model. The relative shortness of the migration timescale is a major issue for understanding planet formation. Since migration is found to be inward for simple disk models, as we will see later, the timescale disparity suggests that a planetary core will fall into the central star before it develops into a gas giant planet. Research on planet migration has concentrated on including additional effects such as gas pressure and on improving the migration rates by means of both analytic theory and multidimensional simulations.

A more detailed analysis reveals that the torque in equation (25) does provide a reasonable estimate for the magnitude of migration rates in gaseous disks in the so-called Type I regime in which a planet does not open a gap in the disk. However, the derivation of the torque in this section is not precise enough to determine whether the migration is inward or outward (i.e., whether T is negative or positive). The density asymmetry about $r = a$ (see equation (14)) typically involves a higher density at smaller r, as in the case of the minimum mass solar nebula. This variation suggests that torques from the inner disk dominate, implying outward migration, as was thought to be the case in early studies. But this conclusion is incorrect. As we will discuss later, inward migration is typically favored, at least for simple disk models. We have not included the effects of gas temperature and gas pressure. A disk with gas pressure propagates density waves launched by the planet. The analysis in this section has only considered effects of material that passes by the planet. In addition, there are effects from material that lies closer to the orbit of the planet (see Fig. 2). This region can also provide torques. We have also assumed that the disk density is undisturbed by presence of the planet. Feedback effects of the disk disturbances and gaps in the disk can have an important influence on migration. Finally, there are other physical effects such as disk turbulence that should be considered. We will consider such effects in subsequent sections.

2.3. Coorbital Torques

Thus far we have considered torques that arise from gas that passes by the planet in the azimuthal direction (the y direction in Fig. 3) with a modest deflection. Gas that resides closer to the planet, $|r-a| \leq 3 R_H$, in the so-called coorbital region, does not pass by the planet. Instead, it follows what are called librating orbits in the corotating frame of the planet as seen in Fig. 2. We will consider librating orbits of particles that reach close to a planet that lies on a fixed circular orbit, the so-called horseshoe orbits. A particle at position o+ in Fig. 2 is in circular motion outside the orbit of the planet. It moves more slowly than the planet and approaches it. The particle in this case is more strongly perturbed by the planet than the more distant particle in Fig. 3. It gets pulled inward by the planet, causing it to change to a circular orbit interior to the planet's orbit once it reaches position i+. Since the particle is now moving faster than the planet, it moves away from it. In the process, the particle has executed a U-turn. At a later time, the particle approaches the planet from behind, at position i–, and the planet pulls the particle outward, causing it to make a second U-turn to position o–. The particle is then back to the initial outer radius that it started on before both encounters. The particle path is closed in the corotating frame of the planet.

With each change in angular momentum of the particle, there is an equal and opposite change in angular momentum of the planet. However, whatever angular momentum is gained by a particle as it changes from the inner radius r_i to outer radius r_o is lost when it later encounters the planet and shifts from r_o to r_i. The reason is that the particle follows a periodic orbit in the frame of the planet, so there is no change in angular momentum over a complete period of its motion. As a result, angular momentum changes do not grow over long timescales. At any instant, the largest angular momentum change possible over any time interval is that acquired in the last particle-planet encounter. Therefore, the time-averaged torque on the planet due to a particle drops to zero over timescales longer than the period of its motion. So although particles on closed horseshoe orbits can come close to the planet, the orbital symmetry limits the torque that they exert on the planet.

For a set of particles, there is a further type of torque cancellation that occurs. Recall that the time period between particle-planet encounters (the libration timescale) varies inversely with the particle-planet orbital separation as $1/x$ (see equation (13)). So even if a set of particles at different radii are initially lined up to encounter the planet at the same time and produce a large torque, they will eventually drift apart in azimuth (phase) and encounter the planet at different times. After a while, at a time when one particle gains angular momentum another will be losing it. This randomization, called "phase mixing," leads to a drop in torque over time. Figure 7 shows the angular momentum evolution of a set of 60 particles that initially lie on the negative x axis within the horseshoe orbit region exterior to the orbit of the planet ($r > a$). The torque (the derivative of the curve) shows both types of cancellations: the time rate of change of angular momentum oscillates due to the orbital symmetry and the amplitude of the oscillations decreases due to phase mixing.

A space-filling continuous set of small particles will be considered to represent coorbital gas. The disk in this case is dissipationless and pressureless, and the planet is on a fixed circular orbit. As is the case for a set of particles described above, the torque that the gas exerts on the planet approaches zero on timescales longer than the time for gas to make successive encounters with the planet.

We now consider the torque caused by gas whose density varies with radius with some imposed initial surface density distribution $\Sigma(r)$. The particle paths in Fig. 2 then become streamlines for the gas flow. Gas loses angular momentum in going from position o+ to i+. This angular momentum is continuously gained by the planet. Similarly, the planet continuously loses angular momentum from the gas that passes from i– to o–. The gas spends a relatively short time in transition between r_i and r_o compared to the time it spends between encounters given by equation (13), where $x = a - r_i$. We consider here the torque exerted on the planet on timescales longer than the transition timescale, but shorter than the timescale between encounters with the planet.

As gas in the coorbital region approaches from inside the orbit of the planet, it carries a mass flux (defined to be positive) that we estimate as

$$\dot{M}_i \sim \Sigma_i |\Omega_i - \Omega_p| r_i w_i \qquad (26)$$

where w_i denotes the radial extent (width) of the coorbital horseshoe region interior to the planet's orbital radius. Quantities Σ_i, Ω_i, and r_i denote the values of the surface density, angular velocity, and orbital radius, respectively, at position i–. This position lies at an intermediate radius within the horseshoe orbit region and interior to the planet's orbit. Gas at this position is executing circular motion, just ahead of an encounter with the planet. Similarly, for coorbital gas outside the orbit of the planet, the mass flux is estimated by

$$\dot{M}_o \sim \Sigma_o |\Omega_o - \Omega_p| r_o w_o \qquad (27)$$

where the subscripts denote the values at a point o+ that lies at an intermediate radius within the horseshoe orbit region outside the orbit of the planet.

Due to the orbital geometry, we have that region widths satisfy $w_i \sim w_o \sim r_o - a \sim a - r_i$. This gas interacts with the planet and undergoes a change in angular momentum per unit mass as it flows from position i– to o–

$$\Delta J = r_o^2 \Omega_o^2 - r_i^2 \Omega_i^2 \sim a \Omega_p^2 w > 0 \qquad (28)$$

where $w = w_o + w_i \sim r_o - r_i$. In going from o+ to i+, the gas undergoes an equal and opposite change in angular momentum, $-\Delta J$. The flows from outside and inside the orbit of the planet impart a torque on the planet. This torque is equal and opposite to the rate of change of angular momentum of gas resulting from the two contributing mass fluxes

$$T_{co} = (\dot{M}_o - \dot{M}_i) \Delta J \qquad (29)$$

$$\sim \dot{M}_o \left(1 - \frac{\dot{M}_i}{\dot{M}_o}\right) a \Omega_p^2 w \qquad (30)$$

We need to be careful in evaluating the ratio of the mass fluxes in the above equation, since departures from unity are critical. From equations (26) and (27), the ratio of the mass fluxes is given by

$$\frac{\dot{M}_i}{\dot{M}_o} = \frac{\Sigma_i}{\Sigma_o} \left(\frac{|\Omega_i - \Omega_p| r_i w_i}{|\Omega_o - \Omega_p| r_o w_o}\right) \qquad (31)$$

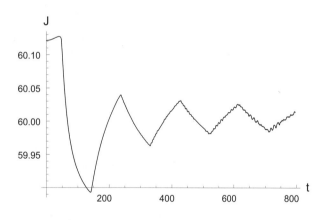

Fig. 7. Total angular momentum for a set of 60 particles, each of mass M, on horseshoe orbits in units of $Ma^2\Omega_p$ as a function of time in units of the planet orbit period in a star-planet system with $M_p = 3 \times 10^{-6}$ M_s (0.3 $M_\oplus$ for a planet that orbits a solar-mass star). The particles start at $t = 0$ distributed between $r = a + R_H/60$ to $r = a + R_H$ (with $R_H \simeq 7 \times 10^{-3} a$) along the star-planet axis 180° from the planet. The changes in angular momentum cause a torque to be exerted on the planet. The torque (the time derivative of J) oscillates and declines in time as angular momentum becomes more constant in time because the particles undergo phase mixing on the libration timescale ~150 planet periods.

Ward (1991) showed that the term in parentheses on the righthand side is equal to B_o/B_i, where $B(r)$ is the Oort constant defined through $2rB = d(r^2\Omega)/dr$ and $B(r) = \Omega(r)/4$ for a Keplerian disk. We then obtain that

$$\frac{\dot{M}_i}{\dot{M}_o} = \frac{\Sigma_i}{\Sigma_o}\frac{B_o}{B_i} \quad (32)$$

Using equation (27), we then approximate $\dot{M}_o$ appearing in the first term on the righthand side of equation (30) as

$$\dot{M}_o \sim \Sigma\Omega_p w^2 \quad (33)$$

Applying equations (30), (32), and (33), we then obtain an expression for the coorbital torque as

$$T_{co} \sim \Sigma\Omega_p^2 aw^3 \left(\frac{\Delta\Sigma}{\Sigma} - \frac{\Delta B}{B}\right) \quad (34)$$

$$\sim \Sigma\Omega_p^2 w^4 \frac{d\log(\Sigma/B)}{d\log r} \quad (35)$$

where $\Delta\Sigma = \Sigma_o - \Sigma_i$ and $\Delta B = B_o - B_i$.

Quantity B/Σ is sometimes called the vortensity, since B is half the vorticity, i.e., half the curl of the unperturbed velocity $r\Omega e_\theta$. The coorbital torque then depends on the gradient of the vortensity. Recall that in this situation, the torque is not permanent. It oscillates and decays to zero, due to phase mixing, as discussed above. The coorbital torque drops to zero or is said to be saturated on long timescales. The gas density in the coorbital region becomes modified over time so that the vortensity becomes constant and the torque approaches zero.

Gas that flows on the closed streamlines shown in Fig. 2 can impart a change in angular momentum on the planet that is at most about equal to the gas angular momentum of the region times its fractional width, w/a (i.e., $\Sigma\Omega_p a^2 w^2$). This angular momentum change is due to the rearrangement of gas within the horseshoe orbit region from some initial state. In this model, gas is trapped within the coorbital region and cannot exchange angular momentum with the large reservoir in the remainder of the disk. Maintaining a coorbital torque over long timescales requires the action of a process that breaks the symmetry of the streamlines and permits exchange with the remainder of the disk. One such process involves the disk turbulent viscosity. The disk viscosity acts to establish a characteristic density distribution, as discussed in section 1. Sufficiently strong turbulence can overcome the effects of phase mixing that saturate the torque. The torque is then given by equation (35), where the density distribution $\Sigma(r)$ is determined by turbulent viscosity in the viscous accretion disk. The vortensity gradient is then maintained at a nonzero value. The angular momentum changes within the coorbital region are transferred to the remainder of the disk through torques associated with frictional stresses caused by disk turbulent viscosity (viscous torques). In this way, a steady-state torque can be exerted on the planet over timescales much longer than the libration timescale $\sim aP/w$.

We compare the coorbital torque with the net torque due to particles outside the coorbital region in a pressureless disk, as we discussed in section 2.2. Taking $w \sim R_H$ in equation (35), we see that the unsaturated coorbital torque [taking $|d\log(B/\Sigma)/d\log r| \sim 1$ and nonzero] is comparable to the torque due to particles outside the coorbital region in equation (18). It can be shown that this is also true for a disk where pressure effects are important, where $H > R_H$. For such a disk, the coorbital torque is generally on the same order as the net torque from the region outside the coorbital zone (equation (25)). The direction of the coorbital torque contribution depends on the sign of the vortensity gradient. For the minimum mass solar nebula model, $\Sigma \propto r^{-3/2}$ and $B \propto r^{-3/2}$. Consequently, the coorbital torque is zero in that case. With more general density distributions of the form

$$\Sigma \propto r^{-\beta} \quad (36)$$

and for smaller values of $\beta < 3/2$, the coorbital torque on the planet is positive.

We have discussed how the disk turbulent viscosity can cause a torque to be exerted in the coorbital region over long timescales. But if the amount of viscosity is too small, the phase mixing process discussed above can dominate and cause the torque to saturate. For the turbulent viscosity to prevent torque saturation, the timescale for turbulence to affect the density in the coorbital region needs to be shorter than the libration timescale on which the torque would drop in a nonviscous disk. This criterion imposes a constraint on the level of viscosity required. So although they are potentially powerful, coorbital torques are not guaranteed to play a role in planet migration in all cases.

Although the coorbital torque is on the same order as the net torque due to material outside the orbital zone, more detailed calculations as described in the next section show it typically does not dominate the total torque, at least for simple disk conditions.

Although the description here is self-consistent, coorbital torques are less well understood than the torques involving noncoorbital material we considered in section 2.2. They are more complicated because they are subject to saturation. The asymmetry in the coorbital region that prevents torque saturation can in principle be produced by effects other than turbulence, such as the migration of the planet itself (see section 3.2). Furthermore, recent simulations suggest that the thermodynamic state of the gas, not considered in this model, can modify the coorbital torque and possibly result in outward migration, as is discussed in section 3.2.

2.4. Type I Migration

The torques arising from gas that lies somewhat inside the orbit of the planet $r < a$ (but outside the coorbital region) act to cause outward migration, while torques arising from gas outside the orbit of the planet act to cause inward migration, as discussed in section 2.2. The net migration torque was estimated in equation (25). But due to the inaccuracy of the approximations, the sign of the torque and its dependence on gas properties, such as the density and temperature distributions, could not be determined. More accurate analytic calculations of the torque involve solving the fluid equations for gas subject to the gravitational perturbing effects of the planet. Both the weakly perturbed region (section 2.2) and coorbital region (section 2.3) are analyzed. The calculations described in this section assume a simple disk model. By a simple model we mean that the disk has smooth disk density distribution, in the absence of the planet, and a smooth and fixed temperature distribution, even in the presence of the planet.

Consider a cylindrical coordinate system centered on the star (r, θ, z). In these calculations, the perturbing potential of a circular orbit planet is expanded in a Fourier series as

$$\Phi(r,\theta,t) = \sum_m \Phi_m(r,a)\cos\left[m(\theta - \Omega_p t)\right] \quad (37)$$

where m is a nonnegative integer, and Ω_p is the orbital frequency of the planet. Such an expansion is possible because the potential is periodic in azimuth and periodic in time with the planet's orbital period. In addition, at fixed radius r (and fixed a), the potential depends only on the relative azimuth between that of a point θ and that of the planet $\Omega_p t$, i.e., $\theta - \Omega_p t$. For each azimuthal wave number m, the gas response is calculated by means of linear theory. There is a torque associated with each m value due to the effects of the density perturbation for each m. The sum of these torques determine the net torque on the planet.

The results show that the gas response is dominated by the effects of resonances (*Goldreich and Tremaine*, 1979, 1980). There are two types of resonances that emerge: Lindblad resonances and corotational resonances. They occur in the regions described in sections 2.2 and 2.3, respectively. Such resonances also arise in the theory of the spiral structure of galaxies and planetary rings.

The gas response to a particular Fourier potential component m in equation (37) is strong at particular radii r_m where the Lindblad resonance condition is satisfied. The Lindblad resonances occur for gas that periodically passes by the planet. They correspond to the so-called mean-motion resonances of particles in celestial mechanics. For a given m value, the forcing frequency experienced by a particle is the time derivative of the argument of the cos function in equation (37) along a particle's path θ(t). The forcing frequency is then given by $m(\Omega(r) - \Omega_p)$. Consider a particle in a circular orbit about a central mass. If the particle is momentarily slightly perturbed, it oscillates radially at a frequency called the epicyclic frequency, often denoted by κ(r). This frequency is the free-oscillation frequency of the particle, analogous to the free-oscillation frequency of a spring in a simple harmonic oscillator. For a Keplerian disk, the radial frequency κ(r) is equal to the circular frequency Ω(r). (This is why noncircular Keplerian orbits are closed ellipses in the inertial frame.) Lindblad resonances occur wherever the absolute value of the forcing frequency matches the free-oscillation frequency. For a Keplerian disk, the Lindblad resonance condition, $|m(\Omega - \Omega_p)| = \kappa$, simplifies to

$$\Omega(r_m) = \frac{m\Omega_p}{m \mp 1} \quad (38)$$

where Ω(r) is given by equation (1).

For each m value there are two Lindblad resonances. An outer Lindblad resonance (OLR) occurs outside the orbital radius of the planet, for which the plus sign is taken in the denominator of equation (38), and an inner Lindblad resonance (ILR), for which the minus sign is taken (see Fig. 8). At each Lindblad resonance, spiral waves are launched that propagate away from the resonance (see Fig. 9). These waves are similar to acoustic waves, but are modified by the disk rotation. One can associate an angular momentum and energy with the waves. They carry energy and angular momentum from the planet, resulting in a torque on the planet. Various processes cause these waves to damp as they propagate. As the waves damp, their angular momentum is transferred to the disk. This situation is reminiscent of the case of ocean waves. They are generated by wind far from land, but undergo final decay when they break at the shore and can cause irreversible changes. Similarly, the waves in disks can modify the underlying disk density distribution Σ(r) and open gaps as they damp, although the disk turbulence can wash out these effects.

At the corotation resonance in a Keplerian disk, the condition is simply that

$$\Omega(r_m) = \Omega_p \quad (39)$$

or $r_m = a$ for all m. That is, the corotation resonance occurs at the orbital radius of the planet. It lies in the region of orbits shown in Fig. 2. The corotation resonance also creates disturbances in the disk, but the disturbances do not propagate. Instead they are evanescent and remain trapped within a radial region whose size is on the order of the disk thickness H.

The analytic calculations determine a torque for each m value at the inner and outer Lindblad resonances and the corotation resonance. The strength of the Lindblad torques increases for resonances that lie closer to the planet. The closer resonances occur for higher m values. But due to effects of pressure, the resonance condition in equation (38) breaks down close to the planet and the torque reaches a maximum value for $m_{cr} \sim r/H$ where the resonance is at a distance on the order of the disk thickness from the planet (see Fig. 8). Three-dimensional effects also weaken the torque close to the planet. The reduction arises because the gas is spread over

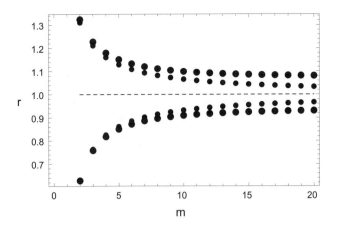

Fig. 8. Radius of a Lindblad resonance in units of a as a function of azimuthal wave number m. Points below (above) the dashed line are for inner (outer) Lindblad resonances. The dashed line is the location of the planet. The two sets of smaller dots are for a Keplerian disk with radii given by equation (38). The resonances get closer to the planet with increasing m. The two sets of larger dots account for azimuthal pressure effects in a disk with thickness to radius ratio H/r = 0.1. With these pressure effects included, the resonances maintain a fixed separation from the planet at high m.

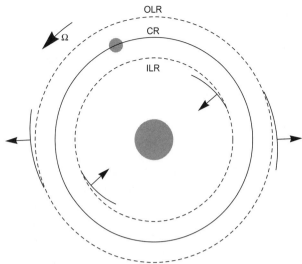

Fig. 9. Schematic of acoustic (pressure) wave propagation in a gas disk involving an inner Lindblad resonance (ILR), outer Lindblad resonance (OLR), and corotation resonance (CR). Spiral waves launched at Lindblad resonances propagate away from the orbit of the planet. The region near the CR, between ILR and OLR, is evanescent (nonpropagating).

the disk thickness H (as was discussed in section 3.2). The critical m value of m_{cr} is sometimes called the torque cutoff. This reduction of torque for $m > m_{cr}$ implies that the sum of the torques over all m values remains finite. In this way, these calculations avoid the singularity that we encountered in equation (17) involving Δr.

The results of these calculations show that planets have a definite tendency to migrate inward (*Ward*, 1997). For a given m value, the torque is stronger at the OLR than the ILR in a disk, even if we ignore the asymmetries caused by density and temperature variations. There are several contributions to this density- and temperature-independent effect. One such contribution can be seen by noticing that for a given m, the OLRs lie slightly closer to the planet's orbital radius than ILRs in equation (38). This asymmetry, as well as other density- and temperature-independent effects, enhances the inward migration effects caused by OLRs over the outward migration effects that are caused by ILRs. With simple disk structures (smooth disk density distributions and smooth, fixed temperature distributions), such effects often dominate the overall torque on the planet for typical density and temperature distributions.

Density and temperature effects can also play an important role in determining the Lindblad torque. The surface density distribution in radius influences the torque balance in at least two ways. It affects the amount of gas at each resonance and it affects the radial pressure force. These two effects typically act in opposition to each other and partially cancel. A declining surface density distribution with radius, as is typical, places more gas at the ILRs than the OLRs.

This effect provides an outward torque contribution, since the ILR torques are enhanced. Pressure effects typically cause the gas at orbital radius close to a to rotate a little more slowly than the planet, as discussed in section 2.1. Although the primary effect here is gravitational and not gas drag as in section 2.1, the gravitational torque on the planet again produces a qualitatively similar effect as drag and acts to cause inward migration. This effect can also be understood in terms of a slightly decreased angular velocity Ω(r) (below the Keplerian rate) in equation (38) that causes both the ILRs and OLRs to move inward. The ILRs are then shifted away from the planet, while the OLRs are shifted toward it. The effect of pressure is to favor the OLRs, which contribute to inward migration.

The temperature distribution in radius also has multiple effects on the Lindblad torque balance. It affects the gas radial pressure force in a similar way that the density distribution does, as just described. The pressure effects of a radially declining temperature then contribute to inward migration. Temperature variations also modify the torque cutoff $m_{cr} \sim a/H$, since disk thickness H depends on the gas sound speed and therefore temperature (see equation (23)). The OLRs, being at a lower temperature (smaller H) than the ILRs, have a higher cutoff m_{cr}. This effect favors the OLRs and its inward torques. Both effects of a declining temperature with radius, also generally expected, contribute to inward migration.

Detailed three-dimensional linear analytic calculations of the Type I migration rates have been carried out by *Tanaka et al.* (2002). They assumed that the gas sound speed is strictly constant in radius. That is, the gas temperature is assumed to be unaffected by the planetary perturbations. For the case of saturated (zero) coorbital corotation torques, where only differential Lindblad torques are involved, the torque on the planet is given by

$$T = -(2.34 - 0.10\,\beta)\,\Sigma(a)\,\Omega(a)^2\,a^4\left(\frac{M_p}{M_s}\right)^2\left(\frac{a}{H}\right)^2 \quad (40)$$

where β is given by equation (36). [A higher sensitivity to density gradients (more negative coefficient of β) was found in the analysis by *Menou and Goodman* (2004).] The torque on the planet resulting from the action of both Lindblad and (unsaturated) coorbital corotation resonances is given by

$$T = -(1.36 + 0.54\,\beta)\,\Sigma(a)\,\Omega(a)^2\,a^4\left(\frac{M_p}{M_s}\right)^2\left(\frac{a}{H}\right)^2 \quad (41)$$

These migration rates are consistent with the estimate in equation (25). Numerical values for migration timescales based on equation (41) are plotted in Fig. 1. As mentioned above, temperature variations can also contribute to planet migration, but are not included in this treatment. Other results [such as the two-dimensional calculation by *Ward* (1997)] suggest that the temperature variation should provide an additional term within the first parenthesis on the righthand sides of equations (40) and (41), i.e., $\sim\chi$, where

$$T(r) \propto r^{-\chi} \quad (42)$$

is the temperature distribution with radius (not to be confused with torque). We typically expect in a disk that $0 < \beta < 1.5$ and that $0.5 \leq \chi < 0.75$. Consequently, the migration is predicted to be inward and on the order of the rate that would be obtained even if density and temperature gradients (β and χ) were ignored. But such gradients may take on different values than assumed here and potentially play an important role in slowing migration, as will be discussed in section 3.1. In addition, some further effects, such as gas entropy gradients, can modify the coorbital torque. Recent studies have suggested that such thermal effects could change Type I migration and may even lead to outward migration (section 3.1) (*Paardekooper and Mellema*, 2006).

In comparing the saturated coorbital torque (equation (40)) to the unsaturated coorbital torque (equation (41)), we see that the effects of the coorbital torque reduce the inward migration rate for $\beta < 1.5$ and nearly vanish for $\beta = 1.5$. This behavior is consistent with equation (35), using the fact that $B \propto r^{-1.5}$. The coorbital torque is then positive for $\beta < 1.5$ and is zero for $\beta = 1.5$.

Nonlinear three-dimensional hydrodynamical calculations have been carried out to test the migration rates, under similar disk conditions (in particular, local isothermality) used to derive the analytic model. Figures 1 and 10 show that the migration rates agree well with the expectations of the theory.

We examine the comparison between simulations and theory in more detail by comparing torque distributions in the disk as a function of disk radius. We define the distribution of torque on the planet per unit disk mass as a function of radius as $dT/dM(r) = 1/(2\pi r \Sigma(r))\,dT/dr(r)$. Figure 11 plots dT/dM (scaled as indicated in the figure) as a function of

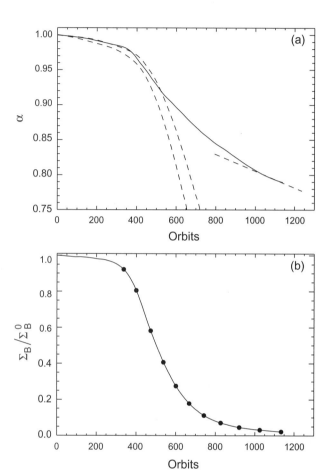

Fig. 10. Migration of a planet orbiting a solar mass star and undergoing growth via gas accretion. The disk parameters are similar to those in Fig. 1. **(a)** The vertical axis is the orbital radius in units of the initial orbital radius 5.2 AU. The horizontal axis is time in units of the initial orbital period 12 yr. The solid curve is the result of three-dimensional hydrodynamical simulations. The lower and upper short dashed curves are based on equations (40) and (41) respectively, applied to a planet of variable mass. The long dashed curve corresponds to migration on the disk viscous timescale. **(b)** Average disk density near the planet relative to the initial value as a function of time. The density is averaged over a band of radial width 2H centered on the orbit of the planet and is normalized by its initial value. The first solid circle marks the time when $M_p = 16.7\,M_\oplus$, subsequent circles occur at integer multiples of $33\,M_\oplus$. The planet initially follows the predictions of Type I migration theory in **(a)** while there is no substantial gap in the disk [no drop in the curve on **(b)**]. After gap opening, the planet follows Type II migration. Obtained from *D'Angelo and Lubow* (2008).

radial distance from the planet based on three-dimensional simulations. The distributions show that the region interior (exterior) to the planet provides a positive (negative) torque on the planet, as predicted in equations (11) and (12) for a particle disk. Also, the integrated total torque is negative, implying inward migration. The theory predicts that the torque density peak and trough occur at distance from the planet r–a = Δr ~ $\mp$H, where the torque cutoff takes effect. For the case plotted in Fig. 11 that adopts H = 0.05r, the predicted locations agree well with the locations of the peaks and troughs in the figure. Furthermore, the torque and torque density should scale with the square of the planet mass. Although the vortensity gradient is not small, since $\Sigma/B \propto r$, we do not find large contributions to the torque density from the coorbital region, as is also expected. As seen in Fig. 11, this expectation is well met for the two cases plotted.

2.5. Type II Migration

The planet's tidal torques cause material interior to the orbit of the planet (not in the coorbital region) to lose angular momentum and material exterior to the orbit of the planet to gain angular momentum, as we saw in section 2.2. The torques then act to clear a gap about the orbit of the planet. As discussed in section 1, Type II migration occurs when the planet mass is sufficiently large that tidal forces cause a gap to clear about the orbit of the planet (*Lin and Papaloizou*, 1986). The tidal torques on the disk interior or exterior to the planet are estimated by equation (11) with Δr ~ H. The tidal torque on the disk that acts to open the gap is then estimated as

$$T_o \sim \Sigma \Omega_p^2 a^4 \left(\frac{M_p}{M_s}\right)^2 \left(\frac{a}{H}\right)^3 \qquad (43)$$

Turbulent viscosity acts to close the gap. The effects of turbulence are often described in terms of a kinematic viscosity ν that is parameterized in the so-called α-disk model as

$$\nu = \alpha c H \qquad (44)$$

where α is a dimensionless number, c is the gas sound speed, and H is the disk thickness, described by equation (23) (*Shakura and Sunyaev*, 1973). The value of the key quantity α is expected to be less than unity because we presume that the turbulent motions are slower than sonic and that the characteristic length scales for the turbulence are smaller than the disk thickness.

The turbulent viscosity provides a torque to close the gap

$$T_c \sim \frac{M_g \Delta J_g}{t_g} \qquad (45)$$

where M_g is the mass supplied by the disk in closing the gap, ΔJ_g is the change in the disk angular momentum per

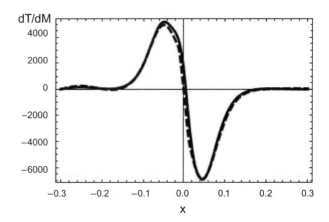

Fig. 11. Scaled torque per unit disk mass on the planet as a function of radial distance from the planet's orbital radius based on three-dimensional simulations. The horizontal scale is in units of a and the vertical scale is in units of $GM_s(M_p/M_s)^2/a$. The solid and long-dashed curves are for 1-$M_\oplus$ and 10-$M_\oplus$ planets, respectively, that orbit a solar-mass star. The disk parameters are H/r = 0.05 and α = 0.004 for both cases. According to linear theory, these two curves should overlap. Torque distributions are averaged over one orbital period. Figure based on *D'Angelo and Lubow* (2008).

unit mass required to close the gap, and t_g is the time for gap closing. We estimate these quantities as

$$M_g \sim \Sigma a w \qquad (46)$$

$$\Delta J_g \sim a \Omega_p w \qquad (47)$$

$$t_g \sim \frac{w^2}{\nu} \qquad (48)$$

where w is the radial extent of the gap, Σ is the disk density just outside the gap, and ν is the disk kinematic turbulent viscosity. The gap-closing timescale estimate t_g is based on the idea that the viscosity acts on a radial diffusion timescale across a distance w, which then implies the quadratic dependence on w. Applying this last set of relations to equation (45), we obtain that

$$T_c \sim \Sigma a^2 \Omega_p \nu \qquad (49)$$

independent of w.

In order to open a gap in the disk, the gap-opening torque must be greater than the gap-closing torque, $T_o \gtrsim T_c$. We then obtain a condition on the planet mass required to open a gap

$$\frac{M_p}{M_s} \gtrsim C_g \left(\frac{\nu}{a^2 \Omega_p}\right)^{1/2} \left(\frac{H}{a}\right)^{3/2} \qquad (50)$$

where C_g is a dimensionless number on the order of unity. *Lin and Papaloizou* (1986) estimated that $C_g = 2\sqrt{10}$. For disk parameters $\alpha = 0.004$ and $H/r = 0.05$, the predicted gap opening at the orbit of Jupiter about a solar-mass star occurs for planets having a mass $M_p \gtrsim 0.2\ M_J$, where M_J is Jupiter's mass. This prediction is in good agreement with the results of three-dimensional numerical simulations (see Fig. 12).

In addition to the above viscous condition, an auxiliary condition for gap opening has been suggested based on the stability of a gap. This condition is to preclude gaps for which steep density gradients would cause an instability that prevents gap opening. This condition, called the thermal condition, is given by the requirement that the Hill radius, given by equation (2), be larger than the disk thickness, $R_H \geq H$ (*Lin and Papaloizou*, 1986). The critical mass for gap opening by this condition is given by

$$\frac{M_p}{M_s} \geq 3 \left(\frac{H}{a} \right)^3 \qquad (51)$$

For $H/r = 0.05$, this condition requires a larger planet mass for gap opening than equation (50) for $\alpha \leq 0.01$. A condition that combines both equations (50) and (51) has been proposed by *Crida et al.* (2006). Even if both the thermal and viscous conditions are satisfied and $R_H \geq H$, a substantial gas flow may occur though the gap and onto the planet (*Artymowicz and Lubow*, 1996; *Lubow and D'Angelo*, 2006).

The migration rate of a planet embedded in a gap is quite different from the Type I case that we have already considered. A planet that opens a gap in a massive disk, a disk whose mass is much greater than the planet's mass, would be expected to move inward, pushed along with the disk accretion inflow. The planet simply communicates the viscous torques across the gap by means of tidal torques that balance them. The Type II migration timescale is then on the order of the disk viscous timescale

$$t_{vis} \sim \frac{a^2}{\nu} \sim \frac{a^2}{\alpha c H} \sim \left(\frac{a}{H} \right)^2 \frac{1}{\alpha \Omega_p} \qquad (52)$$

which is $\sim 10^5$ yr at Jupiter's orbital radius about the Sun for $\alpha = 0.004$, $H = 0.05a$, and $\Omega_p = 2\pi/12$ yr^{-1}. Therefore, the migration timescale can be much longer than the Type I migration timescale for higher-mass planets that open gaps, as is found in simulations ($M_p > 0.1\ M_J$ in Figs. 1 and 10). However, this timescale is still shorter than the observationally inferred global disk depletion timescales $\sim 10^6$–10^7 yr. The actual migration rate may need to be somewhat smaller, in order to explain the abundant population of observed giant exoplanets beyond 1 AU (*Ida and Lin*, 2008).

In practice, the conditions for pure Type II migration are unlikely to be satisfied. The disk mass may not be very large compared to the planet mass and the disk gap may not be fully clear of material. However, simulations have shown that to within factors of a few, the migration timescale matches the viscous timescale, t_{vis}. This result holds over a wide range of parameters, provided that the tidal clearing is substantial and the disk mass is at least comparable to the planet mass (e.g., Fig. 10).

Another way of understanding the Type II torques is to recognize that the distribution of tidal torques per unit disk mass, as seen in Fig. 11, still approximately applies, even if the disk has a deep gap that is not completely clear of material. The disk density through the gap region adjusts so that the planet migration rate is compatible with the evolution of the disk-planet system.

3. OUTSTANDING QUESTIONS

We describe some of the major issues involved with the theory of planet migration. The topics, descriptions, and references are by no means intended to be complete. The purpose is to introduce a few of the questions and some of the suggested solutions.

Planet migration is difficult to calculate for a variety of reasons, as described near the end of section 1. There are technical challenges in determining the properties of migration torques. For example, the coorbital torques are subject to saturation by nonlinear feedback effects. Lindblad torques grow in strength with proximity to the planet. Their contributions depend on the details of how the torques are limited near the planet. For the simple disk structures (e.g., moderate turbulent viscosity, fixed and smooth temperature distributions, smooth density distributions — apart from the density perturbations and gaps produced by the planet), the theory has been

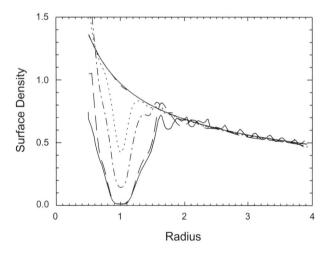

Fig. 12. Azimuthally averaged disk surface density normalized by the unperturbed value at radius a as a function of r in units of a = 5.2 AU. The central star has a solar mass. The disk is simulated in three dimensions with parameters H/r = 0.05 and α = 0.004. The density profiles are for planets with masses of 1 (long-dashed), 0.3 (dot-dashed), 0.1 (dotted), 0.03 (short-dashed), and 0.01 (thin solid) M_J. Only planets with masses $M_p \geq 0.1\ M_J$ produce significant perturbations. The thick solid line is based on a two-dimensional simulation of a 1-M_J planet by *Lubow et al.* (1999). Obtained from *Bate et al* (2003).

verified by various multidimensional nonlinear hydrodynamic simulations (e.g., Figs. 1, 10, and 11). But such simple disk structures may not arise in real systems. Indeed, the problems with shortness of planet migration timescales suggests that more complicated disk structures and physical processes (such as more complicated thermal effects) may be important. The disk structures influence migration, since migration torques are sensitive to gradients of disk properties. But we do not have a complete theory for these disk structures. In addition, migration depends somewhat on the detailed properties of disk turbulence that are not well understood.

3.1. Limiting/Reversing Type I Migration

As seen in Fig. 1, the timescales for Type I migration, based on the model described in section 2, are short compared to gas disk lifetimes of several million years. They become even shorter for a disk having a mass greater than the minimum mass solar nebula. As emphasized earlier, the shortness of the timescales for Type I migration is a serious problem for understanding how planets form and survive in a gaseous disk. To be consistent with the ubiquity of gas giant exoplanets and formation of Jupiter and Saturn, some studies suggest that the inward Type I migration rates must be reduced by more than a factor of 10 (*Alibert et al.,* 2005; *Rice and Armitage,* 2005; *Ida and Lin,* 2008). One major question is whether there are processes that could slow the migration. Several ideas have been proposed. Major uncertainties about them arise due to our lack of knowledge about the detailed structure of the disks.

We briefly discuss below a few of the several suggested mechanisms for slowing Type I migration. We also discuss some processes that may reverse migration, resulting in outward migration.

The fundamental tendency for Type I to be rapid can be difficult to cancel without artificially fine tuning some other effect that provides outward torques. In general, processes that provide outward torques tend to result in either rapid inward or rapid outward migration. However, there are effects that provide an opposing torque that increases as the planet continues migrating (via feedback effects or "traps"). Such processes can naturally maintain halted migration once the inward and outward torques are in balance.

3.1.1. Weak turbulence. For an inwardly migrating planet, there is a dynamical feedback effect associated with the radial motion of the planet that raises the axisymmetric gas surface density within a region interior to the orbit of the planet, and lowers the density within a region exterior to the orbit (*Ward,* 1997), in the same sense as a plow operating in the radial direction. The planet pushes material away from its orbital radius as it migrates. This leads to a pileup of material in the direction of its radial motion (see Fig. 13). The feedback then enhances the positive torques that arise in the inner disk and slows inward migration. For the feedback to exist, the effects of disk turbulence must be sufficiently weak, otherwise the density perturbations are erased by the effects of turbulent diffusion. This feedback grows with planet mass.

Above some critical planet mass M_{cr}, the feedback becomes strong enough that the planet can no longer migrate. When migration is halted, the density perturbations are initially mild and there is no substantial gap (unlike the Type II migration case). There is just a sufficient density asymmetry across the orbit of the planet to change the competition between inward and outward torques away from favoring the inward torques. In subsequent evolution, the planet begins forming a gap.

The value of the critical planet mass M_{cr} depends on several factors such as how far from the planet the density perturbations occur. For disks with $H/r \sim 0.05$ and low turbulent viscosity $\alpha \lesssim 10^{-4}$, the density perturbations produced by shocked density waves in two-dimensional disks cause migration to be sharply reduced at values of $M_{cr} \sim 10\ M_\oplus$ (*Rafikov,* 2002; *Li et al.,* 2009) (see Fig. 14). For disk turbulent viscosity parameter $\alpha \gtrsim 10^{-3}$, Type I migration proceeds with little reduction, since the density perturbations are washed out by the effects of disk turbulence. We have little direct information on what the α value should be. Some indirect evidence based on observed accretion rates onto young stars suggests that the average $\alpha \gtrsim 10^{-4}$. Theory suggests that there may be disk regions of low α in magnetically unstable disks (the so-called dead zones of low ionization) (*Gammie,* 1996).

3.1.2. Disk density/temperature variations. The torque on a planet depends somewhat on differences in the properties of the disk across the planet's orbit, as we have seen in section 2. If the density and temperature smoothly decreases in radius at rates typically expected, the outcome is inward migration (see equation (41) and Fig. 1). We expect this to be generally true across the disk, but there could be regions where the density abruptly changes or where the radial temperature gradient is less negative or is even positive. To slow migration, the effects of these gradients would need to counteract the general trend of inward migration discussed earlier that is due to other asymmetric effects, such as the asymmetries associated with Keplerian rotation. It is possible that rapid changes in disk properties with radius could occur as a consequence of strong radial variations in disk opacity or turbulent viscosity. They can in turn modify the typically adopted gradients of density and temperature that led to our estimated migration timescales. A planet could then experience slowed migration or even be trapped with no further inward migration (e.g., *Menou and Goodman,* 2004; *Masset et al.,* 2006; *Matsumura et al.,* 2007). However, our current knowledge about disk structure is limited.

Disk temperature variations can also be caused by the planet itself. The planet reduces the disk thickness H of gas close to it, as a consequence of its gravity. As a result, gas close to and just outside the orbit of the planet is further exposed to the stellar radiation and experiences additional heating by the star. Gas just inside the orbit of the planet experiences some shadowing. This effect acts to decrease the temperature gradient χ defined in equation (42). Calculations by *Jang-Condell and Sasselov* (2005) have shown that this effect can reduce the Type I migration rates by up to about a factor of 2.

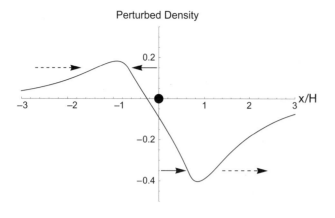

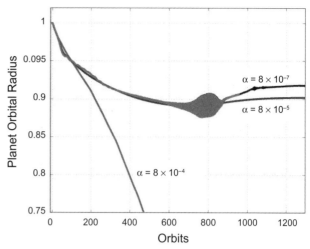

Fig. 13. Schematic of the axisymmetric density perturbation in arbitrary units as a function of radial distance from the planet (shown as dot at the origin) in units of disk thickness H. The solid lines with arrows indicate the velocity perturbation on the gas caused by tidal torques on the planet. These velocities are directed away from the orbit of the planet. The dashed lines with arrows indicate the gas velocity, in the frame of the planet, due to the assumed inward migration of the planet. Interior to the orbit of the planet, the velocities due to tides and migration oppose each other and lead to a pileup of material. Outside the orbit of the planet, the velocities reinforce each other and lead to a density drop.

Fig. 14. Influence of disk viscosity parameter α on the migration of a planet with mass 10 $M_⊕$ in a disk with H/r = 0.035 and mass ~0.1 $M_⊙$ based on two-dimensional simulations. The vertical axis is the orbital radius in units of the initial orbital radius. The horizontal axis is the time in units of the initial planet orbital period. For α = 8 × 10^{-4} the migration follows the Type I rate. At the lower values, the migration halts due to a feedback effect (see Fig. 13). Figure based on *Li et al.* (2009).

3.1.3. Turbulent fluctuations. The effects of disk turbulence are typically described by means of a turbulent viscosity (e.g., equation (44)). The numerical simulations used in Fig. 1 applied this model. Viscosity provides an approximate description of the dynamical effects of turbulent fluctuations that are averaged over the small space and short timescales characteristic of the turbulence. However, the time-dependent, small-scale density fluctuations can give rise to fluctuating random torques on the planet that are not described by viscosity. Unlike the Type I torque that acts continuously in the same direction, the fluctuating torque undergoes changes in direction on timescales characteristic of the turbulence that are short compared with the migration timescale. The fluctuating torque causes the planet to undergo something like a random walk. For the effects of the random walk to completely dominate over Type I migration, the amplitude of the fluctuating torque must be much larger than the Type I torque that acts steadily.

The change in the angular momentum of the planet due to a random torque T_R over some time t is given by $\sim \sqrt{N} T_R t_r$, where $N = t/t_R$ is the number of fluctuations felt by the planet and t_R is the characteristic timescale for the torque fluctuation, perhaps on the order of the orbital period. The planet experiences many torque fluctuations over the migration timescale, $N \gg 1$. The change in the planet angular momentum by the Type I torque T is given by ~Tt. For random torques to dominate, we require $T_R \geq$ and $\sqrt{N}T$. We take t to be the Type I timescale of ~10^5 yr and $t_R \sim 10$ yr, so that $N \sim 10^4$ and we require T_R to be ≥100 T. Torque T_R depends linearly on the planet mass, while the Type I torque T increases quadratically with the planet mass and the Type I migration time t decreases with the inverse of the planet mass. Therefore, the random torque is then more important for lower-mass planets. The nature of the random torque depends on the properties of the disk turbulence, in particular the amplitude T_R and correlation timescale t_R for the fluctuating torques, which are currently not well determined.

Some global simulations and analytic models suggest that turbulent fluctuations arising from a magnetic instability [the magneto-rotational instability (*Balbus and Hawley*, 1991)] are important for migration of lower-mass planets (*Nelson*, 2005; *Johnson et al.*, 2006) (see Fig. 15). Some other simulations, those done for a small region of the disk called "box" simulations, suggest that these fluctuating torques are not effective enough to play an important role in planet migration (*Yang et al.*, 2009). Both the global and box simulations are very computationally demanding and the current results cannot be regarded as definitive. The box simulations have higher resolution than the global ones, since they cover a smaller region. However, they may miss important effects, if they occur on larger scales than the box size. The direct global simulations, such as those in Fig. 15, cover much less time than the Type I migration timescale for a low-mass planet and consequently may not reveal the competing effects of unidirectional Type I torques.

Unlike Type I migration, the survival of a planet within a disk where fluctuating torques dominate is described statistically. Over time, there is a smaller probability that planets survive in the disk. But there is always a nonzero probability of survival.

However, if turbulent fluctuations are important for migration, then the eccentricities of planetesimals are pumped up so highly that collisions between them may result in their destruction rather than accretion (*Ida et al.*, 2008). Therefore, although the turbulent fluctuations may inhibit the infall of planetary cores into the central star by migration, they may inhibit the buildup of the cores necessary for giant planet formation in the core accretion model.

3.1.4. Ordered magnetic field. An ordered magnetic field can affect the nature of disk planet interactions. *Terquem* (2003) analyzed the effects of an ordered (nonrandom) torroidal magnetic field on Type I planet migration. The magnetic field introduces additional resonances, called magnetic resonances, that lie closer to the planet than the Lindblad resonances, whose locations are given by equation (38). As a result, the magnetic resonances can be stronger than the Lindblad ones. As in the case of the Lindblad resonances, a magnetic resonance located interior (exterior) to the orbit of the planet causes outward (inward) planet migration. If the magnetic field strength falls off sufficiently fast with distance from the star, the inner magnetic resonance dominates, and slowed or even outward migration can occur. For example, in the case that the magnetic energy density is comparable to the gas thermal energy density in the disk, an outward torque on the planet can be produced, if the magnetic-to-gas energy density ratio falls off faster than r^{-2}. How this ratio this varies within a disk is not known, although such a variation is quite plausible in some situations. If the magnetic field is responsible for the disk turbulence, then the torroidal field would contain a fluctuating component that complicates the outcome.

3.1.5. Migration driven by nonisothermal effects in the coorbital region. The disk is heated by the central star and by viscous turbulent dissipation. Many studies of disk-planet interactions simplify the disk temperature structure to be locally isothermal. In this approximation the temperature distribution depends on radius in a fixed, prescribed manner. Such a situation can arise if the optical depth of the disk is low enough that it efficiently radiates any excess energy due to compression caused by interactions with the planet. The locally isothermal assumption is frequently applied in numerical simulations and was applied in obtaining equation (41). The behavior in the isothermal limit tends to suggest that coorbital torques do not typically dominate migration (e.g., equation (41)).

But recent work has suggested that nonisothermal behavior could have an important effect on coorbital torques and the overall planet migration rate. The nonisothermal regime has been explored in simulations by *Paardekooper and Mellema* (2006), who found that slowed and even outward migration due to coorbital torques may occur in certain regimes. Recent studies have suggested that a background radial entropy gradient could play a role in determining the corotational torque if the gas undergoes adiabatic changes

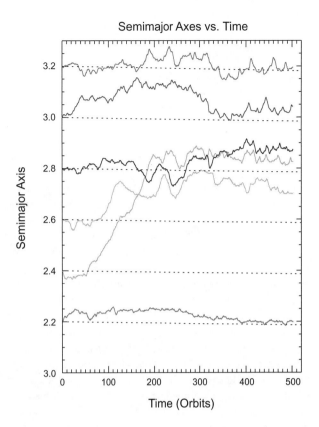

Fig. 15. Orbital radius as a function of time in orbits for 3 $M_\oplus$ planets embedded in a gaseous disk based on simulations. The dotted lines plot the migration of planets in a disk without turbulent fluctuations. The solid lines plot the migration of planets in a disk with turbulent fluctuations due to the MHD turbulence. The random motions are due to the fluctuating torques. Obtained from *Nelson* (2005).

in its interactions with the planet. This effect may provide an additional contribution to the coorbital torque beyond the vortensity gradient that appears in equation (35). The slowing/reversing of migration appears to involve the conditions that the disk have a negative radial entropy gradient, sufficient viscosity to avoid coorbital torque saturation, and a thermal timescale that is long enough for the gas to behave adiabatically as it passes the planet, but short enough to avoid phase mixing of the entropy. The nature of this effect and the conditions required for it to operate is an active area of investigation (e.g., *Baruteau and Masset*, 2008).

3.2. Other Migration Processes Involving a Gaseous Disk

3.2.1. Runaway coorbital migration, Type III migration. The coorbital torque will be saturated (reduced to zero) unless some effect is introduced to break the symmetry of the streamlines, such as turbulent viscosity (see discussion in section 2.2). However, planet migration itself introduces an asymmetry and could therefore act to prevent torque saturation. (The coorbital torque model presented in section 2.3

ignored the effects of planet migration on the horseshoe orbits in Fig. 2.) The coorbital torque for a migrating planet could then depend on the rate of migration. Under some conditions, the coorbital torque could in turn cause faster migration and in turn a stronger torque, resulting in an instability and a fast mode of migration (*Masset and Papaloizou*, 2003). The resulting migration is sometimes referred to as Type III migration. To see how this might operate in more detail, we consider the evolution of gas trapped in the coorbital region (*Artymowicz*, 2004; *Ogilvie and Lubow*, 2006). For a sufficiently fast migrating planet, the topology of the streamlines changes with open streamlines flowing past the planet and closed streamlines containing trapped gas (see Fig. 16). The leading side of the planet contains trapped gas acquired at larger radii, while the gas on the trailing side is ambient material at the local disk density. The density contrast between material on the trailing and leading sides of the planet gives rise to a potentially strong torque. The migration timescales are on the order of the Type I migration timescales. This timescale can be very short for planets that are massive enough to partially open a gap and would otherwise not undergo Type I migration. The migration timescales have been found to be as short as 10–20 orbits in the case of a Saturn-mass planet embedded in a cold and massive disk. A major question centers around the conditions required for this form of migration to be effective and therefore whether planets typically undergo such migration.

3.2.2. Migration of eccentric orbit planets. The analysis of migration in section 2 assumed that planets reside on circular orbits. This assumption is not unreasonable, since there are strong damping effects on eccentricity for a planet that does not open a gap in the disk (*Artymowicz*, 1993). Some eccentricity may be continuously produced by turbulent fluctuations in the gas, as described above, or by interactions with other planets. In general, eccentricity damping is faster than Type I migration. For planets that open a gap, it is possible that they reside on eccentric orbits in the presence of the gaseous disk. In fact, one model for the observed orbital eccentricities of exoplanets attributes the excitation of eccentricities to disk-planet interactions (*Goldreich and Sari*, 2003; *Ogilvie and Lubow*, 2003).

A planet on a sufficiently eccentric orbit embedded in a circular disk can orbit more slowly at apoastron than the exterior gas with which it tidally interacts. Similarly, a planet can orbit more rapidly than the tidally interacting gas at periastron. These angular velocity differences can change the nature of the "friction" between the planet and the disk discussed in section 2.2. For example, at apoastron the more slowly orbiting planet could gain angular momentum from the more rapidly rotating nearby gas that lies outside its orbit. Furthermore, since the planet spends more time at apoastron than periastron, the effects at apoastron could dominate over effects at periastron. It is then possible that outward migration could occur for eccentric orbit planets undergoing Type I migration, assuming such planets could maintain their eccentricities (*Papaloizou*, 2002).

In the case of a planet that opens a gap, simulations sug-

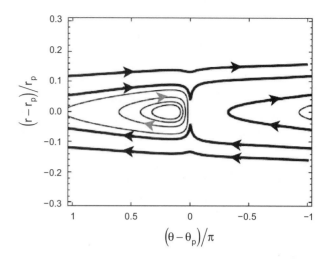

Fig. 16. Coorbital streamlines near an inwardly migrating Saturn-mass planet $M_p = 3 \times 10^{-4}$ M_s located at the origin in the co-moving frame of the planet that orbits a solar-mass star. The inward migration rate is $0.002\Omega_p a$. Angle θ increases to the left. The heavy streamlines are open and pass by the planet. The light streamlines are closed and contain trapped material on the leading side of the planet (streamlines based on *Ogilvie and Lubow*, 2006). The trapped material, acquired at earlier time, is retained from regions further from the star. The open streamlines carry ambient disk material. In contrast to the nonmigrating case in Fig. 2, the asymmetry and the density differences between the trapped (retained) and open (ambient) gas gives rise to a coorbital torque.

gest that slowed or even outward migration may occur as the planet gains eccentricity from disk-planet interactions (*D'Angelo et al.*, 2006). The situation is complicated by the fact that the gaseous disk can gain eccentricity from the planet by a tidal instability (*Lubow*, 1991). For the outward torque to be effective, there needs to be a sufficient difference in the magnitude and/or orientation between the planet and disk eccentricities, so that the planet moves slower than nearby disk gas at apoastron.

3.2.3. Multiplanet migration. Thus far, we have only considered single-planet systems. Of the many planetary systems detected to date by Doppler techniques, more than 10% are found in multiplanet systems (*Butler et al.*, 2006). Systems have been found to have orbits that lie in mutual resonance, typically the 2:1 resonance. The resonant configurations are likely to be the result of convergent migration, migration in which the separation of the orbital radii decreases in time. This process occurs as the outer planet migrates inward faster than the inner planet. Planets can become locked into resonant configurations and migrate together, maintaining the planetary orbital frequency ratio of the resonance (see also the chapter by Fabrycky). The locking can be thought of as a result of trapping the planets within a well of finite depth. Just which resonance the planets become locked into depends on their eccentricities and the relative rate of migration that would occur if they migrated independently. As planets that are initially well separated come closer together, they lock into the first resonance that provides a deep enough potential to trap them against the effects of their convergence. We discuss below some consequences of resonant migration.

To maintain a circular orbit, a migrating planet must experience energy, E, and angular momentum, J, changes that satisfy $\dot{E} = \Omega_p \dot{J}$, where the angular speed of the planet Ω_p varies as the planet migrates. As the planets migrate together in a resonant configuration, their mutual interactions cause deviations from this relation. As a result, their energies and angular momenta evolve in a way that is incompatible with maintaining a circular orbit. Substantial orbital eccentricities can develop as a consequence of migration (*Yu and Tremaine*, 2001; *Lee and Peale*, 2002). In addition, migration can cause a large amplification of an initially small mutual inclination of the planetary orbit planes (*Yu and Tremaine*, 2001; *Thommes and Lissauer*, 2003) (see Fig. 17).

Planetary system GJ876 is a well-studied case in which the planets are in a 2:1 resonance. If the system's measured eccentricities are due to resonant migration, then according to theory (*Lee and Peale*, 2002), the system migrated inward by less than 10%. Such a small amount of locked migration seems unlikely. It is more reasonable to expect that some process limited further eccentricity growth. Disk-planet interactions could have limited the eccentricities and inclinations that could be developed by resonant migration. For a single planet interacting with a disk, theory and simulations suggest that eccentricity is generally damped for planets of low mass, too low to open a gap. A higher-mass (gap-opening) planet can undergo eccentricity growth due to its interaction with the disk. But the level of eccentricity produced is limited and eccentricity damping occurs above that level (*D'Angelo et al.*, 2006). The disk-planet mutual inclination can also be suppressed or limited by the dissipation of disk warps (*Lubow and Ogilvie*, 2001). Recent simulations suggest that disk-planet interactions due to gas interior to the orbit of the inner planet could have limited the eccentricities of multiplanet systems to observed levels (*Crida et al.*, 2008).

But it may be possible under certain circumstances, such as a depleted inner disk, that large eccentricities and orbital inclinations could develop due to such processes.

3.3. Other Forms of Migration

3.3.1. Migration in a planetesimal disk. After the gaseous disk is cleared from the vicinity of the star, after about 10^7 yr, there remains a disk of solid material in the form of low-mass planetesimals. This disk is of much lower mass than the original gaseous disk. But the disk is believed to have caused some migration in the early solar system with important consequences (*Hahn and Malhotra*, 1999; *Tsiganis et al.*, 2005).

There is strong evidence that Neptune migrated outward. This evidence comes from observations of Kuiper belt objects that are resonantly trapped exterior, but not interior, to Neptune's orbit. The detailed dynamics of a planetesimal disk are somewhat different from the case of a gaseous disk, as considered in section 2. The planetesimals behave as a nearly collisionless system of particles. Jupiter is much more massive than the other planets and can easily absorb angular momentum changes in Neptune. As Neptune scatters planetesimals inward and outward, it undergoes angular momentum changes. It is the presence of Jupiter that breaks the symmetry in Neptune's angular momentum changes. Once an inward-scattered planetesimal reaches the orbit of Jupiter, it gets flung out with considerable energy and does not interact again with Neptune. As a result of the loss of inward-scattered particles, Neptune gains angular momentum and migrates outward, while Jupiter loses angular momentum and migrates slightly inward.

A somewhat analogous process occurs in gaseous circumbinary disks, disks that orbit around binary star systems (*Pringle*, 1991). The circumbinary disk gains angular momentum at the expense of the binary. The binary orbit contracts as the disk outwardly expands. A gap-opening planet embedded in a circumbinary disk (or under some

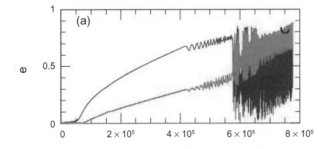
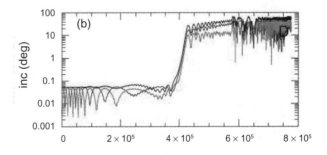

Fig. 17. Evolution of **(a)** orbital eccentricity and **(b)** inclination of a two-planet system in which an inward migration of 10^{-5} AU yr^{-1} is imposed on the outer one. The horizontal axis is in units of years. Both planets have a mass equal to Jupiter's mass. The orbital eccentricities of the planets grow as a result of resonant migration in the 2:1 resonance. The eccentricity of the inner planet grows faster than the outer one. A small initial mutual inclination grows to large values as the eccentricities increase. From *Tommes and Lissauer* (2003).

conditions, a disk that surrounds a star and massive inner planet) would undergo a form of Type II migration that could carry the planet outward (*Martin et al.*, 2007). In the solar system case, the Sun-Jupiter system plays the role of the binary. Viscous torques are the agent for transferring the angular momentum from the binary outward in the gaseous circumbinary disk, while particle torques play the somewhat analogous role in the planetesimal disk.

In a planetesimal disk, another process can operate to cause migration. This process is similar to the runaway coorbital migration (Type III migration) described above, but applied to a collisionless system of particles (*Ida et al.*, 2000). Interactions between the planetesimals and the planet in the planet's coorbital zone can give rise to a migration instability.

3.3.2. Kozai migration. A planet that orbits a star in a binary star system can periodically undergo a temporary large increase in its orbital eccentricity through the a process known as the Kozai effect (see also the chapter by Fabrycky). Similar Kozai cycles can occur in multi-giant-planet systems. The basic idea behind Kozai migration is that the increased eccentricity brings the planet closer to the star where it loses orbital energy through tidal dissipation. In the process, the planet's semimajor axis is reduced and inward migration occurs (*Wu and Murray*, 2003). We describe this in more detail below.

Consider a planet in a low-eccentricity orbit that is well interior to the binary orbit and is initially highly inclined with respect to it. The orbital plane of the planet can be shown to undergo tilt oscillations on a timescale of $\sim P_b^2/P_p$, where P_b is the binary orbital period, P_p is the planet's orbital period, and by assumption $P_b \gg P_p$. Under such conditions, it can be shown that the component of the planet's angular momentum perpendicular to the binary orbit plane (the z component) is approximately conserved, $J_z = M_p\sqrt{GM_s a_p(1-e_p^2)}\cos I$, where a_p and e_p are respectively the semimajor axis and eccentricity of the planet's orbit and I is the inclination of the orbit with respect to the plane of the binary.

The conservation of J_z is easily seen in the case that the binary orbit is circular and the companion star is of low mass compared to the mass of the star about which the planet orbits. On such long timescales $\gg P_b$, the companion star can be considered to be a continuous ring that provides a static potential. In that case, the azimuthal symmetry of the binary potential guarantees that J_z is conserved. By assumption, we have $\cos I \ll 1$ and $e_p \ll 1$ in the initial state of the system. As the planet's orbital plane evolves and passes into alignment with the binary orbital plane, $\cos I \sim 1$, conservation of J_z requires $e_p \sim 1$. In other words, J_z is initially small because of the high inclination of the orbit. When the inclination drops, the orbit must become more eccentric (radial), in order to maintain the same small J_z value. The process then periodically trades high inclination for high eccentricity.

During the times of increased eccentricity, the planet may undergo a close encounter with the central star at periastron distance $a_p(1-e_p)$. During the encounter, the tidal dissipation involving the star and planet results in an energy loss in the orbit of the planet and therefore a decrease in a_p. This process then results in inward planet migration. The energy loss may occur over several oscillations of the orbit plane.

Another requirement for the Kozai process to operate is that the system must be fairly clean of other bodies. The presence of another object could induce a precession that washes out the Kozai effect. Given the special requirements needed for this process to operate, it is not considered to be the most common form of migration. However, there is good evidence that it does operate in some systems (*Takeda and Rasio*, 2005). The Kozai effect can also occur in two-planet systems, where the outer planet plays the role of the binary companion. The process can be robust due to the proximity of outer planet (*Nagasawa et al.*, 2008).

3.4. Techniques Used in Numerical Simulations

Numerical simulations provide an important tool for analyzing planet migration. They can provide important insights in cases where nonlinear and time-dependent effects are difficult to analyze by analytic methods. Simulations, as well as analytical models, depend on a model of appropriate physical processes, such as heating and cooling, the treatment of the disk turbulence (either by a viscosity or detailed modeling of the disk instability that causes the turbulence), and the model of the disk structure. Below we discuss some issues related to the use of simulations for simple (isothermal and α-disk) multidimensional models of disk-planet interactions.

Some powerful grid-based hydrodynamics codes [such as the ZEUS code (*Stone and Norman*, 1992)] have been adapted to the study of disk-planet interactions. In addition, particle codes based on smoothed particle hydrodynamics (SPH) (*Monaghan*, 1992) have sometimes been employed. A systematic comparison between many of the codes has been carried out by *de Val-Borro et al.* (2006). We discuss a few basic points. For planets that open a gap, grid-based codes offer an advantage over particle-based codes. The reason is that the resolution of grid-based codes is determined by the grid spacing, while the resolution of particle-based codes is determined by the particle density. If a planet opens an imperfect gap, the particle density and resolution near the planet is low. Higher resolution occurs where the particle density is higher. On the other hand, SPH well represents regions of a cold disk that are away from a planet. The gas follows the trajectories of SPH particles that are slightly modified by gas pressure and tidal forces. Smoothed particle hydrodynamics provides high resolution in dense regions, such as near the cores of low-mass planets embedded in disks (*Ayliffe and Bate*, 2009).

There have been variable resolution techniques developed for grid-based codes in which the highest resolution is provided in regions near the planet where it is needed, such as nested grid methods (*D'Angelo et al.*, 2002). Such techniques need to provide a means of joining the regions of high and low resolution without introducing artifacts (such as wave reflections) or lowering the overall accuracy of the scheme.

Grid-based codes that simulate disk-planet interactions typically employ numerical devices to improve convergence. For example, the gravitational potential of the planet is often replaced by one that does not diverge near the planet. The potential is limited by introducing a smoothing length, a distance within which the potential does not increase near the planet. Another limitation is that the simulated domain of the disk is typically limited to a region much smaller than the full extent of the disk. Techniques have been developed to ensure that reflections from the boundaries do not occur, e.g., by introducing enhanced wave dissipation near the boundary or approximate outgoing wave boundary conditions.

The timesteps of codes are limited by the Courant condition. Short timesteps often result from the region near the inner boundary of the computational domain (smaller radii) where the disk rotation is fastest. As a result, it is difficult to extend the disk very close to the central star. The FARGO scheme (*Masset*, 2000) is very useful in overcoming this limitation. However, the method is difficult to apply to a variable grid-spaced code.

Convergence is a major issue with these simulations. Ideally, one should demonstrate that the results of the simulations are sufficiently insensitive to the locations of the boundaries, the size of the smoothing length, the size of the timesteps, and the grid resolution. Even the direction of migration can be affected by the size of the potential smoothing length in certain cases. In a two-dimensional simulation, a finite smoothing length ~H provides a means of simulating the reduced effects of planet gravity on a disk of finite thickness. But in the two-dimensional case, the limit of zero smoothing length is unphysical. In practice, testing for convergence is computationally expensive, but can be done for a subset of the models of interest, perhaps over a limited time range. Demonstrating convergence is important for providing reliable results.

4. FUTURE PROSPECTS

The theory of planet migration is intertwined with the theory of planet formation (see chapters by Chambers and by D'Angelo et al.). The timescale for a planet to grow within a disk is a key element in understanding whether planet migration is a major obstacle to planet formation. We have pointed out that the formation timescales for gas giant planets in the core accretion model do present a problem for the simplest planet migration theories. However, alternative migration models, some of which are described in section 3, may be appropriate. Future prospects for resolving this issue depend on advances in the theory of planet formation.

Prospects for making progress in the theory of planet migration rely on including and more accurately representing physical processes in the models, improving computer simulations, and obtaining a better understanding of the structure of planet-forming gaseous disks.

Planet migration in disks is a consequence of the action of Lindblad and corotational resonances, as described in section 2. The theory of Lindblad resonances is better understood. Their linear and nonlinear properties have been analyzed in more detail. The corotational resonances are somewhat more delicate and less well understood. Progress on the theory of planet migration will likely involve further investigations of the role of corotational resonances (e.g., *Paardekooper and Mellema*, 2006).

Improvements to computer capabilities and codes should allow simulations to be carried out with higher resolution over longer timescales. Progress will be made by also including more physical effects. For example, most multidimensional simulations have made only the simplest assumptions about the thermal properties of the disk. The use of a turbulent viscosity is a serious approximation to the effects of turbulence. Directly simulating the instability that produces the turbulence along with planet migration is a major computational challenge (e.g., *Nelson*, 2005).

Some calculations have suggested that higher-mass eccentric orbit planets could undergo slowed or outward migration due to their interactions with a gaseous disk. But we do not know whether such planets have acquired their eccentricities at this early stage. Observations of the eccentricities of young planets would be quite valuable in understanding this issue. It would be useful to know whether planetary eccentricities are determined at early times, suggesting that planet eccentricities are present while the planet is embedded in its gaseous disk.

A major uncertainty in the theory of planet migration is the physical state of the disk. Low-mass planet migration behaves very differently depending on the level of disk turbulence and the structural properties of the disk (section 3). For example, in weakly turbulent disks, feedback effects may halt Type I migration. While in highly turbulent disks, fluctuating torques may dominate over Type I torques. The presence of rapid radial density variations or weakened (or inverted) radial temperature variations can substantially alter migration, since it depends on the competition between torques involving material just inside and outside the orbit of the planet. The better determination of disk properties will likely rely on some combination of improved theory and observations. It is unlikely that theory alone will be able to make much progress along these lines because, for example, the density structure depends on the disk turbulence that is difficult to accurately predict from first principles. Once such difficulty is that magnetically driven turbulence is strongly affected by the abundance and size distribution of dust grains that control the level of disk ionization (e.g., *Salmeron and Wardle*, 2008). The observational determination of disk properties (disk density and temperature as a function of distance from the star) in the inner parts of protostellar disks, where planet formation is expected to occur, is important for such purposes. New telescopes such as ALMA and JWST may be quite valuable in making such determinations.

Acknowledgments. This work was partially supported by NASA grant NNX07AI72G to S.L.. We thank P. Armitage, G.

D'Angelo, and J. Pringle for carefully reading a draft and suggesting improvements. We thank an anonymous reviewer for many helpful suggestions. We also benefited considerably by discussions at the Cambridge University Isaac Newton Institute Program "Dynamics of Discs and Planets."

REFERENCES

Alibert Y., Mousis O., Mordasini C. and Benz W. (2005) New Jupiter and Saturn formation models meet observations. *Astrophys. J., 626*, L57–L60.

Armitage P. J., Livio M., and Pringle J. E. (2001) Episodic accretion in magnetically layered protoplanetary discs. *Mon. Not. R. Astron. Soc., 324*, 705–711.

Artymowicz P. (1993) Disk-satellite interaction via density waves and the eccentricity evolution of bodies embedded in disks. *Astrophys. J., 419*, 166–180.

Artymowicz P. (2004) Migration Type III. In *KITP Conference on Planet Formation,* available online at *online.kitp.ucsb.edu/online/planetf_c04*.

Artymowicz P. and Lubow S. H. (1996) Mass flow through gaps in circumbinary disks. *Astrophys. J. Lett., 467*, L77–L80.

Ayliffe B. A. and Bate M. R. (2009) Gas accretion on to planetary cores: Three-dimensional self-gravitating radiation hydrodynamical calculations. *Mon. Not. R. Astron. Soc., 393*, 49–64.

Balbus S. A. and Hawley J. F. (1991) A powerful local shear instability in weakly magnetized disks. I. Linear analysis. *Astrophys. J., 376*, 214–222.

Baruteau C. and Masset F. (2008) On the corotation torque in a radiatively inefficient disk. *Astrophys. J., 672*, 1054–1067.

Bate M. R., Lubow S. H., Ogilvie G. I., and Miller K. A. (2003) Three-dimensional calculations of high- and low-mass planets embedded in protoplanetary discs. *Mon. Not. R. Astron. Soc., 341*, 213–229.

Bodenheimer P., Hubickyj O., and Lissauer J. J. (2000) Models of the in situ formation of detected extrasolar giant planets. *Icarus, 143*, 2–14.

Butler R. P., Wright J. T., Marcy G. W., Fischer D. A., Vogt S. S., et al. (2006) Catalog of nearby exoplanets. *Astrophys. J., 646*, 505–522.

Crida A., Morbidelli A., and Masset F. (2006) On the width and shape of gaps in protoplanetary disks. *Icarus, 181*, 587–604.

Crida A., andor Z., and Kley W. (2008) Influence of an inner disc on the orbital evolution of massive planets migrating in resonance. *Astron. Astrophys., 483*, 325–337.

Cuzzi J. N. and Weidenschilling S. J. (2006) Particle-gas dynamics and primary accretion. In *Meteorites and the Early Solar System II* (D. S. Lauretta and H. Y. McSween Jr. eds.), pp. 353–381. Univ. of Arizona, Tucson.

D'Angelo G. and Lubow S. H. (2008) Evolution of migrating planets undergoing gas accretion. *Astrophys. J., 685*, 560–583.

D'Angelo G., Henning T., and Kley W. (2002) Nested-grid calculations of disk-planet interaction. *Astron. Astrophys., 385*, 647–670.

D'Angelo G., Lubow S. H., and Bate M. R. (2006) Evolution of giant planets in eccentric disks. *Astrophys. J., 652*, 1698–1714.

de Val-Borro M. et al. (2006) A comparative study of disc-planet interaction. *Mon. Not. R. Astron. Soc., 370*, 529–558.

Gammie C. F. (1996) Layered accretion in T Tauri disks. *Astrophys. J., 457*, 355–362.

Goldreich P. and Sari R. (2003) Eccentricity evolution for planets in gaseous disks. *Astrophys. J., 585*, 1024–1037.

Goldreich P. and Tremaine S. (1979) The excitation of density waves at the lindblad and corotation resonances by an external potential. *Astrophys. J., 233*, 857–871.

Goldreich P. and Tremaine S. (1980) Disk-satellite interactions. *Astrophys. J., 241*, 425–441.

Goldreich P. and Tremaine S. (1982) The dynamics of planetary rings. *Annu. Rev. Astron. Astrophys., 20*, 249–283.

Hahn J. M. and Malhotra R. (1999) Orbital evolution of planets embedded in a planetesimal disk. *Astron. J., 117*, 3041–3053.

Hayashi C. (1981) Structure of the solar nebula, growth and decay of magnetic fields and effects of magnetic and turbulent viscosities on the nebula. *Progr. Theor. Phys. Suppl., 70*, 3553.

Hourigan K. and Ward W. R. (1984) Radial migration of preplanetary material: Implications for the accretion time scale problem. *Icarus, 60*, 29–39.

Ida S. and Lin D. N. C. (2004) Toward a deterministic model of planetary formation. I. A desert in the mass and semimajor axis distributions of extrasolar planets. *Astrophys. J., 604*, 388–413.

Ida S. and Lin D. N. C. (2008) Toward a deterministic model of planetary formation. IV. Effects of type I migration. *Astrophys. J., 673*, 487–501.

Ida S., Bryden G., Lin D. N. C. and Tanaka H. (2000) Orbital migration of Neptune and orbital distribution of trans-Neptunian objects. *Astrophys. J., 534*, 428–445.

Ida S., Guillot T., and Morbidelli A. (2008) Accretion and destruction of planetesimals in turbulent disks. *Astrophys. J., 686*, 1292–1301.

Jang-Condell H. and Sasselov D. D. (2005) Type I planet migration in a non-isothermal disk. *Astrophys. J., 619*, 1123–1131.

Johnson E. T., Goodman J., and Menou K. (2006) Diffusive migration of low-mass protoplanets in turbulent disks. *Astrophys. J., 647*, 1413–1425.

Lee M. H. and Peale S. J. (2002) Dynamics and origin of the 2:1 orbital resonances of the GJ 876 planets. *Astrophys. J., 567*, 596–609.

Li H., Lubow S. H., Li S., and Lin D. N. C. (2009) Type I planet migration in nearly laminar disks. *Astrophys. J. Lett., 690*, L52.

Lin D. N. C. and Papaloizou J. (1979) Tidal torques on accretion discs in binary systems with extreme mass ratios. *Mon. Not. R. Astron. Soc., 186*, 799–812.

Lin D. N. C. and Papaloizou J. C. B. (1986) On the tidal interaction between protoplanets and the protoplanetary disk. III — Orbital migration of protoplanets. *Astrophys. J., 309*, 846–857.

Lodato G. and Rice W. K. M. (2004) Testing the locality of transport in self-gravitating accretion discs. *Mon. Not. R. Astron. Soc., 351*, 630–642.

Lubow S. H. (1991) A model for tidally driven eccentric instabilities in fluid disks. *Astrophys. J., 381*, 259–267.

Lubow S. H. and D'Angelo G. (2006) Gas flow across gaps in protoplanetary disks. *Astrophys. J., 641*, 526–533.

Lubow S. H. and Ogilvie G. I. (2001) Secular interactions between inclined planets and a gaseous disk. *Astrophys. J., 560*, 997–1009.

Lubow S. H., Seibert M., and Artymowicz P. (1999) Disk accretion onto high-mass planets. *Astrophys. J., 526*, 1001–1012.

Lynden-Bell D. and Pringle J. E. (1974) The evolution of viscous discs and the origin of the nebular variables. *Mon. Not. R. Astron. Soc., 168*, 603–637.

Martin R. G., Lubow S. H., Pringle J. E., and Wyatt M. C. (2007) Planetary migration to large radii. *Mon. Not. R. Astron. Soc., 378*, 1589–1600.

Masset F. S. (2000) FARGO: A fast eulerian transport algorithm for differentially rotating disks. *Astron. Astrophys. Suppl., 141,* 165–173.

Masset F. S. and Papaloizou J. C. B. (2003) Runaway migration and the formation of hot Jupiters. *Astrophys. J., 588,* 494–508.

Masset F. S., Morbidelli A., Crida A., and Ferreira J. (2006) Disk surface density transitions as protoplanet traps. *Astrophys. J., 642,* 478–487.

Matsumura S., Pudritz R. E., and Thommes E. W. (2007) Saving planetary systems: Dead zones and planetary migration. *Astrophys. J., 660,* 1609–1623.

Mayor M. and Queloz D. (1995) A Jupiter-mass companion to a solar-type star. *Nature, 378,* 355–359.

Menou K. and Goodman J. (2004) Low-mass protoplanet migration in t tauri disks. *Astrophys. J., 606,* 520–531.

Monaghan J. J. (1992) Smoothed particle hydrodynamics. *Annu. Rev. Astron. Astrophys., 30,* 543–574.

Nagasawa M., Ida S., and Bessho T. (2008) Formation of hot planets by a combination of planet scattering, tidal circularization, and the Kozai mechanism. *Astrophys. J., 678,* 498–508.

Nelson R. P. (2005) On the orbital evolution of low mass protoplanets in turbulent, magnetised disks. *Astron. Astrophys., 443,* 1067–1085.

Ogilvie G. I. and Lubow S. H. (2003) Saturation of the corotation resonance in a gaseous disk. *Astrophys. J., 587,* 398–406.

Ogilvie G. I. and Lubow S. H. (2006) The effect of planetary migration on the corotation resonance. *Mon. Not. R. Astron. Soc., 370,* 784–798.

Paczynski B. (1978) A model of self-gravitating accretion disk. *Acta Astron., 28,* 91–109.

Paardekooper S.-J. and Mellema G. (2006) Halting type I planet migration in non-isothermal disks. *Astron. Astrophys., 459,* L17–L20.

Papaloizou J. C. B. (2002) Global m = 1 modes and migration of protoplanetary cores in eccentric protoplanetary discs. *Astron. Astrophys. 388,* 615–631.

Pringle J. E. (1991) The properties of external accretion discs. *Mon. Not. R. Astron. Soc., 248,* 754–759.

Rafikov R. R. (2002) Planet migration and gap formation by tidally induced shocks. *Astrophys. J., 572,* 566–579.

Rice W. K. M. and Armitage P. J. (2005) Quantifying orbital migration from exoplanet statistics and host metallicities. *Astrophys. J., 630,* 1107–1113.

Salmeron R. and Wardle M. (2008) Magnetorotational instability in protoplanetary discs: The effect of dust grains. *Mon. Not. R. Astron. Soc., 388,* 1223–1238.

Shakura N. I. and Sunyaev R. A. (1973) Black holes in binary systems. Observational appearance. *Astron. Astrophys., 24,* 337–355.

Stone J. M. and Norman M. L. (1992) ZEUS-2D: A radiation magnetohydrodynamics code for astrophysical flows in two space dimensions. I — The hydrodynamic algorithms and tests. *Astrophys. J. Suppl., 80,* 753–790.

Takeda G. and Rasio F. A. (2005) High orbital eccentricities of extrasolar planets induced by the Kozai mechanism. *Astrophys. J., 627,* 1001–1010.

Tanaka H., Takeuchi T., and Ward W. R. (2002) Three-dimensional interaction between a planet and an isothermal gaseous disk. I. Corotation and Lindblad torques and planet migration. *Astrophys. J., 565,* 1257–1274.

Terquem C. E. J. M. L. J. (2003) Stopping inward planetary migration by a tooroidal magnetic field. *Mon. Not. R. Astron. Soc., 341,* 1157–1173.

Terquem C. E. J. M. L. J. (2008) New composite models of partially ionized protoplanetary disks. *Astrophys. J., 689,* 532–538.

Thommes E. W. and Lissauer J. J. (2003) Resonant inclination excitation of migrating giant planets. *Astrophys. J., 597,* 566–580.

Tsiganis K., Gomes R., Morbidelli A., and Levison H. F. (2005) Origin of the orbital architecture of the giant planets of the solar system. *Nature, 435,* 459–461.

Ward W. R. (1991) Horeshoe orbit drag. In *Lunar Planet. Sci. XXII,* pp. 1463–1464. Lunar and Planetary Institute, Houston.

Ward W. R. (1997) Protoplanet migration by nebula tides. *Icarus, 126,* 261–281.

Weidenschilling S. J. (1977) Aerodynamics of solid bodies in the solar nebula. *Mon. Not. R. Astron. Soc., 180,* 57–70.

Wu Y. and Murray N. (2003) Planet migration and binary companions: The case of HD 80606b. *Astrophys. J., 589,* 605–614.

Yang C. C., Mac Low M.-M., and Menou K. (2009) Planetesimal and protoplanet dynamics in a turbulent protoplanetary disk: Ideal unstratified disks. *Astrophys. J., 707,* 1233–1246.

Yu Q. and Tremaine S. (2001) Resonant capture by inward migrating planets. *Astron. J., 121,* 1736–1740.

Part V:
Exoplanet Interiors and Atmospheres

Terrestrial Planet Interiors

C. Sotin
Jet Propulsion Laboratory, California Institute of Technology

J. M. Jackson
California Institute of Technology

S. Seager
Massachusetts Institute of Technology

The discovery and study of exoplanets has always motivated the question of the existence and nature of terrestrial exoplanets, especially habitable planets. Exoplanet mass and radius measurements (yielding average density) are possible for a growing number of exoplanets, including terrestrial planets. The mass and radius provide a constraint for terrestrial planet interior models and, via models, enable interpretation of a planet's bulk composition. This chapter describes the fundamental equations for calculating the interior structure of terrestrial planets (including silicate-rich, iron-rich, and water-rich planets). A detailed description of interior structure models is given, with emphasis on the equation of state. High-pressure experimental measurements are needed to understand the internal structure and evolution of massive terrestrial exoplanets. Interpretation of low-mass planet interiors are fundamentally limited by degeneracies, because there are only two data points per planet: mass and radius. For example, large terrestrial planets having a massive primordial atmosphere of H_2 and He cannot be distinguished from planets having dominant internal H_2O layers based on the mass and radius alone. The occurrence of plate tectonics on exoplanets more massive than Earth is a controversial question, and, although unanswered, this chapter addresses the relationship between plate tectonics and thermal convection. As more and more low-mass exoplanets are being discovered, mass vs. radius statistics will build up. The hope for terrestrial exoplanet mass and radius measurements is that unique populations will emerge from statistics, helping us to understand planet formation and evolution.

1. INTRODUCTION

The search for terrestrial exoplanets is one of the most exciting challenges of the twenty-first century. For the first time astronomers have the chance to uncover a large sample of planets that are predominantly rocky or icy and not the easier-to-detect giant planets, which are composed mostly of H and He. Terrestrial planets are of major scientific significance because they are the planets suitable for life as we know it and amenable for future observational searches for atmospheric biosignature gases. In the conventional sense a habitable planet is one with some surface liquid water, because all life on Earth requires liquid water. In contrast to terrestrial planets, giant and Neptune-sized planets enshrouded by gas envelopes have no solid or liquid surfaces to support life as we know it, and their temperatures just below the deep atmosphere rapidly become too hot for life to exist.

The solar system planets are conveniently divided into two categories: the terrestrial planets (Mercury, Venus, Earth, Mars) located in the inner solar system, and the giant planets (Jupiter, Saturn, Uranus, Neptune) located in the outer solar system. The difference between terrestrial and giant planets is related to the planet's ability to retain a large primordial atmosphere of H and He. Whether or not a planet retains substantial amounts of H and He depends primarily on a planet's mass. *Wuchterl et al.* (2000), for example, propose that a planet less than 15 times the mass of Earth ($M_\oplus$) will be terrestrial in nature, i.e., without significant H and He.

Beyond the terrestrial and giant planet groupings, each category can actually be subdivided into two subsets. The terrestrial planets include both silicate-dominated planets like Earth and Venus, as well as Mercury-like planets, which are enriched in iron (i.e., depleted in silicates). The Mercury-type exoplanet is of interest because finding many close to the host star will help further understand Mercury's formation.

The giant planet category includes not only the H/He-dominated Jupiter and Saturn but also icy planets Uranus and Neptune. The ice giants have a much smaller H-He envelope than Jupiter and Saturn (10–15% by mass). Note that how water-rich Uranus and Neptune are is not accurately known. Although their atmospheres are hydrogen-rich, there is a tradeoff for their interior bulk compositions between ice + an H/He envelope and a combination of rock/iron, much less ice, and a more massive H/He envelope. Extrapolating to icy planets with insignificant H-He envelopes compared to Uranus and Neptune are exoplanets of significant astrobiological interest: ocean planets or water worlds, which could also be called super Ganymedes or super Titans. Under the right atmospheric mass and interior temperature conditions, some of these planets are likely to harbor internal liquid water oceans. Icy exoplanets with thin atmospheres can be considered as large analogs of the large icy satellites of Jupiter (Europa, Ganymede, Callisto) and Saturn (Titan).

The bulk interior composition of an exoplanet can be constrained with planet mass and radius measurements together with planet interior equations. The planet mass is measured via radial velocity measurements (see chapter by Lovis and Fischer) and the planet radius via transit measurements (see chapter by Winn). Some exoplanets are discovered with the radial velocity method and are later found to transit, while others are discovered by the transit technique and followed up for mass measurements via radial velocity.

Conceptually the equations that describe a terrestrial planet interior are the conservation of mass, hydrostatic equilibrium, energy transport, and the equation of state. The most basic assumptions common to most models are that the terrestrial planet interior is formed from three basic materials (iron, silicate, and water) with some mixture and of various phases, and that the planets have differentiated interiors. The new territory for terrestrial exoplanet interior models lies in the equation of state, because terrestrial exoplanets more massive than Earth (super Earths) can have interior pressures much higher. For example, a planet ten times more massive than Earth can have internal pressures three times higher (e.g., *Sotin et al.*, 2007), which implies that new high-pressure mineral phases may exist. Super Earths are loosely defined to be planets between 1 and 10 $M_\oplus$ that are predominantly rocky or icy (see the Appendix of the chapter by Seager and Lissauer for exoplanet definitions).

This chapter describes the concepts and equations needed to understand and model terrestrial planet interiors. Application to the mass-radius relationship for a variety of terrestrial planet types and a detailed presentation of the equation of state for different materials is given. This chapter also focuses on terrestrial planet internal dynamics by providing the equations describing subsolidus convection in planetary mantles and investigating the controversial question of plate tectonics. We conclude with an outlook for the future of characterizing terrestrial exoplanet interiors based on the limited observational data (two data points per planet) and anticipated planet harvests.

2. INTERIOR STRUCTURE

2.1. Bulk Composition and Structure

The minerals that compose the different layers of a planet are determined according to the elements that are present, the pressure, and the temperature. Previous studies (e.g., *Sotin et al.*, 2007, and references therein) have shown that only a limited number of elements are necessary to describe the total mass of the planet: O, Fe, Mg, and Si explain 95% of the mass of Earth. If Ni, S, Al, and Ca are added, then 99.9% of the mass is taken into account (Table 1). The latter three minor elements add a lot of complexities in the system. For example, experimental studies suggest that the oxidation state of iron and exsolution of metallic iron particles from iron-bearing silicates is strongly affected by the presence of Al (*Frost et al.*, 2004) and extreme pressures and temperatures (*Jackson et al.*, 2009). If S is added to Fe, the melting temperature decreases with increasing pressure and the phase relations become very complex (*Chen et al.*, 2008). Although there is considerable debate surrounding the identity of the light element and crystal structure of the Fe-dominant alloy in Earth's core, the effect on total planet mass is small. Based on the available data, adding Ni, S, Al, and Ca to their closest major element results in a less than 1% error in the model-calculated total mass. Therefore, the composition of each layer can be described with four elements: O, Fe, Mg, and Si (see *Sotin et al.*, 2007). In the mantle, Al is equally divided between Mg and Si for charge conservation and Ca is added to Mg. In the core, Ni and S are added to Fe.

Due to Earth's seismically active interior, the average structure of Earth (Fig. 1) is well known from global Earth models (*Dziewonski and Anderson*, 1981). Earth is mainly composed of two layers: the iron-rich core (one-third of the mass) and the silicate mantle (two-thirds of the mass). These two layers differentiated very early in Earth's evolutionary history because the melting temperature of iron alloys is lower than that of silicates and their density is higher. It is envisaged that this differentiation processes occurred into the planetesimals by segregation and into protoplanets by Rayleigh-Taylor instabilities (*Chambers*, 2005). When the protoplanets collided to form the terrestrial planets, their iron cores would have merged. The change in gravitational energy during the accretion processes controls the temperature difference between the core and the mantle (e.g., *Solomon*, 1979).

TABLE 1. Mass fraction of the major elements contained in EH enstatite chondrites (values from *Javoy*, 1995).

Element	Enstatite Model Mass Fraction
O	30.28
Fe	33.39
Si	19.23
Mg	12.21
Total	95.11
Ni	2.02
Ca	1.01
Al	0.93
S	0.85
Total	99.92

This example shows that 8 elements account for more than 99.9% of the total mass.

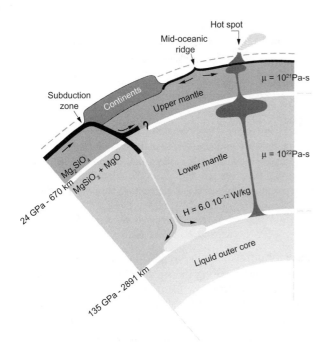

Fig. 1. Interior structure of Earth. Seismic data, laboratory experiments, and numerical simulations are used in order to simulate the interior structure and dynamics of Earth. Pressure and depth of major interfaces are indicated on the left. The description of each layer is described in the text.

The differentiation processes control the amount of iron that is left in the silicate mantle. This explains why the martian mantle is enriched in iron compared to that of Earth. Earth's mantle is divided into an upper mantle and a lower mantle. Partial melting in the mantle migrates to the surface to form the oceanic crust. Similar differentiation processes produce the continental crust. In the simulations for Earth described below, the layers have the following characteristics:

1. *The core.* The solid inner core (radius of 1220 km) is nestled inside a liquid outer core (radius out to 3480 km). These two layers are merged into one single iron-rich core, because the chemical differences between the two layers lead to minimal changes in total planet mass as calculated from the model.

2. *The lower mantle.* Bound by the major seismic discontinuities below (P ~ 135 GPa, radius = 3480 km) and above (P ~ 23 GPa, radius = 5701 km), the lower mantle is composed primarily of ~70% $(Mg,Fe)SiO_3$ perovskite ("pv"), ~20% $(Mg,Fe)O$ periclase, and ~10% $CaSiO_3$ perovskite. We merge $CaSiO_3$-pv into $(Mg,Fe)SiO_3$-pv. We note the likely existence of postperovskite near Earth's core-mantle-boundary (*Murakami et al.*, 2004), and the occurrence of a high-spin to low-spin crossover in $(Mg,Fe)O$ under deep mantle conditions (*Badro et al.*, 2003; *Sturhahn et al.*, 2005). Due to very limited equation of state data under these conditions for low-spin $(Mg,Fe)O$ and postperovskite, we exclude these phases in our model.

3. *The upper mantle.* This layer is mostly composed of olivine, ortho- and clino- pyroxenes, and garnet (e.g., *Vacher et al.*, 1998). By neglecting Ca and Al, the upper mantle is assumed to be simply made of olivine ($[Mg,Fe]_2SiO_4$) and orthopyroxene enstatite ($[Mg,Fe]_2Si_2O_6$). This simplification might imply an overestimate of the planet radii, because (1) garnet is denser than the other silicates, and (2) we neglect the transition zone as a distinct layer containing high-pressure olivine and garnet phases. This neglect introduces only ~0.2% error in the model-calculated total planet mass (see *Sotin et al.*, 2007).

4. *The crust.* With a laterally varying thickness from oceanic (~6 km) to continental (~20 to 60 km), the crust represents about 0.4% of Earth's mass. We therefore neglect the crust when calculating the mass-radius relationship.

In addition, two other planet layers might be considered but are omitted here because they also do not contribute significantly to the total planet mass and radius for Earth considered as an exoplanet: (1) The hydrosphere (oceans), which represents only 2.5×10^{-4} of Earth's mass. The oceans may contain more than 50% of the total H_2O. (2) The atmosphere, which is a very thin layer representing only 10^{-6} of Earth's mass.

The thickness of each layer depends on its elementary composition. The input parameters are the total mass of the planet, the composition of the star (Fe/Si and Mg/Si), and the Mg content of the mantle (Mg# = Mg/[Mg + Fe]). Together with the Fe/Si ratio, the Mg content and total planet mass determines the size of the core. Another approach (e.g., *Valencia et al.*, 2006) is to fix the size of the core. Both approaches give similar results. The Mg# is largely unknown except for Earth (~0.9) and Mars (~0.7). Its value depends on the degree of differentiation of a planet. The more differentiated the planet is, the larger the value of Mg#, because more Fe settles into the planetary core.

The abundances of the rock-forming elements Si and Mg, as well as Fe, can be measured in stellar atmospheres [see Fig. 2 and, e.g., *Huang et al.* (2005)]. The relevant issue for terrestrial planet models is whether or not the range of Si, Mg, and Fe found in stars needs to be considered. It is interesting to find that the variation of Mg, Si, and Fe in solar-type stars is not large enough to cause significant differences in model-calculated planet mass. Assuming a solar or stellar composition for the bulk composition of a planet is still debated (*McDonough and Sun*, 1995). Although CI chondrites have a solar composition, it has been suggested that the composition of Earth may be similar to that of EH enstatite chondrites (*Javoy*, 1995; *Mattern et al.*, 2005). In this case, rocks are more Si-rich and the ratio Mg/Si and Fe/Si are lowered to 0.734 and 0.878, respectively (Table 2). A precise description of the effect of varying Mg, Fe, and Si ratios will be given in section 3.1 but it can be already noted that varying molar ratios Fe/Si and Mg/Si in the range discussed above causes only small differences on computed planetary radii (0.3% on average).

The next step in modeling a terrestrial planet interior is to transform elementary composition into mineralogical com-

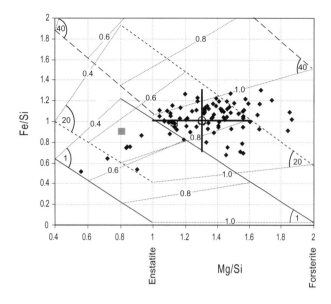

Fig. 2. The composition of the stars with orbiting planets compared to the solar composition (square) and the enstatite endmember composition (filled square). The circle is the average value of the relevant abundance ratios of all the stars in the study and the cross shows the uncertainty. Domains for different values of the core mass fraction are delimited by plain lines (1%), dotted lines (20%), and dashed lines (40%).

TABLE 2. Comparison of the solar (e.g., *Cox et al.*, 1999) and EH enstatite chondrite composition (values are based on Earth values).

	Solar		EH Enstatite Chondrites	
(Fe/Si)	0.977	0.986	0.878	0.909
(Mg/Si)	1.072	1.131	0.734	0.803

Two models are endmember models described in the text. For each composition, the first column indicates the solar ratio and the second the corrected ratio when the minor elements (Ca, Al, Ni) are replaced by the major elements as described in the text.

position for the mantle. If y_{LM} and y_{UM} are the iron content of each phase in the lower and upper mantle, respectively (Table 1), then

$$Mg\# = \left(\frac{Mg}{Mg+Fe}\right)_{silicates} = 1 - y_{LM} = 1 - y_{UM} \quad (1)$$

We assume magnesium and silicon are only found in the upper and lower mantle so that the ratio Mg/Si for each mantle must equal the input ratio. If x_{LM} and x_{UM} are the proportion of perovskite and olivine in the lower and upper mantle (Table 1), respectively, then

$$\left(\frac{Mg}{Si}\right)_{UM} = \left(\frac{Mg}{Si}\right)_{LM} \Leftrightarrow x_{LM} = 1 - x_{UM}/2 \quad (2)$$

For a given value of x_{UM}, the amount of iron is fixed in the mantle with equations (1) and (2). Iron being present in the silicate mantles and in the core, the ratio Fe/Si links the mass (and the size) of the metallic core (M_1) with x_{LM}. But there is only one set (x_{LM} ; M_1) that provides the correct mass for the whole planet. This is illustrated in Fig. 2, where the different lines illustrate different compositions depending on the value of both the Mg# and the relative amount of olivine. For example, if there is no core, the ratio Fe/Si of a pure enstatite mantle is equal to (1-Mg#) whereas it is equal to (2-2xMg#) for a pure olivine (dunite) mantle. Models of planetary differentiation suggest that the formation of the core of a planet during its accretion is the consequence of partial melting caused by the transfer of kinetic energy into heat when the planet becomes large enough. The melting temperature of iron alloys being lower than that of silicates and their density larger, Rayleigh-Taylor instabilities of the dense iron-rich liquids lead to the formation of the iron cores. The larger the planet, the more iron can be segregated into the core since the amount of gravitational energy available is proportional to the mass of the planet. Although the information from the solar system gives only two points (Mars and Earth), the trend is at least observed with smaller Mg# for Mars than for Earth. The upper bound for the metallic core can be calculated if all the iron segregated out of the silicates and migrated to form the core (Fig. 2).

Returning to the composition of stellar atmospheres, measured abundances define a domain bounded by an Fe/Si ratio and Mg/Si ratio between 0.6 and 1.7, and 0.8 and 2, respectively. Some stars are magnesium rich and are close to the forsterite limit while other stars are close to the enstatite limit (*Grasset et al.*, 2009). The average stellar abundance ratios are (Fe/Si = 1.1; Mg/Si = 1.3). By comparison, the Sun is enriched in Si compared to most of the stars. However, the uncertainties on stellar compositions are large. The abundance in a given element X is usually the comparison of the atomic ratio of the amount of this element to the amount of hydrogen (H) abundance with the solar ratio

$$[X] = \log\left(\frac{X}{H}\right)_{\star} - \log\left(\frac{X}{H}\right)_{\odot} \quad (3)$$

The atomic ratios Fe/Si and Mg/Si can therefore be calculated using the following equations

$$\left(\frac{Fe}{Si}\right)_{\star} = \left(\frac{Fe}{Si}\right)_{\odot} 10^{[Fe]-[Si]} \text{ and } \left(\frac{Mg}{Si}\right)_{\star} = \left(\frac{Mg}{Si}\right)_{\odot} 10^{[Mg]-[Si]} \quad (4)$$

The composition of stars hosting exoplanets (*Beirao et al.*, 2005; *Gilli et al.*, 2006) is an example that provides the values of the (Fe/Si) and (Mg/Si) ratios plotted in Fig. 2. These values are compatible with planet core mass fraction between 20% and 40% and Mg# between 0.6 and 0.9 (Fig. 2). The average of the stellar values is (Fe/Si; Mg/Si) = (1.0; 1.3). The solar Fe/Si ratio is similar whereas the solar Mg/Si is smaller (1.1). If an enstatite composition is assumed for the protoplanets that accreted to form Earth (*Javoy*, 1995; *Mattern et al.*, 2005), then the Mg/Si ratio is even smaller and the mantle composition would be very close to the pyroxene end member (no olivine) with a 30% core mass fraction. The exact composition of terrestrial planets compared to that of iron- and rock-forming elements in their host star is debated. The uncertainties on the stellar values of [Fe], [Mg], and [Si] are high, in the range 0.03 to 0.1, meaning that the abundance ratios also span a relatively wide range. The elemental abundance uncertainties can be converted into uncertainties of 0.3 to 0.5 for the Fe/Si and Mg/Si ratios (Fig. 2). These values are on the order of the difference between solar composition and EH composition (Table 2), at least for the Mg/Si ratio. Uncertainty about the iron content is the most critical for determining the mass-radius relationship because the molar mass of iron is much larger that that of Mg and Si.

2.2. Equations

The relation between the mass of a planet (M) and its radius (R) is derived from four equations: mass conservation, energy transport (generally described by an adiabat, together with temperature variations across boundary layers), hydrostatic equilibrium, and the equation of state (the relationship between density, pressure, and temperature for a given material in thermodynamic equilibrium).

The expression for the total mass and density distribution is

$$M = 4\pi \int_0^R r^2 \rho(r) dr \qquad (5)$$

where the density (ρ) depends on composition (section 2.1), temperature, and pressure.

The temperature profile is determined by the equations describing heat transfer. Earth's mantle behaves like a highly viscous fluid over geologic timescales and heat transfer is dominated by subsolidus convection. The temperature of Earth's mantle is at most a few hundreds of degrees lower than melting temperature (*Mosenfelder et al.*, 2009). In more massive planets, the temperature in the silicate mantle is controlled by the competition between the heating due to the decay of long-lived radiogenic elements and the cooling by subsolidus convection that is more and more efficient as temperature increases and viscosity decreases (*Tozer*, 1972). Although there is an extensive literature on convection processes, three-dimensional spherical models of thermal convection in a volumetrically heated fluid with isothermal boundaries are still limited by computer performance (see section 4). Earth's internal dynamics are characterized by hot plumes forming at the core-mantle boundary and cold sheets of oceanic lithosphere recycled into the mantle due to plate tectonics. The relationship between convection and plate tectonics is an active domain in Earth science and is discussed in section 4. For the purpose of modeling the planetary radius and the interior structure, it suffices to model the temperature profile by a large temperature difference at each layer interface and adiabatic temperature gradients in the convective mantle and in the convective core. As discussed in previous papers (*Valencia et al.*, 2006; *Sotin et al.*, 2007), the effect of temperature on the determination of the radius is small. On the other hand, the temperature profile controls the amount of partial melt in the mantle, which in turn affects degassing (*Elkins-Tanton and Seager*, 2008), viscosity and convection (see section 5), and hence planet evolution.

The adiabatic temperature profile is given by

$$\frac{dT}{dP} = \frac{\alpha T}{\rho C_P} = \frac{\gamma T}{\rho \Phi} \qquad (6)$$

with γ and Φ the Grüneisen and the seismic parameters, respectively

$$\gamma = \gamma_0 \left(\frac{\rho_0}{\rho}\right)^q \qquad (7)$$

$$\Phi = \frac{K_S}{\rho} = \frac{dP}{d\rho} \qquad (8)$$

In equations (6)–(8) T, P, ρ, α, C_p, and K_S are temperature (K), pressure (Pa), density (kg/m^3), thermal expansion coefficient (K^{-1}), heat capacity (J/kg/K), and adiabatic bulk modulus (GPa), respectively. The temperature within each layer can be computed using equations (6)–(8). Within the necessary accuracy of predicting compositions of exoplanets, the parameters are relatively well constrained for the candidate phases (Table 3) at pressure values relevant to Earth's case, with some exceptions. By applying equations (6)–(8) to a much larger pressure domain, we expect important deviations of the computed thermal profiles compared to the real profiles. But at the present time, it is not possible to provide an accurate model for planets much more massive than Earth, because very few experimental and theoretical constraints exist for silicates under these conditions (*Mosenfelder et al.*, 2009; *Umemoto et al.*, 2006).

The pressure (P) is computed using the hydrostatic equilibrium equation

$$\frac{dP(r)}{dr} = -\rho(r) g(r) \qquad (9)$$

The remaining equation is the equation of state (EOS), which links density to temperature and pressure. Two different approaches are commonly used in Earth sciences for describing the pressure and temperature dependences of materials (e.g., *Jackson*, 1998). One method introduces the effect of temperature in the parameters that describe the mineral's isothermal EOS and is achieved using the third-order Birch-Murnaghan EOS with the thermal effect incorporated using the mineral's thermal expansion coefficient

$$\begin{cases} P(\rho, T) = \frac{3}{2} K_{T,0}^0 \left[\left(\frac{\rho}{\rho_{T,0}}\right)^{7/3} - \left(\frac{\rho}{\rho_{T,0}}\right)^{5/3} \right] \\ \qquad \left\{ 1 - \frac{3}{4} (4 - K'_{T,0}) \left[\left(\frac{\rho}{\rho_{T,0}}\right)^{2/3} - 1 \right] \right\} \\ K_{T,0}^0 = K_0 + a_P (T - T_0) \\ K'_{T,0} = K'_0 \\ \rho_{T,0} = \rho_0 \exp\left(\int_{300}^{T} \alpha_{T,0} dT\right) \\ \alpha_{T,0} = a_T + b_T \cdot T - c_T \cdot T^{-2} \end{cases} \quad (10)$$

The relation between pressure, temperature, and density is then described using the eight parameters known at ambient pressure, T_0, ρ_0, K_0, $K'_{T,0}$, a_P, a_T, b_T, and c_T, the reference temperature, density, bulk modulus, pressure, and temperature derivatives of bulk modulus, and thermal expansion coefficients respectively (Table 3).

The second approach dissociates static pressure and thermal pressure by implementing the Mie-Grüneisen-Debye formulation

$$\begin{cases} P(\rho, T) = P(\rho, T_0) + \Delta P_{th} \\ P(\rho, T_0) = \frac{3}{2} K_0 \left[\left(\frac{\rho}{\rho_0}\right)^{7/3} - \left(\frac{\rho}{\rho_0}\right)^{5/3} \right] \\ \qquad \left\{ 1 - \frac{3}{4} (4 - K'_0) \left[\left(\frac{\rho}{\rho_0}\right)^{2/3} - 1 \right] \right\} \\ \Delta P_{th} = \left(\frac{\gamma}{V}\right) \left[E(T, \theta_D) - E(T_0, \theta_D) \right] \\ E = 9nRT \left(\frac{T}{\theta_D}\right)^3 \int_0^{\theta_D/T} t^3 dt / (e^t - 1) \\ \theta_D = \theta_{D0} \left(\frac{\rho}{\rho_0}\right)^\gamma \\ \gamma = \gamma_0 \left(\frac{\rho}{\rho_0}\right)^{-q} \end{cases} \quad (11)$$

where T_0, ρ_0, K_0, K'_0, θ_{D0}, n, γ_0, and q are the reference temperature, density, isothermal bulk modulus, pressure derivative of bulk modulus, reference Debye temperature, number of atoms per chemical formula, and scaling exponents, respectively (Table 3).

The third-order Birch-Murnaghan (BM) EOS is usually chosen for the upper mantle where the pressure range is limited to less than 25 GPa. Other EOS can be used, such as the Vinet EOS, which provides the same result as BM and MGD at low pressure, because the parameters entering into these equations are well-constrained from laboratory experiments. The Mie-Grüneisen-Debye (MGD) formulation is preferred for the lower mantle and core, as it permits a self-consistent approximation on the vibrational properties of the phases in the absence of limited experimental data. As noted in *Sotin et al.* (2007), the electronic pressure term should become important at higher pressures, requiring a precise description. Within the temperature range of Earth's lower mantle, the thermal pressure term in the Mie-Grüneisen-Debye formulation provides estimates close to EOS derived from select *ab initio* calculations and shock experiments (*Thompson*, 1990).

In order to solve for pressure, temperature, and density in the interior model, an iterative process is implemented until convergence is attained [see detailed descriptions in *Valencia et al.* (2006) and *Sotin et al.* (2007)]. As noted before, the input parameters are mass, composition (Mg/Si and Fe/Si), and proportion of iron in the mantle (Mg#). The model starts with estimated values for the core. At the end of the iteration process, the Fe/Si ratio is calculated and compared to the stellar value. A new core mass fraction is estimated using the bisection method. The code, as described in *Sotin et al.* (2007), iterates the computation of temperature, density, and pressure profiles until the difference between the computed Fe/Si and the stellar Fe/Si is less than 0.5%.

Super Earths are more massive than Earth and the validity of the EOS at much higher pressures and temperatures is questionable (*Grasset et al.*, 2009; *Valencia et al.*, 2009). The Vinet and MGD formulation appear to be valid up to 200 GPa (e.g., *Seager et al.*, 2007). Above ~200 GPa, electronic pressure becomes an important component that cannot be neglected. At very high pressure (P > 10 TPa), first-principles EOS such as the Thomas-Fermi-Dirac (TFD) formulation can be used (e.g., *Fortney et al.*, 2007; *Seager et al.*, 2007). The pressure at the core-mantle boundary (CMB) of terrestrial planets 5 and 10 times more massive than Earth is equal to 500 GPa and 1 TPa, respectively (*Sotin et al.*, 2007). In this intermediate pressure range, one possibility is to use the ANEOS code, which was developed to process shock experiments (*Thompson et al.*, 1990). The study by *Grasset et al.* (2009) compares the density-pressure curves of iron and forsterite using the MGD, TFD, and ANEOS formulations (Fig. 3). The TFD formulation, as expected, predicts values of densities much too small at low pressure. We emphasize that the TFD theory is for a pressure-ionized gas of noninteracting electrons; it cannot describe chemical bonds. In contrast, the ANEOS seems to fit the MGD at low pressure and the TFD at very high pressure. Therefore, the ANEOS appears to be a good choice in the intermediate pressure range from 0.2 to 10 TPa. However, laboratory experiments approaching this very high pressure range are required to confirm which phases are present at these high pressures and to determine the density of material.

TABLE 3. Parameters describing the equation of state, bulk composition of the layers,
mineralogical transformations, and thermal profiles in the planets (values based on Earth values).

	Variable		Layer 1		Layer 2				Layer 3			
Equation of State / Composition	Equation of state		EOS2		EOS2				EOS1			
	Phases		Iron-rich phase		Perovskite		Periclase		Olivine		Enstatite	
	Layer composition (%)		100		x_2		$1-x_2$		x_3		$1-x_3$	
	Components		Fe	FeS	$MgSiO_3$	$FeSiO_3$	MgO	FeO	Mg_2SiO_4	Fe_2SiO_4	$Mg_2Si_2O_6$	$Fe_2Si_2O_6$
	Phase composition (%)		80	20	$1-y_2$	y_2	$1-y_2$	y_2	$1-y_3$	y_3	$1-y_3$	y_3
	Density at ambient conditions (kg/m³)	ρ_0	8340	4900	4108	5178	3584	5864	3222	4404	3215	4014
	Reference temperature (K)	T_0	300		300		300		300		300	
	Reference bulk modulus (GPa)	K_0	135		254.7		157		128		105.8	
	Pressure derivative of the bulk modulus	K_0'	6,0		4.3		4.0		4.3		8.5	
	Debye temp. (K) α parameter (10^{-5} K^{-1})	θ_0/a_T	474/5.5		736/2.2		936/3.8		757/3.0		710/3.2	
	Gruneisen parameter / α parameter (10^{-8} K^{-2})	γ_0/b_T	1.36/NA		2.23/NA		1.45/NA		1.11/0.74		1.009/NA	
	Power exponent / α parameter (K)	q/c_T	0.91/NA		1.83/NA		3.0/NA		0.54/−0.5		1.0/NA	
	Temperature derivative of bulk modulus (GPa/K)		−0.045		−0.02		−0.019		−0.016		−0.026	
	References		[1],[2]		[3],[7]		[4],[7]		[5]		[6]	
Thermal Profile	Upper temp. drop (K)	ΔT	800		300				1200			
	References		[7]		[7]		[7]					

References: [1] *Uchida et al.* (2001); [2] *Kavner et al.* (2001); [3] *Mosenfelder et al.* (2009) (BM3S model); [4] *Hemley et al.* (1992), *Sinogeikin and Bass* (2000), *Sinogeikin et al.* (2000); [5] *Duffy et al.* (1995), *Bouhifd et al.* [1996]; [6] *Vacher et al.* (1998), *Angel and Jackson* (2002), *Jackson et al.* (2007); [7] *Poirier* (2000); [8] *Irifune* (1987). NA: Not applicable.

2.3. Earth and Solar System Terrestrial Planets

The planets Mars, Venus, and Earth are obviously better known than any terrestrial-like exoplanet. In order to validate the model approach described above, previous studies (e.g., *Valencia et al.,* 2006; *Sotin et al.,* 2007) have compared the observed radius to the computed values obtained with the numerous approximations. For the three solar system planets, the reference elementary composition (Mg/Si and Fe/Si) is the solar composition (Table 2). The difference between these extreme models is surprisingly less than 1% in radius (Table 3), which is quite small. In order to investigate the influence of chemical composition, values of the radius have been computed using a nonsolar composition corresponding to the composition of enstatite chondrites (*Javoy,* 1995, 1999), which represents another end member. In this case the amount of Mg is much smaller, leading to a smaller value of the Mg/Si ratio (Fig. 2). The resulting difference is that planet radius is much less than 1% (*Sotin et al.,* 2007). Similarly, if the value if the Mg# is changed from 0.9 to 0.7, the change in computed planet radius varies by less than 1% (*Sotin et al.,* 2007). Considering the uncertainties in the stellar composi-

tions (Fig. 2) compared to the difference between the solar composition and the terrestrial composition, the stellar composition can be assumed as a good approximation. However, one must note that this does not work for both Mercury and the Moon, which are enriched and depleted in iron, respectively.

For Earth, the locations of the interfaces between layers are well constrained but the composition of the deep layers relies on the assumption that the bulk composition is chondritic. In the results described in Fig. 4, the Fe/Si, Mg/Si, and Mg# are those of Model 1 given in Table 2. The model fit is quite good for Earth's pressure and density, two profiles that are provided by seismic inversion (*Dziewonski and Anderson,* 1981). The value of the model radius is 43 km (0.7%) larger than the value of 6371 km. Considering the assumptions made for this simulation, and in particular the fact that only four elements are taken into account (Si, Mg, Fe, O), the model fit is considered to be quite good. The temperature profile computed for Earth is, as expected, close to the real one since information from Earth was used to calculate the temperature profile. It is interesting to note that the value of Fe/Si that would provide an exact fit to the radius is 1.10 (Table 4), a value that is 12% larger than the solar value.

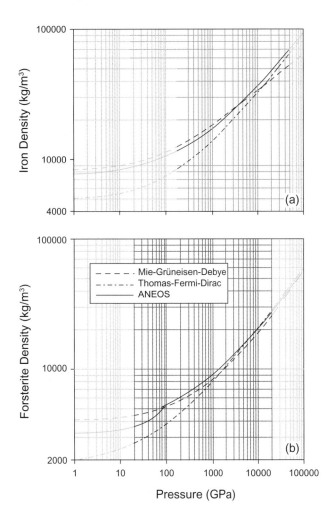

Fig. 3. Comparison between the different equations of state (EOS) described in the text for a pressure range up to 100 TPa. After *Grasset et al.* (2009).

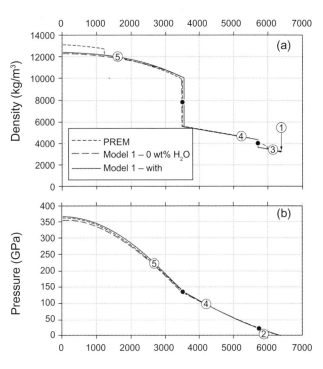

Fig. 4. Comparison between predictions and PREM for Earth with Fe/Si = 0.987, Mg/Si = 1.136, and Mg# = 0.9. The value of the radius is 6414 km, which is 0.6% larger than Earth's mean radius. The computed and measured profiles of both density and pressure cannot be distinguished except in the solid inner core for the density profile. The numbers 1, 2, and 3 refer to the core, lower mantle, and upper mantle, respectively.

The same model has been applied to Mars and Venus. For Venus, the Earth-like model (Mg# = 0.9) gives a radius only 5 km (0.1%) larger than the observed value. It suggests that the Earth-like model applies very well to Venus, an observation that was already made by *Anderson* (1980). For Mars, taking a terrestrial value for the Mg# gives a radius that is 40 km (1.5%) smaller than the measured value (Table 4). However, the measurements made on SNC meteorites suggest that the Mg# for Mars is smaller than Earth's value (*Dreibus and Wänke*, 1985). If one takes Mg# = 0.7 then the radius is only 23 km (0.7%) smaller (Table 4). A smaller value of the Mg# means that less iron differentiated into an iron core, which is in agreement with estimates of the core radius (*Yoder et al.*, 2003). Values of Mg# on the order of 0.75 have been calculated in order to explain the chemical composition of the basaltic meteorites supposed to have come from Mars (*Dreibus and Wänke*, 1985).

Mercury and the Moon, two objects much smaller than Earth, are bodies that had very different accretion histories than Earth, leading to different internal composition. Applying the model developed for Earth, therefore, does not yield accurate radii, giving a radius 12% too large for Mercury and 8% too small for the Moon. There are three scenarios that have been proposed to explain Mercury's high density (*Cameron et al.*, 1988): the equilibrium condensation scenario, planetary evaporation of the crust and part of the mantle, and a major planetary collision. The vaporization scenario (*Fegley and Cameron*, 1987) predicts elementary composition of the surface that will be tested by two forthcoming missions: NASA's MESSENGER mission and ESA's Bepi Colombo mission. It is interesting to note that if Mercury had the same Fe/Si ratio as the other terrestrial planets, its size would be 640 km less than Mars. This suggests that even without vaporization, the planet closest to the Sun would have been quite small. For the Moon, the situation is also different because the Moon formed as Earth had already differentiated into an iron-rich core and a silicate mantle. Therefore the Fe/Si ratio for the Moon is much lower than the solar value. A value of 0.22 for the Fe/Si ratio is found in order for the model to match the exact value of the Moon's radius (Table 3). In this case, the Moon would have an iron core 405 km in radius. Such a value is in agreement with values proposed by *Kuskov and Kronrod* (2001).

TABLE 4. Comparison of the radius observed for the solar system bodies and the predictions given by the models.

Name	Mass/$M_\oplus$	Planetary Radius					Best Fit			
		Measured	Model 1	Model 2	Model 3	Model 4	Model 1	Model 2	Model 3	Model 4
Mercury	0.055	2437	2705	2723	2706	2715	8	8	7.5	7.5
Mars	0.107	3389	3349	3366	3342	3357	0.78	0.84	0.71	0.79
Venus	0.81	6051	6056	6071	6008	6032	0.96	1.03	0.80	0.85
Earth	1	6371	6414	6447	6379	6405	1.10	1.19	0.92	0.99
Moon	0.0123	1738	1600	1642	1591	1621	0.22	0.48	0.30	0.30

The different models correspond to solar composition (models 1 and 2) and EH composition (models 3 and 4). Models 1 and 3 have a value of the Mg# equal to 0.9, whereas models 2 and 4 have a values of the Mg# equal to 0.7. The right part of the table indicates the required value of Fe/Si (Earth-like planet) in order to get the value of the measured radius for each body. Results from *Sotin et al.* (2007).

2.4. Exoplanets: Super Earths, Super Mercuries, and Icy Planets

We now turn to a review of exoplanet interior composition calculations. Although the topic is not as mature as interior structure modeling for solar system terrestrial planets, a number of fundamental issues can still be described.

One goal in exoplanet interior modeling is to provide relationships between mass and radius. The studies by *Valencia et al.* (2006) and *Sotin et al.* (2007) do this for planets up to 10 $M_\oplus$. The mass-radius relationship can be written

$$R/R_\oplus = a\left(M/M_\oplus\right)^b \quad (12)$$

where (a,b) is equal to (1, 0.274) for planets less than 10 $M_\oplus$. Values of different studies are very close to each other (Table 5). One can note that the more massive the planet, the smaller the value of parameter "b" because the compression due to the pressure effect becomes more and more important. Also, *Grasset et al.* (2009) propose a more complex equation where the parameter "b" depends on the mass of the planet. The values that they give are very close to those proposed by *Valencia et al.* (2006) although the EOS and the parameters in those EOS were different. The mass-radius relationship for terrestrial planets can be plotted (Fig. 5). A 10-$M_\oplus$ planet is about twice as large and a 100-$M_\oplus$ planet is only 3.2 times as large (Fig. 5). For an example of a study that considers mass-radius relationships for planet masses >10 $M_\oplus$, see *Seager et al.* (2007). For a description of why mass-radius relationships for exoplanets of different internal composition follow a similar functional form, see the detailed discussion in *Seager et al.* (2007).

Another goal of exoplanet interior modeling (using the modeling approach described in section 2) is to investigate the mass dependence of several planetary characteristics such as mean density, radius of the core, depth of the upper-lower mantle interface, equilibrium heat flux, and other properties. The mean density being the mass divided by the volume, its dependence varies as $(M/M_\oplus)^{1-3*b}$, which is equal to $(M/M_\oplus)^{0.178}$ if one takes b as being equal to 0.274. A 10-$M_\oplus$ planet would have a density 50% larger (Table 6). The surface gravity acceleration (not taking into account the centrifugal acceleration) increases as $(M/M_\oplus)^{1-2*b}$, which is equal to $(M/M_\oplus)^{0.452}$. Therefore the gravity acceleration of a 10-$M_\oplus$ planet is about three times larger. It means that the pressure gradient in such planets is three times larger than on Earth, which has implications for the depth of phase transitions and mineralogical transformations, which are pressure-dependent. Both the core and the mantle thickness increase with increasing mass. On the other hand, the thickness of the upper mantle decreases because of the larger pressure gradient described above (Fig. 6).

The radius of the core depends on the EOS that one takes for iron and its state depends on the iron phase diagram at pressures relevant to the center of exoplanets. There is a lack of experimental data to constrain the EOS of iron and silicates at pressures above 500 GPa (e.g., *Poirier*, 2000; *Valencia et al.*, 2009). Such pressures exist within planets that are only a few Earth masses. For example, the central pressure of a 5-$M_\oplus$ planet is 1700 GPa (*Sotin et al.*, 2007). Therefore more reliable results have to wait for laboratory experiments that will confirm *ab initio* calculations (*Volcadlo et al.*, 1997; *Alfe*, 2009). Several studies have investigated the melting temperature of iron alloys in order to assess the state of Earth's iron core (e.g., *Williams et al.*, 1987; *Anderson and Ahrens*, 1994, 1996; *Boehler*, 1993; *Alfe*, 2009). The melting temperature of Fe-FeS alloys is smaller than that of silicates at the same pressure. It suggests that cores would be liquid (Fig. 7). However, the large uncertainties of the composition of metallic cores (e.g., *Poirier*, 1994) and the lack of experimental data beyond 300 GPa make any conclusion very tentative. The state of the iron core is a critical parameter in order to assess the existence of an internal magnetic field that would protect the atmosphere from bombardment of energetic stellar ions that would efficiently erode it.

Another parameter that is important to consider when simulating the thermal evolution of terrestrial planets is the radioactive heat flux. On Earth, about 50% of the heat flux comes from the decay of the long-lived radioactive elements

TABLE 5. Coefficients a and b of the law $(R/R_⊕) = a(M/M_⊕)^b$ for the different cases explained in the text.

$M/M_⊕$ Range of mass	Terrestrial Planets a	b	Ocean Planets (50 wt%) a	b	Super Mercury a
0.01–1*	1.00	0.306	1.258	0.302	
1–10*	1.00	0.274	1.262	0.275	
1–10†	1.00	0.267–0.272			~0.30
1–100‡	1.01	0.252–0.285	1.253	0.239–0.272	

*Sotin et al. (2007).
†Valencia et al. (2006).
‡Grasset et al. (2009).

such as ^{40}K, ^{235}U, ^{238}U, and ^{232}Th. The other 50% is the cooling of the planet, which depends on the efficiency of mantle convection (see section 3). Assuming that the amount of radioactive elements is proportional to the mass, then the radiogenic heat flux varies as $(M/M_⊕)^{0.452}$ (Fig. 8 and Table 6). The radiogenic heat flux is equal to the radiogenic heating rate divided by the surface area. The heat flux scaling results from the radiogenic heating rate being proportional to planet mass and the planet mass being proportional to $R^{0.274}$.

The existence of icy planets such as more massive versions of Ganymede is very likely. It has been proposed (*Kuchner*, 2003; *Léger et al.*, 2004) that some of these planets could turn into water planets. These planets would have formed in the outer stellar system at a distance where H_2O ice can condense and would have then migrated by a type 2 migration toward their star. The recent discovery of OGLE-2005-BLG-390Lb (*Beaulieu et al.*, 2006) is a possible example.

Calculations for a planet that contains 50% H_2O by mass (Fig. 5) suggest that the radius would be more than 25% larger compared to a planet with an Earth-like interior composition (*Sotin et al.*, 2007). *Grasset et al.* (2009) expanded the calculations and provided relationships for any amount of H_2O. With an H_2/He envelope of about 10% by mass, and assuming that Uranus and Neptune are otherwise composed of water, the Uranus and Neptune H_2O mass fraction would need to be as large as 75% in order to fit the mass and radius. The choice of EOS (equation (11)) for the high-pressure phase of ice is discussed in *Sotin et al.* (2007). It must be noted that they use a value of q = 1 instead of the surprising negative value published by *Fei et al.* (1993) in order to compute adiabatic profiles at pressures larger than 60 GPa. Another implication of the presence of the thick H_2O layer is that the pressure at the interface with the silicates is larger than the pressure at the upper-lower mantle interface. Therefore the upper mantle layer would not exist in such planets.

If the surface temperature becomes larger than 300 K, ice melts and an ocean forms. Because the melting temperature of the low-pressure phase of ice (ice I) decreases with increasing pressure, the depth of the ocean is limited by the melting temperature of the high-pressure phase of ice. *Sotin et al.* (2007) have investigated the thickness of an ocean layer as a function of the mass of the planet. For a H_2O mass fraction of 50%, the thickness varies from 475 km to 50 km for planets 1 and 10 times Earth's mass, respectively. The smaller thickness of the ocean with increasing mass is due to the pressure gradient effect described earlier. Other parameters such as heat flow, surface temperature, solutal convection, latitudinal circulation, superrotation, and others may affect the thickness of the ocean. As noted in *Léger et al.* (2004), for such large amounts of H_2O the liquid ocean is not in contact with the silicate shell. It means that if volcanism exists in the silicate shell, this volcanism will be in contact with high-pressure ice and not with liquid, which gives conditions quite different from those existing on Earth's seafloor.

Application of this model to the large icy moons of Jupiter and Saturn gives very good results, although these moons have masses much lower than Earth's mass. The icy moons Ganymede, Callisto, and Titan have an H_2O mass fraction close to 50% (*Sotin and Tobie*, 2004). If a three-layer structure is assumed (silicate mantle and a layer with the density

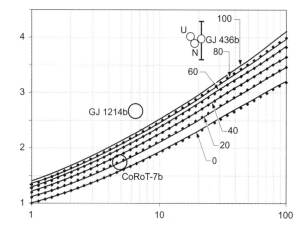

Fig. 5. Mass-radius curves for planets with Earth-like composition and for water planets. The amount of H_2O varies from 0% (almost Earth-like) to 100%. The large icy satellites of Jupiter and Saturn have about 50% water by mass. The two circles correspond to CoRoT-7b and GJ 1214b, the smallest planets for which both mass and radius have been determined; the size of the circle approximately represents the uncertainty in interior composition.

TABLE 6. Values of the core radius, upper mantle thickness, mantle thickness, and
radioactive flux for planets with different masses.

Mass Relative to Earth	Density (kg/m³)	Radius (km)	Core Radius (km)	Upper Mantle Thickness (km)	Mantle Thickness	Radioactive Heat Flux (mW/m²)
1	5515	6422	3074	658	3348	47
2	6239	7866	3698	494	4168	64
3	6706	8838	4104	416	4734	77
6	7587	10730	4870	310	5860	106
10	8309	12308	5486	248	6822	133

of ice VI covered by a layer with the density of ice I), the calculated radii of the large icy moons match the actual radii by better than 1%. *Sotin et al.* (2007) have also applied this model to Uranus and Neptune. These outer planets have a thick hydrogen-rich outer layer that is taken care of by reducing both the radius and the mass by 1 $R_\oplus$ and 2 $M_\oplus$, respectively (*Hubbard et al.*, 1995; *Podolak et al.*, 2000). The curves on Fig. 5 show that if composed predominantly of H, He, and H_2O, these planets must contain more than 50% mass fraction of H_2O. Not shown is that Uranus and Neptune may alternatively have interiors dominated by rock, with much less ice and a more massive H/He envelope.

So far iron-rich, silicate-rich, and icy planets have been described in detail, without regard to how to identify them. In reality, deducing exoplanet interior composition is very difficult. Only two data points are available per planet: mass and radius. Even with perfect measurements there is simply not enough information to uniquely identify a planet's interior composition. Two exceptions are at the density extremes; giant planets of low density must be composed almost entirely of H and He, and any planet of extremely high density must be iron-dominated because iron is the densest cosmically abundant substance. See section 5 for further discussion of interior composition degeneracies.

3. DYNAMICS

The likelihood of plate tectonics on exoplanets larger than Earth can be assessed using either scaling laws or numerical models describing mantle thermal convection. Two papers, *Valencia et al.* (2007) and *O'Neill and Lenardic* (2007), came to opposite conclusions based on scaling laws and numerical calculations, respectively. *Valencia et al.* (2007) conclude that as planetary mass increases, the shear stress available to overcome resistance to plate motion increases while the plate thickness decreases, thereby enhancing plate weakness. These effects contribute favorably to the subduction of the lithosphere, an essential component of plate tectonics. On the other hand, the numerical simulations described by *O'Neill and Lenardic* (2007) suggest that increasing planetary radius acts to decrease the ratio of driving to resisting stresses. They conclude that super-sized Earths are likely to be in an episodic or stagnant lid regime.

Here we review the major concepts and definitions relevant for plate tectonics. The different assumptions and parameters used in each study are first described. The definition of thermal boundary layer and lithosphere is clarified. It is confirmed that scaling laws predict that the more massive the planet, the more likely the occurrence of plate tectonics. No scaling laws exactly describe the numerical calculations reported in the study by *O'Neill and Lenardic* (2007), where the vigor of convection of a fluid heated from both within and below is examined. The lack of information on scaling laws describing velocity and the ratio of driving forces to resistance is emphasized. Finally, simulations of the thermal evolution of terrestrial planets (e.g., *Papuc and Davies*, 2008) are described.

3.1. Mantle Convection Processes in Terrestrial Exoplanets

A planet's internal dynamics are driven by solid-state convective heat transfer in the mantle. Heat sources include radiogenic internal heating due to the decay of the long-lived radioactive elements ^{40}K, ^{232}U, ^{235}U, and ^{242}Th, and the initial heat stored in the planet during accretion and differentiation. The convective processes in the mantle control the thermal evolution of the planet (e.g., *Schubert et al.*, 2001). In a fluid that is heated from within, cooled from the top (cold surface temperature), and heated from below (hot core), cold plumes form at the upper cold thermal boundary layer and hot plumes form at the hot thermal boundary layer, which corresponds to the core-mantle boundary (e.g., Sun et al., 2007). The efficiency of heat transfer is mainly controlled by the mantle viscosity, which depends on a number of parameters including, but not limited to, mineral composition of the mantle, temperature, pressure, and grain size. Laboratory experiments (*Davaille and Jaupart*, 1993) and numerical studies (*Solomatov and Moresi*, 2000; *Grasset and Parmentier*, 1998) have shown the major role of temperature-dependent viscosity. Scaling laws have been derived and then employed to predict the thermal evolution of exoplanets (e.g., *Valencia et al.*, 2007; *Papuc and Davies*, 2008). Scaling laws can also predict the velocity and hence the shear stress acting below the lithosphere. If the shear stress is large enough, it may overcome the resistance of the lithosphere and produce faulting. This would trigger the plate tectonics regime as described in both studies by *O'Neill and Lenardic* (2007) and *Valencia et al.* (2007).

Scaling laws based on the stability of the thermal boundary layers are used to describe the heat that can be removed by convective processes. Laboratory experiments (e.g., *Davaille and Jaupart,* 1993) and numerical simulations (*Grasset and Parmentier,* 1998) have shown that the temperature difference across the upper thermal boundary layer (ΔT_{TBL}) is proportional to a viscous temperature scale (ΔT_η) defined by

$$\Delta T_\eta = \left| \frac{1}{\partial Ln(\eta)/\partial T} \right| \quad (13)$$

where the viscosity η strongly depends on temperature. The viscosity of Earth's mantle can be determined by studying the postglacial rebound and the temperature of the mantle (T_m) is estimated around 1350°C at 80 km depth from studies of mantle rocks sampled by kimberlites (e.g., *Bertrand et al.,*

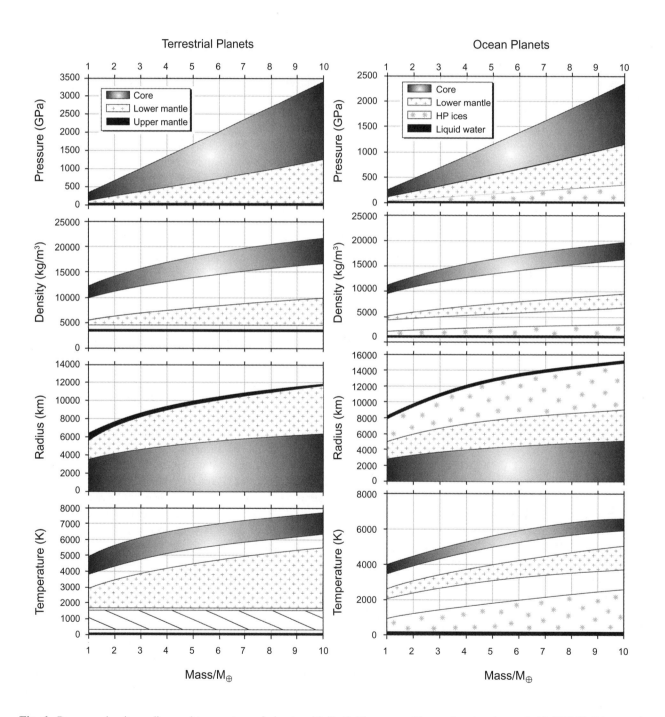

Fig. 6. Pressure, density, radius, and temperature of planets with Earth-like composition and water planets (with 50% H$_2$O by mass). The mass of exoplanets varies from 1 to 10 M$_\oplus$. From *Sotin et al.* (2007).

1986). It seems relevant to use η (1350°C) = 10^{21} Pa s^{-1} as an anchor point for the viscosity law. Laboratory experiments suggest that the deformation rate of a solid is a thermally activated process that can be described by an Arrhenius-type law

$$\eta = A e^{\frac{Q}{RT}} \quad \text{or} \quad \eta = \eta_0 e^{\frac{Q}{R}\left(\frac{1}{T} - \frac{1}{T_0}\right)} \quad (14)$$

where Q is the activation energy (Q = E + P ΔV), R the gas constant, and (η_0, T_0) = (10^{21} Pa s^{-1}; 1350°C). The viscous temperature scale is therefore

$$\Delta T_\eta = \frac{RT_m^2}{Q} \quad (15)$$

Different viscous laws are used to describe how viscosity depends on temperature. *Valencia et al.* (2006) use the expression $\eta = \eta_0(T/T_0)^{-30}$, which gives a viscous temperature scale equal to $T_m/30$. The two viscous temperature scales are equivalent at low temperature but differ significantly at large temperatures with the Arrhenius-type law, giving a viscous temperature scale 15% and 80% larger at a temperature equal to 1350°C and 2250°C, respectively.

As illustrated in Fig. 9, the thermal boundary layer is under the conductive lid. Its thickness δ is controlled by the value of the thermal boundary layer Rayleigh number (Ra_{TBL}) defined as

$$Ra_{TBL} = \frac{\alpha \rho g \Delta T_{TBL} \delta^3}{\kappa \eta(T_m)} \quad (16)$$

where α is the coefficient of thermal expansion, ρ is the density, ΔT_{TBL} = 2.25 ΔT_η = 2.25 RT_m^2/Q, and κ is the thermal diffusivity. The value of the thermal boundary layer Rayleigh number depends on the boundary conditions but numerical simulations and laboratory experiments suggest that a value of 20 is adequate. This value must not be confused with the value of the critical Rayleigh number for convection to occur, which is around 1000. For a given value of the mantle temperature T_m, one can calculate the viscosity (equation (14)), the thickness of the thermal boundary layer (equation (4)), and the temperature across the thermal boundary layer. This provides the heat flux (q) that is transferred by convection

$$q = k \frac{\Delta T_{TBL}}{\delta} \quad (17)$$

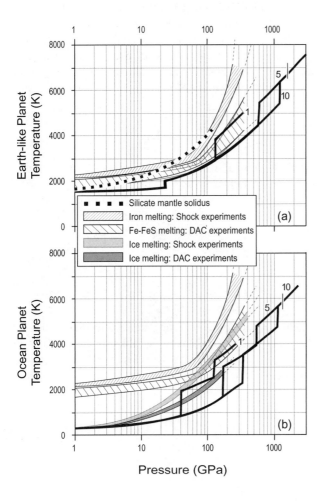

Fig. 7. Comparison between temperature profiles and melting curves of H$_2$O and iron alloys. Melting domains of iron (*Williams et al.*, 1987; *Anderson and Ahrens*, 1994, 1996; *Boehler*, 1993; *Alfe et al.*, 2002) and ice VII (*Mishima and Endo*, 1978; *Fei et al.*, 1993; *Frank et al.*, 2004) are plotted as a function of pressure. Above these domains, the component is always liquid. **(a)** Earth-like planets: Thermal profiles for 1, 5, and 10 $M_\oplus$ are above the melting curve of the FeS component in the core (P,T) domain. **(b)** Ocean-planets: The transition from liquid to ice VII occurs at low pressure (between 1 and 2 GPa depending on the surface temperature). The iron core is probably liquid, but whatever the planetary mass, it is colder and at lower pressure than for the Earth-like case.

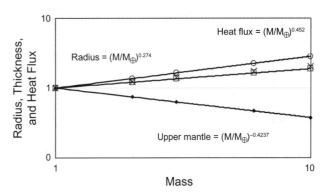

Fig. 8. Equilibrium heat-flux, radius, and thickness of the upper mantle thickness for different values of planet mass. The equilibrium heat flux is the internal radiogenic heating rate divided by the surface area. Scaling relationships are normalized to Earth values.

where k is the thermal conductivity. In order to compare the vigor of convection for different planet sizes, it is assumed that there is equilibrium between internal heating (radiogenic heating) and heat flux. As described above, the surface heat flux varies almost as the square root of mass ($M^{0.452}$). One must note that in this surface heat flux expression, the secular cooling is omitted.

Taking an Earth-like planet without plate tectonics and the concomitant efficient heat removal as a reference, the equilibrium heat flux would be 47 mW/m² , which is the heat removed by convection if the viscosity is equal to 1.7×10^{19} Pa s⁻¹. Such a low viscosity corresponds to a mantle temperature of 1925 K (1650°C). As discussed in *Papuc and Davies* (2008), such a high value of the temperature produces a lot of partial melt in the mantle. This temperature may reflect the present time state of Venus as previous numerical studies suggested (e.g., *Arkani-Hamed*, 1994; *Sotin and Labrosse*, 1999; *Moresi and Solomatov*, 1998).

Then the thickness of the lithosphere (D) can be calculated, assuming continuity of temperature and heat flux at the conductive lid/thermal boundary layer interface (Fig. 9). It is interesting to note that the mass dependence is the same for the thickness of both the thermal boundary layer and the lithosphere thickness ($M^{-0.39}$). The lithosphere thickness decreases with increasing mass because the heat flux (equation (17)) is larger for planets of increasing mass (Fig. 10). The Rayleigh number can be calculated using (b–D) as the thickness of the convective layer where b is the thickness of the mantle

$$Ra = \frac{\alpha \rho g \Delta T_{tbl}(b-D)^3}{\kappa \eta(T_m)} \quad (18)$$

The set of equations (13)–(18) can be used in order to predict the thermal evolution of an exoplanet (*Papuc and Davies*, 2008). The determination of the Rayleigh number (equation (18)) is used to estimate velocities and stresses below the lithosphere and to assess the likelihood of plate tectonics.

3.2. Scaling Laws: Relations Between the Vigor of Convection and Plate Tectonics

The lithosphere breaks when the stresses induced by convection become larger than the yield stress (*Moresi and Solomatov*, 1998). This approach was used by *O'Neill and Lenardic* (2007). In order to compare with the study of *Valencia et al.* (2007), scaling laws for stresses are now derived.

The velocity (u) of the plumes and the horizontal velocity within the boundary layers depend on the vigor of convection (*Schubert et al.*, 2001). Numerical simulations suggest that the velocity scales to the Rayleigh number (equation (18)) as

$$u = 0.12 \frac{\kappa}{(b-D)} Ra^{2/3} \quad (19)$$

Applied to an Earth-like planet without plate tectonics, equation (19) gives a value on the order of 1 cm/yr, which is compatible with values of plate velocities on Earth.

By equilibrating stresses on the lithosphere, the normal stress (σ) applied on the lithosphere by the convection process can be written

$$\sigma = \tau \frac{L}{D} \quad (20)$$

where τ is the convective shear stress affecting the base of the lithosphere on a distance L, which is the width of the convective cell. This length (L) is also the distance between hot plume and cold plume, or half the distance between

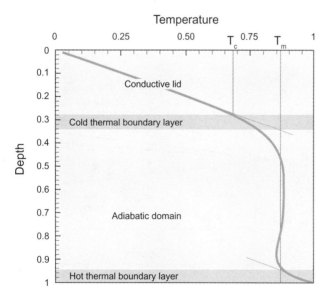

Fig. 9. Schematic view of the temperature profile in a fluid heated from within and from below.

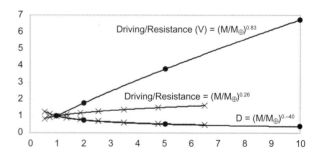

Fig. 10. Ratio of driving forces to resistance (yield strength) for two plate tectonic studies (*O'Neill and Lenardic*, 2007; *Valencia et al.*, 2007). The ratio is much higher in Valencia et al.'s study due to a larger shear stress and a larger distance between plumes. Note that the trend for the lithosphere thickness is similar for both studies.

the cold plumes if no hot plumes are present in the case of convection processes in a volumetrically heated fluid (e.g., *Parmentier et al.*, 1994). The shear stress can be written (*Schubert et al.*, 2001)

$$\tau = \eta \frac{u}{(b-D)/2} \quad (21)$$

Following *Valencia et al.* (2006), the length of the plate (L) can be defined as the time it takes for the thermal boundary layer to reach its thickness δ and it is found to increase with increasing mass

$$L \approx \frac{u\delta^2}{\kappa} \approx M^{0.32} \quad (22)$$

Therefore, one can determine the value of the normal stress, which balances the shear stress, and it is found that it increases with increasing mass

$$\sigma = \tau \frac{L}{\delta} \approx M^{0.86} \quad (23)$$

The convective stress (equation (23)) has to be compared with the yield stress, which increases with increasing mass because of the pressure gradient effect mentioned above. The ratio of these two stresses increases with increasing mass (Fig. 10) as $M^{0.83}$.

The next part of the present study is to scale the numerical experiments of *O'Neill and Lenardic* (2007) (in order to compare with *Valencia et al.*, 2007). First, one must list the hypotheses of their simulations, which include the following:

1. Although the viscosity law is not given in *O'Neill and Lenardic* (2007), they use the model described in the study by *Moresi and Solomatov* (1998), in which the viscous law is the Frank-Kamenetskii approximation described by

$$\eta = \eta_0 e^{-\gamma\left(\frac{T-T_0}{T_1-T_0}\right)} \quad (24)$$

where η_0 is the viscosity at the surface temperature (T_0 = 300 K) and γ is a constant set to 14 (*Moresi and Solomatov*, 1998). Therefore, the viscous temperature scale is equal to $(T_1-T_0)/\gamma$. Since the basal temperature seems to be the same whatever the scaling factor, the viscous temperature scale and therefore the temperature difference in the cold thermal boundary layer is the same and equal to 235 K.

2. The aspect ratio has been set up to 4. In their Fig. 1, the conductive lid regime leads to a much higher internal temperature than that in the plate tectonics regime. The number of cold plumes is equal to 7 and no hot plume can be seen. This is in agreement with the work by *Parmentier and Sotin* (2000) for a volumetrically heated fluid and with the predictions of the scaling laws.

3. The heat flux is set up such that the ratio of internal heating to basal heating is 64–36% for the reference active-lid state. When the convection pattern turns from active lid regime into the conductive lid regime, the amount of heat that can be removed is much less. Little is said in *O'Neill and Lenardic* (2007) about the change in heat flux. However, their Fig. 1 shows that the conductive lid is thick and that the heat flux must quite small.

In the case of mixed heating, one can use the isoviscous scaling laws proposed by *Sotin and Labrosse* (1999)

$$\theta = \frac{\Delta T_{TBL}}{T_1 - T_c} = \frac{T_m - T_c}{T_1 - T_c} = 0.5 + 1.256 H_S^{3/4}/Ra^{1/4} \quad (25)$$

$$q = k\frac{T_c - T_0}{D} = 0.3446k\frac{T_1 - T_c}{b-D}(\theta)^{4/3}(Ra)^{1/3} \quad (26)$$

where

$$Ra = \frac{\alpha\rho g(T_1 - T_c)(b-D)^3}{\kappa\eta(T_m)}$$

and

$$H_S = \frac{H(b-D)^2}{k(T_1-T_c)}$$

with H the volumetric heating rate in W/m³. In order to solve for T_c, T_m, D and q, a fourth equation is needed, which is the temperature difference across the boundary layer given by equation (13). When these equations are used, the heat flux is a factor of 4 lower for the conductive lid regime than that for the active lid regime and the temperature of the convective mantle increases to reach a value larger than 1500 K. The heat flux is smaller than the heat flux necessary to transfer just the radiogenic heating. It results in an increase of the mantle temperature until the equilibrium is reached. The heat flux predicted by this scaling is almost identical to the one predicted by scaling laws for fluids with temperature-dependent viscosity because the viscosity variations in the convective domain are small. When this scaling is applied, the ratio of driving force to resistance still increases with increasing mass (~$M^{0.26}$) but with a much smaller coefficient (Fig. 10).

The fact that the two-dimensional numerical experiments of *O'Neill and Lenardic* (2007) predict the transition from the conductive lid regime into the active lid regime (plate tectonics) at different mass than the predictions of scaling laws is not surprising. First, scaling laws are one-dimensional, whereas the brittle failure of the lithosphere may be easier at either upwelling plumes or downwelling plumes that can only be realistically accounted for by three-dimensional models. Second, the scaling laws describing the horizontal velocity in the thermal boundary layer (equation (19)) are based on simulations for isoviscous fluids. It is therefore important to determine the scaling laws for fluids having complex velocities. Third, the length of the plate plays a major role in the scaling but it is poorly constrained. For example, the terrestrial plates are longer than the proposed scaling (equation (22)) and small-scale convection has been proposed as the oceanic plate thickens (e.g., *Davaille and Jaupart*, 1994).

The conclusion about plate tectonics is that the scaling of the driving to resistance forces is not yet available to determine whether planets more massive than Earth are likely to have plate tectonics. Further simulations are required and the effects of different parameters including the geometry, the thickness of the upper mantle, and the presence of water, the viscosity, and composition of the mantle need to be investigated.

4. RECENT HIGHLIGHTS

Out of the hundreds of known exoplanets, only a few have a mass less than 10 $M_\oplus$ (Fig. 11). The HARPS instrument found a planet with a minimum mass of 1.9 $M_\oplus$ (*Mayor et al.*, 2009). In this GJ 581 system, three other planets were already known to orbit around this M dwarf, including two with a minimum mass less than 10 $M_\oplus$. Another important discovery is the detection of OGLE-2005-390-Lb by gravitational lensing around an M star (*Beaulieu et al.*, 2006). This planet would have a mass equal to $5.5^{+5.5}_{-2.7}$ times Earth's mass. It orbits its star at about 5 AU, a distance where the blackbody temperature is about 50 K. While measurements of both mass and radius is available for almost 100 planets (Fig. 12), only three of these planets have masses much lower than 100 $M_\oplus$ (Figs. 12 and 13): GJ 436b, GJ 1214b, and CoRoT-7b (Table 7). CoRoT-7b lies on the Earth-like composition curve, whereas GJ 436b and GJ 1214b have much lower densities, which make them consistent with ice giants. However, these two planets could instead have H/He envelopes surrounding a predominantly rocky interior.

TABLE 7. Low-mass exoplanets with known radius.

	Mass ($M_\oplus$)	Radius ($R_\oplus$)	Distance to the Star (AU)
CoRoT-7b	4.8 (<11)	1.72	0.017
GJ 436b	22.8	4.92	0.0287
GJ 1214b	6.55	2.68	0.0144

4.1. Exoplanet Internal Structure: The Case of CoRoT-7b

Searching for transiting planets, the Convection Rotation and Transit (CoRoT) telescope (*Rouan et al.*, 1999; *Borde et al.*, 2003) reported a planet with a radius equal to 1.7 $R_\oplus$ (*Léger et al.*, 2009) and a mass lower than 11 $M_\oplus$. This discovery was followed by HARPS observations to refine the value of the planet mass to around 4.8(±0.8) $M_\oplus$ (*Queloz et al.*, 2009). This planet, known as CoRoT-7b, has a period of 0.854 days around a G9V star. Such a planet is on the Earth-like planet mass-radius curve (Fig. 5). If the radius is 1.72 $R_\oplus$, the mass should be 7.24 $M_\oplus$ according to the model for Earth-like planet composition described in section 2. Considering an uncertainty of 0.13 on the radius, the mass uncertainty would be around 0.45 $M_\oplus$ (*Grasset et al.*, 2009). CoRoT-7b has a lower mass than needed for an Earth-like composition, which suggests the presence of low-density material. Due to expected atmospheric escape for a planet in an 0.8-d orbit about a Sun-like star, it is hard to imagine how CoRoT-7b could have retained a thick atmospheric layer of H_2 and He because the planet orbits so close to its host K star: 0.0172 AU (see also the discussion on degeneracies in section 4.2). If there is little to no atmosphere, it would imply the presence of H_2O at depth. To further expand on CoRoT-7b's density, a planet with (mass; radius) = (4.8; 1.72)$_\oplus$ has a density of 5.21. However, an Earth-like composition planet that size would have a density of 7.85 due to the effect of compression from pressure. With these numbers, the surface gravity is equal to 16 m/s².

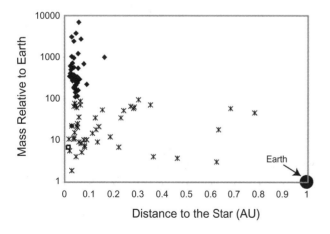

Fig. 11. Mass vs. semimajor axis for all planets observed by transits (filled diamonds) and for the smallest planets observed by any detection method (stars). Only three planets less than 100 $M_\oplus$ have been observed by transit and radial velocity (CoRoT-7b, GJ 1214b, and GJ 436b).

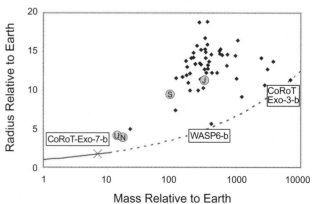

Fig. 12. Mass vs. radius for planets observed by transit.

CoRoT-7b is so close to its star that it must be locked in synchronous rotation. The dayside could have temperatures as high as 2000 K whereas the nightside could be very cold depending on how the atmosphere carries the heat all around the planet. Such a high temperature is quite incompatible with the presence of H_2O. Also, the escape rate of gaseous species must be quite high. Building an interior model for CoRoT-7b is a difficult task if one does not know the atmospheric composition. For illustration, an Earth-like composition planet the size of CoRoT-7b would have an iron core about 5100 km in radius and a silicate mantle about 6000 km thick with a thin upper mantle of only 300 km, which is more than two times thinner than Earth's upper mantle. Such a planet would have a radiogenic heating rate equal to 170 TW (24 TW for Earth)

at t = 4.5 G.y. and 510 TW at t = 0. The equivalent heat flux is equal to 115 and 345 mW/m², respectively. If convection processes can remove more heat than produced, then the planet can cool down. Otherwise it would keep heating up until convection processes are vigorous enough to remove more heat than radiogenic heating (or other internal heat sources). With a surface temperature of 2000 K, it is likely that a magma ocean would be present on the dayside. Also, tidal dissipation may provide an additional heating source. Plate tectonics is difficult to imagine in such a context, because the strong heating implies that the plates may not exist or would be too ductile. CoRoT-7b may be volcanically active due to tidal stresses and may resemble Io, Jupiter's most volcanically active moon.

4.2. Degeneracies: Example of GJ 1214

The above discussion on CoRoT-7b illustrates the issue of degeneracy since it is not possible to determine which light component is present to explain CoRoT-7b's density. The interior composition degeneracy can also be illustrated with GJ 1214b, discovered by *Charbonneau et al.* (2009). The planetary mass and radius are in agreement with a composition of primarily water enshrouded by a H/He envelope. GJ 1214b has a low enough density ($\rho = 1870 \pm 400$ kg m⁻³) that it cannot be composed of rocky and iron material alone (Fig. 13). In fact, the planet cannot be composed of pure water ice (Fig. 5). The planet almost certainly contains a gas component.

Rogers and Seager (2010) have investigated three different possibilities for the interior composition that match the GJ 1214b measured mass and radius. One possibility is a planet with an iron core, silicate mantle, and water outer layer, topped by a thick H/He envelope up to 6.8% of the planet mass. A second possibility is a planet with no interior water, but with an iron core, silicate mantle, and more massive H/He envelope. A third possibility is a water planet that requires at least 47% water by mass and has a massive steam atmosphere.

The identification of the gas component would provide important constraint. Specifically, a measured scale height from transmission spectra (see chapters by Winn, Burrows and Orton, and Meadows and Seager) can in principle distinguish between an envelope or atmosphere dominated by light elements such as H and He and an atmosphere composed of water vapor. The details of the interior composition, however, will likely remain unknown.

For a further description of interior degeneracies and how to quantify them, see, e.g., *Valencia et al.* (2007), *Adams et al.* (2008), *Zeng and Seager* (2008), and *Rogers and Seager* (2010).

4.3. Laboratory Measurements

Although our knowledge of the phase behavior and the corresponding EOS at very high pressures and temperatures is still limited, there have been some recent breakthroughs

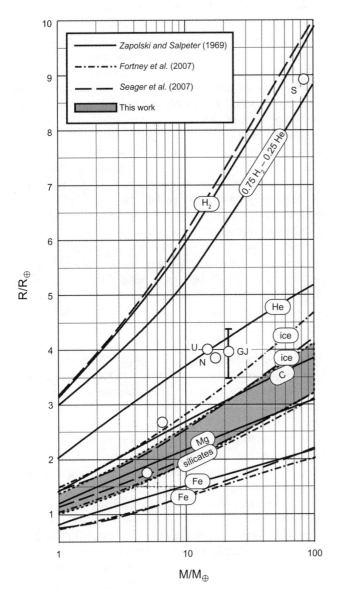

Fig. 13. Mass-radius curves for different types of planets (from *Grasset et al.*, 2009).

that allow us to better characterize the interior structure of terrestrial planets more massive than Earth. Recent progresses in high-pressure experiments, especially at advanced radiation sources, have produced more accurate EOS for candidate materials within such planets (*Angel et al.*, 2009; *Jackson*, 2010). For example, the electronic spin crossover occurring in the Fe^{2+} component of $(Mg,Fe)O$ has been shown to be associated with a density increase of ~3% (*Badro et al.*, 2003; *Lin et al.*, 2005), occurring over a wide temperature interval (*Sturhahn et al.*, 2005). Such an effect has been shown to have dynamical implications (*Bower et al.*, 2009). The volume-compression of $(Mg,Fe)O$ with various iron concentrations have been measured with high precision up to pressures of ~130 GPa, showing complex EOS behavior near the spin transition that cannot be explained by Birch-Murnaghan or Vinet formalisms (*Lin et al.*, 2005; *Fei et al.*, 2007; *Zhuravlev et al.*, 2009). At the core-mantle boundary region in Earth (P ~ 115–135 GPa and T ~ 3300–4300 K), silicate perovskite is likely to transform into postperovskite with a ~1.5% density increase (*Murakami et al.*, 2004). Although this region occupies only ~3% of the total volume of Earth, similar regions inside larger terrestrial planets occupy more than 50% of the total volume (Fig. 6). Experiments approaching such conditions have also shown that there may exist significant postperovskite-like structural variations and chemically heterogeneous phase assemblages (*Murakami et al.*, 2004; *Mao et al.*, 2004; *Oganov et al.*, 2005; *Tschauner et al.*, 2008; *Catalli et al.*, 2009; *Wicks et al.*, 2010). At extremely large compressions, first-principles calculations show that postperovskite disassociates into MgO and SiO_2 at ~10.5 Mbar (1.05 TPa) at 5000 K (*Umemoto et al.*, 2006). This pressure is easily achieved in the mantle of terrestrial planets or icy planets more than 8 $M_\oplus$. In the area of terrestrial-like cores, the isothermal EOS of hcp-structured iron has been measured with similar precision to 197 GPa at 298 K, yielding K_{0T} = 163.4 ± 7.9 GPa, and dK/dP = 5.38 ± 0.16 (*Dewaele et al.*, 2006). Recent progress in *ab initio* calculations using quantum Monte Carlo techniques suggests that hcp-structured iron has a relatively high melting temperature at Earth's inner-core outer-core boundary (P = 330 GPa) of 6900 ± 400 K (*Sola and Alfe*, 2009).

5. FUTURE PROSPECTS

A major limitation for terrestrial planet interior compositional characterization is the fact that only two data points are available per planet: mass and radius. Even with perfect measurements there is simply not enough information to uniquely identify the planet's interior composition. (One exception is iron-dominated planets, because a very dense planet has no other alternative.)

There are two possible paths forward. One path involves observations and interpretation of an exoplanet atmosphere to help break the interior composition degeneracy for a specific exoplanet. For example, an apparent massive water planet, with an interior dominated by water and a thick steam atmosphere, cannot be distinguished from a silicate planet with a thick H/He atmosphere or envelope based on a mass and radius measurement alone. An atmosphere measurement could help discriminate between the two via the scale height: A steam atmosphere would have a much smaller scale height than a H/He atmosphere. A major cautionary note is that the atmospheres of solar system planets do not really correlate with interior composition. Careful work must be done to understand which types of planet interiors can be constrained further with which kind of atmosphere measurements.

A second path forward from the interior composition ambiguity is statistics. With enough planets with a measured mass and radius (perhaps dozens or more), the hope is that specific planet populations in the mass-radius diagram (Fig. 13) will emerge. With distinct planet populations, characteristics of terrestrial planets in general can be identified, even if the actual composition of individual planets cannot. With even more optimism we hope to see distinct exoplanet populations in the mass-radius-period parameter space to be able to form general statistical statements about planet formation and migration.

NASA's Kepler space telescope has presented early results (*Borucki et al.*, 2010) for 300 planet candidates orbiting faint (V = 12–14) host stars. A major challenge for Kepler is follow-up radial velocity for mass measurements and planet verification; many Kepler planet candidate host stars may simply be too faint. Borucki et al. state that half of the 2010 announced planet candidates might be false positives. There is, however, still value in planet radius as a function of period. Planets over a certain size cannot be predominantly rocky, regardless of their mass. A relevant early Kepler result that should stand regardless of the false positive rate is that for periods shorter than 30 days, the majority of exoplanets are found to be Neptune-sized and smaller, and not Jupiter-sized.

The link between thermal convection and plate tectonics is still an active topic in Earth science. When did plate tectonics start? Has it been present since the end of accretion or did it happen much later in the history of Earth? Numerical simulations combined with better knowledge of the physical properties of mantle rocks might help answer these questions. The continuously improving computing power makes possible simulations with smaller and smaller grid sizes, which become appropriate for describing the thermal boundary layers of Earth-like planets. The strongly temperature-dependent characteristics of mantle minerals and rocks, such as viscosity, are also better handled with smaller grid size and therefore a larger number of grid points. Accurate knowledge of the appropriate phase assemblages and their EOS at very high pressures and temperatures is still limited; however, rapid progress is occurring (section 4.3). As described in section 2, the postperovskite phase is present at conditions near Earth's core-mantle boundary. On larger terrestrial planets, this phase or its disassociated phase may be the main solid phase(s) in the silicate mantle.

There are many space telescope concepts to build on the exquisite Kepler photometry. These concepts focus on transiting exoplanet surveys of stars that are much brighter than Kepler's and that are therefore amenable to radial velocity follow-up observations for planet confirmation and mass measurements. The Transiting Exoplanet Survey Satellite (TESS) study, led by P.I. George Ricker, aims to observe over 2 million stars brighter than about I = 13 for transiting planets with periods less than 2 months. Some planets orbiting M stars will be down to Earth size and orbiting in the habitable zone of the host stars. PLAnetary Transits and Oscillations of stars (PLATO) is under study by ESA. The goal is to monitor 100,000 main-sequence stars brighter than about V = 11, using an ensemble of 28 identical small, very-wide-field telescopes, assembled on a single platform and all looking at the same 25° diameter field. Even more exoplanet transit mission concepts are under study, leading us to believe that a large pool of transiting terrestrial planets with measured mass and radius is a very likely possibility in the future.

The search for terrestrial exoplanets is just starting. The last 20 years have shown that planetary systems are very common. The next 20 years may shed light on the uniqueness of Earth-like planets.

Acknowledgments. This work has been carried out at the Jet Propulsion Laboratory-California Institute of Technology, under contract with the National Aeronautics and Space Administration. C.S. acknowledges JPL R&TD support. J.M.J. acknowledges the National Science Foundation (0711542 and CSEDI 0855815) and Caltech for partial support of this work.

REFERENCES

Adams E. R., Seager S., and Elkins-Tanton L. (2008) Ocean planet or thick atmosphere: On the mass-radius relationship for solid exoplanets with massive atmospheres. *Astrophys. J., 673,* 1160–1164.

Alfe D. (2009) Temperature of the inner-core boundary of the Earth: Melting of iron at high pressure from first-principles coexistence simulations. *Phys. Rev. B, 79,* 060101(R).

Alfe D., Price G. D., and Gillan M. J. (2002) Iron under Earth's core conditions: Liquid-state thermodynamics and high-pressure melting curve from *ab-initio* calculations. *Phys. Rev. B., 65,* 165118, DOI: 10.1103/PhysRevB.65.165118.

Anderson D. L. (1980) Tectonics and composition of Venus. *Geophys. Res. Lett., 7,* 101–102.

Anderson W. W. and Ahrens T. J. (1994) An equation of state for liquid iron and implications for the Earth's core. *J. Geophys. Res., 99,* 4273–4284.

Anderson W. W. and Ahrens T. J. (1996) Shock temperature and melting in iron sulfides at core pressures. *J. Geophys. Res., 101(B3),* 5627–5642.

Angel R. J. and Jackson J. M. (2002) Elasticity and equation of state of orthoenstatite, $MgSiO_3$. *Am. Mineral., 87,* 558–561.

Angel R. J., Jackson J. M., Speziale S., and Reichmann H. J. (2009) Elasticity measurements on minerals: A review. *Eur. J. Mineral., 21(3),* 525–550, DOI: 10.1127/0935-1221/2009/0021-1925.

Arkani-Hamed J. (1994) Effects of the core cooling on the internal dynamics and thermal evolution of terrestrial planets. *J. Geophys. Res., 99,* 12109–12119.

Badro J., Fiquet G., Guyot F., Rueff J.-P., Struzhkin V. V., Vanko G., and Monaco G. (2003) Iron partitioning in Earth's lower mantle: Toward a deep lower mantle discontinuity. *Science, 300,* 789–791.

Beaulieu J.-P., Bennett D. P., Fouquet P., Williams A., Dominik M., et al. (2006) Discovery of a cool planet of 5.5 Earth masses through gravitational microlensing. *Nature, 439,* 437–440.

Beirao P., Santos N. C., Israelian G., and Mayor M. (2005) Abundances of Na, Mg and Al for stars with exoplanets. *Astron. Astrophys., 438,* 251–256.

Bertrand P., Sotin C., Mercier J-C., and Takahashi E. (1986) From the simplest chemical system to natural one: Garnet peridotite barometry. *Contrib. Mineral. Petrol., 93,* 168–178.

Boehler R. (1993) Temperatures in the Earth's core from melting-point measurements of iron at high static pressures. *Nature, 363,* 534–536.

Borde P., Rouan D., and Léger A. (2003) Exoplanet detection capability of the CoRoT space mission. *Astron. Astrophys., 405,* 1137–1144.

Borucki W. J., Koch D., Basri G., Batalha N., Brown T., et al. (2010) Kepler planet-detection mission: Introduction and first results. *Science, 327,* 977–980.

Bouhifd M. A., Andrault D., Fiquet G., and Richet P. (1996) Thermal expansion of forsterite up to the melting point. *Geophys. Res. Lett., 23,* 1143–1146.

Bower D. J., Gurnis M., Jackson J. M., and Sturhahn W. (2009) Enhanced convection and fast plumes in the lower mantle induced by the spin transition in ferropericlase. *Geophys. Res. Lett., 36,* L10306, DOI: 10.1029/2009GL037706.

Cameron A. G. W., Fegley B., Benz W., and Slattery W. L. (1988) The strange density of Mercury: Theoretical considerations. In *Mercury* (F. Vilas et al., eds.), pp. 692–708. Univ. of Arizona, Tucson.

Catalli K., Shim S.-H., and Prakapenka V. (2009) Thickness and Clapeyron slope of the post-perovskite boundary. *Nature, 462,* 782–785.

Chambers J. E. (2005) Planet formation. In *Treatise on Geochemistry, Vol. 1: Meteorites, Comets, and Planets* (A. M. Davis et al., eds.), pp. 461–475. Elsevier, Oxford.

Charbonneau D. and 18 colleagues (2009) A super-Earth transiting a nearby low-mass star. *Nature, 462,* 891–894.

Chen B., Li J., and Hauch S. A. (2008) Non-ideal liquidus curve in the Fe-S system and Mercury's snowing core. *Geophys. Res. Lett., 35,* L07201, DOI: 10.1029/2008GL033311.

Cox A. N., ed. (1999) *Allen's Astrophysical Quantities,* 4th edition. Springer-Verlag, New York. 714 pp.

Davaille A. and Jaupart C. (1993) Transient high Rayleigh number thermal convection with large viscosity variations. *J. Fluid Mech., 253,* 141–166.

Davaille A. and Jaupart C. (1994) Onser of thermal convection in fluids with temperature-dependent viscosity — Application of the oceanic mantle. *J. Geophys. Res., 99,* 19853–19866.

Dewaele A., Loubeyre P., Occelli F., Mezouar M., Dorogokupets P. I., and Torrent M. (2006) Quasihydrostatic equation of state of iron above 2 Mbar. *Phys. Rev. Lett., 97,* 215504.

Dreibus G. and Wänke H. (1985) Mars: A volatile rich planet. *Meteoritics, 20,* 367–382.

Duffy T. S., Zha C. S., Downs R. T., Mao H. K., and Hemley R. J. (1995) Elasticity of forsterite to 16 GPa and the composition of the upper mantle. *Nature, 378,* 170–173.

Dziewonski A. M. and Anderson D. L. (1981) Preliminary reference Earth model. *Phys. Earth Planet. Inter., 25,* 297–356.

Elkins-Tanton L. T. and Seager S. (2008) Ranges of atmospheric mass and composition of Super-Earth exoplanets. *Astrophys. J., 685,* 1237–1246.

Fegley B. and Cameron A. G. W. (1987) A vaporization model

for iron/silicate fractionation in the Mercury protoplanet. *Earth Planet. Sci. Lett., 82,* 207–222.

Fei Y., Mao H., and Hemley R. (1993) Thermal expansivity, bulk modulus, and melting curve of H₂O ice VII to 20 GPa. *J. Chem. Phys., 99,* 5369–5373.

Fei Y., Zhang L., Corgne A., Watson H. C., Ricolleau A., Meng Y., and Prakapenka V. (2007) Spin transition and equations of state of (Mg, Fe)O solid solutions. *Geophys. Res. Lett., 34,* L17307.

Fortney J. J., Marley M. S., and Barnes J. W. (2007) Planetary radii across five orders of magnitude in mass and stellar insolation: Application to transits. *Astrophys. J., 659,* 1661–1672.

Frank M. R., Fei Y., and Hu J. (2004) Constraining the equation of state of fluid H₂O to 80 GPa using the melting curve, bulk modulus, and thermal expansivity of ice VII. *Geochim. Cosmochim. Acta, 68,* 2781–2790.

Frost D. J., Liebske C., Langenhorst F., McCammon C. A., Trønnes R. G., and Rubie D. C. (2004) Experimental evidence for the existence of iron-rich metal in the Earth's lower mantle. *Nature, 428,* 409–412.

Gilli G., Israelian G., Ecuvillon A., Santos N. C., and Mayor M. (2006) Abundances of refractory elements in the atmospheres of stars with extrasolar planets. *Astron. Astrophys., 449,* 723–736.

Grasset O. and Parmentier E. M. (1998) Thermal convection in a volumetrically heated, infinite Prandtl number fluid with strongly temperature-dependent viscosity: Implications for planetary evolution. *J. Geophys. Res., 103,* 18171–18181.

Grasset O., Schneider, J., and Sotin C. (2009) A study of the accuracy of mass-radius relationships for silicate-rich and ice-rich planets up to 100 Earth. *Astrophys. J., 693,* 722–733.

Hemley R. J., Stixrude L., Fei Y., and Mao H. K. (1992) Constraints on lower mantle composition from P-V-T measurements of (Fe,Mg)SiO₃-perovskite and (Fe,Mg)O. In *High Pressure Research: Applications to Earth and Planetary Sciences* (Y. Syono and M. H. Manghnani, eds.), pp. 183–189. Geophysical Monograph Series, Vol. 67, American Geophysical Union, Washington, DC.

Hubbard W. B., Podolak M., and Stevenson D. J. (1995) The interior of Neptune. In *Neptune and Triton* (D. P. Cruikshank, ed.), pp. 109–138. Univ. of Arizona, Tucson.

Huang C., Zhao G., Zhang H. W., and Chen Y. Q. (2005) Chemical abundances of 22 extrasolar planet host stars. *Mon. Not. R. Astron. Soc., 363,* 71–78.

Irifune T. (1987) An experimental investigation of the pyroxene-garnet transformation in a pyrolite composition and its bearing on the constitution of the mantle. *Phys. Earth Planet. Inter., 45,* 324–336.

Jackson I. (1998) Elasticity, composition and temperature of the Earth's lower mantle: A reappraisal. *Geophys. J. Intl., 134,* 291–311.

Jackson J. M. (2010) Synchrotron-based spectroscopic techniques: Mössbauer and high-resolution inelastic scattering. In *High-Pressure Crystallography: From Fundamental Phenomena to Technological Applications* (E. Boldyreva and P. Dera, eds.), in press. Springer Science, DOI: 10.1007/978-90-481-9258_5.

Jackson J. M., Sinogeikin S. V., and Bass J. D. (2007) Sound velocities and single-crystal elasticity of orthoenstatite to 1073 K at ambient pressure. *Phys. Earth Planet. Inter., 161,* 1–12, DOI: 10.1016/j.pepi.(2006)11.002.

Jackson J. M., Sturhahn W., Tschauner O., Lerche M., and Fei Y. (2009) Behavior of iron in (Mg,Fe)SiO₃ post-perovskite assemblages at Mbar pressures. *Geophys. Res. Lett., 36,* L10301, DOI: 10.1029/2009GL037815.

Javoy M. (1995) The integral enstatite chondrite model of the Earth. *Geophys. Res. Lett., 22,* 2219–2222.

Javoy M. (1999) Chemical Earth models. *C. R. Acad. Sci. Paris, 329,* 537–555.

Kavner A., Duffy T. S., and Shen G. (2001) Phase stability and density of FeS at high pressures and temperatures: Implications for the interior structure of Mars. *Earth Planet. Sci. Lett., 185,* 25–33.

Kuchner M. J. (2003) Volatile-rich earth-mass planets in the habitable zone. *Astrophys. J. Lett., 596(1),* L105–L108.

Kuskov O. L. and Kronrod V. A. (2001) Core sizes and internal structure of Earth's and Jupiter's satellites. *Icarus, 151,* 204–227.

Léger A., Selsis F., Sotin C., Guillot T., Despois D., Mawet D., Ollivier M., Labèque A., Calette C., Brachet F., Chazelas B., and Lammer H. (2004) A new family of planets? "Ocean-planets." *Icarus, 169,* 499–504.

Léger A., Rouan D., Schneider J., Alonso R., Samuel B., et al. (2009) Transiting exoplanets from the CoRoT space mission: VIII. CoRoT-Exo-7b: the first Super-Earth with measured radius. *Astron. Astrophys., 506,* 287–302.

Lin J.-F., Struzhkin V. V., Jacobsen S. D., Hu M. Y., Chow P., et al. (2005) Spin transition of iron in magnesiowüstite in the Earth's lower mantle. *Nature, 436,* 377–480, DOI: 10.1038/nature03825.

Lunine J. I., Macintosh B., and Peale S. (2009) The detection and characterization of exoplanets. *Phys. Today, May 2009,* 46–51.

Mao W. L., Shen G., Prakapenka V. B., Meng Y., Campbell A. J., et al. (2004) Ferromagnesian postperovskite silicates in the D'' layer of the Earth. *Proc. Natl. Acad. Sci., 101(45),* 15867–15869, DOI: 10.1073/pnas.0407135101.

Mattern E., Matas J., Ricard Y., and Bass J. (2005) Lower mantle composition and temperature from mineral physics and thermodynamic modelling. *Geophys. J. Intl., 160(1),* 973–990, DOI: 10.1111/j.1365-1246X.

Mayor M. and 12 colleagues (2009) The HARPS search for southern extra-solar planets? XVIII. An Earth-mass planet in the GJ 581 planetary system. *Astron. Astrophys., 507,* 487–494.

McDonough W. F. and Sun S.-S. (1995) The composition of the Earth. *Chem. Geol., 120,* 223–253.

Mishima O. and Endo S. (1978) Melting curve of ice VII. *J. Chem. Phys., 68,* 4417–4418.

Moresi L. and Solomatov V. (1998) Mantle convection with a brittle lithosphere: Thoughts on the global tectonic styles of the Earth and Venus. *Geophys. J. Intl., 133,* 669–682.

Mosenfelder J. L., Asimow P. D., Frost D. J., Rubie D. C., and Ahrens T. J. (2009) The MgSiO₃ system at high pressure: Thermodynamic properties of perovskite, postperovskite, and melt from global inversion of shock and static compression data. *J. Geophys. Res., 114(B1),* DOI: 10.1029/2008JB005900.

Murakami M., Hirose K., Kawamura N., Sata N., and Onishi Y. (2004) Post-perovskite phase transition in MgSiO₃. *Science, 304,* 855–858.

Oganov A. R., Martoňák R., Laio A., Raiteri P., Parrinello M. (2005) Anisotropy of Earth's D'' layer and stacking faults in the MgSiO₃ post-perovskite phase. *Nature, 438,* 1142–1144.

O'Neill C. O. and Lenardic A. (2007) Geological consequences of super-sized Earths. *Geophys. Res. Lett., 34,* L19204.

Papuc A. M. and Davies G. F. (2008) The internal activity and thermal evolution of Earth-like planets. *Icarus, 195,* 447–458.

Parmentier E. M. and Sotin C. (2000) 3D numerical experiments on thermal convection in a very viscous fluid: Implications for the dynamics of a thermal boundary layer at high Rayleigh number. *Phys. Fluids, 12,* 609–617.

Parmentier E. M., Sotin C., and Travis B. J. (1994) Turbulent 3D thermal convection in an infinite Prandtl number, volumetrically heated fluid: Implications for mantle dynamics. *Geophys. J. Intl., 116,* 241–251.

Podolak M., Podolak J. I., and Marley M. S. (2000) Further investigations of random models of Uranus and Neptune. *Planet. Space Sci., 48,* 143–151.

Poirier J.-P. (1994) Physical properties of the Earth's core. *C. R. Acad. Sci., 318,* 341–350.

Poirier J.-P. (2000) *Introduction to the Physics of the Earth Interior,* 2nd edition. Cambridge Univ., Cambridge. 264 pp.

Queloz D. and 39 colleagues (2009) The CoRoT-7 planetary system: Two orbiting super-Earths. *Astron. Astrophys., 506,* 303–319.

Rogers L. A. and Seager S. (2010) Three possible origins for the gas layer on GJ 1214b. *Astrophys. J., 716,* 1208–1216.

Rouan D., Baglin A., Barge P., Copet E., Deleuil M., Léger A., Schneider J., Toublanc D., and Vuillemin P. (1999) Searching for exosolar planets with the CoRoT space mission. *Phys. Chem. Earth, 24,* 567–571.

Schubert G., Turcotte D. L., and Olson P. (2001) *Mantle Convection in the Earth and Planets.* Cambridge Univ., Cambridge.

Seager S., Kuchner M., Hier-Majumder C., and Militzer B. (2007) Mass-radius relationships for solid exoplanets. *Astrophys. J., 669,* 1279–1297.

Sinogeikin S. V. and Bass J. D. (2000) Single-crystal elasticity of pyrope and MgO to 20 GPa by Brillouin scattering in the diamond cell. *Phys. Earth Planet. Inter., 120,* 43–62.

Sinogeikin S. V., Jackson J. M., O'Neill B., Palko J. W., and Bass J. D. (2000) Compact high-temperature cell for Brillouin scattering measurements. *Rev. Sci. Instruments, 71(1),* 201–206.

Sola E. and Alfe D. (2009) Melting of iron under Earth's core conditions from diffusion Monte Carlo free energy calculations. *Phys. Rev. Lett., 103(7),* DOI: 10.1103/PhysRevLett.103.078501.

Solomatov V. S. and Moresi L. N. (2000) Scaling of time-dependent stagnant lid convection: Application to small-scale convection on Earth and other terrestrial planets. *J. Geophys. Res., 105,* 21795–21817.

Solomon S. C. (1979) Formation, history and energetics of cores in the terrestrial planets. *Phys. Earth Planet. Inter., 19,* 168–182.

Sotin C. and Labrosse S. (1999) Three-dimensional thermal convection in an iso-viscous, infinite Prandtl number fluid heated from within and from below: applications to the transfer of heat through planetary mantles. *Phys. Earth Planet. Inter., 112,* 171–190.

Sotin C. and Tobie G. (2004) Internal structure and dynamics of the large icy satellites. *C. R. Acad. Sci. Phys., 5,* 769–780.

Sotin C., Grasset O., and Mocquet A. (2007) Mass-radius curve for extrasolar Earth-like planets and ocean planets. *Icarus, 191,* 337–351.

Stevenson D. J. (1999) Life-sustaining planets in interstellar space? *Nature, 400,* 32.

Sturhahn W., Jackson J. M., and Lin J.-F. (2005) The spin state of iron in minerals of Earth's lower mantle. *Geophys. Res. Lett., 32,* L12307, DOI: 10.1029/2005GL022802.

Sun D., Tan E., Helmberger D., and Gurnis M. (2007) Seismological support for the metastable superplume model, sharp features, and phase changes with the lower mantle. *Proc. Natl. Acad. Sci., 104(22),* 9151–9155.

Thompson S. L. (1990) *ANEOS — Analytic Equations of State for Shock Physics Codes.* Sandia Natl. Lab. Doc. SAND89-2951, Albuquerque, New Mexico.

Tozer D. C. (1972) The present thermal state of the terrestrial planets. *Phys. Earth Planet. Inter., 6,* 182–197.

Tschauner O., Kiefer B., Liu H., Sinogeikin S., Somayazulu M., and Luo S.-N. et al. (2008) Possible structural polymorphism in Al-bearing magnesiumsilicate post-perovskite. *Am. Mineral., 93,* 533–539.

Uchida T., Wang Y., Rivers M. L., and Sutton S. R. (2001) Stability field and thermal equation of state of ε-iron determined by synchrotron X-ray diffraction in a multianvil apparatus. *J. Geophys. Res., 106,* 21799–21810.

Umemoto K., Wentzcovitch R. M., and Allen P. B. (2006) Dissociation of $MgSiO_3$ in the cores of gas giants and terrestrial exoplanets. *Science, 311,* 983–986.

Vacher P., Mocquet A., and Sotin C. (1998) Computation of seismic profiles from mineral physics: The importance of the non-olivine components for explaining the 660 km depth discontinuity. *Phys. Earth Planet. Inter., 106,* 275–298.

Valencia D., O'Connel R. J., and Sasselov D. D. (2006) Internal structure of massive terrestrial planets. *Icarus, 181,* 545–554.

Valencia D., O'Connel R. J., and Sasselov D. D. (2007) Inevitability of plate tectonics on Super-Earths. *Astrophys. J. Lett., 670,* L45–L48.

Valencia D., O'Connel R. J., and Sasselov D. D. (2009) The role of high-pressure experiments on determining super-Earth properties. *Astrophys Space Sci., 322,* 135–139.

Volcadlo L., de Wijs G. A., Kresse G., Gillan M., and Price G. D. (1997) First principles calculations on crystalline and liquid iron at Earth's core conditions. *Faraday Discuss., 106,* 205–217.

Wicks J. K., Jackson J. M., and Sturhahn W. (2010) Very low sound velocities in iron-rich (Mg,Fe)O: Implications for the core-mantle boundary region. *Geophys. Res. Lett., 37,* L15304, DOI: 10.1029/2010GL043689.

Williams Q., Jeanloz R., Bass J., Svendsen B., and Ahrens T. J. (1987) The melting curve of iron to 250 gigapascals — A constraint on the temperature at Earth's center. *Science, 236,* 181–184.

Wuchterl G., Guillot T., and Lissauer J. J. (2000) Giant planet formation. In *Protostars and Planets* (V. Mannings et al., eds.), p. 1081. Univ. of Arizona, Tucson.

Yoder C. F., Konopliv A. S., Yuan D. N., Standish E. M., and Folkner W. M. (2003) Fluid core size of Mars from detection of the solar tide. *Science, 300,* 299–303.

Zeng L. and Seager S. (2008) A computational tool to interpret the bulk composition of solid exoplanets based on mass and radius measurements. *Publ. Astron. Soc. Pac., 120,* 983-991.

Zhuravlev K. K., Jackson J. M., Wolf A. S., Wicks J. K., Yan J., and Clark S. M. (2009) Isothermal compression behavior of (Mg,Fe)O using neon as a pressure medium. *Phys. Chem. Minerals, 37,* 465–474, DOI: 10.1007/s00269-009-0347-6.

Giant Planet Interior Structure and Thermal Evolution

Jonathan J. Fortney
University of California, Santa Cruz

Isabelle Baraffe
Ecole Normale Supérieure de Lyon — CRAL

Burkhard Militzer
University of California, Berkeley

We discuss the interior structure and composition of giant planets, and how this structure changes as these planets cool and contract over time. Here we define giant planets as those that have an observable hydrogen-helium envelope that includes Jupiter-like planets, which are predominantly H/He gas, and Neptune-like planets, which are predominantly composed of elements heavier than H/He. We describe the equations of state of planetary materials and the construction of static structural models and thermal evolution models. We apply these models to transiting planets close to their parent stars, as well as directly imaged planets far from their parent stars. Mechanisms that have been postulated to inflate the radii of close-in transiting planets are discussed. We also review knowledge gained from the study of the solar system's giant planets. The frontiers of giant planet physics are discussed with an eye toward future planetary discoveries.

1. INTRODUCTION

The vast majority of planetary mass in the solar system, and indeed the galaxy, is hidden from view in the interiors of giant planets. Beyond the simple accounting of mass, there are many reasons to understand these objects, which cut across several disciplines. Understanding the structure of these planets gives us our best evidence as to the formation mode of giant planets, which tells us much about the planet formation process in general. As we will see, giant planets are vast natural laboratories for simple materials under high pressure in regimes that are not yet accessible to experiment. With the recent rise in number and stunning diversity of giant planets, it is important to understand them as a class of astronomical objects.

We would like to understand basic questions about the structure and composition of giant planets. Are they similar in composition to stars, predominantly hydrogen and helium with only a small mass fraction (~1%) of atoms more massive than helium? If giant planets are enhanced in "heavy elements" relative to stars, are the heavy elements predominantly mixed into the hydrogen-helium (H-He) envelope, or are they mainly found in a central core? If a dense central core exists, how massive is it, what is its state (solid or liquid), and is it distinct or diluted into the above H-He envelope? Can we understand whether a planet's heavy element mass fraction depends on that of its parent star?

Giant planets are natural laboratories of hydrogen and helium in the megabar to gigabar pressure range, at temperatures on the order of 10^4 K. How do hydrogen and helium interact under these extreme conditions? Is the helium distribution within a planet uniform, and what does this tell us about how H and He mix at high pressure? What methods of energy transport are at work in the interiors of giant planets? Can we explain planets' observable properties such as the luminosity and radius at a given age?

The data that we use to shape our understanding of giant planets comes from a variety of sources. Laboratory data on the equation of state (EOS, the pressure density-temperature relation) of hydrogen, helium, warm fluid "ices" such at water, ammonia, and methane; silicate rocks; and iron serve at the initial inputs into models. Importantly, data are only available over a small range of phase space, so that detailed theoretical EOS calculations are critical to understanding the behavior of planetary materials at high pressure and temperature. Within the solar system, spacecraft data on planetary gravitational fields allows us to place constraints on the interior density distribution for Jupiter, Saturn, Uranus, and Neptune. For exoplanets, we often must make do with far simpler information, namely a planet's mass and radius only. For these distant planets, what we lack in detailed knowledge about particular planets, we can make up for in number.

Within six years of the Voyager 2 flyby of Neptune, the encounter that completed our detailed census of the outer solar system, came the stunning discoveries of the giant exoplanet 51 Peg b (*Mayor and Queloz*, 1995) and also the first bona fide brown dwarf, Gliese 229B (*Nakajima et al.*, 1995). We were not yet able to fully understand the structure and evolution of the solar system's planets before we were given a vast array of new planets to understand. In particular, the close-in orbit of 51 Peg b led to immediate questions regarding its history, structure, and fate (*Guillot et al.*, 1996; *Lin et al.*, 1996). Four years later, the first transiting planet,

HD 209458b (*Charbonneau et al.*, 2000; *Henry et al.*, 2000), was found to have an inflated radius of ~1.3 R_{Jup}, confirming that proximity to a parent star can have dramatic effects on planetary evolution (*Guillot et al.*, 1996). The detections of over 50 additional transiting planets (as of August 2009) has conclusively shown that planets with masses greater than that of Saturn are composed predominantly of H/He, as expected. However, a great number of important questions have been raised.

Much further from their parent stars, young luminous gas giant planets are being directly imaged from the ground and from space (*Kalas et al.*, 2008; *Marois et al.*, 2008; *Lagrange et al.*, 2009). For imaged planets, planetary thermal emission is only detected in a few bands, and a planet's mass determination rests entirely on comparisons with thermal evolution models, that aim to predict a planet's luminosity and spectrum with time. However, the luminosity of young planets is not yet confidently understood (*Marley et al.*, 2007; *Chabrier et al.*, 2007).

In this chapter we first outline the fundamental physics and equations for understanding the structure and thermal evolution of giant planets. We next describe the current state of knowledge of the solar system's giant planets. We then discuss current important issues in modeling exoplanets, and how models compare to observations of transiting planets, as well as directly imaged planets. We close with a look at the future science of giant exoplanets.

2. EQUATIONS AND MODEL BUILDING

2.1. Properties of Materials at High Pressure

Materials in the interiors of giant planets are exposed to extreme temperature and pressure conditions reaching ~10000 K and 100–1000 GPa (1–10 Mbar) (see Fig. 1). The characterization of materials' properties under these conditions has been one of the great challenges in experimental and theoretical high pressure physics. Ideally one would recreate such high pressure in the laboratory, characterize the state of matter, and then directly measure the equation of state (EOS) as well as the transport properties needed to model planetary interiors. While a number of key experiments have been performed, for a large part of Jupiter's interior we instead rely on theoretical methods.

There are both static and dynamic methods to reach high pressures in laboratory experiments. The highest pressure in static compression experiments have been reached with diamond anvil cells (*Hemley and Ashcroft*, 1998; *Hemley*, 2000; *Loubeyre et al.*, 2002). One has been able to reach ~400 GPa (4 Mbar), which exceeds the pressure in the center of Earth but is far from the ~4000 GPa at Jupiter's core-envelope boundary. While many diamond anvil cell experiments were performed at room temperature, the combination with laser heating techniques has enabled one to approach some of temperature conditions that exist inside planets.

The challenges in dynamic compression experiments are quite different. Reaching the required pressures is not the primary concern, but instead it is the difficulty in achieving a high enough density. Most dynamic experiments compress the material with a single shock wave. The locus of final ρ–T–P points reached by single shock from one particular initial point is called the Hugoniot. While this method provides direct access to the EOS (*Zeldovich and Raizer*, 1966), the compression ratio rarely exceed values of 4. Instead the material is heated to very high temperatures that exceed those in planetary interiors (*Jeanloz et al.*, 2007; *Militzer and Hubbard*, 2007). Recently, static and dynamic compression techniques have been combined to address this issue (*Eggert et al.*, 2008). By precompressing the sample in a diamond anvil cell before a shock wave was launched, Eggert et al. were able to probe deeper into planetary interiors. Earlier experiments by *Weir et al.* (1996) employed reverberating shock waves to reach high densities and thereby approached the state of metallic hydrogen at high temperature.

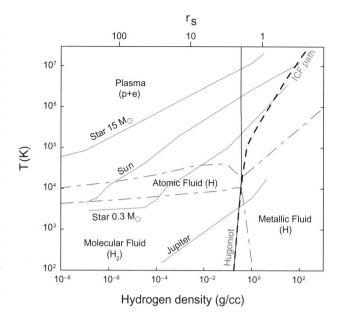

Fig. 1. Density-temperature phase diagram of hot dense hydrogen. The dash-dotted lines separate the molecular, atomic, metallic, and plasma regimes. The solid lines are isentropes for Jupiter and stars with 0.3, 1, and 15 $M_\odot$. Single-shock Hugoniot states as well as the inertial confinement fusion paths are indicated by dashed lines. The thin solid line shows ρ-T conditions of PIMC simulations.

The first laser shock experiments that reached megabar pressures predicted the material to be highly compressible under shock conditions and to reach densities six times higher than the initial state (*da Silva et al.*, 1997; *Collins et al.*, 1998). However, later experiments (*Knudson et al.*, 2001, 2003; *Belov et al.*, 2002; *Boriskov et al.*, 2005) showed smaller compression ratios of about 4.3, which were in good agreement with theoretical predictions (*Lenosky et al.*, 1997; *Militzer and Ceperley*, 2000).

Until very recently, all models for giant planet interiors were based on chemical models (*Ebeling and Richert*, 1985;

Saumon and Chabrier, 1992; Saumon et al., 1995; Juranek and Redmer, 2000) that describe materials as a ensemble of stable molecules, atoms, ions, and free electrons. Approximations are made to characterize their interactions. The free energy of the material is calculated with semianalytical techniques and all other thermodynamic variables are derived from it. Chemical models require very little computer time, can easily cover orders of magnitude in pressure temperature space, and have therefore been applied to numerous star and planet models. Chemical models do not attempt to characterize all the interactions in a many-body system. Molecular hydrogen at high density, for example, is typically approximated by a system of hard spheres where the excitation spectrum of the isolated molecule is modified by a density dependent term. In general, chemical models have difficulties predicting materials' properties in the strongly coupled regime where interaction effects dominated over kinetic effects. At high density and temperature where molecules are no longer stable or near a metal-to-insulator transition, chemical models require input from experiments or other theoretical techniques to fit adjustable parameters.

Recently, progress in the field of theoretical description of planetary materials at high pressure has come from first principles computer simulations. Such methods are based on the fundamental properties of electrons and nuclei and do not contain any parameters that are fit to experimental data. While approximations cannot be avoided altogether to efficiently derive a solution to the many-body Schrödinger equation, such approximations are not specific to a particular material and have been tested for a wide range of materials and different thermodynamic conditions.

First-principles simulation can now routinely study the behavior of hundreds of particles at very different pressure and temperature conditions. Here we summarize three different approaches: path integral Monte Carlo (PIMC), density functional molecular dynamics (DFT-MD), and quantum Monte Carlo (QMC). The challenge of performing accurate simulations has always been to make sure that the approximations, which are often necessary to perform the calculations at all, do not impact the predictions in a significant way. There are fundamental approximations, such as the assumption of simplified functionals in DFT or the nodal approximation in QMC and PIMC simulations with fermions (*Foulkes et al.*, 2001). These approximations can in most cases only be checked by comparison between different methods or with experimental results. There are also controlled approximations, such as using a sufficiently large number of particles, long enough simulations, or a large enough basis set. These approximations can always be verified by investing additional computer time but it is not always possible to perform all tests at all thermodynamic conditions.

PIMC (*Pollock and Ceperley*, 1984; *Ceperley*, 1995) is a finite-temperature quantum simulation method that explicitly constructs paths for electrons and nuclei. All correlation effects are included, which makes PIMC one of the most accurate finite-temperature quantum simulation methods available. The only fundamental approximation required is the fixed node approximation (*Ceperley*, 1991, 1996) that is introduced to treat the fermion sign problem, which arises from the explicit treatment of electrons. The method is very efficient at high temperature and can provide one coherent description of matter reaching up to a fully ionized plasma state. PIMC simulations have been applied to hydrogen (*Pierleoni et al.*, 1994; *Militzer and Ceperley*, 2000, 2001), helium (*Militzer*, 2006, 2009), and their mixtures (*Militzer*, 2005). At low temperature, this method becomes more computationally demanding because the length of the path scales like 1/T. At temperatures below ~5000 K where electronic excitations are not important, it is more efficient to use a ground-state simulation method, discussed next.

Density functional molecular dynamics simulations rely on the Born-Oppenheimer approximation to separate the motion of electrons and nuclei. For a given configuration of nuclei, the instantaneous electron ground-state is derived from density functional theory. Forces are derived and nuclei are propagated using classical molecular dynamics. Excited electronic states can be incorporated by using the Mermin functional (*Mermin*, 1965).

DFT is a mean field approach and approximations are made to treat electronic exchange and correlation effects. Electronic excitation gaps are underestimated in many materials. A more sophisticated description of electronic correlation effects is provided by quantum Monte Carlo (*Foulkes et al.*, 2001), where one uses an ensemble of random walks to project out the ground-state wave function. This method represents the ground-state analog of PIMC. To avoid the fermion sign problem, one also introduces a nodal approximation based on a trial wave function. While most QMC calculations were performed for fixed nuclei, the method has recently been extended to fluids and calculations for fluid hydrogen have been performed (*Pierleoni et al.*, 2004; *Delaney et al.*, 2006).

The EOS of dense hydrogen has been the subject of several DFT-MD studies (*Lenosky et al.*, 1997; *Desjarlais*, 2003; *Bonev et al.*, 2004). Figure 2 compares DFT-MD EOS from *Militzer et al.* (2008) with predictions from free-energy models. Even at the highest densities, one finds significant deviations because no experimental data exist to guide free-energy models. Furthermore, the density is not yet high enough for hydrogen to behave like an ideal Fermi gas.

Of importance for the interiors of giant planets and the generation of their magnetic fields are the properties of the insulator-to-metal transition in dense hydrogen (*Chabrier et al.*, 2006). According to predictions from the best simulation methods currently available, quantum Monte Carlo (*Delaney et al.*, 2006) and DFT-MD (*Vorberger et al.*, 2007), this transition is expected to occur *gradually* in the condition of giant planet interiors. Earlier DFT-MD simulations (*Scandolo*, 2003; *Bonev et al.*, 2004) had predicted a sharp dissociation transition but all these results have now been attributed to inaccuracies in the wave function propagation with the Car-Parinello method (*Car and Parrinello*, 1985). With the more accurate Born-Oppenheimer propagation method, the transition occurs gradually (see discussion in *Militzer and Hubbard*, 2009) but gives rise to a region of negative $\partial P/\partial T|_V$

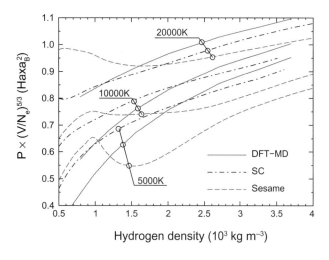
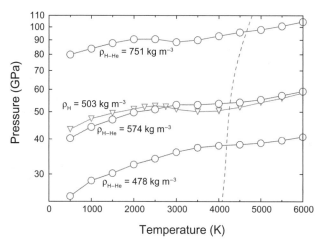

Fig. 2. Comparison of the DFT-MD EOS with the Saumon-Chabrier (SC) and SESAME models. Three isotherms for pure hydrogen are shown in the metallic regime at high pressure. P is scaled by the volume per electron to the power 5/3 to remove most of the density dependence.

Fig. 3. Isochores derived from DFT-MD simulations of H-He mixtures (circles, Y = 0.2466) and pure hydrogen (triangles). Results for mixtures predict a positive Grüneisen parameter, $(\partial P/\partial T)|_V > 0$, along Jupiter's isentrope.

(*Bagnier et al., 2000*; *Vorberger et al., 2007*) that is shown in Fig. 3. This leads to a negative Grüneisen parameter and could introduce a barrier to convection in Jupiter and other giant planets. However, *Militzer et al.* (2008) demonstrated that in a hydrogen-helium mixture, the region of $\partial P/\partial T|_V < 0$ is shifted to lower temperatures than occur in Jupiter.

The question of whether the insulator-to-metal transition in hydrogen is smooth or of first order (plasma phase transition) has been debated for a long time. A series of free-energy models were constructed with a plasma phase transition (*Ebeling et al., 1989*; *Saumon and Chabrier, 1992*), while others do not include one (*Ross, 1998*; *Lyon and Johnson, 1992*). Recent shock wave experiments (*Fortov et al., 2007*) show evidence of an insulator-to-metal transition at temperature and pressure conditions that are consistent with theoretical predictions. In their interpretation, the authors carefully suggest that their data may provide evidence for a first-order phase transition.

Recently *Militzer et al.* (2008) and independently *Nettelmann et al.* (2008) used first-principles simulations to derive an EOS to model Jupiter's interior. While both studies relied on the same simulation technique, DFT-MD, the groups derived very different predictions for the size of Jupiter's heavy element core, the distribution of heavy elements in its mantle, and the temperature profile in its interior. The differences between the two approaches are analyzed in *Militzer and Hubbard* (2009) and will be summarized later in this chapter. While some deviations in the simulation parameters such as the number of particles and their effect on the computed EOS have been identified, the differences in the predictions are mainly due to additional model assumptions, e.g., whether helium and heavy elements are homogeneously distributed throughout the interior

(*Militzer et al., 2008*) or not (*Nettelmann et al., 2008*). Only the deviations in the interior temperature profile is a direct consequence of the computed EOS. While pressure and internal energy can be obtained directly from simulations at constant volume and temperature, the entropy does not follow directly and is typically obtained by thermodynamic integration of the free energy. *Militzer et al.* (2008) and *Nettelmann et al.* (2008) used different methods to compute the free energy but more work is needed to understand which method yields more accurate adiabats.

Two recent papers (*Morales et al., 2009*; *Lorenzen et al., 2009*) provide evidence for the phase separation of the hydrogen-helium mixtures at the interior of Saturn and possibly also of Jupiter. Both papers rely on DFT-MD simulations but *Morales et al.* (2009) used larger and more accurate simulations and employed thermodynamic integration to determine the Gibbs free energy of mixing. *Lorenzen et al.* (2009) used a simplified approach where mixing entropy, which is the most difficult term to calculate, was taken from a noninteracting system of particles. While others have shown that nonideal mixing effects are important (*Vorberger et al., 2007*), the impact of this approximation on the pressure-temperature conditions where the hydrogen and helium start to phase-separate remains to be studied, but it may explain why *Lorenzen et al.* (2009) predict higher phase-separation temperatures than *Morales et al.* (2009).

In addition to hydrogen and helium, water has also been extensively studied experimentally (e.g., *Lee et al., 2006*) and computationally (e.g., *Cavazzoni et al., 1999*; *Schwegler et al., 2001*; *French et al., 2009*). Often water, ammonia, and methane are grouped together as "planetary ices," as a generic phrase for O-, C-, and N-dominated volatiles (mostly H_2O, NH_3, and CH_4), which are likely found in fluid form, not solid, within giant planets. Furthermore, these components are probably not found as intact molecules at high pressure. Water is found from first-principles calculations to dissociate into $H_3O^+ + OH^-$ ion pairs above ~2000 K at 0.3 Mbar (*Cavazzoni et al.*, 1999; *Schwegler et al.*, 2001; *Mattsson and Desjarlais*, 2006), and indeed high electrical conductivities are measured at pressures near 1 Mbar (*Chau et al.*, 2001). Moving to even heavier elements, "rock" refers primarily to silicates (Mg-, Si-, and O-rich compounds) and often includes iron and other "metals" as well (see *Stevenson*, 1985; *Hubbard*, 1984). Uncertainties in the EOSs of heavy elements are generally less important in structure and evolution calculations than those for hydrogen and helium, but they certainly do have important quantitative effects (*Hubbard et al.*, 1991; *Baraffe et al.*, 2008).

This section is only able to give a flavor of the vast array of EOS science going on at the boundary between physics and planetary sciences. We are now in an era where long sought-after advances in experiment and first principles theory are occurring at a fast pace.

2.2. Basic Equations

To a first approximation, a planet can be considered as a nonrotating, nonmagnetic, fluid object where gravity and gas pressure are the two main contributors to forces in the interior (radiation pressure is negligible in planetary interiors). Planetary structure and evolution can thus be described in a spherically symmetric configuration and are governed by the following conservation equations

$$\text{Mass conservation:} \quad \frac{\partial m}{\partial r} = 4\pi r^2 \rho \quad (1)$$

$$\text{Hydrostatic equilibrium:} \quad \frac{\partial P}{\partial r} = -\rho g \quad (2)$$

$$\text{Energy conservation:} \quad \frac{\partial L}{\partial r} = -4\pi r^2 \rho T \frac{\partial S}{\partial t} \quad (3)$$

where L is the intrinsic luminosity, i.e., the net rate of radial energy flowing through a sphere of radius r. The rate of change of the matter entropy is due to the variation of its internal energy and to compression or expansion work, according to the first and second laws of thermodynamics. A complete model requires an additional equation describing the transport of energy in the planet

$$\frac{\partial T}{\partial r} = \frac{\partial P}{\partial r} \frac{T}{P} \nabla \quad (4)$$

where $\nabla = \frac{d \ln T}{d \ln P}$ is the temperature gradient. If energy transport is due to radiation or conduction, it is well described by a diffusion process with

$$\frac{\partial T}{\partial r} = \frac{3}{16\pi acG} \frac{\kappa LP}{mT^4} \quad (5)$$

The total opacity of matter, κ, accounts for radiative and conductive transport and is defined by $\kappa^{-1} = \kappa_{Ross}^{-1} + \kappa_{cond}^{-1}$ where κ_{Ross} and κ_{cond} are the Rosseland mean radiative opacity and the conductive opacity, respectively. Energy transport by conduction in a fluid results from collisions during random motion of particles. For planets essentially composed of H/He, energy transfer is due to electrons in the central ionized part, whereas molecular motion dominates in the outer envelope. Because of the high opacity of H/He matter in planetary interiors, convection is thought to be the main energy transport and the temperature gradient is given by the adiabatic gradient $\nabla_{ad} = \left(\frac{d \ln T}{d \ln P}\right)_S$. If the planet has a core made of heavy material (water or ice, rock, iron), heat transport can be due to convection or conduction (electrons or phonons), depending on the core material, its state (solid or liquid), and the age of the planet.

2.3. Thermal Evolution and Atmospheric Boundary Conditions

Starting from a high-entropy, hot initial state, the luminosity of a planet during its entire evolution is powered by the release of its gravitational E_g and internal E_i energy and is given by (see equation (3))

$$L(t) = -\frac{d}{dt}(E_g + E_i) = -\int_M P \frac{d}{dt}\left(\frac{1}{\rho}\right) dm - \int_M \frac{de}{dt} dm \quad (6)$$

where e is the specific internal energy. The virial theorem, which applies to a self-gravitating gas sphere in hydrostatic equilibrium, relates the thermal energy of a planet (or star) to its gravitational energy as

$$\alpha E_i + E_g = 0 \quad (7)$$

with $\alpha = 2$ for a monoatomic ideal gas or a fully nonrelativistic degenerate gas, and $\alpha = \frac{6}{5}$ for an ideal diatomic gas. Contributions arising from interactions between particles yield corrections to the ideal EOS (see *Guillot*, 2005). The case $\alpha = \frac{6}{5}$ applies to the molecular hydrogen outer regions of a giant planet. Note that the mass fraction involved in these regions, for a Jupiter-like planet, is usually negligible compared to that involved in the central core and the metallic H region.

According to equations (6) and (7), the planet radiates

$$L \propto -\frac{dE_g}{dt} \quad (8)$$

defining a characteristic thermal timescale τ_{KH}

$$\tau_{KH} \sim \frac{E_g}{L} \sim \frac{GM^2}{RL} \quad (9)$$

For a 1 M_{Jup} gaseous planet, with negligible heavy element content, $\tau_{KH} \sim 10^7$ yr at the beginning of its evolution (note that this value is highly uncertain since it depends on the initial state and thus on the details of the planet formation process; see section 4.6) and $\tau_{KH} > 10^{10}$ yr after 1 G.y. [see, e.g., *Baraffe et al.* (2003) for values of R and L at a given age]. The reader must keep in mind that equation (9) is a rough estimate of the characteristic timescale for cooling and contraction of a plane. This timescale can be longer by 1 or 2 orders of magnitude than the value derived from equation (9). Indeed, when degeneracy sets in, a significant fraction of the gravitational energy due to contraction is used to increase the pressure of the (partially) degenerate electrons and the luminosity of the planet is essentially provided by the thermal cooling of the ions (see *Guillot,* 2005).

The rate at which internal heat escapes from a planet depends on its atmospheric surface properties, and thus on the outer boundary conditions connecting inner and atmospheric structures. Put another way, although the interior may be efficiently convecting, the radiative atmosphere serves as the bottleneck for planetary cooling. For objects with cold molecular atmospheres, the traditional Eddington approximation assuming that the effective temperature equals the local temperature at an optical depth $\tau_{Ross} = 2/3$ provides incorrect thermal profiles and large errors on T_{eff} (see *Chabrier and Baraffe,* 2000, and references therein). Modern models for planets incorporate more realistic atmospheric boundary conditions using frequency-dependent atmosphere codes. Inner and outer temperature-pressure profiles must be connected at depths where the atmosphere becomes fully convective, implying an adiabatic thermal profile, and optical depth is greater than 1. The connection is done at a fixed pressure, usually a few bars (*Burrows et al.,* 1997, 2003; *Fortney and Hubbard,* 2003; *Guillot,* 2005) or at a fixed optical depth, usually at $\tau = 100$ (*Chabrier and Baraffe,* 2000). The numerical radius corresponding to the outer boundary conditions provides, to an excellent approximation, the planet's photospheric radius, where the bulk of the flux escapes.

2.4. Effect of Rotation

The solar system giant planets are relatively fast rotators with periods of about 10 h for Jupiter and Saturn, and about 17 h for Neptune and Uranus (see *Guillot,* 2005, and references therein). Rotation modifies the internal structure of a fluid body and yields departures from a spherically symmetric configuration. Its effects can be accounted for using a perturbation theory, which has been extensively developed during the past century for planets and stars (*Tassoul,* 1978; *Zharkov and Trubitsyn,* 1974, 1976, 1978). Abundant literature exists on the application of this theory to models for our solar system giant planets (*Hubbard and Marley,* 1989; *Guillot et al.,* 1994; *Podolak et al.,* 1995). One often refers to the so-called theory of figures, presented in details in the works of *Zharkov and Trubitsyn* (1974, 1976, 1978).

Here we only briefly explain the concept of the theory. The underlying idea is to define a rotational potential V_{rot} such that surfaces of constant P, ρ, and U coincide. $U = V_{rot} + V_{grav}$ is the total potential (rotational + gravitational) and obeys the hydrostatic equilibrium equation $\nabla P = \rho \nabla U$. The solution for the level surfaces (or figures) can be expressed in terms of an expansion in even Legendre polynomials, also called zonal harmonics, P_{2n}. As a consequence of rotation, the gravitational potential departs from a spherically symmetric potential and can be expressed in terms of even Legendre polynomials and gravitational moments J_{2n} (also called zonal gravitational harmonics) as

$$V_{grav} = \frac{GM}{r}\left[1 - \sum_{n=1}^{\infty}\left(\frac{a}{r}\right)^{2n} J_{2n} P_{2n}(\cos\theta)\right] \quad (10)$$

where r is the distance to the planet center, a its equatorial radius, and θ the polar angle to the rotation axis. The gravitational moments can be related to the density profile of the planet (see *Zharkov and Trubitsyn,* 1974) as

$$J_{2n} = -\frac{1}{Ma^{2n}} \iiint \rho(r,\theta) r^{2n} P_{2n}(\cos\theta) dv \quad (11)$$

where dv is a volume element. A measure of J_{2n} by a spacecraft coming close to a planet thus provides constraints on its density profile. This powerful and elegant method was applied to our four giant planets using the determination of J_2, J_4, and J_6 from the trajectories of the space missions Pioneer and Voyager [see *Guillot* (2005) for a review; see also section 3).

As is well known from the theory of figures, the low-order gravitational moments provide constraints on the density/pressure profile in the metallic/molecular hydrogen regions but do not sound the most central regions (*Hubbard,* 1999; see also Fig. 4 of *Guillot,* 2005). Consequently, the presence of a central core — its mass, composition, and structure — can only be *indirectly* inferred from the constraints on the envelope provided by the gravitational moments.

2.5. Mass-Radius Relation

The mass-radius relationship for planets, and more generally for substellar/stellar objects, contains essential information about their main composition and the state of matter in their interior. The fundamental work by *Zapolsky and Salpeter* (1969) is a perfect illustration of this statement. The analysis of cold (zero-temperature) spherical bodies of a given chemical composition and in hydrostatic equilibrium shows the existence of a unique mass radius relation and of a maximum radius R_{max} at a critical mass M_{crit}. The very existence of a maximum radius stems from two competing physical effects characteristic of the state of matter under planetary conditions. The first effect is due to electron degeneracy, which dominates at large masses and yields a mass-radius relationship $R \propto M^{-1/3}$ characteristic of fully degenerate bodies (*Chandrasekhar,* 1939). The second effect stems from the classical electrostatic contribution from ions (Coulomb effects), which yields a mass radius relation $R \propto M^{1/3}$, characteristic of incompressible Earth-like planets.

Zapolsky and Salpeter (1969) find a critical mass of 2.6 M_{Jup} where the radius reaches a maximum value $R_{max} \sim 1 \, R_{Jup}$ for a gaseous H/He planet. The critical mass increases as the heavy element content increases, while R_{max} decreases.

The true mass-radius relationship, derived from models taking into account a realistic equation of state (see section 2.1) yields a smoother dependence of radius with mass, as displayed in Fig. 4. The transition between stars and brown dwarfs marks the onset of electron degeneracy, which inhibits the stabilizing generation of nuclear energy by hydrogen burning. The typical transition mass is $\sim 0.07 \, M_\odot$ (*Burrows et al.*, 2001; *Chabrier and Baraffe*, 2000). Above this transition mass, the nearly classical ideal gas yields a mass-radius relationship $R \propto M$. In the brown dwarf regime the dominant contribution of partially degenerate electrons, balanced by the contribution from ion interactions, yields $R \propto M^{-1/8}$ instead of the steeper relationship for fully degenerate objects. The increasing contribution of Coulomb effects as mass decreases competes with electron degeneracy effects and renders the radius almost constant with mass around the critical mass. The full calculation yields, for gaseous H/He planets, $M_{crit} \sim 3 \, M_{Jup}$, amazingly close to the results based on the simplified approach of *Zapolsky and Salpeter* (1969). Below the critical mass, Coulomb effects slightly dominates over partially degenerate effects, yielding a smooth variation of radius with mass close to the relation $R \propto M^{1/10}$.

3. SOLAR SYSTEM GIANT PLANETS

3.1. Jupiter and Saturn

Jupiter and Saturn, composed primarily of hydrogen and helium, serve as the benchmark planets for our understanding of gas giants. Indeed, these planets serve as calibrators for the structure and cooling theory used for all giant planets and brown dwarfs. The significant strides that have been made in understanding these planets have come from theoretical and experimental work on the equation of state of hydrogen, as well at the spacecraft-measured gravitational moments.

As discussed in section 2, with a basic knowledge of the EOS of hydrogen, helium, and heavier elements, along with observations of Jupiter's mass and radius, one can deduce that Jupiter and Saturn are mostly composed of hydrogen (see, e.g., *Seager et al.*, 2007). The same argument shows that Uranus and Neptune are mostly composed of elements much heavier than hydrogen.

Models of the interior structure of Jupiter in particular were investigated by many authors in the twentieth century, and investigations of Jupiter and Saturn became more frequent with the rise of modern planetary science in the 1950s and 1960s. However, even these initial models were uncertain as to whether the hydrogen is solid or fluid. The path toward our current understanding of giant planets started with the observation by *Low* (1966) that Jupiter emits more mid-infrared flux than it received from the Sun. Soon after,

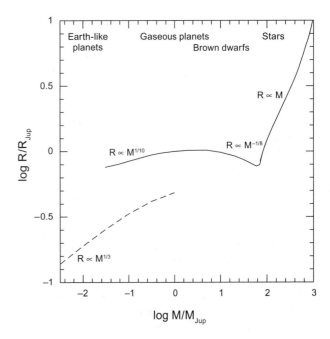

Fig. 4. Characteristic mass-radius relationship for stellar and substellar gaseous H/He objects (solid line) (models from *Baraffe et al.*, 1998, 2003, 2008) and for pure ice planets (dashed line) (models from *Fortney et al.*, 2007).

Hubbard (1968) showed that the observed heat flux could only be carried by convection (as opposed in radiation or conduction), showing that Jupiter's interior is fluid, not solid. Thus began the paradigm of giant planets as hydrogen-dominated, warm, fluid, convective objects.

However, it was clear that Jupiter and Saturn were not composed entirely of hydrogen and helium in solar proportions — they have radii too small for their mass. *Podolak and Cameron* (1974) showed that these planets likely had massive heavy elements cores, which tied their interior structure to a possible formation mechanism (*Perri and Cameron*, 1974; *Mizuno et al.*, 1978). *Stevenson and Salpeter* (1977a,b) performed detailed investigations of phase diagrams and transport properties of fluid hydrogen and helium, putting our understanding of these planets on a firmer theoretical footing. The issues discussed in these papers still reward detailed study. Also, in the early to mid 1970s, the first thermal evolution models of Jupiter were computed (*Grossman et al.*, 1972; *Bodenheimer*, 1974; *Graboske et al.*, 1975) and connections were made between these models and similar kinds of cooling models for fully convective very low mass stars. Much of the early knowledge of giant planet structure and evolution is found in *Hubbard* (1973) and *Zharkov and Trubitsyn* (1978). A consistent finding over the past few decades is that fully adiabatic cooling models of Jupiter can reproduce its current T_{eff} to within a few K.

The situation for Saturn is more complicated. There is a long-standing cooling shortfall in thermal evolution models of Saturn: The planet is $\sim 50\%$ more luminous than one calculates for a 4.5-G.y.-old, adiabatic, well-mixed planet (*Pollack et al.*, 1977; *Stevenson and Salpeter*, 1977a). The most likely explanation is that the He is currently phase separating from the liquid metallic hydrogen (*Stevenson*, 1975), and has been for the past 2–2.5 G.y. This immiscible He should coalesce to form droplets that are denser than the surrounding H/He mixture, and then "rain" down within the planet. This differentiation is a change of gravitational potential energy into thermal energy. Recent models of the evolution of Saturn with the additional energy source indicate that the He may be raining down on top of the core (*Fortney and Hubbard*, 2003). The largest uncertainly in properly including He phase separation into cooling models is the current observed abundance of He in Saturn's atmosphere, which does appear to be depleted in He relative to protosolar abundances, but the error bars are large (*Conrath and Gautier*, 2000). (Note that this "He rain" cannot power the inflated hot Jupiters, as their large radii require interior temperatures warmer than needed for phase separation.) The *in situ* observation that Jupiter's atmosphere is modestly depleted in He as well shows that our understanding of the evolution of both these planets is not yet complete.

Figure 5 shows models of Jupiter's and Saturn's interiors taken from *Guillot* (2005). As Jupiter is 3.3 times more massive than Saturn, a greater fraction of its interior mass is found at high pressure. Therefore, most of the hydrogen mass of Jupiter is fluid metallic, while for Saturn it is fluid molecular, H_2. Jupiter's visible atmosphere is warmer than Saturn's. If their interiors are fully adiabatic, this implies that Jupiter is always warmer at a given pressure than Saturn. Since Jupiter has a larger interior heat content (residual energy leftover from formation), this also means that at a given age, Jupiter is always more luminous than Saturn.

Structural models for Saturn show that it is more centrally condensed than Jupiter. Although significant central condensation is expected due to the compressibility of H/He, detailed models of Saturn indicate even greater central condensation is needed to match its gravitational field. This confirms that a significant fraction of the heavy elements must be in the form of a central dense core. Recent estimates from *Saumon and Guillot* (2004) indicate a core mass of 10–20 $M_\oplus$. The majority of Saturn's heavy elements are within the core. For Jupiter, the situation is less clear cut for several reasons. First, a 10-$M_\oplus$ core would only be 3% of the planet's total mass, so EOS inputs must be accurate to this same percentage for real constraints on the core mass. Second, since Jupiter is more massive than Saturn, a greater fraction of its interior is in the 1–100-Mbar region, which is difficult to model. *Guillot* (1999) and *Saumon and Guillot* (2004) found that Jupiter models without any core were allowed, and that the majority of the planet's heavy elements must be mixed into the H/He envelope. This would be a clear difference compared to Saturn. On the whole Saturn possesses <25 $M_\oplus$ of heavy elements, while Jupiter is <40 $M_\oplus$. Saturn appears to have a relatively greater fraction of heavy elements by mass.

Very recently, two groups have computed new models of the interior of Jupiter, based on first-principles equations of

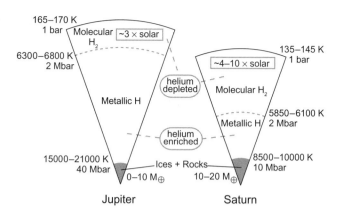

Fig. 5. Interior views of Jupiter and Saturn, from calculations with the *Saumon et al.* (1995) "chemical picture" H/He EOS. Jupiter is more massive, which leads to a greater fraction of its mass in the high-pressure liquid metallic phase. It also has a higher temperature at a given pressure. Interior temperatures are taken from *Guillot* (2005).

state for hydrogen and helium. *Militzer et al.* (2008) simulate the hydrogen-helium mixtures directly, under the pressure-temperature conditions found in the interior of Jupiter. This work predicts a large core of 14–18 $M_\oplus$ for Jupiter, which is in line with estimates for Saturn and suggests that both planets may have formed by core accretion. The paper further predicts a small fraction of planetary ices in Jupiter's envelope, suggesting that the ices were incorporated into the core during formation rather than accreted along with the gas envelope. Jupiter is predicted to have an isentropic and fully convective envelope that is of constant chemical composition. In order to match the observed gravitational moment J_4, the authors suggest that Jupiter may not rotate as a solid body and predicted the existence of deep winds in the interior leading to differential rotation on cylinders. Differential rotation as well as the size of Jupiter's core can potentially be measured by the forthcoming Juno orbiter mission, briefly described below.

Alternatively, *Nettelmann et al.* (2008) compute EOSs for hydrogen, helium, and water separately, and investigate interior models under the assumption of an additive volume rule at constant temperature and pressure. They hypothesize a heavy-element-enriched metallic region and heavy-element-depleted molecular region. They overall find a large abundance of heavy elements within Jupiter, but far less in a distinct core. A further key difference between these two models that has not yet been investigated is interior temperatures — the *Nettelmann et al.* (2008) model predicts higher temperatures than do *Militzer et al.* (2008). This will have important consequences for cooling models of the planet, which will need to be investigated. The *Nettelmann et al.* (2008) and the *Militzer et al.* (2008) approaches are compared in *Militzer and Hubbard* (2009), while Fig. 6 gives a schematic comparison.

The resolution of the H/He phase separation issue in Saturn may only come from an entry probe into Saturn's atmosphere, which could measure the He/H$_2$ ratio *in situ*, as was done for Jupiter. If the He mixing ratios were known precisely in each atmosphere, it is possible that a phase diagram and evolution history combination could be derived that constrains the demixing region of the high-pressure H/He phase diagram (*Fortney and Hubbard*, 2003). In theory, if atmospheric abundances of He, C, N, and O were known, this would be a strong constraint on the properties of the entire H/He envelope. In practice, this has not yet happened for Jupiter because the Galileo Entry Probe was apparently only able to measure a lower limit on the O abundance from water vapor. The forthcoming Juno mission will use microwave spectroscopy to measure the deep H$_2$O and NH$_3$ abundances at pressures near 100 bar (below cloud condensation levels), which should help to resolve this issue for Jupiter. Juno is expected to launch in 2011 and arrive at Jupiter in 2016, with detailed data analysis beginning in 2018.

Juno, which is a low-periapse orbiter, will also exquisitely map the planet's gravitational and magnetic fields. Work by *Hubbard* (1999) has shown that if surface zonal flows extend down to 1000 km depth (P ~ 10 kbar) then this should be observable in the planet's gravity field. This will give us a view of the *mechanics* of the interior of the planet. Does it rotate as a solid body, on cylinders, or something more exotic? Furthermore, the tidal response of Jupiter to the Galilean satellites may also be detectable, which will help to constrain the planet's core mass.

The Juno orbiter will also map Jupiter's magnetic field. The magnetic fields of Jupiter and Saturn are large and predominantly dipolar, with a small tilt from the planet's rotation axis, similar to Earth. This tilt is 9.6° for Jupiter, but less than 1° for Saturn. These fields are consistent with their production via a dynamo mechanism within the liquid metallic region of the interior. Unfortunately, dynamo physics is not understood well enough to place strong constraints on the interiors of Jupiter and Saturn.

3.2. Uranus and Neptune

Uranus and Neptune have not received the same attention that has been paid to Jupiter and Saturn. The neglect of the "ice giants" has been due to relatively less precise observational data, and the complicated picture that the data has revealed. The first stumbling block has been the relatively high density of these planets, compared to Jupiter and Saturn. This high density shows that these planets are *not* predominantly composed of hydrogen and helium. But then what is their composition? Mostly the fluid planetary ices of water, ammonia, and methane? Mostly rock and iron? The mass/radius of these planets can be matched with a very wide range of compositions from the three categories of H/He gas, ice, and rock (e.g., *Podolak et al.*, 1995). A high pressure mixture of rock and H/He can very nicely mimic the pressure-density relation of the ices (*Podolak et al.*, 1991), which means that even gravity field data, which helps to elucidate central condensation, cannot break the ice vs. rock/gas degeneracy. Therefore, modelers often have had

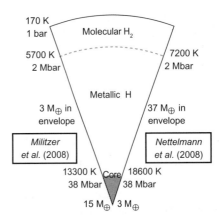

Fig. 6. Possible revised interior views of Jupiter by *Militzer et al.* (2008) and *Nettelmann et al.* (2008). Temperature, pressures, and abundances of heavy elements are labeled.

to resort to cosmogonical constraints, such as an assumed ice-to-rock ratio from protosolar abundance arguments, to constrain structural models. Recent estimates of the H/He mass fraction yield 1–2 M$_\oplus$ in both planets (with an ice-to-rock ratio of 2.5), with a hard upper limit of ~5 M$_\oplus$ in each if only rock and H/He gas are present (*Podolak et al.*, 2000).

Nearly all collected knowledge on the structure and evolution of these planets has been gathered in the *Uranus* and *Neptune and Triton* University of Arizona Space Science Series book chapters by *Podolak et al.* (1991) and *Hubbard et al.* (1995), respectively. These chapters summarize our state of knowledge of Uranus and Neptune as of the early 1990s, after the Voyager 2 encounters. Importantly, novel research is also presented in those chapters that is not found in the more readily available literature. Both chapters are worth detailed study. Since that time, the only more recent update is the work of Marley and Podolak, who investigated Monte Carlo interior models of these planets with a minimum of assumptions regarding interior density (*Marley et al.*, 1995; *Podolak et al.*, 2000). Without assumptions regarding the layering of gas, ices, and rock, these authors perform Monte Carlo studies of the interior density distribution, which uses the gravity field alone to show where density jumps, if any, must occur.

The second major complication with these planets, after composition degeneracy, is the interior heat flow. While fully adiabatic, fully convective thermal evolution models reproduce the current luminosity of Jupiter, and underpredict the luminosity of Saturn; they *overpredict* the luminosity of Neptune and Uranus. The situation for Uranus is especially dramatic, as no intrinsic flux from the planet's interior was detected by Voyager 2. At least two important ideas partially address the heat flow issue. *Hubbard and Macfarlane* (1980) suggested that the absorbed and reradiated stellar flux may be large enough to swamp the intrinsic flux, a smaller component. This would be a larger effect in Uranus than Neptune, since it is closer to the

Sun. This same effect, on a much more dramatic scale, is seen for the hot Jupiters, where the intrinsic flux is unmeasurable, since it is 10^4 times smaller than reradiated absorbed flux. While this effect is certainly real in Uranus and Neptune, it alone cannot explain the low heat flows (*Hubbard et al.,* 1995).

The problem may well be in the assumption that the interior is partitioned into well-defined layers of H/He gas, the fluid ices, and rock. If these distinct layers exist, then convection should be efficient in each layer, and the interior heat should be readily transported to the surface. However, it is well known that composition gradients can readily suppress convection, as a much steeper temperature gradient is needed for convective instability to occur, from the Ledoux criterion. If large regions of the interior of these planets are stably stratified, then stored residual energy from formation will be "locked" into the deep interior, and will only be transported quite slowly. At gigayear ages this would lead to a small intrinsic luminosity. A promising explanation for the reduced heat flow of Uranus and Neptune is that the deep interiors of the planets, which are likely a mix of fluid ices and solid rock, are predominantly stratified, with only the outer approximately one-third of the heavy element interior region freely convecting (*Hubbard et al.,* 1995).

Recently, interior geometry as proposed above was investigated with three-dimensional numerical dynamo models (*Stanley and Bloxham,* 2006). Uranus and Neptune are known to have complex magnetic fields that are nondipolar, nonaxisymmetric, and tilted significantly from their rotation axes. Stanley and Bloxham have found that they can reproduce the major features of these fields with an outer convecting ionic shell, in the outer ~20–40% of the interior region, with the innermost regions not contributing to the dynamo. This is strong evidence that the interiors of these planets are complex and are not fully convective. In addition, structure models that include a pure H/He envelope, a pure icy layer, and an inner rocky core are *not* consistent with the gravity field data of either planet. A possible interior view of each planet is shown in Fig. 7.

Since we will not have additional data to constrain the gravitational moments of these planets, or an entry probe, for perhaps decades, further progress in understanding the structure and evolution of Uranus and Neptune must come from new theoretical ideas. At this time, new thermal evolution models are needed to explore in some detail what regions of the interior are indeed convective. Future work on heat transport in double-diffusive convective regions in planetary interiors (see section 4) should focus on Uranus and Neptune as well as exoplanets.

4. RECENT HIGHLIGHTS: TRANSITING AND DIRECTLY IMAGED PLANETS

4.1. The Observed Mass-Radius Relationship

The determination of mass-radius relationships of exoplanets, using photometric transit and Doppler follow-up techniques, provides an unprecedented opportunity to extend our knowledge on planetary structure and composition. The

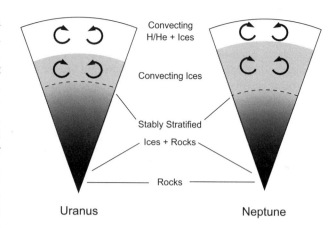

Fig. 7. Possible interior views of Uranus and Neptune. White indicates a composition of predominantly H_2/He gas (with smaller amounts of heavy elements mixed in), solid gray is predominantly ices, and black is predominantly rock. The gray-to-black gradient region in each planet shows where the interior may be statically stable due to composition gradients. Circles with arrow heads indicate convection. Neptune appears to be composed of a greater fraction of heavy elements and it may have a larger freely convective region; however, this and other inferences are uncertain.

twentieth century was marked by the solar system exploration by space missions, revealing the complexity and diversity of giant planets in terms of chemical composition and internal structure. The prolific fishing for transiting exoplanets in the beginning of this twenty-first century, with a catch of over 50 objects, not only confirms this diversity but also raises new questions in the field of planetary science. Two benchmark discoveries illustrate the surprises planet hunters were faced with. The very first transiting planet ever discovered, HD 209458b (*Charbonneau et al.,* 2000; *Henry et al.,* 2000), was found with an abnormally large radius, a puzzling property now shared by a growing fraction of transiting exoplanets. At the other extreme, a Saturn-mass planet, HD 149026b (*Sato et al.,* 2005), was discovered with such a small radius that more than 70 $M_\oplus$ of heavy elements is required to explain its compact structure. This discovery raised in particular new questions on the formation process of planets with such a large amount of heavy material. The diversity in mean density of transiting planets yet discovered is illustrated in Fig. 8.

4.2. Irradiation Effects

The discovery of HD 209458b and the additional transiting exoplanets that followed has opened a new era in giant planet modeling. The modern theory of exoplanet radii starts with models including irradiation effects from the parent star. These effects on planet evolution are accounted for through the coupling between inner structure models and irradiated atmosphere models, following the same method described in section 2.3. Current treatments are based on simplified treatments of the atmosphere, using one-dimensional plane-parallel atmosphere codes. However, they allow one to understand the main effects on planetary evolution. In reality the impinging stellar flux has an angle of incidence that is a function of the latitude and longitude, but in one dimension one attempts to compute a planetwide or dayside average atmosphere profile, using a parameter f, which represents the redistribution factor of the stellar flux over the planet surface (*Baraffe et al.*, 2003; *Burrows et al.*, 2003; *Fortney et al.*, 2006). The incident stellar flux F_{inc} is explicitly included in the solution of the radiative transfer equation and in the computation of the atmospheric structure and is defined by

$$F_{inc} = \frac{f}{4}\left(\frac{R_*}{a}\right)^2 F_* \qquad (12)$$

where R_* and F_* are the stellar radius and flux respectively, and a the orbital separation. The current generation of models often use $f = 1$, corresponding to a stellar flux redistributed over the entire planet's surface, or $f = 2$ if heat is redistributed only over the dayside. Heat redistribution is a complex problem of atmospheric dynamics, depending in particular on the efficiency of winds to redistribute energy from the dayside to the nightside. This question is a challenge for atmospheric circulation modelers (see chapter by Showman et al., as well as *Showman et al.*, 2008). This nascent field is growing rapidly with observational constraints provided by infrared lightcurves obtained with Spitzer, which are starting to provide information on the temperature structure, composition, and dynamics of exoplanet atmospheres.

Although treatments of irradiation effects can differ in the details, with possible refinements accounting for phase and angle dependences of the incident flux (*Barman et al.*, 2005; *Fortney and Marley*, 2007), different models converge toward the same effect on the planet atmosphere and evolutionary properties. Atmospheric thermal profiles are strongly modified by irradiation effects (see chapter by Burrows, as well as *Barman et al.*, 2001; *Sudarsky et al.*, 2003) as illustrated in Fig. 9a. The heating of the outer layers by the incident stellar flux yields to an isothermal layer between the top of the convective zone and the region where the stellar flux is absorbed. The top of the convective zone is displaced toward larger depths, compared to the nonirradiated case. The main effect of the shallower atmospheric pressure-temperature profile is to drastically reduce the heat loss from the planet's interior, which can maintain higher entropy for a longer time (*Guillot et al.*, 1996). Consequently, the gravitational contraction of an irradiated planet is slowed down compared to the nonirradiated counterpart and the upshot is a larger radius at a given age (see Fig. 9b).

The quantitative effect on the planet's radius depends on the planetary mass, parent star properties, and orbital distance. Typical effects on the radius of irradiated Saturn-mass or Jupiter-mass planet located at orbital distances ranging between 0.02 AU and 0.05 AU around a solar-type star are on the order of 10–20% (*Baraffe et al.*, 2003; *Burrows et al.*, 2003; *Chabrier et al.*, 2004; *Arras and Bildsten*, 2006;

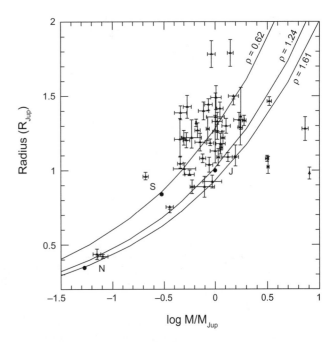

Fig. 8. Mass-radius diagram for the known transiting planets (data are taken from the website of F. Pont, *www.inscience.ch/transits*). Three isodensity curves are shown for the mean densities of Saturn ($\bar{\rho} = 0.62$), Jupiter ($\bar{\rho} = 1.24$), and Neptune ($\bar{\rho} = 1.61$). These three solar system giant planets are also indicated by solid points.

Fortney et al., 2007). A consistent comparison between the theoretical radius and the observed transit radius requires an additional effect due to the thickness of the planet atmosphere (*Baraffe et al.*, 2003; *Burrows et al.*, 2003, 2007). The measured radius is a transit radius at a given wavelength, usually in the optical, where the *slant* optical depth reaches ~1. This is at atmospheric layers above the photosphere (*Hubbard et al.*, 2001; *Burrows et al.*, 2003). The latter region is defined by an averaged normal optical depth $\tau \sim 1$, where the bulk of the flux is emitted outward and which corresponds to the location of the theoretical radius. The atmospheric extension due to the heating of the incident stellar flux can be significant, yielding a measured radius larger than the simple theoretical radius. This effect can add a few percent (up to 10%) to the measured radius (*Baraffe et al.*, 2003; *Burrows et al.*, 2003). Effects of irradiation on both the thermal atmosphere profile and the measured radius must be included for a detailed comparison with observations and can explain some of the less inflated exoplanets. They are, however, insufficient to explain the largest radii of currently known transiting exoplanets, such as TrES-4 with a radius $R = 1.78\ R_{Jup}$ (*Sozzetti et al.*, 2009). This fact points to other mechanisms to inflate close-in planets.

4.3. Determining Transiting Planet Composition

The fraction of planets that are larger than can easily be explained [*Miller et al.* (2009) put his fraction at ~40%] serve

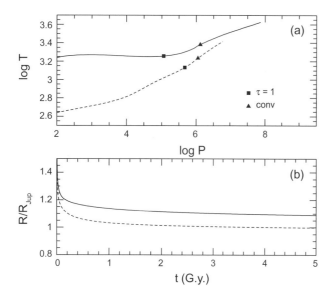

Fig. 9. Effect of irradiation for a planet at 0.05 AU from the Sun. **(a)** T–P profiles of atmosphere models with intrinsic T_{eff} = 1000 K (a high value representative of giant planets at young ages) and surface gravity log g = 3 (cgs). The solid line corresponds to the irradiated model and the dashed line to the nonirradiated model. The locations of the photosphere ($\tau \sim 1$) and of the top of the convective zone are indicated by symbols (models after *Barman et al., 2001*). **(b)** Evolution of the radius with time of a 1 M_{Jup} giant planet. Irradiated case: solid line; nonirradiated case: dashed line (models after *Baraffe et al., 2003*).

as a handicap to the goal of understanding the composition of these planets. We would like to be able to constrain the fraction of planetary mass that is the non-H/He heavy elements, and understand how this varies with planetary mass, stellar mass, stellar metallicity, distance from the parent star, and other factors. Work on understanding the heavy element enrichment of planets has progressed, but must be regarded as incomplete, due to the poorly understood large-radius planets.

It is certainly clear that a number of exoplanets, like our solar system giant planets (see section 3), are enriched in heavy material, as these planets have radii smaller than pure H/He objects. This idea is supported by our current understanding of planet formation via the core-accretion scenario. This model is itself consistent with current observations showing that metal rich environments characterized by the high metallicity of the parent star favors planet formation (*Udry and Santos, 2007*).

The exoplanets with the largest fraction of mass in heavy elements known at the time this chapter is written are WASP-7b, with M_p = 0.96 M_{Jup} and R_p = 0.915 R_{Jup} (*Hellier et al., 2009*), HAT-P-3b with M_p = 0.6 M_{Jup} and R_p = 0.89 R_{Jup} (*Torres et al., 2007*), and HD 149026b with M_p = 0.36 M_{Jup} and R_p = 0.755 R_{Jup} (*Sato et al., 2005*). Their small radii indicate a global mass fraction of heavy material greater than what Jupiter contains (more than 12%) and in the case of HD 149026b, significantly greater than what Saturn contains (more than 30%). This suggests that the presence of a significant amount of heavy material is a property shared by all planets. Evolutionary models must thus take into account such enrichment and many efforts are currently devoted to the construction of a wide range of models with different amounts of heavy elements. Exploration of the effects of materials of different composition is limited by the available equations of state under the conditions of temperature and pressure characteristic of giant planets. As a simple starting point, and given the large uncertainty on the nature of heavy elements and their distribution inside planets, current models often assume that all heavy elements are located in the core and are water/ice, rock and/or iron. A possible interior density profile of HD 149026b with a core of either ice or rock is shown in Fig. 10. Uncertainties in current planetary models due to uncertainties in the available EOS, the distribution of heavy elements, and their chemical composition have been analyzed in detail in *Baraffe et al.* (2008). Depending on the total amount of heavy material and its composition, the radius of an enriched planet can be significantly smaller compared to that of H/He-dominated giant planets. This is illustrated in Fig. 11, which compares the radius evolution with time of planets with different core sizes (in $M_{\oplus}$) and different core compositions.

As noted, the problematic planets are those that are "too large," such that we cannot constrain the heavy element mass in their interiors. However, if the "true" additional energy source or contraction-stalling mechanism affects nearly all hot Jupiters to some degree (which seems likely given that one-third to one-half of planets are too large), then we may be able to take out its effect on planetary radii, and examine the heavy element enrichments of the transiting planets as a collection. This was first done by *Guillot et al.* (2006), who postulated an unspecified additional interior energy source equal to 0.5% of the absorbed incident energy. All planets then required at least some amount of heavy elements to match the measured radii, and they found a correlation between stellar metallicity and planetary heavy element mass. Subsequently, *Burrows et al.* (2007) found a similar correlation, using only enhanced atmospheric opacities for all planets, although this cannot explain the largest-radius planets. Importantly, *Guillot* (2008) find that the stellar metallicity/planetary heavy element correlation continues to hold with a larger number of planets. This appears to fit well with expectations for the core-accretion theory of giant planet formation (*Dodson-Robinson and Bodenheimer, 2009*). Determining whether this and perhaps other relations hold in the future will be an important area of transiting planet research in the future, as the number of detected planets continues to climb.

4.4. Proposed Solutions to the Radius Anomaly

As mentioned in section 3.1, a significant fraction of transiting planets have large radii that cannot be explained by the effect of thermal irradiation from the parent star alone. Other mechanisms are required to explain these inflated planets. Whether these mechanisms are peculiar,

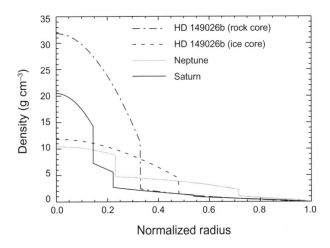

Fig. 10. Interior density as a function of normalized radius for two possible models for HD 149026b compared with Neptune and Saturn. All planet models have been normalized to the radius at which P = 1 bar. The Neptune profile is from *Podolak et al.* (1995) and the Saturn profile is from *Guillot* (1999). The Saturn and Neptune models have a two-layer core of ice overlying rock, but this is only an assumption. For Neptune in particular the interior density profile is uncertain. The two profiles of HD 149026b assume a metallicity of 3× solar in the H/He envelope and a core made entirely of either ice or rock. After *Fortney et al.* (2006).

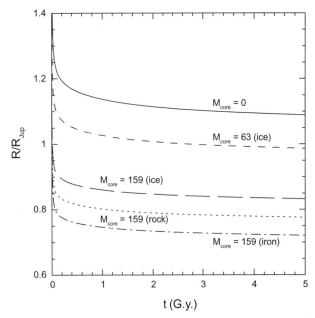

Fig. 11. Effect of heavy element enrichment on the evolution of the radius with time of a 1-M_{Jup} giant planet. Solid line: H/He planet; dashed line: $M_{core} = 63\ M_\oplus$ (ice); long dashed line: $M_{core} = 159\ M_\oplus$ (ice); dotted line: $M_{core} = 159\ M_\oplus$ (rock); dash-dotted line: $M_{core} = 159\ M_\oplus$ (iron).

operating only under specific conditions, for certain planets, or whether basic physics is missing in the modeling of close-in giant planet structure are still open questions. In either scenario the inflation mechanism(s) must be common. As noted, the latter alternative stems from the suspicion that all observed transiting planets may contain a certain amount of heavy material. The denser planetary structure resulting from the presence of heavy material could counteract the effect of the "missing" mechanism, yielding an observed "normal" radius (*Fortney et al.*, 2006).

Many mechanisms have been proposed since the discovery of HD 209458b, but no mechanism has gained the consensus of the community as being clearly important. We describe below the main ideas and comment on their status.

4.4.1. Atmospheric circulation. Based on numerical simulations of atmospheric circulation on hot Jupiters, *Showman and Guillot* (2002) suggested a heating mechanism in the deep interior driven by strong winds blowing on the planetary surface. The idea is that stellar irradiation produces strong day-night temperature contrasts, which drive fast winds. Their simulations produced a downward kinetic energy flux, about 1% of the absorbed stellar incident flux, which dissipates in the deep layers. This mechanism provides an extra source of energy that slows down the planet evolution and can explain the large observed radii (*Guillot and Showman*, 2002; *Chabrier et al.*, 2004).

This attractive and original scenario still requires confirmation. The very existence of the downward kinetic energy flux and its strength strongly depend on the outcome of atmospheric circulation models. Although this field has almost half a century history for solar system planets, it is still in its infancy when applied to hot Jupiters where conditions are very different. Current models show important divergences (see the chapter by Showman et al.) and their robustness to describe the conversion process of heat from the parent star into mechanical energy (i.e., winds) has even been questioned (*Goodman*, 2009). The substantial downward transport of kinetic energy reported by *Showman and Guillot* (2002) has not been found in somewhat similar numerical simulations performed by *Burkert et al.* (2005). Moreover, the physical process that converts kinetic energy into thermal energy in the *Showman and Guillot* (2002) scenario still needs to be identified.

4.4.2. Enhanced atmospheric opacities. Burrows et al. (2007) suggest that abnormally large radii of transiting planets can be explained by invoking enhanced opacities. Enhanced opacities could for instance be due to missing or underestimated opacities in the current generation of model atmospheres. In practice, *Burrows et al.* (2007) implement these higher opacities by simply using supersolar metallicities (up to 10× solar). The main effect of enhanced opacities is to retain internal heat and to slow down the contraction. These authors conclude that a combination of enhanced opacities and the presence of dense cores can provide a general explanation to the observed spread in radii of transiting planets. Interestingly, this work also finds the same correlation as suggested by *Guillot et al.* (2006) between planet core mass and stellar metallicity.

If opacity calculations are indeed correct, then higher opacities could be due only to increased metallicity. This would, however, present several weaknesses. It may be difficult to maintain a substantial fraction of heavy elements in strongly irradiated atmospheres, which are radiative down to deep levels. Perhaps mixing via atmospheric dynamics suffices to keep the atmosphere well mixed (*Showman et al.,* 2009; *Spiegel et al.,* 2009). Moreover, as discussed in *Guillot* (2008), *Burrows et al.* (2007) do not take into account the subsequent increase of molecular weight due to the increase of the planet atmosphere metallicity. This effect counters the effect of enhanced opacities and may even dominate in some cases, yielding the opposite effect on the radius (see discussion in *Guillot,* 2008). Finally, for the most inflated transiting planets yet detected, models with enhanced opacities are unable to reproduce their radii and another additional mechanism is required (*Liu et al.,* 2008).

4.4.3. Tidal effects. Another alternative, first suggested by *Bodenheimer et al.* (2001), is that tidal forces may heat the planet. There has come to be a large body of work in this area. If a planet has an eccentric orbit or a nonsynchronous rotation, internal tidal dissipation within the planet produces an energy source, which can slow down its contraction, or reinflate the planet after a previous contraction phase (see, e.g., *Gu et al.,* 2003). Following this idea, *Liu et al.* (2008) suggest a combination of enhanced atmospheric opacities, the presence of heavy elements and heating due to orbital tidal dissipation, assuming nonzero eccentricities, to explain the most inflated planets. Almost all studies on tidal effects have assumed a tidal equilibrium state, yielding very short timescales for synchronization ($\sim 10^5$–10^6 yr) and circularization ($\sim 10^8$–10^9 yr) compared to the estimated age of known exoplanetary systems. This assumption implies that for eccentricity to be maintained for several gigayears and contribute to the heating, it should be continually excited. In the case of HD 209458b, *Bodenheimer et al.* (2001) suggested the presence of an unseen planetary companion, which could force eccentricity.

Tidal heating due to a finite current eccentricity may explain some of the large transit radii currently observed. It is, however, certainly not the mechanism that could explain all inflated planets. Constraints on the eccentricity of HD 209458b based on the timing of the secondary eclipse (*Deming et al.,* 2005) yield that a nonzero eccentricity is very unlikely. This explanation seems also improbable for TrES-4 based on Spitzer observations by *Knutson et al.* (2009), who can rule out tidal heating at the level required by *Liu et al.* (2008) to explain this planet's bloated size.

More recently, *Jackson et al.* (2008) and *Levrard et al.* (2009) have revisited the tidal stability of exoplanets. Essentially all examined transiting planets have not reached a tidal equilibrium state, implying that they will ultimately fall onto the central star (*Levrard et al.,* 2009). *Jackson et al.* (2009), confirming this view, find that it is the youngest parent stars that tend to harbor the closest-in hot Jupiters, implying a loss of close-in planets with time. More importantly, these works stress that conventional circularization and synchronization timescales, which are widely used in the community, are in most cases not correct. In the *Levrard et al.* (2009) view, nearly circular orbits of planetary transiting systems currently observed may not be due to tidal dissipation. Conversely, observations of nonzero eccentricity would be naturally explained without the need for gravitational interactions from undetected companions.

Ibgui and Burrows (2009) and *Miller et al.* (2009) have coupled a standard second-order tidal evolution theory (e.g., *Jackson et al.,* 2009) to planet structural evolution models for close-in giant planets in single-planet systems to investigate under what circumstances tidal heating by recent eccentricity damping (which leads to an energy surge and radius inflation), together with tidal semimajor axis decay, could inflate these planets. *Miller et al.* (2009) find that this radius inflation can occur for some planets at gigayear ages, perhaps explaining some large radii, but this mechanism is unlikely to explain all inflated radii.

Tidal heating could also be produced by a large obliquity, i.e., the angle between the planetary spin axis and the orbital normal, as suggested by *Winn and Holman* (2005). Since planet obliquity is expected to be rapidly damped by tidal dissipation, one possibility for maintaining a nonzero obliquity is for it to be locked in a Cassini state, i.e., a resonance between spin and orbital precession. However, *Levrard et al.* (2007) show that although the probability of capture in a spin orbit resonance is rather good around 0.5 AU, it decreases dramatically with semimajor axis. Also, *Fabrycky et al.* (2007) rule out possible drivers, such as the presence of a second planet, of a high-obliquity Cassini state for HD 209458b. They conclude that very special configurations are required for obliquity tides to be an important source of heating. On the whole, it thus seems difficult to invoke tidal dissipation as the mechanism that could explain all currently observed inflated exoplanets, although important quantitative effects have been identified. Considerable work in this area continues.

4.4.4. Double diffusive convection. *Chabrier and Baraffe* (2007) suggest that the onset of layered or oscillatory convection, due to the presence of molecular weight gradients, can reduce heat transport in planetary interiors and slow down the contraction, providing an explanation for the large spread in radii of transiting planets. The formation of layers is a characteristic of double-diffusive convection, which may occur in a medium where two substances diffuse at different rates. This is a well-known process in oceans or salty lakes where the two substances are heat and salt. Applied to the interior of planets, in the presence of a compositional gradient, convection can break into convective layers separated by thin diffusive layers. The heat transport efficiency is thus significantly reduced because of the presence of multiple diffusive layers, compared to the case of a fully homogeneous planet where convection is assumed to be fully adiabatic. This process is similar to the so-called semiconvection in stars (*Stevenson,* 1979). Based on a phenomenological approach, and assuming a molecular weight gradient in the most inner part of the planet, *Chabrier and Baraffe* (2007) show that a significant number of diffusive layers strongly reduce the heat escape. They suggest that the composition gradient is inherited from the formation process, during accretion of planetesimals and gas, or due to core erosion. The upshot is a significantly inflated planet compared to its homogeneous and adiabatic counterpart.

The idea that planetary interiors may not be completely homogeneous and convection not fully efficient was already suggested by D. Stevenson for Jupiter (*Stevenson*, 1985) and for Uranus and Neptune (*Hubbard et al.*, 1995) (see section 3). Double-diffusive convection is indeed a well-known process under Earth conditions, which have similarities with those found in the interior of giant planets (for details, see *Chabrier and Baraffe*, 2007). Whether this scenario can explain all inflated transiting planets is still debatable. The key questions are whether the diffusive layers commonly form, and can survive on timescales of gigayears, characteristic of the age of the transiting exoplanets. Development of three-dimensional numerical simulations of this process under planetary conditions could provide clues about its long-term existence in giant planet interiors.

4.5. Hot Neptune Planets

The dream of discovering exoplanets of a few Earth masses is now becoming reality. More than a dozen planets with masses in the Uranus-Neptune range (≤ 20 $M_\oplus$) have been detected by radial velocity surveys (*Udry and Santos*, 2007). Because these light planets are very close to the detection threshold, their discovery suggests that they are rather common. Given their low mass and close orbit, the question of their origin deserves some attention. Formation models based on the core accretion scenario suggest that hot Neptunes are composed of a large heavy material core (ice or rock) and have formed without accumulation of a substantial gaseous envelope (*Alibert et al.*, 2006; *Mordasini et al.*, 2008). Another suggestion is that hot Neptunes could have formed from more massive progenitors and have lost most of their gaseous envelope (*Baraffe et al.*, 2005). The latter idea arises from observational evidences that close-in exoplanets may undergo evaporation processes induced by the high-energy flux of the parent star (*Vidal-Madjar et al.*, 2003), as well as high mass-loss rates found in some early models (e.g., *Lammer et al.*, 2003).

The interpretation of the *Vidal-Madjar et al.* (2003) observations of an extended neutral hydrogen atmosphere around HD 209458b is an active area that is still controversial (*Ben-Jaffel*, 2007, 2008; *Holmström et al.*, 2008; *Vidal-Madjar et al.*, 2008). However, at the same time, the various groups computing models of evaporative mass loss at small orbital distances have been converging to mass-loss rates that yield a total mass loss of only ~1% for HD 209458b over the lifetime of the system (*Yelle*, 2004; *Tian et al.*, 2005; *García Muñoz*, 2007; *Yelle et al.*, 2008; *Murray-Clay et al.*, 2009). If this is correct, than these evaporation processes are too small to significantly affect the evolution of the *currently observed* close-in planets. Small evaporation rates also seem to be consistent with the observed mass

function of exoplanets (*Hubbard et al.*, 2007). Improved statistics in the low-planetary-mass regime, confirmation of the observations of *Vidal-Madjar et al.* (2003), and the extension of similar observations to other transiting planets will shed light on these issues.

Independently of these issues, the interior properties of two hot Neptune exoplanets were recently revealed by the remarkable discovery of the first transiting Neptune-mass planets, GJ 436b (*Gillon et al.*, 2007) and HAT-P-11b (*Bakos et al.*, 2010). Both planets have a radius comparable to that of Neptune, indicating heavy material enrichment greater than 85%, which is the overall heavy element content of Uranus and Neptune (*Guillot*, 2005). Given current uncertainties on planetary interior structure models, only the bulk of heavy elements can be inferred for each, about 20 $M_\oplus$ for a total mass of ~22 $M_\oplus$ for GJ 436b, which is the more well studied of the two (see Fig. 12). Although small in terms of mass, the contribution of the H/He envelope to the total planetary radius is significant. This is illustrated in Fig. 12, where the radii of pure water and pure rocky planets of the same mass are also indicated.

Better determination of the chemical composition of heavy material and its distribution within the planet must await improved EOSs for H/He and heavy elements (*Baraffe et al.*, 2008). These discoveries, however, confirm the large heavy element content that can be expected in exoplanets and support the general picture of planet formation drawn by the core accretion model.

4.6. Young Giant Planets

Giant planet thermal evolution models are being tested at gigayear ages for solar system planets and the transiting planets. It is clear from giant planet formation theories (see the chapter by D'Angelo and Lissauer) that these planets are hot, luminous, and have larger radii at young ages, and they cool and contract inexorably as they age. However, since the planet formation process is not well understood *in detail*, we understand very little about the initial conditions for the planets' subsequent cooling. Since the Kelvin-Helmholtz time is very short at young ages (when the luminosity is high and radius is large), it is expected that giant planets forget their initial conditions quickly. This idea was established with the initial Jupiter cooling models in the 1970s (*Graboske et al.*, 1975; *Bodenheimer*, 1976).

Since our solar system's giant planets are thought to be 4.5 G.y. old, there is little worry about how thermal evolution models of these planets are affected by the unknown initial conditions. The same may not be true for very young planets, however. Since giant planets are considerably brighter at young ages, searches to directly imaged planets now focus on young stars. At long last, these searches are now bearing fruit (*Chauvin et al.*, 2005; *Marois et al.*, 2008; *Kalas et al.*, 2008; *Lagrange et al.*, 2009). It is at ages of a few million years where understanding the initial conditions and early evolution history is particularly important. Traditional evolution models (section 2), which are applied to both giant

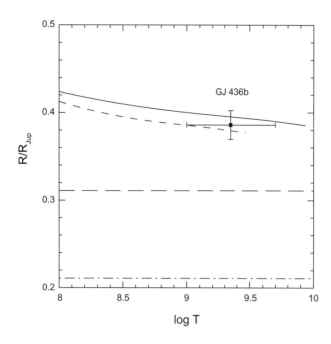

Fig. 12. Evolution of a planet with the mass of GJ 436b (22.6 $M_\oplus$) and different heavy element contents and compositions. The solid line corresponds to a model with a water core of 21 $M_\oplus$ and the dashed line to a rocky core of 19.5 $M_\oplus$ (after the models of *Baraffe et al., 2008*). The long dashed line indicates the radius of a pure water planet and the dash-dotted line corresponds to a pure rocky planet (after the models of *Fortney et al., 2007*).

planets and brown dwarfs, employ an arbitrary starting point. The initial model is large in radius, luminosity, and usually fully adiabatic. The exact choice of the starting model is often thought to be unimportant, if one is interested in following the evolution for ages greater than 1 m.y. (*Burrows et al., 1997*; *Chabrier and Baraffe, 2000*).

Thermal evolution models, when coupled to a grid of model atmospheres, aim to predict the luminosity, radius, T_{eff}, thermal emission spectrum, and reflected spectrum, as a function of time. When a planetary candidate is imaged, often only the apparent magnitude in a few infrared bands are known. If the age of the parent star can be estimated (itself a tricky task) then the observed infrared magnitudes can be compared with calculations of model planets for various masses, to estimate the planet's mass. Recall that mass is not an observable quantity unless some dynamical information is also known. It is not known if these thermal evolution models are accurate at young ages — they are relatively untested, which has been stressed by *Baraffe et al.* (2002) for brown dwarfs and *Marley et al.* (2007) for planets.

Marley et al. (2007) examined the issue of the accuracy of the arbitrary initial conditions (termed a "hot start" by the authors) by using initial conditions for cooling that were not arbitrary, but rather were given by a leading core-accretion planet-formation model (*Hubickyj et al., 2005*). The core-accretion calculation predicts the planetary structure at the end of formation, when the planet has reached its final mass. The *Marley et al.* (2007) cooling models used this initial model for time zero, and subsequent cooling was followed as in previously published models. Figure 13 shows the resulting evolution. The cooling curves are dramatically different, yielding cooler (and smaller) planets. The initial conditions are not quickly "forgotten," meaning that the cooling curves do not overlap with the arbitrary start models for 10^7 to 10^9 years. What this would mean, in principle, is that a mass derived from "hot start" evolutionary tracks would significantly underestimate the true mass of a planet formed by core accretion.

Certainly one must remember that a host of assumptions go into the formation model, so it is unlikely that these new cooling models are quantitatively correct. However, they highlight that much additional work is needed to understand the *energetics* of the planet-formation process. The *Hubickyj et al.* (2005) models yield relatively cold initial models because of an assumption that accreting gas is shocked and readily radiates away this energy during formation. This energy loss directly leads to a low-luminosity starting point for subsequent evolution. Significant additional work on multidimensional accretion must be done, as well as on radiative transfer during the accretion phase, before we can confidently model the early evolution. Thankfully, it appears that detections of young planets are now beginning to progress quickly, which will help to constrain these models.

5. FUTURE PROSPECTS

The future of understanding the structure, composition, and evolution of giant planets is quite promising. Most immediately, and with the biggest impact, will be the detection of more Neptune-class transiting planets, in addition to GJ 436b and HAT-P-11b. Of particular interest will be the mass-radius relation of planets around ~10 $M_\oplus$, since this is estimated to be a boundary between planets that have H/He envelopes and those that do not. A radius that is larger than that calculated for a pure water planet is unambiguous evidence of a H/He envelope. Certainly the diversity of radii for the transiting gas giants has been surprising, and understanding Neptune-class radii as a function of mass and orbital distance will be fascinating.

Several groundbased surveys for transiting planets are now scouring the northern hemisphere for Jupiter-class planets, with some additional attention being paid to the southern hemisphere. The required photometric precision to detect a Jupiter-type transit (a 1% dip in stellar flux) is not difficult to achieve. Furthermore, the detection of HAT-P-11b opens up the possibility that these same groundbased surveys will also be able to detect significant numbers of Neptune-type planets as well. In addition, the implementation of orthogonal transfer array CCDs may allow for groundbased photometric precision that approaches that of spacebased platforms (e.g., *Johnson et al., 2009*), which would significantly help the cause of detecting smaller planets from the ground.

The CoRoT mission, which has already announced five planets, and will continue to search for a few more years, will

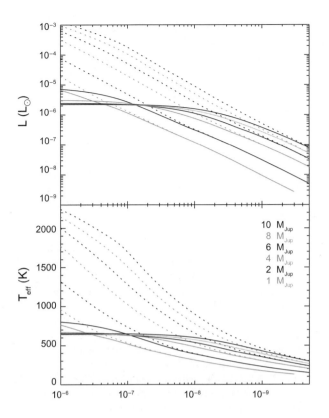

Fig. 13. Thermal evolution of giant planets from 1 to 10 M_{Jup}, adapted from *Marley et al.* (2007). The dotted curves are standard "hot start" models with an arbitrary initial condition, and the solid curves use as an initial condition the core accretion formation models of *Hubickyj et al.* (2005).

be important in adding to our sample size in two ways. The first is of course smaller planets. The second is increasing the sample size of transiting planets at wider orbital separations. CoRoT is surveying particular areas of the sky for 120 days, meaning that 40-day orbits of giant planets can be easily confirmed via 3 detected transits. Around a solar-type star, a 40-day orbit yields a semimajor axis of 0.23 AU, which is a factor of 33 reduction in flux from that intercepted at 0.04 AU. Understanding radii at the largest possible range of incident fluxes may allow us to understand the reason(s) for the large radii of the currently detected planets.

The Kepler mission, which began taking science data in May 2009, is an even more ambitious telescope to detect smaller transiting planets in longer-period orbits. Although its main goal is to ascertain the frequency of Earth-radius planets in Earth-like orbits around Sun-like stars, it will also do the same for larger planets. Kepler will be able to detect multiple transits of planets out to 1 AU, a factor of over 600 reduction in flux from 0.04 AU. In addition, even longer-period transiting giant planets may be detected, if suitable followup is done. Since some mechanisms proposed to explain the large radii of the close-in planets should be significantly muted at larger orbital separations, finding planets farther from their parent stars will likely be the *most important* step in understanding what leads to these large radii. In Fig. 14 we show a specific prediction for the contraction of a 1 M_{Jup} planet over a factor of 250,000 in incident stellar flux. As detailed in *Fortney et al.* (2007), the effects of stellar flux are muted beyond the current group of close-in transiting planets, and planets out to ~1 AU should have radii quite similar to those at ~0.1 AU, in the absence of missing physics for these more distant planets.

The orbital dynamics of particular exoplanets in some systems may give us direct constraints on a planet's interior state. The apsidal precession rate of a planetary orbit is directly proportional to the tidal Love number, k_2. This number parameterizes the internal density distribution of a planet. Authors have recently pointed out instances in which k_2 could be measured or well constrained. These include precession due to a tidal-induced gravitational quadrupole on the planet by its parent star, for very close-in planets (*Ragozzine and Wolf*, 2009), which could be measured as a change in transit shape over time. Another affects transiting planets in multiplanet systems, which now only includes HAT-P-13b (*Bakos et al.*, 2009). For this system a refined measurement of current planetary eccentricity can constrain k_2 as well as Q, its tidal dissipation quality factor (*Batygin et al.*, 2009).

Detections of transiting giant planets at younger ages would be very important as these planets would inform our understanding of contraction with time in the face of intense stellar irradiation. This would shed light on the initial conditions for evolution, postformation, as well as allow us to better understand the nature of the physical process that is causing large radii at gigayear ages. Toward this goal, several transit surveys of open clusters have been performed, but they have not netted any planets to date.

The focus of comparing models to observations is already shifting from *specific* planets to *samples* of planets, and will soon shift to a statistically significant number of planets, with the additional detections from the ground and from space. A good reference for the kind of work, just starting to be done, is *Fressin et al.* (2007), who analyzed in detail the OGLE transiting planet survey. They simulate the OGLE survey — given the properties of the thousands of stars that were monitored, the known planet frequency as a function of stellar mass and of orbital distance — and implement giant planet contraction models, to derive constraints on, for instance, possible separate populations of planets.

The planets directly imaged by *Marois et al.* (2008) and *Kalas et al.* (2008) have fully opened the door to direct imaging, which began yielding planetary candidates a few years ago (e.g., *Chauvin et al.*, 2005). The characterization of these planets will present different challenges compared to the transiting planets. While transiting planets yield accurate masses and radii, the atmospheric characterization is challenging. For directly imaged planets, masses and radii likely cannot be measured, but spectra should be more easily obtained. Spectra can yield the planet's T_{eff}, which can be directly compared to thermal evolution models. However, current techniques are limited to planet-to-star flux ratios

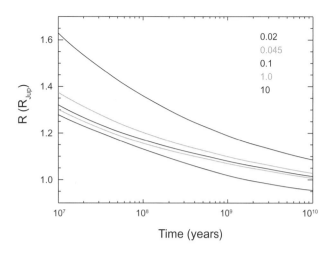

Fig. 14. Contraction of a 1 M_{Jup} planet, with 25 $M_{\oplus}$ of heavy elements (in a core), over time. The planets are placed at 0.02, 0.045, 0.1, 1.0, and 10 AU from a constant luminosity Sun, to show the effect of stellar irradiation on thermal evolution. Adapted from *Fortney et al.* (2007).

of ~10^{-5}. New instruments coming online in the very near future, such as the Gemini Planet Imager (GPI) on Gemini South and Spectro-Polarimetric High-contrast Exoplanet Research (SPHERE) at the VLT will allow for contrasts of ~10^{-6}–10^{-7}. This will allow for the direct imaging and characterization of giant planets at a variety of masses, and also a variety of *ages*, so the first ~1–100 m.y. of giant planet evolution should be relatively well understood via observations.

We are now over a decade into the new era of studying giant planets as a class of astronomical objects. It is the expansion of this class beyond Jupiter, Saturn, Uranus, and Neptune that will enable us to better understand the formation and evolution of these planets. Much like our understanding of the stars is greatly enhanced by studying more than just the Sun, so will our understanding of giant planets grow.

Over the past several years we have seen strange and startling transiting planets with hugely inflated radii, and those with small radii that must include vast interior stores of heavy elements. They have expanded our imaginations regarding the possible structure of giant planets. We are now beginning to see these original oddballs within the continuum of giants planets that extend far beyond what we see in our solar system. Astronomers will do doubt continue to creatively find ways to detect and characterize many more planets in the future.

Acknowledgments. J.J.F. was partially supported by the National Science Foundation and the NASA Outer Planets Research Program.

REFERENCES

Alibert Y., Baraffe I., Benz W., Chabrier G., Mordasini C., Lovis C., Mayor M., Pepe F., Bouchy F., Queloz D., and Udry S. (2006) Formation and structure of the three Neptune-mass planets system around HD 69830. *Astron. Astrophys., 455,* L25–L28

Arras P. and Bildsten L. (2006) Thermal structure and radius evolution of irradiated gas giant planets. *Astrophys. J., 650,* 394–407.

Bagnier S., Blottiau P., and Clarouin J. (2000) Multiscale recursion in dense hydrogen plasmas. *Phys. Rev. E, 61,* 6999–7008.

Bakos G. A., Howard A. W., Noyes R. W., Hartman J., Torres G., Kovacs G., Fischer D. A., Latham D. W., Johnson J. A., Marcy G. W., Sasselov D. D., Stefanik R. P., Sipocz B., Kovacs G., Esquerdo G. A., Pal A., Lazar J., and Papp I. (2009) HAT-P-13b,c: A transiting hot Jupiter with a massive outer companion on an eccentric orbit. *Astrophys. J., 707,* 446–456.

Bakos G. Á., Torres G., Pál A., Hartman J., Kovács G., Noyes R. W., Latham D. W., Sasselov D. D., Sipőcz B., Esquerdo G. A., Fischer D. A., Johnson J. A., Marcy G. W., Butler R. P., Isaacson H., Howard A., Vogt S., Kovács G., Fernandez J., Moór A., Stefanik R. P., Lázár J., Papp I., and Sári P. (2010) HAT-P-11b: A super-Neptune planet transiting a bright K star in the Kepler field. *Astrophys. J., 710,* 1724–1745.

Baraffe I., Chabrier G., Allard F., and Hauschildt P. H. (1998) Evolutionary models for solar metallicity low-mass stars: Mass-magnitude relationships and color-magnitude diagrams. *Astron. Astrophys., 337,* 403–412.

Baraffe I., Chabrier G., Allard F., and Hauschildt P. H. (2002) Evolutionary models for low-mass stars and brown dwarfs: Uncertainties and limits at very young ages. *Astron. Astrophys., 382,* 563–572.

Baraffe I., Chabrier G., Barman T. S., Allard F., and Hauschildt P. H. (2003) Evolutionary models for cool brown dwarfs and extrasolar giant planets. The case of HD 209458. *Astron. Astrophys., 402,* 701–712.

Baraffe I., Chabrier G., Barman T. S., Selsis F., Allard F., and Hauschildt P. H. (2005) Hot-Jupiters and hot-Neptunes: A common origin? *Astron. Astrophys., 436,* L47–L51.

Baraffe I., Chabrier G., and Barman T. (2008) Structure and evolution of super-Earth to super-Jupiter exoplanets. I. Heavy element enrichment in the interior. *Astron. Astrophys., 482,* 315–332.

Barman T. S., Hauschildt P. H., and Allard F. (2001) Irradiated planets. *Astrophys. J., 556,* 885–895.

Barman T. S., Hauschildt P. H., and Allard F. (2005) Phase-dependent properties of extrasolar planet atmospheres. *Astrophys. J., 632,* 1132–1139.

Batygin K., Bodenheimer P., and Laughlin G. (2009) Determination of the interior structure of transiting planets in multiple-planet systems. *Astrophys. J. Lett., 704,* L49–L53.

Belov S. I., Boriskov G. V., Bykov A. I., Il'Kaev R. I., Luk'yanov N. B., Matveev A. Y., Mikhaĭlova O. L., Selemir V. D., Simakov G. V., Trunin R. F., Trusov I. P., Urlin V. D., Fortov V. E., and Shuĭkin A. N. (2002) Shock compression of solid deuterium. *Soviet J. Exp. Theor. Phys. Lett., 76,* 433–435.

Ben-Jaffel L. (2007) Exoplanet HD 209458b: Inflated hydrogen atmosphere but no sign of evaporation. *Astrophys. J. Lett., 671,* L61–L64.

Ben-Jaffel L. (2008) Spectral, spatial, and time properties of the hydrogen nebula around exoplanet HD 209458b. *Astrophys. J., 688,* 1352–1360.

Bodenheimer P. (1974) Calculations of the early evolution of Jupiter. *Icarus, 23,* 319–325.

Bodenheimer P. (1976) Contraction models for the evolution of Jupiter. *Icarus, 29,* 165–171.

Bodenheimer P., Lin D. N. C., and Mardling R. A. (2001) On the tidal inflation of short-period extrasolar planets. *Astrophys. J., 548*, 466–472.

Bonev S. A., Militzer B., and Galli G. (2004) Ab initio simulations of dense liquid deuterium: Comparison with gas-gun shockwave experiments. *Phys. Rev. B, 69*, 014101.

Boriskov G. V., Bykov A. I., Il'Kaev R. I., Selemir V. D., Simakov G. V., Trunin R. F., Urlin V. D., Shuikin A. N., and Nellis W. J. (2005) Shock compression of liquid deuterium up to 10^9 GPa. *Phys. Rev. B, 71*, 092104.

Burkert A., Lin D. N. C., Bodenheimer P. H., Jones C. A., and Yorke H. W. (2005) On the surface heating of synchronously spinning short-period jovian planets. *Astrophys. J., 618*, 512–523.

Burrows A., Marley M., Hubbard W. B., Lunine J. I., Guillot T., Saumon D., Freedman R., Sudarsky D., and Sharp C. (1997) A nongray theory of extrasolar giant planets and brown dwarfs. *Astrophys. J., 491*, 856–875.

Burrows A., Hubbard W. B., Lunine J. I., and Liebert J. (2001) The theory of brown dwarfs and extrasolar giant planets. *Rev. Mod. Phys., 73*, 719–765.

Burrows A., Sudarsky D., and Hubbard W. B. (2003) A theory for the radius of the transiting giant planet HD 209458b. *Astrophys. J., 594*, 545–551.

Burrows A., Hubeny I., Budaj J., and Hubbard W. B. (2007) Possible solutions to the radius anomalies of transiting giant planets. *Astrophys. J., 661*, 502–514.

Car R. and Parrinello M. (1985) Unified approach for molecular dynamics and density-functional theory. *Phys. Rev. Lett., 55*, 2471–2474.

Cavazzoni C., Chiarotti G. L., Scandolo S., Tosatti E., Bernasconi M., and Parrinello M. (1999) Superionic and metallic states of water and ammonia at giant planet conditions. *Science, 283*, 44–46.

Ceperley D. M. (1991) Fermion nodes. *J. Stat. Phys., 63*, 1237–1267.

Ceperley D. M. (1995) Path integrals in the theory of condensed helium. *Rev. Mod. Phys., 67*, 279–355.

Ceperley D. M. (1996) Path integral Monte Carlo methods for fermions. In *Euroconference on Computer Simulation in Condensed Matter Physics Chemistry: Monte Carlo and Molecular Dynamics of Condensed Matter Systems* (K. Binder and G. Ciccotti, eds.), pp. 443–482. Italian Phys. Soc. Conf. Proc. Vol. 49, Editrice Compositori, Bologna, Italy.

Chabrier G. and Baraffe I. (2000) Theory of low-mass stars and substellar objects. *Annu. Rev. Astron. Astrophys., 38*, 337–377.

Chabrier G. and Baraffe I. (2007) Heat transport in giant (exo) planets: A new perspective. *Astrophys. J. Lett., 661*, L81–L84.

Chabrier G., Barman T., Baraffe I., Allard F., and Hauschildt P. H. (2004) The evolution of irradiated planets: Application to transits. *Astrophys. J. Lett., 603*, L53–L56.

Chabrier G., Saumon D., and Potekhin A. Y. (2006) Dense plasmas in astrophysics: From giant planets to neutron stars. *J. Phys. A, 39*, 4411–4419.

Chabrier G., Baraffe I., Selsis F., Barman T. S., Hennebelle P., and Alibert Y. (2007) Gaseous planets, protostars, and young brown dwarfs: Birth and fate. In *Protostars and Planets V* (B. Reipurth et al., eds.), pp. 623–638. Univ. of Arizona, Tucson.

Chandrasekhar S. (1939) *An Introduction to the Study of Stellar Structure.* Univ. of Chicago, Chicago.

Charbonneau D., Brown T. M., Latham D. W., and Mayor M. (2000) Detection of planetary transits across a Sun-like star. *Astrophys. J. Lett., 529*, L45–L48.

Chau R., Mitchell A. C., Minich R. W., and Nellis W. J. (2001) Electrical conductivity of water compressed dynamically to pressures of 70–180 GPa (0.7–1.8 Mbar). *J. Chem. Phys., 114*, 1361–1365.

Chauvin G., Lagrange A.-M., Dumas C., Zuckerman B., Mouillet D., Song I., Beuzit J.-L., and Lowrance P. (2005) Giant planet companion to 2MASSW J1207334-393254. *Astron. Astrophys., 438*, L25–L28.

Collins G. W., da Silva L. B., Celliers P., Gold D. M., Foord M. E., Wallace R. J., Ng A., Weber S. V., Budil K. S., and Cauble R. (1998) Measurements of the equation of state of deuterium at the fluid insulator-metal transition. *Science, 21*, 1178–1181.

Conrath B. J. and Gautier D. (2000) Saturn helium abundance: A reanalysis of Voyager measurements. *Icarus, 144*, 124–134.

da Silva L. B., Celliers P., Collins G. W., Budil K. S., Holmes N. C., Barbee T. W. Jr., Hammel B. A., Kilkenny J. D., Wallace R. J., Ross M., Cauble R., Ng A., and Chiu G. (1997) Absolute equation of state measurements on shocked liquid deuterium up to 200 GPa (2 Mbar). *Phys. Rev. Lett., 78*, 483–486.

Delaney K. T., Pierleoni C., and Ceperley D. M. (2006) Quantum Monte Carlo simulation of the high-pressure molecular-atomic crossover in fluid hydrogen. *Phys. Rev. Lett., 97*, 235702.

Deming D., Seager S., Richardson L. J., and Harrington J. (2005) Detection of thermal emission from an extrasolar planet. *Nature, 434*, 740–743.

Desjarlais M. P. (2003) Density-functional calculations of the liquid deuterium Hugoniot, reshock, and reverberation timing. *Phys. Rev. B, 68*, 064204.

Dodson-Robinson S. E. and Bodenheimer P. (2009) Discovering the growth histories of exoplanets: The Saturn analog HD 149026b. *Astrophys. J. Lett., 695*, L159–L162.

Ebeling W. and Richert W. (1985) Thermodynamic properties of liquid hydrogen metal. *Phys. Stat. Sol., 128*, 467–474.

Ebeling W., Förster A., Kremp D., and Schlanges M. (1989) Ionization kinetics in non-ideal dense plasmas: Ionization fronts. *J. Phys. A, 159*, 285–300.

Eggert J., Brygoo S., Loubeyre P., McWilliams R. S., Celliers P. M., Hicks D. G., Boehly T. R., Jeanloz R., and Collins G. W. (2008) Hugoniot data for helium in the ionization regime. *Phys. Rev. Lett., 100*, 124503.

Fabrycky D. C., Johnson E. T., and Goodman J. (2007) Cassini states with dissipation: Why obliquity tides cannot inflate hot Jupiters. *Astrophys. J., 665*, 754–766.

Fortney J. J. and Hubbard W. B. (2003) Phase separation in giant planets: Inhomogeneous evolution of Saturn. *Icarus, 164*, 228–243.

Fortney J. J. and Marley M. S. (2007) Analysis of Spitzer spectra of irradiated planets: Evidence for water vapor? *Astrophys. J. Lett., 666*, L45–L48.

Fortney J. J., Saumon D., Marley M. S., Lodders K., and Freedman R. S. (2006) Atmosphere, interior, and evolution of the metal-rich transiting planet HD 149026B. *Astrophys. J., 642*, 495–504.

Fortney J. J., Marley M. S., and Barnes J. W. (2007) Planetary radii across five orders of magnitude in mass and stellar insolation: Application to transits. *Astrophys. J., 659*, 1661–1672.

Fortov V. E., Ilkaev R. I., Arinin V. A., Burtzev V. V., Golubev V. A., Iosilevskiy I. L., Khrustalev V. V., Mikhailov A. L., Mochalov M. A., Ternovoi V. Y., and Zhernokletov M. V. (2007) Phase transition in a strongly nonideal deuterium plasma generated by quasi-isentropical compression at megabar pressures. *Phys. Rev. Lett., 99*, 185001.

Foulkes W. M., Mitas L., Needs R. J., and Rajagopal G. (2001) Quantum Monte Carlo simulations of solids. *Rev. Mod. Phys., 73,* 33–83.

French M., Mattsson T. R., Nettelmann N., and Redmer R. (2009) Equation of state and phase diagram of water at ultrahigh pressures as in planetary interiors. *Phys. Rev. B, 79,* 054107.

Fressin F., Guillot T., Morello V., and Pont F. (2007) Interpreting and predicting the yield of transit surveys: giant planets in the OGLE fields. *Astron. Astrophys., 475,* 729–746.

García Muñoz A. (2007) Physical and chemical aeronomy of HD 209458b. *Planet. Space Sci., 55,* 1426–1455.

Gillon M., Pont F., Demory B.-O., Mallmann F., Mayor M., Mazeh T., Queloz D., Shporer A., Udry S., and Vuissoz C. (2007) Detection of transits of the nearby hot Neptune GJ 436 b. *Astron. Astrophys., 472,* L13–L16.

Goodman J. (2009) Thermodynamics of atmospheric circulation on hot Jupiters. *Astrophys. J., 693,* 1645–1649.

Graboske H. C., Olness R. J., Pollack J. B., and Grossman A. S. (1975) The structure and evolution of Jupiter — The fluid contraction stage. *Astrophys. J., 199,* 265–281.

Grossman A. S., Graboske H., Pollack J., Reynolds R., and Summers A. (1972) An evolutionary calculation of Jupiter. *Phys. Earth Planet. Inter., 6,* 91–98.

Gu P.-G., Lin D. N. C., and Bodenheimer P. H. (2003) The effect of tidal inflation instability on the mass and dynamical evolution of extrasolar planets with ultrashort periods. *Astrophys. J., 588,* 509–534.

Guillot T. (1999) A comparison of the interiors of Jupiter and Saturn. *Planet. Space Sci., 47,* 1183–1200.

Guillot T. (2005) The interiors of giant planets: Models and outstanding questions. *Annu. Rev. Earth Planet. Sci., 33,* 493–530.

Guillot T. (2008) The composition of transiting giant extrasolar planets. *Phys. Scripta T, 130,* 014023.

Guillot T., Chabrier G., Morel P., and Gautier D. (1994) Nonadiabatic models of Jupiter and Saturn. *Icarus, 112,* 354–367.

Guillot T., Burrows A., Hubbard W. B., Lunine J. I., and Saumon D. (1996) Giant planets at small orbital distances. *Astrophys. J. Lett., 459,* L35–L38.

Guillot T., Santos N. C., Pont F., Iro N., Melo C., and Ribas I. (2006) A correlation between the heavy element content of transiting extrasolar planets and the metallicity of their parent stars. *Astron. Astrophys., 453,* L21–L24.

Hellier C., Anderson D. R., Gillon M., Lister T. A., Maxted P. F. L., Queloz D., Smalley B., Triaud A. H. M. J., West R. G., Wilson D. M., Alsubai K., Bentley S. J., Cameron A. C., Hebb L., Horne K., Irwin J., Kane S. R., Mayor M., Pepe F., Pollacco D., Skillen I., Udry S., Wheatley P. J., Christian D. J., Enoch R., Haswell C. A., Joshi Y. C., Norton A. J., Parley N., Ryans R., Street R. A., and Todd I. (2009) Wasp-7: A bright transiting-exoplanet system in the southern hemisphere. *Astrophys. J. Lett., 690,* L89–L91.

Hemley R. J. (2000) Effects of high pressure on molecules. *Annu. Rev. Phys. Chem., 51,* 763–800.

Hemley R. J. and Ashcroft N. W. (1998) The revealing role of pressure in the condensed matter sciences. *Phys. Today, 51,* 26–32.

Henry G. W., Marcy G. W., Butler R. P., and Vogt S. S. (2000) A transiting "51 Peg-like" planet. *Astrophys. J. Lett., 529,* L41–L44.

Holmström M., Ekenbäck A., Selsis F., Penz T., Lammer H., and Wurz P. (2008) Energetic neutral atoms as the explanation for the high-velocity hydrogen around HD 209458b. *Nature, 451,* 970–972.

Hubbard W. B. (1968) Thermal structure of Jupiter. *Astrophys. J., 152,* 745–754.

Hubbard W. B. (1973) Interior of Jupiter and Saturn. *Annu. Rev. Earth Planet. Sci., 1,* 85–106.

Hubbard W. B. (1984) *Planetary Interiors.* Van Nostrand Reinhold, New York.

Hubbard W. B. (1999) NOTE: Gravitational signature of Jupiter's deep zonal flows. *Icarus, 137,* 357–359.

Hubbard W. B. and Macfarlane J. J. (1980) Structure and evolution of Uranus and Neptune. *J. Geophys. Res., 85,* 225–234.

Hubbard W. B. and Marley M. S. (1989) Optimized Jupiter, Saturn, and Uranus interior models. *Icarus, 78,* 102–118.

Hubbard W. B., Nellis W. J., Mitchell A. C., Holmes N. C., McCandless P. C., and Limaye S. S. (1991) Interior structure of Neptune — Comparison with Uranus. *Science, 253,* 648–651.

Hubbard W. B., Podolak M., and Stevenson D. J. (1995) The interior of Neptune. In *Neptune and Triton* (D. P. Kruikshank, ed.), pp. 109–140. Univ. of Arizona, Tucson.

Hubbard W. B., Fortney J. J., Lunine J. I., Burrows A., Sudarsky D., and Pinto P. (2001) Theory of extrasolar giant planet transits. *Astrophys. J., 560,* 413–419.

Hubbard W. B., Hattori M. F., Burrows A., and Hubeny I. (2007) A mass function constraint on extrasolar giant planet evaporation rates. *Astrophys. J. Lett., 658,* L59–L62.

Hubickyj O., Bodenheimer P., and Lissauer J. J. (2005) Accretion of the gaseous envelope of Jupiter around a 5–10 Earth-mass core. *Icarus, 179,* 415–431.

Ibgui L. and Burrows A. (2009) Coupled evolution with tides of the radius and orbit of transiting giant planets: General results. *Astrophys. J., 700,* 1921–1932.

Jackson B., Greenberg R., and Barnes R. (2008) Tidal evolution of close-in extrasolar planets. *Astrophys. J., 678,* 1396–1406.

Jackson B., Barnes R., and Greenberg R. (2009) Observational evidence for tidal destruction of exoplanets. *Astrophys. J., 698,* 1357–1366.

Jeanloz R., Celliers P. M., Collins G. W., Eggert J. H., Lee K. K. M., McWilliams R. S., Brygoo S., and Loubeyre P. (2007) Achieving high-density states through shock-wave loading of precompressed samples. *Proc. Natl. Acad. Sci., 104,* 9172–9177.

Johnson J. A., Winn J. N., Cabrera N. E., and Carter J. A. (2009) A smaller radius for the transiting exoplanet WASP-10b. *Astrophys. J. Lett., 692,* L100–L104.

Juranek H. and Redmer R. (2000) Self-consistent fluid variational theory for pressure dissociation in dense hydrogen. *J. Chem. Phys., 112,* 3780–3786.

Kalas P., Graham J. R., Chiang E., Fitzgerald M. P., Clampin M., Kite E. S., Stapelfeldt K., Marois C., and Krist J. (2008) Optical images of an exosolar planet 25 light-years from Earth. *Science, 322,* 1345–1348.

Knudson M. D., Hanson D. L., Bailey J. E., Hall C. A., Asay J. R., and Anderson W. W. (2001) Equation of state measurements in liquid deuterium to 70 GPa. *Phys. Rev. Lett., 87,* 225501.

Knudson M. D., Hanson D. L., Bailey J. E., Hall C. A., and Asay J. R. (2003) Use of a wave reverberation technique to infer the density compression of shocked liquid deuterium to 75 GPa. *Phys. Rev. Lett., 90,* 035505.

Knutson H. A., Charbonneau D., Burrows A., O'Donovan F. T., and Mandushev G. (2009) Detection of a temperature inversion in the broadband infrared emission spectrum of TrES-4. *Astrophys. J., 691,* 866–874.

Lagrange A.-M., Gratadour D., Chauvin G., Fusco T., Ehrenreich D., Mouillet D., Rousset G., Rouan D., Allard F., Gendron É., Charton J., Mugnier L., Rabou P., Montri J. and Lacombe F. (2009) A probable giant planet imaged in the β Pictoris disk. VLT/NaCo deep L'-band imaging. *Astron. Astrophys., 493*, L21–L25.

Lammer H., Selsis F., Ribas I., Guinan E. F., Bauer S. J., and Weiss W. W. (2003) Atmospheric loss of exoplanets resulting from stellar X-ray and extreme-ultraviolet heating. *Astrophys. J. Lett., 598*, L121–L124.

Lee K. K. M., Benedetti L. R., Jeanloz R., Celliers P. M., Eggert J. H., Hicks D. G., Moon S. J., MacKinnon A., da Silva L. B., Bradley D. K., Unites W., Collins G. W., Henry E., Koenig M., Benuzzi-Mounaix A., Pasley J., and Neely D. (2006) Laser-driven shock experiments on precompressed water: Implications for "icy" giant planets. *J. Chem. Phys., 125*, 014701.

Lenosky T. J., Kress J. D., and Collins L. A. (1997) Molecular-dynamics modeling of the Hugoniot of shocked liquid deuterium. *Phys. Rev. B, 56*, 5164–5169.

Levrard B., Correia A. C. M., Chabrier G., Baraffe I., Selsis F., and Laskar J. (2007) Tidal dissipation within hot Jupiters: A new appraisal. *Astron. Astrophys., 462*, L5–L8.

Levrard B., Winisdoerffer C., and Chabrier G. (2009) Falling transiting extrasolar giant planets. *Astrophys. J. Lett., 692*, L9–L13.

Lin D. N. C., Bodenheimer P., and Richardson D. C. (1996) Orbital migration of the planetary companion of 51 Pegasi to its present location. *Nature, 380*, 606–607.

Liu X., Burrows A., and Ibgui L. (2008) Theoretical radii of extrasolar giant planets: The cases of TrES-4, XO-3b, and HAT-P-1b. *Astrophys. J., 687*, 1191–1200.

Lorenzen W., Holst B., and Redmer R. (2009) Demixing of hydrogen and helium at megabar pressures. *Phys. Rev. Lett., 102*, 115701.

Loubeyre P., Occelli F., and LeToullec R. (2002) Optical studies of solid hydrogen to 320 GPa and evidence for black hydrogen. *Nature, 416*, 613–617.

Low F. J. (1966) Observations of Venus, Jupiter, and Saturn at λ20μ. *Astron. J., 71*, 391.

Lyon S. P. and Johnson J. D., eds. (1992) Hydr5251. In *SESAME, the Los Alamos National Laboratory EOS Database*, LANL Report LAUR-92-3407.

Marley M. S., Gómez P., and Podolak M. (1995) Monte Carlo interior models for Uranus and Neptune. *J. Geophys. Res., 100*, 23349–23354.

Marley M. S., Fortney J. J., Hubickyj O., Bodenheimer P., and Lissauer J. J. (2007) On the luminosity of young Jupiters. *Astrophys. J., 655*, 541–549.

Marois C., Macintosh B., Barman T., Zuckerman B., Song I., Patience J., Lafrenière D., and Doyon R. (2008) Direct imaging of multiple planets orbiting the star HR 8799. *Science, 322*, 1348–1352.

Mattsson T. R. and Desjarlais M. P. (2006) Phase diagram and electrical conductivity of high energy-density water from density functional theory. *Phys. Rev. Lett., 97*, 017801.

Mayor M. and Queloz D. (1995) A Jupiter-mass companion to a solar-type star. *Nature, 378*, 355–359.

Mermin N. D. (1965) Thermal properties of the inhomogeneous electron gas. *Phys. Rev., 137*, 1441–1443.

Militzer B. (2005) Hydrogen helium mixtures at high pressure. *J. Low Temp. Phys., 139*, 739–752.

Militzer B. (2006) First principles calculations of shock compressed fluid helium. *Phys. Rev. Lett., 97*, 175501.

Militzer B. (2009) Path integral Monte Carlo and density functional molecular dynamics simulations of hot, dense helium. *Phys. Rev. B, 79*, 155105.

Militzer B. and Ceperley D. M. (2000) Path integral Monte Carlo calculation of the deuterium Hugoniot. *Phys. Rev. Lett., 85*, 1890–1893.

Militzer B. and Ceperley D. M. (2001) Path integral Monte Carlo simulation of the low-density hydrogen plasma. *Phys. Rev. E, 63*, 066404.

Militzer B. and Hubbard W. B. (2007) Implications of shock wave experiments with precompressed materials for giant planet interiors. In *Shock Compression of Condensed Matter* (M. Elert et al., eds.), pp. 1395–1398. AIP Conf. Proc. 955, American Institute of Physics.

Militzer B. and Hubbard W. B. (2009) Comparison of Jupiter interior models derived from first-principles simulations. *Astrophys. Space Sci., 322*, 129–133.

Militzer B., Hubbard W. B., Vorberger J., Tamblyn I., and Bonev S. A. (2008) A massive core in Jupiter predicted from first-principles simulations. *Astrophys. J. Lett., 688*, L45–L48.

Miller N., Fortney J. J., and Jackson B. (2009) Inflating and deflating hot Jupiters: Coupled tidal and thermal evolution of known transiting planets. *Astrophys. J., 702*, 1413–1427.

Mizuno H., Nakazawa K., and Hayashi C. (1978) Instability of a gaseous envelope surrounding a planetary core and formation of giant planets. *Progr. Theor. Phys., 60*, 699–710.

Morales M. A., Schwegler E., Ceperley D., Pierleoni C., Hamel S., and Caspersen K. (2009) Phase separation in hydrogen-helium mixtures at Mbar pressures. *Proc. Natl. Acad. Sci., 106*, 1324–1329.

Mordasini C., Alibert Y., Benz W., and Naef D. (2008) Giant Planet Formation by Core Accretion. In *Extreme Solar Systems* (D. Fischer et al., eds.), pp. 235–242. ASP Conf. Ser. 398, Astronomical Society of the Pacific, San Francisco.

Murray-Clay R. A., Chiang E. I., and Murray N. (2009) Atmospheric escape from hot Jupiters. *Astrophys. J., 693*, 23–42.

Nakajima T., Oppenheimer B. R., Kulkarni S. R., Golimowski D. A., Matthews K., and Durrance S. T. (1995) Discovery of a cool brown dwarf. *Nature, 378*, 463–465.

Nettelmann N., Holst B., Kietzmann A., French M., Redmer R., and Blaschke D. (2008) Ab initio equation of state data for hydrogen, helium, and water and the internal structure of Jupiter. *Astrophys. J., 683*, 1217–1228.

Perri F. and Cameron A. G. W. (1974) Hydrodynamic instability of the solar nebula in the presence of a planetary core. *Icarus, 22*, 416–425.

Pierleoni C., Ceperley D., Bernu B., and Magro W. (1994) Equation of state of the hydrogen plasma by path integral Monte Carlo simulation. *Phys. Rev. Lett., 73*, 2145–2149.

Pierleoni C., Ceperley D. M., and Holzmann M. (2004) Coupled electron-ion Monte Carlo calculations of dense metallic hydrogen. *Phys. Rev. Lett., 93*, 146402.

Podolak M. and Cameron A. G. W. (1974) Models of the giant planets. *Icarus, 22*, 123–148.

Podolak M., Hubbard W. B., and Stevenson D. J. (1991) Models of Uranus' interior and magnetic field. In *Uranus* (J. T. Bergstralh et al., eds.), pp. 29–61. Univ. of Arizona, Tucson.

Podolak M., Weizman A., and Marley M. (1995) Comparative models of Uranus and Neptune. *Planet. Space Sci., 43*, 1517–1522.

Podolak M., Podolak J. I., and Marley M. S. (2000) Further investigations of random models of Uranus and Neptune. *Planet. Space Sci., 48*, 143–151.

Pollock E. and Ceperley D. M. (1984) Simulation of quantum many-body systems by path-integral methods. *Phys. Rev. B, 30*, 2555–2568.

Pollack J. B., Grossman A. S., Moore R., and Graboske H. C. (1977) A calculation of Saturn's gravitational contraction history. *Icarus*, *30*, 111–128.

Ragozzine D. and Wolf A. S. (2009) Probing the interiors of very hot Jupiters using transit light curves. *Astrophys. J.*, *698*, 1778–1794.

Ross M. (1998) Linear-mixing model for shock-compressed liquid deuterium. *Phys. Rev. B*, *58*, 669–677.

Sato B., Fischer D. A., Henry G. W., Laughlin G., Butler R. P., Marcy G. W., Vogt S. S., Bodenheimer P., Ida S., Toyota E., Wolf A., Valenti J. A., Boyd L. J., Johnson J. A., Wright J. T., Ammons M., Robinson S., Strader J., McCarthy C., Tah K. L., and Minniti D. (2005) The N2K Consortium. II. A transiting hot Saturn around HD 149026 with a large dense core. *Astrophys. J.*, *633*, 465–473.

Saumon D. and Chabrier G. (1992) Fluid hydrogen at high density — Pressure ionization. *Phys. Rev. A*, *46*, 2084–2100.

Saumon D. and Guillot T. (2004) Shock compression of deuterium and the interiors of Jupiter and Saturn. *Astrophys. J.*, *609*, 1170–1180.

Saumon D., Chabrier G., and Horn H. M. V. (1995) An equation of state for low-mass stars and giant planets. *Astrophys. J. Suppl.*, *99*, 713–741.

Scandolo S. (2003) Liquid-liquid phase transition in compressed hydrogen from first-principles simulations. *Proc. Natl. Acad. Sci.*, *100*, 3051–3053.

Schwegler E., Galli G., Gygi F., and Hood R. Q. (2001) Dissociation of water under pressure. *Phys. Rev. Lett.*, *87*, 265501.

Seager S., Kuchner M., Hier-Majumder C. A., and Militzer B. (2007) Mass-radius relationships for solid exoplanets. *Astrophys. J.*, *669*, 1279–1297.

Showman A. P. and Guillot T. (2002) Atmospheric circulation and tides of "51 Pegasus b-like" planets. *Astron. Astrophys.*, *385*, 166–180.

Showman A. P., Menou K., and Cho J. Y.-K. (2008) Atmospheric circulation of hot Jupiters: A review of current understanding. In *Extreme Solar Systems* (D. Fischer et al., eds.), pp. 419–441. ASP Conf. Ser. 398, Astronomical Society of the Pacific, San Francisco.

Showman A. P., Fortney J. J., Lian Y., Marley M. S., Freedman R. S., Knutson H. A., and Charbonneau D. (2009) Atmospheric circulation of hot Jupiters: Coupled radiative-dynamical general circulation model simulations of HD 189733b and HD 209458b. *Astrophys. J.*, *699*, 564–584.

Sozzetti A., Torres G., Charbonneau D., Winn J. N., Korzennik S. G., Holman M. J., Latham D. W., Laird J. B., Fernandez J., O'Donovan F. T., Mandushev G., Dunham E., Everett M. E., Esquerdo G. A., Rabus M., Belmonte J. A., Deeg H. J., Brown T. N., Hidas M. G., and Baliber N. (2009) A new spectroscopic and photometric analysis of the transiting planet systems TrES-3 and TrES-4. *Astrophys. J.*, *691*, 1145–1158.

Spiegel D. S., Silverio K., and Burrows A. (2009) Can TiO explain thermal inversions in the upper atmospheres of irradiated giant planets? *Astrophys. J.*, *699*, 1487–1500.

Stanley S. and Bloxham J. (2006) Numerical dynamo models of Uranus' and Neptune's magnetic fields. *Icarus*, *184*, 556–572.

Stevenson D. J. (1975) Thermodynamics and phase separation of dense fully ionized hydrogen-helium fluid mixtures. *Phys. Rev. B*, *12*, 3999–4007.

Stevenson D. J. (1979) Semiconvection as the occasional breaking of weakly amplified internal waves. *Mon. Not. R. Astron. Soc.*, *187*, 129–144.

Stevenson D. J. (1985) Cosmochemistry and structure of the giant planets and their satellites. *Icarus*, *62*, 4–15.

Stevenson D. J. and Salpeter E. E. (1977a) The dynamics and helium distribution in hydrogen-helium fluid planets. *Astrophys. J. Suppl.*, *35*, 239–261.

Stevenson D. J. and Salpeter E. E. (1977b) The phase diagram and transport properties for hydrogen-helium fluid planets. *Astrophys. J. Suppl.*, *35*, 221–237.

Sudarsky D., Burrows A., and Hubeny I. (2003) Theoretical spectra and atmospheres of extrasolar giant planets. *Astrophys. J.*, *588*, 1121–1148.

Tassoul J.-L. (1978) *Theory of Rotating Stars*. Princeton Series in Astrophysics, Princeton Univ., Princeton.

Tian F., Toon O. B., Pavlov A. A., and De Sterck H. (2005) Transonic hydrodynamic escape of hydrogen from extrasolar planetary atmospheres. *Astrophys. J.*, *621*, 1049–1060.

Torres G., Bakos G. Á., Kovács G., Latham D. W., Fernández J. M., Noyes R. W., Esquerdo G. A., Sozzetti A., Fischer D. A., Butler R. P., Marcy G. W., Stefanik R. P., Sasselov D. D., Lázár J., Papp I., and Sári P. (2007) HAT-P-3b: A heavy-element-rich planet transiting a K dwarf star. *Astrophys. J. Lett.*, *666*, L121–L124.

Udry S. and Santos N. C. (2007) Statistical properties of exoplanets. *Annu. Rev. Astron. Astrophys.*, *45*, 397–439.

Vidal-Madjar A., Lecavelier des Etangs A., Désert J.-M., Ballester G. E., Ferlet R., Hébrard G., and Mayor M. (2003) An extended upper atmosphere around the extrasolar planet HD 209458b. *Nature*, *422*, 143–146.

Vidal-Madjar A., Lecavelier des Etangs A., Désert J.-M., Ballester G. E., Ferlet R., Hébrard G., and Mayor M. (2008) Exoplanet HD 209458b (Osiris): Evaporation strengthened. *Astrophys. J. Lett.*, *676*, L57–L60.

Vorberger J., Tamblyn I., Militzer B., and Bonev S. (2007) Hydrogen-helium mixtures in the interiors of giant planets. *Phys. Rev. B*, *75*, 024206.

Weir S., Mitchell A., and Nellis W. (1996) Metallization of fluid molecular hydrogen at 140 GPa (1.4 Mbar). *Phys. Rev. Lett.*, *76*, 1860–1863.

Winn J. N. and Holman M. J. (2005) Obliquity tides on hot Jupiters. *Astrophys. J. Lett.*, *628*, L159–L162.

Yelle R. V. (2004) Aeronomy of extra-solar giant planets at small orbital distances. *Icarus*, *170*, 167–179.

Yelle R., Lammer H., and Ip W.-H. (2008) Aeronomy of extrasolar giant planets. *Space Sci. Rev.*, *139*, 437–451.

Zapolsky H. S. and Salpeter E. E. (1969) The mass-radius relation for cold spheres of low mass. *Astrophys. J.*, *158*, 809–813.

Zeldovich Y. B. and Raizer Y. P. (1966) *Elements of Gas Dynamics and the Classical Theory of Shock Waves*. Academic, New York.

Zharkov V. N. and Trubitsyn V. P. (1974) Determination of the equation of state of the molecular envelopes of Jupiter and Saturn from their gravitational moments. *Icarus*, *21*, 152–156.

Zharkov V. N. and Trubitsyn V. P. (1976) Structure, composition, and gravitational field of Jupiter. In *Jupiter, Studies of the Interior, Atmosphere, Magnetosphere and Satellites*, pp. 133–175. Univ. of Arizona, Tucson.

Zharkov V. N. and Trubitsyn V. P. (1978) *Physics of Planetary Interiors* (W. B. Hubbard, ed. translator). Pachart, Tucson.

Giant Planet Atmospheres

Adam Burrows
Princeton University

Glenn Orton
Jet Propulsion Laboratory, California Institute of Technology

---

Direct measurements of the spectra of giant exoplanets are the keys to determining their physical and chemical nature. The goal of theory is to provide the tools and context with which such data are understood. It is only by putting spectral observations through the sieve of theory that the promise of exoplanet research can be realized. With the new Spitzer and Hubble Space Telescope data of transiting "hot Jupiters," we have now dramatically entered the era of remote sensing. We are probing their atmospheric compositions and temperature profiles, are constraining their atmospheric dynamics, and are investigating their phase lightcurves. Soon, many nontransiting exoplanets with wide separations (analogs of Jupiter) will be imaged and their lightcurves and spectra measured. In this paper, we present the basic physics, chemistry, and spectroscopy necessary to model the current direct detections and to develop the more sophisticated theories for both close-in and wide-separation giant exoplanets that will be needed in the years to come as exoplanet research accelerates into its future.

1. INTRODUCTION

Our understanding of gas giant planets was informed for many decades by remote telescopic observations and *in situ* measurements of Jupiter and Saturn. These detailed investigations provided a fine-grained view of their atmospheric compositions, temperatures, dynamics, and cloud structures. However, they left us with a parochial view of the range of possible orbits, masses, and compositions that has now been shattered by the discovery of extrasolar giant planets (EGPs) in the hundreds. We have found gas giants in orbits from ~0.02 AU to many AU, with masses from below Neptune's to ~10 M_{Jup}, and around stars from M to F dwarfs. The corresponding stellar irradiation fluxes at the planet vary by a factor of ~10^5, and this variation translates into variations in atmosphere temperatures from ~100 K to ~2500 K. With such a range of temperatures and of orbital distances, masses, and ages, atmospheres can have starkly different compositions, can be clear or cloudy, and can evince dramatic day-night contrasts.

One must distinguish imaging of the planet itself by separating the light of planet and star, something that can currently be contemplated only for wide-separation planets, from measurements of the summed light when the orbit is tight and the planet cannot be separately imaged. In the latter case, the planet's light can be a nontrivial fraction of the total, particularly in the infrared (IR). When transiting, such hot Jupiter systems provide an unprecedented opportunity to measure the planet's emissions by the difference in the summed light of planet and star in and out of secondary eclipse and by the phase variation of that sum. Generally, the star itself will not vary with the period of the planet's orbit. Moreover, in a complementary, but different, fashion, the wavelength dependence of the transit depth is now being used to probe the composition of the planet's atmosphere near the terminators. The EGPs, by dint of their mass and luminosity, have been the first discovered, and will serve as stepping stones to the terrestrial exoplanets.

To understand in physical detail the growing bestiary of EGPs requires chemistry to determine compositions, molecular and atomic spectroscopy to derive opacities, radiative transfer to predict spectra, hydrodynamics to constrain atmospheric dynamics and heat redistribution, and cloud physics. In short, global three-dimensional radiation-hydrodynamic general circulation models (GCMs) with multispectral, multiangle, and nonequilibrium chemistry and kinetics will be needed. We are not there yet, but basic treatments have emerged that allow us to interpret and constrain day-night differences, profiles, molecular compositions, and phase lightcurves.

In this chapter we lay out some of the basic elements of any theoretical treatment of the atmospheres, spectra, and lightcurves of EGPs. This theory provides the necessary underpinnings for any progress in EGP studies, a subject that is engaging an increasing fraction of the world's astronomical and planetary science communities. In section 2.1, we summarize the techniques for calculating molecular abundances. We follow in section 2.2 with an explication of general methods for assembling opacity tables. Section 2.3 touches on Rayleigh scattering, and then we continue in section 2.4 with a tutorial on albedos and phase functions. In section 2.5, we explain the nature of the transit radius. Section 2.7 contains a very useful analytic model for the atmospheric thermal profile of EGPs, which is a generalization for irradiated atmospheres of the classic Milne problem. This model incorporates a condition for thermal inversions. Then, in section 3 we summarize lessons learned and knowledge gained from the decades-long study of Jupiter and Saturn. This includes discussions of their spectra, cloud layers, temperatures, and compositions. Having set the stage, we review in section 4.1 the general chemistry and atmospheric character of EGPs as a function of orbital distance and age. This subsection includes a diversion into the putative evolution of Jupiter itself. We follow this in section 4.2 with a few paragraphs on theoretical EGP planet/star flux ratios as a function of wavelength and distance, focusing on wide-separation (>0.2 AU) EGPs. Then, in section 4.3 we present highlights from recent campaigns of direct detection of transiting EGP atmospheres, with an emphasis on secondary eclipse measurements, the compositions inferred, atmospheric temperatures, and thermal inversions. Finally, in section 5 we list some of the outstanding open issues and future prospects in the study of EGP atmospheres and spectra.

2. MODELING CONCEPTS AND EQUATIONS

In this section, we present the core ingredients necessary to construct theories of the atmospheres and spectra of EGPs. These includes general chemistry, opacities, and simple formalisms for the calculation of albedos and phase lightcurves. We discuss the concept of "transit radius" and include an analytic theory for the temperature profiles of irradiated exoplanets. The resulting approximate equations make clear the key role of opacity and its wavelength dependence in determining the character of EGP thermal profiles, in particular in creating thermal inversions when they arise. Such inversions have been inferred for many of the hot Jupiters seen in secondary eclipse and are emerging as one of the most exciting and puzzling features of current exoplanet research.

2.1. Calculation of Atomic and Molecular Abundances in Chemical Equilibrium

Before any opacities or atmospheric models can be calculated, the abundances of a mixture of a large number of species have to be determined for the given temperature and pressure. The assumption of chemical equilibrium is a good starting point, although nonequilibrium kinetics may play a role. Despite this, we present here a straightforward discussion of such calculations. Much of this presentation on abundances and that of section 2.2 on opacities can be found in *Burrows and Sharp* (1999) or *Sharp and Burrows* (2007), to which the reader is referred. In addition, there are excellent reviews and/or accounts in *Lodders* (1999), *Fegley and Lodders* (1994, 2001), *Lodders and Fegley* (2002), and *Sharp and Huebner* (1990).

For a given temperature, pressure, and composition, the equilibrium abundances of the various species can be determined by minimizing the total Gibbs free energy of the system. This requires a knowledge of the free energy of each species as a function of temperature, which is normally obtained from thermodynamic data. At the temperatures for which data are tabulated, least-square fits can be made for a set of polynomials whose highest order is given by

$$\Delta G_{pi}(T) = aT^{-1} + b + cT + dT^2 + eT^3 \quad (1)$$

where a, b, c, d, and e are fitted coefficients, and $\Delta G_{pi}(T)$ is the fitted Gibbs free energy of formation at temperature T of species i in phase p. The polynomials are evaluated at the tabulated points and the deviations from the tabulated values are obtained.

In performing the calculation for a particular temperature, pressure, and composition, the Gibbs free energy for each species is obtained from the database using the fitted coefficients at the temperature required, then the total free energy of the system is minimized to obtain the abundances of the gas-phase species, together with any condensates. The total dimensionless free energy is given by

$$\frac{G(T)}{RT} = \sum_{i=1}^{m} \left[n_{1i} \left\{ \frac{\Delta G_{1i}(T)}{RT} + \ln P + \ln\left(\frac{n_{1i}}{N}\right) \right\} \right] + \frac{1}{RT} \sum_{p=2}^{s+1} \left[n_{p1} \Delta G_{p1}(T) \right] \quad (2)$$

where R is the gas constant ($k_B N_A$), and for the first sum for the gas phase with p = 1, P is the total pressure in atmospheres, N is the number of moles, m is the number of species, n_{1i} is the number of moles of species i, and $\Delta G_{1i}(T)$ is the corresponding free energy of that species. The second sum is over the s condensed phases, which may include multiple phases of the same species, but except at a phase boundary, only one phase of a particular species in a condensed form is assumed present at any time, although one may consider solid or liquid solutions. Consequently, n_{p1} is the number of moles of a condensed species and $\Delta G_{p1}(T)$ is the free energy of that species. Since there is only one species per phase, for convenience we generally set i equal to 1.

A subset of 30 gas-phase species out of nearly 350 gas-phase species are usually the most important and selected for detailed treatment. The species are the neutral atoms, H, He, Li, Na, K, Rb, Cs, Al, Ca, and Fe; the ions, e^-, H^+, and H^-; and the metal hydrides, MgH, CaH, FeH, CrH, and TiH; with the remaining molecules being H_2, N_2, CO, SiO, TiO, VO, CaOH, H_2O, H_2S, NH_3, PH_3, CO_2, and CH_4.

Figure 1 depicts a representative result (here at 1 atm pressure and solar elemental abundances) for the temperature dependence of the equilibrium mixing ratios of water, methane, carbon monoxide, hydrogen sulfide, phosphene, molecular nitrogen, ammonia, TiO, and VO. The last two species may or may not play a role in EGP atmospheres, whereas the others certainly do.

2.2. Calculation of Atomic and Molecular Opacities

The calculation of the absorption cross sections and opacities of molecules is made more difficult than that for atoms by the substantially larger number of transitions and levels involved. Polyatomic species can have hundreds of millions, even billions, of vibrational and rotational lines, multiple electronic states, and a complicating mix of isotopes. Since it is not possible to measure with precision many transitions to determine their oscillator strengths or Einstein A coefficients, ab initio calculations using quantum chemical techniques are frequently necessary. Such calculations can be, and frequently are, calibrated with only a few measurements at selected wavelengths, but the experimental determination of the quantum numbers of the upper and lower states of even a given measured transition can be ambiguous. Moreover, particularly for hot Jupiters, the high temperatures experienced require a knowledge of absorption transitions from excited states, the so-called "hot bands," for which there are rarely measurements. For instance, the methane hot bands and some of the hot bands of water are completely unconstrained by current experiment, despite the fact that both water and methane are important greenhouse gases in Earth's atmosphere.

The situation is made even more difficult by the almost complete absence of calculations or measurements of line broadening coefficients or line profiles. The upshot is that theorists rely on imperfect and by-and-large uncalibrated compilations of theoretical line lists, and very approximate theories for line broadening. Some molecules are done better than others, but even for atoms, for which theory and measurement are rather better, line shapes are not well constrained. In the context of EGPs, this is particularly relevant for the alkali metals.

Despite these drawbacks, a rather sophisticated and extensive database of opacities for the constituents of EGPs has been assembled. In this subsection, taken in part from the review paper by *Sharp and Burrows* (2007), we summarize the techniques and methodologies needed to derive these opacities and use cgs units for specificity. Another useful review is that of *Freedman et al.* (2008).

The calculation of the line strengths for each line of each species depends on the data available for the species being considered, so different methods have to be used. In order to reduce the chances of errors with input data in different forms, it is recommended to convert, if necessary, all the line strengths into the same uniform system, with the best being integrated line strengths in cm^{2}s^{-1} species^{-1}. These depend only on the temperature. The lines should then be broadened into a profile that is dependent on the pressure, then the absorption in cm^{2}species^{-1} across the profile should be computed, summing the contributions from any overlapping profiles. The absorption for each species obtained in this manner depends on the temperature and pressure. The total opacity of the gas is obtained by summing the individual contributions weighted by the corresponding number densities (in cm^{-3}) for each of the species, yielding the total volume opacity in cm^{2}cm^{-3}, i.e., cm^{-1}. However, the total mass opacity in cm^{2}g^{-1} is usually the required result, and is obtained by dividing the volume opacity by the gas mass density.

In its most general LTE (Local Thermodynamic Equilibrium) form, the integrated strength S of a spectral line in cm^{2}s^{-1} species^{-1} is

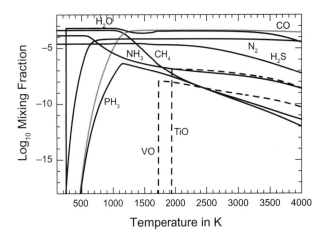

Fig. 1. The log (base 10) of the mixing fraction as a function of temperature at a total gas pressure of 1 atm for the seven molecules shown with solid curves CH$_4$ (black), CO (gray), N$_2$, NH$_3$, H$_2$O, H$_2$S, and PH$_3$, and the two molecules shown with dashed curves TiO and VO. At 4000 K, CO and N$_2$ are the most stable species, containing nearly all the carbon and nitrogen, respectively. With decreasing temperature, CO reacts with H$_2$, forming CH$_4$, which becomes the dominant carbon-bearing species at low temperatures, and N$_2$ reacts with H$_2$, forming NH$_3$, which likewise becomes the dominant nitrogen-bearing species at low temperatures. Except above about 3000 K, H$_2$O is fully associated containing nearly all the available oxygen that is not bound in CO. Below about 1600 K, its abundance temporarily falls slightly due to the condensation of silicates that reduce the available oxygen; however, the mixing fraction of H$_2$O then rises again when CO is converted to CH$_4$, which releases the oxygen tied up in CO. Finally, at 273 K H$_2$O drops effectively to zero due to the condensation of ice. With decreasing temperature, both TiO and VO rise as they associate, then sharply drop to effectively zero when condensates involving Ti and V form. Taken from *Sharp and Burrows* (2007). Available in color at *arxiv.org/abs/0910.0248*.

$$S = \frac{\pi e^2 g_i f_{ij}}{m_e c} \frac{e^{-hcF_i/kT}}{Q(T)} \left[1 - e^{-hc(F_j - F_i)/kT}\right] \quad (3)$$

where g_i and f_{ij} are the statistical weight of the ith energy level and the oscillator strength for a transition from that level to a higher level j; F_i and F_j are the term values (excitation energies) in cm^{-1} of the ith and jth levels participating in the transition; and Q(T) is the partition function of the species at some temperature T. The other symbols have their usual meanings. Note that the first term in equation (3) gives the line strength in cm^{2}s^{-1} absorber^{-1}, the next term with the Boltzmann factor and the partition function converts this to the required line strength, and the last term is the stimulated emission correction factor, where $F_j - F_i$ is the transition frequency in wave numbers, i.e., $\bar{\nu}$ in cm^{-1}. Although monochromatic opacities are frequently displayed as functions of wavelength, it is recommended that all opacity calculations be performed internally in wave numbers, even if some of the input data are given in wavelengths, since most molecular spectroscopic constants and energy levels are expressed in cm^{-1}, and adopting a uniform system of units reduces the chances of error. Note that some of the data available are not expressed in the form of oscillator strengths and statistical weights, collectively given as gf-values, but in other forms that must be converted to the required line strengths.

The general expression for calculating the partition function is given by

$$Q(T) = \sum_{i=1}^{n} g_i e^{-hcF_i/kT} \quad (4)$$

where the summation is performed over the first n levels, whose contributions are required at the highest temperatures of interest. The term value of the lowest level F_1, i.e., the ground state, is zero by construction.

Figure 2 portrays a representative set of absorption cross sections as a function of wavelength in the IR for the vibration-rotation transitions of H_2O, NH_3, and CH_4. Many of the most important absorption bands in EGP atmospheres are shown and these were calculated using the formalism described in this subsection.

2.3. Rayleigh Scattering

Rayleigh scattering is a conservative scattering process by atoms and molecules. Although strong in the ultraviolet/blue, the scattering cross sections quickly fade toward the red region of the spectrum ($\propto \lambda^{-4}$). Rayleigh scattering has little effect on the spectra of isolated brown dwarfs, but irradiated EGPs reflect a nonzero fraction of the incident intensity.

The Rayleigh scattering cross sections are derived from polarizabilities, which are in turn derived from refractive indices. The refractive indices are readily available at 5893Å (NaD) and are assumed not to vary strongly with wavelength. The Rayleigh cross sections are derived via

$$\sigma_{Ray} = \frac{8}{3} \pi k^4 \left(\frac{n-1}{2\pi L_0} \right)^2 \quad (5)$$

where n is the index of refraction, k is the wave number ($2\pi/\lambda$), and L_0 is Loschmidt's number, the number of molecules per cubic centimeter at STP (= 2.687 × 10^{19}). Given the strong inverse dependence on wavelength of equation (5), Rayleigh scattering is most pronounced in the blue and ultraviolet and is ultimately responsible for Earth's blue sky.

2.4. Albedos and Phase Functions

Objects in the solar system, such as planets, asteroids, and moons, are seen and studied in reflected solar light. The brightness of the reflection depends upon the orbital distance, the stellar flux, the reflectivity of the object, the detector angle (the "phase angle"), and the object's radius.

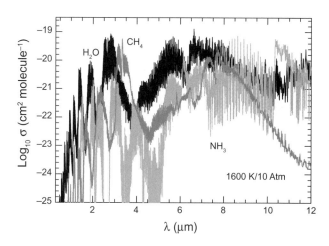

Fig. 2. The log (base 10) of the monochromatic absorption σ in cm^{2}molecule^{-1} as a function wavelength λ in micrometers in the infrared at a temperature of 1600 K and a pressure of 10 atm for the vibration-rotation transitions of H_2O, NH_3, and CH_4. The contribution due to different isotopes is included. For this plot we chose a high enough temperature and pressure to ensure that the lines were sufficiently broadened to suppress very rapid and large-amplitude fluctuations in the absorption cross section that can otherwise be in evidence over short wavelength intervals. In this way, the main band features (which are nevertheless generic for each species) are more easily seen. At significantly lower pressures the broadening of the lines is much smaller and the absorption can change so rapidly in short wavelength intervals that the main features do not show up so clearly. As can be seen here, H_2O has a strong absorption feature just shortward of 3 μm, and CH_4 has a strong peak near of 3.3 μm. In the region of 8 μm to 9 μm all three molecules absorb strongly; however, between about 10.5 μm and 11 μm NH_3 has absorption that is distinctly higher than that of the other two molecules. When the combined opacity is calculated, the individual absorptions must be weighted by the abundances. Taken from *Sharp and Burrows* (2007). Available in color at *arxiv.org/abs/0910.0248*.

The reflectivity, in the guise of an "albedo" (defined below), bears the stamp of the composition of its surface and/or atmosphere, and its wavelength dependence is a distinctive and discriminating signature. Solar-system objects are too cold to emit much in the optical, where the Sun is brightest, but emit in the mid-IR, where the peak of a ~40–800 K black body resides. Hence, there is a simple and obvious separation in their spectra between reflection and emission bumps that allows an unambiguous definition of the albedo and its interpretation as a dimensionless reflectivity bounded by a value of "1" (however, see below), the latter implying full reflection and no absorption.

However, some EGPs, the close-in and transiting variants, are so near their central stars that their surface and atmospheric temperatures can be quite large (~1000–2500 K). Hot planets (due to either proximity or youth) can be self-luminous in the near-IR (and even in the optical). As a result, the reflection and emission components can overlap in wavelength space, and are not so cleanly separated, as they are for solar system objects. The upshot is that the albedo and "reflectivity" might be misnomers, particularly in the near-IR. Nevertheless, the planet/star flux ratio as a function of wavelength is an important probe of EGP atmospheres and has to date been used with profit to diagnose their thermal and compositional character. This is true despite the fact that the associated albedos could be far above 1 at some wavelengths (Note that if the complete radiative transfer solution with stellar irradiation is derived, the concept of an albedo is redundant and unnecessary.) With this caveat in mind and the traditional interpretation intact for planets at greater orbital distances and older ages similar to the Jupiter/Sun pair, we proceed to develop the formalism by which the planet/star flux ratio is calculated, the albedo is defined, and why they are important. In the process we distinguish the geometric albedo (A_g), the spherical albedo (A_s), and the Bond albedo (A_B), and connect them to the planet/star flux ratio, F_p/F_*. We also introduce the "phase function," by which the observed lightcurve is described. Much of the development below is taken from papers by *Sudarsky et al.* (2000, 2005) and *Burrows et al.* (2004). The papers by *Marley et al.* (1999) and *Burrows et al.* (2008b) are also good resources on this general topic.

Planetary phase is a function of the observer-planet-star orientation, and the angle whose vertex lies at the planet is known as the phase angle (α). The formalism for the computation of planetary brightness as a function of phase angle has been presented by numerous authors. Following *Sobolev* (1975), one can relate the planetary latitude (ψ) and longitude (ξ) to the cosine of the angle of incident radiation (μ_0) and the cosine of the angle of emergent radiation (μ) at each point on the planet's surface

$$\mu_0 = \cos\psi \cos(\alpha - \xi) \quad (6)$$

and

$$\mu = \cos\psi \cos\xi \quad (7)$$

where latitude is measured from the orbital plane and longitude is measured from the observer's line of sight. The phase angle is then

$$\alpha = \cos^{-1}\left(\mu\mu_0 - \left[(1-\mu^2)(1-\mu_0^2)\right]^{1/2} \cos\phi\right) \quad (8)$$

where ϕ is the azimuthal angle between the incident and emergent radiation at a point on the planet's surface. The emergent intensity from a given planetary latitude and longitude is

$$I(\mu, \mu_0, \phi) = \mu_0 S \rho(\mu, \mu_0, \phi) \quad (9)$$

where the incident flux on a patch of the planet's surface is $\pi\mu_0 S$, and $\rho(\mu, \mu_0, \phi)$ is the reflection coefficient. In order to compute the energy reflected off the entire planet, one must integrate over the surface of the planet. For a given planetary phase, the energy per time per unit area per unit solid angle received by an observer is

$$E(\alpha) = 2S\frac{R_p^2}{d^2}\int_{\alpha-\pi/2}^{\pi/2}\cos(\alpha-\xi)\cos(\xi)d\xi \times \\ \int_0^{\pi/2}\rho(\mu,\mu_0,\phi)\cos^3\psi d\psi \quad (10)$$

where R_p is the planet's radius and d is the distance to the observer. This quantity is related to the geometric albedo (A_g), the reflectivity of an object at full phase ($\alpha = 0$) relative to that of a perfect Lambert disk [for which $\rho(\mu, \mu_0, \phi) = 1$] of the same radius under the same incident flux, by

$$A_g = \frac{E(0)d^2}{\pi S R^2} \quad (11)$$

where $E(0)$ is $E(\alpha)$ at $\alpha = 0$. A planet in orbit about its central star displays a range of phases, and the planet/star flux ratio is given by

$$\frac{F_p}{F_*} = A_g\left(\frac{R_p}{a}\right)^2 \Phi(\alpha) \quad (12)$$

where $\Phi(\alpha)$ is the classical phase function, which is equal to $E(\alpha)/E(0)$, R_p is the planet's radius, and a is its orbital distance. This formula is one of the core relationships in the study of irradiated and reflecting EGPs.

$\Phi(\alpha)$ is normalized to be 1.0 at full face, thereby defining the geometric albedo, and is a decreasing function of α. For Lambert reflection, an incident ray on a planetary patch emerges uniformly over the exit hemisphere, A_g is 2/3 for purely scattering atmospheres, and $\Phi(\alpha)$ is given by the formula

$$\Phi(\alpha) = \frac{\sin(\alpha) + (\pi - \alpha)\cos(\alpha)}{\pi} \quad (13)$$

However, EGP atmospheres are absorbing and the anisotropy of the single scattering phase function for grains, droplets, or molecules results in non-Lambertian behavior. For instance, back-scattering off cloud particles can introduce an "opposition" effect for which the planet appears anomalously bright at small α. Figure 3 provides some theoretical EGP phase functions taken from *Sudarsky et al.* (2005) in which this effect is clearly seen. Figure 4 depicts the corresponding phase curves for some of the solar system objects. These two figures together suggest a likely range for exoplanets.

Both A_g and $\Phi(\alpha)$ are functions of wavelength, but the wavelength-dependence of A_g is the most extreme. In fact, for cloud-free atmospheres, due to strong absorption by molecular bands, A_g can be as low as 0.03, making such objects very "black." Rayleigh scattering serves to support A_g, but mostly in the blue and UV, where various exotic trace molecules can decrease it. The presence of clouds increases A_g significantly. For instance, at 0.48 μm, Jupiter's geometric albedo is ~0.46 and Saturn's is 0.39 (*Karkoschka*, 1999). However, for orbital distances less than 1.5 AU, we expect the atmospheres of most EGPs to be clear. The albedo is cor-

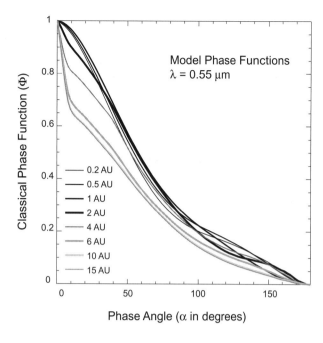

Fig. 3. Theoretical optical phase functions of 1-M_{Jup}, 5-G.y. EGPs ranging in orbital distance from 0.2 AU to 15 AU from a G2V star. Near full phase, the phase functions for our baseline models at larger orbital distances peak most strongly. For the cloud-free EGPs at smaller orbital distances (0.2 AU, 0.5 AU, and 1 AU), the phase functions are more rounded near full phase. Taken from *Sudarsky et al.* (2005). Available in color at *arxiv.org/abs/0910.0248*.

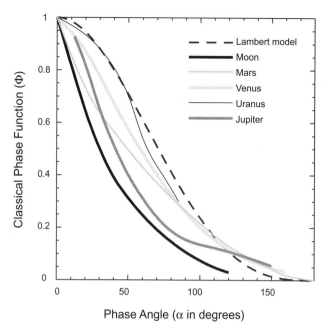

Fig. 4. The measured visual phase functions for a selection of solar system objects. A Lambert scattering phase curve, for which radiation is scattered isotropically off the surface regardless of its angle of incidence, is shown for comparison. The phase functions of the Moon and Mars peak near full phase (the so-called "opposition effect"). A red bandpass Jupiter phase function, taken from *Dyudina et al.* (2004), is also plotted. Taken from *Sudarsky et al.* (2005). Available in color at *arxiv.org/abs/0910.0248*.

respondingly low. As a consequence, the theoretical albedo is very nonmonotonic with distance, ranging in the visible from perhaps ~0.3 at 0.05 AU, to ~0.05 at 0.2 AU, to ~0.4 at 4 AU, to ~0.7 at 15 AU. In the visible (~0.55 μm), the geometric albedo for a hot Jupiter is severely suppressed by NaD at 0.589 μm. Due to a methane feature, the geometric albedo can vary from 0.05 at ~0.6 μm to ~0.4 at 0.625 μm. Hence, variations with wavelength and with orbital distance by factors of 2 to 10 are expected.

$\Phi(\alpha)$ and A_g must be calculated or measured, but the sole dependence of $\Phi(\alpha)$ on α belies the complications introduced by an orbit's inclination angle (i), eccentricity (e), and argument of periastron (ω). Along with the period (P) and an arbitrary zero of time, these are most of the so-called Keplerian elements of an orbit. Figures 5 and 6 in the chapter by Murray and Correia define these orientational and orbital parameters. In the plane of the orbit, the angle between the planet and the periastron/periapse (point of closest approach to the star) at the star is θ. In celestial mechanics, θ is the so-called "true anomaly." For an edge-on orbit (i = 90°), and one for which the line of nodes is perpendicular to the line of sight (longitude of the ascending node, Ω, equals 90°) and parallel to the star-periapse line (ω = 0°), θ is complementary to α (α = 90°– θ). As a result, θ = 0° at α = 90° (greatest elongation) and increases with time. Also, for such an edge-on orbit, α = 0° at superior conjunction. In general

$$\cos(\alpha) = \sin(\theta+\omega)\sin(i)\sin(\Omega) - \cos(\Omega)\cos(\theta+\omega) \quad (14)$$

Since it is common to define the observing coordinate system such that α = 90°, we have the simpler formula

$$\cos(\alpha) = \sin(\theta+\omega)\sin(i) \quad (15)$$

In order to produce a model lightcurve for a planet orbiting its central star, one must relate the planet's orbital angle (θ, the true anomaly), as measured from periapse, to the time (t) in the planet's orbit

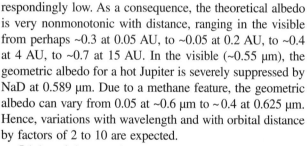

where P is the orbital period and e is the eccentricity. By combining equation (16) with equation (15), we derive the exact phase of any orbit at any time.

For a circular orbit, R is equal to the semimajor axis (a). However, a planet in an eccentric orbit can experience significant variation in R, and therefore stellar irradiation [by a factor of $[\frac{1+e}{1-e}]^2$]. For example, if e = 0.3, the stellar flux varies by ~3.5 along its orbit. For e = 0.6, this variation is a factor of 16! Such eccentricities are by no means rare in the sample of known EGPs. Therefore, in response to a changing stellar flux it is possible for the composition of an EGP atmosphere to change significantly during its orbit, for clouds to appear and disappear, and for there to be lags in the accommodation of a planet's atmosphere to a varying irradiation regime. Ignoring the latter, equations (12) and (14) can be combined with $\Phi(\alpha)$ and the standard Keplerian formula connecting θ and time for an orbit with a given P and e to derive an EGP's lightcurve as a function of wavelength, i, e, Ω, ω, and time. Depending upon orientation and eccentricity, the brightness of an EGP can vary in its orbit not at all (for a face-on EGP in a circular orbit) or quite dramatically (e.g., for highly eccentric orbits at high inclination angles). Since astrometric measurements of stellar wobble induced by EGPs can yield the entire orbit (including inclination), data from the Space Interferometry Mission (SIM) (*Unwin and Shao*, 2000) or Gaia (*Perryman*, 2003) could provide important supplementary data to aid in the interpretation of direct detections of EGPs.

The spherical albedo is the fraction of incident light reflected by a sphere at all angles. For a theoretical object with no absorptive opacity, all incident radiation is scattered, resulting in a spherical albedo of unity. The spherical albedo is related to the geometric albedo by $A_s = qA_g$, where q is the phase integral

$$q = 2\int_0^\pi \Phi(\alpha)\sin\alpha\, d\alpha \qquad (17)$$

For isotropic surface reflection (Lambert reflection) q = 3/2, while for pure Rayleigh scattering q = 4/3. Although not written explicitly, all the above quantities are functions of wavelength.

Van de Hulst (1974) derived a solution for the spherical albedo of a planet covered with a semi-infinite homogeneous cloud layer. Given a single-scattering albedo of σ (= $\sigma_{scat}/\sigma_{total}$) and a scattering asymmetry factor of g = $<\cos\theta>$ (the average cosine of the scattering angle), van de Hulst's expression for the spherical albedo of such an atmosphere is

$$A_s \approx \frac{(1-0.139s)(1-s)}{1+1.170s} \qquad (18)$$

where

$$s = \left[\frac{1-\sigma}{1-\sigma g}\right]^{1/2} \qquad (19)$$

The Henyey-Greenstein single-scattering phase function

$$p(\theta) = \frac{1-g^2}{(1+g^2-2g\cos\theta)^{3/2}} \qquad (20)$$

is frequently used as a fit to the overall scattering phase function, where again g = $<\cos\theta>$. Other phase functions can be used, but their specific angular dependence has been found to be less important than the value of the integral, g, itself. In addition, Rayleigh and cloud particle scattering are both likely to result in significant polarization of the reflected light from an EGP (*Seager et al.*, 2000). The degree of polarization will depend strongly on orbital phase angle and wavelength, and can reach many tens of percent; however, polarization may be difficult to measure. To date, there is no credible evidence for polarized light from any EGP. The degree of polarization in the optical and UV is expected to be largest.

The important Bond albedo, A_B, is the ratio of the total reflected and total incident powers. It is obtained by weighting the spherical albedo by the spectrum of the illuminating source and integrating over all wavelengths

$$A_B = \frac{\int_0^\infty A_{s,\lambda} I_{inc,\lambda}\, d\lambda}{\int_0^\infty I_{inc,\lambda}\, d\lambda} \qquad (21)$$

where the λ subscript signifies that the incident intensity varies with wavelength. With the Bond albedo, one can make a crude estimate of the "effective temperature" of the planet's emission component in response to irradiation. Under the assumption that the incident total power is equal to the emitted power, one derives

$$T_{eff} = T_* \left(\frac{fR_*}{a}\right)^{1/2} (1-A_B)^{1/4} \qquad (22)$$

where T_* is the stellar effective temperature, a is the orbital distance, R_* is the stellar radius, and *f* is a measure of the degree of heat redistribution around the planet. It is equal to 1/4 if the reradiation is isotropic, and 1/2 if the planet reemits only on the dayside and uniformly. Note that T_{eff} is independent of the planet's radius.

Equation (22) is useful in estimating the temperatures achieved by radiating planets, but has conceptual limitations. First, it presumes that the planet is not self-luminous and radiating the residual heat of formation in its core. This assumption is not true for young and/or massive EGPs. It is not true of Jupiter. Second, it encourages the notion that the reflection and emission peaks are well-separated. This is not correct for the hot Jupiters, for which the two components overlap and merge. Third, in radiative transfer theory and equilibrium, the true flux from a planet is the net flux. Without internal heat sources, this is 0 for irradiated planets. Finally, equation (22) is often used to determine the temperature of an atmosphere. However, EGP atmospheres have temperature profiles. In their radiative zones the temperatures can vary by factors of ~3, as can the effective photospheric temperatures in the near- and mid-IR. Hence, one should employ equation (22) to obtain atmospheric temperatures only when very approximate numbers are desired. Importantly, calculating the irradiated planet's spectrum with a T_{eff} derived using equation (22) results in large errors across the entire wavelength range that can severely compromise the interpretation of data.

2.5. The Transit Radius

A transiting planet reveals its radius (R_p) by the magnitude of the diminution in the stellar light during the planet's traverse of the stellar disk. This is the primary eclipse. In fact, it is the ratio of the planet and star radii that is most directly measured, so a knowledge of the star's radius is central to extracting this important quantity. With a radius and a mass (for transiting planets, the inclination must be near 90° and the inclination degeneracy is broken), we can compare with theories of the planet's physical structure and evolution. However, since the effective edge of the planetary disk is determined by the opacity of the atmosphere at the wavelength of observation, the radius of a gas giant is wavelength-dependent. Importantly, the variation with wavelength of the measured radius can serve as an ersatz atmospheric "spectrum." From this spectrum, one can determine the atmosphere's constituent atoms and molecules. Specifically, the apparent radius is larger at wavelengths for which the opacity is larger and smaller at wavelengths for which the opacity is smaller. For example, the planet should be larger in the NaD line at ~5890Å (as in fact was found for HD 209245b) than just outside it. It should be larger near the absorption peaks of the water spectrum for those planets with atmospheric water than in the corresponding troughs.

However, the "transit radius" is not the same as the classical photospheric radius of an atmosphere. The latter is determined by the depth in the atmosphere where $\tau_\nu = 2/3$ in the radial direction. The transit radius is where this same condition obtains along the chord from the star, perpendicular to the radius to the center. At the spherical radius where a light beam experiences $\tau_\nu = 2/3$, τ_ν along the chord can be much larger. Therefore, to achieve the $\tau_\nu = 2/3$ condition along the chord pushes the transit radius (also referred to as an "impact parameter") to larger values. It is only at this greater altitude and lower pressure that the chord optical depth is ~2/3. An important difference between the transit radius spectrum and the actual "emission" spectrum from a classical photosphere (such as is relevant at secondary eclipse) is that in the latter case, if the atmosphere were isothermal the spectrum would be a black body and there would be no composition information. However, the transit radius spectrum always manifests the wavelength dependence of the opacities of its atmospheric constituents, even if the atmosphere were isothermal. This difference can be exploited to maximize the scientific return from the study of a given transiting EGP. In sum, if the atmosphere is extended and if the monochromatic opacity at the measurement wavelength is large, the transit radius and the photosphere radius can be rather different and the distinction should always be kept in mind.

A difficulty in interpreting transit radius spectra is that one is probing the planetary limb, the terminator. This means that when comparing with data models must incorporate profiles on both the dayside and nightside, and at both the equator and poles. In particular, the dayside and the nightside can be at different temperatures and have different compositions. This complication is not always appreciated.

One can estimate the magnitude of the excess of the transit radius over the photospheric radius using a simple exponential atmosphere (see also *Fortney*, 2005). The wavelength-dependent optical depth, τ_{chord}, along a chord followed by the stellar beam through the planet's upper atmosphere is approximately

$$\tau_{chord} \sim \kappa \rho_{ph} H \sqrt{\frac{2\pi R_p}{H}} e^{-\left(\frac{\Delta R_{ch}}{H}\right)} \qquad (23)$$

where κ is the wavelength-dependent opacity, ρ_{ph} is the mass density at the photosphere, ΔR_{ch} is the excess radius over and above the $\tau_{ph} = 2/3$ radius (the radius of the traditional photosphere), and H is the atmospheric density scale height. The latter is given approximately by $kT/\mu g m_p$, where μ is the mean molecular weight, g is the surface gravity, T is some representative atmospheric temperature, and m_p is the proton mass. By definition, and assuming an exponential atmosphere, $\tau_{ph} = \kappa \rho_{ph} H = 2/3$. For τ_{chord} to equal 2/3, this yields

$$\Delta R_{ch} = H \ln \sqrt{\frac{2\pi R_p}{H}} \sim 5H \qquad (24)$$

This excess can be from ~1% to ~10%, depending upon the wavelength, temperature, gravity, and deviations from a strictly exponential profile. It is smallest for high-gravity planets at larger orbital distances, whereas in the UV near Lyman-α, and with a planetary wind, this excess for HD 209458b is measured to be a factor of 2–3.

At times, the transit radius spectrum is called the "transmission spectrum." Since what is measured is the effective area of the planet, not the spectrum of light transmitted through the planetary limb, this is a slight misnomer. One is measuring the atmospheric edge position (relative to some zero-point) as a function of wavelength, and not the light transmitted through the finite extent of the atmosphere. To properly do the latter would require a resolved image of the planet that distinguishes the atmosphere from the opaque central planetary disk. Note also that when the atmospheric opacity is low (and, hence, light is more easily "transmitted"), the dip in the stellar flux (what is actually measured) is smaller, not larger. Nevertheless, this rather pedantic point does not inhibit one from profiting from "transmission spectrum" measurements.

2.6. Analytic Model for the Temperature Profile of an Irradiated Planet

One can derive a model for the temperature profile of an irradiated EGP atmosphere that is a generalization of the classical Milne problem for an isolated atmosphere. This model can incorporate the difference between the opacity to the insolating ("optical") and emitting ("infrared") radiation streams, contains a theory for stratospheres and thermal inversions, and a condition for their emergence. The mathematical development of this analytic theory is taken from *Hubeny et al.* (2003), to which the reader is referred for further details. The paper by *Chevallier et al.* (2007) is also of some considerable utility.

The equation of hydrostatic equilibrium is fundamental in atmospheric theory, and obtains as long as the Mach number of the gas is low. It can be written

$$\frac{dP}{dz} = -g\rho$$
$$\frac{dP}{dm} = g \quad (25)$$

where P is the pressure, g is the acceleration due to gravity (generally assumed constant), ρ is the mass density, z is the altitude, and m is the areal (column) mass density defined by $dm = -\rho dz$. Note that for a constant g, the pressure and the column mass are directly proportional.

If the atmosphere is convective, the temperature gradient follows an adiabat and is given by

$$\frac{d\ln T}{d\ln P} = C_p / R \quad (26)$$

where C_p is the specific heat at constant pressure and R is the gas constant. Given the temperature and pressure at any point in the convective region, together with the specific heat (which can be determined using the composition), the temperature at any atmospheric pressure can be determined. With that knowledge, equation (26) can then be used to determine the altitude scale. Hence, the combination of the equation of hydrostatic equilibrium and equation (26) can be used to determine the adiabatic lapse rate for a given specific heat, pressure, and temperature.

However, for most of a strongly irradiated atmosphere, and in radiative zones in general, radiation carries the energy flux. The radiative transfer equation is written as

$$\mu \frac{dI_{\nu\mu}}{dm} = \chi_\nu \left(I_{\nu\mu} - S_\nu \right) \quad (27)$$

where $I_{\nu\mu}$ is the specific intensity of radiation as a function of frequency, ν, angle μ (the cosine of the angle of propagation with respect to the normal to the surface), and the geometrical coordinate, taken here as the column (areal) mass m. The monochromatic optical depth is defined as $d\tau_\nu = \chi_\nu dm$.

S_ν is the source function, given in LTE by

$$S_\nu = \frac{\kappa_\nu}{\chi_\nu} B_\nu + \frac{\sigma_\nu}{\chi_\nu} J_\nu \quad (28)$$

Here κ_ν is the true absorption coefficient, σ_ν is the scattering coefficient, B_ν is the black body function, J_ν is the zeroth angular moment of the specific intensity, and $\chi_\nu = \kappa_\nu + \sigma_\nu$ is the total absorption coefficient. The zeroth moment is equal to the radiation energy density, divided by 4π. All coefficients are per unit mass. The first moment of the transfer equation is written

$$\frac{dH_\nu}{dm} = \chi_\nu \left(J_\nu - S_\nu \right) \quad (29)$$

which can be rewritten, using equation (28), as

$$\frac{dH_\nu}{dm} = \kappa_\nu \left(J_\nu - B_\nu \right) \quad (30)$$

H_ν is the first angular moment of the specific intensity and is equal to the radiation flux, divided by 4π. Integrating over frequency one obtains

$$\frac{dH}{dm} = \kappa_J J - \kappa_B B \quad (31)$$

where κ_J and κ_B are the absorption and Planck mean opacities, respectively, defined by

$$\kappa_J = \frac{\int_0^\infty \kappa_\nu J_\nu d\nu}{\int_0^\infty J_\nu d\nu} \quad (32)$$

and

$$\kappa_B = \frac{\int_0^\infty \kappa_\nu B_\nu d\nu}{\int_0^\infty B_\nu d\nu} \quad (33)$$

These two opacities are often assumed to be equal. However, one should distinguish them here because the difference between κ_J and κ_B turns out to be crucial in the case of strongly irradiated atmospheres.

The second moment of the transfer equation is

$$\frac{dK_\nu}{dm} = \chi_\nu H_\nu \quad (34)$$

and integrating over frequency one obtains

$$\frac{dK}{dm} = \chi_H H \quad (35)$$

where

$$\chi_H = \frac{\int_0^\infty \chi_\nu H_\nu d\nu}{\int_0^\infty H_\nu d\nu} \quad (36)$$

which is referred to as the flux mean opacity.

Finally, the radiative equilibrium equation is written as

$$\int_0^\infty \kappa_\nu \left(J_\nu - B_\nu \right) d\nu = 0 \quad (37)$$

Using the above mean opacities, this can be rewritten

$$\kappa_J J - \kappa_B B = 0 \quad (38)$$

Substituting equation (38) into (31), one obtains another form of the radiative equilibrium equation

$$\frac{dH}{dm} = 0, \quad \text{or} \quad H = \text{const} \equiv \left(\sigma/4\pi\right)T_{\text{eff}}^4 \qquad (39)$$

where σ is the Stefan-Boltzmann constant.

From equation (38), one has $B = (\kappa_J/\kappa_B)J$, which yields an expression for T through J using the well-known relation $B = \sigma T^4$. To determine J, one uses the solution of the second moment of the transfer equation $K(\tau_H) = H\tau_H = (\sigma/4\pi)T_{\text{eff}}^4 \tau_H$, where τ_H is the optical depth using the flux-mean opacity, and expresses the moment K through J by means of the Eddington factor, $f_K \equiv K/J$. Similarly, one expresses the surface flux through the second Eddington factor, $f_H \equiv H(0)/J(0)$ (see also *Hubeny*, 1990)

$$T^4 = \frac{3}{4}T_{\text{eff}}^4 \frac{\kappa_J}{\kappa_B}\left[\frac{1}{3f_K}\tau_H + \frac{1}{3f_H}\right] + \frac{\kappa_J}{\kappa_B}WT_*^4 \qquad (40)$$

where W is the dilution factor, $(R_*/a)^2$. This solution is exact within LTE and is the generalization of the classical Milne atmosphere solution.

The usual LTE-gray model consists in assuming all the mean opacities to be equal to the Rosseland mean opacity. If one adopts the Eddington approximation ($f_K = 1/3$; $fH = 1/\sqrt{3}$), then one obtains a simple expression

$$T^4 = \frac{3}{4}T_{\text{eff}}^4\left(\tau + 1/\sqrt{3}\right) + WT_*^4 \qquad (41)$$

We will consider the most interesting case, namely strong irradiation, defined by $WT_*^4 \gg T_{\text{eff}}^4$. In this case, the second term in the brackets is negligible, and one may define a penetration depth as the optical depth where the usual thermal part $\left(\propto T_{\text{eff}}^4\right)$ and the irradiation part $\left(\propto WT_*^4\right)$ are nearly equal, to whit

$$\tau_{\text{pen}} = W\left[\frac{T_*}{T_{\text{eff}}}\right]^4 \qquad (42)$$

The behavior of the local temperature in the case of a strict gray model is very simple — it is essentially constant, $T = T_0 \equiv W^{1/4}T_*$ for $\tau < \tau_{\text{pen}}$, and follows the usual distribution $T \propto \tau^{1/4}T_{\text{eff}}$ in deep layers, $\tau > \tau_{\text{pen}}$. In the general case, one has to retain the ratio of the absorption and Planck mean (assuming still that the flux mean opacity is well approximated by the Rosseland mean). In the irradiation-dominated layers ($\tau < \tau_{\text{pen}}$), the temperature is given by

$$T = \gamma W^{1/4}T_* \qquad (43)$$

where

$$\gamma \equiv \left(\kappa_J/\kappa_B\right)^{1/4} \qquad (44)$$

γ is approximately 1 for no or weakly irradiated atmospheres. However, in the case of strong irradiation, γ may differ significantly from unity. Moreover, it may be a strong function of temperature, and, to a lesser extent, of density. This is easily seen by noting that in optically-thin regions, the local mean intensity is essentially equal to twice the irradiation intensity, since the incoming intensity is equal to irradiation intensity, and the outgoing intensity is roughly equal to it as well. The reason is that in order to conserve the total flux when it is much smaller than the partial flux in the inward or outward direction, both fluxes should be almost equal, and so too must the individual specific and mean intensities.

The local temperature in the upper layers is given, using equation (43), as

$$T/T_0 = \gamma(T) \qquad (45)$$

It is clear that if γ exhibits a strongly nonmonotonic behavior in the vicinity of T_0, for instance, if it has a pronounced minimum or maximum there, equation (45) may have two or even more solutions.

This is the origin of thermal inversions, when they occur, in the atmospheres of strongly irradiated EGPs (*Hubeny et al.*, 2003). The essential element is the differential absorption in the optical on the one hand (since this is where most of the irradiating stellar light is found), and in the IR (since this is where most of the emission at the temperatures achieved in the atmospheres of hot Jupiters occurs). As the above formalism makes plain, an inversion is not possible for gray opacities, whatever the degree of irradiation. For strongly irradiated EGPs, if there is not a strong optical absorber at altitude, then stellar optical light is absorbed rather deeply in the atmosphere, near pressures of ~1 bar. This is near and interior to the corresponding emission photospheres in the near-IR. The result is a more-or-less monotonic decrease of temperature with altitude and decreasing pressure. However, if there is a strong optical absorber at altitude, the large value of γ will allow another solution to equation (45) for which there is an inversion. The resulting higher temperatures in the upper atmosphere will result in higher fluxes in the mid-IR (e.g., in the IRAC and MIPS bands of Spitzer) where their photospheres reside. We see signatures of such inversions in the measured spectra of many transiting EGPs, such as HD 209458b, XO-1, TrES-4, and HD 149026b. However, to date we do not know what chemical species is absorbing at altitude in the optical (see section 4.3).

3. LESSONS FROM JUPITER AND SATURN

Jupiter and Saturn are the largest planets in our solar system and serve as initial paradigms for the atmospheres of EGPs. With the largest exoplanets invariably referred to as "super Jupiters," it is instructive to start with an assessment of the properties of "regular Jupiters." Both Jupiter and Saturn have been studied with a variety of remote sensing techniques across a wide spectral range, and these approaches have provided sufficient information to determine physical and chemical properties and their variation in both time and space. In the case of Jupiter, these physical and chemical properties have been bolstered by *in situ* observations made by the Galileo mission atmospheric probe, which both extended and served as a measure of ground truth for remote sensing observations. This subsection will concentrate on properties of their atmospheres and the spectra of their upwelling fluxes in order to provide analogies with exoplanets, which can boast fewer observational constraints.

Observations of Jupiter and Saturn have been made in some detail from the time of Galileo, with serious groundbased observations taking place with the advent of photographic film and spectroscopy. In the second half of the 20th century, NASA spacecraft ventured to both — first with the Pioneers 10 (Jupiter) and 11 (Jupiter and Saturn) flyby spacecraft. Later observations were made with more instrumentation (including an IR spectrometer, and wide and narrow-angle cameras) by Voyagers 1 and 2, which visited both planets, and provided an abundance of information on the dynamics, structure, and composition of both Jupiter and Saturn. The Galileo spacecraft orbited Jupiter and determined more about the dynamics of individual regions on the planet and dropped a direct probe into its atmosphere. Most recently, the Cassini spacecraft obtained substantial information on the atmosphere, flying by Jupiter in a gravity assist on its way to Saturn, which it began orbiting in 2005. Most recently the New Horizons spacecraft flew by Jupiter on its way to Pluto, collecting information on Jupiter's clouds.

These spacecraft observations were complemented by observations from Earth-orbiting platforms beginning with observations of auroral emission from Jupiter by the International Ultraviolet Explorer (IUE). They include ultraviolet through near-IR imaging and spectroscopy by the instrument complement on board the Hubble Space Telescope (HST), disk-averaged spectroscopy by ESA's Infrared Space Observatory (ISO), and even X-ray observations by the Chandra Observatory. Additional spectroscopic and occultation observations were made using instruments onboard NASA's Kuiper Airborne Observatory (KAO).

The ever-improving competence of groundbased observational facilities has substantially increased our knowledge of these atmospheres. In particular, increasing spatial and spectral discrimination became available with larger primary mirrors (minimizing diffraction-limited spatial resolution) and a variety of active optical systems, which have been extremely effective in reducing the blurring due to atmospheric turbulence. These have been accompanied at the smaller-telescope end by a host of amateur Jupiter watchers wielding increasingly sophisticated instrumentation, including multifiltered CCD cameras.

3.1. Spectra

Models for the structure of the atmospheres of Jupiter and Saturn can be understood in terms of the various regions in which energy transfer is by either radiation or convection, similar to stellar atmospheres. To good accuracy, much of the atmospheres of both planets can be approximated as self-gravitating fluids in hydrostatic equilibrium. At depth, the atmospheres of both Jupiter and Saturn are dominated by convective processes, and gases rise and fall at rates faster than they can radiate away energy. Thus, adiabatic conditions hold (equation (26)).

It is important to note the "principal players" in radiative transfer in Jupiter and Saturn. In the visible and near-IR, the spectrum is dominated by CH_4 (methane) absorption, with some additional opacity from the H_2-H_2 collision-induced fundamental and H_2 quadrupole lines at higher resolution. Figure 5 depicts the near-IR spectrum of Jupiter from 1 to 6 μm. For Jupiter, the 5-μm spectrum is not dominated by reflected sunlight, as is the case at the shorter wavelengths, but by thermal emission from depth, as the result of a dearth of gaseous absorption in this region. It is this atmospheric "window" that allows glimpses of Jupiter's composition down to a few bars of pressure, where H_2O vapor was first detected in Jupiter's atmosphere from observations using NASA's KAO, and where high-resolution spectroscopy detected PH_3 and trace constituents such as AsH_3 (arsine) and CO (carbon monoxide) at depth. For colder Saturn, more of this spectral window is composed of reflected sunlight.

At longer wavelengths, in the middle- and far-IR, the spectrum is dominated by thermal emission. Figure 6 shows the brightness temperature spectrum of both Jupiter and Saturn. Rather than flux, these spectra are plotted in brightness temperature, which is defined as the temperature that a blackbody source would need to emit at a particular wavelength in order to match the observed flux.

Plotting a spectrum in brightness temperature rather than directly in flux is a convenient way to display a spectrum over a broad range, such as the ones in Fig. 6, because the flux varies over several orders of magnitude. In addition, plotting the brightness temperature provides a quick way to estimate the depth in the atmosphere from which most of the radiation emerges. For example, the source of a strong emission feature with a high brightness temperature is likely to be emerging from molecular emission very high in the hot stratospheres of either planet. More details can be found in *dePater and Lissauer* (2004), *Irwin* (2003), or *Goody and Yung* (1989).

We also note that Fig. 6 plots its primary spectral scale in cm^{-1} ("wave numbers") rather than wavelength. Historically, this is the result of such spectra being derived from Fourier-transform spectrometers, where inverse of the path length difference of a Michaelson-like interferometer is the native spectral unit. A wave number (cm^{-1}) is equal to the frequency of the radiation in Hertz (s^{-1}) divided by 29.97. Because the frequency or wave number is proportional to quantum energy, such a plot enables molecular lines to be displayed in a manner proportional to the transition energy. The result is often (but not always!) a regular series of line transitions (or groups of line transitions known as manifolds). Some of these are evident in Fig. 6 in the far-IR for NH_3 (ammonia) or PH_3 (phosphine) or near 10 μm for NH_3.

Figure 6 shows that the far-IR, where the bulk of Jupiter's and Saturn's flux emerges, is controlled by the collision-induced absorption of H_2, which varies so slowly that it constitutes a virtual continuum opacity source. Two broad rotational lines are evident, and the far-IR lines of NH_3 in Jupiter and PH_3 in Saturn show up as absorption features

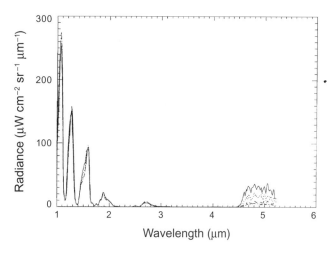

Fig. 5. Near-IR spectrum of Jupiter taken in a relatively "cold" but reflective region of the atmosphere by the Galileo Near-Infrared Mapping Spectrometer (NIMS) in 1996. All radiances except those near 5 µm are sunlight reflected from clouds. Absorption is primarily due to CH_4 gas with some contributions from a H_2 collision-induced fundamental band near 2.2 µm. From *Irwin et al.* (1998).

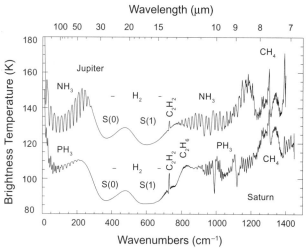

Fig. 6. Middle-IR brightness temperature spectra of Jupiter and Saturn, derived from Voyager and Cassini IR spectrometers. Prominent spectral features are identified. Note that the H_2 collision-induced absorption is a smooth continuum in the spectrum, with the broad rotational transitions S(0) and S(1) identified.

on the long-wavelength "translational" component of the H_2 absorption. When helium (He) collides with H_2, a slightly different spectrum is created than for pure H_2-H_2 absorption, and the translation component is stronger. Fitting this shape accurately provides a means to determine the ratio of He to H_2 in these planets. This part of the spectrum also allows one to determine the ratio of ortho-H_2, where the spins of the hydrogen atoms are parallel, to para-H_2, where the spins are antiparallel (parallel but pointed in opposite directions), because para-H_2 alone is responsible for all even rotational transitions, such as S(0), and ortho-H_2 alone is responsible for all odd rotational transitions, such as S(1). In a quiescent atmosphere, there is an equilibrium between these two species of H_2 that is purely a function of temperature and how their energy levels are populated. In reality, however (particularly in Jupiter), H_2 gas can be transported vertically much faster than this temperature reequilibration can take place, and the para-ortho H_2 ratio varies across the face of the planet. This is useful as an indirect means to track upwelling and downwelling winds. For example, a para-H_2 value that is lower than the value expected from local thermal equilibrium, but greater than 0.25, would indicate rapid upwelling from the deep, warm atmosphere where its value is close to the high-temperature asymptotic value of 25%. However, at the high temperatures of hot-Jupiter atmospheres, transiting or otherwise, para- and ortho-hydrogen are in thermal and statistical equilibrium.

The spectra of Jupiter and Saturn are also filled with discrete transitions, such as the rotational lines of NH_3 and PH_3 in the far-IR. Lines arising from a combination of vibrational and rotational quantum transitions can be identified easily in the mid-IR. Both planets display emission features arising from CH_4 (methane) and from higher-order hydrocarbons, such as C_2H_2 (acetylene) and C_2H_6 (ethane), which are byproducts of the photolysis of methane by ultraviolet radiation in the upper atmosphere. They appear in emission rather than absorption because the stratospheres of both planets are substantially warmer than the upper troposphere. This is the result of warming by sunlight absorbed in the near-IR by CH_4 and small atmospheric particulates. NH_3 and PH_3 can also be seen as components of the mid-IR spectrum; their lines are in absorption because they are most abundant in the troposphere. In fact, ammonia is a condensate — much like water vapor in Earth's atmosphere. It does not appear as strongly in Saturn's spectrum because Saturn is colder than Jupiter, and ammonia condenses out deeper in the atmosphere than in Jupiter and is at the limits of detectability above the H_2 collision-induced continuum. PH_3 lines are, in fact, also detectable in Jupiter's spectrum, but are more difficult to discern among the forest of NH_3 lines in Fig. 6.

The existence of absorption and emission features from H_2 and CH_4 in the thermal spectrum is also useful for determining temperatures. Both molecules are well mixed in the atmosphere, so any changes that take place in their emission from point to point can be attributed to changes of temperature rather than abundance. There are techniques for inverting the radiative-transfer equations for the emitted flux, given observations of thermal emission in regions dominated by H_2 or CH_4 that allow temperatures to be determined (some authors say "retrieved") as a function of altitude. For the H_2 absorption features, the relevant range is in the upper troposphere — around ~100–400 mbar (~0.1–0.4 bar) in atmospheric pressure. For CH_4 emission, the relevant range is the stratosphere at pressures less than ~100 mbar.

Details about molecular transitions in planetary atmospheres can be found in *Goody and Yung* (1989) and *Irwin* (2003). *Irwin et al.* (1998) also discuss in some detail the techniques for retrieving atmospheric temperatures in the giant planets.

3.2. Clouds

The visual appearances of Jupiter and Saturn are strongly affected by clouds. Jupiter's visible atmosphere is famously heterogeneous, and cloud motions and colors have been studied for decades. An extensive review of historical and amateur images is given by *Rogers* (1995). Because light reflected from a generally homogeneous upper-atmospheric haze layer, Voyager and Cassini imaging observations of Jupiter's and Saturn's cloud fields to determine wind velocities confirmed and greatly refined groundbased studies that showed variable zonal (east-west) winds that are variable with latitude, but generally constant in time. Jupiter's wind vectors are both prograde and retrograde with respect to the rotation of the deep interior, ranging from a maximum near 140 m/s and a minimum near –50 m/s. Doppler tracking of the Galileo probe suggests that these zonal winds persist into the atmosphere at least as deep at the 20-bar pressure level. Saturn's winds display a strong jet of higher speeds, reaching 500 m/s near the equator. A drop of Saturn's maximum jet speed has been detected in comparing HST-derived winds with those derived by Voyager imaging, but it is not clear whether this is due to a real change in the wind speed itself or a change of the altitude of the cloud particles being tracked. Meridional (north-south) winds are much smaller and only notable around large, discrete features such as Jupiter's Great Red Spot or smaller vortices. Jupiter's largest and longest lived vortices are all anticylonic (rotating counterclockwise in the southern hemisphere, clockwise in the northern hemisphere). Cyclonic storms are also evident and Galileo observations noted that they are often the source locations for observable lightning; however, their lifetimes are generally measured in weeks rather than months to years.

At near-IR wavelengths, the dominance of gaseous absorption by CH_4 and some H_2 allows tracking of higher-altitude clouds and hazes. Imaging through wavelengths that sample particulates at various levels provides a means of shaping a three-dimensional picture of cloud systems. Stratospheric hazes can be seen at high latitudes that are thought to be products of charged-neutral interactions in Jupiter's polar auroral regions. Images of near-IR reflectivity in Saturn have an overall banded structure that is similar to Jupiter's, but Saturn seldom shows individual discrete features at these wavelengths that are large enough to be detected from Earth.

The 5-µm "spectral windows" for Jupiter and Saturn provide insight into their cloud systems at several bars of atmospheric pressure. The lack of substantial gaseous absorption allows cloud tops to be sensed, and thermal emission begins to dominate most of the 4.8–5.3-µm region for Jupiter and the 5.0–5.3-µm region for Saturn. Analogous to the visible region, Jupiter's 5-µm appearance is spectacularly heterogeneous (Fig. 7), with some regions of the atmosphere nicknamed "5-µm hot spots" ostensibly cloudless down to deep (5 bar atmospheric pressure) and warm (275 K) cloud-top levels. These are mixed with regions of both intermediate and cold clouds across the face of the disk. The warmer regions are highly correlated with areas of darker color in the visible, and the hot spots are correlated with gray areas. This implies that the lighter-colored clouds represent an upper layer of tropospheric clouds with particles thick enough to be optically thick at wavelengths near 5 µm. Such clouds could be related to an ammonia ice condensate ("cirrus") cloud near 1 bar to 500 mbar in pressure, but spectral signatures of ammonia ice have only been seen in particulates undergoing rapid upwelling or regions of turbulent flow to the northwest of the Great Red Spot. This requires that ammonia condensate to be (1) riming nucleation sites made of other materials or (2) particles covered by a coating of other material. The morphology of these clouds can also be used to diagnose upwelling and downwelling motions, just as the para-H_2 distribution discussed earlier. Upwelling gas is wet and capable of forming clouds, but — after reaching dryer and colder altitudes — the condensate has precipitated out and downwelling regions will generally be cloudless because of the dry nature of the gas. Thus the relatively cold 5-µm interior of the Great Red Spot (near the center of Fig. 7) diagnoses a region of upwelling, and its ostensibly warm (actually less cloudy) periphery indicates regions of downwelling.

Saturn's thermal emission at ~5 µm reveals a cloud structure in the several-bar pressure region every bit as heterogeneous as Jupiter's, although one might say it is a negative version. A series of warm zonal bands are interlaced with a cold bands, and discrete regions are dominated not by clear, warm areas as in Jupiter's atmosphere but by cloudy, cold ones. Cassini VIMS observations of this region are similar, but take advantage of the entire spectral window by observing only thermal emison on Saturn's nightside. The persistence of some cloud features that are detectable in the visible can also be found in this spectral region, an indicator that the dynamics creating them persist over a wide vertical range. For example, an irregular band near high northern latitudes known as the "polar hexagon" attests to the presence of meridional waves that perturb the normally uniform zonal flow, and the combination of visible, near-IR, 5-µm, and mid-IR observations shows that this feature persists in both the cloud and thermal fields from atmospheric pressures less than ~100 mbar down to levels with pressures of several bars.

3.3. Temperatures

Temperatures in the atmospheres of Jupiter and Saturn can be determined using several different techniques. For individual locations, occultations of optical light from stars by the atmosphere have been used to determine atmospheric density as a function of altitude and then transformed into temperatures and pressures using the ideal gas law, together with the equation of hydrostatic equilibrium and an estimate

Fig. 7. 4.85-μm image of Jupiter taken at the NASA Infrared Telescope Facility using the facility NSFCam2 instrument on September 24, 2008. The filter used is 0.24 μm wide and illustrates the primarily thermal radiance emerging from Jupiter's multilayered cloud systems, with more radiance emerging from cloud tops in the deepest atmosphere. Jupiter's Great Red Spot can be seen in the lower center of the figure as a region whose periphery is defined by dry, downwelling gas that creates a region relatively clear of clouds.

of the mean molecular weight. A similar approach has been successful using occultations by the atmosphere of coherent spacecraft signals in the radio.

One such result for a Voyager spacecraft radio occultation is shown in Fig. 8 for Saturn, but the Galileo probe at Jupiter provided the most direct measurement of temperatures available for the giant planets (Fig. 8), with direct measurements consistent with a "dry" adiabat (i.e., one without perturbations to the lapse rate from exchanges of latent heat from condensing molecules) from ~400 mbar to ~20 bars (thick, asterisked curve in Fig. 8). At higher levels, before the entry velocity of the probe became subsonic, the deceleration of the probe was Doppler tracked, the density of impeding gas determined, and knowledge of the mean molecular weight was used to determine the temperature profile of Jupiter at that location in the stratosphere (regular asterisked curve in Fig. 8). This profile showed the presence of substantial vertical thermal waves in the stratosphere, which should provide an substantial additional source of mechanical energy for the upper stratosphere.

The most successful approach for mapping temperatures across the disk is to invert observations of thermal emission in spectral regions dominated by well-mixed constituents. As described earlier, spectral regions dominated by collision-induced absorption by well-mixed H_2 and the 7-μm band of CH_4 (Fig. 6) are used in both planets to map temperatures. Figure 8 contains temperatures derived using this technique by the Cassini Composite Infrared Spectrometer (CIRS) instrument for Jupiter and Saturn near the equator and for latitudes of ±30°. This approach has a vertical resolution no better than a scale height, and so it acts as a low-pass filter for temperature variability with altitude. Nonetheless, it provides a good means to derive a range of temperatures over time and space, similar to instrumentation in weather satellites orbiting Earth. The technique has been expanded from spacecraft spectroscopy to a series of mid-IR filtered images of Jupiter and Saturn, which provide similar temperature information over a wide horizontal range without as much vertical information overlap. This has provided a means for groundbased observations to supplement the close-up coverage provided by spacecraft, with spatial resolutions that have often been as good as those available from spacecraft instruments with the deployment of mid-IR imaging instruments on 8-m or larger-class telescopes, which minimize the effects of blurring by diffraction. The entire host of temperature information verifies that the atmosphere is indeed separable into largely convective vs. largely radiative regions, with the convective regions over a variety of different latitudes and bands converging at depth in both planets (Fig. 8).

3.4. Compositions

Over 99.9% of the composition of both planets by volume consists of hydrogen, helium, and methane. A summary of our current knowledge of the compositions of Jupiter and Saturn in the detectable part of the atmosphere is given in Table 1. Their relative abundances, particularly the He/H_2 ratio, are important constraints on models of their formation, evolution, and interiors. For Jupiter, the value shown in Table 1 is the one derived independently by the Galileo Probe Mass Spectrometer (GPMS) and a purpose-built Helium Abundance Detector (HAD). The measured value for the He abundance may be influenced by its ability to form droplets at high pressure and "rain out" of the deep atmosphere toward the center of the planet. Neon (Ne) is soluble in the He drops and appears to be severely depleted in Jupiter with respect to solar abundances. The abundances of Ar (argon), Kr (krypton), and Xe (xenon) determined by the GPMS are all enhanced by ~2.6 with respect to solar values.

CH_4 is present throughout the atmosphere because it represents the simplest stable form of carbon, originally delivered to the planet as methane or other carbon-bearing ices in planetestimals. This is also true of NH_3 for N, H_2O for O, and possibly PH_3 for P, and H_2S for S. Ammonia and water both condense out in the detectable parts of Jupiter's and Saturn's atmospheres; both are responsible for at least part of the atmospheric opacity in the submillimeter through microwave part of the spectrum and, with phosphine, in the 5-μm spectral window. The measurement of O/H via the Galileo GPMS value is considered to be too low because the Galileo probe descended in a 5-μm "hot spot" that also proved to be a region of unusual desiccation. Chemical equilibrium models predict that H_2S should react with NH_3 to form NH_4SH (ammonium hydrosulfate) clouds around

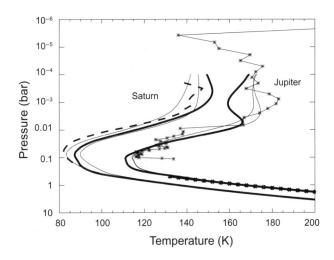

Fig. 8. Temperature profiles for Jupiter and Saturn. Curves in solid lines are derived from Cassini CIRS (infrared) experiment observations, representative of near-equatorial (thick solid) and ±30° latitude (regular solid) curves. Galileo probe Atmospheric Structure Instrument (ASI) results are shown by the regular solid curve and asterisks for the accelerometer data in the "inverted" stratosphere and by the thick solid curve and asterisks for the direct measurements in the troposphere. A Voyager radio occultation curve is shown for Saturn by the dashed curve. Note the warmer temperatures in Jupiter than in Saturn, the "inverted" temperatures in the radiatively controlled atmosphere at pressures lower than ~100 mbar, vertical waves measured by the Galileo ASI accelerometer experiment, and the convergence of the several temperature profiles at different latitudes with depth.

TABLE 1. Compositions of Jupiter and Saturn in volume mixing ratios.

Gas	Jupiter	Saturn
H_2	0.862 ± 0.003	0.882 ± 0.024
He	0.136 ± 0.003	0.113 ± 0.024
CH_4	$2.4 \pm 0.5 \times 10^{-3}$	$4.7 \pm 0.2 \times 10^{-3}$
NH_3	$\leq 7 \times 10^{-4}$	$\leq 1 \times 10^{-4}$
H_2O	$\leq 6.5 \pm 2.9 \times 10^{-5}$	$0.2\text{--}2 \times 10^{-7}$
CO_2		$\leq 3 \times 10^{-10}$
CO	1×10^{-9}	$(1\text{--}3) \times 10^{-9}$
PH_3	$\leq 6 \times 10^{-6}$	$\leq 4.5 \times 10^{-6}$
GeH_4	$\sim 2 \times 10^{-8}$	3.4×10^{-7}
H_2S	$1\text{--}8 \times 10^{-5}$	$<10^{-6}$
C_2H_2	$\leq 3 \times 10^{-8}$	$\leq 1 \times 10^{-7}$
C_2H_6	$\leq 3 \times 10^{-6}$	$\leq 5 \times 10^{-6}$
HCN	$<1 \times 10^{-10}$	$<10^{-9}$
CH_3D	$\sim 2 \times 10^{-8}$	$3.2 \pm 0.2 \times 10^{-7}$
HD	$\sim 1.1 \times 10^{-5}$	$0.7\text{--}1.7 \times 10^{-8}$
Ar	$1.0 \pm 0.4 \times 10^{-5}$	
Ne	$2.3 \pm 0.25 \times 10^{-5}$	

~1–3 bar pressure (between the upper NH_3 cloud and the deeper H_2O cloud layers). Molecules in chemical disequilibrium at the cold upper tropospheric temperatures such as PH_3 and GeH_4 (germane) are present because upwelling from warmer depths takes place more rapidly than molecular decomposition.

Methane is present in the stratosphere and is destroyed by solar ultraviolet radiation, with higher-order hydrocarbons resulting from the chemical recombination process. The most abundant of these are C_2H_2 (acetylene), C_2H_4 (ethylene), and C_2H_6 (ethane), which act as the principal means for stratospheric radiative cooling. Enhanced abundances of hydrocarbons have been reported in polar stratospheric "hot spot" regions, the result of additional auroral-related chemistry. Reports of H_2O line emission attest to the presence of H_2O in the stratosphere, in abundances suggesting infall from the exterior, either from ring particles or from interplanetary ice.

The isotopic $^{13}C/^{12}C$ ratios, determined by the Galileo probe for Jupiter and by the Cassini CIRS experiment for Saturn, are close to the terrestrial value, suggesting a protosolar value with almost no chemical fractionation of carbon isotopes in either atmosphere. The observed D/H ratios, determined from the same set of experiments, also suggest protosolar values. In the case of Saturn, this is inconsistent with interior models, which predict an enhancement of deuterium because of the larger mass of its core.

In addition to the reviews already mentioned, detailed summaries on several topics concerning jovian and saturnian atmospheres, interiors, icy satellites, rings, and magnetospheres are given in *Bagenal et al.* (2004) for Jupiter and *Brown et al.* (2009) for Saturn.

4. SUMMARY OF EXTRASOLAR GIANT PLANET FEATURES AND HIGHLIGHTS

Using the formalisms sketched in section 2, and the current knowledge of jovian planet atmospheres summarized in section 3, one can establish a theoretical framework in which to study EGPs for any age, orbit, mass, and elemental composition. This theoretical edifice is also understood using the simple models provided in section 2.4 and section 2.6, with which one can interpret most of their measured and anticipated properties. In this section, we highlight a few of the interesting ideas and features that have emerged in the last few years concerning EGP evolution, generic planet/star flux ratios, and the distinctive character of the family of hot Jupiters. This summary is in no way comprehensive, but provides the reader with a snapshot of some of the salient facts and issues in this rapidly evolving field.

4.1. Evolution of Extrasolar Giant Planet Atmospheres and Chemistry

Unlike in a star, EGP atmospheric temperatures are sufficiently low that chemistry is of overriding importance. The atmosphere of a gaseous giant planet is the thin outer layer of molecules that controls its absorption and emission spectra and its cooling rate. Molecular hydrogen (H_2) is the dominant constituent, followed by helium. An EGP's effective temperature (T_{eff}) can vary from ~2500 K at birth for the more massive EGPs or in a steady state for the most severely irradiated to ~50 K for the least-massive EGPs in wide orbits after evolving for billions of years. This wide range translates into a wide variety of atmospheric constituents that for a given mass and elemental composition can evolve significantly.

Without any significant internal sources of energy, after formation an EGP gradually cools and shrinks. Its rate of cooling can be altered by stellar irradiation, or when old and light by hydrogen/helium phase separation (*Fortney and Hubbard*, 2004). Jupiter itself is still cooling and its total IR plus optical luminosity is about twice the power absorbed from the Sun. Hence, the temperatures and luminosities achieved are not just functions of mass and composition, but of mass, age, composition, orbital distance, and stellar type.

Since EGPs evolve through atmospheres of various compositions and temperatures, age is a key parameter in their study. A theoretical evolutionary scenario for Jupiter itself can serve to exemplify these transformations. The evolution of other EGPs with different masses and orbits will be different in detail, but for the wide-separation variety not in kind. The description below is based in part on the review by *Burrows* (2005), to which the reader is referred. Another useful review can be found in *Marley et al.* (2007).

At birth, Jupiter had a T_{eff} near 600–1000 K and the appearance of a T dwarf (*Burgasser et al.*, 2002). It had no ammonia or water clouds and, due to the presence of atomic sodium in its hot atmosphere, had a magenta color in the optical (*Burrows et al.*, 2001). Due to the formation and settling to depth of the refractory silicates that condense in the temperature range ~1700–2500 K, its atmosphere was depleted of calcium, aluminum, silicon, iron, and magnesium. (Note that such silicate clouds should exist at depth in the current Jupiter.) Water vapor (steam) was the major molecule containing oxygen, gaseous methane was the major reservoir of carbon, gaseous ammonia and molecular nitrogen were the contexts for nitrogen, and sulfur was found in H_2S. At lower temperatures (<~700K), FeS would be the equilibrium reservoir of sulfur. However, as noted above, earlier in the planet's history when refractory species condensed out and settled gravitationally, the atmosphere was left depleted of most metals, including iron. The result of this "rainout" was that the chemistry simplified and sulfur was in the form of H_2S, as is observed in the current Jupiter. The rainout phenomenon of condensates in the gravitational field of EGPs and solar system giants is a universal feature of their atmospheres (*Fegley and Lodders*, 1994; *Burrows and Sharp*, 1999).

As Jupiter cooled, the layer of alkali metals was buried below the photosphere to higher pressures, but gaseous H_2, H_2O, NH_3, and CH_4 persisted. At a T_{eff} of ~400 K, water condensed in the upper atmosphere and water clouds appeared. This occurred within its first 100 m.y. Within less than a billion years, when T_{eff} reached ~160 K, ammonia clouds emerged on top of the water clouds, and this layering persists to this day (see section 3).

Stellar irradiation retards cloud formation, as does a large EGP mass, which keeps the EGP hotter longer. Proximity to a star also keeps the planet hotter longer, introducing a significant dependence of its chemistry upon orbital distance. Around a G2V star such as the Sun, at 5 G.y. and for an EGP mass of 1.0 M_{Jup}, water clouds form at 1.5 AU, and ammonia clouds form beyond 4.5 AU (*Burrows et al.*, 2004). Jupiter's and Saturn's current effective temperatures are 124.4 K and 95 K, respectively. Jupiter's orbital distance and age are 5.2 AU and 4.6 G.y. The orbital distance, mass, and radius of a coeval Saturn are 9.5 AU, 0.3 M_{Jup}, and 0.85 R_{Jup}. However, as an EGP of whatever mass cools, its atmospheric composition evolves through a similar chemical and condensation sequence. Figure 9 depicts the atmospheric temperature/pressure (T/P) profile for a sequence of 1-M_{Jup}, 5-G.y. models as a function of orbital distance (0.2–15 AU) from a G2V star. As the EGP's orbital distance increases, its atmospheric temperature at a given age and pressure decreases. Superposed on the plot are the H_2O and NH_3 condensation lines. Clearly, a given atmospheric composition and temperature can result from many combinations of orbital distance, planet mass, stellar type, and age. This lends added complexity to the study of EGPs.

The atmospheres of hot Jupiters at orbital distances of ~0.02–0.07 AU from a G, F, or K star are heated and maintained at temperatures of ~1000–2000 K, roughly independent of planet mass and composition. The transiting EGPs discovered to date are examples of such hot objects. At high temperatures, carbon is generally in the form of carbon monoxide, not methane. As a result of this and the rainout of most metals, EGP atmospheric compositions are predominantly H_2, He, H_2O, Na, K, and CO. There are, however, significant day/night differences in composition, temperature, and spectrum that distinguish a hot Jupiter from a lone and isolated planet or star. For instance, on the nightside, carbon might be found in methane, whereas on the dayside it might be in CO. This suggests that nonequilibrium chemistry, where the chemical rates and the dynamical motions compete in determining the composition, might be at play. Exotic general circulation models (GCMs) with credible dynamics, radiative transfer, chemistry, and frictional effects will soon be necessary to understand the equatorial currents, jet streams, day/night differences, terminator chemistry, and global wind dynamics of irradiated EGPs, in particular, and of orbiting, rotating EGPs, in general.

There is another interesting aspect to the orbital-distance dependence of EGP properties and behavior. At the distance of Jupiter, the IR photosphere is close to the radiative-convective boundary, near 0.5 bar. This means that a good fraction of the stellar radiation impinging upon Jupiter from the Sun is absorbed directly in the convective zone, which thereby redistributes this heat more or less uniformly to all latitudes and longitudes of the planet. The upshot of this, and that the luminosity from the core due to the remaining residual heat of formation is comparable to the stellar irradiation, is that Jupiter (and Saturn) emit isotropically in the IR, the effects of cloud banding and structure not withstanding. However, as a gas giant planet moves inward toward its primary star, the pressure level of the photosphere and the pressure level of the dayside radiative-convective zone separate. When the EGP is at ~0.05 AU, and after 1 G.y., the photospheres have moved little, but the radiative-convective boundary is now near a kilobar. Moreover, the internal flux is miniscule compared with the magnitude of the irradiation. The result is that heat is not efficiently redistributed by internal convective motions, but by zonal winds in the radiative atmosphere *around* the planet to the nightside. This dayside/nightside dichotomy is a central feature of hot Jupiters and can lead to severe thermal contrasts as a function of longitude.

Curiously, a star with solar compostion at the edge of the light-hydrogen-burning main sequence ($M_* \sim 75\ M_{Jup}$) has a T_{eff} of ~1700 K. Therefore, an irradiated EGP, with a radius comparable to that of such a star, can be as luminous.

4.2. Planet/Star Flux Ratios of Wide-Separation Extrasolar Giant Planets

The planet/star flux ratio vs. wavelength is the key quantity is the study of orbiting planets. Figure 10, taken from *Burrows et. al.* (2004), depicts orbit-averaged (*Sudarsky et al.*, 2000) such flux ratios from 0.5 μm to 30 μm for a 1-M_{Jup}/5-G.y. EGP in a circular orbit at distances of 0.2 to 15 AU from a G2V star. These models are the same as those depicted in Fig. 9. The water absorption troughs are in evidence throughout. For the closer EGPs at higher atmospheric temperatures, carbon resides in CO and methane features are weak. For those hot Jupiters, the NaD line at 0.589 μm and the corresponding resonance line of K I at 0.77 μm are important absorbers, suppressing flux in the visible bands. Otherwise, the optical flux is increased by Rayleigh scattering of stellar light. As a increases, methane forms and the methane absorption features appear in the optical (most of the waviness seen in Fig. 10 for a ≥ 0.5 AU shortward of 1 μm), at ~3.3 μm, and at ~7.8 μm. At the same time, Na and K disappear from the atmosphere and the fluxes from ~1.5 μm to ~4 μm drop. For all models, the mid-IR fluxes longward of ~4 μm are due to self-emission, not reflection. As Fig. 10 makes clear, for larger orbital distances a separation between a reflection component in the optical and an emission component in the mid-IR appears. This separation into components is not so straightforward for the closer, more massive, or younger family members. For these EGPs, either the large residual heat coming from the core or the severe irradiation buoys the fluxes from 1 to 4 μm. The more massive EGPs, or, for a given mass, the younger EGPs, have larger J-, H-, and K-band fluxes. As a result, these bands are diagnostic of mass and age. For EGPs with large orbital distances, the wavelength range from 1.5 μm to 4 μm between the reflection and emission components may be the least favorable search space, unless the planet is massive or young.

When water or ammonia clouds form, scattering off them enhances the optical fluxes, while absorption by them suppresses fluxes at longer wavelengths in, for example, the 4–5-μm window. Because water and ammonia clouds form

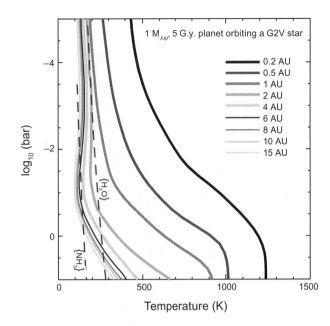

Fig. 9. Profiles of atmospheric temperature (in Kelvin) vs. the logarithm base 10 of the pressure (in bars) for a family of irradiated 1-M_{Jup} EGPs around a G2V star as a function of orbital distance. Note that the pressure is decreasing along the ordinate, which thereby resembles altitude. The orbits are assumed to be circular, the planets are assumed to have a radius of 1 R_{Jup}, the effective temperature of the inner boundary flux is set equal to 100 K, and the orbital separations vary from 0.2 AU to 15 AU. The intercepts with the dashed lines identified with either NH_3 or H_2O denote the positions where the corresponding clouds form. Taken from *Burrows et al.* (2004). Available in color at *arxiv.org/abs/0910.0248*.

in the middle of this distance sequence, the geometric albedo (A_g) is not a monotonic function of a. This effect is demonstrated in Fig. 11, taken from *Sudarsky et al.* (2005), which plots the planet/star flux ratio for a jovian-mass planet orbiting a G2V central star as a function of orbital distance at 0.55 μm, 0.75 μm, 1 μm, and 1.25 μm. Clearly, the planet/star flux ratio does not follow an inverse square law. Clouds can form, evaporate, or be buried as the degree of stellar heating varies with varying distance. This behavior is included in the models of Fig. 10, but its precise manifestations depend upon unknown cloud particle size, composition, and patchiness. As a consequence, direct spectral measurements might constrain cloud properties. As Fig. 10 suggests, the planet/star contrast ratio is better in the mid- to far-IR, particularly at wide separations. For such separations, the contrast ratio in the optical can sink to 10^{-10}.

For the closest-in EGPs (not shown on Fig. 10), such as HD 189733b, HD 209458b, OGLE-TR56b, 51 Peg b, and τ Boo b, the contrast ratio in the optical should be between 10^{-5} and 10^{-6} and is more favorable, although still challenging. However, for these hottest EGPs, alkali metal (sodium and potassium) absorption is expected to dominate in the optical and very near-IR. Sodium has already been detected

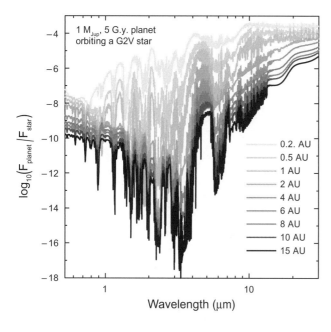

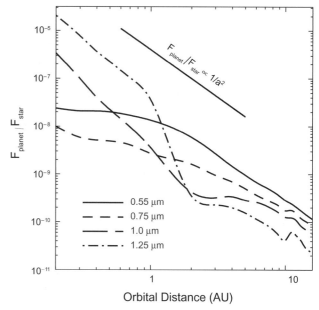

Fig. 10. Theoretical planet to star flux ratios vs. wavelength (in micrometers) from 0.5 μm to 30 μm for a 1-M_{Jup} EGP with an age of 5 G.y. orbiting a G2V main-sequence star similar to the Sun. This figure portrays ratio spectra as a function of orbital distance from 0.2 AU to 15 AU. Zero eccentricity is assumed and the planet spectra have been phase-averaged as described in *Sudarsky et al.* (2003). The associated T/P profiles are given in Fig. 9. Table 1 in *Burrows et al.* (2004) lists the modal radii for the particles in the water and ammonia clouds. Note that the planet/star flux ratio is most favorable in the mid-IR. Taken from *Burrows et al.* (2004). Available in color at *arxiv.org/abs/0910.0248*.

Fig. 11. Planet/star flux ratio as a function of orbital distance at 0.55 μm, 0.75 μm, 1 μm, and 1.25 μm assuming a G2V central star. In each case, the plotted value corresponds to a planet at greatest elongation with an orbital inclination of 80°. Note that the planet/star flux ratios do not follow a simple $1/a^2$ law. Taken from *Sudarsky et al.* (2005).

in the transit spectrum of HD 209458b. As a result, the expectation is that the optical albedos of hot Jupiters should be quite low, perhaps less than ~5%.

Importantly, the transiting EGPs are so hot and their orbits are so favorably inclined that the planet/star flux ratios in the mid-IR, in particular in the Spitzer bands, can range between ~10^{-3} and ~10^{-2}. For these objects, the gap between the reflection and emission components is completely closed. Moreover, the variation in the summed light of planet and star during eclipses and the execution of the orbit can be used to derive the planetary spectrum itself.

4.3. Hot Jupiter Highlights

The discovery of ~70 (and counting) transiting giant planets allows one to address their physical structures by providing simultaneous radius and mass meansurements. However, the proximity of transiting EGPs to their primaries boosts the planet/star flux contrast ratios in the near- and mid-IR to values accessible by Spitzer during secondary eclipse. Such measurements in the IRAC (~3.6, ~4.5, ~5.8, ~8 μm) and MIPS (~24 μm) bands, as well as IRS spectral measurements from ~5 to ~14 μm, are the first *direct* detections of planets outside the solar system. In addition, Spitzer has yielded photometric lightcurves as a function of orbital phase, and revealed or constrained brightness distributions across planet surfaces. This has been done (dramatically so) for HD 189733b at 8 and 24 μm, for HD 149026b and GJ 436b at 8 μm, and, although nontransiting, for both υ And b at 24 μm and HD 179949b at 8 μm. It will soon be done for many more hot Jupiters. Excitingly, Spitzer measurements alone have yielded exoplanet compositions, temperatures, and longitudinal temperature variations that have galvanized the astronomical and planetary communities.

Furthermore, using NICMOS, STIS, and ACS on HST, precision measurements of transit depths as a function of wavelength have enabled astronomers to identify atmospheric compositions at planet terminators. In this way, sodium, water, methane, hydrogen, and carbon monoxide have been inferred. Moreover, Canada's Microvariability and Oscillations of STars (MOST) microsatellite has obtained a stringent upper limit of ~8% to the optical albedo of HD 209458b (*Rowe et al.*, 2008). Hence, in ways unanticipated just a few years ago, these space telescopes are constraining the degree of heat redistribution from the daysides to the nightsides by zonal winds (providing a glimpse of global climate), are signaling the presence of thermal inversions, and are revealing exoplanet chemistry. The new data from Spitzer, HST, and MOST have collectively inaugurated the era of remote sensing of exoplanets and the techniques articulated in this chapter were developed to interpret such data.

As described above, there are many recent highlights in the study of giant exoplanet atmospheres, but we will focus here on only a few examples. They are the spectral and photometric measurements of HD 189733b by *Grillmair et al.* (2008) and *Charbonneau et al.* (2008) and the photometric measurements of HD 209458b by *Knutson et al.* (2008) and *Deming et al.* (2005), along with their interpretations (see also *Barman*, 2008). These objects exemplify the two basic classes of EGP atmospheres that have emerged, those without and with significant thermal inversions at altitude, and are the two best-studied transiting EGPs.

Figure 12a shows a comparison of the HD 189733b data with three theoretical models for the planet/star flux ratio. The best-fit model is consistent with most of the data and assumes solar elemental abundances and only modest heat redistribution to the nightside ($P_n = 0.15$). One of the salient features is the fact that the IRAC 1 (~3.6 μm) to IRAC 2 (~4.5 μm) ratio is greater than 1. This has been interpreted to mean that the temperature profile monotonically decreases outward and that there is no significant inversion. Figure 13a shows such temperature profiles (but for TrES-1), as well as the locations of the effective photospheres in the Spitzer bands. An appreciable inversion would reverse the IRAC 1/IRAC 2 ratio, since the effective photosphere in IRAC 2 is further out and at much lower pressures than that for IRAC 1. In addition, the depression of the IRAC 2 flux is consistent with the presence of carbon monoxide, which has a strong absorption feature there. More important is the comparison between theory and the IRS spectrum of *Grillmair et al.* (2008) between ~5 and ~14 μm. Not only is the slope of the spectrum roughly matched, but there is a ~3-σ detection of a feature just longward of ~6 μm. This is best interpreted as the flux peak due to the opacity window between the P and R branches of the v_2 vibrational bending mode of water vapor (see Fig. 2). Water would explain both the general slope of the IRS spectrum and this peak and together they are now taken to be the best indications of the presence of water (steam) in an EGP atmosphere.

Figure 12b compares theoretical models for the planet/star flux ratio of HD 209458b at secondary eclipse with the corresponding Spitzer photometric data of *Knutson et al.* (2008) (IRAC) and *Deming et al.* (2005) (MIPS). The differences from HD 189733b are illuminating. Here, the IRAC 1/IRAC 2 ratio is less than 1 and the IRAC 3 (~5.8 μm) flux is higher than the IRAC 4 (~8 μm) flux. These features are best interpreted as signatures of a thermal inversion at altitude. There must be an absorber of optical (and/or near-UV) stellar light to create the thermal inversion and a hot outer atmosphere (see Fig. 13b), with the result that the planet/star ratio is much higher in the IRAC 1, 2, and 3 channels than predictions without inversions (e.g., the black curve without an absorber). In fact, centered near IRAC 3, the IRAC data look like an *emission* feature, centered in the broad water band from ~4.5 to ~8 μm, as opposed to the absorption trough expected for an atmosphere with a negative temperature gradient. Emission features are expected in atmospheres with thermal inversions.

HD 189733b and HD 209458b together represent the two classes of hot-Jupiter atmospheres that are emerging to be

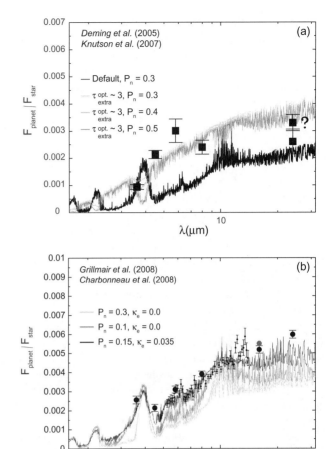

Fig. 12. (a) The planet-star flux ratios at secondary eclipse vs. wavelength for three models of the atmosphere of HD 209458b with inversions and for one model without an extra upper-atmosphere absorber of any kind (lowest curve). The three models with stratospheres have different values of P_n (= 0.3, 0.4, and 0.5), but are otherwise quite similar. (To distinguish the models, note that at a wavelength of ~4 μm the fluxes increase in the order $P_n = 0.5$, 0.4, and 0.3.) The old, default model (generally with lower flux) has a P_n of 0.3. This figure demonstrates that models with an extra upper-atmosphere absorber in the optical and with $P_n \geq 0.35$ fit the data; the old model does not fit. Superposed are the four IRAC points and the MIPS data at 24 μm from *Deming et al.* (2005) (square block on right). Also included, with a question mark beside it, is a tentative update to this 24 μm flux point, provided by Drake Deming (personal communication). (b) Comparison of spectral observations with broadband photometry and theoretical models of the dayside atmosphere of HD 189733b. The small dots with error bars show the mean (unweighted) flux ratio spectra from ~5 to ~14 μm, taken by IRS on Spitzer and published in *Grillmair et al.* (2008). The plotted uncertainties reflect the standard error in the mean in each wavelength bin. The large filled circles show broadband measurements at 3.6, 4.5, 5.8, 8.0, 16, and 24 μm from *Charbonneau et al.* (2008). Shown also are atmospheric model predictions for three values of a dayside-nightside heat redistribution parameter, P_n, and two values for the extra upper-atmosphere opacity, κ_e. The models increase in flux from $P_n = 0.3$, to 0.1, to 0.15 (the latter with a nonzero κ_e) (see *Burrows et al.*, 2008a, for an explanation). See the text for discussion. Available in color at *arxiv.org/abs/0910.0248*.

explained. As Fig. 12 indicates, we can reproduce their basic spectral features, but there remain many anomalies. The fits at ~24 μm are problematic and the detailed shapes of the spectra are not well reproduced. Moreover, the models depend upon the degree of heat redistribution to the nightside (modeled as P_n) (*Burrows et al., 2008a*), which crucially depends upon the unknown zonal flows and global climate. More important, although we know that there are inversions in the atmospheres of HD 209458b, TrES-4, XO-1b, and (likely) HD 149026b, we don't know why. What is the nature of this "extra absorber" in the optical (section 2.6)? With what other planetary properties is the presence or absence of an inversion correlated? Metallicity? Gravity? The substellar flux on the planets TrES-1 and XO-1b are similar, but only the latter shows an inversion. In Fig. 12b, we have introduced an ad hoc optical absorber at altitude merely to determine the opacity (κ_e) needed to reproduce the measurements. Its physical origin is not addressed. Some have suggested it is TiO or VO (*Fortney et al., 2008*). However, these species can easily condense out and should not persist for long in the upper atmosphere. The so-called cold-trap, which operates in the Earth's upper atmosphere to render it dry, should also operate for these metal oxides here (*Spiegel et al., 2009*). Hence, the nature of the extra optical absorber at altitude is currently unknown. One might well ask whether the slight thermal inversion introduced due to zonal heat redistribution at depth can be the culprit (*Hansen, 2008*), but the magnitude of the resulting temperature excursion is slight (≤100 K) and does not approach the ~1000K needed to explain the data. A recent suggestion by *Zahnle et al.* (2009) involves the photolytic production of allotropes of sulfur and HS. However, whether a sulfur allotrope or any sulfur compound achieves the necessary abundance at the requisite upper pressure levels and what its production systematics are with the properties of the planet and star (metallicity and/or stellar spectrum) remains to be seen. The theoretical task is emerging to be complex and challenging.

5. FUTURE PROSPECTS

The theoretical study of exoplanet atmospheres in general, and of EGP atmospheres in particular, has been energized by continuing discoveries and surprises that show no sign of abating. In the short term, the last cold cycle and the upcoming warm cycle(s) of Spitzer will provide more lightcurves and secondary eclipse measurements. HST will continue to yield high-sensitivity transit data. Groundbased extreme adaptive optics systems that will enable high-contrast imaging of exoplanets from under the glare of their parent stars will be developed and perfected. In the next decade, the James Webb Space Telescope (JWST) will constitute a quantum leap in sensitivity and broader wavelength coverage, which together may lay to rest many of the outstanding questions that have emerged during the early heroic phase of EGP atmospheric characterization. These include, but are not limited to:

1. How do zonal flows in EGP atmospheres affect their spectra and phase curves?

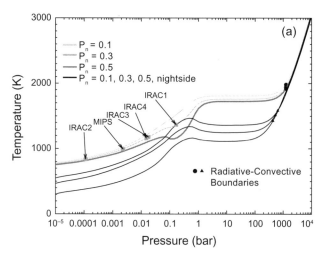

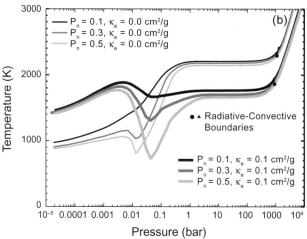

Fig. 13. Dayside (set of higher curves around 10 bar) and nightside (lower set of curves) temperature-pressure profiles for **(a)** TrES-1, as a representative of a transiting EGP without a thermal inversion at altitude, and **(b)** HD 209458b, as a representative of a transiting EGP with such a thermal inversion. These profiles incorporate the external substellar irradiation/flux and an internal flux for the planet corresponding to the temperature of 75 K is included for HD 209458b. See text for a discussion. Figures taken from *Burrows et al.* (2008a), to which the reader is referred for further details. Available in color at *arxiv.org/abs/0910.0248*.

2. What is the magnitude of heat redistribution to the planet's nightside? What is "P_n"?
3. How bright is the nightside?
4. Upon what do the longitudinal positions of the hot and cold spots of strongly irradiated planets depend?
5. What is the depth of heat redistribution?
6. What causes the strong thermal inversions in a subset of the hot Jupiters?
7. Is there a central role for photolysis and nonequilibrium chemistry in EGP atmospheres, particularly in hot Jupiter atmospheres?

8. Can the chemistry and irradiation be out of phase? Is there chemical and thermal hysteresis?

9. How different are the compositions on a planet's dayside and nightside?

10. How different are the temperatures and chemical compositions at the terminators of hot Jupiters during ingress and egress?

11. Could there be significant gaps in our understanding of the dominant chemical constituents of EGP atmospheres?

12. What is the origin of dissipation in planetary atmospheres that stabilizes their zonal flows?

13. What are the signatures of nonzero obliquity and asynchronous spin in the measured lightcurves and spectra?

14. What are the unique features of the lightcurves of EGPs in highly elliptical orbits and can tidal heating contribute to their thermal emissions?

15. What are the thermal and chemical relaxation time constants of EGP atmospheres under the variety of conditions in which they are found?

16. At what orbital distances are the dayside and nightside IR fluxes the same due to efficient heat redistribution?

17. What is the role of metallicity and planet mass in EGP compositions, lightcurves, and spectra?

18. What are the spectral and evolutionary effects of atmospheric clouds and hazes?

19. Do the convective cores of hot Jupiters cool differently on their nightsides and daysides?

Our theories are poised to improve significantly, with credible multidimensional simulations undertaken and nonequilibrium chemistry and cloud formation addressed. Methods are being designed to understand day-night coupling, the degree of heat redistribution, and the pattern of zonal flow. Moreover, with the discovery of more transiting planets such as GJ 436b in the ice giant mass and radius regime, the variety of signatures will multiply. Finally, more wide-separation EGPs will be separately imaged and probed. Given the brilliant nature of its past, whatever the manifold uncertainties, this field is on a trajectory of future growth and excitement that is establishing it as a core discipline of twenty-first century astronomy.

Acknowledgments. The authors would like to acknowledge D. Sudarsky, I. Hubeny, and C. Sharp for past collaborations and much wisdom. They would also like to thank T. Barman and D. Spiegel for careful readings of the manuscript and numerous constructive suggestions. This work was partially supported by NASA via Astrophysics Theory Program grant # NNX08AU16G and with funds provided under JPL/Spitzer Agreements No. 1328092 and 1348668. G.S.O. would like to acknowledge partial support by funds to the Jet Propulsion Laboratory, California Institute of Technology. We are grateful to NASA's Infrared Telescope Facility and JPL collaborators Leigh Fletcher and Padma Yanamandra-Fisher for the image shown in Fig. 7.

REFERENCES

Bagenal F., Dowling T. E., and McKinnon W. B., eds. (2004) *Jupiter: The Planet, Satellites and Magnetosphere.* Cambridge Univ., Cambridge.

Barman T. (2008) On the presence of water and global circulation in the transiting planet HD 189733b. *Astrophys. J. Lett., 676,* L61–L64.

Brown R., Dougherty M., and Krimigis S., eds. (2009) *Saturn After Cassini/Huygens.* Springer, Heidelberg.

Burgasser A., Kirkpatrick J. D., Brown M. E., Reid I. N., Burrows A., et al. (2002) The spectra of T dwarfs. I. Near-infrared data and spectral classification. *Astrophys. J., 564,* 421–451.

Burrows A. (2005) A theoretical look at the direct detection of giant planets outside the solar system. *Nature, 433,* 261–268.

Burrows A. and Sharp C. M. (1999) Chemical equilibrium abundances in brown dwarf and extrasolar giant planet atmospheres. *Astrophys. J., 512,* 843–863.

Burrows A., Hubbard W. B., Lunine J. I., and Liebert J. (2001) The theory of brown dwarfs and extrasolar giant planets. *Rev. Mod. Phys., 73,* 719–765.

Burrows A., Sudarsky D., and Hubeny I. (2004) Spectra and diagnostics for the direct detection of wide-separation extrasolar giant planets. *Astrophys. J., 609,* 407–416.

Burrows A., Budaj J., and Hubeny I. (2008a) Theoretical spectra and light curves of close-in extrasolar giant planets and comparison with data. *Astrophys. J., 678,* 1436–1457.

Burrows A., Ibgui L., and Hubeny I. (2008b) Optical albedo theory of strongly-irradiated giant planets: The case of HD 209458b. *Astrophys. J., 682,* 1277–1282.

Charbonneau D., Knutson H. A., Barman T., Allen L. E., Mayor M., Megeath S. T., Queloz D., and Udry S. (2008) The broadband infrared emission spectrum of the exoplanet HD 189733b. *Astrophys. J., 686,* 1341–1348.

Chevallier L., Pelkowski J. and Rutily B. (2007) Exact results in modeling planetary atmospheres — I. Gray atmospheres. *J. Quant. Spectrosc. Rad. Transfer, 104,* 357–376.

Deming D., Seager S., Richardson L. J., and Harrington J. (2005) Infrared radiation from an extrasolar planet. *Nature, 434,* 740–743.

dePater I. and Lissauer J. J. (2004). *Planetary Sciences.* Cambridge Univ., Cambridge.

Dyudina U. A., Sackett P. D., Bayliss D. D, Seager S., Porco C. C., Throop H. B., and Dones L. (2004) Phase light curves for extrasolar Jupiters and Saturns. *Astrophys. J., 618,* 973–986, astro-ph/0406390.

Fegley B. and Lodders K. (1994) Chemical models of the deep atmospheres of Jupiter and Saturn. *Icarus, 110,* 117–154.

Fegley B.and Lodders K. (2001) Very high temperature chemical equilibrium calculations with the CONDOR code. *Meteoritics & Planet. Sci., 33,* A55.

Fortney J. J. (2005) The effect of condensates on the characterization of transiting planet atmospheres with transmission spectroscopy. *Mon. Not. R. Astron. Soc., 364,* 649–653.

Fortney J. J. and Hubbard W. B. (2004) Effects of helium phase separation on the evolution of extrasolar giant planets. *Astrophys. J., 608,* 1039–1049.

Fortney J. J., Lodders K., Marley M. S., and Freedman R. S. (2008) A unified theory for the atmospheres of the hot and very hot Jupiters: Two classes of irradiated atmospheres. *Astrophys. J., 678,* 1419–1435.

Freedman R. S., Marley M. S., and Lodders K. (2008) Line and mean opacities for ultracool dwarfs and extrasolar planets. *Astrophys. J. Suppl., 678,* 1419–1435.

Goody R. M. and Yung Y. L. (1989). *Atmospheric Radiation: Theoretical Basis, 2nd Edition.* Oxford Univ., New York.

Grillmair C., Burrows A., Charbonneau D., Armus L., Stauffer J.,

Meadows V., van Cleve J., von Braun K., and Levine D. (2008) Strong water absorption in the dayside emission spectrum of the exoplanet HD 189733b. *Nature, 456,* 767–769.

Hansen B. M. S. (2008) On the absorption and redistribution of energy in irradiated planets. *Astrophys. J. Suppl., 179,* 484–508.

Hubeny I. (1990) Vertical structure of accretion disks — A simplified analytical model. *Astrophys. J., 351,* 632.

Hubeny I., Burrows A., and Sudarsky D. (2003) Possible bifurcation in atmospheres of strongly irradiated stars and planets. *Astrophys. J., 594,* 1011–1018.

Irwin P. G. J. (2003) *Giant Planets of Our Solar System: Atmospheres, Composition and Structure.* Praxis, Chichester.

Irwin P. G. J., Weir A. L., Taylor F. W., Lambert A. L., Calcutt S. B., Cameron-Smith P. J., Baines K., Orton G. S., Encrenaz T., and Roos-Serote M. (1998) Cloud structure and atmospheric composition of Jupiter retrieved from Galileo near-infrared mapping spectrometer real-time spectra. *J. Geophys. Res., 103,* 23001–23021.

Karkoschka E. (1999) Methane, ammonia, and temperature measurements of the jovian planets and Titan from CCD spectrophotometry. *Icarus, 133,* 134–146.

Knutson H. A., Charbonneau D., Allen L. E., Burrows A., and Megeath S. T. (2008) The 3.6–8.0 μm broadband emission spectrum of HD 209458b: Evidence for an atmospheric temperature inversion. *Astrophys. J., 673,* 526–531.

Lodders K. (1999) Alkali element chemistry in cool dwarf atmospheres. *Astrophys. J., 519,* 793–801.

Lodders K. and Fegley B. (2002) Atmospheric chemistry in giant planets, brown dwarfs, and low-mass dwarf stars. I. Carbon, nitrogen, and oxygen. *Icarus, 155,* 393–424.

Marley M. S., Gelino C., Stephens D., Lunine J. I., and Freedman R. (1999) Reflected spectra and albedos of extrasolar giant planets. I. Clear and cloudy atmospheres. *Astrophys. J., 513,* 879–893.

Marley M. S., Fortney J., Seager S., and Barman T. (2007) Atmospheres of extrasolar giant planets. In *Protostars and Planets V* (B. Reipurth et al., eds.), pp. 733–747. Univ. of Arizona, Tucson.

Perryman M. A. C. (2003) The GAIA mission. In *GAIA Spectroscopy: Science and Technology* (U. Munari, ed.), p. 3. ASP Conf. Series 298, Astronomical Society of the Pacific, San Francisco.

Rogers J. H. (1995) *The Giant Planet Jupiter.* Cambridge Univ., Cambridge.

Rowe J. F., Matthews J. M., Seager S., et al. (2008) The very low albedo of an extrasolar planet: MOST spacebased photometry of HD 209458. *Astrophys. J., 689,* 1345–1353.

Seager S., Whitney B. A, and Sasselov D. D. (2000) Photometric light curves and polarization of close-in extrasolar giant planets. *Astrophys. J., 540,* 504–520.

Sharp C. M. and Burrows A. (2007) Atomic and molecular opacities for brown dwarf and giant planet atmospheres. *Astrophys. J. Suppl., 168,* 140–166.

Sharp C. M. and Huebner W. F. (1990) Molecular equilibrium with condensation. *Astrophys. J. Suppl., 72,* 417–431.

Sobolev V. V. (1975) *Light Scattering in Planetary Atmospheres.* Pergamon, Oxford.

Spiegel D., Silverio K., and Burrows A. (2009) Can TiO explain thermal inversions in the upper atmospheres of irradiated giant planets? *Astrophys. J., 699,* 1487–1500.

Sudarsky D., Burrows A., and Pinto P. (2000) Albedo and reflection spectra of extrasolar giant planets. *Astrophys. J., 538,* 885–903.

Sudarsky D., Burrows A., and Hubeny I. (2003) Theoretical spectra and atmospheres of extrasolar giant planets. *Astrophys. J., 588,* 1121.

Sudarsky D., Burrows A., Hubeny I., and Li A. (2005) Phase functions and light curves of wide-separation extrasolar giant planets. *Astrophys. J., 627,* 520–533.

Unwin S. C. and Shao M. (2000) Space interferometry mission. In *Interferometry in Optical Astronomy* (P. J. Lena and A. Quirrenbach, eds.), pp. 754–761. *SPIE Proc. Series* 4006.

van de Hulst H. C. (1974) The spherical albedo of a planet covered with a homogeneous cloud layer. *Astron. Astrophys., 35,* 209–214.

Zahnle K., Marley M. S., Lodders K., and Fortney J. J. (2009) Atmospheric sulfur photochemistry on hot Jupiters. *Astrophys. J. Lett., 701,* L20–L24.

Terrestrial Planet Atmospheres and Biosignatures

Victoria Meadows
University of Washington

Sara Seager
Massachusetts Institute of Technology

The search for terrestrial exoplanets — rocky worlds in orbit around stars other than the Sun — is one of humanity's most exciting science goals. The discovery of super Earths, terrestrial planets more massive than Earth, has opened a new era in exoplanet science, confirming the basic idea that our solar system is not the only planetary system to harbor terrestrial planets. Terrestrial exoplanets will expand planetary diversity, with masses and compositions likely very different from those found in our solar system. Most significantly, terrestrial exoplanets have the potential to host habitable environments on or below their solid surfaces, and are the most likely places beyond our solar system to search for signs of life. In the coming decades, instrumentation will be developed to expand our census of terrestrial exoplanets and directly characterize the atmospheres and biosignatures of these worlds. In the meantime, scientific progress in this field is made via extensive photochemical, climate, and radiative transfer modeling of terrestrial planetary environments together with remote sensing studies of solar system terrestrial planets, including Earth. This chapter provides an overview of terrestrial exoplanet atmosphere modeling techniques, a review of the scientific advances to date, and a discussion of outstanding questions and future directions.

1. INTRODUCTION

The field of exoplanet atmosphere characterization stands on a divide. On the one side are the dozens of hot Jupiter-mass (M_{Jup}) exoplanets with published atmosphere observations. On the other side lie the emerging population of super Earths and anticipated Earth-mass ($M_\oplus$) planets, with atmospheres as yet to be characterized. Super Earths are loosely defined to be planets >1 $M_\oplus$ and <10 $M_\oplus$ that are predominantly rocky or icy (see the Appendix of the chapter by Seager and Lissauer for exoplanet definitions). Terrestrial exoplanets are likely to include planets with characteristics akin to the solar system's terrestrial planets, as well as classes of planets not found in our own solar system.

The range of possible exoplanet atmospheric mass and composition has yet to be uncovered theoretically and observationally. As such, there is not yet a definitive categorization of atmosphere types (but see *Seager and Deming*, 2010) (see also section 4.1). Terrestrial exoplanet atmospheres are expected to show a wide diversity, just as the orbits, masses, and radii of known exoplanets do. This atmosphere diversity is likely larger than that seen for giant exoplanets, which retain the gases with which they formed. In contrast, terrestrial planet atmospheres are affected by the planet's initial composition, which is a function of the nature and position of other planets in the system, and by subsequent evolution, which can be substantial.

Terrestrial exoplanets are of significant scientific interest for astrobiology. Terrestrial planets with solid and liquid surfaces are the most likely environments to harbor life beyond our solar system. A habitable exoplanet is conventionally defined as one with some surface liquid water. Water's properties make it an ideal solvent for life, and it is likely easier to remotely detect a surface biosphere than a subsurface one. In contrast to terrestrial planets, giant and Neptune-sized planets enshrouded by gas envelopes have no solid or liquid surfaces to support life, and their temperatures just below the deep atmosphere rapidly become too hot for the stability of life's complex molecules.

Terrestrial exoplanet science is set against a backdrop of groundbreaking recent observational advances and ambitious plans for future telescopic missions. Until the many observational challenges are overcome, the field remains largely dominated by work done to understand the exoplanet context for the terrestrial planets in our own solar system, and theoretical modeling to understand the likely diversity of terrestrial planet characteristics and evolutionary processes.

1.1. The Solar System's Terrestrial Planets

A terrestrial planet is a solid-surfaced planet with a bulk composition dominated by iron and rock. In our solar system, the terrestrial planets are the four planets closest to the Sun, namely, Mercury, Venus, Earth, and Mars (Fig. 1). The bulk composition of these solar system planets is silicate rocks and iron.

1.1.1. Internal composition and magnetic fields. Mercury, Venus, Earth, and Mars have similar internal structures, with

Fig. 1. The solar system's four terrestrial planets to scale, Mercury, Venus, Earth, and Mars. Image credit: NASA.

a metallic core surrounded by a silicate mantle, lithosphere, and crust (Fig. 2) (see the chapter by Sotin et al.). The solar system terrestrial planets, however, have magnetospheres that differ markedly from each other in strength and origin. A planet's magnetosphere, whether intrinsic or induced, is important for habitability as it often serves as the first line of defense against stripping of the planetary atmosphere by the solar wind. Motions in Earth's molten metallic outer core, combined with the planet's relatively rapid rotation, create a dynamo-generated intrinsic magnetic field, protecting our atmosphere from loss via solar wind ablation (*Dehant et al.,* 2007). The remaining terrestrial planets do not have strong intrinsic magnetic fields, either due to loss of internal heating, which froze the metallic cores (as was likely the case for the smaller planets Mercury and Mars), or due to very slow rotation or lack of plate tectonics [as in the case of Venus (*Nimmo,* 2002)]. Mercury has an extremely weak intrinsic magnetic field, and may modify its solar wind interaction via induced currents in the planet's metallic interior (*Slavin et al.,* 2007). Venus supports a weak, variable "bow-shock" magnetosphere, induced by direct interaction between the planet's ionosphere and the solar wind (*Russell,* 1976; *Lipatov,* 1978). Mars' magnetosphere is largely generated by nonuniform remnant magnetization of the planet's crust. In the northern hemisphere, weak crustal magnetic fields allow solar wind interactions with the atmosphere and ionosphere, enhancing the atmospheric loss rate. In the southern hemisphere a stronger crustal field forms localized "magnetospheres" that can stand off the solar wind to altitudes of many hundreds of kilometers above the surface (*Mitchell et al.,* 2001).

1.1.2. Atmospheric composition and evolution. The atmospheric composition of the solar system's terrestrial planets is a function of both their initial volatile inventory and billions of years of evolution. Unlike the gas giant planets, which capture their atmospheres directly from the solar nebula during formation, terrestrial planets rapidly lose their hydrogen-rich primary atmospheres, and acquire secondary atmospheres after formation via outgassing, volcanism, or comet and asteroid impact. A terrestrial planet's atmospheric composition is therefore strongly influenced by the delivery of volatile material to the forming planet. These early atmospheres are then subject to evolutionary processes including photochemistry, interaction and cycling with the planet's solid body, and atmospheric loss processes (*Zahnle et al.,* 2007). Earth's atmosphere has also been heavily modified over billions of years by the presence of the biosphere (*Catling and Kasting,* 2007). Present-day terrestrial planet atmospheres are dominated by oxidized gases in contrast to the hydrogen-dominated jovian planets. CO_2 is the predominant atmospheric gas on both Mars (95%) and Venus (97%), and N_2 (79%) and O_2 (21%) on Earth.

The evolution of a planet's atmosphere is strongly affected by its mass and distance from the Sun. Venus' proximity to the Sun may have triggered a runaway greenhouse phenomenon early in its history, which vaporized its oceans and baked carbon dioxide out of the crust (*Ingersoll,* 1969; *Kasting et al.,* 1984). The ensuing dense atmosphere was subject to photolysis and loss of hydrogen to space, resulting in an irreversible loss of water from the planet's surface. The loss of volatiles from the Venus crust would have shut down the hydrological cycle, and likely also inhibited crustal subduction, shutting down plate tectonics (*Phillips and Hansen,* 1998; *Ragenauer-Lieb et al.,* 2001). The loss of plate tectonics would have inhibited the planet's ability to resequester CO_2 into the planet's interior (*Walker et al.,* 1981). Venus' characteristically dense CO_2 atmosphere and fierce surface temperatures are a direct consequence of this terrestrial planet's distance from the Sun, and Venus may represent a common end state for terrestrial planet evolution. With a planet-wide ocean and ~60 bars of CO_2 locked in the crust, Earth could suffer a similar fate to Venus' as the Sun brightens.

Mars, on the other hand, is sufficiently distant from the Sun that it did not suffer a runaway greenhouse, but instead lost its habitability via loss of its atmosphere [see *Jakosky and Phillips* (2001) for a short review of Mars' volatile and climate history]. Mars' rapid atmospheric loss was due primarily to the planet's small mass. The smaller mass was less effective at gravitational retention of the atmosphere against the impinging solar wind. The smaller mass would have provided insufficient radiogenic heating to keep the planet's core molten much past 0.5–1 G.y., and the subsequent loss of the dynamo-generated magnetic field would have also contributed to atmospheric loss.

Earth is unique in the solar system in maintaining habitability over billions of years of evolution. Geological records suggest that Earth has probably remained habitable over the last 3.7 G.y., and climate models predict that it will likely lose its surface liquid water within the next billion years.

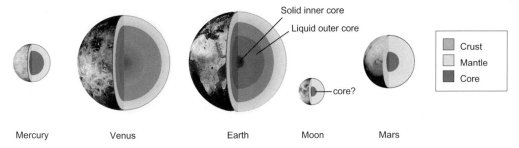

Fig. 2. Internal structure of the solar system's terrestrial planets. Earth and Venus are believed to have similar internal structures. Mercury differs from the other terrestrial planets by having a disproportionately large iron core to silicate mantle ratio. Image credit: NASA.

On the very early Earth, energy from radioactive decay and gravitational accretion would have driven enhanced volcanism and the release of material from the deep Earth. Volcanos predominantly outgas H_2O and CO_2, with traces of N_2 and sulfur gases. Outgassed water would have cooled and condensed to form the ocean, in which the CO_2 and sulfur gases could dissolve, leaving N_2 as the dominant atmospheric gas. Earth has maintained this secondary atmosphere via gravitational retention against escape, and magnetic field protection against solar wind ablation. Earth's atmosphere is vital to its habitability. The atmosphere maintains surface pressure, which retains Earth's oceans, and the relatively dense atmosphere buffers against large day/night temperature changes, and raises the mean surface temperature above the freezing point of water, via the greenhouse mechanism. The atmosphere also contains ozone, which blocks UVC radiation in the range 200–290 nm, and provides a partial shield against DNA-damaging UVB (290–320 nm) radiation. Earth's mass also preserves habitability by providing radiogenic heating to drive plate tectonics over billions of years. Plate tectonics is an integral part of the carbonate-silicate cycle, a planetary process that relies on temperature-dependent sequestration and release of CO_2 to buffer global temperature over geological timescales (*Walker et al.*, 1981).

The early Earth also serves as an explorable example of different types of habitable planets. Although past environments on Earth cannot be observed directly, geological and biological constraints on earlier planetary conditions exist (e.g., *Rye et al.*, 1995, and references therein; *Gaucher et al.*, 2003). Abundant O_2 at atmospheric levels in excess of 10% were not widespread on our planet until ~2.3 b.y. ago (Ga) even though life is believed to have been present on Earth's surface from at least 3.7 Ga (*Catling and Kasting*, 2007) (Fig. 3). Earth prior to 2.3 Ga therefore provides an example of a habitable planet with a biosphere that was not dominated by oxygenic photosynthesis, as it is today (see Fig. 3).

1.2. Remote Sensing Characterization of Exoplanetary Environments

Our only means of studying terrestrial exoplanet environments will be via astronomical remote sensing techniques. Remote sensing studies of solar system planets have made use of photometers and spectrometers on groundbased telescopes, as well as flyby and orbiting spacecraft. Using radiation emitted by or reflected from solar system planets, it is often possible to determine surface temperature and pressure, atmospheric temperature profiles, surface composition and albedo, atmospheric trace gas abundances and distributions, and the presence and nature of clouds and aerosols. Many of these properties have been subsequently corroborated or refined via "ground truthing" from measurements made by *in situ* planetary exploration probes, landers, or rovers.

1.2.1. Observational challenges. Distant terrestrial exoplanets will pose special observational challenges due to the relative faintness of the planet with respect to its parent star. At visible wavelengths an Earth-like planet in the habitable zone, seen in reflected light, would typically be 10^{10} times fainter than its parent star. At mid-infrared wavelengths the planet is seen in emitted radiation, and the contrast ratio between planet and star drops to $\sim 10^7$. Telescopes that hope to detect terrestrial exoplanets must suppress the light from the parent star well above these contrast ratios to provide adequate signal to noise (S/N) for detection, and must simultaneously provide sufficient angular resolution to separate the star and planet so that photons can be collected directly from the planet. The observational challenges will likely lead to very limited S/N on the targets and/or low spectral resolution, at least for the first generation of exoplanet characterization missions.

A fundamental constraint on future terrestrial planet observations is that even the next generation of telescopes will be unable to spatially resolve the planet, seeing it instead as a disk-integrated point source. This limitation has two consequences: no direct constraints on planetary size, and no spatial resolution with which to explore the planetary environment. Without planetary size, our ability to quantify planetary albedo, emissivity, and effective temperature is limited. Planetary size cannot be measured unless the planet is transiting the host star (although observations in reflected and thermal radiation can be used to determine size).

The lack of spatial resolution on the planet's disk will also pose challenges. A single observation will effectively integrate all viewing geometries, solar illuminations, and in-

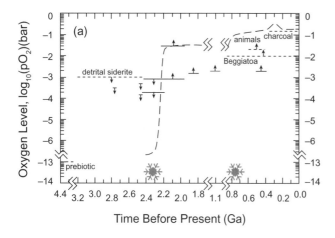

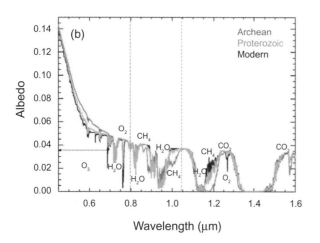

Fig. 3. (a) The history of Earth's atmospheric O_2. The thick dashed line shows a possible evolutionary path for atmospheric O_2 that satisfies geochemical and biological constraints, and modeling results. In the Hadean and Archean periods (prior to 2.4 Ga) the partial pressure of O_2 was extremely small ($\sim 10^{-13}$–10^{-7} bar). Between 2.3 and 2.2 Ga, O_2 rose to at least 10% of the present atmospheric levels. Snowflake symbols indicate the occurrence of episodes of low-latitude glaciation (Snowball Earth events), which are broadly correlated with the atmospheric oxygen history [see Fig. 4.6 of *Catling and Kasting* (2007), and see this reference for more detailed discussion]. (b) Simulated visible-near-infrared spectra of Earth as seen from space during the Archean, Proterozoic, and Modern eras. The Archean differs markedly from the other two spectra due to its low atmospheric oxygen, and therefore ozone abundance, and its relatively high methane and carbon dioxide abundances (*Meadows*, 2006).

homogeneities on the visible disk into a point source, and will not allow us to make specific soundings over clear or cloudy regions, as can be done for planets in our solar system. This will limit our ability to determine surface and atmospheric composition, atmospheric temperature structure, and accurate surface temperatures. The presence of clouds will also truncate the atmospheric column that can be observed, and will limit our ability to sense the surface at some wavelengths. The characteristics of the planet must be disentangled from this disk-integrated view using wavelength or temporal resolution.

1.2.2. Conceptual challenges. Finally, we face a challenge even greater than the observational ones described above. Given the likely diversity of terrestrial exoplanets, there is a very high probability that the planets we attempt to study will have environments that are completely unlike those of the terrestrial planets in our solar system. In these situations, the history of exploration of planets in our solar system warns us that there is a high risk of data misinterpretation. Limited or poor S/N data and limited wavelength coverage may lead us to miss important discriminators, or to seek out and find similarities with known planetary properties, when further measurements may reveal that indeed there are none. We may also be unable to accurately interpret the data because of missing fundamental knowledge of the behavior of atmospheric and surface constituents under the physical and chemical conditions found in the alien environment.

The challenges to remote sensing characterization of terrestrial exoplanets will be addressed with technological advancements in astronomical instrumentation, and an improved understanding of the potentially observable characteristics of terrestrial exoplanet environments via analog observations of solar system planets and theoretical modeling.

2. CONCEPTS AND MODEL DESCRIPTIONS

Theoretical models of planetary environments, extrapolated from our understanding of the solar system's planetary atmospheres and surfaces, allow us to explore the potential range of terrestrial exoplanet environments for different initial compositions and parent stars. Models also allow us to simulate observed planetary characteristics in the disk-average for a range of viewing geometries and solar (stellar) illuminations. The insight gained through modeling can aid in defining instrumentation capabilities for detection, and in minimizing characterization ambiguities from limited observational data. In this section, we provide an overview, focused on remote sensing detectability of terrestrial planet environmental characteristics, and describe the models required to simulate environments and spectra for terrestrial exoplanets.

2.1. Characteristics and Processes in Terrestrial Planet Atmospheres

In this section we briefly describe major components and characteristics of terrestrial planet atmospheres including total atmospheric pressure, atmospheric temperature structure as a function of pressure, atmospheric composition, the effect of clouds and aerosols, and atmospheric escape. Where applicable, we will use examples from terrestrial atmospheres in our solar system to illustrate key points.

2.1.1. Total atmospheric pressure. The atmospheric pressure at the solid surface of a terrestrial planet is predominantly a function of the total mass of the atmosphere and secondarily a function of the mass of the planetary body, which affects the acceleration due to gravity. The global mean surface pressure, P_s, can be determined from the definition of force per unit area

$$P_s = \frac{gm_a}{4\pi R_p^2} \quad (1)$$

where m_a is the total mass of the atmosphere and R_p is the radius of the planet. Although higher-mass planets are more likely to retain more massive atmospheres against progressive loss, this simplistic relationship is often undermined by planetary history and evolutionary processes. For example, Venus is slightly smaller and less massive than Earth, yet retains an atmosphere nearly 100 times as massive as Earth's. This is due to the dramatic evolution of the Venus atmosphere via the runaway greenhouse effect, which likely released a large fraction of the solid planet's CO_2 stores into the atmosphere.

Pressure, P, at any altitude in the atmosphere, z, is caused by the weight of the overlying atmosphere. When compression due to gravity is balanced by gas pressure, the atmosphere is said to be in hydrostatic equilibrium. Assuming hydrostatic equilibrium, and using the ideal gas law, the atmospheric pressure decrease with altitude, z, can be expressed as

$$P = P_0 e^{\left(\frac{-z}{H}\right)} \quad (2)$$

where P_0 is the pressure at $z = 0$ (the surface pressure) and H, the scale height, denotes the altitude range over which the pressure drops by a factor of e. The scale height for an ideal gas is given by

$$H = \frac{kT}{Mg} \quad (3)$$

where k is Boltzmann's constant, T is the temperature of the gas, M is the mean molecular mass of the gas, and g is the acceleration due to gravity. Consequently, scale heights are larger (and atmospheres are "puffier") when the temperature of the atmosphere is higher, or when the mean molecular mass and/or the planet's acceleration due to gravity are lower. A small scale height implies a rapid decrease in pressure with altitude.

2.1.2. Atmospheric temperature structure. The temperature structure of a planet's atmosphere refers to the pattern of change in temperature with altitude above the planetary surface. An atmosphere's temperature structure will depend on the spectral energy distribution of the star irradiating it, as well as physical and chemical processes within the atmosphere that modify opacity and energy transport. Terrestrial planetary atmospheric structure is typically broken up into five main regions: troposphere, stratosphere, mesosphere, thermosphere, and exosphere (Fig. 4). Although this terminology was developed for Earth, it is commonly applied to other planets.

The troposphere starts at the planetary surface and ends at the tropopause, which is typically at a temperature minimum. The very lowest portion of the troposphere (100 m from Earth's surface) is often referred to as "the planetary boundary layer." The troposphere will typically contain the bulk of the planetary atmosphere (80–98% of the total atmospheric mass) and the majority of a planet's clouds and aerosols. For terrestrial atmospheres with some transparency in the visible, the troposphere is heated primarily by the absorption of visible stellar radiation at the planetary surface, which is reradiated at mid-infrared wavelengths, and this energy is transported throughout the troposphere via convection (hence "tropos," or turning). Tropospheres therefore monotonically cool with altitude, and are often at or close to the adiabatic lapse rate. The colder temperatures near the tropopause can act as a "cold trap," ensuring that volatiles such as water condense and rain out, remaining within the troposphere.

The stratosphere starts at the tropopause, and energy is transported via radiation, rather than convection. This stratified region, if present (Venus and Mars both lack stratospheres) has a temperature structure that monotonically increases with altitude. On Earth and Titan this is due to the presence of molecular species that absorb solar UV radiation. On Earth, ozone is formed, and preferentially survives, at water-depleted stratospheric altitudes, and UV absorption by the ozone is responsible for stratospheric heating.

Above the stratopause in Earth's mesosphere, temperature decreases with altitude as O_3 production decreases, and the overlying atmospheric opacity is sufficiently attenuated that CO_2 can radiate efficiently to space, cooling the atmosphere. Earth's atmosphere produces a second temperature minimum at the mesopause. This behavior is confined to Earth and Titan. Other massive atmospheres in our solar system show only one temperature minimum at the tropopause, and a more isothermal mesosphere.

Above the mesopause, in the thermosphere absorption of solar radiation shortward of 0.1 μm causes the temperature to increase with altitude. In Earth's thermosphere UV radiation heats the upper atmosphere primarily via photolysis and ionization of O_2. The heating is also aided by there being too few atoms or molecules available to emit infrared radiation and cool the atmosphere, and by the principle radiators being NO and O, which are less efficient than CO_2 at atmospheric cooling. In contrast, Venus has a thermosphere that is much cooler than Earth's, despite its closer proximity to the Sun and higher incident UV. This is attributed to larger concentrations of CO_2 in its thermosphere, which is a more efficient coolant. Much of a planet's thermosphere is in radiative equilibrium except in the very tenuous upper regions. Here energy is transferred by collisions between particles (conduction), which leads to a near-isothermal temperature structure. Thermospheres can have diurnal temperature swings of many hundreds of Kelvin.

The outer part of a planetary atmosphere is referred to as the exosphere. At its lower-altitude limit, the exobase, a particle's mean free path (the average distance between collisions) is greater than the particle's scale height, and the atmosphere in effect becomes collisionless, allowing the escape of atmospheric molecules to interplanetary space.

2.1.3. Atmospheric composition. The atmospheric composition of a planet is governed by its initial inventory of volatiles, and subsequent evolution via chemical processes

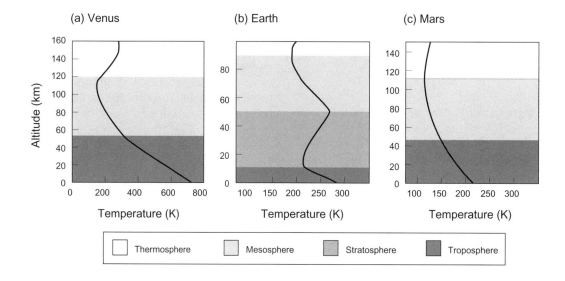

Fig. 4. The atmospheric temperature structure of **(a)** Venus, **(b)** Earth, and **(c)** Mars. Note that Earth's ozone produces a stratosphere, which Venus and Mars do not have. The stratosphere is characterized by rising temperatures from the tropopause near 12 km altitude up to the base of the mesosphere at 50 km altitude. Image credit: F. Bagenal.

(discussed here) and atmospheric loss processes (discussed in section 2.1.6 below).

Chemical and photochemical alteration modify the vertical and latitudinal distribution of atmospheric gases, and photochemistry is dependent on the spectrum of incoming stellar radiation. The vertical distribution of a gas is controlled by its photochemical or geological sources, its transport, and its photochemical loss or reaction with the planetary surface. The balance between these processes can result in very different vertical profiles for constituents of a planetary atmosphere, as shown in Fig. 5. Here we will focus on examples from Earth's atmosphere, where these processes are best understood.

Bulk gases, such as O_2 and N_2 in Earth's atmosphere, are often evenly mixed, meaning that they display the same mixing ratio, e.g., in parts per million by volume (ppmv) throughout the troposphere, stratosphere, and mesosphere. Evenly mixed gases are valuable for temperature retrievals, as the modification in the profile of an absorption band from an evenly mixed species can be interpreted as being due to the atmospheric temperature structure, and not due to changes in species abundance with altitude. However, neither O_2 nor N_2 have strong absorption bands in the mid-infrared, and so temperature structure retrievals are often performed with CO_2. CO_2 is a bulk gas and evenly mixed in the atmospheres of Venus and Mars. It is a trace gas in Earth's atmosphere and yet remains evenly mixed due to its relative unreactivity. Because the majority of its sources are surface sources like volcanism, it shows a small degree of variability in Earth's planetary boundary layer and then remains evenly mixed until it photodissociates at altitudes above 90 km.

The majority of trace gases, however, are not evenly mixed. The vertical distribution of water vapor is strongly affected by its tendency to condense to form clouds and fog, and so it is not evenly mixed below 15 km. The mixing ratio of water vapor is ~1–3% at the surface (10,000–30,000 ppm), with the high end of the range typically found in the tropics, and decreases thoughout the troposphere, reaching a minimum of about 4 ppm (relative humidity of about 1%) at altitudes above 15 km. This is primarily due to the cold trap at the tropopause freezing out and precipitating water. Above the cold trap the abundance of water vapor increases slightly due to the oxidation of methane in the stratosphere. At altitudes above 80 km, water is readily photolyzed, and the mixing ratio drops rapidly with increasing altitude.

Ozone also has an interesting vertical distribution, confined primarily between the tropopause and 40 km altitude, with a maximum in the stratosphere near 17–25 km altitude. O_3 is formed photochemically from oxygen, a process that is simply represented with four reactions called the Chapman cycle. O_2 is first photolyzed by UV radiation at wavelengths shortward of 240 nm. The resulting oxygen atoms quickly react (typically in less than a second) with O_2 to form ozone via a three-body reaction

$$O_2 + O + M \rightarrow O_3 + M \tag{4}$$

where M represents any other molecule, and is typically the abundant N_2 or O_2 in Earth's stratosphere. The M molecule carries away excess energy from the reaction, and the reaction effectively converts solar radiation into thermal heating of the stratosphere. Ozone loss can proceed via the reaction

$$O_3 + O \rightarrow O_2 + O_2 \tag{5}$$

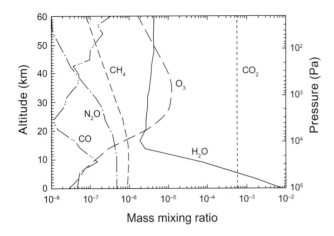

Fig. 5. Mixing ratio (abundance) as a function of altitude for trace gases in Earth's atmosphere showing the evenly mixed trace species CO_2, and others whose concentration changes significantly with altitude (e.g., O_3 and H_2O) due to atmospheric and surface sources and sinks. Image credit: T. Robinson, University of Washington.

This reaction is relatively slow in Earth's atmosphere because the overall abundance of O_3 is relatively low, especially when compared with O_2, the competing reactant. While O_3 can also be lost via reactions with catalysts such as Cl, Br, and OH, the net effect is the overall production of O_3 with incoming solar radiation. The ozone layer forms in a distinct altitude region of the stratosphere, bounded by decreasing atmospheric density above 60 km altitude, which slows the three-body reaction, and at decreasing altitudes by O_2 self-shielding of 240-nm photons.

CH_4 and N_2O are two other gases that are not evenly mixed vertically in Earth's atmosphere. Both have strong sources at the planet's surface from the surface biosphere and in the case of CH_4, from hydrothermal vents. Both constituents remain evenly mixed throughout the troposphere, but are photochemically destroyed in the stratosphere and mesosphere.

Finally, several of the trace gases that are produced by land surface processes such as vegetation, biomass burning, and anthropogenic activities show hemispherically asymmetric distributions. Typically, gases such as CH_4, CO_2, and CO have slightly higher atmospheric concentrations in the land-dominated northern hemisphere than in the ocean-dominated southern hemisphere.

2.1.4. Planetary energy balance and surface temperature. The equilibrium effective temperature of a terrestrial planet is given by the radiative balance created between incoming radiation from the parent star and outgoing radiation emitted by the planet. Equating the radiative equilibrium temperature with the planet's effective temperature assumes that the terrestrial planet is heated via solar radiation, and not significantly heated via internal energy sources such as radioactive decay or tidal energy, or energy of formation or contraction, as would be the case for a jovian planet. For Earth, this assumption is valid as the surface flux of energy due to radioactive decay in the planet's interior is 0.086 W m^{-2}, compared to intercepted solar flux of 340 W m^{-2}. For jovian planets, which have significant internal energy sources, T_{eff} is often a far more complicated function of planetary mass, age, and composition, requiring detailed model calculations to determine.

For a terrestrial planet, the rate at which stellar energy is absorbed is given by $F_\star \pi R^2 (1-A_B)$, where $F_\star$ is the stellar irradiance at the planet's orbit (in W m^{-2}), πR^2 is the area of the planet of radius R intercepting the radiation, and A_B is the albedo of the planet, a measure of the planet's reflectivity, defined as the ratio of the outward to inward flux of stellar radiation. $1-A_B$ is therefore the fraction of radiation absorbed by the planet. Assuming the planet radiates as a blackbody, then the rate of emission of the planet over its entire surface area is given by $4\pi R^2 \sigma T_e^4$, where T_e is the effective temperature of the planet. If we equate the rate at which energy is absorbed by the planet with the rate at which it radiates to space, then we have the energy balance equation

$$\sigma T_e^4 = \frac{F_\star (1-A_B)}{4} \qquad (6)$$

Using this formula to calculate the effective temperature of Earth gives 255 K. This is also the likely average temperature retrieved from mid-infrared observations of Earth averaged over the thermal infrared. This is significantly lower than Earth's average *surface* temperature (288 K) but close to the average temperature of the atmosphere, indicating that the majority of Earth's radiation is lost to space from the atmosphere, rather than from the surface.

The discrepancy between a planet's effective temperature and its surface temperature is a measure of the planet's greenhouse effect. On Earth, this difference is 32 K, but on Venus, the greenhouse effect produces a surface temperature 513 K hotter than the planet's effective temperature. A terrestrial exoplanet's surface temperature then is not simply a function of its size and distance from the parent star, but is largely dependent on the mass and composition of the planetary atmosphere.

2.1.5. Effects of clouds and aerosols. Condensible species in a planet's atmosphere can produce clouds or hazes. Clouds in the troposphere are associated with convection and condensation of a volatile species, like the water ice clouds seen in the atmosphere of Earth or the CO_2 clouds seen on Mars. Hazes can also be formed via photochemistry in both the troposphere and stratosphere. When formed in the stratosphere, photochemical haze layers are often planet-wide. This is the case for the atmospheres of Venus and Titan, which are dominated by sulfuric acid and hydrocarbon hazes respectively.

Clouds are formed by atmospheric gases at altitudes where the temperature drops sufficiently that the partial pressure of the gas exceeds its saturation vapor pressure and it condenses or freezes out. On Earth, water clouds are formed by rising moisture-laden air that condenses, and typically can be found from the surface (fog) up to 18 km altitude, depending on the location of the tropopause. On Mars, both water and CO_2 ice clouds may be present, forming near 10 km and the colder 50 km altitude respectively. On Venus, however, the "cloud" particles are produced photochemically at high altitudes (near 80–90 km) via a UV-driven reaction with sulfur dioxide, water, and atomic oxygen from the photolysis of CO_2. These particles remain small at higher altitudes but can collide and coagulate to form larger particles at lower altitudes. This results in a photochemically produced planet-wide cloud deck from 45 to 70 km altitude with fine haze above it up to 90 km.

Clouds can both reflect and absorb radiation, and can affect the planetary energy balance by producing net cooling or warming effects under different conditions. Perhaps the most intuitive effect of both water and CO_2 clouds is to enhance reflectivity in the visible, which increases the planetary albedo. On Earth, low and middle liquid water clouds are strong visible-light reflectors but poor absorbers in the visible and infrared, and so provide a net cooling effect. However, high ice clouds, which are often thin, allow visible light to reach the surface and absorb more infrared radiation than liquid water, providing a net heating effect. At night, when clouds are not illuminated and cannot enhance the planet's albedo, they instead serve as greenhouse warming agents, by absorbing escaping infrared radiation and reradiating it back toward the surface. Thick CO_2 ice clouds behave slightly differently from low water clouds in that although they also significantly increase the albedo of a planet, CO_2 ice is more efficient at scattering escaping infrared radiation back toward the planetary surface, and so can potentially produce a net warming of the atmosphere, despite its high visible reflectivity. This CO_2 blanketing behavior has been invoked as a potential means of both warming early Mars (*Forget and Pierrehumbert*, 1997) and of extending the outer limits of the habitable zone.

Cloud formation can also affect climate, and atmospheric evolution, by modifying the tropospheric lapse rate (the rate at which temperature decreases with altitude). When a cloud forms, a gas condenses or freezes into the liquid or solid state, thereby releasing latent heat into the surrounding atmosphere. Condensation and latent heat release reduces the rate at which moist air parcels cool as they rise, reducing the atmospheric temperature lapse rate. One consequence of this in a water-rich troposphere is that the tropopause cold trap gets pushed higher into the atmosphere. This is why Earth's tropopause is much higher in the tropics than at the poles. In an extreme case a high tropopause may eventually allow water to rise to an atmospheric region where it is vulnerable to UV photolysis and subsequent escape. This mechanism is the core of the "moist greenhouse" phenomenon, which may ultimately be responsible for Earth's loss of water, and habitability, as the Sun warms.

2.1.6. Atmospheric escape and evolution. Atmospheric escape, the progressive loss of a planet's atmosphere to space, is a long-term process, and is typically most relevant to the compositional and mass evolution of planetary atmospheres over millions to billions of years. For terrestrial planets extremely close to the parent star (hot Earths) atmospheric escape may strip the entire atmosphere in considerably less than a gigayear (*Lammer et al.*, 2009). In our solar system, atmospheric escape is one of the key processes that control the bulk and trace gas composition of a planetary atmosphere.

Atmospheric escape occurs in three stages: upward transportation of a gas from the lower to upper atmosphere, conversion in the upper atmosphere from a molecular to more easily lost atomic or ionic form, and escape to space via a number of possible loss mechanisms. Any one of these stages can be the limiting process for atmospheric escape.

The loss processes can be categorized into three basic types: thermal hydrostatic escape, thermal hydrodynamic escape, and nonthermal escape. Which escape process dominates on a given solar system planet depends on the planet's mass, upper atmosphere composition, distance from the Sun and consequent planetary exospheric temperature, and presence of a planetary magnetic field.

2.1.6.1. *Thermal (hydrostatic) escape:* This process occurs when the thermal velocity of an atom or molecule exceeds the escape velocity of the planet. This escape occurs in the planetary exosphere, the uppermost region of the atmosphere where a particle's mean free path (the average distance between collisions) is greater than the particle's scale height, and the atmosphere in effect becomes collisionless.

The simplest equation describing thermal escape is given as

$$\sqrt{\frac{3k_B T_{exo}}{m}} > \frac{1}{6}\sqrt{\frac{2GM_p}{R_p}} \quad (7)$$

Here, k_B is Boltzmann's constant, T_{exo} is the exospheric temperature, m is molecular or atomic mass, G is the gravitational constant, M_p is the planet mass, and R_p is the planet radius. This simple equation shows that thermal escape is more likely to occur for lower-mass species and hotter exospheres, and will be slowed for larger-mass planets and planets of higher density (with larger M/R ratios).

The thermal velocities of particles of a given mass are not, however, identical, and instead follow a Maxwellian distribution in which the high-energy tail of particles is more likely to escape. Slower atoms and molecules then move to fill the high-velocity tail. This process is called "Jeans escape." The Jeans escape flux can be calculated by integrating the Maxwellian velocity distribution over the upward hemisphere for escape velocities consistent with the exobase temperature.

The Jeans escape flux is typically given in units of $cm^{-2} s^{-1}$ and can be expressed as

$$\Phi_{Jeans} = \frac{n_c}{2\sqrt{\pi}} B \sqrt{\frac{2k_B T_c}{m}} (1 + \lambda_c) e^{-\lambda_c} \quad (8)$$

where n_c is the number density of particles at the exobase, and the subscript c refers to exobase properties. The factor B effectively slows down the escape by accounting for the slow repopulation time of the energetic tail of the Maxwellian distribution, and has a value on the order of 0.5 to 0.8. The escape parameter λ_c is

$$\lambda_c = \frac{GM_p m}{kTr_c} = \frac{r_c}{H} \quad (9)$$

or can also be described by $\lambda_c = E_{esc}/kT_c$, where the escape energy $E_{esc} = 1/2 mv_{esc}^2$. Here M_p is the planetary mass, r_c is the radial distance from the planet center, and G is the universal gravitational constant.

Note that thermal escape requires replenishment of the escaping gas and its velocity distribution, and so is "diffusion limited," depending on the rate of upward diffusion from the lower atmosphere for the species that is escaping.

2.1.6.2. *Hydrodynamic escape:* This is a much faster escape process in which the atmosphere behaves like a dense fluid expanding radially outward. Hydrodynamic escape occurs when the upper atmosphere of a planet is heated by large amounts of stellar EUV radiation such that the particles are so energetic that gravity cannot stop the outward flow. During hydrodynamic escape, heavier elements such as C, N, and O can be carried away with the hydrodynamic flow of hydrogen.

The Jeans escape parameter (λ_c) helps to define whether an atmosphere is in the hydrostatic or hydrodynamic escape regime with a value of one roughly discriminating between the two cases. $\lambda_c \gg 1$ is safely in the Jeans escape regime, and small values of λ_c are necessary but not sufficient to drive hydrodynamic flow. Hydrodynamic escape is limited by the energy available to drive it, but the transition between the hydrostatic and hydrodynamic regime is controlled by adiabatic cooling in the energy budget. When the stellar radiation is not able to overcome the atmospheric cooling rate, a negative temperature gradient (decreasing T with increasing altitude) develops at the top of the atmosphere. This negative temperature gradient is a macroscopic signature of hydrodynamic escape.

2.1.6.3. *Nonthermal escape:* This term generally refers to collisional processes between charged species that produce atoms energetic enough to escape from a planetary atmosphere to space. These collisional processes can enable the escape of heavier atoms such as N, C, and O, whose thermal velocities are too small for thermal escape. In general, a wide variety of processes are included under the nonthermal escape umbrella. Which nonthermal process dominates or even plays an important role depends on many different planet-specific factors, including the exospheric temperature, the presence of a planetary magnetic field, the planet-star separation, the host star type, and especially the planet's escape velocity. For more details on nonthermal escape processes and a summary of numerical models, see the excellent review articles by *Hunten* (1982) and *Shizgal and Arkos* (1996).

Briefly, these processes are:

Charge exchange: Charge exchange involves a collision between a photodissociated and magnetically accelerated ion and an atom, where the excess energy of the ion is transferred to the atom. Charge exchange from the H^+ ion to the H atom is thought to be Earth's dominant hydrogen-loss process, even faster than Jeans escape of hydrogen.

Conversion of photochemical energy into kinetic energy: Electrons created from photoionization, atoms created from molecular photodissociation, and recombination to an excited state can produce atmospheric species that carry excess kinetic energy, allowing them to escape.

Ion escape or "polar wind": Ion escape refers to ions that escape along magnetic field lines that are open to a planetary magnetotail. On Earth, ion escape is an important process in the polar regions.

Sputtering and knockon: Sputtering and knockon occur when atoms or ions impact onto either a solid surface or an upper atmospheric layer. In sputtering, the impacts generate a "backsplash" of escaping atoms, whereas in knockon, the impact generates a forward acceleration that leads to energetic particles heading back out of the planet's atmosphere.

Pickup: Pickup occurs when ions in the planetary atmosphere are "picked up" by the solar-wind-generated interplanetary magnetic field, and accelerated away from the planet. Strong planetary magnetic fields, such as Earth's, protect the atmosphere from escape by deflecting the solar wind. In contrast, in planets that lack a strong magnetic field, such as Venus, the solar wind can penetrate deep into the planet's exospheres, ionizing neutral species and sweeping them away.

2.1.6.4. *Atmospheric evolution via escape processes:* All the above escape processes have influenced the composition and evolution of solar system planet atmospheres in some way (see the review by *Lammer et al.*, 2009), although some of the impacts are still under debate. Thermal escape and possibly hydrodynamic escape may have both been important during Venus' catastrophic runaway greenhouse phase. Interestingly, hydrodynamic escape would have swept up heavier atoms than hydrogen, including deuterium and oxygen, and would have greatly diminished the atmospheric D/H fractionation ratio that is currently used to estimate the water lost from the planet over time (*Kasting and Pollack*, 1983). Atmospheric escape has also heavily modified Mars' atmosphere over time, initially probably via thermal loss processes from this small planet, followed by significant nonthermal loss processes of hydrogen and heavier atoms via processes such as sputtering and pickup once the planetary dynamo and global magnetic field were lost (*Chassefière*, 1996).

Slow thermal escape to space has been important for Earth during its evolution, with the continual loss of light gases from Earth's atmosphere. This loss rate is an important photochemical sink. For example, on Earth, hydrogen is produced via photodissociation of hydrogen compounds (such as volcanic emissions of H_2O and H_2S). Hydrogen escapes via Jeans escape and the nonthermal charge exchange reaction so that there is no net accumulation of hydrogen in the atmosphere. There is some debate, however, as to how much hydrogen Earth had in its early atmosphere and for how long it persisted before thermal escape (*Tian et al.*, 2005a). Some researchers have also postulated that the photodissociation and subsequent hydrogen escape from biogenically produced gases such as CH_4 may have helped to accelerate the overall oxidation of Earth's early atmosphere (*Catling et al.*, 2001). On exoplanets with a higher escape velocity than Earth, such as super Earths, H_2 may accumulate, producing a very different photochemical balance in the planetary atmosphere.

Applying planetary escape processes to our understanding of the lifetimes and compositions of terrestrial exoplanets is challenging, due to the largely unknown properties of exoplanet atmospheres and host stars. The extreme ultraviolet (EUV) radiation from host stars, required to heat the upper atmosphere for thermal or hydrodynamic escape, is difficult to measure, or to predict for earlier stages of the star's history, as is the stellar wind, which plays a significant role in nonthermal escape mechanisms. The evolution of stellar EUV radiation can be approximated using $(t_0/t)^{5/6}$, where t_0 is the present time (*Zahnle and Walker*, 1982; *Ribas et al.*, 2005), which encompasses the very early periods of stellar evolution in which stellar EUV levels can reach 1000 times the "midlife" amount. This is particularly important for terrestrial planets around M stars, for which the peak early activity phase can last from 1 to 8 G.y. An initial atmosphere lost during the high-activity phase could only redevelop during the star's quiet phase when the outgassing rate to replenish the atmosphere is higher than the loss to escape. For calculations of atmospheric lifetimes, the initial outgassed planetary atmospheric mass is largely unconstrained and could span 1–20% of the planet's total mass depending on the planet's initial water inventory (*Elkins-Tanton and Seager*, 2008).

2.2. Remote Sensing of Planetary Atmospheres

Exoplanets will reflect stellar radiation in the visible and near-infrared, and will emit thermal radiation in the near- and mid-infrared. Detecting and measuring this radiation provides a probe of the planet's atmospheric and surface properties. These include the planet's bulk and trace atmospheric composition, its atmospheric and possibly surface temperatures, the presence and nature of clouds, and the behavior of its environment as a function of time.

2.2.1. The bulk atmosphere. To characterize a planet's environment and habitability, the most fundamental properties are the presence of an atmosphere and the atmospheric composition and mass.

The presence of an atmosphere could be determined from characteristic spatial or time-dependent temperature behavior, such as a reduced day-night temperature gradient, or from spectral detection of atmospheric gases. The bulk components of Earth's atmosphere, N_2 and O_2, are relatively difficult to detect remotely, as they have few spectral features, and many of these are quite narrow. In contrast, Earth's trace gases H_2O and CO_2 have many broad spectral features from the UV to the far-infrared (Fig. 6). The CO_2 15-μm band is seen prominently in the mid-infrared spectra of Venus, Earth, and Mars, and additional bands near 1.6 μm and 2.4 μm may be seen in dense atmospheres in which CO_2 is a bulk constituent (e.g., *Segura et al.*, 2007). Water vapor has many strong, broad bands throughout the visible and mid-infrared spectral regions, including bands at 0.94, 1.1, and 6.3 μm. Water also absorbs from 20 to 150 μm, but its presence is difficult to disentangle from temperature effects at low spectral resolution. To detect atmospheres that are reducing, rather than oxidizing, gases such as ethane, which has strong bands near 6.8 and 12.2 μm, and methane, which absorbs in the visible (0.86 μm), near-infrared (1.67 and 2.3 μm) and mid-infrared (7.7 μm), could be searched for.

Super Earth atmospheres detected via transit transmission spectroscopy are likely to be the first confirmed terrestrial exoplanet atmospheres. Transit transmission observes the backlit atmosphere of a planet as it transits the face of its parent star. For this technique, "puffy" atmospheres with large-scale heights are easier to detect (*Miller-Ricci et al.*, 2008). Strong molecular bands, such as those from CO_2 and H_2O, can be optically thick in transmission over many scale heights, so larger scale height atmospheres will produce larger absorption signals in transmission spectra. The largest scale heights are produced by hot, low mean molecular weight atmospheres, such as hydrogen-rich atmospheres, on planets with low surface gravities. In contrast, terrestrial planets with high bulk density and thin atmospheres of high molecular weight will have smaller atmospheric scale heights. Although it will be challenging, *Deming et al.* (2009) have shown that it may be feasible to observe transmission spectra for a subset of super Earths.

Quantifying atmospheric pressure and total atmospheric mass will be extremely difficult due to the spectral invisibility of many bulk atmospheric gases and the presence of clouds, which can shield a troposphere from remote investigation. Atmospheric mass is required to quantify trace gas mixing ratios for studies of atmospheric chemistry, and plays a key role in climate modeling of surface temperature due to greenhouse warming. For relatively transparent atmospheres (e.g., Mars, Earth) radiation escapes from at or near the surface, allowing the total or near-total mass of the atmosphere to be derived from the planet's spectrum. For opaque atmospheres (e.g., Venus or the jovian planets), only the atmospheric mass above a cloud deck or another opaque layer can be determined. Observations at different wavelengths can probe different levels of an atmosphere, thereby providing additional constraints on the vertical variation of pressure.

Atmospheric pressure can be sought using a number of challenging techniques. Modeling the effects of bulk gas broadening of infrared bands of trace absorbers can be used to infer the atmospheric pressure at the emitting level in an atmosphere. Rayleigh scattering, if observed, can also indicate the presence of an atmosphere, but it can be masked by clouds or surface spectral characteristics. Both Venus and Mars do not show a blue Rayleigh scattering tail, because they have strong blue absorbers in their clouds and surface respectively, and the planet-wide Venus clouds truncate the 97-bar atmosphere at a path length of only 30 mbar. The formation of dimers, pairs of molecules held together by van der Waals forces, is pressure sensitive and could also probe atmospheric pressure. For an Earth-like planet, the relative intensities of the oxygen A band and the oxygen dimer $(O_2)_2$

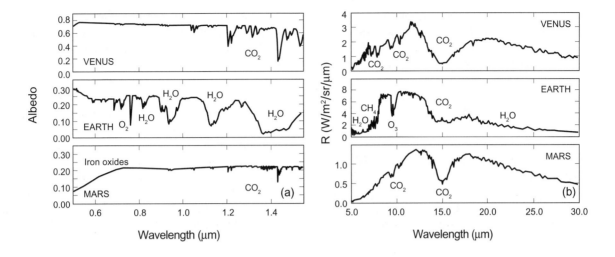

Fig. 6. (a) Visible light and (b) mid-infrared spectra of Venus, Earth, and Mars calculated using a radiative transfer model (*Meadows*, 2006). These spectra show strong absorption from H_2O in Earth's atmosphere, and strong absorption from CO_2 in all three terrestrial atmospheres. Image credit: T. Pyle, Spitzer Science Center.

(0.57–0.60 µm) could be used to constrain atmospheric pressure (*Tinetti et al.*, 2006a).

2.2.2. Surface and atmospheric temperature. The holy grail of habitability is a planetary surface pressure and temperature that is conducive to the existence of surface water in its liquid phase. If liquid water cannot be directly detected, habitability may still be recognized if the planetary surface pressure and temperature can be determined. Surface temperature, however, is arguably one of the most difficult planetary characteristics to directly measure in the disk-average. Thermal-infrared or microwave spectra may return atmospheric and possibly surface temperatures, provided the wavelengths used are relatively free of atmospheric absorption. For planets with cloud cover, the disk-averaged view of the planet will contain a mix of the surface and cloud temperatures, systematically biasing the temperature measurement to be lower than the actual surface temperature.

For an Earth-like terrestrial planet whose atmospheric absorption is dominated by H_2O, CO_2, and O_3 (e.g., Earth, Mars), surface temperature data is best sought in the spectral window between 8 and 12 µm. At these wavelengths, weak water vapor bands and a strong, narrow O_3 band near 9.6 µm are the only significant sources of atmospheric absorption. On either side of the window, strong absorption by H_2O and CO_2 largely preclude surface observations. High water vapor atmospheres (as seen in Earth's tropics) can partially "close" this window, returning a brightness temperature from the atmosphere a kilometer or so above the surface, rather than the actual surface temperature. Atmospheres of different composition may produce a different atmospheric window. In the case of a several-bar CO_2 atmosphere, the atmospheric window is extremely narrow, between 8.5 and 9.0 µm, and CO_2 far-wing absorption precludes detection of the surface temperature even within the atmospheric window (*Segura et al.*, 2007). Finally, surface emissivity must also be estimated to calculate the surface temperature, and this may be difficult to determine.

Because infrared spectral features depend both on constituent abundance and the atmospheric temperature structure, atmospheric temperatures are required to quantify trace gas amounts from thermal radiances, as is the case for Earth's H_2O and O_3. Global-scale constraints on the atmospheric thermal structure are also valuable for studies of the planet's climate and chemical equilibrium. The atmospheric temperature structure can be retrieved from the spectrum of a well-mixed gas with well-characterized absorption features. The CO_2 15-µm band provides the best available constraints on the atmospheric thermal structure for terrestrial planets in our solar system. For planets with H_2-dominated atmospheres, as may be the case for some super Earths, the atmospheric temperature structure could be retrieved from the hydrogen continuum absorption at wavelengths longer than 20 µm.

2.2.3. Surface composition. In determining surface composition, the visible spectral region is often superior to the mid-infrared. This is due in part to the relatively small atmospheric windows to the surface in the mid-infrared range. This problem is compounded by the typical lack of mid-infrared surface spectral features, and the uniformly high mid-infrared emissivity of most Earth surface types (ocean/land/vegetation). One exception is CO_2 ice, which has an abrupt change in emissivity across the 10–12-µm wavelength range, and so might be identified in a disk-average spectrum at mid-infrared wavelengths for a planet with a relatively thin atmosphere, as is the case for Mars (*Tinetti et al.*, 2005).

In contrast, the optical and near-infrared regions of the spectrum display a rich array of spectral features associated with surface composition (Fig. 7) and observations may be able to distinguish between a world dominated by oceans, sand, rock, ice, or clouds. An ice-covered surface may indicate that the planet is not habitable, and determining the composition of the ice could further constrain surface temperature.

2.2.4. Sensitivity to cloud cover. Cloud cover typically results in significant loss of information when attempting to remotely characterize a planet. Cirrus clouds high in Earth's atmosphere obliterate even a strong signal due to ozone at mid-infrared wavelengths, although clouds at lower levels still allow the detection of O_3 (*Des Marais et al.,* 2002). At visible wavelengths, oxygen-rich atmospheres display strong 0.76-μm O_2 A-band absorption, even in the presence of high clouds, because O_2 is evenly mixed. Although the contrast is reduced, the feature may be easier to detect on a cloud-covered planet because the clouds' higher albedo produces a stronger observed signal in both the continuum and depth of the band. However, the cloud truncates the observable column, resulting in a lower limit on the quantification of atmospheric gases.

Although the photochemical hazes that shroud Venus and Titan are opaque at visible wavelengths, they display atmospheric windows at near-infrared wavelengths (*Meadows and Crisp,* 1996; *Smith et al.,* 1996), which allow remote sensing of the lower atmosphere and underlying planetary surface. For Venus, thermal radiation from the hot lower atmosphere and surface escapes through the clouds and can be detected on the nightside of the planet. In the case of Titan, the haze is sufficiently transparent at near-infrared wavelengths that the surface can be detected even when the moon is fully illuminated.

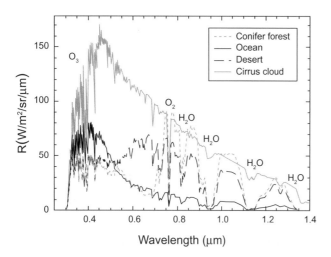

Fig. 7. Surface reflectivity spectra. This diagram shows simulated radiance spectra along lines of sight through Earth's atmosphere over surfaces of different types. In all cases, the incoming spectrum of sunlight was the same, but the radiation reflected back to the observer is quite diverse. This behavior allows us to understand surface composition in remotely sensed data. Image credit: D. Crisp and R. Hasler.

2.2.5. Phase and seasonal variations in spectral features. As with time-variable photometry, monitoring how different spectral features change with time may help us infer spatial information in disk-averaged observations (e.g., the presence of continents or oceans), and may also reveal variations in surface and atmospheric composition that are linked to day-night or seasonal variations (e.g., seasonal ice caps or dust storms). Variations in the retrieved atmospheric temperature or pressure could potentially also yield information about global-scale weather systems, or seasonal variations in the total atmospheric mass, as is the case for Mars. Information on changes in surface characteristics, temperature, and climate over the course of an orbit are important for understanding the potential habitability of a planet.

2.3. Models for Terrestrial Planet Atmospheres

To interpret data taken for terrestrial planets, and to explore the phase space for plausible terrestrial exoplanet environments, planetary scientists often employ computer models to generate environmental parameters and synthetic spectra for the planets under study. For terrestrial planets, these models can be used to explore four major aspects of the planetary environment: atmospheric photochemistry, climate, radiative transfer, and atmospheric escape. While often represented by stand-alone models, these different classes of physical processes are intimately linked in the planetary environment, and are therefore often linked in modeling efforts. For example, photochemical models must also account for the effects of radiative transfer and escape processes on the chemical composition of an atmosphere, and use a vertical temperature structure that is often obtained from a climate model.

Below we describe the inputs, outputs, and fundamental equations for each of these basic types of models.

2.3.1. Photochemical models. Stellar radiation can break the bonds of chemical species (e.g., $H_2O + h\nu \rightarrow H + OH$) driving nonequilibrium photochemical reactions in planetary atmospheres. A photochemical model calculates the spatial distribution and atmospheric concentration of chemical species, and is used to understand the factors that control this distribution. At different regions in the atmosphere, chemical production and destruction, atmospheric transport, and radiative transfer are computed, subject to boundary conditions such as surface fluxes (e.g., from volcanos or biology), surface sinks (e.g., rain, dry deposition), and escape from the atmosphere at the upper boundary.

Input to the photochemical model includes a starting atmospheric composition, the vertical temperature and pressure structure, and the stellar radiation incident at the top of the planetary atmosphere (irradiation). One-dimensional atmospheric photochemistry models calculate the vertical profile of constituents in an atmosphere, while three-dimensional planetary photochemical models calculate the horizontal spatial distribution of constituents as well.

Chemical species concentrations are controlled by four types of processes: surface fluxes emitted by geological or biological sources; photochemical reactions that form or remove species; transport of chemical species away from their point of origin; and loss from the atmosphere via either escape from the planet's gravitational influence, or by deposition to the surface via direct reaction or precipitation. In a chemical model the effect these processes have on atmospheric composition is mathematically parameterized in the form of the continuity equation

$$\delta n/\delta t = -\Delta \cdot (nF) + P - L \qquad (10)$$

which gives the change in number of molecules, n, with time t on a fixed vertical grid as a result of net transportation in and out of a given atmospheric volume ($-\Delta(nF)$) (via diffusion for one-dimensional vertical transport, and via diffusion and winds for the vertical and horizontal transport required for three-dimensional models), addition by sources (production, P), and removal by sinks (losses, L). This equation predicts the distribution of chemical species based on temperature, pressure, and radiation-dependent reaction rates of atmospheric chemical reactions coupled to approximations of the atmospheric dynamics and radiation field.

Given the complexity of the processes being described by the continuity equation, an exact solution is not possible, and instead either analytical approximations are made, or numerical solutions are found. The chemical production and loss of a given species depends on the ensemble of local concentrations of other species that produce or react with that species. The chemical system is specified as a system of coupled ordinary differential equations describing the chemical evolution of the species at each spatial gridpoint in the model and over a given time step (t, t + Δt). This can be integrated numerically, with the quality of the solution depending on the grid resolution, the time step, and the algorithms used for the transport and chemistry.

For model input, the atmospheric structure is typically provided as a list of temperatures and pressures as a function of atmospheric altitude, and may have either been directly or remotely measured (as is the case for planets in our own solar system) or calculated using a climate model (see below). The input stellar spectrum is scaled to be consistent with the planet's distance from its parent star, and its wavelength dependence is also important, as the vast majority of photochemical reactions in planetary atmospheres are driven by the parent star's UV spectrum. Even so, a spectral resolution as coarse as a few nanometers is usually adequate for stellar spectral input to a photochemical model. The photochemical model will include a radiative transfer module to calculate the radiation field at different points in the atmospheric column, based on the atmospheric composition and stellar irradiation. Most importantly, photochemical models require a suite of reaction rates and absorption cross sections that are pertinent and as comprehensive as possible for the atmosphere under consideration. Reaction rates are typically compiled from the available literature, or estimated using molecular modeling techniques (*http://jpldataeval.jpl.nasa.gov*; IUPAC, *http://ww.iupac-kinetic.ch.cam.ac.uk*).

2.3.2. Radiative transfer models. An atmospheric radiative transfer model can be used to calculate the wavelength-dependent radiation field at any point in a planetary atmosphere, including the top of the atmosphere and the planetary surface. To do this, the model resolves the planetary atmosphere into vertical layers and calculates the emitted, absorbed, and scattered radiation into and out of each layer. Knowledge of the radiation field can be used to generate synthetic radiance spectra of the planet and to calculate net heating and cooling rates in a given atmospheric layer, as input to a climate model. Synthetic spectra can be generated for a variety of different viewing and emission angles, stellar zenith and illumination fractions, surface and atmospheric compositions, and atmospheric constituent and temperature distributions. These synthetic spectra can be used to understand the radiation field experienced by life on the surface of the planet, or to provide a simulation of the planet's telescopic appearance.

Input to the model includes the vertical temperature and pressure structure, atmospheric composition, atmospheric component optical properties, and stellar irradiation. Two-stream models (one stream up and down) can provide reliable estimates of hemispherically averaged irradiances, but if accurate descriptions of the angle-dependent radiation field are required (e.g., to understand or predict limb-darkening or phase-dependent behavior on an exoplanet), then multiple-stream models must be used (e.g., *Meadows and Crisp*, 1996).

Radiation in a planetary atmosphere is affected by three processes: emission of radiation, typically thermal radiation, from the atmosphere and surface; absorption of radiation by atmospheric gases and the surface; and the scattering of radiation, both into and out of an atmospheric layer, by the surface, by gases (Rayleigh scattering), and by aerosols (Mie, or other forms of scattering). In a planetary atmosphere, the change in intensity of radiation transmitted through the atmosphere is equal to the difference in intensity between radiation emitted along, or scattered into, the line of sight, and radiation absorbed or scattered out of the line of sight. Considering these processes leads to an equation of transfer of the form

$$\mu dI(\tau,\mu,\phi,\nu)/d\tau = I(\tau,\mu,\phi,\nu) - S(\tau,\mu,\phi,\nu) \qquad (11)$$

(*Goody and Yung*, 1989) where I denotes the radiance, μ is the cosine of the zenith angle, S is the source function, τ is the column-integrated vertical optical depth, ϕ is the azimuth angle, and ν is the frequency (wavenumber) of the radiation. The source function, S, is given by

$$\begin{aligned} S(\tau,\mu,\phi,\nu) = \\ \frac{\omega(\tau,\nu)}{4\pi} \int_0^{2\pi} d\phi' \int_{-1}^{1} d\mu' P(\tau,\mu,\phi,\mu',\phi',\nu) \\ + I(\tau,\mu',\phi',\nu) + [1 - \omega(\tau,\nu)] B[\nu, T(\tau)] \\ + \frac{\omega(\tau,\nu)}{4\pi} F_\odot P(\tau,\mu,\phi,-\mu_\odot,\phi_\odot,\nu) e^{(-\tau/\mu_\odot)} \end{aligned} \qquad (12)$$

Here the terms describe the angle- and wavelength-dependent atmospheric and incident stellar radiation scattered both into and out of the beam, and the thermal source due to Planck emission. Here $\omega(\tau, \nu)$ is the single-scattering albedo, which is the ratio of scattering efficiency to total light extinction (scattering and absorption). $P(\tau, \mu, \phi, \mu', \phi', \nu)$ is the scattering phase function, a function that describes the probability that a photon will be scattered at a given angle to its direction of incidence. Essentially, the total source described by this equation is a weighted sum of thermal emission and scattering from other directions, with the single-scattering albedo controlling the weight given to each.

This equation predicts wavelength- and angle-dependent atmospheric transmission using knowledge of the temperature, pressure, atmospheric composition and optical properties, and incident stellar radiation. Especially when scattering is included, the equation of transfer and its associated source function are solved using numerical methods. One such method is the discrete ordinate method, which can provide accurate solutions to the equation of transfer, but only by considering spectral regions that are sufficiently narrow that the optical properties and source function can be considered as constant within each spectral region (*Stamnes et al.*, 1988). Consequently, the monochromatic equation of transfer must be evaluated at a very large number of wavelengths to resolve rapidly varying gas absorption coefficients. The quality of the solution from a radiative transfer model depends on the number of layers considered for the atmosphere, the grid resolution (i.e., the wavelength resolution used to solve the equation of transfer over rapidly varying gas absorption coefficients), and the accuracy of the input information, which includes laboratory and theoretically derived gas absorption coefficients, absorption line profiles, scattering phase functions, and wavelength-dependent surface reflectivities.

As its fundamental input, a radiative transfer model requires a description of the vertical distribution of atmospheric temperature, pressure, gas species abundances, and aerosols. This atmosphere file can be obtained from measurements and observations of the planet, or it can be generated from the output of a photochemical and/or climate model, allowing calculation of synthetic spectra for either measured or modeled planetary atmospheres.

The model also requires as input the optical properties of the atmospheric gases and aerosols and the surface. Rayleigh scattering is used to describe the scattering properties of gases. Gas absorption and emission characteristics are given using a combination of gas absorption coefficients and a "line profile" that provides atmosphere-dependent information on the width and shape of the molecular line, allowing calculation of the effect of the far wings of a line on the spectrum. Gas absorption coefficients can be obtained from either molecular "line lists," which provide measured and/or calculated wavelength positions and intensities for individual molecular line transitions, or absorption cross-sections, which are often laboratory measured. The most widely used molecular line lists are provided by the HITRAN [High-resolution TRANsmission molecular absorption database (*Rothman et al.*, 2009)] and GEISA (Gestion et Études des Information Spectroscopiques Atmosphériques (*Jacquinet-Housson et al.*, 2005)] databases. Absorption coefficients for a given atmosphere are obtained by summing the total gas optical depth (gas absorption coefficient multiplied by number of molecules) for all molecules that absorb at a given wavelength, including contributions from the far wings of distant lines.

For liquid aerosol droplets such as those in Earth's water clouds or Venus' sulfuric acid clouds, optical properties such as absorption and scattering cross sections and single-scattering albedos can be derived from optical constants (e.g., refractive index, reflectance, transmittance) using a Mie scattering algorithm for spherical particles (e.g., *Wiscombe*, 1980). For nonspherical particles such as ice crystals and mineral dust, which can dominate in dry, cold atmospheres, different algorithms must be used. These algorithms more accurately characterize the nonspherical absorption cross section, which can be enhanced over that of an equivalent volume spherical particle at infrared wavelengths. Similarly, the scattering phase function of nonspherical particles can show more enhanced forward and backscattering behavior in some wavelength regimes when compared to spherical counterparts (*Zuffada and Crisp*, 1997).

The surface optical properties are characterized by an albedo, a, and a surface reflection function $P_a(\mu, \mu')$, where μ and μ' are the incident and scattered angles, respectively. Wavelength-dependent albedo spectra can be obtained for different planetary surface types, including rock and vegetation, from sources such as the U.S. Geological Survey Spectral Library (*http://speclab.cr.usgs.gov*) (*Clark et al.*, 1993). For the surface reflection function, a Lambertian approximation, which assumes an equal probability for an incoming photon being scattered in any direction, is often used. For some flat reflective surfaces, however, most notably liquid water, it is important to take into account specular reflection, where the angle of incidence equals the angle of reflection. "Glint," a spatially localized region of brightness, occurs on water where the local slope provides a direct specular reflection of the Sun. Earth's oceans are not perfectly flat, but do specularly reflect, with the resulting ocean glint comprised of multiple glint spots on the wind-ruffled ocean surface. The shape and size of this glint spot is related to the viewing geometry, the wind-generated water roughness (*Cox and Munk*, 1954a,b) and ocean-air temperature difference (*Shaw and Churnside*, 1997). Probability density functions that predict the distribution of wave slopes as a function of wind speed, such as the Cox-Munk formalism (*Cox and Munk*, 1954a,b), can be incorporated into radiative transfer models to recreate observed ocean glint.

2.3.3. Climate models. A planet's climate is controlled by the amount of stellar radiation intercepted by the planet, and the fraction of that energy that is absorbed. A climate model is used to calculate the thermal structure of terrestrial planet environments by modeling the radiative effects of the atmosphere, clouds, and surface, and the transport of heat via atmospheric and surface processes. One-dimensional climate models describe the globally averaged vertical thermal structure of a planet, where the one dimension refers to altitude.

Energy balance models provide a two-dimensional approximation for planetary climate by including equator-to-pole transport, and calculating the resultant average climate within circular latitudinal zones. Three-dimensional climate models, such as the general circulation models (GCMs), calculate the vertical and horizontal temperature structure for a planetary atmosphere using a three-dimensional wind field to distribute energy around the planet. Three-dimensional GCMs require a large amount of *a priori* knowledge of a planetary environment to produce accurate results, and are best reserved for planets with good observational characterization, including good temporal coverage and some spatial information (see the chapter by Showman et al.).

One-dimensional models cannot model or predict horizontally spatially inhomogeneous characteristics of a planetary environment, but are relatively robust to restricted input information. One-dimensional models, although crudely approximating the planet's atmosphere as a single vertical profile, have been used to successfully characterize the observed globally averaged climates of solar system terrestrial planets and to understand the physical processes governing the vertical structure of the atmosphere/surface system. For initial observations of terrestrial exoplanets, the disk-integrated nature of the data lends itself to analysis using one-dimensional climate models. Input to a one-dimensional climate model includes an atmospheric composition and pressure structure, the stellar radiation incident at the top of the atmosphere, and a starting vertical temperature structure.

The main function of the climate model is to explicitly calculate the solar (stellar) and terrestrial fluxes of radiation, and to determine net radiation at the top and bottom of each atmospheric layer so that temperature structure can be determined. Many planetary processes govern the temperature structure, including stellar radiative heating and thermal radiative cooling, which are calculated using radiative transfer theory. Processes that move heat around the system must also be considered. These include vertical diffusive heat transport within the surface, vertical convective heat transport within the atmosphere, and latent heat transport due to vertical mixing, condensation, evaporation, and precipitation of clouds and rain. All these processes affect how much radiation will be absorbed and emitted by, or transported into or out of, a given atmospheric layer.

The imbalance between the radiation into and out of the layer can be used to calculate the "heating rate" ($h_R(z)$) in that layer (negative heating rates imply cooling). For a "time-marched" equilibrium calculation, the atmospheric profile is initialized with an assumed vertical temperature/pressure structure and gas mixing ratio profile, and the heating rates for a number of layers in the atmosphere are then calculated. The temperature structure as a function of altitude ($T(z)$) is then adjusted in a subsequent time step, Δt, such that

$$T'(z) = T(z) + h_R(z)\Delta t / \rho c_p \qquad (13)$$

where ρ is the atmospheric density and c_p is the specific heat.

At the end of each time step a revised radiative temperature profile is produced. As the time-stepping progresses, the atmospheric lapse rate is calculated, and if it exceeds a critical lapse rate the atmosphere is presumed to be convectively unstable. A degree of vertical mixing sufficient to reestablish the stable lapse rate is performed, and the model proceeds to the next radiative time step. Time-stepping is used to march the atmosphere's response forward until convective adjustment is no longer required and the heating rates for each atmospheric layer approach zero. The atmosphere is then said to be in radiative-convective equilibrium.

As most climate models are based on radiative transfer models, they share many of the same input data, which is described in more detail in the radiative transfer model section above. The principal inputs for a climate model are the vertical distribution of atmospheric gases, and especially those that contribute to atmospheric absorption and emission or radiation, such as CO_2, H_2O, and O_3, and an assumed temperature profile as a function of pressure. This initial T profile can be isothermal, if the temperature structure of the atmosphere is unknown. However, a good guess at the initial profile will likely shorten the time it takes the model to reach equilibrium. Climate models also require specification of cloud altitude and optical properties, wavelength-dependent surface albedo, and the flux and spectral energy distribution of the incoming stellar radiation.

2.3.4. Atmospheric escape models. Actual calculations for hydrodynamic escape are based on the hydrodynamic equations of fluid mechanics. The calculations require as input the solar or stellar extreme UV radiation, and the masses and constituents of planetary upper atmospheres. These inputs can be challenging to obtain or even constrain for early planetary atmospheres and for exoplanets and their host stars. As output, the escape model provides escape fluxes, and the density and temperature structure of the thermosphere and exosphere being modeled. For a description of a numerical model for atmospheric escape, see *Tian et al.* (2005b) (and references therein).

3. EARTH AS AN EXOPLANET

Earth is the only known example of a habitable planet. Both observational and modeling studies have therefore focused on the astronomical detectability of Earth's habitability as a starting point for our search for habitable planets elsewhere.

3.1. Observational Studies

All directly imaged exoplanets will be spatially unresolved to the telescopes studying them. Simultaneous, disk-integrated views of Earth are therefore the best product to understand what characteristics can be derived from spatially averaged, i.e., point source, forms of data. Earth is the *most* studied planet in our solar system. An extensive suite of Earth-observing satellites obtain high temporal and spatial sampling via photometry and spectroscopy at wavelengths from the visible through the mid-infrared. This data, however, is not always useful for compiling an astronomical view of Earth. The bias in Earth-observing satellites toward spatial resolution and detailed regional views are at the expense of global synoptic coverage. To study Earth as an exoplanet, simultaneous disk-averaged views of the rotating Earth for different viewing angles and phases are needed.

Observational studies of Earth as an exoplanet have therefore had to rely not on satellite data but instead primarily on Earthshine measurements and observations of Earth from interplanetary spacecraft en route to other scientific targets.

3.1.1. Earthshine observations. Earthshine is sunlight scattered from Earth that scatters off the Moon (the unlit, but visible portion) and travels back toward Earth. Earthshine can be seen with the naked eye most readily near crescent phases. For centuries this phenomenon has been poetically referred to as "the new Moon with the old Moon in her arms." Earthshine provides a good approximation to a disk-averaged source of Earth light for study because scattered radiation from the rough lunar surface results in a sampling of many regions of Earth.

Earthshine observations of Earth's spectrum were used to identify Earth's principal absorbing species at visible (*Arnold et al.,* 2002; *Woolf et al.,* 2002) and near-infrared (*Turnbull et al.,* 2006) wavelengths. Identified spectral features included water vapor bands at 0.72, 0.82 and 0.94 μm, which confirm the presence of an atmosphere and indicate the presence of surface liquid water. Spectral signatures from the oxygen A band (0.76 μm) and the oxygen B band near 0.69 μm are potential markers for photosynthetic life. The broad ozone Chappuis bands, between 0.5 and 0.7 μm, indicate the likely presence of O_2, and a UV-absorbent gas that could shield the planet's surface from DNA-damaging UVB radiation. Carbon dioxide can be seen in six features from 1.4 to 2.1 μm and methane in numerous weak features between 0.86 and 1.7 μm.

In Earthshine spectra, indicators for atmospheric pressure were also observed. Rayleigh scattering was seen at visible wavelengths, despite partial cloud cover. The presence of Rayleigh scattering is a good indicator of the presence of an atmosphere, but may be difficult to use quantitatively for an accurate pressure derivation in the disk-average due to the different scattering path lengths due to partial or even total cloud coverage.

Earthshine observations have also been used to identify the presence of the vegetation "red edge" in Earth's disk-integrated spectrum. The red edge is a sudden rise in the reflection spectra of photosynthetic vegetation at wavelengths at and longward of 0.7 μm (for overviews, see *Seager et al.,* 2005; *Kiang et al.,* 2007). This rise in reflectivity is typically on the order of 6 to 20 times the reflectivity at 0.5 μm, and depends on vegetation type. Due to water vapor absorption bands centered near 0.7 and 0.72 μm, however, spectra of vegetation taken through an atmospheric column (as would be the case for Earthshine data) will not show a rise until 0.75 μm. The signature of the red edge has been measured in disk-integrated Earthshine data, which showed a small (2–3%) but positive and reproducible correlation between a rise in Earth's reflectivity at wavelengths longward of 0.75 μm and the fraction of Earth's forests that were cloud free at the time of observation (see, e.g., *Montañés-Rodríguez et al.,* 2006; *Arnold,* 2008; and references therein). The corresponding signature of the red edge varies strongly with surface types visible on the observed disk, dropping to zero for views of Earth dominated by the Pacific Ocean. The vegetation red edge may also be more detectable at different observational phases, especially those close to dichotomy (*Tinetti et al.,* 2006b).

Earthshine observations are arguably the most readily obtained observations of disk-averaged Earth light, but their sensitivity can be compromised by the significant data reduction required to remove the reflectivity properties of the Moon's surface, and the "extra" passage of the reflected radiation back through Earth's atmosphere to detection by the telescope. Additionally, Earthshine observations at a given site are geometrically constrained to only a limited range of Earth's surface reflected from the Moon while the telescope is in darkness. Observations from the south pole during a continuous polar night would permit continuous temporal and longitudinal monitoring of Earthshine for several extended periods around the equinoxes (*Briot et al.,* 2010).

3.1.2. Spacecraft observations. Global snapshot observations of Earth can be obtained from robotic spacecraft en route to other solar system targets. The NASA EPOXI mission (the repurposed Deep Impact spacecraft), for example, has made several dedicated observations of Earth (Figs. 8 and 9).

The earliest observational study of Earth as an exoplanet for the purpose of understanding whether or not it was habitable or inhabited was that undertaken by the Galileo spacecraft. Galileo flew by Earth during a gravity assist maneuver on the way to its primary mission at Jupiter (*Sagan et al.,* 1993). The data collected on this flyby included UV and NIR spectra, visible photometry at three separate sites on the planet, and radio signals. Signs of life were found in the data, including gases such as oxygen, at abundances that were unlikely to be photochemically produced; methane in severe thermodynamic disequilibrium; and absorption from chlorophyll. The least potentially ambiguous sign of life, however, was the presence of modulated radio signals that were uniquely attributable to technology.

Other spacecraft have taken observations of Earth either during flyby or en route to another solar system planet. While en route to Mars, the Mars Global Surveyor (MGS) Thermal Emission Spectrometer (TES) took mid-infrared wavelength (5–20 μm) observations of Earth (*Christensen and Pearl,* 1997). More recently, the Mars Express spacecraft, also en route to Mars, observed Earth at near-infrared wavelengths (1–5 μm) with the OMEGA instrument (see, for example, Figs. 9 and 10 in *Tinetti et al.,* 2006a). As yet unpublished Earth data have also been acquired during Earth flybys by MESSENGER and Rosetta and during lookback observations by the Venus Express and LCROSS missions.

The first spaceflight mission to include a dedicated "Earth as an exoplanet" observing program was the EPOXI mission. EPOXI observed the whole disk of Earth from distances of

3.2. Earth Models

Spatially and/or spectrally resolved models of Earth have been developed to better understand Earth's global characteristics, and from vantage points, and at wavelengths, that may not be accessible to existing Earthshine, spacecraft, or Earth-orbiting satellites. These models can serve as theoretical "laboratories" to determine the detectability of signs of habitability and life in disk-integrated spectra of Earth, and to develop and test observational and analysis techniques for exoplanet characterization. Here we divide the models into two basic types: photometric models, which can be used to predict lightcurve behavior, and spectroscopic models, which can produce both lightcurves and higher-resolution spectral simulations.

3.2.1. Photometric models. Photometric models simulate the disk-integrated Earth in different photometric bands to better understand Earth's temporal and photometric variability, and to search for effects with changing solar illumination fraction (phase). These models can be used to generate lightcurves, changes in brightness with time, that occur as a function of Earth's rotation as more or less cloud, land, and oceans rotate into view. Different planetary surfaces have different reflectivities and emissivities, which alter the disk-averaged radiation detected. In the visible, the highly reflective clouds (>60%) dominate Earth's lightcurve, while at mid-infrared wavelengths almost all planetary surfaces have emissivities close to unity, and so the relatively warm oceans and continents dominate the emitted radiation.

Reflectance models capture Earth's spatially and temporally varying reflectance variations without atmospheric absorption and scattering, using bidirectional reflectance distribution functions (BRDFs) for a limited number of surface types, including clouds. These models first demonstrated and quantified the photometric variability of the disk-averaged rotating Earth (*Ford et al.*, 2001). In the absence of clouds this variation is as high as 150% due to highly reflective land surfaces and specular reflectance ("glint") from the ocean, and is reduced to 10–20% in the presence of a realistic cloud fraction (*Ford et al.*, 2001). This is in contrast to photochemically generated haze-covered planets such as Venus and Titan, which show extremely small diurnal reflectivity changes at visible wavelengths. These models indicated that a future terrestrial exoplanet characterization mission with the capability to measure 5% visible variation could potentially detect weather, the rotational period, and seasonal changes in cloud patterns for an Earth-like planet.

Pallé et al. (2008) modeled Earth's photometric variability using three months of cloud data taken from satellite observations. They showed that a hypothetical distant observer could measure Earth's rotation rate even in the presence of clouds, because Earth has a relatively stable cloud pattern. These cloud patterns arise in part because of Earth's continental arrangement and ocean currents. Beyond detecting Earth's rotation rate, *Pallé et al.* (2008) found deviations from the periodic photometric signal, indicative of active weather. From continuous observations of Earth over a few-month period,

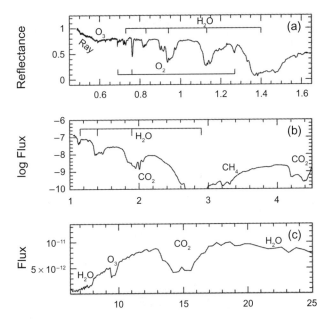

Fig. 8. Earth's observed disk-integrated spectrum. **(a)** Visible wavelength spectrum from Earthshine measurements plotted as normalized reflectance (*Turnbull et al.*, 2006). **(b)** Near-infrared spectrum from NASA's EPOXI mission with flux in units of W m^{-2} μm^{-1} (*Robinson et al.*, 2010). **(c)** Mid-infrared spectrum as observed by Mars Global Surveyor en route to Mars with flux in units of W m^{-2} Hz^{-1} (*Pearl and Christenson*, 1997). Major molecular absorption features are noted, including Rayleigh scattering.

0.18–0.3 AU from both equatorial and polar vantage points, and acquired spatially and temporally resolved visible photometric and near-infrared spectroscopic observations (*Livengood et al.*, 2008). The visible photometry was taken at seven wavebands spanning 300–1000 nm over the course of 24 hours to produce multiwavelength lightcurves of Earth. Analysis of these lightcurves by *Cowan et al.* (2009) revealed that the largest relative changes in the color of Earth over a full rotation occurred at the reddest wavelengths. *Robinson et al.* (2010) attributed the reduction of Rayleigh scattering and therefore the enhanced transparency of the atmosphere (to both absorptive and scattering components) at the longer wavelengths, and to the relatively high albedo of red continents vs. blue oceans. Using a principal component analysis of the multiband lightcurves, *Cowan et al.* (2009) found that 98% of the diurnal color changes in Earth's lightcurves are due to only two dominant eigencolors. The spectral and spatial distributions of the eigencolors correspond to cloud-free continents and oceans, enabling construction of a crude longitudinally averaged map of Earth (Fig. 10). This analysis demonstrated that it might be possible to retrieve spatial information from temporally-resolved multiwavelength photometry of disk-averaged terrestrial planets.

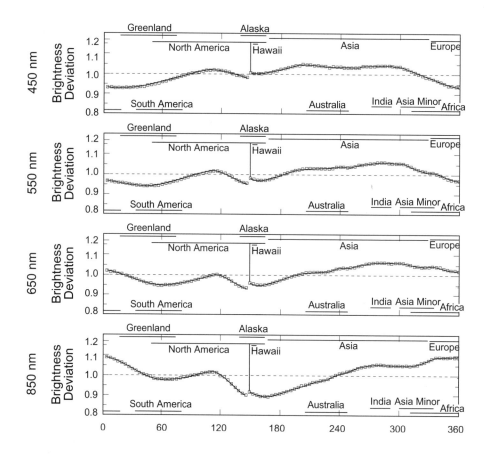

Fig. 9. Lightcurves at visible wavelengths taken by the NASA EPOXI mission reusing the Deep Impact spacecraft. Observations were taken every 15 minutes in broadband filters centered around 450, 550, 650, and 850 nm for a full Earth rotation. The brightness in a specific measurement is divided by the average brightness of the planet over a full rotation. These time-resolved, multiwavelength lightcurves show Earth getting redder over Africa (shorter wavelengths decrease in brightness, longer wavelengths increase in brightness) and bluer over the Pacific. From *Livengood et al.* (2008).

Earth's rotation rate could be extracted, weather identified, and the presence of continents inferred.

Photometric models can also be used to explore the phase-dependent detectability of ocean "glint," a means of directly detecting the presence of liquid water on a planetary surface. *Williams and Gaidos* (2008) used a photometric model to show that a planet with specularly reflecting oceans will appear brighter near crescent phase, and have a higher polarimetric signal, than a planet with isotropically scattering surfaces. This is due to the increasing dominance of the fixed-position glint spot as the illuminated fraction of the planet decreases.

3.2.2. Spectroscopic models. Spectroscopic models simulate the disk-integrated spectrum of Earth. The goal is to understand the effect of different surface types, clouds, phases, and seasons on Earth's spectrum. When validating against or comparing models to real Earth data, the purpose is to aid in identification of atmospheric and surface components, and possibly to retrieve quantitative abundances. Spectroscopic models for analyzing Earthshine data have been used by several Earthshine observers, including *Woolf et al.* (2002), *Turnbull et al.* (2006), and *Montañés-Rodríguez et al.* (2006).

When comparing with datasets that are limited in wavelength, phase, and temporal ranges, then some atmospheric phenomena, including multiple-scattering and phase-dependent effects from clouds, are generally ignored.

If Earth models are to serve as "forward models," those that are used to predict the appearance of the planet for observational and data analysis *planning* purposes over a wide range of viewing angles, solar-illuminations, wavelength ranges, and daily and seasonal timescales, then more rigorous models are appropriate. These models can most accurately capture the scattering and phase dependent effects of clouds, which can dominate the disk-averaged spectrum and can mimic, alter, or mask atmospheric absorption and surface signatures. Perhaps the most comprehensive of these forward models is the three-dimensional spectral model first described in *Tinetti et al.* (2006a,b) and significantly improved via validation with the EPOXI Earth observations by Robinson et al. (personal communication, 2010). This model generates synthetic spectra for a grid of points on the planet using a line-by-line radiative transfer model that includes atmospheric absorption, Cox-Munk specular reflectance from the ocean, self-consistent Rayleigh scattering,

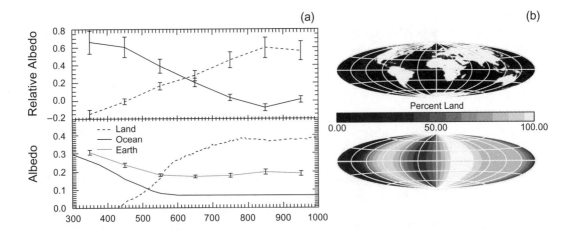

Fig. 10. (a) Composite eigencolors for Earth's change in color over a rotation, compared to example spectra for ocean and land. (b) Map generated using the spectral and spatial distributions of the eigencolors, and compared to the corresponding map of Earth's continents and oceans. From *Cowan et al.* (2009).

and multiple scattering from clouds and aerosols. As input, the model ingests Earth-observing satellite and ground station data that specify surface types, including ice and snow coverage, atmospheric mixing ratios as a function of altitude, and cloud position and optical depth. As output, the model can generate spatially resolved, or disk-averaged, spectra of Earth for arbitrary viewing angle and solar illumination (phase) and over timespans from hours or days to a year. This model can predict the absolute brightness of Earth to within 8% and follow hourly changes in relative brightness to within 3%. Models of this type can be used to generate simulated Earth data to explore the detectability of planetary phenomenon such as seasonal and phase-dependent variability in brightness and spectra, the detectability of ocean glint in the presence of realistically scattering clouds (*Robinson et al., 2010*), and at mid-infrared wavelengths, the accuracy of temperature retrieval in disk-averaged data.

4. SUMMARY OF HIGHLIGHTS AND OUTSTANDING QUESTIONS

The most recent major advances in terrestrial exoplanet science encompass observational and theoretical study of super Earths, a reexamination of the surface habitable zone for terrestrial exoplanets, and theoretical work to understand the interaction between planet and star for stars of different spectral type.

4.1. The Discovery and Characterization of Super Earths

Three different exoplanet discovery techniques — radial velocity, transit, and microlensing — have uncovered a population of over 20 planets with minimum masses between 1.9 and 10 Earth masses ($M_\oplus$), the so-called "super Earths." The term "super Earth" is usually intended for rocky planets but can be applied to <10 $M_\oplus$ planets with icy interiors or significant gas envelopes, which are often also labeled as "mini-Neptunes." Although there is no widely accepted definition of super Earths (see the Appendix in the chapter by Seager and Lissauer), super Earth masses are usually capped at 10 $M_\oplus$, as this is a suggested theoretical limit at which a forming planetary core can entrain sufficient nebular gas to become an ice or gas giant planet. Observations over the next decade should help us understand the dividing mass between different kinds of exoplanets.

The most exciting goal in the search for super Earths is to find a planet orbiting in the star's habitable zone. Yet the vast majority of super Earths are hot Earths with semimajor axes interior to 0.1 AU. This is due to the discovery selection effect that planets close to the star are easier to detect with the radial velocity and transit search methods. This small semimajor axis range includes the lowest (minimum) mass planet found to date, Gl 581e, at M sin i = 1.9 $M_\oplus$ (*Mayor et al., 2009*). There are a handful of super Earths with greater planet-star separations than Gl 581e, including MOA-2007-BLG-192-Lb found at 0.62 AU, close to the orbit of Venus in our system (*Bennett et al., 2008*), and Gl 581d, with a semimajor axis of 0.22 AU, which is interior to the orbit of Mercury in our system, but possibly within the habitable zone for its much cooler M star host (*Mayor et al., 2009*) (see section 3.3). The most distant super Earth found to date is OGLE-05-390Lb, a cool super Earth found at 2.1 AU, detected via gravitational microlensing (*Beaulieu et al., 2006*).

Two super Earths are known to transit their host stars, CoRoT-7b (*Léger et al., 2009*) and GJ 1214b (*Charbonneau et al., 2009*). Transiting exoplanets are the most sought after because the planetary radius and inclination can be measured (see the chapter by Winn). The inclination determination from transit removes the M_p sin i ambiguity in planets detected with the radial velocity method alone. With both size and mass, density can be calculated, which in turn constrains interior composition (see the chapter by Sotin). Both CoRoT-7b and GJ 1214b are fascinating objects because their atmospheres are almost certainly extremely different from those of solar system terrestrial planets.

CoRoT-7b is a 4.80 ± 0.79 $M_\oplus$, 1.68 ± 0.09 $R_\oplus$ exoplanet in a 0.85-day orbit (0.017 AU) about a K0V star (*Léger et al.*, 2009; *Queloz et al.*, 2009). CoRoT-7b's daytime effective temperature is in the range 1800–2600 K. CoRoT-7b is so hot that regardless of its true atmospheric temperature, the atmosphere is anticipated to be not only free of H and He, but possibly C, N, S, and O as well. Its atmosphere would likely be composed of silicates and other refractory molecules such as Ca, Al, and Ti (*Schaefer and Fegley*, 2009), with silicate clouds and rain possible. Unfortunately, due to the planetary system's 150-pc distance from Earth, the related relative faintness of the star (V = 11.7), and the stellar variability, follow-up atmosphere measurements may not be possible. Nevertheless, CoRoT-7b is definitively a terrestrial planet. From the planet's measured mass and radius, and using interior models (see the chapter by Sotin et al.), CoRoT-7b is expected to be dominated by silicates with the presence of some lighter material. CoRoT-7b has a companion planet, at 8.4 ± 0.9 $M_\oplus$ (CoRoT-7c) (*Queloz et al.*, 2009). If gravitational interaction with its companion enables CoRot-7b to maintain an eccentricity greater than only 10^{-5}, possibly driven by a small eccentricity in CoRot-7c's orbit, then tidal heating of the solid body might drive extreme volcanism and resurfacing of the planet (*Barnes et al.*, 2010).

The second known transiting super Earth, GJ 1214b (*Charbonneau et al.*, 2009), is expected to have a very different atmosphere than the super Earth CoRoT-7b. This is because of GJ 1214b's lower bulk density, determined from its relatively large radius (2.68 ± 0.13 $R_\oplus$) for its mass (6.55 ± 0.98 $M_\oplus$), and its lower effective temperature, which is calculated to be 500–558 K assuming an albedo of 0.35 to zero, and atmospheric redistribution of absorbed stellar energy. This lower effective temperature compared to CoRoT-7b is due to GJ 1214 b's larger semimajor axis (0.014 AU, 1.5-day period) about a much cooler star (an M 4.5 V star with an effective temperature of 3026 ± 130 K). The relevant point to GJ 1214b's potential atmospheric composition its that its large radius and low average density (1.87 ± 0.4 g cm^{-3}) implies it very likely has an H/He envelope in the range 0.01–5% by mass (*Rogers and Seager*, 2010); for comparison, Neptune's H/He envelope is 5–15% of the planet mass. There is some possibility GJ 1214b has a massive steam atmosphere (without an H envelope) surrounding a planet made mostly of water (*Rogers and Seager*, 2010). GJ 1214b is therefore more appropriately called a mini-Neptune rather than a super Earth. Regardless of the terminology, the atmosphere is expected to be reduced, with large amounts of water vapor, and also including methane and other reduced molecules. GJ 1214b's proximity to our solar system (13 pc) and its relatively high planet/star radius ratio make the planet more amenable to follow-up observations that could probe the planet's environment.

In the observational study of terrestrial exoplanets we still await the discovery of planets that are more Earth-like in their atmosphere and interior composition, with masses in the 1–2 $M_\oplus$ range and semimajor axes within or near to the star's habitable zone.

4.2. The Nature and Habitability of Super Earths

A recent highlight of super Earth research is the foray into predictions or studies of new kinds of terrestrial planet atmospheres, and the fledgling efforts to put the diversity of exoplanet atmospheres into a single framework. Here we expand upon relevant material in *Seager and Deming* (2010) and *Rogers and Seager* (2010).

Theoretical attempts to evaluate and constrain the possible atmospheric compositions and properties for super Earths used calculations of atmospheres outgassed during planetary accretion, considering bulk compositions drawn from differentiated and/or primitive solar system meteoritic compositions (*Elkins-Tanton and Seager*, 2008; *Schaefer and Fegley*, 2009). Instead of narrowing down possibilities, this work emphasized the large range of possible atmospheric mass and composition of outgassed super Earths even before consideration of atmospheric escape and other forms of atmospheric evolution (e.g., Fig. 11). *Seager and Deming* (2010) identified five categories of possible super Earth atmospheres.

The most straightforward case to categorize is a hot super Earth with no atmosphere. This type of planet will have lost its atmosphere via escape processes. The planet would no longer be able to outgas at a sufficient rate to replenish its atmosphere, possibly due to tidally accelerated outgassing of its volatiles earlier in its history. Tidally locked transiting planets lacking atmospheres may be identified by a hotspot at the substellar point (e.g., *Seager and Deming*, 2009). For planets with atmospheres, the hot spot is likely to be displaced from the substellar point in the direction of the zonal winds, smeared out by atmospheric superrotation, or not detectable at the levels of the atmosphere probed by a particular observational wavelength.

A second category is one of very hot super Earths. Rocky worlds with atmospheric temperatures well over 1500 K may have undergone hydrodynamic escape and lost not only hydrogen but volatile compounds of C, N, O, and S as well. At these temperatures the atmosphere may be composed of silicates enriched in more refractory elements such as Ca, Al, and Ti (*Schaefer and Fegley*, 2009).

A third category of super Earths would have also outgassed their atmospheres, but suffered loss of hydrogen via atmospheric escape, and may in many ways resemble Earth's earliest atmosphere. These atmospheres would have little or no hydrogen, and a dominant CO_2 and water vapor composition. Water-rich super Earths, or "ocean worlds," may have a predominantly water vapor atmosphere (*Kuchner*, 2003; *Léger et al.*, 2004).

A fourth category would be outgassed H-rich atmospheres that have been retained against atmospheric escape. These atmospheres would be dominated by hydrogen, and may include other observable gases consistent with more reducing atmospheres such as H_2O, CH_4, or CO.

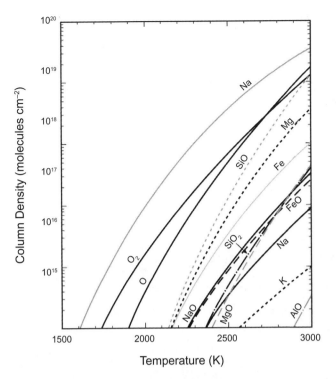

Fig. 11. Atmospheric composition (as column density, molecules cm^{-2}) as a function of temperature for CoRoT-7b, with a surface gravity, g, of 36.2 cm^{-2}, and assuming it has the composition of the bulk silicate Earth. The dominant gas at all temperatures considered is Na, followed by O$_2$ and O. At temperatures in excess of 2200 K, SiO, Mg, and Fe become more prevalent in the atmosphere. From Fig. 2 of *Schaefer and Fegley* (2009).

As a fifth category, very young, cold, or more massive super Earths may display a primordial atmosphere captured from the protoplanetary nebula, which will be dominated by hydrogen and helium. Helium can be quite difficult to detect remotely in any kind of planet (*Hunten and Münch*, 1973).

Work to date on attempting to model super Earth atmospheres has taken several paths. One approach is to consider atmospheres similar to Earth, Venus, or Mars (or their atmospheres in earlier epochs) as the basis for the modeling. Using climate models, assumed greenhouse gas inventories, and different fractional cloud cover, *Selsis et al.* (2007) and *von Bloh et al.* (2007) both found that Gl 581d is more likely to be habitable (i.e., with surface temperatures consistent with liquid water) than Gl 581c, which is likely too close to its parent star for habitability. Other investigators consider atmospheres that radically depart from the terrestrial planets in our solar system. Water planets, like GJ 1214b, may have similar interior compositions to Jupiter's icy moons, which have up to 50% water by mass, and might be anticipated to have massive steam atmospheres (*Kuchner*, 2003; *Léger et al.*, 2004). In a different approach, *Miller-Ricci et al.* (2008) considered GJ 581c and three possibilities relating to hydro-

gen content to determine likely observable characteristics for planets with hydrogen-rich and hydrogen-poor atmospheres (Fig. 12). An exploration of terrestrial planets with geophysical sulfur cycles dominating over carbon cycles is described in *Kaltenegger and Sasselov* (2010). Others have attempted to quantify the atmospheric escape processes and atmospheric lifetime, which is challenging in the face of the unknown initial atmospheric mass and the star's activity history (e.g., *Lammer et al.*, 2007).

Super Earth interior and atmosphere composition will be diverse, as already demonstrated by the similar mass but marked difference in density and likely bulk composition for CoRoT-7b and GJ 1214b. This planetary diversity is determined by the planet's formation, migration, and evolution history.

4.3. Revisiting the Habitable Zone

Interest in attempting to better define the limits of the habitable zone for liquid water on planetary surfaces is a growing scientific area. The habitable zone is the range of orbital distances around a star where a terrestrial planet can maintain liquid water on its surface. A revisit of the habitable zone was spurred by the exciting discovery of planets Gl 581c and Gl 581d, each of which have minimum masses less than 10 $M_\oplus$ and orbit in or very close to the habitable zone of their parent star (*Udry et al.*, 2007). The parameters of the planetary system provided constraints for theoretical explorations of which atmospheric conditions would actually allow the planet to be habitable. *Selsis et al.* (2007) provided a comprehensive discussion of the different atmospheric parameters that affect habitability, including an emphasis on the radiative effect of clouds. Their conclusion, assuming that the planets had masses equal to the minimum masses inferred from radial velocity measurements, were that Gl 581c, near the inner edge of Gl 581's habitable zone, was likely uninhabitable, but that Gl 581d, at the outer edge, potentially could be (Fig. 13). GJ 581d is further from the star than a generous empirical outer limit for the habitable zone based on early Mars' apparent habitability 4 G.y. However, *Selsis et al.* (2007) argue that the outer limit of the habitable zone can be pushed outward from even the habitable Mars limit by assuming that the planet is 100% covered with CO$_2$ clouds. CO$_2$ clouds scatter effectively in the thermal infrared, trapping radiation in the lower atmosphere, and compensating for the increased albedo in the visible (*Forget and Pierrehumbert*, 1997). *Selsis et al.* (2007) also discuss the possibility for an enhanced greenhouse with the addition of a super Earth atmosphere's more reducing gases, such as CH$_4$ and NH$_3$, to the standard terrestrial CO$_2$ and H$_2$O greenhouse mechanism.

Other recently explored aspects of planetary habitability include both the planet's environment (e.g., tidal heating) and the effect of stellar activity on atmospheric photochemistry and erosion. Both effects are likely to be most important for planets orbiting M dwarfs, where the habitable zone is very close to the parent star.

Significant tidal heating, enough to affect the planet's geo-

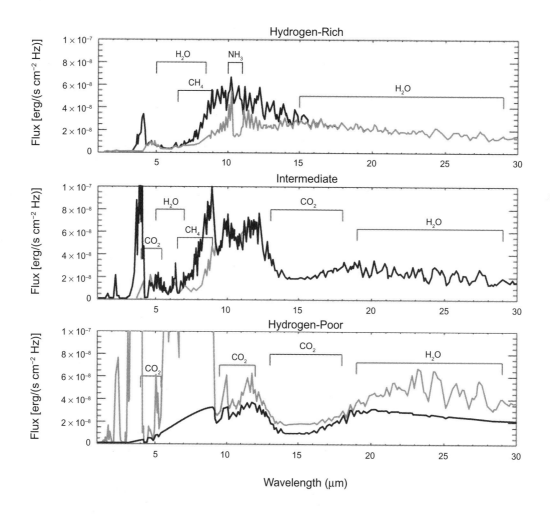

Fig. 12. Emission spectra for three model atmospheres considered for planet Gl 581c. The spectra illustrate atmospheres that range from hydrogen-rich to hydrogen-poor and with and without photolysis and clouds. In the top two panels, the dashed line shows the equilibrium model, and the solid line spectra show a model in which NH_3 and CH_4 are assumed to be photochemically removed. In the lower panel, the dashed line denotes a cloud-free model and the solid line includes a sulfuric acid cloud deck at a temperature of 400K. From *Miller-Ricci et al.* (2008).

logical activity, is a function of both the planet's semimajor axis, which must typically be within 0.15 AU of the parent star, and its eccentricity, which must be nonzero. The habitable zone for M dwarfs contains regions interior to 0.15 AU for M-dwarf masses below 0.4 $M_\odot$. *Barnes et al.* (2009) argue that these close-in planets may experience significant tidal heating via their gravitational interaction with the parent star. Closer to the star, the internal heating from tidal forces could augment the body's radiogenic heating rate to drive intense volcanism, similar to that seen on Io, and possibly preclude habitability. Tidal heating drops off extremely rapidly with distance from the star ($\propto a^{-6.5}$), and farther from the star provides little if no internal heating. Between these two extremes lies a region dubbed "the tidal habitable zone" (*Barnes et al.*, 2009), where tidal heating can potentially augment radiogenic heating to maintain geophysical processes such as plate tectonics.

Stellar activity on the erosion of terrestrial planet atmospheres for planets in the habitable zones of M dwarfs might be critical for a planet's habitability. The long-term enhanced X-ray and extreme ultraviolet flux produced by active M dwarfs can result in thermospheric heating and expansion of the atmosphere. *Tian* (2009) showed that super Earth atmospheres were robust against thermal escape even for M-dwarf extreme ultraviolet enhancements as large as 1000 times that experienced by Earth today. The resultant extended exosphere, however, can make the planetary atmosphere vulnerable to loss via ion pickup. *Lammer et al.* (2007) showed that terrestrial planets with weak magnetic moments at orbital distances ≤0.2 AU could lose tens to hundreds of bars of atmospheric pressure as a result of this mechanism. CO_2-rich atmospheres were also shown to have higher exospheric temperatures and so are at higher risk of loss. Super Earth planets may therefore have an enhanced probability of habitability around M dwarfs because their possibly stronger magnetic moments and higher gravitational acceleration would constrain the expansion of the thermosphere-exosphere and reduce atmospheric escape.

The maintenance of habitability in the face of stellar

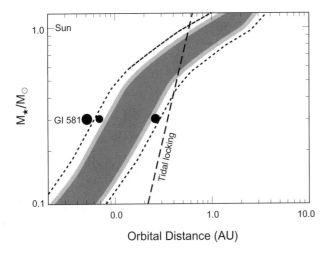

Fig. 13. The 5-G.y. continuously habitable zone for the Gl 581 planetary system. The darker area is defined by the empirical "early Mars" and "recent Venus" criteria. The light gray region gives the theoretical inner (runaway greenhouse) and outer limits with 50% cloudiness, with H_2O and CO_2 clouds, respectively. The dotted boundaries correspond to the extreme theoretical limits, found with a 100% cloud cover, which is unlikely for a planet with an active hydrological cycle. Gl 581d, with a semimajor axis near 0.22 AU, is the most likely planet in this system to be habitable. From *Selsis et al.* (2007).

evolution can be important for exoplanets and includes the evolution of Sun-like stars onto the red giant branch. The climate models used to perform these studies indicated that water-rich super Earths might have the best chance of maintaining habitable environments in the face of significant stellar evolution (*von Bloh et al.*, 2009).

The recent studies summarized in this section have highlighted the diversity of stellar and planetary processes that can affect the range of orbital distances at which a planet is likely to be habitable.

4.4. Earth-Like Planets Around Non-Sun-Like Stars

Recent research focus on the effect of stellar radiation on planetary atmospheric photochemistry has helped to better understand the equilibrium composition and the lifetime of gases in planetary atmospheres irradiated by stars of different spectral type. Calculations typically use atmospheric photochemistry or coupled climate-photochemical models. The models have been used to explore the effects of stellar radiation on an Earth-like atmosphere's ozone layer, and the corresponding surface UV flux, and on the production and destruction of compounds that have biological sources, such as oxygen and biogenic sulfur gases.

Segura et al. (2003) used a coupled climate-photochemical model to simulate Earth-like atmospheres with O_2 concentrations from present atmospheric levels (PAL) down to 10^{-5} PAL for planets in the habitable zones around F, G, and K stars (Fig. 14). Their results showed the effectiveness of oxygenated atmospheres in producing ozone layers that adequately shield the planetary surface in response to UV from all spectral types. This protective effect broke down at O_2 levels below 10^{-2} PAL, however, and planets around all three types of stars were subject to dangerous surface radiation environments at these lower oxygen levels. In a complementary paper, *Segura et al.* (2005) showed that an O_3 layer would also be formed in Earth-like planets orbiting M stars, and could be as astronomically detectable as those for planets around F, G, and K stars. *Segura et al.* (2005) also showed that the stellar spectrum and lowered UV flux greatly increased the lifetime of biosignatures gases such as CH_4 and N_2O, and introduced CH_3Cl as a potential biosignature gas under these conditions. These gases have relatively weak absorption features in Earth's spectrum, but their enhanced atmospheric lifetime around M stars results in stronger features in the disk-averaged planetary spectrum than would be seen if the planet were orbiting a G star like our Sun.

Grenfell et al. (2007) used a similar photochemical climate modeling technique to explore the effect of orbital distance within the habitable zone on the photochemistry of Earth-like planets orbiting F, G, and K stars. They found, perhaps counterintuitively, that O_3 concentrations increased as planets moved further from the star due to slowing of the destruction reaction rate with lower temperatures. The biosignature gases CH_4 and CH_3Cl increased in abundance due to a drop-off in UV-mediated destruction via the water molecule, itself also less abundant further from the star due to temperature-induced removal from the planetary atmosphere via enhanced condensation.

Terrestrial exoplanet atmosphere modeling efforts have helped us to explore the diversity of atmospheric conditions and resulting spectra for even strictly Earth-like planets as they are acted upon by different stellar spectral energy distributions. Perhaps most importantly, these studies allow us to identify the effect of wavelength-dependent changes in stellar spectra on the critical photochemical reactions for production and destruction of relevant species, and to search for trends in photochemical responses to the parent star.

5. BIOSIGNATURES

Another area of research at the forefront of exoplanet atmosphere science is the field of biosignatures. This field identifies global modifications to a planet's environment that can be detected remotely in disk-averaged observations.

From studying Earth we know that these biosignatures can be divided into at least three main classifications: atmospheric gases, surface reflectivity features, and time-dependent changes in either atmospheric gases or surface reflectivity. Surface reflectivity features, such as the "red edge" of vegetation, are discussed in section 3.1.1. Here we concentrate on biosignatures associated with atmospheric gases.

An atmospheric biosignature is a gas whose presence in a planetary atmosphere indicates that the planet likely harbors life. To be classified as a biosignature, a gas must be generated by life (typically as a metabolic byproduct, but

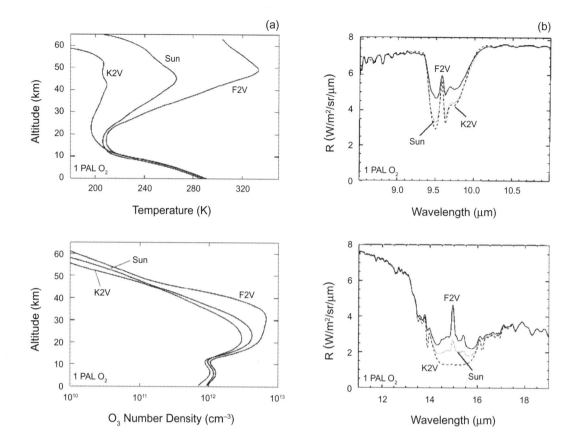

Fig. 14. (a) The vertical temperature structure and ozone mixing ratio for an Earth-like planet in the habitable zone of an F2V, G2V (the Sun), and K2V star. **(b)** Spectra showing the shape and strength of the O_3 9.6-µm and CO_2 bands for Earth-like planets around F, G, and K stars. Note that although the planet around the F2V star produces a "super-ozone" layer due to the star's higher UV output, the resultant stratospheric heating results in a weaker 9.6-µm ozone absorption band feature, due in part to a changed atmospheric temperature gradient. The CO_2 band is also shown to be diagnostic of the planetary temperature structure and the presence of ozone, with ozone-poor cooler stratospheres, such as those around K stars, having a much different CO_2 band shape than hotter F-star planet stratospheres. From *Segura et al.* (2003).

potentially also from biomass degradation or burning), must accumulate in a planetary atmosphere to sufficient levels to be detectable, and must be spectrally active in the wavelength range being observed.

The ideal atmospheric biosignature gas would therefore (1) not exist naturally in the planetary atmosphere at ambient temperatures and pressures, (2) not be created by geophysical processes, (3) not be produced by photochemistry, (4) not be significantly destroyed by photochemistry, and (5) have a strong spectral signature.

Gases that can be produced by life, but are also produced by geophysical or photochemical processes, are potential "false positives" for biosignature gases.

5.1. Redox Biosignatures and Chemical Disequilibrium

The canonical concept for the search for atmospheric biosignatures is to find an atmosphere severely out of thermodynamic equilibrium (*Lederberg*, 1965; *Lovelock*, 1965). All atmospheres are out of equilibrium due to the incoming energy from the star. In defining a biosignature, the question is the degree of disequilibrium, and whether or not it can be explained by abiotic planetary processes alone. Redox reactions are those where electrons are transferred between molecules (and/or compounds) and are central to all life on Earth. Redox reactions are used by life to exploit chemical energy gradients, and are also a key part of metabolism, which includes production of energy-storing materials. Some products of redox reactions end up as "waste," and can accumulate in the atmosphere if photochemical destruction pathways are slow. The active production of these gases by life can drive a detectable chemical disequilibrium in the atmosphere.

Earth's atmospheric oxygen and methane, both produced by life, are out of thermodynamic equilibrium and if detected together, would form an extremely robust biosignature. O_2 makes up 21% of Earth's atmosphere by volume, yet O_2 is highly reactive and therefore will remain in significant quantities in the atmosphere only if it is continually produced (*Catling and Claire*, 2005). On Earth, plants and photosynthetic bacteria generate oxygen as a metabolic byproduct; there are no continuous abiotic sources of large quantities of O_2. Ozone (O_3) is a photolytic product of O_2, generated after O_2 is split up by solar UV radiation. In the mid-infrared spectral range O_3 is more detectable than oxygen, and so often serves as a proxy for the presence of oxygen in an atmosphere. CH_4, which may have been produced in large quantities by methanogenic bacteria on early Earth, is considered a biosignature gas, although CH_4 also has abiotic origins, including hydrothermal activity and water/rock reactions. In a strongly oxidized atmosphere like the modern Earth's, methane has a lifetime of only 8.4 years, and must be continually replenished to maintain its current abundance.

In practice it could be difficult to detect a pair of gases in thermal disequilibrium. Earth has a relatively prominent oxygen absorption feature at 0.76 µm (Fig. 8), whereas methane at present-day levels of 1.6 ppm has only extremely weak spectral features. Early in Earth's history CH_4 may have been present at much higher levels (1000 ppm or even 1%), possibly produced by widespread methanogenic bacteria (*Haqq-Misra et al., 2008,* and references therein). Such high CH_4 concentrations would be easier to detect, but since Earth was not oxygenated until 2.5 Ga, O_2 and CH_4 would not be detectable concurrently (see *Des Marais et al., 2002*). However, in the Proterozoic eon between 2.5 and 0.53 Ga, oxygen concentrations of a few thousand ppm may have coexisted with methane concentrations of 20–100 ppm. A spectroscopic detection of life on Earth may therefore have been easier during these 2 b.y. than in the present Phanerozoic eon (*Segura et al., 2003*).

A simpler, but arguably less robust biosignature is a single gas completely out of chemical equilibrium. Oxygen and ozone are Earth's two most detectable single biosignature gases (*Léger et al., 1993*). Nitrous oxide (N_2O) is also produced by life — albeit in small quantities — during microbial redox reactions, and so could also stand as a single biosignature gas, although like methane, it is also more compelling in the presence of oxygen.

In many cases a single gaseous biosignature could also be made more robust by obtaining time-resolved observations of the gas, which if produced by life, may exhibit seasonal variations in production. Seasonal variations are seen in both Earth's CH_4 and CO_2, at the level of 1–2% or less. However, in the disk-average, such a seasonal variation can only be seen if there is a hemispherical asymmetry in production as viewed by the observer. This could occur if the life that was producing the gas was present predominantly on one hemisphere of the planet. This is actually the case for an equatorial view of the seasonal production of CO_2 from Earth's land-dominated northern hemisphere, when disk-averaged with the out-of-phase, minimal production from the ocean-dominated southern hemisphere.

A major challenge with a using a single biosignature outside the context of redox chemistry is vulnerability to false positives. To avoid false positives we must look at the whole atmospheric context. For example, a high atmospheric oxygen content might indicate a planet with life, or one undergoing a runaway greenhouse with evaporating oceans. When water vapor in the atmosphere is being photodissociated with H escaping to space, O_2 will build up in the atmosphere for a short period of time. In this case, O_2 can be associated with a runaway greenhouse via very saturated water vapor features, since the atmosphere would be filled with water vapor at all altitudes (*Schindler and Kasting,* 2000). Other O_2 and O_3 false positive scenarios (*Selsis et al.,* 2002) are discussed and countered in *Segura et al.* (2007).

5.2. Earth's Minor Biosignature Gases Amplified

Most work to date has focused on extrapolations of exoplanet biosignatures as seen on modern-day Earth (O_2, O_3, N_2O) or early Earth (possibly CH_4) biosignatures. Research into biosignature gases that are produced by metabolisms that are not currently globally significant on Earth is underway. *Pilcher* (2003) suggested that organosulfur compounds, particularly methanethiol (CH_3SH, the sulfur analog of methanol) could be produced in high enough abundance by bacteria, possibly creating a biosignature on other planets. Pilcher emphasized a potential ambiguity in interpreting the 9.6 µm O_3 spectral feature since a CH_3SH feature overlaps with it. It is interesting to note that that CH_3SH and dimethylsulfide (DMS: CH_3SCH_3) (*Pilcher,* 2003) are not produced as redox metabolic byproducts, but instead originate from secondary metabolism occurring in specific organisms.

The reduced UV radiation on quiet M stars, as emphasized in *Segura et al.* (2005), enables longer biosignature gas lifetimes and therefore higher concentrations to accumulate. Segura et al. showed that the Earth-like biosignature gases CH_4, N_2O, and even biogenic methyl chloride (CH_3Cl) have higher concentrations and therefore stronger spectral features on planets orbiting M stars compared to Earth.

5.3. Beyond Terracentrism

The NRC report on "The Limits of Organic Life in Planetary Systems" (*Baross et al.,* 2007) proposed that the conservative requirements for life of liquid water and carbon could be replaced by the more general requirements of a liquid and an environment that can support covalent bonds (especially between hydrogen, carbon, and other atoms). This potentially opens up a new range for habitable planets, namely those beyond the ice line in a "cryogenic" habitable zone where water is frozen but other liquids such as methane and ethane are present. Although no one has yet studied the possibility of exoplanet biosignatures on cold exoplanets, research on life in nonwater liquids (*Bains,* 2004) and the possibilities for methanogenic life in liquid methane on the surface of Titan (*McKay and Smith,* 2005) are a useful start.

5.4. Unique and Nonunique Biosignature Gases

If we use the criteria for an ideal biosignature gas given at the beginning of this section, then Earth's most robust biosignature gas O_2 satisfies all five of the criteria and N_2O satisfies the first three. Even though O_2 and N_2O are also generated by photochemistry, these abiotic production rates are minuscule when compared to the biological sources.

Microbial life on Earth emits a much larger suite of metabolic byproducts than the few we have discussed so far, including H_2, CO_2, N_2, NO, NO_2, H_2S, and SO_2 (e.g., *Madigan et al.*, 2001). Here we consider the feasibility of each as a biosignature.

On Earth, the gas N_2 makes up 78% of our atmosphere, and although it has strong biological production, the question of whether it is a true biosignature is moot because as a homonuclear molecule N_2 has no rotational-vibrational transitions and hence no spectral signature at visible and infrared wavelengths, failing criterion 5.

CO_2, making up about 0.035% of our atmosphere, is demonstrably produced by life, but also has strong geological sources, including volcanism. For Venus and Mars (the only other solar system terrestrial planets with an atmosphere), CO_2 makes up more than 97% of the atmosphere. As such, CO_2 is considered a gas that results from planet formation and evolution, and hence is not a useful biosignature. Similarly, the gases H_2, H_2S, SO_2, NO, and NO_2 are produced on Earth either by volcanos or by photochemistry, making them nonunique biosignatures. On modern Earth these gases have a very low abundance in the atmosphere and hence also lack any detectable spectral signature for a remote observer.

Many exoplanets may simply not have a unique biosignature like Earth's O_2 and N_2O. In the preparation for the search for exoplanet biosignatures, we must aim to consider the more common gases produced both as metabolic byproducts and by either geophysical or photochemical processes. To understand their potential as biosignatures, we have to understand and model the sources and sinks — i.e., model exoplanet environment scenarios to determine under which cases the common gases would be much more abundant than can reasonably be produced by geologic or photochemical processes. In other words, absent a unique biosignature such as the simultaneous presence of $O_2/O_3/CH_4$, the main criterion for a biosignature is that the gas exists in such great quantities that its presence in a planetary atmosphere is well above the amounts that could be produced by any abiotic processes. In such a scenario we may be relatively certain but never 100% positive that we have found signs of life on another planet. Alternatively, we may also be guided toward identifying biosignatures without *a priori* assumptions, by characterizing a planetary environment sufficiently well that we can identify an atmosphere that is out of chemical equilibrium, with likely active biological sources.

6. FUTURE PROSPECTS

The search for terrestrial exoplanets is just beginning. Many exciting opportunities for both modeling advances and observational discoveries await us in the coming decade.

6.1. Atmosphere Models

The field of modeling of exoplanet environments is still largely in its infancy, and is certainly hampered by the relative dearth of observational constraints. Paradoxically, until sufficient data for terrestrial exoplanets are obtained, models are our only means of exploring planetary environment parameter space, assuming that the laws of physics and chemistry are universal.

The main challenge facing terrestrial exoplanet atmosphere models is the ability to develop and run more self-consistent models for planetary atmospheres unlike those found in our solar system. A large fraction of existing climate and photochemical models were specifically developed to model Earth. These models often contain "shortcuts," approximations that are valid for the Earth system, but preclude accurate calculations of the conditions for planets with different atmospheric or cloud compositions, or temperature and pressure structures. In the next five years, versatile, rigorous self-consistent photochemical and climate models that include realistic clouds and hazes will likely be developed to simulate a very wide range of planets with atmospheric conditions very unlike our own. In the longer term, further development will include the coupling of boundary condition models, such as atmospheric escape, with geological outgassing, weathering, ocean, microbial biosphere, and land vegetation models. These interdisciplinary coupled models will allow us to take the first steps toward understanding not just the interplay between exoplanet environmental components to produce a resultant environment, but the processes that drive the evolution of the atmosphere and surface over time.

Parameter space for forward terrestrial atmosphere models is daunting, especially for the ideally complete model described above. A complementary modeling area ripe for signficant progress in the next decade is retrieval techniques. Retrieval techniques depend on inverse models that can be used to quantify planetary properties from observational data. Exoplanet atmosphere retrieval techniques will differ significantly from remote sensing retrieval algorithms currently used by the Earth-observing community, who have the luxury of spatial resolution, and do not have to face the challenge of a spatially inhomogeneous, yet unresolved surface for which the combination of viewing angle, solar illumination, and surface type will not be known. Statistical techniques to deal with what is initially likely to be extremely limited data will also need to be further developed. Model fits to a few data points are likely to be highly degenerate [as are current datasets for hot Jupiters (see *Seager and Deming*, 2010)]. Retrieval techniques to constrain the range of possible values consistent with the data will need to be developed (e.g., *Madhusudhan and Seager*, 2009).

A third atmosphere model area that requires improvement is in model input in the form of molecular line lists. Descriptions of molecular spectral activity, such as line lists and absorption cross-sections, are needed for planetary temperatures and pressures that are far removed from standard temperature and pressure, or even from environments in our own solar system. In particular, super Earths encourage us to better understand molecular behavior, including collisionally induced absorption at higher pressures than are typical for the solar system's terrestrial planets.

The "holy grail" of exoplanets is to identify a habitable planet. Until observation of habitable-zone atmospheres are both possible and plentiful, the question of what constitutes a habitable zone will receive major theoretical focus. Early definitions of the habitable zone, while seminal (*Kasting et al.*, 1993), appear too narrow for today's wide diversity of exoplanets. These early models largely ignored the radiative effect of clouds and aerosols on the limits of the habitable zone, and used blackbodies to characterize stellar irradiation. Revisiting the habitable zone will require new modeling for planets in Earth to super Earth mass ranges and for very different planetary composition. The effects of clouds, aerosols, and ocean feedback on the runaway and moist greenhouse scenarios will need to be remodeled, and an improved understanding of the limits of greenhouse warming (especially from reduced atmospheres) and cloud feedback on the outer limits of the habitable zone also needs to be explored. Additional factors, such as realistic stellar energy distributions, stellar activity, orbital evolution, and tidal heating will also need to be folded into our consideration of what constitutes a habitable zone.

6.2. Anticipated Observations

NASA's Kepler space telescope, launched in March 2009, promises the most significant terrestrial planet discoveries in the short term. Kepler aims to determine the fraction of stars that harbor Earth-sized planets in Earth-like orbits. Knowing whether Earth-sized planets are common or rare will be a major scientific advancement, and will also be an important design driver for future terrestrial planet detection and characterization missions. However, Kepler's deep pencil beam survey means that the planets are too distant and therefore too faint for follow-up observations of atmospheric characteristics.

Atmosphere observations may soon be possible for a potentially habitable planet different from Earth analogs: transiting super Earths orbiting M dwarfs. An M dwarf has a much lower luminosity than the Sun, and the corresponding M-dwarf habitable zone is closer to its star. This greatly increases the probability of seeing the habitable planet in transit as both the geometric probability of detecting the transit and the depth of the transit signal increase. The transit probability for habitable M-dwarf planets is about 2.3–5%, significantly higher than the 0.5% probability of an Earth-Sun analog transit. Similarly, the magnitude of the planet transit signature is proportional to the planet-to-star area ratio and can be up to 100 times stronger for an Earth-sized planet orbiting a low-mass star compared to an Earth-Sun analog. Because of the habitable planet's proximity to the host star, the radial velocity signal is also enhanced, and may be up to 30 times easier to detect than for an Earth-Sun analog. The M-dwarf planet's close-in orbit also results in a period up to 90 times shorter than Earth's, allowing multiple observations a year, to improve the S/N of the detection. A super Earth larger than Earth (and up to about 10 $M_\oplus$ and 1.75 $R_\oplus$) is even easier to detect due to its larger transit signal and mass signature than Earth.

To study atmospheres of the transiting super Earths orbiting M stars, hopes are pinned on the James Webb Space Telescope (JWST), scheduled for launch in 2014. The JWST is a large (6.5 m diameter), infrared-optimized space telescope with instrumentation spanning 0.6–27 µm. With its large mirror and suite of infrared instrumentation, JWST will allow us to study transiting super Earth atmospheres using techniques analogous to those currently being used to study hot Jupiters. Observations to detect CO_2 and H_2O in the thinner atmospheres of potentially habitable terrestrial planets will be challenging when compared to the puffy, large-scale-height atmospheres of hot Jupiters. The observations are also likely to be time consuming, with multiple transits required to build sufficient S/N. One estimate suggests that 85 hours of JWST observations would be required to measure CO_2 to a S/N of 28 for a habitable super Earth at a distance of 22 pc (*Deming et al.*, 2009). JWST's targets will likely be a handful of rare, but highly valuable, transiting super Earths in the habitable zones of nearby low-mass stars. Groundbased surveys such as MEarth (*Nutzmann and Charbonneau*, 2008) and spacebased surveys such as the proposed TESS (*Ricker et al.*, 2010) will search M stars to find the most favorable planet-star systems.

The ultimate goal in terrestrial planet studies is the discovery and characterization of a true Earth analog: an Earth-mass, Earth-sized planet in an Earth-like orbit about a Sun-like star. Direct imaging is the only method to provide numerous targets accessible for atmospheric spectroscopy. The most ambitious mission concept planned to date is a spaceborne telescope that could directly image a terrestrial exoplanet. Direct imaging would allow us to obtain time-resolved photometry and spectra for planets in the habitable zones of their parent stars. These observations would aim to detect signs of habitablility and biosignatures (see the chapter by Seager and Meadows).

Direct imaging of terrestrial exoplanets is incredibly challenging, due to the enormous star/planet flux ratio. For example, Earth is much fainter than the Sun (10^7 fainter at mid-infrared wavelengths and 10^{10} at visible wavelengths). In other words, to observe a distant Earth-analog planet at visible wavelengths, the telescope must not only be sufficiently large for angular separation of the planet and star, but must also be capable of supressing the diffracted light from the star by 10 billion times. This level of suppression is not possible with even the largest of the planned future groundbased telescopes. Although no direct imaging space missions are planned, technology development is ongoing. A collection of related ideas to block out the starlight (at any wavelength) are referred to as "Terrestrial Planet Finder" missions. See the chapter by Traub and Oppenheimer for details on direct imaging missions from space.

A recent development has given renewed promise for a Terrestrial Planet Finder type of mission. The idea is to use the already planned JWST (launch date 2014) together with a novel-shaped external occulter placed tens of thousands of kilometers from the telescope in order to suppress the diffracted starlight (*Cash*, 2006; *Soummer et al.*, 2009). Most of the time the JWST would be functioning as planned, and during this time the starshade would fly across the sky to the next target star. While there are many technical and programmatic concerns for the JWST and external occulter idea, none currently appear to be without solution.

Terrestrial exoplanet science is poised to expand significantly in the coming decade. We anticipate that we will discover Earth-sized planets within the habitable zones of their parent stars, the measurement of transmission and secondary eclipse spectra of super Earth atmospheres, and the development of space missions to find and characterize Earth analogs. Advances in observations will allow us for the first time to undertake the scientific search for habitable worlds and life beyond our solar system. Advances in modeling will allow us to direct and focus our search, and will teach us how to interpret the observations to determine planetary characteristics and recognize signs of life. This is indeed an extremely exciting time, as we are privileged to live in an era that provides the scientific capability to address questions that humanity has been pondering for millenia.

Acknowledgments. This work was partially supported by the NASA Astrobiology Institute.

REFERENCES

Arnold L. (2008) Earthshine observation of vegetation and implication for life detection on other planets: A review of 2001–2006 works. *Space Sci. Rev., 135,* 323–333.

Arnold L., Gillet S., Lardière O., Riaud P., and Schneider J. (2002) A test for the search for life on extrasolar planets. Looking for the terrestrial vegetation signature in the Earthshine spectrum. *Astron. Astrophys., 392,* 231–237.

Bains W. (2004) Many chemistries could be used to build living systems. *Astrobiology, 4,* 137–167.

Barnes R., Jackson B., Greenberg R., and Raymond S. N. (2009) Tidal limits to planetary habitability. *Astrophys. J. Lett., 700,* L30–L33.

Barnes R., Raymond S., Greenberg R., Jackson B., and Kaib N. (2010) CoRoT-7 b: Super-Earth or Super-Io? *Astrophys. J. Lett., 709,* L95–L98.

Baross J. A., Benner S. A., Cody G. D., Copley S. D., Pace N. R., et al. (2007) *The Limits to Organic Life in Planetary Systems.* National Academies, Washington, DC.

Beaulieu J.-P., Bennett D. P., Fouqu P., Williams A., Dominik M., et al. (2006) Discovery of a cool planet of 5.5 Earth masses through gravitational microlensing. *Nature, 439(7075),* 437–440.

Bennett D., Bond I., Udalski A., Sumi T., Abe F., et al. (2008) A low-mass planet with a possible sub-stellar-mass host in microlensing event MOA-2007-BLG-192. *Astrophys. J., 684,* 663–683.

Briot D., Arnold L., Jacquemoud S., Schneider J., Agabi A., et al. (2010) The LUCAS experiment: Spectroscopy of Earthshine in Antarctica for detection of life. *EAS Publ. Ser., 40,* 361–365.

Cash W. (2006) Detection of Earth-like planets around nearby stars using a petal-shaped occulter. *Nature, 442,* 51–53.

Catling D. and Claire M. (2005) How Earth's atmosphere evolved to an oxic state: A status report. *Earth Planet. Sci. Lett., 237,* 1–20.

Catling D. and Kasting J. F. (2007) Planetary atmospheres and life. In *Planets and Life: The Emerging Science of Astrobiology* (W. Sullivan and J. Baross, eds.), pp. 91–116. Cambridge Univ., Cambridge.

Catling D., Zahnle K., and McKay C. (2001), Biogenic methane, hydrogen escape, and the irreversible oxidation of early Earth. *Science, 293,* 839–843.

Charbonneau D., Berta Z., Irwin J., Burke Ch., Nutzman Ph., et al. (2009) A super-Earth transiting a nearby low mass star. *Nature, 462,* 891.

Chassefière E. (1996) Hydrodynamic escape of oxygen from primitive atmospheres: Applications to the cases of Venus and Mars. *Icarus, 124,* 537–552.

Christensen P. and Pearl J. C. (1997) Initial data from the Mars Global Surveyor thermal emission spectrometer experiment: Observations of the Earth. *J. Geophys. Res., 102(E5),* 10875–10880.

Clark R. N., Swayze G. A., Gallagher A. J., King T.V.V., and Calvin W. M. (1993) *The U.S. Geological Survey, Digital Spectral Library: Version 1: 0.2 to 3.0 Microns.* U.S. Geological Survey Open File Report 93-592, 1340 pp.

Cox C. and Munk W. (1954a) Statistics of the sea surface derived from Sun glitter. *J. Marine Res., 13,* 198–227.

Cox C. and Munk W. (1954b) Measurement of the roughness of the sea surface from photographs of the Sun's glitter. *J. Optical Soc. Am., 44,* 838–850.

Cowan N. B., Agol E., Meadows V. S., Robinson T., Livengood T. A., et al. (2009) Alien maps of an ocean-bearing world. *Astrophys. J., 700,* 915–923.

Dehant V., Lammer H., Kulikov Y., Griemeier J. M., Breuer D., et al. (2007) Planetary magnetic dynamo effect on atmospheric protection of early Earth and Mars. *Space Sci. Rev., 129(1),* 279–300.

Deming D., Seager S., Winn J., Miller-Ricci E., Clampin M., et al. (2009) Discovery and characterization of transiting super Earths using an all-sky transit survey and follow-up by the James Webb Space Telescope. *Publ. Astron. Soc. Pac., 121,* 952–967.

Des Marais D. J., Harwit M. O., Jucks K. W., Kasting J. F., Lin D. N. C., et al. (2002) Remote sensing of planetary properties and biosignatures on extrasolar terrestrial planets. *Astrobiology, 2(2),* 153–181.

Elkins-Tanton L. and Seager S. (2008) Ranges of atmospheric mass and composition of super-Earth exoplanets. *Astrophys. J., 685,* 1237–1246.

Ford E. B., Seager S., and Turner E. L. (2001) Characterization of extrasolar terrestrial planets from diurnal photometric variability. *Nature, 412,* 885–887.

Forget F. and Pierrehumbert R. T. (1997) Warming early Mars with carbon dioxide clouds that scatter infrared radiation. *Science, 278,* 1273–1276.

Gaucher E. A., Thomson J. M., Burgan M. F., and Benner S. A. (2003) Inferring the palaeoenvironment of ancient bacteria on the basis of resurrected proteins. *Nature, 425,* 285–288.

Goody R. M. and Yung Y. L. (1989) *Atmospheric Radiation: Theoretical Basis, 2nd edition.* Oxford Univ., Oxford, New York.

Grenfell J. L., Stracke B., von Paris P., Patzer B., Titz R., Segura A., and Rauer H. (2007) The response of atmospheric chemistry on Earthlike planets around F, G and K stars to small variations in orbital distance. *Planet. Space Sci., 55(5),* 661–671.

Haqq-Misra J. D., Domagal-Goldman S. D., Kasting P. J., and Kasting J. F. (2008) A revised, hazy methane greenhouse for the Archean Earth. *Astrobiology, 8,* 1127–1137.

Hunten D. M. (1982) Thermal and nonthermal escape mechanisms for terrestrial bodies. *Planet. Space Sci., 30,* 773–783.

Hunten D. M. and Münch G. (1973) The helium abundance on Jupiter. *Space Sci. Rev., 14,* 433–443.

Ingersoll A. (1969) The runaway greenhouse: A history of water on Venus. *J. Atmos. Sci., 26,* 1191–1198.

Jacquinet-Housson N., Scott N. A., Chédin A., Crepeau L., Garceran K., Armante R., et al. (2005) The 2003 edition of the GEISA/IASI spectroscopic database. *J. Quant. Spectrosc. Radiat. Transfer, 95,* 429–467.

Jakosky B. M. and Phillips R. J. (2001) Mars' volatile and climate history. *Nature, 412,* 237–244.

Kaltenegger L. and Sasselov D. (2010) Detecting planetary geochemical cycles on exoplanets: Atmospheric signatures and the case of SO_2. *Astrophys. J., 708,* 1162–1167.

Kasting J. F. and Pollack J. B. (1983) Loss of water from Venus. I — Hydrodynamic escape of hydrogen. *Icarus, 53,* 479–508.

Kasting J. F., Pollack J. B., and Ackerman T. P. (1984) Response of Earth's atmosphere to increases in solar flux and implications for loss of water from Venus. *Icarus, 57,* 335–355.

Kasting J. F., Whitmire D. P., and Reynolds R. T. (1993) Habitable zones around main sequence stars. *Icarus, 101,* 108–128.

Kiang N. Y., Segura A., Tinetti G., Govindjee, Blankenship R. E., Cohen M., Siefert J., Crisp D., and Meadows V. S. (2007) Spectral signatures of photosynthesis. II. Coevolution with other stars and the atmosphere on extrasolar worlds. *Astrobiology, 7(1),* 252–274.

Kuchner M. J. (2003) Volatile-rich Earth-mass planets in the habitable zone. *Astrophys. J. Lett., 596,* L105–L108.

Lammer H., Lichtenegger H. I. M., Kulikov Y. N., Griessmeier J.-M., Terada N., et al. (2007) Coronal mass ejection (CME) activity of low mass M stars as an important factor for the habitability of terrestrial exoplanets. II. CME-induced ion pick up of Earth-like exoplanets in close-in habitable zones. *Astrobiology, 7,* 185–207.

Lammer H., Kasting J. F., Chassefière E., Johnson R. E., Kulikov Y. N., and Tian F. (2009) Atmospheric escape and evolution of terrestrial planets and satellites. *Space Sci. Rev., 139,* 399–436.

Lederberg J. (1965) Signs of life — Criterion-system of exobiology. *Nature, 207,* 9–13.

Léger A., Pirre M., and Marceau F. J. (1993) Search for primitive life on a distant planet: Relevance of O_2 and O_3 detections. *Astron. Astrophys., 277,* 309.

Léger A., Selsis F., Sotin C., Guillot T., Despois D., et al. (2004) A new family of planets? "Ocean-planets." *Icarus, 159,* 499–504.

Léger A., Rouan D., Schneider J., Barge P., Fridlund M., et al. (2009) Transiting exoplanets from the CoRoT space mission VIII. CoRoT-7b: The first Super-Earth with measured radius. *Astron. Astrophys., 506,* 287–302.

Lipatov A. S. (1978) The induced magnetosphere of Venus. *Cosmic Res., 16(3),* 346–349.

Livengood T. A., A'Hearn M. F., Deming D., Charbonneau D., Hewagama T., Lisse C. M., McFadden L. A., Meadows V. S., Seager S., Wellnitz D. D., and the EPOXI-EPOCh Science Team (2008) EPOXI empirical test of optical characterization of an Earth-like planet. *Bull. Am. Astron. Soc., 40,* 385.

Lovelock J. (1965) A physical basis for life detection experiments. *Nature, 207,* 568–570.

Madhusudhan N. and Seager S. (2009) A temperature and abundance retrieval method for exoplanet atmospheres. *Astrophys. J., 707,* 24–39.

Madigan M. M., Martinko J., and Parker J. (2001) *Brock Biology of Microorganisms.* Prentice Hall, Boston.

Mayor M., Bonfils X., Forveille Th., Delfosse X., Udry S., et al. (2009) The HARPS search for southern extra-solar planets XVIII. An Earth-mass planet in the GJ 581 planetary system. *Astron. Astrophys., 507,* 487–494.

McKay C. P. and Smith H. D. (2005) Possibilities for methanogenic life in liquid methane on the surface of Titan. *Icarus, 178,* 274–276.

Meadows V. S. (2006) Modelling the diversity of extrasolar terrestrial planets. In *Direct Imaging of Exoplanets: Science and Techniques* (C. Aime and F. Vakili, eds.), p. 25–34. IAU Colloquium Series 200, Cambridge Univ., Cambridge.

Meadows V. S. and Crisp D. (1996) Ground-based near-infrared observations of the Venus nightside: The thermal structure and water abundance near the surface. *J. Geophys. Res., 101,* 4595–4622.

Miller-Ricci E., Seager S., and Sasselov D. (2008) The atmospheric signatures of super-Earths: How to distinguish between hydrogen-rich and hydrogen-poor atmospheres. *Astrophys. J., 690,* 1056–1067.

Mitchell D. L., Lin R. P., Mazelle C., Rème H., Cloutier P. A., Connerney J., Acuña M. H., and Ness N. F. (2001) Probing Mars' crustal magnetic field and ionosphere with the MGS Electron Reflectometer. *J. Geophys. Res., 106(E10),* 23419–23428.

Montañés-Rodríguez P., Pallé E., Goode P. R., and Martín-Torres F. J. (2006) Vegetation signature in the observed globally integrated spectrum of Earth considering simultaneous cloud data: Applications for extrasolar planets. *Astrophys. J., 651,* 544–552.

Nimmo F. (2002) Why does Venus lack a magnetic field? *Geology, 30,* 987–990.

Nutzman P. and Charbonneau D. (2008) Design considerations for a ground-based transit search for habitable planets orbiting M dwarfs. *Publ. Astron. Soc. Pac., 120,* 317–327.

Pallé E., Ford E. B., Seager S., Montañés-Rodríguez P., and Vazquez M. (2008) Identifying the rotation rate and the presence of dynamic weather on extrasolar Earth-like planets from photometric observations. *Astrophys. J., 676,* 1319–1329.

Pearl J. C. and Christensen P. R. (1997) Mars Global Surveyor Thermal Emission Spectrometer: Observations of Earth. *Bull. Am. Astron. Soc., 29,* 970.

Phillips R. J. and Hansen V. L. (1998) Geological evolution of Venus: Rises, plains, plumes, and plateaus. *Science, 279,* 1492–1497.

Pilcher C. B. (2003) Biosignatures of early Earths. *Astrobiology, 3,* 471–486.

Queloz D., Bouchy F., Moutou C., Hatzes A., Hebrard G., et al. (2009) The CoRoT-7 planetary system: Two orbiting super-Earths. *Astron. Astrophys., 506,* 303.

Ragenauer-Lieb K., Yuen D., and Branlund J. (2001) The initiation of subduction: Criticalilty by addition of water? *Science, 294,* 578–581.

Ribas I., Guinan E. F., Gudel M., and Audard M. (2005) Evolution of the solar activity over time and effects on planetary atmospheres. I. High-energy irradiances (1–1700 Å). *Astrophys. J., 622,* 680–694.

Ricker G. R., Latham D. W., Vanderspek R. K., Ennico K. A., Bakos G., et al. (2010) Transiting Expoplanet Survey Satellite (TESS). *Bull. Am. Astron. Soc., 42,* 459.

Robinson T. D., Meadows V. S., and Crisp D. (2010) Detecting oceans on extrasolar planets using the glint effect. *Astrophys. J. Lett.,* in press.

Rogers L. A. and Seager S. (2010) A framework for quantifying the degeneracies of exoplanet interior compositions. *Astrophys. J., 712(2)*, 974–991.

Rothman L. S., Gordon I. E., Barbe A., Benner D. C., Bernath P. F., et al. (2009) The HITRAN 2008 molecular spectroscopic database. *J. Quant. Spectrosc. Radiat. Transfer, 110*, 9–10.

Russell C. T. (1976) The magnetosphere of Venus: Evidence for a boundary layer and a magnetotail. *Geophys. Res. Lett., 3(10)*, 589–590.

Rye R., Kuo P. H., and Holland H. D. (1995) Atmospheric carbon dioxide concentrations before 2.2 billion years ago. *Nature, 378*, 603–605.

Sagan C., Thompson W. R., Carlson R., Gurnett D., and Hord C. (1993) A search for life on Earth from the Galileo spacecraft. *Nature, 365*, 715–721.

Schaefer L. and Fegley B. (2009) Chemistry of silicate atmospheres of evaporating super-Earths. *Astrophys. J. Lett., 703(2)*, L113–L117.

Schindler T. L. and Kasting J. F. (2000) Synthetic spectra of simulated terrestrial atmospheres containing possible biomarker gases. *Icarus, 145*, 262–271.

Seager S. and Deming D. (2009) On the method to infer an atmosphere on a tidally locked super Earth exoplanet and upper limits to GJ 876d. *Astrophys. J., 703*, 1884–1889.

Seager S. and Deming D. (2010) Exoplanet atmospheres. *Annu. Rev. Astron. Astrophys., 48*, 631–672.

Seager S., Turner E. L., Schafer J., and Ford E. B. (2005) Vegetation's red edge: A possible spectroscopic biosignature of extraterrestrial plants. *Astrobiology 5*, 372–390.

Segura A., Krelove K., Kasting J. F., Sommerlatt D., Meadows V., Crisp D., Cohen M., and Mlawer E. (2003) Ozone concentrations and ultraviolet fluxes on Earth-like planets around other stars. *Astrobiology, 3*, 689–708.

Segura A., Kasting J. F., Meadows V., Cohen M., Scalo J., Crisp D., Butler R. A. H., and Tinetti G. (2005) Biosignatures from Earth-like planets around M dwarfs. *Astrobiology, 5*, 706–725.

Segura A., Meadows V. S., Kasting J. F., Cohen M., and Crisp D. (2007) Abiotic formation of O_2 and O_3 in high-CO_2 terrestrial atmospheres. *Astron. Astrophys., 472*, 665–679.

Selsis F., Despois D., and Parisot J.-P. (2002) Signature of life on exoplanets: Can Darwin produce false positive detections? *Astron. Astrophys., 388*, 985–1003.

Selsis F., Kasting J. F., Levrard B., Paillet J., Ribas I., and Delfosse X. (2007) Habitable planets around the star Gliese 581? *Astron. Astrophys., 476*, 1373–1387.

Shaw J. A. and Churnside J. H. (1997) Scanning-laser glint measurements of sea-surface slope statistics. *Appl. Opt., 36*, 4202–4213.

Shizgal B. D. and Arkos G. G. (1996) Nonthermal escape of the atmospheres of Venus, Earth, and Mars. *Rev. Geophys., 34*, 483–505.

Slavin J. A., Krimigis S. M., Acuña M. H., Anderson B. J., Baker D. N., et al. (2007) MESSENGER: Exploring Mercury's magnetosphere. *Space Science Rev., 131(1–4)*, 133–160.

Smith P. H., Lemmon M. T., Lorenz R. D., Sromovsky L. A., Caldwell J. J., and Allison M. D. (1996) Titan's surface, revealed by HST imaging. *Icarus, 119(2)*, 336–349.

Soummer R., Cash W., Brown R. A., Jordan I., Roberge A., et al. (2009) A starshade for JWST: Science goals and optimization. In *Techniques and Instrumentation for Detection of Exoplanets IV* (S. B. Shaklan, ed.), p. 74400A. SPIE Conf. Series 7440, Bellingham, Washington.

Stamnes K., Tsay S.-C., Wiscombe W., and Jayaweera K. (1988) Numerically stable algorithm for discrete-ordinate radiative transfer in multiple scattering and emitting layered media. *Appl. Opt., 27*, 2502–2509.

Tian F. (2009) Thermal escape from super Earth atmospheres in the habitable zones of M stars. *Astrophys. J., 703*, 905–909.

Tian F., Toon O. B., Pavlov A. A., and De Sterck H. (2005a) A hydrogen-rich early Earth atmosphere. *Science, 308*, 1014–1017.

Tian F., Toon O. B., Pavlov A. A., and De Sterck H. (2005b) Transonic hydrodynamic escape of hydrogen from extrasolar planetary atmospheres. *Astrophys. J., 621*, 1049–1060.

Tinetti G., Meadows V. S., Crisp D., Fong W., Velusamy T., and Snively H. (2005) Disk-averaged spectra of Mars. *Astrobiology, 5*, 461–482.

Tinetti G., Meadows V. S., Crisp D., Fong W., Fishbein E., Turnbull M., and Bibring J.-P. (2006a) Detectability of planetary characteristics in disk-averaged spectra. I: The Earth model. *Astrobiology, 6*, 34–47.

Tinetti G., Meadows V. S., Crisp D., Kiang N. Y., Kahn B. H., Fishbein E., Velusamy T., and Turnbull M. (2006b) Detectability of planetary characteristics in disk-averaged spectra. II: Synthetic spectra and light-curves of Earth. *Astrobiology, 6*, 881–900.

Turnbull M. C., Traub W. A., Jucks K. W., Woolf N. J., Meyer M. R., Gorlova N., Skrutskie M. F., and Wilson J. C. (2006) Spectrum of a habitable world: Earthshine in the near-infrared. *Astrophys. J., 644*, 551–559.

Udry S., Bonfils X., Delfosse X., Forveille T., Mayor M., et al. (2007) The HARPS search for southern extra-solar planets. XI. Super-Earths (5 and 8 $M_\oplus$) in a 3-planet system. *Astron. Astrophys., 469*, L43–L47.

von Bloh W., Bounama C., Cunmtz M., and Franck S. (2007) The habitability of super-Earths in Gliese 581. *Astron. Astrophys., 476(3)*, 1365–1371.

von Bloh W., Cuntz M., Schröder K.-P., Bounama C., and Franck S. (2009) Habitability of super-Earth planets around other suns: Models including red giant branch evolution. *Astrobiology, 9*, 593–602.

Walker J. C. G., Hays P. B., and Kasting J. F. (1981) A negative feedback mechanism for the long-term stabilization of the Earth's surface temperature. *J. Geophys. Res., 86*, 9776–9782.

Williams D. M. and Gaidos E. (2008) Detecting the glint of starlight on the oceans of distant planets. *Icarus, 195*, 927–937.

Wiscombe W. J. (1980) Improved Mie scattering algorithms. *Appl. Opt., 19*, 1505–1509.

Woolf N. J., Smith P. S., Traub W. A., and Jucks K. W. (2002) The spectrum of Earthshine: A pale blue dot observed from the ground. *Astrophys. J., 574*, 430–433.

Zahnle K. J. and Walker J. C. G. (1982) The evolution of solar ultraviolet luminosity. *Rev. Geophys., 20(2)*, 280–292.

Zahnle K., Arndt N., Cockell C., Halliday A., Nisbet E., Selsis F., and Sleep N. H. (2007) Emergence of a habitable planet. *Space Sci. Rev., 129*, 35–78.

Zuffada C. and Crisp D. (1997) Particle scattering in the resonance regime: Full-wave solution for axisymmetric particles with large aspect ratios. *J. Optical Soc. Am., 14*, 459–469.

Atmospheric Circulation of Exoplanets

Adam P. Showman
University of Arizona

James Y-K. Cho
Queen Mary, University of London

Kristen Menou
Columbia University

We survey the basic principles of atmospheric dynamics relevant to explaining existing and future observations of exoplanets, both gas giant and terrestrial. Given the paucity of data on exoplanet atmospheres, our approach is to emphasize fundamental principles and insights gained from solar system studies that are likely to be generalizable to exoplanets. We begin by presenting the hierarchy of basic equations used in atmospheric dynamics, including the Navier-Stokes, primitive, shallow-water, and two-dimensional nondivergent models. We then survey key concepts in atmospheric dynamics, including the importance of planetary rotation, the concept of balance, and simple scaling arguments to show how turbulent interactions generally produce large-scale east-west banding on rotating planets. We next turn to issues specific to giant planets, including their expected interior and atmospheric thermal structures, the implications for their wind patterns, and mechanisms to pump their east-west jets. Hot Jupiter atmospheric dynamics are given particular attention, as these close-in planets have been the subject of most of the concrete developments in the study of exoplanetary atmospheres. We then turn to the basic elements of circulation on terrestrial planets as inferred from solar system studies, including Hadley cells, jet streams, processes that govern the large-scale horizontal temperature contrasts, and climate, and we discuss how these insights may apply to terrestrial exoplanets. Although exoplanets surely possess a greater diversity of circulation regimes than seen on the planets in our solar system, our guiding philosophy is that the multidecade study of solar system planets reviewed here provides a foundation upon which our understanding of more exotic exoplanetary meteorology must build.

1. INTRODUCTION

The study of atmospheric circulation and climate began hundreds of years ago with attempts to understand the processes that determine the distribution of surface winds on Earth (e.g., *Hadley*, 1735). As theories of Earth's general circulation became more sophisticated (e.g., *Lorenz*, 1967), the characterization of Mars, Venus, Jupiter, and other solar system planets by spacecraft starting in the 1960s demonstrated that the climate and circulation of other atmospheres differ, sometimes radically, from that of Earth. Exoplanets, occupying a far greater range of physical and orbital characteristics than planets in our solar system, likewise plausibly span an even greater diversity of circulation and climate regimes. This diversity provides a motivation for extending the theory of atmospheric circulation beyond our terrestrial experience. Despite continuing questions, our understanding of the circulation of the modern Earth atmosphere is now well developed (see, e.g., *Held*, 2000; *Schneider*, 2006; *Vallis*, 2006), but attempts to unravel the atmospheric dynamics of Venus, Jupiter, and other solar system planets remain ongoing, and the study of atmospheric circulation of exoplanets is in its infancy.

For exoplanets, driving questions fall into several overlapping categories. First, we wish to understand and explain new observations constraining atmospheric structure, such as lightcurves, photometry, and spectra obtained with the Spitzer, Hubble, or James Webb Space Telescopes (JWST), thus helping to characterize specific exoplanets as remote worlds. Second, we wish to extend the theory of atmospheric circulation to the wide range of planetary parameters encompassed by exoplanets. Existing theory was primarily developed for conditions relevant to Earth, and our understanding of how atmospheric circulation depends on atmospheric mass, composition, stellar flux, planetary rotation rate, orbital eccentricity, and other parameters remains rudimentary. Significant progress is possible with theoretical, numerical, and laboratory investigations that span a wider range of planetary parameters. Third, we wish to understand the conditions under which planets are habitable, and answering this question requires addressing the intertwined issues of atmospheric circulation and climate.

What drives atmospheric circulation? Horizontal temperature contrasts imply the existence of horizontal pressure contrasts, which drive winds. The winds in turn push the atmosphere away from radiative equilibrium by transporting heat from hot regions to cold regions (e.g., from the equator to the poles on Earth). This deviation from radiative equilibrium allows net radiative heating and cooling to occur, thus helping to maintain the horizontal temperature and pressure contrasts that drive the winds (see Fig. 1). Spatial contrasts in thermodynamic heating/cooling thus fundamentally drive the circulation, yet it is the existence of the circulation that allows these heating/cooling patterns to exist. (In the absence of a circulation, the atmosphere would relax into a radiative-equilibrium state with a net heating rate of zero.) The atmospheric circulation is thus a coupled radiation-hydrodynamics problem. On Earth, for example (see Fig. 2), the equator and poles are not in radiative equilibrium. The equator is subject to net heating, the poles to net cooling, and it is the mean latitudinal heat transport that is both responsible for and driven by these net imbalances.

The mean climate (e.g., the global-mean surface temperature of a planet) depends foremost on the absorbed stellar flux and the atmosphere's need to reradiate that energy to space. Yet even the global-mean climate is strongly affected by the atmospheric mass, composition, and circulation. On a terrestrial planet, for example, the circulation helps to control the distribution of clouds and surface ice, which in turn determine the planetary albedo and the mean surface temperature. In some cases, a planetary climate can have multiple equilibria (e.g., a warm, ice-free state or a cold, ice-covered "snowball Earth" state), and in such cases the circulation plays an important role in determining the relative stability of these equilibria.

Understanding the atmosphere/climate system is challenging because of its nonlinearity, which involves multiple positive and negative feedbacks between radiation, clouds, dynamics, surface processes, planetary interior, and life (if any). The inherent nonlinearity of fluid motion further implies that even atmospheric-circulation models neglecting the radiative, cloud, and surface/interior components can exhibit a large variety of behaviors.

From the perspective of studying the atmospheric circulation, transiting exoplanets are particularly intriguing because they allow constraints on key planetary attributes that are a prerequisite to characterizing an atmosphere's circulation regime. When combined with Doppler velocity data, transit observations permit a direct measurement of the exoplanet's radius, mass and thus surface gravity. (Note that combining Doppler velocity and transit measurements lifts the mass-inclination degeneracy.) With the additional expectation that close-in exoplanets are tidally locked if on a circular orbit, or pseudo-synchronized if on an eccentric orbit, the planetary rotation rate is thus indirectly known as well. [Pseudo-synchronization refers to a state of tidal synchronization achieved only at periastron passage (=closest approach), as expected from the strong dependence of tides with orbital separation.] Knowledge of the radius, surface gravity, rotation rate, and external irradiation conditions for several exoplanets, together with the availability of direct observational constraints on their emission, absorption, and reflection properties, opens the way for the development of

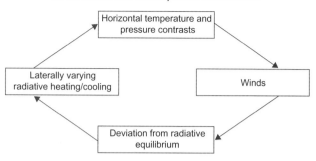

Fig. 1. Atmospheric circulation results from a coupled interaction between radiation and hydrodynamics: Horizontal temperature and pressure contrasts generate winds, which drive the atmosphere away from local radiative equilibrium. This in turn allows the spatially variable thermodynamic (radiative) heating and cooling that maintains the horizontal temperature and pressure contrasts.

comparative atmospheric science beyond the reach of our own solar system.

The need to interpret these astronomical data reliably, by accounting for the effects of atmospheric circulation and understanding its consequences for the resulting planetary emission, absorption, and reflection properties, is the central theme of this chapter. Tidally locked close-in exoplanets, for example, are subject to an unusual situation of permanent day/night radiative forcing, which does not exist in our solar system. (Venus may provide a partial analogy, which has not yet been fully exploited.) To address the new regimes of forcings and responses of these exoplanetary atmospheres, a discussion of fundamental principles of atmospheric fluid dynamics and how they are implemented in multidimensional, coupled radiation-hydrodynamics numerical models of the general circulation model (GCM) type is required.

Contemplating the wide diversity of exoplanets raises a number of fundamental questions. What determines the mean wind speeds, direction, and three-dimensional flow geometry in atmospheres? What controls the equator-to-pole and day-night temperature differences? What controls the frequencies and spatial scales of temporal variability? What role does the circulation play in controlling the mean climate (e.g., global-mean surface temperature, composition) of an atmosphere? How do these answers depend on parameters such as the planetary rotation rate, gravity, atmospheric mass and composition, and stellar flux? And, finally, what are the implications for observations and habitability of exoplanets?

At present, only partial answers to these questions exist (see reviews by *Showman et al.*, 2008b; *Cho*, 2008). With upcoming observations of exoplanets, constraints from solar system atmospheres, and careful theoretical work, significant progress is possible over the next decade. While a rich variety of atmospheric flow behaviors is realized in the solar system alone — and an even wider diversity is possible on exoplanets — the fundamental physical principles obeyed

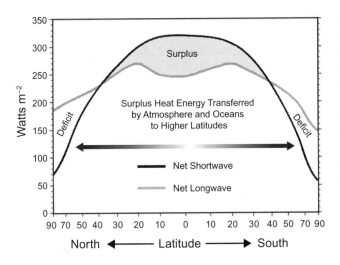

Fig. 2. Earth's energy balance. Earth absorbs more sunlight at the equator than the poles (higher curve, denoted "shortwave"). The Earth also radiates more infrared energy to space at the equator than the poles (lower curve, denoted "longwave"). However, because the atmospheric/oceanic circulation act to mute the latitudinal temperature contrasts (relative to radiative equilibrium), the longwave radiation exhibits less latitudinal variation than the shortwave absorption. Thus, the circulation leads to net heating at the equator and net cooling at the poles, which in turn drives the circulation. Data are an annual-average for 1987 obtained from the NASA Earth Radiation Budget Experiment (ERBE) project. Copyright M. Pidwirny, *www.physicalgeography.net*, used with permission.

by all planetary atmospheres are nonetheless universal. With this unifying notion in mind, this chapter provides a basic description of atmospheric circulation principles developed on the basis of extensive solar system studies and discusses the prospects for using these principles to better understand physical conditions in the atmospheres of remote worlds.

The outline of this chapter is as follows. In section 2, we introduce several of the equation sets that are used to investigate atmospheric circulation at varying levels of complexity. This is followed in section 3 by a tutorial on basic ideas in atmospheric dynamics, including atmospheric energetics, timescale arguments, force balances relevant to the large-scale circulation, the important role of rotation in generating east-west banding, and the role of waves and eddies in shaping the circulation. In section 4, we survey the atmospheric dynamics of giant planets, beginning with generic arguments to constrain the thermal and dynamical structure and proceeding to specific models for understanding the circulation of our "local" giant planets (Jupiter, Saturn, Uranus, Neptune) as well as hot Jupiters and hot Neptunes. (The terms hot Jupiter and hot Neptune refer to giant exoplanets with masses comparable to those of Jupiter and Neptune, respectively, with orbital semimajor axes less than ~0.1 AU, leading to high temperatures.) In section 5, we turn to the climate and circulation of terrestrial exoplanets.

Observational constraints in this area do not yet exist, and so our goal is simply to summarize basic concepts that we expect to become relevant as this field expands over the next decade. This includes a description of climate feedbacks (section 5.1), global circulation regimes (section 5.2), Hadley-cell dynamics (section 5.3), the dynamics of the so-called midlatitude "baroclinic" zones where baroclinic instabilities dominate (section 5.4), the slowly rotating regime relevant to Venus and Titan (section 5.5), and finally a survey of how the circulation responds to the unusual forcing associated with synchronous rotation, extreme obliquities, or extreme orbital eccentricities (section 5.6). The latter topics, while perhaps the most relevant, are the least understood theoretically. In section 6 we summarize recent highlights, both observational and theoretical, and in section 7 we finish with a survey of future prospects.

2. EQUATIONS GOVERNING ATMOSPHERIC CIRCULATION

A wide range of dynamical models has been developed to explore atmospheric circulation and climate. Such models are used to explain observations, understand mechanisms that govern the circulation/climate system, determine the sensitivity of a planet's circulation/climate to changes in parameters, test hypotheses about how the system works, and make predictions.

Developing a good understanding of atmospheric circulation requires the use of a hierarchy of atmospheric fluid dynamics models. Complex models that properly represent the full range of physical processes may be required for detailed predictions or comparisons with observations, but their very complexity can obscure the specific physical mechanisms causing a given phenomenon. In contrast, simpler models contain less physics, but they are easier to diagnose and can often lead to a better understanding of cause and effect in an idealized setting. Whether a given model contains sufficient physics to explain a given phenomenon is a question that can only be answered by exploring a hierarchy of models with a range of complexity. Exploring a hierarchy of models is therefore invaluable because it allows one to determine the minimal set of physical ingredients that are needed to generate a specific atmospheric behavior — insight that typically cannot be obtained from one type of model alone.

The equations governing atmospheric behavior derive from conservation of momentum, mass, and energy for a fluid, which we here assume to be an electrically neutral continuum. For three-dimensional models, where momentum is a three-dimensional vector, this implies five governing equations, which are generally represented as five coupled partial differential equations for the three-dimensional velocity, density, and internal energy per mass (with other thermodynamic state variables determined from density and internal energy by the equation of state). The Navier-Stokes equations, described in section 2.1, constitute the canonical example and provide a complete representation of a continuum, electrically neutral, viscous fluid in three dimensions.

So-called reduced models simplify the dynamics in one or more ways, for example, by reducing specified equations to their leading-order balances. For example, because most atmospheres have large aspect ratios (with characteristic horizontal length scales for the global circulation typically 10–100 times the characteristic vertical scales), the vertical momentum balance is typically close to a local hydrostatic balance, with the local weight of fluid parcels balancing the local vertical pressure gradient [see, e.g., *Holton* (2004, pp. 41–42) or *Vallis* (2006, pp. 80–84) for a derivation]. The primitive equations, described in section 2.2, formalize this fact by replacing the full vertical momentum equation with local vertical hydrostatic balance. Although the system is still governed by five equations, this alteration simplifies the dynamics by removing vertically propagating sound waves, which are unimportant for most meteorological phenomena. It also leads to mathematical simplification, making it easier to obtain analytic and numerical solutions. This is the equation set that forms the basis for most cutting-edge global-scale climate models used for studying atmospheres of solar system planets, although some global-scale high-resolution models now include nonhydrostatic effects.

A further common reduction is to simplify the dynamics to a one-layer model representing (for example) the vertically averaged flow. The most important example is the shallow-water model, described in section 2.3, which governs the behavior of a thin layer of constant-density fluid of variable thickness. This implies three coupled equations for horizontal momentum and mass conservation (governing the evolution of the two horizontal velocity components and the layer thickness) as a function of longitude, latitude, and time. Although highly idealized, the shallow-water model has proven surprisingly successful at capturing a wide range of atmospheric phenomena and has become a time-honored process model in atmospheric dynamics (see, for example, *Pedlosky*, 1987, Chapter 3).

A further reduction results from assuming the fluid layer thickness is constant in the shallow-water model. Given that density is also constant, the mass conservation equation then becomes a statement that horizontal convergence/divergence is zero. This constraint, which has the effect of removing gravity (buoyancy) waves from the system, allows the horizontal velocity components to be represented using a stream function, leading finally to a single governing partial differential equation for the stream function as a function of longitude, latitude, and time. Section 2.4 describes this two-dimensional, nondivergent model. The impressive reduction from five coupled equations in five dependent variables (as for the Navier-Stokes or primitive equations) to one equation in one variable leads to great mathematical simplification, enabling analytic solutions in cases when they are otherwise difficult to obtain. Moreover, the exclusion of buoyancy effects, gravity waves, and vertical structure leads to a conceptual simplification, allowing the exploration of (for example) vortex and jet formation in the most idealized possible setting.

A comparison of results from the full range of models described here provides a path toward identifying the relative roles of acoustic waves, vertical structure, buoyancy effects, and gravity waves in affecting any given meteorological phenomenon of interest. We now present the equations associated with each of these models.

2.1. Navier-Stokes Equations

Let $\mathbf{u} = \mathbf{u}(x, t)$ be velocity at position x and time t, where $x, \mathbf{u} \in \mathbb{R}^3$. If the frictional force per unit area of the fluid is linearly proportional to shear in the fluid, then it is a Newtonian fluid (e.g., *Batchelor*, 1967). Such fluids are described by the Navier-Stokes equations

$$\frac{D\mathbf{u}}{Dt} = -\frac{1}{\rho}\nabla p + \mathbf{f}_b + \frac{1}{\rho}\nabla \cdot \left\{ 2\mu \left[\mathbf{e} - \frac{1}{3}(\nabla \cdot \mathbf{u})\mathbb{I} \right] \right\} \quad (1a)$$

where

$$\frac{D}{Dt} = \frac{\partial}{\partial t} + \mathbf{u} \cdot \nabla \quad (1b)$$

is the material derivative (i.e., the deriative following the motion of a fluid element). Here, ρ is density, p is pressure, $\mathbf{f}_b$ represents various body forces per mass (e.g., gravity and Coriolis), μ is molecular dynamic viscosity, and $\mathbf{e} = 1/2[(\nabla \mathbf{u}) + (\nabla \mathbf{u})^T]$ and $\mathbb{I}$ are the strain-rate and unit tensors, respectively. In equation (1a) the quantity inside the braces is the viscous stress tensor. Here, as in equation (3) below, the average normal viscous stress (bulk viscosity) has been assumed to be zero.

Equation (1a) is closed with the following equations for mass (per unit volume), internal energy (per unit mass), and state

$$\frac{D\rho}{Dt} = -\rho \nabla \cdot \mathbf{u} \quad (2)$$

$$\frac{D\varepsilon}{Dt} = -\frac{p}{\rho}(\nabla \cdot \mathbf{u}) + \frac{2\mu}{\rho}\left[\mathbf{e} : (\nabla \mathbf{u})^T - \frac{1}{3}(\nabla \cdot \mathbf{u})^2 \right] + \frac{1}{\rho}\nabla \cdot (K_T \nabla T) + \mathcal{Q} \quad (3)$$

$$p = p(\rho, T) \quad (4)$$

where $\varepsilon = \varepsilon(T, s)$ is specific internal energy, s is specific entropy, K_T is heat conduction coefficient, T is temperature, and $\mathcal{Q}$ is thermodynamic heating rate per mass. In equation (3), ":" is the scalar-product (i.e., component-wise multiplication) operator for two tensors. Equations (1)–(4) constitute six equations for six independent unknowns, $\{\mathbf{u}, p, \rho, T\}$. Note that for a homogeneous thermodynamic system, which involves a single phase, only two state variables can vary independently; hence, there are only two thermodynamic degrees of freedom for such a system.

The neutral atmosphere is well described by equations (1)–(4) when the characteristic length scale L is much larger than the mean free path of the constituents that make up the atmosphere. Hence, the equations are valid up to heights where ionization is not significant and the continuum hypothesis does not break down. Under normal conditions, the atmosphere behaves like an ideal gas. The parameters μ, K_T, and other physical properties of the fluid depend on T, as well as ρ. When appreciable temperature differences exists in the flow field, these properties must be regarded as a function of position. For large-scale atmosphere applications, however, the terms involving μ in equations (1a) and (3) are small and can be neglected in most cases. The typical boundary conditions are $\mathbf{u} \cdot \mathbf{n} = 0$ at the lower boundary, where $\mathbf{n}$ is the normal to the boundary, and ρ, p → 0 as z → ∞. For local, limited area models, periodic boundary conditions are often used.

2.2. Primitive Equations

On the large scale (to be more precisely quantified below), the motion of an atmosphere is governed by the primitive equations. They read (e.g., *Salby*, 1996)

$$\frac{D\mathbf{v}}{Dt} = -\nabla_p \Phi - f\mathbf{k} \times \mathbf{v} + \mathcal{F} - \mathcal{D} \quad (5a)$$

$$\frac{\partial \Phi}{\partial p} = -\frac{1}{\rho} \quad (5b)$$

$$\frac{\partial \varpi}{\partial p} = -\nabla_p \cdot \mathbf{v} \quad (5c)$$

$$\frac{D\theta}{Dt} = \frac{\theta}{c_p T} \dot{q}_{net} \quad (5d)$$

where

$$\frac{D}{Dt} = \frac{\partial}{\partial t} + \mathbf{v} \cdot \nabla_p + \varpi \frac{\partial}{\partial p} \quad (5e)$$

Note here that p, rather than the geometric height z, is used as the vertical coordinate. This coordinate, which simplifies the gradient term in equation (5a), is common in atmospheric studies; it renders z = z(**x**, p, t) a dependent variable, where now $\mathbf{x} \in \mathbb{R}^2$. In equation (5) **v**(**x**, t) = (u, v) is the (eastward, northward) velocity in a frame rotating with Ω, where Ω is the planetary rotation vector as represented in inertial space. (Cardinal directions are defined here consistent with everyday usage; east is defined to be the prograde direction, that is, the direction in which the planet rotates. North is the direction along which $\Omega \cdot \mathbf{k}$ becomes more positive. The coordinate system is righthanded.) Φ = gz is the geopotential, where g is the gravitational acceleration (assumed to be constant and to include the centrifugal acceleration contribution) (see *Holton*, 2004, pp. 13–14) and z is the height above a fiducial geopotential surface; **k** is the local upward unit vector; $f = 2\Omega \sin \phi$ is the Coriolis parameter, the locally vertical component of the planetary vorticity vector 2Ω; ∇_p is the horizontal gradient on a p-surface;

ϖ = Dp/Dt is the vertical velocity; $\mathcal{F}$ and $\mathcal{D}$ represent the momentum sources and sinks, respectively; $\theta = T(p_{ref}/p)^\kappa$ is the potential temperature, where p_{ref} is a reference pressure and $\kappa = R/c_p$ with R the specific gas constant and c_p the specific heat at constant pressure; and $\dot{q}_{net}$ is the *net diabatic heating rate* (heating minus cooling). [The potential temperature θ is related to the entropy s by $ds = c_p d \ln \theta$. When c_p is constant, this yields $\theta = T(p_{ref}/p)^\kappa$.] Note that $\dot{q}_{net}$ can include not only radiative heating/cooling but latent heating and, at low pressures where the thermal conductivity becomes large, conductive heating. The Newtonian cooling scheme, which relaxes temperature toward a prescribed radiative-equilibrium temperature over a specified radiative time constant, is one simple parameterization of $\dot{q}_{net}$.

The fundamental presumption in the use of equations (5) is that small-scale processes are parameterizable within the framework of large-scale dynamics. Here, by "large" scales, it is typically meant that L ≥ a/10, where a is the planetary radius. By "small" scales, it is meant those scales that are not resolvable numerically by global models, typically ≤a/10. Regions of the atmosphere where small-scale processes are important are often highly concentrated (e.g., fronts and convective updrafts). Their characteristic scales are ≪a/10. Therefore, it is possible that the equation (5) set — as with all the other equation sets discussed in this chapter — leaves out some processes important for large-scale dynamics.

To arrive at equation (5), one begins with equations (1)–(4) in spherical geometry (e.g., *Batchelor*, 1967). Two approximations are then made. These are the "shallow atmosphere" and the "traditional" approximations (e.g., *Salby*, 1996). The first assumes z/a ≪ 1. The second is formally valid in the limit of strong stratification, when the Prandtl ratio (N^2/Ω^2) ≫ 1. Here, N = N(x, z, t) is the Brunt-Väisälä (buoyancy) frequency, the oscillation frequency for an air parcel that is displaced vertically under adiabatic conditions

$$N = \left[g \frac{\partial (\ln \theta)}{\partial z} \right]^{1/2} \quad (6)$$

These approximations allow the Coriolis terms involving vertical velocity to be dropped from equation (1a) and vertical accelerations to be assumed small. The latter is explicitly embodied in equation (5b), the hydrostatic balance, which we discuss further below.

Hydrostatic balance renders the primitive equations valid only when $N^2/\omega^2 \gg 1$, where 2π/ω is the timescale of the motion under consideration. This condition, which is distinct from the Prandtl ratio condition, restricts the vertical length scale of motions to be small compared to the horizontal length scale. Therefore, the hydrostatic balance approximation breaks down in weakly stratified regions (here we refer to the dynamically evolving hydrostatic balance associated with circulation-induced perturbations in pressure and density; the mean background density and pressure — i.e., those that would exist in absence of dynamics — will remain hydrostatically balanced even when the circulation induced perturbations are not). The hydrostatic assumption filters vertically propagating sound waves from the equations.

According to equation (5d), when $\dot{q}_{net} = 0$, individual values of θ are retained by fluid elements as they move with the flow. In this case, equation (5) also admit a dynamically important conserved quantity, the potential vorticity

$$q_{PE} = \left[\frac{(\zeta+f)\mathbf{k}}{\rho}\right] \cdot \nabla\theta \quad (7a)$$

where $\zeta = \mathbf{k} \cdot \nabla \times \mathbf{v}$ is the relative vorticity. This quantity provides the crucial connection between the primitive equations and the physically simpler models that follow. For example, undulations of potential vorticity are often a direct manifestation of Rossby waves, which are represented in all the models presented in this section. The conservation of the potential vorticity q_{PE} following the flow

$$\frac{Dq_{PE}}{Dt} = 0 \quad (7b)$$

and the redistribution of q_{PE} implied by it, is one of the most important properties in atmospheric dynamics.

2.3. Shallow-Water Model

For many applications, equation (5) is too complex and broad in scope. In the absence of observational information to properly constrain the model parameters, reduction of the equations is beneficial. A commonly used approach is to collapse the three-dimensional primitive equations to a two-dimensional, one-layer model. Such reduction allows investigation of horizontal vortex and jet interactions in an idealized setting.

Among the most widely used one-layer models is the shallow-water model. Consider a thin layer of homogeneous (i.e., constant-density) fluid, bounded above by a free surface and below by an impermeable boundary, so that its thickness is h(**x**, t). The dynamics of such a layer is governed by the following equations (e.g., *Pedlosky*, 1987, Chapter 3)

$$\frac{D\mathbf{v}}{Dt} = -g\nabla h - f\mathbf{k} \times \mathbf{v} \quad (8a)$$

$$\frac{Dh}{Dt} = -h\nabla \cdot \mathbf{v} \quad (8b)$$

where

$$\frac{D}{Dt} = \frac{\partial}{\partial t} + \mathbf{v} \cdot \nabla \quad (8c)$$

Forcing and dissipation are not included in equation (8), but they can be added in the usual way. In the absence of forcing and dissipation, the equations preserve the potential vorticity

$$q_{sw} = \frac{\zeta+f}{h} \quad (9)$$

following the flow.

If equation (8) is derived as the vertical mean of the flow of an isentropic atmosphere with a free upper boundary, h must be replaced by h^κ in the geopotential gradient term. If they are derived as a vertical mean of the flow of an isentropic atmosphere between rigid upper and lower boundaries, h must be replaced by $h^{\kappa/(1-\kappa)}$ in the geopotential gradient term. Note that while $\nabla \cdot \mathbf{v} \neq 0$ in equation (8b), $\nabla \cdot \mathbf{u} = 0$, since the layer is homogeneous (i.e., density is constant). Hence, the sound speed $c_s \to \infty$, and the sound waves are filtered out from the system. However, the system does retain gravity waves, which propagate at speed $c_g = \sqrt{gh}$.

The shallow-water equations are widely used as a process model in geophysical fluid dynamics. They are much simpler than the primitive equations, yet they still describe a wealth of phenomena — including vortices, jet streams, Rossby waves, gravity waves, and the interactions between them. Excluded are any processes that depend on the details of the vertical structure — including vertically propagating waves, baroclinic instabilities (see section 3.7), and depth-dependent flow. However, because both rotational and buoyancy processes are included (the latter via the variable layer thickness), the shallow-water model — as well as all the models discussed so far — includes a fundamental length scale called the Rossby radius of deformation (often simply called the deformation radius), which is a natural length scale for a variety of phenomena that depend on both rotation and stratification. In the shallow-water system, this length scale is

$$L_D = \frac{\sqrt{gh}}{f} \quad (10)$$

2.4. Two-Dimensional, Nondivergent Model

This is the simplest useful one-layer model for largescale dynamics. For large-scale weather systems characterized by $U/c_g \ll 1$, we can apply a rigid upper boundary to the shallow-water model, since c_g represents external gravity wave speed in the model. Then, H is large and equation (8b) implies $\nabla \cdot \mathbf{v} \ll 1$. Taking $\nabla \cdot \mathbf{v} = 0$ then gives the two-dimensional nondivergent equation

$$\frac{D\mathbf{v}}{Dt} = -g\nabla h - f\mathbf{k} \times \mathbf{v} \quad (11)$$

where D/D_t is same as in equation (8c). The $\nabla \cdot \mathbf{v} = 0$ restriction on the velocity implies that we can define a stream function, $\psi(\mathbf{x}, t)$, such that

$$\mathbf{v} = \left(-\frac{\partial \psi}{\partial y}, \frac{\partial \psi}{\partial x}\right) \quad (12)$$

Using this definition, equation (11) can finally be written as a single governing equation for the evolution of the stream function

$$\frac{D}{Dt}(\nabla^2\psi + f) = 0 \qquad (13)$$

From equation (13), we see that $q_{2D} = \nabla^2\psi + f$ is the materially conserved potential vorticity for the two-dimensional nondivergent model. This also results simply by letting $h \to$ constant in equation (9).

In addition to the nonlinear vorticity advection, this equation — along with all the other equation sets described in this section — represents the dynamical effects of latitudinally varying Coriolis parameter. This is the so-called "beta effect," where $\beta \equiv df/dy$ is the northward gradient of the Coriolis parameter.

Equation (13) describes Rossby waves, nonlinear advection, and phenomena — such as the formation of zonal jet streams — that require the interaction of all these aspects (see section 3.6). However, it lacks a finite deformation radius ($L_D \to \infty$) and does not possess gravity-wave solutions. Therefore, any phenomena that depend on finite deformation radius, gravity waves, or buoyancy cannot be captured. (These assumptions render the equation valid only for $U/c_g \ll 1$ and $L/L_D \ll 1$.) As a result, the two-dimensional nondivergent model cannot serve as an accurate predictive tool for most applications; however, its very simplicity renders it a valuable process model for investigating jet formation in the simplest possible setting. For a review of examples, see, for example, *Vasavada and Showman* (2005) or *Vallis* (2006).

2.5. Conserved Quantities

Potential vorticity conservation has been emphasized throughout because of its central importance in atmospheric dynamics. There are other useful conserved quantities. For example, the full Navier-Stokes equation gives

$$\frac{D}{Dt}(\mathbf{r} \times \mathbf{u}) = \mathbf{r} \times \left(-\frac{1}{\rho}\nabla p - 2\Omega \times \mathbf{u} - \nabla\Phi^* + \mathbf{F}\right) \qquad (14)$$

where Φ^* is effective geopotential and $\mathbf{F}$ represents any additional forces on the fluid. From this, we obtain the conservation law for specific, axial angular momentum $\mathcal{M}$

$$\frac{D\mathcal{M}}{Dt} = -\frac{1}{\rho}\frac{\partial p}{\partial \lambda} + \mathcal{F}_\lambda \cos\phi \qquad (15a)$$

where

$$\mathcal{M} = (\Omega r \cos\phi + u) r \cos\phi \qquad (15b)$$

Equation (15) relates the material change of $\mathcal{M}$ to the axial components of torques present. For a thin atmosphere, r can be replaced with a.

Equation (1) also gives the material conservation law for the specific total energy E

$$\frac{DE}{Dt} = -\frac{1}{\rho}\nabla \cdot (p\mathbf{u}) + (\dot{Q}_{net} + \mathbf{u} \cdot \mathbf{F}) \qquad (16)$$

where E is the total energy including kinetic, potential, and internal contributions: $E = 1/2\mathbf{u}^2 + \Phi + c_v T$ with c_v the specific heat at constant volume and T the temperature. In flux form, the conservation law is

$$\frac{\partial}{\partial t}(\rho E) + \nabla \cdot [(\rho E + p)\mathbf{u}] = \rho(\dot{Q}_{net} + \mathbf{u} \cdot \mathbf{F}) \qquad (17)$$

As already noted for $\mathcal{M}$, the total energy reduces in the appropriate way for the various simpler physical situations discussed in previous subsections. For example, Φ and $c_v T$ terms do not exist for the two-dimensional nondivergent case. An important issue in the study of atmospheric energetics is the extent to which Φ and $c_v T$ are available to be converted to $1/2\,\mathbf{u}^2$.

3. BASIC CONCEPTS

The equation sets summarized in section 2 describe nonlinear, potentially turbulent flows with many degrees of freedom. Unfortunately, due to the nonlinearity and complexity, analytic solutions rarely exist, and one must resort to solving the equations numerically on a computer. To represent the atmospheric circulation of a particular planet, the chosen equation set is solved numerically with a specified spatial resolution and timestep, subject to appropriate parameter values (e.g., composition, gravity, planetary rotation rate), boundary conditions, and forcing/damping (e.g., prescriptions for heating/cooling and friction).

Such models vary greatly in complexity and numerical method. General circulation models (GCMs) in the solar system studies literature, for example, typically solve the three-dimensional primitive equations with sophisticated representations of radiative transfer, cloud formation, surface/atmosphere interactions, surface ice formation, and (if relevant) oceanic processes. These models are useful for exploring the interaction of dynamics with surface processes, radiation, and climate and are needed for quantitative comparisons with observational records.

However, because of their complexity, numerical simulations with full GCMs are computationally expensive, limiting such simulations to only moderate spatial resolution and making it difficult to broadly survey the relevant parameter space. Even more problematic, because of the inherent complexity of nonlinear fluid dynamics and its possible interactions with radiation and surface processes, it is rarely obvious *why* a given GCM simulation produces the output it does. By itself, the output of a sophisticated three-dimensional model often provides little more fundamental understanding than the observations of the actual atmosphere themselves. To understand how a given atmospheric circulation would vary under different planetary parameters, for example, an understanding of the *mechanisms* shaping the circulation is required. Although careful diagnostics of GCM results can provide important insights into the mechanisms that are at play, a deep mechanistic understanding does not always flow naturally from such simulations.

Rather, obtaining a robust understanding requires a diversity of model types, ranging from simple to complex, in which various processes are turned on and off and the results carefully diagnosed. This is called a modeling hierarchy and its use forms the backbone of forward progress in the field of atmospheric dynamics of Earth and other solar system planets (see, e.g., *Held*, 2005). For example, despite the existence of numerous full GCMs for modern Earth climate, significant advances in our understanding of the *mechanisms* shaping the atmospheric circulation rely heavily on the usage of linear models, simplified one-layer nonlinear models (such as the two-dimensional nondivergent or shallow-water models), and three-dimensional models that do not include the sophisticated treatments of radiation and subgridscale convective processes included in full GCMs. [To illustrate, a summary of the results of such a hierarchy for understanding Jupiter's jet streams can be found in *Vasavada and Showman* (2005).] Even more fundamentally, obtaining understanding requires the development of basic theory that can (at least qualitatively) explain the results of these various models as well as observations of actual atmospheres. One of the major goals in performing simplified models is to aid in the construction of such a theory (see, e.g., *Schneider*, 2006).

Exoplanet GCMs will surely be useful in the coming years. But, as with solar system planets, we expect that a fundamental understanding will require use of a modeling hierarchy as well as basic theory. In this section we survey key concepts in atmospheric dynamics that provide insight into the expected atmospheric circulation regimes. Emphasis is placed on presenting a conceptual understanding and as such we describe not only GCM results but basic theory and the results of highly simplified models as well. Here we focus on basic aspects relevant to both gaseous and terrestrial planets. Detailed presentations of issues specific to giant and terrestrial exoplanets are deferred to section 4 and section 5.

3.1. Energetics of Atmospheric Circulation

Atmospheric circulations involve an energy cycle. Absorption of starlight and emission of infrared energy to space creates potential energy, which is converted to kinetic energy and then lost via friction. Each step in the process involves nonlinearities, and generally the atmosphere self-adjusts so that, in a time mean sense, the conversion rates balance.

What matters for driving the circulation is not the *total* potential energy but rather the *fraction* of the potential energy that can be extracted by adiabatic atmospheric motions. For example, a stably stratified, horizontally uniform atmosphere can contain vast potential energy, but none can be extracted — any adiabatic motions can only *increase* the potential energy of such a state. Uniformly heating the top layers of such an atmosphere would further increase its potential energy but would still preclude an atmospheric circulation.

For convecting atmospheres, creating extractable potential energy requires heating the fluid at lower altitudes than it is cooled. This creates buoyant air parcels (positively buoyant at the bottom, negatively buoyant at the top); vertical motion of these buoyant parcels releases potential energy and drives convection. [To emphasize the importance of the distinction, consider a hot, isolated giant planet. The cooling caused by its radiation to space *decreases* its *total* potential energy, yet (because the cooling occurs near the top) this *increases* the fraction of the remaining potential energy that can be extracted by motions. This is what can allow convection to occur on such objects.] But, most atmospheres are stably stratified, and in this case extractable energy — called available potential energy — only exists when density varies horizontally on isobars (*Peixoto and Oort*, 1992, Chapter 14). In this case, the denser regions can slide laterally and downward underneath the less-dense regions, decreasing the potential energy and creating kinetic energy (winds). Continual generation of available potential energy (required to balance its continual conversion to kinetic energy and loss via friction) requires heating the regions of the atmosphere that are already hot (e.g., the tropics on Earth) and cooling the regions that are already cold (e.g., the poles). For Earth, available potential energy is generated at a global-mean rate of ~2 W m^{-2}, which is ~1% of the global-mean absorbed and radiated flux of 240 W m^{-2} (*Peixoto and Oort*, 1992, pp. 382–385).

The rate of frictional dissipation can affect the mean state, but rigorously representing such friction in models is difficult. For solar system planets, kinetic-energy loss occurs via turbulence, waves, and friction against the surface (if any). Ohmic dissipation may be important in the deep interiors of gas giants (*Kirk and Stevenson*, 1987; *Liu et al.*, 2008), as well as in the upper atmosphere where ionization becomes important. These processes sometimes have length scales much smaller (by up to several orders of magnitude) than can easily be resolved in global, three-dimensional numerical models. In Earth GCMs, such frictional dissipation mechanisms are therefore often *parameterized* by adding to the equations quasiempirical damping terms (e.g., a vertical diffusion to represent turbulent kinetic-energy losses by small-scale shear instabilities and breaking waves). A difficulty is that such prescriptions, while physically motivated, are often nonrigorous and the extent to which they can be extrapolated to other planetary environments is unclear. Perhaps for this reason, models of hot Jupiters published to date do not include such parameterizations of frictional processes (although they all include small-scale viscosity for numerical reasons). [Note that a statistically steady (or quasisteady) state can still occur in such a case; this requires the atmosphere to self-adjust so that the rates of generation of available potential energy and its conversion to kinetic energy become small.] Nevertheless, *Goodman* (2009) has highlighted the possible importance that such processes could play in the hot Jupiter context, and future models of hot Jupiters will surely explore the possible effect that friction may have on the mean states.

Solar system planets offer interesting lessons on the role of friction. Despite absorbing a greater solar flux than any other thick atmosphere in our solar system, Earth's winds are relatively slow, with a mean wind speed of ~20 m s^{-1}. In contrast, Neptune absorbs a solar flux only 0.1% as large, but has wind speeds reaching 400 m s^{-1}. Presumably, Neptune can achieve such fast winds despite its weak radiative forcing because its frictional damping is extremely weak. Qualitatively, this makes sense because Neptune lacks a surface, which is a primary source of frictional drag on Earth. More puzzling is the fact that Neptune has significantly stronger winds than Jupiter (Table 1) despite absorbing only 4% of the solar flux absorbed by Jupiter. Possible explanations are that Jupiter experiences greater frictional damping than Neptune or that it has equilibrated to a state that has relatively slow wind speeds despite weak damping. This is not well understood and argues for humility in efforts to model the circulations of exoplanets.

3.2. Timescale Arguments for the Coupled Radiation Dynamics Problem

The atmospheric circulation represents a coupled radiation hydrodynamics problem. The circulation advects the temperature field and thereby influences the radiation field; in turn, the radiation field (along with atmospheric opacities and surface conditions) determines the atmospheric heating and cooling rates that drive the circulation. Rigorously attacking this problem requires coupled treatment of both radiation and dynamics. However, crude insight into the thermal response of an atmosphere can be obtained with simple timescale arguments. Suppose τ_{advect} is an advection time (e.g., the characteristic time for air to advect across a hemisphere) and τ_{rad} is the radiative time (i.e., the characteristic time for

TABLE 1. Planetary parameters.

Planet	a* (10^3 km)	Rotation period† (Earth days)	Ω (rad s^{-1})	Gravity‡ (m s^{-2})	$F_\star$§ (W m^{-2})	T_e¶ (K)	H_p** (km)	U†† (m s^{-1})	Ro‡‡	L_D/a§§	L_β/a¶¶
Venus	6.05	243	3 × 10^{-7}	8.9	2610	232	5	~20	10	70	7
Earth	6.37	1	7.27 × 10^{-5}	9.82	1370	255	7	~20	0.1	0.3	0.5
Mars	3.396	1.025	7.1 × 10^{-5}	3.7	590	210	11	~20	0.1	0.6	0.6
Titan	2.575	16	4.5 × 10^{-6}	1.4	15	85	18	~20	2	10	3
Jupiter	71.4	0.4	1.7 × 10^{-4}	23.1	50	124	20	~40	0.02	0.03	0.1
Saturn	60.27	0.44	1.65 × 10^{-4}	8.96	15	95	39	~150	0.06	0.03	0.3
Uranus	25.56	0.72	9.7 × 10^{-5}	8.7	3.7	59	25	~100	0.1	0.1	0.4
Neptune	24.76	0.67	1.09 × 10^{-4}	11.1	1.5	59	20	~200	0.1	0.1	0.6
WASP-12b	128	1.09	6.7 × 10^{-5}	11.5	8.8 × 10^6	2500	800	–	0.01–0.3	0.1	0.2–1.5
HD 189733b	81	2.2	3.3 × 10^{-5}	22.7	4.7 × 10^5	1200	200	–	0.03–1	0.3	0.4–3
HD 149026b	47	2.9	2.5 × 10^{-5}	21.9	1.8 × 10^6	1680	280	–	0.06–2	0.8	0.6–4
HD 209458b	94	3.5	2.1 × 10^{-5}	10.2	1.0 × 10^6	1450	520	–	0.04–1	0.4	0.5–3
TrES-2	87	2.4	2.9 × 10^{-5}	21	1.1 × 10^6	1475	260	–	0.03–1	0.3	0.4–3
TrES-4	120	3.5	2.0 × 10^{-5}	7.8	2.5 × 10^6	1825	870	–	0.03–1	0.4	0.4–3
HAT-P-7b	97	2.2	3.3 × 10^{-5}	25	4.7 × 10^6	2130	320	–	0.02–1	0.3	0.4–3
GJ 436b	31	2.6	2.8 × 10^{-5}	9.8	4.3 × 10^4	660	250	–	0.1–3	0.7	0.8–5
HAT-P-2b	68	5.6	1.3 × 10^{-5}	248	9.5 × 10^5	1400	21	–	0.1–3	1	0.8–5
CoRoT-Exo-4b	85	9.2	7.9 × 10^{-6}	13.2	3.0 × 10^5	1080	300	–	0.1–4	1	0.9–5

*Equatorial planetary radius.
†Assumes synchronous rotation for exoplanets.
‡Equatorial gravity at the surface.
§Mean incident stellar flux.
¶Global average blackbody emission temperature, which for exoplanets is calculated from equation (51) assuming zero albedo.
**Pressure scale height, evaluated at temperature T_e.
††Rough estimates of characteristic horizontal wind speed. Estimates for Venus and Titan are in the high-altitude superrotating jet; both planets have weaker winds (few m s^{-1}) in the bottom scale height. In all cases, peak winds exceed the listed values by factors of two or more.
‡‡Rossby number, evaluated in midlatitudes using wind values listed in the table and L ~2000 km for Earth, Mars, and Titan, 6000 km for Venus, and 10^4 km for Jupiter, Saturn, Uranus, and Neptune. For exoplanets, we present a range of possible values evaluated with L = a and winds from 100 to 4000 m s^{-1}.
§§Ratio of Rossby deformation radius to planetary radius, evaluted in midlatitudes with H equal to the pressure scale height and N appropriate for a vertically isothermal temperature profile.
¶¶Ratio of Rhines length (equation (35)) to planetary radius, calculated using the equatorial value of β and the wind speeds listed in the table.

radiation to induce large fractional entropy changes). When $\tau_{rad} \ll \tau_{advect}$, we expect temperature to deviate only slightly from the (spatially varying) radiative equilibrium temperature structure. Because the radiative equilibrium temperature typically varies greatly from dayside to nightside (or from equator to pole), this implies that such a planet would exhibit large fractional temperature contrasts. On the other hand, when $\tau_{rad} \gg \tau_{advect}$, dynamical transport dominates and air will tend to homogenize its entropy, implying that lateral temperature contrasts should be modest.

In estimating the advection time, one must distinguish north-south from east-west advection; east-west advection (relative to the pattern of stellar insolation) will often be dominated by the planetary rotation. For synchronously rotating planets, a characteristic horizontal advection time is

$$\tau_{advect} \sim \frac{a}{U} \quad (18)$$

where U is a characteristic horizontal wind speed. A similarly crude estimate of the radiative time can be obtained by considering a layer of pressure thickness Δp that is slightly out of radiative equilibrium and radiates to space as a blackbody. If the radiative equilibrium temperature is T_{rad} and the actual temperature is $T_{rad} + \Delta T$, with $\Delta T \ll T_{rad}$, then the net flux radiated to space is $4\sigma T_{rad}^3 \Delta T$ and the radiative timescale is (*Showman and Guillot*, 2002; *James*, 1994, pp. 65–66)

$$\tau_{rad} \sim \frac{\Delta p}{g} \frac{c_p}{4\sigma T^3} \quad (19)$$

In deep, optically thick atmospheres where the radiative transport is diffusive, a more appropriate estimate might be a diffusion time, crudely given by $\tau_{rad} \sim H^2/D$, where H is the vertical height of a thermal perturbation and D is the radiative diffusivity.

Showman et al. (2008b) estimated advective and radiative time constants for solar system planets and found that, as expected, planets with $\tau_{rad} \gg \tau_{advect}$ generally have small horizontal temperature contrasts and vice versa.

For hot Jupiters, most models suggest peak wind speeds of ~1–3 km s^{-1} (section 4.3), implying advection times of ~10^5 s based on the peak speed. Equation (19) would then suggest that $\tau_{rad} \ll \tau_{advect}$ at p $\ll$ 1 bar whereas $\tau_{rad} \gg \tau_{advect}$ at p $\gg$ 1 bar. Thus, one might crudely expect large day-night temperature differences at low pressure and small day-night temperature differences at high pressure, with the transition occurring at ~0.1–1 bar. These estimates are generally consistent with the observational inference of *Barman* (2008) and three-dimensional numerical simulations (e.g., *Showman et al.*, 2009; *Dobbs-Dixon and Lin*, 2008) of hot Jupiters — although some uncertainties still exist with modeling and interpretation.

For synchronously rotating terrestrial planets in the habitable zones of M dwarfs, a mean wind speed of 20 m s^{-1} (typical for terrestrial planets in our solar system; see Table 1) would imply an advection time of ~3 Earth days. For a temperature of 300 K, equation (19) would then imply that τ_{rad} is much smaller (greater) than τ_{advect} when the surface pressure is much less (greater) than ~0.2 bar. This argument suggests that synchronously rotating terrestrial exoplanets with a surface pressure much less than ~0.2 bar should develop large day-night temperature differences, whereas if the surface pressure greatly exceeds ~0.2 bar, day-night temperature differences would be modest. As with hot Jupiters, these estimates are consistent with three-dimensional GCM simulations (*Joshi et al.*, 1997), which suggest that this transition should occur at ~0.1 bar. These estimates may have relevance for whether CO_2 atmospheres would collapse due to nightside condensation and hence whether such planets are habitable.

3.3. Basic Force Balances: Importance of Rotation

Planets rotate, and this typically constitutes a dominant factor in shaping the circulation. The importance of rotation can be estimated by performing a scale analysis on the equation of motion. Suppose the circulation has a mean speed U and that we are interested in flows with characteristic length scale L (this might approach a planetary radius for global-scale flows). To order of magnitude, the strength of the acceleration term is U^2/L (namely, U divided by a time L/U to advect fluid across a distance L), while the magnitude of the Coriolis term is fU. The ratio of the acceleration term to the Coriolis term can therefore be represented by the Rossby number

$$Ro \equiv \frac{U}{fL} \quad (20)$$

Whenever Ro $\ll$ 1, the acceleration terms $D\mathbf{v}/Dt$ are weak compared to the Coriolis force per unit mass in the horizontal momentum equation. Because friction is generally weak, the only other term that can balance the horizontal Coriolis force is the pressure-gradient force, which is just $-\nabla_p \Phi$ in pressure coordinates. The resulting balance, called geostrophic balance, is given by

$$fu = -\left(\frac{\partial \Phi}{\partial y}\right)_p \qquad fv = \left(\frac{\partial \Phi}{\partial x}\right)_p \quad (21)$$

where x and y are eastward and northward distance, respectively and the derivatives are evaluated at constant pressure. In our solar system, geostrophic balance holds at large scales in the mid- and high-latitude atmospheres of Earth, Mars, Jupiter, Saturn, Uranus, and Neptune. Rossby numbers range from 0.01 to 0.1 for these rapidly rotating planets, but exceed unity in the stratosphere of Titan and reach ~10 for Venus, implying in the latter case that the Coriolis force plays a less important role in the force balance (Table 1). (Titan's Rossby number is smaller near the surface, where winds are weak.) Note that, even on rapidly rotating planets, horizontal geostrophy breaks down at the equator, where the horizontal Coriolis forces go to zero.

Determining Rossby numbers for exoplanets requires estimates of wind speeds, which are unknown. Some models of hot Jupiter atmospheres have generally suggested peak winds of several kilometers per second near the photosphere, with mean values perhaps a factor of several smaller. (Photosphere is defined here as the approximate pressure at which infrared photons can escape directly to space.) To illustrate the possibilities, Table 1 presents Ro values for several hot Jupiters assuming a range of wind speeds of 100–4000 m s^{-1}. Generally, if mean wind speeds are fast (several kilometers per second), Rossby numbers approach or exceed unity. If mean wind speeds are hundreds of meters per second or less, Rossby numbers should be much less than 1, implying that geostrophy approximately holds. One might thus plausibly expect a situation where the Coriolis force plays an important but not overwhelming role (i.e., Ro ~ 1) near photosphere levels, with the flow transitioning to geostrophy in the interior if winds are weaker there.

Geostrophy implies that, rather than flowing from pressure highs to lows as often occurs in a nonrotating fluid, the primary horizontal wind flows *perpendicular* to the horizontal pressure gradient. Thus, the primary flow does not erase the horizontal pressure contrasts; rather, the Coriolis forces associated with that flow actually help preserve large-scale pressure gradients in a rotating atmosphere. Geostrophy explains why the isobar contours included in most midlatitude weather maps (say of the U.S. or Europe) are so useful: The isobars describe not only the pressure field but the large-scale wind field, which flows along the isobar contours.

In many cases, rotation inhibits the ability of the circulation to equalize horizontal temperature differences. On rapidly rotating planets like the Earth, which is heated by sunlight primarily at low latitudes, the mean horizontal temperature gradients in the troposphere generally point from the poles toward the equator. (The troposphere is the bottommost, optically thick layer of an atmosphere, where temperature decreases with altitude and convection may play an important role; the tropopause defines the top of the troposphere, and the stratosphere refers to the stably stratified, optically thin region overlying the troposphere. In some cases, a stratosphere's temperature may increase with altitude due to absorption of sunlight by gases or aerosols; in other cases, however, the stratosphere's temperature can be nearly constant with altitude.) Integration of the hydrostatic equation (equation 5b) implies that the mean pressure gradients also point north-south. Thus, on a rapidly rotating planet where geostrophy holds and the primary temperature contrast is between equator and pole, the mean midlatitude winds will be east-west rather than north-south — thus limiting the ability of the circulation to homogenize its temperature differences in the north-south direction. We might thus expect that, everything else being equal, a more rapidly rotating planet will harbor a greater equator-to-pole temperature difference.

In rapidly rotating atmospheres, a tight link exists between horizontal temperature contrasts and the vertical gradients of the horizontal wind. This can be shown by taking the derivative with pressure of equation (21) and invoking the hydrostatic balance equation (5b) and the ideal-gas law. We obtain the thermal-wind equation for a shallow atmosphere (*Holton*, 2004, pp. 70–75)

$$f\frac{\partial u}{\partial \ln p} = \frac{\partial (RT)}{\partial y} \qquad f\frac{\partial v}{\partial \ln p} = -\frac{\partial (RT)}{\partial x} \qquad (22)$$

where R is the specific gas constant (i.e., the universal gas constant divided by the molar mass). The equation states that, in a geostrophically balanced atmosphere, north-south temperature gradients must be associated with a vertical gradient in the zonal (east-west) wind, whereas east-west temperature gradients must be associated with a vertical gradient in the meridional (north-south) wind. Given the primarily equatorward-pointing midlatitude horizontal temperature gradient in the tropospheres of Earth and Mars, for example, and given the weak winds at the surface of a terrestrial planet (a result of surface friction), this equation correctly demonstrates that the mean midlatitude winds in the upper troposphere must flow to the east — as observed for the midlatitude tropospheric jet streams on Earth and Mars.

Geostrophy relates the three-dimensional structure of the winds and temperatures at a given time but says nothing about the flow's time evolution. In a rapidly rotating atmosphere, both the temperatures and winds often evolve together, maintaining approximate geostrophic balance as they do so. This time evolution depends on the ageostrophic component of the circulation, which tends to be on the order of Ro smaller than the geostrophic component. The fact that a time-evolving flow maintains approximate geostrophic balance implies that, in a rapidly rotating atmosphere, adjustment mechanisms exist that tend to reestablish geostrophic balance when departures from it occur.

What is the mechanism for establishing and maintaining geostrophy? If a rapidly rotating atmosphere deviates from geostrophic balance, it implies that the horizontal pressure gradient and Coriolis forces only imperfectly cancel, leaving an unbalanced residual force. This force generates a component of ageostrophic motion between pressure highs and lows. The Coriolis force on this ageostrophic motion, and the alteration of the pressure contrasts caused by the ageostrophic wind, act to reestablish geostrophy.

To give a concrete example, imagine an atmosphere with zero winds and a localized circular region of high surface pressure surrounded on all sides by lower surface pressure. This state has an unbalanced pressure-gradient force, which would induce a horizontal acceleration of fluid radially away from the high-pressure region. The horizontal Coriolis force on this outward motion causes a lateral deflection (to the right in the northern hemisphere and left in the southern hemisphere, leading to a vortex surrounding the high-pressure region. [Northern and southern hemispheres are here defined as those where $\Omega \cdot \mathbf{k} > 0$ and < 0, respectively, where Ω and $\mathbf{k}$ are the planetary rotation vector and local vertical (upward) unit vector.] The Coriolis force on this vortical motion points radially toward the high-pressure center, resisting its lateral expansion. This process continues until the inward-pointing Coriolis force balances the outward pressure-gradient force — hence establishing geostrophy and inhibiting further expansion. Although this is an extreme example, radiative heating/cooling, friction, and other forcings gradually push the atmosphere away from geostrophy, and the process described above reestablishes it.

This process, called geostrophic adjustment, tends to occur with a natural length scale comparable to the Rossby radius of deformation, given in a three-dimensional system by

$$L_D = \frac{NH}{f} \quad (23)$$

where N is Brunt-Väisälä frequency (equation (6)), H is the vertical scale of the flow, and f is the Coriolis parameter. Thus, geostrophic adjustment naturally generates large-scale atmospheric flow structures with horizontal sizes comparable to the deformation radius [for a detailed treatment, see *Holton* (2004) or *Vallis* (2006)].

Equation (22), as written, applies to atmospheres that are vertically thin compared to their horizontal dimensions (i.e., shallow atmospheres). However, a nonshallow analog of equation (22) can be obtained by considering the three-dimensional vorticity equation. In a geostrophically balanced fluid where the friction force is weak, the vorticity balance is given by (see *Pedlosky*, 1987, p. 43)

$$2(\Omega \cdot \nabla)\mathbf{u} - 2\Omega(\nabla \cdot \mathbf{u}) = -\frac{\nabla \rho \times \nabla p}{\rho^2} \quad (24)$$

where Ω is the planetary rotation vector and $\mathbf{u}$ is the three-dimensional wind velocity. The term on the right, called the baroclinic term, is nonzero when density varies on constant-pressure surfaces. In the interior of a giant planet, however, convection tends to homogenize the entropy, in which case fractional density variations on isobars are extremely small. Such a fluid, where surfaces of constant p and ρ align, is called a barotropic fluid. In this case the right side of equation (24) can be neglected, leading to the compressible-fluid generalization of the Taylor-Proudman theorem

$$2(\Omega \cdot \nabla)\mathbf{u} - 2\Omega(\nabla \cdot \mathbf{u}) = 0 \quad (25)$$

Consider a Cartesian coordinate system (x_*, y_*, z_*) with the z_* axis parallel to Ω and the x_* and y_* axes lying in the equatorial plane. Equation (25) can then be expressed in component form as

$$\frac{\partial u_*}{\partial z_*} = \frac{\partial v_*}{\partial z_*} = 0 \quad (26)$$

$$\frac{\partial u_*}{\partial x_*} + \frac{\partial v_*}{\partial y_*} = 0 \quad (27)$$

where u_* and v_* are the wind components along the x_* and y_* axes, respectively. These equations state that the wind components parallel to the equatorial plane are independent of direction along the axis perpendicular to the equatorial plane, and moreover that the divergence of the winds in the equatorial plane must be zero. The flow thus exhibits a structure with winds constant on columns, called Taylor columns, that are parallel to the rotation axis. In the context of a giant planet, motion of Taylor columns toward or away from the rotation axis is disallowed because the planetary geometry would force the columns to stretch or contract, causing a nonzero divergence in the equatorial plane — violating equation (27). Rather, the columns are free to move only along latitude circles. The flow then takes the form of concentric cylinders, centered about the rotation axis, which can move in the east-west direction. This columnar flow provides one model for the structure of the winds in the deep interiors of Jupiter, Saturn, Uranus, Neptune, and giant exoplanets (see section 4).

3.4. Other Force Balances: The Case of Slowly Rotating Planets

The pressure-gradient force can be balanced by forces other than the Coriolis force. An important example is cyclostrophic balance, which is a balance between horizontal centrifugal and pressure-gradient force, expressed in its simplest form as

$$\frac{u_t^2}{r} = \frac{1}{\rho}\frac{\partial p}{\partial r} \quad (28)$$

where we are considering circular flow around a central point. Here, u_t is the tangential speed of the circular flow and r is the radius of curvature of the flow. The lefthand side of the equation is the centrifugal force per unit mass and the righthand side is the radial pressure-gradient force per unit mass. Cyclostrophic balance is the force balance that occurs within dust devils and tornados, for example.

In the context of a global-scale planetary circulation, if the primary flow is east-west, the centrifugal force manifests as the curvature term $u^2 \tan\phi/a$, where a is planetary radius and ϕ is latitude (see *Holton*, 2004, pp. 31–38 for a discussion of curvature terms). Cyclostrophic balance can then be written

$$\frac{u^2 \tan\phi}{a} = -\left(\frac{\partial \Phi}{\partial y}\right)_p \quad (29)$$

This is the force balance relevant on Venus, for example, where a strong zonal jet with peak speeds reaching ~100 m s^{-1} dominates the stratospheric circulation (*Gierasch et al.*, 1997). Equation (29) can be differentiated with respect to pressure to yield a cyclostrophic version of the thermal-wind equation

$$\frac{\partial (u^2)}{\partial \ln p} = -\frac{a}{\tan\phi}\frac{\partial (RT)}{\partial y} \qquad (30)$$

where hydrostatic balance and the ideal-gas law have been invoked.

Thus, on a slowly rotating, cyclostrophically balanced planet like Venus, equation (30) would indicate that variation of the zonal winds with height requires latitudinal temperature gradients as occur with geostrophic balance. Note, however, that for zonal winds whose strength increases with height, the latitudinal temperature gradients for cyclostrophic balance point equatorward *regardless* of whether the zonal winds are east or west. This differs from geostrophy, where latitudinal temperature gradients would point equatorward for a zonal wind that becomes more eastward with height but poleward for a zonal wind that becomes more westward with height.

3.5. What Controls Vertical Velocities?

In atmospheres, mean vertical velocities associated with the large-scale circulation tend to be much smaller than horizontal velocities. This results from several factors, as detailed below.

3.5.1. Large aspect ratio. Atmospheres generally have horizontal dimensions greatly exceeding their vertical dimensions, which leads to a strong geometric constraint on vertical motions. To illustrate, suppose the continuity equation can be approximated in three dimensions by the incompressibility condition $\nabla \cdot \mathbf{u} = \partial u/\partial x + \partial v/\partial y + \partial w/\partial z = 0$. To order of magnitude, the horizontal terms can be approximated as U/L and the vertical term can be approximated as W/H, where L and H are the horizontal and vertical scales and U and W are the characteristic magnitudes of the horizontal and vertical wind. This then suggests a vertical velocity of approximately

$$W \sim \frac{H}{L}U \qquad (31)$$

On a typical terrestrial planet, $U \sim 10$ m s^{-1}, $L \sim 1000$ km, and $H \sim 10$ km, which would suggest $W \sim 0.1$ m s^{-1}, 2 orders of magnitude smaller than horizontal velocities. However, this estimate of W is an upper limit, since partial cancellation can occur between $\partial u/\partial x$ and $\partial v/\partial y$, and indeed for most atmospheres equation (31) greatly overestimates their mean vertical velocities. On Earth, for example, mean midlatitude vertical velocities on large scales (~10^3 km) are actually ~10^{-2} m s^{-1} in the troposphere and ~10^{-3} m s^{-1} in the stratosphere, much smaller than suggested by equation (31).

3.5.2. Suppression of vertical motion by rotation. The geostrophic velocity flows perpendicular to the horizontal pressure gradient, which implies that a large fraction of this flow cannot cause horizontal divergence or convergence (as necessary to allow vertical motions). However, the *ageostrophic* component of the flow, which is O(Ro) smaller than the geostrophic component, can cause horizontal convergence/divergence. Thus, to order of magnitude one might expect $W \sim RoU/L$. Considering the geostrophic flow and taking the curl of equation (21), we can show that the horizontal divergence of the horizontal geostrophic wind is

$$\nabla_p \cdot \mathbf{v_G} = \frac{\beta}{f}v = \frac{v}{a\tan\phi} \qquad (32)$$

where in the rightmost expression β/f has been evaluated assuming spherical geometry. To order of magnitude, $v \sim U$. Equation (32) then implies that to order of magnitude the horizontal flow divergence is not U/L but rather U/a. On many planets, the dominant flow structures are smaller than the planetary radius (e.g., L/a ~ 0.1–0.2 for Earth, Jupiter, and Saturn). This implies that the horizontal divergence of the geostrophic flow is ~L/a smaller than the estimate in equation (31). Taken together, these constraints suggest that rapid rotation can suppress vertical velocities by close to an order of magnitude. For Earth parameters, the estimates imply $W \sim 0.01$ m s^{-1}, similar to the mean vertical velocities in Earth's troposphere.

3.5.3. Suppression of vertical motion by stable stratification. Most atmospheres are stably stratified, implying that entropy (potential temperature) increases with height. Adiabatic expansion/contraction in ascending (descending) air would cause temperature at a given height to decrease (increase) over time. In the absence of radiation, such steady flow patterns are unsustainable because they induce density variations that resist the motion — ascending air becomes denser and descending air becomes less dense than the surroundings. Thus, in stably stratified atmospheres, the radiative heating/cooling rate exerts major control over the rate of vertical ascent/decent. Steady vertical motion can only occur as fast as radiation can remove the temperature variations caused by the adiabatic ascent/descent (*Showman et al.*, 2008a), when conduction is negligible. The idea can be quantified by rewriting the thermodynamic energy equation (equation (5d)) in terms of the Brunt-Väisälä frequency

$$\frac{\partial T}{\partial t} + \mathbf{v}\cdot\nabla_p T - \omega\frac{H_p^2 N^2}{Rp} = \frac{\dot{q}_{net}}{c_p} \qquad (33)$$

where H_p is the pressure scale height. In situations where the flow is approximately steady and vertical thermal advection is more important than (or comparable to) horizontal thermal advection, an estimate of the magnitude of the vertical velocity in an overturning circulation can be obtained by equating the right side to the last term on the left side. Converting to velocity expressed in height units, we obtain (*Showman et al.*, 2008a; *Showman and Guillot*, 2002)

$$W \sim \frac{\dot{q}_{net}}{c_p} \frac{R}{H_p N^2} \qquad (34)$$

[Note that oscillatory vertical motions associated with (for example) fast waves can occur adiabatically and are not governed by this constraint.] For Earth's midlatitude troposphere, $H_p \sim 10$ km, $N \sim 0.01$ s^{-1}, and $\dot{q}_{net}/c_p \sim 3 \times 10^{-5}$ K s^{-1}, implying $W \sim 10^{-2}$ m s^{-1}. In the stratosphere, however, the heating rate is lower, and the stable stratification is greater, leading to values of $W \sim 10^{-3}$ m s^{-1}.

For a canonical hot Jupiter with a three-day rotation period, horizontal wind speeds of kilometers per second imply Rossby numbers of ~ 1, and for scale heights of ~ 300 km (Table 1) and horizontal length scales of a planetary radius, one thus estimates RoUH/L ~ 10 m s^{-1}. Likewise, for R = 3700 J kg^{-1}K^{-1}, $H_p \sim 300$ km, $N \approx 0.005$ s^{-1} (appropriate for an isothermal layer on a hot Jupiter with surface gravity of 20 m s^{-1}), and $\dot{q}_{net}/c_p \sim 10^{-2}$K s^{-1} (perhaps appropriate for photospheres of a typical hot Jupiter), equation (34) suggests mean vertical speeds of $W \sim 10$ m s^{-1} (*Showman and Guillot*, 2002; *Showman et al.*, 2008a). Thus, despite the fact that the circulation on hot Jupiters occurs in a radiative zone expected to be stably stratified over most of the planet, vertical overturning can occur near the photosphere over timescales of $H_p/W \sim 10^5$ s. Because the interior of a multi-G.y.-old hot Jupiter transports an intrinsic flux of only ~ 10–100 W m^{-2} (as compared with typically $\sim 10^5$ W m^{-2} in the layer where starlight is absorbed and infrared radiation escapes to space), the net heating $\dot{q}_{net}$ should decrease rapidly with depth. The mean vertical velocities should thus plummet with depth, leading to very long overturning times in the bottom part of the radiative zone.

3.6. Effect of Eddies: Jet Streams and Banding

Although turbulence is a challenging problem, several decades of work in the fields of fluid mechanics and atmosphere/ocean dynamics have led to a basic understanding of turbulent flows and how they interact with planetary rotation. In a three-dimensional turbulent flow, vorticity stretching and straining drives a fluid's kinetic energy to smaller and smaller length scales, where it can finally be removed by viscous dissipation. This process, the so-called forward energy cascade, occurs when turbulence has scales smaller than a fraction of a pressure scale height (~ 10 km for Earth, ~ 200 km for hot Jupiters). On global scales, however, atmospheres are quasi-two-dimensional, with horizontal flow dimensions exceeding vertical flow dimensions by typically factors of ~ 100. In this quasi-two-dimensional regime, vortex stretching is inhibited, and other nonlinear processes, such as vortex merging, assume a more prominent role, forcing energy to undergo an *inverse* cascade from small to large scales (*Vallis*, 2006, pp. 349–361). This process has been well documented in idealized laboratory and numerical experiments (e.g., *Tabeling*, 2002) and helps explain the emergence of large-scale jets and vortices from small-scale turbulent forcing in planetary atmospheres.

Where the Coriolis parameter f is approximately constant, this process generally produces turbulence that is horizontally isotropic, that is, turbulence without any preferred directionality (north-south vs. east-west). This could explain the existence of some quasicircular vortices on Jupiter and Saturn, for example; however, it fails to explain the existence of numerous jet streams on Jupiter, Saturn, Uranus, Neptune, and the terrestrial planets. On the other hand, *Rhines* (1975) realized that the variation of f with latitude leads to anisotropy, causing elongation of structures in the east-west direction relative to the north-south direction. This anisotropy can cause the energy to reorganize into east-west-oriented jet streams with a characteristic latitudinal length scale

$$L_\beta = \pi \left(\frac{U}{\beta}\right)^{1/2} \qquad (35)$$

where U is a characteristic wind speed and $\beta \equiv df/dy$ is the gradient of the Coriolis parameter with northward distance y. This length is called the Rhines scale.

Numerous idealized studies of two-dimensional turbulent flow forced by injection of small-scale turbulence have demonstrated jet formation by this mechanism (e.g., *Williams*, 1978; *Cho and Polvani*, 1996a; *Cho and Polvani*, 1996b; *Huang and Robinson*, 1998; *Marcus et al.*, 2000; see *Vasavada and Showman*, 2005, for a review). Figure 3 illustrates an example. In two-dimensional nondivergent numerical simulations of flow energized by small-scale turbulence, the flow remains isotropic at low rotation rates (Fig. 3, left) but develops banded structure at high rotation rates (Fig. 3, right). As shown in Fig. 4, the number of jets in such simulations increases as the wind speed decreases, qualitatively consistent with equation (35).

The Rhines scale can be interpreted as a transition scale between the regimes of turbulence and Rossby waves, which are a large-scale wave solution to the dynamical equations in the presence of nonzero β. Considering the simplest possible case of an unforced two-dimensional nondivergent fluid (section 2.4), the vorticity equation reads

$$\frac{\partial \zeta}{\partial t} + \mathbf{v} \cdot \nabla \zeta + v\beta = 0 \qquad (36)$$

where ζ is the relative vorticity, $\mathbf{v}$ is the horizontal velocity vector, and v is the northward velocity component. The relative vorticity has a characteristic scale U/L, and so the nonlinear term has a characteristic scale U^2/L^2. The β term has a characteristic scale βU. Comparison between these two terms shows that, for length scales smaller than the Rhines scale, the nonlinear term dominates, implying the dominance of nonlinear vorticity advection. For scales exceeding the Rhines scale, the linear $v\beta$ term dominates over the nonlinear term, and Rossby waves are the primary solutions to the equations. [Adopting a stream function ψ defined by $u = -\partial\psi/\partial y$ and

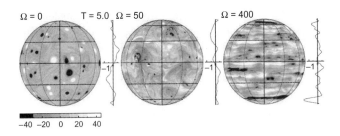

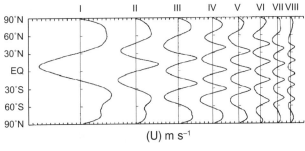

Fig. 3. Relative vorticity for three two-dimensional nondivergent simulations on a sphere initialized from small-scale isotropic turbulence. The three simulations are identical except for the planetary rotation rate, which is zero on the left, intermediate in the middle, and fast on the right. Forcing and large-scale friction are zero, but the simulations contain a hyperviscosity to maintain numerical stability. The final state consists of isotropic turbulence in the nonrotating case but banded flow in the rotating case, a result of the Rhines effect. The spheres are viewed from the equatorial plane, with the rotation axis oriented upward on the page. After *Yoden et al.* (1999), *Hayashi et al.* (2000), *Ishioka et al.* (1999), and *Hayashi et al.* (2007).

Fig. 4. Zonal-mean zonal winds vs. latitude for a series of global, two-dimensional nondivergent simulations performed on a sphere showing the correlation between jet speed and jet width. The simulations are forced by small-scale turbulence and damped by a linear drag. Each simulation uses different forcing and friction parameters and equilibrates to a different mean wind speed, ranging from fast on the left to slow on the right. Simulations with faster jets have fewer jets and vice versa; the jet widths approximately scale with the Rhines length. From *Huang and Robinson* (1998).

$v = \partial\psi/\partial x$, the two-dimensional nondivergent vorticity equation can be written $\partial\nabla^2\psi/\partial t + \beta\partial\psi/\partial x = 0$ when the nonlinear term is neglected. In Cartesian geometry with constant β, this equation has wave solutions with a dispersion relationship $\omega = -\beta k^2/(k^2 + l^2)$ where ω is the oscillation frequency and k and l are the wave numbers (2π over the wavelength) in the eastward and northward directions, respectively. These are Rossby waves, which have a westward phase speed and a frequency that depends on the wave orientation.]

How does jet formation occur at the Rhines scale? In a two-dimensional fluid where turbulence is injected at small scales an upscale energy cascade occurs, but when the turbulent structures reach sizes comparable to the Rhines scale, the transition from nonlinear to quasilinear dynamics prevents the turbulence from easily cascading to scales larger than L_β. [In a two-dimensional model, turbulence behaves in a two-dimensional manner at all scales, but in the atmosphere, flows only tend to behave in two dimensions when the ratio of widths to heights $\gg 1$ (implying horizontal scales exceeding hundreds of kilometers for Earth).] At the Rhines scale, the characteristic Rossby-wave frequency for a typically oriented Rossby wave, $-\beta/\kappa$, roughly matches the turbulent frequency $U\kappa$, where κ is the wavenumber magnitude. (The Rossby wave frequency results from the dispersion relation, ignoring the distinction between eastward and northward wavenumbers; the turbulent frequency is 1 over an advective timescale, L/U.) At larger length scales (smaller wavenumbers) than the Rhines scale, the Rossby-wave oscillation frequency is faster than the turbulence frequency and so wave/turbulent interactions are inefficient at transporting energy into the Rossby-wave regime. This tends to cause a pile-up of energy at a wavelength L_β. As a result, flow structures with a wavelength of L_β often contain more energy than flow structures at any other wavelength.

But why are these dominant structures generally banded jet streams rather than (say) quasicircular vortices? The reason relates to the anisotropy of Rossby waves. In a two-dimensional nondivergent fluid, the Rossby-wave dispersion relation can be written

$$\omega = -\frac{\beta k}{k^2 + l^2} = -\frac{\beta\cos\alpha}{\kappa} \qquad (37)$$

where k and l are the wavenumbers in the eastward and northward directions, $\kappa = \sqrt{k^2 + l^2}$ is the total wavenumber magnitude, and α is the angle between the Rossby-wave propagation direction and east. Equating this frequency to the turbulence frequency $U\kappa$ leads to an anisotropic Rhines wavenumber

$$k_\beta^2 = \frac{\beta}{U}|\cos\alpha| \qquad (38)$$

When plotted on the k-l wavenumber plane, this relationship traces out a dumbbell pattern; k_β approaches β/U when $\alpha \approx 0°$ or $180°$ and approaches zero when $\alpha \approx \pm 90°$. The Rhines scale L_β is thus $\sim\pi(U/\beta)^{1/2}$ for wavevectors with $\alpha \approx 0°$ or $180°$, but it tends to infinity for wave vectors with $\alpha \approx \pm 90°$. Because of this anisotropic barrier, the inverse cascade can successfully drive energy to smaller wavenumbers (larger length scales) along the $\alpha \approx \pm 90°$ axis than along any other direction. Wave vectors with $\alpha \approx \pm 90°$ correspond to cases with $k \ll l$, that is, flow structures whose east-west lengths become unbounded compared to their north-south lengths. The result is usually an east-west banded pattern and jet streams whose latitudinal width is $\sim\pi(U/\beta)^{1/2}$. Although these considerations are heuristic, detailed numerical simulations generally confirm that in two-dimensional flows forced by small-scale isotropic turbulence, the turbulence reorganizes to produce jets with widths close to L_β (Fig. 3) (for a review, see *Vasavada and Showman*, 2005).

In practice, when small-scale forcing is injected into the flow, jet formation does not generally occur as the end product of successive mergers of continually larger and larger vortices but rather by a feedback whereby small eddies interact directly with the large-scale flow and pump up jets (e.g., *Nozawa and Yoden*, 1997; *Huang and Robinson*, 1998). The large-scale background shear associated with jets distorts the eddies, and the interaction pumps momentum upgradient into the jet cores, helping to maintain the jets against friction or other forces.

The above arguments were derived in the context of a two-dimensional nondivergent model, which is the simplest model where Rossby waves can interact with turbulence to form jets. Such a model lacks gravity waves and has an infinite deformation radius L_D (equations (10) and (23)). It is possible to extend the above ideas to one-layer models exhibiting gravity waves and a finite deformation radius, such as the shallow-water model. Doing so suggests that, as long as β is strong, jets dominate when the deformation radius is large (as in the two-dimensional nondivergent model); however, the flow becomes more isotropic (dominated by vortices rather than jets) when the deformation radius is sufficiently smaller than $(U/\beta)^{1/2}$ (e.g., *Okuno and Masuda*, 2003; *Smith*, 2004; *Showman*, 2007).

On a spherical planet, $\beta = 2\Omega a^{-1} \cos \phi$, which implies that a planet should have a number of jet streams roughly given by

$$N_{jet} \sim \left(\frac{2\Omega a}{U}\right)^{1/2} \qquad (39)$$

These considerations do a reasonable job of predicting the latitudinal separations and total number of jets that exist on planets in our solar system. Given the known wind speeds (Table 1) and values of β, the Rhines scale predicts ~20 jet streams on Jupiter/Saturn, 4 jet streams on Uranus and Neptune, and 7 jet streams on Earth. This is similar to the observed numbers (~20 on Jupiter and Saturn, 3 on Uranus and Neptune, and 3 to 7 on Earth depending on how they are defined). [In Earth's troposphere, an instantaneous snapshot typically shows distinct subtropical and high-latitude jets in each hemisphere, with weaker flow between these jets and westward flow at the equator. The jet latitudes exhibit large excursions in longitude and time, however, and when a time and longitude average is performed, only a single local maximum in eastward flow exists in each hemisphere, with westward flow at the equator.] Moreover, the scale-dependent anisotropy — with quasi-isotropic turbulence at small scales and banded flow at large scales — is readily apparent on solar system planets. Figure 5 illustrates an example for Saturn: Small-scale structures (e.g., the cloud-covered vortices that manifest as small dark spots in the figure) are relatively circular, whereas the large-scale structures are banded.

These considerations yield insight into the degree of banding and size of dominant flow structures to be expected on exoplanets. Equation (39) suggests that the number of bands scales as $\Omega^{1/2}$, so rapidly rotating exoplanets will tend to exhibit numerous strongly banded jets, whereas slowly rotating planets will exhibit fewer jets with weaker banding. Moreover, for a given rotation rate, planets with slower winds should exhibit more bands than planets with faster winds. Hot Jupiters are expected to be tidally locked, implying rotation rates of typically a few days. Although wind speeds on hot Jupiters are unknown, various numerical models have obtained (or assumed) speeds of ~0.5–3 km s^{-1} (*Showman and Guillot*, 2002; *Cooper and Showman*, 2005; *Langton and Laughlin*, 2007; *Dobbs-Dixon and Lin*, 2008; *Menou and Rauscher*, 2009; *Cho et al.*, 2003, 2008). Inserting these values into equation (39) implies $N_{jet} \sim 1–2$. Consistent with these arguments, simulations of hot Jupiter atmospheres forced at small scales have obtained typically ~1–3 broad jets (*Cho et al.*, 2003, 2008; *Langton and Laughlin*, 2008a). Thus, due to a combination of their slower rotation and presumed faster wind speeds, hot Jupiters should have only a small number of broad jets — unlike Jupiter and Saturn. If the wind speeds are much slower, and/or rotation rates much greater than assumed here, the number of bands would be larger. [Published three-dimensional models of hot Jupiters forced by day-night heating contrasts also exhibit only a few broad jets (e.g., *Cooper and Showman*, 2005; *Showman et al.*, 2008a, 2009; *Dobbs-Dixon and Lin*, 2008; *Menou and Rauscher*, 2009) but for a different reason. In this case, the jet width may be controlled by the Rossby deformation radius, which is close to the planetary radius for many hot Jupiters. As a result, the day-night heating contrast injects significant energy into the flow directly at the planetary scale, and the β effect can reorganize this global-scale energy into a banded jet pattern. Because the forcing and jet scales are comparable, such a flow could lack an inverse cascade.]

3.7. Role of Eddies in Three-Dimensional Atmospheres

Although atmospheres behave in a quasi-two-dimensional manner on large scales due to large horizontal:vertical aspect ratios, small Rossby numbers, and stable stratification, they are of course three-dimensional, and as a result they can experience both upscale and downscale energy cascades depending on the length scales of the turbulence and other factors. Nevertheless, the basic mechanism discussed above for the interaction of turbulence and the β effect still applies and suggests that even three-dimensional atmospheres can generally exhibit a banded structure with a characteristic length scale close to L_β. Consistent with this idea, numerical simulations show that banding can indeed

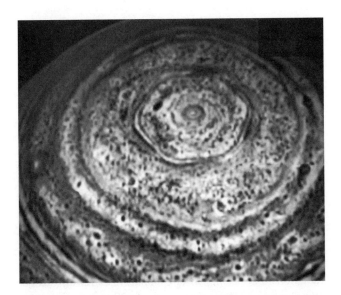

Fig. 5. Saturn's north polar region as imaged by the Cassini Visible and Infrared Mapping Spectrometer at 5-μm wavelengths. At this wavelength, scattered sunlight is negligible; bright regions are cloud-free regions where thermal emission escapes from the deep (~3–5-bar) atmosphere, whereas dark region are covered by thick clouds that block this thermal radiation. Note the scale-dependent anisotropy: Small-scale features tend to be circular, whereas the large-scale features are banded. This phenomenon results directly from the Rhines effect (see text). Photo credit: NASA/VIMS/Bob Brown/Kevin Baines.

occur in three dimensions even when β is the only source of horizontal anisotropy (e.g., *Sayanagi et al., 2008; Lian and Showman,* 2010). Nevertheless, on real planets, banding can also result from anisotropic forcing — such as the fact that solar heating is primarily a function of latitude rather than longitude on rapidly rotating planets like Earth and Mars.

A variety of studies show that, in stably stratified, rapidly rotating atmospheres, the characteristic vertical length scale of the flow is approximately f/N times the characteristic horizontal scale (e.g., *Charney,* 1971; *Dritschel et al.,* 1999; *Reinaud et al.,* 2003; *Haynes,* 2005). For typical horizontal dimensions of large-scale flows, this often implies vertical dimensions of one to several scale heights.

Several processes can generate turbulent eddies that significantly affect the large-scale flow. Convection is particularly important in giant planet interiors and near the surface of terrestrial planets. Shear instabilities can occur when the wind shear is sufficiently large; they transfer energy from the mean flow into turbulence and reduce the shear of the mean flow. A particularly important turbulence-generating process on rapidly rotating planets is baroclinic instability, which is a dynamical instability driven by the extraction of potential energy from a latitudinal temperature contrast. (When two adjacent, stably stratified air columns have differing temperature profiles, potential energy is released when the colder column slides underneath the hotter column. For this reason, baroclinic instability, which draws on this energy source, is sometimes called "slantwise convection.") On non- (or slowly) rotating planets, the presence of a latitudinal temperature contrast would simply cause a direct Hadley-type overturning circulation, which efficiently mutes the thermal contrasts. However, on a rapidly rotating planet, Hadley circulations cannot penetrate to high latitudes (section 5.3), and, in the absence of instabilities and waves, the atmosphere poleward of the Hadley cell would approach a radiative-equilibrium temperature structure, with strong latitudinal temperature gradients and strong zonal winds in thermal-wind balance with the temperature gradients. In three dimensions, this radiative-equilibrium structure can experience baroclinic instabilities, which develop into three-dimensional eddies that push cold polar air equatorward and down and push warm low-latitude air poleward and up. This process lowers the center of mass of the fluid, thereby converting potential energy into kinetic energy and transporting thermal energy between the equator and poles. The fastest instability growth rates occur at length scales comparable to the Rossby deformation radius (equation (23)), and the resulting eddies act as a major driver for the midlatitude jet streams on Earth, Mars, and perhaps Jupiter, Saturn, Uranus, and Neptune.

A formalism for describing the effect of eddies on the mean flow can be achieved by decomposing the flow into zonal-mean components and deviations from the zonal mean. Here, "eddies" are defined as deviations from the zonal-mean flow and can represent the effects of turbulence, waves, and instabilities. Denoting zonal means with overbars and deviations therefrom with primes, we can write $u = \bar{u} + u'$, $v = \bar{v} + v'$, $\theta = \bar{\theta} + \theta'$, and $\omega = \bar{\omega} + \omega'$. Substituting these expressions into the three-dimensional primitive equations (equation (5)) and zonally averaging the resulting equations yields

$$\frac{\partial \bar{\theta}}{\partial t} = \frac{\overline{\theta q}}{Tc_p} - \frac{\partial \overline{(\theta v)}}{\partial y} - \frac{\partial \overline{(\theta \omega)}}{\partial p} \qquad (40)$$

$$\frac{\partial \bar{u}}{\partial t} = f\bar{v} - \frac{\partial \overline{(uv)}}{\partial y} - \frac{\partial \overline{(u\omega)}}{\partial p} + \overline{\mathcal{F}} - \overline{\mathcal{D}} \qquad (41)$$

It can be shown that

$$\overline{uv} = \bar{u}\bar{v} + \overline{u'v'} \qquad (42)$$

and similarly for zonal averages of other quadratic terms. Thus, the terms involving derivatives on the right sides of equations (40)–(41) represent the effect of transport by advection of the mean flow (e.g., $\bar{u}\bar{v}$) and the effect of transport by eddies (e.g., $\overline{u'v'}$). For example, $\overline{u'v'}$ can be thought of as the northward transport of eastward momentum by eddies. If $\overline{u'v'}$ has the same sign as the background jet shear $\partial \bar{u}/\partial y$, then eddies pump momentum up-gradient into the jet cores. If they have opposite signs, the eddies transport momentum down-gradient out of the jet cores. [See *Cho* (2008) for a brief discussion in the exoplanet context.] Molecular diffusion is a down-gradient process, but in many cases eddies can transport momentum up-gradient, strengthening the jets.

To illustrate, consider linear Rossby waves, whose dispersion relation is given by equation (37) in the case of a two-dimensional nondivergent flow. For such waves, it can be shown that the product $\overline{u'v'}$ is opposite in sign to the north-south component of their group velocity (see *Vallis*, 2006, pp. 489–490). Since the group velocity generally points away from the wave source, this implies that Rossby waves cause a flux of eastward momentum *toward* the wave source. The general result is an eastward eddy acceleration at the latitudes of the wave source and a westward eddy acceleration at the latitudes where the waves dissipate or break. This process is a major mechanism by which jet streams can form in planetary atmospheres.

The existence of eddy acceleration and heat transport triggers a mean circulation that has a back-influence on both $\bar{u}$ and $\bar{\theta}$. On a rapidly rotating planet, for example, the eddy accelerations and heat transports tend to push the atmosphere away from geostrophic balance, leading to a mismatch in the mean north-south pressure gradient and Coriolis forces. This unbalanced force drives north-south motion, which, through the mass continuity equation, accompanies vertical motions. This circulation induces advection of momentum and entropy, and moreover causes an east-west Coriolis acceleration $f\bar{v}$, all of which tend to restore the atmosphere toward geostrophic balance [for descriptions of this, see *James* (1994, pp. 100–107) or *Holton* (2004, pp. 313–323)]. Such mean circulations tend to have broad horizontal and vertical extents, providing a mechanism for a localized eddy perturbation to trigger a nonlocal response.

Clearly, jets on planets do not continually accelerate, so any jet acceleration caused by eddies must (on average) be resisted by other terms in the equation. Near the equator (or on slowly rotating planets where Coriolis forces are small), this jet acceleration is generally balanced by a combination of friction (e.g., as represented by $\bar{\mathcal{D}}$ in equation (41)) and large-scale advection. However, in the mid- and high-latitudes of rapidly rotating planets, Coriolis forces are strong, and away from the surface the eddy accelerations are often balanced by the Coriolis force on the mean flow $f\bar{v}$.

When averaged vertically, the east-west accelerations due to the Coriolis forces tend to cancel out (this occurs because mass continuity requires that, in a time average, there be no net north-south mass transport, hence no net east-west Coriolis acceleration), and the vertically integrated force balance (on a terrestrial planet, for example) is between the vertically integrated eddy accelerations and surface friction (e.g., *Peixoto and Oort*, 1992, Chapter 11). This means that regions without active eddy accelerations must have weak time-averaged surface winds, although the time-averaged winds may still be strong aloft. Conversely, the existence of strong time-averaged winds at the surface can only occur at latitudes where eddy accelerations are active. A prime example is the eastward mean surface winds in Earth's midlatitudes (associated with the jet stream), which are enabled by the eddy accelerations from midlatitude baroclinic instabilities.

4. CIRCULATION REGIMES: GIANT PLANETS

4.1. Thermal Structure of Giant Planets: General Considerations

Because of the enormous energy released during planetary accretion, the interiors of giant planets are hot, and due to the expected high opacities in their interiors, this energy is thought to be transported through their interiors into their atmospheres primarily by convection rather than radiation (*Stevenson*, 1991; *Chabrier and Baraffe*, 2000; *Burrows et al.*, 2001; *Guillot et al.*, 2004; *Guillot*, 2005; chapter by Fortney et al.). Jupiter, Saturn, and Neptune emit significantly more energy than they absorb from the Sun (by factors of 1.7, 1.8, and 2.6, respectively, implying interior heat fluxes of ~1–10 W m^{-2}) showing that these planets are still cooling off, presumably by convection, even 4.6 G.y. after their formation. Even for giant exoplanets that lie close to their parent stars, the stellar insolation generally cannot stop the inexorable cooling of the planetary interiors (e.g., *Guillot et al.*, 1996; *Saumon et al.*, 1996; *Burrows et al.*, 2000; *Chabrier et al.*, 2004; *Fortney et al.*, 2007). Thus, we expect that in general the interiors of giant planets will be convective, and hence the interior entropy will be nearly homogenized and the interior temperature and density profiles will lie close to an adiabat. [Compositional gradients that force the interior density structures to deviate from a uniform adiabat could exist in Uranus and some giant exoplanets (*Podolak et al.*, 1991; *Chabrier and Baraffe*, 2007).]

The interiors will not precisely follow an adiabat because convective heat loss generates descending plumes that are colder than the background fluid. We can estimate the deviation from an adiabat as follows. Solar system giant planets are rotationally dominated; as a result, convective velocities are expected to scale as (*Stevenson*, 1979; *Fernando et al.*, 1991)

$$w \sim \left(\frac{F\alpha g}{\Omega \rho c_p} \right)^{1/2} \quad (43)$$

where w is the characteristic magnitude of the vertical velocity, F is the heat flux transported by convection, and α and c_p are the thermal expansivity and specific heat at constant pressure. To order of magnitude, the convected heat flux should be $F \sim \rho w c_p \delta T$, where δT is the characteristic magnitude of the temperature difference between a convective plume and the environment. These equations yield

$$\delta T \sim \left(\frac{F\Omega}{\rho c_p \alpha g} \right)^{1/2} \quad (44)$$

Inserting values for Jupiter's interior ($F \sim 10$ Wm^{-2}, $\Omega = 1.74 \times 10^{-4}$ s^{-1}, $\rho \sim 1000$ k gm^{-3}, $\alpha \sim 10^{-5}$ K^{-1}, $c_p \approx 1.3 \times 10^4$ J kg^{-1} K^{-1}, and $g \approx 20$ m s^{-2}), we obtain $\delta T \sim 10^{-3}$ K, suggesting fractional density perturbations of $\alpha \delta T \sim 10^{-8}$. Thus, deviations from an adiabat in the interior should be extremely small. Even objects with large interior heat fluxes, such as brown dwarfs or young giant planets, will have only modest deviations from an adiabat; for example, the above equations suggest that an isolated 1000-K Jupiter-like planet with a heat flux of 6×10^4 W m^{-2} would experience deviations from an adiabat in its interior of only ~0.1 K.

At sufficiently low pressures in the atmosphere, the opacities become small enough that the outward energy transport transitions from convection to radiation, leading to a temperature profile that in radiative equilibrium is stably stratified and hence suppresses convection. At a minimum, this transition (called the radiative-convective boundary) will occur when the gas becomes optically thin to escaping infrared radiation (at pressures less than ~0.01–1 bar depending on the opacities). For Jupiter, Saturn, Uranus, and Neptune, this transition occurs at pressures somewhat less than 1 bar. However, in the presence of intense stellar irradiation, the absorbed stellar energy greatly exceeds the energy loss from the interior; thus, the mean photospheric temperature only slightly exceeds the temperature that would exist in thermal equilibrium with the star. In this case, cooling of the interior adiabat can only continue by development of a thick, stably stratified layer that penetrates downward from the surface and deepens with increasing age. Thus, for old, heavily irradiated planets, the radiative-convective boundary can instead occur at large optical depth, at pressures of ~100–1000 bar depending on age (see section 4.3 and chapter by Fortney et al.).

To what extent can large horizontal temperature differences develop in the observable atmosphere? This depends greatly on the extent to which the infrared photosphere and radiative-convective boundaries differ in pressure. [We define the infrared photosphere as the pressure at which the bulk of the planet's infrared radiation escapes to space.] On Jupiter, Saturn, Uranus, and Neptune, the infrared photospheres lie at ~300 mbar, close to the pressure of the radiative-convective boundary. Thus, the convective interiors — with their exceedingly small lateral temperature contrasts — effectively outcrop into the layer where radiation streams into space. On such a planet, lateral temperature contrasts in the observable atmosphere should be small (*Ingersoll and Porco*, 1978), and indeed the latitudinal temperature contrasts are typically only ~2–5 K in the upper troposphere and lower stratospheres of our solar system giant planets (*Ingersoll*, 1990).

If instead the radiative-convective boundary lies far deeper than the infrared photosphere — as occurs on hot Jupiters — then large lateral temperature contrasts at the photosphere can potentially develop. Above the radiative convective boundary, convective mixing is inhibited and thus cannot force vertical air columns in different regions to lie along a single temperature profile. Laterally varying absorption of light from the parent star (or other processes) can thus generate different vertical temperature gradients in different regions, potentially leading to significant horizontal temperature contrasts. In reality, such lateral temperature contrasts lead to horizontal pressure gradients, which generate horizontal winds that attempt to reduce the temperature contrasts. The resulting temperature contrasts thus depend on a competition between radiation and dynamics. These processes will be particularly important when strongly uneven external irradiation occurs, as on hot Jupiters.

Several additional processes can produce meteorologically significant temperature perturbations on giant planets. In particular, most giant planets contain trace constituents that are gaseous in the high pressure/temperature conditions of the interiors but condense in the outermost layers. This condensation releases latent heat and changes the mean molecular weight of the air, both of which cause significant density perturbations. On Jupiter and Saturn, ammonia, ammonium hydrosulfide, and water condense at pressures of ~0.5, 2, and 6 bar, causing temperature increases of ~0.2, 0.1, and 2 K, respectively, for solar abundances. For Jupiter, this corresponds to fractional density changes of ~10^{-3} for NH_3 and NH_4SH and ~10^{-2} for H_2O. At deep levels (~2000 K, which occurs at pressures of ~5000 bar on Jupiter), iron, silicates, and various metal oxides and hydrides condense, with a total latent heating of ~1 K and a fractional density perturbation of ~10^{-3}. For Uranus/Neptune, the condensation structure is similar but also includes methane near the tropopause, with a latent heating of ~0.2 K and a fractional density change of ~3×10^{-3} at solar abundance. Conversion between ortho and para forms of the H_2 molecule is also important at temperatures <200 K and produces a fractional density change of ~10^{-2} (*Gierasch and Conrath*, 1985). Note that methane, and plausibly water and other trace constituents, are enhanced over solar abundances by factors of ~3, 7, and 20–40 for Jupiter, Saturn, and Uranus/Neptune, respectively, so the actual density alterations for these planets likely exceed those listed above by these factors.

The key point is that these fractional density perturbations exceed those associated with dry convection in the interior by orders of magnitude; condensation could thus dominate the meteorology in the region where it occurs. Such moist convection could have several effects. First, if the density contrasts become organized on large scales, then significant vertical shear of the horizontal wind can occur (via the thermal-wind equation) that would otherwise not exist in the outermost part of the convection zone. Second, it can generate a background temperature profile that is stably stratified (e.g., *Stevens*, 2005), allowing wave propagation and various other phenomena. Third, the buoyancy produced by the condensation could act as a powerful driver of the circulation. For example, on Jupiter and Saturn, condensation leads to powerful thunderstorms (*Little et al.*, 1999; *Gierasch et al.*, 2000) that are a leading candidate for driving the global-scale jet streams (section 4.2). The cloud formation associated with moist convection can also significantly alter the temperature structure due to its influence in scattering/absorbing radiation.

For giant planets hotter than those in our solar system, the condensation sequence shifts to lower pressures and the more volatile species disappear from the condensation sequence. Important breakpoints occur for giant exoplanets with effective temperatures exceeding ~150 and ~300–500 K, above which ammonia and water, respectively, no longer condense, thus removing the effects of their condensation from the meteorology. The latter breakpoint will be particularly significant due to the overriding dominance of the fractional density change associated with water condensation. Nevertheless, condensation of iron, silicates, and metal oxides and hydrides will collectively constitute an important source of buoyancy production even for hot giant exoplanets, as long as such condensation occurs within the convection zone, where its buoyancy dominates over that associated with dry convection (equation (44)). For hot Jupiters, however, evolution models predict the existence of a quasiisothermal radiative zone extending down to pressures of ~100–1000 bar (*Guillot*, 2005; chapter by Fortney et al.). Latent heating occurring within this zone is less likely to be important, simply because the ~1-K potential temperature change associated with silicate condensation is small compared to the probable vertical and horizontal potential temperature variations associated with the global circulation (section 4.3). For example, in the vertically isothermal radiative zone of a hot Jupiter, a vertical displacement of air by only ~0.5 km would produce a temperature perturbation (relative to the background) of ~1 K, similar to the magnitude of thermal variation caused by condensation.

4.2. Circulation of Giant Planets: General Considerations

We now turn to the global-scale atmospheric circulation. As yet, few observational constraints exist regarding the circulation of exoplanets. Infrared lightcurves of several hot Jupiters indirectly suggest the presence of fast winds able to advect the temperature pattern (*Knutson et al.*, 2007, 2009b; *Cowan et al.*, 2007), and searches for variability in brown dwarfs are being made (e.g., *Morales-Calderón et al.*, 2006; *Goldman et al.*, 2008). But, at present, our local solar system giant planets represent the best proxies to guide our thinking about the circulation on a wide class of rapidly rotating giant exoplanets. We consider basic issues here and take up specific models of hot Jupiters in section 4.3. In the observable atmosphere, the global-scale circulation on Jupiter, Saturn, Uranus, and Neptune is dominated by numerous east-west jet streams (Fig. 6). Jupiter and Saturn each exhibit ~20 jet streams whereas Uranus and Neptune exhibit three jets each. Peak speeds range from ~150 m s^{-1} for Jupiter to over 400 m s^{-1} for Saturn and Neptune. The winds are determined by tracking the motion of small cloud features, which occur at pressures of ~0.5–4 bar, over periods of hours. The jet streams modulate the patterns of ascent and descent, leading to a banded cloud pattern that can be seen in Fig. 6.

4.2.1. Constraints from balance arguments. For giant exoplanets and solar system giant planets alike, the dynamical link between temperatures and winds encapsulated by the thermal-wind equation allows plausible scenarios for the vertical structure of the winds to be identified. The likelihood that giant planets lose their internal energy by convection suggests that the interior entropy is nearly homogenized and hence that the interiors are close to a barotropic state (section 3.3). This would lead to the Taylor-Proudman theorem (equations (25)–(27)), which states that the winds are constant on cylinders parallel to the planetary rotation axis. This scenario, first suggested by *Busse* (1976) for Jupiter, Saturn, Uranus, and Neptune, postulates that the jets observed in the cloud layer would simply represent the intersection with the surface of eastward- and westward-moving Taylor columns that penetrate throughout the molecular envelope.

How closely will the molecular interior of a giant planet follow the Taylor-Proudman theorem? It might not apply if geostrophy does not hold on the length scale of the jets (i.e., Rossby number ≥1), but this is unlikely, at least for the rapidly rotating giant planets of our solar system (*Liu et al.*, 2008). However, for giant exoplanets, geostrophy could break down if the interior wind speeds are sufficiently fast, planetary rotation rate is sufficiently low, the interior heat flux is sufficiently large (so that convective buoyancy forces exceed Coriolis forces associated with the jets, for example), or magnetic effects dominate. Alternately, even if the interior exhibits geostrophic columnar behavior, the interior could exhibit a shear of the wind along the Taylor columns if density variations on isobars in the interior are sufficiently large. To illustrate how this might work, we cast equation (24) in a cylindrical coordinate system and take the longitudinal component, which gives

$$2\Omega \frac{\partial u}{\partial z_*} = -\frac{\nabla \rho \times \nabla p}{\rho^2} \cdot \hat{\lambda} \qquad (45)$$

where u is the zonal wind, z_* is the coordinate parallel to the rotation axis (see section 3.3), and $\hat{\lambda}$ is the unit vector in the longitudinal direction. Suppose that horizontal density variations (on isobars) occur only in latitude. A geometrical argument shows that the magnitude of $\nabla \rho \times \nabla p$ is then $|\nabla p|(\partial \rho / \partial y)_p$, where $(\partial \rho / \partial y)_p$ is the partial derivative of density with northward distance, y, on a surface of constant pressure. The magnitude of the pressure gradient is dominated by the hydrostatic component, $-\rho g$, and thus we can write

$$2\Omega \frac{\partial u}{\partial z_*} \approx \frac{g}{\rho} \left(\frac{\partial \rho}{\partial y} \right)_p \qquad (46)$$

which, to order of magnitude, implies that the magnitude of zonal-wind variation along a Taylor column can be expressed as

$$\Delta u \sim \frac{g}{L\Omega} \left(\frac{\delta \rho}{\rho} \right)_p \Delta z_* \qquad (47)$$

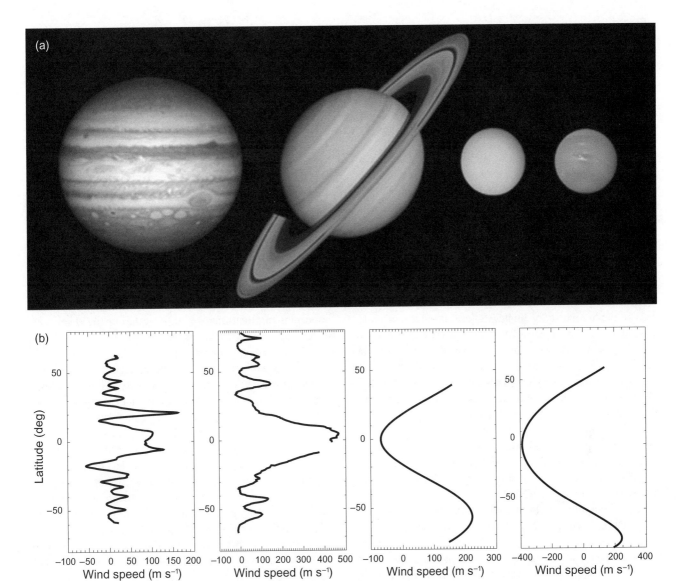

Fig. 6. (a) Jupiter, Saturn, Uranus, and Neptune shown to scale in visible-wavelength images from the Cassini and Voyager spacecraft. (b) Zonal-mean zonal wind profiles obtained from tracking small-scale cloud features over a period of hours. Jupiter and Saturn have an alternating pattern of ~20 east-west jets, whereas Uranus and Neptune have 3 broad jets.

where L is the width of the jets in latitude, $(\delta\rho/\rho)_p$ is the fractional density difference (on isobars) across a jet, and Δz_* is the distance along the Taylor column over which this fractional density difference occurs. In the interior, convection supplies the fractional density contrasts, and as shown in section 4.1, these plausibly have extremely small values of 10^{-6}–10^{-8} below the region where condensation can occur. We can obtain a crude order-of-magnitude estimate for the thermal-wind shear along a Taylor column by assuming that these fractional density differences are organized on the jet scale and coherently extend across the full ~10^4-km vertical length of a Taylor column. Inserting these values into equation (47), along with parameters relevant to a Jupiter-like planet ($g \approx 20$ m s^{-2}, L $\approx$ 5000 km, $\Omega = 1.74 \times$ 10^{-4} s^{-1}) yields $\Delta u \sim 0.001$–0.1 m s^{-1}. If this estimate is correct, then Jupiter-like planets would exhibit only small thermal-wind shear along Taylor columns, and the Taylor-Proudman theorem would hold to good approximation in the molecular interior.

However, the above estimate is crude; if the jet-scale horizontal density contrasts were much larger than the density contrasts associated with convective plumes, for example, then the wind shear along Taylor columns could exceed the values estimated above. The quantitative validity of equation (44), on which the wind-shear estimates are based, also remains uncertain. To attack the problem more rigorously, *Kaspi et al.* (2009) performed full three-dimensional numerical simulations of convection in giant-planet interiors, suggesting that the compressibility may play an important role in allowing the generation of thermal-wind shear within giant planets, especially in the outermost layers of their convection zones. Such wind shear is particularly likely to be important for young and/or massive planets with large interior heat fluxes.

Although fast winds could exist throughout the molecular envelope, Lorentz forces probably act to brake the zonal flows in the underlying metallic region at pressures exceeding ~1–3 Mbar (*Kirk and Stevenson*, 1987; *Grote et al.*, 2000; *Busse*, 2002). The winds in the deep, metallic interior are thus often assumed to be weak. The transition between molecular and metallic occurs gradually, and there exists a wide semiconducting transition zone over which the Taylor-Proudman theorem approximately holds yet the electrical conductivity becomes important. *Liu et al.* (2008) argued that, if the observed jets on Jupiter and Saturn penetrated downward as Taylor columns, the Ohmic dissipation would exceed the luminosity of Jupiter (Saturn) if the jets extended deeper than 95% (85%) of the planetary radius. The observed jets are thus probably shallower than these depths. However, if the jets extended partway through the molecular envelope, terminating within the semiconducting zone where the Ohmic dissipation occurs, the shear at the base of the jets must coexist with lateral density contrasts on isobars via the thermal wind relationship (equation (46)). This is problematic, because, as previously discussed, sufficiently large sources of density contrasts are lacking in the deep interior. *Liu et al.* (2008) therefore argued that the jets must be weak *throughout* the molecular envelope up to shallow levels where alternate buoyancy sources become available (e.g., latent heating and/or transition to a radiative zone).

In the outermost layers of a giant planet, density variations associated with latent heating and/or a transition to a radiative zone allow significant thermal wind-shear to develop. On Jupiter and Saturn, this so-called "baroclinic" layer begins with condensation of iron, silicates, and various metal oxides and hydrides at pressures of $\sim 10^4$ bar and continues with condensation of water, NH_4SH, and ammonia (at ~10, 2, and 0.5 bar, respectively), finally transitioning to the stably stratified stratosphere at ~0.2 bar. Early models showed that the observed jet speeds can plausibly result from thermal-wind shear within this layer assuming the winds in the interior are zero (*Ingersoll and Cuzzi*, 1969). For example, water condensation could cause a fractional density contrast on isobars of up to 10^{-2}, and if these density differences extend vertically over ~100 km (roughly the altitude difference between the water-condensation level and the observed cloud deck), then equation (47) implies $\Delta u \sim 30$ m s^{-1}, similar to the observed speed of Jupiter's midlatitude jets (Fig. 6).

To summarize, current understanding suggests that giant planets should exhibit winds within the convective molecular envelope that exhibit columnar structure parallel to the rotation axis, transitioning in the outermost layers (pressures less than thousands of bars) to a baroclinic zone where horizontal density contrasts associated with latent heating and/or a radiative zone allow the zonal winds to vary with altitude.

4.2.2. Constraints from observations. Few observations yet exist that constrain the deep wind structure in Jupiter, Saturn, Uranus, and Neptune. On Neptune, gravity data from the 1989 Voyager 2 flyby imply that the fast jets observed in the cloud layers are confined to the outermost few percent or less of the planet's mass, corresponding to pressures less than $\sim 10^5$ bar (*Hubbard et al.*, 1991). By comparison, only weak constraints currently exist for Jupiter and Saturn. The Galileo probe, which entered Jupiter's atmosphere at a latitude of ~7°N in 1995, showed that the equatorial jet penetrates to at least 22 bar, roughly 150 km below the visible cloud deck (*Atkinson et al.*, 1997), and indirect inferences suggest that the jets penetrate to at least ~5–10 bar at other latitudes (e.g., *Dowling and Ingersoll*, 1989; *Sánchez-Lavega et al.*, 2008). In 2016, however, NASA's Juno mission will measure Jupiter's gravity field, finally determining whether Jupiter's jets penetrate deeply or are confined to pressures as shallow as thousands of bars or less. These results will have important implications for understanding the deep-wind structure in giant exoplanets generally.

4.2.3. Jet-pumping mechanisms. Two scenarios exist for the mechanisms to pump the zonal jets on the giant planets (for a review see *Vasavada and Showman*, 2005). In the "shallow forcing" scenario, the jets are hypothesized to result from moist convection (e.g., thunderstorms), baroclinic instabilities, or other turbulence-generating processes in the baroclinic layer in the outermost region of the planet. Two-dimensional and shallow-water models of this process, where random turbulence is injected as an imposed forcing (or alternately added as a turbulent initial condition), show success in producing alternating zonal jets through the Rhines effect (section 3.6). For appropriately chosen forcing and damping parameters (or for appropriate initial velocities when the turbulence is added as an initial condition), such models typically exhibit jets of approximately the observed speed and spacing (*Williams*, 1978; *Cho and Polvani*, 1996a; *Huang and Robinson*, 1998; *Scott and Polvani*, 2007; *Showman*, 2007; *Sukoriansky et al.*, 2007). Most models of this type produce an equatorial jet that flows westward rather than eastward as on Jupiter and Saturn, although robust eastward equatorial flow has been obtained in a recent shallow-water study (*Scott and Polvani*, 2008). Three-dimensional models of this process can explicitly resolve the turbulent energy generation and therefore need not inject turbulence by hand. These models likewise exhibit multiple zonal jets via the Rhines effect, in some cases spontaneously producing an eastward equatorial jet as on Jupiter and Saturn (e.g., *Williams*, 1979, 2003; *Lian and Showman*, 2008, 2010; *Schneider and Liu*, 2009). Figure 7, for example, illustrates three-dimensional simulations from *Lian and Showman* (2010) where the circulation is driven by latent heating associated with condensation of water vapor; the condensation produces turbulent eddies that interact with the planetary rotation (i.e., nonzero β) to generate jets. Globes show the zonal wind for a Jupiter case (top row), Saturn case (middle), and a case representing Uranus/Neptune (bottom row). The Jupiter and Saturn cases develop ~20 jets, including an eastward equatorial jet,

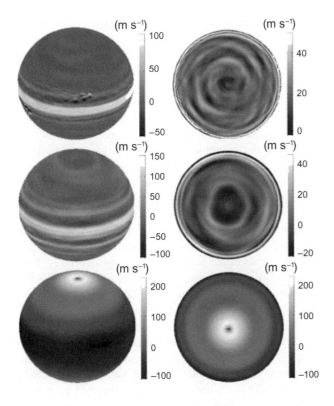

Fig. 7. Illustration of the effect of latent heating on the circulation of a rapidly rotating giant planet. Shows zonal (east-west) wind from three-dimensional atmospheric simulations of Jupiter (top row), Saturn (middle row), and a case representing Uranus/Neptune (bottom row) where the circulation is driven by condensation of water vapor, with an assumed deep abundance of 3, 5, and 30× solar for the Jupiter, Saturn, and Uranus/Neptune cases, respectively. Left column shows an oblique view and right column shows a view looking down over the north pole. Note the development of numerous jets (including superrotating equatorial jets on Jupiter and Saturn), which bear qualitative resemblance to the observed jets in Fig. 6. From *Lian and Showman* (2010).

whereas the Uranus/Neptune case develops a 3-jet structure with a broad westward equatorial flow — qualitatively similar to the observed jet patterns on these planets (Fig. 6). Note that shallow forcing need not imply shallow jets; because the atmosphere's response to a perturbation is nonlocal, deep jets (penetrating many scale heights below the clouds) can result from jet pumping confined to the cloud layer if the frictional damping in the interior is sufficiently weak (*Showman et al.*, 2006; *Lian and Showman*, 2008).

In the "deep forcing" scenario, convection throughout the molecular envelope is hypothesized to drive the jets. To date, most studies of this process involve three-dimensional numerical simulations of convection in a self-gravitating spherical shell excluding the effects of magnetohydrodynamics (e.g., *Aurnou and Olson*, 2001; *Christensen*, 2001, 2002; *Heimpel et al.*, 2005). The boundaries are generally taken as free-slip impermeable spherical surfaces with an outer boundary at the planetary surface and an inner boundary at a radius of 0.5 to 0.9 planetary radii. Most of these studies assume the mean density, thermal expansivity, and other fluid properties are constant with radius. These studies generally produce jets penetrating through the shell as Taylor columns. When the shell is thick (inner radius ≤0.7 of the outer radius), only approximately three to five jets form (*Aurnou and Olson*, 2001; *Christensen*, 2001, 2002; *Aurnou and Heimpel*, 2004), inconsistent with Jupiter and Saturn. Only when the shell has a thickness ~10% or less of the planetary radius do such simulations produce a Jupiter- or Saturn-like profile with ~20 jets (*Heimpel et al.*, 2005). However, on giant planets, the density and thermal expansivity each vary by several orders of magnitude from the photosphere to the deep interior, and this might have a major effect on the dynamics. Only recently has this effect been included in detailed numerical models of the convection (*Glatzmaier et al.*, 2009; *Kaspi et al.*, 2009), and these studies show that the compressibility can exert a significant influence on the jet structure. A difficulty in interpreting all the convective simulations described here is that, to integrate in reasonable times, they must adopt viscosities and heat fluxes orders of magnitude too large.

The fast interior jets in these convection simulations imply the existence of strong horizontal pressure contrasts, which in these models are supported by the impermeable upper and lower boundaries. On a giant planet, however, there is no solid surface to support such pressure variations. If these fast winds transitioned to weak winds within the metallic region, significant lateral density variations would be required via the thermal-wind relationship. As discussed previously, however, there exists no obvious source of such density variations (on isobars) within the deep interior. Resolving this issue will require three-dimensional convection simulations that incorporate magnetohydrodynamics and simulate the entire molecular + metallic interior. Several groups are currently making such efforts, so the next decade should see significant advances in this area.

4.3. Hot Jupiters and Neptunes

Because of their likelihood of transiting their stars, hot Jupiters (giant planets within 0.1 AU of their primary star) remain the best characterized exoplanets and thus far have been the focus of most work on exoplanet atmospheric circulation. Dayside photometry and/or infrared spectra now exist for a variety of hot Jupiters, constraining the dayside temperature structure. For some planets, such as HD 189733b, this suggests a temperature profile that decreases with altitude from ~0.01 to 1 bar (*Charbonneau et al.*, 2008; *Barman*, 2008; *Swain et al.*, 2009), whereas other hot Jupiters, such as HD 209458b, TrES-2, TrES-4, and XO-1b (*Knutson et al.*, 2008, 2009a; *Machalek et al.*, 2008), appear to exhibit a thermal inversion layer (i.e., a hot stratosphere) where temperatures rise above 2000 K.

Infrared lightcurves at 8 and/or 24 μm now exist for HD 189733b, Ups And b, HD 209458b, and several other hot Jupiters (*Knutson et al.*, 2007, 2009b; *Harrington et al.*, 2006; *Cowan et al.*, 2007), placing constraints on the day-night temperature distribution of these planets. Lightcurves for HD 189733b and HD 209458b exhibit nightside brightness temperatures only modestly (~20–30%) cooler than the dayside brightness temperatures, suggesting efficient redistribution of the thermal energy from dayside to nightside. On the other hand, Ups And b and HD 179949b exhibit larger day-night phase variations that suggests large (perhaps >500 K) day-night temperature differences. These and other observations provide sufficient constraints on the dayside temperature structure and day-night temperature distributions to make comparison with detailed atmospheric circulation models a useful exercise. Here we survey the basic dynamical regime and recent efforts to model these objects (see also *Showman et al.*, 2008b; *Cho*, 2008).

The dynamical regime on hot Jupiters differs from that on Jupiter and Saturn in several important ways. First, because of the short spindown times due to their proximity to their stars, hot Jupiters are expected to rotate nearly synchronously (or pseudo-synchronously in the case of hot Jupiters on highly eccentric orbits) with their orbital periods, implying rotation periods of 1–5 Earth days for most hot Jupiters discovered to date. This is 2–12 times slower than Jupiter's 10-h rotation period, implying that, compared to Jupiter, hot Jupiters experience significantly weaker Coriolis forces for a given wind speed. Nevertheless, Coriolis forces can still play a key role: For global-scale flows (length scales L ~ 10^8 m) and wind speeds of 2 km s^{-1} (see below), the Rossby number for a hot Jupiter with a three-day period is ~1, implying a three-way force balance between Coriolis, pressure-gradient, and inertial (i.e., advective) terms in the horizontal equation of motion. Slower winds would imply smaller Rossby numbers, implying greater rotational dominance.

Second, hot Jupiters receive enormous energy fluxes (~10^4–10^6 W m^{-2} on a global average) from their parent stars, in contrast to the ~10 W m^{-2} received by Jupiter and 1 W m^{-2} for Neptune. The resulting high temperatures lead to short radiative time constants of days or less at pressures <1 bar (*Showman and Guillot*, 2002; *Iro et al.*, 2005; *Showman et al.*, 2008a) (see also equation (19)), in contrast to Jupiter where radiative time constants are in years. Even in the presence of fast winds, these short radiative time constants allow the possibility of large fractional day-night temperature differences, particularly on the most heavily irradiated planets — in contrast to Jupiter where temperatures vary horizontally by only a few percent.

Third, over several billion years of evolution, the intense irradiation received by hot Jupiters leads to the development of a nearly isothermal radiative zone extending to pressures of ~100–1000 bar (*Guillot et al.*, 1996; *Guillot*, 2005; chapter by Fortney et al.). The observable weather in hot Jupiters thus occurs not within the convection zone but within the stably stratified radiative zone. Moreover, as described in section 4.1, this separation between the photosphere and the radiative convective boundary may allow large horizontal temperature differences to develop and support significant thermal wind shear. For example, assuming that a lateral temperature contrast of ΔT ~ 300 K extending over several scale heights occurs over a lateral distance of 10^8 m, the resulting thermal-wind shear is Δu ~ 2 km s^{-1}. Thus, balance arguments suggest that the existence of significant horizontal temperature contrasts would naturally require fast wind speeds.

The modest rotation rates, large stable stratifications, and possible fast wind speeds suggest that the Rossby deformation radius and Rhines scale are large on hot Jupiters. On Jupiter, the deformation radius is ~2000 km and the Rhines scale is ~10^4 km (much smaller than the planetary radius of 71,400 km), which helps explain why Jupiter and Saturn have an abundant population of small vortices and numerous zonal jets (Figs. 5 and 6) (*Williams*, 1978; *Cho and Polvani*, 1996a; *Vasavada and Showman*, 2005). In contrast, a hot Jupiter with a rotation period of 3 Earth days and wind speed of 2 km s^{-1} has a deformation radius and Rhines scale comparable to the planetary radius. Thus, dynamical structures such as vortices and jets should be more global in scale than occurs on Jupiter and Saturn (*Menou et al.*, 2003; *Showman and Guillot*, 2002; *Cho et al.*, 2003). This scaling argument is supported by detailed nonlinear numerical simulations of the circulation, which generally show the development of only approximately one to three broad jets (*Showman and Guillot*, 2002; *Cho et al.*, 2003, 2008; *Dobbs-Dixon and Lin*, 2008; *Langton and Laughlin*, 2007, 2008a; *Showman et al.*, 2008a, 2009).

A variety of two-dimensional and three-dimensional models have been used to investigate the atmospheric circulation of hot Jupiters. *Cho et al.* (2003, 2008) performed global numerical simulations of hot Jupiters on circular orbits using the one-layer equivalent barotropic equations (a one-layer model that is mathematically similar to the shallow-water equations described in section 2.3). They initialized the simulations with small-scale balanced turbulence and forced them with a large-scale deflection of the surface to provide a crude representation of the pressure effects of the day-night heating gradient. However, no explicit heating/cooling was included. Together, these effects led to the production of several broad, meandering jets and drifting polar vortices (Fig. 8). The mean wind speed in the final state is to a large degree determined by the mean speed of the initial turbulence, which ranged from 100 to 800 m s^{-1} in their models. At the large (~planetary scale) deformation radius relevant to hot Jupiters, the equatorial jet in the final state can flow either eastward or westward depending on the initial condition and other details. The simulations exhibit significant time variability that, if present on hot Jupiters, would lead to detectable orbit-to-orbit variability in lightcurves and secondary eclipse depths (*Cho et al.*, 2003; *Rauscher et al.*, 2007, 2008).

Langton and Laughlin (2007) performed global, two-dimensional simulations of hot Jupiters on circular orbits using the shallow-water equations with a mass source on the dayside and mass sink on the nightside to parameterize the effects of dayside heating and nightside cooling. When the obliquity is assumed zero and the mass sources/sinks are sufficiently large, their forced flows quickly reach a steady state with wind speeds of ~1 km s⁻¹ and lateral variations on the order of unity in the thickness of the shallow-water layer. On the other hand, *Langton and Laughlin* (2008a) numerically solved the two-dimensional fully compressible equations for the horizontal velocity and temperature of hot Jupiters on eccentric orbits and obtained very turbulent, time-varying flows. Apparently, in this case, the large-scale heating patterns produced hemisphere-scale vortices that were dynamically unstable, leading to the breakdown of these eddies into small-scale turbulence.

Several authors have also performed three-dimensional numerical simulations of hot Jupiters. *Showman and Guillot* (2002), *Cooper and Showman* (2005, 2006), *Showman et al.* (2008a), and *Menou and Rauscher* (2009) performed global simulations with the three-dimensional primitive equations where the dayside heating and nightside cooling was parameterized with a Newtonian heating/cooling scheme, which relaxes the temperatures toward a prescribed radiative-equilibrium temperature profile (hot on the dayside, cold on the nightside) over a prescribed radiative timescale. *Dobbs-Dixon and Lin* (2008) performed simulations with the fully compressible equations in a limited-area domain, consisting of the equatorial and midlatitudes but with the poles cut off. Like the studies listed above, they also adopted a simplified method for forcing their flow, in this case using a radiative diffusion scheme. *Showman et al.* (2009) coupled their global three-dimensional dynamical solver to a state-of-the-art, nongray, cloud-free radiative transfer scheme with opacities calculated assuming local chemical equilibrium (Figs. 9 and 10). The above models all generally obtain wind structures with approximately one to three broad jets with speeds of ~1–4 km s⁻¹. Despite the diversity in modeling approaches, the studies described above agree in several key areas.

1. First, most of the above studies generally produce peak wind speeds similar to within a factor of 2–3, in the range of one to several kilometers per second. (In intercomparing studies, one must be careful to distinguish mean vs. peak speeds and, in the case of three-dimensional models, the pressure level at which those speeds are quoted; such quantities can differ by a factor of several in a single model.) This similarity is not a coincidence but results from the force balances that occur for a global-scale flow in the presence of large fractional temperature differences. Consider the longitudinal force balance at the equator, for example, and suppose the day-night heating gradient produces a day-night temperature difference ΔT_{horiz} that extends vertically over a range of log-pressures $\Delta \ln p$. This temperature difference causes a day-night horizontal pressure-gradient acceleration that, to order of magnitude, can be written $R\Delta T_{horiz}\Delta \ln p/a$, where a is the planetary radius. At high latitudes, this could be balanced by the Coriolis force arising from a north-south flow, but the horizontal Coriolis force is zero at the equator. At the equator, such a force instead

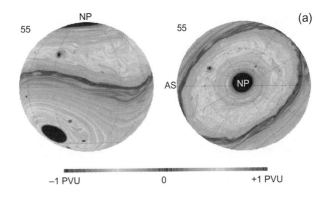

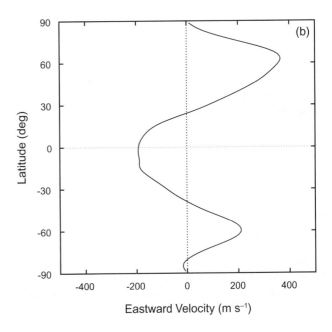

Fig. 8. (a) One-layer, global equivalent barotropic simulation of the hot Jupiter HD 209458b from *Cho et al.* (2003). Globes show equatorial (left) and polar (right) views of the potential vorticity (see section 2); note the development of large polar vortices and small-scale turbulent structure, which result from a combination of the turbulent initial condition and large-scale "topographic" forcing intended to qualitatively represent the day-night heating contrast. (b) Plot shows zonal-mean zonal wind vs. latitude, illustrating the three-jet structure that develops.

tends to cause acceleration of the flow in the east-west direction. Balancing the pressure-gradient acceleration by $\mathbf{v} \cdot \nabla \mathbf{v}$, which to order of magnitude is U^2/a for a global-scale flow, we have

$$U \sim \sqrt{R\Delta T_{horiz}\Delta \ln p} \qquad (48)$$

This should be interpreted as the characteristic variation in zonal wind speed along the equator. For R = 3700 J kg⁻¹K⁻¹, ΔT_{horiz} ~ 400 K, and $\Delta \ln p$ ~ 3 (appropriate to a temperature

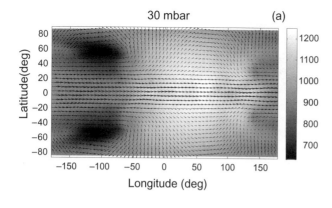
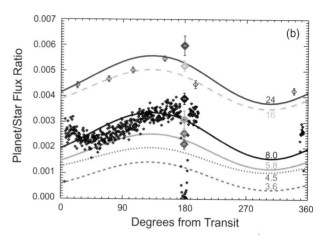

Fig. 9. Results from a three-dimensional simulation of the hot Jupiter HD 189733b from *Showman et al.* (2009). The dynamics are coupled to a realistic representation of cloud-free, nongray radiative transfer assuming solar abundances. (a) Temperature (grayscale, in K) and winds (arrows) at the 30-mbar pressure, which is the approximate level of the mid-IR photosphere. A strong eastward equatorial jet develops that displaces the hottest regions eastward from the substellar point (which lies at 0° longitude, 0° latitude). (b) Lightcurves in Spitzer bandpasses calculated from the simulation (curves; labels show wavelength in micrometers) in comparison to observations (points) from *Knutson et al.* (2007, 2009b), *Charbonneau et al.* (2008), and *Deming et al.* (2006).

difference extending vertically over three scale heights), this yields U ~ 2 km s^{-1}.

Likewise, consider the latitudinal force balance in the midlatitudes. To order of magnitude, the latitudinal pressure-gradient acceleration is again given by $R\Delta T_{horiz}\Delta \ln p/a$, where here ΔT_{horiz} is the latitudinal temperature contrast (e.g., from equator to pole) that extends vertically over $\Delta \ln p$. If Ro $\gg$ 1, this is balanced by the advective acceleration U^2/a, whereas if Ro $\ll$ 1, it would instead be balanced by the Coriolis acceleration fU, where f is the Coriolis parameter (section 3.3). The former case recovers equation (48), whereas the latter case yields

$$U \sim \frac{R\Delta T_{horiz}\Delta \ln p}{fa} \quad (49)$$

Here, U is properly interpreted as the characteristic difference in horizontal wind speed vertically across $\Delta \ln p$. Inserting R = 3700 J kg^{-1} K^{-1}, ΔT_{horiz} ~ 400 K, $\Delta \ln p$ ~ 3, f ~ 2 × 10^{-5} (appropriate in midlatitudes to a hot Jupiter with a three-day period), and a ≈ 10^8 m, we again obtain U ~ 2 km s^{-1}. [The similarity of the numerical estimates from equation (48) and (49) results from the fact that we assumed the same horizontal temperature differences in longitude and latitude and that Ro ~ 1 for the parameter regime explored here.] While not minimizing the real differences in the numerical results obtained in the various studies, the above estimates show that the existence of wind speeds of a few kilometers per second in the various numerical studies is a basic outcome of force balance in the presence of lateral temperature contrasts of hundreds of Kelvins.

2. Second, the various numerical studies all produce a small number of broad jets (approximately one to four). This was first pointed out by *Showman and Guillot* (2002), *Menou et al.* (2003), and *Cho et al.* (2003) and, as described previously, results from the fact that the Rossby deformation radius and Rhines scale are comparable to a planetary radius for conditions relevant to typical hot Jupiters. At least two specific mechanisms seem to be relevant. When the deformation radius is close to the planetary radius, the day-night forcing tends to inject energy at horizontal scales comparable to the planetary radius; the β effect then anisotropizes this energy into zonal jets whose widths are on the order of a planetary radius. Alternately, if the energy injection occurs primarily at small horizontal scales, an inverse cascade can reorganize the energy into planetary-scale jets if the planetary rotation is sufficiently slow and friction sufficiently weak (allowing fast winds and a large Rhines scale). Follow-up studies are consistent with these general expectations (*Cooper and Showman*, 2005; *Cho et al.*, 2008; *Langton and Laughlin*, 2007, 2008a; *Dobbs-Dixon and Lin*, 2008; *Showman et al.*, 2008a, 2009), but additional work to clarify the relative roles of L_D and L_β in setting the jet widths would be beneficial.

Despite these similarities, there remain some key differences in the numerical results obtained by the various groups so far. These include the following:

1. As yet, little agreement exists on whether hot Jupiters should exhibit significant temporal variability on the global scale. The two-dimensional simulations of *Cho et al.* (2003, 2008) and *Langton and Laughlin* (2008a) produce highly turbulent, time-variable flows, whereas the shallow-water study of *Langton and Laughlin* (2007) and most of the

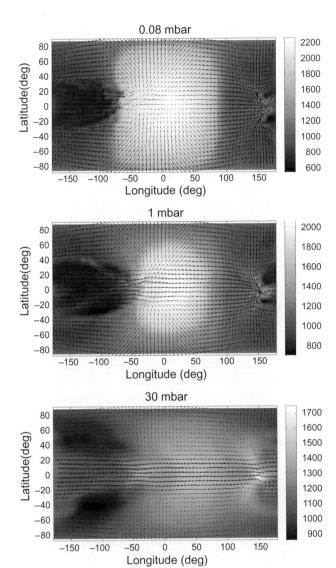

Fig. 10. Three-dimensional simulation of HD 209458b, including realistic cloud-free, nongray radiative transfer, illustrating development of a dayside stratosphere when opacity from visible-absorbing species (in this case TiO and VO) are included. Such stratospheres are relevant to explaining Spitzer data of HD 209458b, TrES-2, TrES-4, XO-1b, and other hot Jupiters. Here, the stratosphere (shown in white) begins at pressures of ~10 mbar and widens with altitude until it covers most of the dayside at pressures of ~0.1 mbar. Panels show temperature (grayscale, in K) and winds (arrows) at the 0.08-, 1-, and 30-mbar levels from top to bottom, respectively. Substellar longitude and latitude are (0°, 0°). From *Showman et al.* (2009).

three-dimensional studies (*Showman and Guillot*, 2002; *Cooper and Showman*, 2005; *Dobbs-Dixon and Lin*, 2008; *Showman et al.*, 2008a, 2009) exhibit relatively steady flow patterns with only modest time variability for zero-obliquity, zero-eccentricity hot Jupiters. The *Cho et al.* (2003, 2008) simulations essentially guarantee a time-variable outcome

because of the turbulent initial condition, although breaking Rossby waves and dynamical instabilities during the simulations also play a large role; the *Langton and Laughlin* (2008a) simulations were initialized from rest and naturally develop turbulence via dynamical instability throughout the course of the simulation.

Time variability can result from several mechanisms, including dynamical (e.g., barotropic or baroclinic) instabilities, large-scale oscillations, and modification of the jet pattern by atmospheric waves. If the jet pattern is dynamically stable (or, more rigorously, if relevant instability growth times are significantly longer than the characteristic atmospheric heating/cooling times), then this mechanism for producing variability would be inhibited; on the other hand, if the jet pattern is unstable (with instability growth rates comparable to or less than the heating/cooling times), the jets will naturally break up and produce a wide range of turbulent eddies. In this regard, it is crucial to carefully distinguish between two-dimensional and three-dimensional models, because they exhibit very different stability criteria (e.g., *Dowling*, 1995). For example, in a two-dimensional, nondivergent model, theoretical and numerical work shows that jets are guaranteed stable only if

$$\frac{\partial^2 u}{\partial y^2} < \beta \qquad (50)$$

a relationship called the barotropic stability criterion. Numerical simulations show that two-dimensional, nondivergent fluids initialized with jets violating equation (50) generally develop instabilities that rob energy from the jets until the jets no longer violate the equation. However, jets in a three-dimensional fluid can in some cases strongly violate equation (50) while remaining stable; Jupiter and Saturn, for example, have several jets where $\partial^2 u/\partial y^2$ exceed β by a factor of 2–3 (*Ingersoll et al.*, 1981; *Sanchez-Lavega et al.*, 2000), and yet the jet pattern has remained almost unchanged over multidecade timescales. This may simply imply that the barotropic stability criterion is irrelevant for a three-dimensional fluid (*Dowling*, 1995; *Vasavada and Showman*, 2005). Thus, it is *a priori* unsurprising that two-dimensional and three-dimensional models would make different predictions for time variability.

However, a recent study by *Menou and Rauscher* (2009) shows that the issue is not only one of two vs. three dimensions. Making straightforward extensions to an Earth model, they performed forced three-dimensional shallow simulations initialized from rest that developed highly turbulent, time variable flows on the global scale, showing that time variability from the emergence of horizontal shear instability may occur in three dimensions under some conditions relevant to hot Jupiters. Like several previous studies, they forced their flows with a Newtonian heating/cooling scheme that generates a day-night temperature difference. However, their forcing setup differs significantly from those of previous studies. They place an impermeable surface at 1 bar and apply their maximum day-night forcing immediately above the surface; the day-night radiative equilibrium temperature difference decreases with altitude and reaches zero near the top of their model. This is the reverse of the setup used in *Cooper and Showman* (2005, 2006) and *Showman et al.* (2008a, 2009) where the forcing amplitude peaks near the top of the domain and decreases with depth, and where the surface is placed significantly deeper than the region of atmospheric heating/cooling to minimize its interaction with the circulation in the observable atmosphere. The differences in the forcing schemes and position of the surface presumably determine whether or not a given three-dimensional simulation develops global-scale instability and time variability. Further work will be necessary to identify the specific forcing conditions that lead to time-variable or quasisteady conditions at the global scale.

It is also worth remembering that hot Jupiters themselves exhibit huge diversity; among the known transiting planets, for example, incident stellar flux varies by a factor of ~150, gravities range over a factor of ~70, and expected rotation rates vary over nearly a factor of 10 (e.g., Table 1). Given this diversity, it is reasonable to expect that some hot Jupiters will exhibit time-variable conditions while others will exhibit relatively steady flow patterns.

2. Another area of difference between models regards the direction of the equatorial jet. The shallow-water and equivalent barotropic models can produce either eastward or westward equatorial jets, depending on the initial conditions and other modeling details (*Cho et al.*, 2003, 2008; *Langton and Laughlin*, 2007, 2008a). In contrast, all the three-dimensional models published to date for synchronously rotating hot Jupiters with zero obliquity and eccentricity have produced robust eastward equatorial jets (*Showman and Guillot*, 2002; *Cooper and Showman*, 2005, 2006; *Showman et al.*, 2008a, 2009; *Dobbs-Dixon and Lin*, 2008; *Menou and Rauscher*, 2009). The mechanisms responsible for generating the eastward jets in these three-dimensional models remain to be diagnosed in detail but presumably involve the global-scale eddies generated by the day-night heating contrast.

To date, most work on hot-Jupiter atmospheric circulation has emphasized planets on circular orbits. However, several transiting hot Jupiters and Neptunes have eccentric orbits, including GJ 436b, HAT-P-2b, HD 17156b, and HD 80606b, whose orbital eccentricities are 0.15, 0.5, 0.67, and 0.93, respectively. When the eccentricity is large, the incident stellar flux varies significantly throughout the orbit — HAT-P-2b, for example, receives ~9 times more flux at periapse than at apoapse, while for HD 80606b, the maximum incident flux exceeds the minimum incident flux by a factor of over 800! These variations dwarf those experienced in our solar system and constitute an uncharted regime of dynamical forcing for a planetary atmosphere. An 8-μm lightcurve of HD 80606b has recently been obtained with the Spitzer Space Telescope during a 30-h interval surrounding periapse passage (*Laughlin et al.*, 2009), which shows the increase in planetary flux that presumably accompanies the flash heating as the planet passes by its star. Two- and three-dimensional numerical simulations suggest that these planets may exhibit dynamic, time-variable flows and show that the planet's emitted infrared flux can peak hours to days after periapse passage (*Langton and Laughlin*, 2008a,b; *Lewis et al.*, 2009).

4.4. Chemistry as a Probe of the Meteorology

Chemistry can provide important constraints on the meteorology of giant planets. On Jupiter, for example, gaseous CO, PH_3, GeH_4, and AsH_3 have been detected in the upper troposphere (p < 5 bar) with mole fractions of ~0.8, 0.2, 0.6, and 0.5–1 ppb, respectively. None of these compounds are thermochemically stable in the upper troposphere — the chemical equilibrium mole fractions are $<10^{-14}$ for AsH_3 and $<10^{-20}$ for the other three species (*Fegley and Lodders*, 1994). However, the equilibrium abundances of all four species rise with depth, exceeding 1 ppb at pressures greater than 500, 400, 160, and 20 bar for CO, PH_3, GeH_4, and AsH_3, respectively. Thus, the high abundances of these species in the upper troposphere appear to result from rapid convective mixing from the deep atmosphere. [Jupiter also has a high stratospheric CO abundance that is inferred to result from exogenous sources (*Bézard et al.*, 2002).] Kinetic reaction times are rapid at depth but plummet with decreasing temperature and pressure, allowing these species to persist in disequilibrium (i.e., "quench") in the low-pressure and temperature conditions of the upper troposphere. Knowledge of the chemical kinetic rate constants allows quantitative estimates to be made of the vertical mixing rate needed to explain the observed abundances. For all four of the species listed above, the required vertical mixing rates yield order-of-magnitude matches with the mixing rate expected from thermal convection at Jupiter's known heat flux (e.g., *Prinn and Barshay*, 1977; *Fegley and Lodders*, 1994; *Bézard et al.*, 2002). Thus, the existence of these disequilibrium species provides evidence that the deep atmosphere is indeed convective.

Given the difficulty of characterizing the meteorology of exoplanets with spatially resolved observations, these types of chemical constraints will likely play an even more important role in our understanding of exoplanets than is the case for Jupiter. Several brown dwarfs, including Gl 229b, Gl 570d, and 2MASS J0415-0935, show evidence for disequilibrium abundances of CO and/or NH_3 (e.g., *Noll et al.*, 1997; *Saumon et al.*, 2000, 2006, 2007). These observations can also plausibly be explained by vertical mixing. For these objects, NH_3 is quenched in the convection zone, but CO is quenched in the overlying radiative zone where vertical mixing must result from some combination of small-scale turbulence (due, for example, to the breaking of vertically propagating gravity waves) and vertical advection due to the large-scale circulation (section 3.7). Current models suggest that such mixing corresponds to an eddy diffusivity in the radiative zone of ~1–10 m^2 s^{-1} for Gl 229b and 10–100 m^2 s^{-1} for 2MASS J0415-0935 (*Saumon et al.*, 2007). These inferences on vertical mixing rate provide constraints on the atmospheric circulation that would be difficult to obtain in any other way.

Several hot Jupiters likewise reside in temperature regimes where CH_4 is the stable form in the observable atmosphere and CO is the stable form at depth, or where CH_4 is the stable form on the nightside and CO is the stable form on the dayside. The abundance and spatial distribution of CO and CH_4 on these objects could thus provide important clues about the wind speeds and other aspects of the meteorology. Methane has been detected via transmission spectroscopy on HD 189733b (*Swain et al.*, 2008) and CO has been indirectly inferred from the shape of the dayside emission spectrum (*Barman*, 2008), although further observations and radiative-transfer models are needed to determine their abundances and spatial variability (if any). *Cooper and Showman* (2006) coupled a simple model for CO/CH_4 interconversion kinetics to their three-dimensional dynamical model of hot Jupiters to demonstrate that CO and CH_4 should be quenched in the observable atmosphere, where thermochemical CO ↔ CH_4 interconversion times are orders of magnitude longer than plausible dynamical times. If so, one would expect CO and CH_4 to have similar abundances on the dayside and nightside of hot Jupiters, in contrast to the chemical-equilibrium situation where CH_4 would be much more prevalent on the nightside (note, however, that their models neglect photochemistry, which could further modify the abundances of both species). For temperatures relevant to objects like HD 209458b and HD 189733b, *Cooper and Showman* (2006) found that the quenching occurs at pressures of ~1–10 bar, and that the quenched abundance depends primarily on the temperatures and vertical velocities in this pressure range.

Thus, future observational determination of CO and CH_4 abundances and comparison with detailed three-dimensional models may allow constraints on the temperatures and vertical velocities in the deep (1–10 bar) atmospheres of hot Jupiters to be obtained. Such constraints probe significantly deeper than the expected visible and infrared photospheres of hot Jupiters and hence would nicely complement the characterization of hot-Jupiter thermal structure via lightcurves and secondary-eclipse spectra. Note also that an analogous story involving N_2 and NH_3 interconversion may occur on even cooler objects (T_{eff} ~700 K "warm" Jupiters).

5. CIRCULATION REGIMES: TERRESTRIAL PLANETS

With current groundbased searches and spacecraft such as NASA's Kepler mission, terrestrial planets are likely to be discovered around main-sequence stars within the next five years, and basic characterization of their atmospheric composition and temperature structure should follow over the subsequent decade. At base, we wish to understand not only whether such planets exist but whether they have atmospheres, what their atmospheric composition, structure and climate may be, and whether they can support life. Addressing these issues will require a consideration of relevant physical and chemical climate feedback as well as the plausible circulation regimes in these planets' atmospheres.

The study of terrestrial exoplanet atmospheres is just beginning, and only a handful of papers have specifically investigated the possible circulation regimes on these objects. However, a vast literature has developed to understand the climate and circulation of Venus, Titan, Mars, and especially Earth, and many of the insights developed for understanding these planets may be generalizable to exoplanets. Our goal here is to provide conceptual and theoretical guidance on the types of climate and circulation processes that exist in terrestrial planet atmospheres and discuss how those processes vary under diverse planetary conditions. In section 5.1 we survey basic issues in climate. The following sections, section 5.2 through section 5.5, address basic circulation regimes on terrestrial planets, emphasizing those aspects (e.g., processes that determine the horizontal temperature contrasts) relevant to future exoplanet observations. Section 5.6 discusses regimes of exotic forcing associated with synchronous rotation, large obliquities, and large orbital eccentricities.

5.1. Climate

Climate can be defined as the mean condition of a planet's atmosphere/ocean system — the temperatures, pressures, winds, humidities, and cloud properties — averaged over time intervals longer than the timescale of typical weather events. The climate on the terrestrial planets results from a wealth of interacting physical, chemical, geological, and (when relevant) biological effects, and even on Earth, understanding the past and present climate has required a multi-decade interdisciplinary research effort. In this section, we provide only a brief sampling of the subject, touching only on the most basic physical processes that help to determine a planet's global-mean conditions.

The mean temperatures at which a planet radiates to space depend primarily on the incident stellar flux and the planetary albedo. Equating emitted infrared energy (assumed blackbody) with absorbed stellar flux yields a planet's global-mean effective temperature at radiative equilibrium

$$T_{eff} = \left[\frac{F_* (1 - A_B)}{4\sigma} \right]^{1/4} \quad (51)$$

where F_* is the incident stellar flux, A_B is the planet's global-mean Bond albedo, and σ is the Stefan-Boltzmann constant. (The Bond albedo is the fraction of light incident upon a planet that is scattered back to space, integrated over all wavelengths and directions.) The factor of 4 results from the fact that the planet intercepts a stellar beam of area πa^2 but radiates infrared from its full surface area of $4\pi a^2$.

Given an incident stellar flux, a variety of atmospheric processes act to determine the planet's surface temperature. First, the circulation and various atmospheric feedback can play a key role in determining the planetary albedo. Bare rock is fairly dark, and for terrestrial planets the albedo is largely determined by the distribution of clouds and surface ice. The Moon, for example, has a Bond albedo of 0.11, yielding an effective temperature of 274 K. In contrast, Venus has a Bond albedo of 0.75 because of its global cloud cover, yielding an effective temperature of 232 K. With partial cloud and ice cover, Earth is an intermediate case, with a Bond albedo of 0.31 and an effective temperature of 255 K. These examples illustrate that the circulation, via its effect on clouds and surface ice, can have a major influence on mean conditions — Venus' effective temperature is less than that of the Earth and Moon despite receiving nearly double the solar flux!

Second, T_{eff} is not the surface temperature but the mean blackbody temperature at which the planet radiates to space. The surface temperatures can significantly exceed T_{eff} through the greenhouse effect. Planets orbiting Sun-like stars receive most of their energy in the visible, but they tend to radiate their energy to space in the infrared. Most gases tend to be relatively transparent in the visible, but H_2O, CO_2, CH_4, and other molecules absorb significantly in the infrared. On a planet like Earth or Venus, a substantial fraction of the sunlight therefore reaches the planetary surface, but infrared emission from the surface is mostly absorbed in the atmosphere and cannot radiate directly to space. In turn, the atmosphere radiates both up (to space) and down (to the surface). The surface therefore receives a double whammy of radiation from both the Sun and the atmosphere; to achieve energy balance, the surface temperature becomes elevated relative to the effective temperature. This is the greenhouse effect. Contrary to popular descriptions, the greenhouse effect should *not* be thought of as a situation where the heat is "trapped" or "cannot escape." Indeed, the planet as a whole resides in a near-balance where infrared emission to space (mostly from the atmosphere when the greenhouse effect is strong) almost equals absorbed sunlight.

To have a significant greenhouse effect, a planet must have a massive-enough atmosphere to experience pressure broadening of the spectral lines. Mars, for example, has ~15 times more CO_2 per area than Earth, yet its greenhouse effect is significantly weaker because its surface pressure is only 6 mbar.

The climate is influenced by numerous feedbacks that can affect the mean state and how the atmosphere responds to perturbations. Some of the more important feedbacks are as follows:

1. Thermal feedback: Increases or decreases in the temperature at the infrared photosphere lead to enhanced or reduced radiation to space, respectively. This is a negative feeback that allows planets to reach a stable equilibrium with absorbed starlight.

2. Ice-albedo feedback: Surface ice can form on planets exhibiting trace gases that can condense to solid form. Because of the brightness of ice and snow, an increase in snow/ice coverage decreases the absorbed starlight, promoting colder conditions and growth of even more ice. Conversely, melting of surface ice increases the absorbed starlight, promoting warmer conditions and continued melting. This is a positive feedback.

Simple models of the ice-albedo feedback show that, over a range of F_* values (depending on the strength of the greenhouse effect), an Earth-like planet can exhibit two stable equilibrium states for a given value of F_*: a warm, ice-poor state with a low albedo and a cold, ice-covered state with a high albedo (e.g., *North et al.*, 1981). Geologic evidence suggests that ~0.6–2.4 G.y. ago Earth experienced several multi-million-year-long glaciations with global or near-global ice cover (dubbed "snowball Earth" events) (*Hoffman and Schrag*, 2002), suggesting that Earth has flipped back and forth between these equilibria. The susceptibility of a planet to entering such a snowball state is highly sensitive to the strength of latitudinal heat transport (*Spiegel et al.*, 2008), thereby linking this feedback to the global circulation.

3. Condensable-greenhouse-gas feedback: When an atmospheric constituent is a greenhouse gas that also exists in condensed form on the surface, the atmosphere can experience a positive feedback that affects the temperature and the distribution of this constituent between the atmosphere and surface. An increase in surface temperature increases the constituent's saturation vapor pressure, hence the atmospheric abundance and therefore the greenhouse effect. A decrease in surface temperature decreases the saturation vapor pressure, reducing the atmospheric abundance and therefore the greenhouse effect. Thermal perturbations are therefore amplified. For Earth, this process occurs with water vapor (Earth's most important greenhouse gas) and is called the "water-vapor feedback" in the Earth climate literature (*Held and Soden*, 2000). The water-vapor feedback will play a major role in determining how Earth responds to anthropogenic increases in carbon dioxide over the next century.

In some circumstances, this positive feedback is so strong that it can trigger a runaway that shifts the atmosphere into a drastically different state. In the case of warming, this would constitute a runaway greenhouse that leads to the complete evaporation/sublimation of the condensable constituent from the surface. If early Venus had oceans, for example, they might have experienced runaway evaporation, leading to a monstrous early water vapor atmosphere (*Ingersoll*, 1969; *Kasting*, 1988) that would have had major effects on subsequent planetary evolution. In the case of cooling, such a runaway would remove most of the condensable constituent from the atmosphere, and if the relevant gas dominates the atmosphere, this could lead to atmospheric collapse. For example, if a Venus-like planet (with its ~500-K greenhouse effect) were moved sufficiently far from the Sun, CO_2 condensation would initiate, potentially collapsing the atmosphere to a Mars-like state with most of the CO_2 condensed on the ground, a cold, thin CO_2 atmosphere in vapor-pressure equilibrium with the surface ice, and minimal greenhouse effect. The CO_2 cloud formation that precedes such a collapse is often used to define the outer edge of the classical habitable zone for terrestrial planets whose greenhouse effect comes primarily from CO_2 (*Kasting et al.*, 1993).

Because CO_2 condensation naturally initiates in the coldest regions (the poles for a rapidly rotating planet; the nightside for a synchronous rotator), the conditions under which atmospheric collapse initiates depend on the atmospheric circulation. Weak equator-to-pole (or day-night) heat transport leads to colder polar (or nightside) temperatures for a given solar flux, promoting atmospheric collapse. These dynamical effects have yet to be fully included in climate models.

4. *Clouds:* Cloud coverage increases the albedo, lessening the absorption of starlight and promoting cooler conditions. On the other hand, because the tropospheric temperatures generally decrease with altitude, cloud tops are typically cooler than the ground and therefore radiate less infrared energy to space, promoting warmer conditions (an effect that depends sensitively on cloud altitude and latitude). These effects compete with each other. By determining the detailed properties of clouds (e.g., fractional coverage, latitudes, altitudes, and size-particle distributions), the atmospheric circulation therefore plays a major role in determining the mean surface temperature.

To the extent that cloud properties depend on atmospheric temperature, clouds can act as a feedback that affects the mean state and amplifies or reduces a thermal perturbation to the climate system. However, because the sensitivity of cloud properties to the global circulation and mean climate is extremely difficult to predict, the net sign of this feedback (positive or negative) remains unknown even for Earth. For exoplanets, clouds could plausibly act as a positive feedback in some cases and negative feedback in others.

5. *Long-term atmosphere/geology feedback:* On geological timescales, terrestrial planets can experience significant exchange of material between the interior and atmosphere. An example that may be particularly relevant for Earth-like (i.e., ocean and continent-bearing) exoplanets is the carbonate-silicate cycle, which can potentially buffer atmospheric CO_2 in a temperature-dependent way that tends to stabilize the atmospheric temperature against variations in solar flux (see, e.g., *Kasting and Catling*, 2003). Such feedback, while beyond the scope of this chapter, may be critical in determining the mean composition and temperature of terrestrial exoplanets.

In addition to the major types of feedback outlined above, there exist dozens of additional interacting physical, chemical, and biological feedback that can influence the mean climate and its sensitivity to perturbations. *Hartmann et al.* (2003) provide a thorough assessment for the Earth's current climate; although the details will be different for other planets, this gives a flavor for the complexity that can be expected.

5.2. Global Circulation Regimes

For atmospheres with radiative time constants greatly exceeding the planet's solar day, the atmosphere cannot respond rapidly to day-night variations in stellar heating and instead responds primarily to the daily-mean insolation, which is a function of latitude. This situation applies to most planets in the solar system, including the (lower) atmospheres of Venus, Earth, Titan, Jupiter, Saturn, Uranus, and Neptune. (Mars is a transitional case, with a radiative time constant ~2–3 times its 24.6-h day.) The resulting patterns of temperature and winds vary little with longitude in comparison to their variation in latitude and height. In such an atmosphere, the primary task of the atmospheric circulation is to transport thermal energy not from day to night but between the equator and the poles. [At low obliquity (<54°), yearly averaged starlight is absorbed primarily at low latitudes, so the circulation transports thermal energy poleward, but at high obliquity (>54°), yearly averaged starlight is absorbed primarily at the poles (*Ward*, 1974), and the yearly averaged energy transport by the circulation is equatorward. At high obliquity, strong seasonal cycles will also occur.] On the other hand, when the atmosphere's radiative time constant is much shorter than the solar day, the day-night heating gradient will be paramount and daynight temperature variations at the equator could potentially rival temperature variations between the equator and poles.

In the next several subsections we survey our understanding of global atmospheric circulations for these two regimes. Because the first situation applies to most solar system atmospheres, it has received the vast majority of work and remains better understood. This regime, which we discuss in sections 5.3–5.5, will apply to planets whose orbits are sufficiently far from their stars that the planets have not despun into a synchronously rotating state; it can also apply to atmosphere-bearing moons of hot Jupiters even at small distances from their stars. The latter regime, surveyed in section 5.6, prevails for planets exhibiting small enough orbital eccentricities and semimajor axes to become synchronously locked to their stars. This situation probably applies to most currently known transiting giant exoplanets and may also apply to terrestrial planets in the habitable zones of M dwarfs, which are the subject of current observational searches. Despite its current relevance, this novel forcing regime has come under investigation only in the past decade and remains incompletely understood due to a lack of solar system analogs.

5.3. Axisymmetric Flows: Hadley Cells

Perhaps the simplest possible idealization of a circulation that transports heat from equator to poles is an axisymmetric circulation — that is, a circulation that is independent of longitude — where hot air rises at the equator, moves poleward aloft, cools, sinks at high latitudes, and returns equatorward at depth (near the surface on a terrestrial planet). Such a circulation is termed a Hadley cell, and was first envisioned by *Hadley* (1735) to explain Earth's trade winds. Most planetary atmospheres in our solar system, including those of Venus, Earth, Mars, Titan, and possibly the giant planets, exhibit Hadley circulations.

Hadley circulations on real planets are of course not truly axisymmetric; on the terrestrial planets, longitudinal variations in topography and thermal properties (e.g., associated with continent-ocean contrasts) induce asymmetry in longitude. Nevertheless, the fundamental idea is that the longitudinal variations are not *crucial* for driving the circulation. This differs from the circulation in midlatitudes, whose longitudinally averaged properties are fundamentally controlled by the existence of nonaxisymmetric baroclinic eddies that are inherently three-dimensional (see section 3.7).

Planetary rotation generally prevents Hadley circulations from extending all the way to the poles. Because of planetary rotation, equatorial air contains considerable angular momentum about the planetary rotation axis; to conserve angular momentum, equatorial air would accelerate to unrealistically high speeds as it approached the pole, a phenomenon that is dynamically inhibited. To illustrate, the specific angular momentum about the rotation axis on a spherical planet is $M = (\Omega a \cos\phi + u)a\cos\phi$, where the first and second terms represent angular momentum due to planetary rotation and winds, respectively (recall that Ω, a, and ϕ are planetary rotation rate, planetary radius, and latitude). If $u = 0$ at the equator, then $M = \Omega a^2$, and an angular-momentum conserving circulation would then exhibit winds of

$$u = \Omega a \frac{\sin^2\phi}{\cos\phi} \quad (52)$$

Given Earth's radius and rotation rate, this equation implies zonal-wind speeds of 134 m s^{-1} at 30° latitude, 700 m s^{-1} at 60° latitude, and 2.7 km s^{-1} at 80° latitude. Such high-latitude wind speeds are unrealistically high and would furthermore be violently unstable to three-dimensional instabilities. On Earth, the actual Hadley circulations extend to ~30° latitude.

The Hadley circulation exerts strong control over the wind structure, latitudinal temperature contrast, and climate. Hadley circulations transport thermal energy by the most efficient means possible, namely straightforward advection of air from one latitude to another. As a result, the latitudinal temperature contrast across a Hadley circulation tends to be modest; the equator-to-pole temperature contrast on a planet will therefore depend strongly on the width of the Hadley cell. Moreover, on planets with condensable gases, Hadley cells exert control over the patterns of cloudiness and rainfall. On Earth, the rising branch of the Hadley circulation leads to cloud formation and abundant rainfall near the equator, helping to explain, for example, the prevalence of tropical rainforests in Southeast Asia/Indonesia, Brazil, and central Africa. (Regional circulations, such as monsoons, also contribute.) On the other hand, because condensation and rainout dehydrates the rising air, the descending branch of the Hadley cell is relatively dry, which explains the abundances of arid climates on Earth at 20–30° latitude, including the deserts of the African Sahara, South Africa, Australia, central Asia, and the southwestern United States. The Hadley cell can also influence the mean cloudiness, hence albedo and thereby the mean surface temperature. Venus's slow rotation rate leads to a global equator-to-pole Hadley cell, with a broad ascending branch in low latitudes. Coupled with the presence of trace condensable gases, this widespread ascent contributes to a near-global cloud layer that helps generate Venus' high Bond albedo of 0.75. Different Hadley cell patterns would presumably cause different cloudiness patterns, different albedos, and therefore different global-mean surface temperatures. On exoplanets, Hadley circulations will also likewise help control latitudinal heat fluxes, equator-to-pole temperature contrasts, and climate.

A variety of studies have been carried out using fully nonlinear, global three-dimensional numerical circulation models to determine the sensitivity of the Hadley cell to the planetary rotation rate and other parameters (e.g., *Hunt*, 1979; *Williams and Holloway*, 1982; *Williams*, 1988a,b; *Del Genio and Suozzo*, 1987; *Navarra and Boccaletti*, 2002; *Walker and Schneider*, 2005, 2006). These studies show that as the rotation rate is decreased the width of the Hadley cell increases, the equator-to-pole heat flux increases, and the equator-to-pole temperature contrast decreases. Figures 11 and 12 illustrate examples from *Navarra and Boccaletti* (2002) and *Del Genio and Suozzo* (1987). For Earth parameters, the circulation exhibits midlatitude eastward jet streams that peak in the upper troposphere (~200 mbar pressure), with weaker wind at the equator (Fig. 11). The Hadley cells extend from the equator to the equatorward flanks of the midlatitude jets. As the rotation rate decreases, the Hadley cells widen and the jets shift poleward. At first, the jet speeds increase with decreasing rotation rate, which results from the fact that as the Hadley cells extend poleward (i.e., closer to the rotation axis) the air can spin up faster (cf. equation (52)). Eventually, once the Hadley cells extend almost to the pole (at rotation periods exceeding ~5–10 days for Earth radius, gravity, and vertical thermal structure), further decreases in rotation rate reduce the midlatitude jet speed.

Perhaps more interestingly for exoplanet observations, these changes in the Hadley cell significantly influence the planetary temperature structure. This is illustrated in Fig. 12 from a series of simulations by *Del Genio and Suozzo* (1987). Because the Hadley cells transport heat extremely efficiently, the temperature remains fairly constant across the width of the Hadley cells. Poleward of the Hadley cells, however, the heat is transported in latitude by baroclinic instabilities (sections 3.7 and 5.4), which are less efficient, so a large latitudinal temperature gradient exists within this so-called "baroclinic zone." The equator-to-pole temperature contrast depends strongly on the width of the Hadley cell.

Despite the value of the three-dimensional circulation models described above, the complexity of these models tends to obscure the physical mechanisms governing the Hadley circulation's strength and latitudinal extent and cannot easily be extrapolated to different planetary parameters. A conceptual theory for the Hadley cell, due to *Held and Hou* (1980), provides considerable insight into Hadley cell dynamics and allows estimates of how, for example, the width of the Hadley cell should scale with planetary size

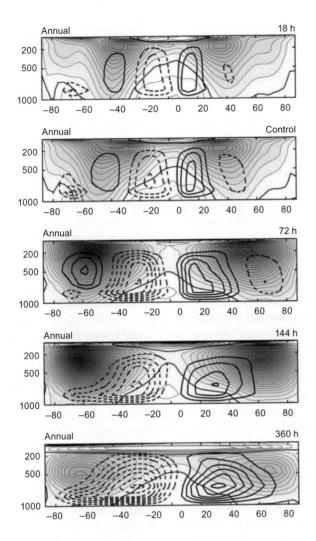

Fig. 11. Zonal-mean circulation vs. latitude in degrees (abscissa) and pressure in mbar (ordinate) in a series of Earth-based GCM experiments from *Navarra and Boccaletti* (2002) where the rotation period is varied from 18 hr (top) to 360 hr (bottom). Grayscale and thin gray contours depict zonal-mean zonal wind, $\bar{u}$, and thick black contours denote stream function of the circulation in the latitude-height plane (the so-called "meridional circulation"), with solid being clockwise and dashed being counterclockwise. The meridional circulation flows parallel to the stream function contours, with greater mass flux when contours are more closely spaced. The two cells closest to the equator correspond to the Hadley cell. As rotation period increases, the jets move poleward and the Hadley cell widens, becoming nearly global at the longest rotation periods.

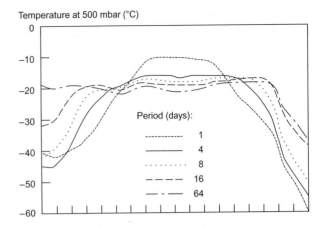

Fig. 12. Zonally averaged temperature in °C vs. latitude at 500 mbar (in the midtroposphere) for a sequence of Earth-like GCM runs that vary the planetary rotation period between 1 and 64 Earth days (labeled in the graph). Latitude runs from 90° on the left to –90° on the right, with the equator at the center of the horizontal axis. The flat region near the equator in each run results from the Hadley cell, which transports thermal energy extremely efficiently and leads to a nearly isothermal equatorial temperature structure. The region of steep temperature gradients at high latitudes is the baroclinic zone, where the temperature structure and latitudinal heat transport are controlled by eddies resulting from baroclinic instability. Note that the width of the Hadley cell increases, and the equator-to-pole temperature difference decreases, as the planetary rotation period is increased. From *Del Genio and Suozzo* (1987).

and rotation rate [see reviews in *James* (1994, pp. 80–92), *Vallis* (2006, pp. 457–466), and *Schneider* (2006)]. Stripped to its basics, the scheme envisions an axisymmetric two-layer model, where the lower layer represents the equatorward flow near the surface and the upper layer represents the poleward flow in the upper troposphere. For simplicity, *Held and Hou* (1980) adopted a basic-state density that is constant with altitude. Absorption of sunlight and loss of heat to space generate a latitudinal temperature contrast that drives the circulation; for concreteness, let us parameterize the radiation as a relaxation toward a radiative-equilibrium potential temperature profile that varies with latitude as $\theta_{rad} = \theta_0 - \Delta\theta_{rad} \sin^2\phi$, where θ_0 is the radiative-equilibrium potential temperature at the equator and $\Delta\theta_{rad}$ is the equator-to-pole difference in radiative-equilibrium potential temperature. If we make the small-angle approximation for simplicity (valid for a Hadley cell that is confined to low latitudes), we can express this as $\theta_{rad} = \theta_0 - \Delta\theta_{rad}\phi^2$.

In the lower layer, we assume that friction against the ground keeps the wind speeds low; in the upper layer, assumed to occur at an altitude H, the flow conserves angular momentum. The upper layer flow is then specified by equation (52), which is just $u = \Omega a \phi^2$ in the small-angle limit. We expect that the upper-layer wind will be in thermal-wind balance with the latitudinal temperature contrast

$$f\frac{\partial u}{\partial z} = f\frac{u}{H} = -\frac{g}{\theta_0}\frac{\partial \theta}{\partial y} \qquad (53)$$

where $\partial u/\partial z$ is simply given by u/H in this two-layer model. [This form differs slightly from equation (22) because equation (53) adopts a constant basic-state density (the so-called "Boussinesq" approximation) whereas equation (22) adopts the compressible ideal-gas equation of state.] Inserting $u = \Omega a \phi^2$ into equation (53), approximating the Coriolis parameter as $f = \beta y$ (where β is treated as constant), and integrating, we obtain a temperature that varies with latitude as

$$\theta = \theta_{equator} - \frac{\Omega^2 \theta_0}{2ga^2H} y^4 \quad (54)$$

where $\theta_{equator}$ is a constant to be determined.

At this point, we introduce two constraints. First, *Held and Hou* (1980) assumed the circulation is energetically closed, i.e., that no net exchange of mass or thermal energy occurs between the Hadley cell and higher latitude circulations. Given an energy equation with radiation parameterized using Newtonian cooling, $d\theta/dt = (\theta_{rad}-\theta)/\tau_{rad}$, where τ_{rad} is a radiative time constant, the assumption that the circulation is steady and closed requires that

$$\int_0^{\phi_H} \theta \, dy = \int_0^{\phi_H} \theta_{rad} dy \quad (55)$$

where we are integrating from the equator to the poleward edge of the Hadley cell, at latitude ϕ_H. Second, temperature must be continuous with latitude at the poleward edge of the Hadley cell. In the axisymmetric model, baroclinic instabilities are suppressed, and the regions poleward of the Hadley cells reside in a state of radiative equilibrium. Thus, θ must equal θ_{rad} at the poleward edge of the cell. Inserting our expressions for θ and θ_{rad} into these two constraints yields a system of two equations for ϕ_H and $\theta_{equator}$. The solution yields a Hadley cell with a latitudinal half-width of

$$\phi_H = \left(\frac{5\Delta\theta_{rad}gH}{3\Omega^2 a^2 \theta_0} \right)^{1/2} \quad (56)$$

in radians. This solution suggests that the width of the Hadley cell scales as the square root of the fractional equator-to-pole radiative-equilibrium temperature difference, the square root of the gravity, the square root of the height of the cell, and inversely with the rotation rate. Inserting Earth annual-mean values ($\Delta\theta_{rad} \approx 70$ K, $\theta = 260$ K, $g = 9.8$ m s^{-2}, H = 15 km, a = 6400 km, and $\Omega = 7.2 \times 10^{-5}$ s^{-1}) yields ~30°.

Redoing this analysis without the small-angle approximation leads to a transcendental equation for ϕ_H (see equation (17) in *Held and Hou*, 1980), which can be solved numerically. Figure 13 (solid curve) illustrates the solution. As expected, ϕ_H ranges from 0° as $\Omega \to \infty$ to 90° as $\Omega \to 0$, and, for planets of Earth radius with Hadley circulations ~10 km tall, bridges these extremes between rotation periods of ~0.5 and 20 d. Although deviations exist, the agreement between the simple Held-Hou model and the three-dimensional GCM simulations are surprisingly good given the simplicity of the Held-Hou model.

Importantly, the Held-Hou model demonstrates that latitudinal confinement of the Hadley cell occurs even in an axisymmetric atmosphere. Thus, the cell's latitudinal confinement does not require (for example) three-dimensional baroclinic or barotropic instabilities associated with the jet at the poleward branch of the cell. Instead, the confinement results from energetics: The twin constraints of angular momentum conservation in the upper branch and thermal-wind balance specify the latitudinal temperature profile in the Hadley circulation (equation (54)). This generates equatorial temperatures colder than (and subtropical temperatures warmer than) the radiative equilibrium, implying radiative heating at the equator and cooling in the subtropics. This properly allows the circulation to transport thermal energy poleward. If the cell extended globally on a rapidly rotating planet, however, the circulation would additionally produce high-latitude temperatures *colder* than the radiative equilibrium temperature, which in steady state would require radiative *heating* at high latitudes. This is thermodynamically impossible given the specified latitudinal dependence of θ_{rad}. The highest latitude to which the cell can extend without encountering this problem is simply given by equation (55).

The model can be generalized to consider a more realistic treatment of radiation than the simplified Newtonian cooling/heating scheme employed by *Held and Hou* (1980). *Caballero et al.* (2008) reworked the scheme using a two-stream, nongray representation of the radiative transfer with parameters appropriate for Earth and Mars. This leads to a prediction for the width of the Hadley cell that differs from equation (56) by a numerical constant on the order of unity.

Although the prediction of Held-Hou-type models for ϕ_H provides important insight, several failures of these models exist. First, the model underpredicts the strength of Earth's Hadley cell (e.g., as characterized by the magnitude of the north-south wind) by about an order of magnitude. This seems to result from the lack of turbulent eddies in axisymmetric models; several studies have shown that turbulent three-dimensional eddies exert stresses that act to strengthen the Hadley cells beyond the predictions of axisymmetric models (e.g., *Kim and Lee,* 2001; *Walker and Schneider,* 2005, 2006; *Schneider,* 2006). [*Held and Hou* (1980)'s original model neglected the seasonal cycle, and it has been suggested that generalization of the Held-Hou axisymmetric model to include seasonal effects could alleviate this failing (*Lindzen and Hou,* 1988). Although this improves the agreement with Earth's observed annual-mean Hadley-cell strength, it predicts large solstice/equinox oscillations in Hadley-cell strength that are lacking in the observed Hadley circulation (*Dima and Wallace,* 2003).] Second, the Hadley cells on Earth and probably Mars are not energetically closed; rather, midlatitude baroclinic eddies transport thermal energy out of the Hadley cell into the polar regions. Third, the poleward-moving upper tropospheric branches of the Hadley cells do not conserve angular momentum — although the zonal wind does become eastward as one moves poleward across the cell, for Earth this increase is a factor of ~2–3 less than predicted by equation (52). Overcoming these failings requires the inclusion of three-dimensional eddies.

Several studies have shown that turbulent eddies in the mid- to high-latitudes — which are neglected in the Held-Hou and other axisymmetric models — can affect the width of the Hadley circulation and alter the parameter dependences suggested by equation (56) (e.g., *Del Genio and Suozzo*, 1987; *Walker and Schneider*, 2005, 2006). Turbulence can produce an acceleration/deceleration of the zonal-mean zonal wind, which breaks the angular-momentum conservation constraint in the upper-level wind, causing u to deviate from equation (52). With a different u(ϕ) profile, the latitudinal dependence of temperature will change (via equation (53)), and hence so will the latitudinal extent of the Hadley cell required to satisfy equation (55). Indeed, within the context of axisymmetric models, the addition of strong drag into the upper-layer flow (parameterizing turbulent mixing with the slower-moving surface air, for example) can lead equation (55) to predict that the Hadley cell should extend to the poles even for Earth's rotation rate (e.g., *Farrell*, 1990).

It could thus be the case that the width of the Hadley cell is strongly controlled by eddies. For example, in the midlatitudes of Earth and Mars, baroclinic eddies generally accelerate the zonal flow eastward in the upper troposphere; in steady state, this is generally counteracted by a westward Coriolis acceleration, which requires an equatorward upper tropospheric flow — backward from the flow direction in the Hadley cell. Such eddy effects can thereby terminate the Hadley cell, forcing its confinement to low latitudes. Based on this idea, *Held* (2000) suggested that the Hadley cell width is determined by the latitude beyond which the troposphere first becomes baroclinically unstable (requiring isentrope slopes to exceed a latitude-dependent critical value). Adopting the horizontal thermal gradient implied by the angular-momentum conserving wind (equation (54)), making the small-angle approximation, and utilizing a common two-layer model of baroclinic instability, this yields (*Held*, 2000)

$$\phi_H \approx \left(\frac{gH\Delta\theta_v}{\Omega^2 a^2 \theta_0} \right)^{1/4} \quad (57)$$

in radians, where $\Delta\theta_v$ is the vertical difference in potential temperature from the surface to the top of the Hadley cell. Note that the predicted dependence of ϕ_H on planetary radius, gravity, rotation rate, and height of the Hadley cell is weaker than predicted by the Held-Hou model. Earth-based GCM simulations suggest that equation (57) may provide a better representation of the parameter dependences (*Frierson et al.*, 2007; *Lu et al.*, 2007; *Korty and Schneider*, 2008). Nevertheless, even discrepancies with equation (57) are expected since the actual zonal wind does not follow the angular-momentum conserving profile (implying that

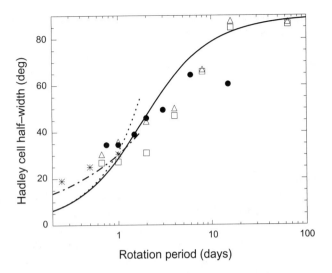

Fig. 13. Latitudinal width of the Hadley cell from a sequence of Earth-like GCM runs that vary the planetary rotation period (symbols) and comparison to the *Held and Hou* (1980) theory (solid curve), the small-angle approximation to the Held-Hou theory from equation (56) (dotted curve), and the *Held* (2000) theory from equation (57) (dashed-dotted curve). GCM results are from *Del Genio and Suozzo* (1987) (squares and triangles, depicting northern and southern hemispheres, respectively), *Navarra and Boccaletti* (2002) (filled circles), and *Korty and Schneider* (2008) (asterisks; only cases adopting $\theta_{rad} \approx 70$ K and their parameter $\Gamma = 0.7$ are shown). Parameters adopted for the curves are g = 9.8 m s^{-2}, H = 15 km, $\Delta\theta_{rad}$ = 70 K, a = 6400 km, θ_0 = 260 K, and $\Delta\theta_v$ = 30 K. In the GCM studies, different authors define the width of the Hadley cell in different ways, so some degree of scatter is inevitable.

the actual thermal gradient will deviate from equation (54)). Substantially more work is needed to generalize these ideas to the full range of conditions relevant for exoplanets.

5.4. High-Latitude Circulations: The Baroclinic Zones

For terrestrial planets heated at the equator and cooled at the poles, several studies suggest that the equator-to-pole heat engine can reside in either of two regimes depending on rotation rate and other parameters (*Del Genio and Suozzo*, 1987). When the rotation period is long, the Hadley cells extend nearly to the poles, dominate the equator-to-pole heat transport, and thereby determine the equator-to-pole temperature gradient. Baroclinic instabilities are suppressed because of the small latitudinal thermal gradient and large Rossby deformation radius (at which baroclinic instabilities have maximal growth rates), which exceeds the planetary size for slow rotators such as Titan and Venus. On the other hand, for rapidly rotating planets like Earth and Mars, the Hadley cells are confined to low latitudes; in the absence of eddies the mid- and high-latitudes would relax into a radiative-equilibrium state, leading to minimal equator-to-pole heat transport and a large equator-to-pole temperature difference. This structure is baroclinically unstable, however, and the resulting baroclinic eddies provide the dominant mechanism for transporting thermal energy from the poleward edges of the Hadley cells (~30° latitude for Earth) to the poles. This baroclinic heat transport significantly reduces the equator-to-pole temperature contrast in the mid- and high-latitudes. For gravity, planetary radius, and heating rates relevant for Earth, the breakpoint between these regimes occurs at a rotation period of ~5–10 d (*Del Genio and Suozzo*, 1987).

On rapidly rotating terrestrial planets like Earth and Mars, then, the equatorial and high-latitude circulations fundamentally differ: The equatorial Hadley cells, while strongly affected by eddies, do not *require* eddies to exist, nor to transport heat poleward. In contrast, the circulation and heat transport at high latitudes (poleward of ~30° latitude on Earth and Mars) fundamentally depend on the existence of eddies. At high latitudes, interactions of eddies with the mean flow controls the latitudinal temperature gradient, latitudinal heat transport, and structure of the jet streams. This range of latitudes is called the baroclinic zone.

The extent to which baroclinic eddies can reduce the equator-to-pole temperature gradient has important implications for the mean climate. Everything else being equal, an Earth-like planet with colder poles will develop more extensive polar ice and be more susceptible to an ice-albedo feedback that triggers a globally glaciated "snowball" state (*Spiegel et al.*, 2008). Likewise, a predominantly CO_2 atmosphere can become susceptible to atmospheric collapse if the polar temperatures become sufficiently cold for CO_2 condensation.

There is thus a desire to understand the extent to which baroclinic eddies can transport heat poleward. General circulation models (GCMs) attack this problem by spatially and temporally resolving the full life of every baroclinic eddy and their effect on the mean state, but this is computationally intensive and often sheds little light on the underlying sensitivity of the process to rotation rate and other parameters.

Two simplified approaches have been advanced that illuminate this issue. Although GCMs for terrestrial exoplanets will surely be needed in the future, simplified approaches can guide our understanding of such GCM results and provide testable hypotheses regarding the dependence of heat transport on planetary parameters. They also may provide guidance for parameterizations of the latitudinal heat transport in simpler energy-balance climate models that do not explicitly attempt to resolve the full dynamics. We devote the rest of this section to discussing these simplified approaches.

The first simplified approach postulates that baroclinic eddies relax the midlatitude thermal structure into a state that is neutrally stable to baroclinic instabilities (*Stone*, 1978), a process called baroclinic adjustment. This idea is analogous to the concept of convective adjustment: When radiation or other processes drive the vertical temperature gradient steeper than an adiabat ($dT/dz < -g/c_p$ for an ideal gas, where z is height), convection ensues and drives the temperature profile toward the adiabat, which is the neutrally stable state for convection. If the convective overturn timescales are much shorter than the radiative timescales, then convection overwhelms the ability of radiation to destabilize the environment, and the temperature profile then deviates only slightly from an adiabat over a wide range of convective heat fluxes (see, e.g., section 4.1). In a similar way, the concept of baroclinic adjustment postulates that when timescales for baroclinic eddy growth are much shorter than the timescale for radiation to create a large equator-to-pole temperature contrast, the eddies will transport thermal energy poleward at just the rate needed to maintain a profile that is neutral to baroclinic instabilities.

This idea was originally developed for a simplified two-layer system, for which baroclinic instability first initiates when the slope of isentropes exceeds ~H/a, where H is a pressure scale height and a is the planetary radius. Comparison of this critical isentrope slope with Earth observations shows impressive agreement poleward of 30°–40° latitude, where baroclinic instabilities are expected to be active (*Stone*, 1978). A substantial literature has subsequently developed to explore the idea further (for a review see *Zurita-Gotor and Lindzen*, 2006). The resulting latitudinal temperature gradient is then approximately H/a times the vertical potential temperature gradient. Interestingly, this theory suggests that otherwise identical planets with different Brunt-Väisälä frequencies (due to differing opacities, vertical heat transport by large-scale eddies, or role for latent heating, all of which can affect the stratification) would exhibit very different midlatitude temperature gradients — the planet with smaller stable stratification (i.e., smaller Brunt-Väisälä frequency; equation (6)) exhibiting a smaller latitudinal temperature gradient. Moreover, although Earth's ocean transports substantial heat between the equator and poles, the theory also suggests that the ocean may only exert a modest influence on the latitudinal temperature gradients in the atmosphere (*Lindzen and Farrell*, 1980a): In the absence of oceanic heat transport, the atmospheric eddies would simply take up the slack to maintain the atmosphere in the baroclinically neutral state. (There could of course be an indirect effect on latitudinal temperature gradients if removing/adding the oceans altered the tropospheric stratification.) Finally, the theory suggests that the latitudinal temperature gradient in the baroclinic zone does not depend on planetary rotation rate except indirectly via the influence of rotation rate on static stability. It is worth emphasizing, however, that even if the latitudinal temperature gradient in the baroclinic zone were constant with rotation rate, the total equator-to-pole temperature difference would still decrease with decreasing rotation rate because, with decreasing rotation rate, the Hadley cell occupies a greater latitude range and the baroclinically adjusted region would be compressed toward the poles (see Fig. 12).

Despite *Stone*'s (1978) encouraging results, several complicating factors exist. First, the timescale separation between radiation and dynamics is much less obvious for baroclinic adjustment than for convective adjustment. In an Earth- or

Mars-like context, the convective instability timescales are ~1 h, the baroclinic instability timescales are days, and the radiation timescale is ~20 d (Earth) or ~2 d (Mars). The ability of baroclinic eddies to adjust the environment to a baroclinically neutral state may thus be marginal, especially for planets with short radiative time constants (such as Mars). Second, the neutrally stable state in the two-layer model — corresponding to an isentrope slope of ~H/a — is an artifact of the vertical discretization in that model. When a multilayer model is used, the critical isentrope slope for initiating baroclinic instability decreases as the number of layers increases, and it approaches zero in a vertically continuous model. One might then wonder why a baroclinic zone can support *any* latitudinal temperature contrast (as it obviously does on Earth and Mars). The reason is that the instability growth timescales are long at small isentropic slopes (*Lindzen and Farrell*, 1980b); the instabilities only develop a substantial ability to affect the mean state when isentrope slopes become steep. In practice, then, the baroclinic eddies are perhaps only able to relax the atmosphere to a state with isentrope slopes of ~H/a rather than something significantly shallower.

The second simplified approach to understanding equator-to-pole temperature contrasts on terrestrial planets seeks to describe the poleward heat transport by baroclinic eddies as a diffusive process (*Held*, 1999). This approach is based on the idea that baroclinic eddies have maximal growth rates at scales close to the Rossby radius of deformation, L_D (see equation (23)), and if the baroclinically unstable zone has a latitudinal width substantially exceeding L_D then there will be scale separation between the mean flow and the eddies. [As pointed out by *Held* (1999), this differs from many instability problems in fluid mechanics, such as shear instability in pipe flow or convection in a fluid driven by a heat flux between two plates, where maximal growth rates occur at length scales comparable to the domain size and no eddy/meanflow scale separation occurs.] The possible existence of a scale separation in the baroclinic instability problem implies that representing the heat transport as a function of *local* mean-flow quantities (such as the mean latitudinal temperature gradient) is a reasonable prospect. Still, caution is warranted, since an inverse energy cascade (section 3.6) could potentially transfer the energy to scales larger than L_D, thereby weakening the scale separation.

The diffusive approach is typically cast in the context of a one-dimensional energy-balance model that seeks to determine the variation with latitude of the zonally averaged surface temperature. This approach has a long history for Earth climate studies (see review in *North et al.*, 1981), solar system planets (e.g., *Hoffert et al.*, 1981), and even in preliminary studies of the climates and habitability of terrestrial exoplanets (*Williams and Kasting*, 1997; *Williams and Pollard*, 2003; *Spiegel et al.*, 2008, 2009). In its simplest form, the governing equation reads

$$c \frac{\partial T}{\partial t} = \nabla \cdot (cD\nabla T) + S(\phi) - I \qquad (58)$$

where T is surface temperature, $D (m^2 s^{-1})$ is the diffusivity associated with heat transport by baroclinic eddies, $S(\phi)$ is the absorbed stellar flux as a function of latitude, and I is the emitted thermal flux, which is a function of temperature. In the simplest possible case, I is represented as a linear function of temperature, I = A + BT (*North et al.*, 1981), where A and B are positive constants. In equation (58), c is a heat capacity (with units $J m^{-2} K^{-1}$) that represents the atmosphere and oceans (if any). The steady-state solution in the absence of transport (D = 0) is T = [S(ϕ)–A]/B. On the other hand, when transport dominates (D → ∞), the solution yields constant T.

The term S(ϕ) includes the effect of latitudinally varying albedo (e.g., due to ice cover), but if albedo is constant with latitude, then S represents the latitudinally varying insolation. To schematically illustrate the effect of heat transport on the equator-to-pole temperature gradient, parameterize S as a constant plus a term proportional to the Legendre polynomial $P_2(\cos \phi)$. In this case, and if D and c are constant, the equation has a steady analytic solution with an equator-to-pole temperature difference given by (*Held*, 1999)

$$\Delta T_{eq-pole} = \frac{\Delta T_{rad}}{1 + 6\frac{cD}{Ba^2}} \qquad (59)$$

where a is the planetary radius and ΔT_{rad} is the equator-to-pole difference in radiative-equilibrium temperature. The equation implies that atmospheric heat transport significantly influences the equator-to-pole temperature difference when the diffusivity exceeds $~Ba^2/(6c)$. For Earth, B ≈ 2 W m^{-2} K^{-1}, a = 6400 km, and c ≈ 10^7 J m^{-2} K^{-1}, suggesting that atmospheric transport becomes important when D ≳ 10^6 m^2 s^{-1}.

The question comes down to how to determine the diffusivity, D. In Earth models, D is typically chosen by tuning the models to match the current climate, and then that value of D is used to explore the regimes of other climates (e.g., *North*, 1975). However, that approach sidesteps the underlying physics and prevents an extrapolation to other planetary environments.

Motivated by an interest in understanding the feedback between baroclinic instabilities and the mean state, substantial effort has been devoted to determining the dependence of the diffusivity on control parameters such as rotation rate and latitudinal thermal gradient (for a review see *Held*, 1999). This work can help guide efforts to understand the temperature distribution on rapidly rotating exoplanets. We expect that the diffusivity will scale as

$$D \approx u_{eddy} L_{eddy} \qquad (60)$$

where u_{eddy} and L_{eddy} are the characteristic velocities and horizontal sizes of the heat-transporting eddies. Because baroclinic eddies will be in near-geostrophic balance (equation (21)) on a rapidly rotating planet, the characteristic eddy velocities, u_{eddy}, will relate to characteristic eddy potential temperature perturbations θ'_{eddy} via the thermal-wind relation, giving $u_{eddy} \sim g\theta'_{eddy}h_{eddy}/(fL_{eddy})$, where h_{eddy} is the characteristic vertical thickness of the eddies. Under the assumption that the ratio of vertical to horizontal scales is $h_{eddy}/L_{eddy} \sim f/N$ (*Charney*, 1971; *Haynes*, 2005), this yields $u_{eddy} \approx g\theta'_{eddy}/(\theta_0 N)$ (where N is Brunt-Väisälä frequency), which simply states that eddies with larger thermal perturbations will also have larger velocity perturbations. Under the assumption that the thermal perturbations scale as $\theta'_{eddy} \approx L_{eddy}\partial\bar{\theta}/\partial y$, we then have

$$D \approx L_{eddy}^2 \frac{g}{N\theta_0}\frac{\partial\bar{\theta}}{\partial y} \qquad (61)$$

Greater thermal gradients and eddy-length scales lead to greater diffusivity and, importantly, the diffusivity depends on the *square* of the eddy size.

Several proposals for the relevant eddy size have been put forward, which we summarize in Table 2. Based on the idea that baroclinic instabilities have the greatest growth rates for lengths comparable to the deformation radius, *Stone* (1972) suggested that L_{eddy} is the deformation radius, NH/f. On the other hand, baroclinic eddies could energize an inverse cascade, potentially causing the dominant heat-transporting eddies to have sizes exceeding L_D. In the limit of this process (*Green*, 1970), the eddies would reach the width of the baroclinic zone, L_{zone} (potentially close to a planetary radius for a planet with a narrow Hadley cell). This simply leads to equation (61) with $L_{eddy} = L_{zone}$. In contrast, *Held and Larichev* (1996) argued that the inverse cascade would produce an eddy scale not on the order of L_{zone} but instead on the order of the Rhines scale, $(u_{eddy}/\beta)^{1/2}$. Finally, *Barry et al.* (2002) used heat-engine arguments to propose that

$$D \approx \left(\frac{ea\dot{q}}{\theta_0}\frac{\partial\bar{\theta}}{\partial y}\right)^{3/5}\left(\frac{2}{\beta}\right)^{4/5} \qquad (62)$$

where $\dot{q}_{net}$ is the net radiative heating/cooling per mass that the eddy fluxes are balancing, a is the planetary radius, and e is a constant of order unity. Equation (62) is probably most robust for the dependences on heating rate $\dot{q}_{net}$ and thermal gradient $\partial\bar{\theta}/\partial y$, which they varied by factors of ~200 and 6, respectively; planetary radius and rotation rates were varied by only ~70%, so the dependences on those parameters should be considered tentative.

As can be seen in Table 2, these proposals have divergent implications for the dependence of diffusivity on background parameters. *Stone*'s (1972) diffusivity is proportional to the latitudinal temperature gradient and, because of the variation of L_D with rotation rate, inversely proportional to the square of the planetary rotation rate. *Green*'s (1970) diffusivity likewise scales with the latitudinal temperature gradient; it contains no explicit dependence on the planetary rotation rate, but a rotation-rate dependence could enter because the Hadley cell shrinks and L_{zone} increases with increasing rotation rate. *Held and Larichev*'s (1996) diffusivity has the same rotation-rate dependence as that of *Stone* (1972), but it has a much stronger depenence on gravity and temperature gradient [g^3 and $(\partial\bar{\theta}/\partial y)^3$]. Moreover, it increases with decreasing static stability as N^{-3}, unlike *Stone*'s diffusivity that scales with N. [The N^{-3} dependence in *Held and Larichev* (1996) gives the impression that the diffusivity becomes unbounded as the atmospheric vertical temperature profile becomes neutrally stable (i.e., as $N \to 0$), but this is misleading. In reality, one expects the latitudinal temperature gradient and potential energy available for driving baroclinic instabilities to decrease with decreasing N. Noting that the slope of isentropes is $m_\theta \equiv (\partial\bar{\theta}/\partial y)(\partial\bar{\theta}/\partial z)^{-1}$, one can reexpress *Held and Larichev*'s (1996) diffusivity as scaling with $m_\theta N^3$. Thus, at constant isentrope slope, the diffusivity properly drops to zero as the vertical temperature profile becomes neutrally stable.] Finally, *Barry et al.*'s (2002) diffusivity suggests a somewhat weaker dependence, scaling with as $(\partial\bar{\theta}/\partial y)^{3/5}$ and $\Omega^{-3/5}$. Note that the rotation-rate dependencies described above should be treated with caution, because altering the rotation rate or gravity could change the circulation in a way that alters the Brunt-Väisälä frequency, leading to additional changes in the diffusivity.

Additional work is needed to determine which of these schemes (if any) is valid over a wide range of parameters relevant to terrestrial exoplanets. Most of the work performed to test the schemes of *Green* (1970), *Stone* (1972), *Held and Larichev* (1996), and others has adopted simplified two-layer quasigeostrophic models (i.e., models where geostrophic balance is imposed as an external constraint) under idealized assumptions such as planar geometry with constant Coriolis parameter (a so-called "*f*-plane"). Be-

TABLE 2. Proposed diffusivities for high-latitude heat transport on terrestrial planets.

Scheme	L_{eddy}	D
Stone (1972)	L_D	$\dfrac{H^2 Ng}{f^2\theta_0}\dfrac{\partial\bar{\theta}}{\partial y}$
Green (1970)	L_{zone}	$L_{zone}^2 \dfrac{g}{N\theta_0}\dfrac{\partial\bar{\theta}}{\partial y}$
Held and Larichev (1996)	L_β	$\dfrac{g^3}{N^3\beta^2\theta_0^3}\left(\dfrac{\partial\bar{\theta}}{\partial y}\right)^3$
Barry et al. (2002)	L_β	$\left(\dfrac{ea\dot{q}_{net}}{\theta_0}\dfrac{\partial\bar{\theta}}{\partial y}\right)^{3/5}\left(\dfrac{2}{\beta}\right)^{4/5}$

cause variation of f with latitude alters the properties of baroclinic instability, schemes developed for constant f may not translate directly into a planetary context. More recently, two-layer planar models with nonzero β have been explored by *Thompson and Young* (2007) and *Zurita-Gotor and Vallis* (2009), while full three-dimensional GCM calculations on a sphere (with ~30 levels) investigating the latitudinal heat flux have been performed by *Barry et al.* (2002) and *Schneider and Walker* (2008) under conditions relevant to Earth. Additional simulations in the spirit of *Barry et al.* (2002) and *Schneider and Walker* (2008), considering a wider range of planetary and atmospheric parameters, can clarify the true sensitivities of the heat transport rates in three-dimensional atmospheres.

5.5. Slowly Rotating Regime

At slow rotation rates, the GCM simulations shown in Figs. 11–12 develop near-global Hadley cells with high-latitude jets, but the wind remains weak at the equator. In contrast, Titan and Venus (with rotation periods of 16 and 243 d, respectively) have robust (~100 m s^{-1}) superrotating winds in their equatorial upper tropospheres. Because the equator is the region of the planet lying farthest from the rotation axis, such a flow contains a local maximum of angular momentum at the equator. Axisymmetric Hadley circulations (section 5.3) cannot produce such superrotation; rather, up-gradient transport of momentum by eddies is required. This process remains poorly understood. One class of models suggests that this transport occurs from high latitudes; for example, the high-latitude jets that result from the Hadley circulation (which can be seen in Fig. 11) experience a large-scale shear instability that pumps eddy momentum toward the equator, generating the equatorial superrotation (*Del Genio and Zhou*, 1996). Alternatively, thermal tide or wave interaction could induce momentum transports that generate the equatorial superrotation (e.g., *Fels and Lindzen*, 1974). Recently, a variety of simplified Venus and Titan GCMs have been developed that show encouraging progress in capturing the superrotation (e.g., *Yamamoto and Takahashi*, 2006; *Richardson et al.*, 2007; *Lee et al.*, 2007; *Herrnstein and Dowling*, 2007).

The Venus/Titan superrotation problem may have important implications for understanding the circulation of synchronously locked exoplanets. All published three-dimensional circulation models of synchronously locked hot Jupiters (section 4.3), and even the few published studies of sychronously locked terrestrial planets (*Joshi et al.*, 1997; *Joshi*, 2003) (see section 5.6), develop robust equatorial superrotation. The day-night heating pattern associated with synchronous locking should generate thermal tides and, in analogy with Venus and Titan, these could be relevant in driving the superrotation in the three-dimensional exoplanet models. Simplified Earth-based two-layer calculations that include longitudinally varying heating, and which also develop equatorial superrotation, seem to support this possibility (*Suarez and Duffy*, 1992; *Saravanan*, 1993).

5.6. Unusual Forcing Regimes

Our discussion of circulation regimes on terrestrial planets has so far largely focused on annual-mean forcing conditions. On Earth, seasonal variations represent relatively modest perturbations around the annual-mean climate due to Earth's 23.5° obliquity. On longer timescales, the secular evolution of Earth's orbital elements is thought to be responsible for paleoclimatic trends, such as ice ages, according to Milankovitch's interpretation (e.g., *Kawamura et al.*, 2007). While it may be surprising to refer to Earth's seasons or ice ages as minor events from a human perspective, it is clear that they constitute rather mild versions of the more diverse astronomical forcing regimes expected to occur on extrasolar worlds.

In principle, terrestrial exoplanets could possess large obliquities (i → 90°) and eccentricities (e → 1). This would result in forcing conditions with substantial seasonal variations around the annual mean, such as order-of magnitude variations in the global insolation over the orbital period at large eccentricities. Even the annual mean climate can be dramatically affected under unusual forcing conditions, for instance at large obliquities when the poles receive more annual-mean insolation than the planetary equator (for i > 54°). Yet another unusual forcing regime occurs if the terrestrial planet is tidally locked to its parent star and thus possesses permanent day- and nightsides much like the hot Jupiters discussed in section 4.3. Relatively few studies of circulation regimes under such unusual forcing conditions have been carried out to date.

Williams and Kasting (1997) and *Spiegel et al.* (2009) investigated the climate of oblique Earth-like planets with simple, diffusive energy-balance models of the type described by equation (58). While seasonal variations were found to be severe at high obliquity, *Williams and Kasting* (1997) concluded that highly oblique planets could nevertheless remain regionally habitable, especially if they possessed thick CO_2-enriched atmospheres with large thermal inertia relative to Earth, as may indeed be expected for terrestrial planets in the outermost regions of a system's habitable zone. *Spiegel et al.* (2009) confirmed these results and highlighted the risks that global glaciation events and partial atmospheric CO_2 collapse constitute for highly oblique terrestrial exoplanets. *Williams and Pollard* (2003) presented a much more detailed set of three-dimensional climate and circulation models for Earth-like planets at various obliquities. Despite large seasonal variations at extreme obliquities, *Williams and Pollard* (2003) found no evidence for any runaway greenhouse effect or global glaciation event in their climate models and thus concluded that Earth-like planets should generally be hospitable to life even at high obliquity.

Williams and Pollard (2002) also studied the climate on eccentric versions of Earth with a combination of detailed three-dimensional climate models and simpler energy-balance models. They showed that the strong variations in insolation occurring at large eccentricities can be rather efficiently buffered by the large thermal inertia of the atmosphere + ocean climate system. These authors thus argued that the annually averaged insolation, which is an increasing function of e, may be the most meaningful quantity to describe in simple terms the climate of eccentric, thermally blanketed Earth-like planets. It is presumably the case that terrestrial planets with reduced thermal inertia from lower atmospheric/oceanic masses would be more strongly affected by such variations in insolation.

It should be emphasized that, despite the existing work on oblique or eccentric versions of the Earth, a fundamental understanding of circulation regimes under such unusual forcing conditions is still largely missing. For instance, variations in the Hadley circulation regime or the baroclinic transport efficiency, expected as forcing conditions vary along the orbit, have not been thoroughly explored. Similarly, the combined effects of a substantial obliquity and eccentricity have been ignored. As the prospects for finding and characterizing exotic versions of Earth improve, these issues should become the subject of increasing scrutiny.

One of the best observational prospects for the next decade is the discovery of tidally locked terrestrial planets around nearby M dwarfs (e.g., *Irwin et al.*, 2009; *Seager et al.*, 2009). In anticipation of such discoveries, *Joshi et al.* (1997) presented a careful investigation of the circulation regime on this class of unusually forced planets, with permanent day and nightsides. The specific focus of their work was to determine the conditions under which CO_2 atmospheric collapse could occur on the cold nightsides of such planets and act as a trap for this important greenhouse gas, even when circulation and heat transport are present. Using a simplified general circulation model, these authors explored the problem's parameter space for various global planetary and atmospheric attributes and included a study of the potential risks posed by violent flares from the stellar host. The unusual circulation regime that emerged from this work was composed of a direct circulation cell at surface levels (i.e., equatorial day-to-night transport with polar return) and a superrotating wind higher up in the atmosphere. While detrimental atmospheric collapse did occur under some combinations of planetary attributes (e.g., for thin atmospheres), the authors concluded that efficient atmospheric heat transport was sufficient to prevent collapse under a variety of plausible scenarios. More recently, *Joshi* (2003) revisited some of these results with a much more detailed climate model with explicit treatments of nongray radiative transfer and a hydrological cycle. It is likely that this interesting circulation regime will be reconsidered in the next few years as the discovery of such exotic worlds is on our horizon.

6. RECENT HIGHLIGHTS

Since the first discovery of giant exoplanets around Sun-like stars via the Doppler velocity technique (*Mayor and Queloz*, 1995; *Marcy and Butler*, 1996), exoplanet research has experienced a spectacular series of observational breakthroughs. The first discovery of a transiting hot Jupiter (*Charbonneau et al.*, 2000; *Henry et al.*, 2000) opened the door to a wide range of clever techniques for characterizing such transiting planets. The drop in flux that occurs during secondary eclipse, when the planet passes behind its star, led to direct infrared detections of the thermal flux from the planet's dayside (*Deming et al.*, 2005; *Charbonneau et al.*, 2005). This was quickly followed by the detection of day-night temperature variations (*Harrington et al.*, 2006), detailed phase curve observations (e.g., *Knutson et al.*, 2007, 2009b; *Cowan et al.*, 2007), infrared spectral and photometric measurements (*Grillmair et al.*, 2007, 2008; *Richardson et al.*, 2007; *Charbonneau et al.*, 2008; *Knutson et al.*, 2008) and a variety of transit spectroscopic constraints (*Tinetti et al.*, 2007; *Swain et al.*, 2008; *Barman*, 2008). Collectively, these observations help to constrain the composition, albedo, three-dimensional temperature structure, and hence the atmospheric circulation regime of hot Jupiters.

The reality of atmospheric circulation on these objects is suggested by exquisite lightcurves of HD 189733b showing modest day-night infrared brightness temperature variations and displacement of the longitudes of minimum/maximum flux from the antistellar/substellar points, presumably the result of an efficient circulation able to distort the temperature pattern (*Knutson et al.*, 2007, 2009b). Dayside photometry also hints at the existence of an atmospheric circulation on this planet (*Barman*, 2008). Nevertheless, other hot Jupiters, such as Ups And b and HD 179949b, apparently exhibit large day-night temperature variations with no discernable displacement of the hot regions from the substellar point (*Harrington et al.*, 2006; *Cowan et al.*, 2007). Secondary-eclipse photometry suggests that some hot Jupiters have dayside temperatures that decrease with altitude, while other hot Jupiters appear to exhibit dayside thermal inversion layers where temperatures increase with altitude. Some authors have suggested that these two issues are linked (*Fortney et al.*, 2008), although substantial additional observations are needed for a robust assessment. Understanding the dependence of infrared phase variations and dayside temperature profiles on planetary orbital and physical properties remains an ongoing observational challenge for the coming decade.

In parallel, this observational vanguard has triggered a growing body of theoretical and modeling studies to investigate plausible atmospheric circulation patterns on hot Jupiters. The models agree on some issues, such as that synchronously rotating hot Jupiters on approximately three-day orbits should exhibit fast winds with only a few broad zonal jets. On the other hand, the models disagree on other issues, such as the details of the three-dimensional flow patterns and the extent to which significant global-scale time variability is likely. The effect of stellar flux, planetary rotation rate, obliquity, atmospheric composition, and orbital eccentricity on the circulation have barely been explored. Further theoretical work and comparison with observations should help move our understanding onto a firmer foundation.

7. FUTURE PROSPECTS

Over the next decade, the observational characterization of hot Jupiters and Neptunes will continue apace, aided especially by the warm Spitzer mission and subsequently JWST. Equally exciting, NASA's Kepler mission and groundbased surveys (e.g., MEarth) (see *Irwin et al., 2009*) may soon lead to the discovery of not only additional super Earths but Earth-sized terrestrial exoplanets, including "hot Earths" as well as terrestrial planets cool enough to lie within the habitable zones of their stars. If sufficiently favorable systems are discovered, follow-up by JWST may allow basic characterization of their spectra and lightcurves (albeit at significant resources per planet), providing constraints on their atmospheric properties including their composition, climate, circulation, and the extent to which they may be habitable. Finally, over the next decade, significant opportunities exist for expanding our understanding of basic dynamical processes by widening the parameters adopted in circulation models beyond those typically explored in the study of solar system planets.

To move the observational characterization of exoplanet atmospheric circulation into the next generation, we recommend full-orbit lightcurves for a variety of hot Jupiters and Neptunes (and eventually terrestrial planets). Target objects should span a range of incident stellar fluxes, planetary masses, and orbital periods and eccentricities (among other parameters) to characterize planetary diversity. To constrain the planetary thermal-energy budgets, lightcurves should sample the blackbody peaks at several wavelengths, including both absorption bands and low-opacity windows as expected from theoretical models; this will provide constraints on the day-night temperature contrasts — and hence day-night heat transport — at a range of atmospheric pressure levels. Dayside infrared spectra obtained from secondary eclipse measurements, while not necessarily diagnostic of the atmospheric circulation in isolation, will provide crucial constraints when combined with infrared lightcurves. By providing constraints on atmospheric composition, transit spectroscopy will enable much better interpretation of lightcurves and secondary-eclipse spectra. Secondary eclipses (or full lightcurves) that are observed repeatedly will provide powerful constraints on global-scale atmospheric variability. Kepler may prove particularly useful in this regard, with its ability to observe hundreds of secondary eclipses of hot Jupiters (such as HAT-P-7b) (*Borucki et al., 2009*) in its field of view. Such an extensive temporal dataset could provide information on the frequency spectrum of atmospheric variability (if any), which would help to identify the mechanisms of variability as well as the background state of the mean circulation. Other novel techniques, such as the possibility of separately measuring transit spectra on the leading and trailing limbs and the possibility of detecting the Doppler shifts associated with planetary rotation and/or winds, should be considered. Specific research questions for the next decade include but are not limited to the following:

1. How warm are the nightsides of hot Jupiters? How does this depend on the stellar flux, atmospheric composition, and other parameters?

2. What are the dynamical mechanisms for shifting the hot and cold points away from the substellar and antistellar points, respectively? Why does HD 189733b appear to have both its hottest and coldest points on the same hemisphere?

3. What is the observational and physical relationship (if any) between the amplitude of the day-night temperature contrast and existence or absence of a dayside temperature inversion that has been inferred on some hot Jupiters?

4. How does the expected atmospheric circulation regime depend on planetary size, rotation rate, gravity, obliquity, atmospheric composition, and incident stellar flux?

5. What are the important dissipation mechanisms in the atmospheres of hot Jupiters? What are the relative roles of turbulence, shocks, and magnetohydrodynamic processes in frictionally braking the flows?

6. How coupled are the atmospheric flows on hot Jupiters to convection in the planetary interior? How are momentum, heat, and constituents transported across the deep radiative zone in hot Jupiters?

7. Does the atmospheric circulation influence the long term evolution of hot Jupiters? Can downward transport of atmospheric energy into the interior help explain the radii of some anomalously large hot Jupiters?

8. What is the mechanism for equatorial superrotation that occurs in most three-dimensional models of synchronously rotating hot Jupiters? Does such superrotation exist on real hot Jupiters?

9. Are the atmospheres of hot Jupiters strongly time variable on the global scale, and if so, what is the mechanism for the variability?

10. How does the circulation regime of planets in highly eccentric orbits differ from that of planets in circular orbits?

11. To what extent can obliquity, rotation rate, atmospheric mass, and atmospheric circulation properties be inferred from disk-integrated spectra or lightcurves?

12. To what extent, if any, does the atmospheric circulation influence the escape of an atmosphere to space?

13. Do the deep molecular envelopes of giant planets differentially rotate?

14. What controls the wind speeds on giant planets? Why are Neptune's winds faster than Jupiter's, and what can this teach us about strongly irradiated planets?

15. What can observations of chemical disequilibrium species tell us about the atmospheric circulation of giant planets and brown dwarfs?

16. What is the role of clouds in affecting the circulation, climate, evolution, and observable properties of exoplanets?

17. For terrestrial planets, how does the atmospheric circulation affect the boundaries of the classical habitable zone?

18. What is the role of the atmospheric circulation in affecting the ice-albedo feedback, atmospheric collapse, runaway greenhouse, and other climate feedback?

19. Can adequate scaling theories be developed to predict day-night or equator-to-pole temperature contrasts as a function of planetary and atmospheric parameters?

20. What is the dynamical role of a surface (i.e., the ground) in the atmospheric circulation? For super Earths, is there a critical atmospheric mass beyond which the surface becomes unimportant and the circulation behaves like that of a giant planet?

Ultimately, unraveling the atmospheric circulation and climate of exoplanets from global-scale observations will be a difficult yet exciting challenge. Degeneracies of interpretation will undoubtedly exist (e.g., multiple circulation patterns explaining a given lightcurve), and forward progress will require not only high-quality observations but a careful exploration of a hierarchy of models so that the nature of these degeneracies can be understood. Despite the challenge, the potential payoff will be the unleashing of planetary meteorology beyond the confines of our solar system, leading to not only an improved understanding of basic circulation mechanisms (and how they may vary with planetary rotation rate, gravity, incident stellar flux, and other parameters) but a glimpse of the actual atmospheric circulations, climate, and habitability of planets orbiting other stars in our neighborhood of the MilkyWay.

Acknowledgments. This paper was supported by NASA Origins grant NNX08AF27G to A.P.S., NASA grants NNG04GN82G and STFC PP/E001858/1 to J.Y-K.C., and NASA contract NNG-06GF55G to K.M.

REFERENCES

Atkinson D. H., Ingersoll A. P., and Seiff A. (1997) Deep zonal winds on Jupiter: Update of Doppler tracking the Galileo probe from the orbiter. *Nature, 388,* 649–650.

Aurnou J. M. and Heimpel M. H. (2004) Zonal jets in rotating convection with mixed mechanical boundary conditions. *Icarus, 169,* 492–498.

Aurnou J. M. and Olson P. L. (2001) Strong zonal winds from thermal convection in a rotating spherical shell. *Geophys. Res. Lett., 28,* 2557–2560.

Barman T. S. (2008) On the presence of water and global circulation in the transiting planet HD 189733b. *Astrophys. J. Lett., 676,* L61–L64.

Barry L., Craig G. C., and Thuburn J. (2002) Poleward heat transport by the atmospheric heat engine. *Nature, 415,* 774–777.

Batchelor G. K. (1967) *An Introduction to Fluid Dynamics.* Cambridge Univ., Cambridge.

Bézard B., Lellouch E., Strobel D., Maillard J.-P., and Drossart P. (2002) Carbon monoxide on Jupiter: Evidence for both internal and external sources. *Icarus, 159,* 95–111.

Borucki W. J., Koch D., Jenkins J., Sasselov D., Gilliland R., et al. (2009) Kepler's optical phase curve of the exoplanet HAT-P-7b. *Nature, 325,* 709.

Burrows A., Guillot T., Hubbard W. B., Marley M. S., Saumon D., Lunine J. I., and Sudarsky D. (2000) On the radii of close-in giant planets. *Astrophys. J. Lett., 534,* L97–L100.

Burrows A., Hubbard W. B., Lunine J. I., and Liebert J. (2001) The theory of brown dwarfs and extrasolar giant planets. *Rev. Mod. Phys., 73,* 719–765.

Busse F. H. (1976) A simple model of convection in the jovian atmosphere. *Icarus, 29,* 255–260.

Busse F. H. (2002) Convective flows in rapidly rotating spheres and their dynamo action. *Phys. Fluids, 14,* 1301–1314.

Caballero R., Pierrehumbert R. T., and Mitchell J. L. (2008) Axisymmetric, nearly inviscid circulations in non-condensing radiative-convective atmospheres. *Q. J. R. Meteorol. Soc., 134,* 1269–1285.

Chabrier G. and Baraffe I. (2000) Theory of low-mass stars and substellar objects. *Annu. Rev. Astron. Astrophys., 38,* 337–377.

Chabrier G. and Baraffe I. (2007) Heat transport in giant (exo) planets: A new perspective. *Astrophys. J. Lett., 661,* L81–L84.

Chabrier G., Barman T., Baraffe I., Allard F., and Hauschildt P. H. (2004) The evolution of irradiated planets: Application to transits. *Astrophys. J. Lett., 603,* L53–L56.

Charbonneau D., Brown T. M., Latham D. W., and Mayor M. (2000) Detection of planetary transits across a Sun-like star. *Astrophys. J. Lett., 529,* L45–L48.

Charbonneau D., Allen L. E., Megeath, S. T., Torres G., Alonso R., et al. (2005) Detection of thermal emission from an extrasolar planet. *Astrophys. J., 626,* 523–529.

Charbonneau D., Knutson H. A., Barman T., Allen L. E., Mayor M., Megeath S. T., Queloz D., and Udry S. (2008) The broadband infrared emission spectrum of the exoplanet HD 189733b. *Astrophys. J., 686,* 1341–1348.

Charney J. G. (1971) Geostrophic turbulence. *J. Atmos. Sci., 28,* 1087–1095.

Cho J. Y.-K. (2008) Atmospheric dynamics of tidally synchronized extrasolar planets. *Philos. Trans. R. Soc. London Ser. A, 366,* 4477–4488.

Cho J. Y.-K. and Polvani L. M. (1996a) The morphogenesis of bands and zonal winds in the atmospheres on the giant outer planets. *Science, 8(1),* 1–12.

Cho J. Y.-K. and Polvani L. M. (1996b) The emergence of jets and vortices in freely evolving, shallow-water turbulence on a sphere. *Phys. Fluids, 8,* 1531–1552.

Cho J. Y.-K., Menou K., Hansen B. M. S., and Seager S. (2003) The changing face of the extrasolar giant planet HD 209458b. *Astrophys. J. Lett., 587,* L117–L120.

Cho J. Y.-K., Menou K., Hansen B. M. S., and Seager S. (2008) Atmospheric circulation of close-in extrasolar giant planets. I. Global, barotropic, adiabatic simulations. *Astrophys. J., 675,* 817–845.

Christensen U. R. (2001) Zonal flow driven by deep convection in the major planets. *Geophys. Res. Lett., 28,* 2553–2556.

Christensen U. R. (2002) Zonal flow driven by strongly supercritical convection in rotating spherical shells. *J. Fluid Mech., 470,* 115–133.

Cooper C. S. and Showman A. P. (2005) Dynamic meteorology at the photosphere of HD 209458b. *Astrophys. J. Lett., 629,* L45–L48.

Cooper C. S. and Showman A. P. (2006) Dynamics and disequilibrium carbon chemistry in hot Jupiter atmospheres, with application to HD 209458b. *Astrophys. J., 649,* 1048–1063.

Cowan N. B., Agol E., and Charbonneau D. (2007) Hot nights on extrasolar planets: Mid-infrared phase variations of hot Jupiters. *Mon. Not. R. Astron. Soc., 379,* 641–646.

Del Genio A. D. and Suozzo R. J. (1987) A comparative study of rapidly and slowly rotating dynamical regimes in a terrestrial general circulation model. *J. Atmos. Sci., 44,* 973–986.

Del Genio A. D. and Zhou W. (1996) Simulations of superrotation on slowly rotating planets: Sensitivity to rotation and initial condition. *Icarus, 120,* 332–343.

Deming D., Seager S., Richardson L. J., and Harrington J. (2005) Infrared radiation from an extrasolar planet. *Nature, 434,* 740–743.

Deming D., Harrington J., Seager S., and Richardson L. J. (2006) Strong infrared emission from the extrasolar planet HD 189733b. *Astrophys. J., 644,* 560–564.

Dima I. M. and Wallace J. M. (2003) On the seasonality of the Hadley cell. *J. Atmos. Sci., 60,* 1522–1527.

Dobbs-Dixon I. and Lin D. N. C. (2008) Atmospheric dynamics of short-period extrasolar gas giant planets. I. Dependence of nightside temperature on opacity. *Astrophys. J., 673,* 513–525.

Dowling T. E. (1995) Dynamics of jovian atmospheres. *Annu. Rev. Fluid Mech., 27,* 293–334.

Dowling T. E. and Ingersoll A. P. (1989) Jupiter's Great Red Spot as a shallow water system. *J. Atmos. Sci., 46,* 3256–3278.

Dritschel D. G., de La Torre Juárez M., and Ambaum M. H. P. (1999) The three-dimensional vortical nature of atmospheric and oceanic turbulent flows. *Phys. Fluids, 11,* 1512–1520.

Farrell B. F. (1990) Equable climate dynamics. *J. Atmos. Sci., 47,* 2986–2995.

Fegley B. J. and Lodders K. (1994) Chemical models of the deep atmospheres of Jupiter and Saturn. *Icarus, 110,* 117–154.

Fels S. B. and Lindzen R. S. (1974) The interaction of thermally excited gravity waves with mean flows. *Geophys. Astrophys. Fluid Dyn., 6,* 149–191.

Fernando H. J. S., Chen R.-R., and Boyer D. L. (1991) Effects of rotation on convective turbulence. *J. Fluid Mech., 228,* 513–547.

Fortney J. J., Marley M. S., and Barnes J. W. (2007) Planetary radii across five orders of magnitude in mass and stellar insolation: Application to transits. *Astrophys. J., 659,* 1661–1672.

Fortney J. J., Lodders K., Marley M. S., and Freedman R. S. (2008) A unified theory for the atmospheres of the hot and very hot Jupiters: Two classes of irradiated atmospheres. *Astrophys. J., 678,* 1419–1435.

Frierson D. M. W., Lu J., and Chen G. (2007) Width of the Hadley cell in simple and comprehensive general circulation models. *Geophys. Res. Lett., 34,* L18804.

Gierasch P. J. and Conrath B. J. (1985) Energy conversion processes in the outer planets. *Recent Advances in Planetary Meteorology* (G. E. Hunt, ed.), pp. 121–146. Cambridge Univ., New York.

Gierasch P. J. et al. (1997) The general circulation of the Venus atmosphere: An sssessment. *Venus II: Geology, Geophysics, Atmosphere, and Solar Wind Environment* (S. W. Bougher et al., eds.), pp. 459–500.

Gierasch P. J., Ingersoll A. P., Banfield D., Ewald S. P., Helfenstein P., et al. (2000) Observation of moist convection in Jupiter's atmosphere. *Nature, 403,* 628–630.

Glatzmaier G. A., Evonukm., and Rogers T. M. (2009) Differential rotation in giant planets maintained by density-stratified turbulent convection. *Geophys. Astrophy. Fluid Dyn., 103,* 31–51.

Goldman B. et al. (2008) CLOUDS search for variability in brown dwarf atmospheres. Infrared spectroscopic time series of L/T transition brown dwarfs. *Astron. Astrophys., 487,* 277–292.

Goodman J. (2009) Thermodynamics of atmospheric circulation on hot Jupiters. *Astrophys. J., 693,* 1645–1649.

Green J. S. A. (1970) Transfer properties of the large-scale eddies and the general circulation of the atmosphere. *Q. J. R. Meteorol. Soc., 96,* 157–185.

Grillmair C. J., Charbonneau D., Burrows A., Armus L., Stauffer J., Meadows V., Van Cleve J., and Levine D. (2007) A Spitzer spectrum of the exoplanet HD 189733b. *Astrophys. J. Lett., 658,* L115–L118.

Grillmair C. J., Burrows A., Charbonneau D., Armus L., Stauffer J., Meadows V., van Cleve J., von Braun K., and Levine D. (2008) Strong water absorption in the dayside emission spectrum of the planet HD 189733b. *Nature, 456,* 767–769.

Grote E., Busse F. H., and Tilgner A. (2000) Regular and chaotic spherical dynamos. *Phys. Earth Planet. Inter., 117,* 259–272.

Guillot T. (2005) The interiors of giant planets: Models and outstanding questions. *Annu. Rev. Earth Planet. Sci., 33,* 493–530.

Guillot T., Burrows A., Hubbard W. B., Lunine J. I., and Saumon D. (1996) Giant planets at small orbital distances. *Astrophys. J. Lett., 459,* L35–L38.

Guillot T., Stevenson D. J., Hubbard W. B., and Saumon D. (2004) The interior of Jupiter. *Jupiter: The Planet, Satellites and Magnetosphere* (F. Bagenal et al., eds.), pp. 35–57. Cambridge Univ., Cambridge.

Hadley G. (1735) Concerning the cause of the general tradewinds. *Philos. Trans., 39,* 58–62.

Harrington J., Hansen B. M., Luszcz S. H., Seager S., Deming D., Menou K., Cho J. Y.-K., and Richardson L. J. (2006) The Phase-dependent infrared brightness of the extrasolar planet υ Andromedae b. *Science, 314,* 623–626.

Hartmann D. L., Betts A. K., Bonan G. B., Branscome L. E., Busalacchi A. J. Jr., et al. (2003) *Understanding Climate Change Feedbacks.* National Research Council of the National Academy of Sciences, The National Academies Press, Washington, DC.

Hayashi Y.-Y., Ishioka K., Yamada M., and Yoden S. (2000) Emergence of circumpolar vortex in two dimensional turbulence on a rotating sphere. In *Proceedings of the IUTAM In Symposium on Developments in Geophysical Turbulence* (R. M. Kerr and Y. Kimura, eds.), pp. 179–192. Fluid Mechanics and Its Applications, Vol. 58, Kluwer, Dordrecht.

Hayashi Y.-Y., Nishizawa S., Takehiro S.-I., Yamada M., Ishioka K., and Yoden S. (2007) Rossby waves and jets in two-dimensional decaying turbulence on a rotating sphere. *J. Atmos. Sci., 64,* 4246–4269.

Haynes P. (2005) Stratospheric dynamics. *Annu. Rev. Fluid Mech., 37,* 263–293.

Heimpel M., Aurnou J., and Wicht J. (2005) Simulation of equatorial and high-latitude jets on Jupiter in a deep convection model. *Nature, 438,* 193–196.

Held I. M. (1999) The macroturbulence of the troposphere. *Tellus, 51A-B,* 59–70.

Held I. M. (2000) The general circulation of the atmosphere. *Paper presented at 2000 Woods Hole Oceanographic Institute Geophysical Fluid Dynamics Program, Woods Hole Oceanographic Institute, Woods Hole, Massachusetts.* Available online at www.whoi.edu/page.do?pid=13076.

Held I. (2005) The gap between simulation and understanding in climate modeling. *Bull. Am. Meteorol. Soc., 86,* 1609–1614.

Held I. M. and Hou A. Y. (1980) Nonlinear axially symmetric circulations in a nearly inviscid atmosphere. *J. Atmos. Sci., 37,* 515–533.

Held I. M. and Larichev V. D. (1996) A scaling theory for horizontally homogeneous, baroclinically unstable flow on a beta plane. *J. Atmos. Sci., 53,* 946–952.

Held I. M. and Soden B. J. (2000) Water vapor feedback anad global warming. *Annu. Rev. Energy Environ., 25,* 441–475.

Henry G. W., Marcy G. W., Butler R. P., and Vogt S. S. (2000) A transiting "51 Peg-like" planet. *Astrophys. J. Lett., 529,* L41–L44.

Herrnstein A., and Dowling T. E. (2007) Effects of topography on the spin-up of a Venus atmospheric model. *J. Geophys. Res.–Planets, 112(E11),* E04S08.

Hoffert M. I., Callegari A. J., Hsieh C. T., and Ziegler W. (1981)

Liquid water on Mars: An energy balance climate model for CO$_2$/H$_2$O atmospheres. *Icarus, 47,* 112–129.

Hoffman P. F. and Schrag D. (2002) Review article: The snowball Earth hypothesis: Testing the limits of global change. *Terra Nova, 14,* 129–155.

Holton J. R. (2004) *An Introduction to Dynamic Meteorology, 4th edition.* Academic, San Diego.

Huang H.-P. and Robinson W. A. (1998) Two-dimensional turbulence and persistent zonal jets in a global barotropic model. *J. Atmos. Sci., 55,* 611–632.

Hubbard W. B., Nellis W. J., Mitchell A. C., Holmes N. C., McCandless P. C., and Limaye S. S. (1991) Interior structure of Neptune — Comparison with Uranus. *Science, 253,* 648–651.

Hunt B. G. (1979) The influence of the Earth's rotation rate on the general circulation of the atmosphere. *J. Atmos. Sci., 36,* 1392–1408.

Ingersoll A. P. (1969) The runaway greenhouse: A history of water on Venus. *J. Atmos. Sci., 26,* 1191–1198.

Ingersoll A. P. (1990) Atmospheric dynamics of the outer planets. *Science, 248,* 308–315.

Ingersoll A. P. and Cuzzi J. N. (1969) Dynamics of Jupiter's cloud bands. *J. Atmos. Sci., 26,* 981–985.

Ingersoll A. P. and Porco C. C. (1978) Solar heating and internal heat flow on Jupiter. *Icarus, 35,* 27–43.

Ingersoll A. P., Beebe R. F., Mitchell J. L., Garneau G. W., Yagi G. M., and Muller J.-P. (1981) Interaction of eddies and mean zonal flow on Jupiter as inferred from Voyager 1 and 2 images. *J. Geophys. Res., 86,* 8733–8743.

Iro N., Bézard B., and Guillot T. (2005) A time-dependent radiative model of HD 209458b. *Astron. Astrophys., 436,* 719–727.

Irwin J., Charbonneau D., Nutzman P., and Falco E. (2009) The MEarth project: Searching for transiting habitable super-Earths around nearby M dwarfs. In *Transiting Planets* (F. Pont et al., eds.), pp. 37–43. IAU Symposium No. 253, Cambridge Univ., Cambridge.

Ishioka K., Yamada M., Hayashi Y.-Y., and Yoden S. (1999) Pattern formation from two-dimensional decaying turbulence on a rotating sphere. *Japan Soc. Fluid Mech.,* Available *online at www.nagare.or.jp/mm/99/ishioka/.*

James I. N. (1994) *Introduction to Circulating Atmospheres.* Cambridge Atmospheric and Space Science Series, Cambridge Univ., Cambridge.

Joshi M. (2003) Climate model studies of synchronously rotating planets. *Astrobiology, 3,* 415–427.

Joshi M. M., Haberle R. M., and Reynolds R. T. (1997) Simulations of the atmospheres of synchronously rotating terrestrial planets orbiting M dwarfs: Conditions for atmospheric collapse and the implications for habitability. *Icarus, 129,* 450–465.

Kaspi Y., Flierl G. R., and Showman A. P. (2009) The deep wind structure of the giant planets: Results from an anelastic general circulation model. *Icarus, 202,* 525–542.

Kasting J. F. (1988) Runaway and moist greenhouse atmospheres and the evolution of Earth and Venus. *Icarus, 74,* 472–494.

Kasting J. F. and Catling D. (2003) Evolution of a habitable planet. *Annu. Rev. Astron. Astrophys., 41,* 429–463.

Kasting J. F., Whitmire D. P., and Reynolds R. T. (1993) Habitable zones around main sequence stars. *Icarus, 101,* 108–128.

Kawamura K. et al. (2007) Northern hemisphere forcing of climatic cycles in Antarctica over the past 360,000 years. *Nature, 448,* 912–916.

Kim H.-K. and Lee S. (2001) Hadley cell dynamics in a primitive equation model. Part II: Nonaxisymmetric flow. *J. Atmos. Sci., 58,* 2859–2871.

Kirk R. L. and Stevenson D. J. (1987) Hydromagnetic constraints on deep zonal flow in the giant planets. *Astrophys. J., 316,* 836–846.

Knutson H. A., Charbonneau D., Allen L. E., Fortney J. J., Agol E., Cowan N. B., Showman A. P., Cooper C. S., and Megeath S. T. (2007) A map of the day-night contrast of the extrasolar planet HD 189733b. *Nature, 447,* 183–186.

Knutson H. A., Charbonneau D., Allen L. E., Burrows A., and Megeath S. T. (2008) The 3.6–8.0 μm broadband emission spectrum of HD 209458b: Evidence for an atmospheric temperature inversion. *Astrophys. J., 673,* 526–531.

Knutson H. A., Charbonneau D., Burrows A., O'Donovan F. T., and Mandushev G. (2009a) Detection of a temperature inversion in the broadband infrared emission spectrum of TrES-4. *Astrophys. J., 691,* 866–874.

Knutson H. A., Charbonneau D., Cowan N. B., Fortney J. J., Showman A. P., Agol E., Henry G. W., Everett M. E., and Allen L. E. (2009b) Multiwavelength constraints on the day-night circulation patterns of HD 189733b. *Astrophys. J., 690,* 822–836.

Korty R. L. and Schneider T. (2008) Extent of Hadley circulations in dry atmospheres. *Geophys. Res. Lett., 35,* L23803.

Langton J. and Laughlin G. (2007) Observational consequences of hydrodynamic flows on hot Jupiters. *Astrophys. J. Lett., 657,* L113–L116.

Langton J. and Laughlin G. (2008a) Hydrodynamic simulations of unevenly irradiated jovian planets. *Astrophys. J., 674,* 1106–1116.

Langton J. and Laughlin G. (2008b) Persistent circumpolar vortices on the extrasolar giant planet HD 37605b. *Astron. Astrophys., 483,* L25–L28.

Laughlin G., Deming D., Langton J., Kasen D., Vogt S., Butler P., Rivera E., and Meschiari S. (2009) Rapid heating of the atmosphere of an extrasolar planet. *Nature, 457,* 562–564.

Lee C., Lewis S. R., and Read P. L. (2007) Superrotation in a Venus general circulation model. *J. Geophys. Res.–Planets, 112(E11).*

Lewis N., Showman A. P., Fortney J. J., and Marley M. S. (2009) Three-dimensional atmospheric dynamics of eccentric extrasolar planets. *Bull. Am. Astron. Soc., 41(1),* 346.01.

Lian Y. and Showman A. P. (2008) Deep jets on gas-giant planets. *Icarus, 194,* 597–615.

Lian Y. and Showman A. P. (2010) Generation of equatorial jets by large-scale latent heating on the giant planets. *Icarus, 207,* 373–393.

Lindzen R. S. and Farrell B. (1980a) The role of the polar regions in climate, and a new parameterization of global heat transport. *Monthly Weather Rev., 108,* 2064–2079.

Lindzen R. S. and Farrell B. (1980b) A simple approximate result for the maximum growth rate of baroclinic instabilities. *J. Atmos. Sci., 37,* 1648–1654.

Lindzen R. S. and Hou A. V. (1988) Hadley circulations for zonally averaged heating centered off the equator. *J. Atmos. Sci., 45,* 2416–2427.

Little B., Anger C. D., Ingersoll A. P., Vasavada A. R., Senske D. A., Breneman H. H., Borucki W. J., and the Galileo SSI Team (1999) Galileo images of lightning on Jupiter. *Icarus, 142,* 306–323.

Liu J., Goldreich P. M. and Stevenson D. J. (2008) Constraints on deep-seated zonal winds inside Jupiter and Saturn. *Icarus, 196,* 653–664.

Lorenz E. N. (1967) *The Nature and Theory of the General Circulation of the Atmosphere.* World Meteorological Organization, Geneva.

Lu J., Vecchi G. A., and Reichler T. (2007) Expansion of the Hadley cell under global warming. *Geophys. Res. Lett., 34,* L06805.

Machalek P., McCullough P. R., Burke C. J., Valenti J. A., Burrows A., and Hora J. L. (2008) Thermal emission of exoplanet XO-1b. *Astrophys. J., 684,* 1427–1432.

Marcus P. S., Kundu T. and Lee C. (2000) Vortex dynamics and zonal flows. *Phys. Plasmas, 7,* 1630–1640.

Marcy G. W. and Butler R. P. (1996) A planetary companion to 70 Virginis. *Astrophys. J. Lett., 464,* L147.

Mayor M. and Queloz D. (1995) A Jupiter-mass companion to a solar-type star. *Nature, 378,* 355.

Menou K. and Rauscher E. (2009) Atmospheric circulation of hot Jupiters: A shallow three-dimensional model. *Astrophys. J., 700,* 887–897.

Menou K., Cho J. Y.-K., Seager S. and Hansen B. M. S. (2003) "Weather" variability of close-in extrasolar giant planets. *Astrophys. J. Lett., 587,* L113–L116.

Morales-Calderón M. et al. (2006) A sensitive search for variability in late L dwarfs: The quest for weather. *Astrophys. J., 653,* 1454–1463.

Navarra A. and Boccaletti G. (2002) Numerical general circulation experiments of sensitivity to Earth rotation rate. *Climate Dynamics, 19,* 467–483.

Noll K. S., Geballe T. R. and Marley M. . (1997) Detection of abundant carbon monoxide in the brown dwarf Gliese 229B. *Astrophys. J. Lett., 489,* L87.

North G. R. (1975) Analytical solution to a simple climate model with diffusive heat transport. *J. Atmos. Sci., 32,* 1301–1307.

North G. R., Cahalan R. F., and Coakley J. A. J. (1981) Energy balance climate models. *Rev. Geophys. Space Phys., 19,* 91–121.

Nozawa T. and Yoden S. (1997) Formation of zonal band structure in forced two-dimensional turbulence on a rotating sphere. *Phys. Fluids, 9,* 2081–2093.

Okuno A. and Masuda A. (2003) Effect of horizontal divergence on the geostrophic turbulence on a beta-plane: Suppression of the Rhines effect. *Phys. Fluids, 15,* 56–65.

Pedlosky J. (1987) *Geophysical Fluid Dynamics,* 2nd edition. Springer-Verlag, New York.

Peixoto J. P. and Oort A. H. (1992) *Physics of Climate.* American Institute of Physics, New York.

Podolak M., Hubbard W. B., and Stevenson D. J. (1991) Models of Uranus' interior and magnetic field. In *Uranus* (J. T. Bergstralh et al., eds.), pp. 29–61. Univ. of Arizona, Tucson.

Prinn R. G. and Barshay S. S. (1977) Carbon monoxide on Jupiter and implications for atmospheric convection. *Science, 198,* 1031–1034.

Rauscher E., Menou K., Cho J. Y.-K., Seager S., and Hansen B. M. S. (2007) Hot Jupiter variability in eclipse depth. *Astrophys. J. Lett., 662,* L115–L118.

Rauscher E., Menou K., Cho J. Y.-K., Seager S., and Hansen B. M. S. (2008) On signatures of atmospheric features in thermal phase curves of hot Jupiters. *Astrophys. J., 681,* 1646–1652.

Reinaud J. N., Dritschel D. G., and Koudella C. R. (2003) The shape of vortices in quasi-geostrophic turbulence. *J. Fluid Mech., 474,* 175–192.

Rhines P. B. (1975) Waves and turbulence on a beta-plane. *J. Fluid Mech., 69,* 417–443.

Richardson L. J., Deming D., Horning K., Seager S., and Harrington J. (2007) A spectrum of an extrasolar planet. *Nature, 445,* 892–895.

Salby M. L. (1996) *Fundamentals of Atmospheric Physics.* Academic, San Diego.

Sanchez-Lavega A., Rojas J. F., and Sada P. V. (2000) Saturn's zonal winds at cloud level. *Icarus, 147,* 405–420.

Sánchez-Lavega A., Orton G. S., Hueso R., García-Melendo E., Pérez-Hoyos S., et al. (2008) Depth of a strong jovian jet from a planetary-scale disturbance driven by storms. *Nature, 451,* 437–440.

Saravanan R. (1993) Equatorial superrotation and maintenance of the general circulation in two-level models. *J. Atmos. Sci., 50,* 1211–1227.

Saumon D., Hubbard W. B., Burrows A., Guillot T., Lunine J. I., and Chabrier G. (1996) A theory of extrasolar giant planets. *Astrophys. J., 460,* 993–1018.

Saumon D., Geballe T. R., Leggett S. K., Marley M. S., Freedman R. S., Lodders K., Fegley B. Jr., and Sengupta S. K. (2000) Molecular abundances in the atmosphere of the T dwarf GL 229B. *Astrophys. J., 541,* 374–389.

Saumon D., Marley M. S., Cushing M. C., Leggett S. K., Roellig T. L., Lodders K., and Freedman R. S. (2006) Ammonia as a tracer of chemical equilibrium in the T7.5 dwarf Gliese 570D. *Astrophys. J., 647,* 552–557.

Saumon D. et al. (2007) Physical parameters of two very cool T dwarfs. *Astrophys. J., 656,* 1136–1149.

Sayanagi K. M., Showman A. P., and Dowling T. E. (2008) The emergence of multiple robust zonal jets from freely evolving, three-dimensional stratified geostrophic turbulence with applications to Jupiter. *J. Atmos. Sci., 65,* 3947–3962.

Schneider T. (2006) The general circulation of the atmosphere. *Annu. Rev. Earth Planet. Sci., 34,* 655–688.

Schneider T. and Liu J. (2009) Formation of jets and equatorial superrotation on Jupiter. *J. Atmos. Sci., 66,* 579–601.

Schneider T. and Walker C. C. (2008) Scaling laws and regime transitions of macroturbulence in dry atmospheres. *J. Atmos. Sci., 65,* 2153–2173.

Scott R. K. and Polvani L. (2007) Forced-dissipative shallow water turbulence on the sphere and the atmospheric circulation of the giant planets. *J. Atmos. Sci, 64,* 3158–3176.

Scott R. K. and Polvani L. M. (2008) Equatorial superrotation in shallow atmospheres. *Geophys. Res. Lett., 35,* L24202.

Seager S., Deming D., and Valenti J. A. (2009) Transiting exoplanets with JWST. In *Astrophysics in the Next Decade* (H. A. Thronson et al., eds.), pp. 123–145. Springer, Dordrecht.

Showman A. P. (2007) Numerical simulations of forced shallow-water turbulence: Effects of moist convection on the large-scale circulation of Jupiter and Saturn. *J. Atmos. Sci., 64,* 3132–3157.

Showman A. P. and Guillot T. (2002) Atmospheric circulation and tides of "51 Pegasus b-like" planets. *Astron. Astrophys., 385,* 166–180.

Showman A. P., Gierasch P. J., and Lian Y. (2006) Deep zonal winds can result from shallow driving in a giant-planet atmosphere. *Icarus, 182,* 513–526.

Showman A. P., Cooper C. S., Fortney J. J., and Marley M. S. (2008a) Atmospheric circulation of hot Jupiters: Three dimen-

sional circulation models of HD 209458b and HD 189733b with simplified forcing. *Astrophys. J., 682,* 559–576.
Showman A. P., Menou K., and Cho J. Y.-K. (2008b) Atmospheric circulation of hot Jupiters: A review of current understanding. In *Extreme Solar Systems* (D. Fischer et al., eds.), pp. 419–441. ASP Conf. Series 398, Astronomical Society of the Pacific, San Francisco.
Showman A. P., Fortney J. J., Lian Y., Marley M. S., Freedman R. S., Knutson H. A., and Charbonneau D. (2009) Atmospheric circulation of hot Jupiters: Coupled radiative-dynamical general circulation model simulations of HD 189733b and HD 209458b. *Astrophys. J., 699,* 564–584.
Smith K. S. (2004) A local model for planetary atmospheres forced by small-scale convection. *J. Atmos. Sci., 61,* 1420–1433.
Spiegel D. S., Menou K., and Scharf C. A. (2008) Habitable climates. *Astrophys. J., 681,* 1609–1623.
Spiegel D. S., Menou K., and Scharf C. A. (2009) Habitable climates: The influence of obliquity. *Astrophys. J., 691,* 596–610.
Stevens B. (2005) Atmospheric moist convection. *Annu. Rev. Earth Planet. Sci., 33,* 605–643.
Stevenson D. J. (1979) Turbulent thermal convection in the presence of rotation and a magnetic field — A heuristic theory. *Geophys. Astrophys. Fluid Dynam., 12,* 139–169.
Stevenson D. J. (1991) The search for brown dwarfs. *Annu. Rev. Astron. Astrophys., 29,* 163–193.
Stone P. H. (1972) A simplified radiative-dynamical model for the static stability of rotating atmospheres. *J. Atmos. Sci., 29,* 406–418.
Stone P. H. (1978) Baroclinic adjustment. *J. Atmos. Sci., 35,* 561–571.
Suarez M. J. and Duffy D. G. (1992) Terrestrial superrotation: A bifurcation of the general circulation. *J. Atmos. Sci., 49,* 1541–1556.
Sukoriansky S., Dikovskaya N. and Galperin B. (2007) On the "arrest" of inverse energy cascade and the Rhines scale. *J. Atmos. Sci., 64,* 3312–3327.
Swain M. R., Vasisht G., and Tinetti G. (2008) The presence of methane in the atmosphere of an extrasolar planet. *Nature, 452,* 329–331.
Swain M. R., Vasisht G., Tinetti G., Bouwman J., Chen P., Yung Y., Deming D., and Deroo P. (2009) Molecular signatures in the near-infrared dayside spectrum of HD 189733b. *Astrophys. J. Lett., 690,* L114–L117.
Tabeling P. (2002) Two-dimensional turbulence: A physicist approach. *Phys. Rept., 362,* 1–62.
Thompson A. F. and Young W. R. (2007) Two-layer baroclinic eddy heat fluxes: Zonal flows and energy balance. *J. Atmos. Sci., 64,* 3214–3231.
Tinetti G. et al. (2007) Water vapour in the atmosphere of a transiting extrasolar planet. *Nature, 448,* 169–171.

Vallis G. K. (2006) *Atmospheric and Oceanic Fluid Dynamics: Fundamentals and Large-Scale Circulation.* Cambridge Univ, Cambridge.
Vasavada A. R. and Showman A. P. (2005) Jovian atmospheric dynamics: An update after Galileo and Cassini. *Rept. Progr. Phys., 68,* 1935–1996.
Walker C. C. and Schneider T. (2005) Response of idealized Hadley circulations to seasonally varying heating. *Geophys. Res. Lett., 32,* L06813.
Walker C. C. and Schneider T. (2006) Eddy influences on Hadley circulations: Simulations with an idealized GCM. *J. Atmos. Sci., 63,* 3333–3350.
Ward W. R. (1974) Climatic variations on Mars. I. Astronomical theory of insolation. *J. Geophys. Res., 79,* 3375–3386.
Williams D. M. and Kasting J. F. (1997) Habitable planets with high obliquities. *Icarus, 129,* 254–267.
Williams D. M. and Pollard D. (2002) Earth-like worlds on eccentric orbits: Excursions beyond the habitable zone. *Intl. J. Astrobiol., 1,* 61–69.
Williams D. M. and Pollard D. (2003) Extraordinary climates of Earth-like planets: Three-dimensional climate simulations at extreme obliquity. *Intl. J. Astrobiol., 2,* 1–19.
Williams G. P. (1978) Planetary circulations. I — Barotropic representation of jovian and terrestrial turbulence. *J. Atmos. Sci., 35,* 1399–1426.
Williams G. P. (1979) Planetary circulations. II — The jovian quasi-geostrophic regime. *J. Atmos. Sci., 36,* 932–968.
Williams G. P. (1988a) The dynamical range of global circulations — I. *Climate Dyn., 2,* 205–260.
Williams G. P. (1988b) The dynamical range of global circulations — II. *Climate Dyn., 3,* 45–84.
Williams G. P. (2003) Jovian dynamics. Part III: Multiple, migrating, and equatorial jets. *J. Atmos. Sci., 60,* 1270–1296.
Williams G. P. and Holloway J. L. (1982) The range and unity of planetary circulations. *Nature, 297,* 295–299.
Yamamoto M. and Takahashi M. (2006) Superrotation maintained by Meridional circulation and waves in a Venus-like AGCM. *J. Atmos. Sci., 63,* 3296–3314.
Yoden S., Ishioka K., Hayashi Y.-Y., and Yamada M. (1999) A further experiment on two-dimensional decaying turbulence on a rotating sphere. *Nuovo Cimento C Geophys. Space Phys. C, 22,* 803–812.
Zurita-Gotor P. and Lindzen R. S. (2006) Theories of baroclinic adjustment and eddy equilibration. *The Global Circulation of the Atmosphere* (T. Schneider and A. H. Sobel, eds.), pp. 22–46. Princeton Univ., Princeton.
Zurita-Gotor P. and Vallis G. K. (2009) Equilibration of baroclinic turbulence in primitive equations and quasigeostrophic models. *J. Atmos. Sci., 66,* 837–863.

Index

Adiabatic temperature profile 379
Advection time 480
Albedo 422, 423
    Bond 116, 423, 425, 500
    geometric 116, 423, 425
    single-scattering 425
    solar system 116
    spherical 423, 425
Aluminum-26/27 301, 309, 310, 311
Andoyer's variables 240, 241
Angrites 311
Angular momentum 305, 331
Antumbra 57
Apodization 128ff
Apsidal lock 229
Asteroid belt 298, 305
    compositional gradient 307
    resonances, *see* Resonances, asteroid belt
Astrobiology 441
Astrocentric coordinates 219, 220
Astrometric detection 6, 157ff
Astrophysical noise 32
Atmospheric dynamics 471ff
    advection time 479
    banding 484ff
    cyclostrophic equilibrium 482, 483
    energetics 478
    frictional dissipation 478
    jet streams 484ff
    Rossby number 481
    shallow-water model 476
    superrotating winds 509
    turbulence 484ff, 505
    vertical velocities 483
Atmospheric thermal tides 252
Atmospheric tides 239, 240, 246ff, 262

Babinet's principle 127
Bacteria
    methanogenic 464, 465
    photosynthetic 464
Banded structures 486ff
Baroclinic instabilities 487, 505ff
Baroclinic zones 505ff
Barotropic stability criterion 497
Barycentric coordinates 19ff, 30, 219, 220
Bayes' theorem 198
Bepi Colombo mission 382
Beryllium-10 310
Biosignature gases 465
Biosignatures 441ff, 463ff
    definition 10, 11
Birch-Murnaghan equation of state 380
Blackbody 62, 113, 271
Bondi radius 328

Brahe, Tycho 6
Brightness temperature 62
Brown dwarfs 157, 202, 319, 403, 412
    desert 202, 319
    thermal structure 489
Brownian motion 298
Bruno, Giordano 3
Brunt-Väisälä frequency 475, 482, 483, 506, 508
Bulk composition 375, 377, 381

C_2H, in disks 276
Calcium-aluminum-rich inclusions (CAIs) 270, 310, 311
Carbon dioxide 276, 442, 443, 445ff, 450, 451, 460, 461, 500, 501, 506, 509, 510
Carbon monoxide (CO) 284
    dissociation 276
    mission 271, 272
    in disks 276, 282
Carbon hydroxide
    dissociation and ionization 276
Cassini mission 429ff, 433
Cassini states 249, 254, 255, 263, 410
Chaotic planetary systems 226ff, 252
Chemical equilibrium 465
Chondrules 300
Chromaticity 140
Chromospheric index 33
Circumstellar envelopes 282
Climate 499
Conservation equations 401
Conservation of mass 376
Contrast 116
Convection 388
Convective stress 389
Cooling timescale 402
Copernicus, Nicolaus 11
Core accretion 203, 207, 270, 319ff, 333
Core-mantle boundary 379
Coriolis forces 475, 477, 480ff, 484, 488, 490, 494, 496, 504, 505
Coronagraph 123ff
    APL 148
    definition 113
    Lyot 129, 138, 148, 149
    phase-mask 148
    star shade 132, 148
CoRoT mission 48, 68, 210, 314, 343, 412, 413
CS disks, *see* Disks, circumstellar dust
Cyanide, in disks 276

Deformable mirror (DM), *see* Mirror, deformable
Descartes, René 269
Deuterium 5, 308
Differential extinction 66
Diffraction
    definition 113
    Fraunhofer 124, 133
    Fresnel 124, 133

Page numbers refer to specific pages on which an index term or concept is discussed. "ff" indicates that the term is also discussed on the following pages.

Disks 347ff
   around solar-type stars 287
   chemistry 275, 306ff
   circumstellar 269ff
   colors 288
   coorbital torque 355ff, 360, 366
   core accretion 334, 341
   corotation radius 335
   debris 269ff, 272, 278, 279
      dust 284
      dynamics 278
      gas 282, 284
      solar-system-like 278
   density/temperature variations 363
   detection limits 286
   diffusion 277
   dispersal 277, 288
   dust settling 300, 301, 340
   dust temperature 274
   dynamical friction 302
   evolution 286
   fragmentation 336, 337, 339
   gap 329, 330, 332, 348, 361, 362
   gas dispersal timescales 334
   gas drag 302, 349, 350
   gas rotation rate 350
   gas temperature 274
   gas-to-dust ratio 271, 279, 281, 282
   gravitational instabilities 300, 301, 334ff, 339
   horizontal shear stress 299
   hydrostatic equilibrium 321, 326
   ice absorption 284
   instability 319, 320, 334, 338, 340ff
   Keplerian shear 323
   lifetimes 279, 285, 288
   linear instability 334
   magnetic field 365
   metallicity 339
   migration rates 304, 349
   molecular hydrogen 281
   oligarchic growth 301, 304
   photoevaporation 277
   polycyclic aromatic hydrocarbons 283
   primordial 271, 272, 279, 280
   protoplanetary 8, 269ff, 272, 273, 297ff, 304, 306, 320ff, 325, 327, 328, 331, 334
      around binary stars 286
      composition 297, 298
      dust density 279
      dynamics 276
      evolution 285
      gas temperature 275
      mass 278
      solids 283
   radial equilibrium 334
   radiative cooling 336ff, 341
   rotation rate 321
   runaway coorbital migration 365
   runaway growth 301
   shear 301
   spiral waves 336
   structure 273ff
   surface density 332
   T Tauri 273, 274
   tidal interactions 304, 333
   torque 350ff
   transitional 271, 272, 279, 280, 287
   turbulence 277, 297, 299, 300, 301, 304, 307, 308, 309, 321, 348, 357, 361, 363, 364, 488
   vertical structure 321, 334, 354
   viscosity 298, 302, 364
   white dwarfs 186
Dispersed fixed-delay interferometers 48
Doppler effect 29ff
Doppler, Christian 27
Dust 297, 298
   accretion 297
   aggregates 300
   circumstellar 271
   dynamics 298
   zodiacal 287
Dust grains 298
   accretion 270
   collisions 299
Dust ring 111
Dynamics, non-Keplerian 217ff

Earth-like planets 11, 243, 375ff, 463
   definition 12
   density 386
   habitable zone 464
   Hadley circulation 505
   heat transport 508
   obliquity effects 509
   ozone 463
   plate tectonics 388
   pressure 386
   radius 386
   seasonal variations 509
   temperature structure 464
   thermal profiles 387
   tidal deformation 243
   zonally averaged temperature 503
   zonal-mean circulation 503
Earth-mass planets 333, 441
   accretion time 270
   migration 304
Earthshine 456ff
Eccentric-orbit planets
   migration 366
Eclipses (see also Occultations) 55, 57
   dimensions 60
   discoveries 63
   duration 57, 61
   light loss 58
Eddington factor 428
Einstein, Albert 79
Einstein A coefficients 421
Einstein delay 179
Einstein radius 96, 196
Einstein ring 80, 82ff, 95, 196
Embryos 297, 303, 311, 323
   lunar-sized 311
   Mars-sized 308, 311

Embryos (continued)
    migration   312, 313
    planetary   308, 320, 322, 324, 333
Energy balance   473
Energy conservation   401, 473
Enstatite   378
Epicurus   3
EPOXI mission   456, 457, 458
Epstein drag   298
Equations of state   376, 379ff, 391, 392, 397ff, 403, 404
Equilibrium heat flux   388
Equilibrium rotation states   261
Etendue   114
Exo-Neptunes
    definition   12
Exoplanets
    atmospheric circulation   471ff
    atmospheric irradiation effects   406ff
    bulk composition   376
    contrast   115, 117
    definition   3, 12
    density measurements   375ff
    direct imaging   7
    discovery rate   40, 64, 65
    discovery selection effects   192ff
    dynamics   15ff
    eccentricity distribution   203ff
    formation signatures   158
    fraction of stars   201
    interior structure   376
    mass determination   158ff, 375ff
    mass distribution   201, 202, 204, 205
    orbital distribution   201ff, 207ff
    orbits, apsidal precession   413
    period determination   158
    pulsars, *see* Pulsar planets
    radius measurements   375ff
    radius, selection effects   195
    satellites   101
    statistical distribution   191ff
    stellar fraction   207
    terrestrial composition   377
    terrestrial planet search   158ff
Exozodi   117, 134ff
Extrasolar giant planets, *see* Giant planets
Extrasolar Planets Encyclopedia   20

Fabry-Pérot cavity   35
Far-infrared, definition   113
Feeding zone   323
Flat fielding   66
Fourier optics   125

Gaia mission   167, 170
Galileo mission   428ff, 456, 492
Gas accretion
    Bondi-type   329
    Hill-type   329, 330
    rates   328, 331
Gas giants, *see* Giant planets
General circulation models   419, 455, 472, 477
Geostrophic equilibrium   480ff, 490

Giant planets   243, 342, 419ff
    atmospheres   419ff, 431, 433ff, 498
    baroclinic layer   492
    chemical equilibrium   420
    chemical evolution   433ff
    composition   397ff, 432
    condensation sequence   490
    cooling timescale   401
    core accretion   270, 325
    definition   12, 113
    effective temperature   425
    evolution   330, 333, 433ff
    formation   270, 288, 319ff
    global-scale atmospheric circulation   490
    interior   397ff
    mass-radius relationship   402ff
    planet/star flux ratio   435, 436
    polarization   425
    radiative-convective boundary   489
    rotation   402
    spectra   419ff
    temperature profiles   426ff, 435, 488
    thermal evolution   397ff, 411, 413, 433ff
    thermal inversions   428
    zonal jets   492
Glint, *see* Specular reflectance
Global circulation models, *see* General circulation models
Granulation   32
Gravitational focusing   302
Gravitational microlensing, *see* Microlensing
Gravitational moments J_{2n} 402
Gravitational potential   241
Gravitational redshift   30
Gravity, surface   60
Great divide   11
Greenhouse effect   447, 500

Habitable planets   99, 191, 375, 441
    definition   12
Habitable zones   27, 117, 227, 259, 319, 443, 451, 459, 461, 467
    continuous, definition   12
    cryogenic   465
    definition   12, 113
    M stars   465
Hadley circulation   487, 501ff
Hafnium-tungsten system   311
Half-lives   310
HCO^+ 284
Heat flux   389
Herschel mission   288, 289
Hertzsprung-Russell diagram   31
Hill radius   227, 228, 303, 305, 323, 328, 329, 350, 352, 354, 362
Hipparcos mission   166, 166, 169
Hot Jupiters   112, 202, 205, 208, 239, 240, 313, 406, 408, 421, 441, 471, 493ff, 497, 498
    atmospheres   409, 419ff, 471, 494, 499
    barotropic simulations   495
    definition   12
    double diffusive convection   410
    energy balance   256

520 *Exoplanets*

Hot Jupiters (continued)
    equatorial jets 498
    frictional processes 478
    geometric albedo 424
    luminosity 425
    observed properties 436
    orbital circularization 256
    period distribution 203
    polar vortices 495
    Rhines scale 494
    Rossby number 484
    spectra 419ff
    spin evolution 250
    thermal inversions 428
    tidal effects 240, 254ff, 262, 410
    wind speeds 480, 481, 486
    zonal winds 494, 495
Hot Neptunes 411, 493ff, 498
Huygens, Christopher 3
Hydrogen cyanide (HCN) 276, 284
Hydrostatic equilibrium 9, 270, 327, 376, 379, 401, 427, 475, 481
Hydroxide (OH) 284

Ice giants 44, 319, 405
Ice line(s), *see* Snow line(s)
Icy planets 376, 383, 384, 400
Icy satellites 384
Individual objects
    μ Arae 45, 209, 229
    μ Arae c 263
    16 Cygni B 228, 230
    2MASS 10415-0935 498
    2MASS J0415-0935 498
    4 Vesta 305, 314
    47 Tucanae 65, 205
    49 Cet 282, 283
    4U 0142+61 184
    51 Ophiuchi 282, 283
    51 Pegasi 4, 28, 40, 41, 160, 172
    51 Pegasi b 158, 347, 397, 435
    55 Cnc 44, 229, 233
    55 Cnc e 45, 263
    61 Vir b 45
    70 Ophiuchi 3
    AU Mic 284
    Barnard's Star 3, 157
    β Pictoris 112, 283
        circumstellar gas 282
        color 284
        gas-to-dust ratios 283
        radius 281
    Callisto 376, 384
    Ceres 314
    Charon 239
    CM Draconis 185
    CoKu Tau/4 272
    Comet Wild 2 314
    CoRoT-7b 390, 459ff
        atmospheric composition 461
        internal structure 390
        mass-orbit relationships 390
        mass-radius relationships 384, 390
        synchronous rotation 391
        temperature 460
        transit 70
    CoRoT-Exo-4b 479
    CS Chu 272
    Earth 347, 381, 473, 478
        aerosols 454
        angular momentum deficit 305
        as an exoplanet 5, 455ff, 467
            angular separations 115
            biosignatures 465
        atmospheric composition 308, 444ff, 481
        atmospheric structure 481
        atmospheric temperature profile 445, 446
        Bond albedo 500
        bulk composition 376
        clouds 447, 448
        differentiation 376, 377
        diurnal color changes 457
        equilibrium rotation states 263
        general circulation model 478
        habitability 442
        Hadley circulation 502
        hydrosphere 377
        internal structure 376, 377, 382, 383, 443
        jet streams 481, 486, 487
        lightcurves, *see* Individual objects, Earth, photometric variability
        orbital eccentricity 305
        ozone 443, 445, 446, 465
        photometric variability 457, 458
        physical parameters 479
        radiogenic elements 383, 384
        Rossby number 480
        secondary atmosphere 443
        source of water 308
        spectral characteristics 119, 120, 451, 458, 459
        spin evolution 252, 253
        thermal emission spectrum 121
        volcanic outgassing 443
        wind speeds 478
    ε Eridani 160
    Europa 376
    Fomalhaut 269, 288
        debris disk 269, 270
    Fomalhaut b 111, 341
    G29-38 186
    Ganymede 376, 384
    GJ 1214b 70, 384, 390, 391, 459ff
    GJ 176b 45, 261, 263
    GJ 436 67
    GJ 436b 44, 45, 390, 411, 412, 436, 479, 498
    GJ 581 43ff, 390
    GJ 581b 45
    GJ 581c 45, 263, 461
    GJ 581d 45, 263, 459, 461
    GJ 581e 45, 261, 263, 459
    GJ 674b 45, 263
    GJ 876 6, 43, 166, 209, 228, 229, 232, 367
    GJ 876b 166
    GJ 876bc 233

Individual objects (continued)
   GJ 876d   45, 263
   Gliese 229b   397, 498
   Gliese 570d   498
   Gliese 581   463
   Gliese 581c   462
   Gliese 581c   461
   Gliese 581d   461, 463
   Gliese 876   342
   HAT-P-11b   45, 411, 412
   HAT-P-13   233
   HAT-P-13b   413
   HAT-P-2b   479, 498
   HAT-P-3b   70, 408
   HAT-P-7b   72, 479, 511
   HD 107146   284
   HD 108874   229
   HD 114762   3, 27, 40, 157
   HD 114762b   157
   HD 11742   4
   HD 118232   282
   HD 128311   229
   HD 1461b   45
   HD 149026   67, 208
   HD 149026b   42, 70, 406, 408, 428, 436
      heavy element abundance   408
      interior structure   409
      physical parameters   479
      thermal inversions   438
   HD 15115   284
   HD 154345   210
   HD 156668b   45
   HD 156846b   21
   HD 158352   282
   HD 160691 = μ Arae, *see* Individual objects, μ Arae
   HD 16417b   45
   HD 166435   197
   HD 17156b   42, 43, 68, 69, 498
   HD 171779   168
   HD 179949b   436, 494, 510
   HD 181327   284
   HD 181433b   45, 74, 263
   HD 189733   42, 67, 69, 70
   HD 189733b   70, 72, 436, 493, 511
      atmospheric dynamics   510
      contrast ratios   435
      infrared lightcurves   493, 494
      lightcurves   496
      physical parameters   479
      planet/star flux ratio   437
      temperature model   496
   HD 190360c   45
   HD 202206   229
   HD 209458   42, 68, 208, 232
   HD 209458b   71, 232, 398, 406, 428
      barotropic simulations   495
      Cassini state   410
      contrast ratios   435
      discovery   63
      infrared lightcurves   493, 494
      mass loss rate   411
      physical parameters   479
      planet/star flux ratio   437
      sodium detection   436
      stratospheres   497
      temperature-pressure curves   438
      temperature-pressure map   497
      thermal inversions   438
      tidal dissipation   257
      tidal evolution   258
      transits   59
   HD 21620   282
   HD 219828b   45
   HD 32297   282
   HD 32297   284
   HD 33636b   166
   HD 40307   44ff
   HD 40307b,c,d   45, 261, 263
   HD 4308b   45
   HD 45364   229
   HD 47186b   45
   HD 4796A   284
   HD 5319   192, 193
   HD 60532   229
   HD 69830   44ff, 202
   HD 69830b   45, 263
   HD 69830c   45, 263
   HD 69830d   45
   HD 72659   199
   HD 73526   229
   HD 7924b   45, 263
   HD 80606b   42, 43, 68, 71, 231, 259, 260, 498
   HD 82943   229
   HD 83443b   22
   HD 92945   284
   HH 30   272, 280
   HR 8799   111, 112, 210, 288
   HR 8799b,c,d   341
   HW Virginis   175, 185
   HW Virginis b   176
   HW Virginis c   176
   Hyperion   240
   Io   391
   ι Draconis   168
   ι Draconis b   166
   Jupiter   340ff, 347, 376, 391, 419, 429, 478, 479
      atmosphere   429, 431, 498
      baroclinic layer   492
      barotropic stability criterion   497
      composition   432ff
      formation   298, 319, 325
      geometric albedo   423
      global-scale atmospheric circulation   490
      Great Red Spot   431, 432
      infrared image   432
      internal structure   397, 400, 403ff, 411
      jet streams   486, 487
      luminosity   425
      magnetic field   405
      mean density   407
      mid-infrared brightness temperature   430
      migration rate   355
      observed properties   428ff

Individual objects, Jupiter (continued)
    physical parameters   479
    radiative-convective boundary   489
    resonance   306
    Rhines scale   494
    Rossby number   480
    rotation   402
    satellites   376, 384
    temperature profiles   431ff, 489
    thermal evolution   434
    tidal dissipation factors   254
    zonal winds   431, 479, 491ff
  Mars   381
    angular momentum deficit   305
    atmosphere   442, 445ff
    cratering   298
    early history   448, 463
    habitability   442, 461
    internal structure   382, 383, 443
    jet streams   481, 487
    magnetosphere   442
    obliquity variations   249
    orbital eccentricity   305
    physical parameters   479
    Rossby number   480
    spectral characteristics   451
    spin evolution   252, 253
  Mercury   240, 381
    angular momentum deficit   305
    cratering   298
    internal structure   382, 383, 443
    magnetic field   442
    orbital eccentricity   305
    spin-orbit resonance   240, 248, 250
  MOA-2003-BLG-53Lb   101, 103
  MOA-2007-BLG-192Lb   103, 207, 459
  MOA-2007-BLG-400Lb   103
  MOA-2008-BLG-310   93
  MOA-2008-BLG-310Lb   102, 103
  Moon   381
    Bond albedo   500
    cratering   298
    formation   297, 305
    internal structure   382, 383, 443
  Neptune   319, 341, 367, 384, 385, 479
    composition   403
    formation   325
    global-scale atmospheric circulation   490
    heat flow   405, 406
    internal structure   397, 405, 406, 409, 411
    jet streams   486, 487
    magnetic field   406
    mean density   407
    physical parameters   479
    radiative-convective boundary   489
    Rossby number   480
    rotation   402
    tidal dissipation factors   254
    zonal winds   491, 492, 493
  OGLE-05-390Lb   459
  OGLE-2003-BLG-235   101, 103
  OGLE-2005-390-Lb   390

OGLE-2005-BLG-071   93, 103
OGLE-2005-BLG-071b   101
OGLE-2005-BLG-071Lb   104
OGLE-2005-BLG-309Lb   101
OGLE-2005-BLG-390Lb   102, 384
OGLE-2005-GLB-169Lb   102
OGLE-2006-BLG-109   102, 103
OGLE-2006-BLG-109Lb   102
OGLE-2007-BLG-368Lb   103
OGLE-TR56b   435
ω Centauri   205
Phobos   239
Pluto   314
PSR 1257+12   4, 40, 229, 230, 232
PSR B1257+12   5, 6, 175, 177, 181ff, 188, 231
PSR B1257+12b   176
PSR B1257+12c   176
PSR B1257+12d   176
PSR B1620-26   175, 178, 182, 183, 184
PSR B1620-26b   176
RS CVn   185
Saturn   340ff, 419, 429
  atmosphere   429, 431
  baroclinic layer   492
  barotropic stability criterion   497
  composition   432, 433
  formation   319
  geometric albedo   423
  global-scale atmospheric circulation   490
  icy satellites   384
  internal structure   397, 400, 403, 404, 409
  jet streams   486, 487
  magnetic field   405
  mean density   407
  mid-infrared brightness temperature   430
  north polar region   486, 487
  observed properties   428ff
  physical parameters   479
  radiative-convective boundary   489
  resonance   306
  Rossby number   480
  rotation   402
  satellites   376
  temperature profiles   431ff
  tidal dissipation factors   254
  zonal winds   431, 491ff
σ Herculis, circumstellar gas   282
τ Boo b, contrast ratios   435
Titan   376, 384
  atmospheric temperature profile   445
  photochemical haze   452
  physical parameters   479
  potential for life   297
  Rossby number   480
  superrotating winds   509
TrES-1   67, 438
TrES-2   479, 493, 497
TrES-4   407, 438, 479, 493, 497
Triton   239
TW Hya   160
υ Andromedae   41, 42, 166, 167, 209, 230
υ Andromedae b   42, 436

Individual objects, υ Andromedae b (continued)
      infrared lightcurves 493, 494
      orbital evolution 225, 226
      temperature variations 510
  υ Andromedae c 166, 167
  υ Andromedae d 167
  Uranus 319, 341, 384, 385
      composition 403
      formation 325
      global-scale atmospheric circulation 490
      heat flow 405, 406
      internal structure 397, 405, 406, 411
      jet streams 486, 487
      magnetic field 406
      physical parameters 479
      radiative-convective boundary 489
      Rossby number 480
      rotation 402
      tidal dissipation factors 254
      zonal winds 491ff
  V391 Pegasi 184
  V391 Pegasi b 176
  VB8 157
  Venus 223, 240, 381, 442
      aerosols 454
      angular momentum deficit 305
      atmosphere 445ff
      atmospheric tides 259
      Bond albedo 500
      current history 463
      early history 500
      equilibrium rotation states 263
      greenhouse effect 447
      internal structure 382, 383, 443
      magnetic field 442
      orbital eccentricity 305
      photochemical haze 452
      physical parameters 479
      Rossby number 480
      spectral characteristics 451
      spin evolution 250, 251
      superrotating winds 509
      tidal evolution 260
      zonal winds 482, 483
  WASP-7b 408
  WASP-12b 479
  XO-1 428
  XO-1b 438, 493, 497
  XO-3b 72
Interferometer 123ff
  definition 113
  nulling 133ff
Interferometric astrometry 163
Interferometry
  groundbased 164ff
  speckle 136ff
International Celestial Reference System (ICRS) 30
International Ultraviolet Explorer (IUE) 429
Interstellar molecular clouds 271, 273
Iodine cell technique 36ff
Iodine-xenon 311
Iron/silicon ratio 378, 379, 381, 382

Iron-60 310
Irradiation effects 408
Isotopes 308, 310

Jacobian coordinates 219
James Webb Space Telescope 11, 151, 288, 438, 467
Jansky, unit of measure 114
Jeans escape, *see* Planets, Jeans escape
Juno mission 405, 492
Jupiter-mass planets 342
  irradiation effects 407, 408
  radius evolution 414

K, *see* Semiamplitude
Kelvin Helmholtz timescale 411
Kepler mission 11, 48, 210, 314, 343, 392, 413, 467, 499, 511
Kepler problem 15ff
Kepler, Johannes 6, 15
Keplerian motion 15ff, 28ff, 348
Kepler's equation 18
Kepler's laws 6, 15ff
  first law 17
  second law 16
  third law 17, 29, 60, 114
Kozai cycles 225, 230, 231, 259, 342
Kozai migration 257, 367
Kuiper belt(s) 270, 272, 278, 289, 298, 314, 321, 333, 367

Lagrange points 228, 323
Lambert scattering 115, 423, 424, 425, 454
Late heavy bombardment 278
Lead-lead method 310
Lidov-Kozai mechanism 258, 259
Lightcurves 496
Limb darkening 59
Lindblad resonance 304, 358, 359, 365
Lithosphere 388
Local thermodynamic equilibrium (LTE) 421
Lomb-Scargle periodogram 192, 198
Love number 218, 233, 244, 413
Luminosity 117

M dwarfs 47
  habitable zone 461, 462, 480, 501
  magnetic fields 32
m sin i, *see* Minimum mass
Magnesium 310, 378, 381
Magnus, Albertus 3
Malmquist bias 197
Mantle thickness 387
Mantle viscosity 386
Mars Global Surveyor mission 456, 457
Mars-mass planets 270
Mass conservation 379, 401, 473
Mass-radius relationships 376, 383, 384, 391
Mature planet, definition 113
Maunder minimum 34
Mercury-like planets 375
MESSENGER mission 314, 382
Metal 297

Metallicity   197, 206
Meteorites   305, 314
   carbonaceous chondrites   307
   chondrule ages   311
   chondrules   308
   CI carbonaceous chondrites   306, 308, 377
   CM carbonaceous chondrites   308
   compared to the Sun   306
   enstatite chondrites   308, 376ff
   HED   311, 314
   iron   307
   ordinary chondrites   307, 308
   SNC   382
   thermal processing   307
Microlensing   7, 79ff, 97, 98, 104, 196
Mid-infrared, definition   113
Mie-Grüneisen-Debye formulation   380
Migration   312, 314, 355, 366
   Type I   8, 203, 304, 312, 332, 342, 348, 350, 355, 357ff, 360, 362, 365, 366, 369
   Type II   8, 203, 304, 332, 342, 348, 360ff
   Type III   365, 366
Milne atmosphere   426ff
Minimum mass   29
Minimum mass solar nebula   279, 297, 323, 336, 355
Mirror, deformable (DM)   140ff
Molecular hydrogen   399
   dissociation   276
   in disks   276, 281
   phase diagram   398
Momentum conservation   473
Multiple-planet systems   209ff, 248

Navier-Stokes equations   474ff
Near-infrared, definition   113
Neptune-like planets   27, 47, 333, 342, 343, 376
Neutron star disks   183
New Horizons mission   314, 429
Newton, Isaac   15, 79
Newtonian cooling   504
Nice model   278
Non-Keplerian perturbations   41
Nova Geminorum   79

Obliquity evolution   245
Occultations (*see also* Eclipses)   55ff
   definition   10
   depth   62
   spectroscopy   62ff
Ocean planets, *see* Water planets
Ohmic dissipation   492
Oligarchic growth   305, 323
Opacities   421
Orbital eccentricity   61
Orbital evolution   217ff, 224, 245, 246
Orbital inclination   60
Orbital migration   342, 343, 350
Orbital period, selection effects   194ff
Orion nebula   281
Ortho- to para- molecular hydrogen ratio   430
Oxygen isotopes   305

Penumbra   57
Periastron advances   222
Phase functions   115, 422ff
Phase-induced amplitude apodization (PIAA)   129
Photodissociation regions   281
Photon flux   114
Photon noise   65, 122, 140, 196
Photons   120ff
Photosphere   426, 481
Pickup ions   449
Pioneer 10 and 11 missions   429
Planck function   62, 113, 271
Planck mean opacity   427
Planet migration   217ff
Planet tectonics (*see also* Plate tectonics)   385
Planet, definition   5
Planetary differentiation   378, 379
Planetary systems, fate   185ff
Planetesimals   297, 301ff, 322
   accretion   270, 301
   atmospheric drag capture   303
   capture   297
   dating   309
   dynamical friction   304
   embryo collisions   305
   feeding zone   303
   formation   270
   formation
      meter-sized barrier   300, 308
      streaming instability   309
      turbulence-driven   308
   gravitational interactions   302
   growth rate   302, 322
   impact strength   301
   largest surviving body   301
   late-stage growth   303, 308, 312
   leftover   270
   Mars-sized   303
   migration   313, 367
   monolithic   301
   oligarchic growth   297, 302, 303
   rubble piles   301
   runaway growth   297, 302, 303
Planets
   atmospheres   10ff, 448, 450
   carbon, definition   12
   completeness corrections in occurrence rates   199
   definition   11, 12
   equilibrium temperatures   447
   formation   7, 8, 311
   free-floating   100, 210
   hydrodynamic escape   449
   interiors   8ff, 385
   ion escape   449
   Jeans escape   448, 449
   mean temperatures   499
   migration   347ff
   nonthermal escape   449
   radius   196
   sputtering   449
   surface temperatures   447

Planets (continued)
    thermal escape   448
Plate tectonics   375, 379, 385, 388, 389
P-mode oscillations   32
Poincaré coordinates   219
Poisson noise, *see* Photon noise
Poisson process   122
Polycyclic aromatic hydrocarbons   283, 284
Poynting-Robertson drag   278, 289
Primordial atmospheres   375
Primordial interstellar grains   272
Proper motion   97
Protoplanetary disks, *see* Disks, protoplanetary
Protoplanets
    growth rate   329, 330
    luminosity   326, 327
    migration   331
Pulsar planets   4, 175ff, 181
Pupil masking   128, 129

Q, *see* Tidal dissipation factor

Radial velocity
    detection threshold   192ff
    measurement   157
    noise   32, 33, 50
    precision   31ff
    selection effects   194
    solar-mass star   29
    technique   6, 27ff
Radiant flux   114
Radiative equilibrium   117, 427, 472, 480
Radiative timescale   480
Radiative transfer equation   427
Radiogenic elements   298, 379, 385
Radius anomaly   403, 408
Ray, definition   113
Rayleigh scattering   422, 425, 435, 450, 457
Rayleigh-Jeans limit   62
Rayleigh-Taylor instabilities   376, 378, 387
Red edge vegetation   11, 119, 456, 463
Red noise   196
Redox reactions   463
Refractory materials   297
Relativistic effects   218
Resonances   217ff, 223, 227
    asteroid belt   306
    corotation   358
    mean-motion   228, 305, 307, 312
    secular   306, 312, 313
    spin-orbit   248
Reynolds number   298
Rhines scale   479, 484, 485, 487, 492, 496, 508
Roche limit   208, 259
Roche lobe   323, 331
Rossby number   479, 480, 482, 486, 494, 496
Rossby waves   476, 477, 484, 485, 486, 488, 497
Rossiter-McLaughlin effect   43, 62ff, 226
Rotation rate evolution   245, 246
Rubidium-87   309
Runaway greenhouse effect   442

Runaway growth   305, 320ff, 322, 323
Russell, Henry Norris   55

Safronov number   208
Safronov, Viktor   314
Saturn-mass planets   349, 407
Scattering, definition   113
Schiaparelli, Giovanni   239, 240
Scintillation noise   66, 140, 144
Seasonal variations   465
Self-luminous planets, definition   113
Semiamplitude   194
    distribution   203
Shear instabilities   487
Shear stress   389
Siderophile elements   308
Silicate grains   297
    in disks   283
    spectral curves   285
    vaporization temperature   280
Simultaneous reference technique   38ff
Snow line(s)   81, 99, 207, 210, 307, 308, 320, 347
    in disks   276
    solar nebula   276
Solar composition   378
Solar cycle   33
Solar nebula (*see also* Solar system,
    protoplanetary disks)   297, 306
    condensation temperatures of minerals   306
    oxidation state   307
    short-lived isotopes   311
Solar system
    as exoplanet system
        blackbody spectra   118
        colors   119
Solar-mass star, migration   360
Space Interferometry Mission (SIM)   170
Specific intensity   113
Speckle, definition   113
Spectral bands   114
Specular reflectance   454, 457
Spitzer Space Telescope   11
Star shade   133
Stardust mission   314
Stars
    Herbig Ae/Be   282, 283
        vaporization temperature   280
    Herbig Be   281
    magnetic fields   32
    occurrence rate, *see* Exoplanets, stellar fraction
    pre-main-sequence   271
    solar-type   287
    T Tauri   277, 279, 281ff
    ZZ Ceti   184
Stellar atmospheres   378
Stellar flux, redistribution factor   407
Stellar jitter   197
Stellar metallicity   205ff, 342
Stellar noise, *see* Radial velocity, noise
Stokes drag   298
Streaming instability   309, 310

Strontium-86/87   309
Struve, Otto   3, 27, 157
Super Earths   9, 11, 27, 44, 47, 209, 210, 239, 240,
    298, 376, 380, 383, 385, 441ff
    atmospheres   450
    atmospheric tides   259
    chacterization   459
    definition   12
    discovery   459
    equilibrium rotation states   262
    habitability   460
    orbiting M dwarfs   467
    spin evolution   250
    tidal effects   240, 259ff
Super Mercuries   383
Supergranulation   32
Symplectic integrators   219, 220
Synchronous rotation   239, 240, 248

Talbot effect   144
Taylor-Proudman theorem   490, 492
Terrestrial exoplanets
    atmospheric temperature profile   445
    characterization   159
    mantle convection   385
    remote sensing   443
Terrestrial planets   308
    atmospheres   441ff, 447, 448, 452ff
    climate models   454
    composition   308
    definition   12, 113
    formation   270, 297ff, 311, 312
    horizontal winds   483
    internal structure   375ff, 441
    magnetic fields   441, 442
    photochemical models   452
    radiative transfer models   453ff
    seasonal spectral variations   452
    solar system   375, 441, 442
    surface composition   451
    surface pressure   451
    surface temperature   451
    tidally locked   510
    vertical winds   483
Terrestrial-mass planets   305
Thermal convection   375
Thermal disequilibrium   465
Thermal evolution models   412
Thermal-wind equation   483, 490, 503
Thomas-Fermi-Dirac formulation   380
Tidal dissipation   204, 222, 223, 225, 239, 244, 245, 413
    Earth-Moon system   253
    hot Jupiters   254
Tidal effects   218, 240, 242, 243
    terrestrial planets   251, 252
Tidal evolution   71ff, 239ff

Tidal habitable zone   462
Tidal interactions   60
Tidal migration   71ff
Tidal stripping   223
Tidal synchronization   72, 472
Timing technique   6, 231
Toomre stability parameter   335
Transit radius   420, 426
Transiting planets
    composition   407
    mass-radius relationship   407
    Neptune-mass   411
    radial velocity discoveries   42ff
    tidally locked   460
Transits   55ff
    contacts   221
    definition   10
    dimensions   60
    discoveries   68
    geometry   221
    mass determination   69
    mass-radius relationship   208
    measurement   65
    period determination   69
    probability   57
    radii determination   69
    selection effects   195
    technique   7
Transit-time variations   232, 233
Transmission spectroscopy   61ff
Tungsten-182   311
Two-body problem   217, 222

Uranium-235/238   309, 310
Uranus-like planets   376

Virial theorem   401
Viscosity   273
Visible, definition   113
Vogel, Hermann   27
Vorticity   476, 477, 482, 485
Voyager missions   429ff

Water planets   384, 387, 460
Water
    dissociation and ionization   276
    in disks   284
Wavefront   121
    definition   113
    sensing   140

Z, *see* Stellar metallicity
Zodiacal disk   117
Zonal gravitational harmonics, *see* Gravitational moments J_{2n}
Zonal winds   485